Ladies in the Laboratory? American and British Women in Science, 1800–1900

A Survey of Their Contributions to Research

Mary R. S. Creese
with Contributions by Thomas M. Creese

The Scarecrow Press, Inc.
Lanham, Md., & London
1998

SCARECROW PRESS, INC.

Published in the United States of America
by Scarecrow Press, Inc.
4720 Boston Way
Lanham, Maryland 20706

4 Pleydell Gardens, Folkestone
Kent CT20 2DN, England

British Library Cataloguing in Publication Information Available

Library of Congress Cataloging-in-Publication Data

Creese, Mary R. S., 1935-
Ladies in the laboratory? : American and British women in science, 1800-1900 : a survey of their contributions to research / Mary R.S. Creese.
p. cm.
Includes bibliographical references and index.
ISBN 0-8108-3287-9 (alk. paper)
1. Women in science—United States—History—19th century.
2. Women in science—Great Britain—History—19th century.
3. Women scientists—United States—Biography. 4. Women scientists Great Britain—Biography. I. Title.
Q141.C69 1998
500'.82'094109034—dc21 97-1125
CIP

ISBN 0-8108-3287-9 (cloth : alk. paper)

™ The paper used in this publication meets the minimum requirements of American National Standard for Information Sciences—Permanence of Paper for Printed Library Materials, ANSI Z39.48-1984.
Manufactured in the United States of America.

To Anna and Cathy, for the pleasure of their company while they were growing up,
and to the memory of Walter George Weir.

Contents

Figures

Preface and Introduction

The topic of "women-in-science," past and present, has received much attention over the last decade or two. Efforts to build up a history describing women's roles in scientific work have taken several approaches. Excluding full biographies, these range from the popular biographical dictionaries and bio-bibliographies (compilations of biographical sketches), to collections of essays on selected groups and broad surveys.[1]

All of these approaches have their own special virtues and their drawbacks. The bio-bibliographies, especially those that focus on particular disciplines, sometimes provide substantial amounts of information (and in addition to being reference works function somewhat as elementary texts offering an introduction to the field), but they can only be a first step toward a coherent, comprehensive story. Broad historical surveys, on the other hand, while essential in mapping out major features in the territory to be covered, also have built-in disadvantages; the reader with a scientific background may find them unsatisfying in their failure to pay sufficiently close attention to individual lives and technical work—the details of how early women scientists managed to enter their fields and what they succeeded in carrying out. Such details, while of considerable interest in any full history of scientific research, are especially important in the women's stories because of the peculiar handicaps faced. Works on selected groups designed to examine particular areas can be very successful; those that focus on a particular theme in women's history, however, have the disadvantage of producing a medley rather than the comprehensive picture to be hoped for from a more systematic approach.

In this survey of the research contributions of nineteenth-century American and British women, I have combined the biographical sketch and historical survey approaches, and attempted, partly by the form of the organization, to overcome some of the drawbacks just mentioned—even at the cost of introducing others. The two-country coverage, even though it has the disadvantage of doubling the length of the work, brings a depth of perspective, missing from a one-country survey, that leads to a more balanced picture of the success, or lack thereof, of each national grouping in each particular field.

The foundation of the study is a bibliography of the scientific journal articles published by about 680 American and British women between 1800 and 1900. This was constructed from entries extracted from the London Royal Society's monumental *Catalogue of Scientific Papers,* 1800–1900, the nineteen-volume index that constitutes the major comprehensive accessible source of scientific journal article titles for the period, covering all fields and all countries. A rich fund of basic data for a survey of this kind, it has remained until now a largely untapped resource.[2] Of the nearly 1,000 women authors found (who produced about 3,400 articles[3]—less than 1 percent of the entries in the *Catalogue*), 41 percent were American and 26 percent British. (Of the rest, 10 percent were Russian and Polish, and the remainder were mainly French, German, Irish, Italian, and Swedish—Figure 0-1.) Organized by field and country, the bibliography clearly offers an excellent basis for a broad comparative survey. The initial plan was to include women of all countries, but that turned out to be too ambitious an undertaking for a single project.

By examining the record of women's research output field by field, it has been possible to establish fairly definitively, in some respects it might be claimed quantitatively, where women were most productive (that is, productive across a wide spectrum of scientific work that includes writing and publication[4]). Figure 0-2 presents the overall picture; Figure 0-3 summarizes American and British data. Patterns and concentrations within fields are also brought out; for instance, distributions across the subfields of chemistry are established, and numbers of women zoologists doing museum taxonomy compared with those who succeeded in entering newer areas such as physiology and cytology.

The two-country coverage reveals in a systematic way, probably for the first time, differences between the two national groups, and, for most fields, why these differences arose. For example, the local circumstances that led to the preeminence of the Americans in zoology, astronomy, and psychology are fairly clear, as are the reasons for the British dominance in geology. The two-country coverage also allows a contrasting of the kind and quality of the work carried out by each national group; in botany, for example, most of the leading Americans were taxonomists and classifiers, while their British counterparts worked in plant genetics, anatomy, and morphology. Quantitative comparative information (numbers of both authors and papers) by field and subfield is provided in the form of graphical summaries following most chapters.

Although the survey covers women who published journal articles within the period 1800–1900, by far the greater number were active during the last two decades of the century, the

a. Authors, by country

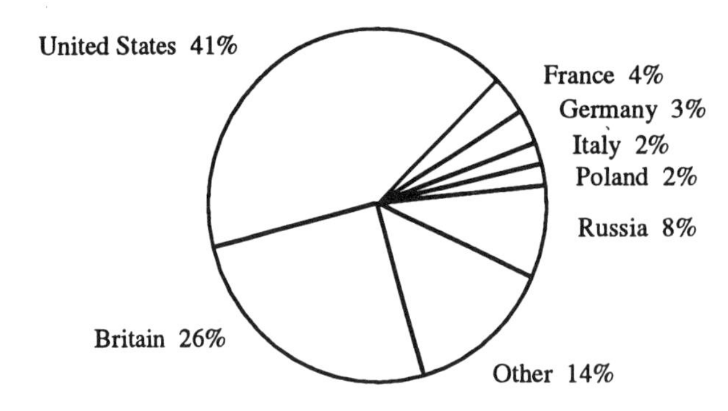

b. Papers, by country

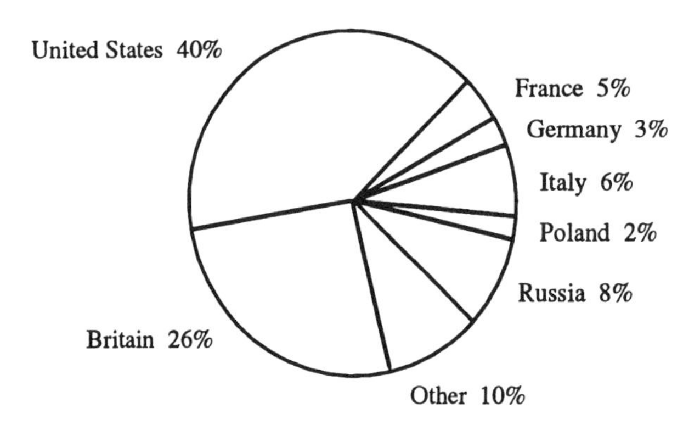

Figure 0-1. Distribution of women authors and their scientific journal publications (all fields) by country, 1800–1900. Data from the Royal Society *Catalogue of Scientific Papers.* The sectors "Other" represent authors and papers from the following countries whose contributions are 1% or less (except as noted): Australia, Austria, Belgium, Canada, China, Denmark, Finland, India, Ireland (2% of authors), Japan, Lithuania, Mexico, Netherlands, New Zealand, Norway, Portugal, Romania, South Africa, Sweden (2% of papers), Switzerland, Yugoslavia, Unidentified (2% of authors).

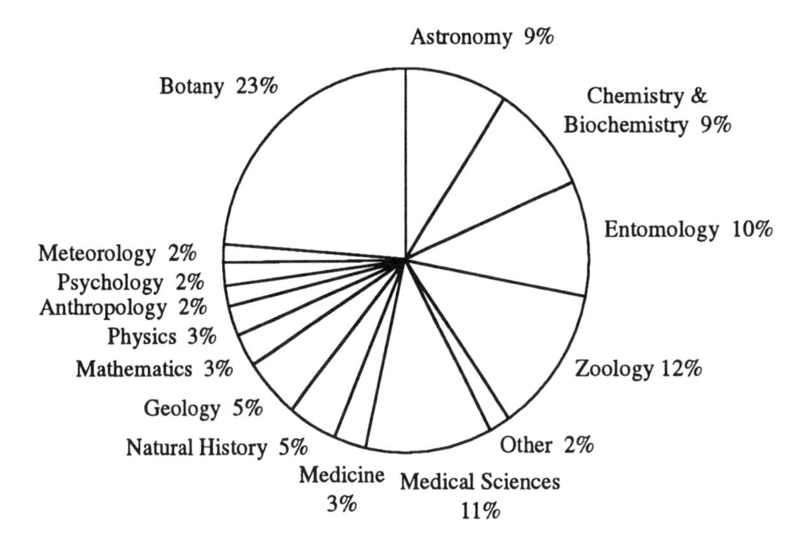

Figure 0-2. Distribution of women's scientific journal publications (all countries) by field, 1800–1900. Data from the Royal Society *Catalogue of Scientific Papers.* The sector "Other" represents publications in the following fields, each less than 1% of the total: General Biology, Geography, Technology.

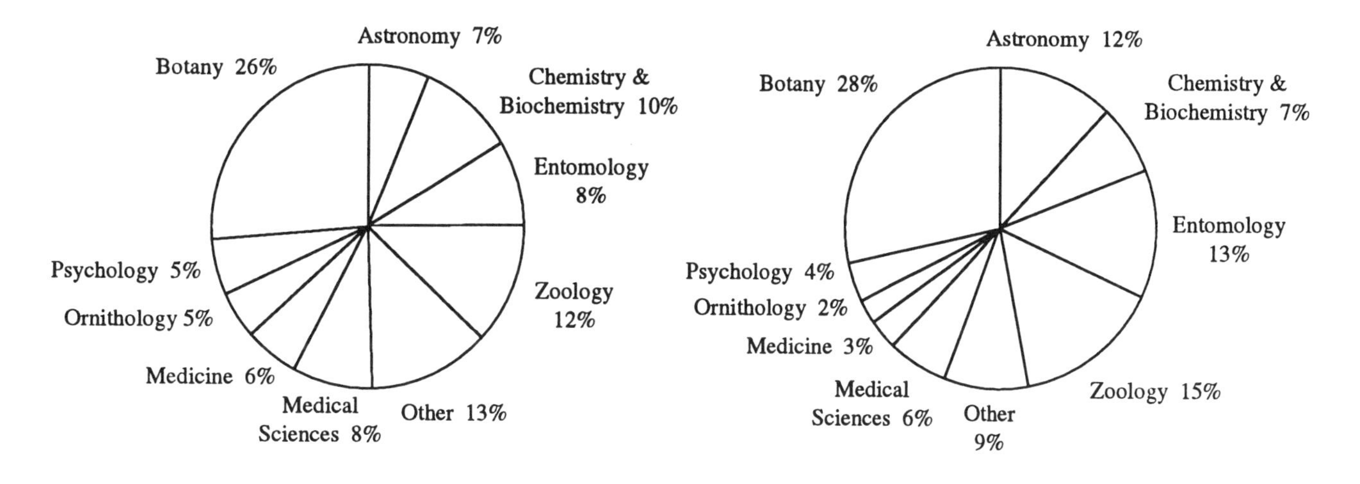

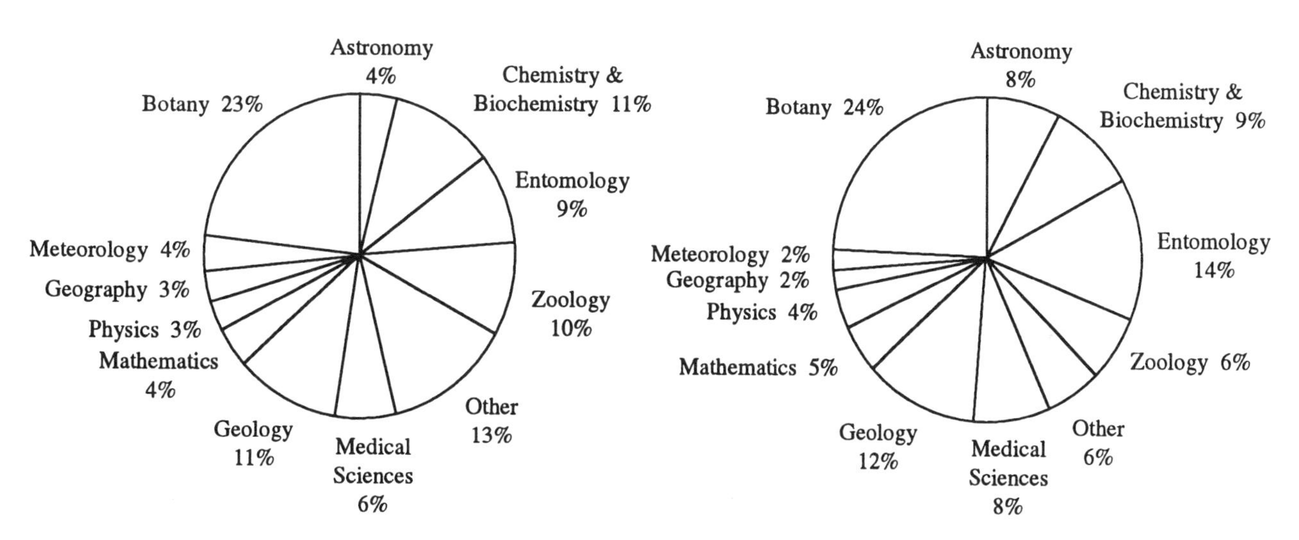

Figure 0-3. Distribution of women authors and their scientific journal publications (United States and Great Britain) by field, 1800–1900. (In a. and c., an author contributing to more than one subfield is counted in each.) Data from the Royal Society *Catalogue of Scientific Papers.* The sectors "Other" represent authors and papers in the following fields, where the contributions are less than 2% (except as noted). United States: Anthropology, General Biology, Conchology, Geography, Geology, Mathematics, Meteorology, Natural History, Physics, Technology. Great Britain: Anthropology, Conchology, Medicine (3% of authors), Natural History (4% of authors), Ornithology (2% of authors), Psychology, Technology.

time when the first of the university-trained women (in many respects a most remarkable group) were beginning their careers. Many of those mentioned were minor characters in the story of women in nineteenth-century scientific research, but the most prominent were among the elite of women scientists of the time. In the British group, for instance, biochemist Ida Smedley MacLean, botanists Ethel Sargant, Edith Saunders, and Margaret Benson, and geologists Maria Ogilvie Gordon and Gertrude Elles were productive research workers who had enviable reputations in their fields.

Overall, this is an attempt to give some definition and shape to a very broad spectrum of work. Perhaps it will point the way toward further comparative analysis likely to be illuminating and productive.

Notes

1. Biobibliographies include the series being brought out by Greenwood Press, so far covering women in anthropology, psychology, mathematics, chemistry/physics, and biology (1987 . . .), and Marilyn Ogilvie's *Women in Science: Antiquity through the Nineteenth Century* (1986). For works on selected groups see, for example, *Uneasy Careers and Intimate Lives* (eds. Pnina Abir-am and Dorinda Outram, 1987)

and *Women of Science: Righting the Record* (eds. Gabriele Kass-Simon and Patricia Farnes, 1990). Broad surveys include Margaret Rossiter's *Women Scientists in America* (1982), Margaret Alic's *Hypatia's Heritage* (1986), and Patricia Phillips's *The Scientific Lady* (1990).

2. Using an author's listing in the *Catalogue* as the criterion for inclusion in the survey keeps the focus on those who contributed original work, that is, on those who by and large had a fair claim to being part of the scientific research community (see also the remarks introducing the bibliography, p. 371). Smaller peripheral groupings such as the early nineteenth-century literary women with interests in science are outside the limits of this study; their contributions were of a different kind, often introductory instructional books for women and young people—see, for instance, Ann Shteir, *Cultivating Women: Cultivating Science* (1996).

3. It should be kept in mind that although the process of producing a count of published papers might at first sight appear mechanical, given an appropriate bibliographical list, it is in fact not so. Choices have to be made and judgment exercised. For instance, should a preliminary note be counted when a full discussion of the same work follows a few months later? Should continuations of reports in consecutive (or almost consecutive) journal issues be considered as one or several publications? Should simultaneous reports of work in more than one country be a single publication or not? Reasonable answers to these questions depend on the details in each case.

4. While this standard may not be wholly applicable in judging the productivity of workers in the early years of the century, it can fairly safely be used for the great majority covered in this survey.

Acknowledgments

Over the course of collecting information for this survey I have had help from close to 100 reference librarians and college and university archivists in Britain and the United States. I am most grateful to them all.

To some I am especially heavily indebted: Ann Phillips, Elisabeth van Houts, and Carola Hicks (Newnham College Cambridge), Kate Perry (Girton College Cambridge), and Carol Bowen (University College London) answered many inquiries over the course of several years with generous patience and unfailing thoroughness. In this country I have relied particularly heavily on Wilma Slaight (Wellesley College), Maida Goodwin and Margery Sly (Smith College), Patricia Albright (Mount Holyoke College), Joseph Svoboda (University of Nebraska), and Elizabeth Sage (University of Chicago). Among the many others who provided me with important material are Gould Colman (Cornell University), Olivia Frederick (Environmental and Community Development Department, Jefferson County, Kentucky), Patricia Harpole (Minnesota Historical Society), Ann Kenne (Iowa State University), Anna Koch (Massachusetts Institute of Technology), and Caroline Rittenhouse (Bryn Mawr College). References to the specific information I received are given in the Notes sections.

I would also like to thank the reference librarians of Anschutz and Watson libraries, University of Kansas, for their help and great patience over several years; in particular, Eleanor Symons helped me find my way into a world of general reference works and library tools with which, as a research chemist, I was quite unfamiliar; Mary Rosenbloom gave me much expert help in the area of women's studies; and the Anschutz staff did wonders unearthing nineteenth-century journals. I am much indebted to the interlibrary loan staff as well; their assistance has been vital to the project.

People whose kindly help and interest have been a special encouragement are Mary Brück (Edinburgh), Ronald Cleevely (Sussex), Bernard Lightman (York University, Toronto), Joan Mason (Open University, United Kingdom), Hugh Torrens (Keele University, United Kingdom), Peter Stevens (Harvard), and Steven Zottoli (Williams College, Massachusetts).

For permission to quote from unpublished archival material, I thank Wellesley College, the University of Nebraska, Kansas State University, the University of Kansas, and David Willcox of Randolph, New Hampshire, and his family. For permission to quote from published materials, I thank Northeastern University Press, the Geological Society (London), the Editorial Board of the Scottish Geological Society, *Geological Magazine* and Cambridge University Press, the Geologists' Association, and the Principal and Fellows of Newnham College Cambridge.

I had critical help in the preparation of the bibliography of scientific journal articles, 1800–1900, from Thomas Creese, my husband. Starting from my lists of about 3,400 article titles, he designed and managed the database from which most of the bibliography was built. He also prepared the graphs included in most chapters, assisted with the material on mathematicians, and provided a great deal of general help and encouragement throughout the course of the work. We were aided in the early stages by support from the University of Kansas General Research Fund (allocation nos. 3693-xx-0038, 3179-xx-0038, 3312-xx-0038). We thank especially Robert Bearse, vice-chancellor for research, for his interest in this project from its beginning. The work on chemists was supported in part by the National Science Foundation, Washington, D.C., Grant No. DIR-8907758.

John Ramirez helped with data entry for the nineteen-century journal article bibliography. The staff of the College of Liberal Arts and Sciences Word Processing Center (University of Kansas), especially Paula Courtney, Pam LeRow, Lynn Porter, and Tonya Elmore, typed the manuscript with great skill and wonderful patience; we are much indebted to them.

Any errors are my responsibility.

Mary Creese
Lawrence, Kansas
May 1997

Part 1

Life Sciences

Chapter 1

BOTANISTS: FROM EARLY EXPLORERS TO PLANT GENETICISTS

Papers in botany make up about 23 percent of the journal publications by women indexed by the Royal Society. Americans contributed almost half (48 percent), British workers another 26 percent, women in France, Belgium, and the Netherlands 7 percent, and Italians 5 percent. Most of the remainder came from women in Sweden, Poland, Ireland, Germany, and Russia.

American women

The botanical contributions of American women constitute about 11 percent of all the scientific papers published by women between 1800 and 1900 (as recorded by the Royal Society). Americans who wrote on botanical topics form the largest author group (119[1]), defined by field and nationality, and make up about 10 percent of women contributing articles in all subjects to nineteenth-century scientific journals. Many of them published only one or two papers, typically reports of undergraduate research or field observations.

The group may be divided into five overlapping but nevertheless distinguishable subdivisions: first, women of the northeast region of the country who were not associated with institutions of higher education, including both amateur investigators and those with formal training in botany (a number were high school science teachers); second, college and university faculty members of the East and Middle West, and women who worked at places closely connected with a college; third, other Midwesterners, many of whom were graduate students who became high school science teachers; fourth, the California botanical explorers and collectors; fifth, United States Department of Agriculture botanists. The first two of these subdivisions (which account for 47 percent of the American authors), together produced about 56 percent of the American botany papers listed by the Royal Society (26 percent and 30 percent, respectively). The large grouping of nonfaculty Midwesterners produced about 23 percent, the Californians, who constituted only 10 percent of the authors, about 17 percent, and the government scientists about 4 percent.

Amateurs and noncollege-faculty women of the Northeast

This subdivision contains a considerable number of authors whose contributions to the botanical literature were very slight. One or two of them were recognized as teachers and educators (rather than botanists) and the stories of a few others are known as well, but many are hardly remembered. For the record, at least names and contributions are mentioned here. Among them are four of the earliest of the American women botanists whose publications are listed in the bibliography, Lucy Millington, Maria Owen, Mary Treat, and Mary Reynolds.[2] Apart from two California explorers and collectors, Rebecca Austin and Caroline Bingham, these four would appear to be the only American women botanists listed in the Royal Society *Catalogue* whose journal articles predate 1880.

LUCY BISHOP MILLINGTON[3] (1825–1900) was well-known in her native state of New York as a competent and observant self-taught naturalist who contributed several short but notable reports to the *New York Botanical Club Bulletin* during the 1870s and early 1880s. Her most important work was her discovery in Warren County, New York, of a parasitic plant that she identified as a mistletoe. Having heard of the ravages of a parasite of this kind on conifers in the Far West, she had set out to investigate what might be the cause of similar trouble in New York State. Her find of the previously unknown dwarf mistletoe, *Arceuthobium pusillum,* she reported to botanist Charles H. Peck of the New York State Museum of Natural History. Peck was then carrying out his own investigation, and within a few weeks of Millington's discovery collected specimens of the plant in other New York locations. He is the author of the species, but in his 1873 report he gave Millington credit for her original discovery.[4]

Born 10 June 1825 in the Adirondack community of New Russia, Essex County, in northeastern New York State, Lucy was the second of the eight children in the family of Lucius and Anne (Sheldon) Bishop. Lucius's father, Elijah Bishop, had founded the hamlet of New Russia in the early 1790s, when he moved west from Vermont. The family prospered; in addition to his farm, two sawmills, and a gristmill, Lucius kept an inn and had a whiskey distillery. His children were capable and talented. Three of them, Annette, Amy, and Bainbridge, became well-known in their time for their artistic work (the Adirondack region was then very popular with landscape painters).

Lucy probably attended the school maintained for the community by her grandfather, but for the most part she was self-taught. In 1846 she married Stokes Potter Millington, a clerk, merchant, and government worker from nearby Elizabethtown.

About ten years later they moved south to Glens Falls in Warren County, where Lucy lived for the next two decades, bringing up three children and carrying out a considerable amount of botanical exploration. She possessed a microscope, a copy of Gray's *Manual of the Botany of the Northern United States* (1848), and one or two other botanical texts. Fungi and ferns were her special interests. The first of her several botanical notes on the dwarf mistletoe appeared in print in 1871, when she was in her mid forties, and shortly after that she published a number of natural history articles in popular magazines. About the same time she began to contribute specimens to the New York State herbarium, a practice she kept up for much of the next three decades.

For a period of about five years, beginning in 1876, she lived in southwestern Michigan, her younger brother Midas Bishop being then in the process of establishing a medical practice there in the community of South Haven. She bought a peach orchard, and was joined by her husband and the younger of her two sons. Unfortunately the agricultural venture was not a success, the trees succumbing to the peach-yellow fungus disease. However, she made a notable botanical contact at that time, forming a friendship with Liberty Hyde Bailey, later a well-known horticulturalist and the first director of the Cornell University Agricultural Experiment Station. A boy of eighteen when Millington arrived in South Haven, Bailey was already much interested in plants, but had never had a chance to talk with a knowledgeable botanist. He sought her out for help with his studies, borrowed her textbooks (the first he had seen), and with her explored and collected in the pinewoods and over the sand dunes near Lake Michigan.

Millington was also interested in birds, and while in South Haven wrote a never-published article, "Birds of Western Michigan."[5]

After the family went back to New York about 1881 she resumed her local botanical exploring and collecting, latterly in the New Russia area to which she returned after the death of her husband. During the 1880s, and especially in the closing years of her life in the 1890s, she kept up a correspondence with Charles Peck, state botanist from 1883. Her rare or unusual finds were forwarded to him, and she sought his opinions on diseases affecting the Bishop-Millington fruit orchard. She died in New Russia, 17 January 1900, at the age of seventy-four. A memorial erected in 1929 in New Russia's public park commemorates the pioneering Bishop family and their efforts for the betterment of the community.

MARIA TALLANT OWEN[6] (1825–1913), Millington's contemporary, is remembered especially for her *Catalogue of Plants Growing without Cultivation in the County of Nantucket, Massachusetts* (1888). Developed from collections she made before 1853, this catalog has come to constitute a valuable early record of the flora (including marine algae) of Nantucket Island. It consists of a substantial annotated list of 787 species and varieties and one form, and provides a baseline against which can be measured subsequent changes in plant distribution resulting from environmental effects within this geographically distinct plant community. The work was painstaking and accurate, Mrs. Owen having regularly consulted specialists, often by correspondence, when the need arose.

She was born on Nantucket, 13 February 1825, the daughter of Nancy (Coffin) and Eben Weld Tallant. Her forebears on both sides were among New England's earliest settlers; through her mother she was related to both Benjamin Franklin and Daniel Webster. She was educated at home and in private schools on the island. Her mother and sisters, all able women, had strong interests in botany and encouraged her study of the subject from an early age. In the 1840s she spent some years teaching in Boston at the Perkins Institution for the Blind, and there met her future husband, Dr. Varillas L. Owen. She returned to Nantucket for a time, ran her own private school, and then taught in the Nantucket High School. In 1853, following her marriage to Owen, she moved to Springfield, Massachusetts, where, for the next fifty years she was remarkably active in cultural, intellectual, and botanical circles. She founded a Shakespeare society and the Springfield Women's Club, as well as the Connecticut Valley Botanical Society and the Springfield Botanical Society. She continued her teaching career, providing instruction in a variety of science subjects, and also French, in private schools in Springfield; Caroline Gray Soule (see chapter 2) was one of her students.

Her *Catalogue* was her principal botanical work. It first appeared in the form of an unannotated list as one section of Edward K. Godfrey's 365-page guidebook *The Island of Nantucket* (1882). She also published several notes on local flora, the last appearing in 1912.[7] Some of her herbarium specimens are still in existence in the Springfield Science Museum. She died in her eighty-ninth year, 8 June 1913, at the home of her daughter, in Plandome, Long Island, New York.

MARY DAVIS TREAT[8] (1830–1923) was by far the most productive of the American women botanists active in the 1870s, as measured by the number of articles she published. Of middle-class background, she was born in Trumansville, Tompkins County, New York, 7 September 1830, the daughter of an itinerant Methodist minister, Isaac M. Davis, and his wife, Eliza. In 1839 the family moved to Ohio. Mary attended public schools and also spent a brief period in an academy for girls. As a young woman she returned to New York State and lived for some time with a sister. In 1861 she married Dr. Joseph Burrel Treat, the son of a Yale-educated Presbyterian minister, an activist in the antislavery cause, lecturer on scientific topics, and a nurse during the Civil War.

After several years in Iowa the Treats moved to New Jersey, and in 1868 settled in the southern part of the state, in the recently founded community of Vineland. Mary, who was then thirty-eight, began her nature studies about that time, taking up her new interest with considerable energy and enthusiasm. Fruit growing was being encouraged in the region, and the attendant problems of insect control soon attracted her attention. Both she and her husband contributed a number of notes and papers to entomological magazines in the late 1860s. Joseph Treat appears to have rather quickly become bored with life in Vineland, however, and moved to New York City. His wife's interests in natural history continued and broadened, and within a year or two she was publishing accounts of her observations of the birds and plants of southern New Jersey. Her first paper on the pine barrens near Vineland appeared in 1870; it was followed by a long series of popular articles in the magazine *Garden and Forest* describing the successive seasonal stages in the plant life of these areas. Her ornithological notes for the most part appeared in such magazines as *Hearth and Home, Harper's,* and *Lippincott's.*

During the three winters of 1876 to 1878 she visited Florida to study the insectivorous plants of the region, particularly their methods of capturing and digesting their prey. Having discovered through Asa Gray of Harvard that Charles Darwin was engaged in similar studies, she began an extensive correspondence with him. Both Gray and Darwin in their writings acknowledged her contributions.[9] While in Florida she spent some time canoeing along the St. Johns River, and there discovered a species of Zephyr lily, named *Zephyranthes treatae* by Harvard botanist Sereno Watson. In addition she found a yellow water lily that was considered extinct by the experts.

Between 1877 and 1890 she brought out a number of books, including *Chapters on Ants* (1877), *Injurious Insects of the Farm and Garden* (1882), *Home Studies in Nature* (1885), *Through a Microscope* (a study of swamp-water microorganisms, coauthored with Samuel Wells and Frederick Leroy Sargent, 1886), *My Garden Pets* (1887), and a critical biography of Asa Gray, *Asa Gray: his Life and Work* (1890). *Injurious Insects,* her principle work in entomology, went through five editions, a chapter on beneficial insects being added in 1887, followed by further additions to three later editions. An attractive volume of 288 pages, illustrated with 163 figures, it was compiled largely from the reports of the early state and federal entomologists, several of whom had published outstanding work over the preceding two decades. In particular, Treat used the 1879 report of John Comstock, entomologist for the United States Department of Agriculture from 1879 until 1881, and the *Annual Reports* of Charles Riley, state entomologist of Missouri from 1868 until 1877. She showed excellent judgment in her choice of materials; Riley's reports, which were quickly recognized as exceptional by entomologists throughout the United States and Europe, were the works that established him as perhaps the foremost entomologist of his time in America. Whole pages of his writings appear verbatim, without quotation marks, in Mrs. Treat's book, and although both she and her editor acknowledged the use of Riley's writings in prefaces, later commentators have wondered why the authorship was not stated as joint. Illustrations were also taken from Riley, without mention of their origin.[10]

Home Studies in Nature, Treat's other book that reached a wide audience, presented material developed from her previously published articles in *Harper's Magazine* and other periodicals. Particularly prominent were her observations on ant behavior and the family life of birds, made in the immediate neighborhood of her home. She gave detailed accounts of her procedures and methods, and all were presented in simple, reassuringly familiar, anthropomorphic terms; undoubtedly the work was intended mainly for women and young people.

Although regarded as being of a rather retiring temperament, from the time she began her nature studies in the late 1860s she nevertheless carried on an extensive correspondence with a great many botanists, entomologists, and nature writers, European as well as American. In addition to Darwin, Gray, and Riley, her contacts included Benjamin Walsh, Illinois state entomologist, Charles Sargent, director of the Arnold Arboretum near Boston, and Theodore Flood, one of the editors of *The Chautauquan* magazine. Among her European corespondents were Gustav Mayr in Vienna and the Swiss naturalist Auguste Forel. Forel named the ant *Aphaenogaster treatae* in her honor, and Mayr named the cynipid oak fig root gall *Belonocnema treatae;* she had discovered the latter on *Quercus virginiana* in Florida. Clearly her careful and accurate observational work was recognized by her contemporaries, and despite the fact that she wrote largely for a popular audience she has an honorable place among the early naturalists of New Jersey.

She lived in Vineland for more than forty-five years. For a time after the death of her husband in 1879 she occupied one half of a double house shared with the C. B. Campbell family. Mrs. Campbell illustrated many of her writings. It was here that she planted a 100-foot-diameter circle of arbor vitae trees, within whose shelter she established the "insect menagerie" where she carried out much of her observational work on insects and spiders. She also had a nearby house called The Cedars, her permanent home, with spacious, carefully planned, and well-tended grounds. She appears to have been comfortably well-off, and throughout the years earned modest sums by her writing; her books yielded royalties, and her magazine articles brought (at least in the 1870s) sums that varied from $7 to $50.[11] Many of the articles were supplied to editors following requests.

A small, cheerful, but not very robust person, she centered her life in her home and community, with occasional traveling; in addition to her three winters in Florida, she made a visit to California. In old age she was severely crippled by a fall; for a time she was cared for by a sister, but the Vineland property was eventually sold and Mrs. Treat was taken to Akron, New York, where she lived with relatives for her remaining years. She died 11 April 1923, in her ninety-third year, and was buried in Siloam Cemetery, Vineland.

By the 1880s and nineties the number of women amateurs making contributions to botanical knowledge in the northeastern United States had increased considerably. A notable center of activity was the Botanical Section of the Rochester Academy of Science, in western New York State. The section was formed in 1881 under the leadership of Mrs. Mary Bosworth Streeter. Its purpose, as then stated, was the systematic study of botany—in particular the collection and identification of plants indigenous to Rochester and the vicinity, and the publication of a complete list of the flora of Monroe County.[12]

FLORENCE BECKWITH[13] (1843–1929), a coauthor of the first list brought out, was one of the eleven charter members of the section, its vice president from 1886 to 1897, and its president for thirty-two years, from 1897 until her death in 1929.

She was also a fellow of the academy. Born in Scottsville, New York, on 7 December 1843, the daughter of Francis Beckwith and his wife Hannah (Goodhue), Florence was descended from New England colonists who had arrived in America in 1635. Her intended husband having been killed in the Civil War, she supported herself through life as editor of *Vick's Illustrated Monthly Magazine,* a popular gardening publication put out by James Vick of Rochester, an English immigrant who had established a seed business in 1860.

Florence was one of the especially energetic members of the academy and added substantially to its pressed-plant herbarium. A careful collector, she made several productive botanical trips to California and Colorado; much of the academy's section on plants from the western United States was donated by her. After retiring she made the work of identifying, mounting, and organizing the herbarium specimens her full-time occupation. She was one of America's first activists in the nature conservation movement, and as early as 1913 took a lead in the effort to establish a wildlife sanctuary at Bergen Swamp, near Rochester. Her most important publication was the 150-page catalog "Plants of Monroe County, N.Y., and adjacent territory," which she coauthored with Mary Macauley and published in the academy's *Proceedings* for 1896. Supplementary lists appeared in 1910 and 1917.[14] She died in Scottsville, New York, on 9 August 1929, at the age of eighty-six.

MARY MACAULEY[15] (1851–1932), Beckwith's close friend over many years, was another of the founding members of the Botanical Section of the Rochester Academy, section secretary for two years and president from 1886 until 1897, the period just before Beckwith accepted the office. She was also a fellow of the academy. The daughter of Robert Macauley and his wife, Jane, immigrants from Northern Ireland, she was born in Rochester and attended Rochester public schools, graduating from the Free Academy in 1872. For the next twenty-one years she taught English and science in public and private schools around Rochester, botany, which in her classes included fieldwork, being her favorite subject. In 1893, at the age of forty-two, she went to Cornell University and graduated with honors five years later. Her coauthored catalog of the plants of Monroe County and the later supplementary lists were probably her only publications, although they represent a creditable amount of botanical work. Like Beckwith, she made important additions to the academy's herbarium, bringing back large collections from frequent collecting trips she made to California, Colorado, Florida, Manitoba, and Tennessee. Failing health and impaired eyesight caused her to give up botanical work in her later years. She died in 1932, the last of the charter members of the academy's Botanical Section.

ANNA H. SEARING[16] of Rochester, New York, who became a corresponding member of the Rochester Academy in 1895, was also active in the Botanical Section and collected material for the herbarium in western and southern regions of the United States. She was one of several botanists who carried out important work in the exploration of the flora of the shore of Lake Ontario. Her short descriptive and observational contributions continued to appear in the academy's *Proceedings* after the turn of the century, and her work on ferns and fungi was of particular interest.

One of the better remembered of the women amateurs active in the northeastern United States in the 1880s is the botanical artist KATE FURBISH[17] (1834–1931). Her name came to the fore in 1976 with the rediscovery of the Furbish lousewort, *Pedicularis furbishiae,* a plant she had discovered ninety years before, but which had subsequently been supposed extinct. The rediscovery became especially memorable because the area where the plant was found, along the banks of the St. John River in northern Maine, was scheduled to be flooded by the Dickey-Lincoln hydroelectric project; completion of the project without a clearly successful transplanting operation therefore meant running foul of the 1973 Endangered Species Act. However, exhaustive searching, with the vigorous assistance of members of the U.S. Army Corps of Engineers, which was to build the dam, finally located more specimens of *Pedicularis furbishiae* in areas downstream from the proposed dam (which in the end was not built).

Kate was born 19 May 1834, in Exeter, New Hampshire, the oldest of the six children and the only daughter of Mary (Lane), of Exeter, and Benjamin Furbish of Wells, Maine. When she was about a year old the family moved to Brunswick, Maine, where they kept a very successful hardware store. A strikingly beautiful girl, she was educated in local schools and later took drawing lessons in Portland and Boston. She also spent a short time in Paris studying art and French literature. Her interest in botany started in childhood. At the age of twelve, with her father's help, she had begun to study native plants, and in 1860, when she was about twenty-six, attended a series of lectures given in Boston by George L. Goodale, later professor of botany at Harvard. Shortly after that, her interest in the subject having considerably deepened, she began what was to become her life's work—collecting, classifying, and recording in watercolor the flora of Maine, a project that has its place among the basic exploratory undertakings then a major part of botanical science in the United States.

After her father's death in 1873, she had adequate independent means, and for the next thirty-five years she traveled the state, county by county, in search of specimens. Her richest finds, which included many plants previously unrecorded in the state, were made during the summers of 1880 and 1881 in the bogs and streams of the then largely untouched wilderness of Aroostook County in northeastern Maine. She wandered alone for the most part, on foot (at times on hands and knees), in rowboats, and on her own improvised rafts. Her work was much respected by the leading professional botanists of her time, including Asa Gray and Sereno Watson of Harvard. Two of her discoveries were named after her, *Aster cordifolius L.,* var. *furbishae* and *Pedicularis furbishiae.* Preferring the artistic medium to the written word, she wrote few journal articles.[18] The major record of her work, her *Illustrated Flora of Maine,* sixteen large folio volumes of about 1,000 exceptionally detailed, delicate, and scientifically accurate watercolor paintings, she presented to Bowdoin College, Maine, in 1908. There they did little more than gather dust for almost seventy years, until discussion of the Furbish lousewort in 1976 aroused public interest and made known the existence of this valuable nineteenth-century botanical-artistic record. Her impressive collection of 4,000 sheets of pressed plants is now housed in the Gray Herbarium at Harvard, and her 182-sheet collection of ferns in the holdings of the Portland Society of Natural History. She was the founder of the Josselyn Botanical Society of Maine, served as its president for a year (1911–1912), and attended its meetings until she was quite old. Her collecting continued until a few years before her death, which occurred 6 December 1931, in Brunswick, Maine. She was ninety-seven.

Six other amateur women botanists from the northeastern region of the United States whose work appeared in the journal literature of the 1880s and 90s are Bessie Putnam, Charlotte Horner, Harriett Paine, Mary Plumer, Mary Saunders, and Martha Flint. Putnam was a regular contributor to botanical and agricultural journals, natural history magazines, and women's periodicals. She also edited sections of educational publications. Born 2 August 1859 in Harmonsburg, Pennsylvania, the daughter of Levi and Elizabeth (Whiting) Putnam, she received an A.M. degree in 1888 from Allegheny College, Meadville, Pennsylvania. Her articles continued to appear until at least the 1920s. Horner, Paine, and Plumer all contributed articles to the *Essex Institute Bulletin* during the 1880s, Horner and Paine on the flora of Massachusetts, Plumer a long article on the dissemination of seeds. Saunders's paper on the flora of colonial days appeared in the same journal in 1895. Horner (1823–1906) was active in the work of the Massachusetts Horticultural Society, which she joined in 1870. Especially noteworthy were her exhibits of the native flora of the state, shown at the society's exhibitions. As a member of the Window Gardening Committee, she encouraged children in growing plants and flowers. She was born in West Boxford, Massachusetts, 5 July 1823 and died in Georgetown, Massachusetts, 18 July 1906. Martha Flint, of Amenia, in eastern New York State, published one paper in the *Botanical Gazette* in 1885. Born in Poughkeepsie, 8 August 1841, she died in 1900.[19]

Included in the grouping of northeastern noncollege-faculty women botanists are three (Keller, Schively, and Williams Wilson) who all taught in girls' schools in Philadelphia and all had notable careers.

IDA KELLER[20] (1866–1932) was a well-known figure in Philadelphia botanical circles. As well as the eight papers listed in the bibliography reporting her original work, she wrote two monographs and a laboratory manual. Her botanical contributions were comparable in quality and quantity to those of several of the college-faculty botanists. The daughter of a Philadelphia physician, William Charles Christian Keller, and his wife Maria Augusta (Cramer), she was born in Darmstadt, Germany, on 11 June 1866, while her parents were there visiting their former home. Following her graduation from the Philadelphia High School for Girls in 1884, she spent two years at the University of Pennsylvania and then a year as an assistant in the herbarium at Bryn Mawr College. She completed her studies in Europe, at the Universities of Leipzig (1887–1889) and Zurich (1889–1890, Ph.D. 1890), and returned to Bryn Mawr as a lecturer. After three years, however, she joined the staff of her old high school and remained there until her retirement in 1930. Following five years as a teacher of chemistry (1893–1898), she served thirty-two years as head of the departments of chemistry and biology. Indeed, both she and her chemist brother, Harry Frederick Keller, had notable careers at about the same time in administration and science teaching in the Philadelphia public schools, Harry joining the staff of the Philadelphia Central High School in 1892 and becoming head of the science departments in 1908. Ida was well regarded by her colleagues, and in 1895 served as vice president of the Association of Colleges and Preparatory Schools of the Middle States.

At a time when women were still being denied membership in many scientific societies, she succeeded in joining several, including the rather exclusive American Society of Naturalists, whose only other female member, the botanist Emily Gregory, also had a Zurich doctorate. Keller belonged to the Philadelphia Academy of Natural Sciences and the Philadelphia Botanical Club, to which she was elected in 1892, the second year of its existence. For many years she was the only woman member of this club; she had the further distinction of being its vice president in 1900.

Her research interests covered various topics in plant morphology and flower fertilization; most of her work appeared in the *Proceedings of the Academy of Natural Sciences of Philadelphia* in the mid to late 1890s, her early years on the staff of the Philadelphia High School for Girls. Her *Handbook of the Flora of Philadelphia and Vicinity* (1905), coauthored by Stewartson Brown, curator of plants at the Philadelphia Academy of Sciences, was published under the auspices of the Philadelphia Botanical Club. A second monograph, *Insects of Philadelphia and Vicinity,* appeared in 1910. She also brought out a chemistry laboratory manual, *Exercises in Chemistry arranged for the Chemical Laboratory of the Girls' High School of Philadelphia* (1894). She died on 10 September 1932, at the age of seventy-six, at her summer home at Aquetong, Bucks County, Pennsylvania.

ADELINE SCHIVELY[21] received a Certificate of Proficiency in biology from the University of Pennsylvania in 1892. Though women were then ineligible for undergraduate degrees, she continued her studies at the graduate level. Her research included a lengthy life history study on *Amphicarpa monoica* (the hog peanut vine). She was awarded a Ph.D. in 1897 and for a time held an honorary fellowship at the university's botanical laboratory. Her later work included morphological studies carried out in collaboration with Ethel Cooke.[22] She taught biology at the Philadelphia Normal School for Girls.

LUCY WILLIAMS WILSON[23] (1864–1937), like Keller, was a prominent figure in the Philadelphia public school system for several decades beginning in the 1890s. Born in St. Albans, Vermont, on 18 August 1864, she was the daughter of Lucy (Crampton) and Samuel Williams, descendent of seventeenth-century English settlers. From 1881, following study at state normal schools in Vermont and Philadelphia, she taught at various institutions in Philadelphia. At the same time, she proceeded with graduate studies at the University of Pennsylvania (Ph.D., 1897); her dissertation research on *Conopholis americana,* the American squawroot, was published the following year (*University of Pennsylvania Publications*). Meanwhile, in 1893, she had married William Powell Wilson, then a faculty member in the botany department at the University of Pennsylvania; they had one son.

Her succession of administrative positions in the Philadelphia school system began in 1893, when she accepted the chairmanship of the departments of biology and geography at the Philadelphia Normal School for Girls, a post she held until 1916. In 1900 she became principal of the Evening High School for Women, and for eighteen years, from 1916 to 1934, she was principal of the South Philadelphia High School for Girls.

From 1891 onward she published a succession of teaching manuals and textbooks, including a number on geography and natural history; many went through several editions. A strong advocate of the Dalton Plan as a teaching method,[24] she made it the trademark of the South Philadelphia High School. The plan largely replaced daily lessons with weeklong (or longer) assignments; these resembled small research projects patterned after university-level work, and were tailored to the individual. Her own research training may well have been the basis of her enthusiasm for the method. She made several visits to Europe, speaking at educational conferences and examining new teaching methods being developed there.

The work brought her wide recognition, including the award of a gold medal at an international exposition in Paris in 1900. As an educational consultant to the Chilean government, she visited South America in 1929. Her monograph, *Chile's New Educational Program,* appeared in 1930; *The New Schools of New Russia,* published as part of Vanguard Press studies of Soviet Russia, came out in 1928, with a Spanish translation following in 1931. Philadelphia recognized her services to its educational institutions in 1934 when it made her the first recipient of its Bok award, a gold medal, and $10,000. For two years from 1934 to 1936, when over seventy, she lectured on education at Temple University in Philadelphia. She died on 3 September 1937, at the age of seventy-three, in Lake Placid, New York.

Five other women science teachers from the northeastern region of the United States, Arnold, Andrews, Hewins, Simons, and Boynton, made modest contributions to the nineteenth-century botanical literature. Each wrote one or two papers, typically reports of undergraduate studies. None of the five went on to a doctoral degree, although Hewins and Boynton both studied at the graduate level.

Isabel Arnold,[25] from Painted Post, New York, studied at Plainfield Seminary, New Jersey, and for four years at Radcliffe College. She then returned to Plainfield as a teacher in the seminary and served as head of the institution from 1906 until it closed in 1919. Her one botanical paper, published in the *Bulletin of the Torrey Botanical Club* in 1888, concerned the flora of her native New York state.

Florence Andrews,[26] the daughter of Mary Amanda (St. John) and Charles Dennison Andrews, was born in Elba, New York, on 25 July 1877. She studied at Cary Collegiate Seminary, Oakfield, New York, and then at Middlebury College, Vermont (A.B., 1900). For ten years, until her marriage to Lucius Atwater in 1910, she taught science in various high schools in New York State. Her one botanical article, published in *Rhodora* in 1900, reported an undergraduate study of the flora around Middlebury.

Nellie Hewins[27] was born on 20 January 1878, in Maspeth, Long Island, New York, and graduated from the Brooklyn Girls' High School in 1895. She then attended Cornell University (B.S., 1898), and two years later took an M.A. at Columbia University. From 1901 onward she taught biology at Newtown High School in New York City, while also continuing graduate work for several years at Columbia and then at New York University. She was especially interested in plant embryology.

Elizabeth Simons[28] (b. 1873) studied biology at the University of Pennsylvania in the late 1890s and carried out her botanical work under the supervision of John Macfarlane, director of the university's botanical garden. She later taught at the Philadelphia High School for Girls where she was a colleague of Ida Keller. She married Eldred E. Jungerich.

Margaret Boynton,[29] the daughter of Mary Whipple (Harwood) and Thomas Cabot Boynton, was born in Lockport, New York, on 1 January 1872. After graduating from Lockport public high school as class valedictorian in 1890 she went to Cornell University (Ph.B., 1895). She then taught science for a year at Northfield Seminary in Massachusetts before returning to Cornell on a graduate fellowship. In 1899 she became assistant to the state entomologist in Albany, New York, but resigned two and a half years later when she married Phineas Windsor. They had three daughters. The family later lived in Urbana, Illinois. Her 1895 paper in the *Botanical Gazette,* on methods of seed dispersal, reported her undergraduate research at Cornell. After her marriage, she limited her scientific work to contributing articles on nature study to various popular periodicals. She was a member of the Illinois Academy of Science.

Botanists on college and university faculties in the east and Middle West and at associated research institutions

Of the eighteen women in this subgroup, twelve taught at eastern women's colleges, three at coeducational institutions in the Midwest, two were closely associated with Columbia University and the New York Botanical Garden, and one worked at the Gray Herbarium at Harvard.

Elizabeth Knight Britton and Anna Murray Vail, the two who worked at Columbia University and the New York Botanical Garden, were the most productive as measured by the number of papers they published in botanical journals. Emily Gregory of Barnard College, New York, and Margaret Ferguson of Wellesley College are also especially noteworthy, although Ferguson, whose research was just beginning around the turn of the century, might well be thought of more as an early twentieth-century botanist than one of the nineteenth, and Gregory had only eleven years of botanical work before her premature death. Both Ferguson and Gregory are distinguished by having interests in the newer branches of the field—plant anatomy, physiology, cytology, and (in Ferguson's case) genetics. Britton and Vail, on the other hand, largely followed the older tradition, and, except for later work in genetics by Vail, published mainly in taxonomy. Four of the eighteen in the subgroup specialized in various areas of cryptogamic botany; in addition to Britton, who became an acknowledged authority in bryology, these were Clara Cummings of Wellesley College, Julia Snow of the University of Michigan and Smith College, and Josephine Tilden of the University of Minnesota, each of whom made notable contributions.

Although the careers of several of the academic women botanists are fairly well-known, others remain quite obscure. Among the almost forgotten is ARMA ANNA SMITH,[30] of Mount Holyoke Seminary, who combined her interest in science with missionary work and spent most of her short teaching career on the staff of the American College for Girls in Scutari, Turkey. She was born on a farm in West Camden, New York, on 12 October 1866, the daughter of Lucy Amanda (Munson) and Samuel Lewis Smith. After attending local schools and the high school in Watertown, New York, she took the classical course at Mount Holyoke, and immediately after graduating in 1891 sailed for Turkey. At the American College for Girls, with which Mount Holyoke then maintained a close relationship, she was mainly responsible for teaching science in various areas from botany to physics, but throughout her three years there taught a Bible class as well. She took great satisfaction in pointing out what she considered the benefits of Christianity to her multiracial audience of girls of many faiths.[31]

Like several American women science students of the period who were able to go to Europe, Smith took advantage of her transatlantic sojourn to spend some time at the University of Zurich. She returned to Mount Holyoke in 1894 and taught Latin and mathematics for one year before going to Cornell, where she studied under George Atkinson, head of the botany department. Her two papers on plant morphology appeared in the *Botanical Gazette* in 1896, the year she received an M.S. degree. Though she went back to her teaching post at Mount Holyoke for a short time in 1897, she left less than a year later to return to her farm home in West Camden to care for her aging father and a sister in poor health. Writing from the relative isolation of the farm six years later, when she was still in her thirties, she remarked that her three years in Scutari were already beginning to seem like a dream.[32] There is no indication that she ever returned to scientific work or teaching.

Two early Mount Holyoke women who made modest contributions to the pre-1900 botanical literature and who are better remembered than Arma Smith are Henrietta Hooker and Alice Carter. These two also have the distinction of being the first women to be awarded Ph.D. degrees in botany by an American university.[33]

HENRIETTA HOOKER[34] (1851–1929) joined the teaching staff at Mount Holyoke Seminary in 1873, and as head of the botany department for twenty years was especially influential in its early development, pushing for the introduction of courses in the newer branches of the field and working for expansion of laboratory space and the acquisition of equipment. She was born in the small town of Gardiner, Maine, on 12 December 1851, the daughter of Eliza Annie (Balletine) and George Washington Hooker. Left an orphan at the age of seven, she spent her childhood with several different relatives, and at sixteen went to work in one of the New England cotton factories. One week of this experience was enough to make her seek the help of her minister in looking for an alternative job. She succeeded in finding a teaching position, first in a district school and then at an academy in West Charleston. Two years later she entered Mount Holyoke, from which she graduated in 1873.

She was invited to stay on as a teacher, first of mathematics and then botany, and she remained at the college for the next thirty-five years, with breaks for graduate study at MIT and the universities of Syracuse, Berlin, and Chicago. Summers at the Woods Hole Marine Biological Laboratory extended her research experience. She received a Ph.D. from Syracuse in 1889, having spent part of that year at the University of Berlin, where she was one of the earliest women to gain access to instruction, albeit informal. Shortly after her return she became head of the botany department at Mount Holyoke, following Lydia Shattuck, who had held the position for the preceding forty-eight years.

Hooker had considerable administrative skill and financial and executive ability; after the loss of the herbarium and plant house in a fire in 1896, she raised much of the money required for building the new facilities. She became a full professor in 1904, when, for the first time, the college's faculty was given academic rank. Her graduate research was on the morphology and embryology of *Cuscuta,* but she does not appear to have published original work in botany after her student days.

A vigorous person with wide interests, she took a prominent part, over a number of years, in organizing and conducting European summer tours for students and teachers, undertakings that must have required considerably more initiative and planning than their equivalent at the present time. She particularly enjoyed northern routes through Norway and Sweden. In 1908, at the age of fifty-seven, she retired from teaching and embarked on projects that she had always wanted to carry out, including building houses and breeding prizewinning Buff Orpington chickens. She was one of the first to import this breed from Britain, and collected many first prizes at poultry shows throughout the country. Also active in local and state politics, she had the distinction of being the first woman delegate elected from South Hadley to a state convention. Mount Holyoke gave her an honorary Sc.D. in 1923, the fiftieth reunion of her class. She died in South Hadley, 13 May 1929, at age seventy-seven.

Hooker's colleague for several years in the late 1880s and her fellow graduate student at Syracuse University was ALICE

CARTER.[35] Born in New York City, on 8 April 1868, Alice was the daughter of Alantha (Pratt) and Samuel F. Carter. She received her undergraduate education at Mount Holyoke Seminary and then took a Ph.D. in botany at Syracuse. Her degree, awarded in 1888, one year before Henrietta Hooker's, would appear to have been the first botany Ph.D. given to a woman by an American university.[36] Following three years of teaching at Mount Holyoke, she went to Cornell University on an American Association of University Women fellowship and received a second graduate degree (M.S. in botany) in 1892. Her thesis research on pollination processes was reported in two papers in the *Botanical Gazette* the same year.

Following her marriage in 1892 to Orator Fuller Cook, a plant scientist whom she had met when they were both students at Syracuse, Alice transferred her botanical interests to work she could more conveniently carry out, namely assisting her husband in his studies on tropical plants for the U.S. Department of Agriculture. She accompanied him on field trips to the Canary Islands and Africa, and the plant collections she made on these expeditions are now in the National Museum, Washington, D.C. She brought out several articles of her own in *Popular Science Monthly* and in botanical journals, including papers on Canary Islands flora and Philippine flora.[37] Her ethnological paper "The aborigines of the Canary Islands" was published in the *American Anthropologist* in 1900. Among her later writings were poems and short stories, which appeared in periodicals, and two plays.[38] She had two sons and two daughters.

At Smith College the hiring of GRACE CHESTER[39] in 1890 marked the establishment of a separate department of botany, and until 1896, when she resigned to spend a year at the University of Berlin, Chester was the department's sole instructor. Under her direction the college herbarium and botanical library were started in 1892.

Born in Sandusky, Ohio, on 11 February 1867, she graduated (B.S.) from Delaware College, Newark, Delaware, in 1885. After a year as a science teacher at Hood College in Frederick, Maryland, she studied botany for two years at Radcliffe College, spending summers at the Marine Biological Laboratory, Woods Hole, Massachusetts. She then taught high school science in Middletown, Connecticut, for two years before joining the Smith faculty. Her 1896 paper in the *Botanical Gazette* described research in plant anatomy carried out at Woods Hole in 1893; a second major paper, also reporting studies in anatomy, appeared in the German Botanical Society's *Berichte* in 1897, following her year at the University of Berlin. She taught biology and teaching methods at the State Normal School in Lowell, Massachusetts, for two years after her return from Germany, but gave up professional work in 1900 when she married George Gow, professor of music at Vassar College (and earlier an instructor at Smith). She had one daughter, Serena Anna.

A second early Smith College faculty member who contributed to the botanical literature was FRANCES SMITH (1870–1948).[40] Smith spent much of her life at the college, first as a student and then as a teacher. Born in Springfield, Massachusetts, on 27 December 1870, she was the daughter of Mary (Stebbins) and George B. Smith. After graduating (A.B., 1893) with a major in classics, she taught high school Latin for two years in Warren, Massachusetts. Her interests turned to botany, however, and she returned to Smith College and took an A.M. in 1900. Immediately after graduating she joined the teaching staff of the botany department as assistant, rising to associate professor by 1911 and professor by 1928. She received a Ph.D. in 1906, following two years of graduate study at the Hull Botanical Laboratory at the University of Chicago (1903–1905). She retired as professor emeritus in 1937.

Although her earliest botanical work, reported in 1899, concerned the distribution of red coloration in New England plants, her major research interest, and the subject of her doctoral dissertation studies, was the morphology and anatomy of the cycads. These members of ancient gymnosperm groups, which look somewhat like palm trees with thick stems and a crown of large compound leaves, have been described as "leftovers" of populations from earlier eras. Smith's two lengthy papers in the *Botanical Gazette* (1907 and 1910) reported work on the morphology of the trunk and on the development of spore-producing and ovule-bearing bodies. Her studies took her frequently to the Tropics, and she often spent her summers traveling in the West Indies, Central America, and along the coast of South America. Part of her sabbatical year, 1928–1929, she spent at the Bishop Museum in Hawaii on an honorary fellowship from Yale University. Her study of Hawaiian ferns was published in 1934.[41] She was a fellow of the AAAS and a member of the Botanical Society of America. Support for the education of Chinese students was one of her special interests; following her death at age seventy-seven (25 May 1948, at Northampton, Massachusetts) a substantial bequest from her estate passed to Ginling University in Nanking.

JULIA SNOW[42] (1863–1927) was a colleague of Frances Smith for twenty-six years, although her most important research, her study of the algae and plankton of Lake Erie, was done before she joined the faculty of Smith College. Born in La Salle, Illinois, on 30 August 1863, she was the daughter of Charlotte (Warner) and Norman Guitio Snow. She studied at the Hungerford Collegiate Institute in Adams, New York, and then at Cornell University (B.S., 1888; M.S., 1889). Following two years as a teacher of botany in preparatory schools in Wisconsin and Indiana, she went on to further study at the University of Zurich on fellowships from the American Association of University Women and the Women's Education Association. She carried out research on algae growth, and was awarded a Zurich Ph.D. in 1893. The next year she joined the faculty of the American College for Girls in Scutari, succeeding Arma Anna Smith of Mount Holyoke. After a summer of further research training at the University of Göttingen in 1895 and a postdoctoral year at Basle University (1896–1897), where she continued her work on algae culturing, she made her way home—the long way round, via Russia and the Orient. The trip was a somewhat unusual adventure for a young woman from Illinois to undertake on her own in the 1890s.

On her return to the United States she became an assistant in botany at the University of Michigan, and after a year was appointed a temporary instructor. She also carried out, over the course of four summers, a microscopical study of the plankton and freshwater algae of Lake Erie, part of a large-scale project in aquatic ecology then under way at the University of Michigan. The project was one of several started in the Midwest shortly before the turn of the century that were of considerable importance in establishing aquatic ecology as an academic discipline in North America. The work at Michigan was directed by Jacob Reighard, professor of zoology, a morphologist who leaned strongly toward the European academic tradition. He had government support in the form of grants provided by the U.S. Fish Commission, a body that generally focused on fairly immediate practical goals, and with which, therefore, he did not always see eye to eye. He had studied the plankton of Lake St. Clair in 1893, that of Lake Michigan in 1894, and planned to continue, with an examination of that of Lake Erie, in 1898–1902.

With her extensive European experience in algae research behind her, Snow, Reighard felt, was uniquely qualified to collaborate in the kind of basic exploratory work he was doing. She carried out extensive studies on the nutrition of algae, experimenting with pure cultures to establish the most important nutrients for each species. She also determined life histories of selected species and examined the roles, beyond that of simply being part of the food chain, that algae played in the lake's ecological balance; one of her findings was that the blue-green algae thrived on the decay products of plankton, which suggested that these algae could convert the waste products of decay into food for crustacea and other zooplankton. Her results were published in the U.S. Fish Commission bulletin of 1902.[43] Much of her methodology and her laboratory modeling was based on similar work then being carried out in Europe about the same time, and her findings were not directly applicable to the immediate practical concerns of the U.S. Fish Commission. Therefore, by 1900 the latter declined to give the work further support, and the $500 per year needed for Snow to continue her laboratory research was not forthcoming.[44]

She spent one more year in the Midwest as head of the department of biology at Rockford College in Illinois, and then in 1901, at the age of thirty-eight, accepted a position at Smith College. She appears to have done no more research, and although she rose from an assistantship to associate professor by 1905 was not promoted further. A quiet and retiring woman, she nonetheless had a particularly broad range of interests and never lost the love of travel that had prompted her round-the-world trip in 1897. On a second such trip she visited unfrequented places in China and India, where the regional art and architecture was her special interest. She died in Northampton, Massachusetts, 24 October 1927, at age sixty-four.

Perhaps the best-remembered of the early women botanists from the northeastern colleges is MARGARET FERGUSON[45] (1863–1951) of Wellesley. An able scientist, she made the most of the opportunities for department building and independent research that her position at Wellesley put within her grasp—opportunities that were few and far between for women of her time.[46] She was born in Orleans, New York, on 20 August 1863, the fourth of the six children of Hannah Maria (Warner) and Robert Bell Ferguson. Robert Ferguson, a farmer in the township of Phelps, New York, was of Scottish descent. From the age of fourteen, Margaret taught in the local public schools, continuing her studies at the same time. In 1885 she graduated from the Genessee Wesleyan Seminary, in Lima, New York, and two years later, at the age of twenty-four, became an assistant school principal. For three years (1888–1891) she studied botany and chemistry as a special student at Wellesley, and then went to Gambier, Ohio, to a position as head of the science department of Harcourt Place Seminary. At the invitation of Susan Hallowell, head of Wellesley's botany department, she returned to Wellesley in 1894 as a member of the teaching staff; she remained at the college for forty-four years.

Following a leave for travel in Europe (1896–1897), she completed her formal education at Cornell University (B.S., 1899; Ph.D., 1901). Her doctoral research on the life history of the white pine (*Pinus strobus*), a particularly detailed and painstaking piece of work, set standards for future plant life history studies. In 1903 she entered an expanded version of her dissertation in a competition set by the Naples Table Association for the best thesis by a woman reporting original scientific work. The prize went to Florence Sabin of the Johns Hopkins Medical School, but Ferguson's entry received honorable mention and was published in the 1904 *Proceedings of the Washington Academy of Sciences.*[47]

She rose rapidly through the academic ranks at Wellesley, becoming associate professor in 1904, and professor and head of the botany department in 1906. She was then forty-three. Her mentor, Susan Hallowell, during her twenty-seven-year tenure as head of the botany department, had always placed strong emphasis on newer teaching methods, including laboratory work, and by so doing had laid the foundations for what was to become one of Wellesley's strongest departments. Under Ferguson's energetic leadership it was further modernized and invigorated, and both the physical facilities and the curriculum were expanded. Planning and fund-raising for what later became the Margaret C. Ferguson Greenhouses depended heavily on her efforts. Her insistence on the necessity of a solid background in chemistry and physics as a prerequisite for botanical research, her emphasis on laboratory work, and her strong commitment to the newer areas of botany—plant physiology, functional morphology, and cytology—made Wellesley one of the best centers in the United States for undergraduate training in botany.[48]

As well as her early work on the physiology of spore germination and her life history studies, Ferguson's investigations concerning Mendelian inheritance in the higher plants were of special note. Much of her work was done on *Petunia.*[49] Her star in the 1910 edition of *American Men of Science,* her election to the vice presidency of the American Microscopical Society in 1914, and to the presidency of the Botanical Society of America in 1929 (she was the first woman to hold that office) attest to her contemporaries' regard for her work. Mount Holyoke College gave her an honorary Sc.D. at its centennial celebrations in 1937. She was a fellow of the AAAS, and in 1943 became a fellow of the New York Academy of Sciences. In 1930, when she was sixty-seven, she was made research professor at Wellesley in recognition of her outstanding contributions to her department.[50] Her studies continued until 1938, when she retired to Seneca Castle, New York. She died of a heart attack in San Diego, California, 28 August 1951, at age eighty-eight.

Notable work in botanical research had been done at Wellesley College before Ferguson's time. CLARA CUMMINGS[51] (1855–1906), whose thirty-year association with Wellesley almost coincided with the first three decades of the existence of the college, was well known for her taxonomic and systematic work on lichens, those organisms formed by a fungus and an alga existing together in symbiotic relationship.

She was born in Plymouth, New Hampshire, on 13 July 1855, the daughter of Elmira and Noah Cummings. Before entering Wellesley in 1876, one year after the college opened, she studied at the New Hampshire State Normal School in Plymouth. She spent one year at the University of Zurich, from 1886 to 1887, but apart from that remained at Wellesley throughout her whole career. Her work on the identification of cryptogamic flora had begun while she was still a student and her talents in the area were early evident. She never took an academic degree, but nevertheless, starting in 1878 as instructor and curator of the Wellesley museum, she rose through the ranks, becoming associate professor in 1886. In 1906, shortly before her death, she was made Hunnewell professor of cryptogamic botany, partly in recognition of the highly specialized work in which she had by then gained an international reputation, and partly to reduce the strain of administrative duties on her deteriorating health.

Her first published work (on mosses and liverworts) was the *Catalogue of Musci and Hepaticae of North America, North of Mexico* (1885), but many of her investigations appeared as appendixes to the publications of other botanists.[52] As chief editor of *Decades of North American Lichens* (1892–1905, 365 numbers issued) and *Lichenes Boreali-Americani* (1894–1905, 280 numbers issued), she was a key figure in the organization of a major system for the distribution of numbered sets of pressed specimens. She also served as associate editor of *Plant World.* Her editorial work and her constant labors on species identification for other lichenologists were among her most valued contributions. She made several extended collecting trips in New England, Alaska (with Grace Cooley), Florida, California, and Colorado, as well as in Italy and Switzerland. Her 1904 study of the lichens of Alaska constituted a substantial addition to the systematics of the group, and is still referred to in North American lichen literature; she reported 217 species, and of these eighty-four were new to Alaska and two were new species.[53] A fellow of the AAAS and a member of several other scientific societies, she served as vice president of the Society of Plant Morphology and Physiology in 1904. She died in Concord, New Hampshire, 28 December 1906, at the age of fifty-one.

Cummings's colleague GRACE COOLEY[54] (1857–1916) taught in the botany department at Wellesley for twenty-one years. She was born in East Hartford, Connecticut, on 26 July 1857. In 1881, at the age of twenty-four, after several years of high school teaching in New Jersey and New York, she enrolled at Wellesley. Although she did not immediately proceed to a degree, she nevertheless held the position of instructor from 1883 to 1896. She carried out summer research at the Woods Hole Marine Biological Laboratory, and in 1893 took an A.M. at Brown University, Rhode Island. Among her publications from this period is a detailed account of the flora of southeastern Alaska, which appeared in the 1892 *Bulletin of the Torrey Botanical Club.* Still a paper of general interest, it reported a trip she and Clara Cummings had taken by coastal freighter along the Alaska panhandle the preceding summer. In the alpine meadows above Juneau, Cooley discovered a new species of buttercup that now bears her name, *Ranunclus cooleyae.* In 1894, following a period of research at the Naples Zoological Station and some time at the University of Zurich, she received a Zurich Ph.D. Her dissertation research on the cellulose content of seeds appeared in the *Memoirs of the Boston Society of Natural History* (1895). Although she became associate professor at Wellesley in 1896, she was not promoted further, possibly because she failed to publish again until 1904, when she brought out two papers on the growing of trees.[55] She was especially active on the Missionary Committee of Wellesley's Christian Association.

In 1904, at the age of forty-seven, she returned to New Jersey, where her teaching career had begun more than twenty-five years earlier. For eleven more years she taught biology in Newark high schools. The welfare of women schoolteachers was her particular concern, and she took a leading part in the formation of the Association of Women Teachers of the Newark High Schools, serving as the organization's first president. She died in Newark, 27 January 1916, at age fifty-eight.

As well as Mount Holyoke, Smith, and Wellesley, a fourth women's college in the northeastern United States, Barnard College in New York City, had on its faculty women scientists who made contributions to late nineteenth-century botanical research.

EMILY GREGORY[56] (1841–1897) established and for eight years ran the department of botany at Barnard. Born in Portage, New York, on 31 December 1841, Gregory was brought up on a farm, and as a girl attended Albion Seminary in Portage. She then taught for a number of years at Dunkirk (Fredonia) Friendship Seminary and elsewhere. At the age of thirty-five she enrolled at Cornell University where she took a B.A. degree in literature (1881), with a minor concentration in botany. Following two years of botany teaching at Smith College she went to Europe for further training. Among the plant anatomists she studied with were Albert Wigand at Marburg (then an old man), Johannes Reinke at Göttingen, Simon Schwendener at Berlin, and Anton de Bary at Strasburg—all of them outstanding botanists. She took her degree at the University of Zurich (1886), submitting a dissertation on "Comparative anatomy of the Filz-like hair-covering of leaf organs." Hers was the first doctorate in botany awarded to an American woman, and probably the first doctorate in any nonmedical field awarded to an American woman by a European university.

On her return to the United States, she accepted a position as a botanist on the Bryn Mawr College faculty. However, serious differences arose between her and Edmund Beecher Wilson, professor of biology, who was in the process of developing a comprehensive undergraduate biology curriculum. She declined to subordinate her botany courses to general biology and after two years was forced out.[57] She spent the next year assisting William Powell Wilson in developing the botanical laboratory at the University of Pennsylvania, and in 1889 offered her services to newly opened Barnard College, which was in the process of establishing a botany department. Though her academic qualifications were excellent and all were eager to appoint her, Barnard could not, within the terms of its agreement with its parent institution, Columbia University, have teaching staff who were not affiliated with the older college. Columbia eventually agreed to accept her as a lecturer on the physiology and anatomy of plants; she thus became the first woman on the Columbia faculty.

The excellent reputation of the botany department at Barnard from its earliest days resulted in large measure from her labors. She spent the summer of 1889 in Berlin, continuing her studies and also buying materials and equipment needed to stock the Barnard laboratory. Most of the needed funds came from private subscriptions. Under her direction the department grew rapidly and Effie Southworth (see below) was taken on as assistant. The first of the herbarium collections was donated by Elizabeth Britton (see below), and the laboratory was expanded.

Gregory's research was in the areas of plant anatomy and physiology, cell structure, and tissue function. Her publications over the eleven-year span of her academic career included, in addition to her original papers, thirty-one review articles (mainly of German works) and eight commentaries on current German research in anatomy and morphology. She frequently spent her summers at academic institutions in Europe and kept well informed on botanical developments there. Her writings did much to introduce new concepts to the American botanical community, which was still to a considerable extent concentrating its efforts in the traditional areas of systematics and taxonomy. Her textbook, *Elements of Plant Anatomy* (1895), is particularly notable. It was the first American botanical book devoted solely to plant anatomy (although earlier works, such as *Gray's Botanical Textbook* in its various revisions, had substantial sections on the subject). Gregory closely followed available German models, such as Julius Sachs's *Lehrbuch der Botanik* (1868), and treated structures in a developmental rather than a comparative manner as an introduction to plant physiology; thus, one-third of her book was given to cell structure and two-thirds to tissue.

She was active in the Torrey Botanical Club in its early years and frequently presented talks at its meetings. From 1888 to 1897 she was associate editor of its *Bulletin*. In 1896 she started the Barnard Botanical Club, an association of alumnae and students of Barnard's botany department. For a short time she served as the club's president.

In her later years she became especially active in religious concerns. Although she had attended church as a child, she had had little inclination toward things spiritual throughout much of her adult life. About the time she moved to New York, however, she underwent a religious awakening, joined a church, and started a Sunday school class of mainly college-age girls. In part she gave up her earlier allegiance to Darwin's theory of natural selection, deciding that science can unravel nature's secrets only to a certain point; beyond that, it seemed to her, fundamental mysteries, which she accepted with humility, must always remain. Her thoughts on the subject were set out in her booklet *A Scientist's Confession of Faith* (1897).[58]

She was promoted to full professor in 1896, but died the following year, at age fifty-five (21 April 1897), in the midst of planning an expansion of the Barnard botanical laboratory coincident with the college's move to larger premises. A small, quiet, reserved woman, she was remembered as an excellent teacher whose interest in her students' work was strong and personal.[59] Barnard established an annual Emily L. Gregory Award for outstanding teaching.

In addition to the faculty members, a few senior undergraduates and graduate students at the Eastern women's colleges published papers in the botanical journals before 1901. Among them, four who worked under Gregory's direction at Barnard, Marion McEwen, Louise Stabler, Anna Pettit, and Louise Dunn, form the most prominent group. McEwen's substantial paper on plant anatomy appeared in the *Bulletin of the Torrey Botanical Club* in 1894, and Stabler's note came out in the same journal in 1891. Anna Pettit, who was awarded a Ph.D. by Columbia University in 1895, published her dissertation research on the underground ripening of the fruit of the peanut plant in a long paper in the *Memoirs of the Torrey Botanical Club* (1893–96). She married G. Lupton Broomell, but died in 1899.[60]

LOUISE DUNN[61] (1875–1902) entered Barnard in 1893 as a special student in botany. In 1897 (the year of Gregory's death) she was awarded an A.B. degree by Columbia University with the highest standing of any Barnard student up to that time. Immediately appointed assistant at Barnard, she continued studies in botany, zoology, and chemistry (A.M., 1899). An enthusiastic and energetic research worker, she spent three summers at the Biological Laboratory at Cold Spring Harbor, Long Island, and one at the Marine Laboratory at Roscoff, Brittany. She presented both her thesis research in plant embryology and work on moss sporophytes at the 1900 meeting of the AAAS. About that time she was promoted to the position of tutor at Barnard. She was an active member of the Barnard Botanical Club, serving as secretary from 1897 to 1898 and as president from 1899 until 1901. A year later (18 December 1902), at age twenty-seven, she died of heart disease.

The research of Gregory's students, particularly at the graduate level, clearly owed much to the close affiliation of Barnard College with Columbia University. The relationship between Radcliffe College and Harvard was comparably close, but only two early botany students from Radcliffe, Mabel Cook and Helen Noyes, made contributions to the journal literature. Cook was born in Dorchester, Massachusetts, on 21 July 1866, and studied at Radcliffe for a year (1886–1887). Her report on the flora of Middlesex County, Massachusetts, appeared in

Rhodora in 1899, and she published a second observational paper, on plants in Maine, two years later. Noyes was from Auburndale, Massachusetts (born on 29 October 1875). She spent two years at Radcliffe, 1896–1898, and later worked at the Alstead School of Natural History in New Hampshire. From there she published a 1900 study of New Hampshire ferns. In 1903 she married Hollis Webster, a botanist with special interests in fungi.[62] Several other students from northeastern colleges who contributed one or two papers to botanical journals and who are known to have gone on to positions as high school teachers were mentioned previously.[63]

At least two late nineteenth-century American women botanists, Alice Rich Northrup and Elizabeth Knight Britton, received their education at the New York Normal School for Women (later Hunter College). Britton became one of the best-known and most productive American women botanists of her time. ALICE NORTHRUP[64] (1864–1922), on the other hand, is hardly remembered, but she nevertheless carried out valuable work in natural history education. She was born in New York City, on 6 March 1864, and studied botany at the Normal School, staying on as an instructor after her graduation in 1883. In 1889 she married John I. Northrup, an instructor in zoology at Columbia University; she was widowed two years later. Her early work included field observational studies of eastern United States and Canadian flora. However, her attention gradually turned to broad aspects of botanical education. She developed a special interest in bringing nature study to New York City schoolchildren, whom she saw as becoming, by the early years of the twentieth century, more and more isolated within the confines of city streets. The School Nature League, organized in 1917, had as its nucleus a group of Hunter College alumnae, former students of Northrup, who tried to bring something of the country to city children through flower shows and nature study in schools. Summer camps for children were another of Northrup's projects. She died at age fifty-eight, killed in a car accident on her way to one of these camps, 6 May 1922.

ELIZABETH KNIGHT BRITTON[65] (1858–1934), the only woman charter member of the Botanical Society of America, was for many years considered the foremost American bryologist. A prime mover in the establishment of the New York Botanical Garden, which was designed to serve American botany as the Royal Botanic Gardens at Kew served British workers, Britton moved in influential social circles where support for such projects was forthcoming. She was a woman of great energy and persistence whose accomplishments in botany were recognized with a star in her listing in *American Men of Science*.[66] She had no graduate degree, and the positions she held after her marriage were somewhat informal and carried no salary; she had no pressing need to earn an income in any case. Her opportunities for professional work were dependent largely on her husband's positions at Columbia University and the New York Botanical Garden, and his status as a widely recognized and respected botanist. In many endeavors she was his full partner, and the results of their joint work were impressive. Her situation, however, was hardly unique, even for a scientist of her standing. Furthermore it had its advantages in that it offered the opportunity for research without incurring the heavy undergraduate teaching loads and administrative burdens that absorbed the time and energy of many of the early faculty members at women's colleges.

Elizabeth was born in New York City, on 9 January 1858, one of the five daughters of Sophie Anne (Compton) and James Knight. Of Scottish-Welsh extraction, Knight was a well-to-do plantation and factory owner in Matanzas, Cuba, where much of Elizabeth's childhood was spent. The rich tropical vegetation of the island undoubtedly stimulated her early interest in plants and natural history. She attended a private school in New York City and then the Normal School. After graduating from the latter at the age of seventeen, she stayed on as a staff member for ten years, assisting in natural science instruction. She joined the Torrey Botanical Club in 1879 and two years later published her first botanical paper, a short note on albinism.

In 1885 she married Nathaniel Britton, also from a well-to-do family, then assistant in geology at Columbia University. He had strong interests in botany also, and switched most of his efforts to that field the following year. Because of her special interest in mosses, a somewhat marginal area of botanical work, Elizabeth became the unofficial curator of the small bryological collection at Columbia. By her own collecting and by buying a number of sizable holdings, she developed the Columbia collection to research standards. Her social connections with wealthy New York families enabled her to raise funds for the acquisitions, the most notable of which was that of the Swiss bryologist, August Jaeger, bought for the then considerable sum of $6,000 in 1893. In 1890 Nathaniel Britton became the first director in chief of the New York Botanical Garden, and for the next thirty-three years he and Elizabeth shared the responsibilities the position entailed. She donated her full-time services to the study and arrangement of the moss herbarium, in which, after 1899, were included the holdings in the Columbia University herbarium. She continued to acquire additional collections, including an important one belonging to the English bryologist William Mitten, bought in 1906. In 1912 she became honorary curator of mosses, a post she held for the rest of her life.

A careful and meticulous observer and an industrious worker, she published seventy papers on the taxonomy of mosses, thirty-one more on ferns and flowering plants, and sixty-one reviews. Work carried out in the late 1880s and the 1890s and published in the two series "Contributions to American Bryology" and "How to Study the Mosses" was particularly important and established her as a leader in this area. Her research on ferns included a revision of the North American species of *Ophioglossum* (1897) and life histories of the curly grass fern and the tropical *Vittaria lineata.*

Despite the lack of an official position at Columbia University, she was major adviser to doctoral students, including Abel Joel Grout, who went on to become, in his turn, the leading bryologist in America. She edited the Torrey Botanical Club's *Bulletin* from 1886 to 1888, was one of the associate editors of

Plant News, and one of the founders (for a time also president) of the Sullivant Moss Society, later the American Bryological Society. In her later career she moved away from research on mosses and, as a founding member of the Wild Flower Preservation Society of America, put considerable effort into nature conservation work. Fifteen species of plants and the moss genus *Bryobrittonia* bear her name. She died on 25 February 1934, in the Bronx, New York City, at age seventy-six. Nathaniel Britton died four months later.

ANNA MURRAY VAIL,[67] a student of Nathaniel Britton at Columbia University, librarian at the New York Botanical Garden from 1900 to 1907, and a productive research worker closely associated with that institution for almost twenty years, is rarely referred to in accounts of early American women botanists. In terms of technical publications authored, however, she ranks rather highly.

Born in New York City on 7 January 1863, she was educated privately, mostly in Europe. During the 1890s she contributed more than twenty substantial papers to North American taxonomic literature, her research on *Asclepias* (the milkweeds), being especially noteworthy. Her fieldwork included extensive exploration in Virginia, some of which was described in a joint report with John Kunkel Small, later head curator at the New York Botanical Garden. For a number of years she assisted Daniel Trembly MacDougal, then also at the New York Botanical Garden, in his studies on mutations. In 1903 she traveled to Europe to compare MacDougal's plants with collections of Hugo de Vries in Amsterdam and with type specimens at the Muséum d'Histoire Naturelle in Paris. She also worked with MacDougal and his collaborator, George Harrison Shull, in a major project to analyze the relationship between various species of the genus *Oenotheras grandiflora.* Their studies were part of current investigations of the competing theories of fast mutation versus slow, continuous, Darwinian evolution; particularly noteworthy were their hybridization experiments to observe the origin of new species by mutation under controlled conditions, investigations they followed by analyses of the stability of the mutants. Vail also carried out joint research on Zygophyllaceae with Per Axel Rhydberg, another worker at the Botanical Garden.[68] She went back to Europe sometime between 1910 and 1921, and lived in France for the next thirty years.

A further noteworthy early worker who was connected with an academic institution in the northeastern region of the United States was the botanical bibliographer MARY ANNA DAY[69] (1852–1924). Day worked at the Gray Herbarium at Harvard for thirty-one years, becoming latterly one of the oldest of the many specialized workers the university employed. Born in Nelson, New Hampshire, on 12 October 1852, she was the daughter of Hannah (Wilson) and Sewell Day. She attended Lancaster Academy, in Lancaster, Massachusetts, from 1869 to 1870, undertaking special studies in botany. For seven years she taught in public schools in Sudbury and Clinton, Massachusetts, and then became an assistant in the Clinton public library.

In 1892, wanting to find a more challenging job, she applied for a position as librarian at Harvard. Despite her almost complete lack of familiarity with technical literature, she was taken on at the herbarium. Her first major assignment was the verification of about 5,000 bibliographical references in a collection of manuscripts by Asa Gray and Sereno Watson, then being prepared for posthumous publication by the herbarium curator W. C. Lane. This exercise undoubtedly gave her a rapid introduction to the botanical literature. Over several years in the late 1890s, she carried out a substantial amount of collecting in Vermont and Massachusetts, bringing back important additions to the herbarium. Her 1899–1900 articles in *Rhodora* on the flora of New England date from this period; her detailed report of the herbaria of New England followed in 1901.[70] She was also the indexer of *Rhodora* for several years.

Her most important work, a project she labored on for more than twenty years, was the preparation of the *Card Index of New Genera, Species and Varieties of American Plants.* The undertaking had been begun by Josephine A. Clark of the U.S. Department of Agriculture, Day's predecessor at the Gray Herbarium. In 1903, after the publication of the first twenty issues (about 28,000 cards), the work was turned over to the herbarium. When completed in November 1923, three months before Day's death, the index contained 170,000 cards. Though she had an assistant for proofreading, most of the indexing she carried out herself, and all was done in the intervals between her regular work. The size of the task may be gauged by the fact that more than 130 scientific serials over many years had to be indexed page by page. Foreign language monographs were also covered since at that time the major part of the botanical literature relating to South America was published in Europe, and North American plants also were often given their first scientific description in the European literature.

Despite a miserably small salary during all her years at Harvard, she managed two trips to Britain and Europe, visiting botanical collections. She shared her home in Cambridge, Massachusetts, first with her sister, Helen Day, and on Helen's death with another sister and a brother-in-law. She died on 24 January 1924, at age seventy-one.

Very few American women botanists whose careers began before the turn of the century held faculty positions at colleges or universities other than the women's colleges in the northeastern region of the country. Among those who did, JOSEPHINE TILDEN[71] (1869–1957) of the University of Minnesota was the most prominent. Tilden was a Midwesterner, born in Davenport, Iowa, on 24 March 1869, the daughter of Eliza Aldrich (Field) and Henry Tilden. She attended Minneapolis Central High School and then studied biological sciences at the University of Minnesota (B.S., 1895; M.S., 1896). From an assistantship in botany awarded her in 1896, she rose through the academic ranks to instructor in 1897, assistant professor in 1903, and professor in 1910 (at age forty-one)—a notable career for a woman scientist at a state university at the time, especially since she never took a doctoral degree. She is the only one of the American women botanists in this study to achieve such early success.

From her undergraduate years onward, Tilden was an outstandingly productive research worker. A student of Conway

MacMillan, founder of the department of botany at the University of Minnesota, she was much influenced also by J. Arthur Harris, a specialist in plant ecology and geographical distribution. Both MacMillan and Harris encouraged particularly her interest in phycology. Her first paper appeared in the *Botanical Gazette* in 1894 and was followed by a steady flow of algae publications. *Minnesota Algae* (1910), her first major monograph, remained a widely used reference work on the blue-green algae for half a century. She traveled extensively, studying algae on the American West Coast, in Polynesia, Australia, New Zealand, and Japan, and early in her career published work on Hawaiian and South Pacific algae, including *Algae Collecting in the Hawaiian Islands* (1902). On American algae especially she became a recognized authority, writing and editing a substantial number of works in the area. Toward the end of her career she took up the subject of the evolutionary development of algae; her last monograph, *The Algae and Their Life Relations: Fundamentals of Phycology,* published by the University of Minnesota Press in 1935 with a second edition in 1937, represented the first American attempt to bring together in one volume a summary of the known characteristics of marine and freshwater algae.

In addition to her original work, she made a noteworthy contribution over the course of many years to the bibliographical literature on algae research. The first issue of her internationally distributed *Index Algarum Universales,* a catalog of all known references to publications on algae, appeared in 1915, and her *Bibliography of the Literature Relating to the Pacific Ocean Algae and to the Fresh Water Algae of the Countries Bordering on the Pacific Ocean* came out in 1920. Her work on the development of standard methods of representing algae in drawings was also of considerable importance. She was an official American delegate to the first three Pan-Pacific Scientific Congresses, held in Hawaii in 1920, in Melbourne and Sydney in 1923, and in Tokyo in 1926. She also attended the First Pan-Pacific Food Conservation Congress in Hawaii in 1924, one of her special research interests being the use of algae as animal food.[72]

A small, cheerful woman whose Minnesota colleagues considered a fine teacher, she communicated her enthusiasm for her subject to generations of students. She took a leading part, along with MacMillan and other faculty members, in founding the Minnesota Seaside Station, a botanical and marine biological station that functioned for a time during the early years of this century in a beautiful setting near Port Renfrew on the southwest coast of Vancouver Island. It offered summer field and laboratory work for small groups of college instructors and high school teachers as well as students. Tilden was subdirector and in addition provided instruction on seaweeds.[73] She also worked to develop the University of Minnesota's field instruction program at Itasca State Park in northwestern Minnesota.

Perhaps her most unusual piece of extra-classroom teaching, however, was the venture she undertook from 1934 to 1935, when she was already sixty-five. On that occasion she led ten graduate students on a round-the-world algae-collecting trip that lasted the entire academic year—a field trip that must surely have ranked as the most memorable in their college careers. They crossed Europe to the Red Sea and went on to Australia, New Zealand, and the Pacific Islands, coming home via California. A point on Stewart Island, New Zealand, was named Minnesota in honor of the expedition.

After retiring in 1937 she lived in a community she founded near Lake Wales in central Florida. She continued to work on a number of writing projects that varied in subject matter from a scholarly treatment of seaweeds to her ideas of good living arrangements for retired people. She died on 15 May 1957, in her eighty-ninth year, at her Florida home, which she called "Ia Ora Na," Tahitian for "friendliness."

Both Susan Nichols and Florence McCormick also held faculty positions at Midwestern institutions, Nichols at Oberlin College, a private, coeducational institution in Ohio, and McCormick at the University of Nebraska. Neither attained prominence in research to the extent that Tilden did, though both published solid work.

SUSAN NICHOLS[74] (1873–1942) was born in Brownsville, Maine, on 12 May 1873, the daughter of the Rev. Charles Lewis Nichols and his wife Anna (Flint). After attending Brownsville schools and Bradford Academy, she went to Cornell University (B.S., 1898). She stayed on for a year as a graduate fellow, working with botanist Willard Rowlee; their study on oak taxonomy appeared as a joint paper in the *Botanical Gazette* in 1900. Following a year in Europe on a fellowship at the American Women's Table at the Naples Zoological Station, she taught science at Houghton Seminary, New York. She returned to graduate studies in 1902, entering the University of Wisconsin (Ph.D., 1904). Her dissertation research was the first of several cytological investigations that she reported at intervals over the next twenty-six years in the *American Journal of Botany* and the *Bulletin of the Torrey Botanical Club;* healing processes in cell structures were her special interest.[75]

Following four years of teaching at a private school for girls in Kentucky (1904–1908), she became an instructor of botany at Oberlin College. She remained there for the next thirty years and concentrated her efforts heavily on teaching, for which she was highly regarded; many of her students kept in close touch with her after they had moved on to professional work. Although she contended with ill health and physical handicap throughout much of her life, she became associate professor in 1911, professor in 1925, and head of the department of botany in 1933. She retired in 1938 and returned to her native state of Maine where two of her sisters still lived. For a time she kept up research on native orchids, begun during her years at Oberlin. She died in Portland, Maine, on 6 December 1942, at age sixty-nine, following a lengthy illness.

FLORENCE MCCORMICK[76] was born in Shippensburg, Pennsylvania, on 21 December 1874, and attended the University of Tennessee (A.B., 1897; M.S., 1900). Her microscopical study on the leaf of *Pinus virginiana,* reported in her one pre-1900 botanical paper, was done in Tennessee. Following six years (1901–1907) as instructor at Winthrop College, in Rock Hill, South Carolina, she obtained an assistantship in botany at the University of Chicago, where she studied from

1909 to 1913, receiving a Ph.D. in plant morphology and physiology in 1914. Her doctoral research was on the liverwort *Symphyogina aspera.* During her two years on the faculty of the University of Nebraska (1914–1916), she held a double appointment, being assistant professor in the department of botany and assistant botanist at the Agricultural Experiment Station. Her work at the station included anatomical investigations on the sweet potato.[77]

Other botanists of the Midwest

Although Tilden, Nichols, and McCormick appear to have been the only women botanists in this overview who held full academic positions in the Midwest, the land-grant colleges and state universities of the region (which had generally accepted women from their beginnings) trained a remarkable number who made modest contributions to the botanical literature as students and research assistants. Iowa College of Agriculture (later Iowa State University), situated on the outskirts of what was then the small, isolated village of Ames, was particularly outstanding in this respect; work by no fewer than eleven of its women students appeared in the scientific journals before the turn of the century, often in the *Proceedings of the Iowa Academy of Sciences.* A paper in the 1881 *American Naturalist* by Ida Twitchell (see *Western botanists,* below) and another by Fanny Perrett in the *Botanical Gazette* the same year, were among the earliest. Both Twitchell and Perrett were students of Charles Bessey, one of the outstanding figures in the history of Midwestern botany, who first taught in Iowa and later in Nebraska. In the 1890s Alice Beach, Mary Rolfs, Emma Sirrine, Emma Pammel, Mary Nichols, Cassie Bigelow, Hannah Thomas, Alice Hess, and Harriet Vandivert studied under Louis Pammel, professor of botany at Iowa College of Agriculture from 1889. Both Beach and Rolfs carried out research on pollination processes. Beach was primarily an entomologist, and in addition to her botanical articles published several papers on Iowa insects (see chapter 2). Sirrine, Emma Pammel, Mary Nichols, Bigelow, and Thomas all had interests in plant morphology. Hess and Vandivert published a joint paper on Iowa fungi. Only Sirrine had a career as a botanist; Rolfs, Pammel, and Nichols became schoolteachers, Bigelow was a businesswoman and Hess a housewife.[78]

EMMA SIRRINE,[79] the daughter of Mary (Meyers) and Jacob Sirrine, was born in Dysart, Iowa, on 13 September 1870. After receiving a B.S. degree in 1894 she stayed on as an assistant in botany (M.S., 1896). Most of her work was on the anatomy of grass leaves, one of Louis Pammel's special interests; three of her studies (on leaf and seed coat structure) appeared in the *Proceedings of the Iowa Academy of Sciences* (1895–1897). After further graduate work at the University of Chicago, she joined the U.S. Department of Agriculture's Bureau of Plant Industry in Washington, D.C., where she was a scientific assistant from 1906 until 1913 and then assistant botanist in charge of the seed testing laboratory. Acknowledged as one of the country's seed experts by the 1920s,[80] she was promoted to associate botanist in 1933. By then she was sixty-three, however, and not far from retirement. Throughout much of the earlier part of her career at the Bureau of Plant Industry she was in frequent correspondence with Louis Pammel, and regularly carried out seed identifications for him from collections made by Iowa farmers. She held memberships in the AAAS and in the Botanical Society of Washington, D.C. In 1937–1938, her last year at the bureau, she served as president of the American Association of Official Seed Analysts (North America). Her retirement was spent in Traer, Iowa.

MARY ROLFS,[81] the daughter of Johanna Maria (Niemeier) and Maas Peter Rolfs, was born in LeClaire, Iowa, on 19 September 1869. After receiving a B.L. degree in 1893 she was a schoolteacher for sixteen years. Her "Notes on pollination" published in 1894 was probably her only scientific paper. She married Herman Schuck in 1908.

EMMA PAMMEL[82] (1874–1904) was born in La Crosse, Wisconsin, on 17 November 1874. She attended the local high school and then enrolled at Iowa College of Agriculture where she studied botany and chemistry (B.S., 1894). Staying on as an assistant in chemistry, she took an M.S. in 1896. Her special interest was the structure of the leaf, and she published two substantial papers in this area in 1896 and 1897, one coauthored by Emma Sirrine. A third paper, reporting bacteriological studies, resulted from collaborative work with Louis Pammel (most likely her brother), whose interests in plant pathology extended to basic bacteriological studies. She taught high school science in Iowa and North Dakota for two years following her graduation, and then married Niels Ebbesen Hansen, professor of horticulture at the University of South Dakota. She died in Brookings, South Dakota, 16 December 1904, at age thirty, leaving two young children.

MARY NICHOLS[83] (1869–1951), a native of Marshall County, Iowa, was the daughter of Lauretta (Hessin) and Benjamin F. Nichols. She received a B.L. from Iowa College of Agriculture in 1891 and an M.S. in botany and geology in 1893. Continuing her studies at Cornell University, she specialized in cryptogamic botany (D.Sc., 1896). Although her early work at Ames included observations on pollination processes, five of her six technical papers published between 1893 and 1896 reported morphological studies. Her investigation of pyrenomycetous fungi development, the subject of her doctoral research, was her most substantial contribution. Like Pammel, she became a high school teacher, first in Des Moines, Iowa, and then in New York City. In 1900 she married John Cox, Jr., an architect; they had two daughters. Her work in education continued after her marriage, though her interests turned to the teaching of younger children. She served as principal of a Friends' private school and later owned and directed the Grayrock Country Home School for Small Children in Chappaqua, New York. She died on 7 October 1951.

CASSIE BIGELOW, the daughter of Sara (Moore) and Daniel A. Bigelow, was born in Ames, Iowa, on 29 July 1874. She took a B.L. degree in 1894. Her short paper on the glands of the hop tree appeared in the *Proceedings of the Iowa Academy of Science* the same year; it was probably her only scientific publication. Apart from three years (1897–1900) that she spent at the Anderson Normal School of Gymnastics in New Haven, Connecticut, she had a long career in business and accounting, first in the Bigelow and Smith store in Ames, then as auditor of Crew's Beggs Company, Pueblo, Colorado, and later with various companies in Puyallup, Washington. She was a member of the Puyallup Women's Business and Professional Club, for a time president of the Women's Chamber of Commerce, and for thirty-four years, from 1917 until she retired in 1951, Puyallup city treasurer.

ALICE HESS, like Sirrine and Pammel Hansen, continued her studies at Iowa College of Agriculture to the graduate level. She took a B.S. in 1899 (with the distinction of being the student with the highest standing in the science course) and an M.S. in 1902. Born in Jones County, Iowa, 12 May 1874, she was the daughter of Esther (Waite) and John Hess. Her major field of interest was probably home economics; she taught sewing for a year at Iowa College of Agriculture, and cooking for a year at Bradley Polytechnic Institute in Peoria, Illinois. She also spent a year at Columbia University. After her marriage in 1904 to fellow Iowa graduate Edgar C. Myers she gave her time to family concerns.[84]

The early prominence of Iowa College of Agriculture in the training of women botanists is perhaps not surprising. The college's botanical laboratory was the first in the United States to offer undergraduate instruction, only Asa Gray's laboratory at Harvard (for graduate work) preceding it. Started about 1873 by Charles Bessey, by the mid-1880s it was among the best teaching laboratories in the country. Well equipped for the time (with twenty compound microscopes and one large binocular, as well as "magnifiers"), it provided senior students with good training in microscopy.[85]

From the late 1880s onward, a number of women carried out botanical investigations while students at colleges and universities in Michigan, Illinois, Minnesota, and Indiana, institutions that could also boast of good laboratory facilities. Many of these women went on to careers as high school teachers and except for one, Katherine Golden of Purdue University, they are not well remembered. Prominent among their research contributions were studies on disease-causing fungi of the region, a poorly understood field at the time, but one that was beginning to be recognized as having considerable economic importance.

Ida Clendenin (1860?–1925), Julia Clifford (1865?–1918), Margaretha Horn, and Etta Knowles were all students at the University of Michigan. Clendenin (B.S., University of Missouri 1886, M.S., Michigan, 1893) published work on plant pathology. She taught biology at the Girls' High School in Brooklyn, New York, from 1895 to 1906 and was especially interested in methods of teaching botany in secondary schools. In 1906 she married Edgar Atchison, and later returned to Missouri. Clifford, a student at Michigan from 1894 to 1896, published two papers on plant anatomy reporting work carried out under Daniel Trembly MacDougal, later of the New York Botanical Gardens. Horn was a native of Westerbergen, Germany. She took a B.S. degree at Kansas State College of Agriculture (later Kansas State University) in 1893 and an M.S. at the University of Michigan in 1896. Her paper reporting studies in plant morphology appeared in the *Botanical Gazette* the same year. She taught botany and zoology in Detroit high schools from 1896 until her death in November 1910. Knowles carried out investigations in both anatomy and plant pathology at the University of Michigan between 1885 and 1889.

Amanda McComb received a Ph.B. degree from the University of Michigan in 1894 and an A.M. in botany from the University of Indiana in 1900; her graduate research was in cytology. Mary Gloss took both her degrees at Northwestern University in Evanston, Illinois (B.S., 1892, M.S., 1897), and her research on ferns was done at the botanical laboratory there. She later taught in high schools in Wisconsin, Michigan, and Illinois. Alice Keener, of Dixon, Illinois, studied at Lake Forest College, Illinois (B.A., 1896), and then taught in Chicago high schools. A short note by her appeared in the *Botanical Gazette* in 1895. She appears to have taken an A.M. at the University of Michigan in 1921. Henrietta Fox and Mary Olson were students at the University of Minnesota in the 1890s. Fox's comprehensive survey of the genus *Cypripedium* Linn was carried out under the guidance of Conway MacMillan. Olson published her studies on a new parasitic species in the *Botanical Gazette* in 1897 and a longer paper in *Minnesota Botanical Studies* in 1899. She later married Harold Melvin Stanford.

Early botany students at Purdue University (Indiana) whose published contributions are listed in the bibliography were Lillian Snyder, Clara Cunningham, and Katherine Golden. Snyder (later Greene), of Crawfords, Indiana, took a B.S. in 1895 and an M.S. in 1896. She was especially interested in the rusts (Uredinea), which, by the 1890s were increasingly being recognized by the U.S. Department of Agriculture as a subject of growing importance because of their effects on forage crops. Three of her papers on Indiana rusts were presented to the Indiana Academy of Sciences between 1896 and 1900. She also carried out bacteriological studies on pear blight. Cunningham (B.S., 1896; M.S., 1897), carried out graduate research on diseases of sugar beet, and also studied the effects of drought on plant tissue. Born in Lafayette, Indiana, in 1876, she spent her professional life as a high school teacher in South Bend and Indianapolis. She married Frank Elbel in 1915.

KATHERINE GOLDEN (1869–1937) is usually thought of as a bacteriologist and food scientist because of her many years of work and remarkable productivity in this area after the turn of the century. Nevertheless, her early studies, done during the fourteen years she spent at Purdue and at the Indiana Agricultural Experiment Station, were in botany.

Born in Stratford, Ontario, on 29 April 1869, she grew up in Lawrence, Massachusetts, and was educated in public schools and at the State Normal School in Salem, Massachusetts, from which she graduated in 1886. At Purdue she took a B.S. in 1890 and stayed on as a graduate student (M.S., 1892). Her main research project was an investigation of yeasts, which led her into the biochemistry of yeast action in bread. However, as assistant botanist at the experiment station she also worked on fungal diseases of sugar beet, reported in a station bulletin in 1891. She was appointed instructor in biology in 1893 and eight years later was promoted to assistant professor. She resigned in 1904 when she married her fellow faculty member, Arville Wayne Bitting. Three years later she and her husband joined the staff of the Bureau of Chemistry, U.S. Department of Agriculture, Katherine as a microanalyst.[86] From 1913 until 1918 they both carried out research for the National Canners' Association, and during the First World War they also worked in the Subsistence Division of the Quartermasters Corps of the U.S. Army. For four years after the war Katherine was employed as bacteriologist with the Glass Container Association.

Most of her research was in the general area of food biochemistry and bacteriology as related to preservation. Her writings on the subject came to more than fifty articles, bulletins, and monographs, many being joint publications with coworkers, particularly her husband. She always worked in partnership with him, although some employers (such as the Quartermaster Corps) preferred that she be formally considered his assistant. Over the years she assembled a unique library of books on gastronomy, which reflected her life's work of research in the field of food science. The collection was later presented to the Library of Congress in her memory. Both she and her husband received honorary doctorates from Purdue University in 1935. She died in San Francisco, at age sixty-eight, on 15 October 1937, just before the publication of her monumental *Gastronomic Bibliography*.[87]

Two early women botanists, Frederica Detmers and Lumina Riddle, took advanced degrees at Ohio State University. FREDERICA (FREDA) DETMERS[88] (1867–1934), the daughter of Heimke (Heeren) and Henry Detmers, was born in Dixon, Lee County, Illinois, on 16 January 1867. Henry Detmers was the founder of the Veterinary College at Ohio State University. Frederica took a B.S. degree in 1887 and an M.S. in 1891, the first in botany given by the university. From 1889 to 1892, she was assistant botanist at the Ohio Agricultural Experiment Station, the first woman to hold a research position of this kind in Ohio. Her M.S. thesis on the rusts of Ohio, *Descriptive Catalogue of the Uredineae of Ohio,* was brought out as a station bulletin in 1892.

She taught science and German in Columbus schools from 1893 to 1906, when she returned to the botany department at the university as a graduate student and instructor. Her Ph.D. dissertation (1912) reported her major three-year investigation of a whole range of factors, ecological, floristic, physiographical, and phytographical, influencing the marsh region of Buckeye Lake; using an approach that was still new in the United States at the time, she combined all of these factors into one comprehensive presentation. She was promoted to assistant professor in 1914, and four years later again joined the staff of the experiment station, first as assistant botanist and then as taxonomist and systematist (1923–1927). Much of her research focused on problems in plant pathology (fungal diseases of vascular plants), and the naturalization and control of weeds. Her last major publication, a comprehensive study on the life history of the Canada thistle, appeared in the station's bulletin for 1927.[89] A charter member of the Ohio Academy of Science, she regularly presented papers at the annual meetings, and she served as vice president in 1918.

In 1927, at the age of sixty, she moved to Los Angeles, taking the job of curator of the herbarium in the botany department at the University of Southern California. She held the post until her death, seven years later, 5 September 1934, in Los Angeles.

LUMINA RIDDLE[90] (1871–1939), received the first Ph.D. degree in botany given to a man or woman by Ohio State University (1905). Born in Woodstock, Champaign County, Ohio, 18 March 1871, Lumina was a descendant of John Cotton, a prominent figure in the history of colonial New England. She entered Ohio State University as an undergraduate in 1894, when she was twenty-three, and took a B.S. in 1897 and an M.S. in 1898. Over the next seven years she alternated periods of further study with teaching in high schools in Michigan and Kansas and at Washburn College in Topeka, Kansas. Throughout the period, she published a creditable number of papers; these included accounts of her studies on algae and protozoa from the Ohio shores of Lake Erie and reports of her masters-level and doctoral research on the embryological life histories of flowering plants in the genera *Alyssum, Elodea, Ranunculus,* and *Staphylea.* In addition she wrote on more general morphological aspects of plants, including fasciation.[91] As a graduate student she frequently presented papers at the annual meetings of the Ohio Academy of Science, and in 1903 served as the group's vice president.

In 1906 she married Bernard Smyth, a widower with three children. A self-taught botanist and schoolteacher, Smyth was also curator of the Kansas State Museum and herbarium. Lumina became assistant curator, and collaborated with her husband on the preparation of a monograph on the flora of Kansas, which appeared in two parts in the *Transactions of the Kansas Academy of Science* as "Catalogue of the flora of Kansas. Pt. 1" (1911), and "Provisional catalogue of the flora of Kansas. Pt. 2—gymnosperms and monocotyls" (1913). Following her husband's death in 1913, she served as museum curator for two years. After that she taught mathematics and teacher training in various Kansas high schools; from 1921 to 1924 she held an assistant professorship in biology at Ottawa University in the small community of Ottawa, Kansas. She went back to Ohio in 1924 and died in Cleveland, on 2 February 1939, at age sixty-seven.

The careers of Riddle and her husband offer an interesting comparison. Bernard Smyth, who acquired his education largely by his own efforts except for a year he managed to spend at the Michigan Normal College in Ypsilanti, for many years held the posts of librarian of the Kansas Academy of Science and curator of the Goss Collection, an important holding of ornithological specimens. He was also professor of botany at Kansas Medical College from 1890 to 1895. Lumina, despite her doctoral degree and creditable output of research, filled in as museum curator for only a short, temporary period and then returned to high school teaching. Of the two, only she is listed in *American Men of Science.*

Like Riddle Smyth, STELLA DENNIS KELLERMAN (1855–1936),[92] carried out botanical work in both Kansas and Ohio, and, although she had no formal institutional affiliation, was associated for several years with Ohio State University. She was born on 25 July 1855 in Amanda, Fairfield County, Ohio, the daughter of Anthony Dennis, a physician who had broad scientific interests. As a child she attended the Academy in nearby Cedar Hill, where one of her teachers was her future husband, William Kellerman. She went on to a female academy, graduating with a "Mistress of Letters." In 1876 she married Kellerman, who was then teaching natural sciences at the State Normal School in Oshkosh, Wisconsin. They spent the two years from 1879 to 1881 in Europe, William Kellerman studying at the universities of Göttingen and Zurich and taking a Zurich doctorate. Having noted the great advantage one of his European professors enjoyed by having a wife who could prepare botanical illustrations, Kellerman arranged drawing lessons for Stella. Shortly after their return to the United States, she began work on the nearly 300 original illustrations used in her husband's *Elements of Botany* (1883), an effort generously acknowledged in the preface. Their collaboration continued after they moved to the Kansas State Agricultural College in 1883; Stella's illustrations appeared in her husband's *Classified List of the Wild Flowers of the Northern United States, with Keys for Analysis and Identification* (1884), and in two jointly authored works, their 1887 paper on Kansas forest trees (see bibliography), and the monograph *Analytical Flora of Kansas* (1888).

Stella's interest in leaf variation, a topic she was to study for more than two decades, began during her years in Kansas as a natural extension of the detailed observational work required in her illustration projects. Her first major contribution in the area, her paper "Evolution in leaves," appeared in the Kansas Academy of Science *Transactions* for 1890. She moved to Columbus, Ohio, the following year, when her husband became chairman of the botany department at Ohio State University, but her leaf variation and evolution studies continued throughout the nineties. Her most notable paper was her broad review of leaf variation published in the *Journal of the Cincinnati Society of Natural History* in 1893. In addition to this and the six other notes listed in the bibliography, she brought out more than twenty short articles on variation in the popular horticultural magazines *Meehan's Monthly* and *Vick's Illustrated Monthly Magazine*.[93]

Included in these writings are her discussions of her ideas on the evolutionary development of corn, arrived at through her interpretation of the morphology of the inflorescences of the plant. She had observed that in abnormal specimens the central stem of the tassel may produce pistillate flowers at its base that mature into grains on small ears. These flowers are surrounded by small spikes bearing functional staminate flowers. She suggested that the abnormal ears represented a reversion to an ancestral type, with bisexual inflorescence, and that gradual evolution through natural selection had produced the modern form in which the staminate flowers had been lost and the pistillate flowers developed. She was the first to point out that the modern ear of the corn plant is the homolog of the central spike of the tassel, an idea that still has relevant implications for modern theories of corn evolution.[94]

In addition to these studies and a considerable amount of practical work and writing on the cultivation of both garden and houseplants, she continued to collaborate with her husband, assisting him in the revision of his textbook, brought out in a second edition in 1897 and further expanded in 1898. She also helped collect specimens for the Ohio State Herbarium (established by Kellerman), and joined in a major study of the nonindigenous vascular plants of Ohio. The latter project, her part in which constituted her major contribution to the knowledge of Ohio flora, was of special importance since it provided baseline data against which all subsequent studies of plant invasions into Ohio have to be compared.[95]

She was one of five women charter members of the Ohio Academy of Science, formed in 1891, and was active in the organization for ten years. She served as vice president in 1894 and 1901, and regularly presented papers at the annual meetings, generally on the topic of morphological variation.

Despite her lack of formal training and her "amateur" status, her botanical work was remarkably wide-ranging and successful, and although her major projects in illustration and her investigations of the local floras of Kansas and Ohio formed part of her husband's work, her studies on morphological variation and her contributions to the culture of garden plants would seem to have been independent studies. Even in the latter, however, she would undoubtedly have benefited considerably from having the help and backing of her husband, a much respected professional. After he died in 1908 (from malaria, on a collecting trip in Guatemala) she did not continue her botanical work. By the late 1890s her interests had begun to shift somewhat anyway, and she was becoming involved in a number of national organizations. For a time she was Ohio delegate to the (Woman's) National Science Club, an organization that met annually in Washington, D.C., throughout the 1890s. One of the club's founding members, she was elected president at its 1897 meeting. In addition, she served as Ohio state chairman of the George Washington Memorial Association, which was then raising funds to establish a national memorial university. She had three children, one of whom, Karl, later became well-known for his botanical work at the U.S. Department of Agriculture. She died in San Diego, California, on 21 July 1936, just before her eighty-first birthday.

Three additional late nineteenth-century women whose botanical contributions are listed in the bibliography did much of their work in Kansas. These were Minnie Reed and Lora Waters, students at the Kansas State College of Agriculture, and A. L. Slosson, who probably had no formal institutional affiliation.[96]

MINNIE REED,[97] whose later work on the edible seaweeds of Hawaii was of special note, took a B.S. degree in 1886 and an M.S. (in botany, horticulture, and domestic economy) in 1893. As an undergraduate she supported herself by teaching in Kansas schools; later she held an assistantship in physical education and taught a calisthenics class for women students. Her botanical work at this period, published mainly in the

Transactions of the Kansas Academy of Science, included several lengthy field studies of Kansas plants, especially ferns and mosses, the subject of her thesis research. She took a second M.S. degree at the University of California in 1899 and then went to Hawaii, where she spent seven years as a science teacher, some of the time at the Kamehameha Boys' School in Honolulu, a manual training school for native boys.

During these years she took up the study of edible seaweeds, then little known in the United States but of great economic value in Japan, which had a virtual monopoly of the supply of agar-agar and seaweed gelatins in the world market. Her research in the area was fairly extensive. Her papers and bulletins included *The Economic Seaweeds of Hawaii and their Food Value,* included in the 1907 reports of the Agricultural Experiment Station, Honolulu, and *The New Ascomycetous Fungi Parasitic on Marine Algae,* which appeared in the University of California Botanical Publications in 1902. She continued her work on algae during a year she spent in Europe on an Association of Collegiate Alumnae Fellowship at the Naples Zoological Station and the University of Berlin. There is little to suggest she did any further research, however. She settled in southern California, and for much of the rest of her life taught practical botany at the Jefferson High School in Los Angeles.

Lora Waters[98] (later Beeler), a contemporary of Minnie Reed at Kansas State Agricultural College, received a B.S. in 1888 and an M.S. in botany and domestic economy in 1893. Her paper on Kansas flora appeared in the *Transactions of the Kansas Academy of Science* in 1896. Until 1898 she was a schoolteacher in Manhattan, Kansas, but following her marriage at about that time she moved to Chicago and gave up professional work.

One of the more remarkable of the early Midwestern women botanists was JANET PERKINS[99] (1853–1933), a native of Lafayette, Indiana, who, over a period of twenty years was engaged in extensive studies of tropical flora from various regions. Throughout much of her research career she was based at the Royal Botanical Museum in Berlin, then directed by Adolf Engler, a leader in the developing field of plant geography and known especially for his work on classification systems of tropical plants. Most of her technical writings were published in German, which may in part explain her omission from studies of early American women botanists.

Born on 20 March 1853, she was the daughter of Jane Rose (Houghteling) and Cyrus Grosvenor Perkins. After receiving her early education at private schools, she went to the University of Wisconsin, where she took a B.S. in 1872, when she was nineteen. She studied further in France and Germany for three years before returning to Chicago, where she was a schoolteacher for the next twenty years. During this period she managed to do a considerable amount of traveling, in Hawaii, California, the Azores, and Europe. In 1895, when she was forty-two, she moved to Germany. She spent three years studying privately at the University of Berlin and a further year at the University of Heidelberg from which she received a Ph.D. in 1899.

Her first lengthy publication reported studies on the systematics of the Monimiaceae, a large and diversified group of woody plants of the tropics. It appeared in 1898 in the *Botanische Jahrbücher,* which Adolf Engler edited. Her first research appointment came in 1902, when she undertook to prepare for the Carnegie Institution of Washington, D.C., a flora of the Philippines. The islands had then only recently become an American possession, following the end of the Spanish-American War, and were attracting considerable attention from American biologists. Perkins's *Fragmenta Flora Philippinae* appeared in 1904. Several other major publications in systematic botany followed in a steady succession. Many resulted from extensive studies carried out on materials coming to the Berlin museum from all over the world. They included *Leguminosae of Puerto Rico* (1907), *Beiträge zur Flora von Bolivia* (1912), works on the plants of Papua (1915), and of southwest Africa (1917). *South-American Monimiaceae* appeared in 1927. These works, and other reports by her on the flora of tropical Africa, were published by the Royal Prussian Academy. She also contributed the sections on the Monimiaceae and the Styracaceae to Engler's *Das Pflanzenreich* (1900–).

Most of these studies were parts of major projects then being carried out by Engler and his coworkers on large-scale intercontinental comparisons, in particular the comparison of floras of tropical Africa and tropical America. Perkins was probably the only early American woman botanist who made substantial contributions to such undertakings. She was a member of several German scientific societies, including the Freie Vereinigung für Systematik Pflanzengeographie, the Deutsche Botanische Gesellschaft, and the Gesellschaft für Erdkünde. She eventually returned to the United States, and died in the Chicago suburb of Hinsdale, 7 July 1933, at the age of eighty.

Among the early women botanists of the central region of the country, one whose colorful personality can still be glimpsed in the fragments of information about her currently available, is LAURA ABBOTT CROSS[100] (1854?–1949). She never explored distant places or contributed to major scientific projects as Perkins did, and she wrote only one botanical paper, albeit the report of a creditable piece of work done as doctoral research at the University of Pennsylvania; Cross, however, stands out as a character.

She grew up on a farm in Clark County, Indiana. In 1878, after her marriage to W. O. Cross, a principal in the Louisville, Kentucky, public school system, she enrolled at Hampton College, a newly established, liberal arts college for women in Louisville. After graduating (A.B.) in 1882, she joined the college faculty, and for several years taught botany and physiology. Cross was a woman with energy and expanding horizons; she supplemented her undergraduate education by taking courses in chemistry and physics offered under the auspices of the Louisville Polytechnic Society, passing the examinations with the highest grade. In 1890, at the age of thirty-six, she went to the University of Pennsylvania and enrolled in graduate study in botany. Her Ph.D. (1894) was the first in botany given to a woman by the University of Pennsylvania.

Returning to Louisville the following year, she bought, for $20,000 and against her husband's strong and vocal opposition, a grand and impressive twenty-seven-room mansion, designed and built before the Civil War for a Louisville judge, Russell Houston. Here she opened a school, offering primary through college preparatory education and boarding accommodation for twenty girls. She herself taught reading, writing, and rhetoric. For forty years she ran the enterprise successfully, even though she never made much money. Then in 1935, having passed her eightieth birthday, she turned her former mansion into a boardinghouse. The income this brought was insufficient to support her, however. Within six years she was evicted, despite a fierce fight on her part, and the house sold at auction for $6,500 to pay delinquent city taxes. She died in a Louisville nursing home in 1949, at the age of ninety-five. The almost one-hundred-year-old mansion, with its marble ballroom, nineteen-foot ceilings, and stained-glass windows, was eventually razed.

SARAH PRICE[101] (1849–1903), a noted botanical explorer and author, was also from Kentucky. The daughter of Maria and Alexander Price, she was born in Bowling Green and lived there all her life. She was educated at St. Agnes Hall, and for some years taught botany and nature study.

From her late thirties until her death, she carried out extensive botanical explorations throughout southern Kentucky, traveling through difficult country by farm wagon, rowboat, or on foot. She put together an extensive herbarium, and, being a skilled artist, made watercolor sketches of more than 1,000 Kentucky plants. This work, and indeed her entire program of botanical investigation in largely unexplored country, is reminiscent of that undertaken by her contemporary, Kate Furbish (see above), whose watercolor record of the flora of Maine is better known than Price's studies. By the early 1890s she began to put her work into print, and her contributions were recognized by a diploma at the Chicago World's Columbian Exposition of 1893. That year, in addition to two papers on cave plants and ferns that appeared in *Garden and Forest,* she published a *Flora of Warren County, Kentucky.* A few years later she brought out two more botanical monographs, *Fern Collector's Handbook and Herbarium* (1897) and *Trees and Shrubs of Kentucky* (1898).

Though primarily a botanist, Price was active in many branches of nature study and maintained contact with a large circle of naturalist colleagues. Her watercolor sketches of Kentucky birds were a significant contribution to the ornithology of the region; she also wrote on the mollusks of Kentucky. Her literary works, *Songs of the Southland* and *Shakespeare, William, 1564–1616,* appeared in the early 1890s. She died in Bowling Green, on 3 July 1903, at age fifty-four.

Probably the only doctoral degree awarded to a woman before the turn of the century by Washington University, St. Louis, Missouri, went to the botanist ANNA ISABELLA MULFORD[102] (1848?–1943). Born in East Orange, New Jersey, she was the daughter of Timothy Mulford, a manufacturer of vehicle wheels. After graduating from the State Normal School in Trenton, New Jersey, she went to Vassar College (A.M., 1886). She stayed on at Vassar as an instructor until 1889, when she began graduate studies at Washington University under the direction of William Trelease, professor of botany and director of the Missouri Botanical Garden.

In the summer and early autumn of 1892 she carried out extensive botanical explorations in the foothills and mountains of central and eastern Idaho and in the sagebrush country to the south. Reporting this field work in the *Botanical Gazette* (1894), she tells how she traveled successively by rail, stage, horseback, and finally on foot, while collecting the 1,100 specimens she found that season. She wrote of seeing Idaho as the meeting place of the Rocky Mountain flora with that typical of California—the sub-Arctic plants of the north meeting the southern cacti. Her doctoral dissertation research, published in the annual report of the Missouri Botanical Garden for 1896, concerned the agaves of the United States, plants of considerable economic importance that are native to the arid and semiarid regions. After receiving her Ph.D. in 1895, Mulford spent three years as an instructor at the Missouri Botanical Garden, where she specialized in programs for high school teachers. After that she taught at McKinley High School in St. Louis for a number of years. She died in East Orange, New Jersey, 16 June 1943, at age ninety-five.

FLORENCE LYON,[103] though she held a post as instructor at the University of Chicago for three years, was also a teacher in Midwestern high schools for much of her life. She was born in Gilbertsville, New York, on 6 March 1860. During the seventeen years between 1879 and 1896, she alternated periods of study at the University of Michigan with teaching in Owosso and Detroit, Michigan. In 1897 she took an S.B. at the University of Chicago and followed this with a Ph.D. (in botany and zoology) in 1901. After a year as instructor of botany at Smith College (1900–1901) she returned to the University of Chicago where she held a similar position until 1904. Her interests were in plant morphology, particularly reproductive organs, and she published several papers in the *Botanical Gazette* between 1901 and 1905.[104] Thereafter she appears to have dropped out of research.[105]

Western botanists

Except for three women who carried out botanical studies as students at the University of California and at Stanford University, the early western women botanists were all explorers and collectors. Included among them are two of the best-known American women botanists of the time, Katharine Curran Brandegee and Alice Eastwood. Curran Brandegee was active from the 1880s onward; most of the others in the group began their studies in the nineties, except for Bingham of Santa Barbara, Plummer Lemmon of Oakland, and Austin, one of the pioneers of botanical exploration in northeastern California.

REBECCA AUSTIN[106] (1832–1919), though her contribution to published scientific work was relatively small, was nevertheless a respected botanist in her time. She was also a colorful figure whose story has something of the romance of the California gold-rush days. A vigorous, capable, and handsome woman, she spent much of her life as a miner's wife in camps in the northern Sierra Nevada. The region was a plant collector's paradise, having an especially beautiful flora then essentially unexplored by botanists.

She was born in Cumberland, Kentucky, on 10 March 1832, into a family by the name of Smith. As a child she lived in Platt County, Missouri, where she managed to begin her education by attending a subscription school during winters. Orphaned at an early age, she moved with foster parents to Illinois, and by the time she was sixteen was teaching school in summer and with the money earned attending Granville Academy in winter. Here she got an introduction to Latin and the natural sciences. She married a physician, A. Leonard, but he died sometime before 1857. Still in her mid-twenties, she went west to Kansas, where a few years later she married James Thomas Austin. With him and her children she moved to California, crossing the mountains on sleds and snowshoes, and arriving at the gold mines on Black Hawk Creek in Plumas County in 1865.

The following year she began collecting and studying plants, a hobby that she kept up for more than thirty years. Her special contributions, most of them first discoveries, concerned the nature of the fluid secreted by the insectivorous plant, *Darlingtonia californica,* the California Pitcher Plant. She both detected the fluid at the bottoms of the closed pitchers and identified it as a secretion of the plant. Her work demonstrated clearly the response of the pitchers to various stimulae, for instance the introduction of fragments of meat, and she recorded detailed observations on the fate of pollinating insects and on the dipterous larvae that live in the pitchers and feed on the captured insects. Her contributions, dealing as they did with both the anatomy and the physiology of the insectivorous plants, were received with much interest; Asa Gray referred to her early reports in his influential *Darwinia* (1876).[107] Also noteworthy was her original work on leaf motion in *Drosera rotundifolia,* which she published in a local county newspaper in 1875 or earlier.

Much of her time was taken up with cooking and washing for miners, and with the various kinds of assistance she was able to provide for the groups of scattered settlers in the region. From her first husband she had learned how to nurse the sick, and her skills were often called upon, there being no other medical help available. Her collecting was done principally near her places of residence. She did not lack opportunities for exploration, however, because her family moved several times in the 1870s, before settling, in 1881, at Davis Creek near the Oregon-California border, where they remained for a number of years. She was the first plant collector in Modoc County, then a frontier wilderness. Shasta, Tehama, Plumas, and Lassen counties of California, and the southern Cascades region of Oregon were other areas she was one of the first botanists to explore. The herbarium specimens she collected laid the foundation for a knowledge of the flora of northeast California, and the letters in her correspondence from Sir Joseph Hooker, Asa Gray, and several other leading botanists of the time attest to the general regard in which her work of observation and her collecting were held. She discovered many new plants; *Lomatium austinae* C. and R., and the strange orchid, *Cephalanthera austinae* (Gray) Hel., were named in her honor. Many of her specimens were distributed to herbaria throughout the United States and Europe. She died in Chico, California, 4 March 1919, just before her eighty-seventh birthday. Her rugged frontier life, which she is said to have found so much to her liking, was probably unmatched by that of any other botanist in this study.

KATHARINE LAYNE (CURRAN BRANDEGEE)[108] (1844–1920) was born in western Tennessee, on 28 October 1844, the oldest of the ten children of Mary (Morris) and Marshall Bolling Layne. During her babyhood the family moved westward, in stages, across the country. Her early years were spent in Carson City, Nevada, then a brawling mining camp and frontier town. About 1853, when she was eight, the family finally settled in the town of Folsom in the Sacramento valley, California. Her upbringing in the "wild west" had a marked effect on her character. Self-reliance in women was possible to a greater extent there than in much of the rest of the country; Katharine grew up expecting to control her own life and was endowed with the psychological strength that made her capable of insisting on reasonable freedom of action.[109]

In 1866, at the age of twenty-two, a handsome, lively, and energetic young woman, she married Hugh Curran of Folsom, an Irish constable. He is said to have been a drunkard, and she left him, but nevertheless continued to help him and paid his debts when he died in 1874. Throughout the years of the marriage, she earned a living as a schoolteacher. A year after she was widowed she enrolled at the Medical Department of the University of California in San Francisco and received an M.D. in 1878. However, women doctors were not very popular in San Francisco, and she found it difficult to establish a practice. Her interests turned more and more to plants, initially those of medicinal value, and, introduced by her friend Hans Herman Behr, a naturalist and physician who taught at the San Francisco College of Pharmacy, she joined the California Academy of Sciences. Thereafter her botanical work gradually increased, much of it done in connection with the academy. In 1883 she became curator of botany at the academy's herbarium.

The time was a lively one in botanical collecting and classification in California, and a great deal of pioneering work was being carried out. Although for many western amateurs (such as Rebecca Austin) their important contacts were still with the better-known East Coast professionals rather than with the academy in San Francisco, there was, nevertheless, a vigorous movement under way to make a break with the dominant eastern establishment (centered at the Gray Herbarium at Harvard) and develop a western school of botany. Western herbaria were being founded and western journals started. Since the University of California did not set up its herbarium until the late 1880s, the position of curator at the academy in San Francisco was one of some importance when Katharine took the job. Further, her eleven-year tenure of office was a remarkable record for a woman at that period.[110] She collected for the herbarium all over the state—in the Sierras, the Coast Range, the valleys and the deserts. Keen also to provide a place for the prompt publication of scientific papers, she established a series of academy bulletins, of which she was acting editor for three years.

In 1889, at the age of forty-five, she married Townshend Stith Brandegee, a Yale-educated civil engineer and botanist, who had made large plant collections during the years of his work on the vast railroad surveys that opened up Colorado to the mining companies in the 1870s and 1880s. He was highly regarded for his extensive knowledge of western flora, and as a botanist had accompanied F. V. Hayden's exploring expedition to southwestern Colorado in 1875. The marriage gave Katharine the financial support that freed her to concentrate entirely on botanical work, and, perhaps of equal importance, the companionship of a close collaborator, whose commitment to collecting and classifying western plants matched hers. Their teamwork over the next thirty years gave impressive results.

Among their early undertakings were the organization of the very successful California Botanical Club and the founding of the journal *Zoe.* The latter, a publication vehicle for both amateur and professional western biologists, came out monthly during its first year (1890–1891) and then four times a year from 1891 to 1894. *Zoe* was really Katharine's publication, a forum for freer discussion and criticism than she found possible in the academy *Bulletin.* However, her control was camouflaged initially, and the first issue of *Zoe* appeared under the editorship of Frank Vaslit. Later numbers were nominally edited by Townshend Brandegee, then by a committee, which included Townshend and Alice Eastwood. By 1893 Eastwood had the formal responsibility. A final volume (5), in which Katharine was openly listed as editor, appeared in 1908.[111]

Although Katharine's papers and notes on California plants, which she had published in the academy *Bulletin* and *Proceedings* during the 1880s, continued to appear throughout the 90s in *Zoe,* the articles in that journal for which she is best remembered were her critical reviews. These essays frequently raised quite a stir, especially when it was realized that the author was female.

One well-known botanist sorely put out by a *Zoe* review of one of his publications (signed "K. B.") was Nathaniel Lord Britton, husband of Elizabeth Britton. The work under review, "A list of state and local floras of the United States and British America," had appeared as a sixty-three-page monograph in volume 5 of the New York Academy of Science's *Annals* in 1889. K. B. confined her critical remarks to Britton's coverage of western lists, but this, she clearly felt, did less than justice to western botany, several important lists having been omitted and the whole being badly organized. Further, Britton's grasp of western geography seems to have been less than perfect: islands off the California coast were associated with the wrong counties, Yellowstone National Park and Big Horn Mountain had apparently undergone displacement, and the classifying of reports of various transcontinental expeditions was confused and inconsistent. Britton's defense of his geographical and organizational unorthodoxies, also printed in *Zoe,* is unconvincing, but he bolstered it with a sexist salvo enlivened with a quotation from *The Taming of the Shrew.*[112] K. B. was uncowed and unrepentant, however. The *Zoe* editorial rebuttal (unsigned) was a somewhat lighthearted, tongue-in-cheek note, which, after touching on a few more of Britton's shortcomings in organization, did acknowledge that Mrs. Brandegee admitted to the truth of one of the grave accusations brought against her—she was female![113]

Katharine's special interest was the work of accurately determining intermediate forms of newly described species. She was much exercised by the rush to establish "new species" then in progress, and indeed considered the trend to be one of the major difficulties facing American botany at the time. Some workers she saw as being single-mindedly bent on creating as many species as possible.[114] One of her most notable reviews discussed at great length the writings of Edward Green, professor of botany at the University of California and a well-known figure among his botanical colleagues; she pointed out very forcefully the areas where, she felt, Green had fallen short in scientific method and accuracy.[115] Even in her early work in the 1880s she had heated arguments over correct species identification, and in many cases she demonstrated that so-called new species were only sub-specifically different. (Her husband joked that her planned, but never completed, flora of California would finally contain only a single species.)

Katharine's lively, ironic style may seem biting by our present restrained and standardized norms of scientific writing, but later western botanists have admired her work and found her criticisms fully justified. Thus Joseph Ewan of the California Academy of Sciences wrote that her reviews fully demonstrated her clear insight and penetrating botanical judgment.[116]

In 1894 she gave up the curatorship of the herbarium (turning it over to Alice Eastwood), and she and Townshend moved to San Diego, where they continued to build their own botanical library and herbarium, and also created a superb botanical garden. Collecting continued, especially in Baja California and Mexico, Townshend going as far south as Durango and Sinaloa. They returned to the San Francisco Bay area in 1906, largely because Katharine found its climate kinder to her increasingly poor health. Townshend became honorary curator of the University of California herbarium in Berkeley, and they both worked there for the rest of their lives. The huge collections they assembled (over 75,000 plants from California, Arizona, Mexico, and Nevada) were rich in type specimens and have remained particularly important in determining range boundaries. All of these materials were donated to the university.

In later years Katharine suffered increasingly from complications arising from diabetes, and she turned more and more to herbarium and library work; field expeditions disrupted her dietary regime (essential before insulin therapy) and activity patterns. The discrimination she felt she had to fight as a woman professional also became more and more of a burden to her. She died in Berkeley on 3 April 1920, at the age of seventy-five, about five years before her husband. Between them the Brandegees described and named approximately forty-five new species of California plants (still currently accepted as good species); many more were based on their collections, though named by others.

In the winter and spring of 1890–1891 ALICE EASTWOOD[117] (1859–1953), until then a high school teacher, plant collector, and aspiring botanical writer from Denver, Colorado, took a tour of California. She spent some time at the California Academy of Sciences with Katharine and Townshend Brandegee; Townshend's work on the Colorado flora she knew well. A year later Katharine Brandegee offered her a $75-per-month assistantship at the academy. Though she had expected to make her life's work the study of the flora of Colorado and the desert plants of Utah, she found the opening attractive; no other city in the American West had an institution comparable to the California Academy of Sciences. The recent death in Denver of a young man who had been her special friend gave her further reason to look for a change of scene. She accepted Katharine's offer and moved permanently to San Francisco. The arrangement with the academy was that Katharine would donate her services if Alice Eastwood were hired as her assistant. When the Brandegees moved to San Diego in 1894, Alice succeeded to the position of curator of the herbarium (and also became, temporarily, editor of *Zoe*).

A Canadian by birth and early upbringing, she was born in Toronto on 19 January 1859, the oldest of the three children of Eliza Jane (Gowdy) and Colin Skinner Eastwood. The family was of English-Irish background, Eliza Gowdy being a first generation immigrant from Northern Ireland; Colin Eastwood was steward of the Toronto Asylum for the insane. Following her mother's death when she was six, Alice was put under the care of an uncle, William Eastwood, a physician who lived in the country and knew some botany. He gave her an introduction to the Latin names of plants. She attended public schools for a time but was then placed in a Catholic boarding institution, the Oshawa Convent. There her interest in botany was further strengthened by a French priest and horticulturalist, Father Pugh, who taught her about gardening while she watched his work of tree grafting.

In 1873, at the age of fourteen, she rejoined her father and one of her brothers, now in Denver, Colorado. Fortunately her opportunities for further plant study continued and expanded; she took a position as nursemaid with a prosperous cattle-ranching family that frequently made trips into the Rocky Mountains, expeditions on which Alice was able to give herself an introduction to the flora of Colorado. After graduating as the class valedictorian of East Denver high school in 1879, she stayed on there as a teacher, spending her summers in botanical exploration and plant collecting in the mountains and arid regions around Denver. Traveling on horseback and on foot, she gathered a rich harvest of material that became the nucleus of the herbarium at the University of Colorado at Boulder. As early as 1881 she went to the East Coast, visited the Gray Herbarium at Harvard, and talked with Asa Gray. She also made the acquaintance of the British naturalist Alfred Russell Wallace, who, on his way home from lecturing in California in 1887, stopped in Denver to see the alpine flora of Colorado. She was his guide and companion on a three-day climb in the region of Grays Peak.

Her ambition was to devote all her time to botanical work. By 1890, thanks to judicious investment of her small savings and a Denver real estate boom, she had a small income from property rentals. She resigned her teaching post, to the great disappointment of her principal, and took the trip to California that determined her later career. Her first monograph, *A Popular Flora of Denver, Colorado* (1893), was written in the period just before she followed Katharine Brandegee as curator at the California Academy of Sciences. It was published at her own expense, and, lacking any marketing experience, she was unable to make it financially successful.

At the academy, in addition to her heavy routine duties, she carried out arduous and extensive fieldwork (always at no expense to the institution), and contributed substantial collections to the herbarium holdings. She had considerable physical stamina, and explored over vast distances and difficult and remote country, on muleback and on foot, across the cattle trails of the Santa Lucia Mountains, along the coast from Pacific Valley to Monterey, and further north in Mendocino, Sonoma, and Napa counties. With her friend and fellow botanist Ida Twitchell Blochmann of Santa Maria, she made horseback trips across the mountains of San Luis Obispo and Santa Barbara counties to the Carrizo Plain. Her monograph, *A Popular Flora of the Pacific Coast,* was published in 1897. She followed it with *A Popular Flora of the Rocky Mountain Region* (1900) and *A Handbook of the Trees of California* (1905).

She was curator of botany at the California Academy of Sciences for fifty-six years, a remarkable tenure by any standard, and one that was particularly outstanding for a woman. Her work was interrupted for six years following the San Francisco earthquake and fire of 1906, which destroyed the academy building and much of the botanical collections. (Irreplaceable type specimens, which she had taken the precaution of segregating from the main collections, as well as the early records of the academy, were saved largely through her personal efforts; her courage and presence of mind at the time of the fire brought her considerable publicity.) From 1906 until 1912 she had no job and no income except for the small sum from her Colorado property rentals. For a short period she worked in the botany department at the University of California, and then visited the Gray Herbarium at Harvard and also the National Herbarium in Washington, D.C. (While at the latter she most likely spent some time with her friend, the Washington-based ornithologist Florence Merriam Bailey—see chapter 4.) In the winter of 1910–1911 she attempted to apply for a paid position at the National Herbarium but found that, as a Canadian citizen, she was ineligible. Deciding to make the most of her unsought free time, she borrowed some money and went to Europe, where she spent ten months studying major British and French botanical collections. An invitation asking her to return to San Francisco and rebuild the academy herbarium awaited her on her return to America in 1912. She accepted, and began the multiyear task of reconstruction in the rebuilt academy, determined to assemble collections of international standard. She was then fifty-three. By the time she retired in 1949 on her ninetieth birthday (receiving on the occasion the title of curator emeritus) she had increased the collections by over 340,000 specimens, despite the fire losses. She also contributed substantially to the rebuilding and development of the botanical library.

During those years of reconstruction a friendship she had formed with Grove Karl Gilbert, one of the leading figures in American geology, deepened to the point where, by 1918, they decided to marry. Gilbert, sixteen years her senior, was then a widower. They planned to provide a home for his infant grandson, whose mother had died. However, Gilbert himself died later that year, and once more, as when she had lost her friend in Denver a quarter of a century earlier, Alice continued without a close tie.[118]

For a time she had assisted in the editing of the western botanical journal *Erythea,* but in 1932, with her assistant John Thomas Howell, she founded and edited *Leaflets of Western Botany.* Indeed, it may be more than coincidence that two of the most outstanding female botanists of the period (Brandegee and Eastwood) both felt sufficiently strongly the constraints imposed by male colleagues that they established their own journals. Eastwood's undertaking got its impetus in part from differences in professional style between her and Willis Linn Jepson, professor of botany at the University of California and editor of *Erythea.* Outgoing and somewhat impatient by nature, she was quite the opposite of the conservative academic, who was reticent, extremely reserved, and cautious in his scientific judgments. Jepson, however, had also been markedly critical of Curran Brandegee, even claiming that the academy's various internal dissensions during the 1880s were brought about by her.[119] Eastwood was probably a little more circumspect overall than the at times caustic Brandegee. In any event, to the credit of both her and Jepson, they appear to have maintained a reasonable working relationship over the years.

Eastwood's publications numbered more than 300. Two-thirds of them appeared after she was fifty, and fifty-five came out after her eightieth birthday. Her work constituted an important contribution to floristic and taxonomic botany, especially of the Rocky Mountains and the California coast region. Several of her papers represented analytical revisions and proposed new species; many were descriptive articles on the occurrence of plants, notes giving range extensions, and so on. Her critical studies were chiefly concerned with west American Liliaceae and the difficult problems among the American representatives of the genera *Lupinus, Arctostaphylos,* and *Castilleja.* She never gave lectures, but her monographs, aimed largely at a general audience, did much to awaken popular interest in western botany.

Despite the fact that she was largely self-taught, she became one of the most knowledgeable of American systematic botanists of her time. She was starred in *American Men of Science,* beginning with the 1906 edition. A fellow of the AAAS and an honorary member of many botanical societies at home and abroad, she was elected a fellow of the California Academy of Sciences in 1928 and unanimously elected an honorary member in 1942. To celebrate her fifty years of service, the academy published volume 25 of the fourth series of its *Proceedings* under the title, "Alice Eastwood Semi-centennial Publications." The Alice Eastwood Grove of California redwood trees bears her name, as do two American genera, *Eastwoodia Brandegee* of the Compositae and *Aliciella Brand.* of the Polemoniaceae. Academy delegate to the International Botanical Congress in Cambridge, England, in 1930, and to the Congress in Amsterdam in 1935, she formed many international contacts over the years and was widely regarded as one of the most cooperative of botanists; she put her collections and data at the disposal of her colleagues and correspondents in many parts of the United States and elsewhere. In 1950 her long years of service to botany were handsomely recognized by the invitation to be honorary president of the Seventh International Botanical Congress in Stockholm. She was then ninety-one years old but made the airplane trip to Sweden on her own. She died at the Stanford University hospital, 30 October 1953, at the age of ninety-four; burial was in her native city of Toronto.

The bibliography also lists botanical work in California by eleven other late nineteenth-century women. Eight of these, Alice Merritt, Ida Twitchell, Luella Engles Trask, Sara Plummer Lemmon, Mrs I. Hagenbuck, Caroline Lord Bingham, Mary Parsons, and Margaret Adamson, were explorers and collectors, though their work (with the exception of Sara Lemmon's) was on a very modest scale compared with that of Brandegee and Eastwood. Many of their publications appeared in *Erythea.* Of the other three, Clara Williams and Edith Byxbee were students at the University of California; Effie McFadden studied botany at Stanford.[120]

ALICE MERRITT[121] was especially interested in the processes of pollination of California wildflowers by insects and published several papers on the subject in 1896–1897. Born in Plymouth, Michigan, on 31 October 1859, she held a degree from the University of Michigan. Until her marriage to Dr. Anstruther Davidson in 1897, she taught botany at the Los Angeles State Normal School (later the University of California, Los Angeles). Her monograph *California Plants in their Homes; a Botanical Reader for Children and Supplement for Use of Teachers* (1898) resulted from her field observations.

IDA TWITCHELL (1854–1931),[122] who was born in Bangor, Maine, on 11 April 1854, was primarily a taxonomist and ethnobotanist. She received a B.S. degree from Iowa Agricultural College in 1878 (a notably early date) and stayed on for at least two years, working in the college herbarium, which was then expanding quickly under the direction of botanist Charles Bessey. Among her special interests were Iowa ferns, her paper on these appearing in the college publication *Aurora* in 1880. She then went west to California and was a schoolteacher for eighteen years in the Santa Maria Valley. In 1888 she married Lazar E. Blochmann with whom she carried out notable work in botanical exploration. They were often joined on their expeditions by Alice Eastwood, who named *Sedum blochmanae* and *Sphacele blochmanae* in Ida's honor. After moving to Berkeley in 1909 she served for a time on the city's board of education and was active in the College Women's Club. She died in Berkeley, 1 August 1931, at age seventy-seven.

LUELLA ENGLES TRASK[123] (1865–1916) was born in Austin, Minnesota, but lived for almost twenty years (1893–1912) on Santa Catalina Island off the coast of southern California. She studied the flora of both Santa Catalina and the neighboring island of San Clemente. Her botanical collections were extensive, and in recognition of her work Alice Eastwood named 5 plants after her: *Aplopappus traskae, Astragalus traskae, Cercocarpus traskae, Eriodictyon traskae* and *Gilia traskae.* She died in San Francisco, 11 November 1916.

SARA PLUMMER LEMMON[124] (1836–1923) probably published only one significant botanical article, an early account of Pacific coast ferns that appeared in 1881 in a weekly newspaper, the *Pacific Rural Press.* It was a most useful piece of documenting and description, however; botanist Joseph Ewan, rediscovering it more than half a century later, realized that it deserved a more permanent place in the scientific literature and reproduced most of it in an article in the *American Midland Naturalist.*[125]

Born in New Gloucester, Maine, in 1836, Sara was the daughter of Elizabeth (Haskell) and Micajah S. Plummer. After studying at the Female College in Worcester, Massachusetts, and the New York State Normal School, she became a teacher in New York City, while at the same time taking courses in chemistry and physics at the Peter Cooper Union College. In 1869, because of poor health, she went to Santa Barbara, California, where an open-air life by the seashore and in the mountains and canyons did much to restore her vigor.

Some time in the 1870s she met a fellow plant enthusiast, John Gill Lemmon, whom she joined on several collecting trips. In 1880, when Sara was forty-four, they were married. A native of Michigan, Lemmon had fought in the Union Army during the Civil War. Having survived a period in the infamous Andersonville Prison in Georgia, he went to California soon after his release to try to regain his health. Although he remained frail and nervous and unable to hold a steady job, he was a daring explorer and enthusiastic collector. Within a short time of his arrival on the West Coast he was sending herbarium specimens to Asa Gray at Harvard and earning a name for himself in the academic world. His marriage to Sara Plummer may well have made exploration easier and more possible for him, because, semi-invalid as he was, he had in her someone who would look after him.

Throughout the 1880s the Lemmon team explored extensively in Arizona and discovered many new plants. Starting in the spring of 1881 from Tucson, then an old adobe town, they made a wedding tour in the nearby Santa Catalina mountains and foothills, rough, arid country where spiny cactus and catclaw bushes made travel extremely difficult. Sara was the first white woman to climb the 9,157-foot summit now known as Mt. Lemmon.

The following year, they continued their explorations, going south and east of Tucson to the Huachuca area, at the time the heart of Apache country. The Apaches were then on the warpath, but the Lemmons proceeded with their collecting, undeterred even by the fact that friends they had expected to stay with had been scalped and their ranch burned. On one occasion they were stopped and searched by an Apache band, but the leader, after carefully examining their baggage, which consisted mainly of plant presses, concluded they were harmless eccentrics and sent them on their way. They ignored the Apache raids on settlers, the nearby outlaw encampments, and the presence of many deserters from armies fighting in a rebellion then under way in Mexico. In 1884 they extended their explorations to northern Arizona, and later they worked through extensive areas of California, from the Mt. Shasta region in the north to the San Bernardino Mountains in the south.

Their permanent residence was in Oakland, California, where they acquired a large frame house that they almost completely filled with plants and books. A wooden sign identified it as the "Lemmon Herbarium." There they put together the sets of specimens sold to scientific institutions to finance their trips. Lack of money, however, was a constant problem. From 1882 to 1892 they held (between them) a salaried position with the California State Board of Forestry. John Lemmon was the official occupant of the post, but it would seem that the responsibilities were shared. Among other things, Sara carried out a considerable amount of artistic work for the board, especially in the period from 1887 to 1891. Particularly skilled with watercolors, she prepared a large collection of paintings of the flora of the Pacific slope; many of these were used to illustrate her husband's handbooks on western trees and ferns. She also contributed a section on forestry for one of his monographs.[126] During her early years in California she was especially interested in marine plants, and is said to have brought out a limited edition of a monograph on the marine algae of the Pacific coast.[127]

During the twenty-eight years of their life together (John Lemmon died in 1908) they amassed collections rich in early records, including type specimens still frequently consulted. These are now housed at the University of California, Berkeley. Each specimen bears the concise label, "J. G. Lemmon & wife," by today's standards a somewhat minimal recognition of Sara as a botanist.

Active in community affairs as well, Sara served on the boards of several organizations, including the California State Red Cross, the California Federation of Women's Clubs (as chairman on forestry), and flora and forestry associations. She was remembered as a pleasant, attractive woman of strong personality. She died in Stockton, California, in 1923, at the age of eighty-seven, outliving her husband by fifteen years. At least six plants bear her name (*Stevia plummerae, Ipomaea plummerae, Euphorbia plummerae, Allium plummerae, Penstemon plummerae,* and *Plummera floribunda*); a great many others have the title, lemmoni.

Botanists in government laboratories

Of the American botanists whose papers are listed in the bibliography, only two would appear to have been appointed to positions in federal government laboratories before 1900. Of these the elder, though not the first to take up her post, was FLORA WAMBAUGH PATTERSON[128] (1847–1928), who joined the Division of Vegetable Pathology at the U.S. Department of Agriculture as a mycologist in 1896.

About that time traditional research in mycology (the study of fungi) was being pushed into the background by an increasing demand for urgent, practical, problem-solving work on diseases of agricultural crops. Fieldwork in plant pathology, with its wide contacts and large-scale problems, was becoming irresistibly attractive to the division's mycologists, and one of the results of this development was the neglect of traditional herbarium work—species identification, specimen care, and systematic indexing of the mycological literature. No male mycologist was willing to stay for very long with this routine work, and the situation had become critical. Patterson, a widow of almost fifty when she started, took on the herbarium job and stuck with it for twenty-seven years. In large measure she owed her appointment to Beverly Thomas Galloway, at that time chief of the Division of Vegetable Pathology. For half a century Galloway was the leading figure in the organization and development of the Bureau of Plant Industry, where he gathered a dedicated and enthusiastic group of associates around him.[129] Patterson took the required Civil Service examination for the herbarium position in 1895, but failed to score among the three highest in the register. When the top man on the list turned down the job, Galloway recommended the appointment of Patterson to Secretary of Agriculture J. Sterling Morton, justifying his choice by pointing out that her training was as good as that of other candidates and that her publications demonstrated more thorough work (with which he was already fairly well acquainted). He was also of the opinion that she would be more loyal to the collections. She was appointed at an annual salary of $1,200, the same as that of women clerical workers at the U.S. Department of Agriculture at the time.

Flora Wambaugh was born in Columbus, Ohio, on 15 September 1847, the daughter of the Rev. A. B. Wambaugh, a Methodist minister, and his wife Sarah (Sells). She was first taught at home by private tutors and then attended Antioch College, from which she graduated in 1860 (A.B.). For a time in the late 1860s she continued her studies at Wesleyan College in Cincinnati, but in 1869 married Edwin Patterson, a steamboat captain. They had two sons. Sometime in the late 1870s Patterson was badly injured in an explosion on his boat, and from then until he died in 1889 was a helpless invalid. Having the responsibility for the support and education of their two sons, Flora completed her A.M. degree at Wesleyan College in 1883. At the age of forty-two, following her husband's death, she began graduate studies in biology at the University of Iowa, where her brother, Eugene Wambaugh, held a faculty position. Her interest in botany started about that time. When her brother moved to Harvard Law School in 1893 she went east with him, hoping to continue her education at Yale. Yale was not yet accepting women for graduate work in botany, however, and so she settled for Radcliffe College, taking special courses in botany. At the same time, she worked as an assistant at the Gray Herbarium at Harvard, and there acquired her special training in mycology and the care of mycological specimens which later impressed Galloway. Her second A.M. degree (1895) was from the University of Iowa. The following year she became assistant pathologist under Galloway in the Division of Vegetable Pathology; in 1901 she was promoted to mycologist in charge of pathological collections and of inspection.

Despite the considerable burden of her routine work, she carried out a creditable amount of research, much of it reported in the succession of U.S. Department of Agriculture bulletins and reports that she authored or coauthored between 1902 and 1925. She kept thoroughly abreast of current theories in her field, and her work was characterized by exactness and attention to detail. Her first research, her graduate work, had concerned the Exoascaceae, a common fungus group that affects both shade and fruit trees. Using materials from Galloway's collections at the U.S. Department of Agriculture as well as collections from Harvard, Cornell, the Missouri Botanic Gardens, and elsewhere, she carried out a survey and prepared a description and classification of the North American species. Her report, published in 1895, included a discussion of the trees affected and an extensive bibliography of published work in the area. Her later studies included investigations of pine blight, rose diseases, and pineapple rot, all problems of considerable economic importance.[130]

A fellow of the AAAS, she was also a member of several other scientific organizations. She retired from government work in 1923 and died five years later, 5 February 1928, at age eighty, in Brooklyn, New York, at the home of her son, Henry Sells Patterson.

EFFIE SOUTHWORTH[131] (1860–1947) was possibly the first woman scientist to be employed by the U.S. Department of Agriculture's Division of Plant Pathology. She worked there for only five years (1887–1892), but her botanical career continued until 1926. Born in North Collins, New York, on 29 October 1860, the daughter of Chloe (Rathbun) and Nathaniel Chester Southworth, she was educated at Allegheny College and the University of Michigan (B.S., 1885). Following one year at Bryn Mawr College on a biology fellowship (1886–1887), she became assistant mycologist in the Division of Plant Pathology in Beverly Galloway's group.

Her research output while with the division was substantial and highly regarded; twelve publications on parasitic fungi appeared in the *Journal of Mycology* or as U.S. Department of Agriculture reports, ten being single-author papers and two joint publications with Galloway.[132] She resigned her position in 1892 to take an assistantship in botany at Barnard College, joining Emily Gregory, whom she had known at Bryn Mawr. Four years later she left Barnard to marry Volney Morgan Spalding, one of the country's leading botanists, whom she had met when she was a student at the University of Michigan. Spalding was then a widower of forty-seven. In 1904 they moved to the Carnegie Institution's Desert Research Laboratory, a facility recently established in Tucson, Arizona, to encourage ecological studies of the Colorado delta region. Here Effie's botanical work continued, some in collaboration with her husband and with Daniel Trembly MacDougal, the laboratory's director; reports of her investigations of water storage in desert plants appeared in 1905 and 1910.[133]

In 1909, because of Volney Spalding's poor health, they moved to Loma Linda in southern California. Shortly after Volney died in 1918, Effie became assistant professor of botany at the University of Southern California; she remained on the faculty there until her retirement in 1926. In 1923, at the age of sixty-three, she took an M.S. degree. She died in 1947, when she was about eighty-seven.

Summary

Despite the fact that about half of the American authors of botanical papers brought out only one recorded pre-1901 journal article, the sheer size of the group makes the overall American contribution a substantial one (Figure 1-1). As noted at the beginning of this chapter, it constitutes about 11 percent of all the scientific papers published by women in the nineteenth century.[134] Figure 1-2, charts a and b summarize numerical data for the distribution of papers and authors by subfield. Fifty-five percent of the publications reported work in the older branches of the discipline—regional studies (exploration) plus classification and taxonomy—with almost 40 percent of the authors contributing to these areas. A little more than half of the remaining publications were in plant anatomy and morphology, contributed by three-tenths of the authors. The work reported in the remaining papers was spread over a number of areas, for the most part general observational studies, pathology, and physiology. About 21 percent of all the publications were in cryptogamic botany.

Some of the main points (by region) to emerge from the survey are the following:

1. The work of women in the northeast region of the country covered a very broad spectrum, ranging from early field observation (much of it done by independent workers) to specialized academic studies in systematics, morphology, and physiology. Several women's colleges, especially Barnard, Mount Holyoke, Smith, and Wellesley, as well as one or two city high schools for girls, provided employment that allowed the possibility of research for a number of well-qualified women botanists; the U.S. Department of Agriculture laboratories and the New York Botanical Garden provided a few additional places. Botanical Garden workers Britton and Vail were particularly productive. Many of the northeastern women were high school teachers and wrote only a single research publication (typically reporting undergraduate studies). As far as postgraduate training was concerned, Cornell University was the most prominent eastern institution.[135]

2. In the Midwest the land-grant colleges and state universities (particularly Iowa Agricultural College and the University of Michigan), with their associated agricultural experiment stations, were especially notable in providing up-to-date botanical training for women students; this survey indicates that they surpassed the eastern women's colleges in the education of female botanists who had at least a modest amount of research experience. Women from both the Midwest and the eastern seaboard took both undergraduate and M.S. degrees at these institutions, and frequently went on to teaching careers in all parts of the country, including the Far West. The research publications resulting from their graduate-level work typically reported laboratory-based studies, often in morphology, physiology, or plant pathology.

3. The published output, primarily in systematics and taxonomy, of the two preeminent women botanical workers in California, Curran Brandegee and Eastwood, was especially notable; their total contributions (original papers, editing, and the assembling of herbarium collections) were such that a strong case could be made for designating these two as the leading American women botanists of the late nineteenth and early twentieth centuries.

Overall in this pre-1900 group perhaps six might be singled out for special note: Curran Brandegee and Eastwood, Tilden, for her pioneering work on algae, and Britton, Perkins, and Vail for their taxonomic studies. (Wellesley College's plant physiologist and geneticist Margaret Ferguson, although discussed here, belongs more to the next generation of botanists; only her earliest report is listed in the bibliography.)

British women

British women form the second largest national group among nineteenth-century women botanists and they contributed about 26 percent of the botanical articles by women indexed in the Royal Society *Catalogue.* Between them, the British and Americans produced three-quarters of the nineteenth-century botanical journal articles written by women.

The British group differs considerably from the American in that more than a quarter of its sixty-four members were active before 1880, sixteen of them before 1850, and one, Agnes Ibbetson, before 1820.[136] Like many of their American counterparts a few decades later, most of the earliest of these workers were concerned with exploration, observation, and classification, still important areas of activity in British botany before 1850; five, Russell and Kirby from the English Midlands, and the algologists, Gifford, Warren, and Merrifield, who worked on the southern coasts, were especially notable. Among the post-1880 British women, there are two prominent subgroups, one associated with London institutions (the British Museum, Kew Gardens Herbarium, and some of the London colleges) and the other with the women's colleges at Cambridge and the Cambridge Botanical Laboratory.

Early workers

AGNES IBBETSON[137] (1757–1823), the earliest of the British women botanists whose work is indexed by the Royal Society, has a place apart in this story. Between 1809 and 1822 she published no less than fifty-eight journal articles, a prodigious number for a woman of her time. Indeed, she would appear to have been not only the most productive woman investigator in the botanical sciences in any western country in the early years of the nineteenth century, but, at least as measured by the Royal Society record, one of the most prolific women of the period publishing in the science journals. Not until the Italian astronomer-physicist Caterina Scarpellini began to put out papers in the 1850s does any other woman come close to matching Ibbetson's output.

She was the daughter of Andrew Thomson, one of a London merchant family, and was possibly related to Thomas Thomson, chemist and founder of *Annals of Philosophy.* After attending a school for young ladies, she passed her time in her early years in purely social activities. She married James Ibbetson, oldest son of the archdeacon of St. Albans, a barrister and writer of legal history; they had no children. After James Ibbetson died in 1790 she lived quietly with a sister in Exmouth, Devon, for another thirty-three years until her death at the age of about sixty-six. Having a comfortable annuity, she occupied herself with neighborhood philanthropic work and in reading and study. History and the sciences, mineralogy, electricity, and most especially botany, were her chosen interests.

About the time she was becoming active in botanical work there was a marked renewal of interest in the internal structure of plants. Originating primarily in the late eighteenth-century technical improvements of the microscope, this interest sparked investigations of many aspects of plant anatomy and life processes, areas that were to dominate botanical research until about 1860.[138] The period was also one of considerable confusion and controversy, however, complicated at times by unchecked speculation about questions in anatomy and morphology, and Ibbetson had her share of erroneous ideas.

Over the course of two and a half decades she investigated a variety of topics in anatomy and physiology, including bud formation and structure, seed formation and growth, flower and leaf structure, and plant motion. Many of her studies involved microscopical examination of plant sections. Her papers, which usually appeared in Nicholson's *Journal of Natural History, Chemistry and the Arts,* or in Tilloch's *Philosophical Magazine,* generally took the form of fairly lengthy and leisurely discussions. Despite her remarkable industry and her long program of direct observations, her arguments relied heavily on conjecture and inference; her interpretations and broad generalizations would seem to have depended to a large extent on the study of groups of structures that were too restricted to adequately test her broader ideas.[139]

She was especially enthusiastic about her theory of the place of origin in plants (especially trees) of flower buds and seeds. One of her fundamental concepts was that the reproductive physiology of plants resembled that of animals in that flower buds and seeds were formed in a special location, the root, and, in the case of trees for example, traveled through the trunk and branches and eventually pushed through the bark. She interpreted various knots and small markings apparent in wood sections as buds en route to their final destination on the outside of the bark. The idea was not wholly illogical as her example of flowering bulbs demonstrated; considering the bulb to be the plant root, she remarked, "Some botanists have accidentally cut a bulb, and found a flower within: this has been the wonder ever since. Is it not extraordinary that no one should have followed this lead, and inquired whether other plants were not formed in the same manner?" And again, after examining some herbaceous plants, she wrote, ". . . the bud appears above the root . . . and then the flower-stalk rises (while increasing the stem) under the bud till the hour of opening and being fructified arrives, as in the *Primula Cyclamen,* etc." To her these observations were "a beautiful series of facts in botanical physiology, which appear to me unanswerable." She had hoped to publish her work in book form, but her ideas were in conflict with those of several of her English colleagues, and "in my application to booksellers I was assured, that after consulting the first botanists, it was decided that no new facts were wanted."[140] Eccentric and naive as her ideas may now seem, they would have been much less so in her time, and, despite the inherent difficulties, Ibbetson's work demonstrates considerable boldness and initiative, as well as perseverance.

Among the more conventional of the early nineteenth-century British women botanical workers was a sizable group who specialized in seaweeds, an area of much popular interest at the time. Outstanding collections assembled by several of them were greatly valued by their male colleagues. However, botanical discoveries were then regularly made known by direct correspondence between workers rather than via published articles (journals being few and very expensive to produce until the late 1830s), and so the work of only three of these women, Gifford, Warren, and Merrifield, is recorded in the journal indexes of the period. In fact Gifford was the last of a chain of lady algologists extending back over a period of more than a century.[141]

ISABELLA GIFFORD[142] (1823?–1891) was the only daughter of Major George St. John Gifford, who had served with Sir John Moore in the Peninsular War and fought at the battle of Coruña. Her mother, also Isabella, was the daughter of John Christie, Esq., of the Brecon district in the south of Wales. Born in Swansea about 1823, she spent her early life in France, the Channel Islands, and Falmouth; the family finally settled in Minehead, on the Bristol Channel, about 1860.

Although she received her basic education from her mother, a talented and cultured woman, Isabella learned her botany on her own. Her first publication, *The Marine Botanist,* appeared in 1840. Attractively written and portable, it was well received; two later editions were brought out (1848, 1853). Her *Introduction to the Study of Algology* was published in 1853, and a short account of the marine flora of Somerset in the *Proceedings of the Somerset Archaeological and Natural History Society* the same year. In addition she provided substantial help to the country's leading expert on marine algae, William Henry Harvey (professor of botany to the Royal Dublin Society), when he was writing his classic *Phycologia Britannica* (1851).

Land plants were also within her province, and she collected over the heights and valleys of Exmoor behind her home and along the north Somerset coast, gaining a broad, general knowledge of the flora of southwest England. A paper she brought out in 1855 discussed rare plants from the neighborhoods of Dunster, Blue Anchor, and Minehead. She had a particular interest in mosses, and frequently sent specimens of rarer ones to colleagues.

A cooperative person, interested in promoting contacts, she actively supported natural history organizations, joining the

Botanical Exchange Club when it started in 1858 and remaining a contributing member for thirteen years. Shortly before her death she joined the London-based Selborne League, an early nature protection society (begun by a group of women), which later developed into the Royal Society for the Protection of Birds. She had hoped to establish a branch at Minehead, but ill health prevented the undertaking. Her correspondence was wide, and she welcomed visits from botanical colleagues who came to consult her at the Minehead home she shared with her mother. Her life on the whole was a quiet one and somewhat isolated, "so much so, that she would count as her 'field day' a long-ago scientific meeting at Dunster, where a paper of hers was read, and her collection of the plants of West Somerset exhibited."[143] In later years rheumatism confined her to house and garden. She died of influenza, 26 December 1891, in her late sixties.

ELIZABETH WARREN[144] (1786–1864), whose paper on the marine algae of Falmouth waters came out in 1849, was a native of Cornwall. She was born in Truro, on 28 April 1786, and for many years lived in Flushing, near Falmouth. A founding member of the Cornwall Polytechnic Society, she took part in its activities over many years as one of its "lady amateurs," serving on the Ladies Committee, often contributing collections of local plants to its exhibitions, occasionally publishing in its *Transactions,* and later becoming a judge for its natural history competitions.

Although seafaring men among her relatives brought her more exotic plants from the East Indies, Hong Kong, the Falkland Islands, and the Crimea, her special interest was the flora of the coastal waters of her own county. When working in Falmouth harbor she discovered the species *Kalymenia dubyi,* until then unknown on British coasts. Her efforts were recognized in 1863, when she received the Polytechnic Society's bronze medal for her collection of British freshwater algae. Somewhat earlier she had contributed material to John Ralfs *British Desmidieae* (1848), and to *Nature-painted British Seaweeds* (1859–1860) by William Grosart Johnstone and Alexander Croall.

Over the course of twenty-five years she corresponded with Sir William Hooker, exchanged live plants and seeds with him, and discussed identification problems. Hooker encouraged her in the preparation of a *Botanical Chart for Schools* (1839), a colored chart, on canvas, setting out classes, orders and genera according to the Linnean System, and incorporating a limited amount of additional information for beginning students. Originally drawn for use by her own relatives, the chart was developed with editorial help from Hooker and produced for general sale. However, despite notable praise from a reviewer (possibly Hooker) who remarked that it "ought to be in the hands of every teacher of youth throughout the kingdom,"[145] the chart was not a financial success. She eventually donated many copies to schools. She died in Flushing on 5 May 1864, at age seventy-eight.

A third early woman algologist whose work is indexed by the Royal Society is MARY WATKINS MERRIFIELD[146] (1804–1889). The daughter of an eminent barrister and conveyancer, Charles Watkins, she was born in Brompton, London, on 15 April 1804. In 1827, she married John Merrifield, also a barrister. They had at least one child, a son, Frederick, who followed his father and grandfather into the legal profession but also became a much respected entomologist and a fellow of the Entomological Society. Mary was perhaps most widely known for her works on art, which she brought out in the 1840s and 1850s before she turned her attention to botany. She was especially interested in Spanish and Italian fresco painting, and made a special study of texts dating from the twelfth to the eighteenth centuries that described methods and procedures used.[147] From there she went on to writing two art handbooks, *Practical Directions for Portrait Painting in Water-colours* (1851), and *Handbook of Light and Shade, with special Reference to Model Drawing* (1855). The latter ran to fourteen editions.

The family had moved to Brighton some time in the 1840s or 1850s, and Merrifield's popular handbook, *Brighton, Past and Present,* appeared in 1857. Her 227-page *Sketch of the Natural History of Brighton and its Vicinity* came out in 1864, and her first botanical paper, "List of marine algae found at Brighton and vicinity," in the *Phytologist* in 1862–1863, when she was in her late fifties. A short appendix to her algae list appeared in the *Journal of Botany* thirteen years later. She published two other botanical papers during the 1870s in the *Journal of the Linnean Society* and in *Nature,* and also assisted several other algologists, including the eminent Swedish worker Jacob George Agardh. Quite late in life she had learned Swedish to be able to correspond with Agardh. Her discussion and summary of his 1866 monograph on the *Monostroma* genus of algae appeared in *Nature* in 1882. This was one of several reviews she published of works on algae, articles for which she became quite well-known among naturalists.[148] She died in Stapleford, Wiltshire, 4 January 1889, in her eighty-fifth year, and is commemorated in the genus *Merrifieldia* (named by Agardh).

Turning now to the early women workers known for their studies of land plants, we have three, Anna Worsley Russell, Margaretta Hopper Riley and Mary Kirby, who were almost contemporary with Merrifield and Gifford. All three did much of their work in the English Midlands; one of them, Worsley Russell, was perhaps the ablest and most outstanding woman field botanist of her time.[149]

The daughter of John Philip Worsley, a sugar refiner, ANNA WORSLEY[150] (1807–1876) was born in Arno's Vale, Bristol, in November 1807. The Worsleys were one of the leading Unitarian families in Bristol and had long taken part in the city's intellectual life. Several of them had strong interests in the sciences, a circumstance that undoubtedly sustained Anna's botanical efforts over the years.

From early in her life she was interested in natural history. At first she concentrated on entomology but later switched to plants. Her first substantial work was a list of the flowering plants of the Bristol area contributed to volume 1 of the *New Botanist's Guide,* published in 1835 by H. C. Watson, an early authority on the geographical distribution of British plants. In 1839 she brought out her thirty-one-page flora of Newbury, *Catalogue of Plants, found in the Neighbourhood of Newbury,* prepared at the request of a relative, Dr. Joseph Bunny. By then she had also begun to study mosses and fungi, the work for which she was best remembered. As an active member of the Botanical Society of London, she donated a number of moss specimens to its holdings. This society, which functioned for twenty years from 1836 to 1856, for the most part drew its membership from field naturalists, including women, interested in collecting and exchanging herbarium specimens. By and large these were people who, because of their social status, could hardly aspire to membership in the country's leading scientific societies (many were from dissenting background, like Anna, or were engaged in reformist politics).[151] Anna continued to take part in the society's annual specimen exchanges until it disbanded.

She studied the flora of several regions of Britain, from the southwest of England to southern Scotland, and regularly exchanged information with other botanists. One of the places she visited frequently was the small village of Langar-cum-Barston near Nottingham, where her brother-in-law Thomas Butler, another botanical enthusiast, was rector. In 1844, at the age of thirty-seven, she married Frederick Russell of Brislington, near Brisol, a fellow Unitarian and a botanist who had accompanied her on plant hunting expeditions for some years. After a time they moved to Kenilworth in Warwickshire, where Anna began a detailed study of the local fungi. Her list of some of the less common species from the Kenilworth neighborhood appeared in the *Journal of Botany* in 1868. She also prepared a remarkable series of more than 730 carefully executed drawings of fungi. She died in Kenilworth at the age of sixty-nine, on 11 November 1876. Her drawings were bequeathed to the British Museum (Natural History), where they are still housed, and her herbarium and a collection of British birds' eggs went to the Birmingham and Midland Institute. She had no children, but one or two of her nieces and nephews continued the family interest in botany.

MARGARETTA HOPPER RILEY[152] (1804–1899) was one of the earliest specialists on ferns in Britain, making several contributions to the field around 1840. She was born in Nottingham, on 4 May 1804. Her family must have been fairly well-to-do, since, as she liked later to tell, her grandfather had entertained Prince Charles Edward, the Young Pretender, on his march south through Nottingham during the 1745 Scottish Jacobite Rebellion. At the age of twenty-one she married John Riley, land agent to the Montague family of Papplewick, a village near Mansfield, and she lived there for the rest of her life.

She had no children, and from shortly after their marriage she and her husband seem to have devoted their spare time to collecting, cultivating, and classifying native British ferns. John Riley was elected to the Botanical Society of London in 1838; his wife most likely joined the following year. Authorship of a paper entitled "On the British genus *Cystea,*" read before the society on 1 November 1839, has been attributed unambiguously to Margaretta. Shortly after that she donated to the society a dried collection of every species and variety of British fern, and about the same time, she and her husband presented to the group a comprehensive monograph on the results of growing every British species side by side.[153] Later in the same year a paper by Margaretta "On growing ferns from seed, with suggestions upon their cultivation and preparing the specimens," was read before the society. Her short note, "*Polypodium Dryopteris* and *calcareum,*" appeared in the *Phytologist* in 1841.

John Riley died suddenly, at the age of fifty, in 1846, leaving an herbarium of about 2,200 sheets and an impressive collection of some 250 species of living ferns. Margaretta lived for another fifty-three years, but although she assisted other fern collectors, she does not appear to have continued her own fern work and disposed of her collections. She remained a keen gardener and student of natural history, and in addition had a wide variety of other interests. When she was fifty she took up watercolor painting. She also wrote poetry, contributed articles on social and religious topics to magazines, brought out a booklet entitled *The Duties of Woman,* and established a local club to encourage thrift. Even in her old age she had a lively interest in history, philosophy, and politics, and was for long one of the notable personalities in her community. Her childhood memories of the Napoleonic Wars were remarkable; she could recall the actual announcement of the 1815 allied victory at Waterloo. She died of bronchitis, 16 July 1899, at the age of ninety-five.

MARY KIRBY[154] (1817–1893), the second of the five children of John Kirby, a Leicestershire businessman of farming background, was born in Leicester on 27 April 1817. Along with her older sister, Sarah, she had the responsibility of taking care of the household from an early age, her mother having died when the children were still young. She attended several schools in Leicester for short periods and later went to lectures on scientific and cultural topics presented by the Leicester Philosophical Society and the Mechanics Institute.

As a child she lived for a time with her grandmother in a country cottage and tended flowers and plants in a garden of her own, but her first extensive botanical work began (when she was probably in her mid-twenties) when she went to Charnwood Forest to recuperate from an illness. A region to the northwest of Leicester quite unlike the surrounding pastoral midlands, Charnwood Forest was a place of great beauty, with wild uplands and rocky summits. Throughout the centuries it had been deforested and reforested again and again, and by Mary's time was mainly scattered moors and heath. She explored the uplands and the surrounding areas thoroughly; then with the cooperation of a local naturalist, the Rev. Andrew Bloxam, she took up the task of writing a flora of Leicestershire.

Although a great number of British local floras were produced during the first half of the century (following the setting up of many popular local amateur societies), at the time only outdated lists were available for Leicestershire. In Bloxam, Mary had a capable and knowledgeable guide. Perpetual curate of Twycross, Leicestershire, for thirty-two years from 1839 to 1871, and later rector of Harborough Magna, Warwickshire, the Rev. Andrew made important collections of almost all plant groups in the region, particularly Leicestershire and the Charnwood Forest area. He corresponded with most of the leading botanists of the time, provided material from which many new species were described (particularly lichens and microfungi), and contributed to accounts of a number of regional floras.[155] Mary acknowledges in her Preface that her *Flora* "would never have been attempted" without Bloxam's encouragement and patient assistance. He had examined every one of the doubtful plants for her, and contributed his list of plants compiled for "Potter's Charnwood Forest." In addition he wrote what was probably the most valuable chapter in the book—that on the difficult genus *Rubi* (the brambles), "Rubus. Lin. Nat. Ord. Rosaceae" (pp. 37–48).[156]

The work took three or four years to complete. A few preliminary copies incorporating alternate blank pages were brought out in 1848 and circulated among local naturalists, with an appeal for assistance. This produced a considerable quantity of additional information. The main edition, which listed 939 plants, appeared two years later, and it seems to have been well-received by the botanical community; Sir William Hooker wrote to Mary expressing his belief in the great value of local studies of the kind it embodied. Bloxam's detailed descriptions of the *Rubi* were especially appreciated.

Although to a considerable extent the product of much collaborative effort (Bloxam contributed numerous short plant descriptions in addition to his chapter on the *Rubi,* and Mary's sister Elizabeth wrote illustrative notes), the *Flora* nevertheless represents a notable amount of organizational and botanical work on Mary's part. Over the course of its preparation she corresponded with a number of the country's botanical experts and published one or two short notes in the *Phytologist* (see bibliography). The work was most likely the only British county flora to appear under a woman's name in the nineteenth century. Short passages from the classics and quotations from the poets adorn many of the sections. This style followed the fashion of somewhat pretentious sentimentality that had overtaken Victorian field botany at the time (not destroying its value but cluttering up its presentation); it was by no means peculiar to Kirby's work. Elizabeth Kirby's notes offered some general historical information about many of the species listed.

Publication of the *Flora,* however, seems to have marked the conclusion of Mary's serious work as a field botanist. Following the death of their father in 1848, she and her sister Elizabeth, possibly to augment the family income, together began to bring out books for children and young people, at first adaptations of classical stories. About the mid 1850s the two sisters moved to Norfolk, living for a time in Wells-upon-Sea, then in Yarmouth, and finally in Norwich. Their hope was that the region would be better than Leicestershire for Elizabeth's fragile health.

They continued to write, and several of their stories were published by London and Edinburgh firms. When contacted by the publisher Thomas Jarrold, they agreed to provide material on plants for his series, *The Observing Eye.* "It was very easy to do, for I had the botanical knowledge at my finger ends; and Elizabeth had such fluency in writing that the sentences seemed to flow from her pen as readily as pearls and diamonds from the mouth of the fairy in the fairy tale." Their work, *Plants of Land and Water,* appeared under joint authorship in 1857. Shortly after that they met the eminent botanist John Lindley, author of *The Vegetable Kingdom* (1846) and a number of other major botanical treatises. Mary later remarked that it was just as well they had finished their plant book for Jarrold before that meeting because "the great doctor's influence was not calculated to encourage any work, except, like his own, the most scientific."[157] Nevertheless, their contributions of popular nature-study works continued, *Caterpillars, Butterflies and Moths* appearing in *The Observing Eye* series in 1860, *Things in the Forest* in 1869, and *Chapters on Trees* in 1873. At the same time, they wrote stories for periodicals (generally after being contacted by publishers who knew that their work was popular) and their manuscripts sold readily.

In 1860, at the age of forty-three, Mary married a clergyman, the Rev. Henry Gregg. Elizabeth bought for him the living of Brooksby Parish, Leicestershire, and she lived with the Greggs until her death thirteen years later. The sisters continued to write together, finding it a totally absorbing occupation that could transport them away to "a world of our own."[158] They coauthored at least twenty-four books. Mary died at Brooksby, 15 October 1893, at the age of seventy-six, having outlived her sister by twenty years and her husband by twelve.

Among the early British women botanists are three, Kent, Twining, and Becker, who made their best contributions in the form of general instructional works.

ELIZABETH KENT[159] published a series of papers in the *Magazine of Natural History* in 1829 and 1830 in which she set out to explain the Linnean system of classification in a way that would "avoid alarming the young student with a crowd of technicalities, at the very outset, unaccompanied by anything to bribe his attention to them." She wanted especially to encourage young ladies to study some area of natural history, stating her considered opinion that some knowledge of the subject would not be a hindrance to the success of girls in society, despite the fact that they were "not only discouraged from the pursuit of natural history, but are very commonly forbidden it."[160]

A stepdaughter of publisher Rowland Hunter, who succeeded to the firm of bookseller Joseph Johnson, Kent had close family connections to the London literary world. Her older sister Marianne was the wife of poet and essayist Leigh Hunt, and she shared in the Hunt social circle that included John Keats, the Shelleys, and others of the early nineteenth-century romantic movement. She herself had some success as an author in the 1820s. Cautious and conservative, she generally published anonymously. Her most popular works were two literary-botanical monographs, *Flora Domestica,* a discussion of growing plants in pots, then a fashionable hobby, and *Sylvan Sketches,* descriptions of common trees and shrubs. In the style for the time, both were generously embellished with poetic allusions, mythological references, and contemporary romantic verse, including some by her brother-in-law.[161] They sold well.

By the late 1820s however, the economic disasters of the post-Napoleonic War period were ruining Kent's family's business. For a time she continued to edit botanical books and papers, but also advertised in the *Times* offering classes in botany for young ladies (her *Magazine of Natural History* papers were probably inspired by teaching experience). Later she worked as a governess and then conducted a small boarding school, but by the late 1860s was living in very straitened circumstances with a nephew's family.

ELIZABETH TWINING[162] (1805–1889), one of the nine children of Elizabeth Mary (Smythies) and Richard Twining, was born into a family of considerable wealth and position. Her father, a fellow of the Royal Society, was a tea merchant, and her grandfather Richard Twining, also a tea-merchant, was one of the directors of the British East India Company; her uncle Thomas Twining joined the Bengal Service of the company, latterly holding resident positions (official, government-agent posts). Elizabeth took up the task of putting her share of the family fortune to good use and was a well-known and influential figure in her time, the promoter of several philanthropic and educational schemes in London. She helped with the founding, in 1849, of Ladies' College, 47 Bedford Square, later Bedford College for Women, where many of those mentioned in this survey received at least part of their education. In addition, while she lived in the family residence at Twickenham, she restored the parish almshouses and established St. John's Hospital. The organization of the first "mothers' meetings" for working-class London women was another of her undertakings. She wrote a number of works dealing with charitable, social, and religious matters, including *Ten Years in a Ragged School* (1857), *Readings for Mothers' Meetings* (1861), and *A Few Words on Social Science to Working People* (1862). A reformer without being a radical, however, she remained skeptical about the crucial issue of women's suffrage.

Her interest in botany was more than passing. She joined the Botanical Society of London in 1839 and also attended British Association meetings. At the 1847 meeting, she presented a paper comparing British and foreign flora. Among several botanical works she published, the most notable was her *Illustrations of the Natural Order of Plants,* brought out in two volumes in 1849 and reissued in a second edition in 1868; the illustration as well as the writing she did herself. Her *Short Lectures on Plants, for Schools and Adult Classes* appeared in 1858, and *The Plant World* in 1866.

LYDIA BECKER[163] (1827–1890), although she had a special interest in botany, was, like Twining, remembered more for her social work than her contributions to science. In contrast to Twining, however, Becker found her great cause in the women's suffrage movement. Born in Manchester on 24 February 1827, she was the oldest of the fifteen children of Mary (Duncuft) and Ernest Hannibal Becker. Becker was a Lancashire businessman of German extraction. From the age of eleven Lydia lived in the country at Altham near Accrington, Lancashire, and developed an interest in flowers and plants. She was educated at home. In the early 1850s she became active in botanical work, helped and encouraged by an uncle and by Manchester physician and chemist John Leigh, who advised her botanical readings; Leigh directed her in particular studies she undertook on plant structure, the area that remained her special interest.

In 1864 she brought out a small volume, *Botany for Novices: A Short Outline of the Natural System of Classification of Plants,* which taught basic principles via a focus on structure. Her one botanical paper, presented at the British Association meeting of 1869 and published in the *Journal of Botany* for that year, discussed her observations of the effect of a parasitic fungus on the structure of *Lychnis dioica* (morning campion). With help from Charles Darwin, with whom she carried on a botanical correspondence over a number of years, she had deduced that the infestation had caused the normally dioecious plant to present a hermaphrodite condition. The paper gave rise to a certain amount of contention at the B.A. meeting, and it was pointed out that if she were right, this was the first known case in the botanical world of a parasitic fungus altering plant development in such a fundamental way. Becker, however, was perfectly capable of defending her position. She pointed out that, if it was the first instance, there was no reason why she should not be the discoverer, and that furthermore she was quite prepared for others to disagree with her—as far as she could see, the B.A.'s botany section was remarkable for everyone disagreeing with everyone else!

In 1865 she moved with her father to Manchester, and there started a ladies' society for the study of science and literature. This venture was not very successful, but the following year, after hearing a public lecture on the subject of women's suffrage, she began to work for the movement—with tremendous energy and enthusiasm. As secretary of the Manchester National Society for Women's Suffrage, she quickly became one of the country's leaders in the cause. An excellent public speaker, clear, lucid, and logical with a forceful command of her subject, she was soon a familiar figure in parliamentary lobbies. From its beginning in 1870 she was the editor of the *Women's Suffrage Journal* and until her death its chief contributor.[164] In 1870 she ran successfully as an independent candidate for the first Manchester School Board, one of the few public administration posts women could then occupy;[165] she was reelected seven times. Her special interest was the overall welfare of female teachers and students; science education for girls and women was a particular concern. Despite all her public work she remained interested throughout her life in botanical research, and for many years attended the annual meetings of the British Association, even going to Canada for the 1884 gathering. About 1890, being in poor health, she went to Aix-les-Bains in southeastern France. She died in Geneva, 18 July 1890, at the age of sixty-three, and was buried there in the St. George cemetery.

Eight other women botanical workers, Edmonds, Hunter, M'Inroy, Scrivenor, Anna Maria Smith, Stackhouse, Walker, and Wright, published papers and notes before 1880 that were indexed by the Royal Society.[166] Of these Stackhouse and Walker would seem to be the best remembered.[167]

EMILY STACKHOUSE (1812–1870) was born in Modbury, Devon, but spent most of her life in Cornwall. The youngest daughter of the Rev. W. Stackhouse, she grew up in a family with strong scientific interests. She was especially attracted to the study of lichens and in 1865 published a list of Musci native to Cornwall in the *Journal of the Royal Institution of Cornwall.* The following year her discussion of rare plants from the Truro district appeared in the same journal; *Rubus radula* (a species of bramble) was first observed in the area by her. She was perhaps best remembered, however, as a botanical artist. Her drawings and woodcuts illustrated Charles Alexander Johns's *A Week at the Lizard* (1839) and also his very popular *Forest Trees of Britain,* which was reissued eleven times, the last edition appearing in 1919. She died in Truro in March 1870.

MRS. A. W. WALKER (née Paton), wife of General James Thomas Walker, collected plants with her husband in Ceylon (Sri Lanka) in the 1830s. Before leaving Britain she had contacted Sir William Hooker, who gave her advice on how to prepare scientifically useful plant sketches. He later dispatched books and scientific equipment to the Walkers, and they sent him live plants and seeds as well as dried specimens for the Kew collections. Mrs. Walker's botanical journal describing an ascent of 7,352-foot Adam's Peak in the south of the island was published in the *Companion to the Botanical Magazine* in 1835, and her "Tour of Ceylon" in *Hooker's Journal of Botany* in 1840. She also contributed pencil sketches of the flora of Ceylon to the works of Robert Wight, a Scottish physician and botanist who published many volumes on the botany of the Indian subcontinent. Her name is commemorated in *Patonia* (dedicated to her by Wight) and in *Liparis walkeriae* Hooker.

The London group

By the late 1880s a few women botanists were forming connections with research institutions in London, especially the British Museum and the Royal Botanic Gardens at Kew. Although unofficial, these connections (frequently long-term) greatly increased the women's possibilities for botanical investigations. Three of those whose careers are sketched here, Lister, Annie Lorrain Smith, and Barton Gepp, specialized in cryptogam taxonomy, an area that, by the 1880s, with the rise of physiology, was increasingly becoming the province of the museum botanist and the amateur rather than the academic.[168]

GULIELMA LISTER[169] (1860–1949), who began to work at the British Museum about this time, was one of those amateurs. A distinguished botanist of professional competence, she became an internationally recognized authority on the Myxomycetes, or Mycetozoa. Born on 28 October 1860 at Leytonstone in Essex, she was one of the seven children and the third of four daughters of Susanna and Arthur Lister. Lister, a London wine merchant, was one of a Quaker family of considerable distinction: Gulielma's grandfather, J. J. Lister, was a physicist and microscopist; Lord Lister, the famous surgeon, was her uncle; of her three brothers, two became noted physicians and one a well-known anatomist. Her mother, the daughter of William Tindall of East Dulwich, was a good artist with formal training.

Gulielma was educated at home except for a year when she was sixteen at Bedford College for Women, London, where she acquired a solid grounding in systematic and structural botany. Her first publication, a study of Alsineae (members of the Pink family) appeared in the *Journal of the Linnean Society* in 1884. Within a few years she joined with her father in the Myxomycetes studies that were to occupy her for the rest of her life.

A fellow of the Royal Society with lifelong interests in many branches of natural history, Arthur Lister came to the study of the Myxomycetes, or slime molds, about 1887, when he was fifty-seven. He chose a fascinating subject to specialize in. For long considered fungi, the Myxomycetes are a small, little-known group (containing only about 700 species), of eukaryotic organisms of worldwide distribution. At certain stages in their life cycle they have attributes usually associated with animals, including the capacity for locomotion; at other stages they produce fruiting bodies and spores resembling those of fungi. The fruiting bodies are tiny and best observed under magnification; intricate, amazingly varied in form and color, and often very beautiful, they have an almost alien and otherworld appearance to the uninitiated.

Accompanied by Gulielma, who made drawings and kept notes, Lister worked over the collections at the British Museum, Kew Gardens, the Paris Natural History Museum, and the University of Strasbourg. Indeed, the cataloging of the British Museum specimens he and Gulielma carried out (at the suggestion of William Carruthers, head of the botany department) marked the beginning of systematic work on these collections.[170] The result of their labors was Arthur Lister's *Monograph of the Mycetozoa,* which appeared in 1894 under the imprint of the British Museum (Natural History); in the preface he fully acknowledged Gulielma's assistance. The work did much to clear up the then confused state of classification of the group. It immediately aroused widespread interest; sales were excellent, and it was quickly reissued in paperback. Further, it brought a great influx of new material to the British Museum, which led to the recognition of new forms and extended knowledge of the geographical distribution of known forms. After her father's death in 1908, Gulielma Lister continued the work, bringing out a second, revised edition of the monograph in 1911. A third edition followed in 1925; it was further enlarged and illustrated with an outstanding collection of her own detailed watercolor sketches of impressive artistic quality. The third edition remains a classic, a key work in the nomenclature and taxonomy of the Myxomycetes.

A remarkable collection of no less than seventy-four carefully and fully indexed manuscript research notebooks, bequeathed to the British Mycological Society and now located in the British Museum (Natural History), provides a continuous record of the day-to-day routine of the Listers' botanical investigations. These went on for almost sixty years, both in Essex and at their seaside home in Lyme Regis, where they spent several months each year. Entries in the notebooks give details of collections examined and the essence of letters received or written; notes were frequently illustrated by careful drawings or watercolor sketches; original letters, reprints, photographs, and other materials were inserted; separate notebooks were reserved for descriptions of special collections or material from particularly important correspondents. Their contacts, as a result of the *Monograph,* were many and worldwide. They included the emperor of Japan, who sent them material and in whose honor Gulielma Lister named a specimen. She greatly prized a gift to her from the emperor of a handsome pair of porcelain vases.

From 1902 onward, for more than three decades, a long succession of papers reported their results, mainly in the *Journal of Botany,* the *Transactions of the British Mycological Society,* and the *Essex Naturalist.* Until Arthur Lister's death authorship was always joint.[171] Included in the reports were a number on Myxomycetes from Switzerland, where the Listers greatly enjoyed working, although Gulielma's real hunting grounds were Epping Forest in Essex and the region around Lyme Regis. She was known as a painstaking and accurate observer, free from bias and preconceived theories. She also kept abreast of studies by others in the area, even learning Polish to follow the contributions of Josef Tomasz Rostafinsky to the systematics of British and European Myxomycetes.

Her association with the British Museum lasted from the late 1880s, when she first visited its collections with her father, until 1939, when the outbreak of the Second World War made travel between Leytonstone and London too difficult for her. For many years until then she was virtually honorary curator of the Myxomycetes, a collection that she had helped to make the most important and complete then in existence. An exhibit of British species in the Botanical Gallery was arranged by her and her father and enriched with her watercolor paintings giving magnified views of typical specimens. She and her friend, the lichenologist Annie Lorrain Smith, were long remembered at the museum for their interest and patience in helping and teaching younger workers.

Gulielma Lister belonged to several botanical organizations, both national and local. In 1903 she joined the British Mycological Society as one of its first 100 members; she gave the organization considerable help in its early years and twice served as president (1912, 1923). Elected a member of the Essex Field Club in 1907, in 1916 she became its first woman president and thereafter remained a permanent vice president. A regular participant in the club's forays, she was long one of its leading spirits. Of special interest to her was the work of its Stratford Museum, to which she contributed a full collection of Epping Forest Myxomycetes and an important conifer specimen collection. When fellowship in the Linnean Society opened to women in 1904 she was one of the first to be elected; she served on the council (1915–1917, 1927–1931) and became a vice president in 1929. An expert field naturalist (with special interests in birds and conifers[172]), she was a strong believer in the benefits to children of regular nature study in the classroom, and gave much time and wise advice to the London School Nature Study Union (founded in 1902), which she chaired for several years. Of attractive personality, gentle and generous, she had a wide circle of friends. Her skill as an artist often brought requests for assistance; she made the colored sketches for her cousin Frederick Janson Hanbury's *Illustrated Monograph of British Hieracia,* several editions of which appeared between 1889 and 1898, and the line drawings for William Dallimore and A. Bruce Jackson's *Handbook of Coniferae* (1923). The latter was an equally successful work, its final edition appearing as late as 1954. She died in her eighty-ninth year, 18 May 1949, at her home in Leytonstone. Collections of her slides and herbarium specimens are now in the British Museum (Natural History) and at the Royal Botanic Gardens, Kew.

Lister's longtime friend, the Scottish botanist ANNIE LORRAIN SMITH[173] (1854–1937), an "unofficial" worker at the museum for over forty years, became the British authority on lichens. Annie was one of the younger children in the large and talented family of the Rev. Walter Smith of Halfmorton in Dumfriesshire, a minister of the Free Church of Scotland. Two of her brothers became professors of philosophy and a third a professor of pathology. After receiving her early education in Edinburgh, she studied French in Orléans and German in Tübingen. She then worked for a number of years as a governess, but about 1888, in her mid-thirties, began to study botany, joining D. H. Scott's classes at the Royal College of Science in South Kensington. Scott frequently took his students to the British Museum (Natural History), and in this way Annie was introduced to the Cryptogamic Department where much of the rest of her working life was spent.

Her first contribution was the remounting for the museum of a recently purchased collection of microscope slides (the de Bary collection), a job that gave her an introduction to microfungi. From there she undertook the arranging of the exhibition stands of microfungi in the museum's Botanical Gallery, a huge task on which she worked part-time until 1908. Gradually building up her knowledge of fungi, she became assistant to William Carruthers, consulting botanist to the Royal Agricultural Society, and spent three years (1899–1901) carrying out seed testing and examining molds and fungi associated with seed germination. Soon after that she became responsible for identifying most of the fungi collections that came into the museum, including materials from tropical East Africa, Angola, and the West Indies. Her reports on these and on new or rare British fungi appeared in the long series of papers she published from about 1897 onward in the *Journal of Botany,* the *Transactions of the British Mycological Society,* and a number of other journals.[174]

In 1906, when the death of James Crombie (a major figure in British lichenology) left unfinished the second volume of his planned two-volume *Monograph of the British Lichens,* Annie Smith agreed to complete her compatriot's project. The undertaking marked the beginning of her special study of lichens, the area in which she concentrated for much of the rest of her career. Although now thought to have been almost all Crombie's work, the second volume of the *Monograph* appeared under her name in 1911; her thoroughly revised second edition of the first volume followed in 1918 and that of the second volume in 1926. Published by the Trustees of the British Museum, the work quickly became, and remained, an important standard; and its appearance gained her wide recognition. Her *Handbook of British Lichens,* sets of well-illustrated descriptive keys based on the monograph, came out in 1921. It was the only available set of keys to all known British lichens until

1952. She was probably best known, however, for her textbook, *Lichens,* which appeared in the Cambridge Botanical Handbooks series in 1921. This work, though in many ways a pioneer volume, was outstanding for its breadth of outlook and the detailed mass of information it contained, by far the most complete treatment of the group to be published to that time. A reviewer, while acknowledging that lichenology was something of a backwater in modern botany, offered the opinion that it would remain a classic.[175] Indeed it did. When reissued in 1975, it was still the standard work in English on the development of the field.[176]

Along with many of the leading naturalists of the time, Continental as well as Irish and British, Annie Smith took part in the 1909–1911 Clare Island Survey. A well-organized intensive field study of the plant and animal life of Clare Island and the surrounding district of west Mayo in western Ireland, the project was initiated by the Irish naturalist Robert Lloyd Praeger. It was one of the earliest and most outstanding large-scale natural history investigations of the time. The resulting reports (a series of no fewer than sixty-seven, brought out promptly by the Royal Irish Academy) constituted a comprehensive record of the plant and animal life of the area; standards were such that these have remained useful reference tools to the present day. Annie Smith's account of the Clare Island lichens appeared in 1911.[177] Her association at this time with Matilda Cullen Knowles (one of a number of notable Irish women biologists who took part in the survey), inspired Knowles to proceed with an extensive and detailed many-year investigation of Irish lichens.[178]

With the staff depletions of the First World War, Annie Smith became acting assistant in the cryptogamic department at the museum and carried responsibility for work on fungi and lichens. Women were still ineligible to join the museum staff, however, and so she remained an unofficial worker until her retirement, her stipend being provided from a special fund.[179] A Civil List pension and an O.B.E. (Officer, Order of the British Empire), the latter honor conferred in 1934, just before her eightieth birthday, came as recognition of her long service to cryptogamic botany.

She held office in a number of national and regional botanical societies. A member of the British Mycological Society from 1899, when she joined as one of the foundation members, she rarely missed the annual meetings over a period of thirty-five years and was twice president (1907, 1917). She strongly advocated the integration of mycology with lichenology, so as to broaden the scope of the former field; the lichen lists published between about 1919 and 1935 in the reports of the British Mycological Society forays (in most of which she participated) remain valuable records of the lichen flora of the time. In 1904 she became one of the first women fellows of the Linnean Society, on whose council she served from 1918 until 1921. The following year she was president of the Botanical Section of the South-Eastern Union of Scientific Societies, where she had been an active and energetic member for many years. She was also a regular and enthusiastic participant in the fungus forays of the Essex Field Club. Specially useful to her colleagues were the abstracts and reviews of recent publications on fungi and lichens that for more than thirty years she prepared for the *Journal of the Royal Microscopical Society.*

An exceptionally vigorous woman with wide interests, Annie Smith was a strong supporter of women's demands for full citizenship and equality of opportunity. Warmhearted and enthusiastic, she was always ready to help students and younger colleagues. She liked to travel, made a trip to the United States, and in 1914 visited Australia with the British Association. Throughout much of her life she shared her home in West Kensington with an older sister, whose death in 1933 depressed her severely. Although she continued to work until the following year, her health was by then failing. She died in London, 7 September 1937, a few weeks before her eighty-third birthday, after three years of illness.

Like Annie Lorrain Smith, ETHEL BARTON[180] (1864–1922) came to join the Botany Department at the British Museum via D. H. Scott's botany classes at the Royal College of Science. Another "unofficial" worker, she spent almost twenty years at the museum and at the Kew herbarium, and was for all practical purposes a member of the staff. Born on 21 August 1864 at Hampton Court Green, Surrey, she grew up in Ticehurst, Sussex, where her family settled in 1872. For a time she went to the same school as her brothers but later was educated at home. She always attributed her breadth of intellectual interests to the fact that she did not attend a girls' school. Following the death of her mother in 1883, the family members scattered, her father going to India and she to Leipzig, where she studied violin and piano for a year and a half. When she returned to England she spent some time with an aunt in Eastbourne and attended classes in botany, studies she continued under Scott when she moved to London in 1889. On the advice of George Murray, then an assistant in the botany department at the British Museum,[181] she made a special study of marine algae. She greatly extended her knowledge of this group in the summer of 1891 when she worked for several months at the Marine Biological Station at Millport, on the island of Cumbrae in the Clyde estuary. In 1902, at the age of thirty-eight, she married Antony Gepp, a staff member of the botany department in the British Museum, and until 1911, when she became seriously ill with tuberculosis, continued to work at the Kew herbarium and the museum. She also brought up two children, a daughter born in 1905 and a son born in 1908.

Throughout the 1890s she contributed a number of papers on various genera of marine algae to the *Journal of Botany* and the *Journal of the Linnean Society,* her contributions on algae from the Cape of Good Hope (1893 and 1896) being especially notable. Her monograph on *Halimeda* appeared in an account of the Siboga Expedition, a Dutch scientific expedition to the East Indies in 1899–1900; a second monograph, on the *Codiaceae,* written in collaboration with her husband, came out in 1911.[182] Her later publications dealt with marine algae from a number of locations, including China, Ceylon, Borneo, the Indian Ocean, the Kermadec Islands, and New South Wales; more than half of these reports were coauthored with her husband.[183]

With the breakdown of her health she had to leave London, and from 1913 until her death on 6 April 1922, she lived in Torquay, Devon. Her name is commemorated in *Ethelia*, established as a subgenus of *Squamariaceae* by her friend the Dutch botanist Anna Weber van Bosse. Other plants named after her are *Caulerpa bartoniae* G. Murray and *Delesseria bartoniae* F. Schmitz. *Lithothamnion geppii* Lemoine and other species were dedicated jointly to her and her husband.

A fourth London woman whose botanical work began in the late 1880s was HENDRINA (RINA) KLAASSEN SCOTT,[184] the daughter of Hendericus Martius Klaassen, F.G.S. The Klaassens were of Dutch extraction. While a student at the Royal College of Science in 1886, Rina Klaassen took D. H. Scott's advanced classes in botany; the following year she married Scott. Her first research, an investigation of the nucleus-like bodies in the cells of the blue-green algae *Oscillaria* and *Tolypothrix*, was reported by Scott in 1888 (with full acknowledgments to Miss H. V. Klaassen[185]). A few years later she collaborated with Ethel Sargant in studies on seedling development, work carried out at the Jodrell Laboratory at Kew. Their joint paper in 1898 on wild Arum seedlings (see bibliography) was the starting point for Sargant's long series of life-history investigations of monocotyledons (see below).

Rina Scott made several contributions in fossil botany, the field in which her husband was one of the leaders at the time; her papers appeared in the *New Phytologist* and the *Annals of Botany*.[186] Also of note is her pioneering work in the development of cinematography for the illustration of flower and plant movements. Her first article on the subject appeared in the *Annals of Botany* in 1903, and she demonstrated this innovative technique to the Linnean Society shortly after she was admitted to the fellowship in 1905.[187]

A person of wide scientific and cultural interests and considerable energy, she followed her husband's research closely, and also assisted him, drawing illustrations for his books and papers, preparing indexes, and providing secretarial help. She regularly accompanied him to meetings of the Linnean Society, the South-Eastern Union of Scientific Societies, and the annual meetings of the British Association. In addition she brought up a family of four daughters and three sons. All three sons died young, however, one in infancy, the second in 1914, when he was still a schoolboy, and the oldest at the British salient at Ypres in 1917, during the First World War. She was an especially fine hostess, and her home in Richmond was a place where botanists often gathered. After D. H. Scott retired from his position at the Jodrell Laboratory in 1906, the family moved to Basingstoke, Hampshire, where they had bought an old farmhouse, East Oakley House. Rina took on a variety of responsibilities on district and parish committees and cultivated an exceptionally fine rose garden. Many botanists, British and foreign, came to visit. She died suddenly on 18 January 1929.

Shortly after the turn of the century MAY RATHBONE[188] (1867?–1960) worked for a time at the British Museum, where she carried out an investigation of the structure of two brown algae, under the general direction of Ethel Gepp and with help from Ethel Sargant. Her report of this work in the *Journal of the Linnean Society* (1904)[189] and a short note in *Nature* (1898) were probably her only botanical publications, although her interest in research continued throughout her life. The daughter of Theodore Rathbone, J.P., of Neston, Cheshire, she was a member of the Liverpool Biological Society for almost half a century, from 1891 until 1940. Sometime before 1908 she qualified as a Licentiate in Medicine and Surgery of the Society of Apothecaries, London,[190] and she spent many years in the capital. For vacations she went to Norway and worked on the preparation of an English-Norwegian glossary of botanical terms and names, still incomplete at the time of her death. Probably through her friendship with Ethel Sargant, Rathbone joined the small group of women who in 1913–1914 contributed to the founding of what became the Botanical Research Fund, later incorporated as a trust for the encouragement of botanical research. Shortly after the First World War she retired to Chipping Camden at the northern edge of the Cotswold Hills in Gloucestershire. She had been elected to fellowship in the Linnean Society in 1908, and at the time of her death at age ninety-three, on 15 November 1960, was one of the oldest members.

The first woman to hold an official position as curator of a natural history museum in London was KATE HALL[191] (1861–1918), curator of the Borough of Stepney Museum in Whitechapel for sixteen years. She was remembered especially for her social and community work in the then thriving Nature Study movement, which sought to bring the subject into schools as a general elementary introduction to biology.

Born in August 1861, she was the daughter of Harry Hall of Newmarket, an animal painter, and his wife Ellen (Paupe). After attending Highfield School in Hendon, Middlesex, she took classes at University College London, beginning in 1881; however, lacking the necessary level of proficiency in Latin, she never received a degree. Her interests in meeting working-class people took her to Toynbee Hall, the settlement house established in Whitechapel in 1885 by Oxford and Cambridge universities. She was appointed curator of the Stepney Museum about 1893, and quickly started a number of innovative programs, including a weekly museum lecture series for local people, which she organized years before a similar program was established at the British Museum.

Perhaps the work for which she was best known was the founding of a second natural history institution designed especially for the benefit of schoolchildren, the Nature Study Museum at St. George-in-the-East. Here the central attractions were a sea anemone tank and an observatory beehive with glass walls through which all the organization and activity of the bee community could be studied. The bees thrived, feeding not only on local garden flowers, but on plunder from convenient sugar ships at the nearby Thames docks. They became famous far beyond the neighborhood; during summer vacations hundreds of London children visited the Nature Study Museum daily, and the hive attracted workingmen from all over the city. As secretary of the Nature Study Union, Hall did much to encourage the introduction of nature study into the regular school curriculum.[192] She spent her leisure hours conducting groups of children, especially parties of boys, on rambles through the parks and open spaces around London, pointing out seasonal changes and teaching the children how to recognize the birds and flowers. She appears to have published only one journal article, an 1891 report of a joint study on the anatomy of *Tmesipteris foresteri,* a South Pacific cryptogam. The work was done under the direction of anatomist A. Vaughan Jennings of Birkbeck College. Her two books, *Nature Rambles in London* (1908) and *Notes on the Natural History of Common British Animals* (1913), were widely read. She became a fellow of the Linnean Society in 1905. In 1909, at the age of forty-eight, she retired from her post at Whitechapel and moved to Lingfield, Surrey, where she died nine years later, on 12 April 1918.

Two of the botanists whose early papers are listed in the bibliography were heads of departments in London colleges for women—Margaret Benson for twenty-nine years at the Royal Holloway College, and Ethel Thomas for eight years at Bedford College.

MARGARET BENSON[193] (1859–1936), one of the notable British paleobotanists of the early decades of this century, was the sixth of the nine children of Edmunda (Bourne) and William Benson. She was born in London on 20 October 1859, but the family moved to Norton House in Hertford when she was twelve. Much of her early education she received at home. Her father, an architect and civil engineer, had a keen interest in the natural sciences and gave all his children a solid introduction to field botany; from her mother, a successful flower painter, she learned how to paint in both watercolors and oils. After a year of classical studies at Newnham College Cambridge (1878–1879), she became assistant mistress at Exeter High School, where she taught until 1887, earning enough money to return to the university. She took a B.Sc. with first-class honours in botany at University College London in 1891. Supported by a Marion Kennedy research studentship she then went on to postgraduate research under the direction of botanist F. W. Oliver of University College. Her study of the embryology of the *Amentiferae,* carried out in part at Newnham College, became a classic in the field. She was awarded a D.Sc. (London) in 1894.

In 1893 she became head of the newly established department of botany at the Royal Holloway College, where she had held the post of senior lecturer in botany and zoology since 1889. Over the next three decades she built a strong department and planned and stocked an excellent botanical garden and a museum and herbarium. Of great value to her in the early planning was a tour of major European laboratories, which she took with her friend and fellow botanist Ethel Sargant during a one-term leave of absence in 1897.

About 1902 she began her studies on fossil plants, work in which she often consulted with D. H. Scott. From then on this was her major research interest. Her first investigations were carried out when the technique of fossil cutting was in its infancy, and she was a pioneer in the preparation of thin sections for microscopic examination. For many years she cut her own sections at a small lapidary bench in a shed in the college grounds using a machine powered by a gas engine. Especially notable were her investigations of the spore-producing bodies and ovules of the Paleozoic seed ferns (pteridosperms), some of the earliest of the seed plants, established as a separate group only in 1904 by Oliver and Scott. Difficult and demanding studies, which involved the analysis and interpretation of intricate structural features often only 2 to 3 millimeters in cross section, these were valued contributions to the understanding of the development of the seed habit, a major event in the evolution of plant life. Also very well received were her studies on early paleozoic herbaceous plants (lycopods) and her work on early examples of the true ferns. She published about twenty original papers (some of which are classics, still cited in accounts of the development of paleobotany), and her findings were soon incorporated into the textbooks.[194]

An enthusiastic traveler and collector, she frequently went to Europe and the Middle East; during a year's sick leave in 1905–1906 she collected in Australia, and in 1914–1915 she visited Australia, Java, and India. Invariably she returned with valuable material for the Royal Holloway botanical collections.

In 1912, in recognition of the standing of her department and her research contributions, the University of London conferred on her the title of university professor. She had been elected a fellow of University College in 1898, and in 1904 became one of the first women fellows of the Linnean Society. For a time she served as a University of London examiner for honors students.

After retiring in 1922 she returned to her family home in Hertford and for several years busied herself with parish and domestic work. She also kept in close touch with her old department, and her happy, genial personality ensured that she was always a welcome visitor. On the occasion of the opening of the college's new botanical laboratories in 1927 her former staff members made her a presentation of her portrait (which remains in the archive collections of Royal Holloway, University of London).

In 1932, at the urging of D. H. Scott, she returned briefly to botanical work and prepared for publication the information she had on the root structure of the Paleozoic seed fern *Heterangium grievii,* material that previously had been only briefly recorded in print by Scott in his *Studies in Fossil Botany.*[195] Although she was then seventy-three, the work reawakened her old enthusiasm and she brought out two additional papers.[196] Her last contribution was made at the Botanical Congress in Amsterdam in 1935. She died suddenly, in Highgate, London, on 20 June 1936, at the age of seventy-six.

ETHEL MILES THOMAS[197] (1876–1944), whose thirty-year teaching career began with her appointment at Bedford College in 1907, was born in Islington, Middlesex (London), on 4 October 1876, the daughter of Mary Emily (Davies) and David Miles Thomas. David Miles Thomas was from Carmarthen, Wales; at the time of Ethel's birth he worked as a tutor. After receiving her early education at home and at Mayo High School, she became a student at University College, where for a time she was president of the women students' union (and where later she was made a fellow). She also attended Sir John Farmer's botany lectures at the Royal College of Science. She was awarded a B.Sc. (London) in 1905.

Her early contacts with Ethel Sargant, whose research assistant she was from 1897 until 1901, and with A. G. (later Sir Arthur) Tansley at University College, awakened her interest in problems in plant anatomy and evolutionary pathways. Following on from the work of the French botanist Léon Guignard on double fertilization in angiosperms, she was the first in Britain to confirm and extend these studies by her observations of the same phenomenon in the marsh-marigold, *Caltha palustris* (1900). Her subsequent work on seedling anatomy, also carried out in close collaboration with Tansley, led her to her theory of the "double leaf-trace." Founded on investigations of vascular structures in the cotyledons of a large number of angiospermous and gymnospermous types, it formed a notable contribution to ideas about evolutionary lines.[198] Her later work, and that of her students, was mainly directed toward the development of a broadly based interpretation of seedling anatomy, a goal she never actually realized, although her publications continued to appear until the mid-1920s.[199]

Ethel Thomas was a distinguished if demanding teacher, and a vigorous and dedicated builder and developer of departments. When she joined Bedford College as assistant lecturer in 1907 a separate department of botany was just being established. As head of this from 1908 until 1916 she directed the move to new premises in Regent's Park in 1913 and developed there the college's Botany Garden. She also initiated planning for a plant physiology laboratory. In 1912, in recognition of her work, she was given the status of Reader in the University of London. She took a London D.Sc. in 1915. However, serious difficulties in outlook between the outspoken and occasionally inflexible Thomas and Bedford College's more cautious principal, Margaret Tuke, led to her dismissal in 1916. The move considerably upset the botany department, Ethel Thomas having by then a very creditable scientific reputation.

During the First World War she served as a Women's Land Army inspector for London and the Home Counties, and also carried out research for the War Office and the Medical Research Committee. After that she was acting head of the Botany Department at University College of South Wales Cardiff for a year (1918–1919), and subsequently, for two years, keeper of the Botany Department at the National Museum of Wales. In 1923 she joined the staff of recently established University College Leicester, where, almost alone, she developed a biological sciences program. Her botany laboratory was the college's first laboratory. She remained head of the biology department at Leicester until she reluctantly accepted retirement in 1937 at age sixty. She was the only one of the early British women botanists in this survey to hold a regular teaching post in a department at a coeducational university.

Prominent in national scientific organizations, she was a life member of the British Association and regularly attended its meetings at home and abroad, visiting Australia in 1914, Canada in 1924, and South Africa in 1929; in 1933 she became vice president of Section K (Botany). In the Linnean Society, of which she became a fellow in 1908, she served on the council from 1910 to 1915. She was also on the Executive Committee of the Imperial Botanical Conference (1924), and served as an examiner for the University of London. Questions affecting the professional status of women were always of interest to her, especially when they concerned opportunities for research; for a time she presided over the Leicester Branch of the British Federation of University Women.

"A striking but complex personality" was how fellow botanist Ellen Delf described her; ambitious, courageous, warmhearted, "yet capable of a steely coldness, hasty at times, yet generous to a fault." Although her tremendous energy and enthusiasm could lead her into difficulties (which she herself was often unaware of), "On the other hand, she brought life and vigour into whatever she undertook . . ." Vivacious and attractive, with dark, lively eyes, she stood out at social gatherings; a reporter at a British Association meeting, having described her as one of Britain's leading women botanists, added, clearly impressed, "she dances like a moonbeam."[200] In 1933 she married Hugh Hyndman, a barrister, whose sudden death the following year was a shock from which she never fully recovered. After retiring she continued botanical work for a time from a research room in Westfield College (London), but by 1940 her health was poor and she was increasingly handicapped by severe memory loss. She nevertheless maintained her independence, declining help and living alone. She died in Woking, Surrey, on 28 August 1944, at the age of sixty-eight.

Thomas, Benson, and Hall all received much of their botanical training at University College London. Two other early women botanists whose papers are listed in the bibliography, Edith Chick and Mary Ewart,[201] are also known to have been students there, and a third, Agnes Fry, collaborated in statistical studies in biology with Karl Pearson of the college's department of applied mathematics, although she does not appear to have been enrolled in classes.

EDITH CHICK[202] (1869–1970), the first woman to hold a Quain studentship in botany at University College, was one of the ten surviving children of Samuel Chick, a businessman in the lace trade, and his wife Emma (Hooley), daughter of a grocer and corn merchant of Macclesfield, Cheshire. Samuel Chick's forbears were West Country people, tenant farmers from the borders of Devon and Somerset. In the early years of the nineteenth century, largely due to the initiative of Samuel's grandmother, Abigail Chick, they had taken up trading in lace, a local cottage-industry product. Samuel went to London as a young man to further the family's business interests, and his children grew up in the capital. Their upbringing was strict, with attendance at family prayers and chapel mandatory and worldly pleasures such as the theater and dancing forbidden. This somewhat restrictive regime doubtless had its benefits; the daughters grew up to be remarkable women. Six of them studied at University College, three, Edith, Harriette, and Frances, concentrating on science subjects.[203]

Edith, the second child and oldest daughter, was born on 29 October 1869. As a young girl she attended Gower Street School, and at age sixteen was expected to join her father in the lace business. She enjoyed school, however, and wanted to become a teacher. After due consideration her father arranged for her to enter Notting Hill High School in 1886. Recently founded by the Girls Public Day School Company, this school was staffed to a large extent by women of the first generation to receive university-level education, people who had trained at the new women's colleges at Cambridge and Oxford or at London colleges. The instruction was of exceptionally high quality. In particular, science teaching was taken seriously, something that was still not the case at many of the best boys' schools of the time, caught as these were by their traditional strong commitment to the classics.

Edith took both Oxford and Cambridge entrance examinations as well as the London matriculation examination, and entered University College in 1889. Her special interests were botany, physics, and mathematics, although she also studied geology and took the Morris Prize in that subject in 1892–1893. She was awarded a B.Sc. in 1894 and received the class prize in botany during her final year. During 1895–1896 and 1898–1899 she continued studies in botany, and then for five years (1899–1904) held the Quain studentship, which provided a stipend of £100 a year, a generous sum for the period.

In addition to a 1900 study on vasular systems carried out under the general direction of Arthur Tansley (see bibliography), she published three other substantial morphological papers, two of them coauthored by Tansley.[204] During her postgraduate years she took an active part in student activities, serving as an officer of the Women's Debating Society in 1896 and as president of the Social Discussion Group in 1898. She was elected a fellow of University College in 1918. In 1903 she married Tansley, who was the outstanding figure in British plant ecology throughout the first half of this century. They had three children, all daughters, who went on to distinguished careers in, respectively, physiology, architecture, and economics. Edith lived to celebrate her hundredth birthday, dying in 1970.

AGNES FRY,[205] in addition to her collaborative report in statistics with Pearson and his coworkers, published two notes on general botanical topics in the *Journal of Botany* and in *Nature* in the 1890s. She was born in Highgate, London, the seventh of the nine children of the distinguished jurist Sir Edward Fry and his wife Mariabella (Hodgkin), both of whom were of Quaker background. Sir Edward was a judge of the High Court, Justice of Appeal, and later a member of the Hague Permanent Arbitration Court. The family name was one that became well-known in Britain from the cocoa and chocolate manufacturing business, established in Bristol by their forbears in the middle of the eighteenth century.

From his boyhood Edward Fry had been interested in the natural sciences, especially botany and geology. He took an active part in his children's education, preparing some of his daughters for examinations and reading geology with them; much of Agnes's botanical work was carried out in collaboration with him. They were the second British father-daughter team to publish a work on the Myxomycetes, their monograph, *The Mycetozoa, and some Questions which they Suggest,* appearing in 1899, five years after the Listers' monograph on the group. This was followed by *The Liverworts, British and Foreign* (1911). Both books were aimed at a popular audience, but offered thoroughly scientific presentations of life history, structure, and phylogeny of the organisms discussed. Agnes continued to collect and study Mycetozoa for a number of years, and was a member of the British Mycological Society almost from its foundation. Her joint work with Karl Pearson about 1900 concerned studies of variability in the vegetable kingdom and the relation of variability to evolutionary theory. An additional note by her on variation appeared in 1902. She also brought out two literary works.[206]

The Cambridge group

Among the British botanists whose work is listed in the bibliography, the largest group associated with a single institution, and in many ways the most interesting and distinguished group, is that formed by seven Cambridge women. These are Bateson, Dawson, Dale, Gowan,[207] and Pertz, all of whom held research studentships or demonstrator positions, Saunders, a senior member of the Newnham College staff, and Sargant, who was a student at Cambridge, although she did most of her later research in her own laboratory.

ANNA BATESON,[208] (1863–1928) collaborated with Sir Francis Darwin on investigations of the effects on plants of artificial and gravitational stimulation, three of their joint papers appearing in 1888 and 1889. She was one of the seven children of Anna and William Henry Bateson. William Bateson was the master of St. John's College; his wife, the daughter of James Aikin of Liverpool, took a prominent part in the early movement for women's education at Cambridge and was a member of the Newnham College Council from 1880 to 1885. All of the Bateson children had exceptional abilities: Mary, a historian and a lecturer at Newnham and fellow of the college, published two well-received monographs, *Medieval England 1066–1350* (1903), and *Borough Customs* (1904); William, who taught botany at Cambridge, was a pioneer in the field of plant genetics. Anna's 1891 study of floral symmetry and recurrent patterns in plant shape was carried out with him.

She was educated by governesses and at a day school in Cambridge, and before entering Newnham in 1882 spent a year in Karlsruhe, Germany. After she had taken both parts of the Natural Sciences Tripos examinations (1884 and 1886, class II in each), she joined Darwin in work in the Botanical Laboratory. She held a Bathurst studentship and an assistant demonstratorship (1887–1889). By 1890, however, having discovered that what she wanted in life was practical, outdoor work, she made up her mind to become a market gardener. This was something of a disappointment to her brother, William, who felt that, the family having escaped from "trade" only two generations back, she should not revert. Nevertheless, after a two-year apprenticeship with a business in South Wales, she bought some land at New Milton, not far from Bournemouth, on the Hampshire coast. For some years she had a hard struggle and much discouragement, but eventually established a solid connection with Bournemouth markets, bought more land, expanded her greenhouse space, and built up a valuable business, probably one of the earliest market garden businesses in Britain to be run by a woman. Until the last months of her life she worked in the garden every day—during the early years she had spent all her time there.

Anna Bateson was an unusual figure for her day in an ordinary English country district. Physically big and strong, she wore boots, overall, and breeches, worked the land like a laborer, and enjoyed her glass of beer and pipe of tobacco. But she was very much interested in her community and active in public affairs, and in time it became clear to those around her that they had in her a person of sound judgment and generous nature, who, though at times gruff or even hard, could be depended on for sound advice in time of difficulty. During the First World War she was asked to be a member of the local Military Service Tribunal, the body that had the heavy responsibility of deciding which men should or should not be exempted from military service. Over the years she also held a number of local and community posts, serving on parish councils, as a Poor Law Guardian, school manager, president of the Women's Institute and district Nursing Association, and as secretary of both the local War Savings Association and the New Forest Women's Suffrage Society. Of the latter she was an especially keen and active supporter. A woman of broad interests, she had a fine appreciation of music and a zest for literature, classical and modern, French as well as English. She knew French and German well, studied Hebrew, and also Arabic, in preparation for a visit to Egypt. She died on 27 May 1928, at Milford, when she was about sixty-five.

ELIZABETH DALE,[209] who was associated with the Cambridge Botanical Laboratory for almost twenty years, worked mainly in plant pathology, the area of special interest of Harry Marshall Ward, professor of botany at Cambridge from 1895 to 1906. She also coauthored a paper on ferns with paleobotanist Sir Albert Seward. Born in Warrington, Lancashire, on 27 March 1868, she was the daughter of John Gallemore Dale, a manufacturing chemist, and his wife Clara (Heys). She received her early education at home from a governess and then went to a private school in Buxton, Derbyshire. In 1887, after studying for a time at Owens College, Manchester, she entered Girton College. She took the Natural Sciences Tripos examinations in 1890 and 1891 (class I, part I; class II, part II).

In 1898 she returned to Cambridge on a two-year Pfeiffer research studentship, and after that stayed on for fourteen years as a research worker in the Botanical Laboratory. For a time she also held a position as assistant demonstrator in botany in the Balfour Laboratory. Much of her research concerned abnormal plant growth, but she also investigated bacterial diseases in plants and soil fungi. She published steadily until about 1914, a number of her reports appearing in *Philosophical Transactions* and the *Annals of Botany*.[210] Her monograph, *The Scenery and Geology of the Peak of Derbyshire,* appeared in 1900. From 1912 until 1914 she was part-time garden steward for Girton, a position she took on full-time from 1914 until 1917. She spent her later years on the Isle of Wight.

MARIA DAWSON[211] was born in London, 6 March 1875. She attended the Roan School in Greenwich and then took a B.Sc. at University College South Wales. In 1897, when she was about twenty-two, she went to Newnham on an 1851 Exhibition science research scholarship. Her most notable work, a pioneering investigation of the nitrogen-fixing nodules on the roots of leguminous plants, was done under the guidance of Marshall Ward. It was reported in three substantial papers in 1900 and 1901.[212] She also carried out two short studies of much general interest (see bibliography). One of these was an analysis of a small fragment from an ancient manuscript brought to England from the lumber room of the synagogue in Old Cairo. A Hebrew inscription put the date of the manuscript at about A.D. 1038. She was able to show that it consisted of flax fibers, glued together with some kind of starch no longer precisely identifiable. The second was an investigation of a specimen of "Indian Soap" sent to the Cambridge laboratory from Canada. She characterized it as a fungal material.

DOROTHEA PERTZ[213] (1859–1939), another worker associated with the botanical laboratory for many years, was born in London in March 1859. She was the daughter of the German scholar Georg Heinrich Pertz, royal librarian in Berlin, and his second wife, Leonore, a daughter of the prominent Scottish geologist Leonard Horner. Dorothea grew up in distinguished scientific circles; one of her aunts, Mary Horner, was the wife of Sir Charles Lyell, a second was married to the botanist Sir Charles Bunbury, and the Horners were close friends of Sir Charles Darwin and his family. Her early years were spent in Berlin; after her father died she lived for a time in Florence, but the family later returned to England, finally settling in Cambridge.

A contemporary of Anna Bateson, she entered Newnham in 1882 and read for the Natural Sciences Tripos examinations with botany as one of her subjects. After passing with class II honors in 1885 she began research, although she also carried out a certain amount of teaching, lecturing at Newnham and conducting correspondence courses. Like Anna Bateson, she collaborated with Sir Francis Darwin on problems in plant physiology, particularly the effects of artificial stimulation. The best known of the Darwin-Pertz joint papers was probably that on the artificial production of rhythm in plants, which appeared in the *Annals of Botany* in 1892.[214] She continued to publish in this area until about 1905, putting out a number of independent notes as well as joint papers.[215] She also collaborated with William Bateson in the late 1890s on some of his early work in plant genetics; their joint paper on inheritance in *Veronica* appeared in 1900.

Although she kept up her work in plant physiology for a time after Darwin retired from his official positions at Cambridge, she began to feel more and more that success in the field was coming to depend on a background in chemistry, physics, and mathematics, a background that she considered she lacked. She therefore looked around for other ways to make useful contributions to botany, and found two, one using her near-native fluency in German, and the other making use of her considerable artistic skill. At the suggestion of Cambridge botanist Frederick Blackman she undertook the job of preparing an index of the papers in plant physiology which had appeared in the early volumes of *Biochemische Zeitschrift* and *Zeitschrift der physiologischen Chemie*—a formidable task that occupied her for a very long period. Before her last illness she had completed the index up to 1935, covering 268 volumes of *Biochemische Zeitschrift* alone, a very considerable and much appreciated contribution to the resources of the Cambridge plant physiology department. She also prepared the illustrations for a long series of papers on floral morphology, put out between 1923 and 1936 by her friend Newnham botanist Edith Rebecca Saunders. This also was a major undertaking, but it allowed Saunders to publish much more work than would otherwise have been possible.

Summing up Pertz's work, Saunders commented that although her friend had enjoyed research, "she had not that quality of mind which seeks its own road towards some goal, being, as a friend expressed it, 'possessed of great intellectual humility in scientific matters'."[216] She was elected a fellow of the Linnean Society in 1905, one of the small group of women admitted in the first months after fellowship was opened to women. In 1932 she received an M.A. degree. During most of her life she was active in social service of one kind or another. After taking training in massage she worked at one of the soldiers' convalescent hospitals in Cambridge throughout the First World War, which must have been a particularly trying time for her with her half-German background. After the war she continued to help some of her former patients who still needed treatment. She did much to encourage the men in occupational therapy, and found the time to teach one Gypsy patient how to read. She died on 6 March 1939, a few days before her eightieth birthday.

Saunders and Sargant, the remaining two Cambridge women botanists whose early papers are listed in the Royal Society index, were without doubt among the most notable women botanists of their time.

EDITH REBECCA SAUNDERS[217] (1865–1945), plant geneticist and morphologist, was born in Brighton, Sussex, on 14 October 1865. She was the daughter of Jane Rebecca (Whitwell) of Islington, Middlesex (London), and John Saunders, a hotel keeper. In 1884, after completing her early education at Handsworth Ladies' College, a school near Birmingham, she entered Newnham as a Birmingham Scholar. She took the Natural Sciences Tripos course, passing the examinations in 1887 and 1888 (class II, part I; class I, part II, botany). She held a Bathurst studentship in 1888–1889 and then was appointed demonstrator in biology at Newnham.

At that time, university lectures were not fully open to women and the Newnham and Girton staff had the responsibility for providing as best they could a considerable amount of the science instruction for women. Edith Saunders and her friend the physiologist Marion Greenwood (see chapter 6) were two of the first staff members to undertake the organization of practical work in biology for women students in the newly converted former chapel known as the Balfour Laboratory. When Greenwood resigned in 1899 to marry biologist George Bidder of Trinity College, Saunders assumed sole responsibility for the direction of this laboratory. For many years she continued to teach much of the practical work in the natural sciences for women students and she was director of studies in natural science at Newnham until 1925. She is said to have cherished happy memories of the strenuous early days when a great deal of work had to be done with very inadequate facilities, but when, at the same time, the knowledge that they were breaking new ground in women's education provided a tremendous stimulus.[218]

Early in her career she became interested in experimental work in evolution, and she was the first person to collaborate with William Bateson on plant breeding experiments. Indeed, she has been called the "mother" of British plant genetics.[219] Her earliest work, on the distinctness of forms and intercrossing processes in *Biscutella laevigata* (a plant in which both dominant and recessive characters can be followed), was published in the 1898 *Proceedings of the Royal Society.* In December 1901 she and Bateson submitted to the Royal Society a report presenting the results of her cross-breeding experiments with *Lychnis, Atropa, Datura,* and *Matthiola.*[220] The data on the first three of these provided simple illustrations of the principles of inheritance first set out by Gregor Mendel in the 1860s (and brought to the fore again by several Continental botanists in 1900).[221] The phenomena presented by Saunders's fourth plant, *Matthiola* (the garden stock), were more complex, and were to occupy her over much of the next twenty-five years. After 1902 she and Bateson were joined by a number of other workers (including Muriel Wheldale and Florence Durham),[222] and the team thus formed began the modern study of plant genetics in Britain.

The work was controversial in that it presented a strong challenge to another vigorous group of investigators in the area of heredity and evolution. These were the biometricians, led by W. F. Raphael Weldon and Karl Pearson, who had already had considerable success with a series of statistical studies of random variation in wild populations. The Weldon-Pearson work depended on the assumption that variation was the product of small continuous changes, and the discontinuity of inheritance of some characteristics demonstrated by the Bateson-Saunders findings thus attacked their basic premise. The often acrimonious debates between the already well-established Biometric School and the Mendelians, as Bateson's group came to be known, went on until Raphael Weldon's death in 1906, after which the controversy was somewhat more muted. Bateson's people had a difficult time gaining acceptance for their new field, even finding themselves shut out of some of the normal channels of communication for a number of years. For a time *Nature* declined to publish their papers.[223] Between about 1918 and 1932, the work of Ronald Aylmer Fisher, J. B. S. Haldane, and the American Sewell Wright, largely reconciled the Bateson-Pearson controversy by interpreting the results of biometry in terms of Mendelian inheritance. This led to the formulation of basic models for evolutionary change, which provided the foundation of modern population genetics.[224]

Saunders's successful analysis of crossbreeding in *Matthiola* presented the most complicated case of genetic interaction then known. *Matthiola* has four independent recessive genes, any of which may make a plant of this normally hairy species glabrous (smooth-skinned). Two of them also make it white. In addition to unraveling the complex interactions that this combination of genetic factors gives rise to, Saunders also described, in connection with doubleness, what, to use later terminology, was the first case of balanced lethals, and the second case of linkage. She reported these and related studies in a series of more than twenty papers, mainly in the period up to 1920.[225] Her selection that year as president of Section K (Botany) of the British Association reflected her position as one of the country's leading figures in plant genetics. In her presidential address she surveyed the development of the field till then, making special reference to the new chromosome theory of heredity put forward by American geneticist Thomas Hunt Morgan. Indeed, her recognition of the importance of cell-level phenomena, and her acceptance of the idea of a connection between phenotypic appearance and chromosome behavior before Bateson came round to that point of view, marked the beginning of a new phase in genetics studies in Britain.[226] She was president of the Genetical Society from 1936 to 1938 and its treasurer for many years.

Her genetics studies involving observations of the hairs on the stems and leaves of plants led her more and more into morphological investigations. Among the thousands of plants she had raised, she had noticed a number with abnormal floral structures. In 1925 she published a paper on the gynoecium (female organs) of the Cruciferae from the point of view of abnormal carpel structures; suggesting that these were derived from four fused carpels, two of which were normally highly modified, she put forward her theory of carpel polymorphism. She considered that the vascular structure of the gynoecium was an important indicator of the evolution of the organ, and after studies of the bundle system of flowers of other families, concluded that carpel polymorphism was present in most of those examined. Her long series of papers on vascular and carpelary structure in a wide range of flowers appeared between about 1922 and 1936, mainly in the *Annals of Botany* and the *New Phytologist*.[227] She summarized her results and theories in a two-volume work, *Floral Morphology* (1937–1939).

Saunders's suggestion that vascular bundles may persist after the floral members that they supplied in ancestral types have been lost in the morphological sense was by no means universally accepted; but she always defended her position stoutly and could point to the existence of many flower structures that were not easily accounted for by other theories then current.[228] By the time of her death the consensus had swung decidedly against her however, particularly on the point of her reluctance to allow sufficiently for evolutionary change in the vascular system.[229] Later commentators have taken a somewhat more favorable view. It is now thought that vestigial vasculation may, in some cases, afford phylogenetically valuable evidence; that is, its study can help to clarify floral structure and improve classification.[230]

In the 1920s and 30s Saunders's theory had a marked effect on morphological thought and did much to stimulate new work on evolutionary aspects of floral structure. Her descriptions of floral vasculature and the accompanying diagrams (executed by Dorothea Pertz) were extremely reliable and very extensive, and her papers were heavily cited by students of floral anatomy for decades. Overall, her work in the area resulted in descriptions of 249 families, 209 dicotyledonous and forty monocotyledonous, a tremendous achievement. Rudolf Schmidt, writing in 1977, commented that she had probably examined more floral material than any other anatomist.[231]

She enjoyed traveling and went with the British Association to Australia, New Zealand, South Africa, China, Japan, and North America. She was one of the first women fellows of the Linnean Society, being elected in March 1905; later she served on the council (1910–1915) and was vice president in 1912–1913. In 1906 she received the Banksian Medal from the Royal Horticultural Society, and in 1925 she became a fellow of that society. As a young woman she had been an alpine climber and throughout her life was an exceptionally fine skater. Tall and handsome, she was remembered as a distinguished figure at botanical meetings. Her intellectual outlook was wide and her conversation lively—"stately in expression and rich in detail."

> On one occasion at an international dinner in Holland at which Royalty was present, she was called upon by the British Ambassador to make an impromptu speech in answer to a somewhat unsuitable reference made by another delegate to the presence of women at the dinner. Miss Saunders rose to the occasion as only she could have done and emphasized briefly the claims of women scientists in simple dignified Biblical English. The impression she made upon the Prince was such that he left the chair, came down to her and asked to have her presented to him. She acknowledged the courtesy with the bow from the waist, which was her lowest obeisance. The Royal personage was charmed and lingered to make her further acquaintance.[232]

She could be a severe critic, but was a loyal friend. In matters of principle she refused to give in. Rather than accept the titular degree when that was finally offered by Cambridge to women in 1923, she preferred to remain (for the purpose of borrowing library books) a resident research student attached to a tutor. "She was essentially humble-minded and never sought recognition for herself, although she was a bonny fighter in genetic controversy and never knew when she was beaten."[233] With the outbreak of the Second World War she gave up research and took on full-time work in the Cambridge offices of the Y.M.C.A., the Women's Voluntary Service, and in the library of the Addenbrooke hospital in Cambridge. She was in the process of resuming her scientific work when she died, 6 June 1945, at the age of seventy-nine, in Cambridge, following a bicycle accident. Unfortunately much of her huge collection of microscope slides and preserved material was destroyed after her death.

ETHEL SARGANT[234] (1863–1918), known among her colleagues especially for her contributions to the study of seedling anatomy and her theories on the origin and relations of monocotyledons, was the first woman ever to preside over a section of the British Association. She was born in London on 28 October 1863, one of the six children of Catherine Emma and Henry Sargant. Sargant was a barrister and conveyancer of Lincoln's Inn; his wife was the daughter of Samuel Beal, M.P. for Derby. Ethel received her early education at the North London Collegiate School, which, under the direction of Frances Mary Buss, was one of the leading girls' schools in England at the time and one that placed a greater than usual emphasis on science subjects. From there she went to Girton College, entering in 1881. She took both parts of the Natural Sciences Tripos examinations (class II, part I, 1884; class III, part II, 1885).

In 1892–1893 she spent a year at the Jodrell Laboratory in Kew Gardens, where she studied under the guidance of D. H. Scott, acquiring experience in research methods, especially plant anatomy. Her first paper, published jointly with Scott in 1893, concerned the physiological anatomy of *Dischidia rafflesiana.* Nearly all the rest of her work was done in her own laboratory, first one built in the grounds of her mother's house at Quarry Hill in Reigate, Surrey,[235] and then one at her own home in Girton Village, Cambridge. In part this was because she was much occupied with the care of both her mother and an invalid sister for considerable periods of time; but she was also reluctant to take on the duties of an academic post, being convinced that, in her case, lecturing and demonstrating would have seriously crippled the faculty for original work. For that she felt she needed enough solitude to be able to concentrate her mind on the problem on hand for long stretches without interruption.

While at the Jodrell Laboratory she had begun studies in cytology as well as anatomy, and these she continued fairly intensely throughout the 1890s. After early work on the question of whether centrosomes exist in the higher plants, she moved on to a general study of oogenesis and spermatogenesis in *Lilium martagon* (the Turk's-cap lily), reported in the *Annals of Botany* in 1896 and 1897. One of her most important results was her demonstration of the synaptic phase in living cells, a point about which there was considerable doubt at the time.

In 1897, with her friend Margaret Benson of Royal Holloway College, she visited several of the leading Continental laboratories, and at Bonn had a memorable meeting with Edward Strasburger, one of the most influential botanists of the period, with whom she had already corresponded. Strasburger's epoch-making research in the late 1870s and 1880s had made clear the process of nuclear fusion in the ovules of gymnosperms and angiosperms, and thereby established modern understanding of reproduction in the higher plants. In 1898 the Russian botanist S. G. Navashin extended this work by uncovering the phenomenon of "double fertilization" in the flowering plants. When Navashin's work first became known in Britain, Sargant took a second look at hand sections of fertilized ovules of *Lilium martagon,* which she had cut four years earlier but hardly glanced at. On examination, she found that they showed double fertilization perfectly. These sections were viewed with much interest at a meeting of the Royal Society in 1899, but she much regretted having missed her chance to discover the phenomenon herself.[236]

Sargant's second area of research was the comparative anatomy of seedlings, also work she started during her year at Kew. Her first paper on the subject (1898), written jointly with Rina Scott, was on the development from seed of *Arum maculatum.* In 1900 she described a new type of transition from stem to root in *Anemarrhena* (Liliaceae), and she followed this two years later with the preliminary statement of her theory on the origin of the seed-leaf in monocotyledons, a theory fully developed in her long memoir in the *Annals of Botany* in 1903.[237]

She had come to this theory after extensive investigations of the anatomical structure of very young monocotyledonous seedlings (at the time an almost untouched field) and comparison of these with certain dicotyledonous seedlings that are exceptional in having a single seed leaf. Her conclusion that in the former, as well as in the latter, the one cotyledon represents a fused pair was contrary to that of the majority of botanists up to that time. She went on to explain the peculiarities of the structure of monocotyledons by the hypothesis that they were essentially geophilous plants originally possessing underground stems, such as bulbs, corms, or rhizomes, as many of them still do. That is, she suggested that monocotyledons arose from some primitive group of dicotyledons via a geophilous habit. The theory offered a satisfactory explanation of the main characteristics of monocotyledons, including the single seed leaf, the usual absence of secondary growth and the sheathing leaf base. Her work was for long cited in accounts of the development of comparative embryology and the origins and relationships of the monocotyledons.[238]

In the course of her studies on seedlings, Sargant also investigated the difficult problems of the anatomy and morphology of the grass embryo. Two papers published jointly with Agnes Arber in 1905 and 1915 reported her findings in this area.[239]

She was elected a fellow of the Linnean Society in 1904 and was the first woman to serve on its council (1906). Her presidential address to the botanical section of the British Association in 1913 (a critical discussion of the progress of botanical embryology since 1870) was an especially memorable one.[240] Although always rather unwilling to be distracted by major lecturing commitments, she gave a course on the ancestry of angiosperms at the University of London in 1907. And from time to time she acted as research adviser to Cambridge students who came to her laboratory in Girton Village where she lived from 1912 on, following the death of her mother. In 1913 she was elected an honorary fellow of Girton College.

Over the years she published about nineteen papers, and her writings, clear, elegant, and vigorous, reflected not only the precision of her observations but also her great enjoyment of English language and literature.[241] Higher education, particularly university education for women, was one of her special interests,[242] and for a period she served as president of the Federation of University Women. During the First World War she gave much time and labor to preparing the Register of University Women for War Work. She died in Sidmouth, on the south coast of Devon, on 16 January 1918, at the age of fifty-four, and was buried in the churchyard of the little church at

Girton. Like Rebecca Saunders, Ethel Sargant had a tremendous zest for research and found keen enjoyment in her botanical work.

Other later contributors

Among the remaining late nineteenth-century women botanists whose papers are listed in the bibliography are four (Auld, Warham, Huie, and Fingland) who were associated with educational institutions, and a larger grouping of independent workers, many of whom contributed one or two papers on local flora to journals which were put out by then flourishing regional clubs and associations.

Both Helen Auld and Amy Warham carried out their botanical investigations as postgraduate students at University College Liverpool, working under the direction of Robert Harvey-Gibson. Warham, who held a B.Sc. from Victoria University, Manchester, taught at the Girls' High School, Pendleton, in 1890–1891 and then lived for a time in St. Andrews. She held a student membership in the Liverpool Biological Society from 1889 until 1896.[243] Lily Huie, from Edinburgh, was a student at Oxford for at least two years between about 1896 and 1898. Her work on cell structure in *Drosera rotundifolia,* carried out in the Oxford physiological laboratory, was reported in two substantial papers in the *Quarterly Journal of Microscopical Science* (see bibliography). She continued work on cell structure and physiology for a number of years after the turn of the century, using facilities in the laboratory of the Royal College of Physicians, Edinburgh, and a microscope in her own home. Her findings were published in the *Proceedings of the Scottish Microscopical Society.*[244] She also brought out a short note in the *Annals of Scottish Natural History* describing observations on weevil life history. Mrs. Fingland published one paper in 1900 on raspberry roots, coauthored by G. F. Scott-Elliot of the botany department of the Royal Technical College (now Strathclyde University), Glasgow.

The larger group of independent workers includes MARIAN OGILVIE-FARQUHARSON[245] (1846–1912), who, although she published only two botanical papers (notes on ferns and mosses from the north of Scotland) and a small *Pocket Guide to British Ferns* (1881), has a special place in the history of British women botanists. Over the course of several years, from 1900 until about 1903, by a determined campaign of repeated petitioning, she almost single-handedly opened fellowship in the Linnean Society to women, overcoming the daunting inertia and resistance of the society's council. This was no small achievement. The Zoological Society and the Royal Entomological Society had admitted women as fellows from their beginnings in 1829 and 1833, respectively, and the Royal Microscopical Society (of which Mrs. Ogilvie-Farquharson was a fellow) had first admitted women, with some limitations, in 1884. However, fellowship in a number of other important national scientific societies, including the Geological Society and the Chemical Society, remained firmly closed to women for another fifteen or twenty years. In April 1900, when she first petitioned for admission to the Linnean Society, the council tried to avoid the question by claiming, unconvincingly, that it could only receive such communications through a fellow. Two months later she resubmitted the petition through Lord Avebury, one of the society's past presidents. But further stalling and prevarication followed, and doubts were raised about the provisions of the society's charter being applicable to women. A barrister's opinion was sought and his verdict was that they were not. Nevertheless she kept up her bombardment of the council, and, with a considerable number of fellows behind her, finally persuaded it to ballot the fellows on the question. A majority being in favor of admitting women, the bylaws were altered and a Supplemental Charter obtained in April 1904. Women were first elected fellows in December of that year.

Marian was born at Privet, Northamptonshire, on 2 July 1846, the oldest daughter of the Rev. J. Nicholas Ridley of Hollington, Hampshire. She received her early education at home, although she also attended classes in London. In 1883 she married Robert F. Ogilvie-Farquharson of Haughton, Aberdeenshire, who had interests in botany and natural history and was president of the Alford (Aberdeenshire) Field Club. Much of the rest of her life was spent in eastern Scotland. Her special interest was the advancement of women's right to participate in professional and especially scientific work. She founded, and for a time presided over, the Scottish Association for Promotion of Women's Public Work. At an international congress in Paris in 1890 she spoke on the position of women in science, and at the Glasgow Exhibition of 1901 she gave an address on the past and future work of women. She did not herself become a fellow of the Linnean Society in 1904, having been blackballed at the first ballot electing women, a turn of events that hurt her deeply.[246] She came forward again as a candidate in March 1908, and that time was elected, but was unable to attend the meeting for formal admission; she was sixty-two and suffering from heart trouble. She died a few years later in Nice, at the age of sixty-five, on 20 April 1912, and was buried in Alford.

Among the modest contributions by other independent women workers of the 1880s and 90s is a note on the plants of the Faeroe Islands by Caroline Birley and Miss L. Copland, which appeared in the *Journal of Botany* in 1891. Birley was best known for her non-botanical works, mainly children's books, published in the 1880s. She had broad interests in the sciences, however, and a collection of zeolites she bequeathed to the British Museum was one of the important additions to its mineral collections made during the early years of this century.[247] Millicent Thomas and M. Buchanan White each published a paper on botanical observations made in Scotland. Thomas's 1900 report of the alpine flora of Clova (in the upper valley of the South Esk) appeared in the *Transactions and Proceedings of the Perthshire Society for Natural Science;* Buchanan White's 1892 note on the fungus *Strobilomyces strobilaceus* found in Perthshire was published in the *Annals of Scottish Natural History.*[248]

The Norfolk botanist and botanical artist Alicia Barnard[249] (1825–1911), although a contemporary of Isabella Gifford and

Mary Kirby, appears to have made her main written contributions in the 1880s, when she was already over sixty. Born on 22 March 1825, she was the grandniece of Sir James Edward Smith, the founder and for forty years president of the Linnean Society. She contributed a list of bryophytes to R. H. Mason's *History of Norfolk* (1884), and the following year published a short note on *Thrincia tuberosa* in the *Transactions of the Norfolk and Norwich Naturalists' Society.* She, her sister Frances, and a brother all held memberships in the Botanical Society of London in the 1840s. She died in Norwich, 1 May 1911.

Ada Selby's list of flowering plants observed in Hertfordshire during the years 1883 to 1888 appeared in the *Transactions of the Hertfordshire Natural History Society and Field Club* in 1886 and 1888. May Roberts contributed two papers on mosses to the *Journal of Botany* in 1896 and 1897. Her observations were made along the upper reaches of the Dovey River in western Wales, near Dinas Mawddwy, in Merioneth County. Miss Dodd's 1894 note on the wildflowers of Mona appeared in the journal of the Isle of Man Natural History and Antiquarian Society.

HELEN SAUNDERS[250] (1830–1914) of South Molton, Devon, published a list of the wild plants of that district and neighboring parishes in the *Report and Transactions of the Devonshire Association* in 1894. Although she received part of her education in France and Germany and spent a few years as a governess in London, she lived much of her life in Devon, giving her time to parish work and writing. She joined the Devonshire Association in 1895 and regularly attended its annual meetings. In 1896 she served on the council and for a time was also a member of the botanical and folklore committees. For more than twenty years she took part in the botanical excursions of the North Devon Botanical Record Committee and was the first to report the presence in England of *Euphrasia minima.* She was also the first to find the Crowberry (*Empetrum nigrum*) in the Exmoor region. As well as her botanical contributions, she published articles on Devonshire folklore and local history.[251] She died 6 October 1914 at age eighty-four.

Also among the later workers were Clarke, Vickers, Layard, Griffiths, the Countess of Selkirk, and Gaye.

GERTRUDE CLARKE[252] (1868?–1929), one of the first women in Britain to take a B.Sc. degree specializing in botany, was the oldest daughter of J. St. Thomas Clarke, a Leicester surgeon. About 1896 she married Dr. Charles Nuttall, also of Leicester. She published a number of semipopular botanical articles,[253] but was best known as the author of the descriptive text for *Wild Flowers as they Grow* (1911), a work illustrated with colored photographs by H. Essenhigh Corke. She and Corke subsequently brought out two similar works, *Trees and how they Grow* (1913), and *Beautiful Flowering Shrubs* (1920). All three of these works ran to several editions, a revised edition of *Flowering Shrubs* appearing as late as 1953. She was a very good public speaker and an excellent organizer; during the First World War she went with the British Red Cross to France and was in charge of recreation huts, work for which she was mentioned in dispatches. She died in St. Albans, Hertfordshire, on 4 May 1929, at the age of sixty-one.

Algologist ANNA VICKERS (1852–1906) was born in Bordeau, on 28 June 1852, published her work in French journals, and died in Roscoff, Brittany, on 1 August 1906. Nevertheless she is included in Ray Desmond's *Dictionary of British and Irish Botanists* (1977) and it would seem likely that her father was British. She published two botanical papers, one on algae of the Canary Islands and the second on algae of Barbados. The former reported fieldwork done during a six-month stay on Grand Canaria in the winter of 1895–1896; the latter recorded studies made during two trips to the West Indies in 1898–1899 and 1902–1903.[254]

She also visited Australia and New Zealand, making the voyage in the company of her parents and two sisters in 1879–1880, when she was twenty-seven. During a stay of six months, the family toured extensively through the hinterlands of Sydney, Melbourne, and Adelaide, explored along the coast of South Australia, and visited Tasmania before going on to New Zealand. Anna was especially interested in the Maori language and presented a short discussion of word derivations (tracing roots to Sanskrit in many cases) in her travel monograph *Voyage en Australie et en Nouvelle-Zélande* (1883). Although botanical topics were not emphasized particularly in this work, she did mention the great variety of exotic algae found along the southern Australian coast between Melbourne and Adelaide. She was also very impressed by the ferns of the region, some immense, some tiny and fragile. Included in the book are many illustrations, sketches made from her photographs, of ferns of both countries and some New Zealand mosses.

Over the course of more than two decades she carried out investigations of marine flora around Roscoff (where there was a marine laboratory, although there is no evidence that she was associated with it), Antibes, Naples, in the Canary Islands, and the West Indies. On her visit to Grand Canaria she discovered thirty-three new species.

When she died at the age of fifty-four, three years after her second trip to the West Indies, a considerable amount of work resulting from this last expedition remained unpublished. The two furthest advanced sections (on *Chlorophyta* and *Phaeophyta*—green and brown algae) of a planned iconography of the marine algae of Barbados were turned over by her family to her friend and colleague Mary Shaw for completion. A lavishly prepared folio-sized volume was brought out in 1908.[255] It contained ninety-three plates of detailed anatomical drawings by Vickers alongside colored representations of the plants drawn by the artists the Mlles. Trottet, the whole augmented with explanatory notes by Shaw. Of the species figured, two of the *Chlorophyta* and three of the *Phaeophyta* were new (Vickers also discovered eight new species of *Florideophyceae,* but these were not covered in the iconography). Shaw had accompanied Vickers on her second Barbados trip, a visit of three and a half months in the winter of 1902–1903, and collected with her along the coasts and coral reefs of the island. She noted in her introduction to the iconography that, in all, Vickers had added twenty-seven new species to the marine flora of the Antilles. Most of the Barbados material went to leading museums in Europe and the United States.

Nina Layard of Ipswich was elected a fellow of the Linnean Society in 1906. She presented observations on several points in plant anatomy at British Association meetings in the 1890s, but her main scientific work was in archaeology (see chapter 14). Frances Griffiths published two papers on botanical topics in the *Proceedings of the Edinburgh Royal Society* in 1888; one, on the effects of light of particular wave lengths on root absorption of minerals and plant growth, was coauthored by her husband, Arthur Bower Griffiths.

Cecely James, Countess of Selkirk,[256] whose note on the fungus causing leaf curl in the peach appeared in the *Journal of the Royal Horticultural Society* in 1900, was the daughter of Sir Philip de Malpas Grey-Egerton, tenth Baronet of Egerton and Oulton in Cheshire, and his wife Anna Elizabeth, daughter of George John Legh, of High Legh, Cheshire. In 1878 she married Dunbar James, sixth Earl of Selkirk, and lived in Kircudbrightshire and London. She died on 10 January 1920.

Selina Gaye[257] was a popular writer, known for her novels, children's stories, translations, and sketches from Hungarian history, which she published over the course of more than thirty years, beginning in 1861. Her paper on arums in the *Midland Naturalist* in 1885 would appear to have been her single botanical article, but she also brought out a natural history work, *The World's Great Farm; Some Account of Nature's Crops and how they Grow* (1894). In the latter, drawing heavily on contemporary travel works, she described plant life in many locations all over the globe, and discussed details of such processes as soil formation, plant growth, seed dispersal, the role of insects and birds, and the effects of man on nature. The work was popular on both sides of the Atlantic and was reissued at least four times, the last edition appearing in 1910. The daughter of the Rev. Charles Hicks Gaye, rector of St. Matthew's Parish, London, Selina had distinguished ancestors, including John Jewell, Bishop of Salisbury in about the middle of the sixteenth century, and the eighteenth-century poet John Gay. Her mother was related to the Reinagles, well-known artists of Hungarian origin. She was educated almost entirely by her father, who taught her several languages, ancient and modern, and Euclidian geometry. She later learned Hungarian on her own.[258]

Summary

The most prominent feature brought out by this overview of British work is the concentration in London and Cambridge of late nineteenth-century botanical research by women. This is hardly surprising, these being the two locations where the bulk of the country's research (particularly in the newer areas of the subject) was being done. Also striking is the prominence of no fewer than seven British women: Saunders, Sargant, Benson, Smith, and Lister were particularly distinguished; Miles Thomas and Barton Gepp stand out as well. It is worth noting that four of these seven were much influenced by D. H. Scott of the Royal College of Science and the Jodrell Laboratory. Sargant and Benson worked especially closely with him, and his lectures were the early inspiration of Smith and Barton Gepp. Indeed contact with workers at the Jodrell Laboratory (probably the leading British botanical laboratory during the last quarter of the century) was most certainly an important stimulus for the women; Scott was one of a number of distinguished workers there. By the 1880s Cambridge also had a well-established botanical laboratory (with close connections to the Jodrell).[259]

Thus, although the leading late nineteenth-century British women had marginal positions in the scientific community in terms of professional posts, their working contacts with the most influential male botanists of the time were close and their research was well received. They left a substantial legacy of contributions, those of Saunders and Sargant in conceptual and interpretative work being particularly notable.

By contrast, the contributions of the women from the earlier part of the century were relatively modest. There were indeed workers of note in the early group. Gifford and Russell, in particular, but also Kirby, Warren, Merrifield, Stackhouse, and a few others were much involved in the botanical efforts of their times and carried out very creditable exploratory studies; Russell's drawings remain valuable and Kirby's *Flora* was a commendable effort; the contributions of Gifford and other women algologists were quickly incorporated into major works in the field. In addition, as productive participants in local clubs and societies that then played a key role in stimulating and focusing regional investigations, the early workers established an extensive background of serious activity by women in the field. Nonetheless, the impressive achievements of several in the post-1880 group, most of whom had the benefit of formal university-level training, undoubtedly constitute the high point of nineteenth-century botanical research by British women, even though their pre-1901 publications represent only their early contributions.

Two-country comparison

Although about twice as many American women as British contributed papers to the nineteenth-century botanical journal literature, both national groups were substantial (119 American and sixty-four British workers, see Figure 1-1), and a comparison brings out a number of similarities and some differences.

The popular local and regional scientific clubs that fostered the work of women amateurs in Britain from the 1830s on had no real counterparts in the United States (much of America being then still a country of settlers and frontiersmen). By the 1870s and 1880s, however, a number of American organizations, such as the Rochester Academy of Science in Rochester, New York, the Boston Society of Natural History, the Essex Institute of Salem, Massachusetts, and the Torrey Botanical Club, had women taking part in their activities and were providing outlets in their journals for women's botanical contributions.[260]

In both countries by the 1880s and 90s one or two large, nonuniversity institutions provided good opportunities for botanical work by a few women—in London the Botanical

Department of the British Museum (Natural History) and the Jodrell Laboratory at Kew Gardens, in San Francisco the herbarium of the California Academy of Sciences, and in New York the New York Botanical Garden. Women who worked in these places were remarkably productive, Smith, Barton Gepp, Lister, and (for a time) Sargant in London, Eastwood and Brandegee in San Francisco, and Britton and Vail in New York.

In both countries also a few women botanists had teaching positions in women's colleges, in Britain in Cambridge and London, in America most often in the small independent women's colleges in the northeastern part of the country, especially Wellesley, Mount Holyoke, Smith, and Barnard. In Britain, however, the closer associations of the women's colleges with larger institutions—Newnham and Girton with Cambridge University, Bedford College and Royal Holloway College with University College London and the Royal College of Science—meant that the British women academics were generally more closely integrated into the national botanical community and had closer regular working links with leading male botanists than was the case with many of their American counterparts (although students at Barnard clearly benefited from the college's link with Columbia University). Thus Margaret Ferguson, Wellesley's most notable early botanist, worked in relative isolation compared with Rebecca Saunders at Newnham College or Margaret Benson at Royal Holloway College.

One of the striking points in the overall comparison between the American and British scenes is the importance in the United States of the coeducational Midwestern universities with their attached agricultural experiment stations. These institutions offered fairly advanced botanical training to a large number of women from the 1870s on, and played an especially prominent role in the 1880s and 90s. Although most of their women botany graduates became high school biology teachers rather than research botanists, as students or student assistants they typically made small but creditable research contributions (often in morphology and plant pathology, areas of ongoing interest to many of the male botanists in the Midwestern universities).[261] There was no comparable botanical research activity by women students at provincial universities in England or at Scottish universities, with work by only three—Auld and Warham at University College Liverpool, and Fingland at the Royal Technical College Glasgow—being identifiable in the Royal Society index.

Comparison of the research subfields of the two groups brings out the importance of cryptogamic botany in women's research in both countries (Figure 1-2, b and d). The British contribution in this area was largely the work of the British Museum women, especially Annie Lorrain Smith and Ethel Barton Gepp; in the United States it was that of Josephine Tilden of the University of Minnesota, Elizabeth Britton of the New York Botanical Garden, and to a somewhat lesser extent

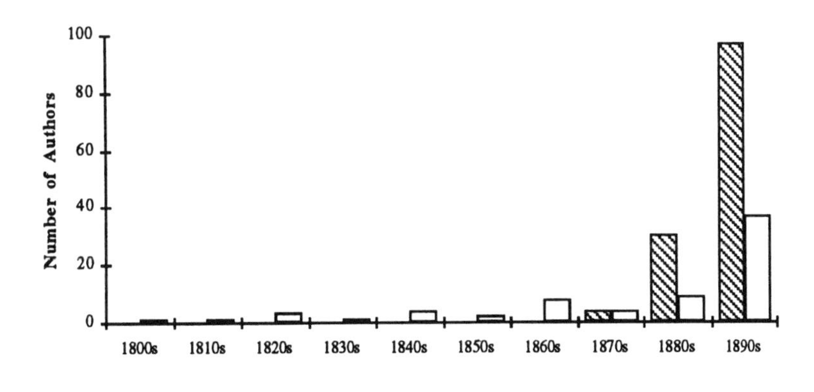

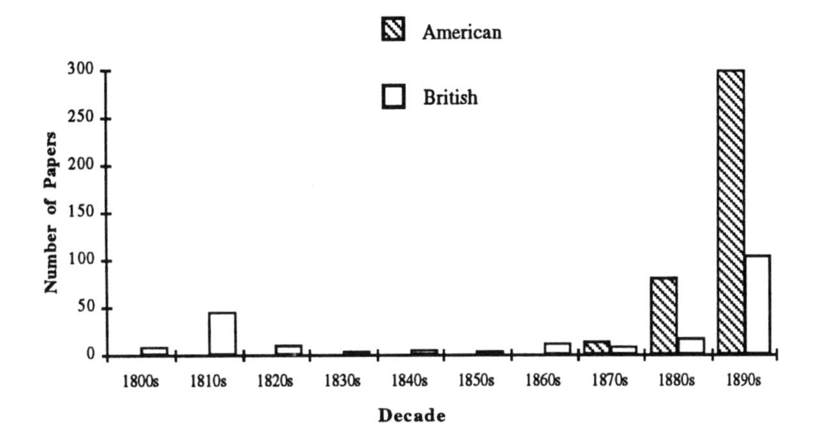

Figure 1-1. American and British authors and papers in botany, by decade, 1800–1900. Data from the Royal Society *Catalogue of Scientific Papers.*

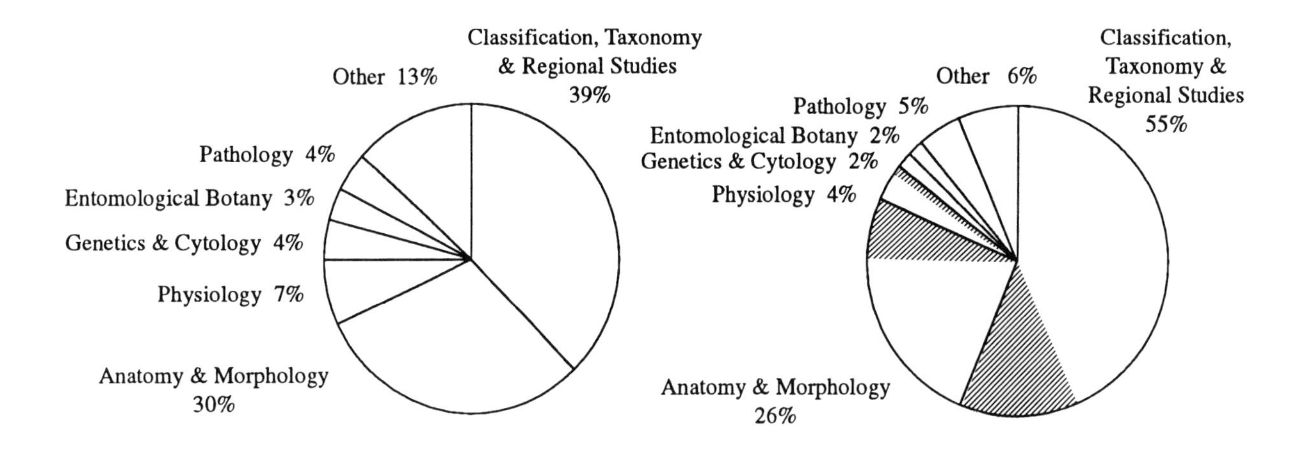

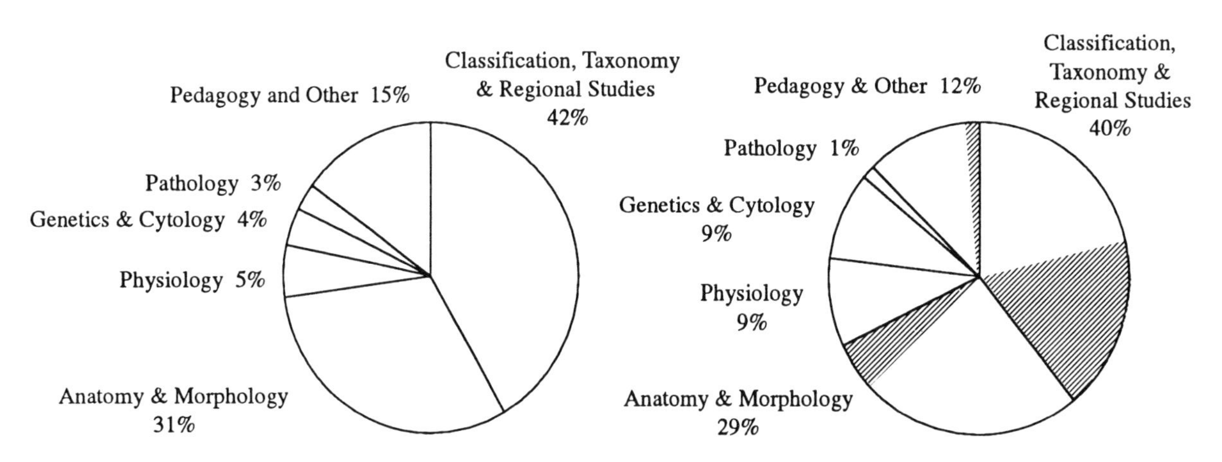

Figure 1-2. Distribution of American and British authors and papers in botany, by subfield, 1800–1900. Shaded sectors represent work in cryptogamic botany. (In a. and c., an author contributing to more than one subfield is counted in each.) Data from the Royal Society *Catalogue of Scientific Papers,* but excluding the 56 early papers in anatomy and physiology by British author Agnes Ibbetson. (If Ibbetson were included, the distribution in chart d. would be 29%, 38%, 16%, 7%, 1%, 9%, clockwise from the top.)

Clara Cummings of Wellesley College; Barton Gepp and Tilden worked on algae, Smith, Britton, and Cummings all specialized in lichens. Alice Eastwood and Katharine Brandegee, the outstanding botanical explorers and classifiers at the California Academy of Sciences in San Francisco, are without British counterparts, work of the kind they were doing on a grand scale in the western United States around the turn of the century having been essentially completed in European countries by that time. Figure 1-2 charts a and b illustrate the notable productivity of the American women classifiers/taxonomists/explorers; though making up less than 40 percent of the authors, they produced 55 percent of the papers. The contributions in plant anatomy, morphology, and genetics of that elite group of British workers—Margaret Benson, Ethel Sargant, Rebecca Saunders, and also Ethel Miles Thomas—are unmatched by those of women botanists of the time in the United States (although work in anatomy and genetics, which Margaret Ferguson was just beginning at Wellesley College around the turn of the century, still awaits analysis and assessment by an interested botanist).

Notes

1. Pre-1900 papers by four American botanists (Detmers, Ferguson, Parsons, and Plummer Lemmon) that are not listed in the Royal Society *Catalogue* have been added to the bibliography, and so the total number of authors listed there is 123. Eight of them (Bunting, De

Graffe Peacock, McNair Wright, Pennington, Annie Slosson, Smith Calvert, Smith Eigenmann, and Thompson) had major interests in other fields and are discussed in the corresponding chapters. Little information has been uncovered about seven (Bacon, Burnett, Merry, Mitchell, Newell, Pease, and Taplin), except that all (with the exceptions of Bacon and possibly Burnett) were from the northeastern region of the country, New York, Massachusetts, and New Hampshire. Alice Bacon was born in Manitowoc, Wisconsin, 13 October 1857, lived in Chicago and in Pasadena, California, in the 1880s, and moved to Brandon, Vermont, in 1890 (JHB, vol. 1, p. 100); she wrote on Vermont orchids. Martha Merry, whose particular interests were fungi, was born in Phoenix, New York, 10 August 1864 (JHB, vol. 2, p. 479).

2. Mary Collins Reynolds was born in Flushing, New York, 30 November 1852, and died in Demorest, Georgia, 23 March 1936. She was especially interested in Florida ferns, on which she published several papers between 1877 and 1881 in the *New York Botanical Club Bulletin* and the *Botanical Gazette* (JHB, vol. 3, p. 149).

3. Liberty H[yde] Bailey, "Lucy Millington," *Torreya,* **39** (1939): 159–63; Beatrice Scheer Smith, "Lucy Millington: nineteenth-century naturalist," *Michigan Audubon News,* **35,** no. 5 (1987): 5, "Lucy Millington: nineteenth-century naturalist. Her notes on Michigan birds," ibid., **35,** no. 6 (1987): 5, and "Lucy Bishop Millington, nineteenth-century botanist: her life and letters to Charles Horton Peck, State Botanist of New York," *Huntia,* **8** (1992): 111–53; information from Essex County Historical Society, Elizabethtown, New York.

4. Charles H. Peck, *New York (State) State Museum, Albany. Twenty-fifth Report* (Albany: New York State Museum, 1873), p. 69.

5. Smith, "Millington . . . Michigan birds".

6. Walter Deane, "Maria L. Owen," *Rhodora,* **16,** (1914): 153–60; Beatrice Scheer Smith, "Maria L. Owen, nineteenth-century Nantucket botanist," ibid., **89,** (1987): 227–39.

7. Owen's post-1900 articles included: "Ferns of Mt. Toby, Massachusetts," *Rhodora,* **3** (1901): 41–3; "The three adventive heaths of Nantucket, Massachusetts," ibid., **10** (1908): 173–9; "*Tillaea* in Nantucket," ibid., **14** (1912): 201–4.

8. John W. Harshberger, *The Botanists of Philadelphia and their Work* (Philadelphia: [Press of T. C. Davis and Son], 1899), pp. 298–302; Harry B. Weiss, "Mrs. Mary Treat, 1830–1923, early New Jersey naturalist," *Proceedings of the New Jersey Historical Society,* **73** (1955): 258–73; Marcia Myers Bonta, *Women in the Field; America's Pioneering Women Naturalists* (College Station, Texas: Texas A&M University Press, 1991), pp. 42–8; Vera Norwood, *Made from this Earth: American Women and Nature* (Chapel Hill, N.C.: University of North Carolina Press, 1993), pp. 42–3. Weiss's paper includes a bibliography of Treat's writings (pp. 270–3); many of these he rediscovered by examination of her correspondence with editors (now in the holdings of the Vineland Historical and Antiquarian Society). A number of the references are incomplete, especially those to ornithology articles in popular magazines from the period predating any indexes to periodical literature; the latter have not been added to the lists of papers by Treat in our bibliography.

9. Darwin, in his *Insectivorous Plants* (New York: D. Appleton, 1897), p. 278, footnote, and p. 281, refers to Mrs. Treat's account of *Drosera filiformis* in her 1873 *American Naturalist* paper. He clearly placed considerable trust in her observational work.

10. See L. O. Howard, "A history of applied entomology," *Smithsonian Miscellaneous Collections,* **84** (1930): 1–545 (whole volume). Considering Riley's strongly held views on questions of authorship, and his extreme reluctance to include the names of coworkers in his own papers and reports, Treat's handling of the authorship of her monograph may well have suited him (see also n. 38, chapter 2). Riley had thought of publishing a popular volume similar to Treat's, but never found the time to do so. He and Treat carried on an extensive entomological correspondence throughout the 1870s, and would appear to have been on good terms. Although on occasion he disagreed with her conclusions, he generally considered her observations reliable and valuable.

11. Weiss, "Treat," pp. 272–3.

12. Florence Beckwith, "Early botanists of Rochester and vicinity and the Botanical Section," *Proceedings of the Rochester Academy of Science,* **5** (1910–18)[1912]: 39–58, especially pp. 39–40.

13. Anna B. Suydam, "Early botanists of Rochester and vicinity, and the Botanical Section, II," ibid., **8** (4), (1941–43): 124–49, on pp. 136–8.

14. Florence Beckwith, Mary E. Macauley, and Milton S. Baxter, "Plants of Monroe County, New York, and adjacent territory. Supplementary list 1910," ibid., **5** (1910–18)[1910]: 1–38, and ". . . Second supplementary list," ibid., **5** (1910–18)[1917]: 59–121.

15. Suydam, "Early botanists," pp. 140–2.

16. *Proceedings of the Rochester Academy of Science,* **2** (1895): 186. Searing is referred to in reports of the meetings of the Botanical Section of the academy as Dr. Searing, but her name does not appear in early *AMD* listings.

17. Lazella Schwarten, "Furbish, Kate," in *NAW,* vol. 1, pp. 686–7; Louise H. Coburn, "Kate Furbish, botanist," *Maine Naturalist,* **4** (1924): 106–9; John Cole, "The woman behind the wildflower that stopped a dam," *Horticulture,* **55,** (1977): 30–5; Richard Saltonstall, Jr., "Of dams and Kate Furbish," *The Living Wilderness,* **40** (January 1977): 42–3.

18. Beyond the five articles listed in the bibliography there is the note, "*Cardamine bellidifolia* in Cumberland County, Maine," *Rhodora,* **3** (1901): 185.

19. For Horner see obituary, *Transactions of the Massachusetts Horticultural Society* (1906): 250; for Putnam, *WWWA,* p. 665 and JHB, vol. 3, p. 116; for Flint, JHB, vol. 1, p. 550. Putnam's post-1900 botanical papers include the following: "In Pennsylvania woods," *American Botanist,* **6** (1904): 62, 63; "Our native Aquilegia," ibid., **19** (1913): 97; "Forest flowers," *American Forest,* **23** (1917): 343–5; "How grandmother kept well," *Nature Magazine,* **5** (1925): 42–4.

20. Witmer Stone, "Ida Augusta Keller (1866–1932)," *Bartonia,* **14** (1932): 59–60; Harshberger, *Botanists of Philadelphia,* pp. 380–2.

21. Ibid., pp. 412–13.

22. Ethel Cooke and Adeline F. Schively, "Observations on the structure and development of *Epiphegus virginiana,*" *Publications of the University of Pennsylvania. Contributions from the Botanical Laboratory,* **2** (1904): 352–98.

23. *NCAB,* vol. 29, pp. 206–7; Sylvester Kohut, Jr., "Wilson, Lucy Langdon Williams," in *Biographical Dictionary of American Educators,* ed. John F. Ohles, 3 vols., (Westport, Conn.: Greenwood Press, 1978), vol. 3, p. 1411; *WWWA,* p. 892; *AMS,* 1910; Edward A. Krug, *The Shaping of the American High School,* 2 vols. (Madison, Wisc.: University of Wisconsin Press, 1972), vol. 2, pp. 166, 196, 199, 224.

24. Ibid., pp. 165–6.

25. Obituary, *NYT,* 5 August 1933, 11: 3 (Arnold died 3 August 1933).

26. *General Catalogue,* Middlebury College, p. 268.

27. *WWWA,* p. 385; JHB, vol. 2, p. 170.

28. JHB, vol. 3, p. 280; Harshberger, *Botanists of Philadelphia,* p. 413.

29. *WWWA,* p. 894.

30. Alumnae records, Mount Holyoke College.

31. Smith was one of many early Mount Holyoke graduates who became teachers in mission fields, fulfilling the hopes and plans of Mary Lyon, Mount Holyoke's founder, one of whose stated principles was the cultivation of a missionary spirit among her students—see Patricia Ruth Hill, *The World their Household* (Ann Arbor, Mich.: University of Michigan Press, 1985), pp. 42–3, 127.

32. Arma Anna Smith to Anna C. Edwards, teacher at Mount Holyoke, letter, dated Aug. 3, 1903 (Mount Holyoke College archives).

33. Walter Crosby Eells, "Earned doctorates for women in the nineteenth century," *American Association of University Professors, Bulletin,* **42** (1956): 644–51.

34. Olive Sprague Cooper and Florence Purlington, "Henrietta Edgecomb Hooker, 1851–1929," *Mount Holyoke Alumnae Quarterly,* **13,** (July 1929): 73–5; Sarah J. Agard, "The Department of Botany. History," ibid., **5** (July 1921): 82–4; *AMS,* 1910.

35. *WWWA,* p. 201; *History of the Fellowships Awarded by the American Association of University Women, 1888–1929. With Vitas of the Fellows,* comp. and ed. Margaret E. Maltby, (Washington D.C.: American Association of University Women, [1929]), p. 37.

36. The choice between Carter and Hooker as the first woman to earn a Ph.D. in botany from an American university is somewhat delicate. Both Eells ("Earned doctorates") and *AMS* (1910) give the year of Hooker's doctorate as 1889, but Cooper and Purlington ("Henrietta Edgecomb Hooker," p. 73) put it a year earlier (1888). Hooker's graduate work at Syracuse might possibly have been completed a year before her degree was formally awarded.

37. See, for instance, "Some Filipino flora," *Plant World,* **4** (1901): 1–5, and "The Dragon Tree of Orotawa," ibid., **4** (1901): 121–4 (see also bibliography).

38. Alice Carter Cook, *Michael; a Playlet of the Time of David* (Boston: Four Seas Company, 1922), and *Komateekay, a One-act Folk Play* (Boston: B. H. Humphries, 1936).

39. Helen A. Choate, "A history of the Botany Department, Smith College. 1875–1950," pp. 11, 12, 16 (manuscript in Smith College archives); faculty files, Smith College archives.

40. Ibid.; *WWWA,* p. 757; *AMS,* 1910–1949.

41. Frances Grace Smith, *Diellia and its Variations* (Honolulu, Hawaii: The Museum, 1934), also published in *Bernice P. Bishop Museum Occasional Papers* (**10,** no. 16). Smith's 1907 and 1910 cycad papers were the following: "Morphology of the trunk and development of the microsporangium of cycads," *Botanical Gazette,* **43** (1907): 187–204; "Development of the ovulate strobilus and young ovule of *Zamia floridana,*" ibid., **50** (1910): 128–41.

42. Mrs. Bernard, "Julia Warner Snow: In Memoriam," *Smith Alumnae Quarterly,* (Nov., 1927): 66; M[artha] Burton Williamson, "Some American women in science," *The Chautauquan,* **28** (1898–99): 161–8, 361–8, 465–73, on p. 470; Stephen Bocking, "Stephen Forbes, Jacob Reighard and the emergence of aquatic ecology in the Great Lakes region," *Journal of the History of Biology,* **23** (1990): 461–98; especially pp. 486–7; *WWWA,* p. 765; *AMS,* 1910–1927.

43. Julia Warner Snow, *The Plankton of Lake Erie, with Special Reference to the Chlorophyceae* (Washington, D.C.: Government Printing Office, 1902, extracted from the U.S. Fish Commission bulletin of 1902, *Contributions to the Biology of the Great Lakes*).

44. Bocking, "Stephen Forbes," pp. 489–90.

45. Ann M. Hirsch and Lisa J. Marroni, "Ferguson, Margaret Clay," in *NAW,* vol. 4, pp. 229–30; Marilyn Bailey Ogilvie, *Women in Science. Antiquity through the Nineteenth Century* (Boston: MIT Press, 1986), pp. 84–5.

46. For an examination of the advantages Wellesley offered early women academics, see Patricia A. Palmieri, "Here was fellowship. A social portrait of academic women at Wellesley College, 1895–1920," *History of Education Quarterly,* **23,** (1983): 195–214. The group Palmieri studied included several of the scientists in this survey—Ferguson and two other botanists (Cummings and Cooley), chemists Roberts and Bragg, zoologist Willcox, mathematician Hayes, psychologist Calkins, and astronomer Whiting. Palmieri argues that at a time when large research universities virtually denied women careers, the Wellesley faculty made a world for themselves in which all had a sense of taking part in a professionally fulfilling undertaking while also sharing a close-knit community life that satisfied their social and intellectual needs.

47. M. Ferguson, "Contribution to the knowledge of the life history of *Pinus* with special reference to sporogenesis, the development of the gametophytes and fertilization," *Proceedings of the Washington Academy of Sciences,* **6** (1904): 1–102. The Naples Table Association was a group of American university women, who, during the 1890s, raised funds for the upkeep of a laboratory table for women investigators at the famed Naples Zoological Station. By the early 1900s the association had enough money to sponsor additional projects promoting scientific research by women, including the contest referred to here.

48. It was also the case that funds for equipment for both teaching and research were scarce, and so money spent on Ferguson's projects left less for other departments (see the discussion of zoologist Mary Willcox, chapter 3).

49. Ferguson's post-1900 publications include the following: "Notes on the development of the pollen tube and fertilization in some species of pines," *Science,* **13** (1901): 668; "The development of the pollen tube and the division of the generative nucleus in certain species of pines," *Annals of Botany,* **15** (1901): 193–224; "The development of the egg and fertilization in *Pinus strobus,*" ibid., **15** (1901): 435–79; "A preliminary study of the germination of the spores of *Agaricus campestris* and other basidiomycetous fungi," *U.S. Department of Agriculture, Bureau of Plant Industry, Bulletin 16* (1902): 1–43; "The spongy tissue of Strasburger," *Science,* **18** (1903): 308–11; "Two embryo-sac mother cells in *Lilium longiflorum,*" *Botanical Gazette,* **43** (1907): 418–19; "Imbedded sexual cells in Polypodiaceae," ibid., **56** (1913): 501–2; "Botany at Wellesley," *Wellesley Alumnae Quarterly,* **8** (May, 1924): 170–83; "A cytological and genetic study of *Petunia.* I," *Bulletin of the Torrey Botanical Club,* **54** (1928): 657–64; "A botanical problem," *Science,* **73** (1931): 193–7; "The morphology of the pollen grains of *Petunia* in relation to hybridity, polyploidy and sterility," *Proceedings of the Sixth International Congress of Genetics,* **2** (1932): 52–3; "To determine genetic ratios when selfing organisms heterozygous for two or more factors," *American Naturalist,* **46** (1932): 91–3; (with A. M. Ottley) "Studies on *Petunia.* III. A redescription and additional discussion of certain species of *Petunia,*" *American Journal of Botany,* **19** (1932): 385–405; (with E. B. Coolidge) "A cytological and genetic study of *Petunia.* IV. Pollen grains and the method of studying them," ibid., **19** (1932): 644–58; "A cytological and a genetic study of *Petunia.* V. The inheritance of color in pollen," *Genetics,* **19** (1934): 394–411; (with B. Hunt) "Studies on *Petunia.* VI. The origin and distribution of color in the anther and in the pollen of *Petunia,*" *Botanical Gazette,* **96** (1934): 342–53.

50. One of Ferguson's less-remembered projects was the management of Wellesley's War Farm, started in the spring of 1918. Twenty

acres of land belonging to the college were farmed by Wellesley students and substantial quantities of vegetables produced. The great success of the War Farm was an important factor in the selection of Wellesley as the site of the training camp for supervisors of agricultural units of the Women's Land Army of America—see Alice Payne Hackett, *Wellesley: Part of the American Story* (New York: E. P. Dutton, 1949), pp. 201–2.

51. Bruce Fink, "A memoir of Clara E. Cummings," *The Bryologist*, **10,** (1907): 37–41; anon., "Miss Clara E. Cummings, Hunnewell Professor of Cryptogamic Botany at Wellesley College," ibid., **10,** (1907): 33–4, reprinted from the *Boston Evening Transcript*, 31 December 1906; Ogilvie, *Women in Science*, pp. 62–3.

52. See, for instance, Clara E. Cummings, lichen list, in L. L. Dame and F. S. Collins, *Flora of Middlesex County, Massachusetts* (Malden: Middlesex Institute, 1888), pp. 165–74; list in Walter Deane, *The Flora of the Blue Hills, Middlesex Fells, Stony Brook and Beaver Brook Reservations of the Metropolitan Park Commission, Massachusetts* (Boston: C. M. Barrows, 1896), pp. 133–6; lichen list in Charles Mohr, "Plant life of Alabama," *Contributions from the United States National Herbarium*, **6** (1901): 1–921, on pp. 267–83; lichen list in E. B. Delabarre, "Report of the Brown-Harvard expedition to Nachvak, Labrador, in the year 1900," *Bulletin of the Geographical Society of Philadelphia*, **3** (1902): 65–212, pp. 196–200.

53. See "The lichens of Alaska," in J. Cardot, Clara E. Cummings, Alexander W. Evans, C. H. Peck, P. A. Saccardo, De Alton Saunders, I. Thériot, and William Trelease, *Cryptogamic Botany, Harriman Alaska Series* (New York: Doubleday Page, 1904), vol. 5, pp. 65–152. For a later reference to this paper see, for example, John Walter Thomson, *Lichens of the Alaskan Arctic Slope* (Toronto: University of Toronto Press, 1979), p. 301.

54. Maxcine Williams, "Dr. Cooley and her buttercup," *Alaska Magazine* (June 1981): 49. Biographical files (faculty), and 7B alumnae biographical files, 1885 (Cooley), Wellesley College archives.

55. Grace E. Cooley, "Ecological notes on the trees of the botanical garden at Naples," *Botanical Gazette*, **38** (1904): 435–45; "Silviculture features of *Larix americana*," *Forestry Quarterly*, **2** (1904): 148–60.

56. Elizabeth G. Britton, "Emily L. Gregory," *Bulletin of the Torrey Botanical Club*, **24** (1897): 221–8; Emanuel D. Rudolph, "Women in nineteenth century American botany; a generally unrecognized constituency," *American Journal of Botany*, **69** (1982): 1346–55, and "The first American plant anatomy book and its author," *American Journal of Botany*, **71**: 5, Pt. 2 (1984): 107–8; Helena L. Jelliffe, "The Barnard Botanical Club. An historical sketch," in anon., *The Barnard Botanical Club* (New York: Irving Press, 1912), pp. 7–10; Rudolf Schmid and Dennis Wm. Stevens, "'Botanical textbooks', an unpublished manuscript (1897) by Emily Lovira Gregory (1841–1897) on plant anatomy textbooks," *Bulletin of the Torrey Botanical Club*, **114** (1987): 307–18; Rudolf Schmidt, "Annotated bibliography of works by and about Emily Lovira Gregory (1841–1897)," ibid., **114** (1987): 319–24.

57. Philip J. Pauly, "The appearance of academic biology in late nineteenth century America," *Journal of the History of Biology*, **17** (1984): 369–97, on p. 382. (Pauly's analysis was based on an examination of E. B. Wilson—M. Carey Thomas correspondence in the Bryn Mawr archives.)

58. Emily Gregory, *A Scientist's Confession of Faith. The Short Story of a Long Journey* [with an introduction by W. H. P. Faunce] (New York: Pusey and Troxell, 1897?, and Philadelphia: American Baptist Publication Society, 1898), especially p. 30.

59. Jelliffe, *Barnard Botanical Club.*

60. For Pettit see JHB, vol. 3, p. 77 and Eells, "Earned doctorates," p. 650.

61. Ada Watterson, "Louise Brisbin Dunn," *Torreya*, **3** (1903): 3–4.

62. For Cook and Noyes see JHB, vol. 1, p. 376 and vol. 3, p. 16, respectively. Cook's second paper was, "A list of plants seen on the island of Monhegan, Maine, June 20–25, 1900," *Rhodora*, **3** (1901): 187–90.

63. Yet others from the Northeast whose minor contributions are indexed in the bibliography include the following: Kate Wilson, Luella Whitney, Mary Frances Peirce, and Anne Townsend. Wilson was a student at Wellesley College from 1888 to 1890. Her short note in the 1890 *Botanical Gazette* was probably her only publication; she died in 1906 (JHB, vol. 3, p. 504). Whitney wrote one paper on the myxomycetes (slime molds) of Vermont. Born 4 May 1875 at Ashburnham, Massachusetts, she did her research as an undergraduate at Middlebury College, Vermont (B.S., 1898). She married Frank Chaffe Dunn in 1903 (JHB, vol. 3, p. 489). Mary Frances Peirce, of Weston, Massachusetts, published a short observational note in 1888 and a second ("Note on *Weigela rosea*") in *Rhodora* in 1908. She appears to have taken an A.B. at Smith College in 1912. Later she lived in Dayton, Ohio (JHB, vol. 3, p. 63). Anne Townsend studied botany under George Atkinson at Cornell University (A.B., 1903). Her anatomical paper that appeared in the *Botanical Gazette* in 1899 was most likely her only botanical publication. She married Charles Howard, an entomologist who worked for a considerable time in Africa (JHB, vol. 3, p. 394).

64. Anon., "Mrs. John I. Northrup," *Natural History*, **22** (1922): 248.

65. John Hendley Barnhart, "The published work of Elizabeth Gertrude Britton," *Bulletin of the Torrey Botanical Club*, **62** (1935): 1–17; anon., *NCAB*, vol. 25, p. 89; William Campbell Steer, "Britton, Elizabeth Gertrude Knight," in *NAW*, vol. 1, pp. 243–4; Nancy G. Slack, "American women botanists," in *Uneasy Careers and Intimate Lives*, eds. Pnina G. Abir-Am and Dorinda Outram (New Brunswick, N.J.: Rutgers University Press, 1987), pp. 95–100; Ogilvie, *Women in Science*, p. 47.

66. Britton was starred in *American Men of Science* starting with the first, 1906, edition. Rossiter (*Women Scientists in America*, pp. 289–90) has pointed out that the significance of Britton's star in later editions is open to question because of the editor's decision to retain this distinction once bestowed, even if the status of former recipients had changed and they were no longer leaders in their fields. During Britton's time there was a gradual shift of emphasis in many American centers of botanical work from traditional taxonomy to newer areas; further, her main interest (the taxonomy of mosses) was hardly a major area of botanical research.

67. S and F, p. 95; JHB, vol. 3, p. 419; *AMS*, 1910–1933; Sharon E. Kingsland, "The battling botanist. Daniel Trembly MacDougal, mutation theory, and the rise of experimental evolutionary biology in America, 1900–1912," *Isis*, **82** (1991): 479–509, on pp. 482, 486.

68. See D. T. MacDougal, assisted by A. M. Vail, G. H. Shull, and J. K. Small, *Mutants and Hybrids of the Oenotheras* (Washington, D.C.: Carnegie Institution, 1905); D. T. MacDougal assisted by A. M. Vail and G. H. Shull, *Mutations, Variations and Relationships of the Oenotheras* (Washington, D.C.: Carnegie Institution, 1907); P. A. Rydberg and A. M. Vail, "*Zygophyllaceae*," *North American Flora*, **25** (1910): 103–16. Vail's other post-1900 papers include the following: "Studies in the Asclepiadaceae. A new species of Vincetoxicum from Chihuahua," *Bulletin of the Torrey Botanical Club*, **28** (1901): 485; "Studies in the Asclepiadaceae," ibid., **29** (1902): 662–8; "Studies in

the Asclepiadaceae. A new species of Vincetoxicum from Alabama," ibid., **30** (1903): 178, 179; "Studies in the Asclepiadaceae. A new species of Asclepias from Kansas and two possible hybrids from New York," ibid., **31** (1904): 457–60; "*Onagra grandiflora* (Ait.), a species to be included in the North American flora," *Torreya,* **5** (1905): 9, 10; "Note on a little-known work on the natural history of the Leeward Islands," *Journal of the New York Botanical Garden,* 7 (1907): 275–9; "Jane Colden, an early New York botanist," *Torreya,* 7 (1907): 21–34. (Colden (1724–1766), the daughter of the surveyor general of the colonial province of New York, is remembered for her work of classification of the plants of the lower Hudson River Valley, carried out in the 1750s. Her brief botanical career ended with her marriage in 1759, but her cataloging was extensive, and, since she had learned from her father the principles of the Linnean system, her work was remarkably precise.)

69. B. L. Robinson, "Miss Day," *Rhodora,* **26** (1924): 41–7.

70. "The herbaria of New England," *Rhodora,* **3** (1901): 67–71, 206–8, 219–22, 240–4, 255–62, 281–8, 285–8. Additional post-1900 articles by Day include, "Bibliography of the botany of the Galapagos Islands," in "Flora of the Galapagos Islands," by B. L. Robinson, et al., *Proceedings of the American Academy of Arts and Sciences,* **38** (1902): 80–2; "*Juncus effusus,* var. *compactus* in New Hampshire," *Rhodora,* **6** (1904): 211; "Additional literature relating to the flora of Louisiana," *Fern Bulletin,* **12** (1904): 54–5; "Botanical writings of the late Alvah Augustus Eaton," *Rhodora,* **10** (1908): 211–14; "Bibliography of the botany of the Galapagos Islands, with additions by Alban Stewart," in A. Stewart, "A botanical survey of the Galapagos Islands," *Proceedings of the California Academy of Sciences,* s. 4, **1** (1911): 246–8.

71. *Who was Who in America,* vol. 3, p. 855; obituary, *Science,* **125** (1957): 1240; anon., "Josephine Elizabeth Tilden. 1869–1957," *Senate Minutes,* University of Minnesota, 1957, pp. 33–4; *AMS,* 1910–1933.

72. Tilden's papers on algae evolution included, "Some hypotheses concerning the phylogeny of the algae," *American Naturalist,* **62** (1928): 137–55, and "A classification of the algae based on evolutionary development, with special reference to pigmentation," *Botanical Gazette,* **95** (1933): 59–77. For her methods of representation see, "Standardization of method for drawing algae for publication," ibid., **95** (1934): 1–19. Her Pan-Pacific Congress presentations were "The study of Pacific Ocean algae," *Proceedings of the Pan-Pacific Science Congress,* **1** (1920): 207–9; "The distribution of marine algae, with special reference to the flora of the Pacific Ocean," *Proceedings of the Third Pan-Pacific Congress,* Tokyo, 1926, pp. 946–53. Her remaining post-1900 journal publications included the following: "*Hydrocoleum holdeni* nom. nov.," *Rhodora,* **3** (1902): 254; "Notes on a collection of algae from Guatemala," *Proceedings of the Biological Society of Washington,* **21** (1908): 153–6; "A phycological examination of fossil red salt from three localities in the southern states," *American Journal of Science,* **19** (1930): 297–304; "The marine and fresh-water algae of China," *Lingnam Science Journal,* 7 (1931): 349–98; (with A. P. Fessenden) "*Bactrophora irregularis,* a new brown alga from Australia," *Bulletin of the Torrey Botanical Club,* **57** (1931): 381–8.

73. Charles E. Bessey, "Life in a seaside summer school," *Popular Science Monthly,* **67** (1905): 80–9.

74. Faculty files, Oberlin College archives; *WWWA,* p. 598; *AMS,* 1910.

75. See Susan P. Nichols, "The nature and origin of the binucleated cells in some Basidiomycetes," *Transactions of the Wisconsin Academy of Sciences, Arts and Letters,* **15,** Pt. 1 (1904): 30–70 (dissertation research); "Methods of healing in some algal cells," *American Journal of Botany,* **9** (1922): 18–27; "The effect of wounds upon the rotations of the protoplasm in the internodes of *Nitella,*" *Bulletin of the Torrey Botanical Club,* **52** (1925): 315–63; "The effect of chloroform upon the rotation in the internodes of *Nitella,*" ibid., **57** (1930): 153–62.

76. *A Record of the Doctors in Botany of the University of Chicago 1897–1916,* by the Doctors in Botany (Chicago: University of Chicago, 1916), p. 63; JHB, vol. 2, p. 421.

77. See Florence A. McCormick, "Some notes on the anatomy of the tuber of *Ipomoea batatas,*" *Botanical Gazette,* **61** (1916): 388–97. Her doctoral research appeared under the title, "A study of *Symphyogina aspera,*" ibid., **58** (1914): 401–18.

78. No information has been collected about Hannah Thomas; Harriet Vandivert was a graduate student at Iowa State College in 1899 and at that time already had a B.Sc. degree. In 1900 she was living in Wichita, Kansas.

79. Alumni records, Iowa State University archives.

80. Letter from Louis Pammel to Emma Sirrine, May 20, 1924, Louis Pammel Correspondence File, Iowa State University archives.

81. Alumni records, Iowa State University archives.

82. Ibid.; anon., "Emma Pammel Hansen, B.S., M.S. (Mrs. N. E. Hansen)," *Proceedings of the Iowa Academy of Sciences,* **12** (1904): xi.

83. Graduate school records, 12/5/636, Cornell University archives; alumni records, Iowa State University archives; *WWWA,* p. 211.

84. Alumni records, Iowa State University archives (for both Bigelow and Hess).

85. J. C. Arthur, "Some botanical laboratories of the United States," *Botanical Gazette,* **10** (1885): 395–406, on pp. 402–3; Richard A. Overfield, *Science with Practice: Charles E. Bessey and the Maturing of American Botany* (Ames, Iowa: Iowa State University Press, 1993), pp. 24, 91.

86. Though their careers are rarely compared, it is worth noting that another successful early woman scientist, Mary Engle Pennington (see chapter 11), joined the Bureau of Chemistry at the U.S. Department of Agriculture just two years before the Bittings. Both Pennington and Katherine Bitting owed their successful entry into the bureau to special circumstances largely: Pennington was backed by her friend Harvey Wiley, bureau chief at the time, and Bitting worked in partnership with her husband. Both women had long careers as food biochemists, and worked closely with various important sectors of the food industry.

87. Information about many of these Midwestern botanists comes mainly from John Hendley Barnhart's *Biographical Notes upon Botanists.* Additional information on Horn was found in the archives at Kansas State University, and dates of death for Clendenin and Clifford in the University of Michigan archives (Clendenin, 11 May 1925, near Mexico, Missouri; Clifford, 18 April 1918, Pasadena, California). For Golden Bitting see, "Mrs. A.W. Bitting, '90," *Purdue Alumnus,* **25** (November, 1937): 14; anon., "The Katherine Golden Bitting collection on gastronomy," *Report of the Librarian of Congress,* 1940, pp. 255–6; Burton Williamson, "American women in science," pp. 471–2; *AMS,* 1910–1933. A few other women from Michigan, Iowa, and neighboring states made minor pre-1901 contributions to the botanical literature, which are listed in the bibliography: Nora Allin (later Mrs. Channing Ellery Dakin) graduated (B.A., 1897) from Iowa State University (later the University of Iowa); she studied botany under Thomas Macbride and published a paper on Iowa fungi in 1898. Emma McGee collected in Iowa and Nebraska. Lora La Mance was from Pineville, Missouri. Fanny Seavey, from Brighton, Illinois, was especially interested in the establishment and organization of botanical gardens. Both Mrs. Preston Lovell and Lucy Osband were from Michigan, Osband being associated with the State Normal

School in Ypsilanti. Rosa Abbott was from Chicago. For Zella Allen Dixon see chapter 16.

88. Ronald L. Stuckey, *Women Botanists of Ohio born before 1900. With Reference Calendars from 1776 to 2028* (Columbus, Ohio: RLS Creations, 1992), pp. 20–1; anon., "Los Angeles, California, Department of Botany of the University of Southern California at Los Angeles," *Chronica Botanica,* **1** (1935): 300–1; *AMS,* 1921–1933.

89. Detmers's post-1901 papers include the following: "Additions to the Ohio flora for 1905–06," *Ohio Naturalist,* **7** (1907): 61; "Annual report on the plants new to the Ohio state list for 1907–08," ibid., **9** (1909): 421–2; "Medicinal plants of Ohio," ibid., **10** (1910): 55–60, 73–85; "A floristic survey of Orchard Island," ibid., **11** (1911): 200–10; "The vascular plants of the cranberry bog in Buckeye Lake," ibid., **11** (1911): 305–6; "A preliminary report on a physiographic study of Buckeye Lake and vicinity," ibid., **12** (1912): 517–32; "An ecological study of Buckeye Lake. A contribution to the phytogeography of Ohio," *Proceedings of the Ohio State Academy of Science,* **5** (1912), Special Paper 19: 1–138; "Two new varieties of *Acer rubrum* L," *Ohio Journal of Science,* **19** (1918–19): 235–7; "Canada thistle, *Cirsium arvense* Tourn. field thistle, creeping thistle," *Ohio Agricultural Experiment Station Bulletin,* 414, (1927): 1–45. One additional paper by Detmers, an 1887 microscopical study on blood corpuscles carried out while she was an undergraduate, is listed in the microscopy section of the bibliography.

90. Stuckey, *Women Botanists,* p. 20; [L. D. Bushnell], obituary, *Transactions of the Kansas Academy of Science,* **42** (1939): 54; information from Ohio State University archives; *AMS,* 1910–1933.

91. See Lumina Cotton Riddle, "Algae from Sandusky Bay," *Ohio Naturalist,* **3** (1902): 317–19; "Fasciation," ibid., **3** (1903): 346–8; "Brush Lake algae," ibid., **5** (1905); 268–9: "Notes on the morphology of *Philotria,*" ibid., **5** (1905): 304–5; "Development of the embryo sac and embryo of *Staphylea trifoliata,*" ibid., **5** (1905): 320–5; "Development of the embryo sac and embryo of *Batrachium longirostris,*" ibid., **5** (1905): 353–63.

92. J. T. Willard, "Stella Kellerman," *The Industrialist,* 12 February 1936 (Kansas State University archives); Ronald L. Stuckey, "Botanical and horticultural contributions of Mrs. William A. Kellerman (Stella Victoria (Dennis) Kellerman), 1855–1936," *Michigan Botanist,* **31** (1992): 123–42.

93. For a list of Stella Kellerman's eighty publications see Stuckey, "Botanical and horticultural contributions," pp. 140–2.

94. Ibid., pp. 135–7; Mrs. W. A. Kellerman, "The evolution of Indian corn," *Annual Report of the Ohio State Academy of Science,* **2** (1894): 32–3 [abstract], "The probable differentiation of the ear and tassel in the Indian corn," *Vick's Illustrated Monthly Magazine,* **18** (184): 29, and "The primitive corn," *Meehan's Monthly,* **5** (1895): 44, 53.

95. William Kellerman and Mrs. W. A. Kellerman, "The non-indigenous flora of Ohio," *Journal of the Columbus Horticultural Society,* **15** (1900): 30–54, and *Bulletin of the Ohio State University,* s. 4, **27** (1900): 1–28.

96. Mrs. A. L. Slosson published a paper on the flora of Kansas in the *Transactions of the Kansas Academy of Science* in 1889 and another on plants of Cherokee County, Texas, in 1890. She was from Leavenworth, Kan.

97. Anon., "Among the alumni," *The Industrialist,* 18 March 1942 (Kansas State University archives); *Annual Catalogue,* Kansas State Agricultural College, 1889–90, 1891–92, 1892–93.

98. Alumni records, Kansas State University archives.

99. JHB, vol. 3, p. 69; *Who was Who in America,* vol. 4, p. 960; *WWWA,* p. 640; obituary, *Science,* **78** (1933): 87.

100. Jean Howerton Coady, "Cross school taught respect for education for over 40 years," *The Courier-Journal,* Louisville, Ky., 21 April 1980 (clipping, University of Pennsylvania archives); Graduate School records, University of Pennsylvania archives; *Ninth Annual Catalogue of the Hampton College, Louisville, Ky., 1886-'87* (Louisville, Ky.: John P. Morton and Co., [1887]), pp. 4, 6; Filson Club: Historical Society, Louisville, Ky.; Eells, "Earned doctorates," p. 650.

101. *Who was Who in America,* vol. 1, p. 995; anon., "Death of Miss Sadie F. Price," *Fern Bulletin,* **11** (1903–04): 85–6.

102. Obituary, *NYT,* 17 June 1943, 21: 5; information from Washington University archives; *WWWA,* p. 583; Eells, "Earned doctorates," p. 651.

103. *Record of Doctors in Botany,* p. 12; S and F, p. 95; *AMS,* 1910–1921. Lyon married Strong Vincent Norton sometime between 1910 and 1916.

104. These post-1900 publications of Lyon included: "Two megasporangia in Selaginella," *Botanical Gazette,* **36** (1901): 308; "The evolution of the sex organs of plants," ibid., **37** (1904): 280–93; "Another seed-like characteristic of Selaginella," ibid., **40** (1905): 73; "The spore coats of Selaginella," ibid., **40** (1905): 285–95.

105. Three botanical papers by yet another Midwestern schoolteacher, Lillien Martin, are listed in the bibliography. Martin taught botany, physics and chemistry in an Indianapolis high school from 1880 until 1889. Her research in chemical botany was carried out (probably during summers) at the Philadelphia College of Pharmacy in the mid 1880s (see also chapter 15).

106. Willis Linn Jepson, "The botanical explorers of California—X," *Madroño,* **2** (1934): 130–3; *Leaflets of Western Botany,* **8** (1957): 84.

107. Asa Gray, *Darwinia. Essays and Reviews Pertaining to Darwinism,* ed. A. Hunter Dupree, (reprint, Cambridge, Mass.: Harvard University Press, 1963), p. 272.

108. Frank S. Crosswhite and Carol D. Crosswhite, "The plant collecting Brandegees, with emphasis on Katharine Brandegee as a liberated woman scientist in early California," *Desert Plants,* (1985): 128–62; Marcus E. Jones, "Katharine Brandegee," *Desert Plant Life* (August, 1932): 41, 51; William Albert Setchell, "Townshend Stith Brandegee and Mary Katharine (Layne) (Curran) Brandegee," *University of California Publications in Botany,* **13,** (1926): 155–78; Hunter Dupree and Marian L. Gade, "Brandegee, Mary Katharine Layne Curran," in *NAW,* vol. 1, pp. 228–9.

109. Crosswhite and Crosswhite, "Plant collecting Brandegees."

110. The California Academy of Sciences was probably the first large scientific institution in the United States to formally encourage female participation, a resolution to that effect having been adopted in 1853. It later became the first to employ women in senior curatorial positions, first Katharine Brandegee and then Alice Eastwood (see F. M. MacFarland, R. C. Miller, and John Thomas Howell, "Biographical sketch of Alice Eastwood," *Proceedings of the California Academy of Sciences,* s. 4, **25** (1948): IX–XIV, especially p. IX). Interestingly enough, Katharine Brandegee wrote that her interest in botany was accidental, and that she would have preferred the study of birds or insects (see Setchell, "Townshend Stith Brandegee and Mary," p. 168).

111. Katharine Brandegee carried out a remarkable amount of western botanical editing in the 1880s and 1890s. Following her work on the Academy of Sciences *Bulletin* in the mid 1880s she edited (as Mary K. Curran) volume 1 of the reestablished academy *Proceedings* (s. 2) in 1888. Volume 2 (1889) was edited (at least nominally) by Townshend Brandegee.

112. "Think you a little din can daunt mine ears . . .", Petruchio's thoughts about Katharina, William Shakespeare, *The Taming of the Shrew,* Act I, Scene 2.

113. Brandegee's review of Britton's monograph appeared in *Zoe,* **1** (1890–91): 286–7; Britton's reply and the *Zoe* response thereto are on pp. 344–6.

114. Katharine Brandegee, "Some sources of error in genera and species," *Zoe,* **5,** (1901): 91–98, especially p. 92.

115. Katharine Brandegee, "The botanical writings of Edward L. Green," ibid, **4** (1893): 63–103.

116. Joseph Ewan, "San Francisco as a Mecca for nineteenth-century naturalists," in *A Century of Progress in the Natural Sciences, 1853–1953,* ed. Edward L. Kessell (San Francisco: California Academy of Sciences, 1955; reprint edns., New York: Arno Press, 1974), pp. 1–63, especially p. 33.

117. John Thomas Howell, "Alice Eastwood: 1859–1953," *Taxon,* **3** (1954): 98–100, and "I remember, when I think . . . ," *Leaflets of Western Botany,* **7** (1954): 153–64; Leroy Abrams, "Alice Eastwood—western botanist," *Pacific Discovery,* **2** (1949): 14–17; Carol Green Wilson, "The Eastwood era at the California Academy of Sciences," *Leaflets of Western Botany,* **7** (1953): 58–64, and *Alice Eastwood's Wonderland: the Adventures of a Botanist* (San Francisco: California Academy of Sciences, 1955); MacFarland, Miller, and Howell, "Biographical sketch of Alice Eastwood;" Joseph Ewan, "Eastwood, Alice," in *NAW,* vol. 4, pp. 216–17; Susanna Bryant Dakin, *The Perennial Adventure. A Tribute to Alice Eastwood: 1859–1953* (San Francisco: California Academy of Sciences, 1954).

118. See Stephen J. Pyne, *Grove Karl Gilbert. A Great Engine of Research* (Austin, Texas: University of Texas Press, 1980), p. 262. Pyne quotes from a 1918 letter, now in the family archives, from Gilbert to his son Arch, in which he spoke of his long relationship with Alice Eastwood and their decision to marry. Alice's attempt to find a position with the National Herbarium in Washington, D.C. about 1910 may well have been influenced by the fact that Gilbert lived in Washington for part of the year.

119. Ewan, "San Francisco as a Mecca," pp. 32–3.

120. About seven of these botanists, Hagenbuck, Adamson, Parsons, Bingham, Williams, Byxbee Nott, and McFadden, only the following fragmentary information has been uncovered. Mrs. I. Hagenbuck worked in San Diego County in southern California. Margaret Adamson and Mary Parsons were from the San Francisco region. Adamson published a note in *Erythea* in 1899, and Parsons a paper on the ferns of Mount Tamalpais in *Zoe* in 1891. Parsons was also the author of the extremely popular monograph *The Wild Flowers of California. Their Names, Haunts and Habits.* First published in 1897 (San Francisco: W. Doxey), it went through many editions and remained in use for more than half a century. Caroline Lord Bingham's botanical observations were made near Santa Barbara (*Botanical Gazette,* 1879 and 1887). Clara Williams (b. 1870) and Edith Byxbee received B.S. degrees from the University of California, in 1897 and 1896, respectively. Both went on to graduate work in cytology at the university botanical laboratory. After receiving an M.S. in 1899, Williams took an M.D. at Johns Hopkins University Medical Department (1902). She spent some time in Serbia, and later practiced in Berkeley (JHB, vol. 3, p. 498, and also the *AMD,* 1916, p. 236). Byxbee received an M.S. degree from the University of California in 1899. She married Charles Palmer Nott in 1898 (JHB, vol. 1, p. 294). Effie McFadden was a student at Stanford in the 1890s.

121. Burton Williamson, "American women in science," p. 473; JHB, vol. 2, p. 479.

122. Alumni records, Iowa State University archives; Ella Dales Cantelow and Herbert Clair Cantelow, "Biographical notes on persons in whose honor Alice Eastwood named native plants," *Leaflets of Western Botany,* **8,** (1957): 86.

123. Ibid., p. 100.

124. *Who's Who in America,* **7** (1912–13) p. 1248; Burton Williamson, "American women in science," pp. 472–3; Ewan, "San Francisco as a Mecca," pp. 23–4; Frank S. Crosswhite, "'J. G. Lemmon & Wife', plant explorers in Arizona, California and Nevada," *Desert Plants,* **1** (1979): 12–21.

125. Joseph Ewan, "Bibliographical Miscellany-V. Sara Allen Plummer Lemmon and her 'Ferns of the Pacific Coast,'" *American Midland Naturalist,* **32** (1944): 513–18.

126. J. G. Lemmon, *How to Tell the Trees and Forest Endowment of the Pacific Slope . . . and also Some Elements of Forestry with Suggestions by Mrs Lemmon. First Series, The Cone Bearers* (Oakland, Calif.: [n.p.] 1902).

127. Burton Williamson, "American women in science," p. 473. No other mention of this work has been discovered.

128. Vera K. Charles, "Mrs. Flora Wambaugh Patterson," *Mycologia,* **21** (1929): 1–4; Beverly T. Galloway, "Flora W. Patterson, 1847–1928," *Phytopathology,* **18** (1928): 877–9; *WWWA,* pp. 626–7; *AMS,* 1910; John A. Stevenson, "Plants, problems and personalities: the genesis of the Bureau of Plant Industry," *Agricultural History,* **28** (1954): 155–62; Gladys L. Baker, "Women in the U.S. Department of Agriculture," ibid., **50** (1976): 190–201.

129. Another of these associates was Emma Sirrine, the Iowa botanist (see above) who joined the Bureau of Plant Industry in 1906 and had charge of the seed testing laboratory from 1913 until 1938. Her career rather closely paralleled that of Patterson, although she does not appear to have published research while a government scientist. A third notable woman research worker in the bureau around that time was Vera K. Charles (A.B., Cornell University) who joined the group in 1903 and published a number of articles in collaboration with Patterson. Charles became widely known as an expert on edible and poisonous mushrooms, and her 1931 bulletin on mushrooms was one of the most widely requested publications of the U.S. Department of Agriculture for more than twenty-five years.

130. See, for instance, "A collection of economic and other fungi prepared for distribution," *U.S. Department of Agriculture, Bureau of Plant Industry, Bulletin 8,* 1902; (with V. K. Charles) "*Septoria spadicea,*" in P. Spaulding, "The present state of the white-pine blights," *U.S. Department of Agriculture, Bureau of Plant Industry,* Circular 35, 4, 1909; (with V. K. Charles and F. J. Veihmeyer) "Some fungus diseases of economic importance. I. Miscellaneous diseases. II. Pineapple rot caused by *Thielaviopsis paradoxa,*" *U.S. Department of Agriculture, Bureau of Plant Industry, Bulletin, 171,* 1910; "*Stemphylium tritici* sp. nov., associated with floret sterility of wheat," *Bulletin of the Torrey Botanical Club,* **37** (1910): 205; (with V. K. Charles) "Mushrooms and other common fungi," *U.S. Department of Agriculture Bulletin 175,* 1915; "Diseases of roses," in F. L. Mulford, "Roses for the home," *U.S. Department of Agriculture Farmers' Bulletin 750,* 1916, revised 1921; (with V. K. Charles) "The occurrence of bamboo smut in America," *Phytopathology,* **6** (1916): 351–6; (with V. K. Charles) "Some common edible and poisonous mushrooms," *U.S. Department of Agriculture Farmers' Bulletin 796,* 1917, revised 1922; "Rose diseases and their prevention," *Florida Fruits and Flowers,* **2** (1925): 37–8.

131. JHB, vol. 3, p. 306; *WWWA,* p. 768; *AMS,* 1910–1949.

132. See bibliography. During this early period in her career Southworth also published, jointly with F. Lamson-Scribner, *The True Grasses,* a translation from the German of the monograph by Eduard Hackel (New York: H. Holt, 1890).

133. Effie Southworth Spalding, "Mechanical adjustment of the Suahara (*Cereus giganteus*) to varying quantities of stored water," *Torrey*

Botanical Club Bulletin, **32** (1905): 57–68; Effie Southworth Spalding and Daniel Trembly MacDougal, *The Water-balance of Succulent Plants* (Washington, D.C.: Carnegie Institution of Washington, 1910).

134. The large number of American women botanists in the 1890s is in large part a reflection of the overall interest in the field nationally; by 1890 botanists made up almost a quarter of the attendance at the annual AAAS meeting (Overfield, *Science with Practice,* p. 96).

135. The University of Zurich was also important in the training of the early women botanists who worked in the northeastern colleges.

136. Botanical work by sixty-six British women is listed in the bibliography, a note by Margaretta Riley and a paper by the ethnologist Mary Kingsley, which do not appear in the Royal Society index, having been added. Riley is of special interest as the first British woman pteridologist (fern specialist); for Kingsley see chapter 14.

137. *DNB,* vol. 10, p. 409; Ann B. Shteir, *Cultivating Women, Cultivating Science* (Baltimore: Johns Hopkins University Press, 1996), pp. 124–35.

138. A. G. Morton, *History of Botanical Science* (London Academic Press, 1981), pp. 364, 366, 368.

139. These conclusions are based on a reading of Ibbetson's later papers only.

140. Agnes Ibbetson, "On the physiology of botany," *Philosophical Magazine,* **56** (1820): 3–9, quotations from pp. 3, 7. Five volumes of Ibbetson's botanical writings and sketches are held in the Botanical Library, British Museum (Natural History). These include three volumes of annotated watercolor sketches of grasses (early work), an illustrated botanical notebook, and her collected papers entitled "Botanical Treatise." In addition, the Linnean Society holds a manuscript, "Phytology," and a collection of Ibbetson's letters to botanist Sir James Smith from whom she had asked assistance in her publishing efforts (Shteir, *Cultivating Women,* p. 256, n. 27). Ibbetson was also interested in soil improvement and agricultural methods, subjects coming increasingly under discussion in her time. Over many years she grew potato crops, experimenting with various manures and lime. She was the only woman whose name appears in the 1814 membership lists of the Bath and West of England Society, an organization interested in agricultural improvement; she contributed at least one paper to its *Correspondence* (ibid., p. 257, n. 53).

141. Anon., obituary for Isabella Gifford, *Journal of Botany,* **30** (1892): 81–3. The writer elaborated on this chain of productive early women seaweed specialists, mentioning no fewer than ten, both British and Irish (p. 81). The earliest was Miss Hutchins of Bantry, who died about 1816 and was commemorated in the genus *Hutchinsia.* Others who made particularly outstanding contributions were Mrs. Griffiths (*Griffithsia*), Mrs. Gatty (*Gattya*), Miss Ball (*Ballia*), Miss Cutler (*Cutleria*), and Miss Warren. Yet others were Miss Poore, Miss Turner, Miss Watt, and Miss White. All of them cooperated with the leading male workers in the field at the time.

142. Ibid.

143. Note attributed to a cousin of Gifford, ibid., p. 83.

144. Isabella Gifford, Memorial, *Journal of Botany,* **3** (1865): 101–3; Frederick Hamilton Davey, *Flora of Cornwall* (Penryn: F. Chegwidden, 1909), p. xliii. (These two sources differ about Warren's date of death, Gifford giving the year as 1863; 1864 would seem the more probable.) See also Shteir, *Cultivating Women,* pp. 186–90.

145. Anon., "Botanical chart for schools," *Annals and Magazine of Natural History,* s. 1, **3,** (1839): 121–2, quote on p. 121.

146. Obituary, *Journal of Botany,* **27** (1889): 160; see also the obituary for Merrifield's son, (J. J. W., "Mr. F. Merrifield,") *Nature,* **113** (1924): 933, and the notice in *Nature,* **39** (1889): 255.

147. Mary Philadelphia Merrifield, *The Art of Fresco Painting, as Practised by the Old Italian and Spanish Masters; with a Preliminary Inquiry into the Nature of the Colours used in Fresco Painting, with Observations and Notes* (Brighton: [n.p.], 1846)—a revised, illustrated edition, with an introduction by A. C. Sewter, was published more than a century later (London: Alec Tiranti, 1952); also, *Original Treatises dating from the Twelfth to the Eighteenth Centuries, on the Arts of Painting in Oil, Miniature, Mosaic and on Glass; of Gilding, Dyeing and the Preparation of Colours and Artificial Gems. Preceded by a General Introduction with Translations, Prefaces and Notes* (London: [n.p.], 1849).

148. See Mary P. Merrifield, "On Monostroma, a genus of algae," *Nature,* **26** (1882): 284–6. The last of Merrifield's review articles, a discussion of monographs on algae published during 1887 and 1888, appeared just after her death—"Recent works on algae," ibid., **39** (1889): 250–2.

149. D. E. Allen, "The botanical family of Samuel Butler," *Journal of the Society for the Bibliography of Natural History,* **9** (1979): 133–6, on p. 134.

150. James E. Bagnall, *The Flora of Warwickshire* (London: Gurney and Jackson, 1891), p. 505; Allen, "Botanical family," and "The women members of the Botanical Society of London, 1836–1856," *British Journal for the History of Science,* **13** (1980): 240–54.

151. Ibid., p. 247.

152. D. E. Allen; "The first woman pteridologist," *Bulletin, British Pteridological Society,* **1** (1978): 247–9.

153. Ibid., p. 247. There is no exact record of authorship of the monograph, various sources attributing it to one or other of the Rileys, but Allen considers there to be little doubt that it was their joint work. A twenty-nine-page octavo volume, *A Catalogue of Ferns, after the Arrangement of Sprengel, to which is added a synoptical table of C.B. Presi's arrangement of genera,* appeared in John Riley's name in 1841. Allen suggests (p. 248) that if this was developed essentially from the 1840 monograph, then Mrs. Riley probably had some claim to being considered a coauthor.

154. Mary Kirby (Mrs. Gregg), *Leaflets from my Life: a Narrative Autobiography* (London: Simpkin and Marshall, 1887); A. R. Horwood and C. W. F. Noel, *The Flora of Leicestershire and Rutland* (London: Oxford University Press), 1933, pp. ccxii–ccxiii.

155. D. L. Hawksworth and M. R. D. Seaward, *Lichenology in the British Isles 1568–1975. An Historical and Bibliographical Survey* (Richmond: Richmond Publishing Co., 1977), pp. 20–1.

156. Mary Kirby, *A Flora of Leicestershire; Comprising the Flowering Plants, and the Ferns Indigenous to the County, Arranged on the Natural System* (London: Hamilton, Adams and Co., 1850), Preface, pp. vi–vii.

157. Kirby, *Leaflets,* pp. 144–5, 147.

158. Ibid., p. 213.

159. Alice M. Coats, *Flowers and their Histories* (London: A. and C. Black, 1968), p. 327 (first published by Hulton Press, 1956); Shteir, *Cultivating Women,* pp. 135–45.

160. Elizabeth Kent, "Considerations on botany, as a study for young people, intended as an introduction to a series of papers illustrative of the Linnean system of plants," *Magazine of Natural History,* **1** (1829): 124–35, quotations from pp. 126, 134. Kent's idea that girls were discouraged from natural history studies at that time was especially conservative.

161. Elizabeth Kent, *Flora Domestica, or the Portable Flower-Garden: with Directions for the Treatment of Plants in Pots; and Illustrations from the Work of the Poets* (London: Taylor and Hessy, 1823); *Sylvan Sketches, or a Companion to the Park and Shrubbery; with Illustrations from the Works of the Poets* (London: Taylor and Hesey), 1825.

162. *DNB,* vol. 19, p. 1315.

163. *DNB,* Supplement, 1968, pp. 159–60; obituary, *Journal of Botany,* **28** (1890): 320; Shteir, *Cultivating Women,* pp. 227–31.

164. Barbara Caine, *Victorian Feminists* (Oxford: Oxford University Press, 1992), many references. Becker's *Women's Suffrage Journal* was an influential publication in its time. It dealt not only with the suffrage question but also with broader issues of the women's movement. The widespread problem of domestic violence and aggravated assault against wives was one to which it devoted considerable attention throughout the 1870s and 1880s (ibid., especially pp. 113, 135–6).

165. Jane Lewis, *Women and Social Action in Victorian and Edwardian England* (Stanford, Calif.: Stanford University Press, 1991), p. 304.

166. Miss Edmonds published a note in the *Journal of Botany* in 1869 on a plant found in the Lake District. Anne Hunter was an honorary member of the Berwickshire Naturalists' Club, the first of its kind in Britain (founded 1831), probably the first to admit women (as honorary members), and an important organization for naturalists for many decades (Hawksworth and Seaward, *Lichenology,* p. 15; see also chapter 4). Anne lived in Coldstream on the river Tweed. Her special interest was fungi, and her two papers in the club's *History* (1842–49, 1868) reported finds near her home. Miss M'Inroy studied mosses near Blair Atholl in Perthshire and near Brechin in Angus, Scotland. She and Mrs. M'Inroy (her mother?) presented their observations to the Botanical Society of Edinburgh (two papers, 1864). Mrs. Scrivenor, of Louth, in eastern Lincolnshire, published an article on procedures for preserving dried flowers (*Journal of Botany,* 1869). Mrs. Anna Maria Smith did her botanical work in the 1870s in the Austrian Tyrol and the mountainous regions to the south and east. She made extensive collections and was one of the principal contributors of Tyrol plants to the Kew herbarium about that time (see Joseph Hooker, *Journal of Botany,* **3** (1874): 209). Her list of the flora of the Fiume region near the head of the Adriatic, then part of Austria, appeared in the journal of the Vienna Zoologisch-Botanische Gesellschaft (1879). Twenty-two years later a supplement to this list was published in Italian (Guido Depoli, "Supplemento allo flora fiumana di Anna Maria Smith," *Rivista Italiana di Scienzi Naturali,* Siena, **21** (1901): 67–73). Mrs. Wright sent two communications to the Edinburgh Botanical Society in 1873, one reporting observations in Shropshire and the other describing plants collected in the south of France. Pre-1880 botanical papers by two additional women, Sarah Bowdich Lee and Elizabeth Hodgson, are listed in the bibliography; Hodgson's botanical survey of Lake Lancashire, a southwestern appendage of the Lake District, was especially notable. For Lee and Hodgson see chapters 4 and 12, respectively.

167. For Stackhouse see Davey, *Flora of Cornwall,* p. liii; Walker is mentioned in H. Trimen, *A Handbook to the Flora of Ceylon,* Pt. V (Delhi: M/S Periodical Experts, 1974), p. 374 (also published in London, 1893–1931); see also Shteir, *Cultivating Women,* pp. 180, 193.

168. Hawksworth and Seaward, *Lichenology,* p. 25. Interestingly, an important early work in the popularization of cryptogamic botany in Britain was written by a woman, Margaret Plues. Her *Rambles in Search of Flowerless Plants* (London: Journal of Horticulture and Cottage Gardener Office) came out in 1864 and was reissued in 1865 (ibid., pp. 16–17).

169. Obituaries, J. Ramsbottom, "Miss Gulielma Lister," *Nature,* **164** (1949): 94, E[lsie] M[aud] W[akefield], "Miss Gulielma Lister," *Transactions of the British Mycological Society,* **33** (1950): 165–6, *Times,* 6 June 1949, 7 and *Essex Naturalist,* **28** (1950): 214; G. C. Ainsworth and Frances L. Balfour-Browne, "Gulielma Lister Centenary," *Nature,* **188** (1960): 362–3; J. J. L[ister], "Arthur Lister, 1830–1908," *Proceedings of the Royal Society,* B, **88** (1915): i–xi; A. B. Rendel, Preface, in Arthur Lister, *A Monograph of the Mycetoza, being a descriptive Catalogue of the Species in the Herbarium of the British Museum* (London: British Museum (Natural History), 2d. ed., revised by Gulielma Lister, 1911); G. C. Ainsworth, "The Lister Notebooks," *Transactions of the British Mycological Society,* **35** (1952): 177–8; G. C. Ainsworth, comp., J. Webster and D. Moore, eds., *Brief Biographies of British Mycologists* (British Mycological Society, 1996).

170. William T. Stearn, *The Natural History Museum at South Kensington. A History of the British Museum* (*Natural History*) (London: Heinemann, 1981), p. 313.

171. See, for instance, the following (with Arthur Lister), "Notes on Mycetozoa," *Journal of Botany,* **40** (1902): 209–13; **42** (1904): 129–40; **43** (1905): 150–6; "Notes on Mycetozoa from Japan," ibid., **42** (1904): 97–9; **44** (1906): 227–30; "Mycetozoa from New Zealand," ibid., **43** (1905): 11–14; "Synopsis of the orders, genera and species of Mycetozoa," ibid., **45** (1907): 176–9; "Notes on Swiss Mycetozoa," ibid., **46** (1908): 216–29. Gulielma Lister's post-1908 single-authored Mycetozoa papers include, "Two new Mycetozoa," ibid., **48** (1910): 73; "Colloderma, a new genus of Mycetozoa," ibid., **48** (1910): 310–12; "Two new species of Mycetozoa," ibid., **49** (1911): 61–2; "Mycetozoa found during the fungus foray in the Forres district, Sept. 12, with the description of new species," *Transactions of the British Mycological Society,* **4** (1912): 38–44; "The past students of the Mycetozoa and their work," ibid., **4** (1912): 44–61 (presidential address); "Notes on Swiss Mycetozoa," *Journal of Botany,* **51** (1913): 95–100; "Notes on the Mycetozoa of Linnaeus," ibid., **51** (1913): 160–4; "Mycetozoa from Arosa, Switzerland," ibid., **52** (1914): 98–104; "A short history of the study of Mycetozoa in Britain, with a list of species recorded from Essex," *Essex Naturalist,* **18** (1914–18): 207–37; "The haunts of the Mycetozoa," ibid., **18** (1914–18): 301–21; "Mycetozoa in Epping Forest," a series of papers and notes that appeared annually in the *Essex Naturalist* from 1912 onward; "Mycetozoa found during the Bettws-y-Coed foray," *Transactions of the British Mycological Society,* **10** (1924): 240–2; "Mycetozoa of the Dublin foray," ibid., **11** (1926): 22–3; "Notes on Irish Mycetozoa," ibid., **11,** (1926): 23–4; "New species of Amaurochaete and some other Mycetozoa," *Journal of Botany,* **64** (1926): 225–7; "Mycetozoa gathered during the Hereford foray," *Transactions of the British Mycological Society,* **12** (1927): 86–7; (with E. M. Wakefield) "The Marlborough foray (list of fungi by G. Lister)," ibid., **93** (1928): 305–16; "A new species of Hemitrichia from Japan," ibid., **14** (1929): 225–7; "Mycetozoa," *Proceedings of the Royal Irish Academy of Science,* **39,** B (1929): 55–7; "Notes on Malayan Mycetozoa," *Journal of Botany,* **69** (1931): 42–3; "New varieties of Mycetozoa from Japan. (*Didymium leonium* var. effusum, *Arcyric pomiformis* var. heterospora, gathered by the Emperor in the Tochigi prefecture, Japan)," ibid., **71** (1933): 220–2.

172. Lister's writings on conifers include, "On conifers grown in suburban gardens," *Essex Naturalist,* **19** (1918–21): 157–70, and "List of flowering plants and conifers in Wanstead Park district," ibid., **27** (1942): 128–36. Among her other articles on subjects other than the Myxomycetes are "Alien plants in south-west Essex," ibid., **16** (1909–11): 330–1; "On some water plants," ibid., **19** (1918–21): 103–15; "The British yellow wagtails," ibid., **19** (1918–21): 152–4; "The water violet in Epping Forest [*Hottonia palustris*]," ibid., **26** (1940): 1–2; "The flora of Wanstead Park district," ibid., **27** (1942): 121–7.

173. Obituaries, Gulielma Lister, *Proceedings of the Linnean Society* (1937–38): 337–9; E. M. Wakefield, *Kew Bulletin* (1937): 442–3,

A. Gepp and A. B. Rendle, "Annie Lorraine Smith, O.B.E.," *Journal of Botany,* **75** (1937): 329–30, and J. Ramsbottom, *Times,* 14 September 1937, 14b; Hawksworth and Seaward, *Lichenology,* pp. 26–7.

174. Smith's post-1900 fungi papers include the following: "Fungi new to Britain," *Transactions of the British Mycological Society,* (1899–1900): 150–8; "Fungi found on farm seeds when tested for germination, with an account of two fungi new to Britain. (Pt. II)," *Journal of the Royal Microscopical Society,* (1901): 613–18. "On some fungi from the West Indies," *Journal of the Linnean Society (Botany),* **35** (1901): 1–19; "Mycorrhiza, the root-fungus," *The South-Eastern Naturalist,* 7 (1902): 9–15; "A disease of the gooseberry," *Journal of Botany,* **41** (1903): 19–23; "New or critical microfungi," ibid., **41** (1903): 257–60; "Diseases of plants due to fungi," *Transactions of the British Mycological Society,* (1903): 55–6; "Notes on fungi recently collected," ibid., (1903): 56–7; (with Carleton Rea) "Fungi new to Britain," ibid., (1902): 31–40; (1903): 59–67; (1905): 127–31; (1906): 167–72; "How the lily is attacked and destroyed by the Botrytis fungus," *Proceedings of the Holmesdale Club,* (1902–5): 78–83; "Recent advances in the study of fungi," *Science Progress in the Twentieth Century,* **1** (1907): 530–7; "Microfungi: a historical sketch" (1907 presidential address), *Transactions of the British Mycological Society,* **3** (1907–11): 18–25; in H. N. Ridley, "On a collection of plants made by H. C. Robinson and L. Wray from Gunong Tahan, Pahang," fungi and lichens by A. L. Smith, *Journal of the Linnean Society (Botany),* **38** (1908): 301–36; "New or rare microfungi," *Transactions of the British Mycological Society,* **3** (1909): 111–23; 220–5; **3,** Pt. 4 (1911): 281–4; **3** (1912): 366–74; (with John Ramsbottom) ibid., **4** (1912): 165–85; "Fungal parasites of lichens," ibid., **3** (1909): 174–8; "*Phaeangella empertri* Boud (in litt.) and some forgotten Discomycetes. A correction," ibid., **4** (1912): 74–6; in Lilian S. Gibbs, "A contribution to the flora and plant formations of Mount Kinabalu and the highlands of British North Borneo," fungi and lichens section by A. L. Smith, *Journal of the Linnean Society* (*Botany*), **42** (1914): 1–240; "The relation of fungi to other organisms" (1917 presidential address), *Transactions of the British Mycological Society,* **6** (1917–19): 17–31; "Hyphomycetes and the rotting of timber," ibid., **6** (1917–19): 54–5; "A drain-blocking fungus," ibid., **6** (1917–19): 262–3.

175. A. H. C., *Journal of Botany,* **59** (1921): 331–3.

176. Hawksworth and Seaward, *Lichenology,* p. 28.

177. "Lichens (Clare Island Survey, Pt. 14)," *Proceedings of the Royal Irish Academy,* **31** (1911): 1–14. Among Smith's other journal articles on lichens are the following: "British Coenogoniaceae," *Journal of Botany,* **44** (1906): 266–8; "Gall formation in Ramalina," ibid., **45** (1907): 344–5; "New localities of rare lichens," ibid., **45** (1907): 345; "New lichens," *Journal of Botany,* **49** (1911): 41–4; "Lichens of the Baslow foray," *Transactions of the British Mycological Society,* **6** (1919): 252; "History of lichens in the British Isles," *South Eastern Naturalist* (1922): 19–35; "Recent work on lichens," *Transactions of the British Mycological Society,* **10** (1925): 133–52; "Lichen dyes," ibid., **11** (1926): 45–58; "Cryptotheciaceae. A family of primitive lichens," ibid., **11** (1926): 189–96; (with M. C. Knowles) "Lichens of the Dublin foray," ibid., **11** (1926): 18–22; "Recent lichen literature," ibid., **12** (1927): 231–75; **15** (1930): 193–235.

178. Matilda Cullen Knowles, *Lichens of Ireland* (London: Williams and Norgate, 1929). See Timothy Collins, "Some Irish women scientists," *UCG Women's Studies Centre Review,* **1** (1992): 39–53, and R. Lloyd Praeger, *Some Irish Naturalists. A Biographical Notebook* (Dundalk: Dungalgan Press, 1949), p. 116. Knowles was a "Temporary Assistant" at the National Museum of Ireland in Dublin from 1907 until her death in 1933.

179. Gepp and Rendle, "Annie Lorrain Smith," p. 330.

180. Obituary, James Britten, "Ethel Sarel Gepp (1864–1922)," *Journal of Botany,* **60** (1922): 193–5.

181. Another woman coworker of George Murray whose paper is listed in the bibliography was Frances Whitting, about whom no biographical information has been collected. Her joint paper with Murray on Atlantic *Peridinaceae,* which would seem to have been her only scientific publication (at least under the name Whitting), appeared in the *Transactions of the Linnean Society* in 1899.

182. E. S. Gepp, *The Genus Halimeda* (Leyden: E. J. Brill, 1901), and Antony Gepp and Ethel Sarel Gepp, *The Codiaceae of the Siboga Expedition, including a Monograph of Flabellarieae and Udoteae,* (Leyden: E. J. Brill, 1911).

183. See, for instance, "Algae. Report on the collection . . . made . . . during the voyage of the 'Southern Cross.' XXI," *Cryptogamia,* London, 1902, pp. 319–20; "List of marine algae collected at the Maldive and Laccadive Islands by J. S. Gardiner," *Journal of the Linnean Society* (*Botany*), **35** (1903): 475–82; "List of marine algae collected by Professor Herdman, at Ceylon, in 1902, with a note on the fructification of *Halimeda,*" *Royal Society Report, Pearl Oyster Fisheries,* Pt. 1, 1903, pp. 163–7; "Chinese marine algae," *Journal of Botany,* **42** (1904): 161–5; "The sporangia of *Halimeda,*" ibid., **42** (1904): 193–7; and the following with Antony Gepp, "Notes on *Penicillus* and *Rhipocephalus,*" ibid., **43** (1905): 1–5; "Antarctic algae," ibid., **43** (1905): 105–9; "Some cryptogams from Christmas Island," ibid, **43** (1905): 337–44; "Some marine algae from New South Wales," ibid., **44** (1906): 249–61; "Marine algae (*Chlorophyceae* and *Phaeophyceae*) and marine phanerogams of the 'Sealark Expedition,' collected by J. Stanley Gardiner," *Transactions of the Linnean Society,* **8,** s. 2. (1908): 163–88; "Marine algae. I. *Phaeophyceae* and *Florideae,*" in *National Antarctic Expedition: Natural History,* vol. 3. London, 1907; "Marine algae of the Kermadecs," *Journal of Botany,* **49** (1911): 17–23. Ethel Gepp's final publication was the marine algae section in Lilian S. Gibbs, "A contribution to the flora and plant formations of Mount Kinabalu and the highlands of British North Borneo," *Journal of the Linnean Society* (*Botany*), **42** (1914): 1–240.

184. Obituaries, F. W. Oliver, *Proceedings of the Linnean Society* (1928–29): 146–7, and anon., "Mrs. D. H. Scott," *Nature,* **123** (1929): 287; *DNB,* Fifth Supplement, 1931–40, pp. 796–7, entry for Scott, Dukinfield Henry.

185. D. H. Scott, "On nuclei in *Oscillaria* and *Tolypothrix,*" *Journal of the Linnean Society* (*Botany*), **24** (1888): 188–192.

186. "On the megaspore of *Lepidostrobus foliaceus,*" *New Phytologist,* **5** (1906): 116–9; "On *Bensonites fusiformis* sp. nov., a fossil associated with *Stauropteris burntislandica,* P. Bertrand, and the sporangia of the latter," *Annals of Botany,* **22** (1908): 683–7; "On Traquairia," ibid., **25** (1911): 459–67.

187. "On the movements of the flowers of *Sparmannia africana,* and their demonstration by means of the kinematograph," *Annals of Botany,* **17** (1903): 762–7. See also, "Animated photographs of plants," *Knowledge and Scientific News,* **1** (1904): 83–6; "Opening of flowers. Use of photographic illustrations," *Gardener's Chronicle,* s. 3, **39,** (1906): 223; "Animated photographs of plants," *Journal of the Royal Horticultural Society,* **32** (1907): 48–51.

188. Obituary, E[llen] M[arion] Delf, *Proceedings of the Linnean Society* (1960–61): 173; membership lists, *Proceedings and Transactions of the Liverpool Biological Society,* 1891–1940.

189. May Rathbone, "Notes on *Myriactis areschougii* and *Coilodesine californica,*" *Journal of the Linnean Society* (*Botany*), **35** (1904): 670–5.

190. This medical qualification licensed a holder to practice, but was somewhat less than a university medical degree.

191. Beatrice Harraden, "Kate Marion Hall," *Proceedings of the Linnean Society* (1917–18): 61–3.

192. Similar programs for schoolchildren were started in the United States—see Alice Rich Northrup (above).

193. Obituaries, E[lizabeth] M[arianne] Blackwell, *Proceedings of the Linnean Society* (1937): 186–9, and "Professor Margaret Benson," *Nature,* **138** (1936): 17; anon., *Chronica Botanica,* **3** (1937) 159–60; *Who was Who,* vol. 3, p. 97; *Newnham College Register 1871–1971,* vol. 1, p. 64.

194. Of special note among Benson's publications were, "*Telangium scottii,* a new species of *Telangium* (Calymmatotheca) showing structure," *Annals of Botany,* **18** (1904): 161–77; "*Miadesmia membranacea,* Bertrand—a new Palaeozoic lycopod with a seed-like structure," *Transactions of the Royal Society,* B, **198** (1908): 409–24; "New observations on *Botryopteris antigua,*" *Annals of Botany,* **25** (1911): 1045–57; "*Cordaites felicis,* sp. nov., a Cordaitean leaf from the Lower Coal Measures of England," ibid., **26** (1912): 201–7; "*Sphaerostoma ovale* (*Conostoma ovale et intermedium,* Williamson), a lower carboniferous ovule from Pettycur, Fifeshire, Scotland," *Transactions of the Royal Society, Edinburgh,* **50** (1914): 1–15 (this paper was cited as late as 1993—see Wilson N. Stewart and Gar W. Rothwell, *Paleobotany and the Evolution of Plants*—Cambridge: Cambridge University Press, 1983, 2d rev. ed. 1993, p. 305); "*Mazocarpon* or the structural *Sigillariostrobus,*" *Annals of Botany,* **32** (1918): 569–89; "*Heterotheca grievii,* the microsporange of *Heterangium grievii,*" *Botanical Gazette,* **74** (1922): 121–42. Included among her earlier works were: "The fructification of *Lyginodendron oldhamium,*" *Annals of Botany,* **16** (1902): 575–6; "A new Lycopodiaceous seed-like organ," *New Phytologist,* **1** (1902): 58–9; "The origin of flowering plants," ibid., **3** (1904): 49–51; "Reforms in cell nomenclature," ibid., **4** (1905): 96; "On the contents of the pollen chamber of a specimen of *Lagenostoma ovoides,*" *Botanical Gazette,* **45** (1908): 409–12; "The sporangiophore—a unit of structure in the Pteridophyta," *New Phytologist,* **7** (1908): 143–9; "*Botrychium lunaria* with two fertile lobes," ibid., **8** (1909): 354; (with Evelyn J. Welsford) "The morphology of the ovule and female flower of *Juglans regia* and of a few allied genera," *Annals of Botany,* **23** (1909): 623–33; "Root parasitism in *Exocarpus,*" ibid., **24** (1910): 667–77.

195. Dukinfield Henry Scott, *Studies in Fossil Botany,* 2 vols. (London: A. and C. Black, 1908–09, 2d edition), vol. 2, p. 411. Scott provided a brief summary of Benson's data in the form of an extract from a letter she had written to him. See also 3d ed. of Scott's *Studies* (1920–23, vol. 2, p.11) and Margaret Benson, "The roots and habit of *Heterangium grievii,*" *Annals of Botany,* **47** (1933): 313–15.

196. "New evidence of isospory in Palaeozoic seed plants," *New Pythologist,* **34** (1935): 92–6; "The ovular apparatus of *Sphenopteridium affine* and *bifidum* and of *Diplopteridium sphenopteridium teilianum* (Walton)," ibid., **34** (1935): 232–44.

197. Obituaries, "Dr. E. N. Miles Thomas," *Nature,* **154** (1944): 481–2, E. M. Delf, *Proceedings of the Linnean Society* (1943–44): 235–6, and *Times,* 1 September 1944, 7; *Who was Who,* vol. 4, p. 1142; E. M. Delf in *Chronica Botanica,* **7** (1942–1943): 48; birth and death certificates, General Register Office for England and Wales; materials from Royal Holloway, University of London, and from University Archives, University of Leicester (including clippings from *Leicester Mercury,* 7 February 1931 and September 1933, and from *Leicester Evening Mail,* 26 May 1933); Student Records, University College London.

198. Thomas's early papers include the following: (with A. G. Tansley) "Root structure in the central cylinder of the hypocotyl," *New Phytologist,* **3** (1904): 104–6; "Some points in the anatomy of *Acrostichum aureum,*" ibid., **4** (1905): 175–89; (with A. G. Tansley) "The phylogenetic value of the vascular structure of spermophytic hypocotyls," *Reports of the British Association* (1906): 761–3; "A theory of the double leaf-trace founded on seedling structure," *New Phytologist,* **6** (1907): 77–91; "Some aspects of 'double fertilization' in plants," *Science Progress,* **1** (1907): 420–6.

199. See Ethel N. Thomas, "Seedling anatomy of Ranales, Rhoeadales and Rosales," *Annals of Botany,* **28** (1914): 695–733; (with A. J. Davey) "Morphology and anatomy of certain pseudo-monocotyledons," *Reports of the British Association* (1914): 578–9; "Observations on the seedling anatomy of the Ebenales," ibid, (1923): 491; "Observations on the seedling anatomy of the genus *Ricinus,*" *Proceedings of the Linnean Society* (1923): 49–50; "The primary vasular system in phanerogams: its character and significance," *Reports of the British Association* (1924): 447.

200. Delf, obituary, p. 236, and anon., *Leicester Evening Mail.*

201. Mary Ewart (b. 30 November 1867) was an undergraduate at University College from 1886 to 1889, (B.Sc., 1889). She continued to work in the college's botanical laboratories for a number of years and published three morphological papers in the *Annals of Botany* and the *Journal of the Linnean Society* between 1892 and 1895 (Student Records, University College London). Sara Agnes Calvert, a contemporary of Ewart, studied at the Royal College of Science in the 1880s and 1890s (B.Sc., 1895). Her two papers on lactiferous tissue (one coauthored by Leonard Boodle) appeared in the first volume of the *Annals of Botany* in 1887.

202. Student Records, University College London; Margaret Tomlinson, *Three Generations in the Honiton Lace Trade. A Family History* (Exeter: Devon Print Group, 1983), especially pp. 64–8; *DNB,* Supplement, 1951–60, pp. 953–4, entry for Tansley, Sir Arthur George.

203. Harriette Chick, later Dame Harriette, went on to an outstanding research career at the Lister Institute (see chapter 6). Frances (1883–1919) took a B.Sc. in 1908, concentrating on chemistry. She carried out postgraduate work with N. T. D. Wilsmore at University College and coauthored two papers with him on polymerization reactions (1908 and 1910); she later moved on to medical and biochemical research, for a time working in the biochemical department at the Lister Institute. She married Sydney Herbert Wood, and had one daughter, but died at the age of thirty-six. Three of the Chick sisters, Elsie (1882–1967), Mary (1873–1938), and Margaret (1876–1963), specialized in language studies at University College (Student Records, University College). Elsie married the Cambridge plant physiologist Frederick Frost Blackman in 1917, and had one son.

204. (with A. G. Tansley) "Notes on the conducting tissue-system in Bryophyta," *Annals of Botany,* **15** (1901): 1–38; "The seedling of *Torreya myristica,*" *New Phytologist,* **2** (1903): 83–91; (with A. G. Tansley) "On the structure of *Schizaea malaccana,*" *Annals of Botany,* **17** (1903): 493–510.

205. See references to Agnes Fry in anon., "The Right Hon. Sir Edward Fry, G.C.B., F.R.S.," *Nature,* **102** (1918): 169–70, and in the *DNB,* Supplement, 1912–21, pp. 200–3, entry for Fry, Sir Edward. See also Agnes Fry, *A Memoir of the Right Honourable Sir Edward Fry, G.C.B.* (London: Oxford University Press, 1921), pp. 61–2.

206. For her 1902 paper see, "Note on variation in leaves of mulberry trees," *Biometrika,* **1** (1902): 258–9. In addition to the biography of her father (n. 205), Agnes Fry published a book of verse, *Winter Sunshine and other Verses* (Leominster: Orphans' Printing Press, 19—?). For a mention of her continuing work on Mycetozoa, which went on until at least 1914, see Gulielma Lister, "A short history of the study of Mycetozoa in Britain, with a list of species recorded from Essex," *Essex Naturalist,* **18** (1914–18): 207–37.

207. Jane Gowan was an "out-student" at Newnham College from 1896 to 1898. She did not take the Tripos examinations but nevertheless collaborated with palaeobotanist Sir Albert Seward in investigations on *Ginkgo biloba* (the maidenhair tree), reported in *Annals of Botany* in 1900. She does not appear to have published any other botanical work (*Newnham College Register 1871–1971,* vol. 1, p. 136).

208. *Newnham College Register 1871–1971,* vol. 1, p. 75; obituaries, B. A. Clough, "Anna Bateson," *Newnham College Roll Letter* (1928): 78–81, and C. Marson, ibid., pp. 81–3. See also *William Bateson, F.R.S. His Essays and Addresses,* ed. B. Bateson (Cambridge: Cambridge University Press, 1928, reprinted by Garland Publishing Inc., New York, 1984), pp. 40–3.

209. *Girton College Register 1869–1946,* vol. 1, p. 43.

210. Elizabeth Dale's post-1900 publications include, "Investigations on the abnormal outgrowths or intumescences on *Hibiscus vitifolius,* Linn. A study in experimental plant pathology," *Philosophical Transactions,* B, **194** (1901): 163–82; "Further investigations on the abnormal outgrowths or intumescences in *Hibiscus vitifolius,* Linn.," *Proceedings of the Royal Society,* **68** (1901): 16–19; "Notes on artificial cultures of *Xylaria,*" *Proceedings of the Cambridge Philosophical Society,* **11** (1901): 100–2; "On the origin, development and morphological nature of the tubers in *Dioscorea sativa,* Linn.," *Annals of Botany,* **15** (1901): 491–500; (with A. C. Seward) "On the structure and affinities of *Dipteris,* with notes on the geological history of the Dipteridinae," *Proceedings of the Royal Society,* **68** (1901): 373–4; "Observations on Gymnoascaceae," *Annals of Botany,* **17** (1903): 571–96; "Further experiments and histological investigations on intumescences, with some observations on nuclear division in pathological tissues," *Proceedings of the Royal Society,* **26** (1905): 587–8, *Philosophical Transaction,* B, **198** (1906): 221–63; "On the morphology and cytology of *Aspergillus repens* de Bary," *Annales Mycologici,* **7** (1909): 215–25; "On the cause of blindness in potato tubers," *Annals of Botany,* **26** (1912): 129–31; "A bacterial disease of potato leaves," ibid., **26** (1912): 133–54; "On the fungi of the soil. I. Sandy loam," *Annales Mycologici,* **10** (1912): 452–77; "On the fungi of the soil. II. Fungi from chalky soil, uncultivated mountain peat and the 'black earth' of the reclaimed Fenland," ibid., **12** (1914): 33–62.

211. *Newnham College Register 1871–1971,* vol. 1, p. 141. The entry gives little information. Dawson was "presumed dead" in 1923.

212. Two of these are listed in the bibliography; the third, entitled "On the economic importance of 'Nitragin'," appeared in *Annals of Botany* in 1901 (**15,** pp. 511–19). The important discovery of symbiotic fixation of nitrogen by leguminous plants had first been reported by German botanists H. Hellriegel and W. Wohlfahrt only thirteen years earlier (Morton, *History of Botanical Science,* p. 420).

213. Obituaries, Agnes Arber, "Miss Dorothea F. M. Pertz," *Nature,* **143** (1939): 590–1, and E. R. Saunders, *Proceedings of the Linnean Society* (1938–39): 245–7; *Newnham College Register 1871–1971,* vol. 1, p. 76.

214. These studies were part of a long series of investigations by Darwin that proved to be an important step on the way to major developments in plant physiology. In collaboration with his father, Charles, Francis Darwin had demonstrated in 1881 that the site of stimulus perception in plants is a localized area in the tip of stem or root and that curvatures due to light or gravity result from the passage of a stimulus from this site to other regions where response takes place. The work was not accepted by the leading plant physiologist of the time, the German botanist Julius Sachs, who had some of the professional's distrust for what seemed to him the Darwins' amateurish methods. Thirty years later the English botanist Peter Boyson-Jensen continued the Darwins' work and demonstrated that the stimulus could be passed through a block of watery gelatin placed between a stimulated excised tip and a stem below, thus showing that the "stimulus" must be a diffusible substance. This started the search that led to the first isolation of a plant hormone (Morton, *History of Botanical Science,* pp. 427–8).

215. Pertz's post-1900 publications include the following: (with Francis Darwin) "On the artificial production of rhythm in plants, with a note on the position of maximum heliotropic stimulation," *Annals of Botany,* **17** (1903): 93–106; "On the distribution of statoliths in Cucurbitaceae," ibid., **18** (1904): 653–4; (with Francis Darwin) "Notes on the statolith theory and geotropism. I. Experiments on the effects of centrifugal force. II. The behaviour of tertiary roots," *Proceedings of the Royal Society,* **73** (1904): 477–90; "The position of maximum geotropic stimulation," *Annals of Botany,* **19** (1905): 569–70; (with Francis Darwin) "On a new method of estimating the aperture of stomata," *Proceedings of the Royal Society,* B, **84** (1912): 136–54.

216. Saunders, obituary, p. 247.

217. Obituaries, H. Godwin, A. R. Clapham, and M. R. Gilson, "Edith Rebecca Saunders, F.L.S.," *New Phytologist,* **45** (1946): 1–3, G[ertrude] L[ilian] Elles and Ethel Shakespear, "Miss E. R. Saunders," *Nature,* **156** (1945): 198, J. McL. Thompson, "Miss E. R. Saunders," ibid., **156** (1945): 198–9, J. B. S. Haldane, "Miss E. R. Saunders," ibid., **156** (1945): 385, H. Hamshaw Thomas, *Proceedings of the Linnean Society* (1947): 75–6, and E[dith] M. C[hrystal], "Edith Rebecca Saunders, F.L.S.," *Newnham College Roll Letter* (January 1946): 41–2; Rudolf Schmidt, "Edith R. Saunders and floral anatomy: bibliography and index to the families she studied," *Botanical Journal of the Linnean Society,* **74** (1977): 179–87; *Newnham College Register 1871–1971,* vol. 1, Staff, p. 7.

218. Godwin, Clapham and Gilson, "Edith Rebecca Saunders," p. 1.

219. Haldane, "E. R. Saunders," p. 385.

220. W. Bateson and Miss E. R. Saunders, *Reports to the Evolution Committee of the Royal Society. Report I* (London: Royal Society, 1902, 160 pp.). In all there were five reports to the Evolution Committee. Reports II, III, and IV (1905, 1906 and 1908) were by Bateson, Saunders and R. C. Punnett. All three of these included sections on Saunders's continuing work on plant crossbreeding, especially on *Matthiola* and sweet peas.

221. Gregor Mendel, "Versuche über Pflanzen-Hybriden," *Mittheilungen der k.k. Mährisch-Schleisischen Gesellschaft zur Beförderung des Akerbaues, der Natur- und Landeskunde in Brünn,* **4** (1865): 3–47; "Ueber einige aus künstliche Befruchtung gewonnen Hieracium-Bastarde," ibid., **8** (1869):26–31. See also Robert Olby, "William Bateson's introduction of Mendelism to England: a reassessment," *British Journal for the History of Science,* **20** (1987): 399–420.

222. Muriel Wheldale, a student at Newnham from 1900 to 1904, acquired an international reputation as a pioneer in plant biochemistry. Most of her working life was spent at Cambridge, her activities divided between the Botany School and the Biochemical Department. She was appointed university lecturer in 1927. Florence Durham, Bateson's sister-in-law, was a student at Girton (NST examinations, 1891, 1892). After six years as a lecturer at Royal Holloway College, London, she returned to Cambridge in 1899 as lecturer in physiology at Newnham. In 1910 she moved with Bateson to the John Innes Horticultural Institute in Merton, and continued her work in genetics for several years (for Wheldale see, Marjory Stephenson, "Muriel Wheldale Onslow (1880–1932)," *Biochemical Journal,* **26** (1932): 915–16; for Durham, *Girton College Register, 1869–1946,* p. 49, and *Girton College Review,* Michaelmas Term, 1949, p. 35).

223. See Robert Olby, "The dimensions of scientific controversy: the Biometric-Mendelian debate," *British Journal for the History of Science,* **22** (1989): 299–320, and R. C. Punnett, "Early days of genetics," *Heredity,* **4** (1950): 1–10.

224. William B. Provine, *The Origins of Theoretical Population Genetics* (Chicago: University of Chicago Press, 1971).

225. Included among Saunders's papers in this area were, "Mendel's theory of heredity," *Report and Transactions of the South-Eastern Union of Scientific Societies,* (1905): 55–61; "Certain complications arising in the cross-breeding of stocks," *Report of the Third International Conference, 1906, on Genetics—Hybridisation, the Cross-breeding of Varieties and General Plant-breeding,* ed. Rev. W. Wilks, M.A., 1906, pp. 143–9; (with W. Bateson and R. C. Punnett) "Further experiments on inheritance in sweet peas and stocks; preliminary account," *Proceedings of the Royal Society,* B., **77** (1905): 236–8; "Double flowers in stocks," *Gardener's Chronicle,* **50,** s. 3 (1911): 324; "Studies in the inheritance of doubleness in flowers. I. *Petunia,*" *Journal of Genetics,* **1** (1910–11): 57–69; "Further experiments on the inheritance of 'doubleness' and other characters in stocks," ibid., **1** (1910–11): 305–76; "On inheritance of a mutation in the common foxglove (*Digitalis purpurea*)," *New Phytologist,* **10** (1911): 57–63; "Double flowers," *Gardener's Chronicle,* **52,** s. 3 (1912): 357–8; "Further contributions to the study of the inheritance of hoariness in stocks (*Matthiola*)," *Proceedings of the Royal Society,* B, **85,** (1912): 540–5; "On the relation of *Linaria alpina* type to its varieties *concolor* and *rosea,*" *New Phytologist,* **11** (1912): 167–9; "Double flowers," *Journal of the Royal Horticultural Society,* **38** (1913): 469–82; "On the mode of inheritance of certain characters in double-throwing stocks. A reply," *Zeitschrift für induktive Abstammungs- und Vererbungslehre,* **10** (1913–14); 297–310; "The results of further breeding experiments with *Petunia,*" *American Naturalist,* **50** (1915): 548–53; "A suggested explanation of the abnormally high records of doubles quoted by growers of stocks (*Matthiola*)," *Journal of Genetics,* **5** (1915–16): 137–43; "On the relation of half-hoariness in *Matthiola* to glabrousness and full hoariness," ibid., **5** (1915–16): 145–58; "Studies in the inheritance of doubleness in flowers. II. *Meconopsis, Althaea* and *Dianthus,*" ibid., **6** (1917): 165–84; "On the occurrence, behavior and origin of a smooth-stemmed form of the common foxglove (*Digitalis purpurea*)," ibid., **7** (1918): 215–28; "Multiple allelomorphs and limiting factors in inheritance in the stock (*Matthiola incana*)," ibid., **10** (1920): 149–78; "Note on the evolution of the double stock (*Matthiola incana*)," ibid., **11** (1921): 69–74; "Further studies on the inheritance in *Matthiola incana.* I. Sap colour and surface character," ibid., **14** (1924): 101–14; "The history, origin and characters of certain interspecific hybrids in *Nolana* and their relation to *Nolana paradoxa,*" ibid., **29** (1934): 387–419.

226. Thomas, obituary, p. 76.

227. See especially, "On carpel polymorphism. I.," *Annals of Botany,* **39** (1925): 122–67, and "Perigyny and carpel polymorphism in some Rosaceae," *New Phytologist,* **24** (1925): 206–22. Saunders's other morphology papers include the following: "The skin-leaf theory of the stem: a consideration of certain anatomico-physiological relations in the spermophyte shoot," *Annals of Botany,* **36** (1922): 135–65; "The bractless inflorescence of the Cruciferae," *New Phytologist,* **22** (1923): 150–6; "A study of *Antirrhinum orontium,*" *Hereditas,* **9** (1927): 17–24; "Illustrations of carpel polymorphism," *New Phytologist,* **27** (1928): 47–60; 175–92; 197–213; **28** (1929): 225–58; **29** (1930): 44–52; 81–95; **30** (1931): 80–118; "On carpel polymorphism," *Annals of Botany,* **45** (1931): 91–110; **46** (1932): 239–88; "The cause of petaloid colouring in 'spetalous' flowers," *Journal of the Linnean Society* (*Botany*), **49** (1933): 199–218; "A study of *Veronica* from the viewpoint of certain floral characters," *Proceedings of the Linnean Society* (1933): 126–8; "On the gynaeceum of *Filipendula ulmaria* Maxim. and *Filipendula hexapetala* Gilib: a correction," *Annals of Botany,* **49** (1935): 848–52; "On certain features of floral construction and arrangement in the Malvaceae," ibid., **50** (1936): 247–82; "The neglect of anatomical evidence in the current solution of problems in systematic botany," *New Phytologist,* **38** (1939): 203–9; "The significance of certain morphological variations of common occurrence in flowers of *Primula,*" ibid., **40** (1941): 64–85. For a complete listing of Saunders's publications on floral anatomy see Schmidt, "Edith R. Saunders," pp. 185–7 (her papers, books, and reports numbered forty-nine; also listed are fourteen abstracts and discussions).

228. Among those with whom Saunders vigorously debated her theories was Agnes Arber (1879–1960), another Newnham-educated botanist, who was, by the 1940s, one of the most distinguished plant morphologists in Britain. (Arber was one of the first women fellows of the Royal Society (1946) and the 1948 recipient of the Linnean Medal). For parts of the Saunders-Arber debate see, for instance, E. R. Saunders, "A reply to comments on the theory of the solid carpel and carpel polymorphism," *New Phytologist,* **25** (1926): 294–306; "On some recent contributions and criticisms dealing with morphology in angiosperms," ibid., **31** (1932): 174–219; "Comments on 'floral anatomy' and its morphological interpretation," ibid., **33** (1934): 127–70.

229. Godwin, Clapham, and Gilson, "Edith Rebecca Saunders," p. 2.

230. See Lincoln Constance, "The systematics of the angiosperms," in Kessel, ed., *Century of Progress,* pp. 405–83, especially pp. 423–4.

231. Schmidt, "Edith R. Saunders," p. 179.

232. Crystal, "Edith Rebecca Saunders," p. 42.

233. Ibid., p. 42. The passages from Crystal are quoted with the permission of the Principal and Fellows of Newnham College, Cambridge.

234. Obituaries, D. H. S[cott], "Miss Ethel Sargant, F.L.S.," *Annals of Botany,* **32** (1918): i–v, A[gnes] A[rber], "Ethel Sargant," *New Phytologist,* **18** (1919): 120–8, E. N. Thomas, "Ethel Sargant," *Proceedings of the Linnean Society* (1918): 41–2, and anon., "Miss Ethel Sargant," *Nature,* **100** (1918): 428; *Who was Who,* vol. 2, p. 983.

235. This laboratory was something of a puzzle to the neighborhood, and caused the local excise officer special concern. He used to pay frequent surprise visits because he suspected that, as soon as his back was turned, Miss Sargant used the small still for the production of distilled water to manufacture illicit spirits (see W. T. Stearn, "Mrs. Agnes Arber," *Taxon,* **9** (1960): 261–3, on p. 263).

236. Arber, "Ethel Sargant," p. 122.

237. "A theory of the origin of the monocotyledons founded on the structure of their seedlings," *Annals of Botany,* **17** (1903): 1–92. Sargant's preliminary statement appeared as, "The origin of the seed-leaf in monocotyledons," *New Phytologist,* **1** (1902): 107–13. The theory was summarized in "The evolution of monocotyledons," *Botanical Gazette,* **37** (1904): 325–45, and "The early history of angiosperms," ibid., **39** (1905): 420–3. It was elaborated further in, "The reconstruction of a race of primitive angiosperms," *Annals of Botany,* **22** (1908): 121–86.

238. See for instance, Constance, "Systematics of the angiosperms," pp. 429–35, 445–51.

239. Ethel Sargant and Agnes Robertson (Mrs. Arber), "The anatomy of the scutellum in *Zea mais,*" *Annals of Botany,* **19** (1905): 115–23; "The comparative morphology of the embryo and seedling in the Gramineae," ibid., **29** (1915): 161–222.

240. Ethel Sargant, "The development of botanical embryology since 1870," *Reports of the British Association* (1913): 692–705.

241. Among Sargant's papers not already mentioned were several on a variety of general topics; see "The adaptations of seedlings to their surroundings," *Report and Transactions of the South-Eastern Union of Scientific Societies,* (1901): 12–15; "The seedlings of geophytes," *The South-Eastern Naturalist,* (1903): 22–6; "The family tree of flowering plants," *Proceedings of the Holmesdale Natural History Club* (1902–05): 50–9; "The native countries of our spring bulbs," ibid., (1910–13): 56–62. In 1910 she was invited by the editors of the Cambridge Botanical Handbooks series to contribute a book on the monocotyledons, but ill health made this impossible and shortly before her death she suggested that Agnes Arber take over the task.

242. Ethel Sargant, "Women and original research," *Frances Mary Buss Schools Jubilee Magazine* (Nov. 1900): 1–8, and "The inheritance of a university," *The Girton Review* (Lent Term, 1901): 1–15.

243. Membership lists, Liverpool Biological Society, 1889–1896, *Proceedings and Transactions of the Liverpool Biological Society.*

244. Lily H. Huie, "Notes on the cells of the so-called hepatico-pancreatic glands of isopods," *Proceedings and Transactions of the Scottish Microscopical Society,* **3** (1901): 85–8; "Metabolism of some ice-bound cells," ibid., **3** (1904): 213–16.

245. B. D. J., *Proceedings of the Linnean Society* (1911–12): 45–6; *Who was Who,* vol. 1, p. 237; D. E. Allen, *The Naturalist in Britain. A Social History* (London: Allen Lane, 1976), p. 169; Andrew Thomas Gage and William Thomas Stearn, *A Bicentenary History of the Linnean Society of London* (London: Academic Press, 1988), pp. 88–93.

246. Of the sixteen women then presented for election (a group that included Margaret Benson, Gulielma Lister, Ethel Sargant, and Annie Lorrain Smith, as well as geologist Maria Ogilvie Gordon, bacteriologist Grace Toynbee Frankland, and zoologist Lilian Gould Veley), Marian Ogilvie-Farquharson alone was rejected—a somewhat churlish performance by those present at this last exclusively male meeting of the society.

247. Stearn, *Natural History Museum,* p. 273. Birley is mentioned in *British Biographical Archive,* microfiche edn, ed. Paul Sieveking (London, 1984), no. 109 (entry from John Foster Kirk, *A Supplement to Allibone's Critical Dictionary of English Literature,* 2 vols., Philadelphia: J. B. Lippincott, 1891). Her books for children included: *We are Seven: a Tale for Children* (London: Darton and Co., 1880), and *My Lady Bountiful: the Tale of Harriet, Duchess of Albans* (London: Walter Smith and Co., 1880).

248. Buchanan White was probably the daughter of Francis Buchanan White, a Perth physician, founder of the *Scottish Naturalist* (later incorporated into the *Annals of Scottish Natural History*) and a fellow of both the Linnean and Entomological Societies.

249. Ray Desmond, *Dictionary of British and Irish Botanists and Horticulturalists* (London: Taylor and Francis), 1977, p. 39; D. E. Allen, "Women members," p. 251.

250. Obituary notices, *Report and Transactions of the Devonshire Association,* **47** (1915): 55–6; *Flora of Devon,* eds. Rev. W. Keble Martin and Gordon T. Fraser (Arbroath: T. Buncle and Co.), 1939, p. 777.

251. See for instance, Helen Saunders, "Devonshire revels," *Report and Transactions of the Devonshire Association,* **28** (1896): 342–50; and "Notes on the history of a North Devon parish, Aissa, Rose Ash," ibid., **32** (1900): 212–28. Her post-1900 botanical papers included, "Botanical notes," ibid., **33** (1901): 469–74; **38** (1906): 491–6; **40** (1908): 303–5; "*Euphrasia minima* Jacq. in England," *Journal of Botany,* **47** (1909): 30; "Double daffodils," ibid., **49** (1911): 62–3.

252. Obituary, *Journal of Botany,* **67** (1929): 183.

253. These included, in addition to the two listed in the bibliography, the following, "Sensation in plant life," *Transactions of the Leicester Literary and Philosophical Society,* **8** (1904): 98–115; "Some rarer agents of pollination," *Knowledge and Scientific News,* **4** (1907): 223–4; "Honey hiding places," ibid., **5** (1908): 96–9; "A plant-life conundrum," ibid., **6** (1909): 90–1.

254. For Vickers's 1896 paper, see bibliography; the second was "Liste des algues marine de la Barbade," *Annales des Sciences Naturelles, Botanique,* s.9, **1,** (1905): 45–66.

255. Anna Vickers, *Phycologia Barbadensis. Iconographia des Algues Marines récoltées à l'Ile Barbade (Antilles): (Chlorophyées et Phéophycées),* avec text explicatif par Mary Helen Shaw (Paris: Librarie des Sciences Naturelles, Paul Klincksieck, 1908), 300 copies printed, at the expense of the author.

256. *Who was Who,* vol. 2, p. 946; *Burke's Peerage and Baronetage,* 105th. ed., 1975, p. 909.

257. *British Biographical Archive,* fiche no. 44, entry from Frances Hays, *Women of the Day, a Biographical Dictionary of Notable Contemporaries,* London: Chatto and Windus, 1885.

258. Papers by four additional women, Eleanor Jex-Blake, Rose Haig Thomas, Ella Tindall, and Miss Hottinger, are also listed in the bibliography. Jex-Blake, who contributed a note on plant variation to *Nature* in 1900, was from Wells, Somersetshire. Mrs. Thomas was a naturalist from Ringwood, Hampshire, who contributed several notes on botanical and zoological subjects to *Nature* during the 1890s. Later papers by her included, "A luminous centipede," *Nature,* **65** (1902): 223, "Swarm of *Velella,*" ibid., **65** (1902): 586, "Bipedal locomotion in lizards," ibid., **66** (1902): 577, and "Parthenogenesis in *Nicotiana,*" *Mendel Journal,* **1** (1909): 5–10. She became a fellow of the Linnean Society in 1913. Tindall published a note in the *Journal of Botany* in 1898 reporting work done near Brauton, North Devon. Miss Hottinger brought out three botanical articles in the *Transactions of the Leicester Literary and Philosophical Society* in the 1890s and a fourth in the same journal in 1905 ("Laurels," n.s., **9,** pp. 35–8). Her paper on the teaching of botany, presented to the Leicester society in 1890, outlined the program of instruction she thought ideal; it suggests that she may have been a teacher.

259. John Gilmour, *British Botanists* (London: Collins, 1944), pp. 44–5.

260. Although there is no indication in the Royal Society index of original papers in nineteenth-century botany being published by American women before the 1870s, at least one well-received work on botanical education appeared in 1829, Almira Lincoln Phelps's *Familiar Lectures on Botany* (see Nancy G. Slack, "Nineteenth-century American women botanists: wives, widows and work," in Abir-am and Outram, eds., *Uneasy Careers,* pp. 77–103, especially p. 77). There were several early botanical clubs for women in the eastern part of the country, including the Female Botanical Society of Wilmington, the Dana Society of Natural History of the Albany Female Academy, and the Philadelphia and Syracuse Botanical Clubs. With the exception of the Syracuse Club, however, these were primarily informal social groups that encouraged private botanical study. The Syracuse Club brought out guides to local flora and fauna and maintained contacts with practicing botanists (Sally Gregory Kohlstedt, "In from the periphery: American women in science, 1830–1880," *Signs,* **4** (1978): 81–96).

261. Very few women held teaching positions at the Midwestern universities, however, and only one, Josephine Tilden, became nationally prominent for research in her field.

Chapter 2

LARGELY LEPIDOPTERISTS: COLLECTORS, CLASSIFIERS, AND STUDENTS OF INSECT DEVELOPMENT: THE ENTOMOLOGISTS

Papers in entomology make up about 10 percent of nineteeth-century articles by women indexed by the Royal Society, and most (about 90 percent) came from the United States and Britain. Thirty-eight Americans[1] and twenty-five British women contributed. Most of the remaining papers came from Germany, Canada, and Ireland.

American women

The American workers fall into three regional groupings: those in the northeast of the country, especially Massachusetts and New York, most of whom were self-taught amateurs,[2] those of the midwest, several of whom worked closely with the early state entomologists of the region, and those of the west and southwest. The northeastern group was the largest, with seventeen women who contributed about 52 percent of the American work; ten midwestern authors produced about 37 percent, and nine from the west and southwest, about 10 percent. Locations of two authors (Slater and Meek), who each wrote one paper, have not been discovered. Except for the writings of Dorothea Dix and Margaretta Morris (a total of six articles) all of the American work appeared after 1870 (Figure 2-1).

The most productive of the early Northeastern women entomologists as measured by the number of papers she published before 1901 was ANNIE TRUMBULL SLOSSON[3] (1838–1926). Born in Stonington, Connecticut, on 18 May 1838, she was the ninth child of Sarah and Gurdon Trumbull. The family was well known in the region. Three of Annie's brothers had notable careers; James, an expert on American Indian languages, taught at Yale, Henry discovered the biblical site of Kadesh Barnea in Palestine, and Gurdon was an outstanding artist. Annie was educated in local schools in Hartford, Connecticut. In 1867 she married Edward Slosson, but they had no children and she was widowed at a relatively early age. Throughout most of her life she spent her winters in Florida, at Miami, Indian Head, and Punta Gorda, her summers near Franconia in the White Mountains of New Hampshire, and the rest of the year in New York City.

Interested in natural history from an early age, she had made a special study of botany in school; in 1884 she published two papers in the *Bulletin of the Torrey Botanical Club*. She was already by then an accomplished writer, well known in literary circles. Her first book, *The China Hunters' Club by the Youngest Member* (a discussion of New England pottery and pottery collectors illustrated by her brother Gurdon), had been published in 1878, and it was followed by several other books and magazine articles describing New England life. A short story, *Fishin' Jimmy* (1889), became popular on both sides of the Atlantic and went through many editions, the last appearing as late as 1932. Much of her early work was written especially for young people, with the aim of encouraging interest in natural history.

Her entomology studies began in 1886, when she was forty-eight. For her fieldwork she was fortunate in having as a companion her brother-in-law, William Cowper Prime, lawyer, journalist, author, and art historian. For nineteen years, until his death in 1905, Prime was her constant source of encouragement for all her entomological undertakings. Although she was known as one of the earliest workers in the Miami region, her most valuable collections were taken mainly in then uncollected areas of the Northeast. The first of her lists of New England insects captured in the alpine region of Mt. Washington appeared in *Entomological News* in 1894.

She was in frequent contact with other entomologists to whom she passed on for further analysis not only many of the specimens she had collected but considerable amounts of biological and field data as well. Among her friends were the Lepidoptera specialists Henry Edwards and A. S. Packard (Packard was a member of the U.S. Entomological Commission and one of the first American authors of entomology textbooks); later she met William T. Davis, remembered for his studies of the natural history of Staten Island. About 100 species were named in her honor—the result of her close cooperation with these men and other leading workers in the field.

She joined the New York Entomological Society at its first meeting in 1892, and was later one of its oldest and most distinguished members. For a time meetings of the society were held at her house on East 23rd Street, New York City. Later, largely thanks to her efforts, accommodation was obtained in the American Museum of Natural History. A prime mover in establishing the society's journal and a frequent contributor to its pages, she was also its major financial backer and a most effective fund-raiser. Money was brought in by specimen auctions to which she regularly contributed the most and those that sold for the highest prices. The reports she published of insect structure and behavior were well-received by her contemporaries, and her articles were valued as enjoyable reading as well.[4] She remained active into her eighties, though after the death of her brother-in-law she slowed down her studies somewhat. She still collected, however, despite periods of ill health, and spent three seasons in the mountains of North Carolina, several in Florida, and many summers at the Delaware Water Gap on the Pennsylvania-New Jersey border. In later years she published her reminiscences of early entomologists she had known and worked with—articles received with much interest by her younger colleagues.[5] She died in her eighty-ninth year, 4 October 1926, at her home in New York City. Her collection of more than 35,000 specimens, left after she had supplied material to virtually every specialist who sent her a request, was given to the American Museum of Natural History shortly before her death.

A second productive Northeastern woman entomologist was CAROLINE SOULE[6] (1855–1920) of Brookline, Massachusetts. The daughter of Maria Goodwin (Gray) and Augustus Lord Soule, Caroline was born in Springfield, Massachusetts; her father was a justice of the Massachusetts Supreme Court. She received much of her education in private schools in Springfield and New York.

Her special interests were heredity and hybridization in Lepidoptera and the rearing of hybrid larvae. She carried out life history studies of several moths. Eighteen of her papers and notes appeared in the journal *Psyche* between 1888 and 1900, three being coauthored by her friend Ida Eliot, with whom she also published the monograph, *Caterpillars and their Moths* (1902).[7] She was a member of the Boston Society of Natural History and of several Boston and Brookline educational associations. A special interest in English literature, also shared with Ida Eliot,[8] inspired several stories and articles that she brought out in educational journals. By her early fifties poor health had caused her to give up all public activities; she appears to have published no more entomological work after 1907.

Both Emily Morton and Maria Smith Fernald also studied Lepidoptera. Morton's work on moth hybrids was of particular interest; Smith Fernald, who worked closely with her husband, the zoologist and entomologist Charles Henry Fernald, became an expert on the Tortricidae (leaf rollers), a large family of moths of considerable economic importance.

EMILY MORTON[9] (1841–1920), the daughter of Edmund Morton and his wife Caroline (Ellison), was born on 3 April 1841, in New Windsor, New York, in a mansion overlooking the Hudson River that had belonged to her mother's family for four generations. Her scientific education was largely self-directed. From early childhood, insects had fascinated her; she studied intensively those few books on entomology that were then available, even reading the state Agricultural Reports for the papers on entomology by Asa Fitch, the State Entomologist for New York.[10] She also had help from relatives and friends who were knowledgeable in natural history. Among these were the prominent entomologists Samuel Scudder, custodian for the Boston Society of Natural History and an editor of *Science,* and A. S. Packard of the U.S. Entomological Commission (also a friend of Annie Slosson).

Included in Morton's early work was a life history of the butterfly *Feniseca tarquinius.* She was the first to investigate the larval stages of this insect and communicated her detailed observations to her friend W. H. Edwards, who published them largely verbatim, but with appropriate acknowledgment of their source, in an 1886 article in the *Canadian Entomologist.* Her specialty, however, was the rearing of moth hybrids, work for which she was widely recognized. She was the first to successfully hybridize the Saturniidae, some of the largest and most colorful of the North American moths. The progeny of the crossings, though infertile, were often spectacular in size, shape, and color. Eight of the hybrids she sold to Baron Rothschild for $40, the only money she ever made from her lifelong work in entomology.[11] Her further contributions included illustrations for the publications of other entomologists, particularly drawings of larval stages of moths and butterflies. She died at her home in New Windsor on 8 January 1920, in her seventy-ninth year. Portions of her Lepidoptera collection went to the American Museum of Natural History and to the Boston Society of Natural History.

MARIA SMITH[12] (1839–1919) was born in Monmouth, Maine, on 24 May 1839, the daughter of Betsy (Torsey) and Ebenezer Smith. She graduated in the first class of the Maine Wesleyan Seminary and Female College, Kent's Hill, Maine, and then stayed on for a time as instructor. In 1862 she married Charles Fernald, just before he left to serve in the Union Navy during the Civil War. They had one son, Henry Torsey Fernald, who later became a noted entomologist and the first professor in the newly established department of entomology at the University of Massachusetts.

Maria first became interested in entomology when her husband, then professor of natural history at Maine State College at Orono, began his studies in that subject in the early 1870s. She collected with him in the fields and woods around Orono, and also by means of a lamp in a window at night or by "sugaring" when moonlight prevented the use of the light. Many of the moths she captured were rare and a few were new species; she was said to have been so skillful in the work that she could take a moth in her hand and transfer it to a cage without disturbing a single one of its scales.[13] Her card catalog of Tortricidae, begun in the late 1870s, was later extended to include insects of all kinds. One section was developed into *A Catalogue of the Coccidae of the World* (360 pp., 1903, also published as *Bulletin 88* of the Hatch Experiment Station of the Massachusetts Agricultural College). It was a work of considerable importance to those investigating scale insects, several species of which are among the most destructive insects known to commercial agriculture. As an experienced cataloger she also regularly helped her husband with various aspects of the bibliographical work he undertook.

The years of Maria Fernald's activity in entomology coincided with a period of rapid development in the field of economic entomology in New England and throughout the country. In 1889, three years after the Fernalds had moved from Maine to the Massachusetts Agricultural College at Amherst, the first of the European gypsy moth plagues occurred, and the following year Massachusetts organized a Gypsy Moth Commission whose technical work was directed by Charles Fernald.[14] Although Maria does not seem to have been directly involved in the subsequent efforts to control the moth, it was she who, thanks to her accurate knowledge of Lepidoptera, identified the caterpillars of the first devastating infestation. Her quick and definitive work made it immediately clear that strenuous efforts would have to be made to check the spread of the species.[15]

Maria was well known in the Amherst community and at the Agricultural College, her home being a center for social and scientific activities. She died 6 October 1919, at age eighty, predeceasing her husband by about two years.

Three women from the Boston area, Anna Hofmann Dimmock, Cora Clarke, and Jennie Arms Sheldon, all made modest contributions to the nineteenth-century entomology journal literature.

ANNA DIMMOCK,[16] the daughter of Ernest Hofmann, was the wife of George Dimmock, a zoologist and entomologist who from 1877 until 1890 edited the entomology journal, *Psyche.* They were married in 1878 and spent some time thereafter in Europe. George Dimmock took a Ph.D. at the University of Leipzig in 1881 and studied for a further year at the Sorbonne. After that he and Anna lived mainly in Boston and were engaged in private research in entomology. Morphological studies by George on the larvae of beetles, in which Anna collaborated, were particularly well received.[17] Her own paper on the nervous system of the larvae of the arthropod *Harpyia,* work carried out in Paris, appeared in *Psyche* in 1882. Later observational studies followed in 1890. The Dimmocks had one child, a daughter, Anna.

CORA CLARKE[18] (1851–1916) had interests in both botany and entomology, though all her pre-1900 publications were in the latter field. Born on 9 February 1851, in Meadville, Pennsylvania, she was the daughter of the Rev. James Freeman Clarke and his wife Anna (Huidekoper). The family moved to the Boston suburb of Jamaica Plain in 1854. Having poor health as a child, Cora received much of her early schooling at home, but later studied at a horticultural school in Newton and at the Bussey Institute in Jamaica Plain, as well as taking classes in botany, biology, and geology at Harvard Summer School.

Her published work included two papers on the larvae of caddis flies, freshwater insects that at the larval stage construct their strange portable "houses" from a great variety of materials. She was also known for her studies of gall gnats and gall wasps (insects of economic importance because their larvae feed on plants), which she was remarkably successful in rearing. Several species were named in her honor. A good photographer, she made excellent photographs of the galls and presented two volumes of these to the Boston Society of Natural History. Her later work included studies of grasses and mosses.[19] She was active in several women's science organizations including the Botany Group of the New England Women's Club, and she assisted in the zoological section of the Society to Encourage Studies at Home, a Boston correspondence school for women. One of the early women fellows of the AAAS (elected in 1884), she was also a council member of the Boston Society of Natural History and a member of the Cambridge Entomological Club. She died in Boston, 2 April 1916, at age sixty-five.

JENNIE ARMS SHELDON,[20] for twenty-five years an assistant to Alpheus Hyatt in the museum of the Boston Society of Natural History and for much of that time a special teacher of biology and geology in the Boston schools, was the author or coauthor of several substantial scientific monographs; these included *Insecta* (with Alpheus Hyatt, 1890), *Concretions of the Champlain Clays of the Connecticut Valley* (1900), *Newly Exposed Geologic Features within the Old "8000 Acre Grant"* (with George Sheldon, 1903), *Guide to the Invertibrata in the Synoptic Collection in the Museum of the Boston Society of Natural History* (1905), and *Observation Lessons on Animals* (studies in zoology for schoolchildren, 1931). Her one nineteenth-century journal publication, a general survey of modes of insect development, appeared as a joint paper with Hyatt in *Psyche* in 1893.

Born in Bellows Falls, Vermont, in 1852, the daughter of Eunice Stratton (Moody) and George Albert Arms, she spent her girlhood in Greenfield, Massachusetts. After attending Greenfield high school and a private school in Boston she went to MIT, entering in 1876, the year when the Woman's Laboratory, run by chemist Ellen Swallow Richards, first opened. Though a member of the MIT class of 1881, she did not take a degree. In 1897, when she was forty-five, she married George Sheldon, a writer and the historian of Deerfield, Massachusetts. Her own historical biography, *The Life of a New England Boy,* describing her father's early years, had been published in 1896. She was a member of the AAAS and the (Woman's) National Science Club, as well as the Boston Society of Natural History.

The best-known contributions to entomology of ADELE FIELDE[21] (1839–1916) were her studies on ant behavior, carried out at the Marine Biological Laboratory at Woods Hole, Massachusetts, just after the turn of the century. She also published several papers in the 1880s recording her observations on insects and microscopic organisms made in Swatow, China. An unusually adventurous, resourceful, and enterprising person, she had an eventful life, with three successive careers, each one of them notable. For twenty-seven years she was a missionary-teacher in Thailand and China; then for twelve years she focused much of her energy on biological research; finally she took up social and political work. In his account of the Woods Hole laboratory and the scientists who worked there, Frank Lillie (for some years the laboratory director) noted that Fielde deserved special mention; he found her "both a notable and gifted worker and a most entertaining and friendly character."[22]

Born in the small town of East Rodman in northern New York State, 30 March 1839, she was the youngest of the five children of Sophia (Tiffany) and Leighton Field. Her parents' financial resources being already overstretched by the time her turn came to go to college, she earned the necessary money herself by teaching for three years. After graduating from the State Normal School at Albany, New York, in 1860, when she was twenty-one, she immediately returned to teaching, first at Watertown and then at Mamaroneck, New York. Four years later she became engaged to Cyrus Chilcott, the brother of her Normal School roommate. Chilcott, a young Baptist minister, was then preparing to go as a missionary to the Chinese community in Bangkok. Adele obtained a post as a missionary teacher, and after a year's delay waiting for a passage on a sailing ship followed her fiancé to the Orient.

Arriving at Hong Kong after a seven-month voyage, she learned that Chilcott had died some months previously. She nevertheless completed her journey to Bangkok and took up her teaching post, remaining there for seven years, until 1872. She spent her spare time during rainy seasons studying Chinese. After a year's leave in America she returned to the Orient in 1873 and began work at the Baptist mission in Swatow, southern China, where she taught for the next twenty years. Much of her effort went into the school she established for Chinese women. A good linguist, she became fluent in the local dialect and in 1883 published a comprehensive *Dictionary of the Swatow Dialect.* It contained 5,442 entries and took her five years to complete. She made the long trip north to Shanghai herself to oversee the printing and carry out the proofreading. Over the next decade she brought out several works in Chinese and yet others about China. They included a Chinese translation of the Book of Genesis, *Pagoda Shadows* (1884), and *A Corner of Cathay* (1894).

Returning to the United States for two years' leave in 1883, she took a course in obstetrics at the Woman's Medical College of Philadelphia, acquiring practical knowledge she wanted to pass on to her Chinese women students in Swatow.[23] In addition, prompted by a growing interest in the theory of evolution, she began studies in biology. There was then no established place in Philadelphia where a woman could obtain instruction in biology, but she managed to persuade several of the scientists associated with the Philadelphia Academy of Sciences to give her laboratory space at the academy and also some tuition. Her teachers included Benjamin Sharp, then professor of zoology at the University of Pennsylvania. Her request for help is said to have prompted reorganization of biology instruction at the University of Pennsylvania and the opening of a course in that area to women, though Fielde herself never studied there.

She returned to China in 1885, but four years later, at age fifty, tendered her resignation as a missionary teacher. Failing health was the ostensible reason, but beyond that she appears to have outgrown many of the dogmas of established religion and so was no longer wholly at ease in the service of the Baptist church. Traveling home via India, the Middle East, Russia, and Europe, she took two years for the journey and made detailed studies of social conditions and customs, as well as political and legal systems, in many of the countries she visited. For the next thirteen years she was steadily employed in her own country as a lecturer, writer, and teacher, making regular appearances before civic and political organizations and scientific societies. She was a popular speaker and was often paid as much as $100 per lecture for a ten-day course.[24] Though an authority on the Orient and Chinese life in particular, her lectures covered a wide range of topics: "The coming revolution in Russia," "Poland and revolution," "Persia in the politics of Europe," "The passage of a race—the Australian aborigines," "The new theory of the origin of the species," "The wonders of ant life," and "The memory of ants" were but a few of her subjects. Most of her writing was in the form of newspaper and magazine articles; her essays and short stories were particularly influential.

Her eight pre-1901 scientific papers listed in the bibliography describe observations made during her second period of residence at Swatow in the 1880s, immediately following her studies at the Philadelphia Academy of Sciences. The communications on aquatic insects and insect larvae were of considerable interest. However, her later studies at the Woods Hole laboratory, where she spent her summers as investigator and lecturer from 1900 until 1907, constitute her most important contributions to entomology. Eighteen papers reporting the results of a long series of studies on ants appeared between 1901 and 1907, mostly in the *Proceedings of the Academy of Natural Sciences of Philadelphia* and the Woods Hole *Biological Bulletin.*[25] She was particularly interested in the olfactory sense of ants and their ability to discriminate between smells. Her experimental methods were innovative, involving the building of artificial nests, which she handled with scrupulous care. Although her theory that each joint of the ant's antenna has a special function in the discrimination of smells was not considered tenable by later workers, Lillie, writing in 1944, noted that probably no one else had made as extensive studies of the discrimination powers of ants. She herself described the work as bringing "variety and delight into my summers."[26] Her scientific publications continued until about 1915, a year before her death. She became a fellow of the AAAS in 1914.

In 1907, at the age of sixty-eight, she moved to Seattle, Washington, perhaps looking for a climate that would alleviate her bronchial disorder. She quickly became very active in state politics, embarking on a vigorous campaign for the passage of a constitutional amendment giving voting rights to women in Washington. When the amendment passed in 1910 she commented that the State of Washington was where she would stay, since there she was in reality an American citizen.[27] An energetic member of a number of civic organizations in the Seattle area, she also worked for Prohibition, although this resulted in her ouster from office as a Trustee of the Seattle Public Library Board. She died after a short illness, on 24 February 1916, at the age of seventy-seven.

The remaining women of the Northeastern group whose papers on entomology are listed in the Royal Society *Catalogue*[28] include two who were the earliest of the American contributors to the field—Dorothea Dix and Margaretta Morris. Dix had strong interests in natural history as a young woman, but she is remembered principally for her pioneering work as a reformer of state institutions for the mentally ill (see chapter 16). Her entomology paper reporting observations on spiders and butterflies appeared in the *American Journal of Science* in 1831. Morris published five entomology papers between 1841 and 1851, a contribution that makes her the most productive American woman naturalist of her time as measured by journal articles published. She and the French naturalist, Madame la Comtesse de Buzelet, who brought out a thirty-five-page "Catalogue des Coléoptères de l'Anjou" in 1852, are perhaps the two most prominent of the women who published on entomology during the middle years of the century—at least as judged by work listed in the Royal Society indexes.

MARGARETTA MORRIS[29] (1797–1867) was born 3 December 1797, probably in Philadelphia. She was the daughter of Ann (Willing) and Luke Morris. From 1812 onward she lived with her widowed mother and an unmarried sister, Elizabeth, in the Germantown suburb of Philadelphia. Later known as the Morris-Littel House, their home stood in the grounds of the historic Witt garden. Elizabeth studied botany, and all three of the Morris ladies attended an evening course of lectures on mineralogy and geology given at the Germantown Academy in the winter of 1820–1821 by Charles Jones Wister. They doubtless also attended a course of botany lectures said to have been given there by the naturalist Thomas Nuttall.

Margaretta Morris knew several of the prominent naturalists of her time, including Nuttall and W. Gambel, for whom she prepared botanical illustrations. Her interests included fungal diseases of plants, but the studies for which she was best known were her investigations, carried out during the 1840s, of the life histories of the seventeen-year locust and an insect she initially thought to be the Hessian fly (*Cecidomyia destructor,* Say, now called *Mayetiola destructor* (Say)). The latter, whose larvae are especially destructive to wheat and barley, is an agricultural pest of major importance. It is now not clear which of the Cecidomiidae Morris was observing, but it was not the Hessian fly, as she herself realized by 1849. Although she presented a detailed life history, her descriptions of the insect itself are not complete enough to allow later workers to positively identify the species. Nevertheless, her writings were received with marked interest, and she was considered something of an authority for a time.[30]

Of the ten Midwestern women entomologists whose work is listed in the bibliography, the first to publish and the most distinguished was MARY MURTFELDT[31] (1839–1913) of Kirkwood, Missouri. The oldest daughter in the family of Esther and Charles W. Murtfeldt, she was born in New York City, 6 August 1839, but grew up on a farm near Rockford, in Ongle County, Illinois, to which her family moved when she was a small child. A serious illness in babyhood left her partially paralyzed, and all her life she walked with crutches. She was educated at home by her parents and a governess until she was twenty, when she went to Rockford College where her sister Augusta was then a student. Ill health caused her to withdraw in 1862 without graduating. Six years later her family moved to St. Louis, Missouri, when her father became corresponding secretary of the Missouri Board of Agriculture and associate editor of Coleman's *Rural World,* then one of the most influential and scientifically oriented agricultural publications in the country. In the same year (1868), Charles Riley, a young English immigrant, scientific writer, artist, and student of entomology, whom Charles Murtfeldt had known professionally in Chicago, became the first state entomologist of Missouri. Riley also moved to St. Louis, and, being then unmarried, he appears to have lived for a time with the Murtfeldts, moving with them in 1870 to the house they built in the suburb of Kirkwood.

Mary Murtfeldt had by then acquired a fairly extensive knowledge of both entomology and botany, and, botanical information being indispensable in working out life histories of new or little-known insects, Riley was quick to secure her services as assistant. Their collaborative study of the pollination of the Yucca plant, in which they demonstrated that the process was carried out by the Pronuba moth, came to be seen as an important contribution to the development of ideas on insect-plant interdependence and evolutionary adaptation. It was of special interest to botanists.[32] After she began working with Riley and had access to his scientific library, Mary concentrated increasingly on entomology. The use of this library, at a time when the general lack of entomological literature was a handicap for even the few professionals in the field, was of critical importance in her early work. For her, crippled and relatively restricted, Riley's presence in her home must have provided an extraordinarily fortunate opportunity for the development of her scientific talents. She is said to have acknowledged that much of her success was due to her close association with him over a period of more than twenty years.[33]

The large Murtfeldt house in Kirkwood, long preserved as one of the fine historic sites of the neighborhood, originally stood on three acres of ground, and though her infirmity prevented her from going out into the countryside, Mary was able, with the help of her family, to build her early entomological collections with specimens caught nearby. Later she managed to interest a number of local boys in assisting; having been instructed on how and where in the fields and woods they might find eggs and larvae, they brought these to her for rearing in her home laboratory.

Over the years she held several positions. As recorded in the 1910 edition of *American Men of Science* these were as follows: assistant to the state entomologist (Riley) from 1876 until 1877 (when Riley went to Washington, D.C. to take the post of entomologist at the U.S. Department of Agriculture); field agent, Division of Entomology, U.S. Department of Agriculture, commissioned to send reports to Riley in Washington, 1880–1893; acting state entomologist, Missouri, 1888–1896; staff contributor in entomology and botany and associate editor for *Farm Progress,* a biweekly journal published by the *St. Louis Republic,* from 1896 (until her death in 1913). Her work with the state would appear to have been done somewhat informally, however, since she was not listed as an employee in the annual reports of the Missouri State Board of Agriculture for the years in question.[34] Her role may have been more that of an adviser or consultant.

Throughout much of the 1870s her major work concerned the investigation of life histories of Missouri insects, especially the Microlepidoptera, studies that she could carry out at home. She described eighteen new species of moths. The greater part of her early work, painstakingly accurate in all observations, was incorporated by Riley into his annual reports on the insects of Missouri.[35] These reports, outstanding in both scientific content and illustration, were greeted with considerable admiration by scientists throughout the United States and Europe (though little appreciated in Missouri at the time[36]) and were a tremendous stimulus to other workers. In discussing Riley's last report, that of 1877, a reviewer for the London *Entomologists' Monthly* remarked that his powers of observation, attention to minute detail, scientific accuracy, and skill with his pencil made him the foremost entomologist of the day.[37] Riley went on to an outstanding career as an economic entomologist and received many honors in both Europe and the United States. Just how much Murtfeldt contributed to the success of his early work is hard to guess, but her obituarist in the *Canadian Entomologist* remarked that she gave Riley "much material assistance" in his Missouri insect life-history studies.[38]

The body of work she published under her own name was substantial (see Figure 2-2, b). After one or two popular articles in the *Little Corporal* magazine and the *Rural New Yorker* in the early 1870s, she brought out more than thirty technical papers in the entomology journals between 1872 and the early years of the present century (see bibliography). She also wrote two monographs, her 130-page manual, *Outlines of Entomology* (1891), and *Stories of Insect Life* (1899). The former was prepared at the request of the State Board of Agriculture and the State Horticultural Society of Missouri to provide farmers and horticulturalists with a basic introduction to entomology in a readable form with a minimum of technical language. Its use as a public school text for the instruction of children from farming districts (who then made up three-quarters of the state's schoolchildren) was also suggested by Board of Agriculture members, the need for the education of future farmers in applied science being already recognized. At the present time, when textbooks in scientific fields proliferate at a remarkable pace, it is easy to overlook the significance of Murtfeldt's *Outlines,* but in 1891 the appearance of a monograph in entomology was still a matter of note. The illustrations were mainly from electrotypes purchased from Riley. In an introduction to the first six chapters, initially printed in the *Twenty-second Annual Report* of the Missouri State Board of Agriculture, Murtfeldt's outstanding reputation among Western entomologists was acknowledged.[39] Her *Stories of Insect Life,* coauthored with Clarence Moores Weed, appeared in 1899 and was reissued in 1901 and 1902. She published little of her botanical work, but her collection of plants from the vicinity of St. Louis was substantial, and S. M. Tracy's "Flora of Missouri" credits Murtfeldt as the collector of many species.[40]

Despite her physical handicap she frequently attended scientific meetings, generally in the company of one of her sisters, and took an active part in the proceedings, either presenting papers or joining in discussions. She was long remembered. As late as the 1930s, Herbert Osborn, recalling the 1883 AAAS meeting in Minneapolis where Murtfeldt presented a paper,[41] mentioned that she and Riley (a particularly striking and handsome man) were conspicuous figures. Osborn described Murtfeldt as a very attractive personality with broad interests, qualities that showed through in her scientific writings.[42]

An honorary member of the St. Louis Academy of Science and a fellow of the AAAS, she was elected vice president of the latter organization's Entomological Club in 1890. Over the years, she took an active interest in the status of women in science and wrote several articles on the subject.[43] Toward the end of her life she gave her large collection of Microlepidoptera to Cornell University. She died in Kirkwood, 23 February 1913, at the age of seventy-four.[44]

The second state in the country to appoint a state entomologist was Illinois, the first appointee (Benjamin Walsh) taking up his post in 1866. Between 1875 and 1882 Cyrus Thomas occupied the position. Although Thomas received no financial support for his investigations beyond his salary (he provided his own facilities and equipment), several assistants were engaged to help him do his share of the work of the U.S. Entomological Commission, of which he was a member. The papers of two of his collaborators, Emily Smith and Nettie Middleton, are listed in the bibliography. Both carried out studies that appeared in Thomas's annual reports, his introductions to which generally acknowledged the work of his assistants. Middleton's lengthy study on butterfly larvae appeared in 1881 as part of the tenth report, and Smith's on the noxious insects of northern Illinois in the seventh report (1878). Smith also published independently her careful and detailed studies on the maple tree pests, the bark louse *Lecanium acericorticis* Fitch and the scale insect, *Pseudococcus aceris;* her work on the latter was especially well regarded. She later studied for a time in Europe, where she met and married a Mr. Pigeon. By the 1920s she was living in Boston.[45]

One of the early women graduates in entomology from Iowa College of Agriculture at Ames was ALICE BEACH[46] (1857–1945), who later held a succession of positions as assistant entomologist at various state agricultural station museums and laboratories. Born on 30 October 1857 in Sumner Hill, New York, she was the daughter of Maria North (Wood) and Isaac Ambrose Beach. She attended schools in Homer, New York, and then studied at the State Normal School in Cortland, New York, graduating in 1878.

For about ten years she taught science in public and private schools, for a time serving as assistant principal of Gregory Institute in Wilmington, North Carolina, and later holding a similar post at Emerson Institute in Mobile, Alabama. After taking her B.S. in 1892 she stayed on at Iowa College of Agriculture as an assistant in the entomological museum while also continuing graduate studies (M.S., 1894). Between 1898 and 1903 she held, successively, positions in the New York State Experiment Station at Geneva (where she had care of the insect collection), at the office of the State Entomologist in Albany, New York, at the State Laboratory of Natural History, University of Illinois (1899–1901), and again with the State Entomologist in Albany (1902–1903). For a time she served as a house director at Iowa College of Agriculture. Her three pre-

1900 publications listed in the bibliography all reported work on Iowa insects during the course of her graduate training (descriptions of new species and varieties of Iowa Thripidae and additions to catalogs and lists). She was a member of the Iowa Academy of Sciences from 1892 until 1915. Later she lived in Boone, Iowa, some fifteen miles west of Ames. She died on 16 August 1945, in her eighty-eight year.

ELIZABETH GIFFORD PECKHAM[47] (1854–1940), arachnologist and entomologist, was known for her very extensive collaborative work with her husband, George Peckham, a teacher of biology, administrator in the public schools in Milwaukee, Wisconsin, and later director of the Milwaukee Public Library.

Born in Milwaukee, 19 December 1854, the daughter of Mary (Child) and Charles Gifford, Elizabeth graduated from Vassar College (A.B., 1876). She married four years later and had three children, but also became very active in her husband's entomological research. Indeed, George Peckham's obituarist wrote that it was not possible to consider Peckham's work separately from his wife's, their collaboration having been complete from their marriage on.[48]

Their investigations were in two areas, the taxonomy of spiders and the psychology of spiders and wasps. In taxonomy they dealt exclusively with the Attidae group, the first of their many papers appearing in the *Transactions of the Wisconsin Academy of Science* in 1885. Further lengthy studies, including investigations of Attidae from Central America and Mexico, followed during the next two decades.

Their observational work on insect psychology, difficult and painstaking research carried out over more than fifteen years, led eventually to the publication in 1898 by the Wisconsin Geological and Natural History Survey of their pioneering study on the instincts and habits of wasps. Enlarged and illustrated, this was republished as the well-received monograph *Wasps Social and Solitary,* a work for long considered a classic in the area of animal psychology. While reporting their entomological observations with strict precision and attention to detail, it also presented a series of vivid word pictures of their working milieu at their summer home in the wooded countryside near Milwaukee—descriptions of days spent in their bean patch, scrambling around on hands and knees to observe the hunting method of the paralyzing wasp *Ammophila urnaria,* or of making an estimate of the amount of food collecting done by a worker in a day by counting wasps from dawn until noon as the insects arrived home with their burdens.[49] Written in a clear, elegant, and simple style, the book was appreciated for its literary value as well as its scientific content.

Elizabeth Peckham was also interested in ornithology and belonged to the Wisconsin Audubon Society. After her husband died in 1914 she appears to have given up the research that had occupied her for more than three decades.[50] She lived for another twenty-six years, dying on 3 January 1940, shortly after her eighty-fifth birthday.

Four other women from the Midwest—Laurene Highfield, Mrs. A. J. Snyder, Hattie Warner, and Bertha Kimball—each contributed one or two papers to the pre-1901 entomology journal literature.[51]

BERTHA KIMBALL[52] took a degree in entomology and horticulture at the Kansas State College of Agriculture in 1890 and an M.S. in botany in 1895. She stayed on for three years as a microscopic draftsman and as a teaching assistant in a variety of subjects including botany, drawing, and physical education for women students, her instructional work in the latter area being shared with botany student Minnie Reed (see chapter 1). Programs of physical training for women were relatively new and innovative undertakings in the mid 90s and the work of the two assistants was well-thought-of by the college. An artist of some talent, Kimball was also a regular contributor of drawings and paintings for the public exhibits organized by the departments of entomology and botany. Her life history study on the insect then known as *Conorhinus sanguisugus,* a member of the Reduviidae (assassin bugs), appeared in the *Transactions of the Kansas Academy of Science* in 1896. In 1898 she married Albert Dickens, a fellow graduate of Kansas State College, who later became professor of horticulture there. She would appear to have given up technical work after her marriage.

A few women in New Mexico, California and Texas made modest contributions to the nineteenth-century entomology journal literature.

Two, Jessie Casad and Wilmatte Porter, worked with Theodore Cockerell of the New Mexico Agricultural College, Las Cruces, and the Agricultural Experiment Station. Casad, Cockerell's student, coauthored two papers on wasps with him (1894, 1895). Porter, a Stanford University graduate, taught high school biology in Las Cruces. Her interest in local plants and animals brought her into contact with Cockerell, then a widower. They were married in 1900 and thereafter she was his constant companion, accompanying him on field trips and collaborating in many of his projects. Her cooperation is said to have greatly increased the effectiveness of Cockerell's own scientific career.[53] A notable number of the many new species he discovered bear the name *wilmattae.* Two of their joint publications are listed in the bibliography, one being a substantial paper that appeared in 1899 on New Mexico bees, their special interest; indeed, Theodore Cockerell's work on wild bees was considered his greatest contribution to entomology.[54] After her marriage, Wilmatte Cockerell taught biology at the state preparatory school associated with the Agricultural College. She was known as an enthusiastic and stimulating teacher who dispensed with standard textbooks and laboratory manuals and followed her own highly effective methods. Her husband often visited her class laboratory and occasionally presented talks. They moved to Boulder, Colorado, in 1903 when Theodore Cockerell became curator of the museum at Colorado College and lecturer at the University of Colorado.

Among the remaining workers from the western region whose papers are listed in the bibliography are three from California—Bertha Chapman, Helen Mills, and Alice Jordan—and one from Texas, Helen King.[55] Chapman and Mills were students at Stanford University in the late 1890s. Chapman studied Mallophaga (fleas) under the direction of Vernon Kellogg, and Mills collaborated with W. A. Snow in an investigation of insect damage to the Monterey pine. Alice Jordan collected near Napa, California; her life history study on the butterfly *Papilio zolicaon* appeared in 1894. Helen King, whose five papers appeared in the *American Naturalist* and *Psyche* between 1878 and 1886, worked near Austin and San Antonio. She was especially interested in phosphorescent insects (fireflies), and also carried out life history studies of several moths, including the bagworm, *Thyridopteryx ephemeraeformis.*

Summary

Regional groupings among early American women entomologists are remarkably well-defined, especially those of the Northeast and the Midwest. With the exception of Fielde, the most notable members of the Northeast group came typically from families that had sufficient financial resources to support their intellectual pursuits, or, like Smith Fernald and Dimmock, they married men with scientific interests who facilitated their work. Several of them found their continuing inspiration for entomological studies in associations with local natural history and entomological clubs and organizations, Slosson with the New York Entomological Society, Clarke, Soule, Arms Sheldon, and Morton with the Boston Society of Natural History, Morris via a connection with the Philadelphia Academy of Sciences. Available obituaries (for Clarke, Morton, and Slosson) and later comments (Mallis on Smith Fernald and Lillie on Fielde) suggest that these early women were accepted fairly comfortably into the larger group of male workers, many of whom were also "amateurs," and their contributions were well-received. Although she worked in the Midwest, Elizabeth Peckham, whose substantial contributions were made in collaboration with her husband, also fits the pattern of independent, "amateur" investigator typical of the Northeast group.

The close association of many of the Midwesterners (Murtfeldt, Middleton, Smith, Beach, Kimball, and Warner) with either early state entomologists or with state agricultural colleges and experiment stations is particularly noticeable, and in part parallels the close link between Midwestern women botanists and the state agricultural colleges. Murtfeldt, the most productive of the Midwesterners with at least thirty-six journal articles and a monograph to her credit by 1901, occupies a distinguished position among all nineteenth-century women entomologists.

British women

Of the twenty-five British women whose papers in entomology were indexed by the Royal Society, only four—Ormerod, Ricardo, Sharpe, and Fountaine—published more than ten papers over the course of their working lives. Ormerod was a well-known agricultural entomologist with many professional connections but no regular institutional attachment, Ricardo and Sharpe were classifiers at the British Museum, and Fountaine an independent collector. Two others—Nicholl and Robson—produced six and seven papers, respectively, and a third worker—Sanders—contributed three substantial articles around the turn of the century. Many, however, appear to have been naturalists and observers who only occasionally sent short notes to entomological journals; about them little information has been uncovered.[56]

One of the best known of the independent observers and collectors was MARGARET FOUNTAINE[57] (1862–1940), who, although she recorded very little of her work in print, over the course of forty years of worldwide collecting acquired a knowledge of tropical butterflies that was probably unique at the time. She also built up a useful collection of about 22,000 specimens, the Fountaine-Neimy collection. Thanks to her now published diaries, a remarkably full account of her life is available.

The story of the reopening of these diaries, which she kept from the age of sixteen until shortly before her death, deserves a mention. In accordance with her will, her butterfly collection and a large, sealed lacquered box were deposited after her death in the Castle Museum, Norwich, with instructions that the box, containing manuscripts, not be opened until 15 April 1978. Shortly after that date the museum officials, in the presence of one of the few surviving relatives of Miss Fountaine, duly broke the seals and revealed the contents. Rather than unpublished notes on Lepidoptera that had been half expected, there were found twelve identical volumes of diaries telling a very private story. Begun 15 April 1878, they run to well over a million words and cover every aspect of Fountaine's life.

She was born on 16 May 1862, the second child and oldest daughter among the eight children of the Rev. John Fountaine and his wife Isabel (Lee Warner). John Fountaine was the rector of the small parish of South Acre in Norfolk, and the family, although not rich, was comfortably off. On both sides they had solid connections with the landed gentry; the Fountaines traced their ancestors back to at least the fourteenth century. The Rev. John appears to have been an easygoing man who devoted much of his time to hunting, fishing, and sailing and did not allow himself to be overwhelmed by his parish duties. His wife, sixteen years his junior, brought up her children very strictly. The two sons were sent to school, but Margaret and her five sisters were taught at home by governesses.

When John Fontaine died in 1878, the rest of the family moved to a house in Norwich, and in due course Margaret and her sisters "came out" and took their part in the social life of the community. Margaret fell in and out of love with a string of young men in fairly rapid succession, ending up with Septimus Hewson, an Irishman, a chorister in Norwich Cathedral and something of a drunkard. Septimus, doubtless recognizing that she belonged to a different social class from his, left her alone, despite the fact that in a suprisingly un-Victorian way she indicated to him how she felt. However, after she let him know that a wealthy uncle had left her an income of £150 a year (enough to live on very comfortably in those days), he changed his mind for a time, and Margaret considered herself engaged. Fortunately Septimus soon left her, and although this was a devastating blow at the time, it was also the event that helped her to break free from the conventions of Victorian society, the restraints of her family, and her mother's exacting demands. She left the cold climate and the "cold men" of her native land, set off in search of blue skies and sunshine, and never really wanted to return to Britain for very long at a time.

Her travels began in 1891, when she was twenty-nine, a year after her break with Hewson. She went first with her sister Florence to Switzerland where she began her butterfly-hunting, carefully identifying her captures from a reference book at the end of each day. She realized very quickly that she had found an occupation she really enjoyed, although she had never had an opportunity to pursue it before. Over the course of the next ten years, with occasional short visits to her family, she continued her traveling and collecting, much of the time in central and southern Europe, and often in the company of a gentleman admirer. She loved the free life and the warmth and sunshine of the Mediterranean lands. Her first publication was a note in the 1897 *Entomologist* on the butterflies of Sicily. A description of her collecting in Greece in 1900 appeared a few years later.[58]

By 1901, when she was thirty-nine, she had moved on to the Middle East, and in Damascus she hired as interpreter and guide Khalil Neimy. Twenty-four, fair-haired, and gray-eyed, despite his Greek-Middle Eastern origins, Neimy had lived in the United States for two years, and spoke English with a strong American accent. They collected together in Syria, Jordan, and the Holy Land. He fell in love with her, and although she at first had a low opinion of him and despised herself for letting the relationship continue, she gradually came to accept him and agreed to marry him. While on a brief visit to England she chanced to find out through friends that he already had a wife. Although extremely disappointed and disillusioned, she nevertheless joined him again. In 1903, now as secret lovers, they went butterfly collecting in Asia Minor and then North Africa.[59] By 1905 they were in Spain, the following year in Corsica, and after that in the Balkans, Khalil in time becoming a skilled, persistent, and knowledgeable collector, at least Margaret's equal in the work.

The winter of 1908–1909 they spent in South Africa, principally because of Khalil's health. They traveled and collected in the Cape, in the new colony of Rhodesia, and in Mozambique. A year later they extended their explorations to America, where they visited Margaret's younger brother Arthur, the black sheep of the Fountaine family, who had been banished to a life of farming in Virginia. This visit to the New World included a trip to the Caribbean, where the collecting was exceptionally good, and then to Costa Rica.[60] India, Ceylon, and Sikkim came next, in 1912, and the following year took them as far as the high passes of the Himalaya, along the Tibetan border.

For three years this whirlwind pace was interrupted when they settled at Myola, inland from Cairns, in North Queensland, Australia. The plan was that they would buy a piece of land, clear it, build a house, and raise horses while Khalil fulfilled residence requirements for British citizenship. But neither knew much about farming, and the monotony and slog of a settler's life was hardly to their tastes. Their unhappiness, exacerbated by periods of seeming mental derangement in Khalil, meant that even then they did not go through with their planned marriage, although he did get his naturalization papers, and, at least as far as Margaret understood the situation, there remained no further legal impediments, his first wife being by then dead.

In 1917 Margaret left by herself for Los Angeles. For a time she explored the mountains and canyons of southern California, often alone, with her butterfly net, as she had wandered through the mountains of south and central Europe years before. She traveled throughout the United States, worked for a while (at twenty-five cents an hour) in a natural history shop in Pasadena, California, to earn some badly needed cash, and continued her collecting. In 1919 she went back across the Pacific, rejoined Khalil, and with him made a brief collecting trip in the South Island of New Zealand. Thereafter, following a few months in London, they took what turned out to be their last joint trip, an expedition to the Far East, to up-country Burma as far north as Mandalay, through Bangkok, and on to the parks of Hong Kong. The journey was completed in the Philippines, where the collecting was spectacular, some of the largest and most beautiful butterflies in the world being found there.[61] Many of them went into their own collection, but in addition they captured specimens to supply to other collectors, several species being much in demand.

West Africa was Margaret's next objective, and she went there on her own in 1926. Shortly after that, she received a commission to collect in the West Indies, and while she was carrying that out Khalil died at his home in Damascus. She missed him sorely after their twenty-six years of friendship, but she kept on traveling and collecting for another thirteen years, with breaks between trips to work with entomologists at the British Museum on classification problems. For a woman in her late sixties and early seventies the list of places she visited during those last years is remarkable, to say the least—South America (along the Amazon and Madeira rivers as far as Bolivia), Cuba, Madagascar, the Great Lakes region of East Africa, Indochina and the East Indies, and finally Trinidad, to her the most beautiful island of the West Indies. There, on 21 April 1940, when she was seventy-eight, she collapsed by the roadside and died soon after. She was buried the next day, in an unmarked grave (according to cemetery records) with several other bodies, at the expense of the local community. In the 1980s, after the publication of the first volume of the diaries, the United Kingdom Women's Club of Trinidad and other British women then living on the island raised money for a commemorative bronze plaque, to be placed on a boulder on a cliff overlooking the Caroni Plain not far from the guest house where she died. The British High Commissioner attended the unveiling ceremony.

Margaret had been a member of the Entomological Society since 1898 and had often attended its meetings when she was in London; on her death her collecting equipment and books went to the youngest member. Her sketchbooks and paintings, some of which appear in excellent reproductions in the published diaries, went to the London Natural History Museum.

Throughout all the years of her friendship with Khalil Neimy the proprieties were maintained by one means or another (in Australia they were "cousins"), and to her English friends she was always a perfectly proper, conventional English lady. Her *Entomologist* obituarist referred briefly to Khalil as "a devoted Syrian courier," who traveled with her during her earlier expeditions. She had few very close friends among entomologists and was clearly very successful in keeping her private life to herself. Although she cannot be classified among the great entomologists, she worked out many butterfly life histories and collected excellent material, much of it in perfect condition because her specimens had often been hatched in cages and killed before they suffered any wear and tear at all. Perhaps more than anything else, however, the story of Margaret Fountaine stands out in the annals of early women naturalists for its sheer color and romance.

A number of the other women observers and collectors published work on Lepidoptera, both foreign and British. Among them was Mary de la Beche Nicholl, whose papers came out between 1897 and 1904. Mrs. Nicholl covered much of the same territory as did Margaret Fountaine in her early travels, visiting Spain, Bulgaria, the Balkans, and Lebanon.[62] Others were Jane Fraser, Edith Wollaston, and Mrs. Samuel Robson. Fraser recorded her observations of butterflies made during her stay on the island of Upolu, Samoa, in the early 1890s. Wollaston was remembered for her collections from St. Helena. The daughter of Joseph Shepherd of Bristol, she married the coleopterist T. Vernon Wollaston and accompanied him on explorations in Madeira, the Canaries, and St. Helena. Although she assisted him with his Coleoptera collecting, her own particular interest was Lepidoptera. Following a long stay on St. Helena, a tiny island tremendously rich in rare insect life, she set herself the task of working out the collections made there. Her study resulted in a lengthy paper on the Lepidoptera of St. Helena (*Annals and Magazine of Natural History,* 1879). She reported ninety-four species, thirty-eight of which were new and which she described in detail. She did little further collecting after her husband's death in 1878, although she lived for another thirty-three years, dying on 23 October 1911.[63] The work by Mrs. Robson on life histories of some butterflies of the Indian subcontinent was carried out at the hill station of Mussooree in northern India in the summer months and at Bankipur, Behar, in the cooler part of the year. All seven of her papers appeared in the Bombay Natural History Society's *Journal* in the early 1890s.

Women who published notes on Lepidoptera they collected within the British Isles include Maude Alderson of Worksop, Nottinghamshire, Mary Kimber of Newbury in Berkshire, Miss R. M. Sotheby of Eastbourne, Mrs. E. C. Bazett of Reading, Miss A. D. Edwards of St. Leonards, Sussex, and Elizabeth Miller of Chelmsford, Sussex. Alderson's list of Lepidoptera from the Worksop neighborhood appeared in 1893. Kimber presented a fourteen-page report on the Macrolepidoptera of her district in the *Transactions of the Newbury Field Club* in 1895, and Sotheby a similar report on the Macrolepidoptera of the Eastbourne area in the *Transactions of the Eastbourne Natural History Society* in 1883. Bazett, Edwards, and Miller each appear to have contributed only one short observational note to the pre-1901 literature. Two collectors in Scotland were Miss M. L. Cottingham and Alice Fowler. Both worked in the western Highlands, Cottingham recording observations on a wide variety of insects in the region bordering the Sound of Jura in Argyllshire, and Fowler working at Inverbroom in western Ross and Cromarty.

Probably the greatest number of papers on Lepidoptera by an early British woman entomologist was produced by EMILY SHARPE of the British Museum. Thirty-seven of her articles, mainly on the museum's African collections, appeared in the journals in the seventeen years between 1890 and 1907, a period during which an immense amount of new, undescribed material was being brought back by travelers and collectors for classification and naming at British and European museums.

Emily was most likely one of the ten daughters of Emily (Burrows) and Richard Bowdler Sharpe. Bowdler Sharpe, an internationally known ornithologist who specialized in taxonomy, was an assistant in the British Museum's Department of Zoology from 1872 until he died in 1909. As noted by William Stearn in his history of the institution,[64] three of Sharpe's daughters (Dora, Daisy, and Sylvia) worked with their father preparing colored plates for his publications. Emily is not mentioned by Stearn, but her contributions were noteworthy, particularly her work on the geographical distribution, taxonomy, and classification of Lepidoptera from what were then British-held territories in East and West Africa. She appears to have stopped publishing in 1907, two years before the death of Bowdler Sharpe.[65]

Another especially productive British Museum worker was GERTRUDE RICARDO.[66] Only her three 1900 papers are listed in the bibliography, but she continued to publish fairly regularly for almost three decades, putting out at least forty-three articles and monographs, many of them very substantial works. Her particular interest was the Tabanidae family, the large-headed and stout-bodied horseflies and deerflies. The collections she studied came from all over the world, from the Arctic to South Africa and from Europe to Australia. The bulk of her work appeared in the *Annals and Magazine of Natural History,* but she also brought out studies in Dutch publications on material from the East Indies, papers in German journals, and a joint monograph in French on African Tabanidae, coauthored with Jacques Surcouf. Her reports on collections in the South African Museum were published in its *Annals;* some of her articles on oriental species appeared in the *Records of the Indian Museum,* Calcutta.[67]

One other woman whose work is listed in the bibliography, Mrs. M. K. Thomas, is also identifiable as a British Museum worker. She brought out four notes during the 1890s on description and classification of beetles and weevils in museum collections.

Except for Eleanor Ormerod, the remaining people (Bewsher, Chawner, Cooper, Pasley, Redmayne, Smee, and Williams) whose work is listed in the bibliography made very modest contributions to the entomological literature.[68] Chawner, however, although she published little, was for many years an active member of the Entomological Society and was remembered for her studies of sawfly larvae.

One of the three children of the Rev. Charles Fox Chawner of Bletchingley, Surrey, and his second wife Frances Sarah (Boulger), ETHEL CHAWNER (1866–1953)[69] was born 21 January 1866. Her work in entomology began after her family moved to Lyndhurst in the New Forest, Hampshire, in 1892. She collected sawfly larvae in the forest and reared them in captivity, through complete life cycles. A number of those she found were previously unknown, including *Selandria serva* F., *Tenthredo perkinsi* Morice, and *Macrophya annulata* Geoffry; several others were unknown until then in the British Isles. Her collections were eventually passed on to entomologist friends who had helped her with identifications, and to the British Museum, although the potential value of the material was greatly reduced because she never labelled her specimens.

Besides her two notes in the *Entomologist* in 1894 and 1895, she published, many years later, a joint paper with A. D. Peacock.[70] Her account of *Hymenoptera phytophaga* in areas of Hampshire and the Isle of Wight appeared in the *Victoria Histories of the Counties of England* series (1900).[71] She also had a special interest in birds, became a life member of the Aviculture Society and studied methods for successfully keeping birds in captivity. During the Second World War she housed the Royal Entomological Society's books on Hymenoptera at her home in Leckford and maintained an efficient postal loan service for the duration. She joined the Entomological Society in 1897, and at the time of her death at age eighty-seven, 16 October 1953, was its fifth most senior member.

Probably the only member of this group of entomologists to work in a university laboratory was Cora Sanders, a student of Edward Poulton, Hope Professor of Zoology at Oxford and a strong proponent of Darwin's theory of evolution. Sanders coauthored a paper with him in 1900 on factors influencing survival in insect populations. She stayed on at Oxford for several years, working in the university museum; her two papers on museum collections of Lepidoptera from Brazil appeared in 1904 and 1908. She married about 1909, becoming Mrs. Hodson, and moved to Melton Mowbray, Leicestershire. Later she was associated with Bablake School, Coventry. She became a fellow of the Linnean Society in 1906.[72]

The best-remembered nineteenth-century British woman entomologist is undoubtedly ELEANOR ORMEROD[73] (1828–1901), author of some fifty-nine journal articles and reports between 1873 and 1900, as well as a great many technical pamphlets and several well-received monographs and handbooks. She was probably the most successful and certainly the most influential woman entomologist of her time in any country. Her specialty was economic entomology, that is, entomology applied to the investigation and control of insects whose depredations affect agricultural crops.

When she came to her work in the latter part of the nineteenth century, British agriculture was going through a particularly troubled period. To the continuing effects of the 1846 repeal of the Corn Laws was being added the increasing stress caused by the industrial revolution's polarization of town and country interests; in addition, the cheap food produced by mechanized farming in North America was beginning to present fierce competition for British farmers. Leaders in agricultural work could see that practices had to be improved, through more and better education at all levels and the introduction of advisory services to farmers. Agricultural entomology, part of this movement, was just beginning to develop. Not many male entomologists had yet shifted their attention far enough away from the traditional areas of collection and classification to consider the opportunities and challenges the new branch of their field presented; and so Ormerod not only had scope for her considerable initiative but filled a national need. By the time of her death she was Britain's "best authority on farm and garden entomology."[74]

The youngest of the ten children of George Ormerod and his wife Sarah, daughter of John Latham, Eleanor was born on 11 May 1828, at Sedbury Park, the family's 800-acre estate near Chepstow, in Gloucester, on the western bank of the Severn, between that river and the Wye. The family had considerable wealth. George Ormerod came from an old Lancashire family, which had held lands in the uplands of the Pennines, on the Lancashire-Yorkshire border, since the early years of the fourteenth century, and subsequently, by judicious marriages, increased these properties. He was magistrate for the counties of Cheshire, Gloucester, and Monmouth, a fellow of the Royal Society and well known as a county historian and the author of *The History of the County Palatine and City of Chester* (1819). His wife's family, the Lathams, traced their descent back to King Edward I and his first wife, Eleanor of Castile; his father-in-law, John Latham, M.D., F.R.S., sometime president of the Royal College of Physicians, was Physician Extraordinary to King George IV.

When not attending to public duties, George Ormerod was much absorbed in his studies and led a somewhat reclusive life. The large family was a close-knit one and Eleanor and her two sisters had few outside social contacts as they grew up. Life was well-regulated. The mother, a kindly woman of considerable artistic talent, taught her daughters the subjects traditionally thought appropriate for girls; Eleanor and her sister Georgiana also had painting lessons during annual visits to London and became good artists. Eleanor studied Latin and modern languages, mainly by herself, and had the freedom of her father's extensive library. Later she was to learn sufficient Russian to be able to correspond with scientific contacts in St. Petersburg. She became interested in natural history when still a child, and a little later helped one of her brothers in his botanical studies, becoming an expert at preparing specimens and learning how to use a good microscope. Her systematic studies in entomology began in 1852, when she was twenty-four. Having chosen beetles as her subject, she obtained a copy of James Stephens's *Manual of British Coleoptera or Beetles* (1839) and supplemented her reading with extensive practical work, dissecting specimens she collected and giving herself a basic foundation in insect anatomy and classification. As her father grew older she became involved in the management of the family estate;

part of this was under cultivation, with the rest being woods and parkland. The work gave her an acquaintance with some of the practical problems of agriculture.

In 1868, at the age of forty, with some sixteen years of serious study behind her, she responded to a request from the Royal Horticultural Society asking for help in the preparation of a collection of insects both useful and harmful in agriculture. The Ormerod estate was excellent collecting territory and over the next few years she gathered together a sizable amount of material for the society. This was done despite the disapproval of some of her family, but with the willing assistance of the farm laborers, who were glad to accept "Miss Eleanor's shillings" for their efforts. The collections were awarded the Horticultural Society's Silver Medal in 1870, the first of several awards that came to her over the course of the succeeding three decades. Two years later she received her next two medals, the University of Moscow Gold Medal and the Moscow Polytechnic Exhibition Silver Medal, for a collection of electrotypes and plaster fruit and flower models entered at the Moscow International Polytechnic Exhibition.

Following the death of George Ormerod in 1873 and the passage of the estate to his oldest son, the family split up. Eleanor and Georgiana first took a house in Torquay, Devon, where they had Latham relatives, but two years later, wanting to be nearer London and Kew Gardens, moved to Isleworth, where they stayed until 1887. Eleanor cultivated the acquaintance of Sir Joseph Hooker, director of Kew, and his wife, and continued her entomological studies in the gardens. Her first technical publications had appeared in the *Journal of the Linnean Society* and the *Entomologist's Monthly Magazine* in 1873. And from then on her articles continued to come out, at the very creditable rate of two or three per year, over the course of the next quarter of a century. At Isleworth she also set up a meteorological station and took daily observations, which she continued after moving to St. Albans in Hertfordshire. The work led to her becoming the first woman fellow of the Meteorological Society (1878).

In 1877, again in response to published requests for help, this time from the editors of agricultural and gardening magazines, Eleanor put out what became the first of her *Annual Reports of Observations of Injurious Insects,* a series that was to continue under her editorship for more than twenty years. Based on the results of a survey she made by correspondence with a great many observers, farmers, horticulturalists, foresters, and members of entomological societies throughout the length and breadth of the country, her first *Notes for Observations of Injurious Insects* was a seven-page pamphlet, printed at her own expense and circulated to her correspondents and to various public agencies. Similar surveys, reporting information on those species that had been most destructive or that had attracted special attention during the year, formed the basis of the subsequent annual reports. These also included brief notes on practical methods for preventing or reducing depredations. Distributed free of charge to contributing observers, they were also sold for one shilling and six pence, a price below the cost of production, and they reached a wide audience.

Finding her own way as far as presentation and format were concerned, she arranged the information in her first four reports under headings of insect names, for instance "6, *Anthomyia ceparum,* Onion fly," or "25, *Abraxas grossulariata,* Magpie moth." However, within a relatively short time she realized more clearly what was better suited to the needs of her agricultural audience, and for their convenience moved somewhat away from traditional entomological form, presenting information under three broad headings: farm crops, orchard and bush fruits, and forest trees. Information about livestock pests was also given, although, with the exception of her work on the ox warble fly, she tended to avoid this area, both because she found the investigations unpleasant and because she thought the subject better left to veterinary surgeons. Illustrations she considered especially important, and after a few years she had these prepared by Horace and E. C. Knight of London, whose work was of excellent quality. A general index to the *Reports* for the years between 1877 and 1898 was prepared by Robert Newstead of Liverpool University.[75] Interestingly, the first 1877 report was occasioned by fears of an imminent invasion of Britain by the Colorado potato beetle, which in the 1860s and seventies was advancing eastward across the United States at an alarming rate. It never did enter Britain. Eleanor Ormerod's own paper on this insect appeared in the *Entomologist* in 1877.

In 1882 she accepted the post of Honorary Consulting Entomologist to the Royal Agricultural Society of England. She did so with hesitation and considerable reluctance because she already had a substantial burden of work. Further, just at the time she took on the responsibility, she was almost run over by a carriage while crossing a London street and suffered a knee injury from which she never fully recovered. From then on she was lame, suffered frequent pain, and was forced to limit her activities. Nevertheless, for the next ten years, in addition to bringing out her own *Annual Report,* she conducted the society's correspondence on injurious insects and prepared the "Reports of the Honorary Consulting Entomologist," published in the society's journal. These were also used by the government's Agricultural Department, with which the society collaborated closely. In 1892, when she was sixty-four, she had her knee operated on, which greatly reduced her pain; but recovery required some reduction in her workload and she resigned from the Royal Agricultural Society. Other reasons as well prompted her resignation. She wrote to her friend Robert Wallace, professor of agriculture and rural economy at the University of Edinburgh as follows:

> I hope and greatly desire to continue all my work, Home, Colonial and publishing; also to act as referee to our Agricultural Journals just as before, but it is much more comfortable working up important points, to having everlastingly to be going over a routine often keeping one from attending to what may be of importance. Who will they get to take my place [at the Royal]? It seems to me a great pity that there is not a properly paid and competent officer for the Board of Agriculture and the R.A.S.E. I am safe in saying this, for I never intend to take office again, not for any amount of money that could be offered, neither do

I mean to do the work of the Government or Society under the polite name of "kindly co-operating!"[76]

Clearly she had had enough of providing regular routine services to an organization—a rather different matter from working independently for the country's agricultural community as a free agent. About this time a paid position for a government entomology adviser was established, and she felt strongly that if substantial assistance were to be provided to him by officers of the Agricultural Society, then this also should be negotiated on a business footing rather than being incorporated, with little acknowledgment, into his reports.[77] Interestingly enough, the society had a paid consulting chemist and a paid consulting botanist, and when Eleanor resigned her honorary position her male replacement was appointed, under a new title of consulting zoologist, at an annual salary of £200.[78] Nevertheless, despite her dissatisfaction and her impatience when dealing with boards and committees, a considerable amount of her work was done in association with the Royal Agricultural Society.

Her public service extended to other spheres as well, notably the legal and educational. As her name became known as an official consulting entomologist she found herself being asked to serve as expert witness, typically in cases of alleged infestation of flour cargos being shipped from America. The critical point to be decided was very often the origin of the infestation. Her examinations of samples often led to such unambiguous conclusions that the cases were settled out of court. But she found the work fairly taxing; the research needed to deal successfully with lawyers' purposely confusing questions, relevant and irrelevant, was inconvenient and wearing.[79]

For a time she was a supernumerary member of the staff of the Royal Agricultural College, Cirencester, and she presented a series of six talks on agricultural entomology there between 1881 and 1884. In 1883 she also gave a lecture before an audience of five hundred students and staff at the Institute of Agriculture at South Kensington on "Insect injuries to farm crops." This was quickly followed by what was clearly seen as a much needed, concentrated course of ten lectures on "Orders of insects," also given at the institute. The subject matter of the course was published the following year as her *Guide to the Methods of Insect Life* (1884), which was later developed into her *Textbook of Agricultural Entomology* (1892).

She was particularly interested in promoting the acceptance of agricultural entomology as a standard academic subject in agricultural education, and for four years (1882–1886) sat on a committee on economic entomology appointed by the Committee of Council on Education. About 1889 she succeeded in having the subject established as an option in an examination administered by the Royal Agricultural Society, and shortly after that as a compulsory examination subject at the Royal Agricultural College. At that time the University of Edinburgh, a leader in agricultural education, decided to establish a chair in economic entomology, and there were those who thought Ormerod would be a suitable candidate (even though three years were still to elapse before women were eligible for undergraduate degrees at Edinburgh). At her suggestion William Fream was appointed. She did, however, act as co-examiner with Fream in economic entomology at Edinburgh from 1896 to 1899, and in 1900, a year before her death, was awarded the first honorary LL.D. that Edinburgh University gave to a woman. When, in 1897, the question of establishing an agricultural lectureship at Oxford arose, she offered the university £100 a year for a period of five years to support such a position, with the important proviso that the subject be accepted as a degree requirement.[80] She felt that instruction in agriculture would greatly benefit those Oxford students, of whom there were many, who were soon to become the great landholders of the country. Despite a similar offer of financial support from the Clothworkers' Company, the university did not establish the lectureship (and £5000 of Ormerod's money eventually went to the University of Edinburgh).

Among her many reports, those that were particularly noteworthy included her 1883 paper on wireworm, an insect widely destructive to grain crops, her observations on the development of the warble fly parasitic on cattle (1885), on the Hessian fly, another scourge of grain crops (1887), on mustard beetles (1887), and on the diamondback moth, a turnip pest (1891). Most of this work was published in the Royal Agricultural Society's journal. To meet the ever-growing requests for help from farmers and stock owners, she also issued, at her own expense, many four-page leaflets dealing with the commonest of Britain's farm pests. The demand for the leaflet on the warble fly, its life history, easy practicable methods of prevention of infestation, and remedy, was such that successive printings totaled 170,000 copies, including 15,000 copies printed by Messrs. Murray of Aberdeen at their own cost. She also handled a very considerable direct correspondence, especially when some "favourite topic" was under discussion. She later recalled once having a run of up to 100 letters a day for a time, with 149 on one day—a situation that required her to get help to keep things in hand. Her regular letter work she estimated at 1,500 a year. Time being a key factor in the problems she was dealing with, she regularly replied, if at all possible, on the day of receipt.[81]

Throughout her work she kept in regular touch with European societies and specialists, including J. Jablonowski of the Hungarian Government Department of Agriculture in Budapest, J. Ritzema Bos in Amsterdam, W. H. Schöyen, the Norwegian State Entomologist in Oslo, Alfred Nalepa in Vienna, Enzio Reuter in Helsinki, and Lindeman in Moscow. Problems discussed with these men over the years ranged from bush fruit and orchard pests to root crop infestations, to the spreading of the Hessian fly. She much enjoyed the contacts and followed the rise and progress of economic entomology in Europe with considerable enthusiasm. Some of her most important connections, however, were with government entomological workers in the British colonial territories and dominions and in the United States. Developments in North America interested her especially, and she was in frequent communication with Charles Bethune, editor of the *Canadian Entomologist,* James Fletcher, Dominion Entomologist in Ottawa, and Charles Riley and Leland Howard, government entomologists in Washington, D.C.

Because of the huge-scale ecological changes brought about by the rapid expansion of agriculture throughout much of North America, the problems of insect devastation that these workers faced were on a scale unknown in Britain, and countermeasures were much further advanced. In particular, chemical sprays were coming into use. Following the advice of the transatlantic experts, Ormerod began a campaign in Britain for the introduction of arsenical sprays (used in the United States from the 1870s) to control fruit tree infestations. Although initially faced with considerable conservative opposition, by about 1890 she had established Paris green as an indispensable insecticide in large orchards; she considered this work her most important accomplishment.[82]

In addition to her 1892 *Textbook of Agricultural Entomology* she brought out a number of well-received monographs, the most important of which were *A Manual of Injurious Insects, with Methods of Prevention and Remedy for their Attacks to Food Crops, Forest Trees and Fruit* (1890), *Handbook of Insects Injurious to Orchard and Bush Fruits, with Means of Prevention and Remedy* (1898), *Notes and Descriptions of a Few Injurious Farm and Fruit Insects of South Africa* (1889, compiled by Ormerod), and *Flies Injurious to Stock . . . with Special Observations on the Ox Warble or Bot Fly* (1900).[83] She also contributed revisions to various parts of William Fream's *Elements of Agriculture* (1892), which, with further periodic revisions, remained a standard textbook for the better part of a century.

From the 1890s she was increasingly troubled by ill health and in her correspondence often mentioned being interrupted in her work by various ailments, including eye trouble. Although she was for many years the center of the national information network for agricultural entomology, and indeed was doing much of the work regularly undertaken by government agencies in other countries, she preferred to function on her own; and thanks to her wealth she could do so effectively. As she grew older, she may also perhaps have developed something of her father's tendency toward reclusiveness. In any case, the crowd and the bustle at the Royal Agricultural Society meetings, where she had to go to present her reports, was not to her taste.[84] Further, she dreaded the way business correspondence piled up during absences: "Does not the collection, all calling 'answer me first,' quite make your heart sink? I cannot face it—it is such a terrible strain, so I stop nearly entirely at home like a limpet on a rock, and keep my work as well as I can in hand."[85] Her lifelong companion was her sister Georgiana, who shared her interest in entomology and who herself occasionally prepared insect diagrams. Georgiana's death in 1896 greatly depressed her. Both sisters were fellows of the Entomological Society, Eleanor since 1876.

Although latterly she was honored by several foreign agricultural societies, and received among other awards the Silver Medal for Applied Entomology of the Société Nationale d'Acclimatation de France (1899), the great event of her last years was the presentation to her of her honorary LL.D. by Edinburgh University (1900). Her state of health by then was such that she had considerable difficulty in making the journey from her home in St. Albans to Edinburgh for the occasion; she had to be accompanied by her lady private secretary and her doctor. But go she did, and took great pleasure in the ceremony. She died in St. Albans, 19 July 1901, at the age of seventy-three, of liver cancer.

Although Ormerod is usually described simply as an "economic entomologist," it is perhaps worth pointing out that the most important part of her work was the collecting, interpreting, and disseminating of scientific information.[86] Like two other nineteenth-century British women scientific writers, her predecessor Mary Somerville and her contemporary, the historian of astronomy Agnes Clerke, she was an excellent expositor of scientific material, skilled at marshaling and condensing information in such a way as to get her message across to her chosen audience. But her work, unlike that of both Somerville and Clerke, was in fact basic research, since the collecting and reporting of current raw data on insect populations constituted a major part of economic entomology. She also carried out a very creditable number of insect development studies.

Footnote

A case has been made recently for explaining Eleanor Ormerod's motivation for her work in terms of an underlying need to acquire the "social legitimacy" and professional prestige that would allow her to challenge prevailing ideas of "feminine purity".[87] Although Ormerod's letters to close relatives (a possible source of insights into her psychology) do not seem to have survived,[88] a large collection of those to her Canadian colleagues and to her friend Robert Wallace have, and many have been included in her autobiography. Admittedly they appear there after passing through editorial hands, which reduces their value, but nevertheless they contain numerous passing general remarks, which, taken as a body, offer a not insignificant glimpse into her thought and character. Neither these letters nor the overall pattern of her life and work would seem to offer any evidence for the existence in her psyche of compelling, nonintellectual needs sufficiently powerful to lead her into sixteen years of serious, independent study in biology followed by another thirty-two years of what, at times, clearly were long spells of grinding work. Whether or not such needs existed we can, of course, never know with certainty. But it seems improbable that a professional commitment such as hers could have been sustained on anything less than a genuine absorption in her subject and the confidence that she was carrying out work worthwhile in itself. Further, unless we are willing to entertain the rather unreasonable supposition that *all* late nineteenth-century women who took up biology did so ultimately to challenge prevailing ideas of feminine purity, there would seem to be no compelling reason for speculating that Ormerod did. Why pick on Eleanor?

Summary

Ormerod is such an outstanding figure in nineteenth-century entomology that she tends to leave all the other women who contributed to the field somewhat in the shade (see Figure 2-2 d). Nevertheless, the contributions of the British Museum classifiers and taxonomists Gertrude Ricardo and Emily Sharpe were also considerable, although these two are largely forgotten figures in the history of nineteenth-century women scientists. The number of British women specializing in Lepidoptera is notable, though hardly surprising, given the inherent attractions of these creatures. The 42 percent of authors whose contributions are labeled "Collectors' Reports" in Figure 2-2 c were almost all lepidopterists.

Two-country comparison

The Royal Society record suggests that fewer British women independent workers were active in entomology than was the case in the United States—at least fewer put their work into print.

Something of a contrast between the British and the Americans lies, as might be expected, in the greater overall American emphasis on applied entomology. There were no equivalents in Britain of the early American state entomologists with whom several of the American women worked, especially those in the Midwest. Likewise, the kinds of opportunities available to Americans for carrying out work in applied entomology as students at agricultural colleges and associated experimental stations were lacking in Britain. The heavier British emphasis on classification and taxonomy, illustrated in Figure 2-2 d, largely reflects the work of British Museum women. Although making up only 12 percent of the British author group, they produced 25 percent of the papers.

Notes

1. Papers by one additional American author (Monks) have been added to the bibliography. Several of those whose entomology papers are listed had their main interests in other fields; for Eigenmann, Hitchcock, Monks, and Treat, see chapters 3, 11, 5 and 1 respectively; for Dix and McNair Wright, see chapter 16.

2. It is important to keep in mind that many of the most outstanding American men entomologists of the late nineteenth and early twentieth centuries were also "amateurs," who made what were major contributions to the field as a hobby. Paid professional positions for entomologists were then few and far between—see Herbert H. Ross, *A Textbook of Entomology*, 3d. ed. (New York: John Wiley, 1965), p. 18.

3. William T. Davis, "Annie Trumbull Slosson," *Journal of the New York Entomological Society*, **34** (1926): 361–4; Veronica

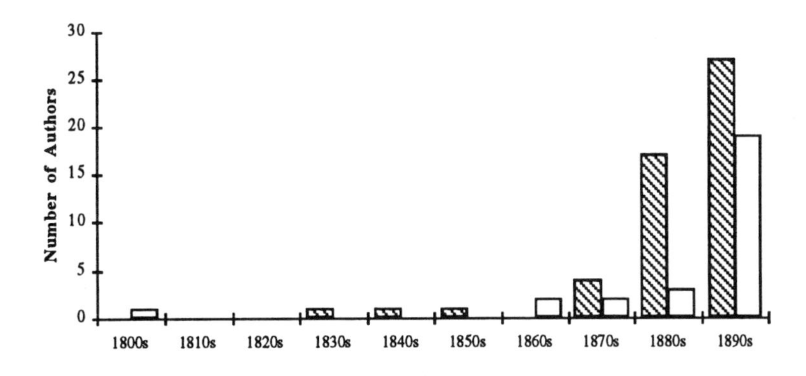

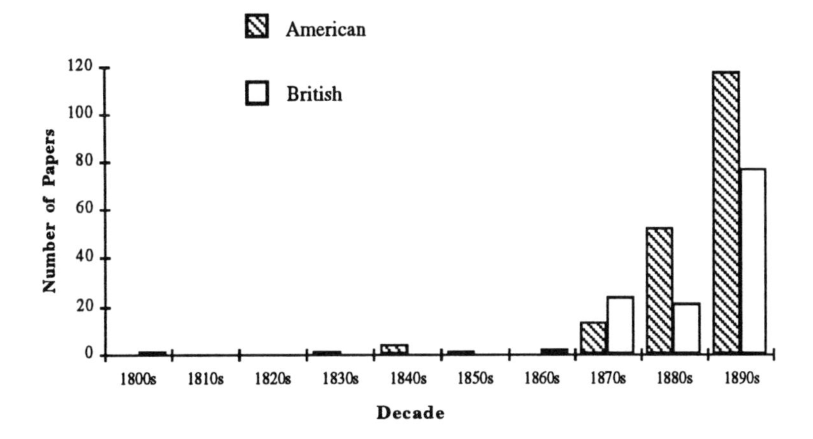

Figure 2-1. American and British authors and papers in entomology, by decade, 1800–1900. Data from the Royal Society *Catalogue of Scientific Papers.*

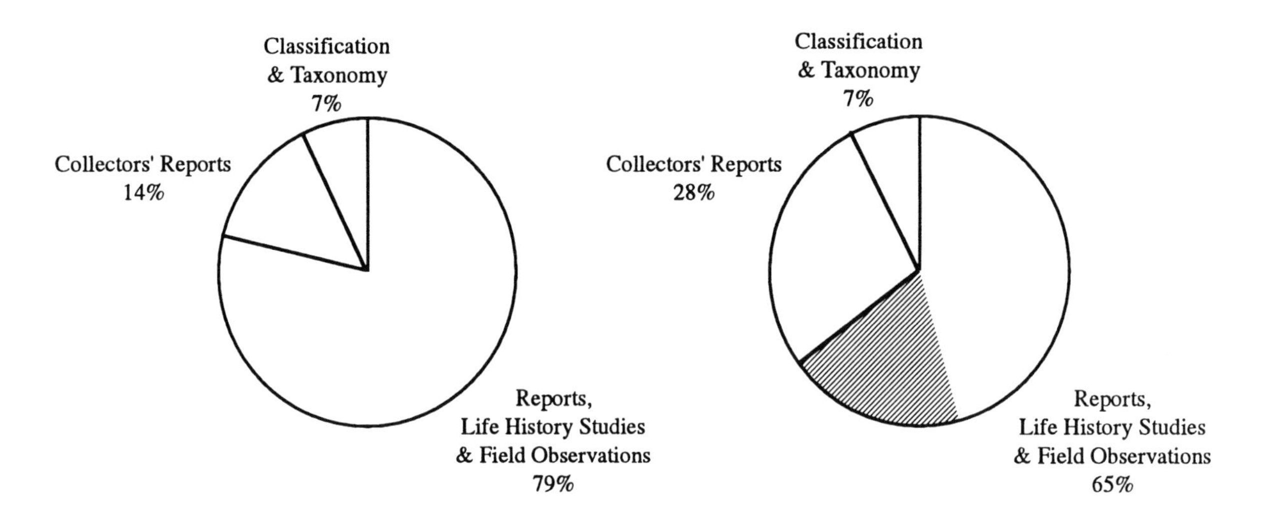

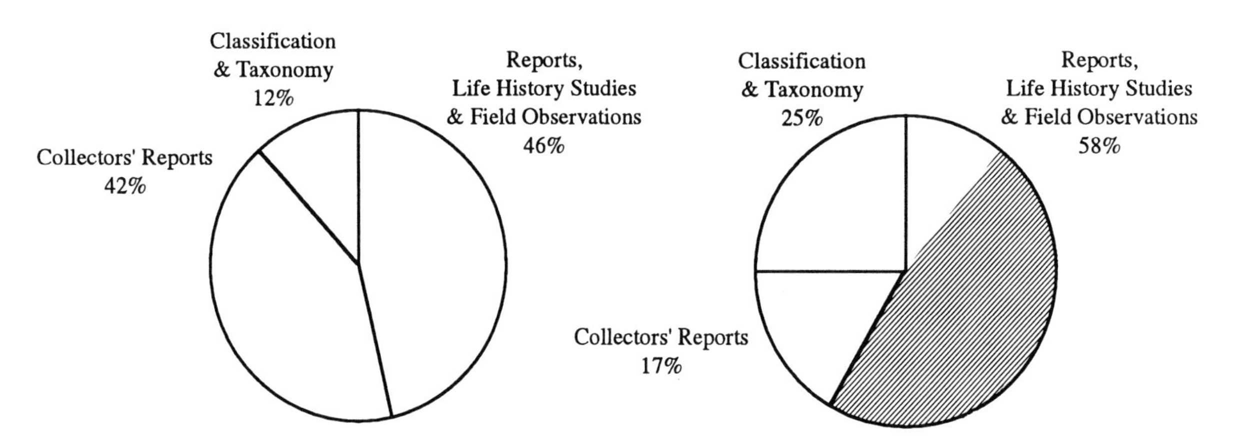

Figure 2-2. Distribution of American and British authors and papers in entomology, by subfield, 1800–1900. (In a. and c., an author contributing to more than one subfield is counted in each.) Shaded sectors represent the contributions of Murtfeldt (19% of the total in b.) and Ormerod (47% of the total in d.). Data from the Royal Society *Catalogue of Scientific Papers.*

Dougherty, "Annie Trumbull Slosson," in Dorothy Feir et al., "Women in ESA: a strong legacy," *American Entomologist,* **36** (1990): 96–127, especially pp. 126–7; Ogilvie, *Women in Science,* p. 185; *WWWA,* p. 752.

4. Davis, "Annie Trumbull Slosson," p. 362.

5. See "A few memories," *Journal of the New York Entomological Society,* **23** (1915): 85–91, and "A few memories. II," ibid., **25** (1917): 93–7. Included among Slosson's post-1900 technical notes were: "On a Florida beach," *Entomological News,* **12** (1901): 10–12; "A successful failure [collecting insects in Florida]," ibid., **12** (1901): 200–3; "Protection of *Chionobas semidae,*" ibid., **12** (1901): 316–17; "Additional lists of insects taken in the Alpine region of Mt. Washington," ibid., **13** (1902): 4–8, 319–21, **17** (1906): 323–6; "Hunting Empids," ibid., **14** (1903): 265–9; "A hunt for *Saldoida* osborn," ibid., **19** (1908): 424–8; "A bit of contemporary history," *Canadian Entomologist,* **40** (1908): 213–19.

6. *Who's Who in America,* vol. 5 (1908–09), p. 1771; *AMS,* 1910, 1921.

7. Soule's moth studies continued for several years after the turn of the century: see for instance, "The 'cocoons' or 'cases' of some burrowing caterpillers," *Psyche,* **9** (1901): 7–8; "Notes on the mating of *Attacus cecropia* and others," ibid., **9** (1901): 224–6; "The inner cocoon of Attacine moths," ibid., **9** (1901): 252; "Mating of *Attacus gloveri,*" ibid., **9** (1901): 255; "The hatching of *Eacles imperialis,*" ibid., **9** (1902): 299–300; "Notes on the hybrids of *Samia cynthia* and *Attacus promethea,*" ibid., **9** (1902): 411–13; "Notes on moths," *Entomological News,* **17** (1906): 395–7; (with Alfred Goldsborough Mayer) "Some reactions of caterpillars and moths," *Journal of Exper-*

imental Zoology, **3** (1906): 415–33; "Some experiments with hybrids," *Psyche,* **14** (1907): 116–17.

8. Ida Eliot, in collaboration with Anna C. Brackett, published a selection of poems, *Poetry for Home and School* (1876).

9. H. H. Newcomb, "Emily L. Morton," *Entomological News,* **28** (1917): 97–101, and **31** (1920): 149–50.

10. Asa Fitch (1809–1878) was the first of the state entomologists in the United States. He was appointed by New York in 1854 to investigate the substantial damage caused to agricultural crops by insects. His fourteen reports made available information on the life histories of many insects and were a great stimulus to workers throughout the country.

11. Newcomb, "Emily L. Morton," (1917), p. 100.

12. E[phriam] P[orter] Felt, "Maria E. Fernald," *Journal of Economic Entomology,* **13** (1920): 153; Arnold Mallis, *American Entomologists* (New Brunswick, N.J.: Rutgers University Press, 1971), pp. 142–7.

13. Mallis, *Entomologists,* pp. 146–7.

14. Gypsy moth eggs were first brought into the United States from France in connection with a study of silk worm diseases. In 1869 adult moths escaped to an area of wasteland near the town of Medford, Massachusetts, where they gradually increased, unnoticed, until in 1889 an unbelievable plague of caterpillars overwhelmed the district. From then until well into the 1900s the insect was an ever-present menace, threatening to wipe out fruit and other trees in the New England states.

15. L. O. Howard, "Address at the dedication of the entomology and zoology building of the Massachusetts Agricultural College," *Science,* **32** (1910): 769–75, on p. 772.

16. See entries for George Dimmock in *NCAB,* vol. 26, pp. 308–9, and *Who was Who in America,* vol. 1 (1897–1942), p. 325.

17. See for instance, George Dimmock, "Circulation of blood in the larva of *Hydrophilus,*" *Psyche,* **3** (1880–1882): 324–6, and "Some glands which open externally on insects," ibid., **3** (1880–1882): 387–99.

18. Anon., "Cora H. Clarke," *Psyche,* **23** (1916): 94; M[artha] Burton Williamson, "Some American women in science," *The Chautauquan,* **28,** (1898–1899): 161–8, 361–8, 465–73, on pp. 367–8; *AMS,* 1910.

19. Cora H. Clarke, "New missionary work. [Preservation of native plants]," *Plant World,* **5** (1902): 81–7; "Mounting mosses," *Bryologist,* **6** (1903): 102–3; "Curbside mosses," ibid., **7** (1904): 74; "A red *Andreaea,*" ibid., **10** (1907): 55; "A suggestion for summer observations," *Rhodora,* **14** (1912): 177–84.

20. Burton Williamson, "American women in science," pp. 366–7; *WWWA,* p. 738; *AMS,* 1910.

21. Helen Norton Stevens, *Memorial Biography of Adele M. Fielde. Humanitarian* (New York: The Fielde Memorial Committee, 1918); P.P.C., *Entomological News,* **27** (1916): 191–2; Frank R. Lillie, *The Woods Hole Marine Biological Laboratory* (Chicago: University of Chicago Press, 1944), p. 143; *WWWA,* p. 290; *AMS,* 1910. (Fielde changed the spelling of her family name, adding the final e; her father had used the form Field.)

22. Lillie, *Woods Hole Laboratory,* p. 143.

23. Fielde's hope was that she might alleviate some of the immense amount of suffering brought about by the fact that trained medical help for women during childbirth was almost nonexistent in China. There were few Chinese women doctors, and male physicians were prohibited from attending births.

24. Stevens, *Memorial Biography,* p. 236.

25. See for instance, Adele M. Fielde, "The study of an ant," *Proceedings of the Academy of Natural Sciences of Philadelphia,* **53** (1901): 425–49; "Further study of an ant," ibid., **53** (1901): 521–44; "Notes on an ant," ibid., **54** (1902): 599–625; "Supplementary notes on an ant," ibid., **55** (1903): 491–5; "Experiments with ants induced to swim," ibid., **55** (1903): 617–24; "Artificial mixed nests of ants," *Biological Bulletin,* **5** (1903): 320–5, "A cause of feud between ants of the same species living in different communities," ibid., **5** (1903): 326–9; "Power of recognition among ants", ibid., **7** (1904): 227–50; "Tenacity of life in ants," ibid., **7** (1904): 300–9; "Observations on the progeny of virgin ants," ibid., **9** (1905): 355–60; "Temperature as a factor in the development of ants," ibid., **9** (1905): 361–7; "The progressive odor of ants," ibid., **10** (1905): 1–16; "Longevity of a velvet ant", ibid., **11** (1906): 265–6.

26. Fielde, quoted by Stevens, *Memorial Biography,* p. 262.

27. Stevens, *Memorial Biography,* p. 319.

28. No biographical information has been discovered about four of the Northeastern group (Aaron, Sargent, Smallwood, and Wadsworth). Mrs. Aaron, of Philadelphia, contributed one note to *Entomological News* in 1897; Annie Bell Sargent's observations on spiders were made in Pennsylvania (1898–1900); Mabel Smallwood worked at Cold Spring Harbor, Long Island, New York, her note on sand fleas appearing in *Science* in 1900 and her longer work on the same topic, "The beach flea: *Talorchestia longicornia,*" as *Cold Spring Harbor Monograph,* No. 1 (1903). She later lived in Chicago, joined the Illinois State Academy of Science when it was organized in 1908, and appears in its membership lists until the early 1930s (*Illinois State Academy of Science Transactions,* 1908–1931; she specified zoology as her interest, but gave no professional affiliation). Mattie Wadsworth lived in Kennebec, Maine; her lists of dragonflies captured in Kennebec County were published between 1890 and 1898.

29. Anon., "Morris, Margaretta Hare," in *Biographical Dictionary of American Science: the Seventeenth through the Nineteenth Centuries,* ed. Clark A. Elliot (Westport, Conn.: Greenwood Press, 1979), pp. 185–6; Jeanette E. Graustein, *Thomas Nuttall, Naturalist. Explorations of America. 1808–1841* (Cambridge, Mass: Harvard University Press, 1967), p. 374.

30. We thank George W. Byers, professor emeritus of entomology, University of Kansas, for a helpful discussion about the Hessian fly and its relatives. It is perhaps worth noting that even the leading naturalists and entomologists of Margaretta Morris's time had little concept of the tremendous number and variety of insects they were dealing with. The errors and uncertainties in Morris's work were by no means unique among workers of the period.

31. Edwin P. Meiners, M.D., "Mary Esther Murtfeldt (1839–1913)," *Lepidopterists' News* (October 1948); obituaries, "Mary Esther Murtfeldt," *Journal of Economic Entomology,* **6** (1913): 288–9, *Canadian Entomologist,* **45** (1913): 157 and Hermann Schwarz, "Miss Mary E. Murtfeldt," *Entomological News,* **24** (1913): 241–2; anon., "Murtfeldt House, 10 Douglass Lane," *Kirkwood Historical Review,* (March 1965): 13; anon., "Brief description of length of individual tours. 10 Douglass Lane," ibid., (December 1981): 34–6; Herbert Osborn, *Fragments of Entomological History: Including Some Personal Recollections of Men and Events* (Columbus, Ohio: privately printed, 1937), pp. 165–6; Burton Williamson, "American women in science," pp. 365–6; Perley Spaulding, "A biographical history of botany at St. Louis. IV," *Popular Science Monthly,* **74** (1909): 240–58, especially pp. 251–2; *WWWA,* p. 589; *AMS,* 1910.

32. This work was published in the *Transactions of the St. Louis Academy of Sciences,* **3** (1868–1877): 55–64, 178–80, 208–10, and also in the *American Naturalist,* **7** (1873): 619–23, under Riley's name only, though Murtfeldt's collaboration was known to those

working in the area. Spaulding ("Biographical history," p. 252) commented that because of her botanical work, as well as her association with Dr. Riley in working out the pollination of *Yucca* and other problems, Murtfeldt deserved mention among prominent St. Louis botanists.

33. Burton Williamson, "American women in science," p. 365.

34. See *Twenty-Second Annual Report of the State Board of Agriculture of the State of Missouri* (1889–1890) (Jefferson City, Mo.: Tribune Printing Co., 1890), issued during the period when Murtfeldt was "acting State Entomologist"; here the secretary of the Board, Levi Chubbuck, refers to her only as "Miss M. E. Murtfeldt, the well-known entomologist of Kirkwood, Mo." (p. 23). In the same report, in a discussion of the question of insect control, Chubbuck states that nothing had been done by the State of Missouri in this matter since Riley's services as state entomologist had been terminated (p. 24). There is no mention of Murtfeldt in the annual Treasurer's Report in this issue either (pp. 27–8). Further, no mention of her appears in the *Twenty-Sixth Annual Report* (for the year 1893) or in the *Twenty-Seventh Annual Report* (for 1894), both of which were issued during the period in which she is reported (in *AMS*) to have been acting state entomologist. In contrast, the *Eleventh Annual Report,* for 1875, and the *Twelfth Annual Report,* for 1876, the last two years Riley was in Missouri, both contain many references to him and reports by him; in addition, his salary ($250 a month) is listed in the Treasurer's Reports (pp. 34–6 in the *Eleventh Annual Report* and pp. 18, 19, 24–5 in the *Twelfth Annual Report*).

35. Charles V. Riley, *Annual Report on the Noxious, Beneficial and other Insects of the State of Missouri,* 9 vols. (Jefferson City, Mo.: Tribune Printing Co., 1869–1877). These came out as appendixes to the *Annual Report of the State Board of Agriculture of the State of Missouri.*

36. Levi Chubbuck, in the *Twenty-Second Annual Report, State Board of Agriculture, Missouri,* 1890, p. 24.

37. See sketch of Charles Riley in *NCAB,* vol. 9, p. 433.

38. Riley's policy of omitting the names of coworkers (Murtfeldt included) from his reports was in keeping with the standard practice of the time. Later, when he was a division chief at the U.S. Department of Agriculture, this policy was challenged by his fellow scientist in the department, the plant pathologist Beverly T. Galloway. Procedures were subsequently changed, and junior workers given credit for their research, but the change was strongly fought by Riley, who believed it would lead to lack of confidence in the work on the part of the public, which was not familiar with the names of the younger scientists (see John A. Stevenson, "Plants, problems and personalities: the genesis of the Bureau of Plant Industry," *Agricultural History,* **28** (1954): 155–62, especially p. 157).

39. Chubbuck, *Twenty-second Annual Report,* p. 182.

40. S. M. Tracy, "Flora of Missouri," *Missouri State Horticultural Society Report* (Appendix), (1885): 1–106.

41. Mary E. Murtfeldt, "Periodicity of *Sabbatia angularis,*" *Proceedings of the AAAS,* **32** (1883): 313 (title only).

42. Osborn, *Fragments,* pp. 165–6.

43. See, for instance, Mary Murtfeldt, "Report on [the] present status of American women in entomology," *Proceedings of the National Science Club,* **3** (1897): 11–14.

44. The two articles from the *Kirkwood Historical Review,* 1965 and 1981 (n. 31), say that all of the Murtfeldt sisters were teachers in St. Louis schools and that their house at 10 Douglass Lane was known as the schoolteachers' house. How the severely handicapped Mary Murtfeldt succeeded in teaching (if she in fact did) in addition to carrying out her entomological work is not discussed.

45. Mallis, *Entomologists,* p. 52.

46. Alumni records, faculty, 1918, University of Illinois, Urbana; letter from Beach to Stephen A. Forbes, University of Illinois, 1 July 1898 (University of Illinois archives, RS 43/1/1–8); alumni records, Iowa State University (see also chapter 1).

47. R[ichard] A[ntony von] Muttkowski, "George Williams Peckham, M.D., LL.D." *Entomological News,* **25** (1914): 145–8; Mallis, *Entomologists,* pp. 348–51; *WWWA,* p. 633; *AMS,* 1910, 1921.

48. Muttkowski, "Peckham," p. 146.

49. George W. Peckham and Elizabeth G. Peckham, *Wasps Social and Solitary* (Boston and New York: Houghton Mifflin, 1905).

50. Elizabeth Peckham's entry in the 1921 edition of *American Men of Science* indicates that she took a Ph.D. at Cornell in 1916. However, there is no mention in the university records of her having done so.

51. See bibliography. The note by Laurene Highfield in the 1894 *American Naturalist* described observations made near Quincy, Illinois; Mrs. Snyder worked in Belvidere, Illinois; Hattie Warner was associated with the Agricultural Experiment Station at the Kentucky State College in Lexington. Her list of the butterflies of Kentucky, the subject of her 1894 paper in the *Canadian Entomologist,* was based on the station collection.

52. Alumni records, Kansas State University archives; Julius Terrass Willard, *History of the Kansas State College of Agriculture and Applied Sciences* (Manhattan, Kan.: Kansas State College Press, 1940), pp. 80–1, 513–14.

53. Mallis, *Entomologists,* pp. 359–61.

54. Another of the Cockerells' joint papers on bees appeared in 1901, "Contributions from the New Mexico Biological Station. IX. On certain genera of bees," *Annals and Magazine of Natural History,* s. 8, 7 (1901): 46–50. Wilmatte Cockerell continued to publish for a few years after the turn of the century; see, for instance, "Eggs of *Arachnis zuni,*" *Entomological News,* **12** (1901): 209; "A trip to the Truchas peaks, New Mexico," *American Naturalist,* **37** (1903): 887–91; "Some aphids associated with ants," *Psyche,* **10** (1903): 216–18; (with Theodore D. A. Cockerell) ["A new Coccid from Beulah"], *Philadelphia Academy of Natural Sciences, Transactions of the American Entomological Society,* **29** (1903): 112–13.

55. Little information has been uncovered about the remaining two workers from the western region, Winnie Harward and Mary Palmer: Harward's observations on insectivorous lizards were made in Albuquerque, New Mexico, in 1900; Palmer published a paper on California trap-door spiders in 1898.

56. For Hopley and Veley, see chapter 3; and for Huie, chapter 1. Papers in entomology by four women not listed by the Royal Society (Alderson, Cottingham, Edwards, and Smee), have been added to the bibliography, giving a total entomology author count there of twenty-nine.

57. Margaret Fountaine, *Love among the Butterflies. The Travels and Adventures of a Victorian Lady,* ed. W. F. Cater (London: Collins, 1980), and *Butterflies and Late Loves. The Further Travels and Adventures of a Victorian Lady,* ed. W. F. Cater (London: Collins, 1986); W.G.S., "Margaret Elizabeth Fountaine," *Entomologist,* 73 (1940): 193–5.

58. "Butterfly hunting in Greece in the year 1900," *Entomologist's Record,* **14** (1902): 29, 35, 64–7.

59. See "A butterfly summer in Asia Minor," *Entomologist,* **37** (1904): 79–84, 105–8, 135–7, 157–9, 184–8, and "Algerian butterflies in the spring and summer of 1904," ibid., **39** (1906): 84–9, 107–9.

60. See "Five months of butterfly collecting in Costa Rica in the summer of 1911," ibid., **46** (1913): 189–95, 214–19, and "Notes on

the life-history of *Papilio demolian* Cram.," *Transactions of the Entomological Society* (1915): 456–8. For an account of Fountaine's South African work see, "Descriptions of some hitherto unknown, or little known larvae and pupae of South African Rhopalocera, with notes on their life-histories," ibid., (1911): 48–61.

61. See "Among the Rhopalocera of the Philippines," *Entomologist,* **58** (1925): 235–9; ibid., **59** (1926): 9–11, 31–4, 53–7. For Fountaine's collecting in southern California see, "List of butterflies taken in the neighbourhood of Los Angeles, California," ibid., **50** (1917): 154–6.

62. Nicholls's post-1900 papers included, "Butterflies of the Lebanon," *Transactions of the Entomological Society* (1901): 75–95, and *Entomologist's Record,* **13** (1901): 169–73, 205–9; "A probably double-brooded *Erebia,*" ibid., **16** (1904): 48–9.

63. For Wollaston see G. T. Bethune-Baker, obituary, ibid., **23** (1911): 325.

64. William T. Stearn, *The Natural History Museum at South Kensington. A History of the British Museum (Natural History)* (London: Heinemann, 1981), pp. 175–7.

65. Among Emily Sharpe's writings not listed in the bibliography are the following: "A list of Lepidoptera collected by Mr. Arthur H. Neumann," in Arthur H. Neumann, *Elephant Hunting in East Africa* (London: Rowland and Ward, 1898), Appendix, pp. 437–47; "A monograph of the genus *Teracolus,*" *Monographiae Entomologicae,* No. 1 (London: Lovell Reeve and Co., 1898); "A list of the Lepidoptera collected by Mr. Ewart S. Grogan in Central Africa," *Annals and Magazine of Natural History,* s. 7, **8** (1901): 278–86; "On the collections of Insecta obtained by Dr. Donaldson Smith in Somaliland," *Entomologist,* **34** (1901), Supplement, 1–8; "Descriptions of three new butterflies from East Africa," ibid., **35** (1901): 40–2; "A list of Lepidoptera collected by Dr. Cuthbert Christy in Nigeria," ibid., **35** (1901): 65–8, 101–7; "Descriptions of two new species of the genus *Planema* from the Uganda Protectorate," ibid., **35** (1901): 135; "On the butterflies collected in Equatorial Africa by Captain Clement Sykes," ibid., **35** (1902): 276–80, 308–11; *A Monograph of the genus Teracolus,* pts. 9 and 10 (1901), pt. 11 (1902) (London: Lovell Reeve and Co., 1914), pp. 101–56; "The butterflies collected in Equatorial Africa by Captain Clement Sykes," *Entomologist,* **36** (1903): 5–8, 36–40; "Description of a new species of the family Lemoniidae," ibid., **36** (1903): 310; "On a new species of butterflies from Equatorial Africa," ibid., **37** (1904): 131–4; "Descriptions of new Lepidoptera from Equatorial Africa," ibid., **38** (1904): 181–3; "Descriptions of new Lycaenidae from Equatorial Africa," ibid., **38** (1904): 202–4; "A list of the Lepidoptera and Heterocera collected in Karamoja between Mounts Elgon and Murosoka," in P. H. G. Powell-Cotton, *In Unknown Africa* (London: Hurst and Blakett, 1904); "Descriptions of two new species of Acraeidae from Entebbe, Uganda," *Annals and Magazine of Natural History,* s. 7, **18** (1906): 75–6; "Description of two new species belonging to the family Nymphalidae," *Entomologist,* **40** (1907): 155.

66. Gertrude Ricardo's family background has not been traced. The following may be relevant: *Burke's Landed Gentry,* 18th ed. (London: Burke's Peerage Ltd., 1972), vol. 3, pp. 762–3, lists a Henry David Ricardo (1833–1873) of Gatcomb Park, Gloucester, who married Ellen, daughter of Ven William Crawley, archdeacon of Monmouth and rector of Bryangwyn. Their oldest daughter was Ellen Gertrude, who died, unmarried, 1 November 1950.

67. Ricardo's post-1900 publications include the following: "Notes on Diptera from South Africa," *Annals and Magazine of Natural History,* s. 7, **7** (1901): 89–110; "Further notes on the Pangoniinae of the family Tabanidae in the British Museum collection," ibid., s. 7, **8** (1901): 286–315, **9** (1902): 336–81, 424–38; "Notes on the smaller genera of the Tabanidae of the family Tabanidae in the British Museum collection," ibid., s. 7, **14** (1904): 349–72; (with F. V. Theobald) "Diptera," in *The Natural History of Sokotra and Abd-el-Kuri,* ed. Henry Ogg Forbes (London: R. H. Porter, 1903); "Notes on the *Tabani* of the palaeoarctic region in the British Museum collection," *Annals and Magazine of Natural History,* s. 7, **16** (1905): 196–202; "Notes on the genus *Haematopota* of the family Tabanidae in the British Museum collection," ibid., s. 7, **18** (1906): 94–127; "Description of a new fly of the family Tabanidae," *Proceedings of the Zoological Society* (1906): 97–8; "Description of some new species of Tabanidae, with notes on some *Haematopota,*" *Annals and Magazine of Natural History,* s. 8, **1** (1908): 54–9; "Descriptions of thirty new species of *Tabani* from Africa and Madagascar," ibid., s. 8, **1** (1908): 268–78, 311–33; "Four new Tabanidae from India and Assam," ibid., s. 8, **3** (1909): 487–90; (with Jacques Surcouf) *Étude Monographique des Tabanides d'Afrique (Groupe des Tabanus)* (Paris: Masson, 1909); "A revision of the genus *Pelecorhynchus* of the family Tabanidae," *Annals and Magazine of Natural History,* s. 8, **5** (1910): 402–9; "A revision of the species *Tabanus* from the oriental region, including notes on species from surrounding countries," *Records of the Indian Museum, Calcutta,* **4** (1911): 111–255; "A revision of the oriental species of the genera of the family Tabanidae other than *Tabanus,*" ibid., **4** (1911): 321–97; "Notes on *Tabani* from the East Indies," *Entomologische Berichte uitgegeven door da Nederlandsche entomologische Vereeniging,* **55** (1912): 347–9; "A revision of the Asilidae of Australia," *Annals and Magazine of Natural History,* s. 8, **9** (1912): 473–88, **10** (1912): 142–60; 350–60; "Dipteren. ii. Notes on species of *Tabanus* from Ceram, Waigiou and Buru," *Bijdragen tot de Dierkunde uitgegeven door het Koninklijk Zoölogisch genootschap Natura Artis Magistra te Amsderdam,* **19** (1913): 70–2; "Tabanidae from Formosa collected by Mr. H. Sauter," *Annales Historico-Naturales Musei Nationalis Hungarici,* **11** (1913): 168–73; "A revision of the Asilidae of Australia (continued)," *Annals and Magazine of Natural History,* s. 8, **11** (1913): 147–66, 409–24, 429–51; "A new species of Tabanidae from the oriental region," ibid., s. 8, **11** (1913): 542–7; "Diptern. ii. The Tabanidae of New Guinea," in Arthur Wickmann, *Nova Guinea, Resultats de l'Expedition Scientifique Néerlandaise à la Nouvelle Guinée* (Leiden: E. J. Brill, 1913), vol. 9, pp. 387–406; "Species of *Tabanus* from Polynesia in the British Museum and in the late Mr. Verrall's collection," *Annals and Magazine of Natural History,* s. 8, **13** (1914): 476–9; "A new species of *Tabanus* from India," ibid., s. 8, **14** (1914): 359–60; "Notes on the Tabanidae of the Australian region," ibid., s. 8, **14** (1914): 387–97; "List of South African Tabanidae (Diptera) in the South African Museum, with descriptions of new species," *Annals of the South African Museum,* **10** (1914): 447–61; "Tabanidae [of Formosa]," *Supplementa entomologica* (Berlin), **3** (1914): 62–5; "Notes on the Tabanidae in the German Entomological Museum," *Archiv für Naturgeschichte,* Abt. A, **80,** Heft 8 (1915): 122–30; "Notes on the Tabanidae of the Australian region (continued)," *Annals and Magazine of Natural History,* s. 8, **15** (1915): 270–91, **16** (1915): 16–40, 259–86; "Notes on a collection of Tabanidae from Hong Kong," *Bulletin of Entomological Research,* **6** (1916): 405–7; "Two new species of *Haematopota* from the Federated Malay States," ibid., **6** (1916): 403–4; "New species of *Haemotopota* from India," *Annals and Magazine of Natural History,* s. 8, **19** (1917): 225–6; "Further notes on the Asilidae of Australia," ibid., s. 9, **1** (1918): 57–60; "Notes on Asilidae: sub-division Asilinae," ibid., s. 9, **3** (1919): 44–79; "A Persian asilid attacking house-flies," *Entomologist's Monthly Magazine,* **56** (1920): 278; "Notes on the Asilinae of the South African and oriental regions," *Annals and Magazine of Natural History,* s. 9, **8** (1921):

175–92, **10** (1922): 36–73; "New species of Asilidae from South Africa," ibid., s. 9; **15** (1925): 234–82; "Stratiomyiidae, Tabanidae and Asilidae," *British Museum* (*Nat. Hist.*) *Bulletin, Insects of Samoa,* **6** (1929): 109–22.

68. Eva Bewsher's observations on bees were made on Mauritius. Charlotte Cooper was from Wormley, Hertfordshire; her 1802 paper "Curious particulars concerning bees" is one of the earliest by a woman listed by the Royal Society. Miss M. S. Pasley was from Windermere, Mary Redmayne contributed her sawfly breeding observations from Litchfield, Elizabeth Smee her studies on Diptera larvae from Wallington, and Juliette Williams her note on "Carniverous caterpillars" from Bickley, Kent.

69. R. B. Benson, obituary, *Entomologist's Monthly Magazine,* **90** (1954): 46; *Burke's Landed Gentry,* vol. 3, p. 182.

70. "Observations on the life histories and habits of *Allantus pallipes* pin., and *Pristiphora pallipes* Lep. (Hym., Tenth.)," *Entomologist,* **56** (1923): 125–8, 179–85.

71. *A History of Hampshire and the Isle of Wight,* 5 vols., (Westminster: A. Constable and Co., 1900–1912).

72. Sanders's post-1900 papers were: "On the Lepidoptera Rhopalocera collected by W. J. Burchell in Brazil, 1825–1830," *Annals and Magazine of Natural History,* s. 7, **13** (1904): 305–23, 356–71, and "The collections of William John Burchell, D.C.L., in the Hope Department, Oxford University Museum. IV. On the Lepidoptera Rhopalocera collected by W.J. Burchell in Brazil, 1825–1830. IV. *Morphinae,*" ibid., s. 8, **1** (1908): 33–42. (Sanders's married name and later places of residence were found in the Linnean Society membership lists, *Proceedings of the Linnean Society,* 1910–1911 and 1914–1915.)

73. *Eleanor Ormerod, LL.D. Economic Entomologist. Autobiography and Correspondence,* ed. Robert Wallace (New York: E. P. Dutton, 1904); Robert Wallace, "Ormerod, Eleanor Anne," *DNB,* Second Supplement, vol. 3, 53–4; obituaries, T. P. N[ewman], "Eleanor A. Ormerod, LL.D.," *Entomologist,* **34** (1901): 235–6, anon., *Entomologist's Monthly Magazine,* **37** (1901): 230; W. F. K., "Miss Eleanor Ormerod," *Nature,* **64** (1901): 330; Virginia Woolf, "Miss Ormerod," *The Dial,* **77** (1924): 467–74; L. O. Howard, "A history of applied entomology," *Smithsonian Miscellaneous Collections,* **84** (1930): 1–545, on pp. 221, 371–2; J. F. McDiarmid Clark, "Eleanor Ormerod (1828–1901) as an economic entomologist: 'Pioneer of purity even more than Paris Green'," *British Journal for the History of Science,* **25** (1992): 431–52.

74. W. F. K., "Eleanor Ormerod."

75. *General Index to Eleanor Ormerod's Annual Reports of Observations of Injurious Insects, 1877–1898* (London: Simpkin and Marshall, 1899).

76. Ormerod to Robert Wallace, August 18, 1892, *Autobiography,* p. 281 (phrase in square brackets added by Wallace).

77. See Clark, "Eleanor Ormerod," pp. 439–40.

78. "Proceedings of the Council. Wednesday, March 1, 1893," *Journal of the Royal Agricultural Society of England,* **4** (1893): xxxvi–xl, on p. xxxviii. There seems little doubt, however, that Ormerod, although she expected fair recognition for her work, was not interested in collecting salary money. When, in the last year of her life, a proposal was made that she be awarded a government pension, she reacted swiftly, pointing out to the persons responsible that such a grant would be quite inappropriate for a person of her wealth. (Ormerod to Robert Wallace, April 1, 1901, *Autobiography,* p. 322). Despite her progressive approach to scientific matters and her national standing in a new field of applied biology, she remained traditional and conservative in much of her thinking; she never, for instance, supported the aims of the women's movement.

79. Ormerod, *Autobiography,* p. 68.

80. Ibid., p. 225. This was not the first attempt at establishing courses in agricultural education at Oxford; the Sibthorpian Professorship of Rural Economy was already in existence, although funds for its support were scarce. Agricultural education courses for candidates for posts in the Indian Civil Service and later the Colonial Service had been established at Oxford in the early 1880s, but these attracted few students—see F. B. Smith, *Florence Nightingale. Reputation and Power* (London: Croom Helm, 1982), p. 148.

81. Ormerod to Robert Wallace (undated), *Autobiography,* pp. 78–9.

82. Difficult as it may be in the 1990s to accept the introduction of arsenical sprays as a major step forward in horticultural practice, it should not be forgotten that it took biologists half a century to fully realize the dangers associated with chemical insecticides. Ormerod followed an American lead in one other area—an odd campaign she took up in 1897 for the control of the house sparrow. Although the evidence, from field observation surveys, that sparrows consumed significant quantities of grain crops was weak, she contrived to ignore this and instead let herself be influenced by American anti-sparrow literature.

83. As she remarked in a letter to Claude Fuller, entomologist to the Department of Agriculture in Natal, all her work was brought out at a financial loss. In her experience, farmers and horticulturalists, although eager for information, were very reluctant to spend money on books. She never asked publishers to take her manuscripts as speculation, but simply had them act as her agents (Ormerod to Fuller, November 5, 1898, *Autobiography,* p. 257).

84. See Ormerod to James Fletcher, February 2, 1891, ibid., p. 208.

85. Ormerod to James Fletcher, September 2, 1889, ibid., p. 198.

86. Ormerod to Robert Wallace, February 25, 1900, ibid., p. 288–9.

87. Clark, "Eleanor Ormerod," p. 432.

88. Wallace, in *Autobiography,* p. 97.

Chapter 3

MUSEUM TAXONOMISTS TO MORPHOLOGISTS AND EMBRYOLOGISTS: WOMEN ZOOLOGISTS

Papers in zoology, including marine zoology, constitute about 12 percent of the pre-1901 articles by women scientists listed in the Royal Society *Catalogue*. Almost half were written by Americans, 15 percent by British workers, and 13 percent by Italians. Smaller but still significant contributions came from women in Russia, Germany, France, Sweden, and Norway.

American women

Among the fifty-five American contributors, three of the most productive (Rathbun, Bush, and Richardson) were museum workers, five (Smith Eigenmann, Sampson Morgan, Foot, Strobell, and Crotty Davenport) did their most significant research without formal institutional positions, fifteen were college faculty members (for the most part in women's colleges), three carried out research as junior instructors at coeducational universities, and one is known to have held a position with the U.S. Department of Agriculture. Most of the others would appear to have been active in zoological research only while they were students or for a short time thereafter; several went on to careers in school teaching or medicine.[1] The museum workers and independent researchers contributed almost half of the American publications; their exceptional productivity was due in large measure to the fact that many of them worked in marine zoology, an area in which there was tremendous collecting activity during the last two decades of the century.

One of the best known of the marine zoologists was MARY JANE RATHBUN[2] (1860–1943), who worked at the U.S. National Museum in Washington, D.C. for more than fifty years. Having entered the field of marine invertebrate taxonomy at this time of intensive collecting by both government and private expeditions, she had unprecedented opportunities. The vast quantities of new and undescribed organisms arriving at the museum generated a huge amount of work in classification, and Rathbun's contributions to the task were substantial. She wrote 158 research papers and monographs over the course of her long career and became an internationally recognized authority on decapod crustacea.

Born in Buffalo, New York, 11 June 1860, she was the youngest of the five children of Charles Howland Rathbun, a descendant of English colonists who came to Massachusetts in the seventeenth century, and his wife Jane (Furey). In 1836 her grandfather Thomas Rathbun, a stone mason, settled in Buffalo where he and a partner successfully operated several large stone quarries; in due course Thomas Rathbun's business passed to his son Charles. Following the death of her mother when she was a year old, Mary, a shy and reserved child, was brought up by an elderly nurse. She was educated in Buffalo public schools, graduating from the Central High School in 1878. Her special skills in English were later to be reflected in the clarity of her scientific writing.

Perhaps the most important influence in the shaping of her early life was her brother Richard, eight years her senior. As children they searched for fossils in the family quarries, and for both of them their childhood fascination with natural history led to long careers as zoologists. In 1881 she accompanied Richard, then a scientific assistant with the U.S. Fish Commission, to the Marine Biological Laboratory at Woods Hole, where she helped him with the sorting and cataloging of the fauna collections being brought in by the commission's survey ships. Richard's superior, Spencer Baird, head of the Fish Commission and curator of invertebrates at the U.S. National Museum, was so impressed with Mary's work that in 1884 he offered her a full-time position as a clerk at the museum. She was assigned to the new division of marine invertebrates and given responsibility for organizing and cataloging museum collections. Two years later, at the age of twenty-six, she was appointed copyist, becoming a regular member of the museum staff with responsibilities for record keeping and cataloging. In 1898 she was formally named second assistant curator, and in 1907 assumed responsibility, as assistant curator, for the marine invertebrate division of which her brother was the formal head. She remained assistant curator until 1914.

For much of that time she worked "alone and unaided" to build up and develop the Division of Marine Invertebrates.[3] The record-keeping system that she introduced not only remained in use for decades but was copied by other divisions within the museum. In addition, the systematic catalog of the thousands of specimens handled by the division was established by her, and the hundreds of cards, made out in her elegant script over the many years before a typist could be employed, constituted the division's core reference materials for decades. In effect she functioned as curator of the marine invertebrate division for much of her career, her brother's time being largely taken up by other duties. By 1914, thanks in large part to her enterprise and energy, the work of the division had increased so much that the help of an assistant had become essential. Over the years she had received a steady government salary that had risen to $1,680 a year by 1910;[4] when the necessary funds were not forthcoming from the government for the support of an assistant, she gave up her own salary for the purpose and Waldo Schmitt was hired. She was then fifty-four. For the next twenty-five years, until her retirement in 1939, she worked as a full-time honorary associate in zoology.

From 1891 on she brought out work at an impressive rate. Her first paper, a study of the genus *Panopeus,* was coauthored by her superior in the marine invertebrate division, James Benedict, but thereafter she published largely without collaborators. Her research centered on recent and fossil decapod crustacea, shrimps, crabs and their relatives, and concerned mainly description, classification, and taxonomy. For many categories she clarified and standardized the classification and was influential in establishing principles of nomenclature.

Among her works are several monumental accounts of marine and freshwater crabs, all of which appeared after 1914, when, with her change of status at the museum, she was able to spend more of her time on research. Included are her comprehensive treatise, *Les Crabes d'Eau Douce, Potamonidae,* published by the Paris Museum,[5] *Crustaceans* (1904), a 337-page monograph reporting on specimens collected by the Harriman Alaska Expedition, and four studies on marine crabs brought out as monographic bulletins by the National Museum.[6] About eighty of her total of 158 publications appeared after she had given up her salaried position. In 1917, when she was fifty-seven, she received a Ph.D. from George Washington University following her presentation of a dissertation reporting one of her comprehensive marine crab studies (*The Grapsoid Crabs of America*). Among her many other investigations, two were especially outstanding as fundamental palaeontological works that provided the descriptive basis for much of the later work in the areas covered; these were *The Fossil Stalk-eyed Crustacea of the Pacific Slope of North America,* published by the U.S. National Museum in 1926, and *The Fossil Crustacea of the Atlantic and the Gulf Coastal Plain,* issued as a special paper by the Geological Society of America in 1935. She also contributed to other works, including L. W. Stephenson's *The Cretaceous Formations of North Carolina* (1923) for which she supplied a chapter on decapod crustacean fossils.

Over the years her advice was frequently sought by American and foreign colleagues in marine zoology and her correspondence was extensive. She received several honors, including an honorary M.A. from the University of Pittsburgh in 1916 and a star in the 1921 edition of *American Men of Science.* She had strong interests in music and the theater, and except during her last four years, when her health failed, she regularly attended the Washington concerts of the Philadelphia and Boston orchestras. She died in Washington, D.C., 4 April 1943, shortly before her eighty-third birthday. Her extensive scientific library went to the museum, and she left $10,000 in memory of her brother to the Smithsonian Institution to further work on Crustacea. The last major publication on which her name is listed as coauthor, *Mesozoic Fossils of the Peruvian Andes,* appeared four years after her death.

Rathbun's fellow marine zoologist, KATHARINE BUSH[7] (1855–1937), the first woman to be awarded a Ph.D. in zoology by Yale University, spent her working life as an assistant in the Peabody Museum at that institution. Her work, like Rathbun's, concerned the classification of materials brought back by the many large scientific expeditions of the time. Over the course of her thirty years at Yale she became a recognized expert on the systematic classification of marine invertebrates, particularly mollusks and marine worms.

The daughter of William Henry Bush, an accountant, and his wife Eliza Ann (Clark), she was born 30 December 1855, in Scranton, Pennsylvania. Shortly after her birth the family moved to Connecticut where her mother's relatives had been prominent citizens for many generations. She was educated in public and private schools in New Haven but never attended a college. In 1879, at the age of twenty-three, she became assistant to Addison Emery Verrill, professor of zoology at Yale, curator of the Peabody Museum and an authority on marine invertebrate zoology.

Her first work consisted of completing catalogs and writing labels; then she progressed to arranging and cataloging collections. In 1883, only four years after she began to work with Verrill, she published her first scientific paper, a catalog of mollusks and echinoderms collected off Labrador. Her publications continued to appear throughout the 1880s and 1890s, with nine more very substantial papers (three of them coauthored with Verrill) on the classification of fauna, particularly mollusks, from the Atlantic coasts. She also assisted in ıevising zoological definitions for the 1890 edition of *Webster's International Dictionary,* though formal recognition for this work went to Verrill.

In 1885 she began taking classes as a special student at the Sheffield Scientific School at Yale, though for the most part the training she received from Verrill substituted for an undergraduate education. She was enrolled in Yale graduate school from 1899 until 1904 and again in 1908–1909, and is thought to have broadened her background by taking nearly all of the available graduate-level courses in biology. Her doctoral dissertation, presented in 1901, concerned the classification of annelids brought back from Alaska by the 1899 Harriman Expedition; in it she described no fewer than three new genera and sixteen new species. A much expanded version of this work (to include fifteen new genera and forty-eight new species) was published in 1905 as part of the reports of the Harriman Expedition.[8] She continued to publish throughout the first decade of this century, her last work, on fauna from Bermuda waters, appearing in 1910, the year Verrill retired from the curatorship of the museum.[9] By then she had for many years been responsible for the general supervision of the specimens kept in alcohol, a substantial fraction of the museum collection. In 1911 and 1912 she worked on enlarging and rearranging the museum exhibits of both vertebrates and invertebrates, but shortly thereafter became ill and was unable to continue the undertaking. The annual report of the museum for 1914 noted that, because of her continuing absence, two additional assistants had to be engaged to take her place.[10]

Katharine Bush led a very quiet life, dedicated almost exclusively to her scientific work. She had several close relatives in the Yale community, however. Her sister Charlotte Eliza, with whom she shared rented rooms in New Haven in the 1880s, was a special student at the Sheffield Scientific School in 1885–1886 and then assistant librarian in the school. Charlotte later married Wesley R. Coe, who became professor of biology at Yale and curator of the Peabody Museum when Verrill retired in 1910. Another sister, Lucy, was a librarian in the museum and assistant to the paleontologist O. C. Marsh from 1883 until 1908.

Unlike Mary Rathbun, who had a steady government salary throughout the first three decades she spent at the U.S. National Museum, Bush, over the course of her career at Yale, was paid by various short-term federal government grants as these became available. For several years in the 1880s and 1890s she was supported, at least in part, by U.S. Fish Commission and U.S. National Museum funds, but whether she was paid at other times is somewhat doubtful.[11] When she left Yale in 1913 or 1914, she became a patient in the Hartford Retreat (later called the Institute for Living), a neuropsychiatric sanitarium. She died there, 19 January 1937, after twenty-three years of poor health, during the last nine of which she was severely incapacitated.

HARRIET RICHARDSON[12] (1874–1958), the third marine zoologist in this study to work in a major East Coast research organization, collaborated in projects at the Smithsonian Institution for several decades, although her most productive period was the twelve years between 1901 and 1913. The daughter of Charles F. E. Richardson and his wife Charlotte Ann (Williamson), she was born in Washington, D.C. She studied at Vassar College, taking an A.B. in 1896 and an A.M. in 1901.

Her first two scientific papers, descriptions of new marine organisms from Alaskan waters, appeared in the *Proceedings of the Washington Biological Society* in 1897. From then until about 1914 she brought out in steady succession some sixty-five papers and monographs reporting work on the classification of marine invertebrates, mainly specimens in collections brought back by various expeditions. Most of her work was published in the *Proceedings of the U.S. National Museum* or as bulletins and reports of the U.S. Fish Commission. Her study entitled "Contributions to the Natural History of the Isopoda," which appeared in the museum *Proceedings* for 1904, was first presented as a Ph.D. dissertation to Columbia University in 1903. A lengthy work, it included sections on isopod crustacea brought back by American expeditions to the western Pacific as well as material in collections from American Pacific coast waters. Notable among her other works are her 727-page *Monograph on the Isopods of North America* (1905), issued as a U.S. National Museum Bulletin, and two memoirs on isopods of the Antarctic collected by two French expeditions.[13] In 1913 the museum brought out a single-volume collection of about twenty of her articles and monographs, many of them major reports, from the ten-year period 1903 to 1913. About that time she married, becoming Mrs. Searle, but she remained associated with the Smithsonian until the early 1950s and continued to publish until at least the early 1920s. A number of her later reports were brought out as sections of works by Dutch and French scientists, analyses of materials (mainly isopod crustacea) brought back from their expeditions.[14] She died 28 March 1958, when she was about eighty-four.

Two other early women marine zoologists were Rosa Smith Eigenmann and Elizabeth Hughes, who both studied for short periods at the University of Indiana with ichthyologist David Starr Jordan and his student, Carl Eigenmann.[15]

ROSA SMITH[16] (1858–1947) was the first prominent woman ichthyologist in America. Although active in scientific work for only thirteen years (1880–1893) and for much of that time unaffiliated with any institution, she published sixteen notes and papers of her own and an additional seventeen papers and monographs as collaborative works. Her taxonomic studies, particularly those coauthored with her husband, Carl Eigenmann, were significant contributions to knowledge of the fishes of South America and western North America.

Born on 7 October 1858 in Monmouth, Illinois, she was the youngest of the nine children of Lucretia (Gray) and Charles Kendall Smith, both of whom were natives of Vermont. Charles Smith was a newspaper printer who founded the *Monmouth Atlas,* the first newspaper in Monmouth County. In 1876 he moved his family to San Diego, California. Rosa was educated at the Point Loma Seminary in San Diego and then at a business college in San Francisco.

Having poor health, she spent as much time as she could outdoors and became interested in natural history, especially fishes she found along the San Diego shoreline. She was the first woman to join the San Diego Society of Natural History, and became its recording secretary and librarian. At a meeting of the society in 1880 she presented a paper on a species of fish she had discovered locally. Her talk was heard by David Starr Jordan, who was at the time studying fishes along the Pacific Coast, and he invited her to join his group for the summer of 1880. Encouraged by Jordan, she then spent two years (1880–1882) studying zoology at the University of Indiana and joined Jordan and a group of his students on a summer trip to Europe. During the five years between 1882 and 1887 she published eleven papers of her own and a monograph coauthored with Joseph Swain, a fellow student of zoology (and later president of the University of Indiana). Most of the papers were short descriptive notes reporting new species of West Coast fishes, typically from the San Diego region; the monograph described a collection of fishes, including five new species, from Johnson Island in the North Pacific. These efforts were well-received, and early in 1887 she was appointed joint curator, with H. F. Lorquin, of fishes, reptiles, and radiates at the California Academy of Sciences.[17]

For a short time in the mid 1880s she worked as a reporter on the staff of the *San Diego Union,* a paper managed by her brother and brother-in-law, but in 1887 she married Carl Eigenmann and after that collaborated closely for several years in his research. Immediately after their marriage, the Eigenmanns spent a year at the Museum of Comparative Zoology at Harvard, where both worked on South American fishes. Rosa also studied cryptogamic botany as a special student, and took some time during the summer of 1888 at Woods Hole, working on U.S. Fish Commission materials. On their return to San Diego she and her husband established a small, private biological laboratory. Between 1888 and 1893 they coauthored sixteen publications, including several substantial monographs reporting the results of their studies at the Harvard Museum. Especially notable among the latter were *A Revision of the South American Nematognathi or Cat-Fishes* (1890) and *A Catalogue of the Freshwater Fishes of South America* (1892). Rosa also continued to work on the taxonomy of the fishes of the San Diego region, publishing four short papers between 1890 and 1893.

In 1891 Carl Eigenmann succeeded David Starr Jordan as professor of zoology at the University of Indiana. The family's return to the Midwest marked the end of Rosa Eigenmann's research; from then on, although she continued to edit her husband's manuscripts, nearly all of her time was taken up with family cares. She had four daughters, the oldest of whom was retarded, and a son who became mentally ill. Carl Eigenmann served as dean of the graduate school at Indiana from 1908 until his retirement in 1926, but during that time his wife took almost no part in university or community life. They returned to California in 1926, one year before Carl's death. Rosa lived to the age of eighty-eight, dying in San Diego, 12 January 1947.

ELIZABETH HUGHES[18] (1853–1935) coauthored three substantial articles in marine zoology in the late 1880s. The granddaughter of David H. Maxwell, the "father of Indiana University," Elizabeth was born on 1 November 1853 in Longport, Indiana. She took an A.B. at the University of Indiana in 1875, and about ten years later returned as a graduate student in zoology (M.S., 1886). Two articles she wrote with Jordan, reviews of the Julidinae and of the genus *Prionotus*, and a third review of North American fishes prepared in collaboration with Carl Eigenmann, appeared in the *Proceedings of the U.S. National Museum* in 1887 and 1888.

Unlike Smith Eigenmann, Hughes published no further work in biology after her student days. For a few years she taught at the Girls' Classical School run by Mary Wright Sewell in Indianapolis, but soon after the turn of the century went to Palo Alto, California, where for a short time she was associated with a girls' preparatory school. She probably came to know about this institution through Jordan, then president of Stanford University. He had given strong support for its founding, despite concern in some quarters that the establishment of such schools would result in too many women seeking to enter Stanford. Hughes was principal of the school, then known as the Harker-Hughes School, during its second year (1903–1904), and her sister Florence Hughes served as business secretary. The institution survived the Depression and many years later merged with the Palo Alto Military Academy. Elizabeth Hughes stayed in Palo Alto only until 1905, however. She died 2 October 1935, at age seventy-one, in El Paso, Texas.

Among the women zoologists who held college faculty positions, those who spent at least part of their careers at Wellesley College form a particularly striking group. Five Wellesley women—Emily Nunn, Mary Willcox, Caroline Thompson, Alice Robertson, and Mary Bowers—all made contributions to the zoological literature, the work of Willcox and Thompson being particularly notable. A sixth Wellesley instructor, Agnes Claypole, although she was on the faculty for only two years, also carried out creditable research while at the college.

From 1878, three years after Wellesley's opening, until 1881, the zoology department was directed by EMILY NUNN (1843–1927).[19] Emily was born in Barton Mills, western Suffolk, England, on 2 July 1843, the third of the eleven children of Miriam Towler (Kendall) and Charles Robert Nunn. Nunns and Kendalls had lived in the region for many generations, the Nunns (farmers and artisans) going back to medieval times. Charles Nunn adopted his wife's Baptist beliefs—a decision which led to the family's moving to the United States in 1852. Settling in northern Ohio, they took up farming, first on rented land near Medina, and later on their own property near Peru, twelve miles south of Norwalk. They were a close-knit, hard-working family, and the children, girls as well as boys, earned money from an early age.

In 1867, Emily, her sister Miriam and younger brother Lucien Lucius (L. L.) moved to nearby Cleveland, where Lucien attended a private academy while the sisters taught in public schools. Two years later Emily joined another sister, Ellen, in Dresden in order to learn German; by 1871 she had returned to the United States and was earning a good income (up to $1,500 a year) teaching German in Chicago.

What awakened Emily Nunn's interest in biology is not recorded, but in 1874, at age thirty, she went to Newnham College Cambridge where she studied for a year and a half under the guidance of pioneering physiologist Michael Foster. During this period she also spent some time at the University of Zurich, unofficially attending zoology lectures. On her return to the United States she became active in Boston area scientific circles, particularly the Boston Society of Natural History (of which she became an associate member in 1881). This brought her to the notice of Wellesley's founder, Henry Durant, then recruiting female faculty members. With Durant's approval, she went to Johns Hopkins University in the autumn of 1877 for further laboratory training, but, dissatisfied at being admitted only to a teachers' class in physiology, left after one semester and went back to Cambridge. Working under the direction of Michael Foster and embryologist Francis Balfour, she joined an ongoing investigation then being pursued by several British physiologists into the specific mode of action in tissue of a variety of metals and salts. Her first paper, a report on changes in frog epidermis brought about by metals, appeared in Volume 1 of the *Journal of Physiology* (1878–1879).

During her first year at Wellesley (1878–1879) she offered a year's course in zoology, and quickly elaborated this to a two-year biology program, with an emphasis on laboratory work and a choice between physiology and comparative anatomy in the second year—at the time an ambitious plan for a small women's college. In the summer of 1879 she continued her research activity, attending (as the only woman) a small Johns Hopkins University program, the Chesapeake Zoological Laboratory at Crisfield, Maryland, and Fort Wool, Virginia. Her microscopical study of the fertilization and early embryology of a small, primitive marine organism, the ctenophore *Mnemiopsis,* was presented at a Hopkins Scientific Association meeting.

In 1881 she was dismissed from Wellesley, largely, it would seem, because of her by then undisguised agnosticism and her occasional eccentric behavior (she was later to undergo periods of treatment for mental illness). She went back to England later that year, joining Thomas Huxley at the Royal College of Science, South Kensington. Her study of the development of the enamel layer in the teeth of vertebrates appeared in a lengthy paper in the *Proceedings of the Royal Society* in 1883.

Despite the shortcomings that made her unacceptable as a women's college faculty member, there is little doubt that Emily Nunn was considered a capable zoologist: with strong recommendations from both Huxley and Michael Foster, she was awarded occupancy of the Cambridge Table at the Naples Zoological Station in 1882, the first woman and the second American to go there. She had hoped to stay for two years but again ran into difficulties, thoughtlessly antagonizing the station's founder, Anton Dohrn; she left after six months. Her report of the station's unique facilities, the materials, specimens and equipment available, and some of the work being done there, appeared in the first volume of *Science* (1883); three years later, in *Century* magazine, she published a fuller, well-illustrated description, explaining the importance of the station for marine biological research at the time and emphasizing its international tone.

In the summer of 1883, while working at Alexander Agassiz's marine biological laboratory at Newport—a forerunner of the Woods Hole laboratory—she met Charles Otis Whitman, then an associate at the Museum of Comparative Zoology, Harvard. Her marriage to Whitman that autumn (at age forty) effectively marked the end of her scientific career.[20] (She was, however, elected to corporate membership in the Boston Society of Natural History in 1884). The elder of her two sons, Francis, was born in 1887, the second, Carroll, in 1890.

Charles Whitman was a notably productive and hard-working zoologist, pioneering in many of his undertakings. Director of the Woods Hole laboratory from 1888 and head of the biology department at the newly opened University of Chicago from 1892, he became more and more immersed in research, administration, teaching and editing. He used much of his own income to support his professional work and also accepted substantial donations from his brother-in-law, L. L. Nunn. (Nunn by then had accumulated considerable wealth from western mining and electrical power transmission projects, and was a strong and generous supporter of several educational undertakings. In recognition of the support he gave Woods Hole he was made a trustee of the laboratory in 1897.) The financial entanglements and Whitman's lack of responsibility for his family's material well-being finally led to bad feeling; by about 1907 Emily and her husband were apart for long periods, and L. L. Nunn had assumed much of the financial responsibility for his sister and her two sons. Emily outlived her husband by seventeen years, dying in Ithaca, New York, 19 October 1927, at age eighty-four.

The major part of the task of establishing a zoology department at Wellesley was undertaken by Emily Nunn's successor, MARY WILLCOX[21] (1856–1953), a notable but largely forgotten malacologist and naturalist, who served on the faculty for twenty-seven years, from 1883 until 1910. Born in Kennebunk, Maine, on 24 April 1856, she was the oldest of the three children of Dr. William Henry Willcox, a congregational minister, and his wife Anne Holmes (Goodenow). The Holmes-Goodenow family was well known in Maine. Mary's great-grandfather, John Holmes, was the first U.S. senator from Maine, whose separation from Massachusetts he did much to secure, and her grandfather, Daniel Goodenow, served as a justice of the Maine Supreme Court. Various members of her family had important connections with Wellesley in its early years: her father served as a trustee from 1886 until 1904, and her aunt, Mrs. Valeria Stone, after whom Stone Hall residence was named, donated substantial funds.

The Willcox family moved to Reading, Massachusetts, soon after Mary's birth, and she attended the State Normal School in Salem. After graduating in 1875 she taught for a year at Frederick Female Seminary (later Hood College) in Frederick, Maryland, and then for two years at Charleston High School in the suburbs of Boston. She spent the summers of 1877 and 1878 at Alexander Agassiz's marine laboratory at Newport, where she clearly impressed Agassiz. He suggested to Henry Durant that she was a suitable candidate for the position of professor of biology at Wellesley, provided she had some additional preparation for the job, preferably study in Europe. Durant accepted Agassiz's suggestion and Willcox accepted the offer of a position. She spent three years (1880–1883) at Newnham College Cambridge, and attended lecture courses by Michael Foster, Francis Balfour, neurologist Charles Édouard Brown-Séquard, and anatomist Joseph Lister (nephew of the famous surgeon Sir Joseph Lister). Laboratory work was of special importance to her and she fitted in a full program. A keen student of languages as well, she spent her vacations in Germany and at the Collège de France in Paris, where she heard lectures by histologist Louis Antoine Ranvier and physiologist Charles Albert François-Frank.

On her return to Wellesley as professor in 1883 she was assigned a zoology laboratory and lecture room fitted up in a basement area under the college gymnasium. During the two years between Emily Nunn's departure in 1881 and Mary Willcox's arrival in 1883, very little course work in zoology had been offered.[22] Willcox, at the age of twenty-seven, faced the task of building a department essentially from the ground up, despite Nunn's preliminary efforts. Financial support for the job was minimal. Her experience at Cambridge, however, had made her aware that good work could still be done in cramped conditions[23] and she began her task at Wellesley with marked energy and enthusiasm.

Well-prepared after three years of transatlantic study, she introduced innovative teaching methods, and placed a strong emphasis on experimental physiology. This area was only beginning to find a place in the regular curricula of American institutions. At most women's colleges courses in descriptive human physiology had been introduced early in the form of lectures on health, or hygiene, taught by the college physician.[24] Willcox, however, had a clear idea of the need to establish physiology at Wellesley as an experimental laboratory science, closely linked to zoology. As a consequence, her department, for a time, was one of the foremost in the country (a circumstance that undoubtedly goes a long way toward accounting for the presence on the early Wellesley faculty of a remarkable collection of productive women zoologists). She wrote as follows about her early course offerings:

> Our course in the anatomy of the cat, introduced as a part of my plan for a good pre-medical course, was copied from one in England and was the first in New England if not in the country. Our course in embryology was far in advance of anything in Harvard or Yale I was told by visitors from those colleges. When I opened [in 1887] a year's laboratory course in physiology Harvard Medical School offered laboratory work in that subject *for one afternoon in the year.* This I know from Professor Porter of the Medical School, who generously helped me in preparing my course.[25]

The course on the anatomy of the cat became famous. Willcox also conducted fieldwork in both freshwater and marine invertebrate zoology. In addition she offered lectures in "philosophical zoology," which included an examination of Darwinian evolutionary theory, a line of thinking viewed with widespread suspicion and indeed aversion in New England society at that time. She was herself a convinced Darwinian, and when at Cambridge had met Charles Darwin socially through her friendship with his daughter-in-law (see below). Her evolution course was a controversial one.

During the 1880s and early 1890s she carried out research at the Newport marine station and the museum of the Boston Society of Natural History. This resulted in the publication of a monograph on the morphology of marine invertebrates in 1892 and a second work, a very successful guidebook to New England birds, in 1895. The following year she brought out three papers on insect anatomy.[26] She was also careful to keep abreast of new developments in her field, returning to Europe in 1889 to attend lectures (as a private student) at the University of Jena.[27]

By the mid 1890s, however, despite her publications and her outstanding and unusual teaching contributions, she felt she needed to get a formal academic degree. (She had not taken the equivalent of one at Cambridge because she had gone there as a special student to study only biology, and lacked the prerequisites required for entering the Tripos examinations.) Accordingly she went to the University of Zurich in 1896. Her dissertation research, an anatomical study of the gastropod *Acmaea fragilis,* Chemnitz, carried out under the direction Arnold Lang, appeared in the *Jenaische Zeitschrift* in 1898. She received a Ph.D., with honors, the same year. She was then forty-two.

After a six-week stay at the Naples Zoological Station, where she worked at the American Woman's Table, Willcox returned to Wellesley and resumed her responsibilities as head of the zoology department. She had in mind the amalgamation of the zoology and botany departments into a broad-based department of biological sciences, a vision very much in line with the ideas of a number of the younger biologists of her time. However, she ran into stiff opposition from Susan Hallowell, head of the botany department and one of the college's most senior faculty members.[28] Hallowell had plans of her own, and was in the process of laying the foundations that would make botany one of Wellesley's strongest and best-equipped departments. Willcox did not press her case, and about 1906, when the ambitious and successful Margaret Ferguson succeeded as head of the botany department, Willcox again agreed to measures that helped the latter at the expense of her own department.[29] Given her early record of introducing some of the most innovative courses in anatomy and physiology offered in the country, the constraints she accepted, though unavoidable in the absence of a compromise with the already strongly established botany department, seem unfortunate. Perhaps, however, the courses she wanted to teach were somewhat too far in advance of the usual curriculum in zoology at the time to be fully acceptable in a small New England women's college; or perhaps she was simply reluctant to become involved in a faculty dispute.

She continued to teach until 1910 and was particularly active in assisting women students to enter medical schools. In addition, for about seven years after her return from Zurich, she kept up her gastropod research and published six more papers dealing with a variety of topics, including anatomy, histology, systematics, and behavior.[30] Her 1906 paper on the anatomy of *Acmaea testudinalis* was introduced as the first part of a projected monograph. However, she published no further scientific work. One might wonder if her differences with the college administration about the direction in which her department should go had not by then sapped much of her enthusiasm and perhaps drained her energy as well. Nevertheless, although her studies on the Acmaeidae family lasted for only eight years, her skill and thoroughness were such that her papers were cited and referred to for decades.

She was a fellow of the AAAS and a member of several zoological societies. She retired in 1910 at age fifty-four because of poor health. Despite increasing difficulties from rheumatism, she remained active for more than three decades in a number of women's organizations. From 1914 until 1919 she was chairman of the Committee on Americanization set up by the Massachusetts Federation of Women's Clubs. Her *Immigrant's Guide to the United States* was published in 1913. In the 1940s, when she was over eighty, she served as director of the League of Women Voters of Newton, Massachusetts. She was especially interested in race relations, and from 1940 until 1945 chaired a committee of the Boston Twentieth Century Association, which was studying the subject. She died on 4 June 1953, at the age of ninety-seven, at her home in Pocasset, on Cape Cod. The lapse of forty-three years between her retirement from academic work and her death meant that few obituaries noting her scientific contributions were written. Her notable research on gastropod anatomy and morphology has been largely overlooked by historians of women in science, despite the 1977 article by David Lindberg, which made clear its importance.

Mary Willcox's reminiscences of her stay at Newnham College are appended below. Her account is of interest not only for the personal story she tells, but because of the prominence of Newnham College in the British parts of this survey. She wrote as follows:

> In the fall of 1880 when the time came for Newnham College to open, Dr. Trumbull, my escort across the Atlantic, took me from Liverpool to Cambridge and established me there. I will try to picture what the college was like in those far-away days, so different from what I saw on my recent visit fifty-five years later.
>
> Although Girton College had been established several years earlier, there seemed to be need of one nearer the University. Professor Henry Sidgwick, who was keenly interested in improving the opportunities for women's education, was the leader of a little group which induced Miss Anne Clough to open Merton House to a few of the hungry women who were to make Newnham College so distinguished. They were older than the ex-school girls who came up in after years, as their brothers did, for further educational advance and they knew better how to value the opportunities for which they had longed; their maturity made them an interesting set. I did not see as much of them as I would have if I had been a resident of South Hall, but I can never regret an assignment which made possible a closer acquaintance with Mrs. Sidgwick.
>
> South Hall being full to overflowing, the next step was a companion building across the street. No second Miss Clough was available but the head of the new North Hall must be a person of distinction. Professor and Mrs. Sidgwick had been married for four years; so, as there were no children, the problem was solved by their coming to live in tiny quarters in North Hall—a bedroom, a small study for Mr. Sidgwick and a large combined living room and office for Mrs. Sidgwick, who became Vice-Principal.
>
> As I arrived, North Hall was just opening; I remember that we students, at our first breakfast, had to carry down the chairs from our bedrooms. The dining hall had at one end a "high table" at which Mrs. Sidgwick sat and to which different students were invited from the four tables which filled the rest of the room. Mr. Sidgwick usually sat at her right and it was a joy to hear his talk, like a mountain stream, full and sparkling, pouring along regardless of the nature of its banks. He had a very serious stammer; I recall the remark of one of his friends, "When the words don't come, you involuntarily finish the sentence to yourself but always find when he completes it that he has made it better than you could." He had a fine vein of humor. I remember when I was visiting at their house afterwards and there had been a little dinner party; after which Mr. Sidgwick leaned against the mantel and, balancing himself on one foot upon the coping about the hearth, kept his wife and me intent as he discussed the mal-adjustment of the universe evidenced by the fact that when your dress shoes begin to be comfortable they cease to be presentable.
>
> As I had come to England for a definite purpose, it seemed wiser to take only the subjects that would further my plan. That meant that I could not enter for a tripos—the examination that crowned and tested the three or four years of work. A preliminary to entrance for the tripos was a successful passing of the "little-go," an examination in classics, mathematics and Paley's Evidences of Christianity. My time was spent entirely on anatomy, biology and physiology. About five years earlier Huxley had succeeded in introducing laboratory work into his course at South Kensington. Dr. Michael Foster, dubbed by Huxley "the Archangel," had been one of his assistants and now was Prelector in Physiology at Cambridge. To him I had written at the suggestion of Mr. Agassiz; he justified his name by his kindness to me. It was under his guidance that Huxley's "Elementary Biology" was used in our little stone-floored laboratory at Newnham. I still quiver with cold as I remember those raw days in the laboratory barely tempered by a little grate fire in one corner.
>
> I had a course in the anatomy of invertebrates under Mr. Frank Balfour, brother of Mrs. Sidgwick, and one in the anatomy of vertebrates under Mr. Joseph Lister, a nephew of Sir Joseph Lister, who hoped to follow in his uncle's footsteps. The days were all alike—lectures and laboratory work, varied by a five-mile bicycle spin through Grantchester and Trumpington or, at the proper season, through the Madingly woods for primroses. As Newnham is about fifteen minutes' walk from the center of Cambridge we were in the midst of open country with hedgerows all about. In the early evenings we had our feeble imitation of the Cambridge "Union," or a colorful dance. Those were the days of "high art," when the vogue among initiates was for the beautiful silks and velvets to be had from Liberty or from the theatrical supply shop of Burnet. Instead of the prevailing fashions we copied the long graceful lines of costumes in old paintings. I do not expect to see again such rich harmonies of color and line as now come before my eyes.
>
> The students in North Hall, as I look back, seem inferior to those I knew later at Wellesley College. I fancy most of the Newnham group intended to become teachers; if so they did not represent as high a social class as our students did. The time had not yet come for women to take up a vocation outside of the home unless they were compelled to do so. Apart from the casual companionships arising from similar courses I made no friends, being, as one of them said of me, "not a girl but a woman." [She was then twenty-four.]
>
> One of my housemates in North Hall was Amy Levy, a young and gifted Jewish girl. I did not see much of her but I recall a rare confidence in which she gave me a picture of her family, overflowing in numbers and cramped into small quarters by limited means. After one year, she did not return. During that year she published a tiny volume of verse, "Xantippe and Other Poems." The title poem was a soliloquy, a wild protest against the limits of a woman's life and against her husband's estimate of her. She had gained some recognition as a poet and all who had known her were shocked some years later when they heard that she had taken her own life . . . [A poem by Amy Levy that was printed in Willcox's manuscript has been omitted.]
>
> The only people who meant much to me were Mrs. Sidgwick and Miss Gladstone. Mrs. Sidgwick I worshipped. She was always remote, as I suppose happily married people are wont to be, and one felt in her the long line of privilege behind her as the sister of Arthur Balfour, then in Parliament and later Prime Minister. She was like an exquisite alabaster vase with the soul

shining through. I think she came to care for me partly because she divined how very much I cared for her, but we never had the personal chitchat common between friends who are equals.

It was during my stay at Newnham that the Society for Psychical Research was formed and we talked often of it and its possibilities. I visited her twice later, once in her Cambridge home in the outskirts and once in her Newnham quarters after she became Principal and had accommodations for guests.

Miss Gladstone was her secretary, and in training to succeed her. She was an open hearted woman, always eager to talk not only about the father whom she adored but about all her relatives and even about family problems. She was younger than Mrs. Sidgwick and, while never forgetting her position, she associated with the students much more on terms of equality. I always felt that knowing her enabled me to understand the hold of Mr. Gladstone on his constituents. She was aloof to anyone whose mind did not go along with hers in her worship of her father, so sure he was right that she had no comprehension of another point of view.

It was a tragedy for her when she was recalled from the Vice Principalship which she adorned to take the place at home of her older sister Mary, a very brilliant woman who had married one of the Hawarden curates years younger than herself. Both she and her husband felt after a few years that he could never have a fair chance so long as he was under Mr. Gladstone's wing and her departure meant that the other daughter must come home to relieve her mother. It meant not merely giving up Newnham but going to a home where she was constantly contrasted with Mary whom her parents and others considered her superior. After the break-up of the family, she took another headship, but it was not successful, and when last I saw her she was living in London alone with her memories.

I must not fail to mention an outside association which meant much to me. A student named Alice Johnson was also doing zoology. She was the fourth daughter and one of eight children of the master of a private school for children of dissenters. No more words are needed to describe the family—intelligent, educated but unpolished, and always a bit assertive—"just as good as Church people"—or shadowed by realizing that other people looked down upon them as "chapel." It was the sort of attitude that had its nearest American parallel in the attitude of mulattoes toward whites.

The Johnsons held open house on Sunday afternoons and through them I became acquainted with a cross section of University life which otherwise I should never have seen. They were not the cream but were worth knowing—youngsters with ideas on all kinds of subjects.

Some of my happiest late afternoons were spent at the home of Horace Darwin. His wife, only a bit older than I and a most sweet woman, had divined that I was lonely and made me free of her house. At the time between tea and dinner I used often to slip down and have a romp with the little Erasmus, then about two years old. I remember calling there later when he must have been eight or nine. He took me out to see his pigeons and showed me the pride of his heart—a beautiful feather. "I had a prettier one," he said, "but I gave it to my sister because she is a girl."

It was through that acquaintance that I had an opportunity of meeting Charles Darwin. He did not come to Cambridge often during term time because he feared the strain of seeing his many friends. During my three years at Newnham, he made one exception to his rule and Mrs. Darwin asked me to dine with them. I was too careless to write an account of it, but I do recall his humorous account of coming into his dining room one night after dinner to find his big sons testing the current statement that a wine glass falling to the floor squarely on its rim would not break. The experiment had not been successful. I remember also a remark that showed his modesty. He had been over to see Mr. Gladstone "and he greeted me just as if he had been one of us." And another answer, "Yes, I have patience."

When he sat down beside me, he asked, as everybody in England had a fashion of doing, if I knew any one in some far away place; he chose Saint Louis. My answer—that Saint Louis was farther from my home in Boston than London was from Constantinople, amused him greatly. And then—it must have been about eight o'clock—his wife said, "Charles, I think you had better go to bed," and he disappeared.[31]

Among the noteworthy group of productive women zoologists who joined the Wellesley faculty during Mary Willcox's tenure was CAROLINE THOMPSON[32] (1869–1921), who became well known for her pioneering work on the biology of social insects, particularly ants and termites. Born in Germantown, Philadelphia, 27 June 1869, she was the daughter of Caroline (Burling) and Lucius P. Thompson. After completing her early education, she taught science and mathematics in private schools for girls in Philadelphia for a time in the early 1890s and then enrolled as an undergraduate in biology at the University of Pennsylvania (B.S., 1898). Her undergraduate research in botany, a morphological study of the internal phloem of *Gelsemium sempervirens* (Carolina yellow jessamine) was published as a contribution from the University Botanical Laboratory in 1897; she presented her findings at the first annual meeting of the Society for Plant Morphology and Physiology the same year. Assisted by a fellowship in zoology, she stayed on at the University of Pennsylvania and began graduate studies under the direction of Thomas Harrison Montgomery, then assistant professor of zoology. Her doctoral research on the microanatomy of nemertean worms (ribbon worms, primitive marine chordates) was published in two papers in the *Zoologischer Anzeiger* in 1900 and in the *Proceedings of the Philadelphia Academy of Natural Sciences* in 1901. The work included investigations of the brain and sense organs and comparison of these with the organs of related species.

Immediately after receiving her Ph.D. in 1901 she became instructor of zoology at Wellesley. She was promoted to associate professor in 1909 and professor in 1916. For some years she continued investigations of the brain and neurological structures in heteronemerteans, but following a suggestion of William Morton Wheeler, professor of entomology at Harvard and a specialist on North American ants, she also began studies of the brains of ants and bees. From there she went on to termites, first investigating the brain structure of the different castes of termite and showing that there was very little differentiation between the brains of different castes and none between the sexes, the only marked difference being in the optic apparatus.

She then took up the question of the origin of castes in termite communities, and in her most important paper, published in 1917, demonstrated that castes cannot be produced by social control of the environment and food supply, contrary to an idea still current at the time. In particular, termite communities cannot manufacture "substitute reproductives" (one of the three castes in termite colonies) by special feeding.[33] Her conclusion that caste is determined by intrinsic genetic factors was an important step forward in the understanding of the termite social system. Indeed, it was said to have "revolutionized the attitude taken by students of termites," an attitude that up until then had been "almost entirely anthropocentric."[34]

From 1917 onward much of her work was done in collaboration with scientists in the Branch of Forest Entomology, Bureau of Entomology, U.S. Department of Agriculture. She regularly used the Woods Hole laboratory facilities, although she also visited the Naples Zoological Station. Two papers published in 1919 and 1920 reported joint research with T. E. Snyder of the bureau on the development of castes and on the phylogenetic origin of castes.[35] She was in the process of extending this line of investigation to an examination of the castes of the honeybee when the work was cut short by her death, at Wellesley, at the age of fifty-two, 5 December 1921. Also left incomplete at the time of her death was a collaborative project on what was planned to be a more or less popular book on termites. She was remembered by her colleagues as an inspiring teacher and an able and original investigator who was not constrained or biased by current fashions but struck out into new areas.

ALICE ROBERTSON,[36] (1859–1922) taught zoology and physiology at Wellesley from 1906 until 1919. Born in Philadelphia, she was of Scottish background, the daughter of Janet (Greaves) and James Robertson. She received her higher education at the University of California, enrolling there when she was already in her late thirties. After being awarded a B.S. in 1898 she stayed on as a graduate student, taking an M.S. in 1899 and a Ph.D. in 1902.

Her graduate research, carried out under the direction of William Emerson Ritter, professor of marine biology at the University of California and director of the San Diego Marine Biological Station, involved a study of the embryology of marine bryozoans (genus *Crisia*). The work appeared in a lengthy paper published in 1903.[37] She remained at Berkeley as an assistant, first in courses of hygiene for women (1902–1904) and then in zoology (1904–1906). Her second major publication on bryozoa appeared in 1905.[38] The work was widely recognized and well regarded. The following year she joined the Wellesley faculty as instructor. An excellent teacher who stressed laboratory work in particular, she was promoted to associate professor in 1909, professor in 1912, and for a time served as head of the zoology department. She resigned in 1919 at the age of sixty because of poor health, and moved to Seattle, Washington. After a year's rest she joined a research group at the University of Washington for a time, and then returned to the University of California, where she worked for a year and a half in the Department of Public Health. She died in California, 22 September 1922. Her *Wellesley College News* obituarist remembered especially her wide intellectual interests, vital concern for all matters political, and "ready Scottish wit."

MARY BOWERS[39] was an instructor in zoology at Wellesley from 1899 until 1908, with two breaks for further study. The daughter of Sarah Abby (Berry) and Roscoe L. Bowers, she was born in Saco, Maine, 2 October 1871. She took a B.L. degree at Smith College in 1895 and three years later received an M.A. from Radcliffe. Her substantial anatomical paper on cranial nerve distribution in the amphibian *Spelerpes bilineatus*, published in the *Proceedings of the American Academy of Arts and Sciences* in 1901, reported research carried out in the late 1890s under the guidance of Edward Mark at the Harvard Museum of Comparative Zoology. Following a year (1898–1899) as demonstrator in zoology and botany at Smith, she became instructor at Wellesley, but at the same time continued graduate studies, spending a year at MIT from 1900 to 1901 and two years at the University of Pennsylvania from 1905 to 1907. Her doctoral research, a histological study on the amphibian *Bufo lentiginosus*, was carried out under the supervision of Edwin Conklin.[40] The University of Pennsylvania awarded her a Ph.D. in 1909. She married Robert William Hall, professor of zoology at Lehigh University in South Bethlehem, Pennsylvania, in 1908; they had one daughter. Mary Hall does not appear to have continued scientific research after her marriage. She lived in South Bethlehem until at least 1949.

Nunn, Willcox, Thompson, Robertson, and Bowers, although the length of their scientific careers varied considerably, all spent the major part of their professional lives on the Wellesley faculty. A sixth Wellesley instructor, Agnes Claypole, held college-level teaching positions at three different institutions during her productive early career. She never remained at any one for more than three years, but succeeded in carrying out notable research and was starred in the 1906 edition of *American Men of Science*.

AGNES CLAYPOLE[41] (1870–1954) was the daughter of Jane (Trotter) and Edward Waller Claypole. She was born in Bristol, England, on 1 January 1870. When she was nine her father accepted a position as professor of geology at Buchtel College (later the University of Akron) in Akron, Ohio, and so the family moved to the United States. Agnes had an identical twin sister, Edith, and as children the two were almost inseparable. They were given their early education at home by their parents (their mother was a writer of scientific articles for the popular press) and then they both attended Buchtel College, graduating with Ph.B. degrees in 1892.

Thereafter the Claypole twins went to Cornell University where both were students of Simon Gage, professor of embryology and histology. Agnes's thesis research (M.S., 1894), a histological study on the enteron of the Cayuga Lake lamprey, appeared in the *Proceedings of the American Microscopical Society* for 1894. Although her sister took a teaching position in the zoology department at Wellesley College after completing her studies at Cornell,[42] Agnes continued graduate work, going to the University of Chicago, where she studied under the direction of Charles Whitman and William Morton Wheeler. She was awarded a Ph.D. in 1896—one of the first in zoology given to a woman by Chicago. Her dissertation research on the embryology and oogenesis of *Anurida maritima*, carried out with

Wheeler's guidance, was published in a long paper in the *Journal of Morphology* in 1898.

Immediately after receiving her degree she joined her sister in the zoology department at Wellesley and assisted there during the two years when Mary Willcox was on leave at the University of Zurich. The two lively and sociable young English sisters, blond and athletic, were well remembered at Wellesley.[43] Agnes's histological studies of embryonic stages in the development of arthropods and the Apterygota (1897 and 1899) were carried out there.

She returned to Cornell in 1898, taking a position as assistant in histology and embryology, but two years later joined her father at the Throop Polytechnic Institute (later the California Institute of Technology) where he was then a faculty member. Starting as an assistant in zoology and geology, she was promoted to instructor the following year. However, she resigned in 1903 when she married Robert Moody, an instructor of anatomy at the University of California, whom she had met at Cornell. For several years thereafter she spent her time on family matters and civic work; for some time she served on the Berkeley Board of Education. In 1918, when she was forty-eight, she again took up teaching and for five years was a lecturer in sociology at Mills College, a women's college in Oakland, California.

Even the presence of these able and notably productive women zoologists among the early faculty at Wellesley and the pioneering curriculum developed by Mary Willcox were not enough to compensate for the zoology department's serious deficiencies in laboratory teaching facilities and equipment, and it was unable, in its early days, to develop a strong undergraduate program.[44] Somewhat in contrast, the department established at Mount Holyoke College by zoologist CORNELIA CLAPP[45] (1849–1934) became especially prominent for its undergraduate teaching, although fewer productive women researchers served on its early faculty.

Cornelia was born in Montague, Massachusetts, 17 March 1849, the oldest of the six children of Eunice Amelia (Slate) and Richard Clapp, both of whom were teachers. The Clapps were of English ancestry, descendants of Richard Clapp, one of the original settlers in Dorchester, Massachusetts.

After graduating from Mount Holyoke Seminary in 1871, Cornelia taught Latin for a year at a boys' school in Andalusia, Pennsylvania, before returning to Mount Holyoke as instructor in mathematics and gymnastics. In 1873, along with her colleague and former teacher the botanist Lydia Shattuck, she attended the Anderson School of Natural History run by Alexander Agassiz on Penikese Island, Massachusetts. This school, one of the predecessors of the Woods Hole Marine Biological Laboratory, opened for Clapp a whole new range of interests. Two years later (in 1875–1876) she taught a laboratory course in zoology at Mount Holyoke, the first one offered there. For a time in the early 1880s she studied with William Sedgwick, professor of biology at MIT, and with Edmund Wilson at Williams College. She also joined in collecting trips in the White Mountains, through the southern states and in Europe. The opening of the Woods Hole laboratory in 1888 gave her regular access to facilities for summer work. That year she took a Ph.B., by examination, at Syracuse University, and followed this with a Ph.D. in 1889. While Mount Holyoke underwent the transition from seminary to college in the early 1890s, she had a three-year leave of absence for study at the University of Chicago under the guidance of Charles Whitman, with whom she had already worked extensively during summers. Her research at Chicago, morphological studies on the toadfish *Batrachus tau* reported in the *Journal of Morphology* in 1899, brought her a second Ph.D., awarded in 1896.

On her return to Mount Holyoke she became head of the zoology department. Then in her mid-forties, vigorous, enthusiastic, and original in her thinking, she quickly gained a reputation for innovative, forward-looking teaching techniques. Her emphasis on field and laboratory work, still somewhat uncommon at the time, made her courses memorable to hundreds of Mount Holyoke students. Further, the department she established maintained its outstanding reputation among zoology departments at women's colleges for many decades, through several generations of her successors.[46] She also kept up her own collecting trips, some of them extensive; the last was to the Far East in 1908–1909.

During her whole career she was closely associated with Woods Hole, where, practical and direct in her approach, she had been a popular member of the community right from the beginning. In addition to continuing a certain amount of research, she was lecturer, librarian, member of the corporation, and for many years trustee—in fact, the only woman on the board of trustees then and for some time after. She had a broad vision of what the Woods Hole program should be like, and always supported Whitman in his advocacy of a combination of instruction and research.

She disliked writing, however, and published very little after her doctoral studies. Nevertheless, these constituted a significant contribution to a current debate on a point in early embryological development, namely the relation of the axis of the embryo to the first cleavage plane. Here she directly and successfully challenged the conclusions of earlier workers by showing that in toadfish eggs the first cleavage plane did not in fact determine the direction of the axis. The work brought her a star in the 1906 edition of *American Men of Science.*

Although her teaching and work of department building were perhaps her most outstanding professional contributions, her prominence at Woods Hole indicates her wide acceptance in the biology community at a time when few women received such recognition. Also notable was her election in 1892 to membership in the American Society of Morphology (forerunner of the American Society of Zoologists).[47] She retired in 1916, at the age of sixty-seven, after more than forty years of teaching, but continued to spend her summers at Woods Hole. Winters she passed in Mount Dora, Florida, where she was active in civic affairs. Mount Holyoke gave her an honorary Sc.D. in 1921, and in 1923 named its new science building Cornelia Clapp Hall. She died at Mount Dora, 31 December 1934, at age eighty-five, of cerebral thrombosis.

LOUISE WALLACE[48] (1867–1968), a younger colleague of Cornelia Clapp at Mount Holyoke Seminary and College for about sixteen years, had a distinguished career in teaching and college administration, which spanned four decades.

The daughter of Elizabeth (Riddle) and William Lockhart Wallace, she was born in Newville, south-central Pennsylvania, on 21 September 1867. Throughout the 1890s she spent her summers at Woods Hole where, under the guidance of Charles Whitman, she undertook morphological work on the same organism that Clapp was studying, *Batrachus tau.* Her first paper appeared in the *Journal of Morphology* in 1893. The same year, she joined the faculty of Mount Holyoke where she taught zoology for three years (1893–1896) while Clapp completed her studies at the University of Chicago. Wallace then served as assistant in zoology at Smith College, and at the same time studied for an A.B. degree, awarded by Mount Holyoke in 1898. She returned to Mount Holyoke as an instructor the following year. Her summer research at Woods Hole continued, under Whitman's direction and with Clapp taking a strong interest in the work; she published her second paper, a study on the toadfish egg, in the *Journal of Morphology* in 1899. In 1901 she spent some time at the Naples Zoological Station, and by 1904 had completed requirements for an A.M. degree at the University of Pennsylvania. Her thesis research on spermatogenesis in the spider, carried out under the supervision of Edwin Conklin, appeared the following year.[49] Shortly after that she was promoted to associate professor at Mount Holyoke. Over the next four years she continued work on spermatogenesis and oogenesis in the spider, and was awarded a Ph.D. by the University of Pennsylvania in 1908.

She remained at Mount Holyoke until 1912, when, at the age of forty-five, she went to the American College for Girls in Scutari, Turkey, as professor of biology. Within a year she became dean of the faculty, the post she held for eleven years, throughout the duration of the First World War and until 1924, when she accepted the position of vice president of the college. Mount Holyoke awarded her an honorary Sc.D. in 1919. In 1927, after fifteen years in the Middle East, she returned to Mount Holyoke as professor of zoology, but she stayed for only a year before joining the faculty at Spelman College (traditionally a school for Negro women) in Atlanta, Georgia. She retired in 1931, when she was sixty-four, and spent some time in Palo Alto, California. However, by the 1940s she had returned to her home state of Pennsylvania and was living in Pittsburgh. She died 21 January 1968, in her one hundred and first year.

Like Wallace, EMILY RAY GREGORY[50] (1863–1946), after some years of teaching in a small Eastern college, in her mid-forties took a position at the American College for Girls in Scutari. The daughter of Mary (Jones) and Henry Duval Gregory, she was born in Philadelphia, on 1 November 1863. She studied music at Wellesley College and graduated from the five-year course with an A.B. in 1885.

Becoming interested in zoology, she continued her education over the next decade while also teaching, first at various girls' schools in Philadelphia and Baltimore, and then at Milwaukee College. During the year 1892–1893 she held a fellowship at the University of Pennsylvania and for three summers worked at Woods Hole under the guidance of Charles Whitman. After being awarded an A.M. degree by the University of Pennsylvania in 1896, she went to the University of Chicago, where, supported by a fellowship and then an assistantship in zoology (1897–1899), she studied with Whitman and William Morton Wheeler. Her research, reported in the *Zoologische Jahrbücher* for 1900, concerned the development of the excretory system of native turtles. In 1899, following the award of her Ph.D., she spent some time at the Naples Zoological Station, and then joined the faculty at Wells College, Aurora, New York, where she taught biology from 1901 until 1909. She then took a two-year appointment to teach biology, physiology, and hygiene at the American College for Girls, and while in Turkey also organized lectures on health and hygiene for Turkish women and for women of the American community. On returning to the United States she taught biology as a temporary instructor at a succession of institutions, including the Municipal University in Akron, Ohio (formerly Buchtel College), Wellesley College, and Sweet Briar College, Virginia. She also continued the kind of community work she had taken up in Turkey, particularly education in preventive medicine. Eugenics and family heredity studies, an area of investigation in which there was then considerable activity, also occupied her attention. From 1918 until 1919 she worked at the War Trade Board, and then, for five years, at the U.S. Treasury Department. She died at age eighty-two, 18 January 1946, at her home in Philadelphia.

Both Anne Barrows and Ethelwyn Foote carried out their zoological research at Smith College, Barrows as a graduate assistant and Foote as an undergraduate.

After receiving her B.L. from Smith in 1897, ANNE BARROWS[51] taught for a year in Providence, Rhode Island, and then held an assistantship in the physiological laboratory at Boston University Medical School. Returning to Smith as an assistant in zoology, she took an A.M. in 1900 and stayed on as an instructor until 1904, with periods of study and research at the University of Munich (1902) and the Naples Zoological Station (1903). Her paper on the anatomy of the respiratory system of the lungless salamander *Desmognathus fusca* (an animal that was also the special research interest of her successor at Smith, Inez Whipple Wilder) appeared in *Anatomischer Anzeiger* in 1900. She married Walter Clarke Seelye in 1904. They had four children, and Anne does not appear to have published further scientific work after her marriage. For a time she served as a school trustee in Worcester, Massachusetts, where the family lived.

ETHELWYN FOOTE,[52] the daughter of Sarah C. (Cole) and Charles Rollin Foote, was born in Williamstown, Massachusetts, on 17 February 1875. After taking her A.B. at Smith College in 1897 she studied zoology at Northwestern University, receiving an A.M. in 1900. As a graduate student she taught science for a year at Ferry Hall in Lake Forest, Illinois. From 1901 until 1903 she was instructor of biology at Pomona College in Claremont, California, and then resumed her studies at the University of California. Her 1897 *Anatomischer Anzeiger* paper, reporting anatomical studies on elasmobranchs (skates) carried out at Smith College, would appear to have been her only technical publication. She married James Stark Bennett in 1907, and they had three children. She and her family later lived in Pasadena, California.

No fewer than ten of the American women whose published work has been classified as zoology in the bibliography are known to have received at least part of their training from biologist Thomas Hunt Morgan, who taught at Bryn Mawr College from 1891 until he moved to Columbia University in 1904. Five of them (King, Peebles, Randolph, Hazen, and Byrnes) went on to faculty positions in colleges or large high schools where they were able to continue research; among the others only one (Sampson) had a productive research career in biology after her student days.

Perhaps the best remembered of these early students of Morgan is HELEN KING[53] (1869–1955), well known in her time for her work on the breeding of superior strains of laboratory rats. Born on 27 September 1869 in Owego, New York, she was the elder of the two daughters of Leonora Louise (Dean) and George A. King. King owned a leather company in Owego. After attending the Owego Free Academy, she went to Vassar College, where she took an A.B. in 1892. She continued her studies, first as an assistant in the Vassar biological laboratory, (1894–1895) and then at Bryn Mawr, where she held fellowships in biology. Her graduate research reflected Morgan's interests at the time, her first paper (1898) reporting work in regeneration and developmental anatomy. She received her Ph.D. in 1899, and stayed on at Bryn Mawr for five years as an assistant in biology, while also teaching science at the Baldwin School for Girls in the town of Bryn Mawr (1899–1907).

Throughout these years she continued to investigate processes of regeneration and embryo development, and by 1908 had published about eight papers in the area. The work brought her a star in the 1906 edition of *American Men of Science.* A fellowship at the University of Pennsylvania gave her the opportunity to work for two years (1906–1908) with Edwin Conklin, whose special interests included cytology and the mechanism of heredity. In 1909 she joined the Wistar Institute of Anatomy and Biology in Philadelphia (an establishment loosely affiliated with the University of Pennsylvania's medical school) as an assistant in anatomy. She was promoted to assistant professor of anatomy in 1913, and in 1927, when she was fifty-eight, became professor of embryology. By the time she retired in 1949, at the age of eighty, she had sat on the institute's advisory board for twenty-four years, had edited its extensive bibliographic service for thirteen years, had been associate editor of the *Journal of Morphology and Physiology* for three years, and had served as vice president of the American Society of Zoologists. In 1932 the Association to Aid Women in Science (formerly the Naples Table Association) awarded its prestigious Ellen Richards Prize jointly to her and astronomer Annie Jump Cannon.

King's successful career at the Wistar Institute resulted from a remarkable meshing of the scientific aspirations of three people, herself, Milton Greenman, and Henry Donaldson, whose separate and quite distinct interests happily fed into one another in such a way that the requirements and goals of all three were met.

The broad objective of Milton Greenman, the anatomist who assumed the directorship of the institute four years before King joined the staff, was the promotion, on a national scale, of cooperation and efficiency in scientific work. Accordingly he reorganized the institute, his base of operations, upgrading it from what had until then been essentially an anatomy and pathology museum, and brought it into the research community as a center for the advancement of biological, and especially anatomical, research. In 1906 he engaged as chief scientist Henry H. Donaldson, a Johns Hopkins-trained neurologist, then a faculty member at the University of Chicago. Donaldson's interests centered on the effects of inherited and environmental factors on the growth and development of the brain and nervous system from birth to maturity; one of his requirements was a baseline standard in the form of a supply of uniformly similar animals that fitted within specified statistical growth and development norms. His biometric model was the albino rat, and he brought with him to Philadelphia several breeding pairs of these animals.

King, despite her Ph.D. and her star in *American Men of Science,* started off at the institute basically as a technician, expected to carry out routine laboratory services. However, she also had the privilege of spending part of her time on her own research interests, which, as a consequence of her work with Conklin at the University of Pennsylvania, now centered on mechanisms of the transmission of inherited characteristics and the effects produced by inbreeding. By carefully designing her projects to take advantage of what she had to do routinely for Donaldson's team (namely maintain a supply of uniformly high-quality rats) she was able to test her own ideas as well. The rat colony, in fact, offered her a means of entering the field of mammalian genetics at a critical time, when geneticists were moving away from the study of amphibians and simple marine invertebrates to models that permitted the exploration of more complex concepts (such as blended inheritance and the effects of the interplay of heredity and environment).

Donaldson's requirement for a supply of "standardized" rats and King's interest in the effects of inbreeding thus neatly complemented each other, and led to the engineering of a superior rat with a narrowed gene pool whose probability of breeding true was high. Director Greenman quickly recognized the tremendous potential of such an animal as standard equipment for biologists, in a sense the living analog of pure chemicals. He saw the production of uniform rats for other scientific laboratories as a way the Wistar Institute could provide valuable service to science, and he seized the opportunity the situation presented to him. The rat colony became an important part of the institute, receiving full administrative support, and the goals of Donaldson, Greenman, and King fitted together into what appears to have been a very satisfactory symbiotic union. Greenman himself devoted much effort to developing good management and husbandry methods; Donaldson's major contribution was the preparation of detailed documentation on the animals, tables of growth and development data that constituted the first such baseline statistics for any animal, including humans.

By the 1930s Wistar rats were used in laboratories worldwide. Breeding and distribution by the institute continued until the 1940s, and the animals were patented under the trade name WISTARAT in 1942.[54] The breeding stock and the rights to the name were sold in the 1960s, but bloodlines survive and more than half the laboratory rats used today have Wistar ancestors. King's achievements in inbreeding and continuous selection for desirable traits producing a narrowed gene pool were unmatched in biological science at the time. It has been suggested that her work might well be considered an approximate forerunner of the process of cloning, the production of exact copies of living organisms.[55]

Her experiments began in 1909 with a series of brother-sister matings of albino rats using carefully selected stock. By 1918 she had examined twenty-five generations. She published complete analyses of the qualities of the inbred animals, including studies of constitutional vigor, fertility, and sex ratios, and demonstrated that they compared favorably with the stock albinos.[56] The findings created something of a sensation in the popular press, where the work was interpreted as implying that King considered incest taboos unnecessary. In response she pointed out that, far from making any general claim that inbreeding is better than outbreeding for maintaining the vigor of a race, she had drawn the limited conclusion that, with careful selection of stock and good environmental conditions, it is not necessarily injurious in rats, even over many generations, and can result in certain benefits, namely the production of strong and healthy genetically homogeneous strains.

She also took up the problem of the domestication of the grey Norway rat, a species not previously raised in captivity. Over the course of three decades of breeding experiments she and a number of collaborators investigated and cataloged the many series of mutations that appeared in this animal. The findings were discussed in about thirty papers, published between 1927 and 1949, the year of her retirement.[57] Included were reports of chocolate-colored rats, curly-haired rats, and "waltzing" rats that traveled in circles; like King's earlier reports, these occasioned much popular interest. The mutant forms provided strains that were used for studying specific characteristics, for example resistance to carcinogenic agents.

Helen King died in her eighty-sixth year, 7 March 1955, at the University of Pennsylvania Hospital in Philadelphia.

Florence Peebles[58] (1874–1956), another student of Morgan at Bryn Mawr in the late 1890s, was the daughter of Elizabeth Southgate (Cummins) and Thomas Chambers Peebles. She was born in Pewee Valley, Kentucky, 3 June 1874. After attending the Girls' Latin School in Baltimore she studied at the Woman's College of Baltimore (B.A., 1895), and then, with a biology scholarship, went on to graduate studies at Bryn Mawr.

Her work was mainly concerned with regeneration in marine animals, although in 1898 she also published a lengthy study on chick embryology. She spent several summers at Woods Hole and in addition carried out research in Europe while holding a Mary E. Garrett European Fellowship at the universities of Munich and Halle in 1898–1899. Bryn Mawr awarded her a Ph.D. in 1900; her dissertation research on regeneration and grafting in hydrozoa was reported in *Archiv für Entwickelungsmechanik der Organismen* the same year. This early work, along with further investigations on regeneration and chick embryology reported in 1902 and 1903,[59] brought her a star in the 1906 edition of *American Men of Science.*

She taught biology at the Woman's College of Baltimore from 1899 until 1906, first as instructor and then as associate professor. Thereafter she taught at Miss Wright's School in Bryn Mawr for six years and at Bryn Mawr College, where she was lecturer and acting head of the biology department in 1913. From 1915 until 1917 she headed the biology department at Sophie Newcomb College, Tulane University, and then she returned to Bryn Mawr for two years as associate professor of biology. She regularly spent summers at Woods Hole, and also had several research and study leaves at the Naples Zoological Station and at various universities in Germany (Bonn, 1905, Würzburg, 1911, Freiburg, 1913), usually obtaining fellowship support. Reports of her work appeared at intervals in American and European journals, mainly in the years just prior to the First World War.[60]

Her work on hydra regeneration was concerned principally with the question of how external influences, particularly those of the immediate environment, affect tissue differentiation and regulation. The conclusions she came to from her early studies, that the characteristics of a cell or organism can be modified by external influences, especially the immediate environment, constituted ideas that were in large measure ahead of her time. Theories similar to those her studies pointed to are now current among environmental biologists, but Peebles's work remained unnoticed for the most part, mainly because she offered no theoretical basis in which to embed her findings and make clear their broader implications.

About 1919, when she was forty-five, she took a break from teaching and spent five years doing experimental farming, followed by three years of travel in the western United States and Europe and a summer of research at the Naples Zoological Station in 1927. She returned to educational work in 1928, taking the position of professor of biological sciences at California Christian College in Los Angeles (later Chapman College), where she set up a department of bacteriology. She managed to remain active in research as well and spent some time at the Scripps Institution of Oceanography in La Jolla, California, where she collaborated with Denis Fox in investigations on marine worms. A paper on regeneration studies published in 1931 reported work begun in Naples in 1927 and completed at the Scripps Institution the following year.[61]

After her "retirement" in 1942 she was associated with Lewis and Clark College in Portland, Oregon, where she established the college's biology laboratory. Indeed, despite her creditable research contributions she was probably best remembered for her teaching and organizational work. In 1946, when she was seventy-two, she finally retired and moved to Pasadena, California, where she was active in community affairs for another ten years, particularly as a counselor for the aged. Her alma mater, Goucher College (formerly the Woman's College of

Baltimore), awarded her an honorary LL.D. in 1954. Two years later, in December 1956, she died at her home in Pasadena, at age eighty-two.

HARRIET RANDOLPH[62] (1856–1927?), the daughter of Mary (Sharpless) and Edward Taylor Randolph, was born in Philadelphia, 27 October 1856. She took an A.B. at Bryn Mawr in 1889 and stayed on for a further year, working under Morgan's direction as a fellow in biology. In 1892 she received a Ph.D. from the University of Zurich, her doctoral research on tube worms (Tubificidae) appearing in the *Jenaische Zeitschrift*. She joined the American Society of Morphology the same year as one of its first three women members.[63]

For the next twenty-one years, until 1913, she taught biology and botany at Bryn Mawr. A substantial paper on regeneration in flatworms and her textbook *Laboratory Directions in General Biology* both appeared in 1897. She later brought out a report on spermatogenesis in the earwig.[64] In 1915, when she was fifty-nine, she appears to have taken graduate classes at the University of California; two years later she continued her studies at Columbia University.

ANNAH HAZEN[65] (1872–1962), although she did not take a degree at Bryn Mawr, studied there for two years (1897–1899) as a graduate fellow. She was born in Hartford, Vermont, on 22 September 1872, the daughter of Abbie (Coleman) and Charles Dana Hazen. After receiving a B.L. degree from Smith College in 1895 she continued her education at Dartmouth College where she studied zoology under William Patten (M.S., 1897). Their joint paper on the morphology and development of *Limulus* (the horseshoe crab), one of Patten's special interests, appeared in the *Journal of Morphology* in 1900. At Bryn Mawr she collaborated in Morgan's regeneration studies, publishing a paper on earthworms in 1899. Her joint article with Morgan on cell structure and development in *Amphioxus* came out in 1900. For a year (1899–1900) she taught biology at New Hampshire Normal School and then served as assistant in zoology for three years at Smith College, where, following further suggestions of Morgan, she continued research on regeneration processes in marine organisms. Three more of her papers appeared in 1902 and 1903.[66] She taught high school biology in Brooklyn for a number of years, but by 1921 was bacteriologist at the U.S. Army Base Hospital at Camp Lee, Virginia. She died in May, 1962, at age ninety.

ESTHER BYRNES[67] (1867–1946), the daughter of Mary (Wilson) and Jacob Fussell Byrnes, was born at Overbrook, Pennsylvania, on 3 November 1867. She graduated from Bryn Mawr (B.A.) in 1891 and spent the next two years as an assistant in the biology department at Vassar College. Returning to Bryn Mawr, she took an A.M. in 1894 and stayed on as an assistant in the biological laboratory until 1897 (Ph.D., 1898). She then became instructor in physiology and biology at the Girls' High School, Brooklyn, New York, a post she held until her retirement in 1932—with one interesting break. For two years (1926 and 1927) she had a leave of absence to tutor the daughters of the Emperor of Japan at Tsuda College, Tokyo.[68] In 1940 she served as director of the Mt. Desert Biological Laboratory (Mt. Desert Island, Maine).

Her early studies, carried out with Morgan at Bryn Mawr and at Woods Hole, concerned limb regeneration in amphibia and maturation and fertilization processes in snails. She continued to work in these areas for a number of years after she left Bryn Mawr, pursuing her investigations in Brooklyn and at the Station for Experimental Evolution, Cold Spring Harbor, Long Island. She also studied the small, freshwater crustacean, cyclops. Her reports of the work, which was well-regarded, appeared at intervals until at least 1919.[69] An especially active member of the New York Science Teachers' Association, she was also a fellow of the New York Academy of Sciences and a member of several other scientific organizations, including the American Society of Naturalists. She died while on vacation in Maine, on 1 September 1946, at age seventy-eight.

The other five Bryn Mawr students of Morgan whose work is listed in the bibliography were Sampson, Smith, Kroeber, Hemenway, and Langenbeck.[70]

LILIAN SAMPSON[71] (1870–1952) was born in the small town of Hallowell, near Augusta, on the Kennebec river in Maine. She was the second of the three daughters of George Sampson, part owner of one of the town's prosperous oilcloth factories, and his wife Isabella, daughter of Thomas Merrick, a well-to-do businessman in the chemical and pharmaceutical trade. Both the Merricks and the Sampsons traced their ancestry back to earliest colonial times, the Sampsons to the Plymouth Colony settlers. The family was hard hit by the tuberculosis epidemics that ravaged the region in the early 1870s, and only Lilian and her elder sister Edith survived. In 1874 they were sent to live with their Merrick grandparents, who by then had moved to Germantown, Pennsylvania. The Germantown family was a large and active one, Thomas and Elizabeth Merrick having brought up thirteen children. Edith and Lilian, only two years apart in age, remained very close throughout their childhood.

After attending Wellesley Preparatory School in Germantown, Lilian, in 1887, followed Edith to Bryn Mawr. As an undergraduate she dutifully obeyed her grandfather's instructions to return home every weekend, and also kept her promise to him that she would practice her violin every day—indeed she played all her life. She took a very solid undergraduate science program, concentrating heavily on mathematics, biology, and physics. E. B. Wilson was one of her zoology professors, and Martha Carey Thomas, later president of the college, her adviser. She graduated with an A.B. in 1891, spent the summer at Woods Hole (the first of many summers she was to spend there), and the following autumn went to the University of Zurich on a Bryn Mawr European Fellowship. Her morphological study on the muscles of the mollusk *Chiton* was begun in Zurich under the direction of Arnold Lang and continued the following year at Bryn Mawr with guidance from Morgan, who had by then succeeded Wilson as professor of zoology. Her first paper, "Die Muskulatur von Chiton," appeared in the *Jenaische Zeitschrift* in 1894, the year she received her M.A. from Bryn Mawr. From then on, throughout the 1890s, she spent her summers at Woods Hole as an independent investigator, working on breeding and development in amphibia.

Her long paper on Anura (frogs and toads) appeared in the *American Naturalist* in 1900.

In 1904 she married Morgan, and the same year they moved to New York City, Morgan having accepted the chair of experimental zoology at Columbia University. Lilian continued to do a limited amount of research in her husband's laboratory for a few years after her marriage and published three papers in embryology and regeneration during the period up to 1906.[72] However, after the birth that year of her first child, her son Howard, she spent most of her time on family matters; one of her major concerns was to relieve her husband of routine business, and make sure that he had the freedom to devote himself to his research. Three more children, Edith, Lilian, and Isabel, were born in the next five years. Lilian took much of the responsibility for their early education herself, giving the two youngest girls their first two years of schooling at home. She also kept up her scientific reading, and followed the progress of her husband's research. The complex task of arranging the move of the family and the Morgan group's laboratory equipment from New York to Woods Hole and back every summer was also hers, and regularly required weeks of planning. Likewise the running of the large summer house they had built at Woods Hole was something of an organizational challenge; its six bedrooms frequently accommodated as many as eighteen people, relatives and graduate students. Lilian was also active in the Woods Hole community; in 1913 she and several other women founded the Summer School Club, later the Children's School of Science, whose activities were to be dancing, singing, and learning about science. She was the first educational chairman, and in 1914 became the science committee chairman.

About 1921, after all her children were in school, the youngest then being ten, she returned to laboratory work, and successfully made the transition from embryology, the line of research she had put aside almost two decades previously, to drosophila genetics, the area in which her husband was then working. (Morgan's genetics research brought him the Nobel prize for Medicine and Physiology in 1933.) She had considered the possibility of studying violin again, but settled for scientific work, although her husband made it clear to her that she would be on her own, in no sense an assistant to him.[73] She was given working space in the Columbia laboratory, but no formal position, and although she was made welcome, her situation was a somewhat awkward one. She took the greatest care to be as unobtrusive as possible, but Morgan was not entirely comfortable with her presence. Although she was undoubtedly well aware of the day-to-day progress of his group's investigations, she was never part of his inner circle of graduate students (which had something of the atmosphere of an exclusive men's club[74]). She had her own area and her own fly stocks, and, being much older than any of the other women in the laboratory and a quiet person not much inclined to talk, she remained relatively isolated. But her skill and commitment became obvious to the students, who soon accepted her.

Her research methods followed the general pattern of the work then going on in the laboratory—observation and genetic deduction, backed by cytological studies. Her most notable contributions to drosophila genetics developed from her discovery and capture, early in 1921, of an unusual mosaic female fly. Mating experiments with this single female produced unexpected and striking results. These she interpreted by suggesting that, instead of the normal pair of X chromosomes, the mosaic female had one "attached-X" chromosome, formed by the joining of two normal X chromosomes end to end, and one Y. The attached-X was inherited and passed on as a single unit. Her subsequent cytological corroboration of this attached-X hypothesis was a striking demonstration of the agreement between breeding results and chromosome behavior, and provided further evidence in support of the still-new theory that chromosomes are the physical basis of inheritance. The work was also a noteworthy contribution to the unraveling in drosophila of the mechanisms of inheritance of sex and parental characteristics, about which very little was then known.[75]

In 1928 the Morgan family moved to Pasadena, following T. H. Morgan's acceptance of an invitation to found a new Division of Biology at the California Institute of Technology. Lilian Morgan continued her drosophila investigations, still working alone without any formal appointment, and she published ten more papers. Her work was frequently cited, clear proof of its continuing value to her successors. The paper she wrote that is most often referred to was her study of the closed-X chromosome, which she discovered in 1933.[76] The closed-X remained an important tool in drosophila genetics, notably in the study of mosaic systems.

All her scientific work was carried out with great thoroughness and persistence, as illustrated especially clearly in her first genetics investigation of the stray mosaic female she chanced upon—an observation that led to the establishment of the exceptional stock, the attached-X, which remained one of the Morgan group's most valuable tools throughout the 1920s. Although she freely discussed her work with other geneticists, she was never part of a professional circle and avoided scientific meetings and the presentation of papers. In 1946, following her husband's death, she was given a research associateship at the California Institute of Technology. She was then seventy-six. The appointment was the only one she ever held and it allowed her to continue her research for another six years. Her last paper was published in 1947.[77]

A forward-looking woman who enjoyed the company of young people, she strongly supported the League of Nations and women's rights causes, although quietly, in the form of financial contributions. She and Tom Morgan kept social life to a minimum, concentrating on work and family. She enjoyed building furniture and camping out. In 1920, when Morgan had taken a sabbatical year at Stanford University, she and their oldest child, Howard, learned to drive a car from an instruction manual and with the other children went camping in Yosemite Park, sleeping on the ground in homemade sleeping bags without a tent. She died at the age of eighty-two, of intestinal cancer, on 6 December 1952, in Pasadena.

AMELIA SMITH[78] (1876–1965) was a graduate student at Bryn Mawr for the two years, 1899–1901. Born in Philadelphia, 23 February 1876, she was the daughter of Mary Allen

(Knight) and Frederick Smith. She studied at the Philadelphia Normal School and then at the University of Pennsylvania, where she took a B.S. in biology in 1899. Her paper on the structure of *Aphyllon uniflorum* was presented at the second annual meeting of the Society for Plant Morphology and Physiology in New York in 1898, when she was still an undergraduate. At Bryn Mawr she carried out research in embryology and earthworm physiology; her note on the spinal cord of the chick embryo appeared in *Anatomischer Anzeiger* in 1899 and her more substantial paper on her earthworm work in the *American Journal of Physiology* three years later.[79] In 1901 she married Philip Powell Calvert, whom she had known as a student at the University of Pennsylvania, and who was later professor of zoology there. Calvert was an authority on the Odonata (dragonflies and damselflies) and Amelia prepared the illustrations for his work on the odonate fauna of Mexico and Central America. They spent the year 1909–1910 in Costa Rica, studying the country's flora and fauna, and subsequently coauthored the book *A Year of Costa Rican Natural History* (1917). Amelia died on 24 December 1965, shortly before her ninetieth birthday.

JOHANNA KROEBER[80] (1880–1969) was born on 8 February 1880. She attended Bryn Mawr in the late 1890s as an undergraduate. Her research, published in the *Biological Bulletin* in 1900, concerned regeneration in *Allolobophora*. After graduating with a B.A. in 1900, she taught for three years at the elementary level in Dr. Sachs's school in New York City and then took a position as teacher of biology at Wadleigh High School, also in New York City. Following her marriage in 1908 to Herman Otto Mosenthal, a physician and later a professor of clinical medicine, she worked for twenty-eight years as a research assistant in biology and paleontology to Henry Fairfield Osborn, curator of vertebrate paleontology at the American Museum of Natural History. She also took courses in biology, geology, and literature at Columbia University and Barnard College, and brought up four children. For many years she served on the boards of the New York Association for the Aid of Crippled Children and the New York Diabetes Association. During the Second World War, although in her sixties, she worked on the American Friends Service Committee and in a factory that sewed aviators' jackets. She died on 13 September 1969, at age eighty-nine.

JOSEPHINE HEMENWAY[81] (1880–1965), who became widely known for her work in pediatrics, was born in Glasgow, Missouri. Having completed her undergraduate education by the age of eighteen, she spent some time as a graduate student with Morgan at Bryn Mawr. Her morphological study of the arthropod *Scutigera* (*Cermatia*) *forceps* (a centipede) was reported in the *Biological Bulletin* in 1900. She took an M.D. at Johns Hopkins Medical School in 1904, interning in Baltimore (and working in areas of the city where she had to have a police escort to get safely to some of her cases). She then served as resident at the Babies' Hospital in New York City under the eminent pediatrician L. Emmett Holt. Holt, not given to complimenting his juniors, is said to have remarked that Hemenway was "the best man I ever had on my staff." She married James Kenyon, a surgeon, in 1911, and they had two daughters.

During the First World War she worked for the War Department, traveling the country and lecturing on disease control at training camps. She had a private practice for a time in New York City and also maintained close affiliations with a number of New York hospitals, including Presbyterian Hospital where she carried out research on growth and obesity. Her widely used book, *Healthy Babies are Happy Babies*, which went through five editions, was first published in 1934.[82] For thirty years she had charge of the baby health department of *Good Housekeeping* magazine, to which she contributed a page on baby care every month. She was a fellow of the New York Academy of Medicine and of the American Medical Association. In 1952, after forty-two years of work in New York City, she retired to Colorado, where her daughters then lived, and settled in the country near Boulder. She died there on 10 January 1965, at age eighty-four.

The Bryn Mawr and Woods Hole students of Thomas Hunt Morgan constituted a substantial fraction of the late nineteenth–early twentieth century American women who made contributions to the zoological research literature. There was also a second distinct group, however, that associated with Radcliffe College and the Museum of Comparative Zoology at Harvard. Six of the early women zoologists whose work is listed in the bibliography (Bowers, Lewis Nickerson, Crotty Davenport, Henchman, Platt, and Perkins[83]) received at least part of their training at these institutions.

Of the six, probably the best remembered now is MARGARET LEWIS NICKERSON (1866–1942).[84] Born in Akron, New York, 8 December 1866, she received her undergraduate education at Smith College (B.A., 1893). She then taught for a year in secondary schools in Cambridge, Massachusetts, and for a further year as instructor of zoology at Smith before taking up graduate studies at Radcliffe in 1895. During her two years there she carried out notable cellular level investigations on the central and peripheral nervous systems of marine worms. The work was begun at Woods Hole in the summer of 1895 at the suggestion of Charles Whitman and continued under the guidance of Edward Mark, director of the zoological laboratory at the Harvard Museum. Her first report appeared in a long paper in *Anatomischer Anzeiger* in 1896 and was followed by several others in American and German journals over the next five years.

After receiving an M.A. from Radcliffe in 1897, she enrolled at the University of Minnesota Medical School, where she was also appointed instructor in histology and embryology. The same year, she married Winfield Scott Nickerson, a fellow medical student and instructor in histology and embryology. For a time she continued research on nerve systems, published in the *Journal of Morphology* in 1901,[85] and her early work was well received, bringing her a star in the 1906 edition of *American Men of Science*. After taking her M.D. in 1904, however, she did little further research, and over the next three decades devoted most of her energies to medical work. From about 1911, when she and her husband separated (although they remained legally married) she served as school physician in the Minneapolis public school system. She also maintained a private practice in the city. In 1936 she retired and moved to Rochester, Minnesota, where she died on 15 August 1942, at age seventy-five.

GERTRUDE CROTTY[86] (1866–1946), the daughter of Millia (Armstrong) and William Crotty, was born in Asequa, Colorado, on 28 February 1866. Her family later moved to Burlington, in east-central Kansas, and she entered the University of Kansas in 1885. After receiving a B.S. degree in 1889, she stayed on for three years as a graduate student and instructor in zoology before going to Radcliffe in 1892. Two years later she married Charles Davenport, then an instructor of zoology at Harvard. She was especially interested in variation in starfish and in turtle embryology; most of her publications reported microscopical studies carried out in the biological laboratory of the Station for Experimental Evolution, Cold Spring Harbor, Long Island, New York, of which her husband became director in 1898. She died on 8 March 1946, shortly after her eightieth birthday, in Upper Nyack, New York.

ANNIE HENCHMAN[87] (1852–1926), was born in Boston, on 22 March 1852, and studied at Radcliffe from 1884 until 1890. She carried out research on the nervous system of the mollusk *Limax maximus* at the Harvard Museum, work reported in the museum's bulletin for 1890–1891; a second short note by her appeared in *Science* in 1897. After that she worked as a teacher and a laboratory assistant in Cambridge, Massachusetts, except for a short time about 1920, presumably after she retired from teaching, when she collaborated with Charles Davenport at the Station for Experimental Evolution.

JULIA PLATT (1857–1935),[88] whose work is still frequently cited in specialist literature on embryology, is one of the more remarkable American women research biologists mentioned in this survey. She was active in her field for only ten years, while carrying out graduate research at a number of institutions, including the Harvard Museum, the universities of Chicago, Freiburg, and Munich, and the Woods Hole and Naples laboratories. Nevertheless, her work in vertebrate embryology was pioneering, and it aroused considerable controversy. Her outstanding contributions, coupled with her almost total absence from general historical accounts of early women biologists, suggest that there is a need to make clear just what she did, even if it means presenting a somewhat technical sketch of her highly specialized work. A few words of background about the initial stages in the development of the vertebrate embryo are perhaps not inopportune by way of introduction.

Fertilization of an ovum triggers into action a wonderfully dynamic system. First the ovum divides producing two daughter cells; they in turn divide and the process continues, rapidly building up a mass of cells, the blastoderm, in which the individual cells are arranged in three layers. The three-layer body then folds in on itself and closes to give a tube structure, the first recognizably vertebrate feature of the embryo. Initially observed in the early nineteenth century and later designated ectoderm (outer layer), mesoderm (middle layer), and endoderm (inner layer), the three layers formed the basis of the "germ-layer theory," one of the fundamental concepts of embryology. According to this theory, the cells in each layer play specified roles in the further development of the embryo, those of the ectoderm giving rise to epidermal and neural structures, those of the mesoderm to skeletal, muscular, and vascular systems, as well as connective tissue, and those of the endoderm to internal organs. The order, neatness, and simplicity of the germ-layer doctrine made it particularly attractive to nineteenth-century embryologists, and despite the fact that the specificity of the layers was beginning to be questioned by a few workers by the early 1870s, the theory remained well-entrenched and strongly supported by the majority.[89] (Its fundamental importance in biological thought is clear, involving as it does the understanding of the initial stages in the development of vertebrate life, and therefore human life.)

Julia Platt was one of the early challengers of the theory. In 1893 and 1894 she published an experimentally well-supported claim that ectodermal rather than mesodermal cells are the source of cartilage in the head structures of amphibian embryos, thus denying the specificity of the germ layers, the central tenet of the doctrine. The controversy to which she offered this substantial contribution was heated, to say the least. Charles Minot, professor of anatomy at Harvard, in a lecture delivered to the New York Pathological Society in 1901, stated categorically that, apart from a few minor unanswered questions, the specific layer origin of every organ in the human body is known. He felt he needed to comment briefly on "A remarkable attempt to upset the doctrine of the germ layers. . . ," namely Julia Platt's work, but he had no specific data to offer that might refute her claim, and merely stated that neither his own investigations nor those of another worker, Hanson Kelly Corning, offered the slightest evidence in her favor. His considered opinion was that "we may, therefore, I think, safely regard this attempt to overthrow the morphological value of the germ layers as unsuccessful. I know of no other attempt of sufficient importance to be even mentioned."[90] Nevertheless, from about 1895 on, a number of eminent anatomists and embryologists began to confirm Platt's interpretations.[91]

Particularly able in her laboratory work and exceptionally acute in her observations, she was free from preconceived notions in her scientific judgments and admirably independent when it came to speaking out and making her ideas known. Indeed, several of her papers fully deserve the description "classic"; few other American women of her generation carried out original experimental work in biology that had an impact comparable to hers.

Born in San Francisco, 14 September 1857, Julia was the daughter of Ellen Loomis (Barlow) and George King Platt. George Platt was a Vermont lawyer and state's attorney from 1840 to 1842. His wife was the daughter of a merchant and real estate dealer in Burlington, Vermont. George died nine days after his daughter's birth. Little has been uncovered about her early years, but she probably spent at least part of her childhood in Burlington, and in any case enrolled at the University of Vermont in 1879, taking the literary-scientific course. During her three years as an undergraduate she lived at home in her mother's house, only a short walk from the university; she graduated with a Ph.B. in 1882, shortly before her twenty-fifth birthday. After the death of her mother the following year, she was on her own, and evidently left with sufficient financial resources to allow her reasonable freedom in planning her immediate future.

By 1887 she had enrolled in classes at Radcliffe College and Harvard Annex, which gave her the opportunity of carrying out laboratory work in the Museum of Comparative Zoology. Her supervisor was the anatomist Howard Ayres, then an instructor at Harvard and Radcliffe (later president of the University of Cincinnati), who specialized in the morphology of the vertebrate head. It was doubtless Ayres who first aroused her interest in this area of investigation. Her study of the order in which segments form at the anterior end in chick embryos (the subject of her first paper, 1889) led her to criticize the conclusions of earlier workers and for some time her findings were contested. But, as with most of her later work, the consensus of opinion eventually swung to her side, and her interpretation came to be accepted as the basis for the designation of the successive stages in chick development.[92]

She spent the summers of 1889 and 1890 at Woods Hole, and the intervening year at Bryn Mawr, where she listened to E. B. Wilson's zoology lectures. At Woods Hole, with the guidance of Charles Whitman, she extended her studies on embryonic head organization, now working on the spiny dogfish. She continued this line of investigation in 1890–1891 at the University of Freiburg, in the laboratory of Robert Ernst Wiedersheim, one of the leading comparative zoologists of his time. In particular, she examined the segmentation anterior to the ear, a region about which there was considerable disagreement, and she made two observations that were to have important effects on the direction of later work in the field. Her first discovery was an additional structure, subsequently referred to as "Platt's vesicle," whose role in head development has not been clarified even today, and whose phylogenetic significance is therefore still not understood; it has recently been noted, however, that Platt's vesicle clearly must be considered in any assessment of the head segment organization in the earliest craniates.[93] Her second major observation, the incomplete divisions between the potential segments of the anterior section of the head and the relationship of these to nerve structure, is likewise an area that is still being investigated.

In the early summer of 1891 she spent some time at the Naples Zoological Station, where she had the American Davis Table. Then she went to Faro, at the northeastern tip of Sicily, to collect *Amphioxus,* which she studied in Wiedersheim's laboratory. Her conclusions about features in the central nervous system in this organism, published in 1892, were controversial and generated much further work, and although some of the questions she raised remain unanswered even now, many of her observations have been confirmed.

She returned to the United States for the winter of 1892–1893, going to the newly opened University of Chicago to work in Whitman's laboratory. It was about this time that she embarked on her best-remembered work, her investigation of the embryo of the aquatic salamander *Necturus* (mud puppy) that led her to the highly controversial conclusion about the ectodermal origin of head cartilage mentioned above. Her preliminary note on this topic, her first attack on the theory of the specificity of the germ layers, appeared in *Anatomischer Anzeiger* in 1893. She continued the work during the winter of 1893–1894 at the University of Munich, where she was permitted to attend lectures as an "exception case" and where she had guidance for her research from Carl Wilhelm von Kupffer, another of the eminent zoologists of the time, well known for his work in comparative anatomy and embryology. Her full paper on her *Necturus* work, a fifty-five-page article in German, appeared in *Archiv für Mikroskopische Anatomie* in 1894. That year, she attended the Eighth Congress, in Strasburg, of the Anatomische Gesellschaft, of which she was a member. She presented two of her *Necturus* slides, and had what was doubtless the most interesting experience of listening to Carl Rabl, professor of anatomy in Prague, speak out strongly in defense of the germ-layer theory in its traditionally accepted form. Although von Kupffer does not appear to have jumped into the fray in 1894, one year later, at the Gesellschaft's Basle meeting, he came to Platt's defense, stating in the keynote address that after looking at her excellent section preparations he had concluded that the evidence was clearly against the specificity of the germ layers.[94]

Platt was to spend another winter at the University of Munich in 1896–1897, working at the Zoological Institute in Richard Hertwig's laboratory, but meanwhile in 1895–1896 she returned to the United States for two semesters at Radcliffe where she joined in Charles Davenport's investigations in experimental morphology; her findings appeared in her last paper, published in the *American Naturalist* in 1899. She also completed studies on the development of the peripheral nervous system, the thyroid gland, and suprapericardial bodies in *Necturus,* work that appeared in two substantial articles in 1896. The following year she went back to Freiburg for more course work and research, and in the spring of 1898 applied for admission as a candidate for a doctoral degree in zoology, submitting a dissertation on the development of the cartilaginous skull in *Necturus.* Her degree (multa cum laude), awarded in 1898, was the third doctorate conferred by the University of Freiburg on a woman.[95]

The receipt of her degree, however, far from ushering in a professional career, marked the end of Platt's scientific work. She was then forty-one. She returned to the United States and for a year searched in vain for a teaching position. Despite her ten years of graduate research under some of the most respected zoologists of the day and her remarkable publications, she was forced to conclude that she would not be able to find the kind of work she wanted.

From 1899 she lived in Pacific Grove, a small community on the promontory marking the southern end of Monterey Bay, on the California coast eighty miles south of San Francisco. Here she built a house near the shore road and occupied herself with community affairs, becoming known as the city's go-getter. Neighborhood improvement, the development of park areas, and beautification projects involving the planting of trees and shrubs were her special care. She did much of the manual labor herself and became a familiar figure in Pacific Grove, regularly pushing her wheelbarrow full of plants and garden tools through the streets on her way to work sites. She was regarded as something of an eccentric in her old-fashioned dress and mannish, gray felt hat, but she could be depended on to take a strong stand on matters that involved the general good. Indeed, on such concerns she would brook no opposition. On one occasion she even removed by force a gate barring access to what had traditionally been public right of way. When the gate was padlocked she filed apart the lock, and when it was then nailed shut she demolished it with her axe, making sure a press photographer was present to record her action! The gate remained open. In 1924 she and a small group of women interested in community affairs secured for Pacific Grove a city charter from the state legislature and a city manager form of government. In 1931, in her seventy-fourth year, she was elected the first woman mayor.

Although there was a research station near her home, the Hopkins Marine Station, there is no indication that she ever worked there or had much contact with station biologists other than an acquaintance with Harold Heath, a Stanford University embryologist. In the 1920s she adopted a teenage boy, Harold, but until then lived alone. She died on 28 May 1935, at age seventy-seven, and was buried at sea in Monterey Bay. Accompanied by a party of city fathers, her body, encased in a wicker basket, was taken by small fishing launch twelve miles out, and committed to the depths; an airplane circling overhead dropped hundreds of roses on the water over her final resting place. Her adopted son placed a memorial plaque on a granite boulder overlooking the sea in one of the parks she had developed. It is still there today. Her house also remains; called "Roserox," it is now a bed-and-breakfast inn.

One might wonder why Julia Platt was unable to find an academic position that would have allowed her to continue the kind of work she so much enjoyed. Several factors were against her, however. One was just the general dearth of such jobs for women, apart from the few opportunities available in the Eastern women's colleges. There was also the unusual, and, to many of her male colleagues, unsettling combination of her considerable scientific accomplishments and her forceful personality. The controversial nature of her published findings (powerful attacks on accepted theory) allied with her independence and boldness in tackling disagreements when she felt she had the evidence to back her up, made her a troublesome person for male colleagues with whom she disagreed. And she had some influential people among her opponents, including Harvard's Charles Minot, who was also the editor of *Science.* Although blunt speech was common enough in scientific argument among men at the time, this style was not so readily accepted in women.[96] In a letter introducing her to Anton Dohrn, director of the Naples Zoological Station, Charles Whitman indicated that Julia was overly given to talk, and he added, perhaps a trifle paternalistically, that he hoped she would not be too "troublesome." She and Dohrn did in fact have a falling out, somewhat to the amusement of Whitman, who reminded Dohrn that he had, after all, warned him.[97] Fortunately Julia and Dohrn were able to resolve their difficulties, but there is little doubt that she had something of a reputation among her colleagues, and not all of them were as willing as Dohrn to accommodate the rougher aspects of her personality. She had to put up with a good deal of annoyance, today it would be called harassment, from men at Woods Hole. Edwin Conklin, in his 1968 reminiscences of early days at that laboratory, recalled one such incident involving Julia when she was collecting hydroids from a floating platform; a large, heavyset woman, she sank the platform well down into the water, and noticing this, Hermon Bumpus, assistant director of the laboratory and the resident practical joker, quietly signaled to a few other men who were standing around that they should all step onto the platform. The latter, as intended, sank under the combined weight, thoroughly soaking Julia in her full skirts. This dunking was an all-too-obvious move by the men to humiliate her, and she soon stopped going to Woods Hole.[98]

Despite her rejection by the American scientific community,[99] Julia Platt's work stands as a fine monument to her. Harold Heath, her California acquaintance, is quoted as saying that she hoped science would go over her findings, "for I know that I am right."[100] Science is slowly doing just that; and for the most part, she was.

In contrast to the students of Morgan and those from Harvard and Radcliffe, KATHARINE FOOT[101] (1852–1944?) had no formal college education. Nevertheless, she did a considerable amount of original work during an independent scientific career of a quarter of a century, beginning in the 1890s, and her early research, carried out at Woods Hole, brought her a star in the 1906 edition of *American Men of Science.*

She was born in Geneva, New York, on 14 October 1852, and received her early education in private schools in New York State. As a young woman in her twenties she was active in the the flourishing New York women's club, Sorosis, which was largely responsible for organizing, in 1873, the Association for the Advancement of Women. The latter group included in its membership many of the most prominent women of the period, among them the future American Indian ethnographer Alice Fletcher (chapter 14). Foot also became interested in Indian peoples, and by the mid 1880s was president of the Washington auxiliary of the Women's National Indian Association, one of the organizations formed about that time by more liberal-minded members of the white community to lobby on behalf of Indians' rights. In the autumn of 1886 she accompanied Alice Fletcher on a three-and-a-half-month trip to Alaska to investigate the situation of Indians in that territory. They traveled across the country to Seattle by train, stopping on the way at Yellowstone Park, and then sailed from Port Townsend, Washington, by schooner, going first to the Aleutian Islands chain and later to the settlements at Sitka and Juneau. The season was advanced, however, the weather poor, and their opportunities for talking with Indians (rather than missionaries) fewer than they had hoped for.

By the 1890s Foot was a resident of Evanston, Illinois, where, until 1897, she lived in the home of Orrington Lunt, a Chicago businessman and a founder and trustee of Northwestern University in Evanston. Lacking any basic training in science, she went to Woods Hole in 1892 "merely as an amusement,"[102] but after her first six weeks in a course on invertebrates, Charles Whitman put her to work on the problem of the maturation and fertilization of the egg of the earthworm, *Allolobophera foetida*. The project was considered a difficult one but she carried it out exceptionally well. Her first paper on the subject appeared in the *Journal of Morphology* in 1894, when she was forty-two, and was followed over the course of more than a decade by at least eleven more, describing cellular level changes during the development of the fertilized egg. Her 1896 lecture on "The centrosomes of the fertilized egg of *Allolobophora foetida*" was the first given by a woman at the Woods Hole laboratory. Diffident and somewhat negative in her personality, she was conservative in her scientific conclusions, and her ideas followed closely the generally accepted theories of the time.

In the spring of 1897 Ella Strobell joined her as an assistant,[103] and from 1899 Strobell's name appears as a coauthor on her papers. These two were among the first workers to use the camera to record stages in the development of fertilized eggs, and the many clear and detailed photomicrographs that illustrated their articles were a significant technical advance over the drawings that until then were the normal method of presenting material seen under the microscope.[104]

The earthworm research led Foot and Strobell directly into current debates on the role of chromosomes in transmitting definite units of hereditary information, particularly those governing sex-linked characteristics. By about 1906 they were investigating the question using hemipterous species, particularly squash bugs, in which certain stages of chromosomal development can be followed relatively clearly. Much of this work, done between 1906 and 1913, appears to have been carried out in a laboratory of their own in New York City. Unfortunately they interpreted their observations as evidence in support of what was soon clearly the loosing side in the debate over the theory of chromosome control of sex-determined characteristics. Nevertheless, they spiritedly defended their conservative position against such heavyweights as T. H. Morgan himself, who by then had swung over to the other side. Writing in 1913 Foot says:

> Our cytological studies have caused us to sympathize with the many investigators who have expressed skepticism of the causal nature of the chromosomes. For several years we have argued that the chromosomes in the forms we have studied show too much variability, both in their morphological and physiological expressions, to justify those theories which obviously demand a rigid compliance to a definite mode of expression. We demonstrated in 1905 that the form and relative size of the chromosomes in *Allolobophora foetida* are inconstant and in every publication since that date we have demonstrated variability in the form, relative size and behaviour of the chromosomes in every form we have studied, and we have consistently argued that such variability attacks the very foundations upon which the popular chromosome speculations of this decade have been built.[105]

In 1914, when Foot was sixty-two and Strobell fifty-two, they went to England and continued work on crossbreeding of Hemiptera with help from entomologist Harry Eltringham of New College Oxford. The results were reported in three long papers that appeared in 1914 and 1915. In all they published eleven papers on Hemiptera breeding studies, the last in the *Biological Bulletin* in 1917. After that Strobell became ill and their research stopped. Foot commemorated their long collaboration by bringing out a collection of their reprints bound together in a single volume.[106]

By the end of the First World War Katharine Foot was living in Paris, where she worked as a volunteer for the American Red Cross. One of the projects she undertook was an investigation of the life cycle of the louse, *Pediculus vestimenti,* and a preliminary examination of possible methods for its control. The work was carried out in part at the École de Médecine, and was made possible by a legacy left by Ella Strobell, which paid the laboratory expenses and allowed Foot to give her time to the project for two years.[107] She lived in London for a period in the 1930s, but by the Second World War had returned to the United States and in 1944 was living in Camden, South Carolina. She was then ninety-two years old. She died some time between 1944 and 1949.

Relatively few of the American women whose published work in zoology is listed in the bibliography had close associations with the growing coeducational universities of the Midwest, South, or West. Among those who did were Merrill at the University of Wisconsin, Langdon and Phelps at the University of Michigan, Sturges at the universities of Michigan and Chicago, Rucker at the University of Texas, and Hartley at Stanford University. None of these went on to careers as zoologists, and only one, Merrill, became a teacher of biology; nevertheless all except Hartley pursued their zoological studies to the master's degree level.

HARRIET MERRILL,[108] a native of Stevens Point, Wisconsin, attended Milwaukee High School and state normal schools. She then taught in Milwaukee for a number of years and in 1888 entered the University of Wisconsin (B.S., 1890). She continued her studies under the guidance of zoologist Edward Asahel Birge and was awarded an M.S. in 1893. During summers and vacations she worked with Charles Whitman, both at Woods Hole (1893) and at the University of Chicago, where she held an honorary fellowship (1894–1896). Her paper on the structure of the microscopic shallow water crustacean *Bunops scutifrons* appeared in the *Transactions of the Wisconsin Academy of Science* in 1893, and a second, shorter article, reporting work on the eyes of the leech, was published the following year. She taught biology in Milwaukee high schools from 1890 until 1899 and then for a year at Downer College, Milwaukee. From 1901 onward she held a position at the Milwaukee Public Museum and also served as a lecturer in public schools. She died on 10 April 1915.

FANNY LANGDON[109] was born in Plymouth, New Hampshire, and received her early education at the New Hampshire State Normal School. She entered the University of Michigan in the early 1890s and took courses from Volney Spalding, professor of botany, and Jacob Reighard, then professor of animal morphology. Summers she spent at Woods Hole. Her long paper on the worm *Lumbricus agricola* appeared in the *Journal of Morphology* in 1895, and a second, on the common sea worm *Nereis virens,* in *Science* two years later. After receiving a B.S. in botany in 1896 and an M.S. the following year, she stayed on at Michigan as instructor in botany (1897–1898) and then as instructor in zoology (1898–1899). She died in Ann Arbor, Michigan, in October 1899.

JESSIE PHELPS[110] was also a student of Jacob Reighard. The daughter of Mary (Irish) and Edwin Phelps, she was born in Pontiac, Michigan, in January 1870. She studied biology at the University of Michigan (B.S., 1894; M.S., 1898). Her two notes on the North American freshwater fish *Amia* (the bowfin), one coauthored by Reighard and S. O. Mast, appeared in *Science* in 1899 and 1900. She later taught hygiene at the State Normal College in Ypsilanti, Michigan.

MARY STURGES,[111] the daughter of Ella (Delafield) and Charles M. Sturges, was born in Chicago in 1869. She took a B.S. in biology at the University of Michigan in 1893 and then went on to graduate work under the guidance of Charles Whitman at the University of Chicago. After receiving an M.S. in 1896 she continued her studies for another two years, carrying out work in cytology at Chicago and Woods Hole. Her paper on the worm *Distomum patellare* was published in 1898 and a cytological note appeared in *Science* the following year. She later worked in the Eugenics Record Office and at Columbia University, where she studied family histories in cancer and undertook research readings in biology.

AUGUSTA RUCKER[112] was born in Paris, Texas, 24 May 1873. She received an A.B. from the University of Texas, Austin, in 1896, and stayed on as a graduate student (A.M., 1899). The following year she was an instructor in the zoology department. Her 1900 paper on *Peripatus eisenii* (a member of the phylum Onyclophora, sister group of the arthropod complex) reported work carried out under the guidance of William Morton Wheeler in the zoological laboratory; she also published a substantial article in 1901 describing a new Texas arachnid, *Koenenia.*[113] These were probably her only zoology publications. She took an M.D. at Johns Hopkins Medical School in 1911, and thereafter worked as a physician, specializing in pediatrics, first at the New York Infirmary for Women and Children and later at the Ruptured and Crippled Hospital. She was still in practice in 1950, by which time she was in her mid seventies.

FLORA HARTLEY (1865–1948)[114] was one of Stanford University's earliest women graduates (A.B., 1895). Born in Yankeetown, Indiana, on 9 July 1865, the daughter of Amanda Angeline (Taylor) and Daniel Hartley, she took a B.S. in education in 1890 at Indiana State Normal School. While an undergraduate at Stanford she worked in the university museum as assistant curator, and later (1898–1900) held an assistantship in bionomics. Her two papers reporting studies in classification and taxonomy appeared in the *Proceedings of the California Academy of Sciences* in 1894 and 1895. She married Charles Wilson Greene, instructor of physiology at Stanford, in 1896. They had three children. In 1909, when she was forty-four, she took an M.S. in home economics at the University of Missouri (where her husband had been professor of physiology since 1900).

A prominent member and an officeholder in the Missouri Federation of Women's Clubs from the early years of this century, she was also a member of the General Federation of Women's Clubs. She chaired the economics section of the national board from 1918 until 1920, and in this capacity worked for passage by the U.S. Congress of a bill to permit the teaching of home economics in high schools. She was best known in Missouri for her extensive work in public welfare and child health care, and it was largely due to her efforts (through the state Federation of Women's Clubs) that hospital facilities for crippled children were established in St. Louis and Kansas City. For a time she served as director of the Missouri Crippled Children Society. During the First World War she was special agent of the U.S. Children's Bureau (1917–1920). Later, as chairman of the public health section of the Missouri Social Welfare Association, she assisted in health programs for schoolchildren throughout the state and lectured extensively on the care and feeding of children.[115]

Among the remaining eight women whose papers in zoology are listed in the bibliography is LOUISE TAYLER, who appears to be the one zoologist in this survey who held a position with the U.S. Department of Agriculture. A student of Edith Claypole at Wellesley College in the mid 1890s, she published a histological study on striped muscle fiber in the *American Monthly Microscopical Journal* in 1897. Her review of the literature on swine kidney worm appeared in the 1900 annual report of the Bureau of Animal Industry, U.S. Department of Agriculture, at which time she was an assistant in the bureau's zoological laboratory. She held B.A. and M.S. degrees.[116]

The importance of Woods Hole in the research work of late nineteenth-century American women zoologists is one of the prominent features in this overview. Not less than twenty (36 percent) of the fifty-five contributors of papers listed in the zoology section of the bibliography are known to have spent at least a summer or two there (or at one of the earlier East Coast marine laboratories). Many did the greater part of their work at Woods Hole, and for several, including some who were particularly outstanding, their experience at that laboratory was their start in biological research. Not surprisingly, Charles Whitman, director of Woods Hole for twenty years, from its opening in 1888 until 1908, was one of the most influential of the men zoologists of the period in fostering women's research efforts. He supervised work of at least ten of the women mentioned here, either at Woods Hole or at the University of Chicago, where he was professor of zoology and department head from 1892 until 1910. Three of these women—Clapp, Gregory, and Agnes Claypole—received Ph.D. degrees from Chicago. Thomas Hunt Morgan and William Morton Wheeler also stand out as people who encouraged women students in original work. Morgan, at Bryn Mawr College and Woods Hole, supervised at least ten in this group, and three of them—King, Peebles, and Byrnes—received doctorates from Bryn Mawr. Wheeler (at, successively, the University of Chicago, the University of Texas, and Harvard) is known to have advised or offered suggestions to at least four.

Six of those discussed here were starred in the first (1906) edition of *American Men of Science,* indicating that they had been ranked among the top 1,000 scientists in the country at the time. Four of these six—Claypole, Clapp, Foot, and Lewis Nickerson—were students of Whitman (though Lewis Nickerson also worked with Edward Mark at Harvard); King and Peebles studied with Morgan. Of the six, only Clapp and King remained prominent beyond the period of their early work. Clapp, although she published little after the turn of the century, was known for her teaching and administrative work at the Woods Hole laboratory for many years; King remained a productive scientist until 1949. The other four, although their stars formally remained in *American Men of Science* because of the editor's policy of not de-starring a scientist once so honored, were in fact no longer numbered among the country's leading scientists by 1910;[117] indeed, Claypole did little research after her marriage in 1903, Lewis Nickerson shifted her energies to medical practice after receiving her M.D. in 1904, and Foot, by about 1906, was pursuing a line in her genetics research that was theoretically unsound. Peebles, although she devoted much time and energy to administration and teaching, continued to publish significant work, but she failed to present her ideas and results fully and forcefully enough to attract attention to their broad significance.

The one other woman in this group to be starred, the marine zoologist Mary Jane Rathbun, received the honor in 1921, by which time she was an established and highly regarded taxonomist, long independent of any mentor and the author of many substantial papers and monographs in her field.[118] Her great productivity, together with that of her fellow marine zoologists Bush, Richardson, and Smith Eigenmann, has a defining influence on the overall picture of American women's publication output in zoology for the period. The heavy preponderance of papers in description and classification (44 percent of the total, as indicated in Figure 3–2 b) in large measure reflects the output of these three women; the fraction is well out of proportion to the number of authors working in the subfield (Figure 3-2 a).

One might wonder at the lack of wider recognition for Willcox and Thompson, both of whom did notable original work. However, Willcox's early training was in England, the director of her doctoral research was Arnold Lang of the University of Zurich, and she did not have an American mentor. Furthermore, apart from her dissertation research, published in German, her outstanding work on gastropod anatomy appeared after the turn of the century, perhaps somewhat late to have brought her recognition in the first edition of *American Men of Science,* which was compiled largely on the basis of information gathered up to 1903. Thompson's best work on ants and termites was done between 1913 and 1918, too late to be considered for recognition in the second edition of *American Men of Science;* she died in 1921, the year of publication of the third edition. There is also the case of the controversial Julia Platt. Platt undoubtedly deserved to be listed, and probably starred, in the first edition of *American Men of Science.*

British women

Pre-1901 papers by British authors make up about 15 percent of the zoology contributions by women indexed in the Royal Society *Catalogue,* in contrast to the 50 percent produced by American workers. There were twenty-six British contributors.[119] The largest subgrouping among them consists of six independent workers, three of whom were active before the 1880s (that is, in the period before university training was open to women); five others worked at University College Liverpool, either as students or independent investigators, four were students at University College London, four were students or staff at Newnham College Cambridge, three were at Oxford, two in Scotland, one at Owen's College Manchester, and one carried out research in the laboratory of the Zoological Gardens, London.

The earliest of the independent workers was JEANNETTE DE VILLEPREUX POWER,[120] wife of James Power. For several years in the 1830s and 40s she lived in Messina, on the northeastern tip of Sicily, but made frequent visits to London, where she had contacts in the naturalist community. She was also a corresponding member of the Gioenian Academy of Natural Science of Catania and well acquainted with a number of Italian scientific men. Her studies of marine life, an area of much popular interest at the time, were carried out during her stay in Sicily and published at intervals between 1837 and 1857 in British and European journals. She also read a number of reports at meetings of both the British Association and the Gioenian Academy.

Her special interest was mollusks. Introducing her 1839 *Magazine of Natural History* paper on the *Argonauta argo,* she explained that she had "for many years past devoted to natural science, and to enriching my cabinet with marine objects, the few hours to be spared from domestic cares." She added that "in fact few are the moments that one of my sex and condition can enjoy study . . .",[121] but she nevertheless succeeded in accomplishing a fair amount. She maintained her own glass aquaria, then recent innovations in natural history studies (though by mid-century a popular fad), and in addition arranged to have government permission to work in a seawater channel flowing through the Messina lazaretto; a sheltered spot in the Messina harbor was another of her work sites. The Sicilian seas were then so very limpid and transparent that observation, she noted, was very easy.

A woman of inventive turn of mind, she designed special marine cages in order to study her animals in their natural habitat, a procedure she recommended to all naturalists. These cages were later called "Gabbiole alla Power" by the Gioenian Academy and "Power Cages" by the London Zoological Society. With the encouragement of Professor Carlo Gemettaro and help from Dr. Anastasio Cocco, a Messina ichthyologist who often accompanied her in her harbor work, she carried out studies on a number of organisms, including investigations of growth and development from the egg stage, and food gathering procedures. She even examined processes of regeneration of excised parts. Although handicapped by a limited background in biology, she made a number of contributions, which were cautiously, but nevertheless fairly happily received by the British naturalist community. Thus, in a footnote to her 1838 *Magazine of Natural History* paper, the editor commented on her contributions as follows:

> . . . this lady's very slight acquaintance with comparative anatomy and physiology renders her observations of less importance than they might otherwise have been, yet there is every reason for believing, that experiments conducted in the manner which she has suggested, may satisfactorily determine whether or not the species of the genus Ocythöe are the true constructors of the shells which they inhabit.[122]

The paper nautilus (*Argonauta argo*), the subject of her main study, was an animal well known since antiquity as a favorite subject of poetic imagery. Two of its arms have membranes associated with them that early writers compared to sails, while its remaining six arms were thought of as oars. Thus equipped the "little navigator" could let itself be wafted along the surface of the sea, like the legendary heros of ancient times. In the 1830s the question of whether this mollusk was the builder of the shell in which it was found or a parasitic occupant was still undecided. Early in her studies Power had observed many specimens of *Argonauta* in her marine cages, and had concluded that the mollusk was indeed the builder of the shell it lived in. Further, the "sails" were, she deduced, organs connected with the construction and repair of the shell.

She communicated these ideas to the Gioenian Academy in 1834 and to the Zoological Society in 1837, but doubts remained among conchologists and many still held to the parasite theory. Therefore in 1838 she undertook further experiments to settle the question. Following a suggestion from Richard Owen (the British naturalist and paleontologist), she cut off one membraneous arm from each of several specimens. Few survived for long after the operation, but one lasted five days, and she was able to observe that it continued to add to its shell on the side where its arms were all present but not on the side that had suffered the amputation. Further, she found that in normal animals, if she broke off pieces of shell at the extremity of the "Great Whorl," the shell was not only repaired but enlarged in a period of four to ten days. In addition, when the shell was broken the membrane was immediately spread over the break and kept there until the repair was successfully completed. These observations, she felt, provided strong evidence against the parasite theory. She also studied the stages in the normal enlargement of the shell, the details of the *Argonauta*'s syphon-powered locomotion methods, and the development of the young.

In one of her last papers, published only in 1857, she reported a long series of observations on the feeding methods of a variety of marine animals, including *Octopus vulgaris*. She was interested in land animals as well, and in a second 1857 paper discussed the habits of martens. Two of these, taken from the forests near Mt. Etna, she had kept as pets. Small, almost fearless predators of the weasel family, they had adapted well to her household and formed a strong attachment to her personally. However, when she left Sicily for a stay in London, entrusting the martens to the care of the Duchess of Belviso, one died and the other escaped—despite the fact that the Duchess's naturalist husband had undertaken to care for them.

Among the other independent contributors of papers on zoological topics were E. F. Staveley and Catherine Hopley.[123] Staveley, an associate of naturalist John Edward Gray, zoology keeper at the British Museum, carried out anatomical studies on spiders and hymenopterous insects in the 1860s. Much of her work on the presence and form of teeth in spiders and hooks on insect wings was communicated to the Zoological Society for her by Gray. Her specimens went into museum collections.

CATHERINE HOPLEY,[124] of Tickenham, was a naturalist and writer, best known in her time for her books and magazine articles that described her experiences during a stay of several years in the United States in the late 1850s and early 60s. She had relatives in some of the northern and midwestern states, including a married sister in Indiana, and she spent some time visiting them and studying local plants and animals. In the spring of 1860, having decided to see something of the southern states as well, she found herself a position as governess to a family in Richmond, Virginia, and so was there at the outbreak of the Civil War. The following year she went south to Florida, where she was governess to the family of the state's governor. She made many friends in Richmond and became very much a Southern sympathizer. Florida, however, she found less to her liking than Virginia. She gave up her post there in the summer of 1862, made her way, with some difficulty, through the military cordon around the blockaded South, and soon after that went home to England.

Her two-volume *Life in the South; from the Commencement of the Civil War,* published anonymously in 1863, offers a lively and colorful account of her stay in the South when it was still a slave-owning society, and a view of the first year of the Civil War from the Southern side. She also wrote a short biography of the Southern general, Thomas ("Stonewall") Jackson, *Stonewall Jackson, a Biographical Sketch and an Outline of his Virginian Campaigns* (1863). Several years later she brought out two children's books based on her American experiences, *Rambles and Adventures in the Wilds of the West* (1872) and *Stories of Red Men* (1880). The former described a summer and autumn in Ohio (and also a winter in Florida), and incorporated a considerable amount of information about American plants, birds, and insects.

Later becoming specially interested in amphibians and reptiles, she worked for a time in the Gardens of the London Zoological Society. She published a note on the glottis of snakes in the *American Naturalist* in 1884 and another on mud fish in the same journal in 1891. Her natural history books for young people, *Snakes: Curiosities and Wonders of Serpent Life,* and *British Reptiles and Batrachians,* appeared in the 1880s. A number of her short notes recording observations on insects appeared in the entomology literature over the years.[125]

The largest subgrouping of early women zoologists associated with a university comprised the five women students or coworkers of William Abbott Herdman, professor of natural history at University College Liverpool from 1881. Herdman organized the Port Erin Marine Biological Station on the Isle of Man and was a leader in the investigation of the fauna of Liverpool Bay and the Irish Sea. Not surprisingly all five Liverpool women—Alice Heath, Fanny Palethorpe, Charlotte Wilson, Laura Thorneley, and Amy Warham—did their research in marine biology. Most of their publications appeared in the *Proceedings and Transactions* of the Liverpool Biological Society. With the possible exception of Heath, all of them were members of the society.

Alice Heath's 1883 paper was the earliest. In it, and in a second (1888) paper she reported anatomical and embryological studies. Fanny Palethorpe and Charlotte Wilson, as students in Herdman's advanced laboratory in 1886, worked on a collection of ascidians from Australian seas. This collection had originally been exhibited by the Australian Museum, Sydney, at the 1883 London Fisheries Exhibition, and was then sent to Herdman for examination. Palethorpe was a student member of the Liverpool Biological Society from 1887 to 1893.

LAURA THORNELEY, an ordinary member of the society for almost half a century, from 1891 to 1940, was a specialist in hydroid fauna. She published her first work in 1894, a supplementary report to a major investigation of material from Liverpool Bay, much of it obtained by dredging expeditions.[126] Her study was carried out at University College and the Port Erin Biological Station. Two years later, following an 1896 Irish expedition to Rockall, she published notes on the hydroids collected in the dredging of the shallow bank surrounding the island. Her later studies included an examination of hydroids from the neighborhood of Rathlin Island and Ballycastle, along the northeast coast of Ireland, and a major investigation of hydroids collected by Herdman on an expedition to Ceylon in 1902, when he went there to investigate the Pearl Oyster Fisheries of the Gulf of Mannar.[127] By 1920 Thorneley was living in Hawkshead in the Lake District.

Amy Warham's anatomical note on *Ascidia virginia* appeared in the Liverpool Biological Society's *Proceedings* for 1893. Three years earlier she had published a botanical paper coauthored with Robert Harvey-Gibson. A student at University College Liverpool in the late 1880s, she received a B.Sc. from Victoria University (of which University College Liverpool was a constituent college at the time). In 1890–1891 she taught at the Girls' High School, Pendleton, and by 1892 was living in St. Andrews.[128]

EDITH PRATT,[129] another early zoologist who studied aquatic organisms, was a student at Owen's College Manchester (Owen's College, later Manchester University, was then also a constituent college of Victoria University). Pratt went to Manchester from Notting Hill High School, London. She took a B.Sc. in 1897 and stayed on for further study under the direction of zoologist Sydney J. Hickson.

One of her first projects was an investigation of the freshwater plankton of Lake Bassenthwaite in the Lake District. At the time, although much was already known about freshwater fauna populations in European lakes, little systematic work had been done in England. Her study, involving the examination of samples collected in Lake Bassenthwaite by Hickson in 1897, was one of the earliest on English lakes. Hickson viewed it as an important beginning, which he hoped would stimulate further work.

Pratt published a lengthy study of the marine fauna of the Falkland Islands in 1898, and a number of papers on the taxonomy, anatomy, and physiology of the marine polyps of the subclass Alcyonaria followed. Among these was her report of an examination of collections brought back by Cambridge zoologist J. Stanley Gardiner from an 1899–1900 dredging expedition to the Maldive and Laccadive archipelagoes (coral formations in the Indian Ocean); under the general supervision of Hickson and with help from Herdman, she attempted a classification, by extremely detailed anatomical study, of various closely related genera into specific groups. Most of her work was done at the zoological laboratory at Owens College, although she also spent some time at the Port Erin Biological Station.

She became an associate of Owens College in 1898, received an M.S. in 1900 and a D.Sc. in 1904. After that, for a short period, she continued her investigations as an honorary research fellow of the University of Manchester, working in Manchester and, for a month in 1905, at the British Association Table at the Naples Zoological Station. Her last published paper was probably that reporting studies on the Pennatulids, the sea pens and sea pansies, one of the most complex groups in the Alcyonaria.[130] Sometime between 1905 and 1909 she married, becoming Mrs. Musgrave.

The four women who carried out work in zoology at University College London were Florence Buchanan, Emma Beck, Margaret Collcutt, and Margaret Robinson.

Buchanan was best remembered for her later work in physiology (see chapter 6), but her early investigations carried out at University College and the Plymouth Marine Laboratory (the first of the publicly funded British marine laboratories) form a significant contribution to the pre-1900 work in zoology by British women. Her studies dealt mainly with the anatomy and taxonomy of polychaete worms, on which she published several papers between 1889 and 1895.

EMMA BECK[131] was enrolled at University College London from 1879 to 1881, although she may have continued to work there informally for a short time after that as well. She studied zoology and comparative anatomy with Ray Lancaster in 1879–1880 and collaborated with him and W. B. S. Benham on a comparative study of the muscular and endoskeletal systems of Limnus and Scorpio. The work was reported in the *Transactions of the Zoological Society* in 1885; Beck's section dealt with the muscular and endoskeletal systems of Scorpio. She does not appear to have continued in scientific work; her other interests were history and architecture.

MARGARET COLLCUTT[132] (b. 1872) enrolled at University College London in 1890, having previously studied at University College Bangor, in North Wales. She ranked equal to the silver medalist in zoology and comparative anatomy in 1893–1894, and received first class in zoology in the Preliminary Science examinations in 1894. Although she remained at University College until 1895, she did not take a degree. Her study of *Hydractinia echinata,* carried out in the college's Zoological Laboratory, appeared in the *Quarterly Journal of Microscopical Science* in 1898.

MARGARET ROBINSON[133] (later Mrs. Browne) worked in the Zoological Laboratory at University College as an undergraduate in the late 1880s. She carried out anatomical studies on decapods under the direction of Walter Weldon. In 1903–1904 and 1908–1909 she returned to University College for further studies.

The zoologist in whose laboratory at University College London women had been accepted since the late 1870s, Ray Lancaster, moved to Oxford in 1891; there he continued to encourage women students. Of the three Oxford-educated women zoologists included here (Kirkaldy, Pollard, and Gould), two carried out their research under his direction.

JANE KIRKALDY,[134] the daughter of W. H. Kirkaldy of Wimbledon, entered Somerville College on an Exhibition Scholarship from Wimbledon High School in 1887. She took first class in the Final Honours School of Natural Sciences in 1891 and stayed on, specializing in zoology under the guidance of Lancaster. Some of her early experimental work was done at University College London in Walter Weldon's laboratory, but later studies were carried out at Oxford—mainly investigations on the anatomy and taxonomy of marine species, including *Amphioxus* and the Branchiostomidae. Two of her papers appeared in the *Quarterly Journal of Microscopical Science* in 1894 and 1895.

She remained at Oxford throughout her career, immersing herself completely, over a period of almost four decades, in teaching and in the organization and administration of the women's colleges, work in which she was considered outstanding. Initially she served as AEW tutor for all women reading natural science,[135] and later she became tutor or lecturer to the various women's societies in turn. To her also came the "delicate and difficult work in organizing almost from the beginning the gradual inclusion of women students in the intricate system of the Natural Sciences School at Oxford," a task she accomplished "with admirable skill."[136] For many years she was on the council of St. Hugh's College.

Her teaching load was always heavy, and she published no more research after 1895. However, she did bring out two books, a 558-page translation from the German of Johan Erik Vesti Boas's *Lehrbuch der Zoologie,* prepared jointly with Miss E.C. Pollard, and her *Introduction to the Study of Biology,* coauthored with I. M. Drummond (who was an Oxford Home Student—probably in the 1890s).[137] Lady Margaret Hall, a sister institution of Somerville College, selected Kirkaldy as one of its four Honorary Jubilee Fellows at its fiftieth jubilee in 1929. The following year, when her health failed, she resigned her positions. She died in a London nursing home on 19 June 1932.

Kirkaldy's colleague E. C. Pollard was also a student of Lancaster. Two of her papers on marine organisms appeared in the *Quarterly Journal of the Microscopical Society* in 1893 and 1894. She held a London B.Sc.

LILIAN GOULD[138] (1861–1936), one of the first women to be admitted to the Linnean Society, was born on 19 February 1861, the youngest daughter of the Rev. J. Nutcombe and his wife Katharine Emma Gould. She went to Somerville College in 1890 on a scholarship, and in 1894 obtained a first class in the Final Honours School of Natural Sciences, with animal morphology as her special subject. While a student she published two papers; one was on color in lepidopterous larvae, an area of special interest to her adviser, zoologist and entomologist Edward Bagnal Poulton, and the other was on the amoeba *Pelomyxa palustris.* She married Victor Herbert Veley, F.R.S., who had also taken first class in the Natural Sciences at Oxford, and collaborated with him in a number of bacteriological investigations. These were of a practical nature. Victor Veley was director and chairman of Baddow Brewery Company; their joint work concerned microorganisms found in spirits, particularly rum. During the First World War, Lilian Veley served as Commandant of a British Red Cross unit in London. Gardening, photography, and the breeding of Siamese cats were her special interests and pleasures. She died on 2 December 1936.

The four Cambridge women whose work is listed in the zoology section of the bibliography were Alice Johnson, Lilian Sheldon, Elizabeth Abbott, and Jessie Sallitt. None of them went on to research careers in biology, but the early embryological and morphological contributions of both Johnson and Sheldon were very creditable. Abbott and Sallitt became schoolteachers, Abbott's career lasting for thirty-six years, Sallitt's for twenty-one.

ALICE JOHNSON[139] (1860–1940) was born in Cambridge on 7 July 1860, the fourth daughter among the eight children of William Henry Farthing Johnson, master of a private school for dissenters, and his wife Harriet (Brimley).[140] She was educated at schools in Cambridge and Dover, and entered Newnham College on a Cambridge Senior Local scholarship in 1878. She held a College scholarship in 1880. After taking class I in part I of the Natural Sciences Tripos examinations in 1881, she stayed on as a research student, holding a Bathurst scholarship in 1882–1883. With the opening of the Balfour Laboratory in 1884, she was appointed its first demonstrator, and held the post of College Demonstrator in animal morphology until 1890.

A student of embryologist Francis Balfour, and after his death in 1882 of his successor Adam Sedgwick, her first research was on the development of the hind limb of the chick. Subsequently, at Sedgwick's suggestion, she undertook an investigation of the sequence of changes that takes place in the blastopore of the newt. Examining the earliest development of the tail end, she made sections of embryos that demonstrated that developmental features were similar to those in vertebrates in general. Her account of the work was presented to the Royal Society in 1884, and her paper was the first by a woman to appear in the society's *Proceedings.* She also looked at the embryological development of the cranial nerves in the newt, a study done in collaboration with Lilian Sheldon, then a Bathurst student of Newnham.

In 1890, at the age of thirty, she gave up her demonstratorship to become private secretary to Eleanor Sidgwick, who succeeded to the office of Principal of Newnham two years later. From then on, for most of the rest of her life, Johnson's work centered on the activities of the Society for Psychical Research, then a flourishing organization in which a number of well-known and influential university people, including Eleanor Sidgwick and her husband, Henry, were much involved (see the discussion of Eleanor Sidgwick in chapter 9). Johnson undertook both investigative work and a great deal of writing and editing. Among the society's early projects in which she assisted were those known as "The Brighton experiments," carried out about 1891. Examinations of the phenomenon of thought transference with hypnotized subjects, they concerned memory in the hypnotic trance; memories were found to come into the mind in successive stages of hypnosis and to recur spontaneously in the corresponding stage when the process was reversed. Johnson became editor of the society's publications in 1899 and remained in that position for seventeen years, until 1916. She was also secretary from 1903 to 1907, research officer for nine years beginning in 1907, and reporter for the society on the Census of Hallucinations from 1900 to 1904. Her original contributions to the society's *Proceedings* throughout the period from 1892 to 1936 were also substantial. In addition she did much of the preparation for publication of the work *Human Personality,* a discussion of psychic phenomena by F. W. H. Myers, left unfinished at the time of his death.[141]

Her ties with Newnham remained close for a number of years; she was secretary to the principal (Mrs. Sidgwick) until 1903, and an associate of the college from 1893 to 1902. She died in Cambridge, 13 January 1940, at age seventy-nine. Although she was well remembered for her "long years of hard, clear-headed work on psychical research,"[142] her early biological studies had been all but forgotten by the time of her death.

LILIAN SHELDON[143] (1862–1942), Johnson's coworker in some of her embryological investigations, was born in Handsworth, near Birmingham, in May 1862. She was the third daughter of the Rev. John Sheldon, vicar of the parish. From Handsworth Ladies College she went to Newnham on a College scholarship in 1880, and took both parts of the Natural Sciences Tripos examinations (1883 and 1884, class II in each). As a Bathurst student from 1885 to 1888, she carried out a substantial amount of research, not only her joint work with Alice Johnson on the embryology of the newt, but anatomical and morphological studies on *Cynthia rustica* and *Peripatus.* Her findings were published in five papers in the *Quarterly Journal of Microscopical Science* between 1888 and 1890. In addition she contributed a section on nemertines (ribbon worms) to the *Cambridge Natural History* series.[144] She remained on the Newnham staff until 1898, as demonstrator in morphology (1892–1893), and lecturer in comparative anatomy (1893–1898). From 1894 to 1906 she was a Newnham associate.

About 1898 she gave up academic work and joined her brother Gilbert, living first in London, and then for a time in Stroud in the Cotswolds. Later they moved to Lympstone near Exmouth (in south Devon), where others of their family had already settled following the father's retirement about 1898. Gilbert Sheldon had been crippled by polio in early childhood, but nevertheless traveled widely. He had special interests in architecture and topography.[145] Lilian joined him in his architectural investigations, and herself published a number of articles on traditional Devonshire buildings—barns, tollhouses, and inns—all well illustrated with many fine photographs. These appeared in the *Transactions* of the Devonshire Association,[146] of which she was a member for many years. During the First World War she worked for the Young Men's Christian Association in Birmingham, where she drove a car. Indeed, she was one of the earliest women drivers in the country. Active in community affairs in Exmouth, she served for a time on the local hospital committee. She died in Exmouth, 6 May 1942, at age eighty.

ELIZABETH ABBOTT[147] (1865–1955), the third of this group of early Cambridge zoologists, was from Pocklington, Yorkshire. The daughter of John D. Abbott, an architect, and his wife Jane (Ellis), she was born on 22 December 1865. She went to Newnham in 1888 from Aske's Haberdasher's School, Hatcham, and held a College scholarship in 1889. In the Natural Sciences Tripos examinations (1891 and 1892) she received class I in part I and class II in part II. Later she took an M.A. at Trinity College Dublin. From 1892 to 1894 she held a Bathurst studentship and carried out research with Hans Friedrich Gadow, whose specialty was vertebrate morphology. Their long joint paper on the evolution of the vertebral column of fishes appeared in *Philosophical Transactions* in 1895. The work involved a study of specimens collected off the coast of Portugal by Gadow. Abbott prepared and examined sections, and in the course of the study noted, and traced, a number of morphological features not previously recognized.

For the nine years between 1894 and 1903 she taught in the City of London School for Girls, and then for three years was headmistress of Northwich High School, while also serving as science lecturer for the London County Council. From 1907 to 1930 she was headmistress of Canton School, Cardiff.

Energetic, enterprising, and deeply interested in all aspects of education, she contributed many articles to educational journals and published several school textbooks and plays, dramatizations for school use of episodes from the classics. The latter included two plays based on Sir Walter Scott's romances *Woodstock* and *Kenilworth,* an episode from Alexander Dumas's *The Three Musketeers* entitled *The Queen's Diamonds,* and a dramatization of Lord Lytton's *The Last Days of Pompeii.* She also brought out an edition of *Much Ado about Nothing.*[148] Active in civic and social organizations as well, after her retirement she was for thirteen years treasurer of the Women's University Settlement. She died on 22 November 1955, a month before her ninetieth birthday.

JESSIE SALLITT[149] (1860–1944) was born 8 August 1860, in Bailiff Bridge, Yorkshire. She went to Newnham from Bradford Girls' Grammar School in 1878, but did not take a Tripos examination. Throughout much of the 1880s and 1890s she taught in a succession of schools (York High School,

1882–1883; King Edward VI High School, Birmingham, 1883–1886; Newcastle High School, 1892–1894; City of London School for Girls, 1894–1896), and from 1896 to 1903 served as headmistress of Yarmouth High School. Her one research paper, published in the *Quarterly Journal of Microscopical Science* in 1884, concerned chlorophyll in infusoria, a topic suggested to her by Ray Lancaster. After she gave up teaching she occupied herself with volunteer social work. She died on 14 December 1944, at age eighty-four.

Another early Newnham College zoologist was IGERNA SOLLAS[150] (1877–1965). Only her 1900 paleontology paper is listed in the bibliography, none of her zoology journal publications having appeared until 1902. However, she taught and was active in research in zoology for about twelve years.

Born on 16 March 1877 at Dawlish in Devon, she was the second daughter of William Johnson Sollas and his first wife, Helen, daughter of William John Coryn of Weston-super-Mare. William Sollas was professor of geology at, successively, the universities of Bristol, Dublin, and Oxford. After receiving her early education at Alexandra School and College, Dublin, Igerna went to Newnham on a Gilchrist scholarship in 1897. She took first class in the Natural Sciences Tripos examinations in both part I (1899) and part II (1901, zoology). After a year (1901–1902) as a Bathurst student and demonstrator at Newnham she became lecturer in zoology, the post she held from 1903 until 1913, with a two-year break (1904–1906) as a college research fellow. She was a Newnham Associate from 1902 to 1913.

Her wide interests involved her in a number of investigations and projects over the course of her scientific career. Much of her research was on marine organisms. While still a student she had contributed an article on sponges to volume 1 (1895) of the *Cambridge Natural History* series, and her journal publications included several papers on sponge structure, polyzoans, and ascidians. She also took part in two major collaborative projects: that with her father on problems in paleontology extended over the course of sixteen years and resulted in a number of substantial joint papers (see chapter 12); the other formed part of a lengthy series of investigations in evolutionary genetics, directed by the Cambridge geneticist William Bateson, in which several of the Newnham staff were closely involved. Sollas carried out cross-breeding experiments with guinea pigs, looking particularly at a dwarf form.[151] In addition, for a time around 1906 to 1908, she assisted in the preparation of the Zoology Section of the *International Catalogue of Scientific Literature,* a monumental index of world scientific literature brought out over the period of 1901 to 1919 by the Royal Society for the International Council.[152]

She lived to the age of eighty-eight, dying in November 1965, but she does not seem to have published further scientific work after 1916. In later life she became interested in the Christian Science movement and contributed occasional articles to Christian Science journals.

Little information has been uncovered about Sophie Fedarb and Nellie Maclagan, the two remaining women whose zoology papers are listed in the bibliography (other than those whose major interests were in other fields). Fedarb's study of earthworms collected in India and the East Indies islands was carried out in the laboratory of the London Zoological Society. She was able to work there thanks to an arrangement with Frank Evers Beddard, naturalist on the 1882–1884 Challenger Expedition and prosector for the Zoological Society, who specialized in earthworms. Four of Fedarb's papers appeared in the 1890s. Maclagan prepared a list of edible British fishes (with their Latin, French, German, and Italian names) for the 1884 *Annual Report* of the Fisheries Board for Scotland.

Work relating to marine life was also carried out by MARION NEWBIGIN (1869–1934). Although Newbigin's later distinguished career in geography is discussed in chapter 13, her early biological contributions deserve mention here. She was in fact one of the noteworthy women marine biologists of the 1890s, with a special talent for bringing her subject to a wide audience. She might well be compared to Rachel Carson, the American biologist who followed in her footsteps half a century later (and is now remembered for her books *The Sea Around Us,* and, most especially, *The Silent Spring*). During the late 1890s Newbigin worked at the Scottish Marine Biological Station at Millport on the island of Cumbrae in the Clyde estuary, by then a flourishing laboratory, focusing its research on the fish and plankton of the Clyde area.[153] Her studies there formed the basis of her extremely popular monograph, *Life by the Seashore: an Introduction to Nature Study* (1901), which became a classic and was to reappear in a number of revised editions over the next three decades.

Much of her research, however, concerned the phenomenon of color in nature, especially the pigments to be found in plants and animals. Starting with an attempt to classify the common plant pigments, work published in the *Transactions* of the Edinburgh Botanical Society in 1896, she went on to animal studies. In particular she investigated the red and yellow pigments of the muscle and ovary of salmon, and the variation of their concentrations in these locations with the season. The work was part of a larger collaborative study, with D. N. Paton of the Edinburgh Royal College of Physicians and several other biologists, on the changes in the metabolism of salmon as they move upriver during seasonal migration. She also investigated color in Crustacea, especially the blue pigment in lobsters. These various studies resulted in a very successful general work, *Colour in Nature: a Study in Biology,* which she brought out in 1898, when she was not yet thirty. Her two early monographs, both of which offered broad, but at the same time solidly scientific discussions of aspects of the natural world, led her directly into her major studies in biogeography.

She taught zoology at the Extra-mural School of Medicine for Women in Edinburgh for at least five years, until 1902, when she became editor of the *Scottish Geographical Magazine,* one of the most influential journals in its field. Even after that, however, she can hardly be said to have turned her back on zoological work, but instead integrated her biological background into many of her geographical studies. Her *Animal Geography* (1913) drew heavily on material she had used in her lectures at the School of Medicine for Women, and her final work, *Plant and Animal Geography* (1936), a classic in biogeography, presented in a broad and general synthesis the ideas on the interactions of organisms with their environments, which she had developed from both her biological and geographical work.

The publications of Power and Staveley, considered along with those of naturalists discussed in chapter 4—Bowdich Lee, Thynne, and Harvey—suggest that interest in zoologically oriented natural history was not lacking among British women in the early and midyears of the century (those who published most likely constituting a small fraction of those active in the area). Given the established importance of biology in British schools and universities by the 1880s,[154] and the fact that the vast majority of women science students were training to become teachers, considerable numbers must have studied zoology at the undergraduate level. And so one might wonder why the early interest did not develop further, and why relatively few women zoology students (compared, for instance, with botanists and geologists) went on to make even modest contributions in the form of graduate research (Pratt and Sollas were notable exceptions). However, as noted in the discussion of British physiologists (chapter 6), access to the necessary laboratory facilities and thus to opportunities for advanced work in the area was limited.

Two-country comparison

Less than half as many British women as Americans published in zoology before 1900, and they wrote little more than a fourth as many papers (see Figure 3-1).

In both countries there was a marked emphasis on marine zoology, a reflection of the dominant position that this field had come to occupy in the biological sciences by the 1880s.[155] The most prominent British women's marine zoology group was that consisting of the five students and assistants of W. A. Herdman at University College Liverpool, although compared with the Americans working in the area their contributions were modest. There were no British counterparts of the very productive American museum workers Rathbun, Bush, and Richardson and the independent Smith Eigenmann, whose combined publications in marine classification and taxonomy constituted about 37 percent of all the zoology papers by the American women.

The widespread ongoing interest in basic embryological and developmental problems is also reflected in women's publications in both countries (Figure 3-2 b and d). Some of their most interesting contributions were in these areas. Notable on the American side was the work of the students of T. H. Morgan and Charles Whitman, and particularly outstanding was that of Julia Platt. Again the British contribution was considerably more modest than the American, but the work of the Cambridge students of Adam Sedgwick (Alice Johnson and Lilian Sheldon), and also that of Elizabeth Abbott, was well thought of.

Although the American museum workers, especially Rathbun, had fairly lengthy and productive careers, it is also the case that, like the British women, several of the most notable of the American zoologists (including Platt and Lewis Nickerson) gave up zoological research after completing their graduate studies. Willcox, Agnes Claypole, and, more especially, Thompson were more successful in pursuing independent research for a period after their student years. King's work in applied genetics at the Wistar Institute in Philadelphia stands by itself, although Sollas's contributions to investigations in evolutinary genetics in the Bateson team in Cambridge should not be overlooked.

There were relatively few college teaching positions open to the British women, the academic careers of only two, Kirkaldy at Oxford and Sollas at Newnham College, being comparable to those of the Americans Willcox, Thompson, Robertson, Clapp, and Wallace at Wellesley and Mount Holyoke. Alice Johnson, Lilian Sheldon, and Marion Newbigin, although they each taught for several years, moved on relatively quickly to other interests.

In addition to the two already-noted circumstances that contributed to the special prominence of the American women (namely a few positions in large museums, and the existence of faculty posts likely to be filled by women in women's colleges), there were two additional factors critical on the American scene. These were the existence of the Woods Hole Marine Biological Laboratory (also important to the women's college faculty members as their summer research facility), and the presence on the Bryn Mawr College faculty of T. H. Morgan. Although there were several marine research laboratories around the British coasts, and women did work at them (Thorneley and Pratt at Port Erin, Buchanan at Plymouth, and Newbigin at Millport), none seems to have played a role in fostering the work of women biologists comparable with that of the Woods Hole laboratory. Indeed the opportunity for learning and research that Woods Hole provided during the twenty years when it was directed by zoologist Charles Whitman was exceptional. Women went there in considerable numbers, summer after summer, from the opening in 1888. For some of those in this survey it provided their introduction to biology; others did the greater part of their best work there.

The presence of the future Nobel Prize-winning biologist Thomas Hunt Morgan on the Bryn Mawr College faculty for the thirteen-year period from 1891 to 1904 likewise had a considerable influence on the emergence of American women as the dominant national group among late nineteenth-century women zoologists. Unlike the male British instructors, all of whom taught men as well as women, Morgan, during his years at Bryn Mawr had women students only; and he put before them (both undergraduates and graduate students) his ideas in embryology and regeneration to work out in the laboratory. The results were remarkable;[156] twenty-two of the papers indexed in the bibliography came from his students, two of whom (King and Peebles) were highly ranked among turn-of-the-century American zoologists.

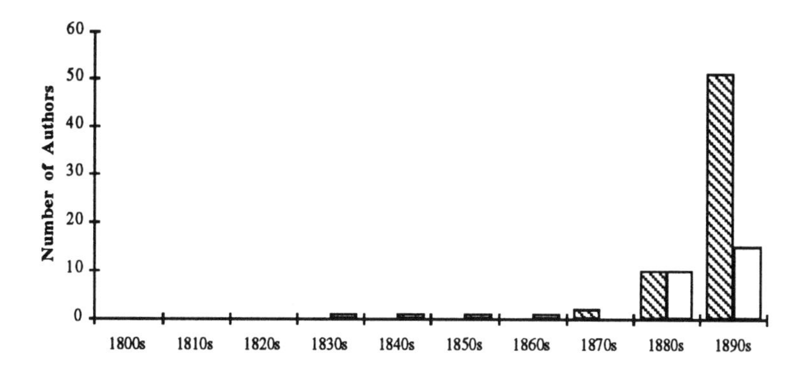

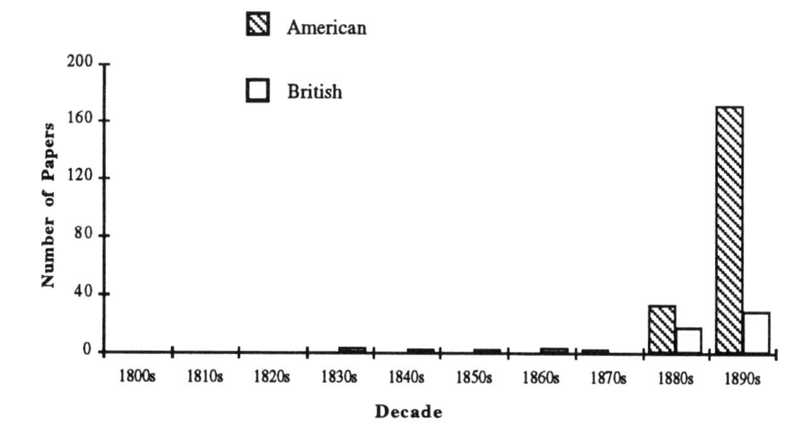

Figure 3-1. American and British authors and papers in zoology, by decade, 1800–1900. Data from the Royal Society *Catalogue of Scientific Papers.*

Notes

1. Six of those whose work is listed in the zoology section of the bibliography also had interests in other fields and are discussed in the corresponding chapters; they are Porter Cockerell (chapter 2), Hitchcock and Pennington (chapter 11), Monks (chapter 7), and Hefferan and Ross (chapter 6).

2. Lucile McCain, obituary, *Science,* **97** (1943): 435–6; Waldo L. Schmitt, obituary, *Journal of the Washington Academy of Sciences,* **33** (1943): 351–2, and "Rathbun, Mary Jane," in *NAW,* vol. 3, pp. 119–21; *WWWA,* p. 672; *AMS,* 1910–1933.

3. McCain, obituary, p. 435.

4. Rossiter, *Women Scientists in America,* p. 58.

5. *Nouvelles Archives,* Paris Museum of Natural History, n. 6 and 8, 1904–1906.

6. These were: *The Grapsoid Crabs of America,* U.S. National Museum Bulletin 97, 1918, 461 pp.; *The Spider Crabs of America,* U.S. National Museum Bulletin 129, 1925, 613 pp.; *The Cancroid Crabs of America of the Families Euryalidae, Portunidae, Atelecyclidae, Cancridae and Xanthidae,* U.S. National Museum Bulletin 152, 1930, 609 pp.; *The Oxystomatous and Allied Crabs of America,* U.S. National Museum Bulletin 166, 1937, 278 pp.

7. Jeanne E. Remington, "Katharine Jeannette Bush: Peabody's mysterious zoologist," *Discovery,* **12** (1977): 3–8; *WWWA,* p. 151; *Who was Who in America,* vol. 1, p. 175; *AMS,* 1910–1933.

8. Katharine Jeannette Bush, "Tubicolous annelids of the tribes Sabellides and Serpulides from the Pacific Ocean," *Smithsonian Institution Publication 1999* (Harriman Alaska Expedition Series, vol. 12 (1905): 167–346). (Bush's dissertation was entitled: "Descriptions of three new genera and sixteen new species belonging to the tribes Sabellides and Serpulides.")

9. See, for example, Katharine J. Bush, "Descriptions of two genera of tubicolous annelids, *Paravermilia* and *Pseudovermilia,* with species from Bermuda referable to them," *American Journal of Science,* s. 4, **23** (1907): 131–6; "Descriptions of new serpulids from Bermuda with notes on known forms from adjacent regions," *Proceedings of the Academy of Natural Sciences, Philadelphia,* **62** (1910): 490–501.

10. Remington, "Bush," pp. 3–4.

11. See Rossiter, *Women Scientists in America,* p. 58.

12. *WWWA,* p. 686; S and F, pp. 361–2; *AMS,* 1910–1955.

13. Harriet Richardson, "Isopodes," in Jean B. A. É. Charcot, *Expédition Antarctique Française (1903–05). Sciences Naturelles: Documents Scientifiques. Crustacés* (Paris, 1906–1908), and in Charcot, *Deuxième Expédition Antarctique Français (1908–10). Sciences Naturelles: Documents Scientifiques* (Paris, 1913).

14. See, for instance, "Description d'un nouveau genre de crustacé isopode de la Nouvelle Zemble et appartenant à la famille des Monnopsidae," *Paris, Muséum National d'Historie Naturelle, Bulletin* (1919): 569–73; "Isopod crustaceans of the Dutch West Indies" [Report of the 1904 expedition], (Utrecht: Universiteit te Utrecht, 1919); "Terrestrial isopoda collected in Java by Dr. Edward Jacobson with descriptions of five new species," *U.S. National Museum, Proceedings,* **60** (1922): 1–7; "Isopodes terrestres," in *Voyage de m. le Baron Maurice de Rothschild en Ethopie et en Afrique Orientale Anglaise . . . Resultats Scientifiques,*

a. American authors

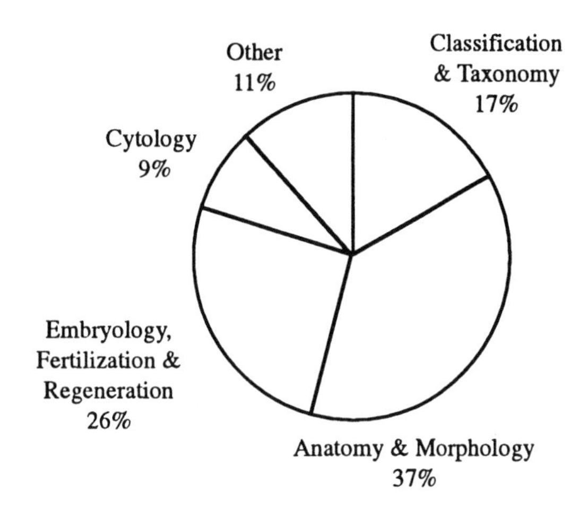

b. American papers

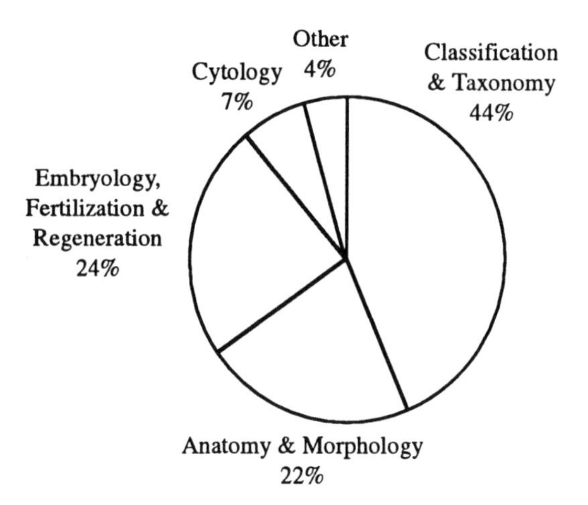

c. British authors

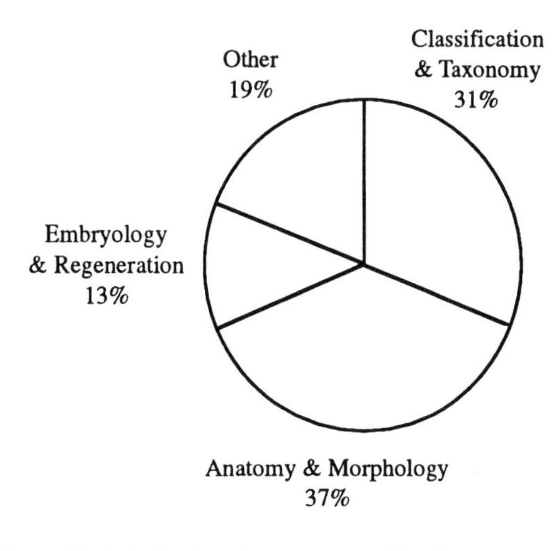

d. British papers

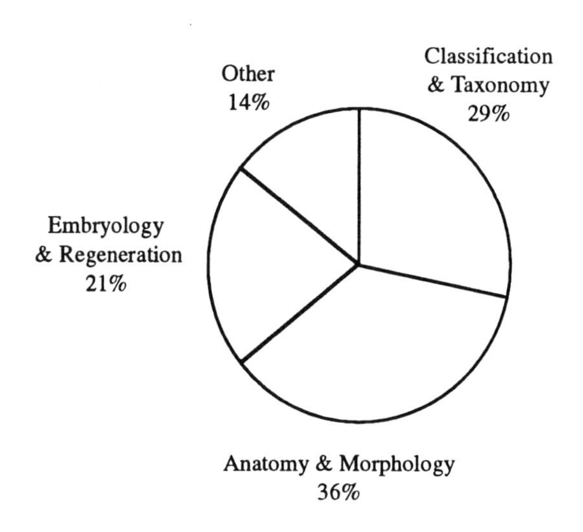

Figure 3-2. Distribution of American and British authors and papers in zoology, by subfield, 1800–1900. (In a. and c., an author contributing to more than one subfield is counted in each.) Data from the Royal Society *Catalogue of Scientific Papers.*

Animaux articulés . . . (Paris: Imprimerie nationale, 1922), pt. 1, pp. 19–34; "Crustacés isopodes terrestres d'eau douce," in *Voyage Zoologique d'Henri Gadeau de Kerville en Syrie* (Paris: J. B. Ballière, 1921–1926), vol. 1, pp. 303–10.

15. A third woman student of marine zoology at the University of Indiana whose work is listed in the bibliography was Jennie Horning. She seems to have authored only one scientific publication, a joint paper with Carl Eigenmann, which appeared in the *Annals of the New York Academy of Sciences* in 1887.

16. Carl L. Hubbs, "Eigenmann, Rosa Smith," in *NAW,* vol. 1, pp. 565–6; M. Burton Williamson, "Some American women in science," *The Chautaquan,* **28** (1898–1899): 161–8, 361–8, 465–73, on pp. 364–5; *WWWA,* p. 271.

17. *Proceedings of the California Academy of Sciences,* **1** (1888): 273.

18. Student Records, University of Indiana archives; information from Palo Alto City Library, Palo Alto, California.

19. Alice Payne Hackett, *Wellesley. Part of the American Story* (New York: E. P. Dutton, 1949), p. 82; Stephen A. Bailey, *L. L. Nunn: a Memoir* (Ithaca, N.Y.: Telluride Association, 1933), pp. 16–17; Linda Tucker and Christiane Groeben, "My life is a thing of the past': the Whitmans in zoology and marriage," in *Creative Couples in Science,* eds., Helena M. Pycior, Nancy G. Slack and Pnina G. Abir-am (New Brunswick, N.J.: Rutgers University Press, 1996), pp. 196–206; Frank R. Lillie, "Biographical sketch of Charles O. Whitman," *Journal of Morphology,* **22** (1911): xv–lxxvii, on pp. xv, xxv.

20. Emily Nunn's difficulties in Naples did not escape notice on her home territory: in the opinion of Bryn Mawr College president Martha Carey Thomas, who considered both Emily and her husband

for positions on the Bryn Mawr faculty, Charles Whitman's early difficulties in finding an academic position were due to his wife, whose "notorious" behavior at Naples had made "all Americans ashamed of her" (Philip J. Pauly, "The appearance of academic biology in late nineteenth century America," *Journal of the History of Biology*, **17** (1984): 369–97, n. 36. Pauly refers to M. Carey Thomas, "Conversations about college organization in 1884," manuscript notebook, Bryn Mawr College archives).

21. Hackett, *Wellesley*, p. 82; David R. Lindberg, "Mary Alice Willcox, 1856–1953," *Annual Report, Western Society of Malacologists*, **10** (1977): 16–17; information from faculty biographical files, Wellesley College archives, including a memorial by Vida Dutton Scudder, "Mary A. Willcox;" *Who was Who in America*, vol. 3, p. 919; *AMS*, 1910–1949.

22. The only course in zoology offered at Wellesley during these two years was one in taxidermy (Hackett, *Wellesley*, p. 82).

23. See Stella Butler, "Centres and peripheries: the development of British physiology, 1870–1914," *Journal of the History of Biology*, **21** (1988): 473–500, especially p. 488.

24. Toby A. Appel, "Physiology in American women's colleges. The rise and decline of a female subculture," *Isis*, **85** (1994): 26–56 (see also chapter 6, n.3).

25. Mary Willcox, undated, typed manuscript, faculty biographical files, Wellesley College archives (original italics).

26. A fourth paper on insect anatomy published from Wellesley in 1896 reported work by Willcox's student, Martha Goddard. Goddard does not appear to have brought out any further work in zoology. Willcox's two monographs were the following: *Notes on the Morphology of Invertebrates* (Boston: [n.p.], 1892, 144 pp.) and *Pocket Guide to the Land Birds of New England* (Boston: Lee and Shepard, 1895, with later editions in 1901 and 1903, 158 pp.) The latter was one of the first guidebooks to use color and coloration as an aid to bird identification; it came out just before Frank Chapman's well-known *Handbook of Birds of Eastern North America* (1895).

27. James C. Albisetti, *Schooling German Girls and Women* (Princeton: Princeton University Press, 1988), p. 224.

28. Willcox is on record as saying, "I hoped that we might cooperate with the department of botany, but as Miss Hallowell rejected my overtures that proved impossible." (Willcox, undated, typed manuscript, Wellesley College archives.) Interestingly, a similar botany-versus-general-biology issue had come up at Bryn Mawr a few years earlier. There the more senior faculty member involved, Johns Hopkins-trained morphologist E. B. Wilson, succeeded in establishing a highly successful comprehensive biology program. In the process of the reorganization the college's botanist, Emily Gregory, who refused to subordinate her courses to general biology, was forced out (see sketch of Emily Gregory, chapter 1).

29. "Somewhere about 1905 or '06, at the urgent desire of [the botany] department I consented to a restriction of ours," (Willcox, undated, typed manuscript). Rossiter (*Women Scientists in America*, p. 24) has pointed out a pattern that, by the 1920s, had evolved for science departments in the Eastern women's colleges. None of these institutions had the financial resources necessary for all-round excellence in the sciences, but most had by then notable programs in one or two fields; for instance, Bryn Mawr was strong in geology and mathematics, Mount Holyoke in zoology, and Wellesley in botany and physics. This spreading out of notable departments developed quite unintentionally as the cumulative result of many unrelated events and circumstances. In the case of Wellesley's botany department, Willcox's acquiescence to the requests of her botanist colleagues, as early as 1906, was of major importance. In fact the Wellesley zoology department suffered a further setback at this time, the newly established department of hygiene and physical education being authorized to offer a course in physiology that rivaled those offered by the zoology department. Instruction in physiology was only returned to the control of the zoology department after more than a decade of contention (Appel, "Physiology in American women's colleges," p. 35, and Hackett, *Wellesley*, pp. 141–2).

30. Willcox's later papers were the following: "A revision of the systematic names employed by writers on the morphology of the Achmaeidae," *Proceedings of the Boston Society of Natural History*, **29** (1899–1901): 217–22; "A parasitic or commensal Oligochaete [*Chaetogaster limnaei* ?] in New England," *American Naturalist*, **35** (1901): 905–9; "Some disputed points in the anatomy of limpets," *Biological Bulletin*, **2** (1901): 331–2 and *Zoologischer Anzeiger*, **24** (1901): 623–4; "Biology of *Acmaea testudinalis* Müller," *American Naturalist*, **39** (1905): 325–33; "Homing of *Fissurella* and *Siphonaria*," *Science*, **22** (1905): 90–1; "Anatomy of *Acmaea testudinalis* (Müller). Part I. Introductory material—external anatomy," *American Naturalist*, **40** (1906): 193–7.

31. Undated manuscript entitled, "Reminiscences of Mary Alice Willcox, a special student of zoology at Newnham College 1880–1883" (Willcox, faculty biographical files, Wellesley College archives). From the context, the account was probably written about 1940. A slightly edited version appears in Ann Phillips, ed., *A Newnham Anthology* (Cambridge: Cambridge University Press, 1979). Permission for the present reproduction has been granted by the Principal and Fellows of Newnham College, Wellesley College, and the Willcox family.

32. Philip P. Calvert, "Caroline Burling Thompson," *Entomological News*, **33** (1922): 62–4; T. E. S[nyder], "Caroline Burling Thompson, 1869–1921," *Science*, **55** (1922): 40–1; *AMS*, 1910, 1921.

33. Thompson's ant, bee, and termite papers include the following: "A comparative study of the brains of three genera of ants, with special reference to the mushroom bodies," *Journal of Comparative Neurology*, **23** (1913): 515–72; "The posterior roots of the mushroom bodies in the worker *Bombos* sp.," ibid., **24** (1914): 283–9; "The brain and frontal gland of the castes of the 'white ant,' *Leucotermes flavipes* Kollar," ibid., **26** (1916): 553–602; "Origin of casts of the common termite, *Leucotermes flavipes*," *Journal of Morphology*, **30** (1917): 83–106.

34. Snyder, "Thompson," p. 41.

35. (with T. E. Snyder) "The question of the phylogenetic origin of termite castes," *Biological Bulletin*, **36** (1919): 115–32, and "The 'third form,' the wingless reproductive type of termites: *Reticulitermes* and *Prorhinotermes*," *Journal of Morphology*, **34** (1920): 591–623. Thompson also published, as single-author papers, "Dual queens in a colony of honey bees," *Science*, **48** (1918): 294–5, and "The development of the castes of nine genera and thirteen species of termites," *Biological Bulletin*, **36** (1919): 379–98.

36. *Who was Who in America*, vol. 1, p. 1041; Helen Goss Thomas, "Dr Alice Robertson—the teacher," *Wellesley Alumnae Quarterly* 7 (Nov., 1922): 23; S. C. Hart, "In Memoriam," *Wellesley College News*, 28 September 1922, p. 8; *AMS*, 1910, 1921.

37. "Embryology and embryonic fission in the genus *Crisia*," *University of California, Publications in Zoology*, **1** (1903): 115–56.

38. "Non-incrusting chilostomatous bryozoa of the west coast of North America," ("Contributions from the laboratory of the Marine Biological Association of San Diego"), *University of California Publications in Zoology*, **2** (1905): 235–323.

39. *WWWA*, p. 355; *AMS*, 1910–1949.

40. Mary A. Bowers, "Histogenesis and histolysis of the intestinal epithelium of *Bufo lentiginosus*," *American Journal of Anatomy*, **9** (1909): 263–79.

41. Ogilvie, *Women in Science*, pp. 57–8; *WWWA*, p. 571; *AMS*, 1910–1949.

42. For Edith Claypole see chapter 6.

43. See Marion E. Hubbard, "Biographical sketch [of Edith Claypole]," in Agnes Claypole Moody and Marion E. Hubbard, *In Memoriam. Edith Jane Claypole* (Berkeley, Calif.: [n.p.], 1915).

44. Marian E. Hubbard, "The plight of our zoology department," *Wellesley Alumnae Magazine*, **12** (1928): 123–30.

45. Charlotte Haywood, "Clapp, Cornelia Maria," in *NAW*, vol. 1, pp. 336–8; obituary, "Cornelia H. Clapp, educator, 85, dies," *NYT*, 2 January 1935, 25; Ogilvie, *Women in Science*, p. 57; Jane Maienschein, "Cornelia Maria Clapp 1849–1934," in *Defining Biology: Lectures from the 1890's*, ed., Jane Maienschein (Cambridge, Mass.: Harvard University Press, 1986), p. 179; *WWWA*, p. 179; *AMS*, 1910–1933.

46. Rossiter, *Women Scientists in America*, pp. 19–20.

47. E. R. Benson and C. E. Quinn, "The American Society of Zoologists, 1889–1989: a century of integrating biological sciences," *American Zoologist*, **30** (1990): 353–96, especially pp. 357, 383.

48. *WWWA*, p. 848; S and F, p. 356; *AMS*, 1910–1955.

49. Louise Baird Wallace, "The spermatogenesis of the spider," *Biological Bulletin*, **8** (1905): 169–84.

50. Obituary, "Miss Emily Ray Gregory," *NYT*, 2 January 1946, 42; *WWWA*, p. 344; *AMS*, 1910–1949 (necrology).

51. *WWWA*, p. 729; Smith College archives.

52. *WWWA*, p. 94.

53. Obituary, "Dr. Helen Dean King, 85, noted zoologist," *NYT*, 10 March 1955, 27; Bonnie Tochter Clause, "The Wistar rat as a right choice: establishing mammalian standards and the ideal of a standardized mammal," *Journal of the History of Biology*, **26** (1993): 329–49; Ogilvie, *Women in Science*, pp. 108–10; *WWWA*, p. 458; *AMS*, 1910–1949.

54. Trademark 396,978, registered 11 August 1942, U.S. Patent Office (Clause, "The Wistar rat," p. 331).

55. Ibid., p. 346.

56. Accounts of King's early work on rat breeding appeared in the literature from about 1911 (see for instance, "The effects of semi-spaying and of semi-castration on the sex ratio of the albino rat (*Mus norvegicus albinus*)," *Journal of Experimental Zoology*, **10** (1911): 381–92; "The sex ratio in hybrid rats," *Biological Bulletin*, **21** (1911): 104–12; "Some anomalies in the gestation of the albino rat," ibid., **24** (1913–4): 377–91). Her main series of reports, published under the general title, "Studies on inbreeding," included the following: "I. The effects of inbreeding on the growth and variability in the body weight of the albino rat," *Journal of Experimental Zoology*, **26** (1918): 1–55; "II. The effects of inbreeding on the fertility and on the constitutional vigor of the albino rat," ibid., **26** (1918): 335–78; "III. The effects of inbreeding with selection, on the sex ratios of the albino rat," ibid., **27** (1918): 1–35; "IV. A further study of the effects of inbreeding on the growth and variability in the body weight of the albino rat," ibid., **29** (1919): 1–54. A sixth paper, published jointly with P. W. Whiting, was entitled, "Ruby-eyed dilute gray, a third allelomorph in the albino series of rat," ibid., **26** (1918): 55–64.

57. Included in King's papers for the period are the following: (with Leo Loeb) "Transplantation and individuality differentials in strains of inbred rats," *American Journal of Pathology*, **3** (1927): 143–67; "Seasonal variations in fertility in the sex ratio of mammals, with special reference to the [Norway] rat," *Archiv für Entwicklungsmechanik der Organismen*, Abth. D, **112** (1929): 61–111; (with H. H. Donaldson) "Life processes and size of the body and organs of the gray Norway rat during 10 generations of captivity. Pt. I. Life processes; Pt. II. Size of the body and organs," *American Anatomical Memoirs*, **14** (1929), 106 pp.; (with Leo Loeb) "Individuality differentials in strains of inbred rats," *Archives of Pathology*, **12** (1931): 203–21; "Birth weight in the gray Norway rat and the factors that influence it," *Anatomical Record*, **63** (1935): 335–54; (with H. H. Donaldson) "On the growth of the eye in three strains of the gray Norway rat," *American Journal of Anatomy*, **60** (1937): 203–29; "Life processes in gray Norway rats during fourteen years in captivity," *American Anatomical Memoirs*, **17** (1939): 1–72; "Labyrinthitis in the rat and a method for its control," *Anatomical Record*, **74** (1939): 215–22; (with Leon F. Whitney) "A second independent occurrence of the curly$_2$ mutation," *Journal of Heredity*, **30** (1939): 211–12; a series of papers, all coauthored by William Ernest Castle, under the general title, "Linkage studies of the rat (*Rattus norvegicus*)," published in the *Proceedings of the National Academy of Sciences*, **26** (1940): 578–80, **27** (1941): 250–4, **27** (1941): 394–8, **30** (1944): 79–82, **34** (1948): 135–6; "Linkage studies of the rat (*Rattus norvegicus*). Shaggy, a new dominant," *Journal of Heredity*, **38** (1947): 341–3; (with C. E. Keeler) "Multiple effects of coat color genes in the Norway rat, with special reference to temperament and domestication," *Journal of Comparative Psychology*, **34** (1942): 241–50; (with Margaret Reed Lewis) "A study of inducement and transplantability of sarcomata in rats," *Growth*, **9** (1945): 155–76; (with P. M. Aptekman and M. R. Lewis) "A method of producing in inbred albino rats a high percentage of immunity from tumors native in their strain," *Journal of Immunology, Virus Research and Experimental Chemotherapy*, **52** (1946): 77–86; (with P. M. Aptekman and M. R. Lewis) "Inactivation of malignant tissue in tumor-immune rats," ibid., **61** (1949): 321–6.

58. *Who was Who in America*, vol. 5, p. 565; Ogilvie, *Women in Science*, p. 145; Maltby, *History of Fellowships*, pp. 50–1; Gabriele Kass-Simon, "Biology is destiny," in *Women of Science. Righting the Record*, eds., G. Kass-Simon and Patricia Farnes, (Bloomington, Ind.: Indiana University Press, 1990), pp. 215–67, on pp. 220–2; *WWWA*, p. 634; *AMS*, 1910–1955.

59. Florence Peebles, "Further experiments in regeneration and grafting of hydroids," *Archiv für Entwicklungsmechanik der Organismen*, **14** (1902): 49–64; "A preliminary note on the position of the primitive streak and its relation to the embryo of the chick," *Biological Bulletin*, **4** (1903): 211–14.

60. See for instance, "The influence of grafting on the polarity of *Tubularia*," *Journal of Experimental Zoology*, **5** (1908): 327–58; "On the interchange of the limbs of the chick by transplantation," *Biological Bulletin*, **20** (1911): 14–18; "Regeneration and regulation in *Paramecium caudatum*," ibid., **23** (1912): 104–12; "Regeneration acöler Plattwürmer. 1. *Aphanastoma diversicolor*," *Bulletin de l'Institut océanographique*, **263** (1913): 1–5; "On some aceolous flatworms from the Gulf of Naples," *Zoologischer Anzeiger*, **43** (1913): 241–4.

61. "Growth regulation in *Tubularia*," *Physiological Zoology*, **4** (1931): 1–36; for the joint work with Fox see, "The structure, functions and general reactions of the marine sipunculid worm *Dendrostoma zostericola*," *Bulletin of the Scripps Institution of Oceanography, Technical Series*, **3**, n. 9 (1933): 201–44.

62. *WWWA*, p. 671; S and F, p. 353; *AMS*, 1910.

63. Benson and Quinn, "American Society of Zoologists," pp. 357, 383.

64. "On the spermatogenesis of the earwig, *Anisolabis maritima*," *Biological Bulletin*, **15** (1908): 111–18.

65. *WWWA*, p. 375; S and F, pp. 259–60; *AMS*, 1910, 1921.

66. "Regeneration in *Hydractinia* and *Podocoryne*," *American Naturalist,* **36** (1902): 193–200; "The regeneration of an oesophagus in the anemone, *Sagartia luciae*," *Archiv für Entwicklungsmechanik der Organismen,* **14** (1902): 592–9; "Regeneration in the anemone, *Sagartia luciae*," ibid., **16** (1903): 365–76.

67. Obituary, "Dr. Esther Byrnes, a science teacher," *NYT,* 5 September 1946, 27; *WWWA,* p. 153; S and F, p. 72; *AMS,* 1910–1949 (necrology).

68. Tsuda College, originally Women's English School, was founded in Tokyo in 1900 by Tsuda Umeko (1864–1929), one of the outstanding figures in the history of women's education in Japan. Among the first Japanese girls to be educated abroad, Umeko spent eleven years (1871–1882) in Washington, D.C., and in the 1890s returned to the United States to study biology at Bryn Mawr (where she collaborated with Morgan in embryological research) and education at Oswego Teachers College. Throughout her life she promoted educational and cultural exchange between the United States and Japan (see Barbara Rose, *Tsuda Umeko and Women's Education in Japan*—New Haven, Conn.: Yale University Press, 1992). Presumably Umeko and Byrnes knew each other from their Bryn Mawr days.

69. See for instance, "Heterogeny and variation in some of the Copepoda of Long Island," *Biological Bulletin,* **5** (1903–1904): 152–68; "On the skeleton of regenerated anterior limbs in the frog," ibid., **7** (1904–1905): 166–9; "Regeneration of the anterior limbs in the tadpoles of frogs," *Archiv für Entwicklungsmechanik der Organismen,* **18** (1904): 171–7; "The regeneration of double tentacles in the head of *Nereis dumerilis*," ibid., **21** (1906): 126–9; "Two transitional stages in the development of *Cyclops signatus* var. coronatus," *Biological Bulletin,* **10** (1905–06): 193–200; "Experiments in breeding as a means of determining some relationships among cyclops," ibid., **37** (1919): 40–8.

70. Clara Langenbeck worked at Woods Hole during the three summers, 1893, 1894, and 1895, and held a graduate fellowship in biology at Bryn Mawr in the winter of 1896–97. Her morphological work on crustaceans appeared in a lengthy paper in the *Journal of Morphology* in 1898.

71. Katherine Keenan, "Lilian Vaughan Morgan (1870–1952): her life and work," *American Zoologist,* **23** (1983): 867–76; Isabel Morgan Mountain, "An introduction to Thomas Hunt Morgan and Lilian Vaughan Morgan," ibid., **23** (1983): 825–7; Ian Shine and Sylvia Wrobel, *Thomas Hunt Morgan. Pioneer of Genetics* (Lexington, Ky.: University of Kentucky Press, 1976), pp. 29, 43–5, 93–107, 126–30; *WWWA,* p. 241; Bryn Mawr College records.

72. "A contribution to the embryology of *Hylodes martinicensis*," *American Journal of Anatomy,* **3** (1904): 473–504; "Incomplete anterior regeneration in the absence of the brain in *Leptoplana littoralis*," *Biological Bulletin,* **9** (1905): 187–93; "Regeneration of grafted pieces of planarians," *Journal of Experimental Zoology,* **3** (1906): 269–94.

73. Keenan, "Lilian Vaughan Morgan," p. 873.

74. Ibid., p. 874. In the early years of the drosophila work, many of Morgan's assistants were women, but from about 1915, when it had become evident that this research was mainstream biology, the fly team was predominantly male, the key members forming an elite group, unusually ambitious, aggressive, and given to working at a fast pace—see Robert E. Kohler, *Lords of the Fly: Drosophila Genetics and the Experimental Life* (Chicago: University of Chicago Press, 1994), pp. 96–7.

75. See Lilian V. Morgan, "Non-cris-cross inheritance in *Drosophila melanogaster*," *Biological Bulletin,* **42** (1922): 267–74. Her procedures and findings were, briefly, as follows: she bred the female mosaic with a black male and the female offspring of these with normal males. The resulting sons she found to have their mothers' Y chromosomes and their fathers' X chromosomes, and therefore to resemble their fathers in X-linked characterstics. Viable daughters had their mothers' attached-X chromosomes and their fathers' Y chromosomes, and resembled their mothers for X-linked characteristics. Of the four genotypes resulting from the crossing of an attached-X female with a normal male, the trisomic X rarely survived, the YY died, the XY was a normal male, and the XXY a normal female. This work was especially valuable as confirmation of Calvin Bridges's theory of sex determination (C. B. Bridges, "Triploid intersexes in *Drosophila melanogaster*," *Science,* **54** (1921): 252–4). The attached-X strain is useful in other ways as well, for instance in providing a means of probing the cytology and physiology of crossing over by maintaining certain combinations of characteristics linked to the X chromosome (since the son always receives the X chromosome of the father, it never lies adjacent to another X chromosome and so can never cross over). The attached-X inheritance pattern became a commonly used demonstration exercise in undergraduate genetics courses (see Keenan, "Lilian Vaughan Morgan," pp. 867–8).

76. "A closed X chromosome in *Drosophila melanogaster*," *Genetics,* **18** (1933): 250–83.

77. Lilian V. Morgan, "A variable phenotype associated with the fourth chromosome of *Drosophila melanogaster* and affected by heterochromatin," *Genetics,* **32** (1947): 200–19. Her other genetics publications not already mentioned include the following: "Polyploidy in *Drosophila melanogaster* with two attached X chromosomes," ibid., **10** (1925): 148–78; "Correlation between shape and behaviour of a chromosome," *Proceedings of the National Academy of Sciences,* **17** (1926): 180–1; (with A. H. Sturtevant, C. B. Bridges, T. H. Morgan, and Ju Chi Li) "Contributions to the genetics of *Drosophila simulans* and *Drosophila melanogaster*," *Carnegie Institute, Washington, Publication 399* (1929): 225–296; "Proof that bar changes to not-bar by unequal crossing over," *Proceedings of the National Academy of Sciences,* **17** (1931): 270–2; "Genetic behaviour of a closed X chromosome of *Drosophila*," *Proceedings of the Sixth International Congress of Genetics* (Ithaca, N.Y.), **2** (1932): 135; "Effects of a compound duplication of the X chromosome of *Drosophila melanogaster*," ibid., **23** (1938): 423–62; "Origin of attached-X chromosomes in *Drosophila melanogaster* and the occurrence of non-disjunction of X's in the male," *American Naturalist,* **72** (1938): 434–46; "A spontaneous somatic exchange between non-homologous chromosomes in *Drosophila melanogaster*," *Genetics,* **24** (1939): 747–52; (with T. H. Morgan and H. Redfield) "Maintenance of a *Drosophila* stock center, in connection with investigations on the germinal material in relation to heredity," *Carnegie Institute, Washington Yearbook,* **42** (1943): 171–4; (with T. H. Morgan and A. H. Sturtevant) "Maintenance of a *Drosophila* stock center, in connection with investigations on the germinal material in relation to heredity," ibid., **44** (1945): 157–60.

78. Anon., "Amelia Smith Calvert," in *American Women 1935–1940. A Composite Biographical Dictionary,* 2 vols., ed., Durwood Howes (Los Angeles: Richard Blank; reprint, Detroit: Gale Research Co., 1981), vol. 1, p. 141; *AMS,* 1910–1944.

79. Amelia C. Smith, "The influence of temperature, odor, light and contact on the movements of the earthworm," *American Journal of Physiology,* **6** (1902): 459–86.

80. Alumnae Records, Bryn Mawr College.

81. Obituary, "Dr Josephine H. Kenyon dies; was a famous baby specialist," *Boulder Daily Camera,* 11 January 1965, 6 (clipping, Bryn Mawr College archives).

82. Josephine Hemenway Kenyon, *Healthy Babies are Happy*

Babies: a Complete Handbook for Modern Mothers (Boston: Little Brown, 1934–1951). The third revised edition (1943) was coauthored by her daughter, Ruth Kenyon Russell. The work also appeared in a paperback ed. (Signet Books) in 1949 and 1950.

83. Helen Perkins worked with Charles Davenport at the Harvard Museum on problems of geotaxis. Their lengthy joint paper appeared in the *Journal of Physiology* in 1897. (For Bowers see the discussion of faculty members of Wellesley College.)

84. Maltby, *History of Fellowships*, p. 39; *American Medical Directory*, 1914–1936; "MN Biography Project File," Minnesota Historical Society; *AMS*, 1910–1933.

85. "Sensory and glandular epidermal organs in *Phascolosoma gouldii*," *Journal of Morphology*, **17** (1901): 381–98.

86. University of Kansas archives; anon., "Necrology: Gertrude Crotty Davenport," *The Graduate Magazine* (University of Kansas), **44** (1946): 28; *WWWA*, p. 230; *AMS*, 1910–1949 (necrology).

87. S and F, pp. 350–1; *AMS*, 1910–1927 (necrology).

88. Student records, University of Vermont; Steven J. Zottoli and Ernst-August Seyfarth, "Julia B. Platt (1857–1935): pioneer comparative embryologist and neuroscientist," *Brain, Behaviour and Evolution*, **43** (1994): 92–106; information from Drew M. Noden, Professor of Anatomy, New York State College of Veterinary Medicine, Cornell University; *WWWA*, p. 649; Ernst Theodor Nauck, *Das Frauenstudium an der Universität Freiburg in Breisgau* (Freiburg im Breisgau: Verlag Eberhard Albert Universitätsbuchhandlung, 1953), p. 31; *AMS*, 1921, 1933.

89. Jane M. Oppenheimer, *Essays in the History of Embryology and Biology* (Cambridge, Mass.: MIT Press, 1967), p. 257.

90. Charles Sedgwick Minot, "The embryological basis of pathology," *Science*, **13** (1901): 481–98, on p. 486. Although dealing mainly with Platt's work, Minot also mentioned observations by Hermann Klaatsch, which reinforced Platt's argument. Minot dismissed Kaatsch, however, as being simply in error.

91. See, for example, Anton Dohrn, "Studien zur Urgeschichte des Wirbelthierkörpers. 22. Weitere Beiträge zur Beurtheilung der Occipitalregion und der Ganglienleiste der Selachier," *Mitteilungen aus der Zoologischen Stazion zu Neapal*, **15** (1902): 555–654; Edwin Goodrich, "On the segmentation of the occipital region of the head in the *Batrachia urodela*," *Proceedings of the Zoological Society* (1911): 101–120; Nils Fritiof Holmgren, "Studies on the head in fishes," *Acta Zoologica*, **21** (1940): 51–267; Drew M. Noden, "The control of avian cephalic neural crest cytodifferentiation. I. Skeletal and connective tissue," *Developmental Biology*, **67** (1978): 296–312; R. Glenn Northcutt, "Ontogeny and phylogeny: a re-evaluation of conceptual relationships and some applications," *Brain, Behaviour and Evolution*, **36** (1990): 116–40; E. H. Gilland, R. Baker, and R. Gould, "Development of neuronal and mesodermal components of the elasmobranch VIth nerve and lateral rectus muscle," *Society for Neuroscience, Abstracts*, **18** (1992): 1303. For a fuller list of articles that contain citations to Platt's work see Zottoli and Seyfarth, "Julia B. Platt."

92. See H. L. Hamilton, *Lillie's Development of the Chick* (New York: Henry Holt and Co., 1952), pp. 148–9. The complex and intricate nature of the process of interpreting microscope sections of embryo tissue should not be overlooked. Additional complications arise from the fact that some of the structures are transitory.

93. Northcutt, "Ontogeny and phylogeny," pp. 129–30.

94. See Zottoli and Seyfarth, "Julia B. Platt," pp. 103–4.

95. Constance Auguste Gelderblom, Hoorn, The Netherlands, received an M.D. from the University of Freiburg in February 1895, and Elizabeth E. Bickford, of Piermont, New Hampshire, a Ph.D. in physiology in December 1895 (Nauck, *Das Frauenstudium*, p. 62).

96. Rossiter (*Women Scientists in America*, pp. 177, 180) has pointed out that personal idiosyncrasies, although regularly overlooked in male candidates for academic positions, were enough to put a woman out of the running—even for consideration for positions in women's colleges where men frequently advised on matters of faculty hiring.

97. See Zottoli and Seyfarth, "Julia B. Platt," p. 101. Their comments on this aspect of Platt's difficulties were based on an examination of the Anton Dohrn correspondence in the archives of the Naples Zoological Station.

98. Edwin G. Conklin, "Early days at Woods Hole," *American Scientist*, **56** (1968): 112–20, especially p. 120, and P[hilip] J. Pauly, "Summer resort and scientific discipline: Woods Hole and the structure of American biology, 1882–1925," in *The American Development of Biology*, eds. R. Rainger, E. R. Benson, and J. Maienscheim (New Brunswick, N.J.: Rutgers University Press, 1988), pp. 121–50, especially p. 134. Julia was not exactly circumspect in her comments herself, however. Conklin mentions that she always referred to him as "that Johns Hopkins gasteropod." She did, all the same, have friends in the American scientific community. Zottoli and Seyfarth mention in particular (p. 101) her cordial relationship with Herbert Neal, whom she knew at Harvard and also during the year he spent in von Kupffer's laboratory in Munich (1896–97). Neal shared her interest in the morphology of the vertebrate head.

99. She was not even listed in *American Men of Science* until 1921, more than two decades after she had given up research, although her work, as we have seen, continued to be discussed in the literature. She was, however, one of the few early women members of the American Society of Morphology, joining at the third annual meeting (1892) at the same time as Cornelia Platt and Harriet Randolph (see Benson and Quinn, "American Society of Zoologists," pp. 357, 383).

100. Platt to Harold Heath, printed in the *Pacific Grove Tribune*, 31 May 1935, and quoted by Zottoli and Seyfarth, "Julia B. Platt," p. 103.

101. Frank R. Lillie, *The Woods Hole Marine Biological Laboratory* (Chicago: University of Chicago Press, 1944), pp. 132, 160; Burton Williamson, "American women in science," p. 469; Maienschein, "Introduction," in *Defining Biology*, p. 40; information from Evanston Historical Society, Evanston, Ill.; Joan Mark, *A Stranger in her Native Land. Alice Fletcher and the American Indians* (Lincoln, Neb.: University of Nebraska Press, 1988), pp. 138, 141–6; *AMS*, 1910–1949 (necrology).

102. Burton Williamson, "American women in science," p. 469.

103. Ella Strobell was born in New York, 26 June 1862, and like Foot was educated in private schools and by tutors. She was Foot's companion and coworker for twenty years, from 1897 until 1917, when her illness brought their scientific collaboration to an end. She died about 1917 or 1918 (*AMS*, 1910; notes in Foot's later technical papers).

104. Among their papers published after 1900 were the following: "Photographs of the egg of *Allolobophora foetida. II.* Further notes on yolk-nucleus and polar rings," *Journal of Morphology*, **17** (1901): 517–54; "The spermatozoa of *Allolobophora foetida*," *American Journal of Anatomy*, **1** (1902): 321–7; "Further notes on the cocoons of *Allolobophora foetida*," *Biological Bulletin*, **3** (1902): 208–13; "The sperm centrosome and aster of *Allolobophora foetida*," *American Journal of Anatomy*, **2** (1903): 365–9; "Prophases and metaphase of the first maturation spindle of *Allolobophora foetida*," ibid., **4** (1905): 199–243.

105. Katharine Foot and E. C. Strobell, "Preliminary note on the results of crossing two hemipterous species with reference to the in-

heritance of an exclusively male character and its bearing on modern chromosome theories," *Biological Bulletin,* **24** (1913): 187–204, on p. 188. See also, "Amitosis in the ovary of *Protenor belfragi* and a study of the chromatin nucleolus," *Archiv für Zellforschung,* 7 (1911): 190–230.

106. Foot and Strobell, *Cytological Studies. 1894–1917* ([n.p.], 1917). Their major 1914–1917 publications were the following: "Preliminary report of crossing two hemipterous species, with reference to the inheritance of a second exclusively male character," *Biological Bulletin,* **27** (1914): 217–36; "Results of crossing *Euschistus variolarius* and *Euschistus servus* with reference to the inheritance of an exclusively male character," *Journal of the Linnean Society. Zoology,* **32** (1914): 337–73; "Results of crossing two hemipterous species, with reference to the inheritance of two exclusively male characters," ibid., **32** (1914): 457–93; "Results of crossing *Euschistus variolarius* and *Euschistus ictericus* with reference to the inheritance of two exclusively male characters," *Biological Bulletin,* **32** (1917): 322–42.

107. Katharine Foot, "Notes on *Pediculus vestimenti,*" *Biological Bulletin,* **39** (1920): 261–79.

108. Membership lists, Wisconsin Academy of Science, *Transactions of the Wisconsin Academy of Science,* vols. 8–17; *The University of Wisconsin. Its History and its Alumni,* ed. R. G. Thwaites (Madison, Wisc.: J. N. Purcell, 1900), p. 733; *University of Wisconsin Alumni Directory 1849–1919,* p. 225.

109. Burton Williamson, "American women in science," p. 468; alumni records, University of Michigan.

110. *WWWA,* p. 644.

111. Ibid., p. 793.

112. *AMS,* 1910–1949.

113. "The Texas *Koenenia* [K. Wheeleri n. sp.]," *American Naturalist,* **35** (1901): 615–30.

114. Alumni records, University of Missouri, Columbia; obituary (clipping, University of Missouri archives) in *Missourian,* Columbia (21 July 1948); *WWWA,* p. 341; *Who was Who in America,* vol. 4, p. 378.

115. Two of Hartley Greene's later papers were the following: "The amount of edible meat in the various cuts from animals of known life history," *Journal of Home Economics,* **2** (1910): 413–24 (thesis research); "The adult woman's challenge to the home economics teacher," ibid., **13** (1921): 193–8.

116. Information about Tayler came from notes in her technical articles. The little uncovered about the remaining seven American women whose papers in zoology are listed in the bibliography is as follows: Isabella Green of Akron, Ohio, took an A.B. at Buchtel College in Akron and then studied zoology under Simon Gage and Burt Wilder at Cornell University (M.S., 1896). Her lengthy 1896 paper reported her graduate research, histological studies on the peritoneal epithelium of several local amphibia. Marguerite Hempstead, who presented a paper on lung development in frogs at the 1900 meeting of the AAAS, was from Meadville, Pennsylvania. Edith Brace was a student of Charles Whitman at the University of Chicago (S.M., 1896). She published a full report of her work on the annelid *Aeolosoma tenebrarum* in the *Journal of Morphology* in 1901 (**17,** 177–84). By 1913 she was living in Rochester, New York (University of Chicago, *Alumni Directory,* 1913). Eva Field carried out her microscopical study of malignant growth in lower animals in Des Moines, Iowa. Annie Mozley, the oldest of the five children of Mabel and Edward Mozley, was born in Pennsylvania, probably in 1857. Her father was an English immigrant, a blacksmith by trade. By 1865 the family was living in Lawrence, Kansas, and Annie attended the University of Kansas as a special student from 1874 until 1877. Her paper on the snake collection in the university museum appeared in the *Transactions of the Kansas Academy of Sciences* in 1878 (Kansas State Census, 1865; University of Kansas archives). Edna Congdon and Gulielma Crocker were students of William Ritter, professor of biology at the University of California. Congdon worked on the anatomy of the freshwater worm *Stenostoma,* and Crocker on the identification of echinoderms and on starfish anatomy. Crocker's collaborative work with Ritter on the rays of the starfish appeared in the reports of the Harriman Alaska Expedition (1900).

117. Rossiter, *Women Scientists in America,* p. 350, n. 15.

118. It is remarkable how few among these early women zoologists received stars for work carried out during their more mature years. In most cases the award recognized work developing directly from the interests of the prominent research directors who supervised their early studies; as would be expected, the areas they worked in were those of major interest at the time.

119. Five of these twenty-six, Florence Buchanan, Agnes Crane, Edith Durham, Marion Newbigin, and Eleanor Ormerod, had major interests in other fields (see, respectively, chapters 6, 12, 14, 13, and 2); Newbigin's work in zoology is also mentioned here.

120. Jeannette Power was born in Juillac, Corrèze, France (Alphonse Rebière, *Les femmes dans la science: notes recueilles,* 2d. ed., Paris: Librairie Nony, 1897, p. 228-9). As well as her journal articles, she published the guidebook *Itinerario della Sicilia* (1839), *Observations ... sur ... plusieures animaux marins et terrestres* (1860), and *Observations sur l'origine des corps météoriques ...* (1867).

121. "Osservazioni fisiche sopra il polpo dell' *Argonauta argo,*" *Magazine of Natural History,* **3** (1839): 101–6, on p. 101.

122. See "Experiments made with a view of ascertaining how far certain marine testaceous animals possess the power of renewing parts which may have been removed," ibid., **2** (1838): 63–5, on p. 65.

123. The remaining three in the subgrouping of six independent workers were Crane, Durham, and Ormerod (see n. 119).

124. *British Biographical Archive,* microfiche edn., ed. Paul Sieveking (London: K. G. Saur, 1984), no. 571, entry from J. F. Kirk, *A Supplement to Allibone's Critical Dictionary of English Literature.*

125. In her *Life in the South* Hopley included a reproduction of the federal passport she had been issued in 1862 when traveling north from Richmond to Baltimore through the lines of the Army of the Potomac. Her personal description included the following pieces of information: age, 30; height, 4 feet 11 inches; complexion, florid; hair, carroty; eyes, hazel; build, robust. If she was in fact thirty then she would have been born about 1832. However, the name she gave in the passport reproduction was Miss Sarah L. Jones (in keeping with the book's being published anonymously), and we cannot be sure that other particulars were not adjusted as well (*Life in the South; from the Commencement of the Civil War.* By, A Blockaded British Subject. 2 vols. London: Chapman and Hall, 1863, vol. 2, p. 381).

126. *Report upon the Fauna of Liverpool Bay and the Neighbouring Seas,* Liverpool Marine Biology Committee, ed. W. A. Herdman (London: Longmans, Green, 1886–).

127. Laura B. Thorneley, "Polyzoa from Ballycastle and Rathlin Island," *Irish Naturalist,* **11** (1902): 161–2; "Report on the Hydroida collected by Professor Herdman at Ceylon in 1902," *Royal Society Report on Pearl Oyster Fisheries,* Pt. ii, 1904, pp. 107–27.

128. The scanty information obtained about the University College Liverpool women came mainly from the membership lists of the Liverpool Biological Society (printed in *Proceedings and Transactions*). Thorneley is mentioned in R. Lloyd Praeger, *Some Irish Naturalists. A Biographical Notebook* (Dundalk: W. Tempest, 1949), p. 197.

129. Student records, the John Rylands University Library, University of Manchester.

130. Pratt's post-1900 publications included the following: "A collection of Polychaeta from the Falkland Islands," *Memoirs of the Manchester Literary and Philosophical Society,* **45** (1901): 1–18; "Some notes on the bipolar theory of the distribution of marine organisms," ibid., **45,** Pt. IV (1901): 1–21; "The mesogloeal cells of *Alcyonium* (preliminary account)," *Zoologischer Anzeiger,* **25** (1902): 545–8; "The Alcyonaria of the Maldives. II. The genera *Sarcophytum, Lobophytum, Sclerophytum* and *Alcyonium,*" in vol. 2, Pt. I (pp. 503–39) of *Fauna and Geography of the Maldive and Laccadive Archipelagos,* ed. John Stanley Gardiner (Cambridge: Cambridge University Press, 1903–1906); "The assimilation and distribution of nutriment in *Alcyoneum digitatum,*" *Reports of the British Association* (1903): 688–9; "Report on some Alcyoniidae collected by Professor Herdman at Ceylon in 1902," *Royal Society Report on Pearl Oyster Fisheries,* Pt. iii, 1905, pp. 247–68; "The digestive organs of the Alcyonaria and their relation to the mesogloeal cell-plexus," *Quarterly Journal of Microscopical Science,* **49** (1906): 327–62. Her last biology paper appeared under the name Edith M. Musgrave (née Pratt), "Experimental observations on the organs of circulation and the powers of locomotion in Pennatulids," *Quarterly Journal of Microscopical Science,* **54** (1909): 443–81.

131. Student Records, University College London. Emma was one of two Beck sisters registered at University College about 1880. The other was Elizabeth, who studied mathematics for five years; she also took classes in history and architecture, as did Emma.

132. Ibid. (University College records indicate that Collcutt received a Natural Science Scholarship for Girton College; however, she does not appear in the Girton records, and it seems fairly certain that she did not go there.)

133. Ibid.

134. Obituary, "Miss J. W. Kirkaldy," *Times,* 22 June 1932, 14f; Vera Brittain, *The Women at Oxford. A Fragment of History* (New York: Macmillan, 1960), pp. 180, 185.

135. The AEW was the Association for the Education of Women, the body responsible for supervising the educational opportunities for women at the Oxford women's halls from 1879 to 1921.

136. Obituary, *Times.*

137. Boas's *Lehrbuch der Zoologie* (1890), was originally published in Dutch in 1888. Kirkaldy and Pollard's English translation appeared in 1896 (London: Sampson, Low and Co.). I. M. Drummond, Kirkaldy's coauthor for *An Introduction to the Study of Biology* (Oxford: Clarendon Press, 1909), later became headmistress of the North London Collegiate School for Girls.

138. Obituary, *Proceedings of the Linnean Society* (1936–1937): 218; *Who Was Who,* vol. 3, p. 2621.

139. *Newnham College Register, 1871–1971,* vol. 1, Staff, p. 65; G.P.B., obituary, "Miss Alice Johnson," *Times,* 10 March 1940, 5c; Ethel Sidgwick, *Mrs. Henry Sidgwick. A Memoir* (London: Sidgwick and Jackson, 1938), various references. Johnson received an M.A. degree in 1928.

140. For a contemporary comment on the Johnson family, interesting for its remarkable frankness, see the reminiscences of American zoologist Mary Willcox above.

141. F. W. H. Myers, *Human Personality, and its Survival of Bodily Death,* ed. Alice Johnson (London: Longmans, 1903). Johnson also wrote a biography of one of her brothers, *George William Johnson, Civil Servant and Social Worker* (Cambridge: privately printed, 1927).

142. Obituary, *Times.*

143. *Newnham College Register, 1871–1971,* vol. 1, Staff, pp. 7–8; obituary, *Reports and Transactions of the Devonshire Association for the Advancement of Science, Literature and Art,* **74** (1942): 38.

144. Lilian Sheldon, "Nemertinea," in *The Cambridge Natural History,* eds. Sydney F. Harmer and Arthur E. Shipley (London: Macmillan, 1896), vol. 2, pp. 97–120.

145. Gilbert Sheldon was particularly remembered for his book on the evolution of roads, *From Trackway to Turnpike. An Illustration from East Devon* (London: Oxford University Press, 1928), but he also published historical novels, short stories, and verse.

146. Lilian Sheldon, "Devon barns," *Reports and Transactions of the Devonshire Association . . . ,* **64** (1932): 389–95; "Devon toll-houses," ibid., **65** (1933): 293–306; "Devon inns," ibid., **69** (1937): 365–90.

147. *Newnham College Register, 1871–1971,* vol. 1, p. 95.

148. Abbott's school textbooks included, *The Science of Everyday Life* (London: G. Bell and Sons, 1918), *Scientific Geography* (Cardiff: Education Publishing Co., 1918) and *Scientific History of the World* (London: G. Bell and Sons, 1920).

149. *Newnham College Register 1871–1971,* vol. 1, p. 66.

150. Ibid., vol. 1, Staff, p. 9; *Dictionary of Scientific Biography,* vol. 12, pp. 519–20, entry for William Johnson Sollas. (See also chapter 12.)

151. Sollas's zoological publications include: "Porifera—Sponges," in *The Cambridge Natural History,* vol. 1 (1895). "On the sponges collected during the Skeat Expedition to the Malay Peninsula, 1899–1900," *Proceedings of the Zoological Society* (1902), ii: 210–12; "On *Hypurgon skeati,* a new genus and species of compound ascidians," *Quarterly Journal of Microscopical Science,* **46** (1903): 729–35; "On *Haddonella topsenti,* gen. et sp. n., the structure and development of its 'pithed fibres'," *Annals and Magazine of Natural History,* s. 7, **12** (1903): 557–62; "On the molluscan *Radula;* its chemical composition and some points in its development," *Quarterly Journal of Microscopical Science,* **51** (1907): 115–36; "On the identification of chitin by its physical constants," *Proceedings of the Royal Society,* B, **79** (1907): 474–81; "The inclusion of foreign bodies by sponges, with a description of a new genus and species *Monaxonida,*" *Annals and Magazine of Natural History,* s. 8, **1** (1908): 395–401; "A new freshwater polyzoon from S. Africa," ibid., s. 8, **2** (1908): 264–73; "Inheritance of colour and of supernumerary mammae in guinea-pigs, with a note on the occurrence of a dwarf form," *Report to the Evolution Committee of the Royal Society, 1902–09,* **5** (1909): 51–79; "Note on parasitic castration in the earthworm, *Lumbricus herculeus,*" *Annals and Magazine of Natural History,* s. 8, **7** (1911): 335–7; "Note on the offspring of a dwarf-bearing strain of guinea pigs," *Journal of Genetics,* **3** (1913–1914): 201–4. As well as her scientific papers, Sollas published a short book, *The Story of Newnham College* (Cambridge: W. Heffer and Sons, 1912).

152. The *International Catalogue* was the successor to the Royal Society's *Catalogue of Scientific Papers,* 1800–1900, on which this survey is based.

153. Sir John Graham Kerr, "The Scottish Marine Biological Association," *Notes and Records of the Royal Society,* **7** (1950): 81–96.

154. Robert E. Kohler, *From Medical Chemistry to Biochemistry. The Making of a Biomedical Discipline* (Cambridge: Cambridge University Press, 1982), p. 42.

155. D. E. Allen, *The Naturalist in Britain. A Social History* (London: Allen Lane, 1976), p. 207.

156. Jane M. Oppenheimer, "Thomas Hunt Morgan as an embryologist: the view from Bryn Mawr," *American Zoologist,* **23** (1983): 845–54, on pp. 852–3.

Chapter 4

NATURALISTS

American ornithologists: Popularizers, occasional contributors, and Florence Merriam Bailey

Less than 2 percent of the journal articles by women listed in the Royal Society *Catalogue* are in ornithology. Twenty-two Americans published thirty-three papers,[1] seven British authors nine. Further contributions came from women in Austria, Canada, Germany, and Ireland.

Of the twenty-two Americans, most either came from or were based in the eastern states. Ten carried out their observations in New England, six in New York, one in Pennsylvania, and one in New Jersey. Both Olive Thorne Miller and Florence Merriam Bailey, the two most prominent members of the group, had close connections with New York, although much of their work was done in other regions of the country. Only two, Jane Hine and Julia McNair Wright, appear to have had permanent residences outside of the eastern region, and McNair Wright too spent her early life in New York.

The earliest American note in ornithology listed in the bibliography, that on hummingbirds by the botanist Lucy Millington, appeared in 1868, but three articles by GRACEANNA LEWIS[2] (1821–1912) appeared shortly thereafter. Graceanna was born on 3 August 1821, the third of the four daughters of Esther (Fussell) and John Lewis, Quaker farmers in Chester County, Pennsylvania. The Lewis family's connection with the region went back some time, John's Welsh ancestor Henry Lewis having arrived there in 1682. The more than 100-acre Lewis farm was land deeded to the family by William Penn. After John died of typhus in 1824, his widow and several of the members of her extended family ran the farm and brought up the four Lewis children. Esther Lewis's discovery of a deposit of hematite on her land, and the arrangement she made to have it mined, brought in some extra money.

Graceanna received her early education from her mother, who had been a teacher before her marriage, and when she was eighteen went for a year to a Quaker school near her home, Kimberton Boarding School. Courses in scientific subjects, including astronomy, botany, and chemistry, were part of the curriculum. In 1842 she got her first job, in a small boarding school opened by her uncle in York, Pennsylvania; she undertook to teach astronomy and botany as well as the basic subjects. When the school closed two years later she taught for a term in a similar school in Phoenixville, but did not particularly enjoy the work and in 1845 returned to the family farm.

Throughout the 1840s and up until the Civil War she and her family, along with many of their Quaker friends, took a strong and unequivocal stand on the antislavery issue, participating both in public abolitionist activities in Pennsylvania and in covert efforts to help escaped slaves. The Lewis farm was one of the stops on the clandestine and illegal Underground Railroad from Maryland and Virginia to Canada, and during the pre-Civil War years the family was constantly providing refuge for slaves. Another of their charitable acts was the adopting of a child of destitute parents, Ellen Bechtel, whom they brought up as their daughter.

Graceanna's interests in natural history probably began when she was still in her twenties. One of her closest friends at the time was Mary Townsend, of Philadelphia, the author of *Life in the Insect World: or, Conversations upon Insects between an Aunt and her Nieces,* published anonymously in 1844. Mary died in 1851, but her friendship had a great influence on Graceanna, who took up the study of birds in part because she wanted to write a work that would be companion to that of her friend. By the early 1860s she had moved to Philadelphia where she was readily accepted into the small group of Quakers who were active in natural history studies. Among them was John Cassin, a specialist in taxonomy and at the time curator of birds at the museum of the Philadelphia Academy of Natural Sciences. Thanks to his help she was able, over the next seven years, to use the academy's library and to study the specimens in its museum. She worked closely with Cassin on the classification of birds collected on many of the major expeditions of the time, including the Pacific railroad explorations of the American West, and she learned his techniques for comparing and contrasting specimens. She may well have had more supervised training than any other American woman naturalist of the period.[3]

For a number of years beginning in the late 1860s she earned money by freelance lecturing, offering instruction in ornithology, zoology, and general natural history in the Pennsylvania and New Jersey area. She also gave short courses of lectures in Rochester, New York,[4] and at Vassar College, and had hopes of obtaining a college teaching position. She twice applied for a professorship at Vassar, with encouragement from astronomer

Maria Mitchell, but both times was passed over in favor of men who had superior academic credentials.

Following these setbacks she took a job at a Friends' school in Philadelphia. However, her dislike of schoolteaching had not diminished since her earlier experience, and within a year she succumbed to an illness that was probably as much psychological as physical. After she recovered, she set about devising various plans to usefully develop the family land, which, after the death of her mother and two of her sisters, she could no longer hope to farm. But none of her plans proved practicable, and in 1883, at the age of sixty-two, she once again turned to schoolteaching, taking a post at the fashionable Foster School for girls in Clifton Springs, New York. Not surprisingly she failed to find the atmosphere congenial, the social outlook of the young ladies she taught being very different from that of the Quaker community she was used to. She resigned after two years and within a short time joined her widowed sister Rachel in Media, Pennsylvania. She lived there for the remaining twenty-seven years of her life, supported in part by a generous allowance from her adopted daughter, Ellen, now married to a relatively well-off Philadelphia lawyer. Ellen also arranged ideal vacations for Graceanna, regularly inviting her to her summer home at Longport on the New Jersey shore, where they both carried out studies on marine life.

Graceanna's first scientific publication was probably her privately printed pamphlet, *Natural History of Birds* (1868), the first part of a projected ten-part general treatise on ornithology, addressed to the general reader. She intended that this work would be not just a presentation of straightforward descriptive material, but a study which would reveal God's design for the ornithological segment of the natural world. The remaining nine parts of the treatise never appeared.

In 1869, thanks to the help of Spencer Baird of the Smithsonian Institution, she obtained her own microscope and immediately started an examination of the fluids in birds' feathers. She wrote describing her observations to F. W. Putnam, then director of the Peabody Academy of Science, Salem, Massachusetts, who read excerpts to the Salem Essex Institute. Another study on birds' feathers, "On the plummage of terns," she reported at the 1869 Salem meeting of the AAAS, where she also presented, in preliminary form, her "Thoughts on the structure of the animal kingdom." Two years later the *American Naturalist* published her papers, "Symmetrical figures in birds' feathers" and "The Lyre bird." The latter, a description of a very rare bird, was based on an examination of specimens in the Philadelphia Academy museum.

Despite the fact that she is now remembered mainly for her work in ornithology, her primary intellectual goal was nothing less than the revealing of the order underlying the complexity of nature. An enthusiastic follower of Louis Agassiz, then probably the most influential proponent in America of the concept of a divine plan in nature, she had an unshakable belief that living forms were not simply the cumulative result of Darwinian struggle for existence against blind forces of nature; on the contrary, there was an orderly progression toward "perfection," which science would eventually make plain. She elaborated her preliminary 1869 presentation on this progression in her twenty-two-page pamphlet *The Development of the Animal Kingdom,* which came out in 1877. The preceding year she had shown her "Chart of the Animal Kingdom" at the Women's Pavilion at the Centennial Exhibition in Philadelphia.[5] It was displayed again, nine years later, at the World's Industrial and Cotton Centennial Exposition in New Orleans. However, although her training in classification and systematics under Cassin at the Philadelphia Academy in the 1860s appears to have been sound enough, she had no effective means of keeping abreast of advances in taxonomy after Cassin died in 1869. She never found anyone else to help her as he had done, and by the 1880s the store of scientific information on which her ideas about classification and her picture of evolutionary trees ultimately had to depend undoubtedly had gaps. Further, after Agassiz's death in 1873, interest in such topics had declined in the scientific community. Graceanna's ideas on evolutionary theory consequently attracted little serious attention.

For some time she continued to carry out a certain amount of work at the Philadelphia Academy museum. In 1883 she read a short paper at the academy, "On the genus *Hyliota,*" discussing three specimens in the museum collection. Even after the turn of the century, when she was in her eighties, she retained her enthusiasm for nature study; two reports of her observations on marine life were presented to the Delaware County Institute of Science in 1909.[6] She was also a competent artist, and was hired to make a series of watercolor paintings of leaves of Pennsylvania trees for the Pennsylvania Forestry Commission. These won a bronze medal at the Chicago Columbian Exposition in 1893. Thereafter she prepared a series of lithographed tree leaf charts that she made available to schools at a cost of fifty cents each.

Although she never joined the AAAS, she became a member of the Philadelphia Academy of Sciences in 1870, after an initial rejection, and in 1881, when the reorganized Rochester Academy of Sciences began to admit women on an equal basis with men, she accepted an honorary (nonresident) membership.[7] She was secretary of the Delaware County Forestry Association for a period in the 1890s. In her later years at Media she joined the Women's Christian Temperance Union, then becoming an influential organization, and she served as secretary of the Media chapter and as Superintendent of Scientific Temperance Instruction for Delaware County. She died of a stroke, 25 February 1912, at age ninety.

OLIVE THORNE MILLER[8] (1831–1918), born ten years after Graceanna Lewis, was a less philosophically ambitious and more conventional and successful naturalist. A prolific and talented writer, she was one of a group that included Florence Merriam Bailey, Sara Hubbard, Bradford Torrey, John Muir, and John Burroughs, all of whom made important contributions to early nature writing in America. Her popular books on birds were much respected, even among many of the professional naturalists of her time.

Olive Thorne Miller (or Olive Thorne) was a pen name, her real name being Harriet Mann Miller. She was born on 25 June 1831, in Auburn, western New York State, the oldest child and only daughter in the family of Mary (Holbrook) and Seth Hunt Mann. Both the Holbrooks and the Manns were old New England families; Captain Benjamin Mann, Harriet's great-grandfather, had commanded a company at Bunker Hill in the Revolutionary War. Seth Mann, a banker, moved his family by stages from New York to Missouri, with stays along the way of three to five years in Ohio, Wisconsin, and Illinois. Their journey to Ohio in the early 1840s via the Mohawk Valley canal, Lake Erie steamer, and finally horse-drawn coach through Ohio, was one that Harriet described for later generations in a children's story, *What Happened to Barbara* (1907). Her formal education suffered many interruptions, but she had five years in a "select" school in a small college town in Ohio. As a girl she considered herself plain, was excessively shy and solitary, and preferred books and story writing to social contact. In 1854, at the age of twenty-three, she married Watts Todd Miller, who was also from a New York family. They settled in Chicago, where Watts worked in a lumber firm, and had four children, born between 1856 and 1868. The younger daughter, Mary Mann, was later to share her mother's interest in ornithology.

For a time Harriet put aside her writing hobby because of family concerns, but by 1870 she had taken it up again. That year the first of her tales for young people was accepted by a magazine, and over the next twelve years she published an incredible amount—over 350 magazine stories and five books, the best known of which was probably *Little Folks in Feathers and Fur, and Others in Neither* (1873). About 1875 the family moved to Brooklyn, New York, and shortly thereafter, at the suggestion of her naturalist friend Sara Hubbard, Harriet took up bird-watching in Prospect Park. From then on her interest in birds grew steadily. Over the course of two decades, from 1883 until 1903, she regularly took summer field trips throughout New England, the Midwest, the mid-Atlantic region, and the West, making carefully recorded observations, which she later wrote up for publication. On one of these trips (1893) she was joined by a fellow bird-watching enthusiast, Florence Merriam (see below). She published eleven books on birds, the first, *Bird Ways,* appearing in 1885, when she was fifty-four. Others included, *In Nesting Time* (1887), *Little Brothers of the Air* (1892), *A Bird Lover in the West* (1894), *Upon the Tree Tops* (1897), *The First Book of Birds* (1899), *The Second Book of Birds; Bird Families* (1901), and *With the Birds in Maine,* her last book of field observations, published in 1903 when she was seventy-two.

She frequently attended meetings of the Linnaean Society in New York City, and was one of the early and enthusiastic women associates of the American Ornithologists' Union. In the late 1890s, along with Florence Merriam, she worked on the union's Committee on Protection of North American Birds. She took great pleasure in becoming a full member of the union in 1901 when women were first allowed to do so, and saw the change as an encouraging recognition of "honest work." By the time of her death she was the organization's oldest member.

All of her work was marked by precision of observation and accuracy of description; her early tendency to anthropomorphize her subjects by applying terms from human child-raising to the activities of parent birds decreased with time. She usually published her shorter articles in popular magazines such as *Atlantic Monthly,* a habit that somewhat annoyed ornithologist William Brewster, who thought she should report new discoveries where ornithologists would have a better chance of keeping track of them. "It is a pity," he complained, "that writers like Mrs. Miller—gifted with rare powers of observation and blessed with abundant opportunities for exercising them—cannot be induced to record at least the more important of their discoveries in some accredited scientific journal, instead of scattering them broadcast over the pages of popular magazines or newspapers, or ambushing them in books with titles such as [*Little Brothers of the Air*]."[9]

She believed strongly in women broadening their sphere of activity beyond traditional family-centered occupations, and was active in several women's clubs in Brooklyn and New York City. As she saw them, these clubs provided an important means for intellectual development to women denied college education. Nature study especially she considered to be an area where women could make valuable contributions. In her later years she even overcame her shyness sufficiently to give public lectures on ornithological topics. She also spoke out strongly for wilderness preservation. In 1904, after her husband's death, she and her daughter Mary moved to southern California,[10] where her elder son and his family were already living. They built a cottage on the outskirts of Los Angeles, in a region whose climate, flowers, and birds Olive Thorne already knew and loved. She continued her writing, producing the last six of her twenty-five books. These were mainly children's stories, but included *The Bird Our Brother* (1908). Though her stories for young people did not join the classics of children's literature, her bird books long retained an honorable place in early American nature writing. She died in California, on 26 December 1918, at age eighty-seven.

Of the American women ornithologists who published in the scientific journals before 1901, the most outstanding was undoubtedly FLORENCE MERRIAM[11] (1863–1948), the first woman fellow of the American Ornithologists' Union. Like Olive Thorne Miller she wrote a number of popular nature books, but she was also the author of two outstanding volumes in western American ornithology.

Florence was born 8 August 1863, in Locust Grove, near Leyden, Lewis County, northern New York State, the youngest of the four children of Caroline (Hart) and Clinton Levi Merriam. Merriam, a Lewis County businessman and banker, was descended from English settlers who came to what is now Concord, Massachusetts, about 1636. As a younger man he had pursued his career in Utica and New York City but he brought his family back to Locust Grove about the time Florence was born. A few years later he was elected Republican representative to the U.S. Congress for the Lewis County district, and served two terms in Washington, D.C. His wife, the daughter of a county court judge and member of the State Assembly, had graduated from Rutgers Female Institute.

Growing up on the large Merriam country estate of "Homewood," Florence developed an interest in natural history at an early age, and was helped and encouraged by her father and her older brother, Hart. She received much of her early education at home, though she also attended school in New York City, where the Merriams often passed the winter. Later she went to a private school in Utica, New York, and then to Smith College as a special student for the four years between 1882 and 1886. At Smith her interest in natural history, particularly ornithology, increased considerably. Along with her fellow student and close friend Fannie Hardy (later Eckstorm—see below), she led student nature walks and started the Smith College Audubon Society, a most successful organization joined by one-third of the college's students. She and Hardy also started a campaign to dissuade the Smith women from wearing hats decorated with the feathers of rare wild birds. Such hats were much in vogue then and for some time after, and Merriam's efforts to educate women about the devastating effect of the millinery trade on wild bird populations were to continue for many years. In 1885 she was nominated for associate membership in the American Ornithologists' Union by her brother Hart, by then chief of the U.S. Biological Survey and one of the founders of the union. She became its first woman associate. Hart was to remain a guiding influence in her ornithological work, providing steady encouragement and sound advice over the years.

By the time she left Smith College, Florence had decided to make writing her life's work. Her first published articles on bird lore appeared in the *Audubon Magazine* in 1887. She continued to study birds and kept extensive field notes. During the winters, when her family was in New York, she worked in one of the city's clubs for working girls, social action groups to which many middle-class women of the time gave their efforts. However, both she and her mother were in poor health—it appears likely that they had tuberculosis—and in 1889 the family spent several months in San Diego County, southern California, at the homestead of her father's brother, Gustavus Merriam. The visit to California not only brought about a great improvement in her health, but was the beginning of her lifelong love of the American West and its birds. Her first book, *Birds Through an Opera Glass,* a collection of her earlier *Audubon Magazine* articles much revised and expanded, appeared in late 1889, when she was twenty-six. Though intended for amateur observers and young people, it was kindly reviewed in *Auk,* and was undoubtedly worthy of the praise (though the reviewer, William Brewster, was a close friend of her brother).

During these early years of her career Florence derived considerable encouragement in her observational work and writing from Olive Thorne Miller, whose reputation as a dependable bird observer and good nature writer was already established. In 1893, following further periods of work in girls' clubs in New York City and a summer of teaching in Hull House, one of the Chicago settlement houses, her health was again in decline, and she joined Miller in a trip west to Utah. They spent the summer together in a small village of Mormon settlers. *My Summer in a Mormon Village* (1894) described her experiences. Though mainly a travel book, it also contained her thoughts about the dreary lives of countrywomen and her opinions of Mormon religious practices. These opinions were considerably toned down between her first draft and the published version; her brother Hart, having read the manuscript, declared that he would never be able to show his face in Utah again if she published the original.[12]

After the Utah summer she proceeded west to California and spent six months at Stanford University, where her brother's friend ichthyologist David Starr Jordan had recently become the first president. About this time she met Alice Eastwood (see chapter 1), the young botanist who was just starting her long and distinguished career at the California Academy of Sciences in San Francisco. They were to be lifelong friends. The following spring, in the nesting season, she went south to her relatives in San Diego County. Her bird studies, made in Twin Oaks Valley where the Merriam homestead was situated, were published in *A-Birding on a Bronco* (1896). She returned home via Arizona, where she did more bird-watching in the San Francisco Mountains near Flagstaff.

Once back in Washington, she settled in her brother's home and became active in several scientific organizations. She joined the Women's National Science Club, was one of the founders of the Audubon Society of the District of Columbia, and, along with Olive Thorne Miller, worked on the American Ornithologists' Union's Committee on the Protection of North American Birds. In addition she did a considerable amount of educational writing on bird protection and economic ornithology, pointing out the value to farmers and fruit growers of insect-destroying birds. First published in magazines such as *Forest and Stream,* her articles were reprinted in book form (*How Birds Affect the Farm and Garden*) in 1896. Her well-illustrated *Birds of Village and Field* (1898) was one of the first popular bird guides produced in America; it was also used as a school textbook.

In 1899, when she was thirty-six, she married Vernon Bailey, a friend of her brother for many years and his colleague at the U.S. Biological Survey. Bailey, from a farming family in Elk River, Minnesota, was one of the survey's outstanding early naturalists. His major undertakings were wide-ranging field studies in the western part of the country. Almost immediately after her marriage Florence joined her husband in his expeditions, and over the course of the next thirty years would accompany him to New Mexico, Texas, California, Arizona, the Pacific Northwest, and the Dakotas. While he carried out his survey assignments she observed birds. The Bailey field trips were very simple, with a minimum of horses and mules, one wagon for provisions, and a man to cook and do the camp chores. They ate the plainest possible food and slept on the ground without watch fires, which Bailey considered unnecessary because he had no fear of animals. Florence enjoyed the outdoor life and her health difficulties apparently disappeared. Her ornithological observations were reported in a long series of papers in *Auk, Bird Lore,* and *Condor.*[13] Her *Handbook of Birds of the Western United States* (1902), the companion volume to Frank M. Chapman's *Handbook of Birds of Eastern North America* (1895), went through many editions and remained the standard work in its area for half a century.

For several summers between about 1909 and 1916, Vernon Bailey carried out fieldwork for the U.S. Biological Survey in North Dakota. Florence accompanied him, and was fascinated by the remoteness and emptiness of the country. Her reports of her observations of the birds of the prairies appeared as the longest series of articles she ever wrote, seventeen papers in *Condor* over a period of five years.[14] Another notable series of papers she published about this time described bird observations made in the autumn of 1914 along the Oregon coast, whose rich population includes forest and mountain birds as well as seabirds.[15] In 1917 the Baileys visited Glacier National Park, and the following year Vernon's *Wild Animals of Glacier National Park* was brought out by the Park Service as a guidebook for visitors; Florence contributed the section on birds.

About this time she was asked by Edward W. Nelson, then chief of the U.S. Biological Survey, to undertake the completion of a work on the bird life of New Mexico started by survey scientist Wells W. Cooke but left unfinished following his sudden death in 1916. She was a logical choice since she knew the western birds perhaps as well as anyone else in the country.[16] The survey appointed her a special assistant so that she could work in its offices, a change of circumstances that gave her great pleasure and made her feel that finally she too had a place in the organization her brother had helped to found. She updated Cooke's data and greatly expanded the overall scope of the project. The result of her efforts, the first comprehensive report of birdlife in the Southwest, was ready for publication by 1919. It was a substantial work and included twenty-four colored plates and many black-and-white drawings and photographs. Publishing costs would be considerable; no government funds were available and no publisher was willing to take the manuscript. After a delay of nine years it was finally brought out in 1928 by the New Mexico Department of Game and Fish under the title *Birds of New Mexico.* Much of the cost was underwritten by private donations. Florence declined to follow the survey's suggestion that Cooke be listed as coauthor, and because of her many original contributions it was agreed that she be sole author with full credit given in the introduction to Cooke for his part.[17]

The work was undoubtedly her most important contribution to the ornithological literature. It and Vernon Bailey's *Mammals of New Mexico,* brought out by the survey in 1931, were landmark publications in western natural history. Florence was duly recognized by her colleagues; the American Ornithologists' Union made her a fellow in 1929 and two years later she became the first woman to receive its Brewster Medal. In 1933 the University of New Mexico gave her an honorary LL.D. She was then seventy years old.

For many years the Baileys made their Washington home a social gathering place for naturalists, their dinner parties being famous in the scientific community. Florence was also an active member of several local ornithological clubs and societies. Through the Washington Audubon Society she organized and taught bird classes for teachers of nature study for more than twenty years. Although she had no children of her own, her early interests in working with young people continued throughout her life, and she served on a number of regional and national associations concerned with assisting children. Both she and her husband also worked in the Boy Scout organization. Her writing continued throughout the 1930s. She contributed the section on birds to her husband's *Cave Life of Kentucky,* published in 1933, and saw her last substantial work, *Among the Birds of Grand Canyon National Park,* brought out by the National Park Service in 1939, when she was seventy-six. She died on 22 September 1948, at age eighty-five, at her home in Washington, D.C., having outlived her husband by six years. Burial was in Locust Grove, New York. *Parus gambeli baileyae,* a chickadee from the higher mountains of southern California, bears her name.

FANNIE HARDY[18] (1865–1946), Florence Merriam's close friend at Smith College in the 1880s, was a well-regarded naturalist and writer and a leading authority on the folk and Indian culture of Maine. Born in Brewer, Maine, on 18 June 1865, she was the oldest of the six children of Emeline (Wheeler) and Manly Hardy. Her forbears on both sides were old Penobscot river families, a Wheeler having been the first settler in Hampden, Maine, and the Hardys having had business interests in lumbering and the fur trade for several generations. Manly Hardy became the most important fur trader in the state, exporting his goods directly to Europe. He had grown up with Indian boys, hunted with them, and learned their language, and he kept close and friendly relations with the Indians of the Penobscot settlement throughout his life. Over a period of about seventy years, until 1900, the Hardys knew most of the Indian guides and the boatmen, hunters, trappers, and head lumbermen in northern and eastern Maine. Manly was also a writer and an authority on Maine birds and animals; his collection of mounted birds was the finest of its time in New England. The vast store of information and experience he gained from a lifetime spent in the wilderness he in large measure passed on to his daughter.

Fannie went to high school in Bangor and then studied for a year at Abbott Academy in Andover, Massachusetts. At Smith College, which she entered in 1885, she was particularly active in nature study groups; along with Florence Merriam she founded the college Audubon Society and led a program of nature walks. Immediately after graduating (A.B.) in 1888 she accompanied her father on a long canoe trip, the first of many such, along the Penobscot lake and river complex into the hill country north and east of Moosehead Lake, and then on through the Nicatowis region. As a child she had learned how to keep full and accurate records, and the notes on animal habits and Indian lore that she brought back from her trips formed the major source she was to draw on in much of her later writing.

For two years, 1889 to 1891, she was superintendent of Brewer schools, the first woman in Maine to hold such a position. However, when she failed to persuade the town to provide money needed for improvements, she resigned and went to Boston, where she spent a year as a reader of scientific manuscripts for the publishing firm of D.C. Heath. In 1893, at age twenty-eight, she married the Rev. Jacob Eckstorm, an Episcopal minister of Norwegian background, who then had pastorates in Oregon. They lived in Oregon City for a year, and then moved to Eastport, Maine. Of their two children, born in 1894 and 1896, only one survived to adulthood. The Rev. Eckstorm died in 1899 and thereafter Fannie returned to Brewer.

She published a number of ornithological articles, beginning in the late 1880s (some in the popular magazine, *Forest and Stream*),[19] but her best-known works were her books. Both a children's text, *The Bird Book,* and a well-received monograph reporting her study of woodpecker life, *The Woodpeckers,* appeared in 1901. She also wrote a number of articles and books on the people of the Maine wilderness. Her accounts of Indian legends came out in the *Atlantic Monthly,* and her book, *The Penobscot Man,* in which she described the life of the river drivers, was published in 1904. In it she spoke out against the milling interests that were threatening Maine's forests and rivers. She also wrote on local Indian languages, history, and handicrafts; her *Handicrafts of the Modern Indians of Maine* appeared in 1932, and *Indian Placenames of the Penobscot Valley and the Maine Coast* in 1941. The latter presented a considerable amount of new material, and established her as a leading authority on the Penobscot Indians.

During the 1920s she made a collection of the ballads of Maine. The work was a major project, painstakingly carried out, and it resulted in the publication of two highly regarded books, *Minstrelsy of Maine: Folk Songs and Ballads of the Woods and Coast* (1927), prepared in collaboration with Mary Winslow Smyth of Elmira College, New York, and *British Ballads from Maine: the Development of Popular Songs, with Text and Airs* (1927), coauthored with Smyth and Phillips Barry. Much of the music in these collections had been in danger of being lost forever; the importance of Fannie Eckstorm's work of preservation was recognized in 1929 by the University of Maine when it awarded her an honorary M.A.

Throughout much of her life she was active in civic and political affairs in the Brewer community and the surrounding region; in 1920 she became the first chairman of the local Women's Republican Committee. One of her special concerns was the development of legal protection of the state's fish and game against outside interests, a cause her father had taken up. She died at age eighty-one, on 31 December 1946.

No information has been uncovered about the other American women whose ornithology papers were indexed by the Royal Society,[20] with the exception of Amelia Watson.[21] The daughter of Sarah (Bolles) and Reed Watson, Amelia was born on 2 March 1856 at East Windsor Hill, Hartford, Connecticut. She was educated in private schools and became a watercolor painter, teaching painting in Martha's Vineyard Summer School for twenty years. Her landscapes of New England and the American South were exhibited in Boston and New York City. Among her other artistic works were the illustrations for Henry Thoreau's *Cape Cod* (1896). Her note in the *Auk,* on the taming of a chipping sparrow, appeared in 1894.

Like many of the American women contributors to nineteenth-century botanical literature, a considerable number of those who wrote about birds before 1901 published only one or two observational notes in the scientific journals. There is no readily available evidence that suggests that they developed their bird-watching hobby into a major, long-term commitment.

Among the four most prominent women discussed here—Graceanna Lewis, Olive Thorne Miller, Florence Merriam Bailey, and Fannie Hardy Eckstorm—Lewis stands somewhat apart. In contrast to the other three, who were field naturalists, her practical experience in ornithology was largely in the areas of taxonomy and classification, and her papers listed in the bibliography resulted from museum work and microscope studies. She had enthusiasm and a wide range of interests, but her major project of revealing the Creator's grand plan for the ornithological segment of the natural world resulted from a line of thought that was already out of date by the time she was writing. On the other hand, Miller, Merriam Bailey, and to a lesser extent Hardy Eckstorm, all made notable contributions to American nature writing and ornithological literature of the late-nineteenth and early twentieth-centuries. Indeed, Merriam Bailey, perhaps the leading authority on birds of the American Southwest by the time she was in her fifties, was a remarkably fine field naturalist and an outstanding writer, able to make serious natural history studies palatable to general audiences. It is worth noting that both Hardy Eckstorm and, most especially, Merriam Bailey had the very considerable advantage of close family relationships with men who were influential in ornithology and natural history; Hardy Eckstorm learned from her naturalist father, and Merriam Bailey was first helped by her brother and then enabled to continue and greatly expand her region of activity by joining her husband's field expeditions.

Along the Pacific coasts: American women conchologists of the 1890s

Papers in conchology constitute less than 1 percent of the articles by women listed in the Royal Society *Catalogue.* About half of them were contributed by workers in the United States and most of the rest by women in Australia and Ireland.

Of the eleven American authors whose papers are listed in the bibliography, one was from Maine (Wentworth),[22] one from Kentucky (Price), and nine (Bradshaw, Campbell, Drake, King, Monks, Shepard, Soper, White, and Woodhead Williamson) from the West Coast. Other than Drake (of Washington State) all the Westerners worked in California. With the exceptions of Monks, Price, Woodhead Williamson, and Shepard[23] these naturalists are not well remembered. All were active during the 1890s, a time when a considerable number of small conchological clubs and summer shell groups, as well as many sections of the Isaac Lea Conchological Chapter of the Aggasiz Association, were in existence throughout the United States and Canada.

Much of the women's published work appeared in the journal *Nautilus*.

MARTHA BURTON WOODHEAD[24] (1843–1922) was born in Leeds, England, on 6 March 1843, and came to the United States about ten years later. She was educated in schools in Ohio and then at Eliot Seminary and Burlington College, Iowa, where she took special courses in philosophy. Little more is known about her early life in the Midwest, except that in 1882 she was editor of the *Enterprise* in Terre Haute, Indiana. She married Charles Wesley Williamson and they had at least two children, Lillian and Estella.

The family moved to Los Angeles in 1887, settling in the area known as University District,[25] close to the recently established University of Southern California. Martha quickly became active in local educational, scientific, and civic organizations. At the urging of Ira More, principal of the Los Angeles State Normal School (later the University of California Los Angeles), she joined the Historical Society of Southern California. She was to serve as one of its vice presidents for twenty years, from 1894 until 1915, except for the year 1899. A member of the national executive committee of the Isaac Lea Conchological Chapter of the Agassiz Association, she held office as the chapter's general secretary from 1893 until 1898.[26] She was a frequent contributor of book reviews to *Nautilus*. In 1894 she ran a summer school course in biology in Long Beach. She also retained her interest in newspaper work and was associated with the Southern California Press Club.

Throughout the 1890s and the succeeding decade she published a remarkable amount of work on West Coast shells, including a forty-one-page annotated list of the shells of the San Pedro Bay area (*Proceedings of the United States National Museum,* 1893). A report of the conchological research activities of the region followed a year later in the *Annual Publications, Historical Society of Southern California*. Her technical articles continued to appear until about 1911[27]; one of her particular interests was the abalone industry of the West Coast and the need for conservation measures.

Education, women's organizations,[28] including the Women's Christian Temperance Union, and the contributions of women to scientific work were among her special interests over many years. Her lengthy, enthusiastic, and well-illustrated three-part paper on American women in science, published in 1898–1899,[29] provides still-useful information about the work of many of her contemporaries. She died on 18 March 1922, in Los Angeles, at age seventy-nine.

IDA SHEPARD[30] (1856–1940), for more than twenty years a curator in the Department of Geology at Stanford University, was another of the early pioneers in West Coast conchology. She was remembered especially for two major works, *The Marine Shells of Puget Sound and Vicinity* (1924), and *The Marine Shells of the West Coast of North America,* brought out in four volumes as Stanford University Publications between 1924 and 1927. The latter was a compilation of descriptions of 2,000 species of mollusks from the West Coast and included many illustrations.

Born in Goshen, Indiana, on 25 November 1856, she was educated at Saline High School in southeastern Michigan and then at the University of Michigan, where, as a special student in science, she studied for a teaching certificate. Her family moved to Long Beach, California, in 1888, and there she began collecting and studying shells. In 1895 she married a fellow shell enthusiast, Tom Shaw Oldroyd; their joint collection became one of the largest privately owned in California.

In 1916 the Oldroyds moved to Stanford to take on the job of cataloging the Hemphill shell collection, recently acquired by the university's geology department. The following year, having sold their own collection to Stanford, they became curators in the geology department, and they continued to work there for the rest of their lives. Through extensive field trips they greatly increased the Stanford collections of both American and foreign shells. Especially notable among their American acquisitions were collections from the Puget Sound area. Ida Oldroyd went to considerable effort to encourage donations to the university and also urged purchases of important holdings, such as the Sarah Mitchell collection of Philippine shells, bought in 1930. She herself, on a round-the-world trip at the time, expedited the shipping of the two tons of shells from Manila to California.

Although Tom Oldroyd died in 1932, Ida remained at Stanford and continued her conchological work. A charter member of the American Conchological Union, she served as its vice president in 1934 and as honorary president from 1935 to 1940. She was one of the early members of the Conchological Club of Southern California and remained an honorary member until her death. She also held corresponding memberships in the Conchological Society of Great Britain and the Peking Natural History Society. As well as bringing out her two major monographs, she made several contributions to *Nautilus* over the years. She died in California, on 9 July 1940, in her eighty-fourth year.

Some early British naturalists

Five of the twenty-two British women naturalists whose papers are listed in the bibliography under the headings conchology, ornithology, and natural history made their contributions before 1860. These were Mrs. Harvey of London, a Dorset naturalist identified only as Miss E. W., Anna Thynne, Jeannette Power, and Sarah Wallis Bowdich Lee. All the others were active during the last two decades of the century.

A cluster of seven who lived in southern Scotland and northern England form something of a subgroup. This includes the three contributors of work in conchology (Janet Carphin of Edinburgh, Mrs. Thew of Warkworth, Northumberland, and Jane Donald Longstaff from Carlisle), Miss Sprague of Edinburgh and Mrs. Brown, who published in the *Transactions* of the Dumfries and Galloway Natural History Society; there are also two ornithologists, C. H. Greet (of Northam, Northumberland) and Eleanor Warrender, both of whom published in the *History* of the Berwickshire Naturalists' Club.

Others mentioned here are Jessie Saxby and Adelaide Traill, bird observers who worked in the Shetlands, islands famous for the richness and variety of their seabird colonies. Saxby was a native of the Shetlands; Traill divided her time between Shetland and Edinburgh. Two other ornithologists worked in the English Midlands—Mrs. Barnard of Rugby, an observer of bird migration, then a new line of study, and Annie Ley of Lutterworth, Leicestershire. Ley also wrote a number of notes to *Nature* on meteorological events of special interest (see chapter 16). Emma Hubbard contributed her notes on the cuckoo from Kew (London). On the south coast were Agnes Crane of Brighton and Frances Gough, a Cornish naturalist whose descriptive notes on more than twenty local varieties of sea anemones appeared in the Penzance Natural History Society's *Transactions.* The remaining three were people who wrote on topics that had interested them when living or traveling abroad, Mrs. E. D. W. Hatch, a traveler in the Rocky Mountains, Miss Martin, who described a stay in the Canaries, and Mary Yate, for several years a resident in army cantonments in India.

Information about most of these women is sketchy at best. Many contributed only a single paper, often a short observational note, to the scientific literature, although one, Donald Longstaff, was well known for her work in paleontology, and a few others, particularly Bowdich Lee, Crane, Power, and Saxby, were fairly well recognized among writers and naturalists of their times.[31]

Miss E. W., one of the earliest workers, studied marine mollusks along the Channel coasts, particularly in the region of Weymouth, Dorset, and communicated her observations to the French naturalist Defrance. The 1822 paper by her listed in the bibliography, perhaps her only publication, is a lengthy extract from a letter to Defrance, who noted that her detailed observations, the result of "le zèle extrêmement remarquable qu'elle met à cette étude," needed to be made available to science; field studies in the area were few at the time and identifications sometimes questionable—E. W.'s were considered reliable by Defrance.

Among the mollusks she described are shipworms (Teredinidae), the piddocks (Pholadidae, other wood borers), surf clams (Mactridae), the heart shells (Cardiidae) and ark shells (Arcidae). She carried out careful anatomical studies, particularly of means of locomotion and of reproductive systems (to determine whether the animals were viviparous or oviparous). She was also interested in establishing varieties of species, especially in the case of shipworms.

ANNA THYNNE, who also worked along the southern coasts, wrote in her 1859 *Annals of Natural History* paper on corals that she had had a special interest over many years in geology and that this led her to look at living corals. With encouragement from the very successful natural history writer Philip Henry Gosse, she collected specimens in Devon and succeeded in maintaining them live in her London home for several years while she observed their development and reproduction. Mrs. Thynne is on record as one of the first to realize (about 1846) that seawater could be kept fresh in a glass bowl by growing seaweeds in it. She did nothing to publicize her discovery, however.[32]

SARAH WALLIS (later BOWDICH and then LEE),[33] (1791–1856), was born on 10 September 1791, the only daughter of John Eglinton Wallis of Colchester, Essex. At the age of twenty-two she married Thomas Edward Bowdich, a naturalist, who, while in the service of the British African Company, carried out a considerable amount of exploration in the Ashanti region, now central Ghana, and played an important role in furthering British settlement on the Gold Coast. Sarah took an enthusiastic interest in her husband's scientific work, especially his botanical and zoological collecting. In 1814, not long after her marriage, she left England alone, intending to join him on "The Coast," but when she arrived she discovered that he had already set out on the return journey. Undeterred by this first disappointment, she went out again the following year, this time accompanying her husband, and remained in West Africa until 1818.

Sarah was the first woman to collect plants systematically in tropical West Africa. Her early observations were recorded in her husband's *Mission from Cape Coast Castle to Ashantie* (1819); she also contributed illustrations to many of his other books on natural history and African travel. The first major work she brought out under her own name, *Taxidermy, or the Art of Collecting, Preparing and Mounting Objects of Natural History,* first appeared in 1820 and went through six editions, the last appearing in 1843.

From about 1819 until 1823 the Bowdiches lived in Paris, where they made a study of the zoological and paleontological collections of Baron Cuvier, then one of Europe's leading men of science. They then returned to West Africa, visiting Madeira and the Cape Verde Islands en route, for what turned out to be Thomas Bowdich's last visit; he died at Bathurst on the Gambia river in 1824. Sarah promptly took up the work of completing and preparing for publication his manuscripts. To his *Excursions in Madeira and Porto Santo, during the Autumn of 1823,* she added three sections, an account of the rest of the voyage and their expedition in Africa up to the time of his death, a description of the British settlements on the Gambia, and an appendix reporting their zoological and botanical work. The latter included translations from accounts in Arabic. Exceptionally well illustrated with zoological and botanical drawings, colored geological sections, depictions of African costumes and more, the work came out early in 1825 and immediately brought Sarah to the attention of London naturalist circles.

She again spent some time in Paris in the late 1820s, renewing her acquaintance with Baron Cuvier and his family, and getting to know a number of the members of the Paris naturalist community. She continued to write, bringing out *The Fresh-Water Fishes of Great Britain* in 1828, well illustrated with her own colored paintings. In addition, she published short, popular accounts in the *Magazine of Natural History* of her experiences and adventures with African animals. Some of these have been listed in the bibliography, including her "Anecdotes of a tamed panther" (1828), which told how she transported a West African panther to England (after having come to know it fairly well in Africa), and presented it to the Duchess of York. During her Paris stay she also wrote for her British audience on current activities in natural history in the

French capital. Her nineteen-page report of Cuvier's presentation to the Paris Academy of Sciences on progress in natural history during 1828 appeared in the *Magazine of Natural History* the following year. And in response to an editorial request, she supplied to the same journal a description of the Muséum National d'Histoire Naturelle in the Paris Jardin des Plantes, pointing out the regrettable fact that there was nothing at all comparable in England at the time.

In 1829 she remarried, becoming Mrs. Robert Lee. She continued to write, bringing out popular books for young people on natural history, generally illustrated by herself, and stories set in Africa.[34] In addition, in 1833, not long after Cuvier's death, she published her 369-page biography, *Memoirs of Baron Cuvier.* In the writing of this she had advice from a number of well-known men of science, including the geographer Alexander von Humboldt and the zoologist Georges Duvernoy. It was a popular work, an American edition and a French translation appearing promptly. In 1854 she was granted a Civil List pension of £50 a year, in recognition of her literary and scientific work. However, she died two years later, at age sixty-five, on 22 September 1856, at Erith, on the east side of London, where she was visiting one of her daughters.

Like Sarah Bowdich Lee, MARY YATE[35] also wrote about her observations on animals in faraway places, although her travels came three-quarters of a century later than Bowdich Lee's. The daughter of the Rev. Conolly McCausland of Bath, she married Lieutenant-Colonel Arthur Campbell Yate in 1895. Yate was an officer in the Indian Army, a member of various learned societies, and the author of many books. Mary's article on keeping polecats as pets appeared in the Bombay Natural History Society's *Journal* in 1898. A lively, nontechnical description, it recounted the animals' habits and activities in and around her bungalow in Chaman, on the northwest frontier of what was then British India. The Yates later lived in Shifnal, Shropshire. They had one son.

JESSIE EDMONDSTON[36] (1842–1940) was from the island of Unst, the most northerly in the Shetland group, and a spot that, in the days of her childhood, was doubtless as remote to many as Chaman. Born at Halligarth, on 30 June 1842, she was the ninth child of Eliza (Macbrair) and Laurence Edmondston. Edmondston was the island doctor and also a naturalist; his wife came from an Edinburgh family.

Jessie had no formal education, but largely taught herself by reading and watching the bird and animal life of her native moors and seashore. She had the guidance of parents who had strong literary and scientific interests, and the inspiration that came from having several talented older brothers and sisters.[37] In 1859, at the age of seventeen, she married Henry Linckmyer Saxby, an English medical student and ornithologist, who, while taking his training at Edinburgh and St. Andrews universities, spent some time as an assistant to Dr. Edmonston in Unst. After qualifying he returned to the island, where in due time he took over his father-in-law's practice. He died at the age of thirty-seven, however, in 1873, leaving Jessie and four young sons. (His highly regarded work, *Birds of Shetland,* was brought out just after his death.)

Even before she was widowed Jessie had taken up the family tradition of writing, and with children to support she turned to literary work as an important source of income. She contributed articles to magazines and daily newspapers and also published a succession of books. Many of these were stories for boys, the most popular of her writings, but she also brought out poems, travel stories, and works on Shetland folklore and the Norse sagas.[38] By the 1890s she was fairly widely known. She gave the inaugural address to the Viking Club in London in 1892, her subject being "Bird omens in Shetland," and for a time she served as the honorary president of the Edinburgh, Orkney and Shetland Literary and Scientific Association. Her two ornithological articles listed in the bibliography appeared in the 1890s, one a note on the feeding habits of the great skua in the *Annals of Scottish Natural History* and the other a longer paper about gulls, which came out in the *Transactions* of the Edinburgh Field Naturalists Society.

Jessie Saxby took an active part in many concerns and issues of the time, including the Home Rule question, the Temperance Movement, and emigration from Scotland to Canada. She visited Canada herself in the interests of women's emigration. A dependable and respected member of the island community, in 1920 she was made a Justice of the Peace. For most of her life she lived at Baltasound, Unst, and died there, on 27 December 1940, at age ninety-eight.

Two-country comparison

Neither ornithology nor conchology was as popular an area of activity with late nineteenth-century British women as it was with their American contemporaries (see Figure 4). Both fields had more than three times as many American as British workers. In ornithology there were no British women of the time whose contributions were comparable with those of the most notable of the American observers and writers, Florence Merriam Bailey and Olive Thorne Miller. Miller's remarkable output of popular bird books, which appeared mainly in the 1880s and 1890s, Bailey's *Handbook of Birds of the Western United States* (1902) and her later major work, *Birds of New Mexico,* have no equivalents in the work of the early British women included here. Spectacular opportunities such as Bailey had in the American West, whose rich bird life was almost unexplored when she joined her husband's field trips to the region in the 1890s, were not so readily available to the British. Aside from the two rather exceptional Americans, however, the two national groups are in some respects not dissimilar. Two-thirds of the Americans were, like the British, contributors of only one or two observational notes or papers to the scientific literature. Their work in ornithology was most likely a spare time hobby rather than a major commitment.

In conchology there was no British counterpart to the prominent cluster of Americans active by the 1890s on the Pacific coast, particularly in southern California. There again there were special opportunities and attractions—a mild climate for seaside work and a rich and little-explored shellfish life. A number of nineteenth-century British women collected and studied fossil shells because of interests in geology, but relatively few would seem to have published observations on current forms.[39]

Of particular interest among the early British naturalists is Sarah Bowdich Lee, a veritable pioneer among women travelers in West Africa, much admired for her energetic and enthusiastic collaboration with her husband in some of the earliest British exploratory ventures in the region. She was clearly an observer, naturalist, and author of considerable talent. A comparison and contrast between her and her successor three-quarters of a century later, the much better known West African traveler Mary Kingsley (see chapter 14), might well be interesting.

Notes

1. Seven additional ornithology papers by Americans have been added to the bibliography, giving a total paper count there of forty. Four of the twenty-two American authors had important interests in other fields—for Furness see chapter 10, for Millington and Treat chapter 1, and for McNair Wright chapter 16.

2. Deborah Jean Warner, *Graceanna Lewis, Scientist and Humanitarian* (Washington, D.C.: Smithsonian Institution Press, 1979); John W. Harshberger, *The Botanists of Philadelphia and their Work* (Philadelphia: T. C. Davis and Son, 1899), pp. 233–6.

3. Warner, *Lewis,* pp. 53–4.

4. The Rochester Society of Natural Sciences is said to have been founded largely because of the interest that Lewis's lectures aroused among women in the area (ibid., p. 82).

5. "Miss Lewis' Chart of the Animal Kingdom," *The New Century for Women,* 4 Nov. 1876, p. 197, cited in Warner, *Lewis,* p. 130 n. 10.

6. Ibid., p. 104.

7. In 1879 Lewis had, on principle, refused an honorary membership in the academy's forerunner, the Rochester Microscopical Society, because that organization did not then admit women as regular members, ibid., pp. 97–9.

8. Florence Merriam Bailey, "Mrs. Olive Thorne Miller," *Auk,* **36** (1919): 163–9, and "Olive Thorne Miller," *Condor,* **21** (1919): 69–73; Robert H. Welker, "Miller, Olive Thorne," in *NAW,* vol. 2, pp. 543–5; *WWWA,* p. 563.

9. William Brewster, "Two corrections," *Auk,* **10** (1893): 365. Brewster had only three months previously published what he thought was a new discovery, only to find that Mrs. Miller had written an account of it three years earlier in the *Atlantic Monthly* and had reported it again in *Little Brothers of the Air.* Miller did send an occasional note to *Science* or *Auk;* in addition to the two listed in the bibliography, we came across "The song of the alder fly-catcher," *Auk,* **19** (1902): 289.

10. Mary Mann Miller, Olive Thorne's second daughter, was born in Chicago in 1859 and was about sixteen when she moved with

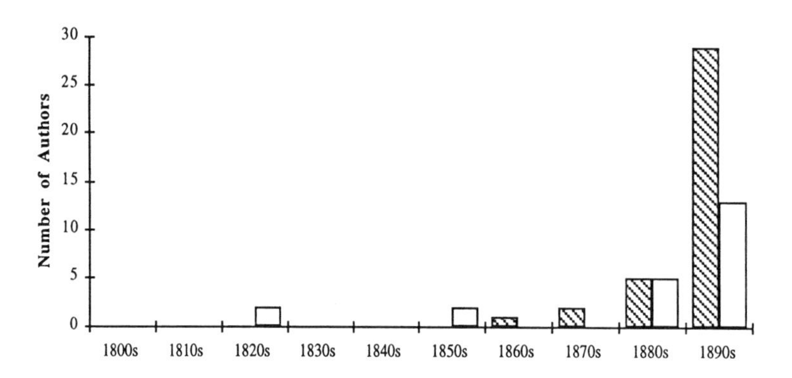

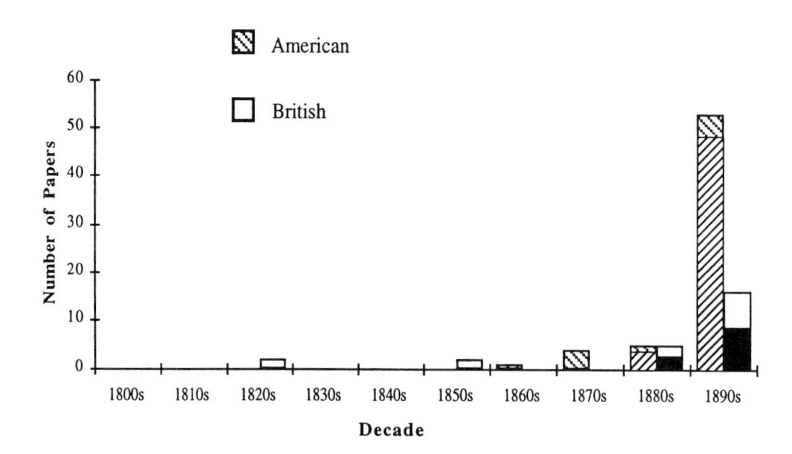

Figure 4. American and British authors and papers in conchology, ornithology, and general natural history, by decade, 1800–1900. Papers in ornithology and conchology constitute 81% of the American papers (back slope) and 48% of the British (solid); the rest are in general natural history. Data from the Royal Society *Catalogue of Scientific Papers.*

her family to Brooklyn. Her paper on the feeding habits of birds appeared in the *Auk* in 1899. However, although she accompanied her mother on some bird observation trips, she does not appear to have followed in her footsteps as a nature writer. After she and her mother moved to Los Angeles in 1904, she became one of the founders of the Bird Lovers' Club of the Southwest Museum, a group that flourished for more than sixty years. Mary is mentioned briefly in biographical sketches of her mother.

11. Harriet Kofalk, *No Woman Tenderfoot: Florence Merriam Bailey, Pioneer Naturalist* (College Station, Tex.: Texas A & M University Press, 1989); Paul H. Oesher, "In Memoriam: Florence Merriam Bailey," *Auk,* **69** (1952): 19–26, and "Bailey, Florence Augusta Merriam," in *NAW,* vol. 1, pp. 82–3.

12. Kofalk, *No Woman Tenderfoot,* pp. 55 and 196 n. 20.

13. See, for instance, the following: "The scissor-tailed flycatcher [*Milvulus forficatus*]," *Condor,* **4** (1902): 30–1; "The Harris hawk [*Parabuteo unicinctus harrisi*] on his nesting ground," ibid., **5** (1903): 66–8; "Additional notes on the birds of the Upper Pecos," *Auk,* **21** (1904): 349–63; "Scott oriole [*Icterus parisorum*], and grey vireo [*Vireo vicinior*] and phoebe [*Sayornis phoebe*] in north-eastern New Mexico," ibid., **21** (1904): 392–3; "Additions to Mitchell's list of the birds of San Miguel County, New Mexico," ibid., **21** (1904): 443–9; "Twelve rock wren nests in New Mexico [*Salpinctes obsoletus*]," *Condor,* **6** (1904): 68–70; "A dusky grouse [*Dendra-gapus obscurus*] and her brood in New Mexico," ibid., **6** (1904): 87–9; "Breeding notes from New Mexico," ibid, **7** (1905): 39–40; "Notes from northern New Mexico," *Auk,* **22** (1905): 316–18; "Nesting sites of the desert sparrow," *Condor,* **8** (1906): 111–12; "A nest of *Empidonax difficilis* in New Mexico," ibid., **8** (1906): 108; "White-throated swifts at Capistrano," ibid., **9** (1907): 169–72; "The palm-leaf oriole [*Icterus nelsoni*]," *Auk,* **27** (1910): 33–5; "Wild life of an alkaline lake," ibid., **27** (1910): 418–27; "An irrigation ranch in the fall migration," *Condor,* **12** (1910): 161–3; "The yellow pines of Mesa del Agua de la Yegua," ibid., **12** (1910): 181–4; "The red-headed woodpecker," *Bird-Lore,* **12** (1910): 86–9; "A drop of four thousand feet," *Auk,* **28** (1911): 219–25; "The oasis of the Llano," *Condor,* **13** (1911): 42–6; "Birds of the cottonwood groves," ibid., **14** (1912): 113–16; "With *Asio* in the greenwood," *Bird-Lore,* **15** (1913): 285–90; "The tufted titmouse [*Baeolophus bicolor*]," ibid., **15** (1913): 394–7.

14. The first four of these papers appeared under the main title, "Characteristic birds of the Dakota prairies." They were subtitled as follows: "I. In the open grassland," *Condor,* **17** (1915): 173–9; "II. Along the lake borders," ibid., **17** (1915): 222–6; "III. Among the sloughs and marshes," ibid., **18** (1916): 14–21; "IV. On the lakes," ibid., **18** (1916): 54–8. The remaining thirteen papers were published between 1918 and 1920 under the general title, "A return to the Dakota lake region," ibid., **20** (1918): 24–37, 64–70, 110–14, 132–7, 170–8; **21** (1919): 3–11, 108–14, 157–62, 189–93, 225–30; **22** (1920): 21–6, 66–72, 103–8.

15. The articles appeared under the general title "Birds of the humid coast," *Condor,* **19** (1917): 8–13, 46–54, 95–101.

16. Oesher, "In memoriam," p. 24.

17. Kofalk, *No Woman Tenderfoot,* pp. 163, 203 n. 3. Kofalk quotes Bailey as saying that she collaborated only with her husband.

18. Janet Wilson James, "Eckstorm, Fannie Pearson Hardy," in *NAW,* vol. 1, pp. 549–51; David C. Smith, "Eckstorm, Fannie Hardy," in *DAB,* supplement 4 (1947), pp. 248–9.

19. Eckstorm's post-1900 articles included, "A description of the adult black merlin (*Falco columbarius suckleyi*)," *Auk,* **19** (1902): 382–5; "The meadowlark in Maine, and other notes," ibid., **26** (1909): 430–2.

20. These were the following: Helen Ball of Worcester, Massachusetts; Abby Bates of Waterville, Maine; Mabel Berry of East Derry, New Hampshire; Caroline Boyce who worked on the New York coast, in the Hudson Valley, and in New Jersey; Mary Bruce of Vermont; Huberta Foote of New York City; Jane Hine of Sedan, Indiana; Mary Hyatt of Sanfordville, New York; Martha Tyler of the Fairbanks Museum, St. Johnsbury, Vermont (she was the museum curator); Ellen Webster of Franklin Falls, New Hampshire; Jennie Whipple of Norwich, Connecticut; Nelly Woodworth of St. Albans, Vermont. Except for Boyce, who wrote in the 1870s, and Whipple, whose paper was contributed in the 1880s, all of these women wrote in the period 1891–1900. Typically each contributed only one or two observational notes to the indexed ornithological literature, most appearing in the *Auk.*

21. *WWWA,* p. 859.

22. D. J. Wentworth (Mrs. Edwin P. Wentworth) published three papers during the 1890s on shells of the coast of Maine. She lived in Portland, Maine, was active in the local branch, Section B, Isaac Lea Chapter of the Agassiz Association, and served as secretary of this section (Marine Shells of the East Coast) in 1896–1897 (*Nautilus,* **10** (1896–1897): 112).

23. Ida Shepard's paper was not indexed by the Royal Society but has been added to the bibliography. For Monks and Price see, respectively, chapters 5 and 1.

24. Information from archives department, University of California Los Angles; *American Malacologists. A National Register of Professional and Amateur Malacologists and Private Shell Collectors,* ed. R. Tucker Abbott (Falls Church, Va.: American Malacologists, 1973–1974), p. 177; *AMS,* 1906–1933 (necrology).

25. Mrs. M. Burton Williamson, "A history of University Town," *Annual Publications, Historical Society of Southern California,* **3,** pt. iii (1895): 17–22, especially p. 21; "Glancing backward," ibid, **11,** pt. ii (1919): 82–90, especially p. 89.

26. *Nautilus,* **8** (1894–1895): 10–11, and **10** (1896–1897): 112.

27. Included among Williamson's post-1900 publications are the following: "How *Potamides* (*Cerithidea*) *californica* Hald. travels," *Nautilus,* **15** (1901): 82–3; "A monograph on *Pecten aequisulcatus,* Cpr. class Pelecypoda," *Bulletin of the Southern California Academy of Sciences,* **1** (1902): 50–61; "On cataloguing a collection of shells," *Nautilus,* **17** (1903): 39–41; "New varieties of *Crepidula rugosa* Nutt. found on *Natica* and on *Norrisia,*" ibid., **19** (1905): 50–1; "Some west American shells, including a new variety of *Corbula luteola* Cpr. and two new varieties of gastropods," *Bulletin of the Southern California Academy of Sciences,* **4** (1905): 118–29; "West American Mitridae—north of Cape St. Lucas, Lower California," *Proceedings of the Biological Society of Washington,* **19** (1906): 193–7; "Abalones and the penal code of California," *Nautilus,* **20** (1906): 85–7; "Note on *Thais* (*Purpura*)," ibid., **25** (1911): 30–1.

28. See *Ladies' Clubs and Societies in Los Angeles in 1892* (reported for the Historical Society of Southern California, Mrs. M. Burton Williamson, comp. and ed., March, 1892—Los Angeles: E. R. King, 1925). This work, published as a monograph three years after Williamson's death by her daughters Lillian A. and Estella M. Williamson, was withdrawn from circulation almost immediately after it appeared. In 1945 the balance of the edition was finally released for sale.

29. M. Burton Williamson, "Some American women in science," *The Chautauquan,* **28** (1898–99): 161–8, 361–8, 465–73.

30. Abbott, *American Malacologists,* p. 139; "Ida Shepard Oldroyd," *Nautilus,* **55** (1942): 140–1.

31. For Donald Longstaff and Crane see chapter 12, for Power,

chapter 3. Only the following has been uncovered about Carphin, Thew, and Sprague: Janet Carphin collected shells at Eyemouth, just north of Berwick-on-Tweed. She was also interested in crustaceans, her article on *Argulus foliaceus* appearing in the *Annals of Scottish Natural History* in 1895. Mrs. Edward Thew, whose work was published by the Berwickshire Naturalists' Club in 1897, assembled collections of shells from the Northumberland coast. Miss Sprague, along with her father, Dr. T. B. Sprague, and her mother, belonged to the Edinburgh Field Naturalists Society. Over the course of several years during the 1890s, with her father's help, she studied microscopic life, collecting mainly from local ponds. For Annie Ley see chapter 16.

32. See Allen, *Naturalist in Britain,* pp. 132–9, for an account of the development of the marine aquarium and the subsequent aquarium craze that swept the country about the middle of the century. Anna Thynne's early observation is mentioned in John E. Taylor, *The Aquarium: its Inhabitants, Structure and Management* (London: Hardwicke and Bogue, 1876), pp. 13–23.

33. "Lee, Sarah," *DNB,* vol. 32, p. 379; "Lee, Sarah (née Wallis)," in Ray Desmond, *Dictionary of British and Irish Botanists and Horticulturalists* (London: Taylor and Francis, 1977), p. 379; anon., "Mrs. Lee," *The Gentleman's Magazine and Historical Review* (1856): 653–4.

34. Among Bowdich Lee's later books were the following: *The Juvenile Album; or, Tales from Far and Near* (1841); *Elements of Natural History, for the Use of Schools . . . ; comprising the Principles of Classification* (1844); *The African Wanderers; or, the Adventures of Carlos and Antonio, embracing Descriptions of the Manners and Customs of the Western Tribes, and the Natural Productions of the Country* (1847); *Adventures in Australia* (1851); *Anecdotes of the Habits and Instincts of Animals* (1852); *British Birds, with Descriptions* (1852); *The Farm and its Scenes* (1852); *Anecdotes of the Habits and Instincts of Birds, Reptiles and Fishes* (1853); *Familiar Natural History* (1853); *Foreign Animals* (1853); *Foreign Birds* (1853); *Trees, Plants and Flowers* (1854); *Playing at Settlers* (1855). All were published in London, the first two by Ackerman and Longmans, respectively, all the rest by Grant and Griffith.

35. *Who's Who,* 1929, pp. 3374–5, entry for Yate, Lieut.-Col. Arthur Campbell.

36. *Who's Who,* 1931, p. 2819; "Edmondston family of Unst, Shetland," *British Biographical Archive,* microfiche edn., no. 361, entry from William Anderson, *The Scottish Nation* (1862).

37. Thomas Edmondston, Jessie's older brother, wrote a very successful *Flora of the Shetland Islands* (Aberdeen: G. Clark, 1845). Two years after it was published he was elected professor of botany at Anderson College, Glasgow University, but he resigned almost immediately, having also been appointed chief naturalist on the round-the-world trip of H.M.S. *Herald*; he was killed in Ecuador, during this expedition. Jessie's sister Eliza was also an author, her *Sketches and Tales of the Shetland Islands* appearing in 1856.

38. Among Jessie Saxby's works were the following: *Lichens from the Old Rock [and other Poems]* (Edinburgh: W. P. Nimmo, 1868); *The One Wee Lassie* (Edinburgh: [n.p.], 1875); *Daala-mist; or Stories from the Shetland Isles* (Edinburgh: T. Gray, 1877); *Folklore from Unst, Shetland* (published in *Leisure Hour,* Jan.-May, 1880); *Breakers Ahead; or, Uncle Jack's Stories of Great Shipwrecks of Recent Times, 1869–1880* (London: T. Nelson and Sons, 1882); *Preston Tower; or, will he no come back again* (Edinburgh and London: Oliphant, Anderson and Ferrier, 1884); *The Lads of Lunda* (London: J. Nisbet, 1887); *The Home of a Naturalist* ([n.p.], 1888); *Lindéman Brothers; or, Shoulder to Shoulder* (London: Sunday School Union, 1888); *Oil on Troubled Waters; a Story of the Shetland Isles* (London: Religious Tract Society, 1888); *The Yarl's Yacht* (London: J. Nisbet, 1889); *West-nor'-west* (London: J. Nisbet, 1890); *Wrecked on the Shetlands* (London: Religious Tract Society, 1890); *Heim-laund and Heim-folk* (Edinburgh: R. and R. Clark, 1892); *Viking Boys* (London: J. Nisbet, 1892); *Lucky Lines; or Won from the Waves* (Edinburgh and London: Oliphant, Anderson and Ferrier, 1893); *Bird Omens in Shetland* [inaugural address to the Viking Club, London, 13 October 1892], with notes on the folklore of the raven and the owl by W. A. Clouston (London: [n.p.], 1893); *The Saga Book of Lunda, wherein is recorded some more of the Adventures of the Viking Boys and their Friends* (London: J. Nisbet, 1893); *Queen of the Isles* (London: W. S. Partridge, 1897); *Shetland Traditional Lore* (Edinburgh: Grant and Murray, 1932).

39. Conchology is not even listed in the index of Patricia Phillips's recent study of British scientific women, *The Scientific Lady. A Social History of Women's Scientific Interests 1520–1918* (New York: St. Martin's Press, 1990). We do know, however, that at least one member of the group discussed here, Jane Donald Longstaff, built up a sizable collection of recent as well as fossil shells, collected worldwide (see chapter 12).

Chapter 5

SOME GENERAL BIOLOGISTS

Papers classified under the broad term biology in this overview constitute less than one-half percent of the material by women indexed in the Royal Society *Catalogue.* However, a number of Americans whose papers are listed in such categories as botany, conchology, physiology, zoology, or some combination of these designated their interest as "biology" in their entries in contemporary biographical dictionaries.[1] Four of them (Bunting, Enteman Key, Monks, and Towle) were high school or college biology teachers; Bunting and Enteman Key took Ph.D. degrees. A fifth woman, Foulke Andrews, who had less academic training than any of the others and carried out her research without formal institutional affiliation, was remarkably productive as measured by the amount of scientific work she published before the turn of the century. Of these five, three received at least part of their training at Bryn Mawr College.

GWENDOLEN FOULKE[2] (1863–1936), the daughter of Julia de Veaux (Powell) and William Parker Foulke, was born at Bala Farm, Chester County, Pennsylvania, on 26 June 1863. She was educated at a private school in Philadelphia and developed an early interest in biology, particularly microscopic freshwater animals and unicellular organisms. By her early twenties she had published several notes and papers in *Science* and in the *Proceedings of the Academy of Natural Sciences, Philadelphia.* In 1888 she spent a year at Bryn Mawr College, taking a special course in biology. She married Ethan Allen Andrews, professor of zoology at Johns Hopkins University, in 1894, and for several years continued her biological work, especially studies on the structure of protoplasm. Two lengthy papers by her appeared in the *Journal of Morphology* and in *Zeitschrift für wissenschaftliche Mikroskopie* in 1897, the year in which her monograph *The Living Substance as Such: and as Organism* was published. Thereafter her interests shifted to animal and child psychology. She had considerable literary and artistic ability and wrote poems, prose fiction, and essays; these were brought out in two volumes by her husband after her death in 1936.[3]

MARTHA BUNTING[4] (1861–1944) was active in scientific work for over three decades, alternating periods of research with high school and college teaching. The daughter of Susan Lloyd (Andrews) and Samuel Bunting, she was born in Philadelphia, on 2 December 1861. After receiving her early education at the Friends' School in Darby, Pennsylvania, she went to Swarthmore College (B.L., 1881). She continued her studies at the University of Pennsylvania under the guidance of pathologist Leo Loeb for three years from 1888 until 1891, and then for another two years, at Bryn Mawr College, where she worked with Thomas Hunt Morgan on the development of the sex cells in *Hydractinia* and *Podocoryne.* Her substantial paper on this research, carried out in part during three summers at the Woods Hole Marine Biological Laboratory, appeared in the *Journal of Morphology* in 1894. She received her Ph.D. the following year.

From 1893 until 1898 she taught biology, first at the Woman's College of Baltimore (later Goucher College) and then at the Girls' High School in Philadelphia. Thereafter she had a further year of study at Columbia University and a summer of research at the Marine Laboratory at Cold Spring Harbor, Long Island. Two papers reporting her investigations of the development of cork tissue appeared in the late 1890s. Carried out at the Woman's College of Baltimore and the Botanical Garden of the University of Pennsylvania, the work was suggested by John Macfarlane, the garden's director. She taught biology at Wadleigh High School in New York City from 1900 until 1910, when, at the age of forty-eight, she returned to research, taking a Carnegie assistantship in physiology with Edward Tyson Reichert in the department of medicine at the University of Pennsylvania. Except for a two-year interruption during the First World War when she carried out war relief work, she continued as a research assistant until about 1920 and remained associated with the university for another decade. Throughout this period she does not appear to have published further under her own name in any of the major scientific journals. She died in Philadelphia, 13 October 1944, in her eighty-third year.

ELIZABETH TOWLE[5] (1876–1959), the daughter of Mary Elizabeth (Ladd) and James Augustus Towle, was born in Painsville, Ohio, on 11 March 1876. She attended private schools and Iowa College (later Grinnell College) in Grinnell, Iowa, before going to Bryn Mawr (A.B., 1898; A.M., 1899). She stayed on at Bryn Mawr for a year as a fellow in biology, carrying out research on limb muscle regeneration in amphibians under the direction of Thomas Hunt Morgan.[6] After two years of high school science teaching in Middletown, Connecticut (1900–1901) and another two as an instructor in biology at Rockford College in Illinois (1901–1903), she returned to graduate studies, first with a one-year fellowship in physiology in the Hull Laboratory at the University of Chicago and then at Columbia University (1904–1906). Her 1904 investigation of the effects of stimuli on the freshwater protozoan *Paramecium,* long a popular organism for experimental work, was carried out under the guidance of physiologist Elias P. Lyon at Chicago.[7] For thirty-four years, from 1907 until 1941, she was head of the science department at the Baldwin School for Girls in Bryn Mawr, Pennsylvania, a school with an outstanding academic reputation. The only variation in this long period of teaching and administration was a two-year leave (1919–1921) when she taught physics at the American College for Girls in Constantinople. Following her retirement in 1941 she worked on various war relief organizations and was active on Bryn Mawr alumnae committees. She died at her home in Haverford, Pennsylvania, on 30 January 1959, at the age of eighty-two.

WILHELMINE ENTEMAN[8] (1872–1955) was born in Hartland, Wisconsin, on 22 February 1872. After taking an A.B. degree at the University of Wisconsin in 1894, she taught high school biology and German in Green Bay, Wisconsin, for four years. She then resumed her studies, going to the University of Chicago where she held an assistantship in zoology from 1899 until 1902. Her work on the anatomy of microscopic freshwater organisms, carried out under the direction Charles Whitman, was reported in four papers and notes published between 1899 and 1901, the year she received her Ph.D.

Over the next few years she taught biology at a succession of institutions. At New Mexico Normal University in 1903–1904, she headed the biology and nature study departments and studied wasps. Her lengthy report on coloration in these insects appeared in 1904.[9] From 1907 to 1909 she was at Belmont College, and then for three years until 1912 taught at Lombard University, Galesburg, Illinois. She married Francis B. Key sometime between 1904 and 1910. Later becoming interested in heredity and eugenics, she worked for two years in the eugenics record office of the Carnegie Institution. After that she held a succession of posts in eugenics research associations and state training schools in Pennsylvania, Washington, D.C., and Michigan.[10] She retired in 1925 and died on 31 January 1955, shortly before her eighty-third birthday.

SARAH MONKS[11] (1847?-1926) taught biology at the Los Angeles State Normal School (later the University of California Los Angeles) for twenty-two years and was active in marine biology research in southern California throughout much of that time. Born in Cold Spring-on-Hudson, New York, she attended Cold Spring schools and in 1867 went to Vassar College, which had opened two years previously. She was remembered as one of the outstanding members of the class of 1871. After receiving an A.B. degree she continued her studies, using facilities at the Philadelphia Academy of Sciences. In 1874 she went to Santa Barbara College in California as a teacher of natural history, but returned to the east the following year. Submitting a thesis on "Coloration in birds," she took an A.M. at Vassar in 1876.

In 1884 she became instructor of biology at the Los Angeles State Normal School and remained there until her retirement in 1906. Her summers were given to scientific work. She had at least one season at the Woods Hole laboratory, and during the 1890s regularly took part in the work of the Marine Biological Laboratory at San Pedro, a West Coast summer school in biology run by the University of California. A considerable amount of collecting was done by dredging in the vicinity, and Monks's special interest was the sorting and classification of echinoderms. She also published work on a variety of biological subjects, including observations on reptiles, amphibians, trap-door spiders, and diatoms.[12] She joined the American Association of Conchologists in 1893. For twenty-five years she made her home in San Pedro, maintaining a workplace on the jetty where she spent most of her time throughout the twenty years of her retirement. She died at San Pedro, on 10 July 1926, at age seventy-nine.

Notes

1. On the other hand, six of the American women who each contributed one or two papers classified in the bibliography as biology had clearly recognizable major interests in more specifically definable areas and are discussed in the corresponding chapters: for Bodington, Fielde, Lewis, and Nunn see, respectively, chapters 15, 2, 4, and 3; for Brown Blackwell and Hinckley see chapter 16.
2. *WWWA,* p. 51; *AMS,* 1910.
3. *The Poems of Richard de Veaux,* pseudonym for Sara Gwendoline Foulke Andrews, (Baltimore, Md.: J. H. Furst, 1938), and *Some Writings of Richard de Veaux* (Ann Arbor, Mich: Edwards Brothers, 1938).
4. *Who Was Who in America,* vol. 4, p. 135; alumni records, Friends Historical Library of Swarthmore College; *WWWA,* p. 145; *AMS,* 1910, 1921.
5. Alumnae records, Bryn Mawr College; *WWWA,* pp. 820–1.
6. Elizabeth W. Towle, "On muscle regeneration in the limbs of *Plethodon,*" *Biological Bulletin,* **2** (1901): 289–99.
7. "A study of the effects of certain stimuli, single and combined, upon *Paramoecium,*" *American Journal of Physiology,* **12** (1904): 220–36.
8. S and F, p. 73; *AMS,* 1910–1949.
9. See *Coloration in Polistes,* publication no. 19, Carnegie Institution, Washington, D.C., 1904, 88 pp.
10. Eugenics and family heredity studies was an area of research that attracted considerable interest in the United States from the late 1870s until the First World War. Much of the fieldwork involved studies of isolated communities, for instance the reconstruction of the descent of rural laboring-class families of apparently "defective" stock. The field-workers in these projects were predominantly middle-class women—see *White Trash. The Eugenic Family Studies, 1877–1919,* ed., Nichole Hahn Rafter (Boston: Northeastern University Press, 1988).
11. *WWWA,* p. 570; M. Burton Williamson, "The Marine Biological Laboratory at San Pedro," *Annual Publication of the Historical Society of Southern California,* **5** (1901): 121–6, on p. 126; *Vassar College Catalogues,* 1867–71; Elizabeth Coffin, "In memoriam. Sarah Preston Monks," *Vassar Quarterly,* **12** (December 1926): 56.
12. Monks's post-1900 papers include: "Variability and autotomy of *Phataria* [echinoderms]," *Proceedings of the Academy of Natural Sciences of Philadelphia,* **56** (1904): 596–600, and "Diatoms," *Bulletin of the Southern California Academy of Sciences,* **7** (1908): 12–17.

Chapter 6

MEDICAL SCIENTISTS: PHYSIOLOGISTS, NEUROLOGISTS, ANATOMISTS, PATHOLOGISTS, AND BACTERIOLOGISTS

Women in the medical sciences form one of the more productive groups participating in scientific research in the late nineteenth century. Thanks in part to an especially strong contribution from Russian women (almost a third of the papers) work in this area constitutes about 11 percent overall of women's publications for the period. Only botanists and zoologists brought out more; and indeed, if papers in clinical medicine had been included with those in the research branches, the medical sciences would rank second only to botany in journal article production (Figure 0-2). In Britain and the United States, work in the medical sciences is somewhat less prominent (compare Figures 0-2 and 0-3[1]).

In the areas of physiology, neurology, anatomy, and pathology (taken together) the largest national contributions came from Britain, the United States, and Russia (each country producing about a quarter of the total), with work by women from Italy, Poland, and Romania also noteworthy. In bacteriology, a field that began to emerge as a distinct discipline (an offshoot of pathology) only in the 1880s[2], the bulk of the papers came from workers in Russia and Italy, but there were also creditable contributions from American and British women.

American women

In the United States four institutions were especially prominent in the training of women research workers in the medical sciences. These were the universities of Michigan, Chicago, and Cornell, and Bryn Mawr College.[3] Of the six women who had especially notable careers, two—physiologist Ida Hyde and neurologist Susanna Phelps Gage—took their undergraduate degrees at Cornell; three others—anatomist/pathologist/bacteriologist Lydia Adams DeWitt, pathologist/toxicologist Alice Hamilton, and pathologist/microscopist/dental surgeon Vida Latham—received medical (or dental) degrees from the University of Michigan, by the 1880s undoubtedly the country's top-ranking institution for the training of women in these areas. The sixth woman, bacteriologist Anna Williams, had an M.D. from the Woman's Medical College of New York.

About half of those whose careers are sketched here worked during at least part of their careers in the developing field of bacteriology, then largely an applied science concerned with the solution of pressing current problems. Typically these women had medical degrees rather than Ph.D.s in a biological science,[4] and they taught or practiced mostly in medical schools and public health laboratories.[5] Just over a quarter of the group might be described as primarily bacteriologists (Figure 6-2 a).

IDA HYDE[6] (1857–1945), perhaps the best remembered of the early women physiologists in the United States, was born in Davenport, Iowa, on 8 September 1857, one of the four children of German immigrants from Würtemberg, Babette (Loewenthal) and Meyer H. (Heidenheimer) Hyde.[7] Meyer Hyde, a merchant, left his family when his children were still young, and their mother supported them by doing mending and cleaning, work she gradually developed into a fairly successful small business. The family moved to Chicago, where Ida went to public schools and had an almost middle-class upbringing until the 1871 Great Fire of Chicago destroyed her home and displaced her mother's customers. From then on she worked for her living, helping to maintain the family and supporting her brother while he completed his schooling and took a degree at the University of Illinois. She was first a milliner's apprentice in a clothing factory, and then one of the factory's buyers and salesladies (experience that she later put to good use by always making her own clothes). In 1881, when attending her brother's graduation at the University of Illinois, she met a number of women students and decided to try to get an education herself, though her mother and brother opposed the idea. She passed the examinations for entrance to the College Preparatory School at the University of Illinois and later was admitted to the first-year class. However, in 1882 she returned to Chicago to nurse her brother through an illness and remained at home for six years, earning a living by teaching in public elementary schools.

Although she was involved in a number of interesting projects in the city schools, including the introduction of science and nature study into the curriculum, her ambition to return to academic work remained strong. In 1888, at the age of thirty-one, she enrolled at Cornell University. She finished the four-year A.B. degree course in three years, and carried out an outstanding undergraduate research project on mammalian heart structure. Reported in the *American Naturalist* in 1891, this work led her to noteworthy discoveries concerning the coronary valves, a subject she was to return to several years later.

In 1892 she received a graduate assistantship for work in biology at Bryn Mawr College, where she studied under the guidance of Jacques Loeb and Thomas Hunt Morgan. Her research was in two areas, respiratory patterns and the mechanism that controls respiration in *Limulus* (the horseshoe crab), and the anatomy and embryology of scyphozoans (jellyfishes). The latter study, which she started at the Woods Hole Marine Biological Laboratory under Morgan's direction, attracted the attention of the German zoologist Alexander Goette, who was working in the same area. When he invited her to continue her research in his laboratory at the University of Strassburg she accepted and went to Germany on an Association of Collegiate Alumnae fellowship (1893–1894).

After completing the study on jellyfish development (reported in *Zeitschrift für wissenschaftliche Zoologie,* 1894), Hyde wanted to proceed to a Ph.D., as Goette had suggested she might. However, the Strassburg faculty declined to accept a female postgraduate student, and so she transferred to the University of Heidelberg, where women had been admitted to a number of courses as auditors since 1891. In 1896, after two years of full-time study as an auditor under the direction of Otto Bütschli, she obtained her degree *multa cum laude.*[8] The same year, after six weeks at the American Davis Table at the Naples Zoological Station, she went to Bern University, where she worked for a short time on problems in muscle physiology under Hugo Kronecker, head of the Physiological Institute.

She returned to the United States in 1897, and thanks to an introduction from Henry P. Bowditch, former dean of the Harvard Medical School, whose acquaintance she had made in Bern, she obtained an Irwin Research Fellowship for work at Radcliffe College and Harvard Medical School. She was then forty. At Harvard she went back to her work on the functioning of the mammalian heart and investigated the contraction of the ventricle in association with increased blood volume when fluid is forced into the organ. Her observation that the coronary blood vessels are compressed by this muscular action, considered one of her most important findings, was reported in volume 1 of the *American Journal of Physiology* (1898). This early cardiovascular research, along with her scyphozoan and *Limulus* studies, was recognized by her election to the American Physiological Society in 1902. She was in fact the only woman until 1913 whose original contributions were considered to meet the society's standards for membership.

Wishing to find a permanent position, she applied for a post at the University of Kansas, a state institution that was then in the process of establishing a medical school and needed a physiologist. Her excellent credentials got her the job. She became assistant professor of zoology in 1898, associate professor of physiology the following year, and in 1905 full professor and head of the physiology department.

Despite a substantial teaching and administrative load while she built up her department and organized the premedical physiology curriculum, she continued her research, both at the University of Kansas and at the Woods Hole laboratory. In the summer of 1904 she worked at the University of Liverpool, one of the leading British centers for marine biology studies. Her interests were broad and she was original and inventive. Over the course of her two decades in Kansas she continued to work on the response of the cardiovascular system to stress, but also put considerable effort into investigations of respiratory processes and the functioning of the nervous system.[9] Following a chance observation that electrolytes in high concentrations affect processes of cell division, she became interested in minute differences in electrical potential within cells. This led her into what were some of the earliest investigations of micro methods for work on single cells. The stimulating microelectrode for intracellular excitation, to the development of which she made notable contributions, was a key tool in the advancement of neurophysiology and the understanding of the functioning of nerves and muscles. An electrolyte-filled glass capillary of tip diameter four to eight micrometers, it was gradually refined by a number of workers over a period of several decades. However, many of the early investigations in which it was used, including Hyde's studies, were subsequently overlooked.[10] Her later work included examinations of the physiological effects of sensory input (such as light and music),[11] and the relationship of diet to health. She also found time to write two textbooks, *Outlines of Experimental Physiology* (1905), and *Laboratory Outlines of Physiology* (1910).

For several years she took summer classes in surgery and clinical medicine at the Rush Medical School in Chicago and at the University of Kansas Medical School, completing most of the requirements for an M.D.,[12] a qualification she felt would help her when dealing with her medical school colleagues. Public health, especially matters affecting children and young people, was of special interest, and she worked to develop projects in this area at Haskell Indian College, a United States government institution in Lawrence, and in local schools. Her program in cooperation with local doctors for medical inspection of schoolchildren for communicable diseases, particularly tuberculosis and spinal meningitis, ran into conservative opposition, however, and the university chancellor, Francis Strong, withdrew his support. Such schemes of "compulsory medicine" were by no means generally acceptable at the time. Hyde's lectures at the university and in Kansas high schools on "social hygiene," were also pioneering in that they included discussion of such matters as the sexual transmission of diseases. For a time in 1918, when she had given up her administrative work at the university, she had a state appointment as chair of the Woman's Committee on Health and Sanitation of the State National Defense Committee.

With her students, a number of whom were women, she encouraged research and set high standards. She was a life member of the Naples Table Association (secretary, 1897–1900), and worked to increase the funding of research tables for women students at the Naples research laboratory (and also at Woods Hole).

Quite possibly the first woman to become full professor and head of a science department in a state university, Hyde achieved much, considering the attitudes of the time. She faced some major difficulties, however, including continual troubles in dealing with administrators and disagreements with the schools of medicine and pharmacy, which supplied many of her students. Her blunt and forceful style, evident in her correspondence with Chancellor Strong, may have helped her to get ahead in her early years, but it must inevitably have had drawbacks in her interpersonal relations with her colleagues. A great frustration to her over a long period was her relatively low salary. Feeling she was being discriminated against in this matter, she several times sent the chancellor strongly worded letters of complaint, which invariably failed to get her what she wanted.[13] From 1915 onward her problems intensified. In that year the university amalgamated her department of physiology (in the College of Liberal Arts and Sciences) with the Medical School physiology department, the merged unit to be run by a committee. The change followed a pattern general at the time, medical schools in a number of institutions taking steps to get the direction of physiology instruction into their own hands.[14] Hyde fought back, wanting to keep control of her own department, but to no effect. By June 1916 a rumor that the university was seeking a new head for its physiology department had reached her. She gave up her administrative work the following year.

For the next three years she seems to have been unsure about what her position at the University of Kansas really was.[15] She had a part-time post for a year, followed by a leave of absence (1919–1920), when she spent some time at the Scripps Institution in La Jolla, California. She then returned to Kansas for one semester, but thereafter went on another leave of absence, which became permanent retirement. In 1922–1923, when she was sixty-five, she took a last trip to Europe, staying for a period in Heidelberg but also doing a considerable amount of traveling. As well as excursions throughout Germany and into Switzerland and Austria, she went to Egypt and even to India. She looked into research possibilities at the University of Heidelberg, but it is unlikely that she carried out any laboratory work.

Within a few years she had settled in California, first in San Diego and later in Berkeley, where she died, on 22 August 1945, shortly before her eighty-eighth birthday. She left $25,000 to endow a woman's international fellowship, administered by the American Association of University Women (the Ida H. Hyde Fellowship), and $2,000 to the University of Kansas to fund a scholarship for women graduate students in biology.

SUSANNA PHELPS[16] (1857–1915), neurologist, was born in Morrisville, New York, 26 December 1857, the daughter of Henry Samuel Phelps, a businessman and former schoolteacher, and his wife Mary (Austin), who had also been a teacher before her marriage. Both the Phelpses and Austins were old New England families. Susanna was first taught at home by her mother and then went to Morrisville Union School and Cazenovia Seminary, in Cazenovia, New York. Her mother wanted her to continue her education at Vassar, but her father was so impressed by accounts of recently opened Cornell University that he persuaded her to go there instead. The first student to enroll in Sage College, the residence hall for women opened in 1875, she received her degree (Ph.B.) in 1880. The following year, she married Simon Henry Gage, assistant professor of histology and embryology.[17] They had one son, Henry Phelps Gage, who was also to become a successful scientist.

Marriage gave Susanna an opportunity to follow her special interests of zoology and comparative anatomy, and she lost no time in making the most of it. Her earliest published investigations, on the physiology of respiration (1885, 1886), were done in collaboration with her husband. She quickly branched out on her own, however; her first independent research, carried out in the late 1880s, concerned the form and relation of the fibers in striated muscles in small animals such as mice and birds, a topic very poorly understood at the time. She demonstrated clearly what the relations of the fibers really were. By the early 1890s she had begun her studies in neurology, particularly the development and morphology of the nervous system in humans. Ten of her twenty-six independent publications were in this area. She had access to several laboratories, including those at Harvard Medical School and Johns Hopkins Medical School. Probably her most important paper was that reporting work (microscopical studies of sections) on the brain and nephric system of a three-week human embryo loaned to her by anatomist Franklin Mall of Johns Hopkins.[18] In 1911 she went to Europe, attended meetings of the Anatomische Gesellschaft in Leipzig and the British Association for the Advancement of Science, and visited several famous laboratories, including those of Camillo Golgi in Italy and Cornelius Ariëns-Kappers in Holland.

She never held a paid professional position and her regular access to adequate laboratory facilities depended on her husband's readiness to let her use his. Having no formal graduate-level training, she must also have depended on him for professional guidance, to an extent equivalent to such training. However, her situation may well have given her a considerable practical advantage over many of her women contemporaries; she had the use of reasonable facilities, the benefit of close interaction with leading male workers in the biological sciences, and was free from the heavy teaching and administrative commitments that absorbed the energies of women scientists who were faculty members in women's colleges. Her fellowship in the AAAS and her star in the 1910 edition of *American Men of Science* attest to her high standing among her contemporaries.

Active in the Cornell University community and concerned with the well-being of the students, she was also interested in educational developments and plans at the national level. In particular, in the late 1890s she worked with the George Washington Memorial Association, of which she was one of the founders, for the establishment of a national Washington Memorial university, a project that was never realized. She died on 5 October 1915, at age fifty-seven, after four years of failing health. Her husband and son gave the Cornell physics department a gift of $10,000 in her memory (she had been the first woman to take laboratory physics at Cornell). Volume 27 (1916–17) of the *Journal of Comparative Neurology* was dedicated to her—a remarkable tribute.

LYDIA ADAMS[19] (1859–1928), remembered especially for her collaboration in pioneering investigations of potential antituberculosis drugs, was born in Flint, Michigan, on 1 February 1859, the second daughter among the three children of Oscar Adams, a Flint attorney, and his wife Elizabeth (Walton). She attended Flint public schools, and then became a teacher herself. At the age of nineteen she married Alton D. DeWitt, also a teacher and a native of Flint. They had two children, born in 1879 and 1880, but she continued to teach, and at the same time advanced her qualifications by completing a two-year course (1884–1886) at the State Normal College in Ypsilanti, graduating with high standing.

She taught in Michigan public schools for another nine years (first in St. Louis, then in South Haven and Portland), but in 1895 enrolled at the University of Michigan, taking a combined medical and science course (M.D., 1898; B.S., 1899). Her research began while she was still an undergraduate, her first two notes appearing in 1897. These reported investigations focused on "muscle spindles," motor and sensory nerve ending structures widespread in vertebrate muscle. The work, carried out with anatomist Carl Huber, gave her an introduction to the field of microscopic neuroanatomy, in which she was to continue very successfully for several years. Having separated from her husband at about the time of her graduation, she stayed on at the University of Michigan as an assistant in pathology. In 1902, at the age of forty-three, she became an instructor, and she remained in that position until 1910, with a year's study leave at the University of Berlin in 1906.

Her research was mainly in two areas, both employing her skills in microanatomy—the nerves involved in the transmission of impulse in the beating of the mammalian heart, and the internal structures of the pancreatic gland.[20] Her work on the pancreas constituted a contribution to the long effort to understand the function of the organ and the etiology of diabetes, an effort that was to bear fruit spectacularly some seventeen years later with the isolation of the hormone insulin and the demonstration of its effectiveness in the treatment of the disease.

The use of pancreatic extracts to treat diabetes had been explored by a number of investigators over the years but without success. Focusing on selected internal structures within the pancreas rather than the whole organ, DeWitt prepared extracts from the cell clusters known as the islets of Langerhans and was able to demonstrate, in a preliminary way, that these islets were production sites of a material essential for carbohydrate metabolism (the disruption of which causes diabetes). Earlier work involving extraction of the whole organ had resulted in a complex mixture in which the material from the islets was essentially swamped by the major secretions of the pancreas, namely digestive enzymes. However, DeWitt had neither the facilities nor the necessary backing to proceed beyond the initial tests reported in her 1906 *Journal of Experimental Medicine* paper. The subsequent isolation and successful testing of the islets of Langerhans extracts by a University of Toronto team led by J. J. R. MacLeod and Frederick Banting brought these workers the 1923 Nobel prize in medicine and physiology.[21]

DeWitt's work in microanatomy brought her election to the Association of American Anatomists in 1902 and a star in *American Men of Science* (1906). Nevertheless, at the University of Michigan she was not accepted into either the Faculty Research Club or the Junior [Faculty] Research Club; membership of these was exclusively male. So, in 1902, she organized the Women's Research Club and for a time served as its president. The club functioned successfully for several decades, encouraging women's research at the University of Michigan.

In 1910 she took the position of instructor in the department of pathology at Washington University, St. Louis, where one of her former Michigan colleagues, leading clinical research scientist George Dock, had recently become professor. At the same time, she became assistant city pathologist and bacteriologist in the St. Louis Department of Health, and from then on put aside her microanatomy studies, moving on to the rapidly developing area of bacteriology. Within two years she and her coworkers had published several notable reports on bacteriological procedures in public health work. Of special note were her studies on tuberculosis and typhoid organisms.[22] The work attracted the attention of Harry Gideon Wells, professor of pathology at the University of Chicago and a man of eclectic viewpoint. He invited her to join his department as assistant professor, with an additional position on the staff of the newly opened Otho S.A. Sprague Memorial Institute. There she became one of the key members of a team that was beginning an investigation of the possibilities for chemical treatment of tuberculosis.

The studies were to be modeled on the classic work of the 1908 winner of the Nobel prize in medicine, Paul Ehrlich, who, starting from an observation made when he was a medical student that dyes stain tissue preferentially, looked for one that had selective affinity for parasitic organisms. Having found a dye that functioned in this way he went on to investigate analogous compounds in which a key grouping of two nitrogen atoms in the dye molecule is replaced by two arsenic atoms. The work resulted in the preparation in 1909 of the compound patented in Germany as Salvarsan, an antisyphilitic drug. Ehrlich's discovery was especially important in the field of chemotherapy because it was the result of a thought-out, logical design; his methodology constituted an early example of procedures common in what is now the well-established subfield of medicinal chemistry.

DeWitt and her colleagues first looked for dyes that would selectively bind with tuberculous lesions, and then, with the help of a number of chemists, modified likely dyes by the incorporation of metal atoms into their molecular structures. The resulting compounds were systematically tested for antituberculosis activity. A lengthy series of papers, most of them under the general title, "Studies in the biochemistry and chemotherapy of tuberculosis" (1913–1926), reported the work of the team (which included several women). The 1923 monograph, *Chemistry of Tuberculosis,* coauthored by Wells, DeWitt, and Esmond Long, presented an overall picture of the major findings.[23] These investigations, extensive and meticulous, did not, in fact, produce an antituberculosis agent, but the procedures developed in the process served as a model for later successful searches for chemotherapeutic compounds.

DeWitt was highly regarded by her colleagues; in 1918 (at the age of fifty-nine and six years after she had joined the Chicago group), she was promoted to associate professor. She served as president of the Chicago Pathological Society in 1924–1925. Recognition also came to her from the University of Michigan, which gave her an honorary A.M. in 1914; this might in part have made up for the Faculty Research Club failing to find a way of admitting her not much more than a decade before. Poor health caused her to retire in 1926. She went to live with her daughter in Winter, Texas, where she died, 10 March 1928, at age sixty-nine.

ALICE HAMILTON[24] (1869–1970), who became an internationally recognized authority in industrial toxicology and pathology, was perhaps the most influential of the American women medical scientists included in this study. The second of the five children of Gertrude (Pond) and Montgomery Hamilton, she was born on 27 February 1869 in New York City in the home of her maternal grandmother, but grew up in Fort Wayne, Indiana. Her home, along with those of her Hamilton grandmother and her uncle, was situated on the large estate acquired by her grandfather, Allen Hamilton, a Scots-Irish immigrant who had successfully invested in land and railroad developments. Much of her early life centered on the activities of her extended family. Since her mother had strong objections to the long hours of the Fort Wayne public schools and her father to the curriculum they offered, she and her three sisters got most of their early education at home. To a large extent they taught themselves, by reading, though they had some formal instruction in Latin and mathematics and help with French and German. At the age of seventeen Alice went for two years to Miss Porter's school in Farmington, Connecticut, and then, since she wanted to be independent and to be able to go out and see the world, she decided on a career in medicine.

Following a year of private tuition in physics and chemistry and another year at a small medical college in Fort Wayne, she was accepted into the Medical Department of the University of Michigan. Soon after she graduated with an M.D. in 1893, at the age of twenty-four, she realized that her inclination was not toward medical practice but rather in the direction of clinical diagnosis, where new methods involving microscopical and chemical analysis were beginning to be used with great success. After internships at hospitals for women and children in Minneapolis and Boston, Hamilton returned to Ann Arbor in 1895 as an assistant in the bacteriological laboratory. The following autumn she went to Europe to round out her studies in bacteriology and pathology. Accompanying her was her older sister, Edith, who, on a Bryn Mawr College fellowship, went to study classics. They succeeded in gaining admittance to classes and laboratories at the Universities of Leipzig and Munich, though various restrictions, imposed because they were women, limited them somewhat: "It was a man's world in every sense," with militarism much in evidence. Alice was also disappointed at finding she learned little new technically. Nevertheless, life in Germany in these days was very pleasant: "At home, if we wanted music we must go to a concert; in Germany we could step out into the park, sit under the trees with a glass of beer . . . and listen to lovely music . . . It was the easiest thing to get off into the country (what a contrast to Chicago!), only a short ride in a fourth-class coach and then fields and mountains and lakes . . . "[25]

Having no job when she returned to the United States, she spent a winter at the Johns Hopkins University Medical School. However, in 1898 she found a position teaching pathology at the Woman's Medical School of Northwestern University in Chicago, and shortly thereafter became a resident of Hull House, the most famous of Chicago's settlement houses. Established around that time by socially conscious people in an attempt to bridge the gap between the comfortably off and the poor, the settlements were places where educated young men and women could live as neighbors of working-class people. The activities of Hull House quickly became the absorbing interest of Hamilton's life. They not only brought her into close contact with the urban poor, but also introduced her to social action groups and the work of trade unions. Her own particular contributions were her baby clinic and the evening classes she ran on preventive medicine and basic hygiene.

When the Woman's Medical School closed in 1902 she took a position as bacteriologist at the new John McCormick Institute for Infectious Diseases. Most of her research involved basic technical studies on vaccine therapy, especially in relation to diphtheria, scarlet fever, and typhoid,[26] but the story of her investigation of one of the immediate practical problems that arose at the time is of much general interest. It centers on the 1902 Chicago typhoid epidemic. The disease was endemic in the city at the time, chlorination of public water supplies having not yet been introduced and the water source (Lake Michigan) being also the dumping place for the city sewage lines. Epidemics were periodic, and that of 1902 was exceptionally bad. Hull House, in the city's Nineteenth Ward, was the center of one of the hardest hit regions, and nobody understood why.

Hamilton knew that the main pumping station that drew water from the lake for the Nineteenth Ward supplied a much larger community, which was not especially badly affected; the milk supply was also the same as in adjacent, less severely stricken wards. Guessing that some local condition was operating around Hull House, she tramped through the neighborhood to find out what it might be. The thing that particularly struck her was the shocking state of the local sewage system, which in that immediate area predated the Great Fire of 1871. There were too few household water closets and their place was taken by outdoor privies; swarms of flies were all-pervasive. Remembering the conclusion drawn during the Spanish-American War, that the combination of open latrines and flies had played an important role in spreading typhoid among American soldiers, she decided that these same factors were worth examining in Chicago's Nineteenth Ward. To put her theory to the test she collected flies from privies and kitchens and demonstrated in the laboratory that they were in fact carriers of typhoid. Her first paper on the subject appeared in 1903,[27] and she later wrote that she was sure she gained more kudos from her investigation of flies and typhoid than from any other piece of work she ever did. Being relatively nontechnical and easily understood, her report caught the attention of the public. Further, it resulted in an inquiry that brought about a complete reorganization of the Chicago Health Department and the appointment of an expert to take charge of tenement house inspection. Sometime later she discovered, to her great mortification, that the flies had had little to do with the excessive number of cases of typhoid in the Nineteenth Ward. The cause was simpler but much more discreditable, and so the Board of Health had not wanted to disclose it: for a three-day period after the epidemic had started, the Nineteenth Ward's local booster pumping station along the main supply line had distributed water affected by a secondary source of contamination. A break in a line at the station had resulted in the direct intake of sewage into water pipes, and some time elapsed before the trouble was discovered and the break repaired.

Though her work in bacteriology continued and she spent some time at the Pasteur Institute in Paris in 1903, Hamilton's interests began to focus more and more on the problems of industrial and occupational diseases. She later wrote,

> Living in a working class quarter, coming in contact with laborers and their wives, I could not fail to hear tales of the dangers workingmen faced, of cases of carbon-monoxide gassing in the great steel mills, of painters disabled by lead palsy, of pneumonia and rheumatism among the men in the stockyards. Illinois then had no legislation providing compensation for accident or disease caused by occupation . . . [28]

When Charles Henderson, professor of sociology at the University of Chicago and an admirer of the German workman's insurance system, persuaded the governor of Illinois to appoint a commission to examine the extent of industrial sickness in Illinois, Hamilton, whose interest in the subject was well known to Henderson, was named one of the five doctors on the commission. The task they faced was one of great complexity, there being not even any information available on what the dangerous occupations in Illinois were. All they could do was begin with trades known to be dangerous and hope that, as they studied them, they would discover others less well known. They had one year for their work, 1910. Hamilton took on the investigation of the lead industries and also served as managing director of the survey as a whole. She wrote in her autobiography,

> It was pioneering, exploration of an unknown field. No young doctor nowadays can hope for work as exciting and rewarding. Everything I discovered was new and most of it was really valuable. I knew nothing of manufacturing processes, but I learned them on the spot, and before long every detail of the Old Dutch Process and the Carter Process of white-lead production was familiar to me, also the roasting of red lead and litharge and the smelting of lead ore and the refining of lead scrap.[29]

While the Illinois study was still in progress she was sent to attend the Fourth International Congress on Occupational Accidents and Diseases in Brussels. She gave a paper on the white-lead industry in the United States, but found herself acutely embarrassed by the fact that the American delegation members were unable to answer any of the questions put to them. Their country had no statistics, no regulations, and no system of compensation; the field of industrial hygiene was simply nonexistent in the United States. In attendance at the Congress, however, was Charles O'Neill, Commissioner of Labor in the U.S. Department of Commerce, and soon after Hamilton returned to Chicago she was asked by him to undertake for the federal government a survey similar to that being done in Illinois, but to cover all the states. With little government backing except papers identifying her as a special investigator for the Bureau of Labor, she would have to discover for herself the methods to be followed, devise her own ways of securing the cooperation she would need from industry, and make her own timetable. No salary would be available but the government would buy her final report. She accepted the offer,

> and never went back to the laboratory. Often I was homesick for the old life but I had long been convinced that it was not in me to be anything more than a fourth-rate bacteriologist. Interesting as I found the subject, and pleasant as I found the life, I was never absorbed in it. Hull-House was more vital to me by far, and I had no scientific imagination, one problem did not suggest another to my mind . . . I never have doubted the wisdom of my decision to give it up and devote myself to work which has been scientific only in part, but human and practical in greater measure.[30]

She was then forty-one, the leading American authority on lead poisoning and one of a very small group of specialists in occupational diseases. The descriptions given in her autobiography of the investigations she carried out for the U.S. Department of Labor make absorbing reading, the technical material being embedded in a vivid picture of the often appalling living conditions of the immigrant communities of exploited manual laborers and their families in the United States in the early years of this century. Her surveys made dramatically clear the high mortality rates for workers in a long list of industries, including lead and the associated enamelware trades, the rubber industry, painting trades, dye works, copper and mercury production, and in explosives and munitions factories. Her work brought about changes in state laws that were landmarks in American industrial safety legislation.

Hamilton reported her findings in over fifty papers in medical and scientific journals and in a number of monographs, the best known of which was probably *Industrial Poisons in the United States* (1925). Clear, fair in presenting all sides of controversial questions, and extensive in its coverage, it was the first American textbook on the subject and a work that further emphasized and strengthened her position of leadership in her field. Many of her monographs came out as government reports; they included the following: *Lead Poisoning in Potteries, Tile Works and Porcelain Enamelled Sanitary Ware Factories* (1912), *Industrial Poisons Used in the Rubber Industry* (1915), *Industrial Poisons Used or Produced in the Manufacture of Explosives* (1917), *Women in the Lead Industries* (1919), *Industrial Poisoning in Making Coal-tar*

Dyes and Dye Intermediates (1921), *Women Workers and Industrial Poisons* (1926), *Industrial Toxicology* (1934), *Recent Changes in the Painters' Trade* (1936), and *Occupational Poisoning in the Viscose Rayon Industry* (1940).

In 1919, after nine years of work for the federal government, she was invited to join Harvard University Medical School, that "stronghold of masculinity against the inroads of women," as assistant professor of industrial medicine. "It seemed incredible at the time, but later on I came to understand it. The Medical School faculty, which was more liberal in [the hiring of women] than the corporation, planned to develop the teaching of preventive medicine and public health more extensively than ever before . . . and I was really about the only candidate available."[31] Much to the relief of some of the members of the corporation and the faculty, she refrained from insisting on her rights to use the Harvard Club, receive her quota of football tickets, or march in graduation processions.

Although much gratified at being chosen for the position, Hamilton found her years on the Harvard faculty were in many respects a disappointment. Her situation was uncertain, since she was given a succession of three-year appointments that in the early years she could not be sure would be renewed. She had a habitual reluctance to put herself forward, particularly if she thought this risked antagonizing others, and she tended to feel inadequate among Harvard's bright young male faculty, some of whom were not above making clear their disdain for a middle-aged woman scientist.[32]

Nevertheless, her contributions to her department and to her field during her years at Harvard were not minor. For a time she continued to work with the Department of Labor and also carried out several studies following up earlier investigations, particularly in the lead industries, where she assessed improvements and monitored changes in procedures and processes. Since her reputation was by then widely established, she was increasingly drawn into consulting work and frequently served as an expert witness in compensation hearings. One of her most important contributions to the research efforts of the Medical School was the securing of financial support for a three-year laboratory and clinical study of lead poisoning. The project, funded by several lead companies, was Harvard's most important investigation of industrial diseases at that time. For more than twenty-five years, beginning in 1919, she was associate editor of the *Journal of Industrial Hygiene* and went to considerable effort to ensure its success. She also took a leading part in organizing national conferences to discuss new industrial hazards, tetraethyl lead in 1925 and radium in 1928.

By the 1920s her work was recognized at the international level. Beginning in 1924, Hamilton served a six-year term on the Health Committee of the League of Nations. She was the only woman on the committee and one of two Americans, the other being Surgeon General Hugh Cumming, head of the Public Health Service. Much of her work concerned the collection and analysis of statistical information on epidemic diseases, and the correlation of that information with environmental conditions. She spent several weeks in the USSR in 1924, having been invited by the Soviet Public Health Service to make a survey of what that country was doing in industrial hygiene. The Institute Obuch in Moscow, the first hospital anywhere devoted to occupational diseases only, particularly impressed her, and she felt somewhat envious of the position of Russian women doctors who seemed to be accepted as equals by their male colleagues. From 1930 until 1932 she was a member of President Hoover's Committee on Social Trends in the United States; at times she voiced strong criticism of the capitalist system and the treatment its labor force received.

Hamilton was never promoted at Harvard and remained assistant professor until her required retirement at the age of sixty-five, when she moved with her sister Margaret to Hadlyme, Massachusetts. For a time she stayed on with the Department of Labor as a special adviser on technical problems in the prevention of industrial diseases, and in this capacity carried out her last survey, an investigation of the hazards caused by the use of carbon disulfide in the rayon industry. She also continued to write and to lecture, teaching on an occasional basis at Tufts University Medical School until 1954, by which time she was eighty-five. From 1944 until 1949 she was president of the National Consumers' League. In addition she found time to write her autobiography, *Exploring the Dangerous Trades,* which appeared in 1943. In 1949 she brought out a revised edition of her textbook *Industrial Toxicology.*

Throughout her life she was vitally interested in political as well as social issues in both America and Europe. She had become a pacificist after seeing conditions in Belgium during the First World War and famine-struck Germany in 1919, but the German invasion of western Europe in 1940 made her rethink her position. During her retirement, when she was in her eighties and nineties, she took a strong public stand against American anticommunism, and in 1963, at the age of ninety-four, signed an open letter calling for an end to the military involvement of the United States in the Vietnam War. Something of a legend in her later years, she received a number of honorary degrees and other distinctions, including the Lasker Award, the National Achievement Award, and selection as New England's "Woman of the Year" in 1957. Mount Holyoke College had given her an honorary Sc.D. in 1926. On her ninetieth birthday, friends and former students established a scholarship and lecture fund in her honor at the Harvard School of Public Health.

Alice Hamilton's long career spanned the development of the field of industrial toxicology in the United States from its beginnings in basic field investigations and diagnostic work to the statistical studies that became an area of major importance by about the middle of this century. Her early reports were pioneering, particularly those recording the procedures and findings of the 1911 Illinois Survey; the latter became the model and key example of what could be accomplished in defining problem areas and pointing the way toward possible solutions. She died at her home in Hadlyme, Massachusetts, on 22 September 1970, of a stroke. She was 101.

A second early woman medical scientist whose work very quickly had major nation-wide impact was ANNA WILLIAMS[33]

(1863–1954), probably the most successful American woman bacteriologist whose career began before 1900. Born in Hackensack, New Jersey, on 17 March 1863, she was the daughter of Jane (Van Saun) and William Williams. After graduating from the New Jersey State Normal School in Trenton in 1883 she started to earn her livelihood as a schoolteacher, like many other independent young women of her time.

However, in 1887, following the death of her infant niece and the severe illness of her sister, she entered the Woman's Medical College of New York. Receiving her M.D. in 1891 she stayed on for three years as instructor in pathology and hygiene. She then worked for a year as a volunteer in the New York City Department of Health diagnostic laboratory, the first municipal laboratory in the country to apply bacteriology to problems in public health. Here her collaborative research with the laboratory's founder, William Hallock Park, resulted in the isolation of the strain of diphtheria bacillus, later known as Park-Williams #8, which was soon to come into worldwide use for the immunization of children. The work was reported in a joint paper that appeared in volume 1 of the *Journal of Experimental Medicine* (1896). Park himself was on vacation when Williams carried out her classic isolation, but he subsequently directed the clinical trials, and because he was the laboratory director he has often been accorded the major credit. Nevertheless, it was largely Williams's volunteer work that formed the basis of New York City's successful campaign against diphtheria, and the program of antitoxin production and mass immunization subsequently developed by her and Park quickly became the prototype for similar programs adopted throughout the United States and in Britain.

In 1896 she visited the Pasteur Institute in Paris, then one of the great centers for bacteriological work, and she brought back a rabies culture that, within two years, permitted large-scale production of rabies vaccine in the United States. Her work on methods for diagnosis of rabies in animals reduced the time needed for the procedure from ten days to a few minutes; the technique she developed in 1905 was not improved upon until 1939.[34]

With these impressive early achievements to her credit, she became the laboratory's assistant director in 1905. She kept the post for almost three decades, until required to retire at age seventy-one in 1934 when a financial squeeze necessitated the retirement of all New York City employees over the age of seventy. During the years she shared the administration with Park, the laboratory was noted for its smooth organization, teamwork, and productivity. The staff was greatly expanded and the additional workers included many women, several of whom coauthored her papers.[35] These numbered over fifty. Following her early work on tuberculosis and rabies, she carried out extensive studies on amoebic infections and on trachoma, especially as these afflictions affected the health of New York City schoolchildren.[36] From 1918 until 1921, while she was a member of the Influenza Commission, her studies centered on questions arising from the 1918 pandemic; she published a number of papers on the bacteriology of the disease.[37] The control of scarlet fever and whooping cough, and the problem of the meningococcus carrier, were among the other subjects she investigated.

In addition to her journal publications, Williams coauthored with Park two very successful books, *Pathogenic Microorganisms including Bacteria and Protozoa: A Practical Manual for Student Physicians and Health Workers*,[38] and *Who's Who among the Microbes* (1929), the latter aimed at a general audience. Her third book, *Streptococci in Relation to Man in Disease and Health*, appeared in 1932.

During the First World War she directed a War Department-sponsored training program at New York University that prepared workers for service in medical laboratories in the United States and Europe. Active in various professional associations as well, she served as president of the Woman's Medical Association in 1915, and was the first woman elected to office in a laboratory section of the American Public Health Association. In private life she took special care to avoid close attachments, believing that, for her, emotional commitments were incompatible with a career.[39] She died on 20 November 1954 in Westwood, New Jersey, in her ninety-second year.

VIDA LATHAM[40] (1866–1958), microscopist, pathologist, and physician-dentist, also had a long and remarkable career over the course of which she received notable professional recognition for her steady output of research contributions. Born at Hindley Hall in Lancashire, on 4 February 1866, she was the youngest of the ten children of John Latham, a physician. She received her early education in Manchester, and then studied at Cambridge for four years (1883–1887).[41] She took a University of London M.Sc. in 1889 and a degree in dentistry (D.D.S.) at the University of Michigan in 1892. While at Michigan she held a position as demonstrator in pathology, bacteriology, and dental anatomy in the Medical School. Continuing her studies at the Woman's Medical College of Northwestern University in Evanston, Illinois, she graduated with an M.D. in 1895. She had come to the United States in search of a better climate because she suffered from tuberculosis of the throat, but at various times she returned to Europe for further study and carried out postgraduate work in tropical medicine in Paris, Berlin, and Hamburg. She later gave lectures in that field at the Woman's Medical College of Pennsylvania—the first woman to do so.

For many years she held the posts of oral surgeon at the Chicago Woman's Hospital and pathologist and dental surgeon at St. Francis Hospital in Evanston. Between 1892 and 1896 she was director of laboratories and professor of histology, bacteriology, and pathology at Northwestern University Dental School. Throughout the 1890s and during the next two decades she also taught at various times at the College of Physicians and Surgeons, Chicago, the American Dental College, and the College of Physicians and Surgeons, Milwaukee.

Although much of her published work appeared in medical and dental journals, she also brought out a considerable number of papers in British and American microscopy journals. Among her pre-1901 publications listed in the bibliography are reports on microscopical methods as well as pathological investigations.[42] For a number of years she directed the microscopical laboratory at the Woman's Medical College of Northwestern University, and she served as associate editor of the *Journal of Microscopy and Natural Science.* She was a fellow of the Royal Microscopical Society, vice president of the American Microscopical Society (1904–1905), and secretary and president (1932) of the State Microscopical Society of Illinois, all notable achievements for a woman scientist of her time. She also held office in the American Medical Association as the first chairman of the Stomatology Section, and was vice president of the first World's Columbian Dental Congress in 1893.

In addition to these undertakings she conducted a successful private practice in medicine and dentistry in Chicago. In 1956, when she was ninety and still practicing, she was selected as "Medical Woman of the Year" by the Chicago branch of the American Medical Women's Association. The following year she was one of a group of four selected by the Chicago Technical Societies Council for their 1957 Merit Award. A woman of many talents, she had interests that ranged from music to stonecutting. She composed a number of musical scores, and while a student at the University of Michigan was organist in Ann Arbor's Methodist Episcopal Church. She died on 17 January 1958, at her home in Chicago, a few weeks before her ninety-second birthday, after a career of more than sixty-five years in the medical sciences.

EDITH CLAYPOLE[43] (1870–1915), another early female contributor to research in pathology and microscopy, was, like Latham, of English background, although her cross-Atlantic transplanting occurred at an earlier age than Latham's. The daughter of Jane (Trotter) and Edward Waller Claypole, she was born in Bristol, on 1 January 1870. When she was nine years old her father accepted a position on the science faculty of Buchtel College (later the University of Akron) in Akron, Ohio, and the family moved to the United States.

Both Edith and her identical twin sister, Agnes, received their early education at home from their mother, a writer of scientific articles for the popular press. They then attended Buchtel College and were awarded Ph.B. degrees in 1892. The following year, they both enrolled at Cornell University, where Edith, under the direction of Simon Gage, undertook a series of microscopical studies on blood. Her M.S. thesis project, an investigation of the blood of the amphibians Necturus and Cryptobranchus, was reported in a long paper in the *Proceedings of the American Microscopical Society* in 1893. Although it had been recognized that there were important differences between mammalian and amphibian blood, few studies had been carried out on the latter. Edith's research focused especially on cell size and fibrin threads, the latter one of the areas of major difference. She was awarded her degree with highest honors in 1893, and her work won a first prize in animal histology from the American Microscopical Society. Her second study on blood, an examination of the response of leukocytes to foreign materials, appeared in the *American Naturalist* in 1894.

For five years (1894–1899) she taught physiology and histology at Wellesley College, serving as acting head of the zoology department from 1896 to 1898 while the head, Mary Willcox, completed doctoral studies at the University of Zurich. Claypole was remembered at Wellesley as an enthusiastic and inspiring teacher. While there, she continued her research, doing some of it at the Woods Hole laboratory during summers (1895, 1896). She published two more microscopical investigations in comparative histology in 1896 and 1897.

In 1899, at age twenty-nine, she returned to Cornell to begin medical studies, supported by an assistantship in the department of physiology. After two years, however, she moved to Pasadena, California, where her family then lived, in order to care for her sick mother. For a year she taught biology at the Throop Polytechnic Institute (later the California Institute of Technology) and then took a job as a pathologist carrying out routine diagnostic work for local physicians and surgeons. In 1902 she went back to medical studies, enrolling at the University of California, Southern Branch, and specializing in pathology. She received an M.D. in 1904. For the next eight years she continued to work in pathology, and acquired an excellent reputation among her colleagues for her exceptional thoroughness and efficiency, particularly in hospital work. In 1912 she accepted the offer of an opportunity to join the department of pathology at the University of California, Berkeley, as an independent research associate. Her sister and her brother-in-law, Robert Moody, then lived in Berkeley, Moody holding a faculty position in anatomy at the University of California.

The facilities the pathology department could provide gave Claypole the chance to take up a line of research in the area of lung pathology that had long interested her, but that until then she had had no chance to pursue fully. Over the next three years she carried out notable work on the differentiation of streptotrichosis from tuberculosis, a study that was probably the most important of her scientific contributions; she developed a simple test to distinguish between these two diseases whose symptoms are very similar.[44] She also collaborated with University of California pathologist and bacteriologist Frederick Gay on animal studies to develop procedures for immunization against typhoid fever. Six of their joint papers appeared in 1913 and 1914.[45] Her premature death at age forty-five in Berkeley, on 27 March 1915, resulted from infection caught during the preparation of typhoid vaccine for the armies in Europe. The Edith J. Claypole Memorial Research Fund in Pathology, established by her friends in her memory, has the aim of encouraging research, preferably by women, in the diagnosis and therapy of infectious diseases.

MARY ROSS[46] (1877–1964) also received much of her early training at Cornell University. She later became a very successful practicing physician, but she took her first doctoral degree in physiology and embryology. Born in Oxford County, Ontario, on 29 January 1877, she came to the United States at the age of twelve. As an undergraduate at Cornell (A.B., 1898) she carried out a substantial microscopical investigation of the air sacks in birds under the guidance of Grant Hopkins, professor of veterinary anatomy. She continued her studies at the University of Pennsylvania, taking an A.M. in 1900, and then returned to Cornell, which awarded her a Ph.D. in 1902. Her research on the development of gastric glands, carried out under the direction of embryologist and histologist Benjamin

Kingsbury, appeared in 1903.[47] She took an M.D. at Johns Hopkins University Medical School in 1907 and then spent two years as resident pathologist at St. Luke's Hospital, New Bedford, Massachusetts. Thereafter, for almost half a century, she practiced in Binghamton, New York. In 1953, in recognition of her outstanding work in infant health care, hospital service, and consulting, she was named "Outstanding General Practitioner of the Year" by the New York State Medical Society. She died on 15 July 1964, at age eighty-seven.

At the University of Chicago three women—Anne Moore, Elizabeth Cooke, and Jeannette Welch—had taken doctoral degrees in physiology by 1901. All were students of Jacques Loeb, a German-educated physiologist/morphologist with an international reputation, who, after a year at Bryn Mawr College, joined the faculty of the newly opened University of Chicago in 1892. A fourth Chicago woman, Mary Hefferan, whose main interests were in bacteriology, received her degree in 1903.

ANNE MOORE[48] (1872–1937), the best remembered of these four, was the daughter of Eugenia (Beery) and Roger Moore. She was born in Wilmington, North Carolina, on 10 August 1872. After attending St. Mary's School in Raleigh, North Carolina, she went to Vassar College (A.B. 1896, A.M. 1897). She stayed on as an assistant until 1898, and then entered the University of Chicago. Her earliest publication reported a study of a new species of Dinophilus, carried out at Woods Hole in the summer of 1898. The first of her papers on the effects on tissue of electrolytes, her major research interest at Chicago, appeared in 1900, and a long paper in the *American Journal of Physiology* two years later discussed her dissertation research.[49] She received her Ph.D. in 1901.

From Chicago Moore went to the State Normal School in San Diego, California (later San Diego State University) as teacher of biology and physiology. Following one year as a substitute, 1901–1902, she held a regular faculty post until 1905. She kept up a modest amount of research in physiology, her study of the response of protoplasmic tissue to light appearing in 1903.[50] She also brought out a textbook, *Physiology of Man and Other Animals* (1909).

By about 1910 she had moved to the New York area, where she carried out an investigation of matters affecting the mentally handicapped.[51] This would appear to have been her last scientific work. Becoming interested in literature and the theater, she worked as a drama critic and later joined the Civic Repertory Theatre, a group started in New York in 1926 by the actress Eva Le Gallienne, whose purpose was to bring serious dramatic works to American audiences. Moore's first book of poems, *Children of God and Winged Things,* was published in 1921, and a second volume, *A Misty Sea,* appeared in 1937. Her *Wayfarers in Toodlume,* a book of short stories, vivid word pictures that dramatize both the beauty and the grimness of human relationships, was published posthumously in 1939. She died on 25 September 1937, at age sixty-five.

ELIZABETH COOKE[52] was an undergraduate during the early 1890s at the University of Michigan (S.B., 1893), where she worked with physiologist William H. Howell on the effects of electrolytes on heart action and on the problem of the causation of the rhythmic heartbeat. Her joint paper with Howell appeared in the *Journal of Physiology* in 1893. As a student of Jacques Loeb at Chicago she studied osmotic properties of muscle; she received a Ph.D. in 1897. Her research activities continued after the turn of the century, when she collaborated with Leo Loeb (Jacques's brother), at the University of Pennsylvania and at the Woods Hole laboratory. Reports of their joint work appeared in 1908 and 1909.[53] By 1911 she was an assistant in experimental therapeutics at the Medical College of Cornell University in New York City.

Jeannette Welch, who received her Ph.D. in 1898,[54] probably published only one paper in physiology, the report of her dissertation research on the correlation of mental and muscular activity, which appeared in volume 1 of the *American Journal of Physiology* in 1898. She died in 1906.

MARY HEFFERAN[55] (1873–1948) was born in Eastmanville, Ottawa County, Michigan, the third child of Emily Amelia (Kent) and Thomas Hefferan. The family later moved to Grand Rapids, Michigan, where Thomas Hefferan was a successful banker. Mary took A.B. and A.M. degrees at Wellesley College (1896, 1898), and in 1899 entered the University of Chicago on a graduate fellowship. Following the award of her Ph.D. in 1903, she stayed on as curator of the bacteriological museum and assistant instructor in the department of bacteriology.

Although she published two early papers reporting morphological studies on the marine worm, *Nereis limbata* (carried out under the direction of Charles Davenport, director of the biological laboratory at Cold Spring Harbor, Long Island), most of her work was in bacteriology. Her dissertation research, supervised by Edwin Oakes Jordan, a specialist in waterborne bacteria, was a broad, comparative study of cultures of a large number of red chromogenic bacteria. The work was aimed at uncovering the general features of relationships and variability across the group, a convenient one for investigation because of the prominent pigment production and the nonpathogenic nature of the organisms. The problem of classifying bacteria into genera and species was a difficult one. Most of those Hefferan looked at had been only briefly described previously (although one of her subgroups consisted of strains of *Bacillus prodigiosus—Serratia marcescens*—an organism known since antiquity for its spectacular appearances on common foodstuffs such as bread; its resemblance to drops of blood often caused consternation). The report of her investigation across the group of morphology, culture, effects on growth media, gas production, range of normal growth and variation, and environmental effects governing pigment color and production, appeared in a lengthy paper in the *Centralblaat für Bakteriolgie* in 1903. Three years later she published an extension of the study, which focused on agglutination characteristics as a classification tool.

She also carried out a joint investigation with Jordan of *Anopheles* (the mosquito) from the Chicago vicinity and western Michigan. Their examination (1905) of such factors as breeding places, egg-laying habits, and the effect of low temperatures on eggs and larvae was prompted by a worldwide upsurge in malaria at the time.

The half-dozen publications she brought out between 1900 and 1906[56] did not result in any promotion, however, and in 1910 she left the University of Chicago, returning to Grand Rapids, where by then her two brothers had followed their father into the banking business. She worked for a time as a bacteriologist in Grand Rapids, but by 1919 had given up scientific work and turned to what she felt were more urgent social and civic problems. She directed the Blodgett Home for Children in Grand Rapids for a number of years, and was remembered as a pioneering social worker. She died on 20 July 1948.

Three other late nineteenth-century medical scientists in this group about whom a modest amount of information is available are Caroline Latimer, Claribel Cone, and Adelaide Peckham. Both Latimer and Cone were from Baltimore and both graduated from the Woman's Medical College of Baltimore. Peckham was for long associated with the Woman's Medical College of Pennsylvania.

CAROLINE LATIMER was born in Baltimore, probably in 1859.[57] She attended a private school, and took her M.D. degree in 1890. A year of postgraduate study at Johns Hopkins Hospital followed, and she then enrolled at Bryn Mawr College where she studied under the guidance of physiologist Joseph Warren. Her paper on salivary gland enzymes, coauthored with Warren, appeared in the *Journal of Experimental Medicine* in 1897. In 1896, after being awarded both A.B. and A.M. degrees by Bryn Mawr, she returned to Johns Hopkins Medical School for two years of research in William Howell's laboratory. Her study of fatigue in cold-blooded animals appeared in 1899. From 1897 until 1899 she also held a teaching position in biology at the Woman's College of Baltimore (afterward Goucher College). Later she became literary assistant to Howard A. Kelly and Lewellys F. Barker, two physicians at Johns Hopkins Hospital; she appears to have retained that post until at least 1920, by which time she would have been about sixty. One of her large projects was assisting in the editing of *Appleton's Medical Dictionary*.

CLARIBEL CONE[58] (1865?–1929) was born in Tennessee at the close of the Civil War, the fifth in a family of thirteen children. She lived in Baltimore from the age of six and attended the Western Female High School, graduating in 1883. After receiving her M.D. in 1890, she won, by competitive examination open to both men and women, an internship in the Philadelphia General Hospital.

Pathology and bacteriology became her major interests, and for ten years from 1893 until 1903 she carried out research at Johns Hopkins Hospital pathological laboratory with pathologist William Henry Welch. Her first paper, on tubercules and tubercule bacilli, appeared in 1899. Throughout this period she also taught at the Woman's Medical College, first as lecturer in hygiene (1893–1895), and then as professor of pathology (1895–1903). She spent the two years from 1903 to 1905 in Europe, at the Pasteur Institute in Paris where she worked with the Russian bacteriologist Ilya Mechnikov, and at the Royal Prussian Institute for Experimental Therapy in Frankfurt, where she was associated with Paul Ehrlich. Her investigation of pathological changes in human epidermal cells was reported in a long paper published in 1907.[59]

Although she rejoined the faculty of the Woman's Medical College in Baltimore for three years from 1907 until 1910, she then returned to Europe, and throughout the First World War lived in Munich. Shortly thereafter she moved to Paris, where, through her friend the American writer Gertrude Stein, she became acquainted with several artists of the modern school who were later to become famous. Her family had considerable wealth, and she and her sister Etta, with whom she shared her home in Paris, acquired a sizable collection of paintings, including works by Matisse, Picasso, Van Gogh, and Cezanne. They also collected fine textiles, jewelry, furniture, and rugs, bought mainly during trips they made to the Orient. All of these were brought back to Baltimore and eventually willed to the Baltimore Museum of Art along with $400,000 to build a wing onto the museum to house the collection.

Cone's work in pathology and bacteriology was well regarded, but she does not appear to have taken up research again after her return from Europe. She was president of the Woman's Medical Society of Maryland from 1925 until 1927 and for many years was prominent in Baltimore social circles. Her Saturday parties, to which came people from every walk of life, have been described as being as near to a "salon" as anything the city of Baltimore had ever known.[60] She died in 1929, when she was in her early sixties.

ADELAIDE WARD PECKHAM[61] (1848–1944), bacteriologist, was born on 31 March 1848, into an old New England family. As a young woman she taught in local schools in New England and in public schools in New York City, but in 1882, at the age of thirty-four, she entered the Medical College of the New York Infirmary for Women and Children. On graduating (M.D.) four years later she went to live with her brother in Philadelphia, started a medical practice, and also took on clinical work, first at the Woman's Hospital of Pennsylvania and later at another smaller hospital in the city where she could treat patients under the age of sixteen.

Her special interest in dealing with children led her to realize the great value of the new science of bacteriology in diagnosing diseases in that section of the population. She therefore changed her career a second time, and, entering the Laboratory of Hygiene at the University of Pennsylvania, studied under John Shaw Billings (then surgeon general of the United States) and Alexander Abbott, director of the laboratory and of the Philadelphia Bureau of Health. Her joint work with Billings, supported over several years by a grant from the Bache Fund of the Smithsonian Institution, resulted in a number of lengthy publications dealing with the effect of environmental conditions on pathogenic colon bacilli. After further training in pathology at Johns Hopkins University in 1898 she returned to the University of Pennsylvania. The following year she took up the positions of superintendent of the recently established bacteriological laboratory at the Woman's Hospital of Pennsylvania, succeeding the Swiss-educated Lithuanian Lydia Rabinovitch (later well-known for her research on tuberculosis at the Koch Chemical Institute in Berlin).

Peckham taught bacteriology at the Woman's Medical College for more than twenty years, through a period of tremendous expansion of her field as its role in the diagnosis and treatment of disease became fully appreciated. After her retirement in 1919, when she was seventy-one, she traveled extensively, even in her eighties taking many trips alone and very successfully through Canada and the western United States. She died in Bloomfield, New Jersey, on 15 May 1944, at age ninety-six.

Along with the work of these turn-of-the-century medical scientists,[62] there is listed in the bibliography a contribution by one American woman, MARY GRIFFITH, who was writing some sixty years earlier. Griffith published two papers on the physiology of vision in 1834 in the London *Philosophical Magazine* and in European journals, and a third on the same topic in the *American Journal of Science* in 1840. Her subject was one of considerable interest to the scientific community of the time (see also the later work of Ladd-Franklin, chapter 15); the phenomenon she reported, the effect of sudden light on the closed and resting eye, immediately aroused the interest of Scottish physicist Sir David Brewster, a pioneer in modern experimental optics, the author of several treatises in that area and the editor of *Philosophical Magazine.* Acknowledging Mrs. Griffith's claim to be the first to report the phenomenon, he went on to repeat her "remarkable experiment," and added his own observations on the subject.[63] Griffith's investigations led her to conclude that the "seat of vision" could not be located in the eye itself.

Mary Griffith lived in Charlies Hope, New Jersey. Widowed when still fairly young and left with a family to support, she took up farming, and in addition became a writer on agricultural subjects. She corresponded frequently with H. A. S. Dearborn, president of the Massachusetts Horticultural Society, and was made an honorary member of the society in 1830. She had broad interests and considerable energy and talent. Her book, *Our Neighbourhood,* published anonymously in 1831, was dedicated to the Horticultural Societies of Pennsylvania and Massachusetts. Written to encourage horticulture and stir interest in rural life and natural history, the work takes the form of a series of letters from a young English settler to his relatives at home, describing his life as an American farmer. A second work, entitled *Camperdown,* also published anonymously, appeared in 1836. It comprised a collection of stories including *Three Hundred Years Hence,* a remarkable utopian novel, almost science fiction, in which Mrs. Griffith described her vision of American society of the future.[64] She died in 1877.

The late nineteenth-century American women who worked in the various branches of the medical sciences covered in this section were on the whole remarkably successful. Those known to have spent the major part of their time on research, or research and college-level teaching, although not a large group compared with botanists or zoologists, include six who had careers that must surely be judged outstanding: Hyde was probably the first woman to become a full professor and head of a science department in a coeducational state university; Hamilton, who achieved international recognition in her field, was the first woman scientist on the Harvard faculty; DeWitt became the first woman associate professor in a science department at the University of Chicago; Gage was a research neurologist of national standing; Williams was a well-known bacteriologist, respected for her outstanding contributions to vaccine therapy; and Latham was duly recognized for her work in microscopy and pathology by election to posts in state and national professional organizations.

That being said, it should also be remembered that the success of Hyde, Hamilton, DeWitt, Williams, and Latham, the five who penetrated most fully into the scientific community, was most likely due in large measure to the fact that their areas of specialization were still new. When Hyde got her position at the University of Kansas in 1898 there were few men with comparable experience and credentials competing against her for this post at what was then a still-developing Midwestern institution. Hamilton's situation was similar when she received her appointment to Harvard's faculty; she was, as she put it, "really about the only candidate available."[65] (It is also the case that, even though they got the jobs, both Hyde and Hamilton had uneasy careers in academe, Hyde eventually being pushed into retirement when suitable male candidates became available for her position, and Hamilton never being promoted during her sixteen years at Harvard.) Latham's specialty of oral pathology and surgery was also a new and developing area, as were the interests of Williams (bacteriology), and DeWitt (bacteriology and chemotherapeutics). Gage, who worked independently without formal institutional backing, simply offered her findings as a contribution to the progress of her field, and largely avoided presenting any challenge to the established scientific community. These provisos notwithstanding, the medical scientists stand out as one of the notable groupings in the American part of this survey.

British women

Seventeen British women contributed to research in the medical sciences. Three of them (Greenwood, Eves, and Alcock) were Cambridge-trained; four (Buchanan, Sowton, Chick, and Toynbee Frankland) had at least part of their training at London colleges; one was a student at Mason College Birmingham, another worked at Bern University, and a third at the Collège de France in Paris. Among the remaining seven, four (Brink, Flemming, Morgan Hoggan, and Stoney) were primarily physicians (see chapter 7), but Morgan Hoggan's twenty-three papers published mainly in the 1870s and 1880s, twenty-two of them coauthored with her husband, form the largest pre-1901 contribution by a British woman to research in the medical sciences (the relatively large fraction of anatomy/morphology papers in the British publications shown in Figure 6-2 d to a large extent represents Morgan's contributions); Newbigin, Saunders, and Gould Veley had major interests in other fields (see chapters 13, 1 and 3, respectively).

Greenwood, Eves and Alcock, postgraduate students and then staff members at Newnham College Cambridge in the 1880s and 90s, belonged to the group of workers led by physiologist Michael Foster, whose laboratory was then in the process of growing and developing into what came to be the famous Cambridge School of Physiology. Already well known internationally, it was quickly becoming established as the leading center for experimental physiology in the English-speaking world.[66]

MARION GREENWOOD[67] (1862–1932), the best remembered and most distinguished of these three early Cambridge women physiologists, was born on 24 August 1862, at Oxenhope in Yorkshire, the daughter of George Greenwood, a businessman and lay preacher, and his wife Agnes (Hamilton). In 1879, at the age of seventeen, she went to Girton College from Bradford High School on a Brown scholarship. She was awarded first class in both parts of the Natural Sciences Tripos Examinations (1882 and 1883, Pt. II, physiology) and stayed on with a Newnham College Bathurst research scholarship.

In 1885 she became demonstrator in Newnham's newly established Balfour Laboratory, and there joined zoologist Alice Johnson, whom she succeeded as head of the laboratory in 1890. For eleven years (1888–1899) she was also a Newnham and Girton lecturer in physiology and botany, and in addition, as head of the Balfour Laboratory for much of this period, she acted as general adviser in scientific studies for women of both colleges. Her friend and colleague the Newnham botanist Rebecca Saunders later recalled (with perhaps a hint of nostalgia)[68] that although the facilities available to the women staff in those days were very inadequate, the stimulus derived from knowing that they were moving into uncharted territory in women's education gave them the energy and enthusiasm needed to take on the necessary work and overcome the many difficulties. When Greenwood resigned her posts in 1899 to marry biologist George Bidder of Trinity College, her duties had to be divided among two or three successors, and although this redistribution doubtless resulted in part from increased student numbers and other changes, it also testifies to her outstanding organizational abilities.

The arrangements in Foster's laboratory, where she carried out her research, were later described by one of her colleagues as follows:

> physiologists and biochemists, still undivorced, habited adjacent rooms to their mutual comfort and benefit. The rooms, in order, down the little dark passage, were the homes of Sheridan Lea, Walter Gaskell, Marion Greenwood, and beyond and through her room, in a cupboard of a place, Langley. Miss Greenwood was in a small passage room, and I shared the one bench with her. No modern Ph.D. aspirant could or would compress his or her activities into the space we were contented with in those days.
>
> At that time women were rare in scientific laboratories and their presence by no means generally acceptable—indeed, that is too mild a phrase. Those whose memories go back so far will recollect how unacceptability not infrequently flamed into hostility. The woman student was rather expected to be eccentric in dress and manner; she was still unplaced, so far as the male in possession was concerned. Miss Greenwood, it so happened, was not only a woman of quite unusual intellectual distinction but she had also great personal charm and a great gift of comradeship . . .
>
> She took her share, and it was a large one, in the government of Newnham and Girton, but I am inclined to think that the best she did for women was just being her gracious and kindly self in those early days of hostility, touched as it was sometimes by a spice of active persecution.[69]

Her first paper, published in 1884, reported histological studies on cells from the gastric secretory granules, an area in which John Langley was heavily involved; likewise her 1890 paper on the effect of the alkaloid nicotine on certain invertebrates reflected Langley's interests. However, the investigations for which she was especially remembered were her own studies on protozoan physiology, especially her work on the role of acid in protozoan digestion published jointly with Rebecca Saunders in 1894, and her study of structural changes in the resting nucleus of protozoa, which appeared in 1896. In the latter she examined the effects of putting a metabolic strain on the organism through restricted diet or partial deprivation of oxygen, and demonstrated that these conditions produced not only gross structural changes but also a change in the distribution and quantity of iron. Her early work brought her Girton's Gamble Prize in 1888, the first time it was awarded.

Although she gave up research when she married (at the age of thirty-seven), and her publications were not numerous,[70] they nevertheless constituted a very solid contribution to the field. As her colleague Hardy remarked, she was "above all an accurate observer,"[71] and she was well thought of by her contemporaries. A story is told of how, on a visit to Sir Michael Foster, the eminent Russian biologist Ilya Mechnikov remarked that he would like to meet the young man M. Greenwood, whose recent papers had impressed him. Sir Michael promised that he should have M. Greenwood as his neighbor at luncheon.[72]

Throughout her life she remained active in the concerns of Girton and Newnham colleges and those of the wider university community. She was well known for her generous hospitality and readiness to provide help whenever it was needed; her house at Cavendish Corner was familiar to many. She served on the governing committee of the Homerton Training College for Teachers and on a number of local and regional organizations, including the Cambridgeshire Voluntary Association for Mental Welfare, and, in 1916–1917, the Women's War Agricultural Committee. Serene and well-balanced, she made an admirable committee chairman and arbitrator. She was also an exceptionally painstaking Poor Law Guardian. A staunch Liberal and a solid supporter of the women's suffrage movement, she worked for the election of women to town and county councils. Her two daughters, Caroline and Anna, were educated at Girton and Newnham, respectively. She died in Cambridge, on 25 September 1932, at the age of seventy.

FLORENCE EVES,[73] from Uxbridge (London), attended the North London Collegiate School and also took classes at University College London before going to Newnham in 1878. She received class I in the Natural Sciences Tripos examinations in 1881 and the same year a University of London B.Sc. with first class honours in botany. She held a Newnham Bathurst research studentship in 1881–1882 and a demonstratorship for the six years 1881 to 1887. Her research on liver enzymes and digestive enzymes, much of it done in collaboration with John Langley, appeared in the *Journal of Physiology* in 1883 and 1885.

After leaving Newnham in 1887 Eves taught for three years at Manchester High School for Girls and then for a year at St. Leonards School, St. Andrews. Thereafter she worked for social reform, concentrating particularly on women's issues. She died on 11 February 1911.

RACHEL ALCOCK[74] (1862–1939) was born in Stockport, Cheshire. She went to Newnham in 1886, after receiving her early education from tutors and at a private school. She took both parts of the Natural Sciences Tripos examinations (Pt. I, class II, 1889; Pt. II, class III, 1890) and stayed on at Newnham with a Bathurst studentship. Her research on digestive processes and on nerve distribution in the primitive fish *Ammocoetes* was supervised by Walter Gaskell, known for his studies on nerve systems in vertebrates. She taught morphology and anatomy to women students in the Balfour Laboratory for a short period in 1891 and again in 1898–1899, filling in after the resignation of zoologist Lilian Sheldon. However, although for several years she continued to serve from time to time as demonstrator in the Balfour Laboratory, by the late 1890s she had become interested in Spanish literature.

She took an M.A. at Liverpool University in 1916, having spent two years (1914–1916) in Madrid, at the Biblioteca Nacional, and in Toledo. The subject of her studies was a late sixteenth-century *comedia, La Famosa Toledana,* by Juan de Quiros, a Toledo *jurado,* a contemporary of Cervantes, and one of the forerunners of Lope de Vega, the father of Spanish national theater. Starting from a handwritten manuscript (which dated back more than 320 years to 1591 and was entirely lacking in both accents and punctuation), she prepared an edition of the play that she published with a long introduction and a substantial section of notes.[75] From 1921 to 1923 she held a Cambridge Mary Bateson research fellowship in Spanish literature; her annotated edition of *Canción de Cuna* (*Cradle Song*), by the early twentieth-century dramatist Gregorio Martínez Sierra, came out in 1923. She died on 2 February 1939, when she was about seventy-seven.

The two early physiologists/neurologists who began their careers at London Colleges, Buchanan and Sowton, both spent somewhat longer periods in scientific research than did the Cambridge women. Both were concerned especially with electrical stimulus of muscle.

FLORENCE BUCHANAN[76] was born on 21 April 1867. She entered University College in 1886, and took a B.Sc., with second class honors in zoology, in 1890. Her earliest published report, a study of the development of the respiratory organs in decapod Crustacea, appeared in 1889. She stayed on as a student at University College until 1892, continuing studies on marine organisms. These included work on polychaete worms collected off the Cornish coast during the summer of 1891 when she occupied a table at the Plymouth Marine Laboratory. Two years later her investigations were extended to polychaetes collected off the coasts of Ireland during a survey conducted by the Royal Dublin Society. By the late 1890s she had moved to Oxford where she began her work on electrical effects in muscle in J. Burdon Sanderson's laboratory. Her first publication in this area appeared in the *Journal of Physiology* in 1899, and a second substantial report followed in 1901. She was awarded a London D.Sc. in 1902 and became a fellow of University College London in 1904. Her research continued at Oxford, where she had laboratory space and the use of apparatus and equipment in the University Museum for a number of years, working expenses being covered largely by Royal Society grants.[77]

The heartbeat and form of the electrocardiogram, and the transmission of reflex impulses were her major areas of interest. She frequently presented her results at meetings of the Physiological Society rather than in full papers, particularly her work on electrical fields related to heartbeat. These included comparative studies in mammals, birds, and reptiles, and changes in the heart's electrical effects between the waking and hibernating states. Her three lengthy reports on reflex impulse in volumes 1 and 5 (1908 and 1912) of the *Quarterly Journal of Experimental Physiology* are especially notable. These were basic studies, detailed investigations of the weak electrical currents associated with voluntary muscular activity, that contributed to the understanding of the nature of the stimuli sent to muscles from the central nervous system. Factors studied included the effects on the reflex contraction of temperature and a variety of drugs (including strychnine and caffeine). Although she began under Burdon Sanderson's direction, after his death she continued the work independently (with discussions from time to time with neurologist Charles Scott Sherrington). By 1910 she was well recognized, and that year received the American Association of Collegiate Alumnae's prize for original scientific research. Soon afterward, her publication rate decreased, and she brought out little after 1913.[78]

S. C. M. SOWTON was active in scientific work throughout the twenty-seven-year period between 1896 and 1923. Until about 1903 most of her research was carried out in association with Augustus Désiré Waller, professor of physiology at St. Mary's Hospital London, and well known for his developmental work in electrocardiography. Sowton's studies focused largely on the electrical effects produced by chemical stimulation. Three of her papers (one joint with Waller) and several of her notes appeared in the Royal Society's *Proceedings* and the *Journal of Physiology* between 1896 and 1903. Some of the research was carried out in Carl Ludwig's famous institute of physiology at Leipzig University, where she worked for three months in the summer of 1897 and again during the spring of 1899.

About 1904 she joined the team led by Charles Scott Sherrington, professor of physiology at the University of Liverpool, whom she had probably met earlier when he taught physiology and pathology in London in the early 1890s. Sherrington became known for his work on the problem of the heartbeat and most especially for his neuron theory of the mode of functioning of the central nervous system (he received the Nobel prize for physiology and medicine in 1932). Sowton's collaborative investigations with him focused particularly on the modification of reflex responses in muscles by electrical stimulus through given nerves, and on the effect of various chemicals on the isolated mammalian heart and other muscular organs. She continued to work with him for a time after he moved to Oxford in 1913, although about 1915 she also collaborated (on heart perfusion experiments) with A. S. and H. G. Leyton at the University of Leeds.

In addition to these purely physiological studies she undertook, from time to time, a number of joint investigations in experimental (physiological) psychology. Her 1906 paper on visual perception reported studies carried out with the distinguished physiologist and experimental psychologist W. G. Smith, then in Sherrington's department but later in Edinburgh where he carried out pioneering work in educational psychology. In the 1920s, she collaborated in two projects in the area of industrial fatigue, a field that, after coming into being via debates over industrial efficiency during the early years of the century, took on a fairly important role during the First World War. Indeed British field investigations pioneered studies in this area. Sowton worked with both Bernard Muscio and Charles Myers, perhaps the two most influential members of the Industrial Fatigue Research Board, a body established by the government in 1918. Her investigations with Muscio, an Australian psychologist who served on the board from 1919 until his return to the University of Sydney in 1922, involved attempts to develop practical fatigue tests with controls for psychological factors; her work with Myers, a physician and industrial psychologist, was on the effects of the menstrual cycle on mental and muscular efficiency.[79]

HARRIETTE CHICK[80] (1875–1977), one of the leading women medical scientists of her generation, had a long and distinguished career at the Lister Institute of Preventive Medicine in London, the largest and most important of the British research institutions of the period carrying out biochemical and bacteriological work. Born in London, on 6 January 1875, she was the fifth child and third daughter among the twelve children (ten of whom survived past childhood) of Emma (Hooley) and Samuel Chick. The Chicks were of West Country origin, most of them having been tenant farmers on the borders of Devon and Somerset. About the beginning of the nineteenth century, largely thanks to the energy and initiative of Samuel's grandmother, Abigail (Tutcher) Chick, they had started a small business in the trading of locally made lace. As a young man Samuel moved to London to further the family's commercial interests, but he kept close ties with Devon and his children regularly spent their summers there. Emma Hooley was from Macclesfield, Cheshire, where her father ran a business in grain and groceries. The Chick children were brought up strictly, with no frivolities allowed and regular attendance at family prayers and chapel required. All seven of the daughters attended Notting Hill High School, a girls' school outstanding for its teaching in the sciences. Subsequently six of these girls, including Harriette, went on to take university degrees—for the time most likely a record achievement for a single family.[81]

Harriette entered University College London in 1894, having previously taken classes at Bedford College, where she held a Reid scholarship. Her capabilities were soon evident. She took the advanced class prize in botany in 1894–1895 and the senior class Gold Medal in the same subject the following year. She also obtained a solid grounding in chemistry, critical for the later biochemical work she was to undertake, and continued studies in this area for another year after receiving her B.Sc. in 1896. By 1898 she had turned her attention to the new field of bacteriology; following a year of study in London she obtained a University College Science Research Scholarship and an 1851 Exhibition Scholarship for further work, first at the Hygienic Institutes of the Universities of Vienna and Munich and then at the new Thompson-Yates Laboratory in Liverpool. In Vienna she studied under Max Gruber, whose school was one of the earliest in bacteriology, and in Liverpool she worked with Sir Rupert William Boyce, professor of pathology, city bacteriologist and member of a royal commission on sewage disposal.

Among the pressing public health needs of the time were sanitary surveys, water analyses, and research into the physiology of infection. Boyce, an energetic Irishman, had an impressive record in these areas, and at his suggestion Harriette undertook an extensive survey of *B. coli commune* contamination in large city water supplies in the region. Although on a smaller scale, the project was not unlike one carried out a decade earlier in Massachusetts in which much of the work was done by the American woman public health chemist Ellen Richards. Harriette Chick's findings appeared in a lengthy Thompson-Yates Laboratory Report for 1900–1901. In 1904, on completion of a study of the function of green algae in polluted water, she received a D.Sc. (London) and the following year she applied for a Jenner Memorial Research Studentship at the Lister Institute. Her application caused something of a stir and called forth a few objections, no woman having been given a fellowship previously. But she was appointed, and stayed at the institute until her retirement in 1945, following which, as an honorary staff member, she still remained in close contact for another twenty-five years.

Her early research at the institute concerned the chemical dynamics on the process of disinfection. In collaboration with Sir Charles Martin, the institute's director, she demonstrated that the killing of bacteria, by either chemical disinfection or heat, proceeded as an orderly reaction governed by established physico-chemical laws. The work showed unequivocally that older theories involving mysterious "vital forces" were no longer tenable. Two long papers in the *Journal of Hygiene* in 1908 discussed the results and the fundamental principles that were involved. The Chick-Martin test for disinfectant efficacy that was then formulated became the standard, replacing the previously used Rideal-Walker test, a procedure subject to serious error because it relied on the testing of bacterial suspensions under idealized conditions (in distilled water). Also part of the investigation was the classic Chick and Martin study, reported in the *Journal of Physiology* between 1910 and 1913, which elucidated the process of protein coagulation by heat. An important step toward the understanding of the structure of proteins, this work pointed the way toward the possibility of regarding these as giant molecules, which, with suitable techniques, could be studied as profitably as simpler ones. The structure of the complex molecular units proteins form in water solution, the viscosity of these solutions, and the effects of neutral salts on the molecular aggregates were other areas studied by Chick and Martin in the period up to 1914.[82]

With the coming of the First World War, however, these studies were put aside for more urgent work, first the testing and bottling of tetanus antitoxin for the army, and then the preparation of agglutinating sera for the diagnosis of typhoid, paratyphoid, and dysentery, then prevalent among allied troops in Belgium and the Middle East. The search for food supplements that would prevent deficiency diseases among the troops was the next major undertaking; Charles Martin, serving with the Medical Corps of Australian army units in the Middle East, had sent a request to the institute that something be found to provide protection from beriberi, the disease caused by B vitamin deficiencies. Work on this problem, which resulted in dried egg and dried yeast supplements being added to soldiers' diets, was followed by further nutritional studies on the B complex. These investigations, as well as studies on the antiscorbutic factor that prevents scurvy (vitamin C), occupied Chick and her colleagues, several of whom were women, up to the end of the war.[83]

By 1919, following severe food shortages in many parts of central and eastern Europe, serious outbreaks there of deficiency diseases began to be reported. Austria was especially badly affected. A preliminary investigation by a combined mission from the Lister Institute and the Medical Research Committee (the forerunner of the Medical Research Council) indicated that the situation presented a unique opportunity for the testing of ideas worked out at the institute during the war years, particularly the relation of nutrition to bone disease. Accordingly, a team headed by Harriette Chick and Elsie Dalyell, with Margaret Hume, H. M. M. Mackay and Hannah Henderson Smith, was dispatched to Vienna, where they spent more than two years working with the staff of the Universitäts Kinderklinik. Their studies clearly demonstrated the relation between nutrition and rickets, dispelling the mystery that had previously surrounded the disease. They further showed that exposure to ultraviolet light or treatment with a fat-soluble substance present in cod-liver oil could cure or prevent the bone damage. This finding constituted an important contribution to public health medicine in that it led to the almost complete disappearance of rickets in the period between the two world wars.

For a number of years after her return to London in 1922 Harriette Chick and her team continued their work on the fat-soluble vitamins (A and D),[84] but about 1926 she returned to her earlier interests, the proteins and the B vitamins. Little was known about the constitution and role of the latter, which were only beginning to come under close scrutiny. She became particularly interested in pellagra, the skin disease caused by deficiency of nicotinamide, a member of the B_2 complex. In 1932 she went to the United States to give a series of lectures on the relation of maize diets to the disease. Extensive work was also done on cereals other than maize, the distribution of the B vitamins in wheat being a topic of special interest.[85]

With the coming of the Second World War, the institute's Division of Nutrition, which Chick then headed, moved to temporary accommodation in Sir Charles Martin's house in Cambridge. A vigorous research program continued and investigations of many aspects of wartime food problems were undertaken. Work carried out on the nutritive value of bread, flour, and potatoes was particularly important; Chick's earlier findings were also used in the planning of the "National Loaf."[86]

She retired in 1945, more than five years after the usual age, but even then kept in touch with the Lister Institute as a member of its governing body and attended annual general meetings until 1975. In addition her writing went on, review articles and joint technical reports appearing in the literature until the mid 1950s. She had served as secretary of the Accessory Food Factors Committee of the Medical Research Council for more than twenty-five years, until 1945, and of the League of Nations Health Section Committee on the Physiological Bases of Nutrition from 1934 to 1937. She was also closely involved with numerous committees on international standardization of units for vitamins and other food factors. Her nutrition work brought her a C.B.E. in 1932 and a D.B.E. in 1949. The division she headed at the Lister Institute had an international reputation, and indeed served as a central meeting place for nutrition scientists, who came there from many countries for brief visits or extended periods of research. She was one of the eleven scientists who in 1941 planned the founding of the Nutrition Society, of which she was made an honorary member eight years later. From 1956 to 1959 she served as the society's president, although then over eighty. In 1974, shortly before her hundredth birthday, the British Nutrition Foundation awarded her its annual prize, and for the occasion she prepared a lecture on her work on rickets in Vienna more than half a century before. The talk was delivered by her friend and former coworker Alice Copping, but she herself provided a lively introduction. Active and interested in news of friends and colleagues until her last days, Dame Harriette died on 9 July 1977 in her one hundred and third year.

GRACE TOYNBEE FRANKLAND[87] (1858–1946) was born in Wimbledon, on 4 December 1858, the youngest daughter among the nine children of Joseph Toynbee, F.R.S., a pioneer ear specialist, and his wife Harriet, the daughter of Nathaniel Holmes. One of her brothers was Arnold Toynbee, later a distinguished historian and philosopher at Oxford. She received her early education at home, then attended Bedford College and also studied for a period in Germany. Bacteriology was her special interest. In 1882 she married Percy Faraday Frankland, a chemist who was also looking into bacteriological problems, especially as these related to public health issues. They had one son.

Grace collaborated in work undertaken in a private analytical laboratory set up by Percy Frankland and his father, Sir Edward Frankland, well-known for his studies on water pollution. Her first publication was a study of microorganisms in air, which appeared as a joint report with her husband in 1887. Further collaborative investigations on microorganisms in water and soil appeared the following year.

The Franklands moved to University College Dundee in 1888 when Percy became professor of chemistry there, and during the next six years Grace continued to work closely with her husband in his bacteriological research. In particular, she assisted in his well-received studies on chemical reactions taking place during fermentation and on the possibility of using fermentation processes to prepare chemically pure substances, an area he was one of the first (after Pasteur) to investigate. Their joint papers on fermentation and on nitrification appeared in 1889 and 1890, and Grace provided a bacteriological appendix to another joint paper on sugar fermentation in 1892. Commenting on the Franklands' scientific collaboration, Percy Frankland's obituarist remarked that "Many women in the past have helped their husbands, but Percy Frankland is the first man to admit it"[88]—probably not an inaccurate claim if only chemists are considered, although the physicians George and Frances Hoggan published many joint papers in physiology and anatomy in the 1870s and 80s.

Grace Frankland most likely came to consider herself more an author and science journalist than a research worker, however. A number of her articles appeared in *Nature* in the 1890s and her monograph, *Bacteria in Daily Life,* was published in 1903. In addition she coauthored two books with her husband, *Microorganisms in Water: their Significance, Identification and Removal* (1894), and the biography *Pasteur* (1898).

In 1894 the family moved to Birmingham, Frankland having accepted the professorship of chemistry at Mason Science College (later the University of Birmingham). They bought land in Westmorland and Yorkshire in 1909 and for a number of years successfully ran sheep farms on the Fells, working there during vacations. They also traveled together extensively throughout Europe, both east and west. Frankland retired from his Birmingham position after the First World War and from then on they lived at Letterawe on Loch Awe, in the western Highlands of Scotland, a region Frankland had known since his childhood. Grace died there in 1946.

Little information has been uncovered about the remaining three women in this group, Gertrude Southall, Julia Divine, and Mrs. Ernest Hart. Southall's 1889 paper on gastric enzymes reported work done as a student at Mason Science College Birmingham. Julia Divine's study of heart action in the tortoise was carried out under the guidance of Hugo Kronecker, in whose institute at Bern University many foreigners worked; Kronecker was well known for his studies on the physiology of muscles, heart, and circulation. Mrs. Ernest Hart's research on fibrin formation and blood corpuscles (their estimation and hemoglobin content) was carried out about 1879 at the Collège de France in Paris. Her adviser was the eminent histologist and anatomist Louis Antoine Ranvier, widely regarded as the founder of experimental histology.[89]

In this group of early British women medical scientists Harriette Chick is unquestionably outstanding; her early collaborative work on the chemistry of disinfection was noteworthy, and her vast output on nutritional research impressive. Her career, however, was only beginning around the turn of the century. Among those who made the bulk of their contributions before 1900, physician Frances Morgan Hoggan is especially notable; her twenty-two articles, mainly on the anatomy and physiology of lymph glands and nerve endings, published in collaboration with her husband between 1875 and 1883, form the largest pre-1901 research contribution to the medical sciences by a woman in either Britain or the United States. Most of the papers were substantial, and the amount of basic experimental work reported is remarkable. Morgan Hoggan might readily have been placed in this section. However, her special position among the first British women doctors and her long and active concern with clinical work suggested her inclusion with the physicians. Among the others Sowton and Buchanan stand out for their very creditable contributions to the knowledge of heart function and nerve structure and action; Greenwood also is notable, despite her relatively short professional career, while Alcock, with her two long papers in the *Journal of Anatomy and Physiology* in 1899, offered a modest but sound contribution to the field.

Nevertheless, considering the fact that physiology held a very prominent position in the biological sciences in late nineteenth-century Britain, and was, by the 1880s, already established as the core subject of general biology in schools and universities,[90] one might wonder why more women did not carry out at least a modest amount of graduate research in the area. At least as far as Cambridge was concerned, however, the facilities available to women for laboratory work were limited. Although their own Balfour Laboratory provided practical instruction in elementary courses, the women depended largely on access to university facilities for advanced work, and these were frequently filled beyond capacity by the male students.[91]

Two-country comparison

In the new field of bacteriology there was one notably successful worker in each country—Anna Williams of the New York City Department of Health diagnostic laboratory and Harriette Chick of the Lister Institute. Their interests were in different areas, Chick's best-known bacteriological studies being directed toward the understanding at the chemical and biochemical level of infection and disease control, while Williams concentrated on public health medicine (epidemiology and immunology). Both women, however, made fine contributions to the solution of urgent problems of the time.

Among the physiologists/anatomists/neurologists the early British women might appear to have been less successful in their professional careers than their American counterparts. None of the British except Greenwood had a formal academic position, and hers was in a women's college. Nevertheless, the two groups are not dissimilar. There were perhaps three Americans whose sustained research in physiology/anatomy/neurology might be compared with that of Morgan, Greenwood, Buchanan, and Sowton. These were physiologist Ida Hyde, neurologist/anatomist Susanna Gage, and anatomist (later bacteriologist) Lydia Adams DeWitt. Gage's substantial output was facilitated by her husband, professor of histology and embryology at Cornell, and she herself had no professional position. On the other hand Hyde held a professorship at the University of

Kansas, and DeWitt had an instructorship at the University of Michigan, which permitted her to do her own research (and later a faculty position at the University of Chicago). The career successes of Hyde and DeWitt are indeed noteworthy. However, it should be remembered that the competition that these two faced from American men was less than that which the British women would have been up against. In the United States the field had not developed to the extent it had in Britain,[92] where Buchanan and Sowton, contemporaries of Hyde and DeWitt, would have had little chance of finding comparable professional positions that permitted independent research. Even for the American women such positions were few.

A subgroup of Americans that would include Moore and Cooke, students at the University of Chicago, Latimer and Towle from Bryn Mawr, and perhaps Schively from the Woman's Medical College of Pennsylvania, equates fairly readily with that formed by the British workers Alcock, Divine, Eves, and Newbigin. These women carried out most of their physiology research as students, or soon after, and later moved on to work in related areas or to other interests. There was no British counterpart of the internationally recognized American industrial disease specialist Alice Hamilton (although physiologist/neurologist S. C. M. Sowton did do some work in this area).

The admission of women into the two national physiological societies followed not dissimilar timetables, neither the Physiological Society nor the American Physiological Society being notably early in accepting female members. The British society lagged well behind its sister organizations the Zoological Society, the Royal Entomological Society, the Royal Microscopical Society, and the Linnean Society, although it did open to women before either the Chemical Society or the Geological Society. Founded in 1876 as a dining club for men working in physiology who were already personal friends, its initial membership was limited to forty.[93] First meetings were in London restaurants, with occasional excursions to such places as the Star and Garter in Richmond, Michael Foster's favorite hostelry. The first scientific meeting was held in 1880 at University College London. Even with the removal of the membership limit in 1884, the dining club character persisted, meetings being held in the refectory of the institution visited. The first mention of a woman being present at any of the society's proceedings is the record of Florence Buchanan's attendance at the lecture part of the Oxford meeting in 1896. She was there as a visitor, her passport being her position as assistant to J. Burdon Sanderson (although the meeting's records included the comment that she had carried out independent work). Ten years later, at the Edinburgh gathering, ladies were present as guests at a full dinner meeting.[94] When, in 1915, women were finally declared eligible for membership on the same terms as men, Buchanan and Sowton were among the six then elected.[95]

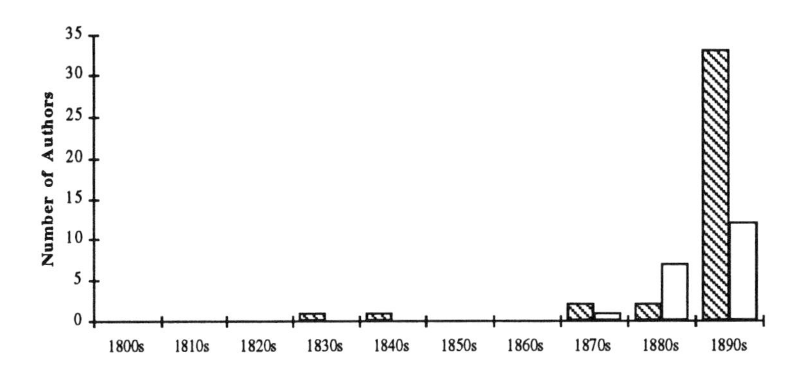

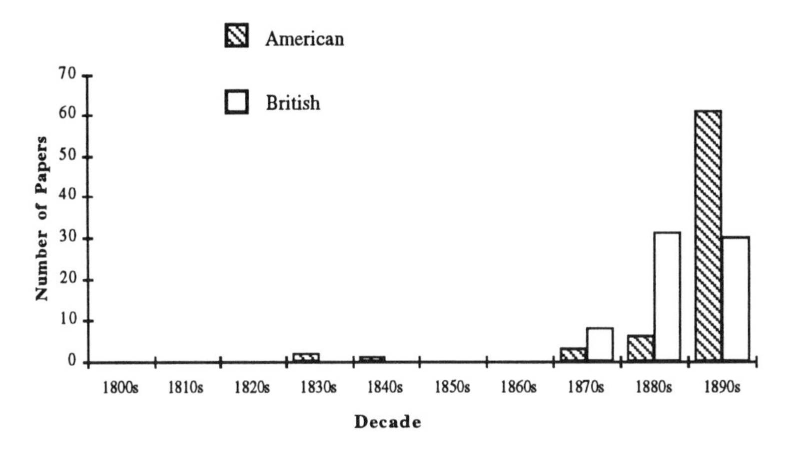

Figure 6-1. American and British authors and papers in medical sciences, by decade, 1800–1900. Data from the Royal Society *Catalogue of Scientific Papers.*

a. American authors

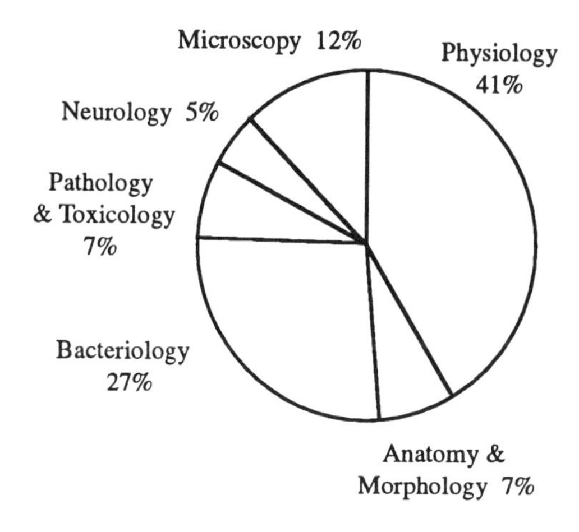

b. American papers

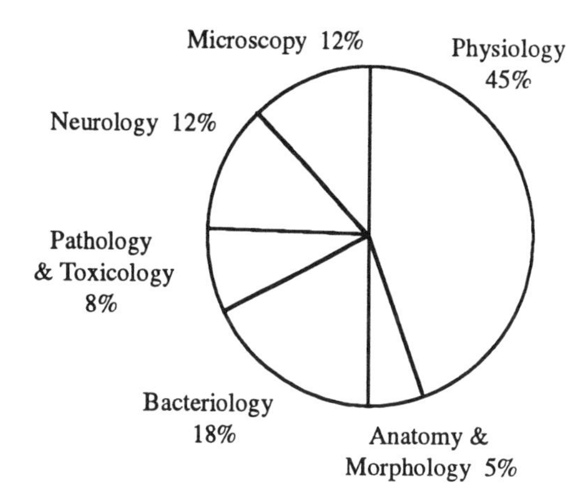

c. British authors

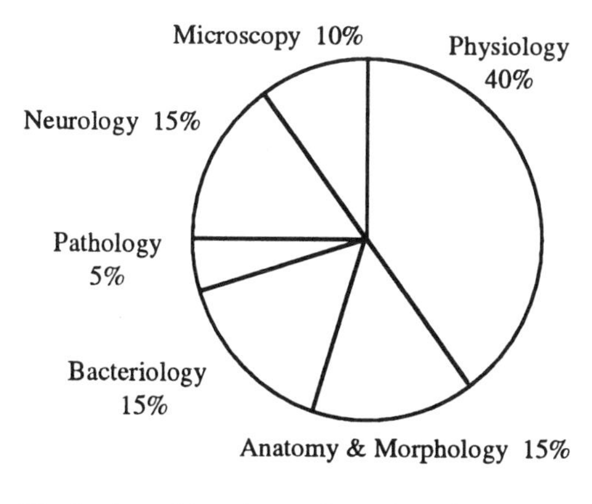

d. British papers

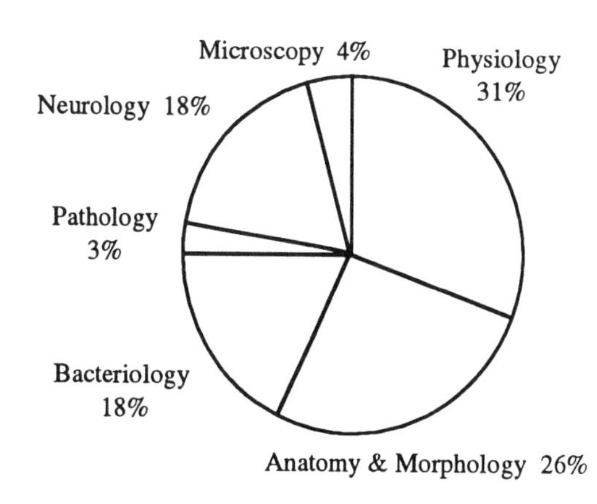

Figure 6-2. Distribution of American and British authors and papers in medical sciences, by subfield, 1800–1900. (In a. and c., an author contributing to more than one subfield is counted in each.) Data from the Royal Society *Catalogue of Scientific Papers.*

The American Physiological Society, established in 1887, set itself stringent eligibility standards and accepted as members only those who had clearly demonstrated their capacity for original research, ongoing at the time of proposal for membership.[96] Not many women had professional positions from which they could compete, the successes, relative to the British, of Hyde and DeWitt notwithstanding. Although Hyde was elected in 1902 (the year DeWitt was elected to the Association of American Anatomists), and the society prided itself on its equal treatment of men and women in the matter of acceptance of members, the actual pattern of admission of women was not very different from that in Britain. After Hyde no women were elected until 1913, and then slowly, one a year on average over the period from 1913 to 1920.[97]

Numerical data on distribution of authors and papers by subfield is summarized in Figure 6-2. Well over half of the American work may be classified in the two areas of physiology (about 45 percent, with Hyde and Gage as the two heaviest contributors) and bacteriology (about 18 percent, with many contributors)—chart b. The fractions of papers in anatomy/morphology and neurology were noticeably greater for the British (the heaviest contributors being physician Hoggan in anatomy and Sowton in neurology)—chart d.[98]

Notes

1. When comparing numbers of publications in such areas as zoology and physiology, however, it should be kept in mind that there is considerable overlap between fields and categorization of papers is frequently a matter of judgment.

2. Russel C. Maulitz, "Pathologists, clinicians and the role of pathophysiology," in *Physiology in the American Context 1850–1940,* ed. Gerald L. Geison (Bethesda, Md.: American Physiological Society, 1987), pp. 209–35, on p. 230.

3. In addition to Bryn Mawr, three other women's colleges (Wellesley, Vassar, and Goucher), are mentioned in this section as places where women were involved in modest amounts of experimental work in physiology. These four colleges were ahead of their time in their willingness to expand their view of physiology beyond lectures on women's health. At Bryn Mawr, physiology was taught in the department of biology; at Wellesley, zoologist Mary Willcox secured its incorporation into her department shortly after she took up her position in 1883; at Vassar and at Goucher, two very able college physicians (Elizabeth Thelberg and Lilian Welsh, respectively), encouraged experimental work (see also Toby A. Appel, "Physiology in American women's colleges. The rise and decline of a female subculture," *Isis,* **85** (1994): 26–56).

4. Several whose papers are listed in medical sciences sections of the bibliography had major interests in other areas. Two (Bryn Mawr-trained Bunting and Towle) were general biologists who went on to careers as high school biology teachers, three (Pennington, Swallow Richards and Hyams) were chemists involved in public health work, five (Gage Day, Stone, Bishop, Mitchell, and Lewi) were primarily practicing physicians; there was also a biochemist (Albro Baker), a zoologist (Nunn Whitman), a botanist (Pammel Hansen) and a psychologist (Carter). For these contributors see chapters corresponding to their areas of major interest.

5. Robert E. Kohler, *From Medical Chemistry to Biochemistry. The Making of a Biomedical Discipline* (Cambridge: Cambridge University Press, 1982), p. 170.

6. Gail Susan Tucker, "Ida Henrietta Hyde: the first woman member of the society," *The Physiologist,* **24** (1981): 1–10, and "Reflections on the life of Ida Henrietta Hyde, 1857–1945; the woman scientist in the twentieth century," *Creative Woman Quarterly,* **1** (1978): 5–8; Ingrith J. Deyrup, "Hyde, Ida Henrietta," in *NAW,* vol. 2, pp. 247–9; Margaret E. Maltby, comp., *History of the Fellowships Awarded by the American Association of University Women, 1888–1929, with the Vitas of the Fellows* (Washington, D.C.: American Association of University Women, [1929]), pp. 14–15; Mary R. S. Creese, "Ida Henrietta Hyde (1857–1945)," in *Women in the Biological Sciences,* eds. L. Grinstein, C. Biermann and R. Rose (Westport, Conn.: Greenwood Press, 1997); Ida Hyde file, University of Kansas archives.

7. The Heidenheimers changed their family name to Hyde soon after their arrival in the United States.

8. Hyde was the third woman to receive a doctorate from the University of Heidelberg, following two Germans, Käthe Windschied, who received a doctorate in modern languages in 1894, and Marie Gernet, who took her degree in mathematics in 1895—see James C. Albisetti, *Schooling German Girls and Women: Secondary and Higher Education in the Nineteenth Century* (Princeton, N.J.: Princeton University Press, 1988), p. 226.

9. Hyde's papers on the nervous system and respiratory processes include the following: "The nervous system of *Gonionema murbachii,*" *Biological Bulletin,* **4** (1902): 40–5; "The nerve distribution in the eye of *Pecten irradians,*" in *Mark Anniversary Volume* (New York: Holt, 1903), pp. 471–82 (invited paper); "Localization of the respiratory center in the skate," *American Journal of Physiology,* **10** (1904): 236–58; "A reflex respiratory center," ibid., **16** (1906): 368–77; "The effect of salt solutions on the respiration, heart beat and blood pressure in the skate," ibid., **23** (1908): 201–13.

10. Gabriele Kass-Simon, "Biology is destiny," in *Women of Science. Righting the Record,* eds. G. Kass-Simon and Patricia Farnes (Bloomington, Ind.: Indiana University Press, 1990), pp. 213–67, on pp. 238, 241–4; Louise H. Marshall, "Instruments, techniques and social units in American neurophysiology, 1870–1950," in Geison, ed., *Physiology in the American Context,* pp. 351–69, on pp. 362–3 and n. 46, p. 369. Hyde described the electrode in a short paper written just before her retirement, "A micro electrode and unicellular stimulation," *Biological Bulletin,* **40** (1921): 130–3. For her early work in this area see, "Differences in electrical potential in developing eggs," *American Journal of Physiology,* **12** (1904): 241–75.

11. Ida H. Hyde and C. Spreier, "The influence of light upon reproduction in *Vorticella,*" *Kansas University Science Bulletin,* **9** (1915): 398–9; Ida H. Hyde and W. Scalopino, "The influence of music on electrocardiograms and blood pressure," *American Journal of Physiology,* **46** (1918): 35–8. Hyde republished the latter study in expanded form in the *Journal of Experimental Psychology* in 1924 (7: 213–24); it won her a prize from the American Psychological Association.

12. Hyde claimed that she completed the requirements for the M.D. but that the degree was withheld because she did part of the final year work *in absentia* (Maltby, *History of Fellowships,* pp. 14–15).

13. Hyde to Chancellor Strong and Board of Regents, 3 May 1903, 20 March 1907, 6 September 1908 (Chancellor's Office, correspondence, faculty, University of Kansas archives). In another letter, of 8 September 1908, Hyde says, "I believe there should be no discrimination, and that it is the desire of the people of the State, that just compensation should be awarded for *service* to the state irrespective of sex." In her last year as head of the department of physiology at the University of Kansas (1916–1917) she was paid $2,600, and she received the same salary the following year. That same year (1917–1918) another woman faculty member, Elizabeth Cade Sprague, head of the home economics department, was also paid $2,600, but many department chairmen, including those in chemistry, physics, astronomy, botany, philosophy, history, and Latin, had salaries of $3,000, 15 percent more than Hyde's. Hyde's successor as chairman of the physiology department, Ole Stoland, was paid $2,500 during his first year as department head (1917–1918), but his salary went up to $2,600 the following year and by 1921–1922 had reached $3,600 (faculty salary cards, 1916–1917 to 1921–1922, and *Minutes, Board of Administration, University of Kansas, 1917–18,* University of Kansas archives).

14. Gieson, ed., *Physiology in the American Context;* Philip J. Pauly, "The appearance of academic biology in late nineteenth century America," *Journal of the History of Biology,* **17** (1984): 369–97.

15. Writing to the chancellor from Woods Hole in June 1918 Hyde says that she infers that she will spend the next year on half salary and asks that he "kindly acquaint me with your decision, pertaining to my relations to the university" (Hyde to Chancellor Strong, 24 June 1918. Other especially pertinent correspondence includes letters from Hyde to Chancellor Strong and Dean Olin Templin, 8 March 1916, and Hyde to Chancellor Strong, 5 June 1916—University of Kansas archives). There seems little doubt that she was being pushed out, and in a rather unfortunate and perhaps somewhat underhanded way.

16. Simon Henry Gage, "Susanna Phelps Gage, Ph.B.," *Journal of Comparative Neurology,* **27** (1916–1917): 5–18; anon., "Memorial to Susanna Phelps Gage," *Science,* **45** (1917): 82–3.

17. Simon Henry Gage was a brother of physician Mary Gage Day (see chapter 7).

18. Susanna Phelps Gage, "A three weeks human embryo with special reference to the brain and nephric system," *American Journal of Anatomy,* **4** (1905): 409–43. Among her other post-1900 papers were the following: "An unusual attitude of the four weeks human embryo. Comparisons with the mouse," *Science,* **17** (1903): 254; "The mesonephros of a three weeks human embryo," *Proceedings of*

the Association of American Anatomists, March 1904, in *American Journal of Anatomy,* **3** (1904): vi; "Total folds of the forebrain. Their origin and development to the third week in the human embryo," ibid., **4** (1905); ix; "Relations of the total folds of the brain tube of human embryos to definitive structure," ibid., **5** (1906): ix-x; "The notochord of the head in human embryos of the third to the twelfth week and comparisons with other vertebrates," *Science,* **24** (1906): 295–6; "Changes in the form of the forebrain of human embryos during the first eight weeks," *Proceedings of the Seventh International Zoological Congress,* 1910, 2 pp.

19. Esmond R. Long, "DeWitt, Lydia Maria Adams," in *NAW,* vol. 1, pp. 468–9; anon., "DeWitt, Lydia M. Adams," *NCAB,* vol. B, 1927, pp. 457–8; Pearl Bliss Cox, "Pioneer women in medicine. Michigan XVI," *Medical Woman's Journal,* **56,** (1949): 48–51, 64, on p. 49; Mary R. S. Creese, "Lydia Maria Adams DeWitt (1859–1928)," in Grinstein, Biermann and Rose, eds., *Women in the Biological Sciences.*

20. Lydia M. DeWitt, "Preliminary report of experimental work and observations on the areas of Langerhans in certain mammals," *Proceedings of the American Association of Anatomists* (1904–1905): viii; "Morphology and physiology of areas of Langerhans in some vertebrates," *Journal of Experimental Medicine,* **8** (1906): 193–239; "Observations on the sinoventricular connecting system of the mammalian heart," *Anatomical Record,* **3** (1909): 475–97; "The pathology of the sinoventricular system or bundle of His," *Physician and Surgeon,* **32** (1910): 145–50. DeWitt also brought out a translation of Johannes Sobotta's *Atlas and Epitome of Human Histology and Microscopic Anatomy* (Philadelphia: W. B. Saunders, 1903).

21. Michael Bliss, *The Discovery of Insulin* (Chicago: University of Chicago Press, 1982).

22. Lydia M. DeWitt, "Report of some experiments on the action of *staphylococcus aureus* on the Klebs-Loeffler bacillus," *Journal of Infectious Diseases,* **10** (1912): 24–35; "A case of generalized infection with a diphtheriod organism," ibid, **10** (1912): 36–42; (with Florence L. Evans) "Laboratory methods of diagnosis of typhoid fever," *Interstate Medical Journal,* **19** (1912): 770–86.

23. Harry Gideon Wells, Lydia M. DeWitt, and Esmond R. Long, *The Chemistry of Tuberculosis; being a Compilation and Critical Review of Existing Knowledge on the Chemistry of the Tubercule Bacillus and its Products* (Baltimore, Md.: Williams and Wilkins, 1923). Among DeWitt's many journal publications are the following: "Preliminary report of experiments in the vital staining of tubercules. Studies in the biochemistry and chemotherapy of tuberculosis IV," *Journal of Infectious Diseases,* **12** (1913): 68–92; "Report on some experimental work on the use of methylene blue and allied dyes in the treatment of tuberculosis. Studies in the biochemistry and chemotherapy of tuberculosis VII," ibid., **13** (1913): 378–403; "Therapeutic use of certain azo dyes in experimentally produced tuberculosis in guinea-pigs. Studies in the biochemistry . . . VIII," ibid., **14** (1914): 498–511; (with Hope Sherman) "Tuberculocidal action of certain chemical disinfectants. Studies in the biochemistry . . . IX," ibid., **15** (1914): 245–56; (with Hope Sherman) "The bactericidal and fungicidal action of copper salts. Studies in the biochemistry . . . XV," ibid., **18** (1916): 368–82; "The present status of tuberculosis chemotherapy," *Journal of Laboratory and Clinical Medicine,* **1** (1915–16): 677–84; "Some derivatives of methylene blue in tuberculosis chemotherapy," *National Association for the Study and Prevention of Tuberculosis. Transactions,* **12** (1916): 257–61; "Gold therapy of tuberculosis," *American Review of Tuberculosis and Pulmonary Diseases,* **1** (1917–1918): 424–30; "The use of gold salts in the treatment of experimental tuberculosis in guinea pigs. Studies in the biochemistry . . . XVIII," *Journal of Infectious Diseases,* **23** (1918): 426–37; (with S. M. Caldwell and Gladys Leavell) "Distribution of gold in animal tissues. Studies in the biochemistry . . . XVII," *Journal of Pharmacology and Experimental Therapeutics,* **11** (1918–1919): 357–77; (with B. Suyenago and H. G. Wells) "The influence of cresote, guaiacol and related substances on the tubercule bacillus and on experimental tuberculosis. Studies in the biochemistry . . . XIX," *Journal of Infectious Diseases,* **27** (1920): 115–35; "Mercury compounds in the chemotherapy of experimental tuberculosis in guinea pigs. I. Studies in the biochemistry . . . XXI," ibid., **28** (1921): 150–69; (with Lauretta Bender) "Blood changes during the progress of experimental tuberculosis in guinea pigs," *National Tuberculosis Association. Transactions,* **18** (1922): 444–9; (with L. Bender) "Hematological studies on experimental tuberculosis in the guinea pig. I. Blood morphology, Arneth counts and coagulation time in normal and untreated tuberculous guinea pigs. Studies in the biochemistry . . . XXVII," *American Review of Tuberculosis,* **8** (1923–1924): 138–62; "The therapeutic and bactericidal value of organic mercurial compounds in experimental tuberculosis in guinea pigs. Studies in the biochemistry . . . XXVIII," ibid., **8** (1923–1924): 234–44; (with L. Bender) "Hematological studies on experimental tuberculosis of the guinea pig. II. The effect of certain drugs on the blood picture in tuberculosis. Studies in the biochemistry . . . XXXI," ibid., **9** (1924–1925): 65–71. (DeWitt's coworker Lauretta Bender (b. 1897) was later known for her work in child psychiatry—see anon., "Medical women of the year," *Journal of the American Medical Women's Association,* **13** (1958): 512–21.)

24. Alice Hamilton, *Exploring the Dangerous Trades. The Autobiography of Alice Hamilton, M.D.* (Boston: Little Brown and Co., 1943); Wilma Ruth Slaight, "Alice Hamilton: first lady of industrial medicine," Ph.D. Dissertation, Case Western Reserve University, 1974; Barbara Sicherman, *Alice Hamilton. A Life in Letters* (Cambridge, Mass.: Harvard University Press, 1984), and "Hamilton, Alice," in *NAW,* vol. 4, pp. 303–6.

25. Hamilton, *Autobiography,* pp. 47, 48.

26. Alice Hamilton, "The toxic action of scarlatinal and pneumonic sera on paramoecia," *Journal of Infectious Diseases,* **1** (1904): 211–28; "The question of virulence among the so-called pseudodiphtheria bacilli," ibid., **1** (1904): 690–713; (with Jessie M. Horton) "Further studies on virulent pseudodiphtheria bacilli," ibid., **3** (1906): 128–47; "The opsonic index and vaccine therapy of pseudodiphtheric otitis," ibid., **4** (1907): 313–25; "On the occurrence of thermostable and simple bactericidal and opsonic substances," ibid., **5** (1908): 570–84; "Surgical scarlatina," *American Journal of Medical Sciences,* **128,** n.s. (1904): 111–29; "Milk and scarlatina," ibid., **130,** n.s. (1905): 879–90; "Experiments in antityphoid inocculation," *Transactions of the Chicago Pathological Society,* **8** (1909–1911): 151–8. Hamilton was also active in research during her four years at the Woman's Medical School—see "A case of heterotopia of the white matter of the medulla oblongata," *American Journal of Anatomy,* **1** (1902): 417–21; "The division of differentiated cells in the central nervous system of the white rat," *Journal of Comparative Neurology, Granville, Ohio,* **11** (1901): 297–320; "The pathology of a case of poliencephalomyelitis," *Journal of Medical Research,* **8** (1902): 11–30.

27. "The fly as a carrier of typhoid; an inquiry into the part played by the common house fly in the recent epidemic of typhoid fever in Chicago," *Journal of the American Medical Association,* **40** (1903): 576–83; see also "The common house fly as a carrier of typhoid," ibid, **42** (1904): 1034; "The role of the house fly and other insects in the spread of infectious diseases," *Illinois Medical Journal,* **9** (1906): 583–7; "Flies and privy vaults," *Transactions of the American Association for the Study and Prevention of Infant Mortality,* **2** (1912): 157–62.

28. Hamilton, *Autobiography,* p. 114.

29. Ibid., p. 121.

30. Ibid., pp. 128–9.

31. Ibid., p. 252.

32. Sicherman, *Alice Hamilton,* pp. 237–8. Hamilton is on record as feeling that she did receive one compliment from a colleague at Harvard: when she asked for a sabbatical leave in 1924 it was granted, but the difficulties it would cause the Medical School were also made plain. There was no one else in the United States whom they could get to teach her course, and Edgar L. Collis (professor of preventive medicine at University College Cardiff, Wales, and a specialist in industry-related diseases) would have to be invited over to substitute for her (ibid., p. 294).

33. Elizabeth D. Robinton, "Williams, Anna Wessels," in *NAW,* vol. 3, pp. 737–9; anon., "Anna W. Williams, scientist, is dead," *NYT,* 21 November 1954, 86:6.

34. Anna W. Williams and Mary M. Louden, "The etiology and diagnosis of hydrophobia," *Journal of Infectious Diseases,* **3** (1906): 452–83. Williams's other rabies papers include: "The diagnosis of rabies," *American Journal of Public Hygiene,* n.s., **4** (1907–1908): 10–15, and *American Public Health Association Report, 1907,* **33,** Pt. ii (1908): 104–9; "Recent studies on rabies," *Women's Medical Journal,* **18** (1908): 113–15; "Cultivation of the rabies organism," *Journal of the American Medical Association,* **61** (1913): 1509–11.

35. Among Williams's women coworkers was Emily Dunning Barringer (1877–1971), the first woman to serve on the staff of a general municipal hospital in New York City and the first woman ambulance surgeon in the United States—Iris Noble, *First Woman Ambulance Surgeon: Emily Barringer* (New York: Julian Messner, 1962).

36. See Anna W. Williams, "Pure cultures of parasitic amoebas on brain-streaked agar," *Proceedings of the Society for Experimental Biology and Medicine,* **8** (1910–1911): 56–8; "Pure cultures of amoebae parasitic in animals," *Journal of Medical Research,* **25** (1911): 263–83; "Significance of the group of haemophilic bacilli in conjunctivitis, especially in that of trachoma," *Proceedings of the New York Pathological Society,* n.s., **12** (1912–1913): 17–29; "Special culture media for pure culture of amoebae found in the intestines of animals," *Transactions of the Fifteenth International Congress on Hygiene and Demography, 1912,* **2** (1913): 43–7; (with G. N. Calkins) "Cultural amoebae, a study in variation," *Journal of Medical Research,* **29** (1913): 43–56; "Cultivation of the rabies organism," *Journal of the American Medical Association,* **61** (1913): 1509–11; "The prevention of trachoma in New York school children," *Archives of Ophthalmology,* **42** (1913): 503; "Relation of 'trachoma inclusions' to certain bacteria growing in intracellular nests," *Proceedings of the New York Pathological Society,* **14** (1914): 28–34; (with others) "A study of trachoma and allied conditions in the public school children of New York City," *Journal of Infectious Disease,* **14** (1914): 261–337; "Causes and treatment of chronic conjunctival infections in childhood," *New York State Journal of Medicine,* **14** (1914): 579–81; (with Anna Von Sholly and Caroline Rosenberg) "Amoebas in the mouths of school children," *Proceedings of the New York Pathological Society,* n.s., **15** (1915): 34–8; (with A. Von Sholly, et al.) "Significance and prevention of amoebic infections in the mouths of children," *Journal of the American Medical Association,* **65** (1915): 2070–3; (with C. Rosenberg) "Purulent conjunctivitis in infants under two months of age: incidence, causes and effects on vision," *Archives of Ophthalmology,* **45** (1916): 155–71 (also published in Spanish, "La conjunctivitis purulenta en niños menores de dos meses," *Crónica medico-quirúrgica, Habana,* **42** (1916): 123–8).

37. Among Williams's influenza papers were the following: "The bacteriology of influenza," *Monthly Bulletin, Department of Health, City of New York,* n.s., **13** (1918): 284–8; "The etiology of influenza," *Proceedings of the New York Pathological Society,* **18** (1918): 83–90; (with W. H. Park) "Studies on the etiology of the pandemic of 1918," *American Journal of Public Health,* **9** (1919): 45–9; "The relationship of *Bacillus influenzae* to the pandemic," *Archives of Pediatrics,* **36** (1919): 107–17; "The bacteriology of the recent pandemic," *Proceedings of the Pathological Society, Philadelphia,* n.s., **22** (1919–1920): 27; (with Olga R. Povitsky) "Growth of *B. influenzae* without the presence of haemoglobin," *Journal of Medical Research,* **42** (1920–1921): 405–17; (with W. H. Park and C. Krumwiede) "Microbal studies on acute respiratory infection with especial consideration of immunological types," *Journal of Immunology,* **6** (1921): 1–4; (with Mary Nevin and Caroline R. Gurley) "Studies on acute respiratory infections. I. Methods of demonstrating micro-organisms, including filterable viruses, from the upper respiratory tract in health, in common colds and in influenza with the object of discovering common strains," ibid., **6** (1921): 5–24.

38. The first edition of this textbook was published by Park; all other editions, from the second (1905) to the eleventh (1939), were coauthored by Williams.

39. Regina M. Morantz-Sanchez, "The many faces of intimacy: professional options and personal choices among nineteenth- and early twentieth-century women physicians," in Pnina Abir-am and Dorinda Outram, eds., *Uneasy Careers and Intimate Lives,* (New Brunswick, N.J.: Rutgers University Press, 1987), pp. 45–59, on p. 55.

40. Helga M. Ruud, M.D., "The Woman's Medical College of Chicago. Records of professors and instructors and alumnae. 1871–1895," *Medical Woman's Journal,* **53,** (1946): 50–4, on p. 53; anon., "Medical women of the year," *Journal of the American Medical Women's Association,* **11** (1956): 434–41, on pp. 435–6; *Who was Who in America,* vol. 4, p. 557; the following obituaries (in Northwestern University archives), "Vida Annette Latham, '92d," *Michigan Alumnus,* June 1937; "Lady doctor, 91, too busy to wed," *Daily News,* 4 February 1957; "Vida Latham, 91, physician, writer, dies," *Chicago Tribune,* 18 January 1958; "Vida Latham, 91, noted doctor-dentist, dead," *Chicago Sun Times,* 18 January 1958; *AMS,* 1933–1955.

41. Latham does not appear to have taken either the Natural Sciences Tripos or the Mathematical Tripos examination at Cambridge, her name being absent from the lists in the *Historical Register of the University of Cambridge to 1910.*

42. Among Latham's later publications, mainly in dental pathology and microscopy, are the following: "Résumé of the histology of dental pulp," *Journal of the American Medical Association,* **39** (1902): 63–74; "Rapid method for examining bacteria in tissues and their staining with haematotoxylin," *Journal of Applied Microscopy,* **6** (1903): 2453–4; "Indications for scientific progress in stomatology," *Journal of the American Medical Association,* **45** (1905): 369–73; "The necessity of a medical education for dentists," *Dental Digest,* **14** (1908): 410–18; "Systemic conditions in relation to oral symptoms and sepsis," *Journal of the American Medical Association,* **55** (1910): 1181–5; "The pathologic findings of some diseases of the teeth and gums," *Dental Summary,* **32** (1912): 819–30; "Anomalies in pulp structure and their relation to clinical work," *Dental Items of Interest,* **39** (1917): 95–118; "The question of correct naming and use of micro reagents," *Medical Woman's Journal,* **29** (1919): 1–4.

43. Agnes Claypole Moody and Marian E. Hubbard, *In Memoriam. Edith Jane Claypole* (Berkeley: [n.p.], 1915); anon., obituary, *Science,* **41** (1915): 527, 754; *NCAB,* vol. 13, 1906, pp. 259–60; Burton Williamson, "Some American women in science," pp. 363–4; *WWWA,* p. 184; *AMS,* 1910.

44. Edith J. Claypole, "On the classification of the streptothrices, particularly in their relation to bacteria," *Journal of Experimental Medicine,* **17** (1913): 99–116; "Human streptotrichosis and its differentiation from tuberculosis," *Archives of Internal medicine,* **14** (1914): 104–19. "Skin reaction in streptothrix infections," *Proceedings of the Society for Experimental Biology and Medicine,* **11** (1913–1914): 41; "A further note on specific hyperleucocytosis in immunized animals," ibid., **11** (1913–1914): 47.

45. F. P. Gay and Edith J. Claypole, "The typhoid carrier state in rabbits as a method of determining the comparative immunizing value of preparations of the typhoid bacillus: studies in typhoid immunization, I," *Archives of Internal Medicine,* **12** (1913): 613–20; "Agglutinability of blood and agar strains of the typhoid bacillus: studies in typhoid immunization, II," ibid., **12** (1913): 621–7; "Induced variations in the agglutinability of *Bacillus typhosus,*" *Journal of the American Medical Association,* **60** (1913): 1141; "Specific and extreme hyperleukocytosis following the infection of *Bacillus typhosus* in immunized rabbits," ibid., **60** (1913): 1590; "Specific hyperleukocytosis: studies in typhoid immunization, IV," *Archives of Internal Medicine,* **14** (1914): 662–70; "An experimental study of methods of prophylactic immunization against typhoid fever: studies in typhoid immunization, V," ibid., **14** (1914): 671–705.

46. Anon., "Mary J. Ross, M.D.," *Journal of the American Medical Women's Association,* **9** (1954): 52; S and F, pp. 148–9; *AMS,* 1910.

47. "The origin and development of the gastric glands of Desmognatus, Amblystoma and pig," *Biological Bulletin,* **4** (1903): 66–95.

48. S and F, pp. 270–1; Faculty Lists, State Normal School, San Diego, 1901–1905; *WWWA,* p. 572; *AMS,* 1910.

49. "On the effects of solutions of various electrolytes and non-conductors upon rigor mortis and heat rigor," *American Journal of Physiology,* 7 (1902): 1–24. A second shorter paper by Moore on the effects of electrolytes appeared in the same volume, pp. 315–19, "On the power of Na_2SO_4 to neutralize the ill effects of NaCl."

50. "Some facts concerning geotropic gatherings of paramoecia," *American Journal of Physiology,* **9** (1903): 238–44.

51. Anne Moore, Ph.D., *The Feeble-minded in New York, a Report prepared for the Public Education Association of New York* (New York: State Charities Aid Association, 1911).

52. University of Chicago alumni directories for 1911 and 1919. Cooke is listed as having died by 1919.

53. Leo Loeb and Elizabeth W. Cooke, "The effect of light on cells in fluorescent solution after addition of potassium cyanide," *Proceedings of the Society for Experimental Biology and Medicine,* **5** (1908): 27–8 (abstract); "The comparative toxicity of sodium chloride and of staining solutions upon the embryo of *Fundulus,*" ibid., **6** (1909): 113–15 (abstract); "Über die Giftigkeit einiger Farbstoff für die Eier von *Asterias* und von *Fundulus,*" *Biochemische Zeitschrift,* **20** (1909): 167–77.

54. University of Chicago alumni directories.

55. S and F, pp. 67–8; *WWWA,* p. 378; *AMS,* 1910–1933; *NCAB,* vol. 16, pp. 153–4, entry for Thomas Hefferan.

56. "A comparative and experimental study of bacilli producing red pigment," *Centralblatt für Bakteriologie,* Abt. 2, **11** (1903): 311–17, 397–404, 456–75, 520–40; "Agglutination and biological relationships in the prodigiosus group," ibid., Abt. 1, **41** (1906): 553–62; (with E. O. Jordan), "Observations on the bionomics of Anopheles," *Journal of Infectious Diseases,* **2** (1905): 56–69 (see also bibliography).

57. S and F, p. 268; *WWWA,* p. 477; *AMS,* 1910. Sources differ about Latimer's date of birth; *AMS* gives 28 March 1860 and the *Register of Graduate Students at Bryn Mawr,* quoted by Siegel and Finley, gives 28 March 1859. Siegel and Finley give Latimer's date of death as between 1930 and 1935.

58. Bertha Tapman Shamer, M.D., "Claribel Cone, M.D." *Journal of the American Medical Women's Association,* **7,** (1952): 431–2; Esther Pohl Lovejoy, *Women Doctors of the World* (New York: Macmillan, 1957), p. 122; Harold J. Abrahams, *The Extinct Medical Schools of Baltimore, Maryland* (Baltimore, Md.: Maryland Historical Society, 1969), especially the Woman's Medical College of Baltimore faculty lists on pp. 106–15, 120.

59. "Zur Kenntnis der Zellveränderungen in der normalen und pathologischen Epidermis des Menschen," *Frankfurter Zeitschrift für Pathologie,* **1** (1907): 37–87.

60. Shamer, "Claribel Cone," p. 432.

61. Rita S. Finkler, "An inspiring visit to a pioneer woman physician," *Medical Woman's Journal,* **48** (1941): 323; obituary, *NYT,* 16 May 1944, 21: 4; Guleilma Fell Alsop, *History of the Woman's Medical College, Philadelphia, Pennsylvania, 1850–1950* (Philadelphia: Lippincott, 1950), pp. 164, 171, 198.

62. Of the remaining late nineteenth-century American women whose papers in the medical sciences are listed in the bibliography but who have not been discussed here, fourteen are mentioned in other chapters (see n. 4). The remaining six are Bertha Ballantyne, Marion Eubank, Louise Katz, Myra Pollard, Mary Schively, and Josephine Kendall. Ballantyne published a joint paper in 1899 with Theodore Hough, then an instructor of biology and physiology at MIT and the Boston Normal School of Gymnastics. Eubank studied with physiologist Winfield Hall at Northwestern University Medical School in the mid 1890s. Katz, who presented a paper on muscle development at the 1900 meeting of the AAAS, was from Ithaca, New York (Cornell University?). Pollard studied at the University of Michigan under physiologist Henry Sewall; their joint paper appeared in the *Journal of Physiology* in 1890. Schively, of Philadelphia, graduated from the Woman's Medical College of Pennsylvania in 1895 and worked with Jacques Loeb at the Woods Hole Laboratory for several summers in the mid 1890s. Kendall, from Boston, took a medical degree at the University of Zurich in 1879, the second American woman to do so. She specialized in ophthalmology (dissertation title, "Über *Herpes corneae*"). Two papers reporting her earlier work in physiology, published jointly with Balthasar Luchsinger, appeared in Pflüger's *Archiv für Physiologie* in 1876 (Hanny Rohner, "Die ersten 30 Jahre des medizinischen Frauenstudiums an der Universität Zürich 1867–1897," *Zürcher Medizingeschichtliche Abhandlungen,* Neue Reihe Nr. 89, Juris Druck & Verlag, Zurich, 1972, p. 81).

63. Sir David Brewster, "On the influence of successive impulses of light upon the retina," *Philosophical Magazine,* **4** (1834): 241–5.

64. A 1950 reprinting of this work has an introduction that records the few known facts about Mary Griffith's life—see Nelson F. Adkins, Introduction, in Mary Griffith, *Three Hundred Years Hence* (Philadelphia: Prime Press, Reprints of Early American Utopian Novels, 1950), pp. 5, 6, especially n. 3, 4. Much of Adkins's information about Griffith came from letters in the *New England Farmer* for 1831, in the holdings of the Massachusetts Horticultural Society. Griffith's earlier work was, *Our Neighbourhood: or, Letters on Horticulture and Natural Phenomena Interspersed with Opinions on Domestic and Moral Economy* (New York: E. Bliss, 1831).

65. Hamilton, *Autobiography,* p. 252.

66. Gerald L. Geison, *Michael Foster and the Cambridge School of Physiology* (Princeton: Princeton University Press, 1978), Introduction, and pp. 309–13; Stella Butler, "Centres and peripheries: the development of British physiology, 1870–1914," *Journal of the History of Biology,* **21** (1988): 473–500.

67. *Newnham College Register, 1871–1971,* vol. 1, Staff, p. 6; obituaries, "Mrs. Bidder. Biological studies at Cambridge," *Times,* 27 September 1932, 9c, ibid., 28 September 1932, 17c, ibid., 29 September 13d, and W[illiam] B[ate] Hardy, "Mrs. G. P. Bidder, *Nature,* **130** (1932): 689–90.

68. See H. Godwin, A. R. Clapham and M. R. Gilson, "Edith Rebecca Saunders, F.L.S.," *New Phytologist,* **45** (1946): 1–3, on p. 1.

69. Hardy, "Mrs. G.P. Bidder," p. 689. Lea, Gaskell and Langley, the men mentioned as Greenwood's laboratory colleagues, along with Hardy himself, were distinguished figures in the early development of the Cambridge School of Physiology (see Geison, *Michael Foster*).

70. In addition to her scientific papers Greenwood published the work, *Domestic Economy in Theory and Practice: a Textbook for Teachers and Students in Training,* coauthored by Florence Baddeley (Cambridge: Cambridge University Press, 1901). The book presented scientific information on such matters as nutritional needs, bacterial contamination of foodstuffs, and general hygiene.

71. Hardy, "Mrs. G.P. Bidder," p. 690.

72. "Mrs. Bidder," *Times,* 27 September.

73. *Newnham College Register 1871–1971,* vol. 1, p. 65.

74. Ibid., p. 37.

75. "*La Famosa Toledana* by Juan de Quiros," *Revue Hispanique,* **41** (1917): 336–562.

76. Student Records, University College London, and information collected from technical papers.

77. No reference to Buchanan was found in the Oxford University Calendar or in lists of members of the Oxford women's colleges.

78. Buchanan's post-1900 papers include the following: "The electrical response of muscle in persistent contraction," *Journal of Physiology,* **27** (1901): 95–160; (with J. Burdon Sanderson) "The Jena researches on the spasm of strychnine," ibid., **28** (1902): xxix–xxxi; "An electrical response to excitation in *Desmodium gyrans,*" ibid., **33** (1905–1906): vii–ix; "Electrical variation accompanying reflex inhibition in skeletal muscle," ibid., **35** (1906–1907): xlii–xlv; "The time taken in passing the synapse in the spinal chord of the frog," *Proceedings of the Royal Society,* B, **79** (1907): 503–4; "Electrical response of muscle in voluntary contraction in man," *Journal of Physiology,* **37** (1908): xlvii–xlviii; "Frequency of heart beat in mouse," ibid., **37** (1908): lxxix–lxxx; "Time taken in transmission of reflex impulses in the spinal chord of the frog," *Quarterly Journal of Experimental Physiology,* **1** (1908): 1–66; "Electrical response of muscle to voluntary reflex and artificial stimulation," ibid., **1** (1908): 211–42; "The frequency of the heartbeat and the form of the electrocardiogram in birds," *Journal of Physiology,* **28** (1909): lxii–lxiv; "On the electrocardiogram, frequency of heartbeat, and respiratory exchange in reptiles," ibid., **39** (1909–1910): xxv–xxvii; "Frequency of heartbeat in sleeping and waking dormouse," ibid., **40** (1910): xlii–xliv; "The significance of the pulse rate in vertebrate animals," [from a lecture delivered to the Oxford University Junior Scientific Club in Nov. 1909], *Science Progress in the Twentieth Century,* **5** (1910–1911): 60–82, reprinted in *Smithsonian Institution Reports, 1910,* (1911): 487–505; "Dissociation of auricles and ventricles in hibernating dormice," *Journal of Physiology,* **42** (1911): xix–xx; "Frequency of the heartbeat in bats and hedgehogs and the occurrence of heart-block in bats," ibid., **42** (1911): xxi–xxii; "The relation of the electrical to the mechanical reflex response in the frog; with a suggestion as to the significance of Loren's active current rhythm in strychnine preparations," *Quarterly Journal of Experimental Physiology,* **5** (1912): 91–130; "Comparison of the wild duck with the tame duck in regard to 0_2-metabolism, heart size and pulse rate," *Journal of Physiology,* **47** (1913): iv–v; "A method of recording the action current of a single spot of skeletal muscle without injuring any other spot," ibid., **64** (1), (1927): ii–iii (this described work done twenty-two years previously).

79. Sowton's post-1900 papers include the following: (with A. D. Waller) "Action of choline, neurine, muscarine and betaine on isolated nerve and heart," *Proceedings of the Royal Society,* B, **72** (1903): 320–45; (with C. S. Sherrington) "On the dosage of isolated mammalian heart by chloroform," *British Medical Journal,* 2 (1904): 162–8; (with C. S. Sherrington) "Relative effects of chloroform upon the heart and upon other muscular organs," ibid., 2 (1905): 181–7; (with C. S. Sherrington) "On the effect of chloroform in conjunction with carbon dioxide on cardiac and other muscles," ibid., 2 (1906): 85–7; (with W. G. Smith) "Observations on spacial contrast and confluence in visual perception," *British Journal of Psychology,* **2** (1906–1907): 196–210; "Some experiments in the testing of tincture of digitalis," *British Medical Journal,* 1 (1908): 310–14; (with B. Moore and F. W. Baker-Young) "A new member of the saponin digitalin group of glucosides," ibid., 2 (1909): 541–2; (with B. Moore, F. W. Baker-Young, and A. Webster) "On the chemistry and biochemical and physiological properties of a sapo-glucoside obtained from the seeds of *Bassia longifolia,*" *Biochemical Journal,* **5** (1910): 94–125; (with R. Magnus) "Zur Elementarwirkung der Digitaliskörper," *Archive für experimentelle Pathologie und Parmakologie,* **63** (1910): 255–62; (with C. S. Sherrington) "Reversal of the reflex effect of an afferent nerve by altering the character of the electrical stimulus applied," *Proceedings of the Royal Society,* B, **83** (1910–1911): 435–46, and *Zeitschrift für allegemeine Physiologie,* **12** (1911): 485–98; (with C. S. Sherrington) "Chloroform and reversal of reflex effect," *Journal of Physiology,* **42** (1911): 383–8; (with C. S. Sherrington) "On reflex inhibition of the knee flexor," *Proceedings of the Royal Society,* B, **84** (1912): 201–14; (with C. S. Sherrington) "Observations on reflex responses to single break shocks," *Journal of Physiology,* **49** (1915): 331–48; (with A. S. Leyton and H. G. Leyton) "Heart perfusion experiments on anaphylaxis," *Proceedings of the Physiological Society* (1915–1916): xiii; (with A. S. Leyton and H. G. Leyton) "On anaphylactic effects as shown in perfusion experiments on the excised heart," *Journal of Physiology,* **1** (1916): 265–84; (by B. Muscio, assisted by S. C. M. Sowton) "Vocational tests and typewriting," *British Journal of Psychology,* **13** (1922–1923): 344–69; (with C. S. Myers) "Two contributions to the experimental study of the menstrual cycle. I. The influence of the menstrual cycle on mental and muscular efficiency," *Medical Research Council, Report 45,* 1928, pp. 1–42. (For a discussion of the field of industrial fatigue, see Richard Gillespie, "Industrial fatigue and the discipline of physiology," in Geison, ed., *Physiology in the American Context* pp. 237–62, and L. S. Hearnshaw, *A Short History of British Psychology 1840–1940* (London: Methuen, 1964), pp. 178, 247, 276).

No reference to Miss Sowton was found in either University of London or University of Liverpool records, but her name appears in lists of younger physiologists (a number of whom became very well known in their fields), who worked with Sherrington at Liverpool and then Oxford—see Judith P. Swanzey, *Reflexes and Motor Integration: Sherrington's Concept of Integrative Action* (Cambridge, Mass.: Harvard University Press, 1969), p. 171, and Ragnar Granit, *Charles Scott Sherrington: an Appraisal* (New York: Doubleday, 1967), p. 97. At least some of Sowton's work was supported by Royal Society grants. She was one of several British women workers (including Marion Greenwood, Christine Tebb, Julia Divine, and Julia Brink) present at the 4th International Congress of Physiologists held in Cambridge in 1898-*Nature,* **58** (1898): 481–6.

80. Margaret Tomlinson, *Three Generations in the Honiton Lace Trade. A Family History* (Exeter: Devon Print Group, 1983); Harri-

ette Chick, Margaret Hume, and Marjorie Macfarlane, *War on Disease. A History of the Lister Institute* (London: Andre Deutsch, 1971), especially pp. 87–92, 124, 147–60; H. M. Sinclair, "Chick, Harriette," *DNB,* 1971–1980 Supplement, pp. 142–3; Student records, University College London; A[lice] M[ary] Copping, "Dame Harriette Chick," *British Journal of Nutrition,* **39** (1978): 3–4.

81. Three of the six Chick sisters who attended University College London became scientists (see sketch of Edith Chick in chapter 1).

82. Harriette Chick's early papers include, "Sterilisierung von Milch durch Wasserstoffsuperoxyd," *Centralblatt für Bakteriologie, Parasitenkunde und Infektionskrankenheit,* Abt. 2, **7** (1901): 705–17 and "A study of the process of nitrification with reference to the purification of sewage," *Proceedings of the Royal Society,* B, **77** (1906): 241–66. Then came the "laws of disinfection" papers, "An investigation of the laws of disinfection," *Journal of Hygiene,* **8** (1908): 92–153; (with C. J. Martin) "The principles involved in the standardization of disinfectants and the influence of organic matter upon germicidal value," ibid., **8** (1908): 654–97; (with C. J. Martin) "A comparison of the power of a germicide emulsified or dissolved, with an interpretation of the superiority of the emulsified form," ibid., **8** (1908): 698–703; "The factors conditioning the velocity of disinfection," *Reports of the Eighth International Congress of Applied Chemistry, Appendix Sect. 6a,* Sect. 11b, **26** (1912): 167–97. Chick's heat coagulation of proteins and solution structure papers were the following: (with C. J. Martin) "The 'heat coagulation' of proteins," *Journal of Physiology,* **40** (1910): 404–30; (with C. J. Martin) "The heat coagulation of proteins. II. The action of hot water on egg albumin and the influence of acid and salts upon reaction velocity," ibid., **43** (1912): 1–27; (with C. J. Martin) "The heat coagulation of proteins. III. The influence of alkali upon reaction velocity," ibid., **45** (1913): 61–9; (with C. J. Martin) "The heat coagulation of proteins. IV. The conditions controlling the agglutination of protein already acted upon by hot water," ibid., **45** (1913): 261–95; (with C.J. Martin) "The viscosity of casein solutions," *Zeitschrift für Chemie und Industrie der Kolloide,* **11** (1913): 102–5; (with C. J. Martin) "The density and solution volume of some proteins," ibid., **11** (1914): 209–26, and *Biochemical Journal,* **7** (1913): 92–6; (with C. J. Martin) "The precipitation of egg-albumin by ammonium sulphate. A contribution to the theory of 'salting out' of proteins," ibid., **7** (1914): 380–98; "The apparent formation of euglobulin from pseudo-globulin and a suggestion as to the relationship between these two proteins in serum," ibid., **8** (1914): 404–20.

83. See the following: (with E. Margaret Hume) "Distribution among foodstuffs (especially those suitable for the rationing of armies) of the substances required for the prevention of beriberi and scurvy," *Journal of the Royal Army Medical Corps,* **29** (1917): 121–59; (with E. M. Hume) "Distribution in wheat, rice and maize grains of the substance, the deficiency of which in a diet causes polyneuritis in birds and beri-beri in man," *Proceedings of the Royal Society,* B, **90** (1917): 44–60; (with E. M. Hume) "Effect of exposure to temperatures at or above 100° upon the substance (vitamin) whose deficiency in a diet causes polyneuritis in birds and beri-beri in man," ibid., B, **90** (1917): 60–8; (with E. M. Hume and Ruth F. Skelton) "The antiscorbutic value of milk in infant feeding," *Lancet,* I (1918): 1–2; (with E. M. Hume and R. F. Skelton) "The relative content of antiscorbutic principle in limes and lemons," ibid., II (1918): 735–8; (with Mabel Rhodes) "The antiscorbutic value of the raw juices of root vegetables with a view to their adoption as an adjunct to the dietary of infants," ibid, II (1918): 774–5; (with E. M. Hume) "The importance of accurate and quantitative measurements in experimental work on nutrition and accessory food factors," *Journal of Biological Chemistry,* **39** (1919): 203–7; (with E. M. Delf) "Antiscorbutic value of dry and germinated seeds," *Biochemical Journal,* 13 (1919): 199–218; (with F. G. Hopkins) "Accessory factors in food," *Lancet,* II (1918): 28–9; (with Mabel E. D. Campbell) "The antiscorbutic and growth promoting value of canned vegetables," ibid, II (1919): 320–2; (with E. M. Hume and R. F. Skelton) "The antiscorbutic value of some Indian dried fruits: (a) tamarind, (b) cocum and (c) mango ('Amchur')," ibid, II (1919): 322–3; *Report on the Present State of Knowledge of Accessory Food Factors* (*Vitamins*), compiled by a committee appointed jointly by the Lister Institute and the Medical Research Council, (London: Medical Research Council, 1919).

84. Publications resulting from the Chick team's work in Vienna and from their continued studies on the fat-soluble vitamins thereafter include, "Role of vitamins in nutrition," *Wiener medizinische Wochenschrift,* **70** (1920): 411–9; (with Elsie J. Dalyell) "Epidemic scurvy among children," *Zeitschrift für Kinderheilkunde,* **26** (1920): 257–69; (with E. J. Dalyell) "Present position of vitamins in clinical medicine," *British Medical Journal,* II (1920): 151–4; (with E. J. Dalyell) "Hunger osteomalacia in Vienna, 1921. I. Its relation to diet," *Lancet,* II (1921): 842–9; (with E. J. Dalyell) "Influence of foods rich in accessory factors in stimulating development in backward children," *British Medical Journal,* II (1921): 1061–6; (with E. J. Dalyell, E. M. Hume, H. M. M. Mackay, H. H. Smith, and Hans Wunberger) "The etiology of rickets in infants: prophylactic and curative observations at the Vienna Universitäts Kinderklinik," *Lancet,* II (1922): 7–11, and *Zeitschrift für Kinderheilkunde,* **34,** (1923): 75–93; (with M. A. Boas) "Influence of diet and management of the cow upon the deposition of calcium in rats receiving a daily ration of the milk in their diet," *Biochemical Journal,* **18** (1924): 433–47; (with Mary Tazellar) "Note upon the effect of the growth of rats, receiving a diet deficient in fat-soluble vitamins, of exposing their environment to the emanation from radium bromide," ibid, **18** (1924): 1346–8; (with E. Mellanby, H. Gray, R. McCarrison, and W. H. Kink) "Discussion of nutritional diseases in animals," *Proceedings of the Royal Society of Medicine,* **17** (1924): 19–30; "Sources of error in the technique employed for the biological assay of fat-soluble vitamins," *Biochemical Journal,* **20** (1926): 119–30; (with H. H. Smith) "Maintenance of a standardized breed of young rats for work upon fat-soluble vitamins, with particular reference to the endowment of the offspring," ibid, **20** (1926): 131–6; (with Margaret Honora Roscoe) "Antirachitic value of fresh spinach," ibid, **20** (1926): 137–52; (with Vladimir Korenchevskii and M. H. Roscoe) "Difference in chemical composition of the skeletons of young rats fed (1) on diets deprived of fat-soluble vitamins, and (2) on a low-phosphorus rachitic diet, compared with those of normally nourished animals of the same age," ibid, **20** (1926): 622–31; (with M. H. Roscoe) "Influence of diet and sunlight upon the amount of vitamin A and vitamin D in the milk afforded by a cow," ibid, **20** (1926): 632–49; (with F. G. Hopkins) "Estimation of vitamin A in cod-liver oil. Comparison between the colorimetric (Rosenheim, Drummond) and the biological methods," *Lancet,* I (1928): 148–50; (with E. M. Hume) "Standardization and determination of vitamin A," *Medical Research Council* (*Britain*), *Special Report,* Ser. no. 202 (1935), 61 pp.

85. (With E. M. Hume) "Production in monkeys of symptoms closely resembling those of pellagra, by prolonged feeding on a diet of low protein content," *Biochemical Journal,* **14** (1920): 135–46; (with M. H. Roscoe) "Dual nature of the water-soluble vitamin B. II. The effect upon young rats of vitamin B_2 deficiency and a method for the biological assay of vitamin B_2," ibid, **22** (1928): 790–9; (with M. H. Roscoe) "A method for the assay of the antineuritic vitamin B, in which the growth of young rats is used as a criterion," ibid., **23** (1929):

493–503; "Effect on vitamin B_2 of treatment with nitrous acid," ibid, **23** (1929): 514–6; (with M. H. Roscoe) "Heat stability of the (anti-dermatitis, 'anti-pellagra') water-soluble vitamin B_2," ibid, **24** (1930): 105–12; (with Alice Mary Copping) "Heat stability of the (anti-dermatitis, 'anti-pellagra') water-soluble vitamin B_2," ibid, **24** (1930): 932–8; (with A. M. Copping) "Alcohol solubility of the anti-dermatitis more heat-stable vitamin B_2 constituent of the vitamin B complex," ibid, 24 (1930): 1744–7; (with A. M. Copping and M. H. Roscoe) "Egg white as a source of the anti-dermatitis vitamin B_2," ibid, 24 (1930): 1748–53; (with A. M. Copping) "The composite nature of the water-soluble vitamin B_1. III. Dietary factors in addition to the anti-neuritic vitamin B_1, and the anti-dermatitis vitamin B_2," ibid, **24** (1930): 1764–79; (with Hester Mary Jackson) "Note on the international standard for the antineuritic vitamin B_1," ibid, **26** (1932): 1223–6; "Current theories of the etiology of pellagra," *Lancet,* II (1933): 341–6; (with A. M. Copping and Constance E. Edgar) "The water-soluble B vitamins. IV. The components of vitamin B_2," *Biochemical Journal,* **29** (1935): 722–34; (with Thomas W. Birch and C. J. Martin) "Experiments with pigs on a pellagra-producing diet," ibid, **31** (1937): 2065–79; (with E. Mellanby and E. M. Hume) "A report on the adoption of a new international reference substance for vitamin B_1 and the definition of the existing unit in terms of this reference substance," *League of Nations, Organisation d'hygiène. Bulletin,* **7** (1938): 942–4; (with Thomas F. Macrae, Archer J. P. Martin, and C. J. Martin) "Curative action of nicotinic acid on pigs suffering from the effects of a diet consisting largely of maize," *Biochemical Journal,* **32** (1938): 10–12; (with T. F. Macrae, A. J. P. Martin and C. J. Martin) "The water-soluble B vitamins other than aneurin (vitamin B_1), riboflavin and nicotinic acid required by the pig," ibid, **32** (1938): 2207–24; "Nutritive value of white flour with vitamin B_1 added, and of whole-meal flour," *Lancet,* II (1940): 511–12; (with T. F. Macrae and Alastair N. Worden) "Relation of skin lesions in the rat to deficiency in the diet of different B_2 vitamins," *Biochemical Journal,* **34** (1940): 580–94; (with M. M. El Sadr and A. N. Worden) "Occurrence of fits of an epileptic nature in rats maintained for long periods on a diet deprived of vitamin B_6," ibid, **34** (1940): 595–600; "Causation of pellagra," *Nutrition Abstracts and Reviews,* **20** (1951): 523–35 (review); (with E. M. Hume) "Recent work on vitamins. The work of the Accessory Food Factors Committee," *British Medical Bulletin,* **12** (1956): 1–2.

86. Chick's publications in these areas include, (with John C. D. Hutchinson and H. M. Jackson) "The biological value of proteins. VI. Further investigation of the balance-sheet method," *Biochemical Journal,* **29** (1935): 1702–11; (with M. A. Boas-Fixen, J. C. D. Hutchinson, and H. M. Jackson) "The biological value of proteins. VII. The influence of variation in the level of protein in the diet, and of heating proteins on its biological value," ibid, **29** (1935): 1712–19; "Nutritive value of milk," *Dairy Industries,* **4** (1939): 166–7; "Nutritive value of the potato," *Chemistry and Industry* (1940): 737–9; (with Philippe Ellinger) "The photosensitizing action of buckwheat (*Fagopyrum esculentum*)," *Journal of Physiology,* **100** (1941): 212–30; "Biological value of proteins contained in wheat flours," *Lancet,* I (1942): 405–8; (with Margery E. M. Cutting) "Nutritive value of the nitrogenous substances in the potato as measured by their capacity to support growth in young rats," ibid, II (1943): 667–9; (with E. B. Slack) "Malted foods for babies. Trials with young rats," ibid, **251** (1946): 601–3; "Nutritive value of proteins contained in wheat flours of different degrees of extraction," *Proceedings of the Nutrition Society,* **4** (1946): 6–9; (with M. E. B. Cutting, C. J. Martin and E. B. Slack) "Digestibility and nutritive value of the nitrogenous constituents of wheat bran," *British Journal of Nutrition,* **1** (1947): 161–82; "Note on methods of determining the nutritive value of proteins," *Chemistry and Industry* (1947): 318–20 (review); (with E. B. Slack) "Nutritive value of proteins contained in wheat flours of different extraction rates," *British Journal of Nutrition,* **2** (1948): 205–13; (with E. B. Slack) "Distribution and nutritive value of the nitrogenous substances in the potato," *Biochemical Journal,* **45** (1949): 211–21; (with E. B. Slack) "Nutritive value of the nitrogenous substances in the potato," *Abstracts and Communications, First International Congress of Biochemistry* (1949): 97–8; "The protein requirement of man," *Pharmazie,* **9** (1954): 452–5 (review).

87. *Who was Who,* vol. 4, p. 405; W. E. Garner, "Percy Faraday Frankland," *Journal of the Chemical Society,* **151** (1948): 1996–2005; *DNB,* vol. 57, pp. 138–9, entry for Toynbee, Joseph.

88. Garner, "Percy Faraday Frankland," p. 1998.

89. One might speculate that Mrs. Hart was the wife of Ernest Abraham Hart (1835–1898), physician, reformer in medicine and public health, medical journalist and editor, and a strong supporter of medical education for women (see the *DNB,* 1901 Supplement, vol. 2, pp. 396–7). Hart's second wife, whom he married in 1872, was Alice, daughter of A. W. Rowlands of Lower Sydenham.

90. Kohler, *Medical Chemistry to Biochemistry,* p. 42.

91. Geison, *Michael Foster,* pp. 306–9. One other notable woman physiologist/biochemist who worked in Foster's laboratory in the 1890s was Christine Tebb (see chapter 11).

92. Kohler, *Medical Chemistry to Biochemistry,* p. 42.

93. Edward Albert Schäfer, *History of the Physiological Society during its First Fifty Years, 1876–1926* (London: Cambridge University Press, 1927), p. 10.

94. Ibid., pp. 114, 135–6.

95. The other four were Ruth F. Skelton, Winifred Cullis (later of the London School of Medicine for Women), Constance L. Terry, and Enid Tribe (London School of Medicine for Women, later Mrs. Oppenheimer, of New York) ibid., p. 154.

96. Eager to establish themselves as a community of *experimental* biologists, the American group considered it necessary to distance itself from the part-time clinicians who taught physiology as purely lecture courses in medical schools (see Pauly, "The appearance of academic biology").

97. Toby A. Appel, "The first quarter century 1887–1912," in *History of the American Physiological Society: the First Century, 1887–1987,* eds. John R. Brobeck, Orr E. Reynolds, Toby A. Appel (Bethesda, Md.: American Physiological Society, 1987), pp. 31–62, on p. 33, and Toby A. Appel, Marie M. Cassidy, M. Elizabeth Tidball, "Women in physiology," ibid., pp. 381–90, Table 1, p. 382.

98. Note that microscopy papers in the medical sciences (12 percent of the American contribution—Figure 6-2 b) are only some of the papers listed in the United States Microscopy section of the bibliography. Others concerned more general biological topics.

Chapter 7

FROM OBSTETRICS TO WAR WORK: PHYSICIANS AND OTHER MEDICAL WOMEN

Women in the medical sciences have been grouped into categories (physiologists, bacteriologists, etc.) according to where the greatest emphasis in their published work appears to fall and according to what could be found out about their overall careers. However, both the spread of interests in individuals and the overlap between categories can be so great that the placing of any one person in a particular grouping is often somewhat arbitrary. Of the nineteenth-century medical women from all countries who published journal articles and who would appear to have been first and foremost physicians, about 40 percent were American, 18 percent Russian, and 16 percent British. The Americans produced about 44 percent of the papers indexed by the Royal Society that have been classed as medicine in the bibliography, and the British 9 percent. Another 20 percent was contributed by Polish and Russian women, and the rest by a number of west European workers.

American women

Of the twenty-seven Americans included here, twenty-two worked mainly in the eastern cities, particularly New York, three spent at least part of their careers in the Midwest, one practiced in San Francisco, and one was a medical missionary in China. Gynecology, children's diseases, and organizational work in fields such as public health were the areas in which they most commonly made their careers. The University of Michigan, the Woman's Medical College of the New York Infirmary for Women and Children, and the Woman's Medical College of Pennsylvania were the three American institutions most prominent in their training. Eight of the Americans—Cushier, Griffith Davis, Putnam Jacobi, Dixon Jones, McNutt, Mergler, Blair Moody, and Mosher—took their M.D. degrees in the 1870s, and of these eight pioneers, four—Putnam Jacobi, Cushier, Mergler, and Mosher—went on to especially notable careers.[1] Nine women (Bryson, Gage Day, Hinds, Moody, Niles, Peckham, Post, Robinovitch, and Sargent) probably all began their professional work in the 1880s. The group of ten who graduated between 1890 and 1900 (Baldwin, Bishop, Bloom, Mitchell Kydd, Lewi, McGee, Mitchell, Blackwell, Sherwood, and Stone) includes three (McGee, Sherwood, and Stone) who made key contributions to the founding and organization of state and national medical organizations.

MARY PUTNAM JACOBI[2] (1842–1906) was undoubtedly one the most prominent of the early women physicians in the United States, especially as measured by the breadth and scope of her research and writings. She was born in London, on 31 August 1842, the oldest of the eleven children of George Palmer Putnam, founder of the publishing firm of G. P. Putnam's Sons, and his wife Victorine (Haven). Both the elder Putnams were New Englanders. The family returned to the United States in 1848 and most of Mary's childhood was spent at Stapleton, on Staten Island. Apart from a year at a private school in Yonkers, New York, and two years (1857–1859) at a public high school in Manhattan, she was educated at home by her mother and by a private tutor who taught her Greek.

Even in her childhood she was something of a leader among her contemporaries. Intellectually precocious, she had considerable literary talent and published her first article in the *Atlantic Monthly* in 1860 when she was seventeen. In 1861, at age nineteen, she enrolled at the New York College of Pharmacy as its first woman student. She graduated from the two-year course in 1863 and then entered the Woman's Medical College of Pennsylvania, an institution opened in 1850 largely as a result of Quaker benevolence and the women's reform movement in the city of Philadelphia. Like most American medical schools of the antebellum period, the training it offered was minimal.[3] Putnam was awarded an M.D. in 1864. Thereafter she worked for short stretches in various hospitals, including the New England Hospital for Women and Children in Boston and soldiers' hospitals in New York City, the Civil War being still in progress. In the spring of 1865 she became engaged to Ferdinand Mayer, a German immigrant, then professor of chemistry at the New York College of Pharmacy; he was probably forty years her senior. The following year, however, after breaking off the relationship with Mayer, she went to Paris, hoping to pursue her medical education at the École de Médecine.

She remained in France for almost five years, experiencing the siege of the city by the Prussian Army in 1870 and the political turmoil that followed. For two years she attended clinics, lectures, and laboratory sessions, and in 1868 was formally admitted (by the Minister of Public Education over the objections of the faculty) to the École de Médecine. She was not the first woman to receive a medical degree from that institution, however, the honor going to Elizabeth Garrett of London, who qualified by passing the final examination in 1870. Nevertheless Putnam, by having an outstanding academic record, helped considerably in winning acceptance for women students. She was awarded a bronze medal for her graduating thesis, completed in 1871. During her stay in France she sent a number of papers on French medical developments to the American *Medical Record* and also contributed articles on French news and politics to American newspapers and magazines. Some of the letters she wrote to her family describing wartime life in Paris were published in the *American Historical Review* in 1917.[4]

Within a relatively short time of her return to the United States in 1871 she was recognized as one of the outstanding physicians in the country. Even among the men, few had better formal training or more clinical experience. She quickly established a successful private practice and combined this with a considerable amount of voluntary hospital work; she also joined the faculty of the Woman's Medical College of the New York Infirmary for Women and Children, the institution opened in 1868 by Elizabeth and Emily Blackwell and Marie Zakrzewska.[5] Here the exceptionally high standards she insisted on initially provoked considerable complaint, but she stayed on as professor of materia medica and therapeutics for eighteen years. She taught or served as consulting physician at several other medical colleges and hospitals in New York City as well. From 1882 until 1885 she lectured on diseases of children at the New York Post-Graduate Medical School. She had a long affiliation with Mt. Sinai Hospital, and, beginning in 1893, she was also visiting physician at St. Mark's Hospital. Two of her major concerns, matters that influenced all her teaching work, were the raising of standards, particularly in the medical education of women, and the overcoming of skepticism about women's ability to practice medicine. In the latter she played a central role. The Association for the Advancement of Medical Education for Women, which she had organized in 1872, later became the Woman's Medical Association of New York City. She served as its president for almost three decades, from 1874 until 1903.

In addition to publishing more than 120 papers in the medical journals on a broad spectrum of topics including pathology, neurology, pediatrics, physiology, and medical education, she wrote a number of books, including one on educational theory, *Physiological Notes on Primary Education and the Study of Language* (1889). She was a pioneer in raising concern for environmental conditions as a health issue, and one of the founders, for a time vice president, of what became the New York Consumers' League. From 1894 onward she worked in the women's suffrage movement.

Short and stocky in build, she did not consider herself good-looking, though she had striking features and penetrating eyes. In 1873 she married Dr. Abraham Jacobi, a German immigrant, and a widower twelve years her senior. Jacobi had taken his medical degree at the University of Bonn in 1851 and came to the United States after escaping imprisonment for revolutionary activities in Germany. He had practiced in New York City since 1853, and by the time of his marriage to Mary Putnam was clinical professor of diseases of infancy and childhood at the College of Physicians and Surgeons and one of the leading pediatricians in America. A considerable number of Mary's professional undertakings after her marriage were carried out in collaboration with him, including the establishment of a permanent pediatric dispensary at Mt. Sinai Hospital and the opening in 1886 of a separate small children's ward at the New York Infirmary for Women and Children. Indeed, it would seem likely that the support and active interest of her husband helped Mary significantly in her medical work.

She received many professional honors, including that of being the first woman admitted to the New York Academy of Medicine. In 1876 she submitted an entry to Harvard's Boylston Prize competition. Her essay on "The question of rest for women during menstruation" attacked the contemporary idea that menstruation in women is a weakness. From careful survey investigations she had concluded that women do not need extra rest while menstruating, and cannot be designated inferior to men simply on the basis of such a requirement. The Boylston Prize had never been given to a woman, but entries were submitted anonymously; Jacobi's having been judged the best, the tradition was broken and she received the prize. A French translation of the essay, "Théorie de la menstruation," was published in the *Revue médico-chirurgicale des maladies des femmes* in 1880.

The Jacobis had three children, of whom only the third, Marjorie, born in 1878, survived past childhood. They lived in New York City most of the year, but spent their summers at Lake George, in upper New York State. Mary's ideal vacation was said to have been to get up at daybreak and write and read for the rest of the day. She died in New York City of a meningeal tumor on 10 June 1906, at the age of sixty-three, following several years of increasingly severe illness.

ELIZABETH CUSHIER[6] (1837–1932), one of Putnam Jacobi's colleagues at the New York Infirmary for Women and Children, graduated from the Woman's Medical College of that institution in 1872. Born on 25 November 1837 into a family of eleven children of whom seven survived to adulthood, she grew up in New York and New Jersey. She received some basic instruction in a private school and later attended public schools, but supplemented her early education with a good deal of self-directed study, in arithmetic, French, and, most especially, English literature. In 1868, when she was thirty-one, the chance reading of an article on a physiological topic awakened her interest in medicine, and she entered a New York homeopathic college for women. After a year she transferred to the Woman's Medical College, and, following her graduation, stayed on at the New York Infirmary for Women and Children, first as an intern and later as staff physician and surgeon specializing in gynecology. She was also particularly interested in normal and pathological histology, and since in the early 1870s no laboratories in the United States accepted women research students in that area, she spent eighteen months carrying out postgraduate study at the University of Zurich.

As a resident physician at the Infirmary for Women and Children, Cushier built up a well-regarded department of gynecological surgery and was widely recognized as the leading woman surgeon of her time. At that period the general academic standards at the New York Woman's Medical College were exceptional for an American institution, and clinical training in Cushier's specialty of gynecology and obstetrics was particularly well organized. In fact no other school in New York City offered its students comparable practical training in these areas.[7] Men's medical colleges then typically made no arrangements for their students to see even one maternity case before graduation, but each senior student at the Woman's Medical College assisted at twelve such cases under the direction of the physician in charge. Cushier's publications, which appeared mainly during the 1880s, generally concerned gynecological procedures.

She became a close friend of Emily Blackwell, who along with her sister Elizabeth, had founded the infirmary in 1857. From 1882 she and Emily shared a house in New York City, where she conducted her eminently successful private practice. Eleven years later they acquired a summer home as well near York Cliffs on the Maine Coast. Both retired from practice in 1900, shortly after the closing of the Woman's Medical College; Cushier was then sixty-two and Blackwell eleven years her senior. After an extended visit to Europe, with eighteen months in Italy and Sicily and a summer in the Austrian Tyrol, they continued to share a home, spending their summers in Maine, until Blackwell's death in 1910. In the years that followed, Cushier, though then in her seventies, felt the lack of an occupation, and with the outbreak of the First World War undertook relief work for France and Belgium and then volunteered her services to the Red Cross. She died on 25 November 1932, on her ninety-fifth birthday.

SARAH MCNUTT[8] (b. 1847?) was also a graduate of the Woman's Medical College of the New York Infirmary for Women and Children. After receiving her M.D. in 1877, she stayed on for over twenty years as assistant to Emily Blackwell, then head of medical work. She also served as attending physician at the infirmary and instructor in the college's department of gynecology. In the early 1880s she held, in addition, a faculty position at the New York Post-Graduate College, where she taught courses on diseases of children and child surgery to both men and women physicians. In 1883 she founded a babies' ward at the Post-Graduate College, the first place where children under the age of two could be admitted to hospital care without their mothers accompanying them. Five years later she and her sister, Julia McNutt, an 1883 graduate of the Woman's College of the New York Infirmary, took a major role in organizing the Babies' Hospital of the City of New York, established under a charter from the State Legislature. During the first year of its operation, the Doctors McNutt were responsible for the medical care this hospital provided, but because of the rapidly increasing demands of the work, which left them little time for their private practices, they later withdrew. Sarah McNutt continued to practice in New York City until at least 1918, latterly limiting her work to gynecology.

MARIE MERGLER[9] (1851–1901), a surgeon and specialist in gynecology, was born in Mainstockheim, Bavaria, on 18 May 1851. She was the youngest of the three children of Henriette (von Ritterhausen) and Francis Mergler. Mergler was a physician who had trained at the University of Wurzburg. The family moved to America and settled in Palatine, Cook County, Illinois, in 1853. Marie attended a public school for a short time, but got most of her early education from her parents, especially her mother.

Though her father encouraged her interest in his medical work, like many young American women of her time who wanted independence, she trained first as a teacher. After graduating from the Cook County Normal School in 1869, she attended the State Normal School in Oswego, New York, where she took the classical course and graduated in 1872. However, after four years on the staff of the Englewood High School in suburban Chicago she came to feel that teaching was for her a less than wholly absorbing occupation. She graduated from the Woman's Hospital Medical College of Chicago (later merged with Northwestern University) as the class valedictorian in 1879, and took a competitive examination for an appointment to the medical staff of the Cook County Insane Asylum at Dunning, Illinois. Though she ranked second she failed to get a job; the work was pronounced unsuitable for a woman doctor. With no immediate prospect of employment, she took instead a year of postgraduate study in pathology and clinical medicine at the University of Zurich (1880–1881). On her return she set up a practice in Chicago and joined the noted Chicago gynecologist William Byford as his surgical assistant at the Woman's Hospital.

The positions she held in Chicago and Cook County medical schools and hospitals during the twenty years from 1881 until her early death in 1901 form an impressive list. In 1882 she accepted the post of attending physician at Cook County Hospital, becoming the second woman on its staff; at the same time, she was professor of materia medica and adjunct professor of gynecology at her alma mater, the Woman's Hospital Medical College of Chicago; in 1886 she became surgeon at the Woman's Hospital, and on the death of Byford in 1890 succeeded him as professor of gynecology at the Woman's Hospital Medical College; in 1895 she became head physician and surgeon at the Chicago Woman's Hospital, and professor of gynecology at the Post-Graduate Medical School of Chicago. Throughout these years medical education for women was a constant concern; from 1899 she served as dean of the faculty of the Woman's Hospital Medical College.

Her papers in the medical journals were chiefly on gynecological topics. She also wrote a monograph, *A Guide to the Study of Gynecology,* published in 1893.

A good diagnostician and an excellent surgeon, remembered by her colleagues for her steady hand, unerring judgment, and unruffled demeanor, Mergler was one of the most distinguished women surgeons in America in the late nineteenth century. She died in Los Angeles of pernicious anemia, at the height of her career, on 17 May 1901, a day before her fiftieth birthday.

Eliza Mosher[10] (1846–1928), the sixth and youngest child in the Quaker farming family of Maria (Sutton) and Augustus Mosher, was born in Cayuga County, western New York State, on 2 October 1846. Among her forbears were Welsh and English settlers who had come to Plymouth Colony, Massachusetts, in the seventeenth century.

She was educated in local schools and in the boarding department of the Friends' Academy in Union Springs, New York. From her girlhood on she was drawn to the care of the sick, and she had considerable experience in nursing, her father, three brothers, and a sister having died of tuberculosis. Her plan to become a doctor initially met with strong opposition, her mother declaring she would as soon shut her up in a lunatic asylum as allow her to study medicine.[11] Nevertheless, in 1869 she entered the New England Hospital for Women and Children in Boston as an untrained doctor's assistant. Her first experience was "on the job." She saw diseased conditions, was present at operations, and assisted at births before she began her studies in the Medical Department of the University of Michigan, which she entered in 1871, one year after it first opened to women. She was enrolled there for four years, spending one of the four at clinics in New York City hospitals in order to gain the wider experience these provided. The strong bent for anatomy that was to mark all her later work came to the fore early in her student career, and in her second year at Michigan she was appointed assistant demonstrator in that subject for the women's class and "quiz master" for the first-year class. On occasion she was even called to demonstrate specimens to the men's class, an unusual assignment for a woman assistant at that time.

Following graduation in 1875 she and her classmate Elizabeth Gerow started a private practice in Poughkeepsie, New York, but within less than two years she accepted the post of resident physician at the new Reformatory Prison for Women at Sherborn, Massachusetts. She was there for two years, organizing and equipping a prison dispensary and hospital, and herself doing all the work in medicine, surgery, obstetrics, and even dentistry that the prison population of 500 women inmates required. The organizational work she accomplished was pioneering. It is unlikely that any woman in the country had taken on an equivalent task before then. The hospital she developed even had a maternity and nursery department, the latter frequently caring for more than sixty infants.

In 1879 she went to Europe for more than a year of postgraduate study at clinics in London and Paris, and on her return expected to go back to private practice. However, at this point she was asked by John D. Long, governor of Massachusetts, and evidently a man of progressive outlook, to take the position of superintendent at the Sherborn reformatory. He felt she was the only woman he could appoint, and that her acceptance was crucial to the advancement of professional women in Massachusetts.[12] She took the job, one that required a commanding personality and considerable personal courage, qualities that, fortunately, she had in good measure. She was remembered as being remarkably successful, capable of dealing quietly and effectively with even the most recalcitrant prisoners, who were at times given to considerable violence. She liked children and understood them well. It is said that she adopted a daughter from among the prisoners at the reformatory, but little is know about the girl.[13]

In 1883 a severe injury to her knee led to her resignation from prison work. For the next seven years she was in constant pain, dependent on crutches and often confined to bed for long stretches of time, until she herself devised an operation that cured the problem. Despite the handicap, during this period she lectured for two semesters at Wellesley College, and along with Dr. Lucy Hall, a fellow Michigan graduate who had been one of her assistants at the Sherborn prison, established a joint practice in Brooklyn, New York. From 1883 until 1887 she and Hall also held jointly the position of resident physician and associate professor of physiology and hygiene at Vassar College, spending alternate semesters in residence. Here Mosher's talents for institutional organization again came to the fore. She started a program of careful physical examination of first-year students, combined with systematic medical record keeping that was probably the first of its kind at a women's college in the United States. She also modernized the physical education program and introduced the divided skirt, a marked improvement in gymnasium dress for women.

Returning to full-time private practice in 1887, she quickly became prominent in medical circles in Brooklyn. She founded the medical training course for mission workers at the Union Missionary Institute in Brooklyn in 1888, and was to lecture there for many years without payment. In 1889 she joined the board of directors of the Chautauqua Summer School for teachers of physical education and for the next two decades served the school as examining physician for women students as well as lecturer on anatomy and hygiene.

In 1896 Mosher once again, with great reluctance, gave up her private practice for a period of administrative work, this time at the University of Michigan, which asked her to be its first dean of women, thereby establishing the position. Simultaneously she was professor of hygiene, resident physician for women students, and director of physical education for women—in all a substantial undertaking. Her faculty appointment was the first for a woman in the undergraduate college of the University of Michigan. Women students had been in residence there for twenty-six years without benefit of the formal supervision of a dean, and some of the requirements Mosher instituted, including participation in the physical education program, took some time to gain acceptance. Her lectures on hygiene covered both personal and community aspects of the subject. She also taught a course on home economics. In 1902 she returned to her Brooklyn practice, but from then on, over the years, continued to teach part-time at a number of educational institutions, including Mount Holyoke College, the Pratt Institute, and Adelphi College in Brooklyn. She was also active in many civic and medical organizations and women's clubs, served as president of the Women's Medical Society of New York, and was senior editor of the *Medical Woman's Journal* for twenty-three years (1905–1928).

She published a number of articles in the medical journals

and in more general magazines. These covered a variety of topics from the health of female prison inmates to an examination of the muscle control of the knee joint; several papers discussed the effects of poor posture in children and young people, a subject in which she was especially interested. Her book, *Health and Happiness: a Message to Girls* (1912), was well-received. She remained in practice until a few months before her death, her career in medicine spanning the remarkable period of fifty-three years; for a time she was probably the oldest active woman physician in America. Her most important contribution is generally considered to have been her pioneering work in the development of health services at educational institutions, including the introduction of systematic record keeping and the promotion of programs in physical education and preventive medicine. She did much to demonstrate the falseness of the claim that higher education is unhealthy for women, although her work also reinforced the idea (perhaps now once more coming to the fore) that women's health requires special attention. She died of pneumonia and cerebral thrombosis, on 16 October 1928, at age eighty-two.

MARY BLAIR MOODY[14] (1837–1919), who for much of her life divided her interests between medical work and natural history, has the distinction of being the first woman member of the Association of American Anatomists. She joined the organization in 1894, her son Robert, then an assistant in pathology at Yale University's medical school, being elected at the same time.

She was born in Barker, New York, on 8 August 1837. After taking an M.D. at the University of Buffalo in 1876 she founded the Women's and Children's Dispensary in Buffalo and served as its senior physician from 1882 until 1886. She published one paper in clinical gynecology in the *Medical Press of Western New York* in 1888 and also wrote book reviews for that journal. An associate editor of the *Buffalo Naturalists' Field Club Bulletin,* she held memberships in the Forestry Association and the Association for the Protection of the Adirondacks. Her note on the singing of birds appeared in *Science* in 1893. She lived for several years in New Haven, Connecticut, spending winters in Pasadena, California, where she studied the local flora, although she does not appear to have published in any of the better-known botanical journals. By 1916 she had moved to Berkeley, California, where her son and daughter-in-law, Agnes Claypole Moody (see chapter 3), then lived.

Readily available biographical information about several of the women physicians who graduated in the 1880s is rather sketchy. However, one whose career was notable though not now widely remembered was MARY GAGE DAY[15] (1857–1935). Mary was born on 20 June 1857 in the farming community of Worcester, Otsego County, central New York State, the seventh of the nine children of Lucy Ann (Grover) and Henry Van Tassel Gage. Her forbears on both sides were of British and Dutch extraction. She received her early education in Otsego County public schools, which had the reputation of being very good at the time, and she later taught for a period in these schools. Like her fellow New Yorker Eliza Mosher, she was strongly impressed at an early age by the toll taken by the frequent epidemics of diphtheria and typhoid, and the ever-present menace of tuberculosis. Medical studies became her ambition. By the mid 1880s she had married Edgar Day and together they went out to Ford County in southwestern Kansas to earn money for her further education. Day joined the laborers on an irrigation project that was to supply the region with water from the Arkansas river; Mary boarded some of the men. They lived the life of prairie pioneers, in a camp of sod houses and tents, with minimal protection from the elements for themselves and their horses. Preserved among her letters to her family is a graphic description of a prairie blizzard in 1886, in which two of Edgar Day's fellow workmen died of exposure after being caught in blinding snow and intense cold.[16]

Before enrolling in the department of medicine at the University of Michigan she spent one year at Cornell University, where her older brother Simon Henry Gage already held a faculty position teaching histology and embryology. Her studies at Cornell in physiology, zoology, and anatomy gave her a year's advanced standing at Michigan and she took her M.D. in 1888. Following further studies at the New York Post-Graduate Medical School and an internship at the Michigan State Institution for Dependent Children, she returned to southwestern Kansas and practiced medicine for six years in the city of Wichita and the surrounding prairies.

Having watched research being carried out at both Cornell and Michigan, she was not slow to take up an original investigation of her own. The problem she looked into, the "loco weed" disease, was a sickness affecting grazing animals east of the Rocky Mountains from Canada to Mexico. For several decades it had been a major concern of American stockmen. Indeed, by the turn of the century losses of cattle and horses were so great that bankruptcies ensued. The symptoms in the afflicted animals included staggering, uncertain gait, strange expression, and emaciation, followed eventually by convulsions and death. The cause was thought by some to be leguminous plants, the chief suspects being *Astragalus mollissimus* and the related genus *Oxytropis lamberti,* which range animals grazing in semiarid regions were known to eat, especially during periods of drought and grass shortage. However, different studies of the effects of ingestion of these plants had given different results, data of various investigators could not be corroborated, and most "locoed" animals lived in an unfavorable environment anyway, where food was often scanty, parasitic infestation frequent, and survival precarious.

Over the course of a year and a half Gage Day carried out a series of experiments involving water extraction of the suspect plants followed by tests of the concentrated extracts on cats and jackrabbits. The test animals developed the same symptoms as the locoed range cattle and horses. Though she succeeded in isolating from the water extract a crystalline material that she considered to be the poison, she had neither the facilities, funds, nor time to pursue the matter further and attempt a chemical identification. She reported the investigation as far as she was able to take it in two papers in the *New York Medical Journal* in 1889. Others were able to repeat her extraction work, but, having assumed that the unknown poison had to be an alkaloid, they dismissed the crystalline material, which was inorganic, as not worth analyzing.

Nevertheless, though severely criticized at the time, partly on the basis of lack of controls in the animal tests, Gage Day's study constituted the only positive proof of the poisonous character of the locoweed until the U.S. Department of Agriculture published a report of its investigation of the problem two decades later. In 1905 the Office of Poisonous Plant Investigations of the Bureau of Plant Industry established a field station at Hugo, Colorado, in cooperation with the Colorado Agricultural Experiment Station, and a thorough, large-scale study was carried out, with fieldwork continued over three years. The results obtained demonstrated conclusively that the locoweed was indeed the source of the poison. The active chemical was subsequently identified as an inorganic salt of barium, and Gage Day's earlier isolation of this material was duly acknowledged.[17] She stayed in Kansas until 1894, when she moved back to her home state of New York and settled in Kingston in the Hudson River valley, where she practiced for the next four decades.[18]

The work for which she was especially remembered in Kingston and Ulster County, New York, was her long public health campaign to wipe out tuberculosis. The slogan she and her colleagues adopted was, "No tuberculosis in New York State by 1920." Early in her practice in Kingston she took the lead in organizing a temporary camp for the tuberculous in the Catskill Mountains, where the newly formulated cure of fresh air day and night, good food, and general medical guidance was made available for the first time to county patients. The Ulster County Committee on Tuberculosis was formed in 1909, and by the following year the camp was replaced by the first permanent county hospital for the tuberculous in New York State. Thanks largely to Gage Day's fund-raising efforts, the original temporary facility was then converted to a summer camp for poor children.

In addition to a private practice and her public health work, she was for many years gynecologist in the Kingston City Hospital, the County Tuberculosis Hospital, and a sanitarium and hospital run by Benedictine Sisters. She also taught at the Benedictine Training School for Nurses and lectured on social hygiene for the New York State Department of Health. During the First World War she served in the Volunteer Medical Service Corps.

Her early observations on chemical-induced insanity in animals on a large scale awakened her interest in mental and psychological processes in general, and effects of drugs on humans in particular. In a 1909 paper on insanity caused by drugs, she pointed out the danger of committing temporarily unbalanced people to insane asylums and argued for the provision of hospital space for their short-term care. Among her other published studies were investigations on the physiological effects of ether anesthesia.[19]

Active in many civic and social organizations in the Kingston area, she was for several years vice chairman of the Ulster County Republican Committee and one of the two representatives from Ulster County on the State Republican Committee. She died in Kissimmee, Florida, on 7 March 1935, at age seventy-seven, following a car accident.

CLARA BLISS HINDS[20] (b. 1851?) trained in the Medical Department of Columbian University (later George Washington University) during the short period between 1884 and 1892 when it saw its way to admitting women students. She was its first woman graduate (M.D., 1887). Her family was prominent in Washington social circles, her father, D. Willard Bliss, being one of the city's well-known physicians. She herself had been one of Washington's most popular debutantes. From an early age she had wanted to become a doctor, but her father had strongly opposed the idea and all applications she submitted to medical schools had been rejected. In 1883, following an unsuccessful marriage that left her with a small daughter to support, she returned to her idea of studying medicine. This time she and three other women applicants succeeded in gaining unofficial admittance to lectures at Columbian's Medical Department. She was then thirty-two. The following year the four women were officially enrolled, and they completed the course, though they were never put forward for internships or residency appointments. Hinds got much of her early clinical experience at the Washington Woman's Clinic, founded in 1890 by Ida Heiberger, and at the Dorothea Dix Dispensary, a hospital for women and children that she and another woman physician ran from 1894 until 1897 when it closed for lack of funds. Though many physicians at the time had difficulty setting up successful private practices, hers appears to have prospered, and was well established after three years.

She joined the Women's Anthropological Society of Washington in 1885, one year after its founding. This group was organized to bring together women who were interested in science in general, and women doctors were some of its most active members; they found that it satisfied their interests and that its aims were in keeping with their view of themselves as an elite professional group.

In 1886 Hinds organized a study section on anthropometry, or child growth. Her interest in the subject stemmed from her study of the work of G. Stanley Hall of Clarke University, a well-known experimental psychologist who specialized in studies of children and adolescents; one of his theories was that a study of children's growth patterns would help explain their later behavior as adults and aid in the curing of disease. Hinds collected data by means of hundreds of questionnaires, distributed to mothers and schoolteachers, asking them to assess growth patterns. Her 1886 paper in the *American Naturalist* reported the findings. The project fitted in well with her feeling that women were better off staying within their prescribed role in scientific work, though in fact she herself attempted to expand that role, especially into areas where she thought women could make special contributions. A second paper she presented before the Women's Anthropological Society in 1890 concerned the effect of regular exercise on women; at the time, she was supplementing her income by serving as the medical director of a women's gymnasium, a place where women could escape for a short time from the severe physical constraints of the corsets then in fashion.[21]

LOUISE ROBINOVITCH (b. 1867) took an MD at the Woman's Medical College of Pennsylvania in 1889. After a period as a resident hospital physician in Philadelphia, she practiced in New York City until 1923, for a time serving as assistant physician at the Insane Asylum on Blackwell's Island. Thereafter, until the late 1930s, she had a practice in Golden, Colorado.

A fellow of the American Medical Association, Robinovitch had continuing interests in research. Her first publication, on methods for reducing fever, appeared in 1892. In the period up to 1911 she carried out a number of studies on the physiological effects of electric currents and procedures for resuscitation after electrocution; her findings appeared in about eleven papers, many in the New York *Journal of Mental Pathology,* and others in European journals. She spent a year at the École de Médecine in Paris in 1906. In the 1920s she coauthored a number of studies on treatments for tuberculosis. She died in the early 1950s.[22]

Little appears to be known about the background of MARY NILES, M.D., who worked as a medical missionary in Canton, China, in the 1890s. The one paper by her listed in the bibliography was published in 1894. It discussed bubonic plague in Canton. Two years earlier, in a letter home to the mission from Canton, she gave a brief account of the kind of work she was engaged in. She regularly saw more than 100 patients a day when on duty at the mission hospital, and in addition performed major operations of many kinds, from breast amputations, to harelip correction, to ovarian cyst removal. She often made house calls, some that involved long journeys by foot, even at night. Not surprisingly her letter included a plea for more medical help.[23]

ANITA NEWCOMB MCGEE[24] (1864–1940), founder of the Army Nurse Corps, was one of the notable members of the post-pioneer group of medical women who graduated in the 1890s. The oldest of the three daughters of Simon Newcomb (later Rear Admiral Newcomb, professor of mathematics and astronomy in the U.S. Navy) and his wife Mary Caroline (Hassler), Anita was born in Washington, D.C., on 4 November 1864. The Newcomb family took an active role in the cultural life of the capital, and the children grew up with a good appreciation of music and the arts. Anita was educated in private schools in the city and then studied abroad for three years (1882–1885), at Newnham College Cambridge and at the University of Geneva, where she took classes in the medical faculty. In 1888, at the age of twenty-four, she married William John McGee, a geologist and anthropologist then on the staff of the U.S. Geological Survey. The Washington wedding was "a brilliant affair."[25]

She was a strong believer in married women having productive work outside the home and soon after the birth of her first child returned to her studies, enrolling in the medical department at Columbian University. During her senior year she was one of the leaders in the women students' campaign to dissuade the medical faculty from closing down its short-lived experiment of accepting women students. Despite a strong effort by many in the Washington community the campaign failed, but those women already enrolled were allowed to finish their course. McGee graduated with honors in 1892. The following year, she was able to get clinical experience with William Johnson, gynecologist at the Central Dispensary and Emergency Hospital. This was a favor not lightly granted, no other woman doctor being accepted at that hospital for another four years.[26] Like Clara Hinds, however, she did most of her clinical work at the Washington Woman's Clinic. Her private practice was only moderately successful and in 1895, shortly after the sudden death of her second child when he was four months old, she closed her office. Although at the time she considered this a temporary break in her career, she never returned to private practice.[27]

In 1893, three years after its founding, she joined the Daughters of the American Revolution, an organization that, in its early days, had a strong appeal to professional women who then made up a large part of its membership. She served on the board of trustees and on numerous committees, and as head of the library she worked to build up what became a valuable collection of materials highlighting the role of women in the Revolution. In 1898, when war with Spain seemed imminent, she drew up a plan for a DAR hospital corps of trained nurses to be available for work in the army or navy, explaining DAR involvement on the basis of women's contributions to the revolutionary cause. Soon after the outbreak of hostilities, congressional authority was given to employ by contract as many nurses, male or female, as might be needed by the military. A committee of DAR members, including the wife of George Sternberg, surgeon general of the armed forces, was organized to establish a female nurse corps. (This aroused considerable resentment among members of the American Red Cross, Clara Barton and other leaders of that organization believing that they should have been given the job.)

As director of the DAR Hospital Corps, McGee had the rank of Acting Assistant Surgeon, and was given the uniform of a regular officer, with jacket, shoulder straps, and a skirt of army cloth.[28] She had charge, under Sternberg, of selecting 1,000 women nurses from the 5,000 who applied to serve, and organizing them into the corps. Despite the fact that the women were given the same training as male nurses, they were restricted to kitchen duty in the army hospitals until typhoid fever and mounting casualties made imperative their employment as nurses. McGee kept close contact with the corps throughout the war, even when they were assigned to duty in the Philippines, and she also traveled throughout the United States organizing DAR relief efforts and distributing supplies to army camps and hospitals. In 1900 she successfully lobbied Congress to create a permanent Army Nurse Corps, and in addition got service-connected benefits for women who had served in the war, including the right of burial in Arlington National Cemetery. Her proposal to establish a federal teaching force of 100 women nurses and a war reserve of 2,000 trained women was not accepted, however. She was awarded the Spanish-American War Medal in acknowledgment of her work, but, having seen the Nurse Corps through its formative stage, she resigned her commission. Her successor, a contract nurse during the war, was appointed the first Superintendent of the Army Nurse Corps.

In 1904, as president of the Association of Spanish-American War Nurses and a representative of the Red Cross of Philadelphia, she took a volunteer group of ex-army nurses to Japan for six months to train Japanese women for military nursing service in the Russo-Japanese War. The group worked on hospital ships and in the main military hospital in Hiroshima. McGee held officer rank in the Japanese army and reported on hospitals throughout Japan, Manchuria, and Korea. For her services the Japanese government awarded her the Imperial Order of the Sacred Crown as well as two Russo-Japanese War medals and a special Japanese Red Cross decoration.

She spent a year as lecturer on hygiene at the University of California in 1911, gave occasional lectures throughout the country, and contributed articles on health topics to magazines, but her work in medicine gradually decreased. She concentrated instead on studies in anthropology and sociology, areas of continuing interest since the early years of her marriage. At that time she had accompanied her husband on a horseback trip through the southwestern United States where he was collecting geological and anthropological materials for the Smithsonian Institution, and this field experience undoubtedly encouraged and stimulated her own investigations. Her first three papers, listed in the bibliography, report work discussed at meetings of the anthropology section of the AAAS in the 1880s and early 1890s. Heredity and cultural evolution patterns in relatively isolated communities and sects, especially those practicing common ownership of property, were her particular interest.

She was a member of the Washington, D.C. Women's Anthropological Society and in 1889, while serving as the recording secretary, helped organize a library of anthropological and sociological works that featured monographs and articles by and about professional women. As a member of the Committee on the Investigation of Directive Forces in Society, she sent surveys to administrators in dozens of cities to find out if they employed women in administrative work and in higher education, but no record of the results of the survey appears to be available. She also used her influence in the men's Anthropological Society of Washington, in which both her husband and her father were officers, to arrange joint meetings at which women anthropologists gave papers. Zelia Nuttall spoke on the "Mexican Calendar System" in 1893, and Alice Fletcher (see chapter 14) gave a series of lectures on folklore in the spring of 1895. Arrangements for the formal merger of the men's and women's societies, under negotiation throughout the 1890s, were completed in 1899 when McGee's husband was president. By then, however, the Anthropological Society was no longer one of the most prestigious organizations of professional men in Washington; membership had declined, and anthropology departments at various universities were replacing the Bureau of Ethnology as centers for specialized work. McGee was elected secretary of the Anthropology Section of the AAAS in 1894, though she resigned before the annual meeting was held. She was again elected secretary for the Detroit meeting in 1897 and this time served as acting chair of the section as well, replacing her husband, who became acting president of the association.[29]

She resigned her membership in the DAR in 1916, rejoined after a lapse of six years, and then resigned again in 1932 without having taken any further part in the government of the organization. By 1929 she had retired to Southern Pines, North Carolina, but she returned to Washington, D.C., in 1938 as a patient in the Barton Health Home. She died on 5 October 1940, in her seventy-sixth year, and was buried with full military honors in Arlington National Cemetery. In 1966 the DAR established the Dr. Anita Newcomb McGee Award, a medal given annually, on the anniversary of the establishment of the Army Nurse Corps, to the U.S. Army "Nurse of the Year."

MARY SHERWOOD[30] (1856–1935), who conducted a practice in the city of Baltimore for almost thirty years, was a leader in the development of preventive medicine and public health care organizations in the United States, particularly those serving children. Born in Ballston Spa, in the Hudson River valley in upper New York State, she came of an old New England family whose founder had come to Boston from Ipswich in 1634. Her father practiced law for a time but then retired to the family farm at Ballston. She studied at the Normal School in Albany, New York, and then for some years alternated periods of schoolteaching with studies at Vassar College, from which she graduated (A.B.) in 1883. For the next three years she taught chemistry, first at Vassar (1883–1885) and after that at Packer Institute, Brooklyn (1885–1886). Having by then decided to study medicine, however, she went to Zurich University, where she was able to get excellent training in the newer branches of the field, especially the developing areas of histology and bacteriology. In 1889, along with her fellow countrywoman Lilian Welsh, she took the first course in bacteriology ever given at Zurich, and one of the first given anywhere. She graduated in 1891, her dissertation research on polyneuritis recurrens, carried out under the eminent internist Professor Eichhorst, being reported in the *Archiv für Pathologische Anatomie und Physiologie* the same year.

On her return to the United States Sherwood went to live in Baltimore, where her brother was finishing graduate studies in economics at Johns Hopkins University. Within a short time she had joined John Kelly of Johns Hopkins Hospital as an assistant in gynecological surgery, first in his private practice and shortly thereafter as a nonresident on his hospital staff. In 1892 she and Lilian Welsh, whose degree was from the Woman's Medical College of Pennsylvania, set up a joint practice in Baltimore, and from then on the lives and work of these two women were closely interwoven. Their early careers called for considerable staying power; patients were few and far between, and the income from their private practice did not even cover office rent. Though both were well-treated by male colleagues who recognized their considerable talents, like other women doctors coming into Baltimore in the 1890s they did not find themselves readily accepted into the city's social circles; a current joke was that humans could be divided into three categories, men, women, and women physicians.[31]

In 1893, when their fortunes were at their lowest ebb, both received offers of attractive positions outside Baltimore, at good salaries for that period and unusual opportunities for practice. But they chose to stay on in the city. Sherwood's optimism and strength of purpose played a significant role in the decision, but there were two other key considerations. One was the fact that in Baltimore they had access to the Johns Hopkins Hospital with its stimulating atmosphere of outstanding research laboratories, libraries, clinics, and lectures; the other was the close connection they had developed with the Evening Dispensary for Working Women and Girls.

This dispensary, the first public clinic for women in Baltimore, was of immense importance not only to working women but also to women doctors. During the eighteen years of its existence, from its opening in 1891, it gave medical care to 22,000 patients and provided extremely valuable clinical experience for no fewer than fifty-two women physicians. Sherwood and Welsh were responsible for the medical work in the dispensary's clinics for a number of years, and for a time also cared for all the obstetrical cases. They worked long hours and were considerably handicapped initially, since they had none of the assistance from the City Health Department, which was later to become routine, no place to which they could send infectious disease patients, and no laboratory facilities for routine diagnostic help. Nevertheless, as was demonstrated by a survey of its work made when it closed in 1910, this small Baltimore dispensary, in addition to its routine patient care, did pioneer service in the development of the public health movement in the United States. The organization of a social service department, a study of birth registration, the first distribution of clean, sterilized milk to sick infants, and a statistical study of deaths from tuberculosis in Baltimore between 1890 and 1900 were but some of the broad, community projects undertaken.

For Sherwood, the dispensary was her introduction to medical sociology and the problems of public health, concerns that occupied much of her time in the latter part of her career. Over the years her service on public health boards and commissions was extensive. She was active on the Maryland Tuberculosis Commission, whose work brought about the founding of the National Tuberculosis Association and in addition led directly to the establishment of public health programs in city and state health departments throughout the country. In the field of child health she became one of the leaders in the Baltimore area. She was influential in the organization of the Bureau of Child Welfare in the City Health Department and later served as its head. She also led the fight for a clean and safe milk supply for Baltimore, a long struggle because of the substantial political protection enjoyed by the dealers involved. By 1912 the National Association for the Prevention of Infant Mortality (later the American Child Health Association) was formed with headquarters in Baltimore. Sherwood was the first chairman of the section on obstetrics.

After medical work, the great cause in her life was higher education for women. It has been said that just the presence in Baltimore in the early 1890s of Sherwood and Welsh, two highly respected women physicians, did much to smooth the path of the first small group of women students admitted to the Johns Hopkins Medical School.[32] For several years Sherwood held staff positions in women's colleges, first at the Woman's Medical College in Philadelphia where she taught pathology, and then at Bryn Mawr College where she lectured on hygiene. Through Welsh, who held a faculty position as college physician at the Woman's College of Baltimore, Sherwood also had a close association with that institution, later serving on its Board of Trustees (1923–1935). In addition she was medical director of the Preparatory School of Bryn Mawr College in Baltimore, a post she occupied until her death. Throughout her career her interest in public affairs in general was very strong and her contributions to the women's suffrage movement substantial. She retired from practice about 1931 and died on 24 May 1935, in her seventy-ninth year, predeceasing Lilian Welsh by three years.

ELLEN STONE[33] (1870–1952), who was superintendent of child hygiene in the City Health Department of Providence, Rhode Island, for twenty-one years, was, like Sherwood, a pioneering worker in the field of child health. The daughter of Alfred Stone, an architect, and his wife Ellen (Putnam), she was born in Providence, on 12 January 1870. She graduated from Radcliffe College (A.B.) in 1895 and received an A.M. from Brown University in Providence the following year. Her physiological study on starfish glands (*American Naturalist,* 1897) was carried out in the physiological chemistry laboratory at Brown.

A strong and outspoken advocate of professional careers for women, she went on to medical studies at the Johns Hopkins University Medical School. After receiving her M.D. in 1900, when she was thirty, she returned to Rhode Island, where she passed the State Board of Medical Registration examinations. Until 1913 she divided her time between private practice in Providence and part-time work in the City Health Department. Thereafter, as superintendent of child hygiene, she initiated a number of public health measures for the protection of children, which, though they were to become routine by the middle of the century, were then pioneering, and made the Providence health department a model in the field. A founder of the Rhode Island Society for Mental Hygiene and one of the early advocates of child guidance clinics, she was particularly interested in uncovering physical causes leading to juvenile delinquency. Among her other accomplishments was the establishment of Providence's first fresh air school for tubercular children. Several of her publications on child health, mostly concerned with children in the public schools, appeared between 1908 and 1916.[34] She retired from professional work in 1934 at age sixty-four, but remained active in civic affairs for a number of years. Throughout much of her life she shared a home with her sister, Esther Stone, an architect. She died on 19 February 1952, a month after her eighty-second birthday.

There remain seven other American women physicians among those whose work is listed in the bibliography who qualified between 1890 and 1900.[35] Four of them—Blackwell, Lewi, Baldwin, and Mitchell Kydd—graduated from the Woman's Medical College of the New York Infirmary for Women and Children. ETHEL BLACKWELL[36] (1870?–1947) was born in Somerville, New Jersey, one of the five children of the Rev. Antoinette Brown Blackwell (see chapter 16) and Samuel Blackwell. Elizabeth Blackwell, founder of the New

York Infirmary for Women and Children, was her aunt. Both Ethel and her sister Edith graduated from MIT in 1891. Ethel went on to graduate work at Johns Hopkins Hospital, and then took an M.D. at the Woman's Medical College. She stayed there for a time as an instructor while also conducting a private practice (shared with an older woman doctor) in New York City. Her two papers on hemoglobin tests, published in the early 1890s, reported investigations carried out at several of the hospitals for women and children in New York City. Following her marriage to Alfred Robinson she gave up medical work but was active in the New York League of Unitarian Women. Following in her mother's footsteps she published two religious works, *Religion of Joy* (1911) and *A Glimpse of God* (1912), a collection of poems. She died at age seventy-seven in Huntington, Long Island, at the home of her son, Horace Robinson, on 1 August 1947.

EMILY LEWI[37] (1867–1946), whose 1895 paper concerned bacteriological investigations of the New York City milk supply, was a successful New York pediatrician and general practitioner for more than fifty years. The daughter of Julia (Seaman) and David Lewi, she was born in New York City and educated at Vassar College, from which she graduated with honors in 1888. After receiving her M.D. from the Woman's Medical College of the New York Infirmary for Women and Children in 1891, she took a post as house physician at the Babies' Hospital of the City of New York, working under the pediatrician L. Emmett Holtz. She also had a long association with the New York Infirmary for Women and Children as attending physician, consulting pediatrician, and lecturer; later she became a member of its board of trustees and an honorary member of the executive committee of its medical board. She died in the infirmary, on 28 February 1946, at age seventy-nine, after a short illness.

HELEN BALDWIN[38] (1865–1946), the daughter of Dr. E. Baldwin and his wife Sarah (Mathewson), was born in Canterbury, Connecticut, on 14 November 1865. After receiving her early education at Thayer Academy, in Braintree, Massachusetts, she studied at Wellesley College (A.B., 1888, with a special certificate in physics and chemistry). She stayed on for a year as laboratory demonstrator in physics, and then enrolled in the Medical Department at the University of Michigan. Better clinical facilities in New York City led her to transfer to the Medical College of the New York Infirmary for Women and Children in her junior year, and she graduated from there in 1892.

After an internship at the New England Hospital for Women and Children in Boston, she spent a year doing postgraduate research at Johns Hopkins University, where she was one of the early women research workers in the new medical school opened in 1893. Following two years at the Philadelphia General Hospital she returned to New York City, and, in addition to setting up a private practice, taught clinical medicine and physiology at the Woman's Medical College of the New York Infirmary for Women and Children. From 1900, after the college had closed, she served in the medical department of the infirmary, becoming head of the department, then consultant, and later president of the medical board. Her experience at Johns Hopkins University had given her a strong interest in research, and for sixteen years, from 1896 until 1912, she carried out a number of investigations in Christian Herter's private biochemical laboratory in New York City. Several of her papers appeared in biochemical journals over the years.[39] By the 1930s, after a career that spanned four decades, she had given up medical work in New York City, except for the care of a few private patients, and had moved to Connecticut. For a time during the Second World War, although by then in her seventies, she served as health officer in her hometown of Canterbury.

In this group of late nineteenth-century American women physicians, the pioneers who began their careers in the 1870s and the first part of the 1880s typically concentrated their efforts on improving medical care for women and children, particularly in the poorer sections of the major cities. Indeed, to a large extent these early women doctors were limited to working with women and children. Most of them received much of their training in women's medical colleges, although the four who were most outstanding—Putman Jacobi, Cushier, Mergler, and Mosher—also studied at European universities (and Mosher's degree was from the University of Michigan). Further, their clinical experience was gained largely in hospitals for women and children attached to women's colleges, and in clinics for working-class women. When they went into private practice their patients came mainly from the same groups. Many of those whose careers began in the late 1880s and in the 1890s were likewise attached to hospitals for women and children, but a few succeeded in entering broader areas of medical work, in particular the public health services, new, low-prestige branches of the profession, that were less attractive to men. But here also the women doctors tended to become involved in undertakings that affected women and children most directly, as in the cases of Sherwood in Maryland and Stone in Rhode Island. Like McGee's work in the U.S. Army Nurse Corps, however, these public health projects were pioneering efforts, involving the development of organizations that, in a relatively short time, became integral parts of the United States medical system.

British women

The opening of medical education to women in Britain came more than two decades after the first women's medical colleges began to graduate female doctors in the United States, and the amount of research contributed by late nineteenth-century British women to the medical journals is considerably less than that of their American counterparts (see Figure 7).

The first two British women to take full medical degrees did so at Continental universities—Elizabeth Garrett in Paris, and Frances Morgan in Zurich, both in 1870.[40] Following the failure, despite a strong fight, of would-be women students to gain entrance to medical schools in London and Edinburgh in the late 1860s and early 1870s, a separate institution, the London School of Medicine for Women, was established in 1874.[41]

The opening of university medical college examinations to women students and their admission to the wards of some hospitals slowly followed. By the 1890s many of the newer English universities and some of the Scottish universities accepted women in mixed classes, but the London School of Medicine for Women remained the major British training institution for women doctors until well into the twentieth century.

About half of those mentioned here are known to have taken medical degrees; among the others were two (Nightingale and Wood) who were influential in early work in public health and nursing.[42]

The earliest physician in the group was the Welsh woman, FRANCES MORGAN[43] (1843–1927). Her M.D. (1870) was the second awarded to a woman by the University of Zurich. She was born on 20 December 1843 in Brecon, south-central Wales, the oldest of the five children of Richard Morgan and his wife Georgiana Catherina, daughter of Captain John Philipps, R.N., of Ystradwrallt, Carmarthenshire. Richard Morgan, a graduate of Jesus College Oxford, was then assistant curate to the vicar of Brecon. After his death in 1851 his widow and children lived for some years in Cowbridge, south Wales, where Frances began her schooling. At the age of ten she was sent to a school in Windsor kept by a friend of the family, and five years later went to Paris, starting what would be, by the time she qualified, very much a cosmopolitan, European education.

After three years of study in Paris and about five more in Düsseldorf, she decided she wanted to take up medicine. Returning to London, she began studies in 1866 with private tutors. Her intention was to take the professional licensing examinations administered by the Society of Apothecaries. The LSA (Licentiate of the Society of Apothecaries) was the only qualification then open to women, but was sufficient to put her name on the Medical Register and allow her to practice. This plan was blocked the following year, however, when the Council of Apothecaries' Hall changed the society's rules to exclude women from the professional examinations. One woman, Elizabeth Garrett, had already slipped through the route they provided and obtained a license—enough was enough.

Fortunately there then opened up what was for Morgan a reasonable alternative. In 1867 the first woman student, the Russian Nadezhda Suslova, received an M.D. from the University of Zurich. That autumn Morgan set out to follow in Suslova's footsteps.[44]

Calm and self-confident, with a cool intelligence and an aristocratic manner, she initially caused a certain amount of amusement mixed with irritation among her fellow students, but they quickly came to regard her as an equal. She worked sixty hours a week at the university (taking, among other extra subjects, a course in Sanskrit) and completed her medical studies in three years. The defense of her dissertation research, a study on progressive muscular dystrophy, took place in March 1870, and the occasion was something of an event in the history of the opening of higher education to women in Europe.

By then there was considerable interest within the university community in medical women, and the examination had to be moved to the Aula, the largest auditorium available, in order to accommodate all who wanted to attend. Even that room was filled. The audience numbered more than 400 and included fifty women who had come to give Morgan their moral support. Her reputation as an exceptionally able student and the suggestion of a disagreement between her and her research director, Anton Biermer, added to the interest. Her findings did indeed differ from material Biermer had already published, and after she had read a summary of her work he launched into a sharp, sometimes angry attack. People in the audience later reported a certain amount of tension building up in the auditorium. But Morgan remained unperturbed, took extensive notes as Biermer spoke, and then responded with a half-hour address in which she explained clearly that the basis of their disagreement lay in the fact that she had made use of British and American research not taken into consideration by Biermer. In the end he was well pleased with his student, and the consensus at the close of that remarkable day in the history of Zurich University was that Morgan's achievements were further strong proof of the success of the great social experiment then quietly under way there. Morgan herself became something of a legend at the university.

The next year she broadened her experience, carrying out postgraduate studies at three of the foremost medical schools in Europe, first in Vienna, where she worked in operative midwifery under the guidance of the surgeon Gustav Braun, and then in Prague and Paris. She returned to London in 1871 and joined forces with the two other women doctors, Elizabeth Blackwell and Elizabeth Garrett, then active in the city. In cooperation with Blackwell she founded the National Health Society, becoming its first honorary secretary, and she was appointed assistant to Garrett at the New Hospital for Women. The latter was a small facility founded by Garrett in 1866 as the St. Mary's Dispensary for Women and Children. As well as supplying desperately needed service to the public, the New Hospital played an important role in providing a certain amount of clinical instruction in midwifery for students at the London School of Medicine for Women. In those days such experience was hard to find, being available at only a few hospitals other than the small facilities run by women; the hostility of the vast majority of British male doctors to the women who were attempting to break into their province was still strong, and particularly so among the members of the Obstetrical Society of London who felt especially threatened by the potential competition. Morgan remained at the New Hospital only until 1875, however, when she resigned over a matter of principle following a disagreement with Garrett.[45]

Although fully trained, Morgan still had no means of having her name placed on the British Medical Register. However, the passage through Parliament in 1876 of the Enabling Bill made it clear that there were no legal impediments to universities examining women for medical degrees, after which their names might be added to the register. The first examining bodies that were persuaded to use this authority were the Irish College of Physicians and Queen's University of Ireland. Morgan presented herself for the examinations in 1877 and in due course received her licence to practice.

In 1874 she married George Hoggan, a Scotsman from a working-class family, who, after many years as an apprentice engineer and then engineer in the Indian Navy, had left the latter service on a small pension and entered the University of Edinburgh to study medicine. While there in the late 1860s he strongly supported the struggle going on for the admission of women to the medical school.

The Hoggans each conducted successful medical practices in London, and in addition carried out a remarkable amount of joint research. This included studies on nerve endings in the skin, and extensive work on the anatomy and physiology of the lymph glands of the skin, muscles, and a number of internal organs, both in the absence of disease and in the presence of cancer and leprosy (see the physiology-anatomy section of the bibliography). George Hoggan also published several single-author papers on changes in the lymphatic system in cancer and eastern leprosy (his interest in the latter disease doubtless arising from his experience in the Far East in the early 1860s), but the bulk of his work was coauthored with his wife. They were probably the first husband-and-wife scientific team to publish a substantial amount of joint research, no fewer than twenty-two of their papers appearing in British, French, and German journals in the short period of eight years between 1875 and 1883. However, toward the end of 1884 George's health broke down. Frances gave up her practice and accompanied him to the warmer climate of the south of France, where he died in 1891.

She returned to Britain, but never went back to medical work, immersing herself instead in social problems that had long been of great concern to her. She was especially interested in the education of the poor and of girls and young women. As early as 1880 she had begun to speak out for the establishment of a system of secondary schools for girls in Wales. Her pamphlet *On the Physical Education of Girls* appeared the same year, and another on *Education for Girls in Wales* in 1882. She also brought out a short work on *The Position of the Mother in the Family, and its Legal and Scientific Aspects* (1884).

She was a strong supporter of the movement then under way to send women doctors and nurses to India, at the time under British rule. Since the 1850s, when missionary reports of the desperate need for medical care for Indian women began to circulate in Britain, modest efforts had been going on to alleviate the situation; by the 1870s Queen Victoria herself was taking an interest in the matter. The widespread prohibition against male doctors attending Indian women meant that only women could fill the need. This gave medical women almost unlimited opportunities, and since their presence posed no threat to any male monopoly they encountered little opposition. Hoggan lobbied in newspapers and via journal articles, visited India twice herself and worked especially to get state support for the women doctors' work.[46] The deplorable conditions in leper colonies in the Middle East also claimed her attention. She and her husband had between them done a creditable amount of basic research on the disease, and she gave both professional advice and material help to authorities in charge of the colonies.

Her interest in the condition of the unfortunate and underprivileged of this world did not stop there. She visited the United States on a number of occasions to study especially the situation of the Negro population, and she frequently expressed great admiration for the achievements of Negro women in their communities, despite the tremendous odds against them. In her pamphlet on *American Negro Women*[47] she drew attention to the fact that in the southern states, after the few years of post-Civil War Reconstruction when Negroes were succeeding in such tasks as setting up public school systems where none had been before, they had then been systematically deprived of their rights by violence and fraud.

With the outbreak of the First World War in 1914, although then over seventy, Morgan Hoggan volunteered her help to the government for hospital administration work. The offer was not accepted. She spent her last years in Brighton, where she died in 1927 at age eighty-three. On 14 March 1970, on the hundredth anniversary of her passing her medical examinations in Zurich, the Brecknock Society commemorated the occasion with a service in Brecon Cathedral.

One of the earliest women to complete her training at the London School of Medicine for Women was EDITH SHOVE.[48] She had begun her medical studies in the early 1870s and in 1874 placed first in the preliminary examinations at Apothecaries' Hall. Three years later, following the passing of the Enabling Bill, the Senate of the University of London voted to admit her to the university medical examinations, part of the requirements for a London medical degree. However, the permission was quickly taken back after more than 200 male medical graduates protested. In 1879, with the formal acceptance of the School of Medicine for Women as one of the schools qualifying for the University of London examinations, Shove finally took the examinations. She was one of the first four women to do so; all passed in the first division. Thereafter she carried out joint research with French physician Charles Rémy on the deterioration of the pancreas in diabetes, reported in *Comptes Rendus* in 1882. The following year she received her degree and was appointed medical officer to the female staff of the Post Office by Postmaster General Henry Fawcett, Elizabeth Garrett's brother-in-law and a strong supporter of the women's movement. She was the first woman doctor to hold a Civil Service appointment, and Fawcett offered her the same pay as men received in the same grade, a precedent that was not always followed in later years.

FLORENCE STONEY[49] (1870–1932), who became known especially for her X-ray work in army hospitals during the First World War, also took her training at the London School of Medicine for Women. She was born in Dublin, the second daughter among the five children of Margaret Sophia, daughter of Robert Storey of Parsonstown (now Birr), County Offaly, and George Johnstone Stoney. George Stoney, whose family had held landed estates in County Clare, was a mathematical physicist, and later, for twenty-five years, Secretary of Queen's University, Dublin. He did much for the improvement of Irish education in general and also was an able advocate of higher education for women. It was in large measure due to his efforts that women were permitted to take examinations in Ireland for legal medical qualification before that privilege was granted them in England or Scotland. Following the death of his wife in 1872 he had the responsibility for the upbringing of his young children; in 1893 he moved the family to London to give his daughters opportunities for higher education not available to them in Dublin at the time.

Florence had delicate health as a child and was taught at home during her early years. She took both the London matriculation examinations and the Cambridge Higher Local examinations, gaining in the latter first class in mathematics and special distinction in botany. She was outstanding as a medical student, taking honors in anatomy, physiology, and materia medica (M.B., London, 1895, M.D., 1898). Some of her studies were carried out at the Royal College of Science (B.S., 1896). Her first publication, an anatomical note, appeared in the *Journal of Anatomy and Physiology* in 1900.

She specialized in radiology, a field then still at the developmental stage, with primitive facilities and apparatus and the question of protection for the operator hardly even considered. After a short time at the Victoria Children's Hospital in Hull, she started X-ray units at both the Royal Free Hospital and the New Hospital for Women in London while also serving as a consultant and building up a general practice (later changed to a specialist practice in radiology). In addition she held the post of demonstrator in anatomy at the London School of Medicine for Women.

At the outbreak of the First World War, having already thirteen years of experience in X-ray work behind her, she offered her services to the government. When these were declined she organized a surgical unit of women doctors and nurses, under the administration of Mrs. St. Clair Stobart and the Women's Imperial Service League, with herself as radiologist and head of the medical staff.

The unit's initial plan to assist with care of the British and Belgian wounded in Brussels had to be abandoned because of the rapid German advance across Belgium. However, by 14 September 1914 they were in Antwerp, where, at the invitation of the Belgian Croix Rouge and the British consul general, they established themselves in a former theater, the Music Hall. Within five days they had converted this to a temporary hospital and their 135 beds were full of British and Belgian casualties. The further German advance was only a matter of time, however, and by October Antwerp came under attack. The hospital was on the Chaussée de Malin, along which the Germans entered the city, and for eighteen hours the building suffered heavy bombardment; the American consul general noted that one of the first buildings to be shelled was the hospital run so successfully by the British women doctors.[50] A basement area was prepared and they remained with their patients, the last British ship leaving Antwerp without them. When all of their wounded had been successfully evacuated, they set out to walk the ten or fifteen miles to the Dutch border. However, their temporary hospital had been situated next to a British ammunition dump, and they were overtaken by three London busses ferrying out the last of the stores. So it came to be that they were conveyed, at maximum possible speed, sitting on ammunition cases, through the empty, shell-cratered streets and past the blazing houses of Antwerp. Fortunately the London drivers were experts; the buses crossed the river Scheldt by a rickety bridge of planks laid across boats twenty minutes before the Belgians blew it up. The Admiralty later awarded Stoney and her corps the 1914 Star.

By 8 November the unit was reestablished at Cherbourg, at the request of the French Croix Rouge, as Anglo-French Hospital No. 2, and Stoney and her staff of seven women doctors and twelve nurses served there until 24 March 1915. This time they were quartered in the sixteenth-century Château Tourlaville, picturesque but hardly ideal for their purposes. However, in the two days between their arrival in Cherbourg and the admission of their first patients they succeeded in organizing it as a temporary hospital. Sanitation was primitive, a bucket system being necessary, and there was no running water except on the ground floor. Water for surgical use had to be carried up two flights and sterilized by boiling over oil stoves. Within a week of their arrival the French had filled seventy-two beds with critically wounded men brought by ship from Dunkirk. The less seriously ill were sent south directly. Most of Stoney's cases were compound fractures, and her X-ray work was invaluable in determining precise locations of shell fragments in the exceptionally septic wounds. With constant practice she also became skilled at distinguishing dead bone from living bone, and found that removal of the former speeded recovery. Of the 120 patients the women's team treated during their four and a half months at Cherbourg only ten died.[51]

By March 1915, with the stream of wounded diverted from the Cherbourg hospitals, Stoney returned to Britain, and was immediately asked to take charge of the X-ray department in Fulham Military Hospital, a 1,000-bed facility. She remained there until 1918, carrying out both the technical work of photographing and the analysis of films. By 1918, 5,000 cases had passed through her hands and she had trained many assistants who then went to other hospitals. With her background in anatomy and her broad experience in general practice, she was able to provide surgeons with especially helpful diagnoses, and other hospitals sent her their difficult cases for bullet and shrapnel location. She was one of the first women doctors to be accepted for full-time work by the War Office, and was for long the only woman member of the large medical staff at Fulham. She had no military rank, however.[52] After the war she was awarded an O.B.E. for her services.

The strain of her four years of military work left her in less than good health and she tired easily. She did not attempt to reestablish herself in London but moved to Bournemouth, where she soon had a large specialist practice. She was also consulting medical officer for the radiology departments of two Bournemouth hospitals, and consulting actinotherapist at the Victoria Cripples Home; for a time she served as president of the Wessex Branch of the British Institute of Radiology. Over the years she contributed a number of original articles to the medical literature, mainly on radiology.[53]

Following her retirement in 1928, when she was fifty-eight, she visited India to study osteomalacia (bone softening) and other diseases of darkness. While there she also advised on the use of ultraviolet light in hospitals. Throughout her life she kept up her early interest in botany and she was a keen gardener. She died at age sixty-two, on 7 October 1932, after a long and painful illness. Her wartime colleagues remembered especially her gay Irish humor.

ALICE CORTHORN,[54] another of the early women doctors trained at the London School of Medicine for Women, spent four years in public service work in India (1898–1902), for which she was awarded the Kaiser-i-Hind Silver Medal. The daughter of Frederick William Corthorn, she was born in London, on 27 June 1859. She received a private education in her early years and then studied at the University of London and the School of Medicine for Women, taking M.B. and B.S. degrees.

Following a two-year spell of duty at Hampstead Fever Hospital and a one-year demonstratorship in physiology at the School of Medicine for Women, she went out to the General Plague Hospital, in Poona, Bombay Province, as special medical officer on Plague Duty. She was there during the epidemic of 1900–1901, when 868 patients were admitted to the hospital. A number of her reports and papers appearing in the period from 1899 to 1902 concerned her observations on plague in both the human and animal populations.

After returning to Britain she spent a year as senior clinical assistant at the New Hospital for Women. About 1904 she became medical officer for several London schools for girls. Wide public health issues remained among her continuing concerns and she was prominent in the national movement then under way to establish an efficient School Medical Service.[55] Her interests in veterinary medicine also continued and she published at least two accounts of treatment of partial paralysis in dogs, some of the work being done in collaboration with veterinary colleagues.[56] She was a fellow of the Royal Society of Medicine.

Only minimal information has been uncovered about E. E. Flemming, Julia Brinck, and Margaret Sharpe, three other early London medical women whose papers are listed in the bibliography. Flemming's anatomical studies, carried out in the mid 1890s, used specimens obtained from the London School of Medicine for Women and the New Hospital for Women. Brinck took her M.D. in 1886 at Bern University,[57] where she studied with Hugo Kronecker, head of the Physiological Institute. Her three papers on muscle nutrition and synthetic processes in cells appeared in German and British journals in the late 1880s. She received a British Medical Association research award in 1890, the first given to a woman.[58] Margaret Sharpe was a radiologist; her three papers on treatments using X-rays and static electricity were published between 1900 and 1907.[59]

There remain to be mentioned two women who were prominent in the late nineteenth-century development of nursing in Britain—one was Florence Nightingale and the other was Catherine Wood. A single paper by each is listed in the bibliography, Nightingale's on "Village sanitation in India," and Wood's on "The progress of nursing in the Victorian era."[60] While Nightingale is a major figure in the social history of Victorian Britain, Wood is not so well remembered. Nevertheless, she published a considerable amount during the fifteen years between 1878 and 1893, both on nursing principles and practice and on the related field of preparation of foods for the sick and the very young.[61] Her 1893 report on *Boards of Guardians and Nurses* was especially noteworthy.

CATHERINE WOOD began her nursing career about 1863,[62] when, promoted to a large extent by Nightingale, the movement to improve nursing and organize it into a profession was in its early stages. Her first position was that of ward sister in the Children's Hospital, Great Ormond Street, London. From there she moved on to the post of Lady Superintendent of Cromwell House, Highgate, and its hospital adjunct (1869–1878). She then returned to Great Ormond Street for ten years, before joining with Mrs. Howard March to found the Alexandra Hospital in Bloomsbury. In 1889 she founded the Nurses' Hostel; she also served as managing director of the Nurses' Hostel Co., Ltd. Active in the wider public sphere as well, from 1892 to 1895 she was engaged as Special Commissioner of the *British Medical Journal* in an inquiry into the condition of workhouse infirmaries in England and Ireland. General education on hygiene and nursing was another of her concerns, and she gave popular lectures under the auspices of various county councils. Later she lived in Hartfield, Sussex. She died on 14 June 1930.

The long life and career of FLORENCE NIGHTINGALE (1820–1910) has been the subject of so many works that a full listing of them would cover several pages.[63] The great two-volume biography by Sir Edward Cook, *The Life of Florence Nightingale* (1913), undoubtedly remains one of the best and most useful, and many of the later works are necessarily retellings of Cook's story. Lytton Strachey's essay in his *Eminent Victorians: Cardinal Manning, Florence Nightingale, Dr. Arnold, General Gordon* (1918), is well known; Margaret Goldsmith's admirable *Florence Nightingale: the Woman and the Legend* (1937), somewhat less so. Later works include Elspeth Huxley's 1975 *Florence Nightingale*, which has a collection of particularly fine illustrations. However, of the more recent discussions of Nightingale's career, one of the most interesting is F. B. Smith's *Florence Nightingale: Reputation and Power* (1982). Not the least noteworthy of the points Smith emphasizes is the remarkable use Nightingale made of statistics, then a new science, to establish herself as a reliable authority in sanitation reform and public health matters, then areas of great activity and rapidly changing standards. In this way she very successfully entered into the national political scene of the day.

Florence was the daughter of Fanny and William Edward Nightingale. William Nightingale had been born into the family of William Shore, of Sheffield, but changed his name when he inherited the extensive Derbyshire estates of Lea Hurst and Woodend from his mother's uncle, Peter Nightingale. He was a quiet man, liberal in outlook, Cambridge-educated, and possessed of very considerable wealth. His wife was the daugher of William Smith, M.P. Shortly after their marriage in 1818, the Nightingales spent three years in Italy, following the fashion for Continental travel prevailing among wealthy English families in that period just after the Napoleonic wars. Their second daughter was born in Florence, 12 May 1820, and named after her birthplace. The family returned to Britain when Florence was a year old, taking up residence at Lea Hurst. But that location was too remote, and the house (with its fifteen bedrooms) too small to suit them. In 1825 they bought Embley Park, on the edge of the New Forest, in Hampshire. After that they spent most of the year there, within fairly easy reach of relatives who lived near Winchester and in Surrey. For the summer months they moved to Lea Hurst, and for "the Season" to London. Mrs. Nightingale was one of ten children, and Florence and her elder sister Parthenope grew up with close contacts to a large, close-knit, and energetic extended family of aunts, uncles, and first cousins.

In their earliest years the Nightingale sisters were taught by governesses, but when Florence was twelve their father took over their instruction. He was a lively but demanding teacher, fluent in Italian. French, Latin, Greek, mathematics, history, and composition were included in the course of study he supervised, and the girls received a far better education than was customary for the time.

Intellectually able, good-looking, and attractive, Florence by the age of sixteen had nevertheless developed the idea that the usual life patterns for young women of her circle did not offer any path she could follow. The answer to the question of what she *could* do was not, however, apparent. Meanwhile she went on taking part in the family's round of social activities, including two years of Continental travel when she was in her late teens, with invitations to balls given by grand dukes, visits to the opera in Florence, and introductions to the social and intellectual elite of Paris. At home again by 1839 she was presented at Court, and in the next few years had two proposals of marriage from eminently suitable young men. All this failed to change her reluctance to settle for a conventional life, however, and she gradually made her way toward something that would satisfy her ambition for independence. In 1851, despite her family's objections, she went for three months to the one place she knew of that professed to provide practical training in nursing—the Institution of Deaconesses at Kaiserswerth, just north of Düsseldorf on the Rhine. This small establishment staffed by Lutheran evangelical deaconesses brought together by Pastor Theodor Fliedner, functioned as a hospital, penitentiary, and school. It had only a minimum to offer in the way of nursing instruction, but in those days of abysmally poor conditions in public hospitals, when nurses were recruited from the most depressed classes of society, the Kaiserswerth women did provide an example of dedication and discipline.

In 1853, when she was thirty-three, Florence secured her first (although unsalaried) job, that of resident superintendent of a newly reorganized charitable institution, the Invalid Gentlewomen's Institution in Upper Harley Street, London. Despite her mother's objections, her father made her a very generous allowance of £500 a year, and one of her aunts arranged an apartment for her in Pall Mall, giving her a place of her own to withdraw to when she wanted. There is some doubt about the effectiveness of her administrative work at Harley Street, but as a strong-willed member of one of the country's wealthiest families she had considerable influence with the ladies and gentlemen on the committees she had to deal with. Her year as superintendent was a valuable first experience, allowing her to try out her ability to effectively establish her control over an organization.

She quickly came to feel that the running of the institution was too limited a sphere of action for her. The issue of hospital reform was in the air, and a number of her close friends were involved. King's College Hospital was to be rebuilt, and her name had been suggested as a possible superintendent of nurses. By the autumn of 1854 the ambitious Florence was busy with plans for recruiting farmers' daughters as nurse trainees, to be instructed along Kaiserswerth lines. But more momentous events were to impinge on her life. The Crimean War had broken out that same year, and early in October, following the Battle of the Alma, the first major encounter in which the British were involved, disquieting news reached people at home. The dispatches of William Howard Russell, the *Times* war correspondent, described in graphic detail the outcome of the total inadequacy of the army medical services to cope with the sick and wounded. There was little new in such military mismanagement and the resulting suffering among the common soldiers, but the widespread public reporting was unprecedented, and consequently something had to be done. Two volunteers came forward immediately with offers to the War Office to supply groups of nurses. But Florence, recognizing an unsurpassable opportunity when she saw it, decided that she, and she alone, would head a nursing expedition. Moving via her Hampshire neighbor, Lord Palmerston, the Home Secretary, she immediately sought backing at the highest government level. She succeeded admirably, and was quickly on her way to Scutari, the city across the Bosporus from Constantinople where the main British military hospitals were located. With her she took thirty-eight nurses and the formally granted right to work in the hospitals under the chief medical officer. Her contingent of thirty-eight constituted all who could be recruited at the time who were considered both acceptable and capable of doing the job; many were from religious houses.

At Scutari she quickly diagnosed the major weakness in the army system, namely the fact that the medical officers were controlled by their military colleagues. The latter made all the key decisions, where the hospitals were to be sited, which men could be spared for orderly duty (generally not the able-bodied), and so on. She had little direct influence with the top officers on the spot, but once again used her private connections with influential people in London to put in train what she saw as the necessary changes. At the same time, however, she and her women constantly prodded the doctors to exert themselves and the orderlies to improve their work. Standards of cleanliness were raised, the food vastly improved, and efficiency increased. Perhaps most important of all for the men was the fact that she had brought with her stores of food and clothing bought with private funds to which she had access. Death rates fell dramatically. By the spring of 1855, half a year after she arrived in Scutari, mortality in the hospitals had dropped from 43 percent to just over 2 percent, while the rates in the French hospitals during the same period remained high, finally topping the worst British rates.

Although it was on this undeniably spectacular achievement that her public reputation was built, Nightingale's real, long-term effectiveness lay elsewhere. Her administrative work at Scutari was questionable at best,[64] and most of the actual nursing continued, necessarily, to be done by male orderlies. However, her outrage at the muddle and inefficiency that confronted her, and her personal ambition to change the whole system, coupled with her political finesse and her direct private connections to people with executive power at home, had remarkable results. Her suggestion for setting up a hospital food supply system based on the French model was accepted by the War Office, and her scheme for a revision of the duties and pay of army medical officers adopted in a Royal Warrent of September 1855.[65] She also pushed for a school of army medicine, and a small one was started at Scutari.

She returned to Britain in July 1856, four months after the end of the war. At the age of thirty-six she was a national heroine, recognized by the sovereign and admired by many important people. Further, she had every intention of continuing the work she had begun—namely reforming the health and sanitation standards of the British army. Having already exposed many of the blunders of the military administration and the War Office, she had made herself more than a few enemies. But several factors were on her side, not the least being the moral justification for her work and her power to call up a tremendous popular emotional response. Of the 94,000 men sent to the Crimea, a staggering 46 percent were lost through disease alone. Furthermore, the tragedy of needless deaths through disease continued in every barracks and army hospital even in peace time. She equated the army's neglect of the situation with nothing less than murder, and vowed to fight the cause of the victims, "lying by their thousands in their forgotten graves."

Despite a strong rebuttal by the military authorities to the accusations launched against them and a number of other setbacks, she succeeded in getting a Royal Commission of Enquiry set up, one consisting mainly of people of her own choosing (she could not, as a woman, expect to sit on such a commission herself, and in any case preferred to work through others). Her 830-page, privately printed *Notes on Matters Affecting the Health, Efficiency, and Hospital Administration of the British Army . . . Presented by Request to the Secretary of State for War* was absorbed into the commission's report. It proposed nothing short of a complete reorganization of the army medical and supply departments, with the arguments for the changes based on clearly presented collections of army health statistics. Four subcommissions were established to oversee the recommended reforms. Many improvements were made, the money spent on army barracks increasing by more than a factor of three between 1854 and 1859. By 1864 every barracks in England had a water supply with one basin for every ten men and a bath for every 100.[66] The changes had an almost immediate effect on the death rate, which, for the army at home, fell from 17 per 1,000 in 1855 to 8.9 (only slightly above the average for comparable civilian age groups) in 1865.

Nightingale's most effective tactic in all her work of sanitary reform was her use of statistics, particularly her condensed presentations of vast amounts of data using colored graphs, bar graphs, and polar area charts (which she invented). The science of statistics was a new one at the time, the field having been essentially founded in its modern form by the Belgian astronomer Adolphe-Jacques Quételet, who organized his country's central statistical bureau in 1841. Quételet demonstrated that, although the study of social data could not be used to predict individual behavior, it did shed considerable light on behavior patterns in populations, and it could be used to make reasonable estimates of future trends. Such ideas were highly controversial, people who believed in free will and individual responsibility regarding them as abhorrent. But Nightingale knew and admired Quételet's work. When she returned to England from Scutari in 1856 she enlisted the help of William Farr, a doctor already expert in the collection and use of epidemiological statistics, and under Farr's guidance produced dramatic graphical presentations illustrating both the impact of disease on the army and the effects of improved sanitary conditions.[67]

About 1858, after she had finished her *Notes on the Army,* she turned her attention to the question of sanitary reform for the army in India. The subcontinent was much in the public mind after the 1857 Mutiny. The latter event had made clear the fact that British rule would have to be maintained by British troops, and it was becoming equally clear that the only effective way of doing this was to reduce the heavy toll of disease. With Farr and Dr. John Sutherland, one of the first Board of Health Sanitary Inspectors, whom she had known in Scutari, Nightingale began to study the sickness and mortality records of the India Office. In addition, in consultation with her medical colleagues, she devised an extensive questionnaire that she sent out to 200 British stations across India, requesting detailed information on their sanitary conditions and arrangements. By 1859 she had secured the establishment of another Royal Commission, the Sanitary Commission for India. It issued its report four years later, and the improvements that followed in the condition of barracks, public sanitation measures, and protection against cholera reduced the mortality rate among British soldiers in India from 69 per 1,000 in 1863 to 18 per 1,000 in 1873.

Nightingale's other great achievement was the establishing of a training scheme for nurses, paid for by the Nightingale Fund. Amounting to almost £44,000, this money had been raised as a thank-you offering for her work in Scutari, and put at her disposal when she arrived back in Britain in 1856. In 1860 the Nightingale Training School for nurse probationers was opened in the recently rebuilt St. Thomas's Hospital. Applicants were selected carefully and the one-year training program was exacting; discipline in work and general conduct was strict, moral standards being one of Nightingale's major concerns (to banish once and for all the early nineteenth-century stereotype of the drunken and slovenly public hospital nurse). She herself monitored the monthly performance reports of each probationer throughout the first twenty years of the program. Although the overall output of nurses was small (100 in the first decade and about 1,900 between 1860 and 1903), the Nightingales, as they were known, set standards of careful work, cleanliness, concern for patients, and confidence in their own skills that were soon taken up, not only throughout Britain, but on the Continent, in the United States and throughout the British Empire. These women rapidly rose to matronships and positions of authority in hospitals, and very effectively spread an ideal of good nursing. Nightingale's efforts thus raised the public standing of the profession within the medical community to a level that, but for her, it would have taken much longer to reach.

Her period of greatest activity was the decade between the late 1850s and the late 1860s, a time when movements for reform were beginning in many areas of British life, including prisons, schools and universities. A number of well-to-do women were active in these efforts, but none of them matched Nightingale for boldness and ambition, and in the sheer audacity with which she used the government itself to carry forward the plans she had worked out. Her success in contriving what were virtually her own handpicked select committees and Royal Commissions, and in managing the evidence presented to these bodies, was unmatched. At the same time, she avoided public controversy by working behind the scenes. No other woman reformer of her time challenged entrenched male power bases of the kind she took on (the War Office, the military elite, and the India Office), and none had the spectacular success she had in such a fundamental matter as the saving of the lives of ordinary people.

About issues that affected women particularly Nightingale remained somewhat cool, her real interests being in movements over which she could keep firm control. On the critical point of women's suffrage she was always skeptical, and although in public she could hardly afford to withhold her nominal support, she was evasive on specific questions and generally pleaded lack of time or poor health as a reason for avoiding direct involvement. As the influential daughter of a very wealthy family with direct personal access to those with executive power, she put through her agenda very well without the vote. Indeed, she had a firm belief in the leadership capacities of the talented few, and considered an increase in the voting population more as a hindrance to society's progress than the reverse.[68] She never saw much reason to cooperate with the early women doctors either; she and Elizabeth Garrett found themselves on opposite sides on a number of issues. She did have a lifelong friendship with Elizabeth Blackwell, who shared her general outlook on a number of medical matters. However, in 1858–1859, when Blackwell proposed establishing a hospital for women as the setting for a combined training program for both women nurses and women doctors, Nightingale declined to give her support, much less any financial help from the then untouched Nightingale Fund.[69] She took the establishment view when it came to the training of women physicians, seeing them as likely to be of little benefit to society; in her opinion well-trained nurses were a more urgent need. Further, a joint program for both nursing and medical students would have meant a serious dilution of her control over the undertaking, something she could not willingly accept.

Much of her writing was incorporated into Royal commission reports, including her important *Notes on the Army* (1858) and her *Observations on the Sanitary State of the Army in India* (1861). However, she also brought out a number of works independently. Especially influential were *How People may Live and not Die in India* (1863—a shortened version of *Observations*) and *Notes on Nursing: What it is and What it is Not* (1860). The latter was probably her most popular book. Even at the relatively high price of five shillings it quickly sold 15,000 copies, went on to several revisions and editions, including an American one, and was translated into other languages. Condensed and forceful in style, and illustrated with discussions of her personal experiences, it emphasized above all Nightingale's two cardinal principles—hygiene, and careful concern for the patient.[70]

By the 1870s, when she was in her fifties, she was no longer taking up major causes whose pursuance involved the exercise of political muscle. From time to time she spoke out strongly on issues that concerned her particularly, such as the supplying of women nurses for the British campaign in the Sudan from 1882 to 1885, or the design of hospitals and women's colleges. Until 1901, when her health failed, she kept up an enormous correspondence, both with former colleagues and with friends and relatives.

A number of honors came to her in her final years, including the Order of Merit (the first given to a woman), bestowed by King Edward VII in November 1907. By then, however, she was too feeble to attend the presentation ceremony and indeed too senile to understand what was happening. The following year she was awarded the freedom of the City of London, becoming the second woman on whom the distinction was conferred. The badge of honor of the Norwegian Red Cross Society was presented to her in 1910—she had already received the German order of the Cross of Merit and the French gold medal of Secours aux Blessés Militaires. She died at her home in Park Lane, London, on 13 August 1910, at age ninety. Burial in Westminster Abbey was proposed, but by her own choice her grave was beside those of her parents at East Wellow, near Embley. Six sergeants of the Guards carried her coffin to the churchyard. Memorial services were held in St. Paul's Cathedral and Liverpool Cathedral. Her life, begun in 1820 in the reign of George IV, had spanned more than the whole Victorian era.

Summary

Though a small group this is a fairly diverse one. Nightingale, one of the prominent figures of her time, occupies a most distinguished position, but it is perhaps worth emphasizing that her most far-reaching achievements (her pioneering use of applied statistics notwithstanding) depended largely on her political skills rather than on medical, technical, or scientific work. The most productive research worker in the group was Morgan, in partnership with her husband, but both Corthorn and Stoney also published a number of original studies. Stoney's four years in military hospitals and many more in specialist practice, Corthorn's public health work in India and later in Britain, and Shove's Civil Service career were all notable professional achievements for women doctors of the period.

Other points that stand out in this overview are the following: most of the physicians in the group had connections, either as students or staff members, with the London School of Medicine for Women or its attached hospital; two people, Stoney and Sharpe (following a pattern seen in other fields of women moving into new, less-populated areas) specialized in radiology; work in India was a common interest for at least three of the group besides Corthorn (Morgan and Stoney both went out for short visits, and Nightingale worked on health issues for the army there). For sheer drama, the wartime experiences of Stoney are memorable, to say the least—one wishes she had written an autobiography. Corthorn's India experience, about which only the sketchiest information appears to have survived, must surely have been remarkable also. Nightingale's story is already an established part of the history of her time.

Two-country comparison

The small size of the British group compared with the corresponding American one (see Figure 7) limits the possibility of a useful comparison between the two, but a few common features are perhaps worth summarizing.

In both countries many of the early women physicians were trained in women's medical colleges and then worked in hospitals and clinics for women and children. Prominent in the United States were such institutions as the Woman's Medical College of Pennsylvania and the Woman's Medical College of the New York Infirmary for Women and Children. In Britain there was the London School of Medicine for Women, which remained the most important institution for the medical training of women in England until well after the turn of the century. The British had no equivalent of the University of Michigan (a strong and forward-looking medical school), which accepted women students from 1870.[71] For women of both countries basic training and also postgraduate experience in Europe remained important for many years.[72]

The great significance to British women of the almost unlimited opportunities for medical work in India that began during the 1880s (and is reflected in even this small sample) had no close parallel in the United States. Considerable numbers of American women were active as medical missionaries in the Far East, including India, from about 1869, and some of them had full medical degrees from reputable medical colleges. The central focus of the Protestant women's foreign mission societies in which most of them served, however, was evangelical;[73] consequently their medical undertakings did not match in scope and complexity that of the British women in India, assisted as the latter were by government backing.[74]

One might be tempted to draw a parallel between Anita McGee's work of organizing the U.S. Army Nurse Corps during

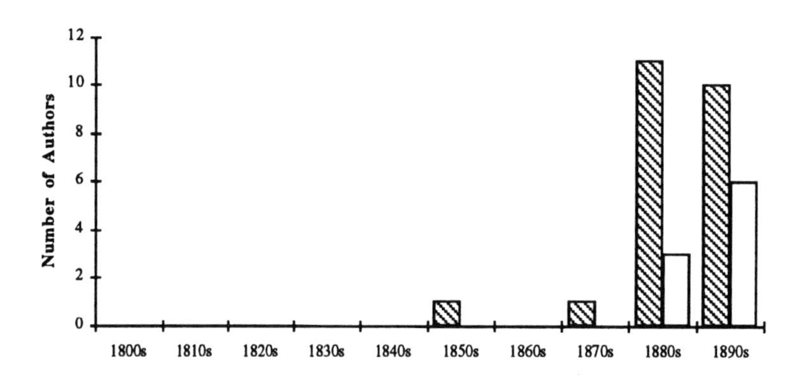

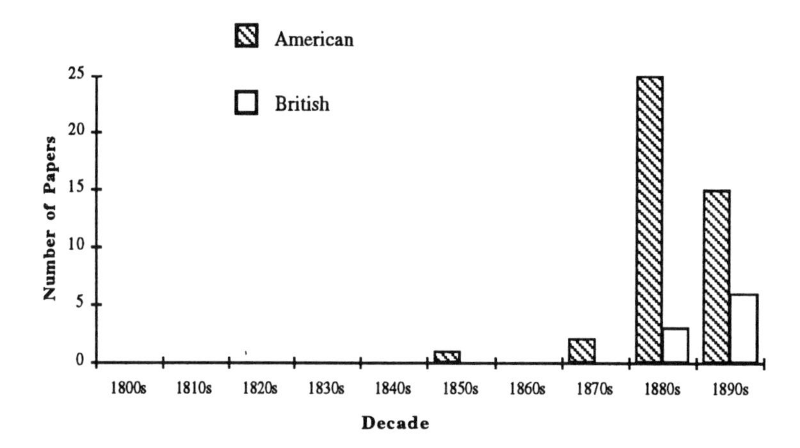

Figure 7. American and British authors and papers in medicine, by decade, 1800–1900. Data from the Royal Society *Catalogue of Scientific Papers.*

the Spanish-American War of the late 1890s and Florence Nightingale's campaign to reform the medical organization of the British army more than forty years previously. McGee's work, however (the founding of a permanent female Army Nurse Corps), was much more narrowly focused than Nightingale's broad and ambitious plans for army and public health reform.

Notes

1. Little information has been collected about two of the women in this early subgroup, Josephine Griffith Davis and Mary Dixon Jones. Both graduated from the Woman's Medical College of Pennsylvania, Griffith Davis in 1877 and Dixon Jones in 1875. Griffith Davis practiced in New York City and later in Hasbrouck Heights, New Jersey. She remained professionally active until at least 1916. Dixon Jones practiced in New York City at least until 1906. Her two papers in gynecology listed in the bibliography reported work done at the Woman's Hospital in Brooklyn in the 1890s—see *Polk's Medical Register of the United States and Canada,* 6th. rev. ed. (Detroit: R. L. Polk and Co., 1900) and *AMD,* 1906, 1914, 1916.

2. Roy Lubove, "Jacobi, Mary Corinna Putnam," in *NAW,* vol. 2, pp. 263–5; Esther Pohl Lovejoy, *Women Doctors of the World* (New York: Macmillan, 1957), pp. 73–5; Rhoda Truax, *The Doctors Jacobi* (Boston: Little, Brown, 1952); John H. Talbott, *A Biographical History of Medicine: Excerpts and Essays on the Men and their Work* (New York: Grune and Stratton, 1990), entry for Abraham Jacobi, pp. 1119–21; Victor Robinson, M.D., "Mary Putnam Jacobi," *Medical Life,* **35** (1928): 334–353.

3. See Thomas Neville Bonner, *To the Ends of the Earth. Women's Search for Education in Medicine* (Cambridge, Mass.: Harvard University Press, 1992), especially pp. 18–24.

4. "Documents," *American Historical Review,* **22** (1916–1917): 836–41. These letters of Putnam to her parents, written in August, September, and December 1870, were vigorous, informed, and thoughtful commentaries on the situation in France; her pro-Republic and anti-Empire stance is very clear.

5. Elizabeth Blackwell was America's first woman physician (and also Britain's—see n. 40). In 1847 she had been admitted, as an "experiment" to Geneva Medical College, in rural New York State. After she graduated in 1849, however, Geneva thought better of its experiment and once more closed its doors to women. Elizabeth's sister Emily and Marie Zakrzewska graduated from the medical department of Western Reserve College in Cleveland in 1854 and 1856, respectively.

6. Sarah J. McNutt, M.D., "Medical women, yesterday and today," *Medical Record,* **94,** (1918): 135–9; "Autobiography of Elizabeth Cushier," ed., Elizabeth B. Thelberg, in Kate Campbell Hurd-Mead, *Medical Women of America: a Short History of the Pioneer Medical Women of America and of a Few of Their Colleagues in England* (New York: Froben Press, 1933), pp. 85–95.

7. Annie Sturges Daniel, M.D., "'A cautious experiment.' The history of the New York Infirmary for Women and Children and the Woman's Medical College of the New York Infirmary; also its pioneer founders, 1853–1893. Excerpts from college catalogues, 1883–1889, inclusive," *Medical Woman's Journal,* **48** (1941): 364–71, especially p. 370.

8. McNutt, "Medical women;" *AMD,* 1906–1925.

9. Helga M. Ruud, M.D., "The Women's Medical College of Chicago. Records of professors and instructors and alumnae. 1871–1895," *Medical Woman's Journal,* **53** (1946): 50–4, on p. 51; Thomas Neville Bonner, "Mergler, Marie Josepha," in *NAW,* vol. 2, pp. 523–30; Cecyle S. Neidle, *America's Immigrant Women* (Boston: Twayne Publishers, 1975), pp. 202–3.

10. Ruth B. Bordin, "Mosher, Eliza Maria," in *NAW,* vol. 2 pp. 587–9; Jessie Hubbell Bancroft, "Eliza M. Mosher, M.D.," *Medical Woman's Journal,* **32** (1925): 122–9.

11. Ibid., p. 122.

12. Ibid., p. 124.

13. Regina M. Morantz-Sanchez, "The many faces of intimacy: professional options and personal choices among nineteenth- and early twentieth-century women physicians," in *Uneasy Careers and Intimate Lives,* eds. Pnina Abir-Am and Dorinda Outram, (New Brunswick, N.J.: Rutgers University Press, 1987), pp. 45–59, on p. 56.

14. *AMD,* 1906, 1916; *Polk's Medical Directory,* 1900; *AMS,* 1910.

15. Simon Henry Gage and Asa Franklin Gage, *Mary Gage Day, M.D. A Memorial Tribute* (Mohawk, N.Y.: Sun Co., 1935); *WWWA,* p. 235; C. Dwight Marsh, "The loco-weed disease of the plains," *U.S. Department of Agriculture, Bureau of Plant Industry,* Bulletin 112 (1909), 116 pp. especially pp. 5, 22, 34, 115.

16. Gage and Gage, *Mary Gage Day,* pp. 18–21.

17. Marsh, "Loco-weed disease," pp. 22, 34.

18. Although Edgar Day's name appears in Wichita City Directories along with his wife's until 1894, from then on he is not mentioned in accounts of her life and work. Her entry in *WWWA* does not even mention their marriage.

19. See "Some studies of the blood before and after etherization by the drop method," *American Journal of Surgery,* **36** (1922): 53–7; "Blood before and after ether anaesthesia," ibid., **38** (1924): 105–10; (with Mark O'Meara) "Evaluation of risk and status of anaesthesia," *Current Researches in Anaesthesia and Analgesia,* **10** (1931): 233–6. Gage Day's papers on drug effects and mental processes were, "Acute toxic insanity due to drugs," *New York State Journal of Medicine,* **9** (1909): 38–41, and "Influences of mental activities on vascular processes," *Journal of Comparative and Physiological Psychology,* **3** (1923): 333–77.

20. Gloria Moldow, *Women Doctors in Gilded-Age Washington* (Chicago: University of Illinois Press, 1987), pp. 29, 57, 79, 90–1, 122–3, 149–51.

21. Ibid., pp. 123 and 205, n. 25 (for the use of the gymnasium, Hinds charged adults $10 a year and children $5).

22. *AMD,* 1906–1950. Robinovitch is incorrectly listed in the Royal Society *Catalogue* as Rabinovitch. For examples of her post-1900 publications see "Electric sleep: an experimental study with an electric current of low tension; illustrated with cardiac and respiratory tracings," *Journal of Mental Pathology,* **7** (1905–1906): 172–7; "Méthode de rappel à la vie des animaux en syncope chloroformique et des animaux en mort apparante causée par l'électrocution . . .," *Comptes Rendus des Séances et Mémoires de la Société de Biologie,* **64** (1908): 167–9; (with G. W. Stiles, Jr.) "A chemical basis for the treatment of tuberculosis; experiments on the action of steapsin and insulin on tubercule bacilli," *American Review of Tuberculosis,* **9** (1924–25); 587–612, and *Chemical News* (London), **130** (1925); 66–70.

23. See letter labeled, "What one noble woman is doing," from Mary W. Niles, M.D., *Medical Missionary Record,* **7** (1892): 157. Only the following information has been collected about the remaining five women of this group graduating in the 1880s—Bryson, Post, Peckham, Sargent, and Cameron Moody: Louise Fiske Bryson, whose

two papers appeared in 1889 and 1890, graduated from the Hahnemann Medical College of Chicago in 1887. She later practiced in New York City. Sarah Post (M.D., 1882, Woman's Medical College of the New York Infirmary for Women and Children) later worked as physician in the infirmary's outpatient department. In addition to her two 1884 papers on diagnostic methods, her pre-1900 publications included a textbook, *Massage, a Primer for Nurses* (1890), and *Psychical Experiences, a Diary* (1893). Grace Peckham also worked at the New York Infirmary for Women and Children, serving as assistant resident physician there in the 1880s. Her four papers (1883–1891) covered a variety of topics from gynecology to nervous disease. Elizabeth Sargent (M.D., 1880, Cooper Medical College, San Francisco—later the Medical Department of University College of San Francisco) practiced in San Francisco, specializing in ophthalmology. Kate Cameron Moody was a dentist, who practiced for a number of years in Mendota, Illinois. Her 1885 paper was read before the Illinois State Dental Society. In 1893 she and her husband, Joseph Moody, also a dentist, moved to Los Angeles, where they built up a large practice. Joseph Moody served as president of the Southern California Dental Association and taught in the Dental Department of the University of Southern California. (Bryson, Post, and Sargent are listed in Polk's *Medical Directory* (1900); some information about Cameron Moody came from an obituary for Joseph Moody in *Annual Publications, Historical Society of Southern California,* **8** (1909–1910): 138–9).

24. Florence E. Oblensky, "Anita Newcomb McGee," *Military Medicine,* **133** (1968): 397–400; Moldow, *Women Doctors,* pp. 90, 122, 124–5, 152–3; *WWWA,* p. 519; *AMS,* 1910–1933.

25. Oblensky, "McGee," p. 397.

26. Moldow, *Women Doctors,* p. 122.

27. McGee had all along experienced difficulty in finding satisfactory child-care arrangements, and often had to ask her mother to look after her children. These problems and the loss of her son appear to have sapped her enthusiasm to continue in private practice (ibid., pp. 124–5).

28. Anon., "History as it is made," *The Chautauquan,* **28,** (1898–1899): 186.

29. Anita Newcomb McGee, "Women officers of the Association for the Advancement of Science," *Science,* **59** (1924): 577. McGee's resignation from the post of secretary of the Anthropology Section in 1894 was probably occasioned by the birth of her second child.

30. Lilian Welsh, M.D., LL.D., *Reminiscences of Thirty Years in Baltimore* (Baltimore, Md.: Norman, Remington Co., 1925); Hanny Rohner, "Die ersten 30 Jahre des medizinischen Frauenstudiums an der Universität Zürich 1867–1897," *Zürcher Medizingeschichtliche Abhandlungen,* Neue Reihe Nr. 89, Juris Druck & Verlag, Zurich, 1972, 96 pp., on p. 83; Florence R. Sabin, "Doctor Mary Sherwood," *Goucher Alumnae Quarterly* (July 1935): 9–13; *WWWA,* p. 742; *AMD,* 1906–1931.

31. The joke is attributed to Dr. Osler, of the Johns Hopkins Hospital (Welsh, *Reminiscences,* p. 44).

32. Sabin, "Mary Sherwood," p. 10.

33. Alumni Graduate Record Form, Brown University; unidentified clipping from a Providence, Rhode Island, newspaper (19 February 1952), Christine Dunlap Farnham Archives, Brown University; *WWWA,* p. 787.

34. Ellen A. Stone, "A fresh air school," *Journal of Out Door Life,* May 1908; "Medical inspection in the public schools of Providence," *Providence Medical Journal,* **12** (1911): 215–17; "The midwives of Rhode Island," ibid., (1912); "Medical supervision of school children in Rhode Island," ibid., (1913); "Health supervision of school children in Rhode Island," *Bulletin of State Board of Health of Rhode Island,* (1915); "A survey of cripples in Rhode Island," *American Journal of Care for Cripples,* **3** (1916): 113–17.

35. About four of these physicians—Bishop, Bloom, Mitchell Kydd, and Mitchell—only the following information has been collected: Frances Bishop (b. 1864) graduated from the University of Michigan Medical Department in 1893 and later practiced in St. Louis, Missouri. Her 1893 paper describing a method for staining tubercle bacilli reported work done in the laboratory of clinical medicine at Michigan (see microscopy section of bibliography). Selina Bloom (b. 1866) graduated from the Laura Memorial Woman's Medical College of Cincinnati in 1892. She specialized in ophthalmology, became a fellow of the American Medical Association, and practiced in Brooklyn, New York, until at least the 1940s. Her two papers listed in the bibliography discussed studies in ophthalmology carried out in the Physiological Institute at the University of Leipzig about 1898. Mary Mitchell Kydd (M.D., 1890, Woman's Medical College of the New York Infirmary) was a coworker of Mary Putnam Jacobi in the 1890s. She practiced until at least 1916 in Jersey City, New Jersey. Charlotte Mitchell, whose 1900 bacteriological paper with Charles Richet appeared in the *Mémoires de la Société de Biologie,* was probably the same Mitchell who graduated from the Woman's Medical College of Pennsylvania and took her licence in 1904. She later practiced in Philadelphia. (See *AMD,* various editions, 1906–1942, for entries on Bishop, Bloom, Mitchell Kydd, and Mitchell.) Papers by three other women—White, Crotty, and Willard—are also listed in the medicine section of the bibliography. Victoria White was a coworker of Mary Putnam Jacobi about 1880. She does not appear in the medical directory sources. For Crotty and Willard see chapters 3 and 16, respectively.

36. Obituary, *NYT,* 2 August 1947, 13: 3; Elinor Rice Hays, *Those Extraordinary Blackwells* (New York: Harcourt, Brace and World, 1967), pp. 215, 258–9, 278–9, 300.

37. Obituary, *NYT,* 2 March 1946, 13: 7; *WWWA,* p. 487.

38. Anon., "Dr. Helen Baldwin," *Medical Woman's Journal,* **47** (1940): 372–3; anon., "In Memoriam: Helen Baldwin, M.D.," *Journal of the American Medical Women's Association,* **1** (1946): 61; *WWWA,* pp. 70–1.

39. See, for instance, Helen Baldwin, "Acetonuria following chloroform and ether anaesthesia," *Journal of Biological Chemistry,* **1** (1906): 239–50; "Changes in the bile occurring in some infectious diseases," ibid., **4** (1908): 213–20; "Influence of lactic acid ferments on intestinal putrefaction in a healthy individual," ibid., **7** (1909): 37–48.

40. The first woman whose name appeared on the British Medical Register (1859) and who was thus entitled to practice was Elizabeth Blackwell. A native of Bristol, Blackwell grew up in the United States and took an M.D. in 1849 at Geneva Medical College, New York (see n. 5). Elizabeth Garrett, having qualified by passing the Society of Apothecaries' examinations in 1865 (before the society altered its regulations to exclude women), was the next to have her name added (1866). Blackwell and Garrett remained the only women on the Medical Register until 1877.

41. See, Catriona Blake, *The Charge of the Parasols. Women's Entry to the Medical Profession* (London: The Women's Press, 1990), and Bonner, *Ends of the Earth,* chapter 6.

42. The medicine section of the bibliography lists papers by four authors—Fowke, Owen, Sime, and Frankland—whose contributions were on somewhat general topics. Fowke's report on the condition of pauper children boarded out by the state was presented at an international medical congress in 1891. Owen and Sime each published short notes in *Nature*; Owen, who was from London, reported her observations of "unconscious bias in walking" in 1884. Sime wrote about the

work of her brother, Professor George Wilson of Edinburgh, on a method of testing for color blindness (his contributions to the subject had been substantial). For Frankland see chapter 6.

43. Onfel Thomas, *Frances Elizabeth Hoggan. 1843–1927* (Newport, Monmouthshire: R. H. Johns, 1971); Bonner, *Ends of the Earth,* pp. 38–9; E[nid] Moberly Bell, *Storming the Citadel: the Rise of the Woman Doctor* (London: Constable and Co., 1953), p. 144; *British Biographical Archive,* microfiche edn., no. 788, entry from Frances Hays, *Women of the Day* (London: Chatto and Windus, 1885).

44. Suslova and Morgan were the first two of a group known as the Zurich Seven, young women from four countries, who, by their early success opened the way for the acceptance of women students in recognized universities. The other five were Louisa Atkins from England, a second Russian, Mariia Bokova, a Scot, Elizabeth Walker, an American, Susan Dimmock, and a Swiss, Marie Vögtlin (Bonner, *Ends of the Earth,* pp. 31, 40).

45. The disagreement arose because of Garrett's determination that all professional work, including surgery, must be done by women, although neither she nor Morgan had yet had extensive surgical experience. When the question arose of performing an ovariotomy, then considered a very risky, almost experimental operation, the hospital committee decided that it should not be done in the hospital lest the death of the patient damage its reputation. Garrett was nevertheless determined to proceed, and Morgan resigned in protest. (Garrett actually performed the operation in a specially prepared room in a private house, with Sir Thomas Smith of St. Bartholomew's Hospital present as an observer. All went well, and within a few years, as Garrett's confidence increased and more women became qualified as surgeons, the situation changed entirely—see Moberly Bell, *Storming the Citadel,* p. 144.)

46. See Frances Elizabeth Hoggan, *Medical Women for India* (Bristol: J. W. Arrowsmith, 1882). The Indian medical schools opened to women without much fuss. Indeed, one of the most distinguished medical women of her generation, Dr. Mary Bird Scharlieb (wife of a British barrister in Madras), graduated from the Madras Medical School in 1878. About seven years later she founded the Royal Victoria Hospital for Women in Madras, although she later returned to London. The low standard of education among Indian girls and young women and the restrictions of caste and the purdah meant that the number of female Indian students admitted to the medical schools was low, but a start was made. In Britain the London School of Medicine for Women did what it could, helping in the selection of suitable candidates and assisting in their training (Moberly Bell, *Storming the Citadel,* Chapter 7; Blake, *Charge of the Parasols,* pp. 175–7; G. Jones, "Women and eugenics in Britain: the case of Mary Scharlieb, Elizabeth Sloan Chesser and Stella Browne," *Annals of Science,* **52** (1995): 481–502).

47. Frances Elizabeth Hoggan, *American Negro Women during their First Fifty Years of Freedom* (London: Personal Rights Association, 1913?). A second short article by Hoggan on the same topic was, "The Negro problem in relation to white women," in *Papers on Interracial Problems,* Universal Races Congress, I. London, 1911.

48. Moberly Bell, *Storming the Citadel,* pp. 99, 103, 178; Blake, *Charge of the Parasols,* pp. 166, 187, 191.

49. *Who was Who,* vol. 3, p. 1299; Barbara McLaren, *Women of the War* (New York: George H. Doran, 1918), pp. 53–8; "Dr. Florence Stoney," *Times,* 8 October 1932, 14c; Dr. Mabel Ramsay, "Dr. Florence Stoney," ibid., 12 October, 1932, 7c; *DNB,* 2d Supplement, vol. 3, pp. 429–31, entry for Stoney, George Johnstone.

50. *Times,* obituary, 8 October 1932.

51. Mabel L. Ramsey and Florence A. Stoney, "Anglo-French Hospital, No. 2, Château Tourlaville, Cherbourg," *British Medical Journal,* **1** (1915): 966–9. In the setting up of the unit's portable X-ray apparatus at Cherbourg, Stoney had the assistance of her sister Edith, a wrangler in the Cambridge Mathematical Tripos, later a lecturer in physics at the London School of Medicine for Women. They generated their own electricity from a stream running through the château grounds. In 1915 Edith Stoney resigned her lectureship to join the Scottish Women's Hospital, Girton and Newnham College Unit, as engineer and radiographer at the hospital the group established at Troyes in France. She then served with the same unit in the Tent Hospital sent in September 1915 with the French Expeditionary Army to Gevgheli, Serbia, retreating with the army to Salonika. By 1917 she was back with a hospital in France, where she served for another two years. After the war she was lecturer in physics in the Women's Department of King's College.

52. Formal equality for women doctors in the British army did not come until 1950. From then on they were commissioned into the Royal Army Medical Corps in exactly the same way as their male colleagues and given the same rank titles. However, during the Second World War there was no practical distinction between men and women army doctors, except that women were barred from serving in the front line or in units immediately behind the front line, and they did not formally hold the King's Commission (Moberly Bell, *Storming the Citadel,* pp. 186–8).

53. Stoney's post-1900 journal articles include the following: "Chronic congestion treated by electricity," *Archives of the Roentgen Ray,* **10** (1905–1906): 325–9; "On the results of treating exothalmic goitre with X-rays," *British Medical Journal,* 1 (1912): 476–80; "The 'soldier's heart' and its relation to thyroidism," ibid., 1 (1916): 706; "On the connection between 'soldier's heart' and hyperthyroidism," *Lancet,* **1** (1916): 777–80; "Fibroid uterus treatment by X-rays," *British Medical Journal,* 2 (1917): 723; "Two skiagrams," *Archives of Radiology and Electrotherapy,* **24** (1919–1920): 352.

54. *British Biographical Archive,* microfiche edn., no. 268 (entry from *The Medical Who's Who, 1914*).

55. See Alice M. Corthorn, "Medical examination of girls in secondary schools," in *Medical Examination of Schools and Scholars,* ed. Theophilus Nicholas Kelynack (London: P. S. King and Son, 1910), pp. 78–94.

56. Alice M. Corthorn and F. Hobday, "An interesting case of hemiplegia in the dog," *Veterinary Journal,* **13** (1906): 78–81; "An interesting case of hemiplegia in the dog due to a tumor of the spinal chord: with appended note by F.E. Batten," ibid, **14** (1907): 155–9. Corthorn's other post-1900 articles include the following: "Albuminuria in plague," *British Medical Journal,* 2 (1901): 671–2; "Innoculation in the incubation stage of plague," ibid, 1 (1902): 198–9; "The coagulating power of the blood in plague," ibid., 1 (1902): 1143–4.

57. Mélanie Lipinska, *Histoire des Femmes Médecins depuis l'Antiquité jusqu'à nos Jours* (Paris: Librairie G. Jacques et Cie, 1900), p. 411.

58. Henry T. Butlin, "Introductory address on research in medicine and women in research. Delivered at the London School of Medicine for Women," *British Medical Journal,* 2 (1911): 835–7. Brink's papers, and also those by Flemming, are listed in the physiology-anatomy section of the bibliography.

59. Sharpe's two post-1900 papers were: "A plea for static electricity," *Archives of the Roentgen Ray,* **11** (1906–1907): 221, and "A case of osteoarthritis treated with X-rays," ibid, **12** (1907): 65–7.

60. A paper by Mary Kingsley on "Nursing in West Africa," not listed by the Royal Society, has been added to the bibliography (for Kingsley see chapter 14).

61. Catherine Jane Wood, *A Handbook of Nursing for the Home and Hospital, with a Glossary of the Most Common Medical Terms* (London: Cassell, 1878); *Food and Cookery for Infants and Invalids* (International Health Exhibition of 1884 Handbooks, London: Clowes, 1884); *A Handbook for the Nursing of Sick Children, with a Few Hints on their Management* (London: Cassell, 1889); *Boards of Guardians and Nurses* (London: Kegan Paul, Trench and Trübner, 1893); *Cottage Lectures on Home Nursing* (London: Christian Knowledge Society, 1893).

62. *Who was Who,* vol. 3, p. 1481.

63. See, for instance, the bibliographies in Monica E. Baly, *Florence Nightingale and the Nursing Legacy* (London: Croom Helm, 1986), pp. 226–9, and in *Florence Nightingale and her Era. A Collection of New Scholarship,* eds. Vern. L. Bullough, Bonnie Bullough, and Marietta P. Stanton (New York: Garland Publishing, Inc., 1990), pp. 324–65.

64. See F[rancis] B[arrymore] Smith, *Florence Nightingale. Reputation and Power* (London: Croom Helm, 1982), pp. 34–6.

65. Ibid., pp. 62–3.

66. Ibid., p. 97.

67. I. Bernard Cohen, "Florence Nightingale," *Scientific American,* **250** (1984): 128–37.

68. Smith, *Florence Nightingale,* pp. 190–1.

69. Lois A. Monteiro, "On separate roads: Florence Nightingale and Elizabeth Blackwell," *Signs,* **9** (1983): 520–33.

70. For a guide to Nightingale's vast output see W. J. Bishop, *Bio-bibliography of Florence Nightingale,* completed by Sue Goldie (London: Dawsons, 1962, 160 pp.), and Sue Goldie and W. J. Bishop, *A Calendar of the Letters of Florence Nightingale* (Oxford: Oxford Microform Publications for the Wellcome Institute for the History of Medicine, 1983).

71. By 1894 coeducation had overtaken segregation by sex in American medical education (Bonner, *Ends of the Earth,* p. 149).

72. The opening to women of Swiss universities in the 1860s and 70s and the possibility of medical training in Paris from about the same time were critical initially. The acceptance of women as auditors in many of the German state universities by the 1890s (and as formally enrolled students shortly after the turn of the century) meant that by the time of the First World War even conservative Germany had surpassed Britain and the United States in the training of women doctors (ibid., pp. 118–19).

73. Patricia Ruth Hill, *The World Their Household,* (Ann Arbor, Mich.: University of Michigan Press, 1985), especially pp. 44, 126–30.

74. Moberly Bell, *Storming the Citadel,* pp. 111–25.

Part 2

Mathematical, Physical, and Earth Sciences

Chapter 8

MATHEMATICIANS AND STATISTICIANS, MAINLY OF THE 1890S, BUT REMEMBERING MARY SOMERVILLE

Most of the nineteenth-century contributions by women to research in mathematics, including applied mathematics and statistics, came from British, American, and Russian workers. The Royal Society *Catalogue* indexes forty-two papers by fourteen British women,[1] fifteen by six Americans, and ten by two Russians. The remaining work came mainly from women in Italy (seven papers), France (six papers), and The Netherlands (five papers). Overall, papers in mathematics constitute about 2.5 percent of the nineteenth-century papers by women indexed by the Royal Society.

American women

Of the American authors whose work is listed in the bibliography,[2] five were from the Midwest or received their undergraduate education there, MacKinnon and Growe at the University of Kansas, Winston at the University of Wisconsin,[3] Hayes at Oberlin College, Ohio, and Schottenfels at Northwestern University in Illinois. The others, Wood and Ladd, were from the eastern seaboard. Wood graduated from Smith College, Ladd from Vassar. Although Ladd's work in mathematical logic in the late 1870s and early 1880s was noteworthy, much of her later research was in optics and experimental psychology. In the latter field she became an acknowledged expert and for her contributions was starred in *American Men of Science* (see chapter 15). Four of the Americans held Ph.D. degrees. Winston's was from the University of Göttingen; Wood, Ladd, and MacKinnon received theirs from Yale, Johns Hopkins, and Cornell universities respectively.

ANNIE MACKINNON[4] (1868–1940) was Canadian by birth and parentage. Born in Woodstock, Oxford County, Ontario, on 1 June 1868, she was one of the five children of Annie (Gilbert) and Malcolm MacKinnon. The family moved to the small town of Concordia in north-central Kansas when she was about two years old and her early education was at local schools. In 1884, at the age of sixteen, she enrolled in preparatory courses at the University of Kansas at Lawrence and the following year entered as a first-year student. After taking a B.S. with Phi Beta Kappa honors in 1889 she stayed on as a graduate student in English and mathematics (M.S., 1891), while also teaching mathematics at Lawrence High School (1890–1892). In 1893 she entered Cornell University's graduate school to study mathematics and mathematical physics. An outstanding student, she received her Ph.D. in 1894, having held a mathematics fellowship during her final year. Her dissertation research in algebraic geometry appeared in the *Annals of Mathematics* (1894) in a long article entitled, "Concomitant binary forms in terms of the roots."

Supported by fellowships from the Association of Collegiate Alumnae (now the AAUW) and the Women's Education Association of Boston, she then spent two years at the University of Göttingen, joining her fellow Americans Mary Winston (see below) and Margaret Maltby (see chapter 9) who had gone there the previous year. As a student of mathematics professor Felix Klein, MacKinnon took an active part in his famous seminars, speaking five times over the course of her four semesters in Göttingen, twice at the seminar on analysis and three times at the number theory seminar.[5]

On her return to the United States in 1896 she became professor of mathematics at Wells College in Aurora, New York, where she taught for five years until her marriage to Edward Fitch in 1901. She does not appear to have continued work in mathematics after that, but became active in the League of Women Voters and in college life at Hamilton College in Clinton, New York, where Edward Fitch was professor of Greek and for a time dean. She died in Clinton, on 12 September 1940, at age seventy-two.

BESSIE GROWE[6] was born on 14 September 1874 in Kentucky, and grew up in Frankfurt, North Dakota. As an undergraduate at the University of Kansas she concentrated on mathematics (A.B., 1897). Her short paper on algebraic geometry appeared in the *Kansas University Quarterly* the same year; a further year of graduate-level work (1898–1899) led to a second note in the same area, published in the *Bulletin of the American Mathematical Society* in 1901. She later became a schoolteacher in Lawrence, Kansas.

MARY WINSTON[7] (1869–1959) was the first American woman to receive a Ph.D. in mathematics from a European university, passing the required examinations, magna cum laude, at Göttingen in 1896. Her 1895 note in *Mathematische Annalen* on hypergeometric functions, the only article she published before 1900, was based on a talk she gave in one of Felix Klein's seminars in 1894 (her talk given the previous semester was the first by a woman at these seminars).

She was born in the town of Forreston, in northern Illinois, on 7 August 1869, the fourth of seven surviving children of Thomas Winston, a country doctor, and his wife Caroline (Mumford). She received much of her early instruction from her mother. At age fifteen (along with an older brother) she was sent to the University of Wisconsin. After graduating (A.B., 1889), she taught mathematics for two years at Downer College in Fox Lake, Wisconsin. She then held a graduate fellowship (1891–1892) at Bryn Mawr College, where she studied under British mathematician, Charlotte Scott (see below).[8] The following year, while at the recently opened University of Chicago as a courtesy fellow, she was introduced to Felix Klein, who was visiting Chicago for the Congress of Mathematicians held in connection with the 1893 World's Columbian Exposition. With his encouragement, Winston, and also Margaret Maltby, a physics student at MIT, went to Göttingen in the autumn of 1893. Along with Grace Chisholm from Girton College Cambridge, these two were the first women officially admitted to a Prussian university, though it was specified that they were "exceptional cases"[9]. Winston held an Association of Collegiate Alumnae fellowship during her last year at Göttingen (1895–1896).

She worked in Klein's group on applications of the theory of analytic functions of a complex variable, one of his major areas of interest at the time. Her dissertation research was an examination of solutions of Lamé's equation with applications to the mechanics of the spherical pendulum and to the theory of the top. Although she never presented the work as a journal article, it was cited by Klein in a monograph brought out in 1897.[10]

Winston's three years in Germany were happy ones. She made close friends with Grace Chisholm and with German families. On winter weekends she often joined a party traveling the thirty or forty miles by train to the Hartz Mountains for hiking trips through the forest paths; in summer she went south to the Türingen Forest, visiting, among other places, the Wartburg, the eleventh-century fortress castle near Eisenach where Martin Luther hid from his enemies in 1521–1522. She returned to the United States in 1896 without publishing her dissertation (a requirement for getting her degree). Unable to bring it out in this country, she had it printed in Göttingen the following year, whereupon the degree was duly granted.

She taught mathematics and elementary German at the high school in the town of St. Joseph in western Missouri during the year 1896–1897, and from there applied for a faculty position at Kansas State Agricultural College in Manhattan, Kansas, which then had only one instructor of mathematics, Josephine Harper. Winston's appointment occasioned substantial disagreement among the members of the Board of Regents, but she was hired (at a salary of $1,450 per year) as professor of mathematics and head of her department,[11] which consisted of herself and Josephine Harper. Among the letters of recommendation supporting her application was one from Henry Byron Newson, professor of mathematics at the neighboring University of Kansas in Lawrence. He was warm in his praise: "With Miss Winston in your faculty your College could boast of a mathematician that ranks with the best anywhere in the country . . ."[12]

The college's faculty of about twenty-three, which was heavily weighted on the side of teachers of applied science, included three other women besides Winston and Harper; two taught home economics and one was an instructor of English. Teaching loads were heavy, faculty turnover high, and in the mathematics department two years of courses at high school level had to be provided. As well as her departmental work, Winston did her share of the community service then expected of faculty at an agricultural college. Her talks at Farmers' Institutes covered such topics as "The German and American school systems" and "The republican part of Germany—her universities." She went to considerable effort to encourage students from country districts in the study of mathematics, and during her first autumn at the college raised $50 for a prize for the student from the country or small-town school who placed first in the college entrance examination in arithmetic and algebra. She taught German as well as mathematics, her class in the former subject being one of the most popular in the college.[13] In 1900, however, when she resigned her position to marry Henry Newson, this period of college-level teaching came to a close. A short entry in the college's *Faculty Records* for June 1900 showed the high regard in which she was held by her colleagues and their regret at losing her.[14]

She did not publish any original work in mathematics after her doctoral research, but in 1902 her translation from the German of the printed version of David Hilbert's celebrated address to the 1900 International Congress of Mathematicians came out in the *Bulletin of the American Mathematical Society.* In this famous lecture, Hilbert outlined what he considered to be the most important problems in mathematics at the turn of the century; Winston's translation appeared in print once again, seventy-four years later, in the proceedings of a 1975 symposium on the developments that had followed from the problems Hilbert presented in 1900.[15]

In 1910, at the age of forty-one, she was left a widow, without a pension and with three children, the youngest only three months old. Though there were openings for mathematics teachers at the University of Kansas about this time she was not considered for a position.[16] Three years later she became assistant professor at Washburn College in Topeka, Kansas; though the job was poorly paid, it allowed her to spend her weekends with her children (looked after by her parents in Lawrence).

Always interested in broader aspects of education, she was active in the Kansas Association of Teachers of Mathematics and chaired the organization in 1915. In 1921, however, after opposing Washburn's president on an academic freedom issue involving the dismissal of a faculty member, she resigned her position. Returning to her home state of Illinois, she spent the rest of her career at Eureka College, near Peoria, retiring in 1942 when she was almost seventy-three. During her later years she and one of her sisters started a chapter of the AAUW in Eureka; international relations became her special interest. Following her death at the age of ninety (5 December 1959) an annual lecture fund was established by her children—the Mary Winston Newson Memorial Lecture on International Relations, at Eureka College.

IDA SCHOTTENFELS[17] (1869–1942) graduated from Northwestern University (A.B., 1892) and then studied at Yale and the University of Chicago (Ph.M., 1896). Thereafter she taught mathematics for five years in Chicago high schools and appears to have been appointed instructor of mathematics at the New York Normal School (later Hunter College) about 1901. For the year 1908–1909 she was the James McLaughlin Professor of Mathematics at Hamline University in St. Paul, Minnesota.

She regularly attended meetings of the American Mathematical Society, going to at least one almost every year between 1900 and 1915. Her continuing interest was the theory of finite groups; she also reported working on analysis, notably complex function theory. She published two papers in 1900 but generally presented her ideas at meetings.[18] Her activity in mathematical research continued until the early 1930s, though by then she attended meetings only occasionally. She died on 11 March 1942, at age seventy-two.

RUTH WOOD[19] (1875–1939), the daughter of Kate (Pond) and S. Eugene Wood, was born in Pawtucket, Rhode Island, on 29 January 1875. After receiving her early education in local schools, she went to Smith College (B.L., 1898). She continued her studies at Yale University (Ph.D., 1901) and then taught mathematics for a year at Mount Holyoke College before returning to Smith as instructor of mathematics in 1902. Following a year of study leave at Göttingen University in 1908–1909 she was promoted to associate professor. In 1914 she became full professor and she served as department head from 1922 until 1928. Her research interests were in non-Euclidian geometry but except for two papers presented at meetings of the American Mathematical Society (1900 and 1902)[20] she brought out little. She died on 5 May 1939, at the age of sixty-four.

ELLEN HAYES[21] (1851–1930), born in the farming community of Granville in southern Ohio, on 23 September 1851, was the oldest of the six children of Ruth (Wolcott) and Charles Coleman Hayes. Charles Hayes was a tanner by trade.

Growing up in a pioneer Midwestern town with forward-looking parents who imposed few restrictions, Ellen developed considerable independence of outlook. She learned to read from her mother, a former schoolteacher, and from age seven to sixteen attended a one-room public school. She then taught in a similar school for about five years to accumulate some money, and in 1872 entered the preparatory department of Oberlin College in northern Ohio. After graduating in 1878 (A.B., with a concentration in science and mathematics), she spent a year as principal of the women's department at Adrien College in Michigan. In 1879 she joined the faculty of newly opened Wellesley College as teacher of mathematics. Promoted to assistant professor in 1882 and associate the following year, she became full professor and head of her department in 1888. In 1897 she was given the position of professor of applied mathematics.

Primarily a teacher, she published few original scientific articles; only her three notes in *Science* (1893 and 1896) appear in the bibliography. She had a special interest in the practical applications of mathematics and taught the college's courses in mathematical physics (including mechanics and thermodynamics) as well as mathematical astronomy. In 1887–1888 she spent some time at the Leander McCormick Observatory at the University of Virginia where she calculated the orbit of the newly discovered Minor Planet 267. Her 1904 articles reported comet orbit calculations and a mathematical analysis of the annual path described by the shadow of a plummet bead.[22] Her major published works were her textbooks, all of which stressed applications. *Lessons on Higher Algebra* appeared in 1891 and was reissued, "with an appendix on the nature of mathematical reasoning," in 1894. It was followed by *Elementary Trigonometry* (1896), *Algebra for High Schools and Colleges* (1897), and *Calculus with Applications; an Introduction to the Mathematical Treatment of Science* (1900). She discussed her ideas in the broader area of the theory of knowledge and reasoning in a general work that came out after her retirement, *How do You Know? A Handbook of Evidence and Inference* (1923).

She was one of the original group of Wellesley faculty members, joining the staff only four years after the college opened. Even in her student days, when she worked her way through Oberlin, she had already been something of a nonconformist, rebelling against life's inequalities, rejecting organized religion in a time when that was a risky position to take, and in a great many day-to-day matters following her own ideas rather than going along with current fashions and trends. As a fellow teacher, some of her colleagues found her "difficult," and indeed her special position of professor of applied mathematics appears to have been created mainly to give her a somewhat separate sphere of activity and so reduce friction between her and other mathematics staff members.

A committed and outspoken socialist and an active member of the Socialist Party, she stood as the party candidate for secretary of state for Massachusetts in 1912, the first woman to be nominated for a state office; she did remarkably well, getting more votes than any other socialist candidate. The same year, one of unprecedentedly violent labor disturbances in the New England cloth mills, she and her like-minded colleague Vida Scudder were invited to speak at a meeting sponsored by the Progressive Women's club of Lawrence, Massachusetts, where the unionized woollen mill workers were on strike. The speeches (on the subject of temporarily evacuating workers' children as a safety precaution) were hardly inflammatory, but the appearance of the two Wellesley women at the meeting made headlines in the local newspapers and conservative voices in Boston advocated their immediate dismissal from the college. This provoked a certain amount of debate among faculty and administrators, but the college community closed ranks and little action was taken against the offenders.[23]

Though her academic credentials were equal to or better than those of a number of the other early faculty appointees at Wellesley[24] and she set high standards in her teaching, as the years passed her lack of advanced training began to be felt and probably increased her difficulties with younger colleagues. She retired in 1916 at the age of sixty-five, after thirty-seven years on the faculty, and stayed on in a house she had built near the college. The Wellesley trustees did not appoint her professor emeritus.

Even in her later years, her interests in women's rights and her activities on behalf of left-wing social causes continued undiminished. During the early 1920s, when whipped-up anti-Red hysteria was sweeping the country, she worked to raise money for Russian children orphaned by the Revolution; in 1927, when she was seventy-six, she was arrested for demonstrating in the Sacco-Vanzetti affair, and became one of the defendants in the American Civil Liberties Union's successful challenge of the legality of the arrests. Throughout these years she brought out a monthly magazine, *The Relay,* designed to help society's less fortunate; it covered current events and included educational and scientific articles. One of her last projects was helping to organize the Vineyard Shore School for Women Workers in Industry, in West Park, New York. She moved to West Park shortly before the opening of the school in 1929 and taught there informally for a short time.

During her retirement she also continued her nonmathematical writings, begun while she was still at Wellesley with *Two Comrades* (1912) and *Letters to a College Girl* (1909), a work in which she stressed the need for girls to study science and mathematics. *Wild Turkeys and Tallow Candles* (1920) was a description of frontier life in Ohio around 1850; it undoubtedly drew much from her close relationship during her childhood with her farmer grandparents. Her historical novel *The Sycamore Trail* appeared in 1929.

She held memberships in the AAAS and the Astronomical and Astrophysical Society of America, and was a founding member of the History of Science Society. She died in West Park, New York, on 27 October 1930, just after her seventy-ninth birthday.

Except for Ida Schottenfels, whose work in research was persistent if not particularly distinguished, none of these six early women mathematicians remained active in original work for very long after their student years. Hayes, whose career began in the 1870s, almost twenty years before the others, was certainly energetic and productive, but she wrote textbooks. Nevertheless, she and Wood, both of whom held faculty positions at small but respected East Coast women's colleges, had what to all appearances were reasonably satisfactory academic careers, becoming full professors without undue delay. MacKinnon and Winston, on the other hand, despite their advanced training, were less successful professionally. Indeed MacKinnon left mathematics on her marriage, after five years of teaching at a small college, and Winston, though she held a position at a coeducational state college early in her career, likewise gave this up on marriage and did not succeed in finding a comparable post again; afterward she taught at smaller, less well-known colleges. Schottenfels, despite her sustained interest in research, was also a teacher in small colleges.

The career of Christine Ladd (1847–1930), the remaining American woman mathematician whose work is listed in the bibliography, is discussed in chapter 15, but the following sketch briefly indicates the nature of her early contributions to mathematics and symbolic logic made before she switched much of her effort to psychology.

In 1877 and 1878, Ladd, an early Vassar College graduate who had continued her studies at Washington and Jefferson College in Washington, Pennsylvania, and also informally at Harvard, published three short papers in the newly founded American journal, *The Analyst,* as well as a number of solutions to mathematical questions in that journal and in the London *Educational Times.* Her special request for admission to Johns Hopkins University graduate school was granted in 1878, largely thanks to this published work, and she completed the requirements for a Ph.D. by 1882. Her degree was awarded at the University's semicentennial celebration in 1926. The London Mathematical Society, somewhat more flexible than Johns Hopkins, elected her one of its women members in 1881.

Ladd's research was carried out under the guidance of logician and psychologist Charles Peirce, then on the Johns Hopkins faculty. His work on the algebraic treatment of what he called "relatives," now part of set theory, he and Ladd applied to symbolic logic. In her dissertation she extended this to the classical syllogism, introducing more modern calculational (algebraic) notation that allowed a new analysis of this ancient subject and indicated conclusions that helped remove a logical stumbling block of very long standing. (The general rule she developed provided an efficient way of testing the validity of any syllogism.) Ladd's study is considered "the first adequate treatment of classical syllogism."[25] Entitled "On the algebra of logic," it was published in *Studies in Logic by Members of the Johns Hopkins University* (1883), which Pierce edited.[26]

Interestingly, the work of Pierce and Ladd on "relatives" is again attracting attention; Zellweger's studies on increasing the flexibility and adaptability of logical notation bring out the significance of their early application of symmetry considerations to propositional logic, pointing to the possibility of higher levels of abstraction.[27]

Ladd lectured in logic and psychology at Johns Hopkins University from 1904 to 1909 and then taught the same subjects at Columbia University from 1910 until 1930. She continued to publish work in logic as well as making her extensive contributions to psychology and the theory of color vision.[28] However, in her later years, although she fully recognized that common logic had to adopt mathematical symbolism, she failed to appreciate the importance of the new ideas being developed by logicians Bertrand Russell, Gottlob Frege, and Giuseppe Peano.[29] The recognition, both national and international, she received came in the main for her research in the theory of color vision.

British women

The forty-two papers in mathematics, applied mathematics, and statistics by fourteen British women form the largest national contribution by women to these fields in the nineteenth century.[30] Eight of the British authors listed in the bibliography are known to have been educated at Cambridge, including the three who were most productive mathematically—Charlotte Scott, Grace Chisholm Young, and Hilda Hudson. Four were students at London colleges. Those who carried out work in statistics were collaborators of Karl Pearson at University College London. All were late nineteenth- early twentieth-century women, except for two contributors from the earlier part of the century, Mary Somerville (whose journal publications were in physics and astronomy) and Ada Lovelace.

CHARLOTTE SCOTT (1858–1931),[31] the earliest of the Cambridge women, spent much of her life in the United States and was the first woman to be recognized in the American mathematical community. For forty years she taught at Bryn Mawr College in Pennsylvania, where the mathematics graduate program she built was at the time the only significant one in a women's college in America. The sixteen papers by her listed in the bibliography constitute the largest pre-1901 contribution to original research by a woman mathematician working in the United States.

She was born in Lincoln, on 8 June 1858, the second of the seven children of the Rev. Caleb Scott and his wife Eliza Ann (Exley). Her father, a Congregational minister, was from 1865 principal of the Lancashire Independent College, Whalley Range, Manchester a training college for nonconformist ministers; he later became minister at City Temple, London. Charlotte was taught at home by tutors until she went to Girton College in 1876 on a Goldsmith's Company scholarship.

The position of women students at Cambridge being then still marginal, they could only take the Mathematical Tripos examinations, the final baccalaureate honors examinations, on an informal basis. The fact that in 1880 Scott was ranked equal (unofficially) to the eighth male student in order of merit in these examinations caused quite a stir, even at the national level; higher mathematics was then generally considered to be beyond the mental capacity of women. As the story is told, when the results were read out at the Senate House, the name of the male eighth "wrangler" was drowned by shouts of "Scott of Girton!" Scott's achievement, and similar successes by three women in other subjects in 1880, had far-reaching effects and led directly to the formal opening of the Tripos examinations to women the following year. Thereafter the names and positions of successful women candidates were recorded in the University Calendar, although many years were to pass before women were awarded Cambridge degrees.[32] Scott stayed on at Girton as resident lecturer in mathematics until 1884, studying under the guidance of the eminent algebraist Arthur Cayley; she took a London B.Sc. (first class honours) in 1882 and a London D.Sc. in 1885. She was probably the first woman member of the London Mathematical Society, being admitted in 1881. About this time she was also very active in social work, an extremely important part of life in her family, and ran a home for working girls in Fulham, London.[33] Shortly after receiving her doctorate, however, she accepted an associate professorship at recently established Bryn Mawr College, and her "American adventure"[34] began.

Of the original faculty of eight assembled by Martha Carey Thomas, then dean and later president of Bryn Mawr College, twenty-seven-year-old Scott was the only mathematician. Over the course of her long career at Bryn Mawr, she built a strong department and officially advised seven Ph.D. students.[35] Her influence on the overall development of the college was also considerable, perhaps second only to that of Thomas. As a member of innumerable committees and boards, she was involved in a whole range of administrative work, including faculty selection and the setting of undergraduate admission standards (demanding in mathematics), course offerings, and mathematical prerequisites. Securing funding for a good mathematics library was one of her continuing concerns.

She was an exceptionally capable and gifted teacher, and although she had little time for the lazy she did her utmost to help less able students who worked hard. Her sense of moral responsibility was a strong one, and not to be subordinated to administrative convenience. In cases of hardship or handicap she spared no effort to provide whatever assistance she could. There are records of her setting out her opinion over no less than seven pages on behalf of a student who had been dismissed because of a crippling illness; she argued that it was nothing short of cruel to deprive the girl of the possibility of intellectual work when she was already limited in other activities.[36] Despite an initial agreement that her lecturing responsibilities would be decreased as other commitments became greater, her load of ten to eleven hours a week remained unchanged over thirty years. In her lectures to advanced students Scott brought a certain elegance and excitement; she gave them a sense of mathematical style.

During her early years at Bryn Mawr, unhappy with the time and effort she was obliged to expend on entrance examinations, she started a move to establish a national test. The College Entrance Examination Board began work in 1901; Scott was the chief examiner in mathematics in 1902 and 1903, and the standards she set then changed little over more than eight decades.

She was very active in the mathematical community at large, holding memberships in both the London and Edinburgh Mathematical Societies, the Deutsche Mathematiker-Vereinigung, and the Circolo Matematico di Palermo, as well as honorary membership in the Amsterdam Mathematical Society. She went to Europe almost every year to attend international meetings and made special efforts to persuade leading mathematicians to visit the United States. Her twenty-two-page report on the 1900 International Congress of Mathematicians in Paris appeared in the *Bulletin of the American Mathematical Society* that year. She arranged for every one of her doctoral students to spend at least one year at some mathematical center in Europe. In the United States she was influential in the early development of the New York Mathematical Society, joining in 1891 and from the beginning contributing to its *Bulletin*. The only woman listed as an officer when the society was reorganized as the American Mathematical Society, she served on the first council (1894). From 1895 to 1897 and from 1899 to 1901 she was again a council member, and in 1905–1906 was one of the two vice presidents, the first woman to hold that office.[37] For twenty-seven years, beginning in 1899, her name appears on the title page of the *American Journal of Mathematics* as one of the two or three mathematicians who cooperated in the preparation of the journal with the editors, first Simon Newcomb, later Scott's close friend Frank Morley. Ranked fourteenth among the top ninety-three mathematicians in America about the turn of the century, she was starred in the 1906 edition of *American Men of Science.*

Scott was a geometer, her research being in the developing field of algebraic geometry, to which she brought "the full resources of pure geometrical reasoning. She was also an enthusiastic searcher and propounder of new ideas and an interpreter of the work of others, adding simplifications and extensions of her own."[38] In the first of her interpretative papers she set out "as a matter of pure pedagogic interest" how more advanced concepts can be made accessible to students; "my contention is not that every step in a rigorous proof can be presented under the guise of elementary mathematics, but that it is quite possible to develop the theory so as to be intelligible and interesting to average students at a much earlier stage then is customary."[39] Her favorite topic was the problem of the analysis of singularities and intersections of plane algebraic curves. Thus in two of her early papers (1892 and 1893—see bibliography) she examined the process of separating from each other the complex ways in which curves cross, join, or fold, so that a figure may be reconstructed to present fewer of these complexities at any one point.[40]

She was an outstanding expositor, and in addition to her more than twenty papers[41] she wrote a much-respected textbook, *An Introductory Account of Certain Modern Ideas and Methods in Plane Analytical Geometry.* The first edition appeared in 1894 and later editions in 1924 and 1961.[42] The term "analytical geometry" did not, in 1894, refer to the material now taught in high schools under that name, but rather to introductory projective algebraic geometry, which presumed a background of Cartesian geometry and differential calculus. The subject was taught at Bryn Mawr as a graduate course. Because of the very rapid changes the field was undergoing during the later years of her career, Scott herself gave up the use of her textbook some time before her retirement and long before many others did.

Even with all her varied interests and activities, she put down no real roots in the United States, and after retiring at the age of sixty-seven returned to Cambridge, where she lived for the rest of her life. Indeed, it would seem that her experience in the United States was perhaps a somewhat lonely one. She had been surprised by the fact that Martha Carey Thomas made a point of maintaining her distance and discouraged her from dropping in for informal conversations during her early years at Bryn Mawr. When she moved from a campus apartment to a house, she persuaded a cousin to come over from England to share it with her and take over domestic responsibilities; occasional visits from her father and one of her brothers helped to keep family ties close. From her late forties on, following a severe attack of rheumatoid arthritis, her health was unpredictable and she suffered increasingly from deafness. Even during her last years of teaching, however, despite the deafness, she lectured "perfectly well,"[43] depending on graduate students to answer questions. On the advice of her doctor she increased her outdoor exercise, taking golf lessons and becoming a reasonably proficient player. She also gardened, and year after year produced a marvelous floral display that lasted from spring to autumn. A new variety of chrysanthemum she developed brought her a prize. In 1922 her former students and seventy members of the American Mathematical Society met at Bryn Mawr for a dinner in her honor. The British logician Alfred North Whitehead, who made a special trip to the United States for the occasion, gave the featured address. During her final years in Cambridge she continued to work on mathematics and in her garden, even though by then very handicapped. She died in Cambridge, on 8 November 1931, at age seventy-three.

PHILIPPA FAWCETT[44] (1868–1948), the young woman who became a legend at Newnham College for her exceptionally fine showing in the Mathematical Tripos examinations of 1890, went to Cambridge eleven years after Charlotte Scott. She was later to teach mathematics at Newnham for ten years before going on to an outstanding career in public education administration.

Born in London, on 4 April 1868, Philippa was the only child of Millicent and Henry Fawcett. Despite the handicap of complete blindness, Henry Fawcett was a fellow of Trinity Hall Cambridge, professor of Political Economy, and in addition a Liberal member of Parliament for Brighton. Later he served as postmaster general in Gladstone's government. His wife, the daughter of Newson Garrett of Aldeburgh on the Suffolk coast, was one of the country's most active feminists, from the 1860s until 1918 a leader of the campaigns for the political emancipation of women and later recognized with a D.B.E. for her public work.

Philippa was taught at home by her mother for a time, then attended Clapham High School. She also took classes at Bedford College for Women and at University College London, before going to Newnham in 1887 as a Winkworth scholar. She had been well-prepared in mathematics, with extra tuition during her high school years from one of her father's Cambridge colleagues. To those who knew her work, her success in placing above the senior wrangler in the 1890 Mathematical Tripos was hardly a great surprise.[45] It was, however, an event without parallel in the history of the university and a landmark in the story of higher education for women; despite modest advances since Scott's time there still persisted the rather widely held belief that women were incapable of the kind of abstract thought that advanced mathematics required.

When the news broke, telegrams of congratulation from supporters of women's education, from leaders of the women's suffrage movement and from some of the country's best-known political figures showered upon Philippa's mother "like snowflakes in a storm."[46] The excitement of the scene at Cambridge when the results were announced was captured in a letter written by Philippa's cousin Marion Cowell to her mother. Marion, her sister Christina, and grandfather Newson Garrett, having got wind of expected good news, had gone up for the occasion:

It was a most exciting scene in the Senate this morning. Christina and I got seats in the gallery, and Grandpapa remained below. The gallery was crowded with girls and a few men, and the floor of the building was thronged by undergraduates as tightly packed as they could be. The lists were read from the gallery and we heard splendidly. All the men's names were read first, the Senior Wrangler was much cheered. There was a good deal of shouting and cheering throughout; at last the man who had been reading shouted "Women." The undergraduates yelled "Ladies," and for some minutes there was a great uproar. A fearfully agitating moment for Philippa it must have been; the examiner, of course, could not attempt to read the names until there was a lull. Again and again he raised his cap, but would not say "ladies" instead of "women", and quite right, I think. He signalled with his hand for the men to keep quiet, but he had to wait some time. At last he read Philippa's name, and announced that she was "above the Senior Wrangler". There was a great and prolonged cheering; many of the men turned towards Philippa, who was sitting in the gallery with Miss Clough, and waved their hats. When the examiner went on with the other names there were cries of "Read Miss Fawcett's name again", but no attention was paid to this. I don't think any other women's names were heard, for the men were making such a tremendous noise; the examiner shouted the other names, but I could not even detect his voice in the noise. We made our way round to Philippa to congratulate her, and then I went over to Grandpapa. Miss Gladstone was with him. She was, of course, tremendously delighted. A great many people were there to cheer and congratulate Philippa when she came down into the hall. The Master of Trinity and Mrs. Butler went up into the gallery to speak to her. Grandpapa was standing at the bottom of the stairs waiting for Philippa. He was a good bit upset. I entreated him not to upset Philippa, and he said he wouldn't. He pressed something into her hand—a cheque, I fancy. [It was really a ring.] She was very composed. A great many of the Dons came to shake hands with her. The undergraduates made way for her to pass through the hall and then they all followed her, cheering, and I saw her no more.[47]

The day's festivities were rounded off with fireworks and a bonfire on the grounds of Newnham College, the women being joined by undergraduates from neighboring Selwyn College.[48]

Philippa stayed on at Cambridge as a Marion Kennedy student in 1891, taking class I in part II of the Mathematical Tripos, a much less competitive test than part I. She then joined the Newnham staff as lecturer in mathematics. She published two research papers in the early 1890s, an article on hydrodynamics appearing in the *Journal of Pure and Applied Mathematics* in 1893 and a study of the electrical properties of mixtures of hydrogen and nitrogen (done at the suggestion of physicist J. J. Thompson) in the *Proceedings of the Royal Society* the following year.

In 1901, however, the course of her life changed. That summer she accompanied her mother on a trip to South Africa, where Millicent Fawcett, by then a prominent figure in British public life and national politics, was to investigate on behalf of the British government conditions in camps that accommodated Boer women and children. These people had been uprooted from their homes by the British military as a result of the guerrilla war that had followed the major campaigns of the Boer War. Mortality rates in the camps were high because of epidemic diseases. Mrs. Fawcett's team of five, which included two women doctors and a nurse, went out on a troopship, and once in South Africa traveled over vast distances, on a special train provided by the Cape government, visiting camps in Natal, the Orange River Colony, and the Transvaal.

As far as Philippa Fawcett's story is concerned, one of the most important aspects of this tour of investigation was the discovery that camp schools set up by the authorities were, almost without exception, a great success. Organized education had not been provided by the Boer government for the population of the Transvaal, and both adults and children grasped the opportunity the camp schools provided; basic English and arithmetic were particularly popular subjects with the adults. After her trip Philippa returned to her teaching position at Newnham for a year, but in 1902 she applied for and obtained a government post in South Africa and went back as private secretary to the acting director of education in the Transvaal. She was concerned mainly with the development of a system of farm schools to meet the needs of the scattered Boer population throughout the veld. To work successfully with people with whom the British had so recently been at war required considerable diplomatic skill, understanding, and genuine goodwill. She had the advantage of already having a point of contact with Boer farmers in outlying regions through her mother, whose work on their behalf had won their confidence, and the three years she spent in South Africa were very fruitful. She came to be regarded as one of the founders of the country's elementary education system.

When she returned to London in 1905 she became chief assistant to the London County Council's director of education. The time was one of tremendous change and reorganization in British primary and secondary education. Following the establishment in 1902 of local education authorities, the London council was in the process of starting its own secondary schools; much of the planning involved was Philippa's work. She was appointed assistant education officer for higher education in 1920. Two teacher training colleges, Avery Hill and Furzedown, were started during her tenure of office. At a time of difficulty she acted as temporary principal of Avery Hill. Her most outstanding work, however, is said to have been the establishing of cordial working relations between the London County Council, London University, and the governing bodies of already-existing secondary schools, doubtless another task requiring considerable diplomatic skill and patience.

As a young woman she was very fond of sports, especially fencing, tennis, and hockey; she played on the Newnham hockey team in 1891.[49] Her links with Newnham lasted throughout much of her life; she was an associate from 1893 to 1906 and from 1907 to 1922, an associate fellow from 1917 to 1919, and a member of the council from 1905 to 1915. She held an M.A. from Trinity College Dublin, and she became a fellow of University College London in 1918. In 1934 she retired from the London County Council. She died on 10 June 1948, at her home in London, at the age of eighty.

Grace Chisholm[50] (1868–1944), the first woman to take a doctoral degree in mathematics by normal course attendance and examination at a Prussian university, was one of a very few women mathematicians of her generation to achieve an international reputation.

She was born in London, on 15 March 1868, the youngest of the four children of Anna Louisa (Bell) and Henry William Chisholm. Henry Chisholm was a Civil Service assistant in the government's Exchequer Bill Department where, as senior clerk, he produced a number of important parliamentary documents. He later had charge of the Weights and Measures Department and for a time served as the first and only Warden of the Standards. When Grace was five he bought a large house with a farm and garden near Haslemere in Surrey, and so she grew up as a country child. In her early years she was troubled by sleep walking and the family doctor ruled that she should have no lessons except those she asked for. She never remembered being taught to read and write, or a time when she could not play the piano and speak French. Much of her time was spent with her father in his carpentry workshop. When she was ten, her sleeping difficulties having decreased, she began formal lessons with a governess; at seventeen she passed the Cambridge Senior Examination. For four years after that she followed a not uncommon course for young, middle-class women of the time and took up charitable work, visiting the poor in some of the worst parts of London. She would have liked to study medicine, but since her mother objected she decided to try mathematics at Cambridge instead. She won a Sir Francis Goldschmid scholarship to Girton College and with her father's encouragement went to Cambridge in 1889.

The great excitement at the end of Chisholm's first year was the spectacular success of Philippa Fawcett in the Mathematical Tripos examinations, without doubt an inspiring example for the younger women students. Chisholm herself took first class in 1892, placing between the twenty-third and twenty-fourth wranglers, a very satisfactory performance, even if it did not match Fawcett's achievement. Her tutor in her final year was William Henry Young, her future husband; although he was known as an effective teacher, she had until then avoided him because of his reputation as a quite ruthless and at times intimidating crammer. Immediately after the results came out, she and her friend and fellow mathematics student Isabel Maddison (see below) responded to a challenge to try the Final Honours School of Mathematics at Oxford.[51] By special arrangement they took the examinations unofficially; Chisholm obtained a first and Maddison a second, although the results were not entered into the university records. In fact, Grace Chisholm's result was the best for that year at Oxford, and she became the first person, man or woman, to obtain a first in mathematics at both Oxford and Cambridge. She returned to Cambridge for the academic year 1892–1893 and took part II of the Mathematical Tripos. After that there was no obvious way forward; Cambridge fellowships were not open to women and her application for a Cornell University scholarship was rejected.

However, as noted above in the discussion of American Mary Winston, it so happened that about then first steps were being contemplated in the experiment of opening Prussian universities to women (as auditors), and that the leading mathematician at Göttingen, Felix Klein, strongly supported the idea of higher education for well-qualified women. Accordingly, in 1893, Chisholm went to Göttingen, submitted the appropriate applications to the Prussian Ministry of Culture in Berlin, and received official permission to attend lectures (as a special case).

She studied astronomy and physics as well as mathematics but concentrated on the latter, working under Klein's supervision on a dissertation on the application of his new theory of groups to spherical trigonometry. Although women at some German universities about this time were not well received by the men students, there seems to have been no difficulty of that sort at Göttingen, where relations between men and women students were friendly and natural.[52] Early in 1895 the question of Chisholm's actually taking a degree came up. Klein, who supported her strongly, felt that although admission to the examination and the doctorate was a matter to be decided by the faculty, an application to the government was in order. She therefore paid a visit to the Ministry of Culture in Berlin where all was approved without difficulty.

Her oral examination, taken in April 1895, was an event that involved considerable formality. She left a fascinating description of it in a letter to her father.[53] Following the announcement of the date, she was provided with a list of the names and addresses of all the professors who were to be present on the occasion, and on each of these she was expected to pay a formal call. On the day of the examination, arrangements were made for the traditional carriage with coachman in livery to take her to the designated place by 5:30 P.M. However, confused instructions to a maid and the coachman's natural expectation that he was to pick up a *gentleman* led to his driving away without her. She set out on foot, got lost in a succession of the little town's squares and triangles, but arrived only five minutes late, even if very hot. Some half dozen professors awaited her, including the acting dean of the faculty Professor Kliehorn, who functioned as committee chairman in charge of the evening's proceedings. All were seated around a long table covered by a dark green cloth. Chisholm was waved to her place at one end. Klein opened the two-hour examination and concentrated on geometry, where she was very confident; in differential equations he led her on "so gently" that things went well there too. Voigt (professor of physics) followed Klein and then Schur, the astronomer, after which an almost exhausted Chisholm was asked to step out to the porter's room. Within a few minutes she was brought back and told that she had passed magna cum laude. Congratulations all round and the presentation of a wonderful bouquet of roses and carnations from Professor and Frau Kliehorn followed; the event was a gala one; Klein especially was delighted with the success of his first female student.[54]

She sent a copy of her printed dissertation[55] to William Young at Cambridge and continued an intermittent correspondence with him after her return to England. She also accepted his suggestion that they collaborate on a book on astronomy. The following spring he proposed marriage, and although she hesitated for a time to risk the mathematical career she wanted by taking on family responsibilities, the wedding took place in June 1896.

Their plans were that Young would continue his tutoring and other university work at Cambridge and Grace would continue her research. She published her first journal article, based on a lecture she had given in the astronomy seminar in Göttingen, in the *Monthly Notices of the Royal Astronomical Society* in 1897. About the same time, Young's ideas about his career changed entirely; he suggested to Grace that he give up teaching and the money it brought and that they go to the Continent, live on savings and investments, and do research. The mathematician G. H. Hardy, writing after Young's death in 1942, remarked that he had never been able to fathom Young's career pattern; after more than ten years of teaching drudgery, in a mathematical atmosphere at Cambridge that in those days was not conducive to research, Young had abruptly rearranged his whole professional outlook. From that point on he began to produce an astonishing amount of original work, becoming "one of the most profound and original of English mathematicians of the last fifty years."[56] The obvious possibility of a link between the change in Young's activities and his wife's earlier experience at Göttingen suggests that the idea of going there was quite likely to have come from Grace. In any event, she was glad to return. The move, in the autumn of 1897, shortly after the birth of their first child, Frank, was not one that their families and friends approved of; people found it hard to understand why any reasonable person would voluntarily give up a promising position at Cambridge, of all places, for no position at all on the Continent.

After a year in Göttingen, where they lived in a small house close to the Kleins, whom they got to know very well, they moved on for a time to Italy to study geometry with Corrado Segre and other Italian mathematicians in Turin. Grace submitted an article to the Turin Academy in 1898–1899 discussing a development of her dissertation on spherical trigonometry. By the autumn of 1899 they were back in Göttingen, where they had their family home until 1908, although for much of the time Young spent most of the academic year in Britain, financial necessity requiring that he resume salaried work of some kind.

Until about 1900 mathematical research for both of them was still relatively unfocused. They were unsure which branch of the subject to concentrate on. During the summer of that year, however, Klein advised them to look into the new field of set theory, developed by Georg Cantor mainly in the 1880s and at the time being further elaborated by Arthur Schönflies. Klein's suggestion had momentous consequences. Over the next twenty-five years the Youngs' contributions to the subject, especially the applications of set theory to problems in mathematical analysis, put them among the leaders in the field. They continued to publish papers in geometry as well, but their outstanding contributions were set theoretical.

As their work developed, the original roles in their partnership reversed. From their Turin days Grace had begun to feel that her husband's mathematical insight was superior to hers. With the beginning of their work on set theory she stepped back for a period from the role of research mathematician and took on that of assistant, although she was an assistant who was quite capable of making original contributions of her own. To Young she was absolutely essential, because it was she who refined into rigorous theorems and results the flood of ideas he poured out to her. In his discussion of William Young's work, G. H. Hardy remarked that "Two features of it stand out on every page, intense energy and a profusion of original ideas. Indeed it is obvious to any reader that Young had a superabundance of ideas, far too many for any one man to work out exhaustively. One feels that he should have been a professor at Göttingen or Princeton, surrounded by research pupils eager to explore every bypath to the end. It may be that he had hardly the temperament or the patience to lead a school in this way, but he never had a chance to try."[57]

Young's one helper was his amazingly energetic wife. In the history of mathematics there have been few partnerships that lasted as long as the Youngs'. During their twenty-five years of research, over the course of which were published three books and at least 220 papers, the great bulk of the original ideas came from Young but very often were developed, elaborated, and revised in detail by Grace, who wrote the final manuscripts while he went on to his next papers.[58] When they were together, mathematics dominated their conversation and when they were apart the discussions continued by letter. Much of their work, especially in the years up to about 1906, was published under Young's name only because they felt that joint papers would bring neither of them the benefit that single-authored work would bring to him.[59] Having a growing family (the last of their six children was born in 1908) Grace was considered not to be in a position to undertake a public career, but her husband still had hopes of obtaining a chair of mathematics at a British university and so had everything to gain from a research reputation. Their most important book, *The Theory of Sets of Points* (1906), was nevertheless published under both names. It was the first textbook in English on a subject now considered basic in mathematical studies, although at the time it was too far removed from tradition for mathematicians to appreciate the need for it.

In 1908 the family moved to Geneva, in part because of the growing unease felt by British nationals in Germany in those years before the First World War. Despite his prodigious output of solid research, Young had failed in all his attempts to get a professorship (except for a low-paid, part-time position at Liverpool), and money to support the family remained in short supply. In 1913, however, he finally secured a professorship at the University of Calcutta, at the very good annual salary of £1,000 although the work involved only a few months of actual residence in Calcutta during the year.

At about the same time, Grace went back to writing papers on her own. Starting with a 1914 article in *Acta Mathematica*,[60] she produced a series of papers that won her a permanent place in the history of the development of the modern theory of real functions. Her special topic was the theory of differentiation, particularly the problem of what a function is like near points where it fails to have a derivative, work begun jointly with her husband in the period from 1909 to 1912.[61] One of her most important papers was her long essay in the *Quarterly Journal* in 1916, which had won her the Gamble Prize at Cambridge the year before. Her more complete statement on the behavior of functions at the edge of differentiability appeared in the London Mathematical Society's *Proceedings* in 1916. What has come to be called the Denjoy-Saks-Young theorem (reflecting her part in its discovery) is one of the fundamental results of the differential calculus. It is still a useful tool for investigating the differentiability of functions.[62] During this period she was also working on two books with her husband, including one on integration and a second edition of their *Theory of Sets of Points*.[63]

The seven years the Youngs spent in Geneva, 1908 to 1915, were especially productive mathematically. At that time also Grace Young was continuing medical studies begun in Göttingen. She was able to do all this in part because throughout much of the period when her children were small one of her sisters-in-law lived with the family. Until 1902 Ethel Young had provided help with the household, and from 1903 to 1913 Mary Ann, a children's nurse by training, to a large extent took over the role of conventional mother for the Young children. In 1913 they moved to Lausanne, and with the coming of the war could obtain only a limited amount of daily help. This meant that Grace could not leave home for the few months of hospital residence she needed to complete her medical degree, and so, although she passed all the examinations, she never qualified as a doctor. She continued to publish mathematics until the late 1920s when her health began to decline. Young also retired at about this time, partly because he felt acutely the loss of her help in checking and reworking his manuscripts.

A woman of tremendous energy and enthusiasm for life, Grace Young had remarkably varied talents, and her interests ranged well beyond mathematics and medicine. She knew more than six languages, was an excellent pianist, a good artist, and a serious student of the Elizabethan period.[64] Perhaps she lost valuable time she might have used to develop her own mathematical thinking during the long period when she worked at refining and revising her husband's ideas. But she also undoubtedly gained from this collaboration, benefiting from the constant stimulus that came from working closely with a person of Young's mathematical ability. When she finally returned to independent research, she was well armed with tools and techniques the joint work had made her familiar with.[65] Her final years were spent in England at the home of her daughter Janet in Croydon, where she helped to look after Janet's two children. She died on 29 March 1944, two weeks after her seventy-sixth birthday.[66]

Of the Youngs' six children, two, Rosalind Cecilia and Laurence Chisholm, became mathematicians. Laurence's career was especially notable. After distinguished work at Cambridge in the early 1930s he was appointed, at the age of thirty-three, to the chair of mathematics at the University of Cape Town. From 1949 until 1975 he was professor of mathematics at the University of Wisconsin.[67] Cecilia (later Mrs. Tanner) also held fellowships at Cambridge for a time, and in 1933, after an unsuccessful bid to obtain a faculty position at the University of Michigan, became assistant lecturer at Imperial College London; she remained on the teaching staff there until her retirement in 1967. Of the other children, one, Janet, became a physician, another, Patrick, took a doctoral degree in chemistry at Oxford and joined Imperial Chemical Industries, and a third, Helen (Leni), studied mathematics for a time at Bryn Mawr but gave it up after she married in 1929. Frank, the oldest, joined the Royal Flying Corps immediately after graduating from the École d'Ingénieurs in Lausanne in 1916; he was killed in action at the Somme the following year.

The third of the notably productive Cambridge-educated mathematicians included here is HILDA HUDSON (1881–1965),[68] who was born in Cambridge, on 11 June 1881. She came of a mathematically distinguished family. Her father, William Henry Hoar Hudson, third wrangler at Cambridge in 1861 and then a fellow of St. John's College, was later professor of mathematics at both King's College and Queen's College (Harley Street), London. Her older brother Ronald was senior wrangler in 1898, and her sister Winifred Mary was bracketed with the eighth wrangler in 1900. She herself tied with the seventh wrangler in 1903. Her mother, Mary Watson, daughter of Robert Turnbull of Hackness, Yorkshire, was one of the first Newnham College students.

Her published work in mathematics dates from 1911, with the exception of a short note offering a simplified proof in Euclidean geometry (see bibliography). This appeared in *Nature* in 1891 when she was a child of ten, and presumably was studying the subject with her father. William Hudson, whose wife died in 1882, most likely had a strong influence on his children's early life and encouraged their interests in mathematics. Among his writings are several monographs on the teaching of mathematics, including geometry and elementary algebra; he was particularly interested in the early stages of mathematical instruction. A strong supporter of higher education for women, he sat on the Newnham College Council for many years.

Hilda entered Newnham from Clapham High School (London) on a Gilchrist scholarship in 1900 and took both parts of the Mathematical Tripos (class 1.3, part II, 1904). Following a year at the University of Berlin (1904–1905) she returned to Newnham as lecturer in mathematics (1905–1910). From 1910 to 1913, she held a Newnham Associate's Research Fellowship and during this period spent some months at Bryn Mawr with Charlotte Scott. Over the years her connections with Newnham remained close (associate, 1911–1926). Brought up on the old-style Tripos, which included a paper on Newton's *Principia,* she knew her planetary physics and was a competent practical astronomer; for a time she had charge of the Newnham observatory.

In 1913 she was appointed lecturer at West Ham Municipal Technical Institute, where, along with a colleague, she prepared students for University of London degrees. Although not remembered as a particularly successful teacher for the average student, she could be inspiring for the dedicated and conscientious. In 1916 she worked for a time with Lieutenant-Colonel Sir Ronald Ross on an ongoing government-funded project of his in epidemiological statistics. Their lengthy joint paper appeared in the Royal Society's *Proceedings* for 1917.[69] That year, she gave up teaching and took a Civil Service post to carry out defense-related research in aeronautical engineering for the Air Ministry. Her considerable success in this very different line of work illustrates her versatility as a mathematician. Mathematical modeling as related to airplane design was then a new field; Hudson headed a subdivision working on problems of aircraft strength and performance.[70] Continuing in this area until 1921, she spent two years as technical assistant with Parnell and Company of Bristol but then retired from salaried work in order to write the book for which she is remembered, *Cremona Transformations in Plane and Space* (1927).

Like Charlotte Scott, Hudson was a geometer. Most of her work centered on the study of Cremona space transformation, an analytical technique for examining the geometry of algebraic surfaces and plane curves. Although now displaced by the development of powerful tools of abstract algebra, it was a challenging and popular branch of investigation in her time. Her studies were notable particularly because her methods, largely those of analytical geometry, were basically elementary, but her powerful, "almost uncanny" geometrical intuition led her to solutions of quite difficult problems and her contributions were substantial.[71] Of special note was the early work she carried out during her three years as a Newnham Associates' fellow, which resulted in a series of ten papers published between 1911 and 1913.[72] From her retirement (when she was forty) until 1928 was also a particularly productive time. Over the course of about six years she brought out or presented six papers as well as her book, her magnum opus.[73] In the latter she put forward a unified account of the major elements of the field of Cremona transformations, supplemented with a comprehensive bibliography of publications covering almost seven decades of research in the area. It was a substantial treatise and much quoted. Among her other professional contributions were many reviews, generally of advanced texts published in French or German; these appeared regularly in the *Mathematical Gazette* over a number of years.

She was devoutly Christian, and a powerful driving force in her mathematical work was her belief that the search for mathematical "truth" was a seeking after the glory of God. Her ideas about the close relationship of mathematics to theology were set out in considerable detail in her paper, "Mathematics and eternity."[74] As she grew older she became increasingly preoccupied with religion and her concept of the spiritual world. After her book came out in 1927 she virtually gave up research and devoted her time to the Student Christian Movement, serving as the finance secretary of its Auxiliary Movement from 1927 until 1939.

A distinguished mathematician, she held both M.A. and Sc.D. degrees from Trinity College Dublin, the doctorate being awarded in 1913. She was one of the few women of her time to sit on the council of the London Mathematical Society, serving from 1917 until at least 1925. Her war work brought her an O.B.E. in 1919. A small woman, with bright eyes behind thick-lensed spectacles and a face that had character, she enjoyed hockey and swimming when young. She lived a simple, almost austere life and paid little attention to material comforts. Thoughtful and fair-minded, during her time in the Civil Service she fought for better pay and working conditions for the people in her subdivision. Early onset of arthritis crippled her severely, and toward the end she moved to St. Mary's Convent and Nursing Home, run by Anglican Sisters in Chiswick, London. She died there, on 26 November 1965, at the age of eighty-four.

The remaining four Cambridge-trained women mathematicians whose papers are listed in the bibliography are Frances Hardcastle, Isabel Maddison, Mildred Barwell, and Mary Beeton, all Girtonians.

FRANCES HARDCASTLE[75] (1866–1941) was the elder daughter of Henry Hardcastle, barrister, of the Inner Temple, and his wife Maria Sophia, daughter of the astronomer Sir John Herschel. She was born at Writtle in Chelmsford, Essex, on 13 August 1866. Educated at home in her early years, she went to Girton in 1888 and took both parts of the Mathematical Tripos course (class II, part I, 1891; class II, part II, 1892). The following two years she spent in the United States, first as an honorary fellow in mathematics at the newly opened University of Chicago, and then as a fellow at Bryn Mawr, where she studied with Charlotte Scott. Her translation from the German of Felix Klein's influential text, *On Riemann's Theory of Algebraic Functions and Integrals* (1893), was completed at Bryn Mawr. She returned to Cambridge as a postgraduate student in 1895 and in 1902 and 1906 held Pfeiffer fellowships. She was awarded an M.A. by Trinity College Dublin in 1905.

Another student of algebraic geometry, Hardcastle published two substantial papers on point-group theory in 1898 (see bibliography). In addition, she prepared for the British Association's "Reports on the State of Science" series a "Report on the Theory of Point-groups," which came out in four parts between 1900 and 1904.[76] Among her other contributions to scientific work were the drawings in Arthur Berry's *Short History of Astronomy* (1899); she also checked most of the numerical calculations in that monograph. From about 1907, however, she gave much of her time to the women's suffrage movement. As joint honorary secretary of the National Union she served on the committee that framed its new constitution in 1907–1908. She then became honorary secretary of the North East Federation of Women's Suffrage Societies. She died in Cambridge, on 26 December 1941, at age seventy-five.

ISABEL MADDISON[77] (1869–1950), Grace Chisholm's close friend at Girton College, spent most of her working life in the United States. She was born in Reading, on 12 April 1869. Before going to Girton as an Exhibition Scholar in 1889, she spent four years at the University College of South Wales and Monmouthshire, where she concentrated on mathematics. Taking class I in the Mathematical Tripos examinations in 1892, she placed equal to the twenty-seventh wrangler. Immediately afterward she and Chisholm sat the Final Honours

School examinations at Oxford, where she was awarded a second class.

The following year, after taking a London honours B.Sc., she received a Bryn Mawr Resident Mathematical Fellowship and began graduate studies under Charlotte Scott. She spent the year 1894–1895 at Göttingen as a Bryn Mawr Mary E. Garrett European fellow, and was there when Grace Chisholm took her oral examination. Bryn Mawr awarded her a Ph.D. in 1896. Her two substantial papers, reporting her graduate research in algebraic geometry and in differential equations, appeared in the *Quarterly Journal of Pure and Applied Mathematics* in 1893 and 1896. She also prepared for the American Mathematical Society an English translation of an address given by Felix Klein at a meeting of the Royal Academy of Sciences at Göttingen in 1895. However, aside from book reviews and a brief note on the history of the map-coloring problem, she published nothing further in her field.[78]

She became reader in mathematics in 1896 and associate professor in 1904, but much of her time from 1896 on was spent in administration. Despite shyness and a certain amount of regret at leaving mathematics, she was very efficient in the work. She held the posts of secretary and then assistant to the president from 1896 to 1926; in 1910, when she became in addition recording dean of the college, she gave up her mathematics professorship.

She was especially interested in making information on European universities readily available to American students, and in 1896 brought out a *Handbook of British, Continental and Canadian Universities, with Special Mention of the Courses open to Women.* There was a real need for such a work at the time; a supplement followed in 1897 and a second edition in 1899. Apart from a year of study leave in 1905 at Trinity College Dublin (from which she received a B.A.) she remained at Bryn Mawr until her retirement in 1926. She died at her home in Wayne, Pennsylvania, on 22 October 1950, at the age of eighty-one. Among her bequests was $10,000 to Bryn Mawr in memory of Martha Carey Thomas, to be used as a pension fund for nonfaculty members of the college staff.

MILDRED BARWELL[79] (1873–1978), who went on to a mathematics teaching career of more than twenty years, was a student at Girton from 1892 to 1896. Born in London, on 5 November 1873, she was the daughter of Richard Barwell, a surgeon, and his wife M. D. Shuttleworth. She received her early education at home and at the Baker Street High School. After taking both parts of the Mathematical Tripos (part I, class II, 1895; part II, class III, 1896) she taught high school mathematics in London for three years and then returned to Girton for a year as assistant resident lecturer in mathematics. From 1900 to 1902 she taught at Dulwich College (London), and thereafter, until her retirement in 1919, was a lecturer in mathematics at Alexandra College Dublin. She took an M.A. at Trinity College Dublin in 1905.

The paper on conformal mapping (a geometrical branch of the theory of functions), which she read before the London Mathematical Society in 1898 (see bibliography), would appear to have been her only technical article, although she later published suggestions about mathematics teaching. Particularly interesting is her paper read before the Mathematical Association in 1913 when she was lecturer at the teacher training department of Alexandra College. It discussed the incorporation of some of the early history of mathematics into high school and even elementary teaching as an effective way to stimulate children's interest.[80]

Later she lived in Bideford, Devon. When she died on 1 March 1978 in her one hundred and fifth year she was the then oldest living Girtonian.

MARY BEETON[81] (1876–1966) was the daughter of Henry Ramé Beeton, a London businessman, later of Reading, and his wife Elizabeth (Dibley). She was born in London, on 4 March 1876. After attending a private school in Surrey and then St. Leonards School, St. Andrews, she entered Girton in 1894. She took class II in part I of the Mathematical Tripos in 1897 and an M.A. at Trinity College Dublin in 1906. From 1897 until 1919 she was much occupied in social work, but in the period up to about 1905 she also collaborated with Karl Pearson of University College London on a number of his statistical investigations of problems in evolutionary biology.[82] From 1912 on she became more and more involved in educational organization, serving for seven years (1912–1919) on the London County Council Education Committee and for six years (1923–1929) on the Benenden School Council. She taught for a time in the early 1920s, and in 1927 founded Halstead Place Preparatory School (for both boys and girls) in Sevenoaks, Kent, where she was joint headmistress until 1932. Her ties with Girton remained close; she was a member of Girton College and of the council from 1918 until 1923. She died in April 1966, at her home in Horley, Surrey, just after her ninetieth birthday.

Nearly all the London-trained women whose papers in mathematics or statistics are listed in the bibliography had associations with Bedford College for Women and University College. The earliest and the most distinguished was SOPHIE WILLOCK BRYANT[83] (1850–1922), probably the first woman to take a degree in mathematics at London University and also the first to receive a London doctorate.

She was born at Sandymount, near Dublin, on 15 February 1850, the third of the six children of the Rev. William Alexander Willock, a mathematician and a fellow of Trinity College Dublin, and his wife, a daughter of J. P. Morris of Skreen Castle. In 1852 Willock gave up his fellowship and accepted the benefice of Ballymony, County Cork, in the south of Ireland. He later moved north to Cleenish, on Loch Erne, in County Fermanagh.

Sophie never went to school. Her father taught her mathematics and philosophy, the two subjects that remained her major academic interests throughout her life. She also read widely in his extensive library and occasional governesses gave her lessons in French and German. Throughout her childhood she was frequently taken on long walking trips in the mountains of Donegal, in the extreme northwest of Ireland, experiences that gave her a lasting love of the hills. When she was thirteen her father died and the family moved to London. Three years later, at the early age of sixteen, she took the Senior Cambridge Local Examination, which had only just been opened to girls. She gained distinction in mathematics, the only girl to do so, and an Arnott scholarship to Bedford College. In 1869 she married a physician, Dr. William Hicks Bryant of Plymouth, but was widowed the following year at the age of twenty. Her strong showing in the Cambridge examination had been noted by Frances Mary Buss, the founder and first headmistress of the North London Collegiate School for Girls, and in 1875 Frances Buss offered her a part-time position, teaching mathematics and German. This was quickly changed to a full-time position as teacher of mathematics. Sophie's bright, engaging personality, intellectual powers, strength of purpose, and commitment to making girls think made her an outstanding teacher. Having taken it upon herself to challenge the conventional view that hers was not a subject for girls, she proved her point by sending a succession of her pupils to study mathematics at Girton College. She enrolled at London University in 1879, a year after degrees were first opened to women, and in 1881 took a B.Sc., with first-class honors in mental and moral science and second-class in mathematics. This she accomplished without giving up her teaching position and while caring for her mother and nieces at the same time. Her D.Sc., in mental and moral science, was awarded in 1884.[84]

From 1878 on she published frequently. A number of her early articles dealt with mathematical and statistical problems (see bibliography), but starting in the late 1880s she became increasingly concerned with education and educational psychology, areas in which she brought out several works. Her 1886 paper in the Anthropological Institute's *Journal* on the testing of schoolchildren is considered a noteworthy contribution to the development of early educational research.[85]

From about 1894 she was much involved in the planning and organization that was to bring about a complete change in the British educational system. Her appointment that year to the Royal Commission for Secondary Education chaired by Lord Bryce was an event of considerable importance in the development of the movement to provide education for girls and women; it reflected the fact that by the 1890s girls' education was finally being recognized as a matter of public concern. In 1900 she became a member of the Consultative Committee of the Board of Education, and the same year took her seat on the first senate of the University of London.[86] As a member of that body she spoke out strongly for teacher training programs and worked toward the founding of the London Day Training College. The establishing of the latter in 1902 was a landmark in British education (it was reconstituted in 1932 as the University of London, Institute of Education).[87] From 1908 until 1914 Bryant served on the London Educational Committee, and for many years was on the executive committee of the Association of Headmistresses, serving as president from 1903 to 1905. She was also one of the ten founding members (the only woman) of the British Psychological Society established in 1901; she served on its first committee. All of these undertakings added up to a tremendous burden of committee work.

After the retirement of Frances Buss in 1895 she became headmistress of the North London Collegiate School, where the challenge she faced, following in the footsteps of that distinguished pioneer in the drive for education for girls and women, was by no means minor. She was a most worthy successor to Frances Buss, however, and under her leadership the school underwent further development and modernization; she did much to improve its curriculum in both science and the humanities, making it the model for day schools for girls in England.[88] One of her special interests was the revitalizing of religious and moral instruction, subjects for which she wrote a number of textbooks.[89] Her work in adult education was extensive also. For a time she traveled to Cambridge once a week to lecture to students at the Cambridge Training College for teachers. In addition, she was a frequent and very popular speaker at Toynbee Hall, the settlement house in Whitechapel run by students from Cambridge and Oxford universities; there she lectured to groups of teachers and adult students on history, literature, and education. The Institute of Hampstead Garden Suburb, which opened in 1909 and over the years had more than 1,000 enrolled adult students, was another organization she strongly supported.

Active in politics and social concerns, she played a prominent part in the women's suffrage movement both locally in London and nationally; along with Millicent Fawcett, Emily Davies, and Frances Balfour, she was one of the four leaders of the demonstrations of the National Union of Suffrage Societies. A Gladstonian Liberal, she worked through the Women's Liberal Federation and the United Irish League for the cause of home rule for Ireland. One of her greatest pleasures was receiving the degree of doctor of literature, honoris causa, from Trinity College Dublin in 1904 when degrees were first opened to women. Although she had left Ireland at the age of thirteen she remained Irish all her life. She backed the movement for higher education for Irish women by herself making sure of the full involvement in it from the earliest stages of the convent-school nuns. Three of her books were about Ireland and its people—*Celtic Ireland* (1889), *The Genius of the Gael* (1913), and a book on the Brehan laws, *Liberty, Order and Law under Native Irish Rule,* completed just before her death and published in 1923.

Very much one of the "new generation" of independent women of the 1890s, Sophie Bryant was thoroughly progressive in all aspects of her life. Physical freedom was of tremendous importance to her; for exercise she rowed on the Thames and during summers bicycled in Ireland. Indeed, she was one of the first London women to ride a bicycle, demonstrating it to her pupils by riding around the school gymnasium. She climbed frequently in Wales, the Lake District, and the Alps; in 1895 she went up the Matterhorn. Her death came in the French Alps in 1922, when she was seventy-two; having planned to walk from Montanvert to Chamonix on 15 August of that year, she set off alone, somewhat behind her companions, and appears to have left the main path and met with an accident. Her body was found two weeks later on August 28.

Although less well-remembered in feminist writings than the very earliest of the British pioneers in late nineteenth-century education for girls and women,[90] Bryant, one of the first of the women educators to have had the benefits, both social and intellectual, of a background of advanced university training, was a most distinguished figure in the educational circles of her time.

ALICE LEE[91] (1858–1939) entered Bedford College in 1876, about ten years after Sophie Willock. In 1879 she matriculated at London University, placing at the head of the group of nine women students entering that year. She took a B.Sc. in 1884, the first Bedford College student to do so, and a B.A. the following year. From then until her resignation because of ill health in 1916, a period of more than thirty years, she remained at Bedford College, first as an assistant in mathematics and physics and later as lecturer in applied mathematics. For two or three years in the 1890s she was also a resident staff member, getting room and board in exchange for undertaking supervisory duties in the student residence.

All of her research was carried out in collaboration with Karl Pearson, professor of applied mathematics and mechanics at University College and one of the most important statisticians of all time. She first met him in 1892, when he publicly criticized the academic standards of Bedford College. She wrote to him in defense of her college, pointing out the difficulties it was then facing as it went through the process of being upgraded from the mixed secondary-tertiary institution it then was to university level; the transition was made even more complicated by the fact that Bedford College had a considerable weight of tradition behind it, having been founded more than four decades earlier (in 1849) as one of the country's pioneer institutions in women's education.

Pearson's interests and activities covered a wide range of social, political, and scientific concerns, including Darwinism, socialism, the women's movement, eugenics, and evolutionary theory. Alice Lee began attending his statistics lectures at University College about 1895. She became especially interested in his applications of statistical methods to problems in evolutionary biology, and under his direction studied for an advanced degree, taking as her research topic an investigation of variation in cranial capacity in humans and its correlation with intellectual ability. The work had, of course, direct connections with the feminist cause.

One of the central tenets of the science of craniology, a field in which there were still many active workers in the closing years of the nineteenth century, was that, just as muscle power increased with size, so did brain power; that is, skull capacity was a measure of mental ability, and consequently males, who generally had larger heads than females, were mentally superior. The theory had strong support from antifeminist anthropologists, including influential members of the Anthropological Institute, who saw the women's rights movement of the time as a potentially serious threat to the established "natural" social order. By 1906, however, the work of Pearson and his biometrics students effectively demolished the whole enterprise of measuring intelligence by skull dimension, and with it the underlying rationale for research in the field of craniology.[92] Alice Lee was one of the two students of Pearson who did much of this work (the other was Marie Lewenz).

Lee's first paper on the subject, "A study of the correlation of the human skull," published in 1901, clearly demonstrated that there was no correlation between skull size and intelligence. Having established formulas for cranial capacity that enabled her to calculate the latter from anatomical measurements, she examined three groups and ranked the individuals within them in order of decreasing skull size. Her groups were women students at Bedford College, male faculty at University College, and a collection of distinguished male anatomists. Individuals in the lists were identified by name and it so happened that some of the most eminent anatomists had cranial capacities lower than those of some of the Bedford College women. She completed her dissertation in 1899, but the findings, not surprisingly, stirred up considerable controversy and the work was not immediately accepted by the examiners; one of the latter was an anatomist with a low ranking in the skull-capacity table. Intervention by Pearson was necessary before Lee finally got her Ph.D. in 1901. The following year the Pearson group put out two further papers in which the criticisms leveled at Lee's original findings were very forcefully answered.[93] Indeed, the group's superior mathematical techniques and high scientific standards ensured that there were no effective challenges; the work very quickly had considerable impact.

Lee held a paid position in Pearson's laboratory until 1907[94] when she gave it up because of poor health, although she continued to work on a voluntary basis for the next twenty years. She kept on her lectureship at Bedford College until 1916 and carried out the laboratory work in her spare time. Much of it was done with calculating machines outdated even by the standards of the day. Her research on the statistical analysis of within-species variation, a branch of evolutionary biology in which the Pearson school for a number of years was extremely active, continued steadily until about 1910, a succession of papers appearing in *Biometrika* from 1902 on[95] (see also the discussion of the work of E. R. Saunders in this area, chapter 1). Lee also contributed to the preparation of tabulated functions much used by statisticians and biologists during her time. Her earliest work on such tables appeared in the *Reports* of the British Association in 1896 and 1899, later papers in *Biometrika* between 1914 and 1927.[96] Throughout the First World War, along with all the Pearson laboratory staff, she carried out statistical work for the government. From 1916 until 1918, this included the calculation of shell trajectories and of range tables of all kinds for the Anti-Aircraft Experimental Section of the Munitions Inventions Department. She also worked on special computing projects for the Admiralty.

Since the Bedford College pension scheme had started too late for her to join and her salary had always been "women's wages," she had little to live on after she gave up paid employment. In 1923 Pearson and Margaret Tuke, principal of Bedford College from 1907 until 1930, petitioned the Home Office for a Civil List pension for her of £70 a year, Pearson emphasizing her very considerable research contributions and her "services to the cause of scientific work."[97] She died in 1939, when she was about eighty-one.

CICELY FAWCETT[98] (1869–1937) was also an assistant of Karl Pearson for a time, coauthoring statistical papers with him and others in 1894, 1901, and 1902, and helping Alice Lee in the preparation of statistical tables. The daughter of W. M. Fawcett, a London barrister, she entered Bedford College in 1886, when she was seventeen. She became an associate in 1894 on receiving her B.Sc. Her teaching activity was limited to three years (1895–1898) as a voluntary assistant for first-year physics classes, and her subsequent long association with the college was as a member of its governing committees. She served on the council from 1913 to 1937, on the Finance Committee for twenty years, and as a Reid Trustee for forty years; in 1937 she became a governor. Active in Hampstead civic affairs as well, she served on the borough council for three years (1919–1922) and for longer periods on committees of other organizations. She died on 15 June 1937, at age sixty-eight.

Work by one other late nineteenth-century British woman mathematician is listed in the bibliography, the 1900 paper by the self-taught geometer ALICIA BOOLE STOTT[99] (1860–1940). Alicia (Alice) was the third of the five daughters of Mary and George Boole. Boole, one of the leading British mathematicians of the early nineteenth century, is remembered especially for his development of the algebra of logic and for his books, *An Investigation of the Laws of Thought on which are Founded the Mathematical Theories of Logic and Probabilities* (1854), and *The Calculus of Finite Differences* (1860). The former, a classic work of nineteenth-century mathematics, was one of the most influential books of the period. Born into an impecunious Lincolnshire farming and trading family and largely self-taught, Boole became professor of mathematics at newly established Queen's College Cork in 1849. His wife was the daughter of the Rev. Thomas Everest of Wickwar, Gloucestershire, and the great-niece of Lieutenant-Colonel Sir George Everest, Superintendent of the Great Trigonometrical Survey of India and the man after whom Mt. Everest was named.

Alicia was born on 8 June 1860 in the family home on Castle Road, outside Cork. Her father died suddenly when she was four, leaving the family almost destitute. She had an unhappy childhood, spent partly with her maternal grandmother in England and partly with her great-uncle, John Ryall, in Cork. When she was about thirteen she rejoined her mother and sisters, now in London. They were still poor, Mrs. Boole supporting her family largely on her Civil List pension of £100 a year, and they lived in a dark, uncomfortable house. Beyond study of the first two books of Euclid, Alicia had little formal mathematical education. When she was about eighteen, however, she was given the chance to amuse herself with a large collection of small wooden cubes that her future brother-in-law, Charles Howard Hinton, was using to develop a geometrical method of building models of the three-dimensional cross sections of figures of dimension four. Fascinated by the possibilities, she began experiments of piling the cubes into various shapes, and in the process developed a remarkably clear understanding of four-dimensional geometry. She found for herself that there are six regular "polytopes" (that is, convex, regular solids) of dimension four, and, using purely Euclidian ruler and compass construction and synthetic methods (because she knew no analytical geometry), she went on to build cardboard models of their three-dimensional cross sections. All of this work, which was amazingly original for a girl of twenty who was almost entirely self-taught, was carried out purely for interest without any thought of publication.

In 1890 she married Walter Stott, an actuary, and for the next ten years gave her time to bringing up her two children and running her household on a very small income; her relaxations were music and natural history (she was a member of the London Zoological Society). About 1900, however, Stott drew his wife's attention to some publications by Peiter Schoute of the University of Groningen, in which were described the cross sections of the regular four-dimensional polytopes. She wrote to Schoute, telling him that she had already determined the whole sequence of sections, and that the middle section, for each polytope, agreed with his results. She also sent him photographs of her complete set of cardboard models. Schoute insisted that she publish her results,[100] and with great enthusiasm asked when he might come over to England to work with her. This began a collaboration that lasted until Schoute's death in 1913. They spent several summer vacations at Hever, in Kent, working together at the house of Ethel Everest, Alicia Stott's cousin. Stott's exceptional powers of geometrical visualization complemented Schoute's more conventional methods and they were an ideal team. Her second paper appeared in 1910.[101] In July 1914, at the tercentenary celebrations of the University of Groningen, she was given an honorary doctorate and her models were placed on exhibition.

In 1930, at the age of seventy, after a break of more than sixteen years, she again took up mathematical work, collaborating with geometer H. S. M. Coxeter, to whom she had been introduced by her nephew Geoffrey Taylor. She and Coxeter investigated Gosset's four-dimensional polytope, rediscovered about that time by Coxeter. She made models of its sections and pointed out that the vertices of the Gosset polytope lie on the edges of another polytope, dividing them according to the "golden section."[102]

Although the field of polytope investigation is not one in which great interest has been taken since Coxeter's work and has so far failed to produce further developments, Stott's contributions are well worth noting. And clearly she much enjoyed what to her was pure hobby. Considering how far she got even without the benefits of formal instruction, one can but wonder what she might have done had she had guidance during her early years from someone like her father. She died at the age of eighty, on 17 December 1940, in a Catholic nursing home, at the height of the bombing of London during the Second World War. Her son Leonard Boole Stott studied medicine and became a pioneer in the treatment and control of tuberculosis, work for which he was later appointed an O.B.E.[103]

There remain in this collection two women whose published work on mathematical topics appeared in the first half of the nineteenth century, Mary Fairfax Somerville and Ada Byron, Countess of Lovelace. Both are mentioned in a number of general works on early women scientists and both are the subjects of fairly recent full-length biographies.[104] Somerville was a well-known and much-recognized figure in the scientific circles of her time, Lovelace considerably less so. Although neither was an original investigator in the way that nearly all of the later women were, both Lovelace and Somerville nevertheless stand out for having been engaged in expository mathematical writing fifty years before most of the others mentioned here began their work.

ADA BYRON, COUNTESS OF LOVELACE,[105] (1815–1852) was the only legitimate child of the poet George Gordon Byron, sixth Baron Byron, by his wife Anne Isabella, daughter of Sir Ralph Milbanke Noel, baronet. The fact that Ada figures in the third canto of *Childe Harold's Pilgrimage,* the autobiographical epic that established Byron's celebrity, makes her one of the romantic figures of the nineteenth century. She also studied mathematics and became a friend of mathematician Charles Babbage, who was at the time experimenting with the design and building of the earliest mechanical computing machines. In 1843, at Babbage's suggestion, she published an article on one of these machines. Because this article constitutes the first and probably the best extensive, illustrated explanation of what such a machine could do, namely receive, store, and manipulate data supplied to it in numerical and symbolic form, Ada Lovelace also has a modest place in the history of computing.[106]

Among the biographical studies of Lady Lovelace that are now available, the one by Dorothy Stein is especially interesting. After a careful perusal of original sources, Stein persuasively argues that over the decades a considerable amount of myth has grown up about the aristocratic Lady Lovelace and her contribution to early work in computer programming.[107]

She was born in London, on 10 December 1815. Within a month of her birth her mother left her abusive husband and soon after obtained a legal separation. Lady Byron kept her daughter (mainly because Byron never pressed his right to retain the child) and also the income and property that had come to her through the marriage settlement—quite an achievement for the time. Throughout her somewhat disjointed upbringing, Ada was taught by her mother and by a succession of governesses and tutors. She became an excellent horsewoman, and, like her mother, had serious interests in science and mathematics, although she did not begin to study these subjects very persistently until about 1834 when she was already nineteen. The following year she married William King, eighth Baron King and Baron Ockham, lord lieutenant of Surrey, created Earl of Lovelace in 1838.

Among her friends at the time was Mary Somerville, who regularly helped her with her mathematics and often accompanied her to social gatherings at Charles Babbage's house. About 1840 she began to study calculus by correspondence with mathematician Augustus de Morgan, whose wife was a friend of her mother. De Morgan, then professor of mathematics at London University, was a gifted and patient teacher and Ada a competent student. At least for a time, she applied herself with enthusiasm.

By 1842 she had progressed far enough to be able to undertake the translation for which she is known—an English version of an article by an Italian engineer, Luigi Menabrea, on Babbage's proposed "Analytical Engine." The original article had resulted from a series of lectures and discussions about his engine that Babbage held in Turin in 1840. It was published in French in the *Bibliothèque Universelle de Genève* in 1842. Ada was asked to prepare the translation by a family friend, the physicist Charles Wheatstone, who solicited English versions of significant foreign technical papers for the recently founded journal, *Taylor's Scientific Memoirs.* Menabrea's account dealt not with the mechanical details of the machine, but rather with its functional organization and mathematical operation; charts (that is, programs) showing punch-card instructions that the machine was to follow in performing its operations provided illustrations. Babbage himself had been ill in the late autumn and winter of 1842 when Ada probably began her translation, but he saw it before it was delivered to the publisher, and suggested that she append some notes of her own that would further illustrate the potentialities of the machine. The Notes that she produced were lengthier than the original paper and resulted from extensive consultation between her and Babbage. She had access to all his relevant papers,[108] in particular the large collection of examples (programs) he had worked out, and did not herself make any original mathematical contribution; but she was clearly a successful expositor, and although she found the work very taxing it gave her great satisfaction. Further, the favorable reception of her paper encouraged her to continue scientific work.

Her interests were wide-ranging. About 1844 she wrote to the chemist Michael Faraday asking if she could study with him. When he gently turned aside the request on the grounds of his poor health she approached Andrew Crosse, a gentleman scientist who lived not far from one of the Lovelace estates. Crosse had done work with voltaic cells, and Lady Lovelace, by then caught up in the subject of mesmerism, a popular interest at the time, wanted to carry out some experiments in physiology, especially an examination of nerve tissue. She even had a vague plan about writing a "Calculus of the nervous system." These ideas were very soon abandoned, however, in favor of studying German. Her only other published work took the form of a note added to an 1848 publication of her husband on the effect of climate and weather on agricultural crop production.[109] She attempted to express "atmospheric heat" available to crops in mathematical terms.

The story of Ada Lovelace's final years makes distressing reading.[110] About 1850 she took up betting on horse races. It seems likely that she wanted money to help her friend (probably lover) John Crosse, the son of Andrew Crosse. Eventually her losses and consequent financial difficulties became substantial. Furthermore, her health, never reliable, now went into a rapid decline. She died in London, on 27 November 1852, at age thirty-six, of cervical cancer, following weeks of intense suffering. She had had three children, but had taken only a limited interest in them; they were brought up in large part by her mother.[111] In the scientific world, a programming language, "Ada," developed by the United States Department of Defense about 1980, commemorates her name.

The mathematical contributions of Mary Fairfax Somerville were of a different order from those of Lady Lovelace. A skillful organizer and expositor who had a good knowledge of the branches of mathematics and mathematical astronomy being pioneered in early nineteenth-century France, Somerville did notable service in introducing these new developments to a relatively wide English-speaking audience. Her achievement was a remarkable one, especially for a woman of her time.

MARY FAIRFAX[112] (1780–1872) was the fifth of the seven children of Lieutenant (later Vice Admiral) William George Fairfax, an English naval officer, and his second wife, Margaret, daughter of Samuel Charters, Solicitor of Customs for Scotland. She was born on 26 December 1780, at the home of her mother's sister, Martha Charters Somerville, in Jedburgh in the Teviot valley in southern Scotland. She grew up in Burntisland, then a small Fifeshire port town, opposite Edinburgh, on the northern shore of the Firth of Forth. Apart from a year she spent at the age of ten at a boarding school at Musselburgh (a town on the eastern outskirts of Edinburgh) she had no formal education as a child. Her uncle Thomas Somerville of Jedburgh gave her some instruction in Latin, but for the most part she taught herself, reading what was available in the small family library. She also listened to her younger brother's geometry lessons and persuaded his tutor to get for her copies of Euclid's *Elements of Geometry* and John Bonnycastle's *Algebra.* These she had to study in secret, however, her father having forbidden her to read mathematics; he was afraid it might do her injury and preferred that she spend her time on music, painting, and needlework, pursuits in which she developed considerable skill.

In 1804, at the age of twenty-four, she married her cousin, Samuel Greig, a captain in the Russian navy. He was the son of Admiral Sir Samuel Greig, who, along with four other British naval officers, had been sent to Russia in 1763 to assist Catherine II in the reorganization of the Russian navy. Greig senior remained in Russia all his life, having been rewarded by Catherine for his work with large estates in Livonia. Since the Fairfaxes declined to allow their daughter to go to Russia, her husband gave up active service and became commissioner of the Russian navy and Russian consul for Britain in London. Mary Greig found life in the capital rather cramped after the freedom and independence she had known in Burntisland, but she busied herself with French lessons and continued her mathematical studies. She got little encouragement from her husband, however; he had a low opinion of women's mental capabilities. Their first child, a son, whom they named Woronzow after the Russian ambassador Sir Simon Woronzow, was born in 1805, and a second son arrived shortly before the death of Samuel Greig in 1807. After that Mary and her two infants returned to the Fairfax family home.

The status of widowhood and a small inheritance from her husband brought her considerable independence, and she felt free to spend time not taken up by her family in the mathematical studies she so enjoyed. She managed to read works on plane and spherical trigonometry, conic sections and astronomy, and attempted Newton's *Principia.* Sociable, vivacious, good-looking (she was called the Rose of Jedburgh), and of solid family background, she moved freely in Edinburgh social and intellectual circles, and she made the acquaintance of John Playfair, the leading figure in mathematics and natural philosophy at Edinburgh University. He helped her over difficulties in her current reading (Laplace's *Mécanique céleste*), and, more importantly, put her in touch with one of his former protégés, William Wallace, a Fifeshire man then professor of mathematics at the Royal Military College at Great Marlow (later at Sandhurst), near London. Wallace, an excellent teacher, provided instruction by correspondence until Mary advanced to the point where he suggested that she get on-the-spot tuition from his brother John, also a mathematics teacher. John Wallace took her through the early volumes of *Mécanique céleste.*

In 1812, when she was thirty-one, she remarried. Her second husband was also her first cousin, William Somerville, son of Thomas Somerville, an army doctor, ten years her senior. Of a generally liberal outlook, he strongly and actively encouraged his wife's studies in mathematics and astronomy. Between 1813 and 1815 they lived in Edinburgh, Somerville being head of the Army Medical Department in Scotland. The location offered certain advantages as far as Mary's mathematical education was concerned. The advances in the new branches of mathematics developing rapidly during the early years of the nineteenth century in France were being assimilated more quickly and more widely in Scottish intellectual circles than in England,[113] and under the direction of the Wallaces Mary became familiar with the new work, especially analysis. She also assembled a small but very good mathematical library of up-to-date French texts.[114]

In 1816 the family, which now included two small daughters,[115] moved to London, following a change in William Somerville's army position. With introductions from an Edinburgh friend, the geologist Leonard Horner (later Warden of London University), the Somervilles quickly made their way into London scientific circles. Among their friends were the Swiss physician and chemist Alexander Marcet and his wife Jane (who had recently brought out a popular instructional work in chemistry), the chemist and mineralogist William Wollaston, the astronomer-physicist Henry Kater, physician and physicist Thomas Young, and Henry Warburton and William Blake, both of whom were interested in geology. William Somerville, although by no means a practitioner of science, was an enthusiastic friend of science; he was elected a fellow of not only the Geological and Linnean Societies but also the Royal Society.[116] His wife systematically continued her studies, still getting help from William Wallace. Much of her instruction, however, came in the form of an informal apprenticeship. Wollaston, Kater, Young, and other eminent London men, among them some of the best scientists in England at the time, accepted her as their protégée because of her serious attitude to her studies and her obvious talent and enthusiasm. Informal but nevertheless critical discussions of scientific principles and findings, direction of readings, and a chance to use apparatus such as the telescope and the goniometer, all part of a common pattern of scientific training at the time, were her route to scientific competence. Close and frequent social interaction with scientific men, typically at evening gatherings in drawing rooms, took the place of classroom and laboratory instruction.

In 1826 she published her first scientific paper, a report of her study "On the magnetizing power of the more refrangible solar rays." The idea that there was a connection between light and magnetism had been circulating for some time, although nobody had produced a convincing demonstration of any such phenomenon. By 1826 the subject was attracting a lot of attention. Her article, in which she reported permanently magnetizing a steel needle by exposing it to ultraviolet light, was one of eleven dealing with aspects of magnetism that appeared in *Philosophical Transactions* that year. It attracted immediate notice, both in Britain and on the Continent; her experiment was repeated by Andreas von Baumgartner, professor of physics in Vienna, who claimed similar results.[117] Half a year later, with the collaboration of her friend the astronomer Sir John Herschel, Somerville repeated the investigation, and much to her distress and embarrassment concluded that she had been mistaken in her initial deductions. The 1826 paper was important all the same because it established her as a person actively involved in scientific work rather than only a student and friend of science.

The following year she was approached, through her husband, by Henry Brougham, a prominent and influential figure in the more liberal and progressive of London's political and educational circles, who suggested that she prepare a condensed English version of Laplace's *Mécanique céleste* (1798–1827). The request was hardly a minor one. The work in question was a five-volume technical treatise in applied mathematical analysis in which the French mathematician P. S. Laplace set out his nebular hypothesis of the solar system, that is, his theory of its evolution from an incandescent gas rotating about an axis.[118] The publication of this work, one of the great guidebooks for nineteenth-century theoretical astronomers, is generally considered to mark the high point in the Newtonian theory of gravitation—the completion of the gravitational part of Newton's *Principia.* Laplace demonstrated that the solar system could be regarded as stable and that it did not require occasional divine adjustment to keep it in place. In the course of his investigations he also developed a number of important mathematical concepts (for instance the Laplacian operator), which became standard tools in mathematical physics.

The mathematical background required for the preparation of an English translation of Laplace's work was considerable, the concepts involved being new and difficult. Brougham, however, had recently established the Society for the Diffusion of Useful Knowledge through which he planned to bring educational books to the public at low cost;[119] included in the list of works he wanted to publish were Newton's *Principia* and Laplace's *Mécanique céleste.* Some translating of Laplace's texts had already been done, but this covered only the first two volumes.[120] What Brougham wanted from Mary Somerville was a translation suitable for a fairly wide audience, accompanied by a discussion of the substance of the work, an examination of how the conclusions had been reached and an explanation of the mathematics used. He had known the Somervilles since their Edinburgh days, was well aware of Mary's mathematical studies, and knew that she had been a keen student of Laplace for almost two decades. He and William Somerville between them persuaded her to see what she could do. She pointed out that, because of the amount of fairly advanced mathematics that it must necessarily include, any work she produced would never be very popular, and that furthermore she would have to add a considerable amount of introductory material. But Brougham accepted these points, and, as she put it, "Thus suddenly and unexpectedly the whole character and course of my future life was changed."[121]

During the three years she worked on this huge project (1827–1830) she consulted from time to time with British scientists, especially Augustus de Morgan, at that time professor of mathematics at the University of London, her friend Sir John Herschel, and mathematician Charles Babbage, but she had considerable difficulty getting her hands on the mathematical material she needed for background. What was available in London at the time was limited. When the work was completed it covered the first four of Laplace's five volumes, and Brougham decided that it was far too long for his sixpenny series. He also felt, however, that it was exceptionally interesting, and suggested that Herschel read through the entire manuscript. The latter appears to have had few doubts about its quality and considered that its publication would be an important and much needed first step in bringing French studies in both mathematical analysis and theoretical astronomy to English readers. With Herschel's recommendation, the publisher John Murray accepted the manuscript. A friend and fellow Scot, Murray had already been contacted by William Somerville when the arrangement for publication with Brougham fell through.

The work appeared in 1831 under the title *The Mechanism of the Heavens,* and although it received one or two rather critical notices, most reviews were overwhelmingly favorable, indeed enthusiastic, in both Britain and France.[122] The following year the preface was published separately as a *Preliminary Dissertation to the Mechanism of the Heavens.* Somerville herself felt that this lengthy and largely nonmathematical introduction was her most important contribution. It placed celestial mechanics in a wider context, and, at a time when elementary texts were in short supply, served to ease the reader's entry into the very technical work that followed. It was immediately pirated by Carey and Lea of Philadelphia. No American edition of the *Mechanism* itself was ever published.[123] Sales in Britain were slow, although by 1837 it was made a textbook for advanced students at Cambridge, guaranteeing a small but steady demand. (Viewed from the longer perspective, this adoption of the book by Cambridge perhaps gives Somerville a modest place among those who helped forward the early development of the nineteenth-century Cambridge school of mathematical physics.)

Public recognition of the importance of Somerville's work came quickly. The Royal Society commissioned a bust of her for its apartments, the Geological Society saluted her through her husband by electing him a member of the council in 1832, and Cambridge University invited her and her husband to be its official guests for a week during which they were entertained as celebrities.

Her first book was no sooner completed than she began a second, *On the Connection of the Physical Sciences,* which developed out of her thoughts about the overlap and interdependence of several of the fields she had found herself dealing with in her first work. What properly constituted the physical sciences was a matter of considerable debate in the first half of the nineteenth century. Much of the material Somerville covered would now come under two headings, astronomy (theoretical and descriptive) and traditional physics, which she subdivided into electricity and magnetism, heat, light and sound, matter in its elementary particles, and the forces that hold these particles together. Under heat she included meteorology and physical geography, and under matter she dealt with topics now thought of as physical chemistry. Chemistry, in its late nineteenth- or twentieth-century form, was omitted from her picture.[124] She collected a considerable amount of material during a stay in Paris in 1832–1833. Among her French contacts were Ampère and Becquerel, both of whom provided information on electromagnetism and other electrical phenomena. In England also she consulted several well-known men of science, some of them the best in their fields, including Brougham, Michael Faraday, Henry Holland, Charles Lyell, and the Cambridge physicist-mathematician William Whewell. Faraday especially, worked with her extensively.

The book, published by Murray in 1834, was a well-researched and lucidly written account of the state of the physical sciences at the time, an up-to-date research report, as it were, backed by concise introductions to basic concepts. All was presented in straightforward prose, although the underlying mathematics, some of it difficult material, was also offered in the form of extensive notes, and an understanding in the reader of first principles was implicitly assumed. By the form of its organization it provided definition and shape to a broad spectrum of scientific work, emphasizing clear interconnections between physical phenomena, unifying ideas from many branches, making guiding concepts more apparent, and so pointing the way forward for future work. It was an immediate success. In America a pirated edition appeared within a few months, brought out by Key and Biddle of Philadelphia, and sizable extracts, especially from the sections on electricity and magnetism, were reprinted in other forms.[125] French, German, and Italian translations also appeared. A second edition, which included Faraday's very latest work and the addition of many new diagrams and figures, appeared in 1835 and sold very well. A third, incorporating more new material, came out the following year and sold even better. By then the book was an established scientific classic and a best-seller on Murray's list; 1837 saw the appearance of a fourth edition. Thus for a number of years it functioned to some extent as an annual report of the progress of physics and astronomy.[126]

In 1835 Mary Somerville received solid public recognition for her contributions to British science in the form of a Civil List pension of £200 a year, a large sum for a scientific person.[127] Three years later her husband became seriously ill and for his sake they decided to go to Rome for the winter. She was then fifty-eight, and showing signs of weariness after more than ten years of challenging writing. William Somerville was sixty-seven. The family left for Italy in September 1838 and except for two short visits never returned to Britain. They had no settled home, but moved around, living for varying periods of time in Florence, Rome, Bologna, and other Italian cities, apparently without being seriously affected by the continuing major political changes taking place in the country during those years. Surprisingly enough, Mrs. Somerville continued her scientific writing, although by then she had only occasional direct contact with people actually engaged in scientific work. In all, nine editions of *The Connection of the Physical Sciences* appeared during her lifetime (four of them after she had left England) and a tenth in 1877, five years after her death. Fifteen thousand copies of the book were sold over a period of four decades.

In 1848 she published a third work, her two-volume *Physical Geography,* the first English-language textbook in the field. Perhaps if she had brought out the first volume of this when it was completed in 1842 the work would have achieved more lasting fame, since then it would have preceded Alexander von Humboldt's acclaimed *Kosmos,* the first volume of which appeared in 1845.[128] Nevertheless *Physical Geography* was immediately successful; it became her most popular work, running to seven editions between 1848 and 1877, with sales of some 16,000 copies. In geographical writing it was pioneering in that it broke away from a common pattern of simply describing the distribution of geographical phenomena country by country; instead it presented a unified picture of the planet, the interrelationships between natural phenomena, the distribution patterns in the animal and vegetable kingdoms and the causes underlying these distributions. By putting aside constraints of political boundaries, Somerville anticipated the regional approach to geographical studies. The work came out at an important time in the development of the field in Britain. The Natural Sciences Tripos established at Cambridge University in 1848 included geology, of which physical geography was a part; in 1850 Oxford recognized physical geography as a separate optional subject in the new School of Natural Science and Somerville's *Physical Geography* appeared in the book lists.[129] It remained in standard geography reference lists for many years.

Her fourth book, *On Molecular and Microscopic Science,* appeared in 1867, when she was in her eighty-ninth year. Consisting of two large volumes, it described the constituents of matter and the structure of microscopic plants and animals. To a large extent its science was out-of-date, but it was kindly received out of deference to its distinguished author.[130] A last work, autobiographical (*Personal Recollections*), was finished just before her death and edited and published by her daughter Martha shortly after. She died on 29 November 1872, in her ninety-second year, and is buried in the English cemetery in Naples. Somerville Hall, one of the first women's colleges at Oxford, commemorates her name; there was also established at Oxford a Mary Somerville scholarship for women in mathematics. Much of her library went to Girton College Cambridge.

Somerville was widely recognized and her honors were many. The Royal Irish Academy accorded her membership in 1834, and the Royal Astronomical Society elected her an honorary fellow in 1835, Caroline Herschel, then eighty-five, being elected at the same time. The two were the first women fellows of this society and the only ones for some years thereafter. In 1869 she received the Patron's Gold Medal of the Royal Geographical Society, and the Victor Emmanuel Gold Medal of the Geographical Society of Florence. The latter society was one of eleven Italian and two American societies that gave her honorary memberships during her later years. She herself, despite the indubitable influence of her books over a broad range of subjects, considered that her intellectual capacities were limited. She saw herself as having intelligence and powers of perseverance that enabled her to analyze, bring together and evaluate the work of others, and above all to explain; but originality was a different matter. Too bound by the attitudes and coventions of her time and too lacking in self-confidence, she accepted the idea that creative imagination in scientific work was something women did not possess. One might regret that she never made contact with Sophie Germain (1776–1831), the French woman mathematician of about her time who did make original contributions (in the areas of number theory and the theory of elasticity). Germain knew and worked with both Lagrange and Legendre, who in turn knew Somerville.

Concluding remarks

Somerville's story (along with that of Lovelace) is placed as something of a lengthy footnote to this account of nineteenth-century British women mathematicians, despite the fact that she unquestionably occupies a foremost place in the history of women of science of the early nineteenth century. Her achievements in scientific exposition in no fewer than four areas—mathematics, physics, astronomy, and the earth sciences—were exceptional, and the recognition she received in her time was unmatched for a woman and well earned. Her contributions, however, were of a different kind from those of the later women, both the Cambridge group and those working in London. Essentially all the later women, at least for some of their careers, were part of the scientific academic community and were involved in original research in mathematics or statistics in areas of then current interest (also among them were two, Sophie Bryant and Philippa Fawcett, who had outstanding public careers at the national level in educational reorganization).

The women who collaborated with Karl Pearson (Lee, Cicely Fawcett, and Beeton), although they were assistants rather than independent research workers, were nevertheless members of what was probably the leading statistical group of the period; Lee especially was remarkably productive. While some of the statistical techniques developed by the Pearson laboratory team have, not surprisingly, been superceded[131] by more efficient or more discriminating methods, others have become central to the analytical arsenal. The work to which Lee made very creditable contributions, including the preparation of tabulated functions, propelled statistics into an important place in science and constituted an essential phase in the emergence and development of dependent areas, such as population genetics.

The interests of the Cambridge women were likewise very much focused on areas where there was great current activity.[132] The fact that four of the Cambridge-trained women—Scott, Maddison, Hardcastle, and Hudson—were algebraic geometers is no coincidence, this branch of mathematics being one where a great deal of research was being done toward the end of the nineteenth century.[133] Cambridge mathematician Arthur Cayley, a heavy contributor to the development of algebraic geometry (and a supporter of higher education for women) was Scott's postgraduate adviser. Maddison, and probably also Hardcastle, attended his lectures. Charlotte Scott's role as mentor for younger British women students of algebraic geometry, even though she was on the other side of the Atlantic, is interesting, and a strong indication of the reputation she had in her field; Maddison, Hardcastle, and Hudson all worked with her as postgraduate students.

One of the most remarkable things overall about this group is simply its size, in particular the size of the distinguished Cambridge-educated subsection. No other university in any country trained as many women who published original work in mathematics in the latter years of the nineteenth century as did Cambridge. How did this come about? The rigors of the selection methods for entrance to Newnham and Girton Colleges combined with four years of closely directed tuition and character building under difficult conditions were doubtless key factors; the ever-present challenge of the Mathematical Tripos was almost certainly another. Whatever the details of the processes that produced the phenomenon, the fact remains that a striking number of women of outstanding ability and great strength of character came out of late nineteenth-century Cambridge.[134] Among them were some of the pioneers of work by women in both mathematics research (Scott, Chisholm, and Hudson) and in education (Philippa Fawcett, Scott, and Barwell).

Two-country comparison

In comparing these early British women mathematicians with their American counterparts, one is struck by the fact that, although several of the American women took Ph.D. degrees or had doctoral-level training (Ladd, MacKinnon, Winston, and Wood), none of them (with the possible exception of Ladd) succeeded in carrying out significant original work in mathematics during their subsequent careers (Winston and Wood concentrated on undergraduate teaching, Ladd and MacKinnon moved to other interests—although Ladd did continue to bring out work in symbolic logic from time to time). Thus, although a number of slightly later American women [for instance Mary Sinclair (1878–1955), Anna Pell (1883–1966) and Mildred Sanderson (1889–1914)], might be compared with Hudson, there was no American with a mathematics research career beginning before the turn of the century who might be thought of as the counterpart of Chisholm Young. Further, there were no pre-1901 contributions by American women to statistics, and no early Americans trained in mathematics had careers in education at the national level as influential as those of Bryant and Philippa Fawcett in Britain. The one exception on the American scene was Cambridge-trained Charlotte Scott. Was Scott extraordinarily lucky in finding a special place where research in mathematics could be done? Was the close connection she maintained with the European mathematical community critical? Or did she have qualities that gave her a special advantage? Probably all these factors were important. With significant work in higher mathematics by American men only beginning in the late 1870s, it is perhaps not surprising that women failed to make much headway until after the turn of the century.[135]

Numerical data for the two groups are summarized in Figure 8.

Notes

1. Among the British women are two, Charlotte Scott and Isabel Maddison, who spent the greater part of their careers at Bryn Mawr College and who could have been classed as contributors to American mathematics. Here both have been placed in the British section with the other Cambridge-trained women mathematicians of the period, in part to emphasize fully the prominent role of the Cambridge women's colleges in early mathematics education for women.

2. A pre-1900 paper by one additional American (Winston) that is not listed in the Royal Society *Catalogue* has been included in the bibliography, giving totals there of seven authors and sixteen papers.

3. The state universities of Kansas and Wisconsin were particularly prominent among American coeducational institutions for early undergraduate education of women in mathematics (see Judy Green and Jeanne Laduke, "Contributors to American mathematics. An overview and selection," in *Women of Science. Righting the Record,* eds. G. Kass-Simon and Patricia Farnes (Bloomington, Ind.: Indiana University Press, 1990), p. 142, n. 12).

4. Anon., *The Graduate Magazine, University of Kansas,* **2** (1903–1904): 147; obituary, including a letter from A. R. MacKinnon, ibid., **39** (1940): 25; Margaret E. Maltby, comp., *History of the Fellowships awarded by the American Association of University Women. 1888–1929. With Vitas of the Fellows* (Washington, D.C.: American Association of University Women, [1929]), pp. 15–16; Graduate

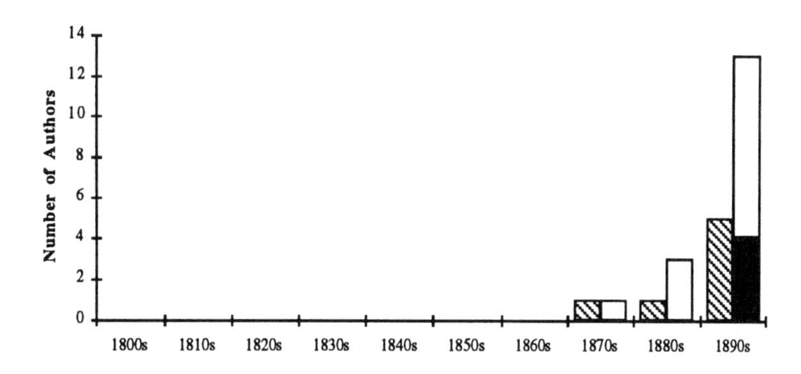

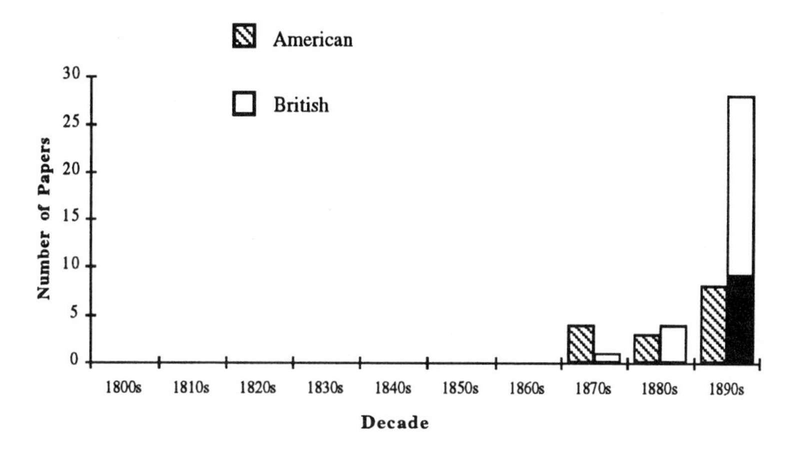

Figure 8. American and British authors and papers in mathematics and statistics, by decade, 1800–1900. (Statistics portion solid black.) Data from the Royal Society *Catalogue of Scientific Papers.*

School Records and Deceased Alumnae Records, Cornell University archives.

5. Titles of MacKinnon's seminar talks were, "Further examples of differentiability of functions;" "Calculus of variations;" "The rational points associated with the conic $ax^2-by=c$;" "Classifying types in the field $\sqrt{-m}$;" "On Smith's curve"—see Karen Hunger Parshall and David E. Rowe, *The Emergence of the American Mathematical Research Community, 1876–1900; J.J. Sylvester, Felix Klein, and E.H. Moore* (Providence, R.I.: American Mathematical Society, London Mathematical Society, 1994), p. 257.

6. Student records, University of Kansas archives; U.S. Decennial Census, 1900 (Kansas).

7. Green and Laduke, "Contributors to American mathematics," pp. 127–9; Maltby, comp., *History of the Fellowships,* p. 16; Betsy S. Whitman, "Mary Frances Winston Newson," in *Women of Mathematics. A Biobibliographic Sourcebook,* eds. Louise S. Grinstein and Paul J. Campbell (Westport, Conn.: Greenwood Press, 1987), pp. 161–4, and "An American woman in Göttingen," *Mathematical Intelligencer,* **15,** (1993): 60–2; *WWWA,* p. 596; *AMS,* 1910–1921.

8. Scott described Winston as a very diligent student with an excellent grasp of her subject even if her abilities were not striking (Whitman, "Winston Newson," p. 162).

9. M. L. Cartwright, "Grace Chisholm Young," *Journal of the London Mathematical Society,* **19** (1944): 185–92, on. p. 186. Officials in the Prussian Ministry of Education were at the time eager to make a limited experiment with the admission of foreign women students to their universities before Prussian women were accepted. Friedrich Althoff, the powerful official in charge of university affairs, asked Klein to seek out, while he was in Chicago at the Columbian Exposition, women interested in doctoral study at Göttingen—see James C. Albisetti, *Schooling German Girls and Women. Secondary and Higher Education in the Nineteenth Century* (Princeton, N.J.: Princeton University Press, 1988), p. 227. Another American, Lillien Martin, studied experimental psychology at Göttingen from 1894 to 1898, but her status as a student would appear to have been less formal than that of Maltby and Winston and she did not receive a degree at that time (see chapter 15).

10. Felix Klein, *The Mathematical Theory of the Top* (New York: Charles Scribner's Sons, 1897), p. 35, n. Winston's dissertation was entitled, *Ueber den Hermite'schen Fall der Lamé'schen Differentialgleichung* (Göttingen: W. Fr. Kästner, 1897).

11. [President] Thomas E. Will Papers, B-1/F-2, B-1/F-7, Kansas State University archives.

12. Letter printed in part in *The Industrialist,* 15 July 1897, p. 166 (Kansas State University archives).

13. Ibid., 15 November 1897, p. 219, and 23 January 1900, p. 237.

14. "We recognise in Miss Mary F. Winston, Professor of Mathematics in the Kansas State Agricultural College, one whose technical acquirements are rarely excelled, and whose logical mind, clear sense of justice, earnestness of character, sympathetic nature and personal charm have combined to make her one of the most efficient and popular teachers of this institution, and one of the most valuable counselors of this Faculty.

We desire to express our deep and sincere regret at her departure . . ." (*Faculty Records, D,* 16 June 1900, p. 184, Kansas State University archives). Quotations made with the permission of Kansas State University.

15. *Mathematical Developments Arising from Hilbert Problems,* ed. Felix E. Browder (Providence, R.I.: American Mathematical Society, 1976), pp. 1–34. Winston's translation originally appeared as "Mathematical problems," *Bulletin of the American Mathematical Society,* **8** (1902): 437–79. While at Göttingen she had studied with Hilbert, one of the major figures in late nineteenth- and early-twentieth century mathematics.

16. This was possibly because of nepotism rules then in force, which prevented close relatives from holding posts in the same institution. Immediately following Henry Newson's death, Mary's sister, Alice Winston, had joined the faculty of the English department at the University of Kansas.

17. *Alumni Directory,* 1910, 1913, 1919, University of Chicago; *Hamline University Bulletin, 1908–09;* S and F, pp. 216–17. Schottenfels is reported to have been chairman of the mathematics department at Adrian College, Michigan, in 1913 (S and F, p. 217), but a search by archivist Bridgette A. Woodall of currently available records at Shipman library, Adrian College, failed to uncover any mention of her. Hunter College did not respond to a request for information about Schottenfels.

18. The fact that her papers are often recorded "by title only" raises the question of whether she actually spoke at very many of those meetings.

19. Smith College archives; *WWWA,* p. 900; *AMS,* 1910–1944.

20. For Wood's 1900 presentation see bibliography. Her 1902 paper, entitled "Non-Euclidian displacements and symmetry transformations," was reported in the *Bulletin of the American Mathematical Society,* **8,** n.s. (1902): 368, 370.

21. Ann Moskol, "Ellen Amanda Hayes (1851–1930)," in *Women of Mathematics,* Grinstein and Campbell, eds., pp. 62–6; Alice Payne Hackett, *Wellesley. Part of the American Story* (New York: E. P. Dutton, 1949), pp. 184–8; *WWWA,* p. 374; *AMS,* 1910.

22. Ellen Hayes, "Comet a 1904," *Science,* **19** (1904): 833–4, and "The path of the shadow of a plummet bead, *Popular Astronomy,* **16** (1908): 279–86.

23. The women faculty at Wellesley in the early years of the century included several who held and expressed strong social and political views. In addition to Hayes and Scudder there were Sarah Whiting, professor of astronomy and physics, an outspoken prohibitionist, Mary Calkins, professor of psychology, a strong advocate of human rights, and Emily Balch, professor of economics and a committed pacifist. Balch was the only faculty member of the period who was actually dismissed by the trustees for her outspoken views and outside interests. After she left Wellesley she spent much of the rest of her life in Geneva working with the Women's International League for Peace and Freedom; in 1946 she shared the Nobel Peace Prize (see Hackett, *Wellesley,* pp. 190–1).

24. Several important faculty positions at Wellesley were initially filled by women who did not have college degrees. These included Mary Horton (Greek), Frances Lord (Latin), and Susan Hallowell (botany)—ibid., pp. 41–3.

25. Evert W. Beth, "Hundred years of symbolic logic," *Dialectica,* **1** (1947): 337 (see also Green and Laduke, "Contributors to American mathematics," p. 122, and Judy Green, "Christine Ladd-Franklin (1847–1930)," in *Women of Mathematics,* Grinstein and Campbell, eds., pp. 121–8, on p. 126).

26. See also Ladd-Franklin's 1890 paper in *Mind* (bibliography).

27. Shea Zellweger, "Sign-creation and man-sign engineering," *Semiotica,* **38** (1982): 17–54, especially p. 46; "Notation, relational iconicity, and rethinking the propositional calculus" (summary), in *Program Abstracts, Eighth International Congress of Logic, Methodology, and Philosophy of Science,* Moscow, USSR, August 17–22, 1987 (vol. 1, pp. 376–9). Ladd's work in the area was later extended by her student Eugene Shen ("The Ladd-Franklin formula in logic: the antisyl-

ogism," *Mind,* n.s. **36** (1927): 54–60, and "The 'complete-scheme' of propositions," *Psyche,* **9** (1929): 48–59).

28. Ladd-Franklin's post-1900 publications in logic include the following: "The reduction to absurdity of the ordinary treatment of the syllogism," *Science,* **13** (1901): 574–6; "Some points in minor logic," *Journal of Philosophy, Psychology, and Scientific Methods,* **1** (1904): 13–15; "Minor logic," ibid., **1** (1904): 494–6; "The foundations of philosophy: explicit primitives," ibid., **8** (1911): 708–13; "Explicit primitives again: a reply to Professor Fite," ibid., **9** (1912): 580–5; "Implication and existence in logic," *Philosophical Review,* **21** (1912): 641–5; "The antilogism—an emendation," *Journal of Philosophy, Psychology and Scientific Methods,* **10** (1913): 49–50; "Symbol logic and Bertrand Russell," *Philosophical Review,* **27** (1918): 177–8; "Bertrand Russell and symbol logic," *Bulletin of the American Mathematical Society,* **25** (1918): 59–60; "The antilogism," *Psyche* (1927): 100–3; "The antilogism," *Mind,* n.s. **37** (1928): 532–4; "Some questions in logic," *Journal of Philosophy,* **25** (1928): 700. She also wrote (jointly with Edward V. Huntington, assistant professor of mathematics at Harvard) the article on "Symbolic logic" in the 1905 edn. of *Encyclopedia Americana,* vol. 9, (New York, 1905). It was reprinted in subsequent edns. (1934, 1941, 1952).

29. Green, "Ladd-Franklin," p. 126.

30. Three of the fourteen British authors—Ayrton, Fry, and Whiteley—had major interests in other areas (see chapters 9, 1, and 11, respectively). Pre-1900 papers by two additional authors (Hudson and Lovelace) not listed by the Royal Society have been added to the bibliography, as well as an additional paper by Bryant.

31. Marguerite Lehr, "Scott, Charlotte Angas," in *NAW,* vol. 3, pp. 249–50; F. S. Macaulay, "Dr. Charlotte Angas Scott," *Journal of the London Mathematical Society,* **7** (1932): 230–40; Patricia Clark Kenschaft, "Charlotte Angas Scott (1858–1931)," in *Women of Mathematics,* Grinstein and Campbell, eds., pp. 193–203, "Charlotte Angus Scott, 1858–1931," *Newsletter—Association for Women in Mathematics,* 7 (1977): 9–10, and ibid., **8** (1978): 11–12; Green and Laduke, "Contributors to American mathematics," pp. 124–7.

32. Cambridge degrees were not fully opened to women until 1948, although titular degrees were awarded from 1923.

33. Evelyn Sharp, *Hertha Ayrton, 1854–1923. A Memoir* (London: Edward Arnold, 1926), p. 93.

34. Scott's years on the Bryn Mawr faculty were thus characterized by her longtime friend Frank Morley, (Macaulay, "Charlotte Angas Scott," p. 231). Morley studied mathematics at Cambridge in the early 1880s, became professor of mathematics at Haverford College (not far from Bryn Mawr) in 1888, and later moved to Johns Hopkins in Baltimore, where Scott often visited him and his family.

35. These students were Ruth Gentry (1894), Isabel Maddison (1896), Virginia Ragsdale (1904), Louise Duffield Cummings (1914), Mary Gertrude Haseman (1916), Bird Margaret Turner (1920), and Marguerite Lehr (1925) (Green and Laduke, "Contributors to American mathematics," p. 143, n. 39). During Scott's time as department head Bryn Mawr ranked third in the country in the number of mathematics doctoral degrees conferred on women (nine); only the University of Chicago and Cornell, both much bigger institutions, gave more (Kenschaft, "Scott" (1987), p. 201).

36. Kenschaft, "Scott" (1977), p. 11.

37. Seventy years were to pass before another woman served as vice president of the American Mathematical Society; Mary W. Gray was elected in 1976, and several other women have held the office since then. In 1983 Julia B. Robinson became the first woman president.

38. Macaulay, "Charlotte Angas Scott," p. 232.

39. C. A. Scott, "On Cayley's theory of the absolute," *Bulletin of the American Mathematical Society,* **3** (1897): 235–46, on p. 235. Among Scott's other notable interpretative papers are, "On von Staudt's Geometrie der Lage," *Mathematical Gazette,* **1** (1900): 307–14, 323–31, 363–70, and "On a recent method for dealing with the intersections of plane curves," *Transactions of the American Mathematical Society,* **3** (1902): 216–63.

40. For an example see Green and Laduke, "Contributors to American mathematics," p. 125.

41. Among Scott's post-1900 papers not already mentioned are, "Note on the geometrical treatment of conics," *Annals of Mathematics,* **2** (1901): 64–72; "On the circuits of plane curves," *Transactions of the American Mathematical Society,* **3** (1902): 388–98; "Note on the real inflections of plane curves," ibid., **3,** (1902): 399–400; "Elementary treatment of conics by means of the regulus," *Bulletin of the American Mathematical Society,* **12** (1905): 1–7; "Note on regular polygons," *Annals of Mathematics,* s.2, **8** (1907): 127–34; "Higher singularities of plane algebraic curves," *Proceedings of the Cambridge Philosophical Society,* **23** (1926): 206–32.

42. The 1961 edition, with corrections and additions, appeared as *Projective Methods in Plane Analytical Geometry* (New York: Chelsea Publishing Co.).

43. See Kenschaft, "Scott" (1977), p. 11.

44. Millicent Garrett Fawcett, *What I Remember* (London: T. Fisher Unwin., 1924), pp. 137–45, 149–74; anon., "Miss P. G. Fawcett. A pioneer in education," *Times,* 12 June 1948, 6e; Sir Fabian Ware, "Miss Philippa Fawcett," ibid., 16 June 1948, 6d; Mrs. C. L. Marson, ibid., 22 June 1948, 7e; Barbara Caine, *Victorian Feminists* (Oxford: Oxford University Press, 1992), pp. 206–8, 213–14; *Newnham College Register, 1871–1971,* vol. 1, Staff, p. 7; *Who was Who,* vol. 4, p. 376; A. T. C. Pratt, *People of the Period,* 2 vols. (London: N. Beeman, 1897), reproduced in *British Biographical Archive,* microfiche edn., ed. Paul Sieveking (London: K. G. Saur, 1984), no. 393.

45. The title "Senior Wrangler" at Cambridge designated the top student of the year in the grueling baccalaureate honours examinations in mathematics. It was the highest undergraduate honour the university could bestow, and for a student the most famous mathematical honor in the country—perhaps the ultimate appraisal of scientific ability. Philippa Fawcett's mother recorded in her memoirs that she had been told that the examiners considered 1890 to have been a "strong year," but that Philippa was 400 marks (13 percentage points) ahead of the senior. "She was ahead on all papers except two, so that the examiners were sure that her place had no element of accident in it. . ." (Fawcett, *What I Remember,* p. 141).

46. Ibid., p. 140.

47. See Fawcett, ibid., pp. 143–4. Miss Clough was the principal of Newnham and Miss Gladstone, a daughter of the statesman, one of the vice principals. The insertion in square brackets was made by Fawcett.

48. Although there is little question about the general jubilation at Cambridge at the time, it is also the case that from about then on (1890) resentment against women competing with men began to grow (see Rita McWilliams-Tullberg, *Women at Cambridge. A Men's University—Though of a Mixed Type* (London: Victor Gollancz Ltd., 1975), p. 102). McWilliams-Tullberg also pointed out (p. 103) the similar pattern of events that had followed the attempts of Elizabeth Garrett to get a medical education in the 1860s. When she first attended classes at Middlesex Hospital she was treated well by her male fellow students, but when it became clear to them that she was an able and ambitious student their attitude changed markedly (Elizabeth Garrett was Philippa Fawcett's aunt).

49. A fine picture of this team, with Philippa Fawcett and geologist Gertrude Lilian Elles (see chapter 12) seated together in the front row, appears opposite p. 50 in *A Newnham Anthology,* ed. Ann Phillips (Cambridge: Cambridge University Press, 1979). Indeed, a fascinating sequence of photographs of Philippa Fawcett exists spread through a number of available works, beginning with a portrait taken about age twelve (plate 11 in Caine, *Victorian Feminists*), progressing to a student photo (Fawcett, *What I Remember,* opposite p. 140), to a group photo of the Newnham College staff in 1896 (McWilliams-Tullberg, *Women at Cambridge,* opposite p. 160). She was a slender, pleasant-faced, good-looking young woman, with an appearance of quiet confidence.

50. Cartwright, "Grace Chisholm Young;" I. Grattan-Guiness, "A mathematical union: William Henry and Grace Chisholm Young," *Annals of Science,* **29** (1972): 105–86; Sylvia M. Wiegand, "Grace Chisholm Young (1868–1944)," in *Women of Mathematics,* Grinstein and Campbell, eds., pp. 247–54.

51. The challenge was possibly made by Chisholm's older brother Hugh, who studied classics at Oxford (Grattan-Guiness, "A mathematical union," p. 121).

52. Albisetti, *Schooling German Girls and Women,* p. 234.

53. Grattan-Guiness, "A mathematical union," pp. 127–9.

54. Although Sofia Kovalevskaia had received a Göttingen Ph.D. in mathematics twenty-one years earlier (1874) it had been awarded to her in absentia, without the oral examination and public defense. Kovalevskaia's friend, the chemist Iuli'ia Lermontova, did successfully take the oral doctoral examination at Göttingen in 1874, but she had carried out her studies and research, by special permission, at Heidelberg and Berlin. Thus Chisholm was the first woman student to complete the whole doctoral course at Göttingen.

55. G. E. Chisholm, *Algebraisch-gruppentheoretische Untersuchungen zur sphärischen Trigonometrie* (Göttingen, 1895).

56. G. H. Hardy, "William Henry Young," *Journal of the London Mathematical Society,* **17** (1942): 218–37, on p. 218.

57. Hardy, "Young," p. 233.

58. In a footnote to one of his 1914 papers Young says, "Various circumstances have prevented me from composing the present paper myself. The substance of it only was given to my wife, who has kindly put it into form. The careful elaboration of the argument is due to her" (W. H. Young, "On integration with respect to a function of bounded variation," *Proceedings of the London Mathematical Society* (2) **13** (1914): 109–150, p. 110).

59. This is stated very explicitly in a letter from Young to his wife quoted by Grattan-Guiness, "A mathematical union," pp. 141–2.

60. "A note on derivates and differential coefficients," *Acta Mathematica,* **37** (1914): 141–54.

61. See the following, all of which were coauthored with W. H. Young: "On derivates and the theorem of the mean," *Quarterly Journal of Pure and Applied Mathematics,* **40** (1909): 1–26; "An additional note on derivates and the theorem of the mean," ibid., **40** (1909): 144–5; "Discontinuous functions continuous with respect to every straight line," ibid., **41** (1910): 87–93; "On the determination of a semi-continuous function from a countable set of values," *Proceedings of the London Mathematical Society,* (2), **8** (1910): 330–9; "On the existence of a differential coefficient," ibid., (2), **9** (1911): 325–35; "On the theorem of Riesz-Fischer," *Quarterly Journal of Pure and Applied Mathematics,* **44** (1912): 49–88.

62. See Wiegand, "Chisholm Young," pp. 250–1, for a discussion of the Denjoy-Saks-Young theorem. See also G. C. Young, "On infinite derivates," *Quarterly Journal of Pure and Applied Mathematics,* **47** (1916): 127–75, and "On the derivates of a function," *Proceedings of the London Mathematical Society* (2), **15** (1916): 360–84. Grace Young's other papers in this area included "Sur les courbes sans tangente," *Enseignement mathématique,* **17** (1915): 348; "Sur les nombres dérivés d'une fonction," *Comptes Rendus,* **162** (1916): 380–2; "On the partial derivates of a function of many variables," *Proceedings of the London Mathematical Society,* (2), **20** (1922): 182–8.

63. The Youngs never published their book on integration. However, their son Laurence brought out an expository monograph, *The Theory of Integration* (Cambridge Tracts in Mathematics and Mathematical Physics) in 1927. Laurence Young noted in his preface (p. vii) that he had much encouragement from his father and "constant assistance" from his mother.

64. She spent about five years, beginning in 1929, on a remarkable historical and literary project—a novel patterned after the style of Sir Walter Scott, whose writing she greatly admired. Entitled *The Crown of England,* the work was anchored in a great deal of research into Elizabethan figures in original sources, first in Geneva and then in London. The main theme was the relationship between Ferdinando Stanley, the elder son of the fourth Earl of Derby, and Alice, the daughter of Sir John Spencer, set among the intrigues of the Elizabethan court of the 1570s. The book ran to almost 400 pages and was rather too scholarly for general readership, however, with many quotations from contemporary documents in original languages. No London publisher would accept it, and although improvements were planned they were not carried out; so far it remains unpublished.

65. Grace Young's single-author papers not already listed include, "On the form of a certain Jordan curve," *Quarterly Journal of Pure and Applied Mathematics,* **37** (1904): 87–91; "A note on a theorem of Riemann's," *Messenger of Mathematics,* **49,** (1919–1920): 73–8; "Démonstration du lemme de Lebesque sans l'emploi des nombres de Cantor," *Bulletin des sciences mathématiques* (2), **43** (1919): 245–7; "On the solution of a pair of simultaneous Diophantine equations connected with the nuptial number of Plato," *Proceedings of the London Mathematical Society,* (2), **23** (1925): 27–44; "Pythagore, comment a-t-il trouvé son théorème?," *Enseignement mathématique,* **25** (1926): 248–55; "On functions possessing differentials," *Fundamenta Mathematicae,* **14** (1929): 61–94; "A time-honoured mystery, from the Meno of Plato," *O Instituto* (Coimbra, Portugal, Imprensa da Universidade, 1929), 78, 24 pp. Papers published jointly with W. H. Young and not already listed include, "Note on Bertini's transformation of a curve into one possessing only nodes," *Atti della Reale Accademia delle Scienze di Torino,* **42** (1906): 82–6; "On the reduction of sets of intervals," *Proceedings of the London Mathematical Society,* (2), **14** (1915): 111–30; "Sur la frontière normale d'une région ou d'un ensemble," *Comptes Rendus,* **163** (1916): 509–11; "On the internal structure of a set of points in space of any number of dimensions," *Proceedings of the London Mathematical Society,* (2), **16** (1918): 337–51; "On the inherently crystalline structure of a function of any number of variables," ibid., (2), **17** (1918): 1–16; "On the discontinuities of monotone functions of several variables," ibid., (2), **22** (1924): 124–42. Grace Young also wrote three educational books for children, one of which, *The First Book of Geometry* (London: J. M. Dent, 1905), was coauthored with W. H. Young (a German translation, *Der kleine Geometer,* appeared in 1908). The other two were, *Bimbo: a Real Little Story for Jill and Molly* (London: J. M. Dent, 1905), and *Bimbo and the Frogs: another Real Story* (London: J. M. Dent, 1907); both were on elementary biology and probably had their origins in her medical studies.

66. William Young died in Lausanne, 7 July 1942. He had been separated from his family by circumstances arising unexpectedly from the outbreak of the Second World War.

67. The Young-Chisholm mathematical talent appeared in the next generation as well, Laurence Young's daughter Sylvia Wiegand, whose article on Grace Young is cited here, became a member of the mathematics faculty at the University of Nebraska.

68. J. G. Semple, "Hilda Phoebe Hudson," *Bulletin of the London Mathematical Society,* **1** (1969): 357–9; M. D. K., "Hilda Phoebe Hudson, 1881–1965," *Newnham College Roll Letter* (1966): 53–4; *Newnham College Register 1871–1971,* vol. 1, Staff, p. 10; *Who was Who,* vol. 1, pp. 357–8, entry for William Henry Hoar Hudson; information from Sister Winifred, S.S.M., St. Mary's Convent and Nursing Home, Chiswick, London.

69. (With Sir Ronald Ross) "An application of the theory of probabilities to the study of a priori pathometry," Pts. II and III, *Proceedings of the Royal Society,* A (650), **93** (1917): 212–40.

70. See H. P. Hudson, "The strength of laterally loaded struts," *The Aeroplane, Aeronautical Engineering Supplement,* **18** (1920): 1178–80; "Incidence wires," *Aeronautical Journal,* **24** (1920): 505–16.

71. Semple, "Hilda Phoebe Hudson," p. 358. One might wonder if Hudson's exceptional geometrical intuition had its origins in her early childhood introduction to Euclidian geometry. Her interests in that area remained strong and in 1916 she published a small monograph, *Ruler and Compasses* (London: Longman's, Green—Longman's Modern Mathematical Series; reissued as part of the compendium, *Squaring the Circle and other Monographs*—New York: Chelsea Publishing Co., 1953). An exposition of the range and limitations of ruler and compass constructions, it included a considerable amount of "elegant geometry" (Semple, p. 357). Hudson also published a paper on "Euclidian constructions" in *Scientia,* Nov. 1921, pp. 346–54.

72. "On the 3–3 birational transformation in three dimensions," *Proceedings of the London Mathematical Society* (2), **9** (1911): 51–6, and **10** (1912): 15–47; "On fundamental points in Cremona space transformations," *Annali di matematica pura ed applicata,* Milano, s. 3, **19** (1912): 45–56; "On cubic birational space transformations," *American Journal of Mathematics,* **34** (1912): 203–10; "On the product of two quadro-quadric space transformations," ibid., **35** (1913): 183–8; "On pinch points," *Quarterly Journal of Pure and Applied Mathematics,* **44** (1913): 161–6; "Curves of simple contact on algebraic surfaces," *Mathematische Annalen,* **73** (1913): 73–85; "Curves of contact of any order on algebraic surfaces," *Proceedings of the London Mathematical Society,* (2), **11** (1913): 389–410; "On binodes and nodal curves," *Proceedings of the International Congress of Mathematicians* (Cambridge), **2** (1913): 118–21; "On the composition of Cremona space transformations," *Rendiconti del Circolo matematico,* Palermo, **35** (1913): 386–8.

73. Hudson's later papers include "The Cremona transformations of a certain plane sextic," *Proceedings of the London Mathematical Society,* (2), **15** (1916): 385–400; "Plane homaloidal families of general degree," *Proceedings of the London Mathematical Society* (2), **22** (1924): 223–47; "Cech's transformations," ibid., (2) **23** (1924): xxvii-xxix; "Double invariant points and curves of Cremona plane transformations," *Rendiconti del Circolo matematico,* Palermo, **50** (1926): 219–28; "Linear dependence of the Schur quadrics of a cubic surface," *Journal of the London Mathematical Society,* **1** (1926): 146–7; (with T. L. Wren) "Involutory point-pairs in the quadro-quadric Cremona space transformation," *Proceedings of the London Mathematical Society,* (2), **24** (1926): xxviii-xxix; "Incidence relations for Cremona space transformations," *Proceedings of the London Mathematical Society,* (2), **26** (1927): 453–69, with an addendum in the *Journal of the London Mathematical Society,* **3** (1928): 3. She also supplied an article on "Analytical geometry, curve and surface," for the 14th edn. (1929) of the *Encyclopaedia Britannica.* One chapter of her book, *Cremona Transformations,* was written in collaboration with her young friend Grace Sadd, who died before the project was completed.

74. "Mathematics and eternity," *Mathematical Gazette,* **12** (1925): 265–70. Hudson's explicitly stated linking of religion and mathematics reflects an outlook more typical of the earlier part of the nineteenth century than her own time. For another example of the importance of religious thinking in the shaping of productive scientific research see George Cantor, *Michael Faraday, Sandemanian and Scientist: A Study of Science and Religion in the Nineteenth Century* (London: Macmillan, 1991).

75. *Girton College Register, 1869–1946,* pp. 50–1.

76. For Part I (1900) see bibliography; the remaining three parts appeared in the *Reports of the British Association,* 1902, pp. 81–93; 1903, pp. 64–77; 1904, pp. 20–9.

77. Anon., "Isabel Maddison," *Bryn Mawr Alumnae Bulletin,* Winter 1951, p. 14; "Vita," Ph.D. Dissertation, and other notes from Bryn Mawr College archives; Betsy S. Whitman, "Ada Isabel Maddison (1869–1950)," in *Women of Mathematics,* Grinstein and Campbell, eds., pp. 144–6.

78. Maddison's translation and note were the following: "The arithmetizing of mathematics" by Felix Klein, translation, *Bulletin of the American Mathematical Society,* **2** (1896): 241–8; "Note on the history of the map-coloring problem," ibid., **3** (1897): 257.

79. *Girton College Register 1869–1946,* vol. 1, p. 639.

80. M. E. Barwell, "The advisability of including some instruction in the school course on the history of mathematics," *Mathematical Gazette,* 7 (1913–1914): 72–8, discussion, pp. 78–9.

81. *Girton College Register 1869–1946,* vol. 1, p. 85.

82. Beeton's post-1900 statistical papers included: (with Karl Pearson) "On the inheritance of the duration of life, and on the intensity of natural selection in man," *Biometrika,* **1** (1902): 50–89, and contributions to projects reported by Pearson under the headings, "Cooperative investigations on plants. II. Variation and correlation in lesser Celandine from divers localities," ibid., **2** (1903): 145–64, and "Cooperative investigations on plants. III. On inheritance in the Shirley poppy. Second memoir," ibid., **4** (1905–1906): 394–426.

83. M. H. W., "Dr. Sophie Bryant," *Nature,* **110** (1922): 458–9; Eleanor Doorley, in *The North London Collegiate School, 1850–1950. A Hundred Years of Girls' Education,* ed. R. M. Scrimgeour (London: Oxford University Press, 1950), pp. 71–91; anon., "Memorial," *Times,* 29 August 1922, 7f; Sheila Fletcher, "Bryant, Sophie," *DNB,* "Missing Persons" vol., 1993, pp. 97–8.

84. The story is told of how one of her examiners, especially impressed by her papers, wrote a note to his colleague saying "There's a very good man in." The other, who knew Mrs. Bryant, replied, "Your man's a woman!" (M. H. W., "Sophie Bryant," p. 458).

85. L. S. Hearnshaw, *A Short History of British Psychology 1840–1940* (London: Methuen, 1964), p. 261. Bryant's writings on educational theory include, "The curriculum of a school for girls of the first grade," *International Conference on Education,* London, 1884, vol. 4, pp. 337–62; *Over-work: from the Teacher's Point of View. With Special Reference to the Work in Schools for Girls* (London: F. Hodgson, 1885); *Educational Ends; or, the Ideal of Personal Development* (London: Longman's, Green, 1887).

86. Bryant was the second woman to sit on the senate of the University of London, the first being Dame Emily Penrose, principal of the Royal Holloway College for Women. Bryant sat on the External side as a representative of Convocation in the Faculty of Science (Margaret J. Tuke, *A History of Bedford College for Women 1849–1937* (London: Oxford University Press, 1939), p. 194, n. 4).

87. Hearnshaw, *Short History,* pp. 254–7.

88. K. Anderson, in *North London Collegiate School,* Scrimgeour, ed., p. 30.

89. These included, *The Teaching of Christ on Life and Conduct* (London: S. Sonnenschein, 1898); *The Teaching of Morality in the Family and School* (London: S. Sonnenschein, 1900); *How to Read the Bible in the Twentieth Century* (London: J. M. Dent, 1918); *Moral and Religious Education* (London: E. Arnold, 1920).

90. These early pioneers were, especially, Frances Buss, Dorothea Beale, and Emily Davies (see, for instance, Caine, *Victorian Feminists*).

91. Rosaleen Love, "'Alice in eugenics-land': feminism and eugenics in the scientific careers of Alice Lee and Ethel Elderton," *Annals of Science,* **36** (1979): 145–8; Tuke, *History of Bedford College,* pp. 129, 167.

92. Elizabeth Fee, "Nineteenth-century craniology. The study of the female skull," *Bulletin of the History of Medicine,* **53** (1979): 415–33, especially pp. 420, 429.

93. Karl Pearson, "On the correlation of intellectual ability with the size and shape of the head," *Proceedings of the Royal Society,* **69** (1902): 333–42; Karl Pearson, Alice Lee and Marie Lewenz, "On the correlation of mental and physical characters in man. Pt. II," ibid., **71** (1902): 106–14.

94. Pearson paid his women assistants rather poorly, £100 a year for full-time work, at a time when even an assistant mistress in a public secondary school earned on average £118. £250 was paid to a male research fellow in the laboratory (Love, "'Alice in eugenics-land'," p. 157). Although he was well aware that some of the women he employed had better academic credentials than the men, and he felt that their work was at the very least equal to that of the men, Pearson considered that £100 was a reasonable salary for a woman who lived at home in London with her family. Paid jobs for women in scientific fields were then almost nonexistent (except for schoolteaching). Hence, although it might be argued that Pearson was exploiting his women assistants, giving them low pay for tedious work, it is also the case that he provided scarce opportunities for a few women graduates, many of whom were extremely eager to put their training to some use. This justification is at best questionable; it is doubtless more acceptable in the context of those early days of women's struggle to enter the scientific professions than it might seem now.

95. These papers by Lee include the following: "Dr. Ludwig on variation and correlation in plants," *Biometrika,* **1** (1902): 316–19; (with Cicely D. Fawcett et al.) "A second study of the variation and correlation of the human skull, with reference to the 'Naqada Crania'," ibid., **1** (1902): 408–67; "The law of ancestral heredity. Appendix III. On inheritance (great-grandparents and great-great-grandparents and offspring) in thoroughbred race-horses," ibid., **2** (1903): 234–6; (with S. Jacob and Karl Pearson) "Craniological notes. III. Preliminary note on inter-racial characters and their correlation in man," ibid., **2** (1903): 347–56; (with Karl Pearson) "On the laws of inheritance in man. I. Inheritance of physical characters," ibid., **2** (1903): 357–462; (with Amy Barrington and Karl Pearson) "On inheritance of coat-colour in the greyhound," ibid., **3** (1904): 245–98; (with J. Blackman and Karl Pearson, "A study of the biometric constants of English brain-weights, and their relationships to external physical measurements," ibid., **4** (1905–1906): 124–60; (with Karl Pearson et al.) "Co-operative investigations on plants. III. On inheritance in the Shirley Poppy. Second memoir," ibid., **4** (1905–1906): 394–426; (with Alexandra Wright and Karl Pearson) "A co-operative study of queens, drones and workers in *Vespa vulgaris,*" ibid., **5** (1906–1907): 407–22; (with Karl Pearson) "On the generalized probable error in multiple normal correlation," ibid., **6** (1908–1909): 59–68; (with Ethel M. Elderton and Karl Pearson) "On the correlation of death rates," *Journal of the Royal Statistical Society,* **73** (1910): 534–9.

96. For the 1896 and 1899 reports see Karl Pearson, "Calculation of the $G(r,\nu)$-integrals. Preliminary report of the committee, consisting of Rev. Robert Harley, A. R. Forsyth, J. W. L. Glaisher, L. Lodge and Karl Pearson," "Reports on the State of Science," *British Association Reports* (1896): 70–82, appendix (pp. 75–82), "Tables of χ-functions, (χ_1, χ_3, χ_5 and χ_7)," calculated by Miss A. Lee, G. U. Yule, C. E. Cullis, and Karl Pearson; Karl Pearson, "Tables of the $G(r,\nu)$-integrals. Report of the committee consisting of Rev. Robert Harley, A. R. Forsyth, J. W. L. Glaisher, L. Lodge and Karl Pearson," ibid., (1899): 65–120, appendix (pp. 77–120), "Table of $F(r,\nu)$ and $H(r,\nu)$ functions" by Miss Alice Lee. Lee's later contributions include the following: "Table of Gaussian 'tail' functions; when the 'tail' is larger than the body," *Biometrika,* **10** (1914–1915): 208–14; "Further supplementary tables for determining high correlations from tetrachoric groupings," ibid., **11** (1915–1917): 284–91; (with H. E. Soper, A. W. Young, B. M. Cave, and Karl Pearson) "On the distribution of the correlation coefficient in small samples. Appendix II to the papers of 'Student' and R. A. Fisher," ibid., **11** (1915–1917): 388–413; (with Karl Pearson) "Table of the first twenty tetrachoric functions to seven decimal places," ibid., **17** (1925): 343–54; "Supplementary tables for determining correlation from tetrachoric groupings," ibid., **19** (1927): 354–404.

97. *Times,* 18 July 1924, 11f.

98. Tuke, *A History of Bedford College,* pp. 303–4; *Times,* 17 June 1937, 1a.

99. Desmond MacHale, *George Boole. His Life and Work* (Dublin: Boole Press, 1985), pp. 158, 261–3; H. S. M. Coxeter, *Regular Polytopes* (New York: Pitman Publishing Corporation, 1948), pp. 163–4, 210, 258–9, and "Alicia Boole Stott (1860–1940)," in *Women of Mathematics,* Grinstein and Campbell, eds., pp. 220–4.

100. In her 1900 paper (see bibliography) she introduced into English the term *polytope,* probably coined by Hoppe in 1882 (Coxeter, *Regular Polytopes,* p. ix).

101. "Geometrical deduction of semiregular from regular polytopes and space fillings," *Verhandelingen der Koninklijke Akademie van Wetenschappen te Amsterdam* (eerste sectie), **11,** No. 1, (1910), 24 pp. In this paper Stott suggested the idea of "partial truncation" and the related processes of *expansion* and *contraction,* for the representation of which she provided a compact notation. These two processes led her to the discovery of a great variety of uniform polytopes.

102. H. S. M. Coxeter, "Wythoff's construction for uniform polytopes," *Proceedings of the London Mathematical Society,* (2), **38** (1934): 327–39, especially p. 338.

103. Alicia Stott's sister Lucy Everest Boole also published scientific work—see chapter 11.

104. Elizabeth Chambers Patterson, *Mary Somerville and the Cultivation of Science, 1815–1840* (The Hague: Martinus Nijhoff Publishers, 1983); Dorothy Stein, *Ada: A Life and a Legacy* (Cambridge, Mass.: MIT Press, 1985); Joan Baum, *The Calculating Passion of Ada Byron* (Hamden, Conn.: Shoe String Press, 1986). A somewhat earlier study of Lovelace is Doris Langley Moore's, *Ada, Countess of Lovelace: Byron's Legitimate Daughter* (London: John Murray, 1977). For sketches in general works see Marilyn Bailey Ogilvie, *Women in Science. Antiquity through the Nineteenth Century* (Cambridge, Mass.: MIT Press, 1986), pp. 49–50, 161–6; Margaret Alic, *Hypatia's Heritage: a History of Women in Science from Antiquity to the late Nineteenth Century* (London: The Women's Press, 1986), pp. 157–63,

181–90; Patricia Phillips, *The Scientific Lady. A Social History of Women's Scientific Interests, 1520–1918* (New York: St. Martin's Press, 1990), pp. 111–15, 207–9.

105. Stein, *Ada;* Baum, *Calculating Passion;* Anthony Hyman, "Byron, (Augusta) Ada, Countess of Lovelace," *DNB,* "Missing Persons" Supplement, 1993, p. 110.

106. See for instance Jerry Martin Rosenberg, *The Computer Prophets* (New York: Macmillan, 1969); Christopher Evans, *The Making of the Micro* (London: Gollancz, 1981); *Charles Babbage and his Calculating Engines,* eds. Philip Morrison and Emily Morrison (New York: Dover, 1961); Anthony Hyman, *Charles Babbage: Pioneer of the Computer* (Oxford: Oxford University Press, 1982); Velma R. Huskey and Harry D. Huskey, "Lady Lovelace and Charles Babbage," *Annals of the History of Computing,* **2** (1980): 229–329, and "Ada, Countess of Lovelace, and her contribution to computing," *Abacus,* **1** (1984): 22–9.

107. Stein's sources included the Lovelace Papers and Somerville Papers, Bodleian Library, Oxford University, manuscripts in the British Library, and a large collection of other documents. Her picture of Ada Lovelace's life and achievements is a rich and complex one, involving a large cast of characters about whom much information is available. The present sketch offers the briefest summary of Ada's career, in part because Stein's argument is a convincing one—despite her talents, wide interests, great enthusiasm, and ambitious hopes and plans, Ada Lovelace's actual achievements were modest.

108. Stein, *Ada,* p. 92; Hyman, "Byron, (Augusta) Ada."

109. Earl of Lovelace, "On climate in connection with husbandry, with reference to a work entitled 'Cours d'Agriculture, par le Comte de Gasparin . . . ,'" *Journal of the Royal Agricultural Society,* **9** (1848): 311–40.

110. Stein, *Ada,* chapters 6 and 7, pp. 207–80.

111. The Lovelaces' daughter Annabella (1837–1917) married Wilfred Scawen Blunt, an Arab scholar. Lady Anne Blunt (as she was known) learned Arabic and spent two years with her husband in the Middle East. They traveled on horseback, without guide or caravan, living in a tent, and wearing Arab dress for the most part. Her diaries, careful descriptions of long journeys made between 1877 and 1879, were published shortly after that. See Lady Anne Blunt, Baroness Wentworth, *Bedouin Tribes of the Euphrates,* edited, "With Some Account of the Arabs and their Horses," by W. S. B[lunt], 2 vols. (London: John Murray, 1879), and also, *A Pilgrimage to Nejd, the Cradle of the Arab Race. A Visit to the Court of the Arab Emir, and "Our Persian Campaign,"* 2 vols. (London: John Murray, 1881). These works quickly became standard sources of information on the region and are classic travel books as well (see Jane Robinson, *Wayward Women. A Guide to Women Travellers*—Oxford: Oxford University Press, 1990, pp. 6–7). Lady Anne was a member of the Royal Asiatic Society. As an author she has more real claim to fame than her more celebrated mother.

112. Patterson, *Mary Somerville;* Martha Somerville, ed., *Personal Recollections from Early Life to Old Age of Mary Somerville, with Selections from her Correspondance* (Boston: Roberts Brothers, 1876); Marie Sanderson, "Mary Somerville: her work in physical geography," *Geographical Review,* **64** (1974): 410–20; Marguerita Oughton, "Mary Somerville, 1780–1872," in *Geographers. Biobibliographical Studies,* vol. 2, eds. T. W. Freeman and Philippe Pinchemel (London and New York: Mansell, 1978), pp. 109–11; *DNB,* vol. 53, pp. 254–5; Geoffrey Sutton and Sung Kyu Kim, "Mary Fairfax Greig Somerville (1780–1872)," in *Women in Chemistry and Physics. A Biobibliographic Sourcebook,* eds. Louise S. Grinstein, Rose K. Rose and Miriam H. Rafailovich (Westport, Conn.: Greenwood Press, 1993), pp. 538–46; Mary T. Brück, "Mary Somerville, mathematician and astronomer of underused talents," *Journal of the British Astronomical Association,* **104** (1996): 201–6. (Somerville's journal articles are listed in the physics and astronomy sections of the bibliography.)

113. Patterson, *Mary Somerville,* p. 9. The relatively rapid introduction of the new French mathematics into early nineteenth-century Scotland was doubtless due in part to the long-standing Franco-Scottish cultural links, but it was speeded by a more immediate phenomenon, the presence in Scotland during the Napoleonic wars of many French prisoners of war. Among them were officers trained in French technical schools who were willing to teach mathematics to interested Scots.

114. Somerville listed these in her autobiography (pp. 79–80). They included the following: Silvestre François Lacroix's three-volume *Traité du calcul différentiel et du calcul intégral* (*Traité des différences et des séries*) (1797–1800), his *Traité de calcul différentiel et de calcul intégral* (1802), and his *Élémens d'algèbre* (1808); Jean Baptiste Biot's *Essai de géometrie analytique* (1805), and *Traité élémentaire d'astronomie physique* (1805); Adrien Marie Legendre's *Éléments de géométrie* (1802), a book that remained influential until the 1880s; Siméon Denis Poisson's two-volume *Traité de mécanique* (1811); Comte Joseph Louis Lagrange's *Traité de la resolution des équations numériques de tous les degrés* (1808) and his *Théorie des fonctions analytiques concernant les principles du calcul différentiel* (1795); Comte Gaspard Monge's *Application de l'analyse à la géométrie, a l'usage de l'école impériale polytechnique* (1807); Louis Benjamin Francoeur's *Traité de mécanique élémentaire* (1801), and his two-volume *Cour complet de mathématiques pures* (second edn. 1819); Marquis Pierre Simon de Laplace's five-volume *Traité de mécanique céleste,* published between 1798 and 1827, and his *Théorie analytique des probabilités* (1812); Jean François Callet's *Tables portatives de logarithmes* (1795).

115. The daughters were Margaret and Martha, born in 1813 and 1815, respectively. Two boys in the family, Woronzow Greig and James Craig Somerville, William's illegitimate son born in South Africa about 1800, remained in Edinburgh to continue their schooling. The younger of the two Greig boys had died in 1814, at the age of nine. The Somervilles also had a son, born in 1815, who died in infancy. Their last child, Mary Charlotte, was born in 1817. Margaret died in 1823.

116. Only some of the fellows of scientific societies at that period and for some time after were practicing scientists—see H. Lyons, *The Royal Society, 1600–1940* (Cambridge: Cambridge University Press, 1944), p. 341. Many had serious interests in science and some joined mainly for social reasons. To a large extent the scientific societies then functioned as gentlemen's clubs, a point that is perhaps not irrelevant in any consideration of their later reluctance to admit women scientists as members.

117. A. Baumgartner, "Untersuchung über Magnetisirung des Eisens durch das Licht, nebst neuen Versuchen über denselben Gegenstand," *Zeitschrift für Physik, Mathematik und verwandte Wissenschaften,* **1** (1826): 263–81. A French translation entitled, "Sur l'animantation de l'acier par la lumière blanche directe du soleil," appeared in *Annales de Chimie et de Physique* (**33,** pp. 333–5) the same year. Baumgartner, who was the editor of the *Zeitschrift,* reported a better and more efficient way of producing the magnetization described by Mrs. Somerville; his method involved dividing the steel needle into alternating polished and oxide-coated zones.

118. The idea behind this hypothesis was not entirely original to Laplace. It had been sketched in qualitative form by Thomas Wright and Immanuel Kant, but Laplace developed and quantified it in his classic work—see Carl B. Boyer and Uta C. Merzbach, *A History of Mathematics* (New York: John Wiley, 1989), p. 550.

119. J. N. Hays, "Science and Brougham's Society," *Annals of Science,* **20** (1964): 227–41.

120. Rev. John Toplis, *A Treatise upon Analytical Mechanics; being the First Book of the Mécanique céleste of P. S. Laplace* (translation and explanatory notes) (Nottingham: H. Barnett, printer, 1814); Rev. Henry H. Harte, *A Treatise of Celestial Mechanics* (translation with explanatory notes, covering vols. 1 and 2 of the original), 2 vols. (Dublin: printed for R. Milliken, 1822–1827).

121. Somerville, *Personal Recollections,* p. 163.

122. Mary Somerville had met a number of French scientists and mathematicians, including Laplace, Biot, and Poisson, during a trip the Somervilles made to Paris in 1817. For press reviews of her book see *Literary Gazette,* **778** (1831): 806–7; [Thomas Galloway], *Edinburgh Review,* **55** (1832): 1–25; *Monthly Review,* n.s., **1** (1832): 134–41; [J. F. W. Herschel], *Quarterly Review,* **47** (1832): 537–59; J. B. Biot, *Journal des Savants* (1832): 28–32; *Athenaeum,* **221** (1832): 43–4 (an unfavorable notice).

123. This is probably related to the fact that an English translation of *Mécanique céleste* was brought out by the American, Nathaniel Bowditch (Boston: Hillard, Gray, Little, and Wilkins, 1829–1839). The appearance of the first volume of this work in 1829, shortly before Mrs. Somerville's manuscript was completed, did not cause her undue anxiety (Patterson, *Mary Somerville,* p. 74). After going through Bowditch's commentaries she concluded that her approach was very different from his, and that the publication of his work should not seriously affect plans for hers. Interestingly enough, only Bowditch's translation is mentioned in Boyer and Merzbach's *History of Mathematics* (p. 623).

124. Her group of subjects corresponded closely to what constituted *la physique* in contemporary French science. By the late eighteenth century this was seen as a branch of science occupying a position between mathematics and chemistry, and separate from both of these. Subjects taught by a *professeur de physique* would typically include properties of matter, heat, light, sound, and electricity, that is, what we would now call traditional physics. Meteorology was also considered (at least by some scientists) as part of a unified view of physics, weather phenomena being seen as subject to the laws of physics (see Maurice Crosland, *Gay-Lussac, Scientist and Bourgeois* (Cambridge: Cambridge University Press, 1978), pp. 117, 126–7).

125. One of these extracts was entitled *Mrs. Somerville, Electro-magnetism. History of Davenport Invention of the Application of Electro-magnetism to Machinery, with Remarks . . . by Professor Silliman* (New York: G. and C. Carville, 1837).

126. Taken along with *The Mechanism of the Heavens,* the early editions of *The Connection of the Physical Sciences* provide a valuable record of the changing planetary theories of the 1830s—see Morton Grosser, *The Discovery of Neptune* (Cambridge, Mass.: Harvard University Press, 1962), pp. 52–4, 79. In 1831, when the *Mechanism* appeared, Somerville had no doubts about what constituted the planetary system. There were seven planets, the farthest out being Uranus, and among the tables she presented were some describing the orbit of the latter. By 1834, when the first edition of the *Connection* came out, the status of Uranus was decidedly less clear and Mrs. Somerville declined to discuss it; following her tables of Jupiter and Saturn she simply stated that tables for the minor planets were commendably accurate, considering they had been discovered so recently. By the third edition (1836) she was willing to mention Uranus again, although she avoided taking risks, and merely noted that while tables of Jupiter and Saturn agreed with the most up-to-date observations, those of Uranus were already inaccurate. However, even the cautious Mrs. Somerville felt able to put into print the suggestion that the cause of the discrepancies might well be the existence of another, further out, planet, and that the observed perturbations of Uranus might be used in the future to calculate the mass and orbit of this unknown planet. These remarks appear to have led the young Cambridge mathematician John Couch Adams to begin, in the early 1840s, the calculations that disclosed the existence of Neptune. (The French astronomer-mathematician Urbain Leverrier was working independently on the same problem at the same time and published his results first—Adams and Leverrier are now considered the codiscovers of Neptune.)

127. Two years later another £100 was added, although this was more because of the Somervilles' serious financial difficulties than as further recognition of her work. It did, however, make Mrs. Somerville's pension equal to the highest awarded to any scientific person at that time (Patterson, *Mary Somerville,* pp. 160–1).

128. Von Humboldt, one of the founders of geography as a science, brought out his massive five-volume treatise, *Kosmos,* between 1845 and 1862. In it he presented (in popular language) a description of the entire physical world, from galaxies to the geography of mosses. When Somerville's work appeared he wrote to her, congratulating her especially on her discussion of the geography of plants and animals (Somerville, *Personal Recollections,* p. 288).

129. During the 1880s and 1890s it was also on the book lists at Manchester University (Oughton, "Mary Somerville," p. 111).

130. Somerville published three additional scientific articles not already mentioned here. One of these was a thirty-eight page exposition on Halley's Comet in the *Quarterly Review* in 1835, based on two articles that appeared in German the same year. The other two (1836 and 1845) reported her own studies of sunlight and its effects on particular substances (see bibliography).

131. See R. A. Fisher, "On the interpretation of χ^2 from contingency tables and the calculation of P," *Journal of the Royal Statistical Society,* **85** (1922): 87–94, for a correction and revision in Pearson's influential Theory of Sampling work. Pearson himself never fully accepted Fisher's revision and persisted in holding to his original view of what has since become known as "degrees of freedom."

132. There is a contrast here between mathematics and chemistry, where some of the most outstanding early British women workers quickly abandoned research in mainstream areas and moved to the emerging (low-prestige) subfield of biochemistry (Mary R. S. Creese, "British women of the nineteenth and early twentieth centuries who contributed to research in the chemical sciences," *British Journal for the History of Science,* **24** (1991): 275–305, on pp. 295–7).

133. Boyer and Merzbach, *History of Mathematics,* chapter 24, especially pp. 615–16.

134. Roy MacLeod and Russell Moseley, "Fathers and daughters: reflections on women, science and Victorian Cambridge," *History of Education,* **8** (1979): 321–33; Creese, "British women in the chemical sciences"; McWilliams-Tullberg, *Women at Cambridge,* especially pp. 25, 143, 222.

135. Judith V. Grabiner, "Mathematics in America: the first hundred years," in *The Bicentennial Tribute to American Mathematics 1776–1976,* ed. Dalton Tarwater (Mathematical Association of America, 1977), pp. 9–24. One of the important events in the story of the opening of mathematical training for American women was the establishment of the University of Chicago in 1892. It soon became the leading American university in advanced mathematics education, and in the period from 1908 to 1939 it conferred more mathematics Ph.D. degrees on women than did any other American institution (Green and Laduke, "Contributors to American mathematics," p. 129).

Chapter 9

RIPPLE-MARKS IN THE SAND, IMAGES ON THE SCREEN, UNIT STANDARDIZATION: STUDIES BY WOMEN IN PHYSICS

Articles in physics make up about 3 percent of the research papers by women indexed by the Royal Society. Contributions came mainly from British and American workers and from Marie Curie in France. British women produced about 40 percent of the total, the Americans and Mme. Curie about 20 and 18 percent, respectively. Smaller contributions came from women from Australia, Canada,[1] Germany, Italy, and Sweden.

American women

The American work was largely student research, and none of the Americans remained active in original investigations for very long after receiving their advanced degrees, though one, Margaret Maltby, carried out notable work during her brief period of postdoctoral studies, and a second, Isabelle Stone, kept up research for several years. Five of the eight women who produced most of the papers studied at Cornell University.[2] Of the remaining three, one (Stone) took a Ph.D. at the University of Chicago, one (Maltby) a Ph.D. at Göttingen University after an S.B. at MIT, and the third (Sabine) carried out research at MIT without taking a degree.[3]

Four of those who were students at Cornell (Baldwin, Spencer, Bruère, and Crehore) were involved in projects in light and optics; the fifth, Noyes, did research on elasticity. Most likely all five were students of the British-born experimental physicist, Edward L. Nichols, professor of physics at Cornell and founder of the *Physical Review,* in which many of their papers were published. Nichols's special interests included radiation, optics, and photometry.

CAROLINE BALDWIN,[4] the daughter of Fannie (Willard) and Alfred Baldwin, was born in San Francisco, on 30 June 1869. She received her early education at public schools in Santa Cruz and from her mother. At the University of California, where she did advanced work in science and mathematics, she was the first woman to graduate from the department of mechanics (B.S., 1892), placing third in the class. She received an Sc.D. from Cornell in 1895, the only doctoral degree in physics awarded to a woman by that institution before the turn of the century.[5] Her 1896 paper in the *Physical Review* describing her photographic study of arc spectra was most likely her only journal publication. She taught physics at the California School of Mechanical Arts in San Francisco from 1895 until 1900, and coauthored with Arthur Merrill the textbook, *Physics Course of the California School of Mechanical Arts,* published in 1898. That year she married Charles Theobald Morrison; she appears to have given up professional work after the birth of her first child in 1900.

MARY NOYES[6] (1855–1936) received her undergraduate education at the University of Iowa (Ph.B., 1881) and stayed on for three years of graduate study, taking an A.M. in 1884. She held a staff position at Lake Erie College in Painsville, Ohio, teaching mathematics, physics, and astronomy from 1885 until 1900,[7] while also continuing her education. Cornell awarded her an M.S. in 1894. Her thesis project on Young's modulus of elasticity, supervised by Nichols, was reported in the *Physical Review* in 1895. An extension of this investigation, carried out under the guidance of Frank Whitman at Western Reserve University in Cleveland, resulted in a second paper in the *Physical Review* in 1896. She received her Ph.D. from Western Reserve the same year. After 1900 she was an instructor in mathematics, physics, and astronomy at the Minneapolis Academy. On retiring she went to Pasadena, California, where she died on 13 September 1936.

MARY SPENCER,[8] the daughter of Henrietta (Elam) and William Brainerd Spencer, was born in Vidalia, Louisiana, on 27 January 1869. She graduated with a B.A. from Newcomb College, New Orleans, in 1892, and then took an A.M. degree at Cornell (1895). Returning to Newcomb College as a faculty member, she taught physics from 1895 until 1901, thereafter becoming professor of mathematics. Her joint paper with Edward Nichols on the influence of temperature on the transparency of solutions, published in the *Physical Review* in 1895, was probably her only research publication.

Alice Bruère (born on 5 October 1863) received her B.S. degree from Cornell in 1895. Two papers reporting her work in optics appeared in the *Physical Review* in 1896 and 1898. Mary Crehore took an M.S. degree in 1894. Her collaborative work with Nichols on changes with time in the spectrum of light from incandescent lime was reported in the *Physical Review* the following year.[9]

ISABELLE STONE,[10] the daughter of Harriet (Leonard) and Leander Stone, was born in Chicago, on 18 October 1870. She took an A.B. at Wellesley College in 1890 and continued her education at the University of Chicago (M.S., 1896; Ph.D., 1897). Her doctorate was the first in physics awarded to a woman by Chicago.[11] Her paper on the electrical resistance of thin films appeared in the *Physical Review* in 1898. She continued her studies on the properties of thin films for several years, carrying out the later work in the laboratories of Columbia University.[12]

Following a year of teaching at the Preparatory School of Bryn Mawr College in Baltimore (1897–1898), she became instructor in physics at Vassar College, a post she held until 1906. She and her sister Harriet Stone, also a graduate of Wellesley College and the University of Chicago, then moved to Italy and together established the Misses Stones' School for American Girls in Rome. This they successfully managed until 1914, when the outbreak of the First World War caused their return to the United States. Throughout much of the war and until 1923 Isabelle Stone was head of the physics department at Sweet Briar College, Sweet Briar, Virginia. She and her sister then opened another school for girls, this time in Washington, D.C. It appears to have run until at least 1944, by which time both women were in their seventies. Stone was a fellow of the AAAS and one of the two women who were charter members of the American Physical Society, having attended the founding meeting in 1899.[13]

Probably the most distinguished of the early American women physicists was MARGARET MALTBY[14] (1860–1944). Born on 10 December 1860 on a family farm in Bristolville, Ohio, she was the youngest of the three daughters of Lydia Jane (Brockway) and Edmund Maltby, both of whom traced their ancestry to seventeenth-century colonists in Connecticut. She received her undergraduate education at Oberlin College, enrolling in 1878 after a year in the preparatory department and graduating (B.A.) in 1882. She had taken a broad selection of courses, including several in the sciences, and she also had considerable artistic talent, which she developed further with a year's study at the Art Student's League in New York City. In 1883 she returned to Ohio where she taught for four years in high schools in Wellington and Massillon.

She became increasingly interested in science, however, and in 1887 went to MIT as a special student in physics (S.B., 1891; M.A., from Oberlin College, the same year). From 1889 she also held a position as instructor at Wellesley College. Continuing her studies at MIT, she carried out work on sound vibrations and pitch with Charles Cross; their joint paper appeared in the *Proceedings of the American Academy of Arts and Sciences* in 1893. The same year, having received an MIT graduate European fellowship, she went to Göttingen University. She thus became, along with the two mathematics students Mary Winston and Grace Chisholm, one of the first three women to be officially enrolled (albeit as "exceptional cases") at a Prussian university.[15] She completed the requirements for the doctorate in two years, receiving her Ph.D., the first awarded to an American woman by Göttingen University, in 1895. Her dissertation research on the measurement of resistance in electrolytic solutions appeared in the *Zeitschrift für Physikalische Chemie* in 1895. An Association of Collegiate Alumnae fellowship enabled her to continue her stay at Göttingen for a further year, and a second paper, on the measurement of the period of electrical oscillations, appeared in *Annalen der Physik und Chemie* in 1897.

In 1896 she returned to Wellesley College, to a one-year temporary position as associate professor and head of the physics department while the chairman, Sarah Frances Whiting, was on leave. She then found a job in her home state of Ohio teaching physics and mathematics at Lake Erie College, but within a year was presented with an opportunity for further research in Germany with Friedrich Kohlrausch, president of the Physikalisch-Technische Reichsanstalt in the Berlin suburb of Charlottenburg. Kohlrausch, who had met her in Göttingen and been impressed by the quality of her work, offered her a one-year post as his private assistant. She joined him in his ongoing investigations of the electrical conductivity of salt solutions, a long program of precise work, carried out painstakingly over the course of several decades, which resulted in substantial contributions to modern theories of electrolytic dissociation and solution structure. During her year at Charlottenburg Maltby investigated the conductivities of aqueous solutions of alkali chlorides and nitrates; the work was reported in a preliminary communication in 1899 and in a long joint paper the following year in the *Wissenschaftliche Abhandlungen der Physikalisch-Technischen Reichsanstalt.*

On her return to the United States in 1899 she worked for a year on problems in theoretical physics with Arthur Webster, then at Clark University in Worcester, Massachusetts, and in 1900 obtained a position at Barnard College of Columbia University as instructor of chemistry and head of the department. Two years later she transferred to Barnard's physics department as adjunct professor, and she remained in that position, with the same salary, for seven years. When she herself raised the question of advancement, then dean and later president of Barnard Virginia Gildersleeve was able to persuade the Columbia University physicists to support Maltby's promotion, first to assistant professor in 1910 (when she was almost fifty) and then to associate professor in 1913. The delay seems to have been largely the result of oversight rather than of active opposition.[16] She remained associate professor and chairman of Barnard's physics department until she retired in 1931 at age seventy.

A conscientious and painstaking teacher, she demanded hard work from students and was herself always willing to provide extra help when needed. Her strong commitment to teaching, coupled with her administrative duties, left her little time for research, although she kept abreast of developments in her field and regularly attended Columbia physics department weekly seminars. Her work in Göttingen and that with Kohlrausch at Charlottenburg had established her reputation, however, and for that early research she received a star in the 1906 edition of *American Men of Science.*

Throughout the years one of her special interests was providing help for capable women to carry out graduate and postdoctoral study. From 1912 until 1929 she gave much time to work on the fellowship committee of the American Association of University Women (AAUW), serving as chairman from

1913 until 1924. Her leadership was instrumental in establishing standards and putting the process for selecting scholarship recipients onto a sounder basis. She also compiled a detailed history of the fellowships,[17] and was honored by having one established in her name in 1926. Another of her particular interests over the years was music; shortly before her retirement she presented what was probably one of the first courses ever taught in the United States on the physics of music, the branch of the field in which she had done her earliest research (with Cross at MIT).

She was very fond of children and in 1901, when she was forty-one, adopted Philip Randolph Meyer, the orphaned son of a near friend. The boy's company meant a great deal to her, and when she traveled during sabbaticals and in connection with her AAUW work she often took Philip with her. Their trips took them not only to Canada and the western United States, but to several countries in Western Europe and North Africa. Philip, and later his wife and three children, remained very close to her throughout the rest of her life.[18] Until crippled by arthritis during her last years she remained active, regularly going to musical performances and continuing to travel. She died in the Columbia-Presbyterian Medical Center in New York City, on 3 May 1944, at age eighty-three.

British women

British women published the largest number of nineteenth-century physics journal articles of any national group. Five of the British had close associations with Cambridge—Ayrton, Fawcett, Klaassen, Everett (whose pre-1901 publications were in astronomy), and Sidgwick. Two others, Bryant and Chambers, received much of their undergraduate education at Durham College of Science and University College Liverpool respectively.[19]

HERTHA MARKS AYRTON[20] (1854–1923) has the notable distinction of being the first woman, and is still the only woman, to have received the Royal Society's Hughes Medal for original work in the physical sciences. Further, she was the first woman to be proposed for fellowship in the Royal Society (in 1902), although she was not accepted, it being quickly and conveniently determined that married women were not eligible under the legal terms of the society's charter. No other women candidates were put forward until the 1940s.

She was born in Portsea, Hampshire, at the home of her maternal grandparents, on 28 April 1854, the third of eight children of a Jewish refugee from Poland, Levi Marks, and his wife Alice Theresa, daughter of Joseph Moss, a Portsea glass merchant. Phoebe Sarah were her given names, but she later changed these to Hertha. Her earliest years were spent in the town of Petworth, Sussex, some thirty miles inland from Portsea, where her father earned a living as a clockmaker-jeweler. Later the family moved to Portsea. They had little money to spare, and there was even less after Levi Marks died in 1861.

Despite her poverty Alice Marks believed that her daughters had to be educated. When Sarah was nine she was sent to London to live with her aunt, Marion Moss Hartog, who ran a school. Mrs. Hartog's family was a talented one and several of her children were to go on to distinguished careers in science and the arts; she saw to it that her niece received a good basic education. When she was sixteen Sarah took a post as resident governess in a family with three children in order to send the money she earned to her mother. Within a few years, however, she moved out on her own and earned her living by offering private tuition, mainly in mathematics. At the same time she studied for the Cambridge University Examination for Women, and passed in 1874. She entered Girton College in 1876, after a delay of two years spent looking after her ailing younger sister.

Although she had failed to win a scholarship, she was helped financially by Barbara Bodichon, one of the cofounders of Girton (and a woman of considerable wealth), whom she had met some years previously and come to know closely. Her studies were interrupted for a year by illness and her preparation was less than adequate, but she wrote the Mathematical Tripos examinations in 1881. Her marked tendency to want to push ahead to new applications without taking time to thoroughly assimilate basic material told against her—as did her preoccupation with the support of her mother and sister. Much to her disappointment she placed in the third class. Nevertheless, while a student she produced her first invention, a line divider, which she patented in 1884. Although the instrument was not initially received with much enthusiasm, in 1885 she was invited to read a paper on its uses before the Physical Society. She took an active part in student life at Girton, joining the Choral Society and founding the college's fire brigade, doubtless very important in those days when student rooms were lit by candles and oil lamps.

She returned to London in 1881, taking a post as mathematics teacher at Kensington High School. She never functioned well within the constraints of classroom teaching, however, and soon went back to private tutoring, work she found much more congenial. Successful in attracting plenty of pupils, she not only made enough money to repay various loans she had taken while at Girton, but was able to help her mother as well. Contacts through her Hartog cousins, her Girton connections and the families of her pupils brought her a full social life, and she came to know a number of prominent people. She also continued her education, and, again thanks to help from Barbara Bodichon, was able to attend evening classes at Finsbury Technical College, where she took classes from William Ayrton, professor of physics and electrical engineering. In 1885 she married Ayrton, a widower with one daughter.

A man of liberal outlook (his first wife, Matilda Chaplin, had been one of the country's pioneers in the push for medical education for women), Ayrton encouraged Hertha in her scientific pursuits. He pressed her to follow her own lines of research whenever possible, in case whatever success she had be attributed to him. Some years elapsed, however, and her daughter Barbara Bodichon Ayrton was born before Hertha found her way into the studies for which she became known. Much of the work was, in fact, a direct continuation of her husband's. Her way was smoothed considerably by a legacy left her in 1891 by Barbara Bodichon, sufficient to allow her to employ a housekeeper and give her energies to research.

In 1893, at the Electrical Congress at the Chicago World's Exposition, William Ayrton read a condensed version of his results on the causes of the variation of potential difference in the electric arc. These results included additional data, obtained for him by Hertha and his assistants after he had set out on the transatlantic voyage and forwarded to him by mail. When he accidentally lost the one and only copy of his paper he had no inclination to reconstitute the results and rewrite, but Hertha, now seriously interested in the subject, took up the project.

Working at the Central Technical College, South Kensington, where Ayrton was then professor of applied physics, she first repeated several of his experiments and then went on to carry out her own full investigation of the electric arc. The work included a careful and painstaking examination of the problem of how to get a constant potential difference between the carbon electrodes and thus a steady current—a problem she solved. In 1895 she agreed to write a series of articles for the *Electrician;* twelve articles in all, these reported much of her arc work. Of special note was a paper she read, on invitation, to the Institution of Electrical Engineers three years later, the first paper presented to the group by a woman. It discussed the hissing of the electric arc; Hertha showed that this phenomenon, and the resulting illumination loss, was caused by the presence of oxygen in the electrode crater, and could be prevented by the exclusion of air from the electrode housing. The paper was well-received and led to her being elected to membership of the institution (in 1899), an event that caused quite a stir in the scientific community. She was the first woman to be so recognized, and was still the only female full member at the time of her death in 1923.

By the late 1890s she had become well known as an expert on the electric arc. She presided over the physical science section of the International Congress of Women in London in 1899, gave a paper at the International Electrical Congress in Paris in 1900,[21] and had her paper, "The mechanism of the electric arc" read before the Royal Society in 1901. Her book, *The Electric Arc* (1902), presented a complete, fully documented account of the subject from the time of Sir Humphry Davy (1778–1829); especially notable for lucidity and precision in data presentation, it was, furthermore, the only such account to date. She dedicated it to her friend and benefactress, Barbara Bodichon.

Although Hertha's electric arc work is always acknowledged as her own and independent of her husband, he took more than a friendly interest, from which she clearly benefited. Writing to her stepdaughter in the summer of 1901 she commented that everything was going on as usual—Ayrton was "boiling down my paper [the one read to the Royal Society] to the last degree of compression," but she was taking care to work along with him and expected to have a full discussion of the subject soon.[22]

About 1901 she put aside her electrical research for a time and took up the problem of wave ripple formation in sand. The project resulted from observations made during a stay at Margate on the Kent coast, where she had gone with her husband who was trying to overcome insomnia and the health difficulties it led to. Using glass tank model systems that could be mechanically rocked to produce stationary waves of different wavelength, she showed that sand grains are moved by small vortices that form within standing waves, and that these vortices (whose presence she first demonstrated with black pepper grains) shift about the ripple crests in response to changing direction of the oscillating wave currents. Not until almost half a century later were these early observations of Ayrton's fitted into a general theoretical framework of sediment transportation; she herself lacked the necessary background to uncover the general principles involved, but her work is still respected and the area remains a subject of research.[23] The paper she read before the Royal Society in 1904, "The origin and growth of ripple-marks," was the first delivered to that venerable body by a woman. Her presentation included a demonstration (using rocked glass vessels of different sizes), of two distinct sand structures, ripples, and sandbars, produced by stationary waves of different wavelength. Her award of the society's Hughes Medal in 1906 recognized her combined contributions on sand ripple formation and the electric arc.

By about 1904 she had gone back to work on the electric arc, joining her husband in an investigation he had undertaken for the Admiralty on the carbons in searchlight projectors. William Ayrton's health was failing, and as he became less able to do the work more and more devolved on Hertha. The project was very much a practical one, involving the determination of length, position, size, shaping, and composition of the electrodes for efficient operation of the light and elimination of the persistent problems of roaring and erratic wandering. Nevertheless, the results were of considerable significance technologically, since they led directly to the standardization of types and sizes of carbon electrodes for arc lamps.[24] The reports submitted to the Admiralty between 1904 and 1908 were officially treated as William Ayrton's, although Hertha assisted in the work presented in the first three and was responsible for all of that in the fourth, which, on her husband's insistence, went to the Admiralty over her name.

Following completion of the searchlight work, she was asked by a film company to look into the problem of producing a steady arc for cinema lighting and thus reduce flicker. This she did, patenting her redesigned carbon arc in 1913.[25]

After her husband's death in 1908 she became even more immersed in her work, moving her laboratory from the top floor of her house to the main ground floor drawing room, doubtless thereby increasing her working space considerably. She continued her investigations of stationary waves and ripple effects for at least another decade, delivering papers to the Royal Society on the topic in 1908 and 1911, although the second of these presentations did not appear in print.[26]

Her general interest in vortices in fluids led directly to her invention of a hand-operated device for deflecting waves of poison gas from the trenches in the First World War, the Ayrton fan. She had great difficulty getting the War Office to consider this invention, but thanks to a contact made with David Lloyd George (then Minister of Munitions) was eventually able to have tests carried out at the school of Military Engineering. These were followed by trials at the front. Contrary to the expectations of the experts, the device was found to be remarkably effective, but bureaucratic delays and obstruction (against which she was obliged to keep up a continual and exhausting struggle) prevented its distribution in quantity to the troops until November 1916, and even then the necessary instruction in its use was seldom provided. Little is now remembered about the Ayrton fan, but at a lecture Hertha gave in 1920 she read noteworthy comments from a letter written from the front in August 1917 by a Major Gillespie, of the Royal Field Artillery. Gillespie attributed the escape of his battery during a gas attack very largely to the Ayrton fans. Even though the gas hung about for days afterward, by judicious flapping at frequent intervals his men were able to keep their quarters fairly free of it.[27] Effectiveness in clearing the dugouts depended on the quickness of the stroke and the angle at which the fan was held, these being critical in the creation of the needed vortices.

In the design and building of the fan and much of the equipment for its testing on a small scale, Hertha's enterprise, originality, and mechanical skills were clearly evident. Her ability to build effective technical devices was remarkable, from her first invention of the line divider made while she was still a student. Even in her childhood she was exceptionally skillful with her hands (sewing clothes for her brothers when she was no more than six). She always thought of her mother as the great influence in her life, but her interest in things mechanical may well have been awakened by her clockmaker father who died when she was seven.

Something of a radical from her student days on, she was a strong advocate of equality for women, encouraged first by her mother's liberal views on education for girls and later by her husband's support. For many years she was active in the women's movement, helping such leaders as Millicent Garrett Fawcett, Elizabeth Garrett Anderson, and Christabel Pankhurst; and her efforts in this sphere did not always sit well with the more socially conservative of her scientific colleagues. She joined the Women's Social and Political Union in 1906 and took part in marches and militant demonstrations, including the great suffrage processions of 1910 and 1911. In some of these marches many of the women, like Ayrton no longer young, were subjected to considerable police brutality. In 1912 her daughter Barbara, who shared her views on women's rights, was imprisoned for the cause; Hertha was extremely proud of her.[28]

As a member of the British Federation of University Women (of which she was a vice president), she was closely associated with the founding in 1919 of the International Federation of University Women. She was also one of the original members of the National Union of Scientific Workers, formed in 1920 with membership open to both men and women. Marie Curie was one of her friends, and indeed as women physicists and wives (later widows) of physicists the two had a fair amount in common. Hertha first met the Curies in 1903 when they attended a Royal Institution Friday Evening Discourse at which Pierre delivered a lecture on the Curies' discovery of radium. Nine years later, Marie, using her maiden name, stayed with Hertha for two months when she needed a place from which to escape the passing scandal concerning her relationship with her colleague Paul Langevin. The following year she and her daughters joined Hertha for a summer vacation, but further visits were prevented by the outbreak of the First World War.

A colorful and striking personality, even if considered somewhat eccentric by some, Hertha Ayrton had talent and self-confidence, and her scientific and technical achievements were considerable. Her laboratory work continued even during her last years, one of her interests being the adapting of the Ayrton fan for various kinds of workplace ventilation; she left much research in progress. Her ties with her daughter and stepdaughter and their families remained very close, and she took great pleasure in the company of her grandchildren. She died in her seventieth year, on 26 August 1923, at North Lancing, Sussex, while on a visit to a friend.

HELEN KLAASSEN[29] (1865–1951) studied at Newnham from 1886 to 1890, and held a College studentship for the two years from 1889 to 1891. She did not take the Tripos examinations, but nevertheless was assistant demonstrator in physics from 1891 to 1896 and after that lecturer for five years until 1901. Her joint note on magnetic transformers with James Ewing, professor of applied mechanics, appeared in the *Electrician* in 1892, and a long article with Ewing on the magnetic qualities of iron followed two years later in *Philosophical Transactions.* She brought out two more research papers, one on solution conductivity (1892) and the other on phase change (1897). In 1902 she moved to London and for a number of years served as manager of various London County Council schools. She was also extremely active in voluntary community work, one of the founders, and later vice president, of the Croydon Nursing Service, a Poor Law Guardian in Camberwell and a member of relief committees. In 1914 she returned to physics, carrying out war-related work in Leeds until 1917, and then for a further year or two at the University of Birmingham. She died on 31 May 1951.

ALICE EVERETT[30] (1865–1949), Klaassen's contemporary at Cambridge, was one of the six children of the English physicist and mathematician Joseph David Everett of Suffolk and his Scottish wife, Jessie, daughter of the Rev. A. Fraser. She was born in Glasgow, on 15 May 1865, her father at the time having a teaching post in mathematics at Glasgow University. The family moved to Belfast in 1867 when Joseph Everett became professor of natural philosophy at Queen's College.

Alice attended a private school in Belfast, and then continued her education at Methodist College, a coeducational day school with a good academic reputation. From there she went to Queen's College, Belfast, which had opened its lectures to women students in 1882. She entered in 1884, taking first place in the first-year scholarship examination in science, although women were not eligible for scholarships for another eleven years. She enrolled at Girton in 1886, with a College studentship, and took the Mathematical Tripos examinations in 1889 (class II, part I). While at Cambridge she also sat the

Royal University of Ireland's B.A. degree examinations in mathematics and mathematical physics, passing with honours in 1887; two years later she was awarded an M.A.

From 1890 until 1895 she held a post as "supernumerary" computer at the Royal Observatory, Greenwich.[31] Initially she worked in the transit department with the observatory's chief assistant, H. H. Turner, making observations with the Transit Circle (taking regular turns with the night work) and carrying out the reducing. In 1892 she was assigned to the Astrographic Catalogue (Carte du Ciel) photographic project, a huge joint international undertaking of several observatories, which had as its goal the surveying of the sky photographically and the cataloging of all stars brighter than 11 magnitude. Under the supervision of assistant George Criswick she was involved in every aspect of the work, the night observing with the astrograph, photographic developing, printing, and making the plate measurements. The formal report (*Astrographic Catalogue 1900.0. Greenwich Section,* vol. 1, 1904) records her contributions.

Both she and her Cambridge friend Annie Russell, who had joined the Greenwich staff in 1891, were proposed for election to fellowship in the Royal Astronomical Society in 1892, but neither gained the necessary two-thirds of the votes cast by the all-male membership. They both became very active members of the British Astronomical Association, however, Everett being appointed secretary in 1893 and Russell editor of the association's journal the following year. Everett contributed a number of notes to the journal in 1891 and 1892, including her observations of Nova Aurigae (1892), work carried out with the Greenwich instruments during her leisure time.

In 1895, after five years at Greenwich, she obtained a post as scientific assistant at the Astrophysical Observatory at Potsdam, at the time the leading European institution for astrophysical research. Her position was a temporary three-year one, filling in for one of the observatory's astronomers called away for military service. She was well-qualified for the work, her assignment being a continuation of the kind of research she had been doing at Greenwich on the Astrographic Catalogue project. About this time she also published two notes in the Royal Astronomical Society's *Monthly Notices* on orientations of orbits of binary stars.

Paid jobs in astronomy were far from plentiful in those day. Everett had applied for the vacant post of assistant at the Dunsink Observatory (Trinity College Dublin) in 1892, but despite excellent testimonials from the Astronomer Royal she failed to get the job. After her time at Potsdam she visited the United States for a year, joining Mary Whitney of Vassar College in Poughkeepsie, New York, for observational work on comets and minor planets. Their two joint papers appeared in the *Astronomical Journal* in 1900. From Vassar she applied for a post at the Lick Observatory in California, but although the director (James Keeler) would have been glad to hire her as assistant astronomer in his stellar spectroscopic program, he was unable to raise funds for a salary.

Following her return to Britain in 1900 she collaborated with her father, by then retired and living in London, in a big project of translating and editing an important new work by Heinrich Hovestadt on the scientific and industrial applications of Jena glass.[32] From then on her interests turned more and more to the mathematics and physics of optical systems, an area in which she expanded her knowledge and skills by taking a practical course in optics. In 1902 her father communicated to the Physical Society on her behalf a short report of work on zonal aberrations of lenses that she carried out at the Davy-Faraday Laboratory of the Royal Institution. The paper was the first by a woman in the society's *Proceedings*.[33]

She did no further work in astronomy, and indeed there is no record of her having had any regular paid employment between 1899 and about 1916. The coming of the First World War, however, brought new job opportunities for women. Alice spent a year working in the optical laboratory of the firm of Hilgers in London, and in 1917, at the age of fifty-two, joined the National Physical Laboratory at Teddington, Middlesex, as a junior assistant in the optical department. Two years later she became an assistant, and held that post until she retired in 1925, when she was sixty. As one of a team of thirteen scientists carrying out research on photometry, spectrophotometry, and the design of optical instruments, she worked mostly on theoretical problems involving calculations of aberrations in lens and mirror systems; her findings were reported in nine papers and notes on geometrical optics, mainly in the *Philosophical Magazine,* between 1919 and 1927.[34]

Even after her formal retirement her interest in new technical developments continued. Preparing herself by means of three years of evening classes (1926–1928), involving mainly practical work in wireless and high-frequency and alternating current technology at the Regent Street Polytechnic in London, she went on to two years of radio communications research at City and Guilds College.[35] About this time she joined the Institute of Radio Engineers as an associate member. The completely new field of television was the area that particularly caught her interest.

The first demonstration in Britain (and probably in the world) of a television image had been given in London by the Scottish engineer John Logie Baird in January 1926. In Baird's audience of forty invited guests, mainly members of the Royal Institution, were two women scientists. Their names are unrecorded, but even if Alice Everett was not one of them, she undoubtedly knew people who saw the demonstration. When in 1927 the Television Society for the promotion of television research was established, she became one of the 325 foundation fellows (of whom only four or five were women). She was to remain an active member for more than twenty years.

In 1933 she suggested improvements for the Baird system scanning equipment, that is, the apparatus that creates the television picture (by translating the scene before the camera into tiny elements according to their light value). Baird's system scanned by mechanical means, using mirrors spaced round a rotating drum.[36] Everett's improvements involved combining two or more drums to carry more mirrors without a corresponding increase in the size of the apparatus, and also the use of more highly reflecting metals for the mirrors. She and the Baird Company applied for a patent for the modifications in

1933.[37] However, in 1935 the British Broadcasting Corporation (the first in the world to provide a regular television service) tried out both the Baird system and the rival Marconi-EMI system. The latter was chosen, since it scanned electronically with an electron beam and produced much higher definition with less flickering. Thus the Everett-Baird patent was abandoned.

Everett continued to work with the Television Society, translating foreign-language publications for its library index system. In 1938 she was awarded a Civil List pension of £100 a year in recognition of her own and her father's contributions to science (she had no pension from the National Physical Laboratory, having worked there for less than the minimum qualifying period of ten years). She died in London, on 29 July 1949, at the age of eighty-four. Her library of technical books was left to the Television Society.

ELEANOR BALFOUR SIDGWICK[38] (1845–1936), another of the women in this group associated with Cambridge, would not have considered herself a physicist; indeed, she was only active in physics research in her "spare time" for a short period of three or four years in the early 1880s. Nevertheless her contributions, reported in three major papers coauthored with her brother-in-law Lord Rayleigh, were noteworthy.

Eleanor was the oldest daughter and the oldest of the eight surviving children of Lady Blanche Harriet, daughter of James Gascoigne Cecil, second Marquis of Salisbury, and James Maitland Balfour. She was born on 11 March 1845 at Wittingehame, East Lothian. The Balfours were an old Scottish family from the county of Fife; James's father had bought the East Lothian property with a fortune made in India. They had in addition a second estate, Strathconan, in the Highlands, and a London house in St. James's Place. The family had close connections with the leading political figures of the time, Lady Blanche Balfour's younger brother Robert Cecil, third Marquis of Salisbury, serving for a period as prime minister. Later her son Arthur Balfour, Eleanor's brother, was to hold the same office.

Eleanor, along with the other children, received much of her early education from her mother, who gave her what was to be a lasting interest in mathematics. French or Swiss governesses provided instruction in French, German, and Italian, all of which Eleanor learned to speak fluently. After the death of her father, when she was eleven, her mother's health was poor, and as the oldest daughter she took on many household responsibilities. As a girl she moved in the best society of the day and had the customary continental travel, either with a governess or with her mother, who regularly went to Bad Schwalbach, near Wiesbaden, for a cure. At the age of eighteen she had her first London season, which included presentation to Queen Victoria.

By the late 1860s, when her brother Arthur was an undergraduate at Cambridge, she acted as vacation hostess for his friends. Among those young men was John Strutt, later Lord Rayleigh, who married her sister Evelyn in 1871. The Cambridge students brought to the Balfour household many of the latest ideas in science and the "modern studies" that were receiving increasing attention at the time. Eleanor continued her own reading at a leisurely pace, taking advantage of whatever opportunities came her way. In the winter of 1872, during a boat trip with her sister and brother-in-law up the Nile river, she and John Strutt spent their mornings doing mathematics at the cabin table. Most of her time, however, was given to Balfour family matters (including the tutoring of her younger brothers for their Cambridge entrance examinations, particularly in mathematics). She alternately ran the family home at Wittingehame and looked after a London house acquired by her brother Arthur, whose political career was just beginning.

Much in the air at the time was the question of education for girls and women, the two Cambridge women's colleges having just been organized. Eleanor was actively interested in the movement from the beginning, and though never a pioneer, was one of those who provided solid support, in the form of money, both for scholarships and for the foundation launched by Henry Sidgwick and Anne Clough to establish a women's residence at Newnham. As soon as Newnham Hall opened, she joined Anne Clough (the first principal) as secretary. In addition she was preparing to take the Cambridge Higher Local Examinations, although she also gave a considerable amount of her time to helping women students in mathematics. In 1876, however, she married Sidgwick (by then praelector in moral and political philosophy at Cambridge), and gave up her plan to proceed with formal studies. She considered that marriage entailed too many other responsibilities. From then on she was much occupied with the administration of Newnham, which remained one of Henry Sidgwick's major concerns. Nevertheless, for a few years in the early 1880s, she had what for her was a delightful change from her main concerns and joined a laboratory research project in physics.

In 1879 Lord Rayleigh (John Strutt) had succeeded Clerk Maxwell in the new chair of experimental physics, and he and his wife moved to Cambridge. Eleanor Sidgwick quickly found herself drawn into her brother-in-law's work, first some private projects and then a major undertaking involving the accurate determination of the e.m.f. of Clark cells and the electrochemical equivalent of silver. This was part of a big program set up by Rayleigh to redetermine the three electrical standards of measurement—the ohm, ampere, and volt. It was tedious, demanding work requiring great precision, but she much enjoyed its basic, practical nature. The project began in the summer of 1880 in the "magnetic" room on the ground floor of the Cambridge laboratory, afterward famous for J. J. Thomson's electron experiments. Eleanor assisted Rayleigh and Arthur Schuster in taking the observations, herself carried out and checked most of the computations involved, and kept the laboratory notebooks. When Schuster moved to the physics department at the University of Manchester (where about ten years later he carried out notable early work on cathode-ray particles), his part of the work passed to Eleanor, and some of hers went to various volunteers, including her sister. The results, which virtually established modern standards for commercial electrical units, were published as three long joint papers (Rayleigh and Sidgwick) in *Philosophical Transactions* (1884, 1885) and in shorter notes in the Royal Society's *Proceedings.*

Other claims on her time did not diminish, however, and in

1880, in the midst of this physics interlude, she was asked to take the post of vice principal for two years in the newly completed North Hall extension of Newnham residence accommodation. Helen Gladstone, a daughter of the prime minister, became her secretary.[39] Her substantial material support of Newnham continued as well; in 1884 she and her sister Alice contributed the bulk of the funds needed for the purchase of a disused building, originally a chapel, which, with the help of further donations, was fitted up as an undergraduate biology laboratory. Known as the Balfour Laboratory, it commemorated Eleanor Sidgwick's younger brother Francis, killed in an accident on Mt. Blanc in 1882, but already one of Cambridge's outstanding young biologists.

Newnham continued to expand, and even after completing her term as vice principal she was much occupied with administrative and fund-raising work. On the death of Anne Clough in 1892 she was offered, and accepted, the principalship. By then a well-known figure in educational circles, she was appointed in 1894 to the Royal Commission on Secondary Education chaired by Lord Bryce (on which also served Sophie Bryant—see chapter 8). The commission was an influential one; its recommendations formed the foundation of the administrative structure of the English secondary school system for both boys and girls for decades to come.

Her considerable services to education brought her recognition on the national level in the form of four honorary degrees—from Victoria University of Manchester and from the universities of Birmingham, St. Andrews and Edinburgh. Locally she served on the Cambridge County Council Education Committee and on the administrative board of the Cambridge Secondary Training College. She remained principal of Newnham until 1911. Throughout her tenure she put special effort into establishing scholarship funds for poor students, research scholarships, and a staff pension scheme. Even after retiring she continued to act as Newnham treasurer, a post she held from 1876 to 1919, and she also served as president of Newnham Council. Her special report on *Higher Mathematics for Women* was published by the board of education in 1912.

For many years, beginning in the early 1870s, she was interested in spiritualism and psychic research. To a late twentieth-century reader this might well seem an anomalous interest for someone of Eleanor Sidgwick's background. But in fact there was a widespread turning to spiritualism at the time. Interest in the movement was part of society's response to the uncertainties and upheavals caused by the serious challenge nineteenth-century science posed to traditional religious values.[40] Many intellectuals from the upper and upper-middle classes were caught up in the debate, among them a number of eminent scientists, including physicists Sir William Crookes, J. J. Thompson, and Lord Rayleigh. Eleanor's husband, Henry Sidgwick, was one of the leaders of the Cambridge psychical research group, having taken up the subject in the course of his own search for a basis for religious conviction. Eleanor was much involved in the Society for Psychical Research, an organization founded in 1882 (with her husband as first president) to examine alleged psychic phenomena fairly and without preconceived bias. She contributed extensively to the society's *Proceedings* and *Journal,* and provided assistance in the editing of these publications. Although her many investigations generally resulted in nothing more than the uncovering of fraud, her studies of thought transference via trance mediums led her to more positive conclusions; later she seems to have felt that telepathy might be a reality under certain circumstances. She continued to work with the society after her husband's death in 1900, served as president in 1908–1909, and went regularly to its London meetings after she retired. The entry on spiritualism in the ninth edition of the *Encyclopaedia Britannica* was supplied by her.[41]

Throughout most of her life, and especially during the 1890s and the early years of the present century, she was closely involved in the ups and downs in the political life of her brother Arthur; he in his turn had strong sympathy for the causes she supported, particularly women's education and the extension of the franchise. In 1916 she moved to her brother Gerald's house near Woking in Surrey, where she lived for another twenty years, her retirement enlivened by the events in the lives of the extended Balfour family and by an active correspondence, especially with Newnham friends. She died on 10 February 1936, a month before her ninety-first birthday.

Little information has been collected about either of the remaining two British women physicists whose papers are listed in the Royal Society *Catalogue*—Ella Bryant and Jesse Chambers.[42] Bryant's one paper was a substantial study on the properties of metals. She entered Durham College of Science Newcastle in 1888 from Newcastle High School, and was awarded a B.Sc. in 1892 (second-class honours in experimental physics). Her study of the stability under heat of various kinds of boilerplate was begun under the direction of Henry Stroud in Newcastle, and continued at the Royal College of Science London, with supervision from the metallurgical engineer Sir William Chandler Roberts-Austin, then professor at the Royal School of Mines. After a lapse of some years she returned to formal studies, and was enrolled in the School of Hygiene at University College London in 1908–1909.

Jesse Chambers took classes in physics and zoology-comparative anatomy at University College London as a private student in the early 1880s and received a third-class certificate in experimental physics at the intermediate science level in 1883. She then attended Mason College Birmingham and University College Liverpool, taking a B.Sc. in 1885 (second-class honours in experimental physics). Her four papers on the polarization of light came out in the *Electrician* and *Philosophical Magazine* in 1886.

Ayrton, Everett, and to a lesser extent Klaassen, the three most notable in this small collection of early women workers in physics, are virtually a subsection of the Cambridge-trained early women mathematicians—as remarkable a group of pioneers as any discussed in this survey (see chapter 8). They might be thought of as the applied mathematicians of the group. Sidgwick, although her physics contributions were noteworthy, was an educationist and administrator rather than a research worker or teacher.

Two-country comparison

A comparison of British and American nineteenth-century women physicists is of marginal significance because both national groups are so small (Figure 9). However, the following points might be noted: most of the nine Americans who studied physics, including the most notable, Margaret Maltby, became teachers in high schools or small colleges. Consequently their combined research output was modest, not even Maltby publishing after her student years and brief postdoctoral interlude. There was no contemporary American equivalent of the enterprising and ambitious Ayrton, and there appears to be nothing to suggest that any of the Americans whose careers in physics began before the turn of the century were involved in technical research during the First World War, as were Ayrton, Klaassen, and Everett (and also a number of British women in other scientific fields). Few women anywhere had the good fortune enjoyed by Sidgwick that enabled her to make a serious and satisfying contribution to an important practical project largely as recreation.

Notes

1. The Canadian physicist Elizabeth Laird (1874–1969) often appears in accounts of work by American women in physics because she was for many years professor of physics and department head at Mount Holyoke College. After an undergraduate education at the University of Toronto, Laird took a Ph.D. in mathematics and physics at Bryn Mawr College (1901). She taught for a year at Ontario Ladies' College before going to Mount Holyoke. She was active in research, despite a substantial administrative load, and spent several productive study and research leaves in British and European laboratories. Her most important work was probably her spark discharge studies, published in *Annalen der Physik* in 1914. After retiring from Mount Holyoke in 1940, she returned to Canada and worked throughout the Second World War with the Canadian National Research Council on a radar project and then with the Ontario Cancer Research Foundation (Katherine Russell Sopka, "Women physicists in past generations," in *Making Contributions: an Historical Overview of Women's Role in Physics,* ed. Barbara Lotze (College Park, Md.: American Association of Physics Teachers, 1984), pp. 13–14; *WWWA,* p. 470).

2. A ninth American woman to publish studies in physics in the nineteenth century was Eunice Foote of Seneca Falls, New York, who joined her husband, Elisha Foote, in experimenting on the heat conducting capacity of gases. Her short paper in the *American Journal of Science* in 1856 directly followed a somewhat lengthier one on the same general topic by her husband, who also published an account of his investigations in the London *Philosophical Magazine* in 1857. Using the same equipment as that described in their initial reports (airtight glass tubes that could be evacuated with "an ordinary air pump"), Eunice extended her observations of the properties of gases to a study of effects produced by compression and expansion. These investigations were reported in the *American Journal of Science* in 1857 and were also discussed at the Montreal meeting of the AAAS the same year. Elisha Foote was a member of the AAAS, having joined

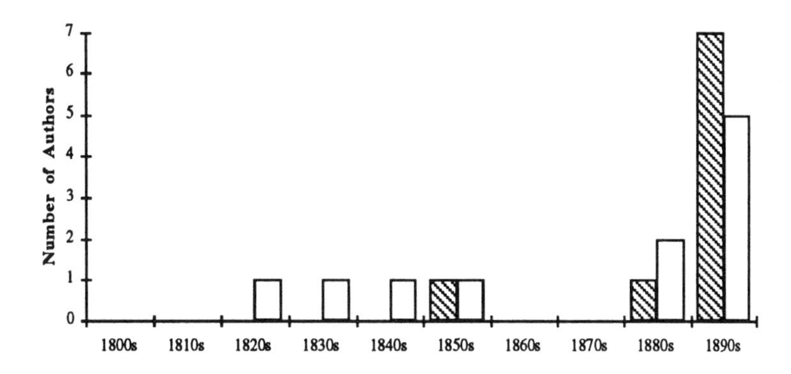

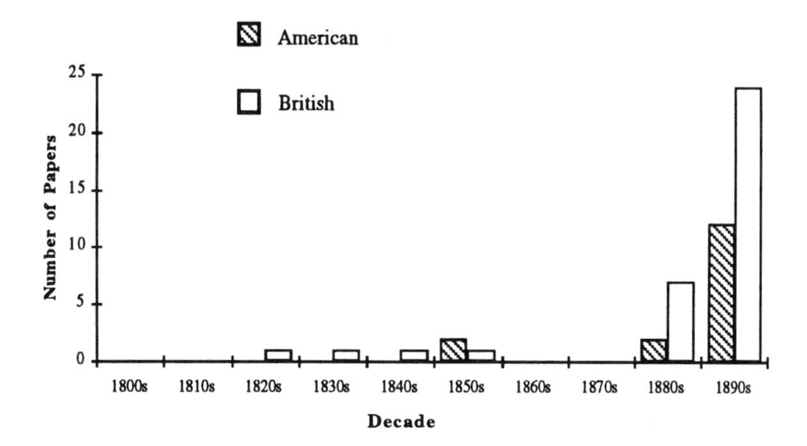

Figure 9. American and British authors and papers in physics by decade, 1800–1900. Data from the Royal Society *Catalogue of Scientific Papers.*

at the Albany meeting in 1856 (see membership lists in the *Proceedings of the American Association for the Advancement of Science,* **10** (1857): xxx, xlvi).

3. Annie Sabine carried out investigations on microphone currents in the Rogers Laboratory of Physics at MIT under the guidance of Charles Cross, the laboratory director. The work was reported in two papers published in 1889.

4. *WWWA,* p. 578.

5. Walter Crosby Eells, "Earned doctorates for women in the nineteenth century," *American Association of University Professors, Bulletin,* **42** (1956): 644–51.

6. S and F, p. 285.

7. During 1897–1898 she taught physics at Mount Holyoke College. There was a close connection between Mount Holyoke and Lake Erie College, the latter being described as a "daughter college" in the *One Hundred Year Biographical Directory of Mount Holyoke College 1837–1937* (South Hadley, Mass.: Alumnae Association of Mount Holyoke College, 1937), p. 7.

8. *WWWA,* p. 770.

9. Dates of birth and/or graduation for Bruère, Crehore, and Spencer came from Cornell University archives.

10. *WWWA,* p. 788; *AMS,* 1910–1944; S and F, pp. 288–9. The sources are inconsistent about Stone's date of birth; *WWWA* reports it as 1870 and *AMS* as 1868.

11. Eells, "Earned doctorates," p. 649.

12. See, for instance, Isabelle Stone, "Color in platinum films," *Physical Review,* **21** (1905): 27–40.

13. The other woman charter member was Marcia Keith, then chairman of the department of physics at Mount Holyoke College. Keith did not have an advanced degree, though she had studied for a year at the University of Berlin (S and F, p. 285).

14. Agnes Townsend Wiebusch, "Maltby, Margaret Eliza," in *NAW,* vol. 2, pp. 487–8; E. Scott Barr, "Maltby, Margaret Eliza," *American Journal of Physics,* **28** (1960): 474–5; Margaret E. Maltby, comp., *History of the Fellowships awarded by the American Association of University Women. 1888–1929. With Vitas of the Fellows* (Washington, D.C.: American Association of University Women, [1929]); Shirley W. Harrison, "Margaret Eliza Maltby (1860–1944)," in *Women in Chemistry and Physics, A Biobibliographic Sourcebook,* eds. Louise S. Grinstein, Rose K. Rose and Miriam H. Rafailovich, (Westport, Conn.: Greenwood Press, 1993), pp. 354–60; *WWWA,* pp. 535–6.

15. See n. 9, chapter 8.

16. Rossiter, *Women Scientists in America,* p. 366, n. 28.

17. Maltby, comp., *History of the Fellowships.*

18. Maltby greatly valued family life; one of her students is quoted as having said that the most memorable advice Maltby gave her was not to forgo marriage for a career (Sopka, "Women physicists," p. 13).

19. For Philippa Fawcett and also Mary Somerville (whose three early papers on effects of solar radiation are listed under physics) see chapter 8. Reports of observations by two additional British women—Hubbard and Crosse—are listed in the physics section of the bibliography. Emma Hubbard, whose note on forces generated by projectiles appeared in *Nature* in 1894, wrote from Kew, London. Cornelia Crosse, the daughter of Captain Berkeley of Exeter, was the second wife of Andrew Crosse (1784–1855) of Fyne Court, Somersetshire. Crosse was a country gentleman with sustained scientific interests who made a special study of voltaic cells and electrolytic plating, subjects on which he published a number of reports. Cornelia Crosse's remarks to the British Association (*Reports,* 1855) took the form of a reply to criticisms of an earlier study by her husband. The latter involved experiments made with electrolytic cells, which Andrew Crosse had discussed before the association just prior to his death. The point in question, namely the possibility of Crosse's having used impure metals as electrodes, his widow settled to her own satisfaction, by repeating his experiments with materials checked for purity; she claimed that she had been able to duplicate his results. The Crosses were friends of Ada Lovelace (see chapter 8). Cornelia brought out a collection of her husband's works, *Memorials, Scientific and Literary, of Andrew Crosse, the Electrician* (London: Longman Brown, 1857), and also a 2-volume memoir, *Red-Letter Days of my Life* (London: Bentley, 1892)—see *Burke's Landed Gentry,* 17th ed. (London, 1952), p. 1138, and Stein, *Ada,* especially pp. 142–3.

20. Evelyn Sharp, *Hertha Ayrton 1854–1923. A Memoir* (London: Edward Arnold, 1926); Joan Mason, "Hertha Ayrton (1854–1923) and the admission of women to the Royal Society of London," *Notes and Records, Royal Society of London,* **45** (1991): 201–20; Glenis Moore, "Hertha Ayrton—first lady of the IEE," *Electronics and Power,* **32** (1986): 583–5; obituaries, Henry E. Armstrong, *Nature,* **112** (1923): 800–1, A. P. Trotter, ibid., **113** (1924): 48–9, anon., "Mrs. Hertha Ayrton. A distinguished woman scientist," *Times,* 28 August 1923, 11e.

21. Ayrton's Paris paper, entitled, "L'intensité lumineuse de l'arc à courants continus," was described by the physicist-mathematician Sylvanus P. Thompson, F.R.S., president of the Institution of Electrical Engineers, as the most remarkable report read at the congress (Mason, "Hertha Ayrton," p. 214).

22. See Sharp, *Hertha Ayrton,* p. 150.

23. For further details see Mason, "Hertha Aryton," pp. 205–6.

24. Trotter, obituary, p. 48.

25. H. Ayrton, British Patents Nos. 1,775 (22 January 1913) and 22,319 (3 October 1913), "A negative carbon electrode, applicable to projection apparatus." U.S. patent No. 1,115,480 (3 November 1913), "Negative arc light carbon."

26. Ayrton's post-1900 journal articles included the following: "The mechanism of the electric arc," *Proceedings of the Royal Society,* A, **68** (1901): 410–11, and *Philosophical Transactions,* A, **199** (1902): 299–336; "Note on electric charging and discharging at a distance," *Nature,* **65** (1901): 390; "Origin and growth of ripple-marks," *Proceedings of Royal Society,* A, **74** (1905): 565–6; "On the non-periodic or residual motion of water in stationary waves," ibid., A, **80** (1908): 252–60; "Local differences of pressure near an obstacle in oscillating water," ibid., A, **91** (1915): 405–10; "Primary and secondary vortices in oscillating fluids: their connection with skin friction," ibid., A, **113** (1926): 44–5 (posth.).

27. "Anti-gas fans in the trenches. Mrs. Ayrton on War Office delay," *Times,* 6 January 1920, 14b (report of an address by Ayrton to the Science Teachers' Association at the Educational Associations Conference, University College London, 5 January 1920). Her main scientific article on the fan, "New method of driving of poisonous gases," appeared in the *Proceedings of the Royal Society,* A, **96** (1919): 249–56.

28. Barbara Ayrton (later Mrs. Gould) was for many years active in social and political work. She was chairman of the Labour Party from 1939 to 1940 and Labour member of Parliament for the constituency of Hendon North from 1945 until her death in 1950 (Mason, "Hertha Ayrton," p. 219, n. 41).

29. *Newnham College Register,* vol. 1, 1871–1923, Staff, p. 8.

30. *Girton College Register, 1869–1946,* p. 37, and other information from Girton College archives, including material collected by Barbara Strachan; Mary T. Brück, "Alice Everett and Annie Russell Maunder, torch bearing women astronomers," *Irish Astronomical*

Journal, **21** (1994): 281–90, "Alice Everett. Early RTS pioneer," *Television* (Dec-Jan 1994): 4, and "Lady computers at Greenwich in the early 1890s," *Quarterly Journal of the Royal Astronomical Society,* **36** (1995): 83–95; *DNB,* 2d Supplement, vol. 1, pp. 638–9, entry for Everett, Joseph David.

31. The Greenwich observatory carried out an experiment during the first half of the 1890s of employing women in temporary positions as computers. Before that the work had been done by boys in their teens, just out of school, who put in very long hours for low wages. Women were accepted only if they had college-level training, and so were better educated than the usual computer. They were still poorly paid, however; Everett's friend Annie Russell, who joined the staff in 1891, received £4 a month, barely enough to live on (Brück, "Everett and Maunder," p. 282). The experiment was discontinued in 1896 (A. J. Meadows, *Greenwich Observatory. Volume 2: Recent History, 1836–1975* (London: Taylor and Francis, 1975), p. 14, and Brück, "Lady computers").

32. Heinrich Hovestadt, *Jena Glass and its Scientific and Industrial Applications,* trs. and eds. J. D. Everett and Alice Everett (London: Macmillan, 1902). See also, Alice Everett, "The Jena glass works, with special reference to astronomical objectives," *Journal of the British Astronomical Association,* **13** (1903): 275–7. A short note by Everett on the Potsdam Observatory appeared in the same issue (pp. 107–8). About this time she was also assisting in the preparation of physics and mathematics entries for the Royal Society's *Catalogue of Scientific Papers.*

33. See Alice Everett, "Photographs of cross-sections of hollow pencils formed by oblique transmission through the annulus of a lens,"*Proceedings of the Physical Society,* **18** (1903): 376, *Journal of the British Astronomical Association,* **13** (1903): 74–5.

34. These were the following: "Note on a paper 'Some generalised forms of an optical equation,'" *Transactions of the Optical Society,* **20** (1919): 203; "On proofs of elementary theorems of oblique refraction," *Philosophical Magazine,* **38** (1919): 480–3; "Note on a simple property of a refracted ray," ibid., **39** (1920): 223–4; "On a projective theorem of Lippich's in geometrical optics. (With a note on the equations of the projection of a straight line on a plane)," ibid., **40** (1920): 113–28; "The unit surfaces of the Cooke and Tessar photographic lenses," *Proceedings of the Physical Society,* **35** (1923): 55–66; "Unit curves of a photographic lens," *Philosophical Magazine,* **46** (1923): 565–7; "Unit magnification of a glass ball," ibid., **46** (1923): 450–4; "Formulae for oblique focal distance in terms of magnification," ibid., **47** (1924): 864–73; "The tangent lens gauge generalised," ibid, **4** (1927): 720–1. Everett also published a note on peculiar instances of light reflection in *Nature,* **90** (1913): 570–1 ("The halo in the rice field and the spectre of the Brocken").

35. City and Guilds College is a constituent college, and in effect the engineering division, of Imperial College of Science, Technology and Medicine of London University.

36. [Alice Everett?], "A new home televisor," *Television* [*Electronic Engineering*], **5** (1932): 183, 223.

37. A. Everett and the Baird Television Company, "A new television drum," Patent application no. 1905/33, January 30, 1933.

38. Ethel Sidgwick, *Mrs. Henry Sidgwick: a Memoir by her Niece* (London: Sidgwick and Jackson, 1936); obituaries, *Times,* 12 February 1936, 16b, and 13 February, 1936, 17b ("Mrs. Sidgwick. Psychical research"); Blanche E. Dugdale, "Sidgwick, Eleanor Mildred," *DNB,* Supplement, 1931–1940, pp. 811–12.

39. For an American student's thoughtful and quite detailed account of living in North Hall, 1880–1883, years that included the period of Sidgwick's vice principalship, see the reminiscences of Alice Willcox in chapter 3.

40. Janet Oppenheim, *The Other World: Spiritualism and Psychical Research in England, 1850–1914* (Cambridge: Cambridge University Press, 1985).

41. Although one might feel tempted to pass over Eleanor Sidgwick's contribution to psychical research as being of less interest and importance than her other considerable achievements, it was not considered minor by her contemporaries. Her obituarist (*Times,* 12 Feb.) paid special attention to her psychical work, and in fact stated that it claimed perhaps even more of her time and effort than educational matters.

42. Information on both Chambers and Bryant came from the *List of Graduates, University of London,* Records Office, University College London.

Chapter 10

OBSERVERS, "COMPUTERS," INTERPRETERS, AND POPULARIZERS: WOMEN IN ASTRONOMY

Articles in astronomy constitute about 9 percent of all the scientific journal publications by women indexed by the Royal Society. Workers in the United States and Britain were the heaviest contributors, producing, respectively, about 52 percent and 21 percent of the output. Another 12 percent came from France, and the remainder from Italy, Russia, and Germany.

American women

Two-thirds of the thirty-one American contributors were either faculty members at women's colleges (Vassar, Mount Holyoke, Smith, and Wellesley), or were employed as assistants at large observatories, particularly that at Harvard. Several of the others whose work is listed in the bibliography published one or two papers as undergraduates but then either stopped doing astronomy or joined computing groups at observatories where their research was published under the name of the institution's director. A few appear to have been observers and amateur astronomers unaffiliated with any institution.

The earliest published work of an American woman in astronomy is the 1847 report by MARIA MITCHELL[1] (1818–1889) of her sighting of a new comet, which appeared in the *Monthly Notices of the Royal Astronomical Society.* For her observation she received a gold medal, the prize offered by the King of Denmark to the first discoverer of a new comet invisible to the naked eye at the time of discovery. The award was the first for astronomical research to an American by a foreign government, and it brought Mitchell to the attention of workers on both sides of the Atlantic. She became the first woman member of the American Academy of Arts and Sciences in 1848 (despite protests from the eminent botanist Asa Gray), and two years later was elected to the AAAS, also as its first woman member. In 1865 she became professor of astronomy at newly founded Vassar College.

The third of ten children in the Quaker family of Lydia (Coleman) and William Mitchell, Maria was born on 1 August 1818 on Nantucket Island. William Mitchell earned a living as a cooper, schoolteacher and bank cashier; he was also a respected amateur astronomer, the second American to spot Halley's comet during its 1835 return, and he had connections to professional scientists at the Harvard Observatory and the Nautical Almanac Office. Maria was educated at island schools where astronomy, a subject of great practical importance to the local seafaring people, was included in the curriculum. Her formal education was over by the time she was sixteen, but being especially interested in mathematics she continued to study on her own, reading a number of works by important British and European scientists of the time, including Gauss's book on orbits, and Airy's on gravitation.[2] From her father she learned how to use telescopes and navigational instruments, and she helped him in the observational work on star transits that he carried out for the U.S. Coast Survey.

In 1836, at the age of eighteen, Maria became librarian at the new Nantucket Athenaeum. She and her father kept up their program of observations, setting up their telescope on clear nights on the rooftop of the bank where he was employed. When she discovered her comet in 1847 she went on to the not-inconsiderable task in these precomputer days of calculating its orbit (which was hyperbolic). An offer of paying astronomical work quickly followed her discovery, a job as a computer for the Nautical Almanac Office. This involved the calculation of data required for the preparation of the astronomical tables for mariners published annually as the *American Ephemeris and Nautical Almanac.* She was the only woman on the Almanac Office staff; the work brought her an annual salary of $300. In 1857 she had a chance to travel to Britain and the Continent (as chaperon of the daughter of a wealthy Chicago banker), and was able to talk with astronomers George Airy and Sir John Herschel, as well as Mary Somerville and others. She was by then a well-known figure, and on her return a group of Boston women presented her with a five-inch Alvan Clark refractor. She observed with this instrument for a number of years, first on Nantucket and then at Lynn, Massachusetts, where she and her father took up residence after the death of her mother in 1861. Two further notes by her appeared in the literature about this time, one on "Minima of Algol" in 1858, and another on double stars in 1861. Four years later, when she was forty-seven, she accepted the professorship at Vassar (taking her father to live with her at the college until his death in 1869).

A condition of her acceptance of the Vassar position was that the college provide an observatory equipped with a telescope that would make possible serious scientific work. A basically good instrument was indeed provided, but technical problems arose, first with the objective lens and then with the mounting, and the college administration was reluctant to spend money for replacement parts and adjustments. Consequently she was unable to pursue her comet and asteroid position measurements and throughout the 1860s and 1870s concentrated mainly on observations of surface features of Jupiter and Saturn. Four of her papers reporting this work and offering some speculation on the nature of the two planets and their satellites appeared in the *American Journal of Science* in the 1870s. She concluded that Jupiter is composed largely of cloud (of which we see only the uppermost layers), rather than being a solid body with a thin cloud layer as then believed. She also started a program of daily sunspot observations and observed several solar eclipses, traveling to Denver for the notable one of 1878. Explanations for the origin of sunspots and the changes they undergo, as well as her popular accounts of other astronomical work, appeared in *Hours at Home* and in *Atlantic* and *Century* magazines. By 1887, one year before her retirement, the college had finally carried out the needed repairs to the mounting of its telescope, and after that measurements of comet and asteroid positions quickly became a major part of the Vassar research effort.

During her early days at the college Mitchell put considerable effort into trying to persuade the administration to make the salaries of its women staff comparable with those of the men (hers was about a third that paid to male professors[3]). Her attempts were unsuccessful, however, and she soon gave up and concentrated on her own department, teaching and the work of acquiring needed equipment absorbing much of her time and energy. As a professor she inspired awe and respect; she had little patience with slow students who were unable to satisfy the mathematical prerequisites she required for her astronomy classes. Undemonstrative of manner and at times blunt of speech, she was nevertheless deeply concerned about the welfare of all her students, and many of those in her advanced classes came to have a deep and lasting attachment to her. Further, the program of rigorous training in astronomy that she established at Vassar served as a model for other women's colleges. She was also active in the wider movement then afoot to help women gain an education, especially in science, and find employment outside of teaching. In 1873 she helped found the Association for the Advancement of Women, and she served as its president (1874–1876) and then as chairman of its committee on science (1876–1878). Another of her undertakings while on the Vassar faculty was the editing of the astronomical column of the *Scientific American.*

She received two honorary LL.D. degrees, one from Hanover College, Indiana, in 1853 and another from Columbia University in 1887. Rutgers Female College gave her an honorary Ph.D. in 1870. When she retired in 1888 at the age of seventy, after twenty-three years at Vassar, she was offered a home at the observatory for the rest of her life. However, she chose to return to Lynn, Massachusetts. She died on 28 June 1889, and is buried on Nantucket Island. A crater on the moon was named after her, and a society in her memory established by Vassar alumnae and Nantucket residents, the Maria Mitchell Association of Nantucket.

Her birthplace, the Mitchell House, was opened to the public in 1904. Four years later a small observatory, the Maria Mitchell Observatory, was built next to it and her own Clark refractor installed in the dome. In 1912 Edward Pickering and Annie Jump Cannon of the Harvard Observatory established a research program there—photographic studies of asteroids and variable stars. A resident astronomer, Margaret Harwood, was appointed, and a year later a seven-and-a-half-inch Cooke wide-field photographic refractor installed. The program is ongoing (with undergraduate volunteers working during summers), and the observatory now has a collection of several thousand variable star photographs. The addition of a new wing provides space for a darkroom and computers. There is also access to more powerful computers at the Harvard-Smithsonian Center for Astrophysics, and data from an automatic photoelectric telescope at the Mount Hopkins Observatory in Arizona augments the observational program. Maria Mitchell would surely be well pleased.

MARY WHITNEY[4] (1847–1921), who followed Maria Mitchell as professor of astronomy at Vassar College, was one of the notably productive late nineteenth-century American women astronomers, as measured by the number of journal articles and notes she published before 1901. Having a good command of mathematics she concentrated on calculating orbits of comets, but also observed minor planets and variable stars. Under her leadership the Vassar program in astronomy became outstanding among science programs at women's colleges.

Born on 11 September 1847 in Waltham, Massachusetts, she was the second of the five children of Mary (Crehore) and John Whitney. Whitney was a prosperous real estate dealer whose ancestors had settled in the colonies in the seventeenth century. Both parents encouraged their childrens' educational pursuits. After she graduated from Waltham high school in 1864, Mary spent a year at the Swedenborgian Academy in Waltham and then entered newly opened Vassar College, where she took classes from Maria Mitchell. Having come in with advanced standing, she graduated with an A.B. in 1868. She then returned to Waltham to be with her mother, her father having died suddenly the preceding year, at about the same time as her older brother was drowned in a shipwreck.

Mary nevertheless continued her studies, and was able to accompany Mitchell and a student group to Burlington, Iowa, to observe the solar eclipse of 1869. Further, having received special permission, following a request by Mitchell, she attended Harvard mathematician Benjamin Peirce's lectures on quaternions in 1869–1870. The following year she listened to his advanced lectures on celestial mechanics as one of only three students found qualified to take the course. After some practical work at the Dearborn Observatory in Chicago, where she spent several months, she was awarded an A.M. degree by Vassar (1872). Two years later she and her family, including her sister who had just completed her undergraduate course at Vassar, went to Zurich. While her sister attended the University of Zurich's medical school, Mary studied mathematics and celestial mechanics, at the same time greatly enjoying the stimulation of the international student community.

The years immediately following her return from Switzerland in 1876 were a period of considerable frustration for her; she found no satisfying way of using her education and training. A teaching position in Waltham high school absorbed only some of her time and energy. So she began an extensive study of the grasses in the area and was also active in the Vassar Alumnae Society. In 1881, when Maria Mitchell's health began to fail, Mary Whitney returned to Vassar as Mitchell's private assistant, sharing the teaching and keeping up the routine work of making observations for the time service provided by the observatory. Her position was a somewhat difficult one, since, being formally only a graduate student, she was unable to start any of the research on which she was eager to begin. In 1887 she was granted leave to get experience in practical work at Harvard in Edward Pickering's laboratory, but within a few months was recalled to Vassar to replace Mitchell, who had decided she must resign.

Whitney became professor of astronomy in 1889, when she was forty-one, very much at the height of her powers and eager to take up what she regarded as the challenge of demonstrating that women were capable of carrying out sustained research of a mathematical nature. She was elected secretary of the Vassar faculty the same year, and received that honor every year she was at the college. Despite a heavy teaching load she lost little time in making herself familiar with the twelve-inch equatorial telescope, by then in working order, and began observations. Her first big project was a collaborative effort, carried out with the Harvard observatory and Mary Byrd of Smith College, to determine the longitude of the new Smith College Observatory. She also began work on double stars and comets, which led to several publications in the early 1890s. About this time, however, both her mother and her sister became seriously ill, and so she brought them to live with her at the college. Much taken up with their care, she greatly reduced her research effort for several years, and only in 1894, after the deaths of both, did she return fully to astronomical work. At that time, having just enough private financial resources at her disposal to hire an assistant, she engaged Caroline Furness, an 1891 Vassar graduate.

In the decade between 1891 and 1901 Whitney published about forty-six articles and observation reports, many in the *Astronomical Journal;* seventeen from the period of 1895 to 1901 were coauthored by Furness. Among these the most important contributions to the field were probably the comet observations carried out in the mid 1890s. A cooperative link with Columbia University was established about 1896, an important step, since it brought Vassar into closer contact with the national astronomical community. The immediate joint project undertaken was the measurement and reduction of a collection of stellar photographs (the Rutherfurd photographs) held by Harold Jacoby of Columbia. Having limited funds at his disposal for getting the work completed, Jacoby proposed that a section of it be carried out by Furness as doctoral research.[5] Furthermore, having a need for computers and finding that young men were not very eager to take the work, he opened the way for Vassar graduates to join the Columbia computing group.

For some time after the turn of the century the Vassar research effort continued to focus on observations of minor planets and comets, but by 1904 Whitney had begun to concentrate on the examination of variable stars. Between 1901 and 1911 she published about twenty-six papers, eight of them coauthored by Furness.[6] In 1901 she and Furness toured European observatories, visiting several in Germany and Austria, and talking with astronomers Dorothea Klumpke in Paris and Lady Margaret and Sir William Huggins in London.

She was known as a competent investigator as well as an outstanding teacher who trained a notable number of future women astronomers. At first all her courses were mathematics-based as Maria Mitchell's had been, but with the growth of the college came a need to provide a class in descriptive astronomy as well. The one she offered became an especially popular elective. She was a founding member of the American Astronomical Society (1899), a fellow of the AAAS, and, as a strong promoter of education for women, a member of the Association for the Advancement of Women. In 1907 she became president of the Maria Mitchell Association. She was also an enthusiastic observer of bird life around the Vassar campus and led student Sunday-morning bird walks. Illness followed by partial paralysis caused her to retire in 1910. She died in Waltham, on 20 January 1921, at age seventy-three.

CAROLINE FURNESS[7] (1869–1936), Mary Whitney's successor as professor of astronomy and director of the Vassar College Observatory, and the second in the line of academic descendants of Maria Mitchell, was born in Cleveland, Ohio, on 24 June 1869. Her parents, Caroline Sarah (Baker) and Henry Benjamin Furness, were of New England background. Henry Furness was a high school science teacher and he encouraged his daughter's early interests in plant and bird life. She spent much of her girlhood in Cincinnati, went to Vassar in 1887 as a scholarship student, and studied astronomy under Mary Whitney. After graduating in 1891 she followed her father's wishes and took up high school teaching, first in West Winfield, Connecticut, and then in Columbus, Ohio (where she took courses in mathematics at Ohio State University).

However, her real interest was further work in astronomy, and after three years she returned to Vassar as Mary Whitney's private assistant. With Whitney she began the highly productive program of comet and minor planet observations, which was to last for almost a decade. She spent her summers doing graduate work at various institutions, including MIT and the University of Chicago's Yerkes Observatory. In 1896, under the direction of Harold Jacoby of Columbia, she began work on the measurement and reduction of a collection of the Rutherfurd stellar photographs. The initial part of the research, presented to Columbia as a Ph.D. dissertation, resulted in the first of the regular publications issued by Vassar College Observatory, "Catalogue of stars within one degree of the North Pole, and the optical distortion of the Helsingfors astrophotographic telescope, deduced from photographic measures" (1900). She was appointed instructor at Vassar in 1903. Over the next few years she completed the measurement and reduction of all the Rutherfurd plates she had been assigned. The work was brought out in 1905 (with the aid of a Carnegie Institution grant) as publication number 2 of Vassar College Observatory, "Catalogue of stars within two degrees of the North Pole." In 1908 she spent a semester working in the astrophysical laboratory at Groningen in The Netherlands, acquiring valuable practical experience. Thereafter her professional advancement proceeded steadily; she was promoted to associate professor in 1911, became director of the Vassar Observatory in 1915, and the following year, at the age of forty-seven, was appointed to the Alumnae Maria Mitchell Chair of Astronomy. Election to fellowship in the Royal Astronomical Society came in 1922.

Her independent study of variable stars began about 1909. A summer at the Harvard observatory in 1911 gave her the opportunity of using instruments not available at Vassar. The work resulted in a series of papers in American and German journals and her highly acclaimed textbook, *Introduction to the Study of Variable Stars* (1915). Among her other publications were reports of micrometric observations of minor planets and comets, observations of the solar eclipse of 1925 (which was total at the Vassar observatory), computations of comet orbits, and work in the history of astronomy.[8] She also edited *Observations of Variable Stars made during the Years 1901–1912 under the Direction of Mary W. Whitney* (1913), and brought out her textbook, *Manual for Practical Exercises in Astronomy* (1933).

During a sabbatical leave (1918–1919) she took a trip around the world, visiting many astronomers and observatories. One of her special interests was the status of women in other countries and their opportunities for education; her article, "Impressions of Japanese women," appeared in the *Vassar Quarterly* in 1920 and the paper, "Medical opportunities for women in Japan" in the *New York Medical Journal* in 1919. She was active in women's groups in New York, and for many years presided over the local branch of the National Alliance of Unitarian Women. She directed the Vassar Observatory until the summer of 1935, when illness overtook her. Shortly before then, however, she made a car trip across the United States, visiting observatories and national parks along the way, and spending some time at Mount Wilson Observatory in Pasadena, California. She died in a New York hospital, on 9 February 1936, at age sixty-six.

At Mount Holyoke Seminary, which had been established as a school for girls in 1837 and was reorganized as a college-level institution in the early 1890s, instruction in elementary astronomy actually predated the Vassar program. ELISABETH BARDWELL (1831–1899)[9], who had graduated from Mount Holyoke in 1866, taught the subject from then until her death thirty-three years later. Up until 1881 only a six-inch telescope was available to her, but in that year the school acquired an eight-inch equatorial instrument, a three-inch meridian circle, and a building to house them, the John Payson Williston Observatory. Bardwell was director of the observatory from its opening until 1896. In its early years the Mount Holyoke program benefited from the assistance of Charles Young, a noted astronomer and talented teacher who was at Dartmouth College in New Hampshire from 1866 until 1877, when he moved to Princeton. He lectured at Mount Holyoke every year from 1869 until 1908, and Bardwell spent a year at Dartmouth in 1873–1874. During the 1880s and 1890s she published four short observational notes in *Popular Astronomy* and the *Sidereal Messenger;* one, co-authored by her colleague Isabella Mack,[10] reported the solar eclipse of 1885 observed from the college observatory. She died on 28 May 1899, at age sixty-seven.

Bardwell's successor at Mount Holyoke was ANNE YOUNG[11] (1871–1961), the niece of astronomer Charles Young. Born in Bloomington, Wisconsin, on 2 January 1871, Anne was the daughter of the Rev. Albert Adams Young and his wife Mary (Sewell). She took a B.L. degree at Carlton College in Minnesota in 1892 and then taught mathematics for three years at Whitman College in Walla Walla, Washington, before returning to Carlton for graduate work in astronomy (M.S., 1897). She published her first four astronomical papers in 1897 and 1898, one, reporting the ephemeris of a comet, being co-authored by Herbert Wilson, professor of astronomy at Carlton. For a year she was principal of the high school in St. Charles, Illinois, and at the same time continued graduate studies for a semester at the University of Chicago. In 1899 she became director of the John Payson Williston Observatory and head of the astronomy department at Mount Holyoke. She had the title of professor from 1904, when for the first time Mount Holyoke's faculty were given academic rank.

A competent and enthusiastic observer, Young carried out extensive work on variable stars, exchanging data with Edward Pickering of the Harvard observatory. However, much of her early work, undertaken with the limited equipment available to her during her first years at Mount Holyoke, involved sunspot observations. She kept up this work throughout the thirty-seven years she was at the college, and from 1907 her daily observations were regularly forwarded to Zurich as part of a global cooperative project.[12] For several months in 1902 she carried out graduate research at the University of Chicago's Yerkes Observatory, and in 1905–1906 spent a year at Columbia University. Her doctoral dissertation, published in 1906, presented her analysis of the double stellar clusters in Perseus based on measurements made on Rutherfurd stellar photographic plates.[13] She was one of eight American astronomers (and the only woman) who in 1911 joined together to form the American Association of Variable Star Observers. From 1922 until 1924 she was president of this association.[14] With the acquisition of better measuring equipment in 1912 she began an extensive project on photographic photometry, using plates borrowed from Yerkes.

An outstanding teacher, she was especially interested in developing laboratory methods for teaching elementary astronomy and in promoting general interest in the field among the public. Her program of open nights at the Mount Holyoke Observatory and her regular monthly column on astronomy in the *Springfield Republican* were important parts of her wider teaching effort. She also revised and edited two textbooks on basic astronomy written by her uncle Charles Young, *Lessons in Astronomy, including Uranology: a Brief Introductory Course without Mathematics* (1918), and *The Elements of Astronomy: a Textbook* (1919). She was a member of the AAAS and of several British and American astronomical societies.

Anne Young was remembered as an attractive woman with a kind and gentle manner. She served on the Mount Holyoke faculty for thirty-seven years, retiring in 1936 when she was sixty-five. Thereafter she lived with her sister Elizabeth in a settlement for elderly relatives of missionaries in Claremont, California, where she died on 15 August 1961, at age ninety.

MARY BYRD[15] (1849–1934), director of the Smith College Observatory for nineteen years, was, like Young, a dedicated teacher who put much effort into pioneering methods of teaching elementary astronomy as a laboratory science. She was born in LeRoy, Michigan, on 15 November 1849, the second of five children of the Rev. John Huntington Byrd, a Congregational minister, and his wife Elizabeth Adelaide (Lowe), both of whom were descendants of families prominent in church and government in early New England. Mary's early childhood was spent in Ohio, but in 1855, when she was six years old, her father, a staunch abolitionist, moved his family west to Kansas Territory. Two years later he was farming on the federal government military reservation of Fort Leavenworth, where his family lived in a section of the barracks accommodation provided.

Mary's first opportunity for formal schooling came when she was ten, and within four years she was helping to teach younger children. After attending Leavenworth high school she spent a year and a half at Oberlin College in Ohio and several years as a schoolteacher, between intervals of returning home to assist her family through periods of illness and other difficulties. Entering the University of Michigan as a third-year student, she graduated (A.B.) in 1878 at the age of twenty-nine. Then followed three years as principal of the high school in Wabash, Indiana, and a year of work at Harvard Observatory, after which, in 1883, she went to Carlton College, Minnesota, as an assistant in the Goodsell Observatory and teacher of mathematics and astronomy. She also had the responsibility, under William Wallace Payne, for sending out by wire from Goodsell twice-daily time signals across Minnesota and a wide region farther to the west. This time service, begun by Payne in 1877 in response to demand by railway companies for accurate time measurements, was by the early 1880s one of the largest and most useful of such services in the United States.

In 1887, when Smith College opened a new observatory, she became its first director. Although she carried a heavy teaching load and had less-than-adequate instruments available to her, she was able to complete several series of investigations. Her earliest was the determination of the exact latitude and longitude of the new observatory, a project carried out with the cooperation of both the Harvard observatory and Mary Whitney at Vassar. The results appeared as a joint publication with Whitney in the *Annals* of the Harvard observatory in 1893; she later presented the work as a Ph.D. dissertation (entitled, "Determination of the longitude of the Smith College Observatory") to Carlton College; her degree was awarded in 1904.

Her special interest, and probably her most valuable contribution to astronomical research, was the fixing of comet positions. She published her first paper on this work in the *Astronomical Journal* in 1895, and followed it with five others in the next decade.[16]

She had strong opinions on the superiority of the laboratory method for teaching astronomy, which would seem to have come in part from her work of mounting, testing, and adjusting the instruments at the Smith observatory. The teaching methods she developed resulted in the writing of many articles and two textbooks; her *Laboratory Manual in Astronomy* appeared in 1899.

In 1906, when she was fifty-seven, and probably near the height of her academic career, she resigned her position in protest over Smith's acceptance of money from the Carnegie and Rockefeller Foundations, a move that she saw as leading inevitably to loss of freedom for the college and the beginning of an imposition of conservative capitalist values on what she had hoped would be an intellectually free academic institution. She returned to the farm in the gently rolling valley of the Wakarusa river near Lawrence, Kansas, which her family had bought in the 1870s and where her mother and two sisters, Abbie and Alice, still worked.

The challenges of teaching elementary astronomy remained an absorbing interest, however, and over the next nine years she completed her series of seven articles on "Astronomy in the high school," begun in 1903. Her second textbook, *First Observations in Astronomy,* appeared in 1913. For a year in 1913–1914 she taught astronomy at the New York Normal School (later Hunter College). Her reflections on the special problems of teaching the subject in the city and on the value of astronomy in the curriculum of normal schools appeared in two papers in 1913 and 1915.[17] By 1917, from her home in Kansas, she was directing a correspondence course for amateurs, organized by the Society for Practical Astronomy, and until about 1921, when she was in her early seventies, she continued to publish suggestions and lesson plans for the study of astronomy at beginning levels.[18] She died at her farm home on 13 July 1934, at age eighty-four, after a two-year illness. A short obituary in a Lawrence newspaper said little of her career, merely recording the fact that for twenty years she had been "director of astronomy at Smith College."[19] A collection of her astronomy journals went to the University of Kansas.

Although Wellesley's Whitin Observatory was not opened until 1900, astronomy was first taught at the college in 1880 by physics professor SARAH WHITING[20] (1847–1927), who joined the faculty when the college opened in 1876. Born on 23 August 1847 in Wyoming, New York, she was the daughter of Elizabeth (Comstock) and Joel Whiting. Whiting, a direct descendant in the eighth generation of Elder Brewster of the *Mayflower,* was a science teacher and principal in a succession of academies in New York State. He taught Sarah classics, mathematics, and physics; while she was still a child she often helped him set up demonstrations for his classes in the latter subject.

After taking a B.A. degree in 1865, at the age of eighteen, at Ingham University in LeRoy, New York, Sarah taught classics and mathematics for nine years, first at Ingham and then at Brooklyn Heights Seminary. Her acceptance of an invitation from Henry Durant, Wellesley's founder, to become professor of physics at Wellesley in 1876 entailed a substantial amount of further training. The subject was to be taught as a laboratory course, although at that time such a treatment had not been attempted at any American college except MIT. Accordingly, for the first two years of her appointment, an arrangement was made for her to attend the MIT undergraduate laboratory classes pioneered by Edward Pickering, then physics professor at the institute. In addition, to familiarize herself with instruments and procedures, she went the rounds of physics research laboratories in several of the major eastern colleges, an experience that, though educational, she found nerve-racking; the presence of a woman in such places was most unusual at the time. The teaching laboratory she opened at Wellesley in 1878 was the second of its kind in the country; it was well-equipped with good apparatus bought by Durant.

In 1879 she was invited by Pickering, by then director of the Harvard observatory, to observe techniques of physics applied to astronomy, particularly the investigation of solar spectra; the following year she introduced a course in elementary astronomy at Wellesley. Having no equipment except a celestial globe and a portable four-inch telescope, she had to be innovative in her teaching methods, and even devised daytime laboratory exercises using photographs and catalogs of celestial objects. Her textbook *Daytime and Evening Exercises in Astronomy* was published in 1912.[21]

She visited Europe frequently, and used her sabbatical leaves (in Berlin in 1889 and in Edinburgh in 1896–1897) to keep up with the latest developments in physics and astronomy. Edinburgh University having opened to women in 1892, she was able to enroll in classes taught by some of the outstanding physicists of the time, making her year there particularly profitable. Then and during later visits to Britain she often spent time with astronomers Sir William and Lady Margaret Huggins at their private observatory at Tulse Hill in London. Her warm friendship and long correspondence with Lady Huggins, who was much interested in Whiting's work in promoting women's education in astronomy, ultimately resulted in Wellesley College receiving many valuable books and a number of astronomical instruments as a bequest from Lady Huggins's estate.[22]

The opening in 1900 of the Whitin Observatory, built with funds provided by Wellesley trustee Mrs. John Whitin, made possible the beginning of a research program at the college. The new equipment included a twelve-inch refracting telescope with spectroscope and photometer attachment as well as a transit instrument. In addition, there was a spectroscopic laboratory, the whole making good training in stellar spectrum photography possible for Wellesley students.[23] Although herself a teacher rather than a research scientist, Whiting trained several women who went on to outstanding original work; one of these was Annie Jump Cannon (B.S., 1894, M.A., 1907, Wellesley), who, starting with an assistantship, eventually received one of the first appointments given a woman by the Harvard Corporation, becoming William Cranch Bond Astronomer in 1938.

Sarah Whiting became a fellow of the AAAS in 1883, one of the few women scientists of the time to be so honored; she joined the American Physical Society soon after its founding in 1899. In 1905 Tufts University gave her an honorary D.Sc. During her forty years on the Wellesley faculty she lived on the campus, sharing her home with a sister. After 1906 they were in Observatory House, built adjacent to the observatory. Of a strongly religious outlook, she supported the Wellesley College Christian Association and its missionary programs; she sought to inspire her students to become women of influence in their communities as well as scholars. She retired from the physics department in 1912 and from the directorship of the observatory in 1916, when she was sixty-nine. Her last years were spent with her sister and a former Wellesley colleague in Wilbraham, Massachusetts, where she occupied herself with gardening, presenting talks, and teaching Bible classes. She died in Wilbraham, on 12 September 1927, shortly after her eightieth birthday.

CHARLOTTE WILLARD[24] (1860–1930) carried out her research in astronomy at the Goodsell Observatory at Carlton College. Born in September 1860, she graduated from Smith College in 1883 and then taught for four years at a succession of seminaries and colleges before becoming instructor of mathematics and observatory assistant at Carlton. Her first publications, a series that appeared in the *Sidereal Messenger* in 1890 and 1891, reported sunspot observations. Like Mary Byrd, one of her predecessors at Goodsell, she had the responsibility of sending out from the observatory the twice-daily time signals, work she discussed in a paper in *Popular Astronomy* in 1894. She also coedited the first volumes of this journal (1893–1895) with William Payne, who published it from the Goodsell Observatory. In 1895 she left Carlton to join the faculty of the American College for Girls in Turkey. With a two-year break from 1899 to 1901, she taught there at least until the First World War. Apart from a joint paper on sunspot observations that appeared as a publication of the Goodsell Observatory in 1901,[25] she does not appear to have contributed further to the astronomical literature. She died in October 1930, at age seventy.

The first of the large observatories to hire women to carry out routine calculations and make the measurements required in stellar spectrum photography was that at Harvard College. Edward Pickering, the very successful observatory director from 1877 until his death in 1919, was largely responsible for employing the women computers, some of whom began with no scientific training, although others had college degrees. Their willingness to work for lower salaries than men generally found acceptable was a major reason for Pickering's decision to hire them—he knew a bargain when he saw it.[26] But it was also the case that he found them to be patient and painstaking workers, despite the unvarying and laborious nature of observatory computing. In addition, he appears to have been glad to be able to offer women a chance to carry out scientific work, and, apart from the poor wages, he generally treated them reasonably and made an effort to see that their labors were to some degree recognized.

The first of the three Harvard women astronomers whose

publications are listed in the bibliography was ANNA WINLOCK[27] (1857–1904). Born in Cambridge, Massachusetts, on 15 September 1857, she was the elder daughter of Isabella (Lane) and Joseph Winlock. Her forbears on both sides had settled in Virginia in pre-revolutionary times, later moving west to Kentucky. Joseph Winlock was director of the Harvard observatory from 1866 until 1875, and so Anna had a certain amount of familiarity with astronomical work from childhood on; in 1869, when she was twelve, she accompanied her father on a solar eclipse expedition to Kentucky. She was educated in Cambridge schools and was especially good at mathematics and Greek, the latter an uncommon subject of study for girls at the time. Her talents in mathematics were quickly put to practical use; following the sudden death of her father shortly after her graduation from high school in 1875, she took a position at the observatory as one of its first female employees. Although she was considered for the directorship of the Smith College observatory when that opened twelve years later, the post went to Mary Byrd. Winlock remained at Harvard for her entire career.

Following the installation of a meridian circle in 1870, Harvard had joined a number of European observatories in a project of preparing a comprehensive star catalog that would list accurate places for most of the stars in the northern heavens from magnitudes 1 to 9. The part of the sky to be included in the catalog was divided, by circles parallel to the celestial equator, into fourteen zones. As assistant to astronomer William Rogers, Winlock worked on the Cambridge zone, which extended from the fiftieth to the fifty-fifth parallel of north declination. Her friend Mary Byrd later commented that it was perhaps fortunate that she had not realized when she started what a long and demanding job she was undertaking. More than 26,000 observations were made of 8,627 stars. The amount of computing required was enormous, and each calculation had to be completely accurate and dependable. The Cambridge zone became part of the catalog of the Astronomische Gesellschaft, which included more than 100,000 stars and for a long time provided fundamental data for precision work in astronomy. Winlock began working on the project while she was still a schoolgirl; before it was finished, twenty years later, she had developed "no small power as a mathematical astronomer."[28]

Besides the Cambridge zone work she collaborated in a number of other observatory projects, including the preparation of a table of positions of variable stars in clusters, computations of asteroid paths, and an exhaustive study of stars close to the north and south poles. The latter, published in volume 18 of the *Annals* of the Harvard College Observatory, at the time constituted the most complete catalog of stars near the poles yet made. Most of the rest of her research was likewise included in *Annals* reports rather than being published as journal articles under her own name. She died in Cambridge, Massachusetts, on 4 January 1904, at age forty-six, after a short illness.

WILLIAMINA FLEMING[29] (1857–1911), who began doing clerical work and simple computing at the Harvard observatory in 1881, remained there for three decades, rising to a position of considerable responsibility and becoming one of the best-known women astronomers of her time.

The daughter of Mary (Walker) and Robert Stevens, she was born in Dundee, Scotland, on 15 May 1857. Her father was a skilled craftsman, a wood-carver and gilder who made picture frames and worked in gold leaf. He also experimented with photography and is said to have introduced the daguerreotype process to Dundee. On her mother's side she was a descendant of the seventeenth-century Jacobite soldier John Graham of Claverhouse, Viscount Dundee ("Bonnie Dundee"), famous in Scottish song and story. A few generations later, her great-grandfather, Captain Walker of the Seventy-ninth Highlanders, served under Sir John Moore in the Peninsular War and died at the battle of Coruña. Williamina received her only formal education in Dundee public schools. After her father died, when she was fourteen, she worked as a student teacher, helping to support her family. In 1877 she married James Orr Fleming and the following year emigrated with him to Boston. They separated in 1879, however, and she took a job as a maid to the family of Edward Pickering. She returned briefly to Scotland for the birth of her son in the autumn of 1879, but then rejoined the Pickering household; within two years she had been given a position on the observatory staff.

Pickering had recently embarked on a large-scale program of classification of the stars according to their spectral lines, that is, the line pattern obtained by dispersing their light by passing it through an objective prism. Some of the earliest successful photographs of stellar spectra had been made about 1872 by Henry Draper, a wealthy New York physician, photographer, and amateur astronomer; by 1880 the combination of photography and spectroscopy was revolutionizing the field by making possible the production of an unprecedented amount of new data. After Draper's death in 1882 his widow made a substantial contribution to the Harvard observatory in his memory; this Henry Draper Memorial largely funded Pickering's spectral photography program. The work called for large numbers of capable assistants for the labor-intensive examination of thousands of photographic plates and measurement of the spectral lines they contained. Efficiency in the huge undertaking demanded careful organization and routine division of labor as in a factory. Having found that women functioned satisfactorily as computers, Pickering employed a considerable number from 1881 on.

By 1886 (when she was still less than thirty) Fleming had acquired considerable experience and had taken on major responsibilities in the detailed management of the stellar classification program. An extremely capable administrator, she had by then overall responsibility for the examination, classification, indexing, care, and storage of the Harvard photographic plate collection. She also did a considerable amount of the editorial work on the observatory publications, including the *Annals,* and in addition to all that supervised the women's work and interviewed new women applicants.

Her earliest research was published in the observatory *Annals* under Pickering's name only, but her key role in the preparation of the important *Draper Catalogue of Stellar Spectra*

(*Annals,* 1890, volumes 26 and 27) was duly noted in the introduction. Indeed, the *Draper Catalogue* was in large part Fleming's work. Her contributions to it were nothing short of amazing. In the course of its preparation she examined (in much greater detail than anything previously attempted) 633 photographic plates containing spectra of about 100 stars each; she measured the brightness and wavelengths of lines of 28,266 spectra, and selected over 10,000 of these spectra with sufficient brightness and clarity to permit classification into categories according to the complexity of lines and bands. From this work she concluded that the classification system used by Pickering until then was inadequate for the description of the star classes she could distinguish. The "Pickering-Fleming System" of classification she went on to devise contained fifteen categories, designated by using most of the letters from A to Q. These categories were differentiated largely by the intensity of hydrogen lines, a generalization of obvious significance, but of the kind that is only too easy to miss when thousands of observations are being dealt with. Containing spectral classes of 10,351 northern stars, Fleming's *Draper Catalogue* was for long the best available comprehensive comparative study of the broad features of stellar spectra. Indeed, in the century since it appeared it has remained a fundamental source for work on the evolution of stars and the structure of the universe.[30]

The examination of the vast numbers of photographs that the *Draper Catalogue* study and her later work entailed provided a great opportunity for discovery, one that Fleming and her staff were not slow to take up—with most remarkable results. Of particular importance were her discoveries of long-period variable stars, on which she published, in the decade of the 1890s alone, some forty articles under her own name in German and American journals. Her 113-page "Photographic study of variable stars" (*Annals,* **47,** pt. 2), published in 1912 after her death, listed 222 she had discovered, and for each of these was provided a selected list of comparison stars, on the basis of which magnitude (brightness variation) could be determined. Commenting on this study in 1912, Oxford astronomer H. H. Turner noted that many astronomers are proud of having discovered one variable and that "the discovery of 222 . . . is an achievement bordering on the marvellous."[31] In addition to her work on the variable stars, she also made a special study of stars with peculiar spectra. Of the 107 Wolf-Rayet stars known at the time of her death, ninety-four were discovered by her (characterized by their broad emission lines, those stars were later considered to be hot, massive bodies ejecting a shell of material); of the twenty-eight then known novae (stars that show a sudden outburst of radiation) she found ten. Her list, "Stars having peculiar spectra" (*Annals,* **56,** 1912) was a substantial contribution to the area. In 1910 alone she found twenty-one new variables, one star of the fourth type, two of the fifth type, four of the sixth, one star with a unique spectrum, and four gaseous nebulae (of 107 nebulae discovered at Harvard between 1888 and 1907 she found fifty-two).[32]

In 1898 Harvard Corporation appointed her curator of astronomical photographs, the first corporation appointment given to a woman. By 1910 nearly 200,000 photographic plates (from both the Harvard and the Arequipa, Peru, observatories) had been examined and cataloged under her supervision. Administrative, clerical, and a great deal of editorial work gradually encroached more and more into her research time, however. For thirty years she worked a nine-hour day, under considerable pressure, took no vacations, and throughout was paid "women's wages." In 1900 that amounted to $1,500 a year, an income on which she could barely run her household and support her son through his undergraduate years at MIT, and one that compared rather unfavorably with the $2,500 paid to some of the male assistants.[33] Nevertheless, she was totally committed to the success of the stellar photographic program and devoted to Pickering.

She received many honors, including election, in 1906, to honorary membership in the Royal Astronomical Society, a distinction bestowed up to that time on only five other women (Mary Somerville, Caroline Herschel, Anne Sheepshanks, Margaret Huggins, and Agnes Clerke). Although she was passed over for the Bruce Medal (an important award administered by the Astronomical Society of the Pacific), the Sociedad Astronomica de Mexico awarded her its Gold Medal in 1911, along with honorary membership in the society. Likewise the Société Astronomique de France gave her an honorary membership. An active and enthusiastic founding member of the Astronomical and Astrophysical Society of America, and the only honorary fellow of Wellesley College, she was starred in the 1906 edition of *American Men of Science.* Altogether these were remarkable achievements for a nineteenth-century woman whose formal education was over by the time she was fourteen, and who began her independent life as a single-parent maidservant. She was indeed a worthy descendant of Bonnie Dundee!

An excellent seamstress and cook and a fine hostess, Williamina was a handsome woman with a friendly, sparkling manner, although she also knew how to keep strict discipline among her assistants. One of her special enjoyments was a Harvard-Yale football game. She died of pneumonia, in Boston, on 21 May 1911, at age fifty-four, after some years of poor health, very possibly brought on by overwork. She left enough unpublished material to fill several volumes of the *Annals.* Annie Jump Cannon succeeded her to the curatorship of the Harvard astronomical photographs, although not immediately to a corporation appointment. Her son, Edward Pickering Fleming, who graduated from MIT in 1901 as a mining engineer and metallurgist, spent some time in Chile as chief metallurgist for a large copper company; later he worked in Utah.

ANTONIA MAURY[34] (1866–1952), the second of the outstanding women astromoners at the Harvard observatory and one of the most talented and original of Pickering's early women assistants, joined the staff in 1888. Born in Cold Spring-on-Hudson, New York, on 21 March 1866, she was the elder daughter of the Rev. Mytton Maury and his wife Virginia (Draper). The family was one of distinction. Antonia's mother, a sister of Henry Draper of the Harvard Draper Memorial, was a daughter of the historian and pioneer physicist, John William Draper. Two generations further back her maternal great-grandmother, Carlotta Joaquina de Paiva Pereira, a Portuguese noblewoman, had served at the court of Emperor Dom Pedro I of Brazil. Her father's people were French Huguenots, the families de la Fontaine and Maury, who settled in Virginia about the beginning of the eighteenth century; her cousin Matthew Fontaine Maury (1806–73), was well-known for his work in geography and hydrography.

Along with her brother and her younger sister Carlotta (see chapter 12) Antonia was educated at home until in her teens. Much of the instruction was provided by her father, an Episcopalian minister with strong interests in geography, who edited several of Matthew Maury's geographical treatises. She went on to study at Vassar, where she was much influenced by Maria Mitchell, and graduated with a B.A. in 1887, having earned honors in astronomy and physics. A year later she joined Mrs. Fleming's group, despite Pickering's hesitation about employing a Vassar graduate for routine computing and stellar classification at a pay of twenty-five cents an hour.

Her first work was the determination of the period of the spectroscopic binary star, Zeta Ursae Majoris, the first of its kind, which Pickering had discovered in 1887. Consisting of two very close stars whose spectral lines shift as they orbit each other, a binary is identifiable only through spectroscopic analysis. With Maury's help Pickering correctly interpreted the spectrum. In 1889 Maury herself discovered the second such binary, Beta Aurigae, and determined its period. Binaries were to remain a subject of continuing interest to her throughout her life.

Her major responsibility when she joined the Harvard group, however, was the preparation of a catalog of selected bright northern stars, whose spectra were to be examined in even more detail than in Fleming's much larger-scale project. A more powerful 11-inch instrument, the Draper telescope, fitted with additional prisms, was used to take the spectral photographs, and when these were enlarged and studied under the microscope they showed an unprecedented complexity of structure. Maury decided that the single spectral sequence of fifteen groups devised by Fleming was insufficient to describe the details of the spectra she was observing, and so she replaced it with a system of twenty-two groups (representing a true sequence based on descending temperature), each characterized by a particular collection of spectral lines. Within each of these groups, however, stars having the same line pattern and color might display differences in line width and intensity, and Maury accordingly introduced three further subdivisions that took these features into account. She emphasized particularly her subdivision "c", in which hydrogen and helium lines were narrow and sharply defined and calcium lines especially intense. The stars that gave rise to these spectra had, she considered, some special intrinsic property.

A meticulous and painstaking worker, gifted with considerable scientific insight, Maury inevitably made slow progress. Pickering, who strongly felt the need of keeping to a definite schedule, was less than enthusiastic about the special effort needed to make reliable width and sharpness measurements. Further, he disagreed with her about the necessity for the more complex system. The emotional stress caused by Pickering's impatience (and perhaps her own boredom with the tedious nature of the work) slowed her down even more; by 1892 she had had enough and left the observatory without completing the study. For the next four years, until her catalog was eventually completed, she worked only occasionally on the project, though continually urged by Pickering to do so. Nevertheless, she also was eager to finish the work, in part for her own reputation and in part in honor of her uncle, Henry Draper, whose legacy had funded the project. Her catalog, finally published in the observatory *Annals* for 1897, was based on an examination of 4,800 photographs. It presented detailed analyses of 681 bright northern stars, classified according to the system she had devised. That issue of the *Annals* was the first to have a woman's name on the title page—one of her conditions for finishing the work.

A study of the spectra of 1,122 bright southern stars, carried out by Annie Jump Cannon and published three years later, completed this section of the Henry Draper Memorial project. Although retaining the essential features of Maury's basic twenty-two-group classification sequence, Cannon dropped the sharpness-of-line criteria and her resulting simplified system was so convenient to use that it was adopted in 1910 by the International Solar Union as the official classification system. Maury's more subtle system seemed about to be forgotten. In 1905, however, the Danish astrophysicist Ejnar Hertzsprung, while studying the problem of the real luminosities of stars, had discovered that among red stars some were very bright (giants) and some very faint (dwarfs); he concluded that these two groups (giants and dwarfs) should show differences in their spectra. In all the catalogs published until then, only Maury's showed the distinction he was searching for. All those stars in her subdivision "c" (those whose spectra were characterized by narrow and sharply defined hydrogen and helium lines plus intense calcium lines) were Hertzsprung's bright giants.

Maury's work was later seen as one of the important advances in observational astronomy that fed directly into the early development of theories of star evolution. The giant red stars, according to later theory, are at a stage in their evolution at which all produce spectra with the sharp, narrow lines she had distinguished in her catalog. Although Pickering continued to feel that the Harvard spectra were of inadequate quality to justify using them to make the distinctions Hertzsprung was claiming, Maury's contribution to spectra analysis was fully recognized in 1922 when the International Astronomical Union modified its official classification system (the Cannon system) to include the prefixing of the letter c to a spectral type to indicate narrow, sharply defined lines. In 1943 her work was acknowledged by her own countrymen when the American Astronomical Society gave her the Annie Jump Cannon award for this early contribution.

Her analyses of spectroscopic binary stars and calculations of their periods of revolution continued intermittently over the years, although she stayed away from Harvard for more than a decade after the publication of her catalog. In an effort to understand the behavior of the component stars of the very complex binary Beta Lyrae, she carried out a detailed investigation of some 300 spectra. Her treatise on the work, her other major contribution to the field, appeared in the observatory *Annals* for 1933.[35] One of her successors at the Harvard observatory, Cecilia Payne-Gaposchkin, is said to have remarked that Maury had a flair for identifying tough problems.[36]

Throughout the 1890s and into the next decade she held teaching positions, first at the Gilman School in Cambridge, Massachusetts (1891–1894), and then at Miss Mason's school in Tarrytown, New York. She also lectured on astronomy to both professional and popular audiences, giving four talks at Cornell in 1899 and others in New York City and elsewhere. In 1908 Pickering gave her an excellent recommendation for an adjunct professorship in physics and astronomy,[37] but she chose instead to return to her work on spectroscopic binaries and remained at or near Cambridge for the rest of her life. Most of her later work was carried out as a research associate at the Harvard observatory between 1918 and her semiretirement in 1935. Pickering died in 1919 and Maury got on well with Harlow Shapley, observatory director from 1921 until 1954. At some time during the 1920s she had taken charge of the Henry Draper Memorial Museum (originally the Draper Observatory) at Hastings-on-Hudson, and after 1935 spent much of her time there as curator. Until her official retirement from the Harvard observatory in 1948 she made annual visits to examine current spectra of Beta Lyrae.

Although somewhat dour in personality, Maury was a woman of broad scientific interests, a good all-round naturalist, an enthusiastic ornithologist, and a conservationist who fought for the preservation of the western redwoods when they were endangered by wartime lumber needs. She died on 8 January 1952, in a hospital in Dobbs Ferry, New York, in her eighty-sixth year.

Several women astronomical computers working at observatories other than Harvard published reports in the astronomical journals before 1901. One of the more productive was ALICE LAMB,[38] of Madison, Wisconsin, who studied at the University of Wisconsin in the early 1880s (B.L., M.L., 1885). While working as an assistant at the university's Washburn Observatory in Madison from 1885 until 1887, she had responsibility for the time service and also collaborated in the regular research program. Much of her work, including observations of minor planets and binary stars, was reported in the *Publications of the Washburn Observatory* in the late 1880s; seven papers are listed in the bibliography. In 1887 she married Milton Updegraff, a fellow graduate of the University of Wisconsin, who became astronomer at the Washburn Observatory in 1886. While with him in Cordova, Argentina, where he served as second astronomer at the Argentine National Observatory from 1887 to 1890, Alice had responsibility for the observatory's time services; she also collaborated in observations and reductions of southern stars. After their return to the United States, Milton Updegraff was professor of astronomy at the University of Missouri, Columbia, for nine years. They moved to the Naval Observatory in Washington, D.C., when he became professor of mathematics in the U.S. Navy, and subsequently lived at a succession of locations according to his navy assignments. Alice does not appear to have published any work in the astronomical journals after 1890. She had three daughters.

MARGARETTA PALMER[39] (1862–1924), whose doctoral dissertation research was published in the first volume of the *Transactions* of the Yale University observatory, worked at Yale for thirty years. Born in Branford, Connecticut, she studied at Vassar where she was introduced to astronomy by Maria Mitchell. After graduating with an A.B. in 1887, she stayed on for two years as an instructor and assistant. She then went to Yale to continue her studies under astronomer William Lewis Elkin, and enrolled in the graduate school when it first opened in 1892. She received a Ph.D. in mathematics in 1894, following completion of her redetermination of the orbit of the comet discovered by Maria Mitchell forty-seven years previously. Hers was the first doctorate awarded to a woman in the United States for a dissertation on an astronomical topic. Remaining at the observatory, she became a research assistant in 1912 and was known particularly for her comet orbit calculations. She also compiled an extensive catalog of references to positions that had been determined for the stars in the *Bonner Durchmusterung.* This compilation remained unpublished, being superceded by the *Geschichte des Fixsternhimmels* (Deutsche Academie der Wissenschaften, 1922–1952), but Palmer made references freely available on request. She died suddenly at her home in New Haven, Connecticut, on 30 January 1924.

Gertrude Wentworth[40] graduated from Boston University and later collaborated with Seth Chandler, an amateur astronomer who was an expert on calculating the orbits of planets. Several of her calculations appeared in 1893 in the *Astronomical Journal,* which Chandler edited.

ELIZABETH BROWN[41] worked for twenty years as a computer in the Nautical Almanac Office in Washington, D.C., calculating the ephemeris of the sun and other tables for the almanac. Born in Front Royal, Virginia, she was the daughter Major Victor M. Brown and his wife Mary (Jacobs). She received a B.S. degree from Columbian (later George Washington) University in Washington, D.C., and then proceeded, with the special permission of the faculty and largely with the help of Simon Newcomb, to graduate studies in mathematics at Johns Hopkins University. In 1888 she married Arthur Powell Davis, chief engineer of the U.S. Reclamation Service. They had four daughters. In addition to her work on the Nautical Almanac she assisted Newcomb in the preparation of tables of sun and planet positions. Her two articles listed in the bibliography (1888 and 1892) reported calculations of comet orbits.

Both Flora Harpham and Mary Tarbox were assistants at Columbia University, although the publications by them listed in the bibliography report earlier student work. FLORA HARPHAM (1860–1925),[42] who was from Minneapolis, entered Carlton College as a second-year student in 1886, at the age of twenty-five, and graduated with a B.A. in 1888. Continuing her studies at Carlton in 1891–1892 (M.A., 1892), she collaborated with Herbert Couper Wilson, director of the Goodsell Observatory, on computations of the elements of comets. After a short time at the University of Chicago she spent three years as assistant to Mary Byrd at Smith College, where she taught astronomy and mechanics and worked with Byrd on the adjustment of the observatory's new twelve-inch equatorial instrument. Her two short papers published in *Popular Astronomy* in the mid 90s reported studies done at Smith.

In 1896, on the recommendation of Mary Whitney of Vassar, she was hired by Harold Jacoby as a computer at Columbia's observatory. She remained there for twelve years, and had charge (as chief computer) of the star cluster section of the Rutherfurd photographs project until the work was completed. She also prepared the results for publication. As contributions from the observatory of Columbia University, the work appeared under Jacoby's name although introductions stated that the calculations were made by Harpham, generally with the assistance of Mary Tarbox, Eudora Magill (who had previously studied at Swarthmore College), and Helen Lee Davis.[43] The results from one part of the project, "Catalog of the stars about 27 Cygni as deduced from the Rutherfurd photographic measures," Harpham submitted to Carlton College as a doctoral dissertation. Her degree, awarded in 1908, was the last of the six Ph.D.s (all in astronomy) that the college ever awarded. She resigned from her position at Columbia in 1908 (she was then about forty-eight). She died on 10 February 1925, at age sixty-four.

The studies on minor planets by MARY TARBOX,[44] carried out at Vassar College under the guidance of Mary Whitney, appeared in the *Astronomical Journal* in 1898. Born in Pomfret, New York, she graduated from Vassar in 1896 and stayed on for a further year as a graduate scholar. In 1898 she joined Harold Jacoby's computing staff at Columbia where for four years she assisted Flora Harpham in the measurement and reduction of Rutherfurd photographs.[45] From 1902 until her marriage in 1905 to Le Roy Nathan Babbitt, she taught in Kemper Hall, Kenosha, Wisconsin. She later lived in Dobbs Ferry, New York.

MARY WAGNER[46] was another student of Whitney's. Their joint papers on observations of a comet appeared in the *Astronomical Journal* in 1893 and 1894. Wagner had been a schoolteacher in Minneapolis for five years before going to Vassar in about 1891. Following two years of study with Whitney, she applied to Edward Pickering for a job at the Harvard observatory. The only position available at the time involved part clerical work and part routine computing at a salary of $500 a year, which was little more than half of what she might have earned as a high school teacher. Very much wanting the experience of working in the observatory, she took the job, but within three months had to return to Minneapolis to care for her seriously ill mother. Although she had little liking for the monotony of computing and could barely live on the pay, she would probably have returned to Harvard had an opportunity arisen. In its absence she enrolled at the University of Minnesota to study zoology (B.A., 1897). She managed to continue some astronomical work as well; a paper by her in the *Astronomical Journal* for 1898 reported observations made at the University of Minnesota observatory. She taught high school science and mathematics from 1897 until 1902, when she returned to Poughkeepsie (the location of Vassar College). However, having some years previously concluded that she could not afford the luxury of doing astronomy, she turned to something that would bring her a living and opened an inn. This she ran for the next twenty years.

Work by twelve more American women astronomers or students of astronomy is listed in the bibliography.[47] They include Adelaide Hobe of the Lick Observatory, the very productive Rose O'Halloran of San Francisco, Hepsa Silliman, and Ellinor Davidson.

ADELAIDE HOBE took a B.S. degree at the University of California in 1899, her joint note on the elements and ephemeris of a comet appearing the same year. Another paper by her, published jointly with Armin Otto Leuschner, director of the students' observatory, came out in 1902. She was an assistant at the Lick Observatory for more than twenty years, until 1922, when, promotion being refused her, she found another position.[48]

In the five years between 1895 and 1900 ROSE O'HALLORAN[49] brought out twenty-five papers and notes on a variety of topics including variable stars, meteors, sunspots, and eclipses, most appearing in the *Publications of the Astronomical Society of the Pacific* and *Popular Astronomy.* Her observational work continued after the turn of the century, with the publication of at least twenty-nine more articles in the period from 1900 to 1910, including regular contributions of notes on variable stars. She remained a member of the Astronomical Society of the Pacific until the 1920s, although her last paper was probably her report on Halley's comet observed from San Francisco and the Lick Observatory in 1910.[50]

Except for Maria Mitchell's first papers, the reports by Hepsa Silliman and Ellinor Davidson are the earliest listed in the bibliography by women contributors working in America. Hepsa Silliman was from New York but published her account of meteors, a worldwide survey that made use of both British and American listings and reports, in the *Edinburgh New Philosophical Journal* in 1862. This paper had been preceded by a short monograph on the same topic published in New York in 1859.[51] Mrs. Silliman is commemorated by the Silliman Foundation at Yale, established in 1883 by a gift of $80,000 from her children. The fund supports a memorial lecture series.

Ellinor Davidson was the wife of George Davidson, an English-born geographer and astronomer, who taught geodesy and astronomy at the University of California.[52] She accompanied him when he led the American expedition to Japan to study the transit of Venus in 1874; her paper in the *American Journal of Science* the following year described observations she made at Nagasaki. An earlier article, published jointly with her husband, reported observations of meteors made in 1869 at Santa Barbara, California.

The important stellar catalogs produced by the three early Harvard women astronomers Fleming, Maury, and Winlock undoubtedly constitute the major contributions by women to turn-of-the-century astronomical research in the United States. Although their work had its genesis in the grand investigational programs of observatory director Pickering and his male colleagues, American and European, it well deserves the recognition it is now beginning to receive.[53] Painstakingly accumulated compilations of fundamental astronomical data, the catalogs have remained of lasting value, constituting key sources in astronomical research. Maury, as noted, went somewhat further than Winlock and Fleming, persisting in presenting interpretations of her own that did not meet with Pickering's approval. Happily she lived to see her convictions vindicated. (If Pickering had embraced her ideas the major credit might well have gone to him.)

One of the notable points emerging from this overview is the key role of Pickering in the careers of so many of the early American women astronomers. Mitchell at Vassar and Bardwell at Mount Holyoke were both well established in their careers before Pickering became director of the Harvard observatory, but almost all of the other early women astronomers on the faculties of the eastern women's colleges at some point came under his influence, and benefited considerably from training and research experience at the Harvard observatory. Thus Mary Whitney spent a short period there in 1887 just before she succeeded Mitchell at Vassar, Caroline Furness kept close contact with Pickering throughout her work on variable stars, and Anne Young regularly exchanged information with him on the same subject. Before going on to teaching and further training at Carlton College, Mary Byrd spent a year at the Harvard observatory. Sarah Whiting of Wellesley had perhaps the closest association with Pickering, having not only spent some time at the Harvard observatory in 1879 but having acquired her introduction to laboratory physics as a special student of Pickering at MIT a few years earlier.

Of the early programs in astronomy at the women's colleges, that at Vassar has long been considered to have been especially outstanding. The Mitchell-Whitney-Furness succession in the early observatory directorships constituted a period of more than half a century of energetic work in teaching and observing. Vassar astronomy graduates were very well thought of, and many found jobs in observatories throughout the country around the turn of the century and in the years immediately after. The practical programs at Vassar and at the other women's colleges followed the pattern general throughout the rest of the country at the time, consisting mainly of observational work on comet positions, asteroids, sunspots, and the brightness of variable stars (Figure 10–2, a and b), all of which fed into the larger data compilation projects of the period. Anything more ambitious would have been surprising, considering the limited equipment available and the heavy teaching loads the women faculty carried.

In addition to Harvard and the women's colleges, one other East Coast institution, Columbia University, stands out as having had a notable role in the training and careers of early women astronomers. As was the case at Harvard, this came about because the observatory had a large number of stellar photographic plates that required measurement and reduction. Both Furness and Young were awarded Ph.D. degrees by Columbia (their dissertation research being analyses of sections of the Rutherfurd plate collection),[54] and by the mid 1890s women computers were being hired, Tarbox from Vassar and Harpham from Carlton College among them.

With the notable exception of Carlton College, institutions outside the northeast region of the country were not prominent in the training or employment of the early women astronomers whose work appears in the bibliography. A few women studied astronomy at the universities of Wisconsin, Colorado, and California, and the University of Chicago provided observatory experience, Whitney working at the Dearborn Observatory, and Furness, Young, and Harpham as volunteers at Yerkes. At Carlton, however, Byrd, Harpham, Willard, and Young all received graduate training and acquired experience as assistants; further, Byrd and Harpham were awarded Carlton Ph.D. degrees.

British women

The Royal Society *Catalogue* lists astronomy papers by ten women who either were British or spent most of their lives in Britain. Two of them, Proctor and Clerke, were primarily scientific writers and expositors, Proctor a popularizer of astronomy and Clerke a historian; a third, Brown, was a solar astronomer and meteorologist, and a fourth, Huggins, carried out work in astrophysics and stellar spectroscopy. Among the others, one was a mathematician (Chisholm); another, Everett, did much of her work in mathematical physics, although at the beginning of her career she spent several years at each of two major observatories—the Royal Observatory, Greenwich, and the Royal Astrophysical Observatory, Potsdam—where she carried out both observational and calculational work on a major star-mapping project, the *Astrographic Catalogue.* Little information is available about the last four—Ashley, Fry, Jones, and Lehmann—whose published work was in any case minor.[55] Probably only Chisholm and Everett had university-level educations; Brown, Clerke, Huggins, and Proctor were self-taught or learned from close relatives.

ELIZABETH BROWN[56] (1830–1899), a well-known figure among amateur British astronomers during the last two decades of the nineteenth century, was best remembered for her detailed and dependable reports of sunspots, prepared daily over the course of many years. The elder of two daughters of Jemimah and Thomas Crowther Brown, she was born on 6 August 1830 at Further Barton, Cirencester, a small market town on the eastern edge of the Cotswolds in Gloucestershire. The Brown family most likely moved into the county in the late eighteenth century when Elizabeth's grandfather established their wine business there. Prosperous and public spirited, Thomas Brown, throughout a long life, gave strong support to all town improvement schemes; he also had strong interests in the sciences, was an active amateur meteorologist, and a fellow of the Geological Society.

Elizabeth was taught at home by a governess and later studied on her own, persistently and widely, reading both literature and science and making herself a very proficient artist. Her mother died some time before she reached the age of eleven and she was very close to her father; he taught her how to watch the sky for meteorological information and how to use a small hand telescope that allowed her to distinguish Saturn's rings and Jupiter's satellites. From 1871 it was she who carried out the daily rainfall recordings for the Royal Meteorological Society, which her father had made for the preceding twenty-six years. For a time she also recorded temperature, deep-well measurements, and thunderstorm activity. Beginning in 1881 her notes on aurorae and other atmospheric effects appeared at intervals, mainly in *Nature.* She was one of the few early women fellows of the Royal Meteorological Society, being elected in 1893—an honor she greatly valued.

Soon after the death in 1883 of her father, whom she had taken care of in his old age, she became publicly active in astronomical work, reading a paper, "Observations of proper motion in certain sunspots," before the then very active and flourishing Liverpool Astronomical Society. This group, which welcomed the participation of women as members, asked her to become director of its Solar Section, with responsibilities for making sunspot observations herself and collecting data from other observers. Her reports appeared regularly in the society's publications over the next six years. Her own work was carried out in one of two observatories (the other housed meteorological instruments) built in the grounds of her home. She had a 3½-inch equatorially mounted refractor and a 6½-inch reflector.

About this time she made two long, somewhat arduous expeditions (with the charge of presenting reports to the Liverpool society) to observe solar eclipses. The first was to Russia in 1887 and the second to Trinidad in 1889. Only the latter was successful scientifically, but she wrote vivid accounts of both.

In Pursuit of a Shadow (1887) described her steamer and rail journey with a young woman friend, "L." (her cousin, a Miss Jeffreys), in stages across Scandinavia and Russia to Kineshma on the Volga, some 200 miles northeast of Moscow. There she and two other English astronomers, Ralph Copeland and Father Stephen Perry, representatives of the Royal Astronomical Society, were guests of Russian astronomer Fëdor Bredichin at his summer home—"a large, rambling building . . . a delightful, romantic house, like no other we had ever seen." However, the great efforts made to transport and set up telescopes, cameras, and other instruments came to naught, clouds obscuring the view at the crucial time. Her second book, *Caught in the Tropics* (1890), described her voyage by Royal Mail steamer *Tagus* to Trinidad, again with L., and their five-week stay on the island, whose colorful tropical life fascinated her. Thanks to an introduction to the acting British administrator, the Hon. Henry Fowler, she was able to install her telescope in a large-windowed tower in the village of Prince's Town, some forty miles south of Port of Spain by the island's small railway. As at Kineshma, weather caused difficulties, and only in the last seconds before the eclipse became total did the clouds clear, letting her witness the breathtaking view—"the silvery light of the corona encircling the death-like blackness of the moon's orb . . ." Even then there was an element of disappointment, however, there being no streamers or solar prominences, only an appearance of rose color on one side, and one beam of rosy light on the other.

About 1890 she joined enthusiastically in a move to organize a metropolitan-based amateur group which, like the Liverpool society, would accept women. From the founding later that year of the new association, the British Astronomical Association, she took a prominent part in its activities. Further, she was offered the directorship of the Solar Section, thus becoming, ex officio, a member of council, where she was joined by her friends Agnes Clerke and Margaret Huggins, also charter members of the association. From then on she gave most of her time to the work of the Solar Section, although she also assisted other sections and at various times made observations on the Moon and on variable and colored stars.

Her special interest remained the daily registering of sunspots, laborious, demanding work requiring considerable artistic skill; her drawings were highly regarded for their precision, detail, and accuracy. She was a regular contributor of papers, notes, and letters to the association's *Journal* until shortly before her death (see bibliography), and also wrote the seven Annual Reports of the Solar Section published in the association's *Memoirs* during the 1890s. Each of these presented a daily calendar of sunspots and faculae (bright spots), and a "Sun-spot Ledger" giving a life history of each spot. Important contributions, they assured her a place among the leading amateur astronomers of her time.

She held memberships in the Astronomical Society of the Pacific, the Astronomical Society of France, and the Astronomical Society of Wales. In 1892, along with two other women (Cambridge-educated Alice Everett and Annie Russell), she was proposed for fellowship in the Royal Astronomical Society, whose meetings she frequently attended. None of the three women received the required two-thirds of the votes cast, however.

Calm, gentle, and unassuming by nature, a member of the Society of Friends and regular attender of the small Cirencester meeting, her life in the country was quiet and simple. After the death of her father her only companion was her sister Jemima, although she had a large circle of friends. In general she avoided interruptions to her daily routine of observing, drawing, writing, and rainfall recording. She had an active correspondence with astronomers overseas as well as in Britain, however, and continued to enjoy travel. In addition to the two eclipse expeditions described, she made one more—to Vadsö in northern Norway in 1896, with a large British Astronomical Association party on the ss. *Norse King.* She also attended the Montreal meeting of the British Association in 1884 and afterward traveled extensively in Canada and the United States. Shorter expeditions took her to Scotland, Ireland, and Spain. She recorded landscapes and wildflowers, a special interest, in color sketches, and described her experiences in vivid letters to her sister.

Never very robust, she died suddenly, at the age of sixty-eight, at her home, on 5 March 1899, after a week of mild illness and in the midst of preparations for a fourth eclipse journey. She left her astronomical observatory and much of its contents, together with the then very considerable sum of £1,000, to the British Astronomical Association.

MARY PROCTOR[57] (1862–1944), although she did relatively little observational work, became well-known in the

1890s in both Britain and America for her popular lectures and writings on astronomy. Born in Dublin, in 1862, she was one of the six children of Mary (Mills) and Richard Anthony Proctor. Richard Proctor, the Cambridge-educated son of an English solicitor, was a mathematics teacher, amateur astronomer, and a fellow of the Royal Astronomical Society. His wife was Irish. About 1863, when the family moved to London, Proctor started out on the career that led to his becoming one of the best-known English-language authors in astronomy, then a subject of tremendous popular appeal. Mary was very close to her father (who had lost his oldest son in 1863) and indeed while still a child became his constant companion. Although she studied for a time at the College of Preceptors, most of her education was provided by her father; his stories of the legends and myths of the stars had fascinated her from her earliest years. She looked after his library and helped him with proofreading from the time she was fourteen.

In 1881, two years after her mother died, her father married an American widow, Mrs. Robert Crawley, from St. Joseph, Missouri, and the family moved to St. Joseph. About this time Mary began a serious study of writing under her father's guidance, and produced a series of articles on comparative mythology. After his death in 1887 she went on writing on mythology and astronomy, and contributed articles to *Science, Scientific American,* and *Knowledge,* the latter a London scientific weekly her father had founded in 1881 and thereafter edited. Her series on "Evenings with the Stars" appeared in *Popular Astronomy* in 1897. For her material she contacted the leading American astronomers of the time, including Edward Pickering at Harvard, who frequently provided information and pictures to illustrate her work. She regularly traveled to observe important celestial events, which she then reported in the astronomy journals. Her account of the 1896 eclipse of the sun viewed from off Stött island along the northern coast of Norway appeared in *Popular Astronomy* the following year. She observed the 1900 eclipse from Norfolk, Virginia, and that of 1905 from Burgos, Spain.[58]

In 1893, following in her father's footsteps, she came to prominence as a public lecturer. Engaged to give a series of six talks on astronomy at the World's Columbian Exposition in Chicago that year, she had expected to speak to children. When a mixed audience presented itself for her first lecture she gave an impromptu talk, which was received with great enthusiasm and given good press coverage in the major Chicago newspapers. From then on she was a popular speaker. As well as having a regular contract with the New York Board of Education, she lectured extensively throughout North America and in Britain. Astronomical topics were then extremely popular, not only at meetings of middle-class literary and scientific societies but also (in Britain) at talks given in working men's institutes.

Overall her most important contributions were probably her books. She was sometimes known as "the children's astronomer," and her books for young people were particularly successful. *Stories of Starland* (1898) was used as a supplementary reader in New York schools. *Giant Sun and his Family* appeared in 1906 and *Half Hours with the Summer Stars* in 1911. By about 1920 she had returned to England and was living in London. She continued to write, and in her sixties and seventies published at an impressive rate; her *Children's Book of the Heavens* and *Evenings with the Stars,* both of which first appeared in 1924, were especially popular.[59] She was elected a member of the AAAS in 1898 and held memberships in American, British, and Mexican astronomical societies.

AGNES CLERKE[60] (1842–1907), although she also considered herself a popularizer of astronomy, wrote for a more mature and informed audience. Like Mary Somerville, her predecessor in the area of astronomical description and scholarly commentary, she was especially successful in condensing current research findings, and her writings were used by specialists in the field. She was also popular with a wide reading public, and indeed her books continued to appear in secondhand bookstores a century after their first publication. Although Irish by birth, her literary and astronomical work was done largely after she had settled permanently in London, and her scientific contacts were for the most part with British and American workers.

The younger daughter among the three children of Catherine Mary and John William Clerke, she was born on 20 January 1842, in the small country town of Skibbereen in County Cork, on the Irish south coast. Her father, the son of the local doctor, was a classical scholar with a degree from Trinity College Dublin. He managed the Provincial Bank in Skibbereen but had wide interests in the sciences, especially chemistry and astronomy. Her mother was one of the seven children of Rickard Deasy, a wealthy brewer from the neighboring town of Clonakilty; educated in a good convent school in Cork, Mrs. Clerke was an accomplished musician, playing both the harp and the piano. Agnes and her sister were given a solid, basic education by their parents. The usual occupations for girls of those days, sketching and needlework, appear to have been replaced by serious study in mathematics and the classics; Agnes, a somewhat frail child, found her greatest enjoyment in music and books. The home environment was one in which scientific pursuits were an important part of life, and she soon developed a special interest in astronomy. Her enthusiasm for the subject was undoubtedly reinforced by the presence in the garden of her father's four-inch aperture telescope, which made possible an occasional glimpse of Saturn's rings or Jupiter's satellites. By the time she was eleven she was reading such works as Sir John Herschel's *Outlines of Astronomy.*

The family moved to Dublin in 1861 when John Clerke took the post of registrar at the court of his brother-in-law, the distinguished lawyer R. M. Deasy, just then appointed Baron of the Irish Exchequer. At about the same time, Agnes's brother Aubrey entered Trinity College Dublin, where he concentrated on mathematics and science, including astronomy.

Starting in 1867, because of Agnes's delicate health, she and her mother and sister began to spend their winters in Italy, and for about four years the two younger women lived there year-round. These were times of great social and political change for Italy, culminating in 1870 in the unification of most of the peninsula's smaller duchies and kingdoms into the modern state. The upheavals do not appear to have disrupted the

Clerkes' activities, however. They lived mainly in Florence, with summers in Bagni di Lucca on the western slopes of the northern Apennines. With their lively and wide-ranging interests they made good use of the extensive resources of the Florence city libraries, freely accessible to the public. Agnes carried out a special study of renaissance science and the work of Galileo. She also resumed her music lessons, begun earlier in Dublin, and became an excellent pianist with a large repertory. She particularly enjoyed playing Chopin.

Both sisters returned to Britain in 1877 and joined the rest of the family in London, where the parents had recently settled on the father's retirement. The same year also saw the appearance of Agnes's first published work, "Brigandage in Sicily," an account of the rise of the Mafia, and "Copernicus in Italy," a review discussion of pre-Copernican Italian astronomy. Both these papers came out in the literary quarterly, the *Edinburgh Review,* whose editor Henry Reeve, recognizing the quality of Agnes's writing, encouraged her to continue. Over the years Reeve and his wife became her close friends, and the *Edinburgh* published about fifty of her articles (two contributions a year), including her first astronomical writing, a review essay on "The chemistry of the stars," which came out in 1880.[61]

About this time she started her first major project, a work that appeared in 1885 under the title, *A Popular History of Astronomy during the Nineteenth Century.* It ran to almost 500 pages and was extremely influential, the first historical survey of the field to be published since Robert Grant's 1852 *History of Physical Astronomy.* To some extent it was an updating and continuation of Grant's work, which had covered the field up to the introduction of the spectroscope; but Clerke successfully concentrated on the newer areas of photographic and spectroscopic research. Although nonmathematical to make the material more accessible to ordinary readers, the work was broadly inclusive, notably complete in its references to original reports and had the added interest of many biographical sketches. Further, it was written with style and lucidity.

Reviews by practicing astronomers, while they might express regret about Clerke's obvious lack of experience in observing and about her giving space to old, speculative theories that are perhaps better consigned to science history's dustbin, by and large were very favorable. Robert Ball, writing in *Nature,* declared the book to be a "most admirable work [which] fills a widely felt want."[62] It reached a wide audience, revised and updated editions appearing in 1887, 1893, and 1902; a German translation, *Geschichte der Astronomie während des neuzehnten Jahrhunderts,* was brought out in Berlin in 1889. It was well received in America also; Edward Holden, director of Washburn Observatory at the University of Wisconsin and later of Lick Observatory, recommended the first edition in *Publications of the Astronomical Society of the Pacific,* which he edited. Holden had supplied Clerke with copies of a number of his publications in 1883, although by then her manuscript was well advanced.[63] Although the first edition was unillustrated, following Grant's style, the one that came out in 1893 included a number of famous pictures, among them Hale's photographs of solar prominences and the Hugginses' 1892 spectrum of Nova Aurigae. The work remains of considerable value even now as a dependable secondary source on the history of nineteenth-century astronomy.

In the late 1880s Clerke was presented with an opportunity for getting a little practical experience, being invited by Sir David Gill, director of the Cape Observatory, to spend some time in Cape Town, South Africa. She had met Gill and his wife when they visited London in 1887. Gill, like her reviewers, was favorably impressed by her *History,* but felt that her future writing would benefit from some familiarity with the problems and procedures of observational work. Being an especially enthusiastic and warmhearted man, he suggested the visit. A great deal of work was then under way at the Cape Observatory on the mapping of the southern stars, and over the course of her two-month stay Clerke was able to watch its progress. With assistance from Gill and his secretary, she worked with a seven-inch equatorial telescope, the first time in her life that she had used an astronomical instrument. The results of her observations on about a dozen red, variable, and emission-line stars were published in the *Observatory* in 1888.

On her return to London she began her second major work, *The System of the Stars* (1890, second edition 1905), a careful and exhaustive survey of current investigations, discoveries and unsolved problems in sidereal astronomy. The project had initially been suggested to her shortly after her *History* came out by J. N. (later Sir Norman) Lockyer, the discoverer of helium, at that time attached to the British Museum; Lockyer was one of the first astronomers she came to know. Her book started off with a discussion of varieties of individual stars, double, red, variable, and so on; later chapters covered topics such as star clusters and nebulae. Unlike the *History,* it included many illustrations and much numerical data. David Gill read and supplied comments on chapter drafts, and Edward Holden gave her written material in addition to photographs taken at the Lick Observatory. Again reviews, by and large, were favorable.[64] She had considerable skill and judgment in extracting the essential from the vast amount of astronomical research that was being reported in the last quarter of the nineteenth century, and so the work of collecting, correlating, digesting, and condensing that she carried out was of considerable service to both research workers and the interested public. In 1892 the Royal Institution acknowledged her contributions with the award of its 100 guinea Actonian Prize for science writing. Although she tended to avoid the use of technical terms, and did not hold back from somewhat dramatic descriptions of the wonders of God's creation, her style, on the whole, was relatively objective and impersonal by the standards of her time.

Once she had the manuscript of this second book in the printer's hands she took a two-month cruise of the Baltic on a yacht owned by a friend of her friend Henry Reeve of the *Edinburgh Review.* This vacation gave her an opportunity to visit the observatories at Copenhagen and Stockholm, where she met astronomers Carl Fredrick Pechüle and K. P. T. Bohlin. The yacht also made a two-week call at St. Petersburg, but she missed visiting the Pulkova Observatory; its astronomer Hermann Struve, to whom she had a letter of introduction, was away, and she was too shy to present herself without the customary formality.

About this time the possibility arose of an appointment at the Royal Observatory, Greenwich. However, the offer as finally made was a "supernumerary computership" at £8 a month, without any definitely stated arrangement for observational work. This she did not find overwhelmingly attractive, and in view of the additional problem of Greenwich Park being considered less than safe for ladies at night, she let pass the opportunity in favor of staying with her literary work—albeit with a certain amount of regret.[65] There was one other professional post for which she might have been considered, that of professor of astronomy at Vassar College. Soon after the retirement of Maria Mitchell in 1888, Clerke's friend Edward Holden wished to recommend her as Mitchell's successor. However, as she later wrote to him,[66] even had the offer been made, and had she felt equal to the challenge involved, she could never have considered leaving her parents to take up the position.

Three more books followed her *System, The Herschels and Modern Astronomy* (biographies of William, Caroline, and John Herschel, 1895), *The Concise Knowledge of Astronomy* (with J. E. Gore and A. Fowler, 1898), and *Problems in Astrophysics* (1903). The latter, her third and last major work, she considered her magnum opus.[67] Whereas her *System* had incorporated much advice and information received from both Gill and Holden, *Problems in Astrophysics* was entirely her own project. It set out her views on worthwhile future research on the sun, stars, and nebulae (planetary work was postponed for another volume). Although her major emphasis was on solar physics and the physics of the stars and nebulae, she also discussed current developments in pure physics that had yet to be applied to stellar spectroscopy and theories of stellar evolution. The topics she discussed make an impressive list for a self-educated woman of her time to have undertaken; they included the temperature of the solar photosphere, Clerk Maxwell's theory of heat, Herzian waves, and Zeeman's observations of the effects of magnetic forces on spectra. In the area of spectroscopy she set out recent theories of Lockyer and Henry Deslandres; her treatment of the solar corona included a discussion of Huggins's electrical theory. A review in *Nature* by the journal's assistant editor R. A. Gregory rather harshly criticized her for meddling in areas Gregory felt were better left to working spectroscopists if errors and misstatements were to be avoided (and he included a list of the latter); but even he felt he had to acknowledge that, overall, the survey was a "great service to astronomers."[68] By and large the work was well received, other reviews in both British and American journals being favorable. Further acknowledgment of its value lies in the fact that immediately after its publication the Royal Astronomical Society finally made Clerke an honorary member. She was then sixty-one, and eighteen years had gone by since the publication of the first edition of her influential *History.* Her friend Lady Margaret Huggins was elected at the same time, these two following Mary Somerville, Caroline Herschel, and Anne Sheepshanks as the only women members of the organization; women were not admitted to full fellowship until 1916.

Clerke wrote two additional books, a literary work, *Familiar Studies in Homer* (1892), and *Modern Cosmogenies* (1905). She contributed about 149 biographical sketches of British and Irish astronomers and other scientists to the *Dictionary of National Biography* (1885–1901), and articles on astronomy and astronomers to the *Encyclopaedia Britannica* and other encyclopedias. Her major biographical essays in the ninth edition of the *Britannica* included those on Alexander von Humboldt, Huygens, Kepler, Lagrange, Laplace, and Lavoisier; that on Laplace was notable since her treatment of his work was mathematical rather than descriptive.[69] Several of these essays were published again in the revised eleventh edition of the *Britannica* (1911); to the latter she also contributed the main article on the history of astronomy, as well as a number of new biographies of astronomers, including those on Hipparchus, Copernicus, and Tycho Brahe. Many of her articles were used, somewhat abridged, in editions of the *Britannica* as late as that of 1961.

While remarkably broad in her general scientific outlook, Clerke was at the same time a conservative Catholic with strong religious views. Living as she did at a time when English society was becoming increasingly secularized, she undertook, during the course of her work of popularizing astronomy, to point out to her readers the awe-inspiring order and design of the divinely created universe—clear proof of the power and glory of God. She fully accepted limitations imposed on science by religion; for her some matters were beyond scientific logic and better left to the Higher Authority.

Despite her pronounced shyness she became a familiar figure among British astronomers, and was frequently present at meetings of the Royal Institution and the Royal Astronomical Society. She had attended sessions of the latter as a visitor, even before being elected an honorary member; although she avoided public speaking, her opinions about new developments in celestial physics were not infrequently sought. She had been a member of the Liverpool Astronomical Society since 1885, and along with her brother joined the British Astronomical Association when it was started in 1890. Somewhat reluctantly she allowed herself to be elected a charter member of its council, served for three years, and was reelected in 1896. During the 1890s her name appeared frequently in the pages of the association's *Journal* over short comments and critical notes on recently published works; many of these were summaries of articles she had brought out in *Knowledge, Nature, The Observatory,* and other magazines. On Edward Holden's recommendation she was elected the first non-American corresponding member of the Astronomical Society of the Pacific. She joined the Astronomical and Physical Society of Canada as a corresponding member when it was founded in 1890.

Generally considered to be fairly cautious in her written presentations and unbiased in her judgments, she provided, over a period of twenty years, a succession of summaries and pictures of progress in astronomy to date, and thus helped to point the way toward what might follow. American astronomers especially appreciated what they saw as her fair estimations of their work.

Her lifelong companion was her sister Ellen (see chapter 13), a kindred soul who also had wide-ranging interests in music, literature, and the sciences. Their brother Aubrey, who shared their Kensington house, was a chancery barrister. Agnes died suddenly at the age of sixty-four, of pneumonia, on 20 January 1907, outliving her sister by less than a year.

MARGARET MURRAY HUGGINS[70] (1848–1915), a close friend of the Clerke sisters for more than twenty years, was also Irish-born. As the collaborator of her husband, the eminent astrophysicist William Huggins, she took part in pioneering developmental work in stellar spectroscopy. William Huggins was an established and highly respected research worker long before their marriage, but Margaret brought her own considerable skills and talents to the partnership.

The daughter of Helen (Lindsay) and John Majoribanks Murray, Margaret was born in Dublin in 1848. Both John and Helen Murray belonged to Scottish families, John's father, Robert, having in 1825 exchanged his post as accountant at the Bank of Scotland in Inverness for that of manager of the Cork office of the newly established Provincial Bank of Ireland. (The Irish banking system was at the time being reformed along Scottish lines.) The family moved to Dublin in 1839 and by 1845 Robert Murray was chief officer of the bank in Ireland. He sent both his sons back to Scotland for their early schooling. The elder, Margaret's father John, then trained as a solicitor at King's Inns Dublin. John's wife, Helen, was the daughter of Robert Lindsay of Taree, near the Scottish east coast town of Arbroath. She died in 1857, when Margaret was nine, and a few years later John Murray remarried.

Although Margaret attended a private school in Brighton for a time, she got much of her early education at home. She spent a lot of time with her grandfather, and with his help learned to recognize the constellations. By the time she was ten she began making observations with her own homemade instruments. A few years later, with the aid of a dark lantern and a small telescope, she was making systematic observations of sunspots and drawings of constellations. She studied on her own, going through various astronomical works, including writings of Sir John Herschel. She also learned the elements of physics and chemistry, taught herself the principles of photography, then a fashionable hobby for ladies interested in art, and, inspired by magazine articles, constructed a spectroscope that allowed her to detect the main Fraunhofer lines in the sun's spectrum. Through Howard Grubb of the famous Dublin firm of Grubb, builders of high-quality astronomical instruments, she met William Huggins. They were married in 1875 when she was twenty-seven; William was fifty-one.

From the early 1860s, working in his private observatory and laboratory at his home at Tulse Hill, London, Huggins had been engaged in the analysis of star spectra, research that he pioneered. Although he had tried photography using the early wet collodion process in the 1860s, his work was largely done by eye comparison for the next ten years. At about the time of his marriage he was switching to the systematic use of photographic methods using the new, much improved dry gelatine plates. Margaret immediately joined in this work, guiding the telescope during long exposures, a maneuvre necessary to keep the image of the star under observation within the narrow slit of the spectroscope. She was a skillful worker, and her considerable success in photographic manipulations gave her great satisfaction. She also did most of the plate development. There were no precedents to follow in this work—exposure times, plate type, developers, all had to be tried out, methods modified and new skills acquired. From 1875 the regular entries in the Tulse Hill Observatory Diaries (the continuous record of the Hugginses' investigations, started by William in 1856) are in Margaret's script and illustrated by her very fine drawings.

The Tulse Hill instruments were such that, by combining photographic with visible records, the Hugginses could produce spectra that covered the whole optical range. They were the first to observe, in the spectrum of Vega, the important set of lines present in the spectrum of atomic hydrogen known as the Balmer series. Their first joint papers (1889) reported studies on the photographic spectra of the planets Uranus and Saturn, and the visible and photographic spectra of the Orion nebula. In the latter they discovered green lines and two violet lines that did not coincide with those of any known terrestrial element. Many years later these were identified as originating from ionized oxygen. They also investigated the spectra of Wolf-Rayet stars, measuring the wavelengths of their (then unidentified) bright emission lines, and they made visual observations, as well as recording photographic spectra, on the nova of 1892, Nova Aurigae.

Margaret developed great accuracy in plate measurement and particularly good judgement in arranging plates in sequences representing stellar development (work of the kind also being carried out very successfully by some of the women astronomers at the Harvard Observatory about the same time). In particular she used series of hydrogen lines in the ultraviolet spectra of white stars to arrange these, and other types, in ordered series. Her major published work was the outstanding *Atlas of Representative Stellar Spectra* (1899), of which she was joint author and for which she prepared the photographic enlargements. Its appearance marked the conclusion of the Hugginses' observational studies. Their joint investigations had gone on over a period of some twenty-two years, during which they were in the forefront of astronomical spectroscopy. By the 1890s, however, recognizing that there were younger workers in the field with advantages of better locations and better climates, they switched their efforts to laboratory spectroscopy.

One of their projects in this second area of study was an examination of the spectra of calcium and magnesium, two of the prominent elements in the sun's spectrum. Their investigation of the dependence of relative line strengths on the physical condition of the source constituted one of the earliest illustrations of the Saha Law (the latter, formulated only in 1920, says that the strengths of an element's characteristic spectral lines depend on the temperature of the source; hence a star's spectrum depends largely on the temperature of its surface layers). Shortly after the turn of the century, influenced by the work of Pierre and Marie Curie, they became interested in radioactivity, and looked at ways of examining radioactive radiation spectroscopically. Several of their papers on the subject appeared between 1903 and 1906.[71]

In addition to the many scientific papers she coauthored with her husband,[72] Margaret carried out a considerable amount of editing, including the preparation of the quarto volume of *The Scientific Papers of William Huggins* (1909). The joint publications constituted a significant fraction of the body of work that brought William his knighthood in 1897 and the Order of Merit in 1902. Margaret also contributed two articles on old astronomical instruments (the armillary sphere and the astrolabe) to the eleventh edition of the *Encyclopaedia Britannica.* Her scientific writing was recognized by the Royal Institution with the award of its Actonian Prize (jointly with her husband).

A striking, attractive, and vigorous personality and an engaging conversationalist, she had wide interests, not the least of which were music and painting. As well as accompanying her husband on piano or organ when he played his Stradivarius, she collaborated with him in the study of ancient violins. Her monograph on the Brescian master-violin maker Giovanni Paolo Maggini, published in 1892, provides a vivid picture of life in sixteenth-century Brescia (Italy).[73] Her artistic work included both watercolors and the pen-and-ink reproductions of old astronomical drawings, which she used, following the bygone tradition, to decorate initial letters of first words of chapters in her works. The restoration of medieval art objects was another special interest; her 1894 paper on the astrolabe in *Astronomy and Astrophysics* was written in part because of the appeal the instrument's intrinsic beauty and historical interest had for her.

She was much interested in education and keen that children should be encouraged to study science; for a time she served as one of the managers of a group of Board schools. Through her friend Sarah Whiting, professor of physics and astronomy at Wellesley College and an occasional visitor to Tulse Hill, she followed developments in women's education in the United States, especially science education. In her own career she evidently had a satisfactory working arrangement with her husband, although she is known to have been considerably more liberal than he on questions of the formal and professional recognition of women as scientists. Her membership in the Royal Astronomical Society at only the honorary level (which came to her in 1903) may perhaps have seemed a reasonable compromise to them. Sir William, although he acknowledged his wife's contributions to their joint scientific work, was nevertheless a man of conservative outlook who did not look with favor on the opening of the major national scientific societies to women.[74] Margaret Huggins was a founding member of the more open British Astronomical Association, and served on its first council.

After Sir William's death in 1910 she moved to a small apartment in Chelsea, and although in poor health took up the task of sorting through the Tulse Hill records and instruments. Arrangements had already been made for the transfer to the University Observatory at Cambridge of the main instruments that Sir William had had on long-term loan from the Royal Society. A considerable amount remained, however, including the Observatory Diaries. These, along with a number of smaller instruments and art treasures, Margaret gave to Wellesley College.

For the last five years of her life she had a Civil List pension of £100 (Sir William had held one of £150 from 1890), bestowed in recognition of her services to astronomy. At the time of her death at age sixty-six, on 24 March 1915, she was starting to write an account of her husband's life and work, a project she declined to delegate to anyone else; her preliminary notes were later used in a privately printed memoir of Sir William.[75]

The prominence in this ten-member British group of women with connections to Ireland is remarkable. Clerke and Huggins were Irish by birth and upbringing, although both had left the island by their mid-twenties. Physicist-astronomer Alice Everett, although of British parentage, lived in Belfast from the age of two until she was nineteen. Mary Proctor might also be included in the British/Irish subgrouping, having been born in Dublin of an Irish mother.

This pronounced interest among young women of Irish background in astronomical work can most likely be regarded as part of the prevailing and widespread fashion of enlightened interest in the sciences in Ireland during the latter part of the nineteenth century—a particularly brilliant period for Irish astronomy that is only beginning to be widely recognized among historians of science.[76] Churchmen, wealthy landed gentry, and academics throughout the island founded and equipped many observatories and made substantial contributions to international astronomical research. Women as well as men were caught up in this trend of the time.

Two-country comparison

Reasons for the relatively low number of British women compared with Americans (Figure 10-1) publishing original work in astronomy in the 1890s are not hard to find. As noted by Kidwell,[77] there were no British equivalents of the astronomy departments of the American women's colleges (Vassar, Mount Holyoke, Smith, and Wellesley) that provided at least a few teaching jobs for women astronomers, some instruments for observational work, and, at the undergraduate level, a basic introduction to the field.[78] Aspiring British women astronomers could, and did, study mathematics and physics, but the necessary subsequent training and experience were hard to acquire. In the United States significant numbers of women found computer positions at such places as the Harvard and Columbia observatories, but not many had even those kinds of opportunities in Britain. Among those mentioned here only Everett worked in a major British observatory before the turn of the century; Clerke's short visit to the Cape observatory was the direct result of her contribution as a historian of astronomy. In the absence of ready entrance to the relatively few existing British institutional facilities, opportunities for women were in private observatories, access to which generally depended on a family relationship, as in Huggins's case, or enough money to have one's own (Brown's case). However, rather than regarding this situation as simply a handicap faced by British women in particular, it is perhaps more illuminating to see it in its wider context, that is, as part of a long-standing overall national pattern that held throughout much of the nineteenth century. During this period observation-based astronomical research in Britain was to a large extent the province of amateurs who had private financial resources, and it was not supported by generous government funding (as in Germany), or by academic corporations (as at Harvard).[79] Thus the work of Margaret Huggins and Elizabeth Brown (in private observatories) fits very readily into the established general pattern.

That said, it is nevertheless the case that a few women (not included in this overview) were in fact working as assistants and computers at British and colonial institutional observatories before the turn of the century. Among them were Elizabeth Kent, assistant to the government astronomer at the Madras observatory for twenty-three years from 1873 to 1896 (and in addition meteorological reporter from 1884 to 1896), Annie Russell, computer at the Greenwich observatory in the early 1890s (at the same time as Alice Everett), Edith Bellamy, whose work at the Oxford University observatory began in 1899, and Mary Orr. Orr, wife of John Evershed, director of the Kodaikanal observatory in southern India, brought out a short monograph, *Southern Stars, a Guide to the Constellations Visible in the Southern Hemisphere* (1896); it resulted from observations she made during a five-year stay in Australia before her marriage. Her later (post-1900) publications included a number of joint papers with her husband in the *Memoirs of the Kodaikanal Observatory.*

It might be noted that, even with the more favorable circumstances in the United States, a large fraction of the pre-1900 journal articles published by American women reported minor studies made at small women's colleges with limited equipment. Further, although large, well-equipped observatories such as Harvard and Columbia did employ women, the majority of them functioned as technicians, doing routine calculations and making routine measurements on photographic plates, "factory-style," with the results of their labors being published under the observatory director's name, and with little prospect of moving on to more interesting or better-paid work. Only three of the large group of Harvard women workers brought out publications listed in the main pre-1901 scientific indexes under their own names—Anna Winlock, Antonia Maury, and Williamina Fleming; and in Fleming's case (although she also published her own papers) her contributions to the *Draper Catalogue,* her major pre-1900 work, were acknowledged in the introduction only, the catalog itself coming out under director Pickering's name.[80]

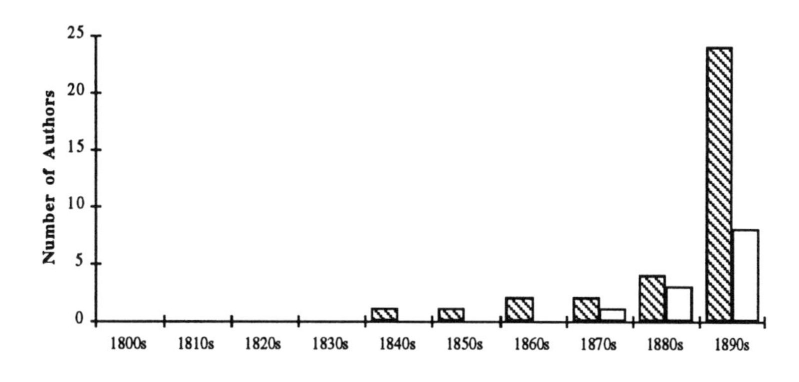

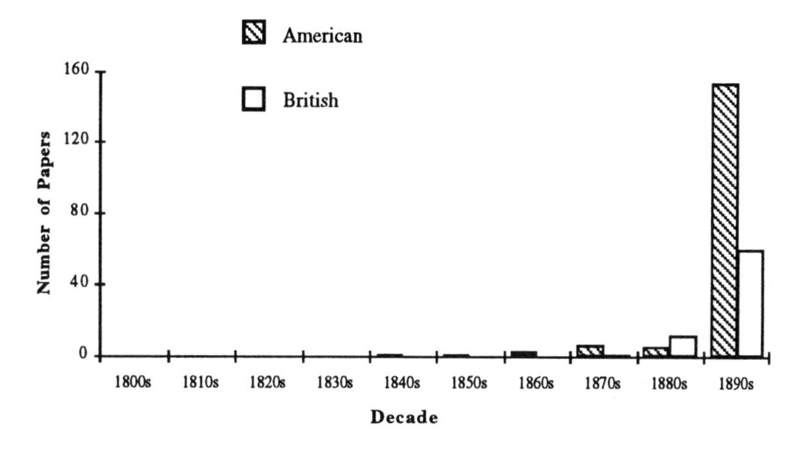

Figure 10-1. American and British authors and papers in astronomy, by decade, 1800–1900. Data from the Royal Society *Catalogue of Scientific Papers.*

On the British side, two women, Margaret Huggins and Alice Everett, were engaged in large projects for the preparation of important stellar catalogs comparable to those on which the Harvard women worked, and although the body of research they produced was much less than the truly remarkable output of the Americans, nevertheless Huggins's joint *Atlas of Representative Stellar Spectra* and Everett's five years of work on the *Astrographic Catalogue* in the 1890s (recorded in the published report),[81] were creditable contributions to major projects. Clearly British women were not entirely out of the picture compared with the Americans, despite their relative lack of opportunity. Among the British contributions, the writings of Agnes Clerke, even if in a nonresearch area, are remarkable. Clerke's broad grasp of the field, ability to pick out major trends and important developments, and her steady work of exposition, make her one of the leading science writers of her time.

Figure 10-2 summarizes author and paper distribution by subfield. In chart b, the American work, the sectors labeled "Comets & Minor Planets" and "Sun & Moon" represent mainly papers by astronomers at the women's colleges; the publications on variable stars by Fleming at Harvard and the notes by Californian observer O'Halloran constitute much of the work represented in the sector labeled "Stars;" the "Stellar Catalogues & Stellar Spectra" sector represents the work of the Harvard women. The large sector designated "Stars" in the British paper subfields (chart d) to a considerable extent represents the writings of Proctor and, more especially, Clerke; that designated "Stellar Spectra" represents the work of Huggins.

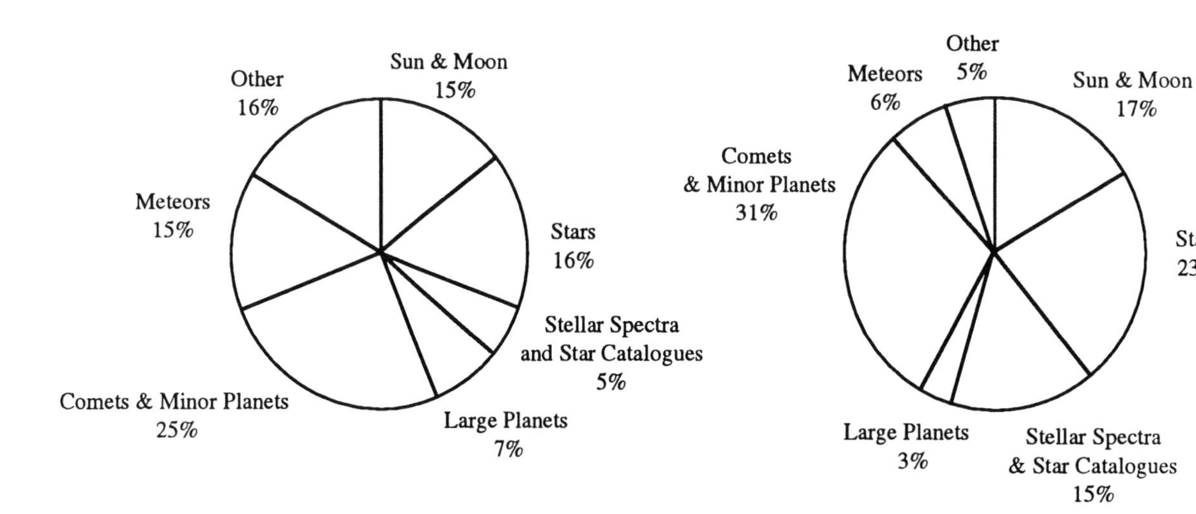

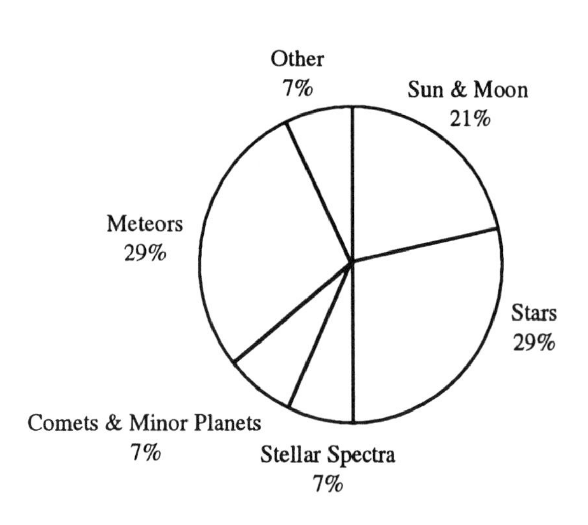

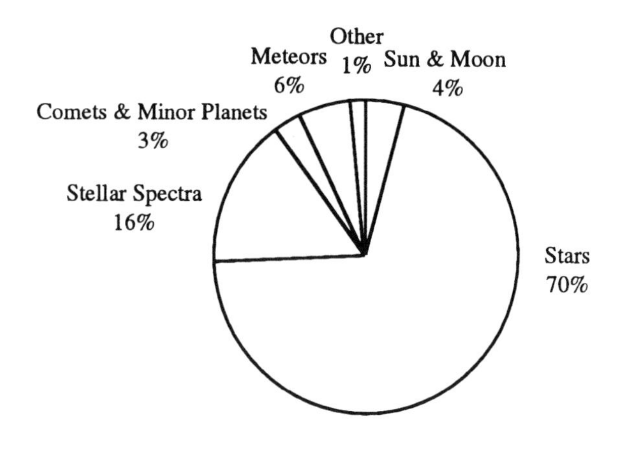

Figure 10-2. Distribution of American and British authors and papers in astronomy by subfield, 1800–1900. (In a. and c., an author contributing to more than one subfield is counted in each.) Data from the Royal Society *Catalogue of Scientific Papers*.

Notes

1. Jane Opalko, "Maria Mitchell's haunting legacy," *Sky and Telescope,* **32** (1992): 505–7; Andrea K. Dobson and Katherine Bracher, "A historical introduction to women in astronomy," *Mercury,* **21** (1992): 4–15; Peggy Aldrich Kidwell, "Three women of American astronomy," *American Scientist,* **78** (1990): 244–51; Pamela E. Mack, "Straying from their orbits. Women in astronomy in America," in *Women of Science. Righting the Record,* eds. G. Kass-Simon and Patricia Farnes (Bloomington, Ind.: Indiana University Press, 1990), pp. 72–116, on pp. 75–9; Frances Fisher Wood, "Sketch of Maria Mitchell," in *What America Owes to Women,* ed. Lydia Hoyt Farmer (Buffalo, N.Y.: Charles Wells Moulton, 1893), pp. 264–70.

2. Karl Friedrich Gauss, *Theoria Motus Corporum Coelestium in Sectionibus Conicus Solem Ambientium* (Hamburg: Perthes et Besser, 1809); George Biddell Airy, *Gravitation: an Elementary Explanation of the Principal Perturbations in the Solar System* (London: C. Knight, 1834).

3. The Vassar administration justified the low salary ($800) it paid Mitchell on the grounds that she still retained her position as a computer of *American Ephemeris* tables and her government salary for this work ($300). Even then her total annual income was considerably less than the $2,500 the college paid some of its male professors (John Lankford and Rickey L. Slavings, "Gender and science: women in American astronomy, 1859–1940," *Physics Today,* **43,** n. 3 (1990): 58–65).

4. Mack, "Straying from their orbits," pp. 79–80; Ogilvie, *Women in Science,* pp. 175–6; Caroline E. Furness, "Mary W. Whitney," *Popular Astronomy,* **30** (1922): 597–608, and **31** (1923): 25–35; *WWWA,* p. 879.

5. The Rutherfurd star cluster photographs comprised several hundred photographic plates turned over to Columbia University in 1890 by astronomer Lewis Rutherfurd when he himself was no longer able to analyze them. Subsections of the collection provided the raw material for the Ph.D. dissertations of three of the women astronomers in this study (Furness, Young, and Harpham).

6. Whitney's post-1900 papers include the following: "Observations of Nova Persei at Poughkeepsie," *Popular Astronomy,* **9** (1901): 220–1; "Observations of Nova Persei," ibid., **9** (1901): 574; "Observations of minor planet 1901 GV made at the Vassar College Observatory, Poughkeepsie, New York," *Astronomische Nachrichten,* **157** (1902): 147–8; "Observations of Nova Persei (Ch. 1226)," ibid., **157** (1902): 387–8; "Measurement of astronomical photographs. Grant No. 23" (Washington, D.C.: Carnegie Institution Year Book, No. 2, 1903), 1904, xxiii; "Comparison stars for the Algol variable 4. 1903," *Popular Astronomy,* **11** (1903): 428–9; "The determination of solar motion," ibid., **12** (1904): 226–30, 311–18; "Observations of new variables," *Astronomische Nachrichten,* **169** (1905): 279–84; "Observations of variables," ibid., **176** (1907): 65–8; "Maxima of long-period variables," *Astronomical Journal,* **25** (1908): 83; "Maxima and minima of long-period variables," *Astronomische Nachrichten,* **178** (1908): 317–20; "Maxima and minima of long-period variables during 1906–07," *Astronomical Journal,* **25** (1908): 198; "SS Cygni," *Astronomische Nachrichten,* **180** (1909): 47–8; "New variable 43. 1909 Draconis," ibid., **183** (1910): 47–8; "Variable star 43. 1909 Draconis," ibid., **183** (1910): 207–8; "New variable in Aquila," *Astronomical Journal,* **26** (1911): 74; "Occultations of stars by the moon, observed at the Vassar College Observatory with the 12-inch telescope," ibid., **26** (1911): 126; "Tail of Halley's Comet," ibid., **26** (1911): 140. Papers from the same period coauthored by Whitney and Furness include: "Observations of minor planets and comets made with the 12-inch telescope of the Vassar College Observatory," ibid., **21** (1901): 116, 160; "Observations of minor planets, made at the Vassar College Observatory," ibid., **22** (1902): 104, and *Astronomische Nachrichten,* **159** (1902): 45–8; "Observations of Comet b 1902 (Perrine) made at the Vassar College Observatory," *Astronomical Journal,* **22** (1902): 195; "Observations of minor planets, made at the Vassar College Observatory," ibid., **23** (1903): 73–4; "Observations of minor planets and comets," ibid., **24** (1905): 153–4; "Observations of comets and minor planets made at the Vassar College Observatory," ibid., **25** (1908): 92; "Observations of comets and minor planets made with the 12-inch equatorial at the Vassar College Observatory," ibid., **25** (1907): 160; "Observations of long-period variables," ibid., **26** (1911): 44, 103–4, and *Astronomische Nachrichten,* **181** (1909): 91–4.

7. Maud W. Makemson, "Caroline Ellen Furness," *Publications of the Astronomical Society of the Pacific,* **48** (1936): 96–100; *Who was Who in America,* vol. 1, p. 433; *WWWA,* pp. 311–12; *AMS,* 1910–1933.

8. Furness's post-1900 papers, in addition to those coauthored by Whitney, include the following: "A photographic catalogue of the north polar stars," *Popular Astronomy,* **9** (1901): 1–7; (with Emma Phoebe Waterman) "Definitive orbit of comet 1886 III," *Astronomische Abhandlungen,* **14** (1908): 27–35; "Var. SX. Draconis," *Astronomische Nachrichten,* **187** (1911): 383–4; "Observations of minor planets and comets," ibid., **189** (1911): 59–62; (with Psyche Rebecca Sutton) "Observations of long-period variables," *Astronomical Journal,* **27** (1911): 24–8; "Observations of comet c 1911 (Brooks) and comet f 1911 (Quenisset) made at the Vassar College Observatory," ibid., **27** (1912): 58; "Observations of variable stars made at Vassar College Observatory," *Popular Astronomy,* **20** (1912): 645–52; "Notes on variable stars," *Astronomische Nachrichten,* **195** (1913): 359–60; "Maxima and minima of variable stars," *Astronomical Journal,* **27** (1913): 191–3; "Observations of long-period variables," ibid., **29** (1915): 47–8; "New variable star in Cygnus," ibid, **30** (1917): 181; "Use of the stereo-comparator in determining proper motions," *Popular Astronomy,* **30** (1922): 158; "The colour of the eclipse of 1927, June 29," *The Observatory,* **51** (1928): 201; "The total solar eclipse of August 1932," *Popular Astronomy,* **36** (1928): 232; (with Irma J. Courtice) "The equation of time," ibid., **38** (1930): 579–83; "Exceptional phenomena of Jupiter's satellites," *The Observatory,* **55** (1932): 114–15; "The longitude of the Vassar College Observatory," *Publication No. 4, Vassar College Observatory,* Poughkeepsie, N.Y., 1934, pp. 1–22.

9. Mount Holyoke College archives; Mack, "Straying from their orbits," pp. 81–2; obituary, *Science,* **9** (1899): 821–2.

10. Isabella Mack was born in Manchester, New Hampshire, in 1851 and graduated from Mount Holyoke Seminary in 1875. She taught mathematics and physics at Mount Holyoke for eleven years (1875–1886) and then returned to Manchester to care for her parents. She became an invalid because of a heart condition and died in Manchester in 1916 (Mount Holyoke College archives).

11. Mount Holyoke College archives; Helen Sawyer Hogg, "Anne Sewell Young," *Quarterly Journal of the Royal Astronomical Society,* **3** (1962): 355–7; Ogilvie, *Women in Science,* p. 178; Mack, "Straying from their orbits," p. 82; *WWWA,* p. 912; *AMS,* 1910–1955.

12. Young's sunspot observational data appeared frequently in *Popular Astronomy* from 1902 until 1936. See "Sunspot observations, 1900–01," *Popular Astronomy,* **10** (1902): 167; "Sunspot observations, 1902–03," ibid., **12** (1904): 214, and "Résume of sunspot observations at Mount Holyoke College," ibid., **18** (1910): 128; **19** (1911): 121; **21** (1913): 115; **24** (1916): 134; **35** (1927): 187; **36** (1928): 138; **38** (1930): 123; **40** (1932): 121; **41** (1933): 97; **42** (1934): 348; **43** (1935): 132; **44** (1936): 111.

13. Anne Sewell Young, "Rutherfurd photographs of the stellar clusters h and χ Persei." (New York: Contributions from the Observatory of Columbia University, no. 24, 1906), pp. 297–368 (a short vita is included on p. 368).

14. Young's contributions on variable stars include the following: (with Louise Freeland Jenkins) "Note on the proper motions of certain long-period variable stars," *Popular Astronomy,* **28** (1920): 10; (with Alice H. Farnsworth) "Proper motions of certain long-period variables," *Astronomical Journal,* **35** (1924): 180, **39** (1929): 47–8. Her other post-1900 publications include: "Nova Persei," *Popular Astronomy,* **9** (1901): 357–8; "On the density of the solar nebula," *Astrophysical Journal,* **13** (1901): 338–43; "Transit of Jupiter's fourth satellite," *Popular Astronomy,* **11** (1903): 574; "Leonid meteors," ibid., **12** (1904): 683; "Longitude of the John Payson Williston Observatory," ibid., **38** (1930): 218.

15. Carlton College archives; Louise Barber Hoblit, "Mary E. Byrd," *Popular Astronomy,* **42** (1934): 496–8; anon., "Miss Mary E. Byrd's resignation," ibid., **14** (1906): 447–8; *WWWA,* p. 152; *AMS,* 1910–1933; *Census of the United States* (1860, 1870, 1910) for Leavenworth and Douglas Counties, Kansas (Washington, D.C.: Government Printing Office).

16. Mary E. Byrd, "Observations of comet b 1900 (Borelly-Brooks) made at the Smith College Observatory, Northampton, Massachusetts, with the 11-inch refractometer and filar micrometer," *Astronomical Journal,* **21** (1901): 115; "Observations of comet b 1902 (Perrine) made with the 11-inch equatorial at the Smith College Observatory, Northampton, Massachusetts," ibid., **23** (1903): 15; "Observations of comet a 1903 (Giacobini) made with the 11-inch equatorial at the Smith College Observatory, Northampton, Massachusetts," ibid., **23** (1903): 127–8; "Observations of comets made with the 11-inch equatorial at the Smith College Observatory, Northampton, Massachusetts," ibid., **24** (1905): 188; (with Harriet W. Bigelow), "Observations of comets," *Astronomische Nachrichten,* **169** (1905): 191–2, **173** (1906): 75–8.

17. "Astronomy and the normal school," *Popular Astronomy,* **21** (1913): 2–4; "Astronomical teaching in the city," ibid., **23** (1915): 154–9, 230–2. Byrd's articles on high school teaching were the following: "Astronomy in the high school," ibid., **11** (1903): 550–2; **12** (1904): 24–7; **12** (1904): 199–202; **13** (1905): 545–9; **15** (1907): 227–37; **21** (1913): 74–8; **23** (1915): 546–53.

18. "Outline of a laboratory course in elementary astronomy," ibid., **14** (1906): 294–8; "A plea for elementary astronomy," ibid., **15** (107): 25–9; "First study of heavenly bodies, lessons I-VII," ibid., **28** (1920): 33–7, 81–6, 143–8, 199–204, 275–81, 314–20, 476–81; "First study of heavenly bodies," ibid., **29** (1921): 102–8. For a note concerning the correspondence course Byrd directed see "A class for amateurs," ibid., **25** (1917): 211.

19. Anon., "Miss Mary Emma Byrd," *Lawrence Daily Journal World,* 30 July 1934, 2, 4.

20. Oglvie, *Women in Science,* pp. 174–5; Gladys A. Anslow, *NAW,* vol. 3, pp. 593–5; Annie J. Cannon, "Sarah Frances Whiting," *Science,* **66** (1927): 417–18, and *Popular Astronomy,* **35** (1927): 539–45; *WWWA,* p. 877.

21. See also Sarah F. Whiting, "Use of graphs in teaching astronomy," *Popular Astronomy,* **13** (1905): 185–90; "Use of drawings in orthographic projection and of globes in teaching astronomy," ibid., **13** (1905): 235–40; "Photographs in teaching astronomy," ibid., **13** (1905): 430–4; "A pedagogical suggestion for teachers of astronomy," ibid., **20** (1912): 156–60; "A solar planisphere," *Publications of the Astronomical and Astrophysical Society of America,* **1** (1910): 254.

22. Sarah F. Whiting, "Priceless accessions to Whitin Observatory, Wellesley College," *Popular Astronomy,* **22** (1914): 487–92, and "The Tulse Hill Observatory Diaries," ibid., **25** (1917): 158–63.

23. See Sarah F. Whiting, "Spectroscopic work for classes in astronomy," ibid., **13** (1905): 387–91.

24. *WWWA,* p. 885; Anne P. McKenney, "What women have done for astronomy in the United States," *Popular Astronomy,* **12** (1904): 171–82, on pp. 174–5; Smith College archives.

25. William W. Payne, H. C. Wilson, Charlotte R. Willard, and A. G. Sivaslian, "Observations of sunspots and measures of solar photographs taken in the years 1889–1892," *Publications of the Goodsell Observatory, Carlton College,* **3** (1901), 84 pp.

26. Women computers at Harvard were generally paid 25 cents an hour (sometimes increased to 30 or 35 cents) at least until 1926. They worked seven hours a day, six days a week, giving them an annual income of about $550, with one month paid vacation a year for full-time employees. Men were paid at the same rate for computing, but few applied for the jobs. Men who did nighttime viewing and other mechanical work received (in 1874) $800 per year (see Mack, "Straying from their orbits," p. 89, and Bessie Zaban Jones and Lyle Gifford Boyd, *The Harvard College Observatory. The First Four Directorships, 1839–1919* (Cambridge, Mass.: Harvard University Press, 1971), p. 390). Women considered the observatory jobs desirable because of the relatively good conditions and fairly pleasant work. (Even for those who had college training, few alternative jobs except schoolteaching were available.) In the period up to 1900 twenty-six women had been hired by the Harvard observatory; the female staff that year numbered nineteen, and it remained large after the turn of the century (Mack, "Straying from their orbits," pp. 88–9).

27. Oglivie, *Women in Science,* pp. 177–8; Mary E. Byrd, "Anna Winlock," *Popular Astronomy,* **12** (1904): 254–8.

28. Byrd, "Anna Winlock," p. 255.

29. Jones and Boyd, *Harvard College Observatory,* pp. 392–5; William E. Rolston, "Mrs. W. P. Fleming," *Nature,* **86** (1911): 453–4; Rossiter, *Women Scientists in America,* pp. 55–7; Dorrit Hoffleit, "Fleming, Williamina Paton Stevens," in *NAW,* vol. 1, pp. 628–30; H. H. Turner, obituary, *Royal Astronomical Society Monthly Notices,* **72** (1912): 261–4; Joseph L. Spradley, "The industrious Mrs. Fleming," *Astronomy,* **18** (1990): 48–51.

30. Rolston, "Mrs. W. P. Fleming;" Spradley, "Industrious Mrs. Fleming."

31. Turner, obituary, p. 262. Fleming's situation in many ways presents a parallel to that in which her contemporaries, the marine zoologists Mary Jane Rathbun and Katharine Bush, found themselves. In both astronomy and marine biology at that time, vast quantities of basic research data (stellar photographs in one case and specimens brought back by big collecting expeditions in the other) were being produced very rapidly. Discoveries almost inevitably awaited the examiners.

32. Nearly all of Fleming's later work was published in the Harvard observatory *Annals* except for a few shorter papers; see, for instance, "Some peculiar spectra," *Publications of the Astrophysical Society of America,* **1** (1910): 240–1, and "The spectrum of a meteor," ibid., **1** (1910): 337. Substantial posthumous works not already mentioned are: "Photographic observations of variable stars during the years 1886–1905, forming a part of the Henry Draper Memorial," *Annals of the Observatory of Harvard College,* **47** (1912): 115–280, and "Spectra and photographic magnitudes of stars in standard regions," ibid., **71** (1917): 27–45.

33. Rossiter, *Women Scientists in America,* p. 57.

34. Jones and Boyd, "Harvard College Observatory," pp. 236–8, 395–400; Ogilvie, *Women in Science,* pp. 130–1; Dorrit Hoffleit,

"Antonia C. Maury," *Sky and Telescope,* **11** (1952): 106, and "Maury, Antonia Caetana de Paiva Pereira," in *NAW,* vol. 4, pp. 464–6; Owen Gingerich, "Maury, Antonia Caetana de Paiva Pereira," *DSB,* vol. 9, pp. 194–5; *AMS,* 1910–1933; Chester A. Reeds, "Memorial to Carlotta Joaquina Maury," *Proceedings of the Geological Society of America* (1939): 157–68.

35. Antonia Maury, "The spectral changes of Beta Lyrae," *Annals of the Astronomical Observatory of Harvard College,* **84,** pt. 8 (1933). Part 6 of the same volume reported the results of her work on the spectroscopic binaries, μ Scorpii and V Puppis.

36. Dobson and Bracher, "Historical Introduction," p. 12.

37. Jones and Boyd, *Harvard College Observatory,* p. 399. The institution to which Maury was being recommended was not specified.

38. *The University of Wisconsin. Its History and its Alumni,* ed., R. G. Thwaites (Madison, Wisc.: J. N. Purcell, 1900), pp. 753, 722; McKenney, "What women have done," pp. 173–4; T. J. J. See, "Brief biographical notice of Professor Milton Updegraff, 1861–1938," *Publications of the Astronomical Society of the Pacific,* **50** (1938): 332–4.

39. Obituary, *NYT,* 31 January 1924, 15:5; *WWWA,* p. 619; M. Burton-Williamson, "Some American women in science," *The Chautauquan,* **28** (1898–99): 161–8, 361–8, 465–73, on p. 162; Dorit Hoffleit, *The Education of American Women Astronomers before 1960* (Cambridge, Mass.: American Association of Variable Star Observers, 1994), p. 10.

40. McKenney, "What women have done," p. 175.

41. Ibid., p. 175; *WWWA,* p. 231.

42. McKenney, "What women have done," p. 179; Carlton College archives; Columbiana files, Columbia University Libraries.

43. See, for instance, Harold Jacoby, "The Rutherfurd photographic measures of the group of the Pleiades," *Contributions from the Observatory of Columbia University,* no. 17, 1901; "Catalogue of 287 stars near the south pole and optical distortion of the Cape of Good Hope astrophotographic telescope," ibid., no. 19, 1902; "Rutherfurd photographs of stars surrounding 59 Cygni," ibid., no. 25, 1908; all of these acknowledge Harpham's work as chief computer. She also published a short historical note under her own name in 1905, "Notes on angle measuring instruments," *Popular Astronomy,* **13,** pp. 149–54.

44. McKenney, "What women have done," p. 177; *WWWA,* p. 62.

45. The introductory notes to nos. 17 and 19 of *Contributions from the Observatory of Columbia University* (see n. 43) contain acknowledgments of Tarbox's work.

46. Mack, "Straying from their orbits," pp. 89–91.

47. For Ellen Hayes, whose note on orbital eccentricity appeared in *Science* in 1893, see chapter 8. About seven of these women (Iliff, Waterbury, Howe, Sanborn, Drake, Sparks, and Moulton) very little information has been collected. Each contributed only one or two papers to the pre-1901 astronomy journal literature. Edna Iliff, Lottie Waterbury, and Fannie Shattuck Howe were all associated with the Chamberlin Observatory in Denver. Iliff and Waterbury, whose joint paper on groups of time stars appeared in *Popular Astronomy* in 1894, were students at the University of Colorado. Fannie Howe was the wife of Herbert Howe, the director of the Chamberlin Observatory. Her 1895 paper in the *Astronomical Journal* reported her observations of the transit of Mercury made at the observatory in 1894. Alice Sanborn was a student at the University of Wisconsin and worked at the Washburn Observatory in the 1880s. Eleanor Drake in 1899 published a description of a meteor observed from the balcony of a private house at Morgan Park, Cook County, Illinois; a similar paper by Marion Sparks, which appeared in *Popular Astronomy* in 1898, recorded meteor observations made in Urbana, Illinois. The 1898 and 1900 *Popular Astronomy* papers of Mary Etta Moulton record observations made in India (in the absence of evidence to indicate that she was British, she has been placed with the Americans since she published in an American journal).

48. Student records, University of California, Berkeley; Lankford and Slavings, "Gender and science," p. 60. Hobe's 1902 paper with Leuschner was "Elements of asteroid 1900 GA and ephemeris for the opposition of 1901–02," *Astronomische Nachrichten,* **157,** pp. 145–8.

49. McKenney, "What women have done," p. 176. O'Halloran's name appears in the membership lists of the Astronomical Society of the Pacific until the early 1920s, but only her private address in San Francisco is given and no institutional affiliation mentioned. She was born in San Francisco and seems to have done most of her work there.

50. O'Halloran's post-1900 papers include the following: "Fluctuations of Nova Persei," *Publications of the Astronomical Society of the Pacific,* **13** (1901): 116–7; "Observations of the variable stars W Lyrae and U^3 Cygni," ibid., **13** (1901): 220–2; "Observations of Nova Persei, 1901," *Popular Astronomy,* **9** (1901): 354–5; "Variable stars," ibid., **10** (1902): 271–3; "The milky way as it appears to observers of the autumn heavens," ibid, **10** (1902): 372–6; "The eclipsed moon in a four-inch telescope," ibid., **10** (1902): 551; "Observations of variable stars," *Publications of the Astronomical Society of the Pacific,* **14** (1902): 95–7; "Observations of the eclipsed moon, October 16, 1902," ibid., **14,** (1902) 188–9; "Maxima of two variables," *Popular Astronomy,* **11** (1903): 52–3; "W Lyrae, V Hydrae, S Ursae Minoris, T Ursae Majoris," ibid., **11** (1903): 216–18; "Stars that periodically glow and fade," ibid., **11** (1903): 294–7; "Observations of variable stars," ibid., **11** (1903): 399–401; "The great sunspot," ibid., **11** (1903): 579–80; "Some details of the recent solar eclipse," ibid., **12** (1904): 27–32; "Variable star notes," ibid., **12** (1904): 496–7; "Variable stars," *Publications of the Astronomical Society of the Pacific,* **16** (1904): 101–3; "Variable star notes," ibid., **16** (1904): 207–10; "Development of the recent large sunspot," ibid., **17** (1905): 20–1; "Variable star notes," ibid., **17** (1905): 14–15, 91–3; **18** (1906): 50–3; "The sunspot maximum of 1905," *Popular Astronomy,* **14** (1906): 368–71; "Light curves of Mira and W Lyrae," ibid., **15** (1907): 95–9; "Variable star notes," ibid., **15** (1907): 381–2; 512–14; "Some recent studies of the solar surface," ibid., **16** (1908): 484–7; "Comet a 1910," ibid., **18** (1910): 181; "Halley's comet," ibid., **18** (1910): 183; "Observations of Halley's comet," ibid., **18** (1910): 453–6.

51. Mrs. G. S. Silliman (Mrs. Hepsa Ely Silliman), *On the Origin of Aerolites* (New York: Wm. C. Bryant, 1859).

52. *World Who's Who in Science. From Antiquity to the Present,* ed. Allen G. Debus (Chicago: Marquis Who's Who, 1968), p. 415.

53. See, for instance, Spradley, "Industrious Mrs. Fleming," and Dobson and Bracher, "Historical introduction."

54. At Harvard, on the other hand, there was no possibility of a woman astronomer presenting part of her research to fulfill doctoral degree requirements since graduate degrees were not awarded to women. Although after 1902 a woman could be granted a Radcliffe doctorate, none of the early Harvard women astronomers received one.

55. For Chisholm and Everett see, respectively, chapters 8 and 9. A pre-1900 astronomy article by one additional British woman, Mary Somerville (see chapter 8) has been added to the bibliography. Mary Ashley, Isabel Fry, Grace Jones, and Susanna Lehmann each wrote only one article indexed by the Royal Society, those by Fry, Jones, and Lehmann being short reports of meteor sightings published as notes in *Nature.* Isabel Fry was probably the daughter of Mariabella (Hodgkin) and Sir Edward Fry, and thus the sister of Agnes Fry (see chapter 1). Grace Jones was from Kensworth in Bedfordshire, some

thirty miles northwest of London, and Susanna Lehmann from Sheffield. Ashley's article on an eclipse of the moon appeared in the *Observatory Magazine* in 1878; she also contributed drawings of lunar features, accompanied by detailed descriptions, to an 1880 issue of the *Selenographical Journal.* She was from Bath, and well enough known in her time as an observer of the lunar surface to be listed briefly in A. Rebière's *Les femmes dans la science, notes recueilles* (Paris: Nony et Cie, 2d ed., 1897).

56. Obituaries, A. S. D. M[aunder], *Observatory,* **22** (1899): 171–2, "In memoriam. Elizabeth Brown, Fellow of the Royal Meteorological Society," *Journal of the British Astronomical Association,* **9** (1898–1899): 214–15, *Quarterly Journal of the Royal Meteorological Society,* **26** (1900): 214–15, *Times,* 16 March 1899, 10b, *Wiltshire and Gloucestershire Standard,* 11 March 1899, *The Friend,* London, **39** (1899): 172, *Friends Quarterly Examiner* (33) 1899: 383 and *Illustrated London News,* 25 March 1899; *Dictionary of Quaker Biography* (unpublished typed manuscript, Library of the Society of Friends, London); information from Gloucester Library, Gloucester; Peggy Aldrich Kidwell, "Women astronomers in Britain, 1780–1930," *Isis,* **75** (1984): 534–46; Kenneth Weitzenhoffer, "A forgotten astronomer," *Astronomy,* **20** (1992): 13–14. The two short quotations are from an 1898 edition of Brown's books, *In Pursuit of a Shadow (Two Eclipse Journeys); Part I, 1887, Russia; Caught in the Tropics, Part II, 1889, The West Indies* (Cirencester: Baily and Woods), pp. 80, 200. I thank Dr. Mary Brück for sending me a list of Elizabeth Brown's articles in the publications of the Liverpool Astronomical Society (see bibliography).

57. Anon., "Proctor, Mary," *National Cyclopaedia of American Biography,* **9,** 1907, pp. 282–3; *WWWA,* p. 663; Jones and Boyd, *Harvard College Observatory,* pp. 416–17; Burton-Williamson, "Some American women in science," pp. 162–3; *AMS,* 1910, 1921.

58. Among Proctor's post-1900 journal articles reporting these and other events were the following: "Borelly's comet," *Scientific American,* **89** (1903): 135; "Eclipse expedition in 1905," *Popular Astronomy,* **12** (1904): 468–9; "Total eclipses of the sun," *Journal of the British Astronomical Association,* **24** (1914): 349–52; "K Crusis," ibid. **25,** (1915): 193–4. She also published descriptions of observatories: "The Amherst College Observatory," *Scientific American, Supp.,* **56** (1903); "A solar observatory in New Zealand," *Knowledge and Scientific News,* n.s. **10,** (1913): 364; "The Cawthorn Observatory," *Journal of the British Astronomical Association,* **24** (1914): 349–52.

59. Proctor's other books include the following: *Legends of the Stars* (London: G. G. Harrap, 1922); *Legends of the Sun and Moon* (London: G. G. Harrap, 1926); *The Romance of Comets* (New York, London: Harper, 1926); *The Romance of the Sun* (New York, London: Harper, 1927); *The Romance of the Moon* (New York, London: Harper, 1928); *The Romance of the Planets* (New York, London: Harper, 1929); *Wonders of the Sky* (New York, London: F. Warne, 1932); (with A. C. D. Crommelin) *Comets; their Nature, Origin, and Place in the Science of Astronomy* (London: Technical Press, 1937); *Our Stars Month by Month* (New York, London: F. Warne, 1937); *Everyman's Astronomy* (London: J. Gifford, 1939).

60. Margaret Lindsay Huggins, "Agnes Mary Clerke," *Astrophysical Journal,* **25,** (1907): 226–30, *Royal Astronomical Society Monthly Notices,* **67** (1907): 230–1, and *Agnes Mary Clerke and Ellen Mary Clerke. An Appreciation* (privately printed, 1907); other obituaries, Hector MacPherson Jr., *Popular Astronomy,* **15** (1907): 165–8, T. J. J. See, ibid., **15** (1907): 323–6, *Times,* 22 January 1907, 12d; H. P. H., "Clerke, Agnes Mary (1842–1907)," *DNB,* 2d Supplement, vol. 1, pp. 371–2; M. T. Brück, "Companions in astronomy. Margaret Lindsay Huggins and Agnes Mary Clerke," *Irish Astronomical Journal,* **20** (1991): 70–7, "Ellen and Agnes Clerke of Skibbereen, scholars and writers," *Seanchas Chairbre (Chronicles of Carbery),* **3** (1993): 23–43, and "Agnes Mary Clerke, chronicler of astronomy," *Quarterly Journal of the Royal Astronomical Society,* **35** (1994): 59–79.

61. [Agnes Clerke], "The chemistry of the stars," *Edinburgh Review,* **152** (1880): 408–33 (a discussion of six recent books).

62. Robert S. Ball, "Astronomy during the nineteenth century," *Nature,* **33** (1886): 313–14, on p. 314.

63. Clerke also received expert guidance in the preparation of her first edition from Ralph Copeland, then chief astronomer at Lord Crawford's private observatory at Dun Echt, Aberdeenshire, and later Third Astronomer Royal for Scotland.

64. Bernard Lightman (York University, Toronto) has carried out a detailed examination of contemporary reviews of Clerke's work. I thank him for a preprint of his article, "Constructing Victorian heavens: Agnes Clerke and gendered astronomy."

65. Clerke's worry about safety if she were to work at the Greenwich observatory was not unjustified. The buildings (old, overcrowded, and constantly undergoing renovation throughout the 1890s) were situated in a public park in an area of growing population. Attempted burglaries and even, in 1894, an anarchist attempt to blow up the observatory, were among the problems that had to be dealt with. By 1913 the observatory was the alleged target of a militant suffragette attack, and for a couple of years police guards were mounted to ward this off. Had Clerke taken the post offered she would have been the observatory's first female computer. At the time, computing was regularly done by boys, who started as soon as they left school at the age of fifteen or sixteen. Women computers were tried as an experiment, in the early 1890s, but the trial stopped in 1896—A. J. Meadows, *Greenwich Observatory, Volume 2, Recent History, 1836–1975* (London: Taylor and Francis, 1975), pp. 14–16, and M. T. Brück, "Lady computers at Greenwich in the early 1890s," *Quarterly Journal of the Royal Astronomical Society,* **36** (1995): 83–95 (see also the sketch of Alice Everett, chapter 9).

66. See Brück, "Agnes Mary Clerke," p. 71.

67. Ibid., p. 73.

68. R. A. Gregory, "The spectroscope in astronomy," *Nature,* **68** (1903): 338–41, especially p. 341.

69. Brück ("Agnes Mary Clerke," p. 63) has pointed out that this article on Laplace probably had its origin in a prize essay written by Agnes Clerke's brother Aubrey when he was a student at Trinity College Dublin. Clerke's biographical work on Laplace, along with other early assessments by Rousse Ball and Augustus de Morgan, was amended and extended by mathematician Karl Pearson about 1929 ("Laplace, being extracts from lectures delivered by Karl Pearson," *Biometrika,* **21** (1929): 202–16). Pearson had a poor opinion of Clerke's history of science writing. However, being an early twentieth-century man of science, very much the intellectual, and a "free thinker" in religious matters, he may well have been thoroughly irritated by Clerke's habit of mixing the spiritual with the scientific, and for falling short, on occasion, in her grasp of technical detail.

70. Obituaries, H. F. Newall, *Royal Astronomical Society Monthly Notices,* **76** (1916): 278–82, Sarah F. Whiting, "Lady Huggins," *Science,* **41** (1915): 852–5, and "Lady Huggins. Her work in Astronomy," *Times,* 25 March, 1915, 10d; Sarah Whiting, "The Tulse Hill Observatory Diaries," *Popular Astronomy,* **25** (1917): 158–63; Brück, "Companions in astronomy;" M. T. Brück and I. Elliott, "The family background of Lady Huggins (Margaret Lindsay Murray)," *Irish Astronomical Journal,* **20** (1992): 210–11; *Who was Who,* vol. 1, p. 357.

71. "On the spectrum of the spontaneous luminous radiation of radium at ordinary temperatures," *Proceedings of the Royal Society,* A, **72**

(1903): 196–9, 409–13, *Astrophysical Journal,* **18** (1903): 151–5; "Further observations on the spectrum of the spontaneous luminous radiation of radium at ordinary temperatures," ibid., **18** (1903): 390–5; "On the spectrum of the spontaneous luminous radiation of radium at ordinary temperatures. Part III. Radiation in hydrogen," *Proceedings of the Royal Society,* A, **76** (1905): 488–92; "On the spectrum of the spontaneous luminous radiation of radium at ordinary temperatures. Part IV. Extension of the glow," ibid., **77** (1906): 130–1.

72. The Hugginses' last joint technical paper was probably, "Note on the interpretation of the spectra of the components of double stars showing contrasted colours," *Astrophysical Journal,* **25** (1907): 65–6.

73. Margaret L. Huggins, comp. and ed., *Gio. Paolo Maggini, his Life and Work,* from material collected by William Ebsworth Hill and his sons Wm., Arthur and Alfred Hill (London: W. E. Hill and Sons, 1892).

74. See Joan Mason, "Hertha Ayrton (1854–1923) and the admission of women to the Royal Society of London," *Notes and Records, Royal Society of London,* **45** (1991): 201–20 on pp. 215–16.

75. John Montefiore, *A Sketch of the Life of Sir William Huggins, K.C.B., O.M.* [Begun by J. Montefiore from material collected by Lady Huggins and completed by Charles Eaten Mills and C. F. Brooke] (Richmond, Surrey: "Times" Printing Works, 1936).

76. See *Science in Ireland 1800–1930: Tradition and Reform* (Proceedings of an International Symposium held at Trinity College Dublin, March 1986), eds. John R. Nudds, Norman D. McMillan, Denis L. Weaire, Susan M. P. McKenna Lawlor (Dublin: Trinity College, 1988), especially the chapter by McKenna Lawlor, "Astronomy in Ireland from the late 18th to the end of the 19th century," pp. 85–96.

77. Kidwell, "Women astronomers in Britain."

78. There was, however, a telescope at Newnham College in use in the 1890s and at least for a time after the turn of the century. Physicist Edith Anne Stoney (Newnham, 1890–1894, Mathematics Tripos), sister of Florence Stoney (see chapter 7), was custodian of the instrument in 1894–1895, and mathematician Hilda Hudson (Newnham, 1900–1904, see chapter 8) held the same post for a period some time later.

79. Allan Chapman, "The Victorian amateur astronomer: William Lassell, John Leech, and their worlds," in *Yearbook of Astronomy* (London: Sidgwick and Jackson, 1994), pp. 159–77.

80. Because reports of findings made at an observatory were regularly published under the director's name only (and often in monograph rather than serial form), this survey may perhaps serve women in astronomical research less well than those in some other fields. However, the practice of using the senior author's name only was hardly restricted to astronomy.

81. W. H. M. Christie, *Astrographic Catalogue 1900.0. Greenwich Section,* vol. 1. (Edinburgh: H. M. Stationery Office, 1904).

Chapter 11

CHEMISTS AND BIOCHEMISTS: LAB. SPACE FOR ASSISTANTS ONLY?

Papers in the chemical sciences make up about 9 percent of the articles by women indexed in the Royal Society *Catalogue.* The largest contributions came from the United States (about 31 percent), Russia (about 26 percent), and Britain (about 26 percent); the remainder came mainly from Poland, Germany, Sweden, and Austria.

American women

Journal articles by forty-four Americans are listed by the Royal Society.[1] Most of the work was carried out at institutions along the eastern seaboard, especially the Massachusetts Institute of Technology (MIT), Yale University, and the Philadelphia College of Pharmacy. Papers from these three places together constitute about 40 percent of the American nineteenth-century contribution. Three women, Ellen Swallow Richards, an instructor at MIT, Martha Austin Phelps, a research worker at Yale, and Helen Abbott Michael, who had access to the laboratories of the Philadelphia College of Pharmacy (and later to a private laboratory), were especially productive. The University of Pennsylvania in Philadelphia also provided advanced training for early women chemists, and five of its women students published papers before 1901. With the exception of Bryn Mawr College, little chemical research came from any American women's college before the turn of the century.[2] In the Midwest the University of Nebraska stands out for the professional activities of the women chemists associated with it or its adjunct, the state Agricultural Experiment Station. On the West Coast only Stanford University had any women chemistry students who published research papers before 1901. Almost half of the women's papers reported work in analytical chemistry (see Figure 11–2, b), often mineral analyses and methods of separation, a field in which there was much activity in the United States at the time.

At MIT, women's activities in chemistry centered around ELLEN SWALLOW RICHARDS[3] (1842–1911), the most prominent American woman chemist of the late nineteenth century and one of only two women chemists starred in *American Men of Science* before 1943. The first woman student at MIT, its first woman instructor, a pioneering investigator in public health research, and later the founder of the discipline of home economics, Richards was one of the most noteworthy of America's early women workers in the sciences.

She was born in Dunstable, Massachusetts, on 3 December 1842, the only child of Fanny Gould (Taylor) and Peter Swallow. Both Swallow, a farmer and schoolteacher, and his wife were descendants of English settlers who had come to Massachusetts in the mid-seventeenth century. Ellen was taught by her parents until she was sixteen and then spent four years at Westford Academy, in Westford, Massachusetts. By the time she was twenty-five she had succeeded in earning enough money by tutoring and housecleaning to enable her to enter newly opened Vassar College. She graduated with an A.B. in 1870 and expected thereafter to teach astronomy in Argentina, having been engaged for the work by the Argentine government. When the arrangement fell through she turned to her other special undergraduate interest, chemistry.

The year 1870 was still a decade and a half before the first American woman to take a doctoral degree in chemistry, Rachel Holloway Lloyd, went to Switzerland to do so, and study in Europe was probably far beyond Ellen Swallow's means in any case. In post-Civil War America opportunities for women to pursue advanced study in any of the sciences were almost nonexistent, but, fully conscious of pushing forward into untrodden territory, she applied to MIT, which had opened to men in 1866. She was admitted, after some delay, as a non-fee-paying special student and as a candidate for a second undergraduate degree—sixteen years were to elapse before MIT's first graduate degree was awarded. Though she was unaware of it at the time, the principal reason for her non-fee-paying status was that it left the institute free to disown her as one of its students should the need arise. She graduated in 1873, receiving the first bachelor of science degree awarded to a woman by an American university. The same year, Vassar College gave her an A.M., following her submission of a thesis reporting the estimation of vanadium in an iron ore deposit.

She was an especially skillful and painstaking analytical chemist, and during her undergraduate years at MIT and for two more thereafter she worked as an assistant with two faculty members who were extensively involved in the application of chemistry to the solution of immediate practical problems. These men were William R. Nichols, a public health chemist, and John Morse Ordway, a former industrial chemist who had kept a large consulting practice. In both the public health research and the industrial problem-solving investigations in which she collaborated, the work was pioneering, requiring considerable skill and originality in execution and breadth of outlook in the interpretation of results. The efforts of Nichols, Ordway, and Swallow were not unimportant in establishing the influence of MIT in the industrial world, at a time when few manufacturing or metallurgical establishments employed chemists of their own, or even consulted them, except in extremis; routine scientific monitoring of industrial performance and professional scientific help in devising technical improvements were still in the future. Ellen Swallow went on to make good use of the first-class training the work gave her.

In 1875 she married Robert Hallowell Richards, chairman of the mining engineering department at MIT, and over the next several decades she carried out a number of analytical investigations directly related to her husband's work. Some of these analyses had noteworthy consequences. Her examination of samples of the Canadian Coppercliffe lode showed that this ore contained 5 percent nickel, missed in previous assays; Coppercliffe very soon became the center of a large nickel industry. She was also responsible for working out improvements that saved the Calumet Mills Company on Lake Superior $200,000 to $300,000 per year. In 1879 she became the first woman member of the American Institute of Mining and Metallurgical Engineers.

The economic security her marriage brought allowed her to pursue her ambition of furthering women's education at MIT. She volunteered her services, including her successful fundraising efforts, to establish and conduct a Woman's Laboratory. By 1879 she was an assistant instructor (without pay). Three years later four of her women students had met the requirements for, and been granted, MIT degrees. To its credit, considering the attitudes of the time, the institute then opened to women on the same basis as men and the pioneering Woman's Laboratory was closed.

Soon after that, Richards finally received an official appointment on the MIT faculty, becoming, at age forty-four, instructor in sanitary chemistry. This field was largely concerned with water analysis, an area in which major pioneering work was then taking place in both Europe and America. Pollution from municipal sewage and industrial wastes and the concomitant public health issues were only partially understood at the time, and analysis techniques and standards were still being developed. In 1887 the Massachusetts State Board of Health began a survey of the inland waters of the state on what was then an unprecedented scale. MIT's William Nichols, who was the leading American expert on water analysis, had died the year before; his successor, Thomas Drown, was given formal charge of the survey, but Drown acknowledged that the undisputed success of the huge Massachusetts project was due largely to Ellen Richards's experience in the area and to the very considerable time and effort she devoted to the work. In 1897, two years after Drown left MIT to become president of Lehigh University, the project was transferred to state laboratories; by then Richards had carried out the analysis of more than 18,000 samples. The survey results led to the establishment of the first state water quality standards in the United States and also brought about the development of the first modern sewage treatment plant in the country.

In 1890 MIT started a program of sanitary engineering, the first in the United States, and as instructor in sanitary chemistry, the position she held until her death in 1911, Richards taught the principles of water, sewage, and air analysis. Her well-received textbook, *Air, Water and Food from a Sanitary Standpoint* (1900), coauthored with her fellow faculty member Alpheus G. Woodman, was based largely on this sanitary engineering course. The work focused largely on analytical methods and the interpretation of results, but also outlined her general ideas about state and municipal responsibility for the maintenance of clean air, the supplying of safe water and the control of food quality. Many of these ideas seem well ahead of what we have achieved even today, almost a century later.

Throughout the 1880s and 1890s her industrial consulting work was also substantial. Particularly noteworthy were studies she carried out as chemist for the Manufacturers' Mutual Fire Insurance Company on the volatility of the machine lubricating oils then in use in large quantity in the textile factories of Massachusetts. Her investigation of the fire and explosion hazard these presented over a range of conditions led to changes that dramatically reduced risk to life and property, and consequently lowered insurance costs as well.

She continued to publish steadily in scientific journals, putting out about seventeen pre-1901 articles (some of them coauthored by her students) on analytical methods and technological procedures in both inorganic and bacteriological chemistry. She also regularly attended the annual meetings of the American Chemical Society and the AAAS, and frequently presented papers at these gatherings. For several years in the late 1890s she was one of a team of eleven MIT faculty members who produced the annual *Review of American Chemical Research*,[4] the forerunner of *Chemical Abstracts.* Her main responsibility was the preparation of the section on sanitary chemistry, which involved abstracting an enormous number of reports of state health boards and federal government agencies.

However, as early as the 1880s Richards was well aware that, despite her extensive involvement in the Massachusetts water survey and other projects, her opportunities for professional development and advancement in chemistry were limited. So she gradually turned to other areas where she felt she had something to offer. Her public health work had made her increasingly conscious of then barely recognized dangers from air and water pollution and adulterated foodstuffs in a society rapidly becoming more and more urban and overcrowded. Nutrition research and the setting up of dietary standards became special concerns, and from there she was drawn into the task of organizing the field of home economics.

At the time work in this area was being carried out by a number of loosely connected groups, including faculty members at land-grant colleges, teachers in public schools, urban leaders, and those concerned with nutrition and the pure food question. Richards put her very considerable energy and talents into the huge task of setting up formal associations, writing necessary handbooks, and providing advice and training for both students and professionals, thereby coordinating the field and giving it form and structure. She was, indeed, "consulting engineer" to the home economics movement.[5] In 1908 she became the first president of the American Home Economics Association, and she was instrumental in founding the association's *Journal of Home Economics.* Although their story is now largely a forgotten piece of American social history, it is nevertheless the case that the first practitioners of this field Ellen Richards did so much to develop played an important part in improving social conditions and the country's living standards in the early years of this century. Further, even if home economics generally had low prestige at academic institutions, it nevertheless filled a second and not unimportant role in that it provided employment opportunities for considerable numbers of educated women, especially those trained in chemistry, at a time when such opportunities were very scarce.

Ellen Richards died of heart disease at the age of sixty-eight, at her home in Jamaica Plain, Boston, on 30 March 1911, following a short period of poor health. At MIT she was remembered by colleagues for her open, cordial friendliness and her readiness to offer generous assistance in their work. The institute now has an endowed professorship in her name.

Evelyn Walton, Margaret Cheney, Alice Palmer, Helen Cooley, Isabel Hyams, Elizabeth Mason, Charlotte Bragg, and Marion Talbot[6], the other women whose work in the chemistry laboratories at MIT resulted in publications listed in the bibliography, were probably all students of Richards, though only Walton, Cheney, and Palmer entered the institute while the Woman's Laboratory still operated.

EVELYN WALTON graduated in 1881 and the following year became the third wife of Richards's friend and colleague John Ordway, then a widower of fifty-nine. He had supervised her thesis research, studies on the crystalline state, reported in her 1881 paper; she had also worked as his research assistant. They moved to New Orleans in 1884 when Ordway became director of manual training at Tulane University.

Neither MARGARET CHENEY nor ALICE PALMER (1857–1929) took degrees at MIT. Cheney, a Boston woman from an old New England family, first attended classes at the institute in 1876. The following year, she coauthored a paper with Richards on a new method for the determination of nickel in mineral analysis, which became a classic in its field. The status of women students at MIT being still somewhat uncertain at that time, she did not immediately proceed toward a degree but spent some time traveling in Europe and California. She returned to the Woman's Laboratory in 1880 and was expected to graduate in 1882, but died in September of that year. A reading room for the women students at MIT opened two years later was named the Margaret Cheney Room. Though renovated and relocated more than once, the room kept the name. Palmer also entered MIT in 1876, and continued to take classes in chemistry until 1883. Her joint publication with Richards on tannin estimation appeared in 1878. She returned to MIT for a further year of study in 1888, but does not appear to have published further work in chemistry.

HELEN COOLEY (b. 1859) trained as a teacher at Oswego Normal School, Oswego, New York, and then taught for a time in Little Britain, in the lower Hudson valley. She entered MIT in 1885, and, having already taken chemistry classes at Harvard Summer School, graduated with an S.B. two years later. Her paper on methods for testing indigo dyes appeared in the *Journal of Analytical Chemistry* in 1888. Within a few years her interests had switched to medicine, and in 1895 she took an M.D. at the New York Medical College for Women. By 1906 she had married John B. Palmer (also a physician), and was dean of the Medical College for Women and professor of ophthalmology and gynecology, while at the same time serving as an assistant surgeon at the Ophthalmic Hospital in New York City. In 1913 she moved to Los Angeles and practiced ophthalmology and otology there until at least 1918. Although she returned to Cambridge, Massachusetts, for a time in the 1920s, she spent most of her retirement in San Marino, California.

Both ISABEL HYAMS (1865–1932) and ELIZABETH MASON (1863–1935) collaborated with Richards over periods of several years. Hyams took an S.B. degree at MIT in 1888 and later joined Richards, presumably as a volunteer assistant, in investigations of bacterial pollution of water supplies. They published their findings in four joint papers between 1898 and 1904.[7] Hyams was remembered in Boston as a philanthropist associated with the Boston Tuberculosis Association and the Boston Sanitorium. She also founded the Louisa M. Alcott Club of South End, Boston. For twenty-seven years she worked as a clerk. Elizabeth Mason, who graduated from Smith College (B.A.) in 1876, studied chemistry at MIT in 1888–1889 and 1890–1893 between periods of schoolteaching in Hingham, Massachusetts, and Yonkers, New York. After receiving her S.B. in 1893 she spent three years as private assistant to Richards. Her one journal publication (1894) reported their joint investigation of the effect of heat on gluten. Mason joined the faculty of Smith College in 1896 and taught chemistry there for thirty-five years, retiring in 1931.[8]

CHARLOTTE BRAGG[9] (1863–1957) became a faculty member at Wellesley College. She was born in Milford, Massachusetts, on 17 August 1863, the daughter of Sarah (Kimball) and Arial Bragg. After graduating from Holliston High School and Wilbraham Academy she taught for five years (1881–1886) in public schools and in a private school in Chattanooga, Tennessee. She entered MIT in 1886, at the age of twenty-three, and took her S.B. in 1890. The same year, she joined the Wellesley faculty as instructor in chemistry and geology. Seven years later she was promoted to associate professor.

All her time and energy went into her teaching, which was outstanding; over the course of her thirty-nine-year career at least 1,500 students went through her classes—an impressive record for a chemistry teacher at a small women's college at the time. She never went on to an advanced degree, and although this meant that she was not promoted beyond associate professor, she nevertheless considered it best for the college as a whole that she pass on opportunities for advanced study and research to younger colleagues. She felt they were better able than she to benefit from such experience. Her one paper, coauthored with Ellen Richards in 1890, reported an undergraduate research project in analytical chemistry. She also wrote a laboratory manual. After retiring in 1930, at the age of sixty-seven, she learned braille and worked for twelve years as a Red Cross volunteer translating books for the blind. When a national drive to collect scrap metals for the war effort was at its height in 1942, she made headlines in Boston newspapers by insisting that the 100-year-old wrought iron fence that had bordered her family cemetery plot be taken for recasting. She died at Melrose, Massachusetts, on 1 September 1957, at the age of ninety-four, and was buried in the Bragg Cemetery at Holliston.

With the exception of an 1874 paper on the analysis of pharmaceutical materials by Kate Crane of the University of Michigan and an 1875 note in the *American Chemist* by Mary Reed of the Worcester Free Institute, the three papers in analytical chemistry published by Ellen Swallow in 1875 were the first in chemistry by an American woman.[10] By the 1880s, however, the second American woman to carry out early chemical research had begun her short but remarkably productive career. HELEN ABBOTT MICHAEL[11] (1857–1904) published thirteen papers in plant chemistry and four in organic synthesis and structure studies during the thirteen years from 1883 to 1896. Her work in botanical chemistry led her to the novel theory that pathways of plant evolution may be traced by chemical "markers" present in plant tissue, these markers being secondary chemical compounds, rather than the basic metabolites present in all species. Though her early enunciation of this concept was for many years overlooked, her original contribution has since been acknowledged.[12]

She was born in Philadelphia, on 23 December 1857, the youngest child of Caroline (Montelius) and James Abbott. The family being well off she received her early education from private tutors; later she studied music composition at the University of Pennsylvania and in Paris, becoming an accomplished pianist. Her interest in science was awakened in the early 1880s when she came across a copy of Hermann von Helmholtz's *Handbuch der Physiologisches Optik* (1867). From the physiology of vision she moved on to studies in general zoology and then to medicine, entering the Woman's Medical College of Pennsylvania in 1882. However, injuries caused by a fall during her second year made her withdraw. During her recuperation she turned to private research, and from 1884 to 1887 worked on various topics in plant chemistry under the guidance of Henry Trimble, professor of analytical chemistry at the Philadelphia College of Pharmacy and an expert on the tannins.[13] Scientific work was relatively well-supported in Philadelphia in the 1880s; there were a number of thriving societies as well as the colleges and institutions of the area. Abbott's educational and social background gave her fairly ready access to these societies and much of her research was presented at public lectures they sponsored. She is said to have been an excellent speaker. Her theory of biochemical markers in plant taxonomy, first published in the *Botanical Gazette* in 1886, was developed in a lecture delivered in 1887 at the U.S. National Museum in Washington, D.C. This talk was considered especially noteworthy, both on scientific grounds and as an example of what a young woman might by then aspire to in scientific work. (In Abbott's case, however, her financial independence was an important contributing factor to the scientific success that came her way; she was hardly an average young woman.)

In 1887 she went to Europe to investigate opportunities for women's higher education. She visited laboratories in Britain, Germany, Sweden, and Switzerland. Several of the botanists and chemists she met already knew her work, and, equipped with letters of introduction from Samuel P. Langley, secretary of the Smithsonian Institute, she was well received. On her return to the United States she went to Tufts College in Boston to study chemistry under Arthur Michael, one of the country's leading organic chemists. She married Michael, a man who shared her broad cultural interests, in 1888. Following a trip around the world, they settled for a year in Worcester, Massachusetts, Arthur Michael becoming director of the chemical laboratory at Clark University. However, he found the position uncongenial and in 1891 he and Helen moved to Bonchurch, on the south coast of the Isle of Wight. They lived there for four years, Arthur setting up a private laboratory and continuing his research. Helen Michael's four substantial papers in synthetic organic chemistry, two of them coauthored by Jean Jeanprêtre, resulted from work carried out in this laboratory.

After the Michaels returned to the United States in 1895 (Arthur to his old position at Tufts College) Helen remained active in chemistry for a period. She looked into the stereochemistry of sugar molecules, a relatively new area of investigation at the time, and presented a long review of synthetic work on sugars before the Franklin Institute in 1895. However, other interests soon claimed her attention, and she became increasingly involved in Boston literary circles. Her home for a time was a center for discussions that ranged over a wide field of cultural topics—science, art, philosophy, and religion. Then in 1903, having returned after an interval of almost twenty years to her early interest in medicine, she took an M.D. at Tufts Medical School. She practiced for a short time in a free hospital for the poor that she herself set up. However, she died of influenza, on 29 November 1904, at the age of forty-six. Her standing among her scientific contemporaries (and perhaps also her social position) is reflected in her membership in a number of societies, including the AAAS and two Philadelphia groups, the Academy of Natural Sciences and the American Philosophical Society.

Although many women students graduated from the Philadelphia College of Pharmacy before 1900, few of them contributed original research to the chemical literature of the period. BERTHA DE GRAFFE (1870–1948)[14] is the only one whose work is listed in the bibliography. Born on 27 September 1870, she grew up in Albany, New York. From 1891 until 1895 she served her apprenticeship in the pharmacy retail store of Otto de Keiffer in Philadelphia while also studying at the College of Pharmacy. She graduated with a Ph.G. (Graduate in Pharmacy) in 1896, her thesis research on the tannins of medicinally important *Ericaceae* appearing in the *American Journal of Pharmacy* the same year. Having a strong interest in chemistry, she stayed on at the college for another year, taking a special chemistry course offered by Abbott's mentor Henry Trimble. In 1900 she married Trimble's successor, Josiah Comegys Peacock, and for the next eighteen years assisted him in the management of their pharmacy retail store. After the business was sold in 1918 she returned to research and collaborated with her husband in several projects, including investigations in dispensing practice and plant analysis. A succession of her papers, coauthored by her husband, appeared in the pharmaceutical journals, especially the *American Journal of Pharmacy,* throughout the 1920s. Her early interest in the tannins continued and her studies on plants containing these compounds were particularly noteworthy.[15] In 1930 the Philadelphia College of Pharmacy recognized her contributions to the field by awarding her an honorary M.Sc. in pharmacy. (Her husband had received the same honor in 1921.) She died on 21 March 1948, in her seventy-eighth year.

By the last decade of the nineteenth century Yale University and the University of Pennsylvania in Philadelphia each had a number of women doing research in chemistry. Four of those at Yale (Austin, Roberts, Fairbanks, and Morris) were students of Frank Austin Gooch, and one (Albro) worked with physiologist Russell H. Chittenden; those at the University of Pennsylvania were all students of Edgar Fahs Smith.[16]

MARTHA AUSTIN[17] (1870–1933) was one of the notably productive early women research chemists in America with sixteen substantial papers to her credit, published over the course of her ten-year scientific career. Born on 13 February 1870 in Georgia, Vermont, she was the daughter of Ann Eliza (Wilson) and George Austin, farmers. After attending Easthampton high school in Easthampton, Massachusetts, she went to Smith College (B.S., 1892). She taught high school science in New York and Massachusetts for five years and in 1896 entered Yale graduate school, where she studied under Gooch in the Kent Chemical Laboratory. Her Ph.D. was awarded in 1898. She stayed on in Gooch's laboratory for another year and then took a post as assistant chemist at the Rhode Island Experiment Station. From 1901 until 1904, when she married Isaac King Phelps, then an instructor in chemistry at Yale, she taught physics and chemistry at Wilson College. Marriage gave her a chance to return to research, however, and for a time she collaborated in her husband's work at Yale. Five joint Phelps papers on the synthesis of organic esters appeared in the *American Journal of Science* between 1907 and 1908. Martha and her husband moved to Washington, D.C., in 1908, and for a year she worked as research chemist at the Bureau of Standards. However, she does not appear to have been professionally active after 1909.

Her early analytical research at Yale, which resulted in four single-author papers as well as joint publications with Gooch, concerned the development of practical and accurate procedures for metal estimations and separations, work of considerable importance to the mining and industrial interests of the time. Several of the papers reported investigations directed toward solving problems in the gravimetric estimation of magnesium and manganese. One especially elegant single-author article describing ideal conditions for the estimation of arsenic as ammonium magnesium arsenate appeared in 1900; the method she then described differs little from that which became a standard and was to remain in textbooks of quantitative analysis for decades. She also collaborated with Gooch in some particularly exacting analyses of prehistoric bronzes collected in Turkestan by the explorer and Harvard geologist, Raphael Pumpelly. Considerable expertise was required in this work, the need to minimize damage to the specimens severely restricting the amount of material to be spared for the analyses. Her results were published in the first volume of Pumpelly's *Explorations in Turkestan. Prehistoric Civilizations of Anau* (1904).

The Phelpses lived in Washington for fifteen years, Isaac working for a time at George Washington University and then with the U.S. Department of Agriculture. Martha was active in a number of academic women's clubs. In 1923 they moved to Massachusetts and shortly after that to New Haven, Connecticut, where Martha died on 15 March 1933, at age sixty-three.

Of the early women chemists who received their graduate training at Yale, perhaps the best remembered is CHARLOTTE FITCH ROBERTS[18] (1859–1917) who taught at Wellesley College for thirty-seven years, except for short absences for study and research. Her doctorate was the first in chemistry given to a woman by Yale.

Born on 13 February 1859 in New York City, Charlotte was the daughter of Mary (Hart) and Horace Roberts. Most of her childhood was spent in Greenfield, Massachusetts. She entered Wellesley College in 1876, one year after it opened. On graduating in 1880 she stayed on, becoming an assistant in the chemistry department and rising to instructor by 1882. Following a year of advanced work at Cambridge University, where she studied under the eminent Sir James Dewar, she was promoted to associate professor. Six years later she went to Yale for graduate study and received a Ph.D. in 1894.

While at Yale, in addition to laboratory research in analytical chemistry, she undertook an investigation, both broad and critical, of the background and current state of what was then the new field of stereochemistry, that branch of chemistry which deals with the three-dimensional structure of molecules. She had probably been introduced to the subject by Dewar at Cambridge. Her survey, which became her doctoral dissertation, she subsequently published as a monograph, *The Development and Present Aspects of Stereo-chemistry* (1896). Frank Gooch described it as the best and clearest introduction in English that he knew of to the background and principles of the field.[19] For a number of years it was used as an advanced textbook.

Roberts returned to Wellesley from Yale as full professor and head of her department. She was then thirty-five. Though her teaching and administrative work left little time for research, a year's leave in 1899–1900 gave her the opportunity to work in Berlin in the laboratory of J. H. van't Hoff, one of the leaders in the development of stereochemical theory; a second sabbatical leave in 1905–1906 saw her again in Europe, auditing lectures at various universities in England, Germany, and Switzerland. She had by then become interested in the life and work of the sixteenth-century Swiss alchemist and physician Paracelsus, often considered the father of modern experimental and observational science. Her last leave (1912–1913) she also spent in Europe, doing extended research on Paracelsus and the later alchemists and their influence on the development of chemistry. This historical work was never published.

Roberts died suddenly, at the age of fifty-eight, of a cerebral hemorrhage on 5 December 1917, at Wellesley. Remembered by her colleagues especially for her buoyant and fun-loving spirit, she was one of the most distinguished members of Wellesley's famous class of 1880, the second to graduate from the college and one that provided three future faculty members. She had made the college her home for most of her adult life, years that almost coincided with the first four decades of Wellesley's existence.[20] An endowed professorship at Wellesley commemorates her name.

CHARLOTTE FAIRBANKS,[21] the second woman doctoral student in chemistry at Yale University, also taught at Wellesley, although only for a short time before going on to medical studies. Her three papers, all in analytical chemistry, resulted from her work in Gooch's laboratory. She was born in St. Johnsbury, Vermont, in 1871. After graduating from Smith College (B.A., 1894) she proceeded directly to graduate study at Yale and received a Ph.D. in 1896, at the age of twenty-five. Her three papers were on analytical methods, two of them coauthored by Frank Gooch. She held a chemistry fellowship at Bryn Mawr College for a year, and in 1897 became instructor at Wellesley. Two years later she entered the Woman's Medical College of Pennsylvania; her M.D. was awarded in 1902. She worked in Philadelphia for six years and then returned to St. Johnsbury, where she practiced until the early 1930s.

The University of Pennsylvania awarded doctorates in chemistry to four women between 1894 and 1900.[22] Three of them—Fanny Hitchcock, Mary Pennington, and Elizabeth Atkinson—had studied biology as special students prior to their graduate work, Atkinson and Pennington each receiving a Certificate of Proficiency, the bachelor's degree equivalent then granted to women by the University of Pennsylvania. The fourth woman, Lily Kollock, held a B.A. from the Woman's College of Baltimore (later Goucher College).

FANNY HITCHCOCK[23] (1851–1936), the first woman to receive a doctoral degree in a scientific field from the University of Pennsylvania, was born in Brooklyn, New York, on 7 February 1851. She was the daughter of Elizabeth and Julius S. Hitchcock. In 1890 she enrolled at the University of Pennsylvania as a special student in biology. She had already studied at Columbia University and had several publications to her credit—two papers in entomology and four notes on fish osteology published in the late 1880s. After three years of study in the graduate department of the University of Pennsylvania, to which she transferred in 1891, and submission of a dissertation on analytical methods, she was awarded a Ph.D. in chemistry (1894). After a year of further work in chemistry at the University of Berlin she returned to the University of Pennsylvania as the first director of the Women's Graduate Department, established in 1892. Ten years later, when the university first admitted women as regular undergraduates, she became director of Women Students, the position she held for nineteen years.

Having adequate private financial resources, she maintained a laboratory at her home in Philadelphia and another at her country residence in Warwick, New York. She carried out a number of scientific investigations, both chemical and biological, including analytical studies on iron ores, the development of reafforestation techniques, and work in plant pathology, but she does not appear to have reported results in any of the main scientific journals. On her retirement in 1921, when she was seventy, she gave her Philadelphia house, her technical library, and her laboratory equipment to the University of Pennsylvania physics department and moved to her Warwick residence. She died there on 25 September 1936, at age eighty-five.

In 1895, one year after Hitchcock was awarded her degree, Smith's second woman doctoral student, MARY PENNINGTON[24] (1872–1952), received her Ph.D. Among American women whose scientific careers began in the nineteenth century, Pennington was probably the only one trained as a chemist who succeeded in making her way fully into the mainstream national scientific community, becoming first a U.S. Department of Agriculture biochemist and later a very successful private consultant to the food storage and transportation industries.

Born in Nashville, Tennessee, on 8 October 1872, she was the elder daughter of Henry Pennington, a Southerner, and his wife Sarah (Molony), who came from an old Pennsylvania Quaker family. The Penningtons moved to Philadelphia shortly after Mary's birth and she always considered herself a Philadelphian. She received her early education in city schools.

After being awarded her doctorate at the age of twenty-three she stayed on at the University of Pennsylvania for two years as a fellow in chemical botany. In 1897, continuing to further broaden and develop her experience beyond her basic training in analytical chemistry, she went to Yale on a one-year fellowship in physiological chemistry. When she came back to Philadelphia she entered into a variety of concurrent and related undertakings involving work in biological chemistry. For eight years, until 1906, she lectured at the Woman's Medical College of Pennsylvania and also directed its clinical laboratory. In 1901 she opened a private laboratory, the Philadelphia Clinical Laboratory. As partners in the undertaking, she engaged Elizabeth Atkinson, whom she had known as a fellow chemistry student at the University of Pennsylvania, and Evelyn Quintard St. John, a physician at the Woman's Medical College. The laboratory operated until 1907, performing chemical and bacteriological analyses for doctors and hospitals. For three years starting in 1904 Pennington was bacteriologist with the Philadelphia Bureau of Health, a position that led immediately to her first major investigation, a rigorous examination of the city's milk supply, then at times subject to distinct fluctuation in quality.[25] The system of dairy herd inspection she devised and the standards for producers and processors she developed brought about major improvements in the quality and safety of the dairy products sold in Philadelphia. Furthermore, her standards were later adopted by health boards throughout much of the United States.

From about 1907 until the end of the First World War she worked with the U.S. Department of Agriculture. Starting as a bacteriological chemist at the Bureau of Chemistry, she became, in 1908, chief of the newly established Food Research Laboratory, set up to implement the 1906 Pure Food and Drug Act. Her appointment to a position at that level, in an era when few women scientists could aspire to anything beyond assistants' posts in government agencies, was facilitated by Harvey W. Wiley, an old family friend and director of the Bureau of Chemistry until 1912. Impressed by her bacteriological work, Wiley wanted her to head the new laboratory. She took the required civil service examination, and without her knowledge Wiley changed the name on her papers to M. E. Pennington, thus obscuring the fact that she was female. She had the highest score of all the applicants for the post and was given the job, despite misgivings in some official quarters when her identity was discovered.

During the eleven years of her direction, the work of the Food Research Laboratory expanded enormously, necessitating a staff increase to fifty-five from the original four (which had included Evelyn St. John and Pennington's sister, Helen M. P. Betts). Pennington herself remained active in technical work, despite heavy administrative duties and much time spent in government-food industry negotiations. Over forty of her articles appeared in scientific journals and as U.S. Department of Agriculture bulletins between 1905 and 1919, many coauthored by members of her staff.[26] The laboratory became a center for pioneering work on the fundamental principles governing bacterial decomposition of foodstuffs. Basic research was supplemented by field trials; standards were established and practical methods developed for the preservation and storage of perishables, especially poultry, eggs, dairy products, and fish. The work, which earned Pennington a star in the 1910 edition of *American Men of Science,* led to the adoption of totally new techniques in the food warehousing and distribution industries in the United States, including the use of insulated railway refrigerator cars and careful temperature control in the transportation of eggs, poultry, and other perishables.

The new standards and procedures Pennington's team developed were strongly opposed by both industry and the general public when they were first introduced. However, she was a very successful public relations agent, and the clear proof she presented for the superiority of cold-stored foodstuffs largely succeeded in overcoming longstanding prejudices. Always avoiding an antagonistic approach, she took care to work constructively with business interests. Her collaboration with the frozen egg industry, then beset by severe technical problems, she found especially satisfactory: over the course of two years, using a combination of chemistry, bacteriology, psychology, and general sanitation, applied in a judicious and practical way, she was able to pull the industry, as she put it, "out of the mire."[27]

Pennington's teacher E. F. Smith was known for the excellent training he gave his students in industrial management. His emphasis on general principles rather than on specific techniques produced a breadth of background that permitted flexibility, while at the same time he instilled caution, ambition, and dedication. These principles would seem to have been especially well received and put to good use in Pennington's case. A nationally recognized expert on the preservation of foodstuffs, she was the official United States delegate to international congresses on refrigeration in Europe in 1908 and 1910 and at Chicago and Washington in 1913. Her work with the Perishable Products Division of the U.S. Food Administration during World War I brought her, in 1919, the award of the Notable Service Medal from Herbert Hoover, U.S. food administrator.

Soon after the end of the war, by then well known in the business world, she resigned her government post and became director of research and development in the American Balsa Company, a New York firm that manufactured insulating materials for low-temperature installations. The position paid her twice her former government salary. In 1922, when she was fifty, she opened a private consulting business in New York City, which she ran very successfully for thirty years. Her research in preservation techniques continued and she became especially interested in frozen foods. Following on from her wartime government work on the development of effective insulation for railroad refrigeration cars, she moved into the design and construction of refrigerated warehouses and household and industrial freezers and refrigerators. For almost three decades, beginning in the early 1920s, the results of her research appeared as government and refrigeration industry reports and as papers in refrigeration engineering journals. Nearly all of her more than forty articles during this period were single-authored. From 1923 until 1931 she directed the Household Refrigeration Bureau of the National Association of Ice Industries. A member of several professional societies, she joined the American Society of Refrigerating Engineers in 1920 as its first woman member and was elected a fellow in 1947. In 1940 the American Chemical Society awarded her its Garvan Medal, an honor it gives to a woman chemist whose work it considers outstanding.

Trained initially as an analytical and inorganic chemist, Pennington was, for two-thirds of her professional life, an entrepreneur very much in the male tradition. Her success in mainstream technological research at a time when women's opportunities for scientific work in the United States were extremely limited is remarkable. Although it clearly depended on exceptional abilities, energy, and dedication, she could also be considered lucky in having as a friend Harvey Wiley, who was in a position to encourage and facilitate her early work. Subsequently, as chief of the U.S. Department of Agriculture's Food Research Laboratory, she was able to establish her reputation in the business and scientific communities. It is perhaps also significant that her area of specialization, bacteriological chemistry related to food preservation, had close connections to the field of home economics, almost exclusively a women's sphere of activity. Pennington has, however, a distinguished place among early women engineers in America as well as among women chemists. And her success in engineering, although based on the outstanding reputation she had built as a government scientist, was not dependent on a male mentor; neither can it be attributed to the field of refrigeration engineering being a professional niche for women.

A serene and quiet woman, known for her "Quaker calm," she maintained strong family ties and had many close friendships. She was a lifelong member of the Society of Friends. Professional associates remembered her as a warm friend and stimulating companion. She died of a heart attack on 27 December 1952, at age eighty, after a fall in her New York City apartment.

ELIZABETH ATKINSON,[28] one of Pennington's partners in their Philadelphia Clinical Laboratory, managed the finances of the business and carried out the nonbiological work, especially analyses for metals. She was born in Wilmington, Delaware, on 7 October 1867, the daughter of Wilmer Atkinson. Like Pennington she came from a Quaker background. After receiving a Certificate of Proficiency in biology from the University of Pennsylvania in 1892, she stayed on unofficially taking undergraduate classes in chemistry. Awarded a Certificate of Proficiency in that field also (1895), and encouraged by E. F. Smith to continue her studies, she took a Ph.D. in 1898. Her two papers on metal separations, one coauthored by Smith, appeared in the *Journal of The American Chemical Society* in 1895 and 1898. She joined the staff of the Woman's Medical College of Pennsylvania in 1899, becoming assistant demonstrator in chemistry, a position she held for three years. She and Pennington maintained their joint business from 1901 until 1907 when Pennington moved to the U.S. Department of Agriculture. Thereafter Atkinson appears to have left scientific work.

The fourth woman to receive a Ph.D. degree in chemistry from the University of Pennsylvania before the turn of the century was LILY KOLLOCK[29] (1873–1951). Her series of eight papers in electrochemistry, describing her collaborative work with E. F. Smith between 1899 and 1910, places her among the small number of American women who were notably productive in chemical research during the first decade of this century.

Lily was born in Philadelphia on 27 April 1873, the daughter of Matthew Henry Kollock, a Virginian, and his wife, Katherine (McElaney), of Philadelphia. After attending girls' schools in Philadelphia and Baltimore she studied at the Woman's College of Baltimore (A.B., 1895). Receiving her Ph.D. in 1899, she spent the following year as instructor in chemistry at Vassar College and then returned to the University of Pennsylvania for a year as an honorary fellow in chemistry. From 1901 until 1907 she was head of the department of physics and chemistry at the Girls' High School, Louisville, Kentucky, and after that, for three years, dean of women at the University of Illinois, a position she does not seem to have found wholly congenial.[30] Throughout these nine years of teaching and administration she went back to the University of Pennsylvania during summers, and continued her collaborative research with E. F. Smith on the development of applications of the rotating anode and the mercury cathode in the estimation of metals in chemical analyses.[31] Clearly research was a major and continuing interest in her life, and her preoccupation with chemistry may have been viewed as reducing her commitment to administrative work.

In 1911 she married Louis John Paetow, professor of history at the University of Illinois, and shortly thereafter moved to Berkeley, California, her husband having taken a position at the University of California. She had two children, born in 1912 and 1916. Following the death of her husband in 1928, she worked for two years as assistant to the dean of women at the University of California. Later she lived for a time in the town of Ross, Marin County, California. She died on 15 December 1951, at age seventy-eight.

GERTRUDE PEIRCE,[32] the remaining woman chemist from the University of Pennsylvania whose work is listed in the bibliography, was one of a group of nine women who entered as special students at the undergraduate level in 1876. Peirce came to study chemistry with Frederick A. Genth, but E. F. Smith, then a laboratory assistant, was her principal instructor. Her single publication, on the nitration of *meta*-chlorosalicylic acid, coauthored by Smith, appeared in volume 1 of the *American Chemical Journal* (1879–1880). She received a Certificate of Proficiency in 1878 but did not proceed to a graduate degree. Comments she wrote after her graduation provide an interesting glimpse of the situation that women chemistry students encountered at that time: they regularly spent seven or eight hours a day in the laboratory, working side by side with the male students, but the young men never exchanged a single word with them. Resentment still ran high at women being allowed to attend men's classes.[33]

Among the Eastern women's colleges, Bryn Mawr was the most notable for the pre-1901 research done by its chemistry students, five papers reporting work by Mary Breed and Margaret MacDonald appearing in the 1890s. For almost four decades, beginning in the mid 1880s, Bryn Mawr's chemistry department had strong connections with Johns Hopkins University; a succession of Hopkins-trained men chemists, known especially for their effective undergraduate teaching and their contributions to the development of smaller departments, joined the faculty at Bryn Mawr, making it, by the 1890s, one of the notable centers in the country for the training of women in organic chemistry. Both Breed's adviser, Edward H. Keiser, and Elmer P. Kohler, MacDonald's adviser, had taken their degrees at Johns Hopkins. Kohler's students especially, both graduate and undergraduate, were prominent for their research contributions during the early years of the present century.[34]

MARY BREED[35] (1870–1949) was remembered especially for her later administrative work at the Margaret Morrison College of the Carnegie Institute of Technology in Pittsburgh, Pennsylvania. Born in Pittsburgh on 15 September 1870, she was the daughter of Cornelia (Bidwell) and Henry Atwell Breed. Among her ancestors was the theologian Jonathan Edwards, who in 1758–1759 served as the third president of the College of New Jersey (later Princeton University). After early schooling in Pittsburgh she went to Bryn Mawr and graduated (A.B.) in 1894. Two of her papers in analytical and inorganic chemistry appeared in 1893 and 1894, one, on the atomic weight of palladium, coauthored with Edward Keiser. She spent the year 1894–1895 on a Bryn Mawr European Fellowship at the University of Heidelberg, where she worked as a private student in the laboratory of the well-known German chemist Victor Meyer. She was probably the first American woman to study chemistry at Heidelberg. On her return to Bryn Mawr she was awarded an A.M. (1895) and six years later a Ph.D. Her doctoral research on polybasic acids was reported in the Bryn Mawr College Monograph Series in 1901.

Already an assistant professor of chemistry at the University of Indiana, Breed became dean of women there in 1901. From 1906 to 1912 she was adviser of women at the University of Missouri, and for a year thereafter head of St. Timothy's School in Catonsville, Missouri. She returned to her hometown of Pittsburgh in 1913, accepting the deanship at the Morrison Branch of the Carnegie Institute, and for sixteen years guided the development of the branch from what was primarily a trade school into a college. Retiring in 1929 she spent several years traveling before settling in Pittsburgh, where she died on 15 September 1949, on her seventy-ninth birthday.

MARGARET MACDONALD[36] (1870–1960), who received her Ph.D degree from Bryn Mawr in 1902, taught and carried out research at a number of schools and research institutions over the course of three decades. Born in Albemarle County, Virginia, on 21 February 1870, she was the daughter of Isabelle Plunket (Mackay) and Abram Addams MacDonald. She studied for two years as a special student at Pennsylvania State College (1893–1895) and then at Mount Holyoke College, from which she graduated (B.S.) in 1898. At Bryn Mawr she worked with Elmer Kohler on investigations of organo-sulfur compounds and syntheses via organo-metallic agents, coauthoring two papers in 1899. For a year (1901–1902) she taught chemistry and physics at Ashville College, and after receiving her Ph.D. taught chemistry "wherever I could get a job"[37]—at the State Normal School in Trenton, New Jersey (1902–1904), as a substitute at Vassar College (1904–1905), and at the School of Agriculture at the Pennsylvania State College (1907–1919). She also spent two years (1905–1907) as assistant chemist at the Delaware Agricultural Experiment Station.

At the Pennsylvania State College her work shifted in the general direction of home economics, as did that of many women chemists of the period (see, for instance, Richards and Bouton), and within a few years she became professor of food and nutrition. She resigned from this position in 1919, and for the last fifteen years of her professional life concentrated on research, both basic and applied. She was remarkably productive. Following a year at the University of Illinois she held an associateship at the School of Hygiene and Public Health at Johns Hopkins University (1920–1924), where she worked on yeast growth and the B vitamins. Five of her papers on this research, two coauthored by Elmer Verner McCollum, were published in the *Journal of Biological Chemistry* and the *American Journal of Hygiene*.[38] In 1924 she became biochemist at the Agricultural Experiment Station at the University of Tennessee in Knoxville. Most of her work there was collaborative, carried out with station colleagues and faculty members in the Department of Home Economic Research. A major project was the investigation of techniques for the removal of natural contaminants from milk products, reported in experiment station circulars and in papers in the *Journal of Home Economics.* Though she officially retired in 1932 at the age of sixty-two, she spent the next three years as a visiting research fellow in organic chemistry and nutrition at Iowa State College; there she again took up her yeast studies and published a report in 1935.[39]

She was a charter member of the East Tennessee Section of the American Chemical Society and a fellow of the AAAS. A council member in the Pennsylvania State College Alumni Association from its establishment, she was instrumental in establishing a loan fund for women students and in securing the college's membership in the American Association of University Women. Her last years were passed in the town of State College, Pennsylvania, where she died on 13 July 1960, at the age of ninety, one of the community's oldest residents.

At Cornell University two women students, Lillian Balcom and Martha Doan, each coauthored papers that appeared in chemical journals before 1900. Balcom was a student of William Ridgely Orndorff, who established a strong program of research in organic chemistry at Cornell. She collaborated in his studies on aldehyde polymerization, their joint paper appearing in the *American Chemical Journal* in 1894.

MARTHA DOAN[40] (1872–1960) was the first woman to receive a doctoral degree in chemistry from Cornell. She went on to a long career of teaching and administration, much of the latter part of which was in the Midwest. Born in Westfield, Indiana, on 6 June 1872, she was the daughter of Phoebe MacPherson (Lindley) and Abel Doan, farmers of Quaker background. Both the Doans and the Lindleys were prominent members of the Westfield community; the home of her grandfather Aaron Lindley had been a station on the Underground Railroad, organized to assist the passage of slaves escaping from the southern states prior to the Civil War.

From the Union High School in Westfield, Martha went to Purdue University, where she took a B.S. degree in 1891, when she was nineteen. A further year of study at Earlham College, in Richmond, Indiana, brought her a B.L. in English literature, and the following year she went back to Purdue and took her third degree, an M.S. She and her sister Mary then entered Cornell, Mary to study English literature. Martha, supported by a Cornell University Sage fellowship, worked toward an advanced degree in chemistry. Her research on compounds of thallium was carried out under the guidance of Louis Munroe Dennis, remembered for his extensive investigations on the rare earths. The work was reported in a joint paper in the *Journal of the American Chemical Society* in 1896; she received her Sc.D. the same year. Her "Index to the Literature of Thallium, 1861–1896" appeared in the *Miscellaneous Collections* of the Smithsonian Institution in 1899. Almost three decades later she brought out a third publication, a translation from the German of Fritz Paneth's *Radio Elements as Indicators and other Selected Topics in Inorganic Chemistry* (1928).

Her first job after graduating was at the Manual Training High School, Indianapolis, where she taught until 1900 when she went to Vassar College as instructor in chemistry. She was at Vassar for fourteen years, for much of the time head resident of Raymond House. She returned to Westfield in 1914 to spend time with her ailing parents, and from then on remained in the Midwest. In 1915 she joined the faculty of Earlham College, where she was concurrently professor of chemistry and dean of women. She later held similar positions at Iowa Wesleyan College until her retirement in 1937.

An energetic and effective teacher and administrator, she was also remembered especially for her erudition and her strict regard for etiquette and the social proprieties. There is no indication of any difficulty during her deanships comparable to that experienced by Lily Kollock and her successors at the University of Illinois. (In Doan's case the possession of a doctoral degree in one of the sciences would not seem to have seriously impaired her administrative abilities or reduced her capacity to deal effectively with women undergraduates!) Throughout her career she was vitally concerned with the development of educational institutions that offered strong programs in teaching, research, and university extension work, a broad combination and one to which considerable importance was attached in the Midwestern institutions where she worked during her mature years. Welcoming opportunities to expand her administrative experience, she served on several occasions as assistant or acting dean of women at the Universities of Wisconsin and Indiana during summer sessions. She was a member of the National Association of Deans of Women and a fellow of the Indiana Academy of Science. Purdue University gave her an honorary doctorate in 1950, and in 1952 Earlham College presented her with an Alumnae Citation award.

Throughout the years of her retirement in rural Westfield she was very active in community affairs, particularly the Friends Meeting and the Western Yearly Meeting. She died at age eighty-seven in Franklin, Indiana, on 15 April 1960. A Westfield village park bears her name—the Martha Doan Park. Her family home, built in 1857, was restored after her death and moved to the Prairie Connor Farm Pioneer Village near Indianapolis, where it now stands.

At the University of Virginia two women, CORA WALKER and FANNIE LITTLETON, carried out chemical research in the 1890s. Both were students of John William Mallet, an Irish chemist who spent much of his career at the University of Virginia. Among Mallet's many interests were mineral and metal analyses, including the analysis of worked metals from bygone communities. Walker's 1895 paper, published in the London Chemical Society's *Transactions,* reported her investigation of silver ornaments from pre-Spanish conquest Inca graves in Peru, brought back to the University of Virginia by J. Lawton Taylor, a mining engineer and former Virginia student. The paper includes some interesting speculation on early Peruvian metallurgical techniques and gives references to a number of old Spanish authors who wrote on that subject.

Two of Littleton's three papers were also concerned with metals, though hers were obtained from less exotic sources, her 1895 work on the structure of silver amalgam being done with silver from the United States Mint in Philadelphia. Littleton was born in Farmville, Virginia, on 10 January 1869, the daughter of the Rev. Oscar Littleton and his wife Alice (Bernard). After graduating from the State Normal School in Farmville at the age of twenty, she taught for three years at Martha Washington College, in Abington, Virginia. From 1893 until 1902 she was an instructor of chemistry and physics at the Farmville State Normal School, and throughout much of this period attended summer schools and studied chemistry as a private student of Mallet at the University of Virginia. She took a B.S. degree at Cornell University in 1900. In 1902 she married Linus Ward Kline, instructor of psychology and pedagogy at the State Normal School. She later lived in Duluth, Minnesota (where her husband taught at the State Normal School), and became active in church work and YWCA administration.[41]

There were several Eastern institutions at which no more than one woman chemist carried out research reported in the chemical literature before 1901. From Columbia University and Barnard College, Harriet Winfield Gibson and Hermann T. Vulté published a joint paper in the *Journal of the American Chemical Society* in 1900. It was the first of a series of three substantial publications they coauthored on the chemistry of oils and soaps. Gibson received her Ph.D. degree from Columbia University in 1899.[42] Delia Stickney, an 1889 MIT graduate who taught chemistry for many years at the English High School in Boston, published a note in the *American Chemical Journal* in 1896 on observations, made with her students in the school laboratory, on the reduction of copper sulphide by heat. The 1875 paper in the *American Chemist* by Mary F. Reed, an assistant in chemistry at the Worcester Free Institute of Industrial Science, was one of the earliest published by an American woman chemist. Reed attended summer school courses in both botany and chemistry at Harvard in 1876 and 1877 and at that time was teaching physical science, presumably in a high school, in Philadelphia.

An 1897 paper by ELLEN P. COOK, who taught chemistry at Smith College, reported research on optical rotation done at the University of Berlin. Cook was born in Ripon, Wisconsin, on 21 June 1865, the daughter of Martha (Smith) and Elisha W. Cook. After early studies at Ripon College she went to Smith, where she took a B.S. in 1893 while also working as assistant in the chemistry department. She stayed on as instructor, had a year's leave for advanced study in Berlin (1897–1898) and continued graduate work at Columbia University. There she studied under the direction of organic chemist Marston Taylor Bogert, collaborating in research on syntheses in the quinazoline series of compounds; her joint paper with Bogert appeared in 1906. She received an A.M. degree the same year and was promoted to associate professor at Smith College. Her manual, *Laboratory Experiments in Organic Chemistry,* came out in 1915.[43]

Finally, among the women chemists who trained at institutions in the eastern region of the country, there is RACHEL HOLLOWAY LLOYD[44] (1839–1900), who slowly, over the course of eight years, got the equivalent of an undergraduate science education at Harvard Summer School. She later took a Ph.D. at the University of Zurich and was most likely the first American woman to receive a doctorate in chemistry. She is best remembered for her work on a large-scale research and development project critical to the establishment of the sugar beet industry in Nebraska, work she carried out while on the faculty of the University of Nebraska. Like Martha Doan, she could well be considered a Midwestern chemist, although, with the exception of her doctoral dissertation research, her journal publications resulted from her joint work with Charles Frederic Mabery at Harvard in the early 1880s.

Lloyd's life during childhood and early womanhood was marked by one tragedy after another, and her courage and steadfastness in continuing on to her educational goals and her ambitious career plans are remarkable. Her Nebraska colleagues later described her as an energetic, attractive, and cultured woman of strong personality and broad interests. She was especially liked and admired by her students, who found in her a patient and sympathetic adviser.

Born in the small town of Flushing, in eastern Ohio on 26 January 1839, she was the second daughter in the Quaker farming family of Abby (Taber) and Robert S. Holloway. Of the four Holloway children she was the only one who survived past infancy. Her mother died when she was five and her father when she was twelve, leaving her with her stepmother, Deborah (Smart) Holloway. She acquired some early education at the Friends School in Flushing, and in 1859, at the age of twenty, married Franklin Lloyd, a Philadelphia chemist also of Quaker background. Little more than six years later Franklin Lloyd was dead; their two children had predeceased him, both dying in infancy. Then in poor health herself, Rachel spent some time in Europe. When she came back she turned, of necessity, to teaching as a means of earning a livelihood, her first position being at the Chestnut Street Female Seminary in Philadelphia.

Starting in 1875 she spent six summers during the following eight years at Harvard, which in 1874 had begun to offer summer courses of intensive instruction especially designed for college and high school teachers. She studied botany during two summers and chemistry all six, specializing in advanced analysis and organic chemistry and collaborating with Charles Mabery on a research project on the synthesis of substituted acrylic acids. The work was reported in three joint papers published between 1881 and 1884 in the *American Chemical Journal.* Meanwhile, in 1880, she had moved to a second teaching position in Clifton Springs, New York. Three years later, having by then several sessions of summer school work to her credit, she became professor of chemistry at the newly opened Louisville School of Pharmacy for Women, in Louisville, Kentucky, and at the same time instructor of chemistry at Hampton College, a small liberal arts college for women opened in Louisville in 1877.[45]

Lloyd's ambitions reached beyond teaching in a small college, however, and in 1885 she resigned her Louisville positions and went to the University of Zurich, which had been formally accepting women students since the 1860s. Having already considerable research experience behind her, she completed the degree requirements in two years and received her Ph.D. in 1887. She was then forty-eight. Her dissertation research on the high-temperature conversion of phenols to aromatic amines, carried out under the direction of Victor Merz, was published in the German Chemical Society's *Berichte* in 1887. She had joined the society the previous year, one of the first American women to do so.[46] The spring and early summer of 1887 she spent in London at the South Kensington School of Science and the Royal School of Mines. From there she accepted a one-year position as acting associate professor of chemistry at the University of Nebraska, joining Henry Hudson Nicholson (a fellow chemistry student at Harvard in the summers of 1880 and 1883), who had proposed her appointment to the Nebraska faculty.[47]

Pressed by large increases in student numbers, the university had opened a new chemical laboratory in 1886. Teaching loads were heavy. Lloyd also had responsibilities as assistant chemist at the Nebraska Agricultural Experiment Station, which was staffed by the university faculty. There she joined Nicholson in a full-scale examination of the economic feasibility of sugar beet cultivation in Nebraska, an extensive many-year undertaking.

The late 1880s marked the beginning of a period of rapid development in Nebraska's agricultural industries and also the beginning of federal support for experimental work at the agricultural station. Lloyd's preliminary investigations in 1888 on the sugar content of a sample of Nebraska-grown beets produced encouraging results. She and Nicholson then organized and directed a larger-scale experiment, involving the determination of the sugar content of beets from many test plots throughout the state. Much of the laborious analytical work she carried out herself. The results fully demonstrated the potential of the industry in Nebraska. The state's first two sugar factories started operating in 1890 and 1891. In 1892 the university opened a Sugar School, one of only two in the country, and the only one dealing with beet-sugar technology; the four-month course prepared students for the analytical work of the factory laboratories and in addition provided instruction in beet culture.[48] Sugar beet became one of Nebraska's major agricultural industries; by 1988, 100 years after Lloyd carried out her preliminary work, the value of the crop was almost $57 million.

Failing health made Lloyd give up her position at the experiment station in 1891, although she continued to teach at the university until 1894. She was a very popular professor, who, although she conducted her share of the more advanced courses, frequently taught both lecture and laboratory sections of the larger, lower-level classes, which often accommodated about eighty students. This gave her contact with most of the undergraduates who came through her department. She also took pains to meet students in less formal settings, being active in a number of their clubs and organizations.

Following her resignation she returned to Philadelphia, where she had Lloyd relatives. She died in Beverly, New Jersey, on 7 March 1900, aged sixty-one. Despite her disappointingly short career, her success in securing a senior appointment on the chemistry faculty of a state university in 1887, and her major role in an extensive and far-reaching agricultural research and development project before the turn of the century give her a special place in the history of women chemists in America.

Rachel Lloyd was the first of a succession of women chemists who worked at the University of Nebraska or the Agricultural Experiment Station from the 1880s onward. The second was ROSA BOUTON[49] (1860–1951), who joined the chemistry department as a junior instructor one year after Lloyd arrived, and who, beginning with a course in "domestic chemistry" in 1894, went on to an outstanding career as the builder of the university's School of Domestic Science.

Rosa was born on 19 December 1860 in Albany, near the present town of Sabetha in northeast Kansas, shortly after her parents, Fanny (Waldo) and Eli F. Bouton, arrived in the region as pioneer settlers from New York State. Eli Bouton was well-educated for the time and worked as a schoolteacher for a number of years. Following the death of his wife in 1867 he married again. Rosa grew up in a family of five children. She studied at the State Normal School in Peru, Nebraska, taking a two-year teaching certificate in 1881.

While she was there her considerable abilities were noticed by Henry Hudson Nicholson, who taught at the Normal School until he became professor of chemistry at the University of Nebraska in 1882. In 1888 Nicholson invited her to join the expanding chemistry department at Lincoln. She continued her studies while helping with instructional work and took a B.S. degree in 1891 and an A.M. in 1893. From 1891 she held a regular faculty position. Joining the American Chemical Society in 1893, she became especially active in the work of the Nebraska local section, frequently presenting papers, either accounts of her own work or reviews of current research. She was elected to the section's executive committee in 1897.[50] Her one journal publication appeared in the *American Chemical Journal* in 1898—a paper on phenylglutaric acids coauthored by her fellow faculty member Samuel Avery.[51]

In 1894, on a tour of chemical laboratories on the Eastern seaboard including those at MIT, she visited Ellen Swallow Richards, then much involved in the organization of home economics as a university-level discipline. Bouton's first course in "domestic chemistry," which included instruction in the qualitative analysis of foodstuffs, detection of adulterants, and methods of domestic sanitation, was offered the same year. For the next four years her time was divided between this new branch of applied chemistry and her regular classes in standard analytical methods.

In 1898 the university, responding to pressure from alumnae groups, decided to establish a separate department of domestic science, and Bouton was informed that she had permission to proceed with the task of organizing it. She was given an initial appropriation of $15 to equip a room with the necessary cupboards, a table, sink, stove, and supplies. (She also had the support of Nicholson who arranged a small amount of credit from the chemistry department.) All of the necessary teaching was to be done by her alone since no funds were allocated for additional instructors—"money was harder to get than blood out of dry bones."[52] She spent the summer of 1898 in Boston, at her own expense, learning enough of the basic material of the new field to begin the undertaking; her new department opened with eleven students in the fall of the same year. In 1900, her student enrollment having meanwhile increased in the two-year period to forty-one, she became director of Nebraska's School of Domestic Science, and in 1906 she expanded the program from a two-year course to a full four-year college course, which included teacher training. Severe lack of funding for staff continued, which meant that she often had to teach new branches of the field herself initially until specialists could be taken on. A new building was provided for the school in 1908 and much of the planning and architectural design was done by her, after careful and extensive studies of established domestic science facilities at other universities (again carried out at her own expense). By 1912 enrollment had increased to 310 students, and the program became well known for its high professional standards and for the large number of distinguished leaders it trained.

Through her articles in *The Nebraska Teacher* and *The Nebraska Farmer* and her extensive work with schools throughout the state, farmers' institutes, and the state Home Economics Association, Bouton was well known in Nebraska. She was also prominent in the national Home Economics Association and well liked in her own department. She therefore received a considerable shock when, in 1912, she got a request from the university's Board of Regents to step down from the headship of her school and make way for "more effective leadership."[53] She was then fifty-one.

Declining to stay on temporarily in a lower position (she felt such a course would not be in the best interests of the school), she went to California and spent some time with her invalid stepmother. For a few years she ran a specialty bakery in San Diego, but wartime scarcities made it unprofitable. About 1918, though nearing sixty, she returned to educational work and became home economics teacher in a high school in Winkelman, a small Arizona mining community. After one term she was asked by the Arizona Agricultural Extension Service to be home demonstration agent for the three northeast counties of the state, Coconino, Apache, and Navajo, a region covering almost 40,000 square miles. For the next fourteen years, despite the wild country, few roads, and long distances between small settlements, she carried out her work of community instruction and when the need arose acted as practical nurse and general counselor as well. Her advice to grow vegetables to vary the common "beans and biscuit" diet markedly improved the nutrition of the region's population. In 1935, at the age of seventy-four, she retired and returned to California, where she lived in rather straitened circumstances with her sister May Bouton and a widowed sister-in-law. Over the years she kept in touch with the home economics department at the University of Nebraska and took great satisfaction in its growth and development. She died in Pomona, California, on 15 February 1951, shortly after her ninetieth birthday, following a long illness.

MARY LOUISE FOSSLER[54] (1868–1952), a third Nebraska woman chemist whose research appears in the nineteenth-century journal literature, was born in Lima, western Ohio, on 14 September 1868. After receiving a B.S. degree from the University of Nebraska in 1894 she was high school principal for two years in Weeping Water, a small Nebraska town about thirty miles east of Lincoln. She then returned to the university and continued studies in chemistry, taking an A.M. in 1898. Her joint paper with Samuel Avery on substituted phenylglutaric acids appeared in the *American Chemical Journal* the same year. She became an instructor in the chemistry department in 1899 and was promoted to assistant professor of chemistry and associate professor of physiological chemistry in 1908.

Fossler came from a family of German immigrants, and both she and her uncle Lawrence Fossler (professor of German at the University of Nebraska), along with several other faculty members, were accused of disloyalty and pro-German sentiment when America entered the First World War. Charges against her were later dropped and she was not, like a number of others at the University of Nebraska and elsewhere, dismissed from her post. Except for a year's leave to take classes at Northwestern University medical school in 1902, she taught at Nebraska until 1919, when, at age fifty-one, she moved to the University of Southern California. Her interests had gradually shifted from chemistry to the biological sciences; in California she taught biochemistry and zoology until her retirement in 1937. She died in Pasadena, California, in January 1952, at age eighty-three.

Elsewhere in the Midwest at the universities of Cincinnati, Chicago, Kansas, Kentucky, and Michigan, and at Iowa Agricultural College a few women were active in chemical research before the turn of the century. At the University of Cincinnati Mary Owens Hooker and Helena Stallo studied under Frank Wigglesworth Clarke in the early 1880s. Owens Hooker coauthored three short papers with Clarke on the constitution of complex inorganic salts, and Stallo one in the same area. After receiving her B.S. degree, Owens Hooker continued her studies in London, where she investigated molecular rearrangements in organic systems under the guidance of Francis Robert Japp at the Royal College of Science. The work was reported in a joint paper published in 1885.

At the University of Chicago, ELIZABETH JEFFREYS[55] (1864–1940) received her Ph.D. in 1898; it was the only doctorate in chemistry the institution conferred on a woman before the turn of the century. Born in Liberty, Ohio, on 9 July 1864, Elizabeth was the daughter of Elizabeth (Rees) and John William Jeffreys. She graduated from Oberlin College (A.B.) in 1887 (and eight years later was awarded an honorary B.Ph.). After some years of high school teaching she enrolled in graduate work at Chicago, the first of the more than twenty women doctoral students of chemist Joseph Stieglitz. Her elegant paper in the *American Chemical Journal* (1899, first published in German in 1897) reported a modification of the Hofmann reaction for the preparation of primary amines, an important synthetic route in organic chemistry discovered by the German chemist A. W. Hofmann sixteen years earlier. Jeffreys' modification produced good yields of higher molecular weight amines, not readily obtainable by the original procedure; her method was given in standard textbooks for decades.[56] Her professional career of more than thirty years was in teaching and administration, however, much of the time at Pritchett College in Glasgow, Missouri, where she became head of the science department in 1905 and president of the college in 1919. For a time she also served as superintendent of Glasgow schools. She retired in 1922, and died on 14 March 1940, when she was seventy-five, in Hubbard, Ohio, at the home of her sister.

Mary Eva Clarke, a student of Joseph Hoeing Kastle at the University of Kentucky in Lexington, took a B.S. degree in 1897, an M.S. in 1898, and a B.S. in education in 1910. She coauthored one paper in inorganic chemistry with Kastle in 1899 and their collaborative research continued after the turn of the century, with three joint papers on topics in both inorganic chemistry and biochemistry appearing in the period up to 1903.[57] She taught high school science in Lexington.

Mary Rice (later Perkins) at the University of Kansas and Lola Ann Placeway at Iowa Agricultural College each coauthored a paper in analytical chemistry in the 1890s. Placeway was a student of Alfred Allen Bennett. Rice worked with Edgar Bailey, and took an M.A. degree in chemistry and mathematics at Kansas in 1896. She had previously taken both a B.A. (1887) and a pharmacy degree (1888). She lived in Lawrence, Kansas, at least until 1928.[58]

Two women students at the University of Michigan (Harriet Bills and Katherine Crane) authored or coauthored work that was published in the 1870s. Harriet Bills was from Boston. Her joint paper (1879), with physiological chemist Victor Vaughan reported her study of the lime content of eggs and eggshell before and after incubation. Katherine Crane[59] was an 1874 graduate of the University of Michigan College of Pharmacy. Her article on methods for establishing the identity and purity of oils, which appeared in the *American Journal of Pharmacy* that year, has the distinction of being, in all probability, the first journal publication by an American woman in the chemical sciences. It predated the earliest published work of Ellen Richards by a year. Crane married Otis Coe Johnson, a fellow graduate of the University of Michigan College of Pharmacy and later a faculty member at the University of Michigan.

A third woman who carried out chemical research at the University of Michigan before 1900 was LAURA LINTON[60] (1853–1915). Her 1894 paper was most likely the first full paper by a woman to appear in the *Journal of the American Chemical Society*. Articles published in that journal in its early years varied noticeably in quality, but Linton's thirteen-page report, "On the technical analysis of asphaltum," stands out for both solid content and clear presentation. It and a second related communication by Linton that appeared in 1896 were received with marked interest by the chemical community. An extensive discussion followed the initial presentation of the 1896 paper at the Cleveland meeting of the American Chemical Society in December 1895; the commercial importance of asphalt had already been recognized, not only as a hydrocarbon source but also as material for improving the then common unpaved roads.

Like Rachel Lloyd and Martha Doan, Linton came from a Midwestern Quaker farming background. She was born in Mahoning County, Ohio, on 8 April 1853, the oldest of the four children of Christina (Craven) and Joseph Wildman Linton. After farming in Ohio, Pennsylvania, and New Jersey, the Lintons settled in Wabasha County in southern Minnesota. Laura trained as a teacher, graduating from Winona Normal School in 1872 at the age of nineteen. She then entered the Academic Department of the University of Minnesota at Minneapolis and took a B.S. degree in 1879. Chemistry was her special interest, and, being known as a careful and conscientious student, she was asked during her senior year to help in a research project. Her assignment was the analysis of a translucent green mineral, collected by faculty member Stephen Farnum Peckham and one of his colleagues during geological survey work along the northern shore of Lake Michigan. The material turned out to be a new variety of the silicate thomsonite. Though Linton was not a coauthor of the paper reporting the analysis, the mineral was given the name lintonite in her honor.

Following her graduation she taught for a year at the high school in Lake City, Minnesota. She was then invited by Stephen Peckham to assist him in the preparation of a comprehensive report on the nation's oil industry for the 1880 United States Census. Peckham's 301-page monograph, *Production, Technology and Uses of Petroleum and its Products,* the work of two years, could almost be described as a classic. It encompassed every aspect of the industry. For the first time in an American publication, a thorough evaluation of the available scientific literature on petroleum chemistry and geology was presented, a considerable amount of material from French and German sources being incorporated. Translations into English were prepared by Linton. She also drew many of the technical illustrations.

In 1882 she registered at MIT and most likely studied in the Woman's Laboratory under Ellen Richards, where she appears to have carried out a research project on the dyeing of silk fibers. She did not take a degree, however, and the following year, at age thirty, went to Lombard University in Galesburg, Illinois, as Conger Professor of Natural Science. By 1884 she was back in Minneapolis as head of the science department in Minneapolis Central High School, a position she held for ten years, while helping to pay the educational expenses of a younger brother and sister.

Beginning in 1894, for a brief but productive period, she returned to research, first in California and after that at the University of Michigan. Once again she was recruited for the work by her friend Stephen Peckham. By then a recognized petroleum expert with a distinguished record in refining technology, Peckham was employed by the Union Oil Company of California for a few months in 1894 in an attempt to improve the performance of the company's sole refinery (that at Santa Paulo) and bring it through a difficult period.[61] While there he continued his ongoing research projects on asphalt. Laura's detailed report of her findings on the composition of twenty-three asphalt samples collected worldwide was published in 1894. As well as being the first full paper by a woman in the American Chemical Society's *Journal,* it was probably the first reporting work by an American woman chemist in an industrial laboratory. The research was concluded at the University of Michigan, both Peckham and Linton having returned to Ann Arbor by 1895. Her third and final paper (1896), reporting analytical work on Trinidad asphalt, was coauthored by Peckham, who later quoted her results at great length in his 1909 technical monograph, *Solid Bitumens.*

Linton then went back to Minneapolis and, at the age of forty-three, enrolled in the College of Medicine at the University of Minnesota. Like her contemporaries Ellen Richards, Mary Pennington, Margaret MacDonald, and Lily Kollock, she doubtless recognized the limited possibilities open to women for career advancement in her field. She may also have had positive reasons for her change of profession. Interest in medicine was strong in her family, both her brother, William Linton, and her sister, Sarah Linton Phelps, having already completed medical school, partly thanks to Laura's financial help. Both were by then practicing in Minneapolis. Perhaps she had simply been waiting her turn for medical training. In any event, Laura received her M.D. in 1900 and immediately thereafter joined the staff of the State Mental Hospital in Rochester, Minnesota. An energetic and innovative doctor and administrator, she rose to the position of assistant superintendent in charge of the women's wards. The program of needlework and handicrafts she introduced for her patients was one of the institution's first attempts at occupational therapy. Further, as one of the heads of the nurses' training school, she taught a course on dietary principles and cooking methods at a time long before such material was included in standard nurses' training. She remained on the hospital staff until her death on 1 April 1915, shortly before her sixty-second birthday.

Very little chemical research was published by women studying at institutions in the western region of the country until after the turn of the century. The only pre-1901 papers would appear to be those by two students at Stanford University, which opened in 1891. Minnie Yoder and Lillian Ray[62] each coauthored one paper in inorganic chemistry and physical chemistry, respectively, but neither pursued a career in the field. Yoder was from Iowa and graduated from the West Des Moines High School in 1890. She received an A.B. from Stanford in 1895 and married Howard C. Lucas the same year. Later she lived in Olympia, Washington. Ray (b. 1873) took an A.B. degree with a concentration in physiology at Stanford in 1897, an A.M. (Stanford) in 1901, and an M.D. at Johns Hopkins Medical School in 1908. Thereafter she practiced in Los Angeles. She married Edwin Albion Titcomb in 1920 and had given up private practice by 1925, taking instead a position as physician at the University of California, Los Angeles, where she remained until at least 1942.

No American woman whose professional work began before the turn of the century can be said to have had a notable career in mainstream research in an established branch of chemistry. Those whose overall careers were the most outstanding moved to new areas, Pennington to bacteriology and food chemistry and then to refrigeration engineering, Richards to public health chemistry and then to home economics, Bouton to home economics, MacDonald to biochemistry, nutrition, and research related to home economics. Abbott Michael's original ideas, which she did not go on to test or develop extensively, concerned topics at the borderline between botany and biochemistry. Lloyd was already close to fifty before she was able to qualify for and find a satisfactory academic position; when she did it was at a recently established Midwestern state university at a point in its development when it was still unlikely to attract the country's leading male chemists. Further, her subsequent work was in agricultural research. Both Austin Phelps and Kollock, the two notably productive women research chemists of their time who each succeeded in working for about a decade in established branches of the field, did so largely by contributing to men's projects. The early women chemists in this study who were faculty members at women's colleges (Roberts, Bragg, Mason, and Cook) carried out very little research compared with their women colleagues in a number of other fields (for instance, astronomy and psychology). In fact, except during study leaves at better-financed and better-equipped institutions, they attempted little laboratory research. It is most remarkable that at the one women's college (Bryn Mawr) where a vigorous research program was built up before the turn of the century, the faculty members responsible were men.

The "marginalization" of women in science is particularly evident among these early chemists. By the 1890s, advanced, graduate-level training in the field was becoming available to American women, but beyond that, opportunities were limited in the extreme, professional positions were few and far between, and women chemists failed to initiate research programs even where it might have been hoped they could do so (in the women's colleges).[63] Turn-of-the-century chemical research in the United States was very much a profession for men. The combination of the craft tradition of the field and its relationship to the country's mining and petroleum interests would appear to have already set in place the conservative outlook that for long made it one of the most difficult branches of scientific work for women to penetrate at anything beyond low-level assistant positions, whatever their academic qualifications.

British women

Papers in the chemical sciences by thirty nineteenth-century British women are indexed in the Royal Society *Catalogue*.[64] Those thirty form the second largest national group in the discipline. Most of them trained at Cambridge or at one of the London colleges, especially Bedford College for Women, University College and the Royal College of Science (later part of Imperial College). Some had undergraduate-level education in London and went on to further work at Cambridge; others studied first at Cambridge (Newnham or Girton Colleges) and then did postgraduate research in London. A few were students at other colleges, including University College Bristol, Yorkshire College of Science (later part of the University of Leeds), Mason Science College (later absorbed into the University of Birmingham), University College Dundee, and Somerville Hall Oxford.

The first of the British women chemists to produce a sustained output of research over an extended period was EMILY ASTON.[65] Born in 1866, she received her early education at Queen's College.[66] She then went to Bedford College, where she carried out her first original work in chemistry, crystallization studies done under the direction of Spencer Pickering and reported in a joint paper in the *Transactions* of the Chemical Society in 1886. She entered University College in 1885, studied geology and mathematics as well as chemistry, and was awarded a B.Sc. (first class honors, geology; third class, chemistry) in 1889.

For the next ten years she continued to take classes at University College, concentrating mainly on analytical chemistry and collaborating in research projects with several of the men chemists on the staff as well as with geologist Thomas George Bonney. Her publications, most of them joint, covered a wide range of topics including mineral analyses, atomic weight determinations (carried out with Sir William Ramsay), and organic structure studies (with John Norman Collie). Perhaps the most noteworthy were four papers on molecular surface energy, coauthored with Ramsay between 1893 and 1902.[67] In the late 1890s she spent some time in Switzerland and France, carrying out research with Philippe Auguste Guye at the University of Geneva on optical rotation, and with Paul Dutoit at the Sorbonne on electrolytic conductivity and molecular association. Her name appears in the chemical literature on fourteen publications in the sixteen years between 1886 and 1902, after which she disappears from the records and presumably dropped out of research. Although she was clearly working only as an assistant on other people's projects, Aston's published output is remarkable for a woman chemist of the period.

Others who collaborated in research carried out at University College in the 1890s were Dorothy Marshall, also a student of Ramsay, Lucy Hall, and Winifred Judson.

DOROTHY MARSHALL was born in 1868 and entered Bedford College in 1886. Two years later she enrolled at University College, where, in addition to chemistry, she studied physics and electrical technology; she graduated with a B.Sc. (third class honors, chemistry) in 1891. A conscientious student, she won several awards, taking three silver medals, in analytical, organic, and general chemistry in 1888–1889, and a prize the following year in philosophy and logic. In 1889 she held a Tuffnell scholarship. As a postgraduate student at University College until at least 1894, she took classes in analytical chemistry and carried out research with Ramsay on the latent heat of evaporation. Her four substantial papers in this area, one coauthored with Ramsay and one with Ernest Howard Griffiths, appeared in 1896 and 1897.

Lucy Hall (b. 12 November 1873) took a B.Sc. at University College in 1896 and collaborated with J. Norman Collie in the preparation of organic nitro compounds and amines. Their joint paper appeared in the *Transactions* of the Chemical Society in 1898. Shortly after that she worked for a time with Friedrich Stanley Kipping, professor of chemistry at University College Nottingham, on further synthetic studies in organic systems.[68] Winifred Judson collaborated with J. W. Walker on reaction rate studies under the general direction of Ramsay. She held a B.Sc.

CLARE DE BRERETON EVANS, who held both B.Sc. and D.Sc. degrees, was a postgraduate student in the 1890s at the Central Technical College, London. She studied aromatic amines, her main paper appearing in the *Transactions* of the Chemical Society in 1897. Later she became lecturer in chemistry at the London School of Medicine for Women, but continued research, working at University College where she undertook a search for trace elements in residues from thorium minerals being examined by Ramsay. She also worked on the closely related matter of improving techniques for weighing extremely small quantities of materials.[69] She joined the German Chemical Society as an associate member in 1897.

Three of the London women whose papers are listed in the bibliography, Halcrow, Dougal, and Whiteley, were associated with the Royal College of Science, South Kensington. The 1880 paper by Lucy Halcrow was probably the first coauthored by a woman to appear in the *Transactions* of the Chemical Society. As a teacher in training, Halcrow took classes in chemistry and physics at the Royal College of Science in 1880–1881,[70] while also assisting Sir Edward Frankland in an investigation of the oxidizing effect of air on peaty water. The study was part of Frankland's many-year examination of river pollution, extremely important public health work for which he was knighted in 1897.

The first publications of MARGARET DOUIE DOUGAL were a series of four articles on the teaching of chemistry, which appeared in *Chemical News* in 1893, but she also carried out research in inorganic chemistry under Thomas Edward Thorpe. Their joint investigations of mixed salts of chromium appeared in the Chemical Society's *Transactions* in 1896.

Dougal's most interesting research from a general point of view was her analysis of a specimen of early iron of Scottish manufacture. In 1887 there had appeared in the *Transactions of the Inverness Scientific Society* (v. 3) a paper by W. Ivison Macadam, in which he described the remains of what may well have been some of the first ironworks in Scotland. The work sites were in somewhat wild and isolated country in Western Ross, in the northwest Highlands, along the shores of Loch Maree and on the banks of streams that flowed into and out of the loch. The area had an adequate supply of timber and running water, primary requirements for the industry as it then operated. Workings near Fasagh on the northeast shore of the loch dated from the late sixteenth century, and a similar site at Letterewe, lower down the loch, was known to have been established in 1609. Raw material had initially been bog ore, found in the immediate vicinity, although later ore was imported from Cumberland; the necessary charcoal was produced from the timber from the surrounding hills. Still in place in the 1880s were the remains of buildings, sluices, sand beds, and other vestiges of what was once an important enterprise and one that had been in operation over a considerable period. There was also a cemetery where immigrant English workers had been buried (Cladh nan Sasunnach). In 1892 Thorpe visited the area and brought back to South Kensington two iron bars. One of these was tested at a nearby engineering laboratory for its behavior under stress and its fracturing characteristics; Margaret Dougal determined its chemical composition. The results of these investigations were quite remarkable, and they provide an interesting comment on the quality of sixteenth-century Scottish craftsmanship in the iron manufacturing business; all the tests, physical, chemical, and mechanical, clearly demonstrated "that the average characteristic of even the best merchant bar of today, despite the enormous advance in metallurgical skill . . . is but little if at all superior to that of the product of three centuries ago, much of which was derived from relatively poor ores smelted by comparatively primitive methods." Dougal further noted that "the fibre is very fine and delicate, and resembles the Swedish iron used by English smiths in the seventeenth and eighteenth centuries."[71]

For a period of fifteen years, until 1909, Margaret Dougal worked as a compiler and indexer, preparing *A Collective Index of the Transactions, Proceedings and Abstracts of the Chemical Society.* The idea for such an index, covering material from 1873 on, was Thorpe's, but almost all the work was carried out by Mrs. Dougal. The first two volumes, covering the twenty years up to 1892, took her upward of five years to produce. After she had completed this labor Sir James Dewar made a point of congratulating her in his presidential address to the Chemical Society. Emphasizing the great service the work had already been and would continue to be to members, he went on to note that, although the task had been of unexpected length and complexity because of lack of system in the previous annual indexing, Margaret Dougal's compilation was "an example of thoroughness and accuracy to her successors."[72] A whole staff of indexers was forthwith engaged for the job of annual indexing. Margaret Dougal continued the collective indexing for the period up to 1902, thus preparing the first three volumes of what remains an invaluable collective series. It is the only one in English devoted to the chemical literature, British and foreign, of the late nineteenth and early years of the twentieth century, the period immediately preceding that covered by *Chemical Abstracts.*

MARTHA WHITELEY[73] (1866–1956), a member of the staff of the Royal College of Science (and later Imperial College) for thirty-one years, was widely known as the editor of the fourth edition of the classic eleven-volume work *Thorpe's Dictionary of Applied Chemistry* (1937–1956), which can be found occupying an impressive two feet of shelf space in the reference section of many university science libraries.

The second daughter of Mary (Bargh) and William Sedgwick Whiteley, Martha was born on 11 November 1866 in Hammersmith, London. After attending Kensington High School she went to the Royal Holloway College for Women and graduated with a B.Sc. in 1890. For the next eleven years she taught science, first at Wimbleton High School (1891–1900) and later at St. Gabriel's Training College, London (1900–1902). In 1898, at the age of thirty-two, she took up postgraduate studies at the Royal College of Science and was awarded a D.Sc. (London) in 1902. The following year she was invited by Sir William Tilden, professor of chemistry, to join the college's academic staff. She became assistant in 1904, demonstrator in 1908, and in 1914, when she was forty-eight, was promoted to lecturer. From 1920 until her retirement fourteen years later she held the position of assistant professor.[74]

With the outbreak of the First World War the chemists at the Royal College of Science, under the leadership of Sir Jocelyn Thorpe, professor of organic chemistry, turned to urgently needed work for the government. Before 1914 Britain had depended heavily on being able to import a variety of chemicals of German manufacture. When the supply of these was suddenly cut off crash programs were set up to try to make good the deficit. Whiteley joined Thorpe in critical synthetic work on drugs needed for army hospitals, primarily phenacetin, novocaine, and β-eucaine, all until then imported. She also worked with Thorpe, a member of the Trench Warfare Committee, on lacrymatory and vesicant gases (tear and blistering gases) for military use. For these services she was appointed an O.B.E. in 1920. Two years previously she had been elected a fellow of the Royal Institute of Chemistry.

Although not an outstandingly productive chemist as measured by the number of journal articles she published, Whiteley's contributions constitute a creditable body of work. She was concerned mainly with the amides and oximes of dicarboxylic acids and related cyclic ureides, including derivatives of caffeine and barbituric acid. Tautomerism in oximes was a particular interest. Much of her work in these areas was published in the period up to 1909, but she returned to the same topics after the war, and throughout the 1920s put out several more substantial papers, coauthored by various collaborators, including two of her women students, Edith Usherwood and Dorothy Yapp.[75]

While Whiteley's work on amides and on tautomerism in oximes undoubtedly developed from Sir Jocelyn Thorpe's studies, these being areas for which he was especially well known, she clearly did not function as his research assistant. Although his junior colleague who worked closely with him in many projects, she had her own research program. Indeed she stands out as probably the only British woman chemist in an established, nonbiological branch of the field whose career began before 1900, who does not fall into the category of "female assistant."

As well as her research papers she also coauthored (with Thorpe) a comprehensive, 241-page laboratory manual.[76] An up-to-date compilation of methods and procedures in organic analysis, the work was welcomed by the chemical community as a valuable addition to the then very meager literature in the area. To a large extent it recorded the experience gained by twenty years' teaching in the organic laboratories of the Royal College of Science, a period during which Whiteley had directed the advanced practical course.

An able teacher, she was remembered by former research students and junior colleagues as a steady and dependable source of inspiration and guidance. She was especially interested in helping and encouraging women students, and her influence in securing the acceptance of women at Imperial College was considerable. She helped with the organizing in 1912 of the Imperial College Women's Association, presided over its activities for many years, and continued to attend its meetings long after she retired. She was also an active member of the British Federation of University Women and held one of the first fellowships it awarded.

Whiteley's role in organizing and coordinating the long struggle for the admission of women to the Chemical Society was especially notable, she and her friend Ida Smedley MacLean being leaders in this cause over a period of about two decades. The question of membership for women had first been raised as early as 1880, but a considerable change in general outlook had to take place (a change that resulted in part from the First World War) before conservative opposition in the society's council was overcome.[77] In 1920, when women were finally accepted, forty years after they had initially presented their case, Whiteley was among the first to be formally admitted. Eight years later she became the first woman to be elected to the society's council, on which she served from 1928 to 1931.

After retiring in 1934 she continued to collaborate with Thorpe, preparing revisions and additions for the *Dictionary of Applied Chemistry.* First brought out over the three years between 1890 and 1893 by Thomas Edward Thorpe (also professor for a time at the Royal College of Science, but not a close relative of Sir Jocelyn), this massive compilation had already been revised and supplemented; the last volume of its third edition came out in 1927, two years after the death of Thomas Edward. By the early 1930s a fourth edition was being planned. The first volume of this appeared in 1937 under the joint editorship of J. F. Thorpe and M. A. Whiteley. Following Sir Jocelyn's death in 1940, just as volume four came out, Whiteley became editor in chief. She carried out her responsibilities in accommodation provided for her at Imperial College or at her home. The latter was destroyed by Second World War German bombing, but her editing and proofreading continued, for a time at the University Chemical Laboratory at Cambridge and later in her South Kensington flat. She was assisted by a distinguished four-man editorial board (I. M. Heilbron and H. J. Eméleus of Imperial College, A. R. Todd, Cambridge, and H. W. Melville, Aberdeen) and an assistant editor, A. J. E. Welch of Imperial College. The last of the planned eleven volumes appeared in 1954 and an index was added in 1956.

In 1945, in recognition of her services to chemistry in general and especially to Imperial College, she was made an honorary fellow of the college, a distinction she took special pride in. Although she never married she had strong family ties; her recreations she listed as "domestic and social duties." She died at home in South Kensington, on 24 May 1956, in her ninetieth year.

Three more London-trained women, Farrer, Buchanan and Boole, wrote papers listed in the bibliography. Farrer attended Bedford College;[78] Buchanan and Boole were students at the London School of Pharmacy. The first woman who had attended lectures at the latter institution (in 1862) was Elizabeth Garrett, the first British-trained woman physician. She, however, had done so without being formally admitted by the Pharmaceutical Society's council. Ten years later she approached the council with a request that a group of women students be allowed to take classes, and although initially denied, permission was granted within a month or so (on the condition that the women enter the building where lectures were given through a different door from that used by the male students). By the late 1870s women were allowed to take laboratory classes as well, compete for school prizes, and become members of the Pharmaceutical Society. Margaret Buchanan was the first woman student to receive a council award. After passing the Minor examination in 1886 with a certificate of honor for botany and materia medica, she went on to the Major examination the following year and won the council's Silver Medal in the Pereira Medal Competition. Her short paper on reactions of cyanic acid was presented to the Pharmacy Students Association in 1888.[79]

LUCY BOOLE[80] (1862–1905) was probably the first woman in England to formally carry out research in pharmacy. She was the fourth of five daughters of George Boole, professor of mathematics at Queen's College Cork, and his wife Mary, daughter of the Rev. Thomas Everest of Wickwar, Gloucestershire.[81] Lucy was born on 5 August 1862 in the Booles' home on the outskirts of Cork. Her father died suddenly when she was two years old, leaving the family almost destitute. They returned to England and Mary Boole brought up her five children on a Civil List pension of £100 a year, supplemented with small earnings (for a time she held the position of librarian and residence supervisor at Queen's College).

Lucy had little formal education in her early years, but passed the Major examination at the School of Pharmacy in 1888 and immediately after became research assistant to Wyndham Dunstan, then professor of chemistry with the Pharmaceutical Society. Her 1889 paper with Dunstan, reporting analysis procedures for tartar emetic, was the first co-authored by a woman to be read before the Pharmaceutical Society. In it she pointed out serious technical and practical difficulties in the pharmacopeial gravimetric method of assay then used, and proposed an alternative, volumetric method. Her suggestion came in for strong criticism initially, but was included in the 1898 British Pharmaceopeia and remained the official method of assay until modified in 1963. Despite her minimal formal training, she became lecturer in chemistry at the London School of Medicine for Women and also ran the small chemistry laboratory there. Her students, however, had to supplement what she was able to provide by going to evening lectures at Birkbeck College.[82] At least until the mid 1890s she remained active in research, working at the Pharmaceutical Society laboratory. Her paper on the blistering constituent of croton oil appeared in 1895. She was later elected a fellow of the Institute of Chemistry. The only one of the five Boole sisters who did not marry, she lived with her mother in a semidetached house near Notting Hill, London. Rumor had it, however, that mother and daughter did not get along particularly well.[83] Lucy died in 1905, in her early forties.

Seven of the early women chemists and biochemists whose work is listed in the Royal Society *Catalogue* received at least part of their training at Cambridge. These were Earp, Field, Gostling, Newton, Sedgwick, Smedley, and Tebb.

ANNIE EARP[84] (1860–1949), the daughter of an old Derbyshire family, was born in Melbourne, in the south of the county, on 12 November 1860. Educated at private schools in Derby and Dover, she entered Newnham College in 1885 and took class I in part I of the Natural Sciences Tripos examinations in 1888. She stayed on for another year on a College scholarship and was awarded class III in part II (1889). Thereafter she taught in high schools for much of the next three decades, except for a break of seven years just after the turn of the century when she served as a social worker and Poor Law Guardian in the London area. Her early teaching posts were also in London, but from 1909 until she retired in 1922 she worked mainly in Birmingham and nearby Nuneaton. Her one chemistry publication, a joint paper with M. M. Pattison Muir in the *Philosophical Magazine* in 1893, described her research on the determination of physical constants for a series of sulfur compounds. She died on 18 August 1949, in her eighty-ninth year.

ELEANOR FIELD[85] (Newnham, 1884–1888) received class I in part I of the Natural Sciences Tripos examinations in 1887. She later took an M.A. degree. From 1889 to 1890 she was an assistant demonstrator in chemistry at Newnham and from 1891 to 1893 held a Bathurst studentship, carrying out research under Pattison Muir. Her two papers, published in the Chemical Society's *Transactions* in 1892 and 1893, reported work on the constitution of salts precipated from solutions containing mixtures of several inorganic compounds. After leaving Cambridge she held a post as assistant mistress for two years (1893–1895) at Liverpool College for Girls and then became lecturer in chemistry at Royal Holloway College for Women. She remained there for nineteen years, until 1914, rising to senior staff lecturer. After retiring she moved to Brno, Czechoslovakia, where she died on 17 November 1932.

MILDRED GOSTLING[86] (1873–1962), the daughter of George James Gostling, a dental surgeon and pharmaceutical chemist, and his wife Sarah Abicail (Aldrich), was born in Stowmarket, Suffolk, on 15 December 1873. She studied at Bedford College and Royal Holloway College, taking a London B.Sc. before going to Newnham in 1898 as a Bathurst student and assistant in the chemistry laboratory. Her research with H. J. H. Fenton on the action of acids on carbohydrates, particularly cellulose, was reported in four papers, as well as a number of notes, between 1898 and 1903. During the year 1901–1902 she held a demonstratorship in chemistry at Royal Holloway College. In 1903 she married a fellow chemist, William Hobson Mills, a fellow at Jesus College Cambridge in the late 1890s and from 1902 head of the chemical department at the Northern Polytechnic Institute, North London. For a time after her marriage Mildred remained active in research, collaborating with her husband on the preparation of dinaphthanthracene derivatives.[87] In 1912 they returned to Cambridge, where Mills took up a position at the university. They had three daughters and a son. Mildred died on 19 February 1962, at age eighty-eight.

LUCY NEWTON[88] (1867–1908) was born in Scarborough, on the Yorkshire coast, on 15 September 1867. She attended schools in Scarborough and Nottingham and went to Newnham in 1886. In the Natural Sciences Tripos examinations of 1889 she was awarded class II (pt. I). She stayed on at Newnham for another year, and then became assistant demonstrator in the chemistry laboratory (1891–1893?). For a short time in the mid 1890s she carried out research in analytical chemistry with Emily Aston at University College London. She died at the age of forty-one, in Malta, on 29 March 1908.

ANNIE SEDGWICK[89] was the elder daughter of Lieutenant-Colonel William Sedgwick of Godalming, Surrey. A student at Girton (Natural Sciences Tripos examinations, class II, part I, 1892, class II, part II, 1893), she collaborated with Cambridge chemist Siegfried Ruhemann in work on organic esters, publishing a joint paper with him in 1895. She stayed on at Cambridge for a time as assistant preparator in physiology in the Balfour laboratory and then continued her chemistry studies at University College London, where she carried out research on pyridine derivatives with J. Norman Collie. In 1897 she married James Walker, then professor of chemistry at University College Dundee and after 1908 at the University of Edinburgh. Apart from a paper on malonic and succinic acids, coauthored with her husband in 1905,[90] she does not seem to have published further work in chemistry. She had one son. In 1921 she became Lady Walker when her husband was knighted for his contributions to industrial and academic chemistry.

The most outstanding of the early Cambridge chemistry students included here, one who went on to a distinguished career in biochemical research, was IDA SMEDLEY[91] (1877–1944). She was born in Birmingham on 14 June 1877, the second of the two daughters of William T. Smedley, a chartered accountant, businessman, and philanthropist, and his wife Annie Elizabeth (Duckworth), daughter of a Liverpool coffee merchant. The parents, both gifted and forward-looking people, gave their daughters an exceptional upbringing for the time, encouraging independence of thought and action from their early years and opening up for their leisure interests a world of cultural and artistic activities.[92]

Ida was taught at home by her mother until she was nine and then went to King Edward VI High School for Girls in Birmingham. At the time, this school was probably the best girls' school in England for science teaching. Headed by Edith Creak, one of the Newnham College pioneers of the early 1870s, it had a science staff made up of Newnham-trained women who in turn sent a succession of their best students to their old college. The first of these, and, it has been said, the most brilliant, was Ida Smedley.[93] While still a schoolgirl she took classes (then men's) in physiology in the medical department of Mason College Birmingham. She sat both London and Cambridge university entrance examinations and was awarded a three-year Gilchrist scholarship for an honours certificate with four distinctions in the Cambridge examinations (the Higher Local Examinations). Entering Newnham in 1896, she took the Natural Sciences Tripos course and obtained class I in part I in 1898 and class II in part II (chemistry and physiology) the following year.

From 1900 to 1903 she held a Bathurst studentship and did postgraduate research at the Central Technical College London under Henry Edward Armstrong, a man who, although often outspoken in his criticisms, was known for helping younger chemists whom he thought showed promise; his emphasis on sound experimental work had a lasting influence on Smedley. For her work with him on benzylaniline sulfonic acids, which she submitted in 1905 as a dissertation for a London D.Sc. degree, she received a supporting grant of £10 from the Chemical Society's Research Fund, the first such grant made to a woman.[94] She returned to Newnham in 1903 as demonstrator in chemistry, but also continued research at the Davy-Faraday Research Laboratory at the Royal Institution, London, carrying out work on the preparation of colorless derivatives of fluorene with a view to understanding the nature of the changes involved in going from colored to colorless compounds in this system.

In 1906 she became assistant lecturer in the chemistry department at the University of Manchester, the first woman to be appointed to the staff there. She stayed for four years, and during that time kept up an ambitious and productive program of research, publishing several substantial papers on the refractive power of unsaturated, aliphatic hydrocarbons, with particular emphasis on the effect of conjugation on refractivity. She also worked on the relationship between chemical constitution and optical properties in aromatic diketones.[95]

In 1910, having been awarded one of the first Beit Memorial Research Fellowships, she went to the Lister Institute of Preventive Medicine in London, and there began (initially with Sir Alfred Harden) the work on the chemistry, metabolism, and biosynthesis of fats that was to continue for the rest of her life. She became a regular member of the institute's staff in 1932. In 1913, when she was thirty-six, she married Hugh MacLean who had joined Harden's group about the time she did. MacLean later became professor of medicine in the University of London at St. Thomas's Hospital.

Like most of her fellow chemists, during the First World War she worked on research critical to the war effort. Along with chemist Chaim Weizmann, director of the Admiralty research laboratories from 1916 to 1919 (later the first president of Israel), she investigated the problem of large-scale production of acetone from starch by fermentation. The project was eminently successful. In 1918 she became a fellow of the Institute of Chemistry.

Her most important research contributions were her investigations of pathways of synthesis and methods of oxidation of long-chain fatty acids in vitro, and the basic metabolic processes whereby fat is synthesised from carbohydrate in living organisms—the latter an extremely complex and difficult topic.

In the area of in vitro fat synthesis she was one of the earliest to build compounds containing unbranched chains of eight or more carbon atoms starting from smaller carbon units. This was an important step toward the understanding of long-chain fatty acid synthesis in nature since early laboratory work had produced only branched chains, rare in living systems. The work of her team on the oxidation (with hydrogen peroxide) of long-chain fatty acids was considered of particular interest because of her demonstration of the marked rate-enhancing effect of copper ions.

On the biochemical side she and her coworkers confirmed and extended in new directions earlier work by George and Mildred Burr on the effects of fat deficiency in rats. Her studies contributed substantially to knowledge of the essential nature in animal metabolism of certain highly unsaturated fats, of the function of these compounds in the animal body, and of the need in the diet for certain starting materials (in particular linoleic acid) for their synthesis. One of her final projects was establishing the structure (with an extremely small quantity of material) of one of the key fatty acids, archidonic acid.[96]

Her book, *The Metabolism of Fat* (1943), completed shortly before her death despite serious health difficulties, summarized her views on the field in which she had become an established authority. A concise survey, it appeared in Methuen's series of Monographs on Biological Subjects. At the time, there was a considerable lack of comprehensive up-to-date monographs on biochemical topics, and the work was intended to fill the gap in the field of fat metabolism. Space limitations prevented presentation of a detailed picture, and she drew criticism from a reviewer for devoting too much attention to current theories of fat metabolism, which were not supported by experimental evidence, but in many respects she succeeded in giving a clear picture of the broad outlines of the field at the time.[97] The book remained in use for at least a quarter of a century, although within a few years of her death the subject of fat metabolism was undergoing considerable change as a result of the introduction of a great many new techniques. The latter included spectroscopic analysis, the use of isotopically labeled precursors, and gas-liquid and thin-layer chromatography; coupled with the development of new methods for the total synthesis of unsaturated fatty acids and some elucidation of the enzyme control of the processes involved, these innovations brought about a fairly rapid increase in chemists' knowledge of the field. Nevertheless Smedley MacLean's work continued to be cited in discussions of the subject at least until the 1960s.[98]

Her interests were by no means restricted to scientific research. She was a founder and the first secretary of the British Federation of University Women, started while she was at Manchester, and she later served as secretary and president (1929–1935). The energy and breadth of vision she brought to this organizational work was considerable; she took a leading part in expanding the movement into the International Federation (an undertaking carried out between the two world wars) and was especially interested in the endowment of international fellowships for research by women.[99] The opening to women of membership in learned societies was another cause she strongly and actively supported. From the early years of the century, she and Martha Whiteley led in the work of organizing support for the admission of women to the Chemical Society. Their skill and persistence did much to keep the cause alive through years of disheartening failure.[100] In 1920, when fellowship was finally opened to women, Smedley MacLean was the first to be formally admitted. She was a member of the society's council from 1931 to 1934. Her concern for the needs of professional women did not stop there; she and her sister, Constance Smedley Armfield, were among those who founded the London Lyceum Club, an organization for professional women that later had branches in a number of other countries. From 1941 until 1944 she served on the Cambridge University Women's Appointments Board.

Vigorous and sympathetic in personality, she was remembered as being an inspiration to the many young students (several of whom were women) who worked with her over the years. She and her husband had two children, a son born in 1914, and a daughter, Barbara Duckworth MacLean, born three years later. She died on 2 March 1944 at the age of sixty-six, after two years of failing health.

CHRISTINE TEBB[101] (1868–1953), although her contributions were considerably less than those of Smedley MacLean, was another early Cambridge-trained biochemist of note. Born on 11 September 1868, she was the daughter of Mary (Scott) and William Tebb of Rede Hall, Barstow. In 1882, after receiving her early education at home and at a private school, she entered Bedford College. She went to Girton College in 1887 and took the Natural Sciences Tripos examinations in 1890 and 1891, gaining class I in both parts and specializing in physiology in part II. From 1891 until 1893 she was laboratory assistant to Newnham physiologist Marion Greenwood (see chapter 6). She also held a Bathurst studentship and carried out research in Sir Michael Foster's physiology laboratory on the enzymatic conversion of maltose to dextrose; the work was reported in her paper in the *Journal of Physiology* in 1894.

After moving to William Halliburton's chemical physiology laboratory at King's College London in the late 1890s, she took up, at Halliburton's suggestion, a series of investigations on the structure of protein fibers in connective tissue, work that brought her into a controversy about the nature of the protein reticulin. Her views, and those of the German chemist Sigfried whose work she had attempted to repeat, were not reconciled until much later, when laboratory techniques for protein tissue analysis and theories of physical chemistry had advanced sufficiently to be able to deal with the complexities involved.[102] From about 1906 until 1910 she collaborated with Otto Rosenheim on the problem of whether the crystalline material "protagon," isolated from the brain, was or was not a single chemical compound. Rosenheim, a German chemist who had come to Britain in 1894 to get away from anti-Semitism in his own country, was then lecturer in physiological chemistry at King's College. He and Tebb coauthored nine papers and notes on the "protagon" problem and closely related topics between 1907 and 1910,[103] and they concluded that "protagon" was a mixture. Heated controversy followed in the biochemical literature, but the Rosenheim-Tebb findings stood the test of time.

Tebb married Rosenheim in 1910. She continued her research at King's College until about 1916, carrying out work on cholesterol supported by grants from the Royal Society. For a few years around 1920 both she and her husband had a period of quasi-retirement at their Hampstead Garden City home; Otto Rosenheim spent much of his time on his rock garden and they went on annual plant-hunting excursions to the Alps and Mediterranean islands. In 1923 they both returned to research at the laboratories of the Medical Research Council in Hampstead, taking up the problem of the constitution of spermine, a base present in various animal tissues. This project continued until 1927, and, with the collaboration of H. W. Dudley and W. W. Starling, resulted in not only the structure determination but also the synthesis of both spermine and a new base, spermidine. Christine's name appears on only the first of the spermine papers (1924) and this was probably her last published scientific work.[104]

She lived for almost three more decades, dying on 19 February 1953, at age eighty-four, after two years of severe illness during which her husband nursed her himself, practically unaided. He disliked publicity and there seem to have been no obituaries for his wife. Her estate of £82,635, subject to the life interest of first her husband and then her two sisters, was left to the Royal Society, the income to be used for biochemical research in Britain by people of outstanding ability irrespective of sex or nationality.[105]

The most notable research contributions by early women chemists who received their training at colleges other than those in London and Cambridge came from two women at University College Bristol, Emily Fortey and Katherine Williams. Although her productive research career lasted for only seven years, EMILY FORTEY[106] is especially noteworthy. Between 1896 and 1903 she authored or coauthored twelve substantial papers and several shorter notes. She was a student at Bristol in 1892–1893, held a Chemical scholarship in 1895, and received a London B.Sc. (with second class honours in chemistry and third class in experimental physics) in 1896. The same year, she was awarded an Associateship of University College Bristol and an 1851 Exhibition Science Research scholarship.

After some early work in photochemistry, published in 1896, she began a five-year collaboration on problems in fractional distillation with Sydney Young, then a member of the Bristol chemistry staff. One of her first major papers (published in 1898 under her name only) described the fractionation of crude petroleum samples from American and Galician deposits. A laborious project carried out at Owen's College Manchester, this work, in conjunction with that of the Russian chemist Vladimir Markovnikov,[107] successfully demonstrated that the cyclohexane fractions from American, Galician, and Caucasian deposits were identical and identical to the synthetic product. Between 1899 and 1903, when she seems to have dropped out of research, Fortey published seven joint papers with Young on fractional distillation. This work not only led quickly to important practical applications,[108] but produced noteworthy theoretical advances and formed a major part of the background research for Young's well-received monograph, *Fractional Distillation* (1903).

KATHERINE WILLIAMS[109] (1848?–1917) became a student at University College Bristol in 1877 at the age of twenty-nine, and took classes there until 1885, concentrating mainly on chemistry. She sat the Cambridge Higher Local Examinations (1881–1884) but did not proceed directly to any degree, taking classes only part-time in 1884–1885. She became an Associate of University College Bristol in 1886, and in 1910, when she would have been about sixty-two, was awarded a B.Sc. by research. For a time in the early 1880s she collaborated with Sir William Ramsay, then principal at University College Bristol, in his early work on atmospheric gases. He is said to have suggested that she repeat the Cavendish experiment on air,[110] but she chose something easier, the determination of oxygen dissolved in water. She also began a series of studies on the chemical composition of foods, and published at intervals in this area over the next three decades, until shortly before her death in January 1917.[111]

At Yorkshire College of Science in Leeds three women students of Thomas Edward Thorpe (Johnston, Kennedy, and Passavant) published minor research contributions in 1881. Margaret Johnston and Catherine Kennedy carried out an analysis of the mineral content of the waters at Boston Spa, Thorp Arch, work that formed part of Thorpe's investigation of the mineral springs of Yorkshire. These waters, although known and used for 300 years, were then for the first time being examined by methods capable of giving relatively complete quantitative information about their mineral content. Neither Johnston nor Kennedy seem to have published any later work in chemistry. Laura Passavant held a Brown Scholarship at Yorkshire College. She collaborated with Thorpe in studies on the relationship between molecular weight and specific gravity, her work being part of a series of investigations in this area in which Thorpe was at the time engaged.

There remain seven British women whose papers in the chemical sciences are listed in the bibliography: E. Johnston, Lloyd, Seward McKillop, Newbigin, Green, Walter, and Marshall.[112]

ETTA JOHNSTON was a student at University College Dundee in the late 1880s. Her joint work with Thomas Carnelley, professor of chemistry at the University of Aberdeen, on the relation between atomic weight and the physiological effect of an element, particularly its toxicity, appeared in a short note in *Nature* in 1890. She also collaborated with Carnelley in some of the public health investigations in the Aberdeen-Dundee area for which he became known and which are still of historical interest.[113] Johnston's studies involved an examination of material commonly used as a sound barrier between floors in dwelling houses in Scotland and a number of other European countries at that time. Known in Scotland as "deafening" material, it was in theory a mixture of coarse mortar and smith's ashes. The work done by Johnston in Dundee, however, showed that in the housing of the poorer classes its composition varied considerably, and that it always contained a sizable amount of organic material. Carnelley had previously shown that air quality, as measured by microorganism concentration, was worst in the houses of the poor; furthermore, the worse the air quality the lower the average age of death. By removing the "deafening" material in test cases Johnston and Carnelly were able to demonstrate that it, in fact, was the source of much of the microorganism contamination.

EMILY LLOYD was a student at Mason Science College Birmingham. Her one chemical paper, a report of osmotic pressure studies, was read before the Mason College Chemistry Society in 1889. Lloyd has the distinction of being the first woman associate of the Institute of Chemistry. In 1892 she applied to take the required examination to qualify as a public analyst (to practice in connection with the administration of the Sale of Food and Drugs Act). The admission of women candidates to this examination had not been contemplated by the

Institute's council, but Lloyd's application having been received, she was admitted and in due course passed.[114]

MARGARET SEWARD (later MCKILLOP)[115] entered Somerville Hall Oxford in 1881. Three years later (the first year the examination was open to women) she sat the Final Honours School of Mathematics, gaining a Second. The following year (1885) she was awarded a First in chemistry final examinations and was immediately appointed Natural Science Tutor for Somerville Hall. Her postgraduate research with William H. Pendlebury on the effect of temperature on reaction rate appeared in the *Proceedings of the Royal Society* in 1889. It brought her a brief mention in Partington's well-known *History of Chemistry;* she is, in fact, the only one of the late nineteenth-century British women chemists included here to be so distinguished.[116] She was later lecturer in chemistry, first for a short period at Royal Holloway College, and then from 1897 until 1912 in the Women's Department at King's College London, where she also held the position of tutor in the faculty of science. From 1910 until 1913 she served on the Delegacy of King's College, the college's governing body.

With the exception of Martha Whiteley, the early British women who worked in traditional, established areas of chemical research did so as assistants to men chemists.[117] Consequently their work, to all intents and purposes, has been absorbed into that of the men; even the two who were most productive in the pre-1900 period, Aston and Fortey, are missing from the story of the research of the time, despite Aston's being the first nineteenth-century British woman to collaborate in notable chemical research projects over on extended period and Fortey's very creditable joint work with Sydney Young. Only Whiteley succeeded in finding a place as an independent worker in an established area of chemistry and remaining active in research and technical writing throughout her long career. Of all the early British women chemists included here, the one who was most successful in original work was Smedley MacLean, and she moved out of established branches of chemistry and into the developing field of biochemistry; there she became a leader in her area.

Two-country comparison

In many respects the work and career patterns of late nineteenth-century British and American women in chemical research are similar. However, no American woman in a traditional branch of chemistry, whose work in research began before 1900 had a career in a major coeducational institution that was as successful as Martha Whiteley's, and no American woman chemist of the time went on to produce a body of original work in basic biochemistry comparable to that which Ida

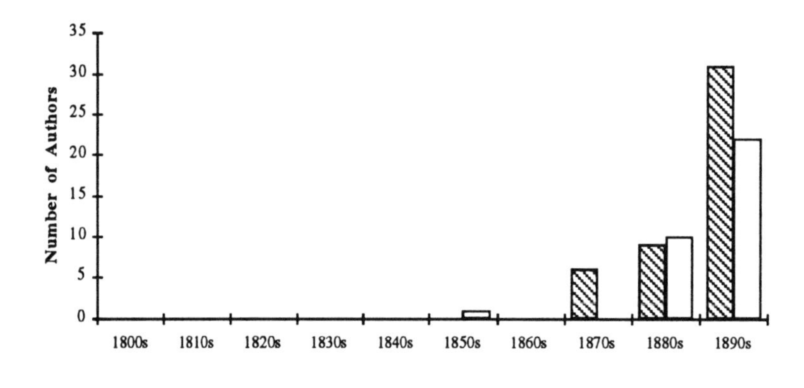

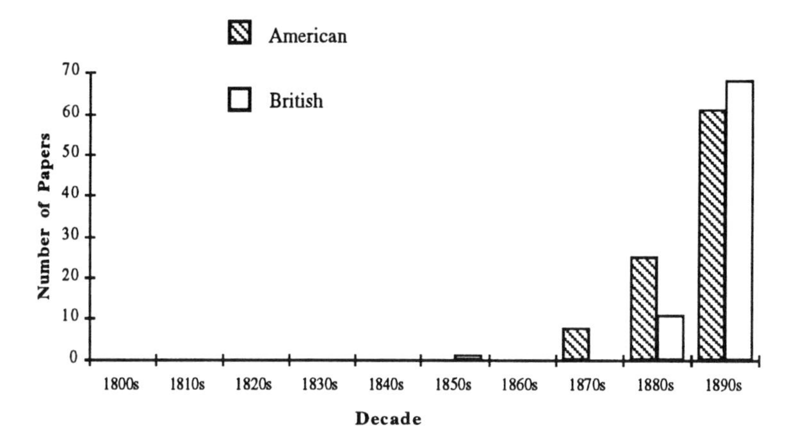

Figure 11-1. American and British authors and papers in chemistry and biochemistry, by decade, 1800–1900. Data from the Royal Society *Catalogue of Scientific Papers.*

a. American authors

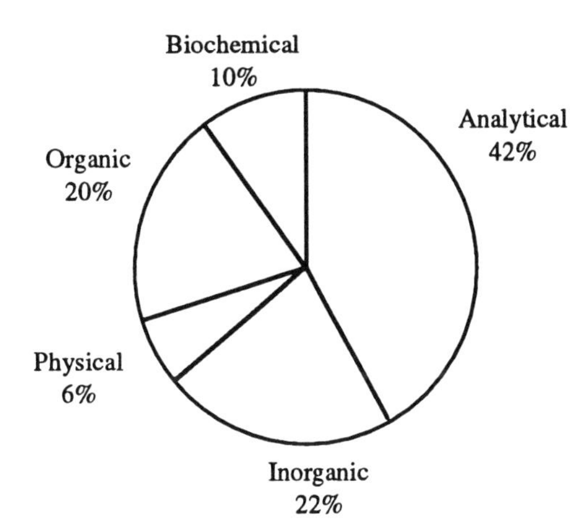

b. American papers

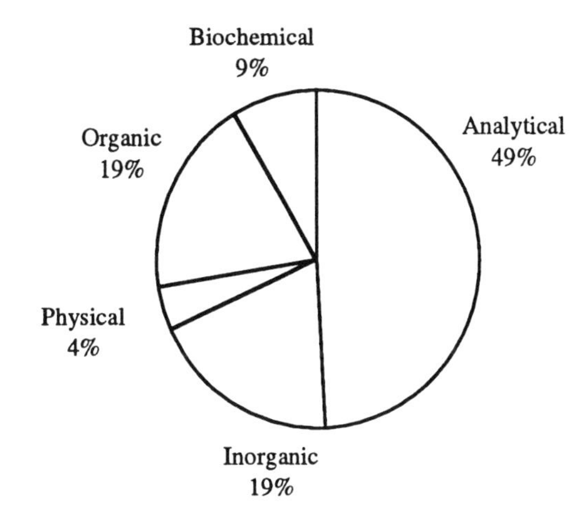

c. British authors

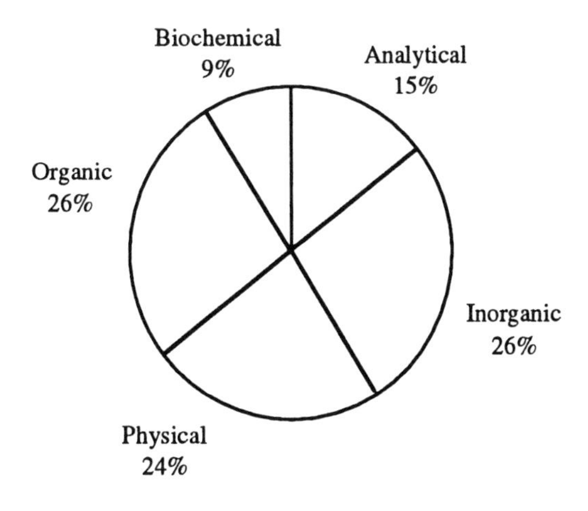

d. British papers

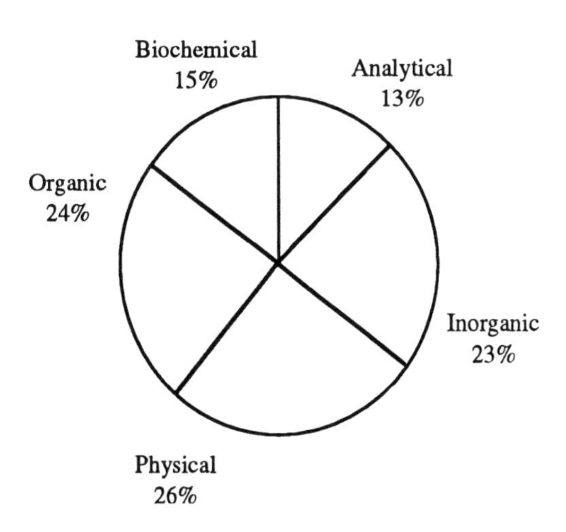

Figure 11-2. Distribution of American and British authors and papers in chemistry and biochemistry, by subfield, 1800–1900. (In a. and c., an author contributing to more than one subfield is counted in each.) Data from the Royal Society *Catalogue of Scientific Papers.*

Smedley MacLean and her collaborators brought out between about 1910 and 1940. Mary Pennington's contributions, though notable, were very much in applied chemistry, narrowly dedicated to the solution of technical and practical problems rather than being concerned with fundamental biochemical processes. Like most of the British, American women who published in mainstream areas of the discipline did so as assistants to men chemists. Those whose careers were most successful moved to new areas such as public health chemistry, bacteriological chemistry, and food chemistry; a considerable number left the field altogether, going into academic administration, home economics, or medicine. Although chemistry at the time was a relatively popular subject with women science students in both countries at the postgraduate as well as the undergraduate level, job opportunities for women chemists were few. On both sides of the Atlantic, the profession was one largely reserved for men.

The graphs (Figures 11-1 and 11-2) summarize numerical data for the two national groups. Figure 11-2 brings out the concentration of the American women in analytical chemistry (a reflection of the overall national focus at the time); in contrast, the British were spread across all the major branches of the field.

Notes

1. The increase in author count by two (forty-two to forty-four) from the number given in a previous communication (Mary R. S. Creese and Thomas M. Creese, "British women who contributed to re-

search in the geological sciences in the nineteenth century," *British Journal for the History of Science,* **27** (1994): 23–54, on p. 24) results from the reclassification of the work of Barker and Bills from physiology to chemistry. The figure of 147 for the number of chemistry papers by American women given in the 1994 article might be considered high. The number depends on what criteria are used for deciding whether or not an article should be included in the count. The choice is especially troublesome with nineteenth-century chemists (particularly Americans and Russians) because of their common habit of publishing a given piece of research in more than one journal—in the case of the Americans more than one American journal or an American and a German journal. The figure 147 was a count of all the relevant entries in the bibliography, irrespective of the virtual duplication of some; for the inter-field comparisons being made in the 1994 article the numbers used are consistent. A more discriminating count, crediting one paper per topic reported, and so perhaps more appropriate in the present study, reduces the number considerably. With, in addition, the reclassification of some borderline papers in plant chemistry to botany and bacteriological chemistry to bacteriology, the present count of American chemistry papers is about ninety-five. A similar pruning operation on the British chemistry papers would reduce the British count (chemistry and biochemistry) to about eighty-four. The Americans remain the group making the largest contribution, although the non-duplicating count significantly reduces their lead (see also n.64 below). Two additional American authors (Bragg and Mason) and nine additional papers not indexed by the Royal Society have been included in the bibliography. Three authors (Gage Day, Keller, and Talbot) had major interests in other fields (see, respectively, chapters 7, 1, and 16).

2. Although Mount Holyoke College had a reputation as an early leader in the training of women chemists at the undergraduate level, even by the first decade of the present century its faculty and students were making only very modest contributions to the research literature.

3. Robert Clarke, *Ellen Swallow: the Woman who Founded Ecology* (Chicago: Follett Publishing Co., 1973); Caroline L. Hunt, *The Life of Ellen H. Richards* (Boston: Whitcomb and Barrows, 1912, 1918); George Rosen, "Ellen H. Richards (1842–1911), sanitary chemist and pioneer of professional equality for women in health science," *American Journal of Public Health,* **64** (1974): 816–19; Patricia H. Hynes, "Ellen Swallow, Lois Gibbs and Rachel Carson: catalysts of the American environmental movement," *Women's Studies International Forum,* **8** (1985): 291–8, Special Issue, *Selections from the Second International Interdisciplinary Congress on Women,* Gronigen, The Netherlands, 17–21 April 1984; obituaries, Isabel Bevier, "Mrs. Richards' relation to the Home Economics movement," *Journal of Home Economics* **3,** (1911): 214–16, ibid., **3** (1911): 370–88, and R[ossiter] W. R[aymond], *Monthly Bulletin, American Institute of Mining Engineers,* n. 58 (1911): xxviii-xxx; Mary R. S. Creese and Thomas M. Creese, "Ellen Henrietta Swallow Richards (1842–1911)," in *Women in Chemistry and Physics: A Bibliographic Sourcebook,* eds., Louise S. Grinstein, Rose K. Rose and Miriam H. Rafailovich (Westport, Conn.: Greenwood Press, 1993), pp. 515–25.

4. See, for example, *Review of American Chemical Research,* **4** (1898), ed. Arthur A. Noyes (Easton, Pa.: Chemical Publishing Co., 1898).

5. Bevier, "Mrs. Richards'," p. 215. Richards published about thirty papers and monographs on topics in home economics. Several were largely works in applied elementary chemistry, such as *The Chemistry of Cooking and Cleaning: a Manual for Housekeepers* (Boston: Estes and Lauriat, several edns., 1882–1916), *Food Materials and their Adulterations* (Boston: Estes and Lauriat, several edns., 1885–1914), *Home Sanitation: a Manual for Housekeepers* (Boston: Tickner and Co., several edns., 1887–1911), *The Art of Right Living* (Boston: Whitcomb and Barrows, several edns., 1904–1911), and *Sanitation in Daily Life* (Boston: Whitcomb and Barrow, several edns., 1907–1919). Further home economics publications by Richards are listed in Creese and Creese, "Ellen Richards," Grinstein, Rose, and Rafailovich, eds., *Women in Chemistry and Physics,* pp. 522–4.

6. For Talbot see chapter 16.

7. The three post-1900 Richards and Hyams papers were: "Notes on *Oscillaria prolifica* (Greville). First paper: Life history," *Technology Quarterly,* **14** (1901): 302–10; "Notes on *Oscillaria prolifica* (Greville). Second paper: Chemical composition," ibid., **15** (1902): 308–15; "Notes on *Oscillaria prolifica* (Greville). Third paper: Coloring matters," ibid., **17** (1904): 270–6.

8. Information about Margaret Cheney, Alice Palmer and Helen Cooley came from the Institute archives, MIT. For Cooley see also the *AMD,* 1906–1936. For Hyams and Mason see Marilynn A. Bever, "The Women of MIT, 1871–1941: who they were, what they achieved," B.S. thesis, Massachusetts Institute of Technology, 1976, pp. 91, 122. Mason is also listed in *WWWA,* p. 547 (date of retirement from Smith College archives). Dates of birth and death for Hyams are 22 March 1865–17 February 1932; for Mason, 30 January 1863–11 September 1935.

9. Helen S. French, "Charlotte Almira Bragg. Department of Chemistry, 1890–1929," *Wellesley Alumnae Magazine,* (June 1929): 287–8; obituaries, "Prof. Charlotte A. Bragg, 94, dies in her home in Melrose," *Boston Sunday Herald,* 1 September 1957, *NYT,* 2 September 1957, 13: 5; anon., "Marblehead woman, 80, gives fence for scrap," *Boston Herald,* 9 September, 1942; Wellesley College archives, faculty biographical file; Mary R. S. Creese, "Charlotte Roberts and Charlotte Bragg—early chemists at Wellesley College," *New England Association of Chemistry Teachers Journal,* **11,** no. 1 (1992): 20–3.

10. In each of the three countries from which came the bulk of women's work in nineteenth-century chemical research (Britain, Russia, and the United States) the first papers by women appeared within the relatively short time span, 1870–1880. In Russia Anna Volkova's first work appeared in 1870, Lidi'ia Zeseman's and Adelheid Lukanina's in 1873, and Iuli'ia Lermontova's in 1874. In Britain the first paper from a modern chemical laboratory authored or coauthored by a woman was Lucy Halcrow's joint publication with Edward Frankland, which appeared in 1880.

11. D. H. Dole, "Biographical sketch," in Helen Abbott Michael, *Studies in Plant Chemistry and Literary Papers* (Cambridge, Mass.: Riverside Press, 1907), pp. 3–107; *Who was Who in America,* vol. 1, p. 836; Ann Tracey Tarbell and D. Stanley Tarbell, "Helen Abbott Michael: pioneer in plant chemistry," *Journal of Chemical Education,* **59** (1982): 548–9; K. Thomas Finley and Patricia J. Siegel, "Helen Cecilia DeSilver Abbott Michael (1857–1904)," in Grinstein, Rose and Rafailovich, eds., *Women in Chemistry and Physics,* pp. 405–9; Mary R. S. Creese, "Helen Abbott Michael," in *American Chemists and Chemical Engineers,* vol. 2, eds. Wyndham D. Miles and Robert F. Gould (Guilford, Conn.: Gould Books, 1994), pp. 187–8.

12. See for instance R. E. Alston and B. L. Turner, *Biochemical Systematics* (Englewood Cliffs, N.J.: Prentice Hall, 1963), pp. 45–6, 67, and R. Darnley Gibbs, *Chemo-taxonomy of the Flowering Plants,* 4 vols. (Montreal: McGill-Queens University Press, 1974), vol. 1, pp. 11–12. Interest in the application of chemistry to systematics goes back to the early years of the nineteenth century, but relatively little work was carried out in the area until after 1910; the field is one of considerable complexity. Alston and Turner (p. 46) pointed out that, though the idea of the importance of secondary compounds in systematics is generally considered to be a fairly recent theory, in reality

it goes back to Abbott's 1886 paper. The theory was reformulated, without reference to Abbott, by Erdtmann in the 1950s (see H. Erdtmann, "Organic chemistry and conifer taxonomy," in *Perspectives in Organic Chemistry,* ed. Sir Alexander Todd (New York: Wiley Interscience, 1956), pp. 453–94). Gibbs (*Chemo-taxonomy,* p. 11) includes Abbott among the modern pioneers of chemotaxonomy.

13. Abbott was by no means the only woman associated with the Philadelphia College of Pharmacy at that time. The institution had admitted women to its classes as auditors since 1876, when Clara Marshall, later dean of the Woman's Medical College of Philadelphia, attended lectures; Susan Hayhurst, the first woman to graduate, received her degree in 1884. She later headed the pharmaceutical department of the Woman's Hospital of Philadelphia (see *The First Century of the Philadelphia College of Pharmacy: 1821–1921,* ed. Joseph W. England (Philadelphia: Philadelphia College of Pharmacy and Science, 1922), pp. 163–4).

14. England, *First Century,* pp. 424–5, 569. Student records, Philadelphia College of Pharmacy archives.

15. See, for instance, Josiah C. and Bertha L. De G. Peacock, "The tannin of wild-cherry bark," *American Journal of Pharmacy,* **95** (1923): 613–23; "The tannin of *Rhus glabra,*" ibid., **97** (1925): 463–71; "Does cascara sagrada contain a tannin?" ibid., **98** (1926): 395–410; "Further study of the tannin of *Geranium maculatum,*" ibid., **100** (1928): 548–57.

16. Only the following information about Julia Morris and Alice Albro has been collected: Morris, whose 1900 paper in analytical chemistry with Frank Gooch appears to have been her only publication, was born in Utica, New York, in 1876, the daughter of Emily (Stevens) and Samuel H. Morris. After taking a B.S. at Smith College (1898) she had one year of graduate study at Yale (1899–1900). In 1904 she married Nellis Barnes Foster, a physician and biochemist associated with Columbia University and the New York Hospital. She later became active in social service work (student records, Yale University archives; *WWWA,* p. 302). Alice Albro (Mrs. Albert Barker) received her Ph.D. in chemistry from Yale in 1898 (Walter Crosby Eells, "Earned doctorates for women in the nineteenth century," *American Association of University Professors, Bulletin,* **42** (1956): 644–51); her graduate work was in biochemistry and physiology and she coauthored two papers (one in physiology) in 1898 and 1899.

17. *Bulletin of Smith College: Alumnae Biographical Register Issue, 1871–1935*; obituary, anon., in Smith College archives; *WWWA,* p. 644; *AMS,* 1921. Mary R. S. Creese, "Martha Austin Phelps," in Miles and Gould, eds., *American Chemists,* pp. 223–4.

18. Anon., "Charlotte Fitch Roberts," *Wellesley College News,* 13 December 1917; *Wellesley Alumnae Magazine,* (February 1950): 192; Ellen L. Burrell, "Charlotte Fitch Roberts," *Wellesley Alumnae Quarterly,* **2,** (January 1918): 80–1; *WWWA,* p. 691; *AMS,* 1910; Creese, "Charlotte Roberts and Charlotte Bragg."

19. For a discussion of Roberts's book see Mary R. S. Creese and Thomas M. Creese, "Charlotte Roberts and her textbook on stereochemistry (1896)," *Bulletin for the History of Chemistry,* no. 15/16 (1994): 31–6.

20. See n. 46, chapter 1, for remarks on the closely shared professional and social lives of the Wellesley faculty community at this period.

21. *WWWA,* p. 282; *AMD,* 1904–1931.

22. The University of Pennsylvania and Yale between them gave eight of the thirteen chemistry Ph.D. degrees that American universities awarded to women in the period up to 1900 (Eells, "Earned doctorates").

23. Lisa Mae Robinson, "The electrochemical school of Edgar Fahs Smith, 1878–1913," Ph.D. dissertation, University of Pennsylvania, 1986, pp. 217–20; *AMS,* 1910.

24. Barbara Heggie, "Ice Woman," *New Yorker,* **17** (10 November 1940): 23–30; Ethel Echternach Bishop, "Mary Engle Pennington: 1872–1952," in *American Chemists and Chemical Engineers,* ed. Wyndham D. Miles (Washington, D.C.: American Chemical Society, 1976), pp. 386–7; Ogilvie, *Women in Science,* pp. 146–7; Anna Pierce, "Mary Engle Pennington: an appreciation," *Chemical and Engineering News,* **18** (10 November 1940): 941–2; Lisa Mae Robinson, "Regulating what we eat: Mary Engle Pennington and the Food Research Laboratory," *Agricultural History,* **64** (1990): 143–53; and "Electrochemical school," pp. 220–5 and 231–3; Vivian Wiser, "Pennington, Mary Engle," in *NAW,* vol. 4, pp. 532–4; Mary R. S. Creese and Thomas M. Creese, "Mary Engle Pennington (1872–1952)" in *Women in Chemistry and Physics,* Grinstein, Rose, and Rafailovich, eds. pp. 461–9.

25. See M. E. Pennington and J. S. McClintock, "A preliminary report on the pasteurized and clean milk of Philadelphia," *American Journal of the Medical Sciences,* **130,** (1905): 140–9. See also the following: (with E. Q. St. John) "The relative rate of growth of milk bacteria in raw and pasteurized clean milk," *Journal of Infectious Diseases,* **4** (1907): 647–56; (with Georgiana Walter) "A bacteriological study of commercial ice cream," *New York Medical Journal,* **86** (1907): 1013–18; "Bacterial growth and chemical changes in milk kept at low temperatures," *Journal of Biological Chemistry,* **4** (1908): 353–94; (with E. L. Roberts) "The significance of leucocytes and streptococci in the production of high grade milk," *Journal of Infectious Diseases,* **5** (1908): 72–84.

26. Among Pennington's especially notable publications from this period of her career were the following: "Chemical and bacteriological study of fresh eggs," *Journal of Biological Chemistry,* **7** (1910): 109–32; "A scientific study of the deterioration of poultry during marketing," *Proceedings of the Pathological Society of Philadelphia,* **14,** n.s. (1911): 66–71; (with H. C. Pierce) "The effect of the present method of handling eggs on the industry and the product," *U.S. Department of Agriculture, Year Book, 1910* (Washington, D.C., 1911), pp. 461–76; "The hygienic and economic results of refrigeration in the conservation of poultry," *American Journal of Public Health,* **2** (1912): 840–8; (with J. S. Hepburn) "The occurrence and permanence of lipase in the fat of the common foul (*Gallus domesticus*)," *Journal of the American Chemical Society,* **34** (1912): 210–22; (with J. S. Hepburn) "Studies on chicken fat. III. Influence of temperature on lipolysis of esters. IV. The hydrolysis of chicken fat by means of lipase," *U.S. Department of Agriculture, Bureau of Chemistry, Circular 103* (1913), pp. 1–12; (with J. S. Hepburn, E. Q. St. John, E. Witmer, M. O. Stafford, and J. I. Burrell) "Bacterial and enzymic changes in milk and cream at 0°," *Journal of Biological Chemistry,* **16** (1913–1914): 331–63; (with M. K. Jenkins, E. Q. St. John, and W. B. Hicks) *A Bacteriological and Chemical Study of Commercial Eggs in the Producing Districts of the Central West* (Washington, D.C.: U.S. Government Printing Office, 1914); (with E. D. Clark) "Refrigeration, transportation and conservation of poultry and fish products," *Journal of Sociologic Medicine,* **16** (1915): 272–305. Additional publications by Pennington are listed in Creese and Creese, "Mary Engle Pennington," pp. 467–8.

27. Pennington, quoted by Pierce, "Mary Engle Pennington," p. 941.

28. Robinson, "Electrochemical school," pp. 229–30.

29. Ibid., pp. 226–7; alumni records, faculty (RS 26/4/1), University of Illinois, Urbana; alumnae records, Goucher College archives; "Directory of Officers and Students," university archives, University of California, Berkeley; Mary R. S. Creese, "Lily Gavit Kollock Paetow," in Miles and Gould, eds., *American Chemists,* p. 208.

30. Kollock was one of a succession of "academic" deans of women at the University of Illinois in the early years of this century who apparently had difficulties meeting the expectations of the university's higher-ranking administrators. One of her successors, the physicist Fanny Cook Gates, left in 1918 after a two-year tenure, when gossip hinted at her possible addiction to drugs. The president of the university declared he would never again hire for the job a woman with a Ph.D. in the sciences. (Maynard Britchford, archivist at the University of Illinois, provided information about Kollock's difficulties as dean. See also Rossiter, *Women Scientists in America,* p. 71.)

31. Kollock's post-1900 papers include the following, all coauthored by E. F. Smith: "The electrolytic method applied to uranium," *Journal of the American Chemical Society,* **23** (1901): 607–9; "Electrolytic estimation of molybdenum," ibid., **23** (1901): 699–71; (with G. H. West) "Use of the rotating anode in electroanalysis," ibid., **26** (1904): 1595–615; "Use of the rotating anode and mercury cathode in electroanalysis," ibid., **27** (1905) 1255–69, 1527–49; "The effect of sulphuric acid on the deposition of metals when using a mercury cathode and rotating anode," ibid., **29** (1907): 797–806; "The determination of zinc with the aid of the mercury cathode and rotating anode," *Transactions of the American Electrochemical Society,* **14** (1908): 59–64; "Estimation of indium with the use of a mercury cathode," *Journal of the American Chemical Society,* **32** (1910): 1248–50.

32. Robinson, "Electrochemical school," pp. 213–14.

33. Ibid., p. 214 (from a letter of Peirce's in the University of Pennsylvania archives).

34. Included among them were Marie Reimer, later professor of chemistry at Barnard College in New York City, Gertrude Heritage, who coauthored a series of creditable papers with Kohler beginning in 1905, Ruth Johnstin, and Annie Macleod. For remarks on the connection between the Bryn Mawr and Hopkins chemistry departments see Dean Stanley Tarbell and Ann Tracy Tarbell, *Essays on the History of Chemistry in the United States, 1875–1955* (Nashville, Tenn.: Folio Publishers, 1986), pp. 35, 83.

35. Obituary, "Miss Mary Breed, retired educator," *NYT,* 16 September 1949, 28: 3; *WWWA,* p. 124.

36. Anon., "Here and there with the girls", *Penn State Alumni News,* October 1931, 18–19; anon., obituary, *Centre Daily Times,* State College Pennsylvania, 14 July 1960 (Pennsylvania State University archives); *WWWA,* p. 516; *AMS,* 1921–1949.

37. MacDonald, quoted in anon., "Here and there," p. 19.

38. See the following: (with E. V. McCollum) "The cultivation of yeast in solutions of purified nutrients," *Journal of Biological Chemistry,* **45** (1921): 307–11; (with E. V. McCollum) "The 'bios' of Wildiers and the cultivation of yeast," ibid., **46** (1921): 525–7; "The synthesis of water-soluble B by yeasts grown in solutions of purified nutrients," ibid., **54** (1922): 243–8; "The synthesis of 'bios' by yeast grown in a solution of purified nutrients," ibid., **56** (1923): 489–99; "Multiplication of yeasts in solutions of purified nutrients," *American Journal of Hygiene,* **5** (1925): 622–34. A sixth note, "A plea for the retention of the term 'bios'," appeared in *Science,* **63** (1926): 187. MacDonald's coauthor, E. V. McCollum, known for his work in nutritional chemistry and especially for studies on the vitamins, had several women research associates while he was at Johns Hopkins University and over the years collaborated with a number of women home economists (see Rossiter, *Women Scientists in America,* p. 209).

39. Margaret B. MacDonald, "The mitogenetic effect on yeast of oligodynamic radiations from metals," *Iowa State College Journal of Science,* **9** (1935): 587–95. Her work in Tennessee was reported in the following articles: (with Esther M. Crawford) "The removal of onion and garlic flavor and odor from milk," *Journal of Home Economics,* **19** (1927): 65–9, and *University of Tennessee Agricultural Experiment Station, Circular 14* (1927); (with E. M. Crawford and Frances A. Briggs) "The effect of mineral oil treatment on the composition of milk," *Journal of Home Economics,* **22** (1930): 213–8; (with Adelaide Glaser) "The removal of the bitter flavor from 'bitterweed' cream," *University of Tennessee Agricultural Experiment Station, Circular 26* (1929); (with Norabelle D. Weathers) "The bitter principle of bitterweed," *University of Tennessee Agricultural Experiment Station, Annual Report* (1931): 34. MacDonald took out the patent, "Purifying milk and milk products" (U.S. Patent No. 1,644,842), in 1927.

40. Deceased alumni records, file 41/2/877, and graduate school records, file 12/5/636, Cornell University archives; information from Earlham College archives, including a short article "Rambling 'Round," by Joe Adams (*Indianapolis Star*?), April 1949, and obituaries in the *Palladium,* Richmond, Indiana, 17 April 1960, the *Evening Star,* Franklin, Indiana, 16 April 1960 and *Quaker Life,* 28 May 1960; personal letter from Francis D. Hole, professor emeritus, University of Wisconsin, nephew of Miss Doan; Mary R. S. Creese, "Martha Doan," in Miles and Gould, eds., *American Chemists,* pp. 63–4.

41. For Littleton see *WWWA,* p. 642, Kline entry.

42. Eells, "Earned doctorates," p. 649. Gibson's two post-1900 papers were, "The nature and properties of corn oil. II," *Journal of the American Chemical Society,* **23** (1901): 1–8; "Metallic soaps from linseed oil and their solubility in certain hydrocarbons," ibid., **24** (1902): 215–22.

43. For Cook see *WWWA,* p. 201; her 1906 paper (with M. T. Bogert) was "Quinazolines. XVIII. Synthesis of 6-nitro-4-keto-2-methyldihydroquinazolines from 5-nitro-acetylanthranil and primary amines," *Journal of the American Chemical Society,* **28** (1906): 1449–1454; Reed's name appears in student lists in the *General Catalogue,* "Summer Courses of Instruction in Chemistry, Botany and Geology," Harvard University, 1876, p. 154, 1877, p. 189.

44. Charles F. Mabery, "Professor Rachel Lloyd," *Journal of American Chemical Society,* **23** (1901): 84; Ann T. Tarbell and D. Stanley Tarbell, "Dr. Rachel Lloyd (1839–1900): American chemist," *Journal of Chemical Education,* **59** (1982): 743–4; Robert N. Manley, *Centennial History of the University of Nebraska,* (Lincoln, Neb.: University of Nebraska Press, 1969), pp. 111, 130, 140; Board of Regents Papers, 1887, 1892, 1893, University of Nebraska archives; student lists, *General Catalogue,* "Summer Courses of Instruction," Harvard, 1875–1883; *Catalogue, 1883–84,* Louisville School of Pharmacy for Women (Louisville, Ky.: Historical Society, Filson Club, [n.d.]); 1885 Louisville City Directory (Caron's); Olin E. Holloway, *Genealogy of the Holloway Families* (Knightstown, Ind.: [n.p.], 1927), pp. 215–16; William Wade Hinshaw, *Encyclopedia of American Quaker Genealogy,* comp. T. W. Marshall (Ann Arbor, Mich., 1936–), vol. 4, p. 533; Mary R. S. Creese, "Rachel Abbie Holloway Lloyd," in Miles and Gould, eds., *American Chemists,* pp. 167–9. For a fuller account of Lloyd's life and work see Mary R. S. Creese and Thomas M. Creese, "Rachel Lloyd, early Nebraska chemist," *Bulletin for the History of Chemistry,* no. 17/18 (1995): 9–14; we are especially grateful to Howard Ephraim Stratton and Lewis Stratton of Flushing, Ohio, and to Margaret L. Glavanis of St. Clairsville Public Library, St. Clairsville, Ohio, for information about Lloyd's early family background.

45. It is perhaps no coincidence that Lydia D. Hampton, president and faculty member of Hampton College, was a fellow chemistry student of Lloyd at Harvard in the summer of 1880. (For Hampton see *WWWA,* p. 210, entry for Lydia Hampton Cowling.) The Harvard Summer School would seem, to some extent, to have served science instructors as a national meeting place and employment information center. (Hampton College awarded Lloyd an A.M. degree, a fact rarely

noted in accounts of her career, although it is recorded in University of Nebraska faculty lists—Board of Regents Papers, April–June, 1893.)

46. Helen Abbott also joined the German Chemical Society in 1886. Lloyd joined the American Chemical Society in 1891, becoming its first woman member, except for Rachel Bodley, who had been given her (largely honorary) membership at the formation of the society in 1874.

47. Nicholson, then in charge of chemistry instruction at the University of Nebraska, had nominated Lloyd for a regular position as associate professor of analytical chemistry. However, the faculty failed to approve the unconditional appointment by a clear majority and her initial position was temporary (Board of Regents Papers, 1887). She became full professor the following year.

48. A report on the school's first year was summarized in *Science,* **19,** (1892): 324. Lloyd reported her analytical results in three articles, all entitled "Experiments in the culture of the sugar beet in Nebraska," 1 April 1890, 15 April 1891, and 1 March 1892, on file in the archives of the University of Nebraska. These formed a part of a series of publications on sugar beet put out by the Nebraska Agricultural Experiment Station.

49. Manley, *Centennial History,* pp. 131, 133, 143, 177, 178; Ernest Douglas, "Miss Rosa Bouton makes good at sixty then retires at seventy-four," *Sunday Journal and Star,* Phoenix, Arizona, 21 April 1935; obituary in an unidentified Nebraska newspaper, 5 March 1951 (University of Nebraska archives); obituary of Eli Bouton, *Sabetha Herald,* Sabetha, Kansas, November 1905 (Mary Cotton Public Library, Sabetha, Kansas); faculty files, Bouton, Rosa, 1912, 2/9/3, and Papers of M. Fedde, Correspondence, 1929–1970, 9/1/10, University of Nebraska archives; Mary R. S. Creese, "Rosa Bouton," in Miles and Gould, eds., *American Chemists,* pp. 26–7.

50. The professional activities of early women chemists in the Nebraska Section of the American Chemical Society are recorded in the society's *Proceedings*—see *Journal of the American Chemical Society* for the years 1891–1910. Throughout this period more women were active in the Nebraska Section than in any other local section in the country. Besides Rachel Lloyd, Rosa Bouton, and Mary Fossler, the three discussed in this survey, Eva Sullivan, Mariette Gray, Mildred Parks, Mariel C. Gere, Mabel A. Hartzell, Mamie Short, and Rachel Corr all contributed to the activities of the section in the years just after the turn of the century. (Parks, Gere, and Corr studied chemistry at the University of Nebraska at the graduate level and received A.M. degrees. Gray, Hartzell, and Short took B.S. degrees.) The prominence of so many early women chemists and chemistry students at Lincoln is remarkable. Employment opportunities offered them by the Agricultural Experiment Station were important, but the influence of Rachel Lloyd and Rosa Bouton, the first women chemists on the faculty, appears to have had a considerable effect in encouraging the active participation of younger women at section meetings. Like Lloyd, Bouton was remembered especially for her concern about the welfare and advancement of her students; indeed, some of them completed their university education thanks only to timely loans from her.

51. All of the journal publications of Nebraska women chemists up to 1910 (except for those of Rachel Lloyd) were coauthored by Samuel Avery, who had himself been a student of Lloyd. He was clearly willing to encourage women's research and his presence on the chemistry faculty from the late 1890s until he became chancellor of the university in 1909 may well have been an additional factor in the prominence of Nebraska women chemists of the period.

52. Anon., "How a department grew. Domestic Science beginning at State University. Work of Miss Rosa Bouton" (manuscript in University of Nebraska archives).

53. Chancellor Avery to Rosa Bouton, June 15, 1912 (file 2/9/3, University of Nebraska archives). Avery, Bouton's old friend and fellow chemist, at the time chancellor, was much distressed by the turn of events, but could do little to ward off the blow that hit Bouton like "a thunderbolt out of a clear sky" (Bouton to Chancellor Avery, June 18, 1912, file 2/9/3). Quotations from archival material are made with the permission of the University of Nebraska.

54. Manley, *Centennial History,* pp. 214, 220, 221, 223; *AMS,* 1910–1949; *Nebraska Alumnus* (February 1952): 10.

55. Obituaries, *NYT,* 15 March 1940, 23: 6, and *Hubbard Eagle,* 21 March 1940 (clipping in Hubbard Public Library, Hubbard, Ohio); *Alumni Directory of the University of Chicago, 1919*; alumni records, Oberlin College archives.

56. See, for instance, Louis F. Fieser and Mary Fieser, *Organic Chemistry* (London: George A. Harrap and Co., 1953), p. 229.

57. Joseph Hoeing Kastle and Mary E. Clarke, "Decomposition of hydrogen peroxide by various substances at high temperature," *American Chemical Journal,* **26** (1901): 518–26; "Cyanogen iodide as an indicator for acids," ibid, **30** (1903): 87–96; "Occurrence of invertase in plants," ibid., **30** (1903): 422–7. Information about Clarke came from student records, University of Kentucky archives.

58. Student records, University of Kansas archives.

59. *Alumni Catalogue of the University of Michigan, 1837–1921,* p. 620.

60. Jean C. Dahlberg, "Laura A. Linton and lintonite," *Minnesota History,* **38,** (1962): 21–3, "A woman to remember," *Lapidary Journal,* **29** (October 1976): 1732–6; obituaries in the *Rochester Post and Record,* 9 April 1915, 1, and the *Minneapolis Journal,* 2 April 1915 (clippings from the Minnesota Historical Society, St. Paul, Minnesota, and the Olmsted County Historical Society, Rochester, Minnesota); student records, University of Michigan archives; alumni records, University of Minnesota archives; student records, Institute archives, MIT; *Catalogue of the Officers and Students of Lombard University, Galesburg, Illinois, 1884*; private communication from Robert J. Endecavageh, Unocal Corporation, Los Angeles, California; Mary R. S. Creese and Thomas M. Creese, "Laura Alberta Linton: an American chemist," *Bulletin for the History of Chemistry,* no. 8 (1990): 15–18; Mary R. S. Creese, "Laura Alberta Linton," in Miles and Gould, eds., *American Chemists,* pp. 165–6 (the reference in this sketch to a Union Oil Company laboratory in Santa Clara, California, is wrong; it should read Santa Paulo).

61. Edger Wesley Owen, *Trek of the Oil Finders: a History of Exploration for Petroleum* (Tulsa, Okla.: American Association of Petroleum Geologists, 1975), p. 165.

62. Stanford University *Alumni Directory,* 1932, pp. 453, 716. Ray (later Titcomb) is also listed in the *AMD,* 1914–1958, which designates her as having retired by 1950.

63. Within another decade the first research papers from Marie Reimer and her group at Barnard College appeared (see, for example, Marie Reimer, "Reactions of organic magnesium compounds with cinnamylidene ester," *American Chemical Journal,* **38** (1907): 227–37; this was followed by a second paper in 1908 developing the same topic and coauthored by Grace Potter Reynolds). By the second decade of the present century, research was being carried out by a women's group headed by Emma Perry Carr at Mount Holyoke College. Even by the 1930s, however, productive women chemists like Carr had only limited success in obtaining the necessary financial support for their projects (see also Rossiter, *Women Scientists in America,* p. 174).

64. The contribution by Margaret Johnston and Catherine Kennedy listed in the bibliography is a subsection of a paper indexed in the Royal Society *Catalogue* under the name of T. E. Thorpe. The

decrease in author count by three from that given in previous communications (Creese and Creese, "British women . . . geological sciences . . . " and Mary R. S. Creese, "British women of the nineteenth and early twentieth centuries who contributed to research in the chemical sciences," *British Journal for the History of Science,* **24** (1991): 275–305) results from the reclassification of papers by Emma Hooker, Eleanor Sidgwick, and Mary Somerville from chemistry into technology, physics, and physics, respectively (see also n. 1).

65. University of London General Register of graduates and undergraduates, 1900.

66. Queen's College was the first of the colleges for ladies in London. It opened in Harley Street in 1848 and provided education for girls over the age of twelve. By the 1860s it had a science curriculum that was outstanding for the time, and was successfully preparing its students for university entrance examinations. It had close ties with King's College, several of whose male faculty were also on its staff—see Rita McWilliams-Tullberg, *Women at Cambridge. A Men's University—though of a Mixed Type* (London: Victor Gollancz, 1975), pp. 19, 31, Patricia Phillips, *The Scientific Lady. A Social History of Women's Scientific Interests 1520–1918* (New York: St. Martin's Press, 1990), p. 239, and Elaine Kaye, *A History of Queen's College* (London: Chatto and Windus, 1972).

67. Three of these papers are listed in the bibliography; the third, "Molecular surface-energy of some mixtures of liquids," appeared in the *Transactions of the Royal Irish Academy,* **32** (1902): 93–100.

68. Some information about Marshall and Hall came from the records of University College (University of London General Register, Pt. III, 1900, and fee entry books).

69. See Clare de Brereton Evans, "Traces of a new tin-group element in thorianite," *Journal of the Chemical Society, Transactions* **93** (1908): 666–8; (with Otto Brille) "The use of the new micro-balance for the determination of electrochemical equivalents and for the measurement of densities of solids," ibid., **93** (1908): 1442–6.

70. Student records, Imperial College of Science, Technology and Medicine, London.

71. Margaret Douie Dougal, "A specimen of early Scottish iron," *Journal of the Chemical Society, Transactions,* **65** (1894): 744–50, quotations from p. 750.

72. Sir James Dewar, Presidential Address, ibid., **75** (1899): 1168.

73. A. A. Eldridge, "Martha Annie Whiteley 1866–1956," *Proceedings of the Chemical Society* (1957): 182–3; *Who was Who,* vol. 5, p. 1162; Miss M. Gosset, "Dr. M.A. Whiteley," *Times,* 22 June 1956, 13f; anon. obituaries, "Great woman chemist," *Yorkshire Evening News,* 26 May 1956, "Invented Tear Gas," *Star* (London), 25 May 1956, "Dr. Martha Whiteley," *Times,* 26 May 1956, 10a; Mary R. S. Creese, "Martha Annie Whiteley (1866–1956): chemist and editor," *Bulletin for the History of Chemistry,* no. 20 (1997): 42–5. See also G. A. R. Kon, "Sir Jocelyn Field Thorpe," obituary, *Journal of the Chemical Society,* **144** (1941): 444–64.

74. The term "assistant professor" should not be confused with the identical American designation. The position held by Whiteley was one peculiar to the constituent colleges of Imperial College and not a London University appointment, although after negotiations it was recognized by the university. It designated a senior faculty rank below professor. In the 1940s the title was changed to reader (information from Anne Barrett, College Archivist, Imperial College of Science, Technology and Medicine).

75. Whiteley's post-1900 papers include the following: "The action of barium hydroxide on dimethylvioluric acid," *Journal of the Chemical Society, Transactions,* **83** (1903): 18–23; "The oxime of mesoxamide and some allied compounds. Part II. Di-substituted derivatives," ibid., **83** (1903): 24–45; "The oxime of mesoxamide (isonitrosomalonamide) and some allied compounds. Part III. Tetra-substituted derivatives," ibid., *Proceedings* (1904): 92–3; "Studies in the barbituric acid series, I. 1:3-Diphenylbarbituric acid and some coloured derivatives," ibid., *Transactions,* **91** (1907): 1330–50; "Studies in the barbituric acid series. Pt. II. 1:3-Diphenyl-2-thiobarbituric acid and some coloured derivatives," ibid., *Proceedings* (1909): 121–3; (with J. V. Backes and R. W. West) "Quantitative reduction by hydriodic acid of hydrogenated malonyl derivatives," ibid., *Transactions* **119** (1921): 359–79; (with A. G. Rendall) "Oxime of mesoxamide (iso-nitrosomalonamide) and some allied compounds. The ethers of isonitrosomalonanilide, isonitrosomalondimethylamide and iso-nitrosomalondibenzylamide," ibid., **121** (1922): 2110–9; (with Edith H. Usherwood) "Oxime of mesoxamide (isonitrosomalonamide) and some allied compounds. Ring formation in the tetrasubstituted series," ibid., **123** (1923): 1069–89; (with Arthur Plowman) "Oxime of mesoxamide (isonitrosomalonamide) and some allied compounds. Structural and stereoisomerism in the methyl ethers of the *p*-tolyl derivatives," ibid., **125** (1924): 587–604; (with Dorothy Yapp) "Reaction of diazonium salts and malonyldiurethan," ibid., **130** (1927): 521–8. In addition to her chemical work, Whiteley also collaborated during her student years in biostatistical research being carried out by Karl Pearson and his team at University College. Her 1900 paper in this area is listed in the bibliography; a second (with Marie Lewenz) was, "Data for the problem of evolution in man. A second study of variability and correlation of the hand," *Biometrika,* **1** (1902): 345–60.

76. Jocelyn Field Thorpe and Martha Annie Whiteley, *A Student's Manual of Organic Chemical Analysis, Qualitative and Quantitative* (London: Longman's, Green, 1925), reviewed in *Nature,* **116** (1925): 707–8.

77. Joan Mason, "A forty years' war," *Chemistry in Britain,* **27** (1991): 233–8.

78. E. M. Farrer appears to have published only one paper, a note on the hydration of salts coauthored with Spencer Pickering of Bedford College, which appeared in *Chemical News* in 1886.

79. E. J. Shellard, "Some early women research workers in British pharmacy. 1886–1912," *Pharmaceutical Historian: Newsletter for the British Society for the History of Pharmacy,* **12** (1982): 2–3.

80. Ibid.; Desmond MacHale, *George Boole. His Life and Work* (Dublin: Boole Press Ltd., 1985), pp. 158, 252–3, 265–6.

81. The Booles were a remarkable family. Lucy's older sister Alicia made noteworthy contributions to mathematics (see Alicia Boole Stott, chapter 8). A younger sister Ethel Lilian (later Mrs. Voynich) became a successful novelist; her most famous work, *The Gadfly* (1897), an anti-Catholic story set in the revolutionary upheavals of mid-nineteenth-century Italy, was translated into thirty languages and was especially popular in Russia and eastern Europe.

82. L. Martindale, *A Woman Surgeon* (London: Victor Gollancz, 1951), p. 32.

83. MacHale, *George Boole,* p. 265.

84. *Newnham College Register 1871–1971,* vol. 1, p. 85.

85. Ibid., pp. 81–2.

86. Ibid., p. 146; entry for Mills, William Hobson, *DNB,* Supplement, 1951–1960, pp. 739–40.

87. William Hobson Mills and Mildred Mills, "The synthetical production of derivatives of dinaphthanthracene," *Proceedings of the Chemical Society,* **28** (1912): 242; *Journal of the Chemical Society, Transactions* **101** (1912): 2194–2208.

88. *Newnham College Register 1871–1971,* vol. 1, p. 90.

89. See the entry for Walker, Sir James, *DNB,* Supplement,

1931–1940, pp. 886–7. Sedgwick's NST dates are from *The Historical Register of the University of Cambridge . . . to the year 1910,* ed. J. R. Tanner (Cambridge: University Press, 1971), pp. 766, 768.

90. James Walker and Annie Purcell Walker, "Tetraethylsuccinic acid," *Journal of the Chemical Society, Transactions,* **87** (1905): 961–7.

91. Obituaries, M. A. Whiteley, "Ida Smedley MacLean. 1877–1944," ibid., **149** Pt. I (1946): 65–7, Leslie C. A. Nunn, *Nature,* **154** (1944): 110 and M[ary] E. de R. E[pps], "Ida Smedley MacLean," *Newnham College Roll Letter* (January, 1945): 50–1; *Newnham College Register 1871–1971,* vol. 1., p. 138; Harriette Chick, Margaret Hume, and Majorie Macfarlane, *War on Disease. History of the Lister Institute* (London: André Deutsch, 1971), pp. 164–6; Creese, "British women . . . chemical sciences," pp. 282–4; Mary R. S. Creese, "MacLean, Ida Smedley," in *DNB,* "Missing Persons" vol., 1993, pp. 433–4.

92. Ida Smedley was an accomplished pianist and an actress of considerable talent, although she had little time for music or theater after her student days. Her sister, Constance Smedley Armfield, became a theatrical producer and a writer. Constance worked for a time in California and then in England (Winifred I. Vardy, comp., *King Edward VI High School for Girls Birmingham, 1883–1925* (London: Ernest Benn, 1928), p. 140).

93. Epps, "Ida Smedley MacLean," p. 50. Other early women chemists and biochemists who were educated at King Edward VI High School and Newnham College were Mary Beatrice Thomas, Hilda Jane Hartle, Muriel Wheldale, Edith Willcock, and Annie Homer. None of these published journal articles before 1900 (and so are not listed in the bibliography) but all took the Cambridge NST examinations in the period between 1897 and 1905. Thomas and Hartle went on to notable careers as educators. Thomas ran the chemical laboratory at Girton College until she retired in 1935 and became well-known for the exceptional quality of her teaching; Hartle was lecturer at Homerton College Cambridge for seventeen years and then principal of Brighton Municipal Training College for another twenty-one. Wheldale spent almost all her career in the Botany School and the Biochemical Department at Cambridge, where her lectures on plant biochemistry formed an important unit in the teaching of advanced botany. She became university lecturer in biochemistry in 1927, one of the first women to receive such an appointment; four years later, shortly before her death, she published the first volume of her *Principles of Plant Biochemistry.* Homer, who was awarded a D.Sc. in 1910, carried out early research in physical-organic chemistry at Cambridge (NST, 1904, 1905, class I in both parts, chemistry in pt. II). She then went on to work in biochemistry at the Lister Institute in London, where during the First World War she did critical research in antisera production, an area in which she published a substantial number of papers. Willcock's work, in biochemistry, was carried out at Cambridge with W. B. Hardy and F. G. Hopkins (see Creese, "British women . . . chemical sciences").

94. Smedley received similar awards of £10 and £15 in 1909 and 1910, respectively (Research Fund Income and Expenditure Accounts, in the reports of Annual General Meetings, *Journal of the Chemical Society, Transactions,* **83** (1903): 635; **97** (1910): 659; **99** (1911): 585).

95. For Smedley's work on fluorene derivatives see, "Studies on the origin of colour; derivatives of fluorene," ibid., **87** (1905): 1249–55. She had begun her work on color at the Central Technical College under Armstrong who had a special interest in the topic (see his proposal for a quininoid theory of color, "The origin of colour and the constitution of colouring matters. The yellow colour of 2:3-hydroxynaphthoic acid," *Proceedings of the Chemical Society,* **4** (1888): 27–31; ibid., **8** (1892): 101–4). Smedley's other papers on constitution and optical properties included, "The refractive power of diphenylhexatriene and allied hydrocarbons," *Journal of the Chemical Society, Transactions,* **93** (1908): 372–84; "The relation between the chemical constitution and optical properties of the aromatic α- and γ-diketones," ibid., **95** (1909): 218–31; "The relative influence of ketonic and ethenoid linking on refractive power," ibid., **97** (1910): 1475–84; "The constitution of the β-diketones," ibid., **97** (1910): 1484–94.

96. For a more detailed account of Smedley MacLean's research, see the quotation from L. C. A. Nunn in Whiteley, "Ida Smedley MacLean," p. 66. Her main papers are the following: "Condensation of crotonaldehyde," *Journal of the Chemical Society, Transactions,* **99** (1911): 1627–33; "Biochemical synthesis of fatty acids from carbohydrates," *Proceedings of the Physiological Society, Journal of Physiology,* **45** (1912): xxv-xxvii; (with Eva Lubrzynska) "The biochemical synthesis of the fatty acids," *Biochemical Journal,* **7** (1913): 364–74, and "Condensation of aromatic aldehydes with pyruvic acid," ibid., **7** (1913): 375–9; (with Ethel Mary Thomas) "The nature of yeast fat," ibid., **14** (1920): 483–93; (with E. M. Thomas) "Observations on abnormal iodine values, with special reference to the sterols and resins," ibid., **15** (1921): 319–33; "The conditions influencing the formation of fat by the yeast cell," ibid., **16** (1922): 370–9; (with Dorothy Hoffert) "Carbohydrate and fat metabolism in yeast," ibid., **17** (1923): 720–41; (with D. Hoffert) "The carbohydrate and fat metabolism of yeast. Pt. II. The influence of phosphates in the storage of fat and carbohydrate in the cell," ibid., **18** (1924): 1273–8; (with Ethel Marjorie Luce) "The presence of vitamin A in yeast fat," ibid., **19** (1925): 47–51; (with D. Hoffert) "The carbohydrate and fat metabolism of yeast. Pt. III. The nature of the intermediate stages," ibid., **20** (1926): 343–57; (with Charles Gaspard Daubney) "The carbohydrate and fat metabolism of yeast. IV. The nature of the phospholipins," ibid., **21** (1927): 373–85; "The isolation of a second sterol from yeast fat," ibid., **22** (1928): 22–6; (with Eleanor Margaret Hume and Hannah Henderson Smith) "The examination of yeast fat for the presence of vitamins A and D before irradiation and of vitamin D after irradiation," ibid., **22** (1928): 27–33; (with Rachel Anne McAnally) "Note on the storage of carbohydrate and fat by *Saccharomyces frohberg* when incubated in sugar solutions," ibid., **28** (1934): 495–8; (with R. A. McAnally) "The synthesis of reserve carbohydrate by yeast," ibid., **29** (1935): 2236–41, **31** (1937): 72–80; (with Leslie Dundonald MacLeod) "The carbohydrate and fat metabolism of yeast," ibid., **32** (1938): 1571–82; (with Marion Alice Battie) "The catalytic action of cupric salts in promoting the oxidation of fatty acids by hydrogen peroxide," ibid., **23** (1929): 593–9; (with Margaret Sarah Beavan Pearce) "The oxidation of oleic acid by means of hydrogen peroxide with and without the addition of copper sulfate: a possible analogy with its oxidation *in vivo,*" ibid., **25** (1931): 1252–66; (with M. S. B. Pearce) "The oxidation of palmitic acid, by means of hydrogen dioxide in the presence of a cupric salt. Preliminary communication," ibid., **28** (1934): 486–94; (with Annie Phyllis Ponsford) "The oxidation of the fatty dibasic acids and of laevulic acid by hydrogen dioxide in presence of a cupric salt," ibid., **28** (1934): 892–7; (with Robert Owen Jones) "The oxidation of phenyl derivatives of fatty acids with hydrogen peroxide in the presence of copper," ibid., **29** (1935): 1877–80; (with Leslie Charles Alfred Nunn) "The oxidation products of the unsaturated acids of linseed oil," ibid., **29** (1935): 2742–5; (with E. A. Hume, L. C. A. Nunn, and H. H. Smith) "Studies of the essential unsaturated fatty acids in their relation to the fat-deficiency disease of rats," ibid., **32** (1938): 2162–77; (with L. C. A. Nunn) "The nature of the fatty acids stored by the liver in the fat-deficiency diseases of rats," ibid., **32** (1938): 2178–84; (with E. A. Hume, L. C. A. Nunn and H. H. Smith) "Fat-deficiency disease of rats. The relative curative

potencies of methyl linoleate and methyl arachidonate," ibid., **34** (1940): 879–83; (with L. C. A. Nunn) "Fat deficiency disease of rats. The effect of doses of arachidonate and linoleate on fat metabolism with a note on the estimation of arachidonic acid," ibid., **34** (1940): 884–902; (with E. M. Hume) "The storage of fat in fat-starved rats," ibid., **35** (1941): 990–5; (with E. M. Hume) "Fat deficiency disease of rats. The influence of tumor growth on the storage of fat and of polyunsaturated acids in the fat-starved rat," ibid., **35** (1941): 996–1002; (with L. C. A. Nunn) "Fat-deficiency disease of rats. The relation of the essential unsaturated acids to tumor formation in the albino rat on normal diet," ibid., **35** (1941): 983–9; (with Doris Elaine Dolby and L. C. A. Nunn) "The constitution of arachidonic acid (preliminary communication)," ibid., **34** (1940): 1422–6; (with C. L. Arcus) "Structure of arachidonic and linoleic acids," ibid., **37** (1943): 1–6. The work of the Burrs referred to was the following: George O. Burr and Mildred M. Burr, "A new deficiency disease produced by the rigid exclusion of fat from the diet," *Journal of Biological Chemistry,* **82** (1929): 345–67, and "On the nature and rôle of the fatty acids essential in nutrition," ibid., **86** (1930): 587–621.

97. A. Kleinzeller, review, *Nature,* **153** (1944): 510.

98. See for instance, T. P. Hilditch and P. N. Williams, *The Chemical Constitution of Natural Fats,* 4th ed. (New York: Wiley, 1964), especially pp. 528, 553–4, 571, 573 (n.1), 609.

99. Smedley's early work on fat metabolism was recognized in 1913 by the award of the Ellen Richards Prize, given by the American Association of University Women to the woman, from any country, judged to have made the most outstanding contribution of the year to scientific knowledge.

100. Mason, "Forty years' war." Interestingly enough, Smedley MacLean's postgraduate adviser, H. E. Armstrong, from his position on the Chemical Society's council for many decades, was one of the most powerful and vocal opponents of the admission of women.

101. *Girton College Register 1869–1946,* vol. 1, p. 47; H. King, "Sigmund Otto Rosenheim, 1871–1955," *Biographical Memoirs of Fellows of the Royal Society,* **8** (1956): 257–67; *Who was Who,* vol. 5, p. 946 (entry for Otto Rosenheim).

102. See John T. Edsall, "The development of the physical chemistry of proteins, 1898–1940," in *The Origins of Biochemistry: A Retrospect on Proteins,* eds. P. R. Srinivasan, Joseph S. Fruton and J. T. Edsall (New York: New York Academy of Sciences, 1978), pp. 53–73, especially pp. 54–5. Tebb's 1898 and 1899 protein papers are listed in the bibliography; others by her in this area include: "The chemistry of reticular tissue," *Journal of Physiology,* **27** (1902): 463–72; "Precipitation of proteins by alcohol and other reagents," ibid., **30** (1903): 25–38.

103. The Rosenheim-Tebb papers include "The non-existence of 'protagon' as a definite chemical compound," *Journal of Physiology,* **36** (1907): 1–16; "Further proofs of the non-existence of 'protagon' as a definite chemical compound," ibid., **37** (1908): i-iv; "Optical activity of 'protagon'; a new physical phenomenon observed in connection with the optical activity of so-called 'protagon'," ibid., **37** (1908): 341–7, 348–54; "So-called 'protagon'," *Quarterly Journal of Experimental Physiology,* **1** (1908): 297–304; "The non-existence of 'protagon' as a definite chemical compound," ibid., **2** (1909): 317–33; "Lipoids of the brain. I. Sphingo-myelin," *Journal of Physiology,* **38** (1909): li-liii; "Lipoids of the adrenal," ibid., **38** (1909): liv-lvi; "Lipoids of the brain. II. A new method for the preparation of the galactosides and of sphingomyelin," ibid., **41** (1910): i–ii; "Die Nicht-Existenz des sogenannten 'Protagons' im Gehirn," *Biochemische Zeitschrift,* **25** (1910): 151–60.

104. Otto Rosenheim, H. W. Dudley and M. C. Rosenheim, "The chemical constitution of spermine. I. The isolation of spermine from animal tissues and the preparation of its salts," *Biochemical Journal,* **18** (1924): 1263–72. Christine Rosenheim's cholesterol papers were, "The cholesterol of the brain. II. The presence of oxycholesterol and its esters," ibid., **8** (1914): 74–81; "The cholesterol of the brain. III. Cholesterol contents of human and animal brain," ibid., **8** (1914): 82–3; "A new colour reaction for 'oxycholesterol'," ibid., **10** (1916): 176–82.

105. *Times,* 5 October 1953, 10e.

106. Student records, University of Bristol.

107. V. V. Markovnikov, ["Recherches sur les composés cycliques de la série de l'hexamethylene"], *Journal of the Russian Physico-Chemical Society,* **30** (1898): 59–90, 151–95; *Bulletin de la Société Chimique de Paris,* **20** (1898): 851–6; [*Liebig's*] *Annalen der Chemie,* **301** (1898): 154–202, **302** (1898): 1–42.

108. The technique used in the final stages of the commercial production of water-free "absolute" alcohol was based on one of Fortey and Young's 1902 observations on the behavior of mixed liquids (E. E. Turner and Margaret Harris, *Organic Chemistry* (London: Longmans, Green and Co., 1952), pp. 60–1; Emily Fortey is referred to here, though only by last name). Fortey's post-1900 papers with Young include: "The properties of the mixtures of the lower alcohols with water," *Journal of the Chemical Society, Transactions,* **81** (1902): 717–39; "The properties of the mixtures of the lower alcohols with benzene, and with benzene and water," ibid., **81** (1902): 739–52; "Fractional distillation as a method of quantitative analysis," ibid., **81** (1902): 752–68; "Vapour pressures and specific volumes of isopropylisobutyrate," ibid., **81** (1902): 783–6; "The vapour pressures and boiling points of mixed liquids. Part II," ibid., **83** (1903): 45–68.

109. Student records and council minutes, University of Bristol. Date of death from Sir William Tilden's, *Sir William Ramsay, K.C.B., F.R.S.* (London: Macmillan, 1918), p. 84.

110. Morris W. Travers, F.R.S., *A Life of Sir William Ramsay, K.C.B., F.R.S.* (London: Edward Arnold, 1956), p. 69. If indeed Katherine Williams did pass up the opportunity to take a fresh look at the Cavendish experiment under Ramsay's guidance, she must surely have wondered in later years if she had made a good choice! In 1785 Henry Cavendish investigated the composition of the atmosphere by passing an electric spark through air to which additional (excess) oxygen had been added and subsequently removing the compound of oxygen and nitrogen thus formed (plus the residual oxygen), by absorption into solutions kept in contact with the gases. Only a small bubble of gas remained, and at the time the experiment was interpreted as demonstrating that air consisted of oxygen and nitrogen. When, in 1894, Ramsay and the physicist Lord Rayleigh again prepared this residual gas bubble and examined it spectroscopically, they were able to show that it was not nitrogen but a new element—the inert gas argon. Ramsay and his collaborators went on to discover the other inert gases. He received the Nobel prize for the work in 1904.

111. Williams's later technical papers were, "The chemical composition of cooked vegetable foods," *Journal of Industrial and Engineering Chemistry,* **5** (1913): 653–6, and "Losses and other chemical changes in boiling vegetables," *Chemical News,* **123** (1916): 145–7. Some of her work was reprinted in American journals: "The chemical composition of cooked vegetable foods," *Journal of the American Chemical Society,* **26** (1904): 244–52; "The chemical composition of cooked vegetable foods, Part II," ibid., **29** (1907): 574–82; "Science for the home; losses and other chemical changes in boiling vegetables," *Scientific American* (*Supplement*), **81** (1916): 362.

112. Marion Newbigin, although she published several biochemical papers in the 1890s, was best known for her work in biology and geog-

raphy (see chapters 3 and 13). No information has been uncovered about either Wilhelmina Green or Lavinia Edna Walter. Green's analysis of tea leaves, published in *Chemical News* in 1885, and Walter's note on derivatives of sulfanilic acid in the Chemical Society's *Proceedings* for 1896 would seem to have been their only chemical publications. Margaret Marshall's paper on the effects of animal and vegetable matter on minerals appeared in the *Proceedings* of the Edinburgh Royal Society in 1851. At the time there was much interest in Britain in the process of soil formation by rock breakdown, an area pioneered by the German chemist Justus von Liebig (1803–1873)—see also the sketch of Elizabeth Hodgson (chapter 12). Marshall's paper was one of the few nineteenth-century journal articles by a British woman on a chemical topic that was published before 1880. (An 1846 paper by Mary Somerville on the effects of light of a particular wavelength on vegetable juices, although also classifiable as chemistry, is listed in the physics section of the bibliography. A report by the London woman Elizabeth Fulhame of her study of the reduction of metallic salts, probably the first notable independent research in chemistry by a woman in the modern era, appeared before 1800 and so is outside the time frame of this survey—see J. F. Coindet, "De l'ouvrage de Mme. Fulhame, intitulée: An Essai on Combustion, etc.," *Annales de Chimie,* **26** (1798): 58–85, and also Creese, "British women . . . chemical sciences," p. 277.

113. H. E. R. and P. P. B., "Thomas Carnelley," *Nature,* **42** (1890): 522–3.

114. Richard B. Pilcher, *The Institute of Chemistry of Great Britain and Ireland. History of the Institute 1877–1914* (London: Council of the Institute of Chemistry, 1914), pp. 113–14.

115. Information from King's College London archives; Vera Brittain, *The Women at Oxford: a Fragment of History* (New York: Macmillan, 1960), pp. 68, 82.

116. J. R. Partington, *A History of Chemistry* (London: Macmillan, 1964, 4 vols.), vol. 4, p. 587.

117. We know, however, that about 40 percent of the thirty early women chemists mentioned here held professional positions in which they used their scientific training (Boole and de Brereton Evans at the London School of Medicine for Women, Field at the Royal Holloway College, Whiteley at the Royal College of Science, Smedley MacLean at the Lister Institute, Newbigin at the School of Medicine for Women in Edinburgh, Seward McKillop at the Royal Holloway College and the Women's Department of King's College London, Tebb-Rosenheim as a research worker at King's College, Williams—possibly as a research worker only—at University College Bristol, Dougal as a compiler for the Chemical Society, Earp and Halcrow as schoolteachers). Further, the estimate of 40 percent is very likely low, since a number of others whose careers have not been followed (such as Emily Lloyd, the first woman associate of the Institute of Chemistry) may well have gone on to salaried positions, including schoolteaching, the most common occupation for women with university training.

Chapter 12

MIDWESTERN CAVES, GRAPTOLITE EVOLUTION, TYROLEAN TECTONICS: STUDIES BY EARLY WOMEN GEOLOGISTS

Geology papers make up about 5 percent of the publications by women indexed by the Royal Society. The largest fraction came from British workers and a further notable contribution was made by a small group of Russians. Most of the rest was produced by women in the United States, Ireland, Italy, Sweden, and France.

American women

Pre-1901 papers by seven Americans—Bascom, Fielde, Fleming, Hayes, Klem, Owen, and Smith—are listed in the index, and articles by two more, Maury and Barbour, who also published before the turn of the century, have been added to the bibliography. Only three of the Americans—Owen, Bascom, and Maury—brought out a notable amount of original work in geology.[1]

LUELLA OWEN[2] (1852–1932) was born in St. Joseph, Missouri, and lived there all her life. Though neither an academic nor employed by government agencies, as were both Bascom and Maury, Owen nevertheless carried out significant early studies in Midwestern geology. She was recognized with a number of honors, including fellowship in both the AAAS and the American Geographical Society, and membership, as the only woman, in the French Société de Spéléologie. She also had strong interests in physical geography and archaeology, two areas that meshed well with her geological work, but the bulk of her writings concerned her speleological investigations and her studies of loess formations. Her classic monograph, *Cave Regions of the Ozarks and the Black Hills* (1898), although written for a general audience, remained the major work on Missouri speleology for almost fifty years, and her papers in the French journal *Spelunca* were instrumental in bringing to the attention of Europeans the great cave formations of North America. The fact that she has been forgotten, omitted from biographical reference works on scientists other than *American Men of Science,* and from accounts of women scientists, is probably in large measure a reflection of her complete lack of institutional connections.

Owen came of distinguished ancestry. Her father, James Alfred Owen, a prominent Missouri lawyer, traced his descent from a line of Welsh barons. The founder of the American branch of the family arrived in Maryland in 1684, and succeeding generations went west to Kentucky, where James Owen was born. In 1847, as a young man of twenty-five, he moved to northwest Missouri, and in 1848 married Agnes Jeanette, daughter of James Cargill, owner of the first gristmill in St. Joseph. The Cargills were a well-to-do family of Scottish background. Of the five surviving children of James and Agnes, three—Luella, Juliette, and Mary—did notable scholarly work, Luella in geology, Juliette in ornithology and botany, and Mary in African American and native American folklore.[3]

Luella was born on 8 September 1852. As a child in the years just before the Civil War, she spent much of her time exploring the caves in the bluffs along the Missouri River on which St. Joseph is situated. Although these activities perforce came to a halt with the occupation of the town by the Union Army (the Owens were known to be Southern sympathizers), her childhood games of exploration undoubtedly awakened the interest that later developed into her many-year study of cave formations. Along with her sisters, she was educated at home by her mother during her early years, and after the war went to the newly opened St. Joseph high school, from which she graduated in 1872 as the class valedictorian.

Although both Mary and Juliette Owen each spent a year or two at Vassar College, Luella does not appear to have done so, and her postsecondary education was largely self-directed. In 1873 she met the pioneer speleologist the Rev. Horace Carter Hovey, an amateur geologist and Presbyterian and Congregational minister, who then had a church in Kansas City. She accompanied him on a number of expeditions to explore the river bluff caves she had known as a child. Hovey introduced her to several scientific publications including the *Proceedings* of the AAAS. About this time she also made contact by correspondence with three noted Midwest geologists, George Frederick Wright of Oberlin College, James E. Todd, then at Tabor College and later state geologist of South Dakota, and Newton Horace Winchell, state geologist of Minnesota and founder of the journal *American Geologist.* Both Wright and Todd worked on the recent geology of the Midwest, Wright specializing in glacial deposits and Todd in the Pleistocene of Kansas and neighboring states, including the loess deposits of Kansas, Nebraska, Missouri, and the Dakotas. Luella's friendships with these three men put her in touch with work in mainstream American geology, and she began to attend meetings, publish articles, and gradually widen her circle of geologist friends, people with whom she was to keep in close touch, by regular and frequent correspondence, over the course of many decades. At her father's suggestion, her first writings were published anonymously or under a pen name, now lost. She traveled much, becoming familiar with the whole Missouri river basin, and following the advice of Winchell and Todd made a special study of loess deposits.

After her father died in 1890 she was free to enter fully into speleological work,[4] which, though always a major interest, she had carefully curtailed during his lifetime to spare him anxiety. There being an estimated 3,500 caves in Missouri, as well as several celebrated caverns, she had ample scope for her investigations close at hand, though she also explored in other regions of the country. Her accomplishments, which would have been notable with twentieth-century equipment, seem even more impressive when it is remembered that she used rope-and-bucket entrance methods and worked by candlelight. The first paper to appear under her own name was "Cavernes américaines," published in *Spelunca* in 1896. In it she discussed Mammoth Cave, Wind Cave, Luray, Cave of the Winds, and the prehistoric Indian cliff dwellings in the Southwest. She followed with "Marble Cave (Missouri) et Wind Cave (Dakota)" in 1897, and descriptions of Crystal Cave and Ha Ha Tonka (Missouri) in 1898 and 1899.

Her monograph, *Cave Regions of the Ozarks and the Black Hills,* was written in part to present her own theory of speleogenesis, the key element of which was her suggestion that the initial event in cave formation, the opening of underground fissures, results from volcanic activity. (The problem of what starts cave formation was one over which speleologists argued for decades after Owen's time.[5]) The first narrow openings would be filled with groundwater, and, she postulated, enlarged by geyser action to form chambers and passageways. Subsequently, when geyser activity waned and finally ceased altogether, water levels would decline. Thus her specific example, the cave region of the South Dakota Black Hills, would, at an earlier stage in its evolution, have presented an appearance comparable to the present-day Yellowstone Park geyser basin, an area she had studied earlier. Although she lacked exact knowledge of geyser morphology, a problem of considerable complexity, she went so far as to speculate on the performance and plume height of Wind Cave and Crystal Cave "geysers," and inside Wind Cave identified three "geyser cones" that she considered to be evidence of declining geyser activity in the water-filled late stages of speleogenesis. Her theory had a fatal flaw, however; limestone, the rock of which the Black Hills are formed, possesses the somewhat unusual property of being less soluble in hot water than in cold, and consequently would not readily have been hollowed out in the way she suggested. It is now generally agreed that most of these limestone caves were formed by artesian groundwater flow; but Luella's theory was bold and in good accord with observations she had made.

About 1900, sensing that she was becoming too old for the strenuous work of spelunking, she went back to her study of loess deposits. But before doing so she took a trip around the world as a working member of the American Geographical Society, this membership entitling her to travel without passport. With a woman companion, and equipped with letters of introduction to geologists and geographers in many countries, she crossed the Pacific, visited the Philippines and China, and returned home via Europe, visiting en route the regions whose geological formations, especially loess desposits, were of particular interest to her.

Her studies on Midwestern loess were published over a period of twenty-five years, between 1901 and 1926, the last article appearing six years before her death. Several papers were first presented at AAAS meetings or at geographical congresses, and the work received considerable attention. Not infrequently in her investigations she made use of cuttings that were then being made through the soft loess formations during road and railway construction. As well as detailed reports of the fossil contents of the beds, she presented her ideas about the mode of their deposition. Theories of aeolian and aqueous origin were then still under debate; Luella came down on the side of the now accepted aeolian theory.[6] After her round-the-world trip she was known outside the country; on at least one occasion the Chinese government sent samples of loess to her for examination. The event is said to have prompted the St. Joseph *News Press* to claim that Miss Owen was the only woman geologist recognized by that government.[7]

Her publications in geographical journals dealt with the effects and potential uses of rivers. Here again, as with her geological work, her interest sprang from local studies. In 1908 she read a paper at the Ninth Geographical Congress in Geneva, correctly forecasting the future importance of the Missouri river as a direct trade route between the American Midwest and Europe. First published in the influential *Scottish Geographical Magazine,* this article was reprinted in the United States *Congressional Record* of 28 January 1913. Her discussion of the extent and effects of flooding along the Missouri and its tributaries between 1844 and 1908 came out in the British *Geographical Journal* in 1910; a third paper, presented at an AAAS meeting in 1912, focused on her particular interest in the changes in land form brought about by river denudation and the relation of this phenomenon to soil conservation.[8] She attended another international geographical congress in Rome in 1913, shortly before the outbreak of the First World War, though she does not appear to have presented a lecture on this occasion.

Archaeology, Luella's other special interest, was one she shared with her mentors Winchell and Wright, both of whom carried out work on early man in North America. When two human skeletons were found by workmen under twenty feet of undisturbed soil near Lansing, Kansas, in 1902, Luella was one of the first geologists at the excavation site. Shortly after that, at George Wright's invitation, she prepared an article on this somewhat controversial find for the journal *Bibliotheca Sacra* (which Wright then edited).[9]

She was a competent artist, and several of her family portraits in watercolor still exist in museum collections in Missouri. During her later years she became interested in genealogy and worked out full family histories for both the Owens and the Cargills (her mother's people). Most of her life was spent in the large house built on the outskirts of St. Joseph by her father in 1859. Following her mother's death in 1911, she and her sisters Mary and Juliette shared the home. She died there on 31 May 1932, at age seventy-nine. After the death eleven years later of Juliette, the last of the three sisters, the fine old house was renovated and maintained for a time, but it was later razed and the space used for a car sales lot.

FLORENCE BASCOM[10] (1862–1945) was America's first "professional" woman geologist. A forceful personality with irreproachable academic credentials herself, she became the doughty and determined builder of a college department recognized for its graduate-level teaching, and her influence reached out to many in the succeeding generation of American women geologists. Her career stands in sharp contrast to Luella Owen's.

Born on 14 July 1862, in Williamstown, Massachusetts, she was the youngest of the three surviving children of John Bascom and his second wife, Emma (Curtiss), a descendant of the early Plymouth Colony settler, Miles Standish. John Bascom, a cultured and upright if somewhat austere man who had a strong influence on his youngest daughter's early life, was professor of philosophy at Williams College. In 1874, when Florence was twelve years old, he became president of the University of Wisconsin, then a small institution with only ten faculty members, and the family moved to Madison. Florence graduated from the local high school before her fifteenth birthday and immediately entered the University of Wisconsin. She was an excellent student with a keen interest in a wide range of subjects; in 1882, at the age of twenty, she received two degrees—an A.B. and a B.L. A visit she made about this time with her father to Mammoth Cave in Kentucky appears to have awakened her interest in geology. She continued her studies, concentrating on science courses, and took a B.S. in 1884. Thereafter she began a research project throughout which she worked closely with Charles van Hise, then a doctoral student at the University of Wisconsin and later one of the leading structural geologists in America. This early association focused her scientific interests and to a large extent led her to the study of structural geology and petrology. The latter was a new field at the time, with few practitioners outside of Germany and little descriptive literature other than original papers in German.

After receiving her M.A. in 1887 she taught for two years at Rockford College, in Illinois. In September 1890 she applied for admission as a graduate student to the geology department at Johns Hopkins University in Baltimore in order to study under George Williams, the "father" of American petrology. Women were not yet being regularly admitted to graduate studies at Johns Hopkins; all the same, since she had been able to make a convincing case that she could not obtain equivalent instruction elsewhere, she was accepted by a special vote of the university executive committee in April 1891 as a special, non-fee-paying student.[11] Although isolated from her fellow students to a large extent—she had to listen to lectures from behind a screen—she nevertheless completed her studies and carried out her assigned fieldwork, an investigation of the formations in South Mountain, the long chain of ridges straddling the border between Pennsylvania and Maryland. By the use of new microscopical techniques, she demonstrated that these formations, previously considered to be sedimentary deposits, were in fact altered volcanics that had originated as ancient lava flows. Publication of her study in the first volume of the *Journal of Geology* in 1893 brought her to the notice of the small group of petrographers then working in the United States. Her Ph.D., awarded the same year, was the first given by Johns Hopkins to a woman; it was also the first doctorate in geology awarded to a woman by an American university, and the only one before the turn of the century.

She then joined the faculty at Ohio State University, thanks to help from her father's friend Edward Orton.[12] Two years later she succeeded in obtaining a position at Bryn Mawr College, where she was to remain for the rest of her career.

Although Bryn Mawr from its founding had put high value on research productivity and Bascom had shown great promise in that direction, her prospects for establishing a research-oriented department there were less than good. Geology was considered a subject unlikely to have wide appeal to women, and both space and money for development were in short supply. Bascom, however, had ambition and drive. Her classroom and laboratory space gradually expanded. She bought, traded, and begged specimens and equipment, and by 1899 was teaching petrography, having acquired one good petrographic microscope and a collection of mineral and rock slides. By 1901, her sixth year at Bryn Mawr, her single course in geology had developed into a "major" and some of her early students had stayed on to become her first graduate students. By 1903 she had an assistant (Benjamin Miller[13]) to teach paleontology and stratigraphy. She became full professor in 1906.

In 1896 she was appointed assistant geologist with the U.S. Geological Survey, the first woman to obtain such a position. Once again she was most likely assisted in getting the appointment by Edward Orton, who collaborated in U.S. Geological Survey projects in addition to his work in Ohio. In due course she was promoted to geologist, and assigned the Piedmont region in Maryland, Pennsylvania, and parts of Delaware and New Jersey as her area of investigation. For many years she spent her summers mapping the crystalline schists and gneisses of this region, traveling on foot or on horseback, working from dawn until dusk and insisting on the same routine for her students, graduate and undergraduate. In the winters she studied thin sections of the rocks, made her evaluations, and wrote her reports. The results, her most important contributions to geological research, appeared as a series of comprehensive reports (generally coauthored by her survey colleagues) in U.S. Geological Survey folios and bulletins for Philadelphia (1909), Trenton (1909), Elkton-Wilmington (1920), Quakertown-Doylestown (1931), and Honeybrook-Phoenixville (1938).[14] She became an authority on the complex, highly metamorphosed, crystalline rocks of the Piedmont area between the Susquehanna and Delaware rivers in southeastern Pennsylvania and neighboring states, and her folios on this region remained for half a century the basis on which later research was built. Her original designation of the terms "Wissahickon gneiss" for the exposures along Wissahickon Creek, and "Bryn Mawr gravel" for surface deposits near Bryn Mawr were adopted generally; these formations were traced widely through the eastern United States.

Well regarded by her colleagues nationally, and starred in *American Men of Science* beginning with the 1906 edition, she had been elected a fellow of the Geological Society of America in 1894, became its first woman councillor in 1924, and its vice president in 1930. From 1896 to 1905 she served as associate editor of the *American Geologist.* In 1907 she had a sabbatical year in Victor Goldschmidt's laboratory in Heidelberg, where she studied the latest techniques in applied crystallography.

Her career at Bryn Mawr, however, was not without serious difficulties at times. Tensions caused by the inevitable conflict between her growing demands for support for her graduate program and the college's limited resources gradually increased, and by 1911 the president, M. Carey Thomas, threatened to eliminate the geology department entirely. Bascom, preoccupied at the time by the death of her father, was unable to meet the challenge this move posed and it was her student, Eleanora Bliss (later Knopf), who came forward and took action. Bliss wrote to the college trustees and former geology students and initiated a campaign that raised $750,000 for the college and saved the department.[15]

Although the number of geology graduates the Bryn Mawr program produced was relatively small (thirty-five women with bachelor degrees, sixteen with M.A.s, and ten with Ph.D.s in the first forty-seven years of its existence[16]), these few graduates constituted a sizable fraction of American women geologists of the period. Hence the importance of Bascom's department, for both undergraduate and graduate training, was greater than student numbers might at a casual glance suggest. Indeed, the department's success was remarkable for a small women's college. Several of Bascom's students went on to notable professional work in geology, and four of them—Ida Ogilvie, Julia Gardner, Eleanora Bliss Knopf, and Anna Jonas Stoss—became fellows of the Geological Society of America. (Ogilvie, who was known for her contributions to glaciology, built up a geology department at Barnard College in New York City as Bascom had done at Bryn Mawr. Gardner, Bliss Knopf, and Jonas Stoss, perhaps benefiting from Bascom's influence, all obtained positions on the U.S. Geological Survey.)

Bascom became something of a legend at Bryn Mawr, where she was known as an inspiring if very demanding teacher; her advanced classes were especially popular. She had a forceful and uncompromising manner, said to have been even caustic at times, which may well have contributed to the substantial and somewhat bitter professional controversies that arose later between her and some of her best students, particularly Jonas Stoss and Bliss Knopf. But it is also the case that these students recalled with gratitude the rigor of the early training she gave them.[17] Her toughness and stamina, both physical and emotional, were doubtless essential for the task of department building she set herself (at times in the face of stiff administrative opposition) and for carrying out her extensive program of demanding fieldwork.

After retiring in 1928 she moved to Washington, D.C., where she was occupied for several years preparing final survey reports. She had planned to spend her last years at her farm on Hoosac Mountain in western Massachusetts, where she had taken a vacation every year with her dog and her riding horse. However, she continued to work until her health failed in the late 1930s. She died of a cerebral hemorrhage in Northampton State Hospital, Massachusetts, on 18 June 1945, a month before her eighty-third birthday.

CARLOTTA MAURY[18] (1874–1938), the third notable early woman geologist included here, was the youngest surviving child of the Rev. Mytton Maury and his wife Virginia (Draper). She was born on 6 January 1874 in Hastings-on-Hudson. Her family was one of some distinction. Her maternal grandfather was the historian and pioneer physicist, John William Draper, a son of Carlotta Joaquina de Paiva Pereira, a member of the Portuguese nobility and one of the ladies of the court of Emperor Dom Pedro I of Brazil. On her father's side she was descended from the French Huguenot families de la Fontaine and Maury who settled in Virginia about the beginning of the eighteenth century. Her great-grandfather James Maury was the first consul from America to Liverpool, and her cousin, Matthew Fontaine Maury (1806–1873), was well known as a geographer and hydrographer.

Along with her brother John and her elder sister Antonia (see chapter 10), she was educated at home until she was in her teens. Much of the instruction was provided by her father, an Episcopalian minister with strong interests in geography and the editor of several of Matthew Fontaine Maury's geographical treatises. He took his children for daily walks and found illustrations for geography and natural history lessons everywhere around him. Carlotta also learned Latin, French, and German at home. She later studied at Radcliffe College and both Columbia and Cornell universities, with a year at the Sorbonne in 1898–1899. Cornell awarded her a Ph.B. degree in 1896 and a Ph.D. in 1902, when she was twenty-eight.

Her first publication, a translation and summary of a long and important article in Spanish by Carlos Sapper, "the father of Central American geology," on the geology of Chiapas, Tabasco, and the Peninsula of Yucatan, appeared in the *Journal of Geology* in 1896. It was coauthored by her adviser Gilbert Dennison Harris, then assistant professor of paleontology at Cornell, editor and publisher of *Bulletin of American Paleontology* and well known for his excellence as a teacher and paleontologist. She followed the translation with the monograph *Chautauqua Lake Shells* (1898). Her dissertation research at Cornell, supported by a Schuyler fellowship in geology, was an investigation of fossils in Oligocene rocks in both Europe and the United States,[19] the first of a number of such cross-Atlantic comparative studies she was to undertake.

For two years, 1904 to 1906, she held a lectureship in geology at Barnard College and an assistantship in paleontology at Columbia University, where she worked with A. N. Grabau, remembered for his influential work on the classification of sedimentary rocks.

In 1907 she returned to field studies, joining her former adviser G. D. Harris in his ongoing investigations of the oil-bearing strata of the Texas and Louisiana Gulf Coast region. As well as teaching at Cornell, Harris was state geologist of Louisiana from 1898 to 1909, a period of tremendous excitement and activity in oil exploration in the Gulf region; the discovery of oil and the sinking of what turned out to be gusher wells near Beaumont, Texas, in 1901 was a major event in the history of the American petroleum industry. Most of Harris's assistants on the Louisiana Survey were his former students, and all of them went on to distinguished careers. Two with whom Maury collaborated closely were Arthur Clifford Veatch, for a time a member of the U.S. Geological Survey and then chief of various international oil explorations, and Leopold Reinecke, a South African geologist. Maury herself was probably the first professional American woman geologist to work on a state geological survey.

The studies of the Harris team provided the first substantial geological knowledge of the Texas-Louisiana oil-productive areas, and indeed the bulk of the paleontological information for the oil-bearing strata of the region available up to 1911 was produced by them.[20] Starting about 1908 they carried out systematic collecting and examination of cuttings from the drilling wells in the important Jennings and Caddo fields and subjected these to microscopic examination. Using well records and samples collected by Reinecke from wells in the Jennings field, Maury carried out the paleontological examinations and produced a structure map for a large area. The Harris-Maury subsurface structure analysis of the Jennings region (based on Maury's paleontological findings) was reported in a U.S. Geological Survey bulletin in 1910; although details were subsequently questioned by geologists who had the advantage of more advanced technology, petroleum geology historian Edgar Owen, sixty-five years later, noted that no new interpretations had been offered and that the pioneering work of the Harris team still held its own.[21] Another substantial report, coauthored by Harris, Maury, and Reinecke, appeared in 1908. It concerned primarily the Louisiana salt domes, the principal type of structure with which oil and gas are associated in the region, but it included a comprehensive review of the occurrence of salt in all parts of the world and was probably the first American work on salt domes that offered a serious examination of the European literature.[22]

Maury's next field study was in the Caribbean and South America. In 1910 she joined a group of about six geologists led by her colleague A. C. Veatch. Veatch was employed by the General Asphalt Company to examine its promising properties in Trinidad and also a concession it held at the Guanoco asphalt lake in Venezuela. As paleontologist on the expedition Maury was able to make a notable contribution. Her discovery in Trinidad of Old Eocene beds with fossil faunas related to those of Alabama and the Pernambuco region of Brazil constituted the first discovery of Old Eocene in the entire Caribbean and northern South American region. Edgar Owen noted that the contributions made by Veatch and his coworkers in Venezuela and Trinidad constituted a major step in South American oil exploration at the period.[23]

Returning to teaching in 1912, Maury spent three years in Cape Colony, South Africa, as professor of geology and zoology at Huguenot College, a women's college founded in 1874 by members of the Dutch Reformed Church but staffed and run by American women, many of them from Eastern women's colleges. (One might wonder if her interest in South Africa did not stem in part from her acquaintance with Leopold Reinecke.) Then in 1916, holding a Sarah Berliner research fellowship from Cornell University, she organized and led the Maury Expedition to the Dominican Republic. Its purpose was to determine exact stratigraphic sequences in the Miocene and Oligocene horizons in the region, to which Santo Domingo was judged to be the key, and to differentiate (for the first time in the Antillean area) between these two horizons. The expedition's stratigraphic results, in the form of type sections and descriptions of fossils, including many new species, were published in two substantial bulletins in 1917 and in a shorter paper in the *Journal of Geology* the following year.[24]

In 1910, with by then considerable experience behind her and several substantial publications to her credit, she became consulting paleontologist and stratigrapher to the Royal Dutch Shell Petroleum Company, Venezuela Division, a position she held for more than twenty years. From 1914 until 1937 she also served as an official paleontologist with the Servico Geologico e Mineralogico do Brasil. The latter appointment she particularly valued, perhaps because of her Brazilian ancestry. She specialized in Antillean, Venezuelan, and Brazilian fossil faunas, and was regarded as an extremely efficient and energetic worker, who, once she took on an investigation, proceeded with speed and precision and "put her whole soul into the work."[25] Having no pressing need to accept every project offered her, she always avoided routine work and took on only investigations that presented a special opportunity to do something she considered exceptional. She wrote easily and skillfully and hired the best draftsmen and engravers to prepare her illustrations.

Her detailed Caribbean and South American stratigraphical and paleontological studies appeared throughout the 1920s and until 1937 in a steady succession of lengthy monographs and bulletins published by the Servico Geologico e Mineralogico do Brasil and in American scientific journals. Excluding sixteen confidential reports on Venezuelan stratigraphy for the Royal Dutch Shell Petroleum Company, her publications numbered more than forty. Her many studies of Recent and Tertiary Mollusca from Florida, Louisiana, and Texas, as well as Puerto Rico, the Virgin Islands, and Brazil were particularly valued; also much respected were her co-relations of Cretaceous fauna and formations of northern South America and the Caribbean with those of the United States and Europe. Her last report, on the Pliocene fossils of Acre, Brazil, a work she considered her best, appeared in 1937, shortly before her death.[26] In the same year she was honored by being elected a corresponding member of the Brazilian Academy of Sciences, an acknowledgment of her outstanding contributions to the geology of Brazil. She was also a fellow of the Geological Society of America, of the American Geographical Society, and of the AAAS.

An exceptionally able paleontologist, Maury was also remembered as a pleasant person who "made enduring friends wherever she went."[27] To a large extent she succeeded in ignoring prejudice against women in professional work thanks to her acknowledged technical abilities, although doubtless her substantial private financial resources increased her freedom of action. She was paid honoraria by the Brazilian government for her reports and monographs, but undertook the initial expenses of their preparation herself. From 1923 onward her home base was a small apartment in Yonkers, New York City; she regularly used the library resources of the American Museum of Natural History and its large collection of invertebrate specimens, living and fossil.

Recreation for her was travel, and in addition to her scientific expeditions she took a round-the-world cruise in her later years. She delighted in the exotic, enjoying the brilliance of the birds in the tropical forests of Venezuela (where she lived for almost a year) and the glory of the mountains, especially the Himalayas and the South African Drakensbergs (where she once gave a memorable party at an ancient cave site). She died on 3 January 1938, three days before her sixty-fourth birthday.

Papers by Carrie Barbour, Mary Fleming, and Mary Klem are also listed in the bibliography.[28]

CARRIE BARBOUR[29] (1862?–1942), the daughter of Adeline (Hinckley) and Samuel Williamson Barbour, grew up on a farm near Oxford, Ohio. In the mid-1880s she attended Oxford Female College and then, despite a special interest in natural history, studied ceramics and wood carving in Cincinnati. After teaching art for two years (1890–1892) at Iowa College (later Grinnell College) in Grinnell, Iowa, she accepted a position at the University of Nebraska, where her brother Erwin Barbour had recently taken up appointments as professor of geology and state geologist.

Soon after arriving in Lincoln Carrie began to assist her brother in the work of the state museum, of which he became curator in 1892. Within a year she had resigned her position in the art department in order to spend her time on the paleontological work of the museum and on graduate studies in biology. In 1893 she was made assistant curator, a post she held for almost half a century. She quickly became involved in the work of the Morrill Geological Expeditions of the University of Nebraska and the State Geological Survey, both of which were organized largely by her brother. In 1897, on the sixth Morrill Expedition, she collected fossils in the Bad Lands of South Dakota, the Black Hills, and the Daemonelix beds of Nebraska and Wyoming. Similar work followed, including a summer of collecting for the survey in 1899, when she and an assistant procured more than 20,000 of the (mainly) commoner species from Nebraska Carboniferous and Permian formations. Some were new to the state and three or four never previously described.[30] Although she brought out a note in the *Publications of the Nebraska Academy of Science* in 1901,[31] she does not appear to have published in technical journals thereafter.

She became assistant professor of paleontology and taught the subject to Nebraska undergraduates for a quarter of a century; her total time on the faculty came to forty-nine years. However, her major contribution to geological research was in the museum where she did much of the "working out" and identification of fossils. By the time of her death at age eighty in June 1942, the Nebraska state museum was one of the most outstanding in the Midwest; her efforts added not a little to its early success.

Of the three women discussed most fully here—Owen, Bascom, and Maury—Bascom alone has received serious attention from students of American women's work in science—in part, no doubt, because her career at Bryn Mawr College is amply documented in the college archives. Comparatively little has been written about Maury, a more elusive subject, and Owen is rarely, if ever, mentioned. Yet Maury's research output was at least the equivalent of Bascom's in quantity and quality. Furthermore, although a noticeable number of American women paleontologists were later to find positions in oil company laboratories, Maury's career as consulting paleontologist and stratigrapher to a large European oil company from 1910 constitutes a remarkably early example of female penetration into a male-dominated profession. Owen's scientific contributions, although modest compared with those of Maury and Bascom, were still very creditable for an independent "amateur" based in a small and relatively remote Midwestern town.

These three form an interesting trio—Bascom the women's college professor of solid, middle-class, New-England background, Maury the aristocratic, independent, freelance scientist with the cosmopolitan lifestyle not atypical of an able petroleum geologist by her time, and Owen the self-taught but cultivated and outward-looking Midwesterner, like Maury a colorful personality and independent spirit. Bascom clearly holds pride of place as the pioneer of work by American women in geology in the late nineteenth century, being not only the first to get a Ph.D. in the field, but going on to a long and distinguished career. On the other hand, her life, shaped as it was by the restrictions of academe and the assignments of the U.S. Geological Survey, would now seem relatively conventional; indeed, by her later years she had become an establishment figure compared with the two free agents, Maury and Owen.

British women

The contribution of nineteenth-century British women to geological work was remarkable. In most fields American workers considerably outnumbered British and produced many more publications, but geology was an exception. Here the British dominated the field by a wide margin; some 60 percent of the geology papers by women indexed in the Royal Society *Catalogue* were by British workers, the largest fraction for any national group within any field covered by the Royal Society.[32]

These early British women geologists fall into two fairly distinct subgroups, the self- or privately taught "amateurs," and the university-trained women who became prominent during the last fifteen years of the century. Each subgroup produced a comparable number of papers, although not unexpectedly the contributions of the later women were, in general, considerably more weighty.

A number of talented and industrious wife-assistants of

some of the famous British men geologists of the early nineteenth century made contributions to their husbands' work. However, although very effective in the tasks they undertook, which often included technical drawing and the preparation of geological landscape sketches, by and large they confined themselves to background roles as helpers and general assistants, as did their American counterparts.[33]

Significant independent work by British women in nineteenth-century geology probably begins with ETHELDRED BENETT[34] (1776–1845). Born at Tisbury, Wiltshire, Etheldred was the daughter of Thomas Benett, a country squire. Her interest in collecting and studying fossils, a popular and fashionable hobby with the middle and upper classes at the time, was most likely encouraged by her half-brother-in-law, Aylmer Bourke Lambert. Lambert, a fellow of both the Royal Society and the Linnean Society and an able all-round naturalist, was a keen collector of fossils and minerals. The countryside immediately surrounding Tisbury presented especially fine collecting opportunities, and by about 1810 at the latest Etheldred had established her own sizable fossil cabinet. Being unmarried, financially independent, and free from major family commitments, she had time to devote to the work; within a few years her collection was extensive.

Many of the specimens she discovered were illustrated in James Sowerby's *Mineral Conchology of Great Britain* (1812–1820), a major work in which the greater part of the numerous specimens presented were new to science, accurately figured, and given Linnean names. Sowerby also published (without telling her or obtaining her permission) a bed-by-bed description she had given him of the Upper Chicksgrove quarry at Tisbury. Based on materials collected by the quarry foreman, John Mountague, it appeared in 1816, and was one of the earliest such quarry sections published in Britain. She had presented a preliminary copy to the Geological Society in 1815, but the later published version contained her detailed information on the fossil contents of each bed.

She corresponded with a number of the prominent men geologists, including William Buckland, Gideon Mantell, and George Greenough, to whose *Geological Map of England and Wales* (Geological Society, 1819) she contributed stratigraphic data on the Wiltshire region. By 1818 she had plans to publish a stratigraphic catalog of the species in her own collection, now numbering several thousand specimens, mainly from the Jurassic-Cretaceous strata of Wiltshire, and widely known to geologists both in Britain and on the Continent. However, her brother's entry into politics and his election in 1819 as M.P. for South Wiltshire for a time divided her interests. Her work finally appeared in 1831 as a section in the third volume of Sir Richard Colt Hoare's *History of Modern Wiltshire;* she also brought it out separately, the same year, as *A Catalogue of the Organic Remains of the County of Wilts.*[35] Despite the fact that she had only a small number of copies printed for private distribution, Benett's list was widely circulated and made available both to her geologist friends and to museums and other institutions (she had also at her disposal a generous supply of off-prints from the Hoare publication). Containing as it did a number of new taxa that she was the first to illustrate, the list was of considerable importance, and it was frequently cited in the geology literature for many decades.[36]

Benett gave many of her best specimens to individual geologists and to public museums throughout the country when she felt this would benefit science. Some of her material went to a museum in St. Petersburg, and in recognition of the importance of her gift, the tsar, assuming from her first name that she was a man, had the University of St. Petersburg give her an honorary doctorate of civil law. Her main collection has an interesting story. After her death in 1845 it was divided and sold, the major part being acquired by an American, Thomas Bellerby Wilson, of Wilmington, Delaware. Wilson was known to have donated his collection to the Philadelphia Academy of Natural Sciences between 1848 and 1852, but over the years it disappeared, and was for long considered lost (this was probably due in part to the fact that for several decades around the turn of the century there was no formal curatorial staff for the academy's paleontological collections). Not until about 1980, 135 years after Benett's death, was this lost collection, comprising some 9,400 specimens of almost 3,000 species, rediscovered at the academy. Including as it does many type specimens, its significance is very considerable.[37] Benett's *Catalogue* had also fallen into obscurity over the years. However, because of its role in questions of nomenclature priority, it has recently been reexamined in detail, in conjunction with studies on the rediscovered collection[38]—a fine reacknowledgment of the achievements of its author.

Etheldred Benett was remembered as becoming something of an eccentric as she grew older. She paid scant attention to ladies' fashions, and when driving to Warminister in her old gig with her manservant, wore a coachman's greatcoat with many capes to it and an old-style bonnet that all but hid her face. She was a pioneer among women paleontologists and perhaps the most distinguished of the early women workers in geology in Britain. Her remarkable initiative and skill and her resulting achievements in collecting, biostratigraphy, systematics, and publication of original work were unmatched for a woman of her time.[39] She died at her home, Norton House, not far from Heytesbury, Wiltshire, on 11 January 1845.[40]

From the same period as Benett, but better remembered because of the spectacular character of her fossil finds, was MARY ANNING[41] (1799–1847) of Lyme Regis in Dorsetshire. She and her fellow townswomen the three Philpot sisters—Margaret, Mary, and Elizabeth—collected extensively from the rich local lias beds, the Philpots assembling an extensive museum of fossil fishes.

One of the two surviving children of a Lyme carpenter, Richard Anning, and his wife Mary (Moore) of Blandford, Dorset, Mary was born on 21 May 1799. Her formal education probably went no further than learning to read and write at the parish school. However, as a small child she accompanied her father and her older brother Joseph on searches for fossils, "curiosities," which were sold to eke out the family income. Richard Anning evidently taught his little daughter how to use tools; she not only developed remarkable skill in finding fossils, but learned how to extract them.

Following the father's death in 1812 the Anning family existed on parish relief for a number of years, and Mary and her brother, with guidance from their mother, persevered with their fossil hunting. Their first spectacular find, a fish-lizard later named *Ichthyosaurus,* located in 1811–1812, was sold to a local collector for £23. As their skills gradually increased they became well-known as fossil hunters. By the 1820s they had made several more important finds (including another ichthyosaurus, *Ichthyosaurus communis,* 1821). Mary, then a young woman in her twenties, did most of the work in the family fossil business, her brother having another occupation. Her 1824 find of an almost complete skeleton of the little-known *Plesiosaurus,* identified by geologist William Conybeare, was a major discovery that firmly established her reputation as an outstanding fossilist. She sold the specimen for at least £100. Important discoveries continued throughout the 1820s. They included the following: *Pterodactylus* (1828, described by Buckland), the first British example of a fossil flying reptile; a second complete plesiosaurus (1829), sold to the British Museum for 100 guineas; the fossil fish *Squaloraja* (1829), an important specimen marking the transition between sharks and rays; yet another ichthyosaurus, and also the *Plesiosaurus macrocephalus* (both 1830), the latter sold for 200 guineas.[42]

Less spectacular than the giant reptiles but also of great importance to collectors were her invertebrate fossil discoveries, including many cephalopods complete with fossil ink bags. Many of these were found during her very productive period in the late 1820s. About this time she also worked with William Buckland in the new area of coprology, the identification of fossil fecal deposits.

She was undoubtedly one of the most notable fossil finders of the early nineteenth century.[43] For almost thirty years, until her health broke down about 1845, she was one of the country's expert fieldworkers, and her discoveries were critical contributions to the emerging fossil record on which Buckland, Conybeare, Henry De la Beche, and other leading figures of the "golden age" of British geology based their new theories. Her services to science were eventually acknowledged by a small annuity, which Buckland succeeded in obtaining for her.

The fact that she left so little in the way of written records of her scientific activities has led to her being largely ignored in accounts of the development of geology, however. Her work has been incorporated into that of the men with whom she collaborated so closely; the ichthyosaurus and plesiosaurus finds have become simply Conybeare's discoveries.[44] A short note concerning Hybodus fossils, written in 1839 to Edward Charlesworth, editor of the *Magazine of Natural History,* was perhaps the only piece of her writing that was published during her lifetime; little-known letters to her geologist friends, or more often to their wives, are most likely the only other extant records of her work made by her.

She was more than just a skillful field technician, however, since over the years, thanks to her native abilities and her friendships with geologists, she built up a very fair knowledge of the bone structure and classification of fossil saurians and fishes.[45] Like Etheldred Benett, she was known to geologists in Europe as well as in Britain ("I am well known throughout the whole of Europe"[46]), and she was quite aware of the significance and importance of the specimens she was discovering, often before the male geologists had examined them. In 1846 she was elected the first honorary member of the new Dorset County Museum in Dorchester.

She died in Lyme on 9 March 1847, at age forty-seven, of breast cancer. An obituary notice by Henry De la Beche appeared in the Geological Society's *Quarterly Journal* in 1848,[47] a notable tribute since she was not, of course, a fellow of the society; two years later the fellows further expressed their appreciation of her work by placing a stained-glass window to her memory in the Lyme church. A portrait of her, executed about 1842, hangs in the Natural History Museum. However, only two British species were named after her in her lifetime, the fish *Acrodus anningiae* (1841) and *Belenostomus anningiae* (1844), both by Louis Agassiz. In 1927 the Karroo reptile genus *Anningia* was named by the South African paleontologist Robert Broom.

Numerous accounts of Anning and her discoveries have been published from about 1857 on, both in popular journals and books. A considerable amount of myth has evolved and some confusion arisen about the details of her life's work. To some extent her story had been kept alive through children's encyclopedias and books for young people, often presenting picturesque but somewhat fictionalized accounts of her early years;[48] less attention has been given to her later technical achievements, but recent work by Torrens does much to redress the balance.[49]

Another early woman fossil collector whose contributions were especially important was ELIZABETH ANDERSON GRAY[50] (1831–1924), of Ayrshire, in southwest Scotland. Over a period of more than half a century she collected from Lower Paleozoic strata, mainly conglomerates and limestones exceptionally rich in faunal remains, in the neighborhood of Girvan. The material she amassed was used by workers in both geology and paleontology, including Charles Lapworth, who, in his work on the "Girvan succession," made extensive reference to Elizabeth's collections in his stratigraphical correlations; specialists in paleontology such as Thomas Davidson, Jane Donald Longstaff (see below), and Frederick Richard Cowper Reed based important contributions on materials she supplied to them.[51] She was one of the foremost Scottish fossil collectors of the late nineteenth and early twentieth centuries.

The younger daughter of Thomas Anderson, an innkeeper, and his wife Mary (Young), she was born at Burns' Arms Inn, Alloway, on 21 March 1831. When she was five her family moved to the Girvan district, her father taking up farming. She had two years of formal education at schools in Girvan and Glasgow, but by the age of sixteen returned to the farmhouse. Her father was a keen naturalist, interested in fossils, and she soon joined in his hobby. In 1856 she married Robert Gray, of Glasgow, an amateur ornithologist who came to Girvan to study coastal birds. Although she moved to Glasgow, she and Gray returned to Girvan every summer, Gray also becoming an enthusiastic fossil collector. He exhibited collections of their finds at meetings of the Glasgow Natural History Society from the early 1860s on, and reports of their joint work appear, under his name only, in the society's *Proceedings*.[52] In 1866 the first of the Gray collections was presented to the Hunterian Museum at Glasgow University. Three years later Elizabeth was able to attend a recently instituted course of lectures on geology for ladies at the university, an opportunity which brought her a better understanding of the scientific significance of the fossils she was collecting.

The Grays had six children, and since each one, by the age of five or six, began to take part in the family summer occupation of fossil hunting, their collections increased rapidly. They moved to Edinburgh in 1874, when Robert, a banker, accepted a position at the head office of the Bank of Scotland. He was soon prominent in Edinburgh scientific societies, and local geologists became aware of Mrs. Gray's work. About 1880 the Keeper of Geology at the Royal Scottish Museum, feeling that Elizabeth might describe (in print) her own fossils, offered to teach her how to do this. But she declined, feeling that she would do better to continue to provide material for specialists; in any case her preferred work was in the field, discovering and collecting.

Robert Gray died in 1887, but Elizabeth, with the help of her four daughters, especially Alice, continued fossil collecting at Girvan for another thirty-seven years. The importance of her work is underscored by the fact that she was awarded the Murchison Fund by the Geological Society in 1903, in recognition of "her great services to geological science."[53] She was elected an honorary member of the Geological Society of Glasgow, and became a fellow of the Royal Physical Society of Edinburgh.

Her methods were meticulous, each fossil she found being carefully labeled with locality and horizon, and special care being taken to keep part and counterpart together. Probably her early contacts with specialists, Davidson as early as 1860 and Lapworth by the seventies, impressed on her the need for precise records. The amount of material collected by the family was immense. When one collection was sold to an institution another was started. After Elizabeth's death the work was continued by her daughters until Alice died in 1942.

Elizabeth went to great effort to get her materials described by competent specialists. This was not always an easy task, a fair amount of persistence being sometimes required to persuade one of the few and often already overburdened experts to undertake the time-consuming labor entailed. She also succeeded in getting her specimens safely housed in good museums. One of her major collections was sold to the British Museum in 1920 for the sum of £2,250 (which she used to establish a fund for her unmarried daughters). Although she chose to confine her activities to collecting, Elizabeth acquired considerable knowledge of the nature and significance of her fossil species and their interrelationships, knowledge that inevitably inspired her to go on to extend her lists, establish occurrences, find larger and better specimens, and discover rare and new species. Her work on the fossil faunas of the Lower Paleozoic rocks of the Girvan district constituted a substantial contribution to British geology, and her collections, rich in type specimens, are still considered essential for studies on Ordovician faunas.[54] She died in Edinburgh, on 11 February 1924, shortly before her ninety-third birthday. Many Paleozoic fossils either bear her family name or a place-name from the Girvan area.

The earliest geological paper by a woman listed by the Royal Society is a report by MARIA DUNDAS GRAHAM[55] (1785–1842), which appeared in the first volume of the Geological Society's *Transactions* in 1824. Maria was the daughter of George Dundas, rear admiral and commissioner of the admiralty. She was born at Papcastle near Crockermouth. In 1809, on her first long voyage to the East, she met Captain Thomas Graham, and married him shortly after they landed in India. Their wedding journey round the coast, with frequent stops to visit temples and ruins inland, was described in her *Journal of Residence in India* (1812) and *Letters on India* (1814). In 1821 she accompanied her husband on his own ship, the *Doris,* bound for Brazil and then Valparaiso, Chile. Captain Graham never reached Chile, however, since he died of fever off Cape Horn. When his widow arrived in Valparaiso she was lent a cottage by friends, and, staying on in the British community for over a year, experienced firsthand the great Chilean earthquake of 1822. She went home the following year, again around Cape Horn, with a call at Rio de Janeiro. After a few months in England to see through publication two books about her Chile and Brazil experiences, she went back to Brazil to take up a post as governess to Donna Maria, daughter of the Prince of Portugal and Emperor of Brazil.

Her diary account of the 1822 earthquake was of considerable interest to the geological community, being one of the earliest detailed and accurate reports of the effects of major seismic activity ever published.[56] Of special note were her observations of the rise in the level of the coastline, since the idea of uplift of major blocks of the land surface as the result of seismic activity was far from being universally accepted at the time. The vice president of the Geological Society, Henry Warburton, considered the report of sufficient interest to request from her an extract for inclusion in his society's *Transactions.* A few years later Charles Lyell quoted from this extract in his celebrated *Principles of Geology*,[57] and following the publication of the latter work George Greenough, in a presidential address to the Geological Society, vigorously attacked Graham and her report. Quite unwilling to accept the possibility of substantial uplift of a coastline along an extent of many miles, Greenough subjected Mrs. Graham's account to what he called a "rigid examination." This consisted largely of an attempt to ridicule her ability to draw trustworthy conclusions about scientific matters by questioning the veracity of almost all her observations. Mrs. Graham, by then back in London and married to the artist Sir Augustus Callcott, promptly sent to the members of the Geological Society a suitable reply, in which she very adequately demonstrated the errors and inconsistencies in Greenough's "rigid examination."[58]

Commenting on Graham's work a century later, Charles Davison in his book *The Founders of Modern Seismology* came down on her side of the argument. He felt that her attention to detail was such that full confidence could be placed in her observations of coastal elevation, even though the extent of the coast affected remained in doubt because she had not herself examined it.[59] Graham was quick to acknowledge that she was no geologist, and in fact felt complete "indifference to all theories connected with [geology]."[60] Nevertheless, her 1824 report concerned no quiet backwater, but touched on major questions and central issues in the theoretical development of the earth sciences at the time.

During the late 1820s she and her husband traveled in Europe and she wrote a number of books on Continental art and architecture, several of which were illustrated by Sir Augustus.[61] She also brought out translations from the French of historical works and wrote a considerable number of moral tales for children. Her best-known work was her celebrated *Little Arthur's History of England,* first published in 1835 but revised and reissued many times, a century edition, taking the history up to the accession of King Edward VIII, being brought out in 1936. She died in London, on 28 November 1842, at age fifty-seven, after several years as an invalid.

Yet another British woman who carried out notable early work in geology was BARBARA, MARCHIONESS OF HASTINGS[62] (1810–1858). She is remembered especially for assembling the Hastings Collection, several thousand fossil vertebrate specimens from locations in Britain and the Continent, now in the British Museum. Although much of the material was bought from dealers and collectors, the marchioness herself collected the large section comprising fossils from Hordle Cliff on the Hampshire coast.

The only child of Henry Edward Yelverton, nineteenth Baron Grey de Ruthyn, and his wife, Anna Maria, daughter of William Kelham, she was born at Brandon House, Warwickshire, on 29 May 1810. A year later, on the death of her father, she became Baroness Grey de Ruthyn in her own right. In 1831 she married George Augustus Rawdon Hastings, second Marquis of Hastings. They had two sons and four daughters. Widowed in 1844, she remarried a year later, her second husband being Captain Hastings Reginald Henry, R.N., who later took the name of Yelverton; he rose to the rank of admiral. From this second marriage there was one child, a daughter.

In 1845 the family settled at Efford House, on a large estate between Milford and Lymington on the Hampshire coast. The region had been well known to paleontologists as a rich hunting ground for many years, and fairly recently mammalian fossils had been discovered there.[63] The marchioness, already a keen collector, quickly began to build up a large holding. Most of her material was obtained from exposed strata in the sea cliffs, the extracting usually being done by her assistant, Henry Keeping of Milford. Occasionally she herself helped. She kept careful records of locations of finds, and over the course of six years worked out the detailed stratigraphy of the area. Existing accounts were inadequate, and the situation along the coast was continually changing because of sea erosion and cliff falls, which obscured the details of the stratigraphy from the short-term observer.

The marchioness's extended observations resulted in her production of a colored, scale-drawn section of the beds on which she based her main papers (see bibliography), and by means of which she recorded the stratigraphic locations of her finds. She housed the specimens in a private museum at Efford House, and herself carried out the work of cleaning, readjusting, and restoring. Much of the material, including turtle shells and crocodile skulls, was extremely fragile and crumbly, and her skill, patience, and neatness were exemplary. Two of the turtle shells and a skull were exhibited at the British Association meeting at Oxford in 1847, at which she gave a short paper. On that occasion the paleontologist Sir Richard Owen proposed that her skull discoveries be named *Crocodilus hastingsiae* in her honor. As she stated in her 1853 paper,[64] the main purpose of her work was to help build a comprehensive picture of tertiary stratigraphy, which depended on local observations of the kind she had made. She was remarkably successful; her papers, the first accurate accounts of Hordle and Beacon Cliffs, have remained among the best.[65] Further, the combination of these writings and her fossil vertebrate collection, the largest ever made from the area, constituted an exceptionally fine contribution to the geological knowledge of the locality.

In 1851 she sold her whole collection to the British Museum for £300. The crocodile skulls and some of the turtle shells were later put on permanent exhibition in the public galleries. She died in Rome on 19 November 1858, at age forty-eight, while traveling to visit her husband who was with the British Mediterranean Fleet near Malta. Burial was in the English Cemetery at Rome.

During the 1860s and 1870s work by four other women amateurs—Elizabeth Hodgson, Charlotte Eyton, Elizabeth Carne, and Agnes Crane—began to appear in the geological literature. All of these women were middle-class, in comfortable circumstances; Eyton, Carne, and Crane had fathers who were well-known naturalists who shared their interests with their daughters.

ELIZABETH HODGSON[66] (1813?–1877) lived in Ulverston, in the northern part of Lancashire, then known as Lake Lancashire, which abuts the Lake District. Her publications dealt largely with the recent and glacial geology of her home area.

Having over many years collected rock samples from the Cumbrian Mountains, she became interested in the pattern of Ice Age transportation of granite debris from these mountains to adjacent regions. A considerable amount of work had already been done on the subject during the 1840s and 50s by a number of eminent geologists, who had traced Cumberland granite erratics to the Cheshire plains and even more distant parts of the country. Elizabeth, an enthusiastic and energetic field-worker, concentrated her attention on the details of her immediate vicinity in North Lancashire and west and northwest Cumberland. She followed direction of ice flow from the higher ground to coastal lowlands by identification of rock type and examination of glacial striae. Well-informed in basic mineralogy and up-to-date on current literature on her subject, she knew the views on glacial geology of experts such as Sir Archibald Geikie and Sir Charles Lyell; she also took care to avoid working in isolation and discussed her findings by correspondence with other geologists, including Adam Sedgwick and people at the British Museum. Several of her papers appeared in the *Geologist* and the *Geological Magazine* between 1864 and 1870. In addition to those listed by the Royal Society she brought out at least one other article on glacial drift, a paper that appeared in 1866 in a local journal, the *North Lonsdale Magazine.*

Her studies of glacial deposits led her to an interest in soil formation by rock disintegration in general. The investigation of the effects of such agents as carbonic acid and decaying vegetable matter on rock breakdown was a relatively new field at the time, pioneered by the German chemist Justus von Liebig, whose views aroused much interest in Britain. Elizabeth, who was familiar with von Liebig's work, set out her ideas on the weathering of the widespread deposits of Carboniferous limestone of her region in her 1867 *Geological Magazine* article.

She had at least one sister, who drew technical illustrations for her work. Among her women scientific friends was Isabella Gifford, the Somerset botanist and naturalist (see chapter 1), who identified for her collections of shells from coastline deposits she worked on. Elizabeth herself was also a competent botanist. Her home region was one of considerable variety in its plant life, and since little systematic botanical work was being done there at the time, her detailed and comprehensive account of the area's flora, which appeared in the *Journal of Botany* in 1874, was of considerable interest. She died in Ulverston, on 26 December 1877, at age sixty-four.

CHARLOTTE EYTON[67] (1839?–1917) was the oldest daughter among the seven surviving children of Thomas Campbell Eyton, of Eyton Hall, near Wellington in Shropshire, and his wife, Elizabeth Frances, daughter and coheiress of Robert Slaney, M.P. for Shrewsbury. Thomas Eyton was a naturalist of considerable standing, author of books on birds and reports on marine life, and a friend of such leading figures in the scientific world as Charles Darwin and Richard Owen.

Throughout the 1860s, when Charlotte was a young woman in her twenties, she carried out a considerable amount of exploration of the recent (Ice Age) geology of her home region of Shropshire, parts of nearby Cheshire, and the north coast of Wales. She kept abreast of current geological literature in her area of interest, reading the journals of regional societies such as the Rugby School of Natural History and the Literary and Philosophical Society of Manchester in which local and regional studies by amateurs working in the English midlands were often published.[68]

Several of her papers appeared in the *Geological Magazine* between 1866 and 1870. For the most part they were concerned with the complex drift deposits (gravels, sands, and clays) of the glacial period, and a number of land features resulting from Ice Age events. Her 1867 paper, "On an old lake-basin in Shropshire," described her investigations of recent deposits in the region around Shrewsbury once occupied by an extensive Ice Age lake (later given the name Lake Lapworth); an 1870 paper discussed the widely occurring blue clay beds of the Cheshire and Shropshire plain. Her monograph, *Notes on the Geology of North Shropshire,* appeared in 1869. She also studied the area known as the Wrekin, the striking isolated hill mass of ancient rocks to the southeast of Shrewsbury. Her book, *The Rocks of the Wrekin, and what is Written upon Them,* was brought out anonymously in 1862. A more general work, *By Flood and by Fell, or Causes of Change, Organic and Inorganic in the Material World* (1872), would appear to have been her last scientific publication. One might wonder if the aging of her naturalist father and his death in 1880 might have affected her work. She died on 6 June 1917, at age seventy-eight.

ELIZABETH CARNE[69] (1817–1873) was the fifth daughter of Joseph Carne and his wife, Mary, daughter of William Thomas of Kidwelley. She was born at Rivière House, in the parish of Phillack, Cornwall, on 16 December 1817. Her father, a fellow of the Royal Society well known for his work on the geology of Cornwall, was the manager of the Cornish Copper Company's smelting works at Hayle on St. Ives Bay, on the north Cornish coast. He later moved his family to Penzance when he joined his father's banking business. On his death in 1858 Elizabeth became head of the Batten, Carne and Carne Bank and inherited a considerable fortune at the same time. This she used to found schools in Penzance and other thinly populated districts of Cornwall. She also built a museum near Penzance to exhibit the large collection of rare ores and minerals her father had assembled in the course of his work with the Copper Company.

Elizabeth's best-known writings were works on general and religious topics, at least one of which appeared under the pseudonym John Altrayd Wittitterly,[70] but her interest in the geology and mineralogy of Cornwall was considerable. Three papers reporting her observations and conclusions about land uplift events and other processes that had influenced the formation and shaping of the Cornish peninsula were published in the *Transactions of the Royal Geological Society of Cornwall,* two of them posthumously. She was very knowledgeable about the mineral composition of the local granites, had some familiarity with the character and origin of the surrounding slates and other metamorphic rocks, and had studied the mineral veins, probably under her father's guidance. However, although it had been recognized that the granites originated in an upwelling of molten rock and the metamorphics in concommitant heating and subsequent cooling processes, a clear picture of the major earth-building events that formed the Cornish landscape had not yet been worked out. And so Elizabeth's geological writing, in which she included a fair amount of speculation, appears somewhat rambling to the modern reader. She died on 7 September 1873, aged fifty-five, at Penzance.

AGNES CRANE,[71] although a contemporary of the first generation of university-trained geologists, was another member of the amateur group. She was born at Thorney Estate, Cambridgeshire, on 9 April 1852, the only daughter of Jane (Turnell) and Edward Crane. Crane was a fellow of the Geological Society and chairman of the Brighton Museum subcommittee. Agnes received her education in private schools in London and after that spent much of her time traveling and writing. In 1881 she accompanied her father on a long tour of both east and west coastal regions of North America; in 1884–1885 they visited Spain, Cuba, and the southern United States. She also traveled extensively throughout the British Isles, Central Europe, and Norway.

From her early twenties Agnes was active in scientific work, her first article, based on studies carried out in the British Museum, appearing in the *Geological Magazine* in 1877. Her particular interest was recent and fossil Brachiopoda, and much of her work was done in collaboration with Thomas Davidson, the specialist in this area with whom Elizabeth Gray cooperated. Agnes assisted in the preparation for publication of Davidson's posthumous memoir on recent Brachiopoda,[72] and also published several articles of her own, mainly in the *Geological Magazine* and the Brighton Natural History Society's *Proceedings.* In 1893 she presented a paper on the evolution of brachiopods at the women's sessions of the World's Congress of Geology, organized as part of the Chicago Columbian Exposition.

Beginning in the late 1880s, with the appearance of the first of the university-trained women geologists, the aspect of British women's activity in geology changes; the substantial contributions of several of the most outstanding members of the later group to a large extent eclipse the work of the "amateurs" of the older tradition. Among the twenty-five women whose geological work is listed in the bibliography, thirteen are known to have had university training. Of these thirteen, three were students at University College London and nine at Cambridge; the other was associated with Oxford. Two of those from University College—Ogilvie Gordon and Raisin—were especially outstanding and received notable honors and awards.

MARIA OGILVIE GORDON[73] (1864–1939), the author of nineteen technical articles by 1900, was the most productive woman field geologist of the late nineteenth century. She was born in the village of Monymusk, in the valley of the Don, Aberdeenshire, on 30 April 1864, the elder daughter among the seven children (five boys and two girls) of the Rev. Alexander Ogilvie and his wife Maria Matilda (Nicol). Alexander Ogilvie was teacher in the Monymusk school. In 1872 he became headmaster of Robert Gordon's Hospital, Aberdeen, which he led for three decades through the crucial period of its development from a monastic institution to one of the best schools in Scotland for secondary and technical education. A man of great strength and resolution, he passed on these qualities in ample measure to his elder daughter.

At the age of nine Maria went to Ladies College (Merchant Company Schools) in Edinburgh, where she remained until she was eighteen, becoming head girl and dux medallist (that is, the top student, academically). Her summers were spent at the family country home in the hills along the river Dee, near Balmoral, and she and her older brother Francis, who also became well-known for his work in geology, often hiked and climbed together. On leaving school in 1882 she entered the Royal Academy of Music in London as a student of piano and composition, and before the end of her first year had been chosen as pianist to accompany the academy's orchestra at public performances. Despite her considerable talent she soon gave up music, however, took the London matriculation examination, and returned to Edinburgh to attend Heriot-Watt College, where she worked toward a science degree.[74] Going back to London in 1889 she enrolled at University College, where she attended Ray Lancaster's classes in zoology. She took a B.Sc. the following year, specializing in geology, botany, and zoology, and was awarded the Gold Medal for zoology and comparative anatomy.

She attempted to continue her studies in Germany, despite the fact that women were not at the time admitted to German universities. Her first choice was the University of Berlin. However, she was unable to find a place there, notwithstanding efforts on her behalf by an influential friend (most probably Baron Ferdinand von Richthofen, professor of physical geography—see below). At Munich she fared better; although mineralogist Paul Heinrich Groth declined to admit her to his laboratory, paleontologist Karl von Zittel accepted her as a private student, as did zoologist Richard Hertwig. She spent four years in Munich (1891–1895) specializing in recent and fossil corals.[75] Her London D.Sc. (1893) was the first in geology awarded to a woman, and the Ph.D. in geology she received from the University of Munich in 1900 was the first given a woman by that institution. She passed the Munich examinations with distinction in zoology, geology, and paleontology.

From her student days on, Oglivie's special interest was the tectonic structure of the South Tyrol, the remote and isolated mountainous region, very complex geologically, to the south of the present Austro-Italian frontier. She was introduced to the area in 1891 by von Richthofen and his wife, who that summer took her there on a monthlong excursion. Captivated by the wild beauty and grandeur of the country she wrote of it as follows:

> Precipitous rocks, generally of a creamy or rose-tinted crystalline dolomite, rise to great heights above the green-swelling passes and grazing land, or sometimes descend at once into deep, gorge-like valleys. The artistic sense scarcely knows which to love most—the romantic region of fir-wood and stream and human habitation, or the wild solitariness of the rocks beyond. Villages are perched mid-way between mountain and ravine, looking in some of the narrower valleys as if a push would throw them into the gap below . . . The barrenness of the dolomite mountains is such that even chamois rarely frequent their clefts and tablelands; snow caps them during nine months out of twelve, and is perpetual on the highest summits.[76]

Von Richthofen had carried out extensive field studies of the area some twenty-five years earlier; he suggested that Maria investigate the geology of the St. Cassian, Cortina d'Ampezzo, and Schluderbach districts in detail. She began in the autumn of 1891, and proceeded with an initial program of two years of very difficult and complicated fieldwork. She herself considered that she was ill prepared when she started.[77] Furthermore, she had no teacher for this first fieldwork, no help or guidance beyond von Richthofen's first directing her to the area. She had to find her own way:

> . . . when I began my field work, I was not under the eye of any Professor. There was no one to include me in his official round of visits among the young geologists in the field, and to subject my maps and sections to tough criticism on the ground.

> The lack of supervision at the outset was undoubtedly a serious handicap.
>
> In point of fact, 17 years passed before I received the first visit of an experienced geologist in the field. That was from Professor [Friedrich August] Rothpletz of Munich, who came with his class of Alpine students for a few days tour in my district [in 1908].[78]

Her study of the Wengen and St. Cassian strata, the basis of her D.Sc. dissertation, was published in the Geological Society's *Quarterly Journal* in 1893. Three major papers reporting her work on coral in the South Tyrol dolomites and on recent and fossil corals followed during the next four years. These were acknowledged as the most complete and comprehensive examination of corals to that time.[79]

Her first three years of fieldwork were particularly challenging, with long daily marches to get to and from her work sites. However, the construction in 1894 of a twenty-mile stretch of driving road giving access to the Sella Massif, and the building of a German and Austrian Alpine Club shelter hut on a mountain terrace in the massif itself somewhat reduced the difficulties.

In 1895 she married Dr. (later Professor) John Gordon, an Aberdeen physician, but nevertheless continued her work in the Tyrol. In the period up to 1913 she brought out on average a paper a year, including extensive studies of the southern and western regions of the Sella Massif, Monzoni and Upper Fassa, and Langkofel.

A number of German and Austrian geologists had worked in the region before her time. In the 1860s von Richthofen had suggested that the puzzlingly irregular dispositions of the limestone masses, the dominant landscape features in the region, might be explained in terms of Darwin's theory of coral reefs. This "Reef Theory" was elaborated during the following decade by Mojsisovics von Mojsvár of the Austrian Geological Survey and had become widely accepted. However, from the late 1870s on, knowledge of the effects of earth-crust movement accumulated rapidly; theories underwent considerable change, and by the 1890s it was possible to entertain the idea that the peculiar forms of the Dolomite massifs in the South Tyrol were not unchanged original structures (coral reefs), but of secondary origin, the results of partial denudation of a mountain system formed by crust movements.

From her early paleontological investigations of the so-called coral reefs, Ogilvie Gordon concluded that these were not in fact coralline. Corals occurred only as relatively small blocks within the limestone matrix, and the predominant fossil remains were those of plant algae. However, although her first three years of work convinced her that the coral-reef theory was no longer tenable, the larger question of finding a satisfactory explanation for the peculiar physical features of the region remained. A number of studies had shown that terrain similar to that in the South Tyrol resulted from complex overthrusting and cross-faulting of previously formed folds. But even these processes failed to adequately explain two features in particular that were typical of the Dolomites—the circular and elliptical mountain shapes with steep, precipitous walls, and the basin-shaped or C-shaped depressions within the massifs themselves containing twisted masses of rock.

In a long paper in the 1899 *Quarterly Journal of the Geological Society* she put forward a possible explanation for these features in terms of a twisting that occurs when overthrusting takes place. She termed the phenomenon "crust-torsion," and demonstrated that in the two series of folding movements that had taken place in the Dolomites, at different periods and with pressure acting in different directions, a kind of "interference" had come into play that had a tendency to cause a turning movement of faulted mountain masses.[80] This demonstration was probably her most important contribution, and it remained work for which she was particularly remembered.

By about 1914, following more than two decades of work in the Dolomites, she had prepared for publication a two-volume monograph of her collected results, covering the stratigraphy, tectonics, and paleontology of the South Tyrol. However, the outbreak of the First World War made her put aside all scientific work for civic responsibilities in Britain (food and clothing distribution and hospital support). Her essentially completed manuscript, together with the engravings for her maps and diagrams, was left in Munich, and in the upheaval that followed these were lost without trace. Her rewritten version, *Das Grödener-, Fassa- und Enneberggebiet- in den Südtiroler Dolomiten. Geologische Beschreibung mit besonder Berücksichtigung der Überschiebungs-erscheinungen,* was brought out by the Vienna Geologische Bundesanstalt in 1927. Described as, "A monument in the field of Alpine geology,"[81] it was outstanding for it comprehensiveness; it became one of the basic references for future work in the region. The sections on tectonics and stratigraphy, illustrated with an exceptionally fine collection of maps, sections, and photographs, were the most notable, but the paleontological findings, which included accurately determined new localities and horizons for various algae and bryozoa, were also of special value.

In 1928 the University of Innsbruck awarded her the diploma of honorary membership in recognition of her work, and the Geological Survey of Austria nominated her an honorary correspondent. Three years later she was elected an honorary member of the Vienna Geological Society (and for a time she was its only female honorary member); the Natural History Museum of Trento also made her an honorary member. In 1932 the London Geological Society awarded her its Lyell Medal; she had been elected a fellow in 1919 when the first eight women were admitted.

She published over thirty original works.[82] Another monograph, her geological guide to the western Dolomites, *Geologisches Wanderbuch der westlichen Dolomiten,* appeared in 1928. Her translation of von Zittel's *History of Geology and Palaeontology* (1901), prepared for the London publisher Walter Scott's Contemporary Science series, was a further important and widely used contribution to the field. Presenting von Zittel's work to English-speaking audiences in slightly condensed form (at his request), the translation came out just two years after the original German edition.

From her school days on, Ogilvie Gordon was an exceptional person, whose energy, ambition, and intellectual qualities marked her as a leader. She overcame opposition from many quarters, family, universities, and scientific circles, in what was a pioneering struggle for a scientific education and the right to follow the path of her own choosing. After her marriage in 1895 she somehow found time not only for family responsibilities (she had two daughters and a son) and geological research, but for social work on national women's action groups as well. She held a number of offices, including those of honorary president of both the Associated Women's Friendly Society and the National Womens Citizens' Associations, president of the National Council of Women of Great Britain and Ireland (1916–1920), and chairman of the Mothercraft and Child Welfare Exhibitions Committee (1919–1921). Her early work on behalf of young people included the preparation of the book *A Handbook of Employments Specially Prepared for the Use of Boys and Girls on Entering the Trades, Industries and Professions* (1908). Shortly after the latter came out she edited *The Health of Nations: Compiled from Special Reports of the National Council of Women,* published by the International Council of Women in 1910.

Although an ardent Scot who took pains to point out that she was in no sense English, after her husband died in 1919 she moved with her children to London, where two of her brothers, Francis, a geologist and educational administrator, and William, a medical doctor, had already been settled for a number of years. She soon became heavily involved in social and civic affairs, serving as a Justice of the Peace and later Chairman of the Marylebone Court of Justice; in fact she was the first woman chair of a London borough court. Herself a strong supporter of the Liberal Party, she worked hard to interest women in politics. She was a good public speaker and could hold a crowd; several times she ran as a Liberal candidate in constituencies where she had no illusions about winning, but where she felt that the opposition she could rally to the more favored party was valuable. She was also active in international women's organizations, serving as honorary corresponding secretary and later vice president of the International Council of Women, and carrying out much of the work of organizing the Council for the Representation of Women in the League of Nations in 1919. These responsibilities entailed a considerable amount of travel to European centers and visits to the United States and Canada as well. For her services she was created a D.B.E. in 1935; the University of Edinburgh gave her an honorary LL.D. the same year. By the 1930s she was a well-known figure in many circles. In 1937, at the invitation of the New South Wales government, she attended the one hundred and fiftieth anniversary of the founding of the penal colony of New South Wales.

All this public work was a considerable burden, however, and throughout much of her life she was torn between the demands it made on her time and her hopes and plans for research. As early as 1908 she found herself having to set aside her scientific work for months at a stretch because of other commitments:

> But from August 1908 to August 1909 my duties as Honorary Corresponding Secretary to the International Council of Women were exceptionally heavy; two large International meetings in Switzerland and in Canada fell within that period, and it is only now, when I have retired from the Secretaryship that I have been able to write an account of my geological researches of 1907 and 1908.[83]

Remarks in many of her later letters to her friend and colleague, the Austrian geologist Julius Pia, reflect the constant conflict: she complained of being overwhelmed with outside work (1922), of getting nothing done because of having to attend meetings (1931), of having no time for any scientific work (1934).[84] Since her geological research was done on the Continent rather than in Britain she also felt a certain amount of professional isolation; ". . . it was a lonely furrow that I ploughed in my field-work abroad. A Britisher—and a woman at that—strayed into a remote and mountainous frontier territory between Austria and Italy, a region destined to be fought over, inch by inch, in the Great War."[85] In her early years she had received help and encouragement from several British geologists (usually fellow Scots, or people with Scottish connections[86]), but her later contacts were largely with the German and Austrian workers whom she had come to know during her student days in Munich and later during the course of her work. Indeed, she felt that she had been awarded the Lyell Medal mainly because of the previous generous treatment given her by the Vienna Geological Society.

Pia described her as a talented observer with a quick mind, an intuitive grasp of her subject and a tremendous enthusiasm for fieldwork. This enthusiasm gave her a tendency to branch off along promising leads as she came to them before she had completed the undertaking already on hand, but it was also the driving force in her work. Two years before her death she wrote of what a joy the work had been, and of the days when she expected "discovery at every corner" as the happiest of her life.[87] In general scientific outlook she was very much a person of the nineteenth century, especially in that she expected to be able to find full and definitive answers to geological questions and was unwilling to leave interpretations for the future.

She died in London on 24 June 1939, at the age of seventy-five; her ashes were buried in Allenvale Cemetery, Aberdeen. The list of mourners attending her memorial service is impressive, and a fitting testimony to the reach and importance of her years of dedicated public service. Included were official representatives of a great many national and international organizations, among them the International Council of Women, the National Council of Women, the National Council of Women of Victoria, Australia, the British Federation of University Women, the India Social Service Group, the National Women's Citizen Association, the National Council for the Unmarried Mother and her Child, the Peace and Disarmament Committee, and the Central Empire Bureau for Women.

CATHERINE RAISIN[88] (1855–1945) was another pioneer among nineteenth-century women scientists, the first to take geology classes at University College, and the second to be awarded a London D.Sc. in geology. She was born in Camden New Town on 24 April 1855, the daughter of Daniel Francis Raisin, a pannierman at the Inner Temple, and his wife Sarah

Catherine (Woodgate). As a child she attended the North London Collegiate School for Girls, opened in Camden in 1850 by Frances Mary Buss. In 1872, at the age of seventeen, she took the University College General Examination for Women, and then attended classes in political economy at University College while also teaching at her old school. Three years later, following the opening to women of a number of science classes at University College, she took John Morris's class in geology and followed that up with his class in mineralogy (only to discover later that it had not been officially open to women). She also studied botany, taking a Special Certificate in that subject in 1877. The next year, when University of London degrees were opened to women, she started to prepare for a B.Sc. She attended Thomas Huxley's lectures in zoology at the Royal School of Mines, but her particular interest was microscopic petrology, taught by the Rev. Thomas Bonney, a well-known figure in late nineteenth-century British geology.

After receiving her degree in 1884 (with second class honours in geology and third class in zoology) she stayed on at University College as Bonney's voluntary assistant. In 1886 she joined the staff of Bedford College as demonstrator in botany and four years later succeeded Grenville A. J. Cole as head of the geology department, becoming the first woman to hold a senior teaching position in the college and one of only two until 1914.[89] She was also the first woman to head a geology department in a British university; she held the appointment for thirty years, during a very formative period in the development of Bedford College. From 1907 she was Norton Sumner Lecturer in geology. She received her D.Sc. in 1898, and in 1902 became a fellow of University College. Her award of the Lyell Fund in 1893 marked the first such recognition of a woman geologist by the London Geological Society. The honor was accepted for her by Bonney since women were not at that time admitted to meetings of the society.

Throughout the 1890s she published a considerable amount of research, some of it in collaboration with Bonney. An outstanding petrographer with a grasp of the optical properties of minerals that was unusual at the time, she was an excellent microscopist who used the most up-to-date techniques available. Her first research, a study of the metamorphic rocks of South Devon published in the Geological Society's *Quarterly Journal* in 1887, was notable as one of the early attempts to recognize and map metamorphic facies. A long series of studies, both petrographical and stratigraphical, dealt with rocks from the western regions of the British Isles, the Ardennes and Switzerland, and collections from the Himalayas and several regions of Africa. She was particularly interested in petrogenetic theory, especially ideas on granetization, but was best known for her careful and detailed investigation of the microscopic structure of the serpentines; of special interest was her joint paper with Bonney in 1905. A 1903 paper on the microstructures in chert was also notable.[90]

Although department head at Bedford College until 1920, she published no further research after 1905, her time presumably going more and more into administration and teaching. For the latter she was well-known, her lectures on petrographic provinces being especially memorable; they were accompanied by practical work on rock specimens of her own collecting from many of the classic areas of Europe and North America.

Catherine Raisin was also a "character." Perhaps some of her style of operation had its origin in her early teaching experience at the North London Collegiate School; one of her obituarists offered the following picture:

> She ruled absolutely in her own sphere and dealt summarily with opposition. All unauthorised perambulation in the Geology Department at Bedford College was quickly stopped, the teaching given by her assistants was carefully scrutinised and in committee she was a doughty, and, to the weaker brethren, an intimidating fighter. She was extremely kind to her students and was affectionately known among them as "The Raisin" and (surreptitiously) "The Sultana." Ever solicitous for their welfare and good name she has been known in the field to immediately take her class out of a quarry upon the arrival of a male party from another college.[91]

When fellowship in the Geological Society was eventually opened to women in 1919, Raisin was one of the early women elected. She was then sixty-four, however, and close to retirement. She belonged to the Geologists' Association, which had never excluded women, for sixty-seven years, was the first woman to serve on its council, and later its most senior member. She became a fellow of the Linnean Society in 1906 soon after it opened to women. Like Ogilvie Gordon she had always taken an active interest in women's welfare and advancement. As an undergraduate in the late 1870s she had given a considerable amount of time to the founding and organizing of a discussion group for women, the Somerville Club, which, when it opened in 1880, already had an enrollment of 1,000 members, a most remarkable number. Under her leadership, first as honorary secretary and later as chairman, this club continued until 1887, when, other educational amenities for women having in the meantime opened up, it was no longer needed. She was also an active supporter of the original Lyceum Club, and took a special interest in its overseas branches. After her retirement in 1920 she gave much of her time to voluntary work in women's action groups. She died of cancer in Ash Priors Nursing Home, Cheltenham, Gloucestershire, on 12 July 1945, at the age of ninety.

One other early woman geologist who was associated with London colleges was MARY FOLEY.[92] She took a B.Sc. with honours in geology at University College in 1891 and received the Morris Prize in geology the same year. Like Raisin she was a student of Thomas Bonney. Her short paper on enclosures of glass in basalt appeared in 1896, and a second on museum collections in 1901.[93] She joined the Geologists' Association in 1892 and served on its council from 1897 to 1900 and again from 1909 to 1912. For two years (1901–1903) she was also the Excursion Secretary. Although she held a post at the University of London for many years she published no further geological research. Much of her time was given to social work related to the welfare and education of women. She died on 22 October 1925.

Among the early Cambridge women geologists, four stand out in particular. These were Gertrude Elles, Ethel Wood,

Ethel Skeat, and Margaret Crosfield, students at Newnham College in the early 1890s. All of them studied under the guidance of Thomas McKenny Hughes and John Edward Marr, two of the leading geologists in Britain at the time.

Of the four, GERTRUDE ELLES[94] (1872–1960) was undoubtedly the most distinguished. Born in Wimbledon on 8 October 1872, she was one of the six children of Jamison Elles, a Scottish businessman who imported Chinese goods, and his English wife, Mary (Chesney). Life during her childhood was divided between London and Scotland, and her great love of the Highlands was early established. From Wimbledon High School she entered Newnham with a College Scholarship in 1891. She took the Natural Sciences Tripos examinations in 1894 and 1895, gaining class II in part I and class I (geology) in part II.

Following a suggestion of John Marr, she and Ethel Wood undertook, while still students, a piece of fieldwork on the Drygill Shales of the Lake District, the results of which were published in the *Geological Magazine* in 1895. This was the first of her studies on the graptolite zones of the Lower Paleozoic. With a Bathurst studentship from Newnham she spent a year in Sweden, a classic region for the study of late Cambrian and early Ordovician formations; she worked with Edvard Holm, Carl Wiman, and especially Sven Törnquist at Lund, all of them eminent figures in Paleozoic geology. After returning to Britain in 1896 she carried out three very substantial studies on the interpretation of graptolite beds in the Welsh Borderland region and the Lake District, studies that extended the earlier epoch-making investigations of Charles Lapworth. Her friend Ethel Wood had joined Lapworth as a research assistant at the University of Birmingham in 1896, and shortly thereafter she and Wood together began the preparation of *British Graptolites,* a monograph that was produced over the next two decades under the general editorship of Lapworth, whose conception it was.[95] Elles wrote the descriptive text and Wood prepared the illustrations.

Graptolite fauna are of critical importance in Lower Paleozoic stratigraphy; for the Ordovician and Silurian periods they have been used as the zonal markers of choice because their characteristic evolutionary pattern along a few dominant trends permits determination of the age of the rocks, at least approximately, by the morphological nature of the fauna. The fossils themselves look like little more than pencil marks on a piece of shale but their study is enormously complex. When Elles and Wood began their work on the monograph the nomenclature of the group was extremely confused; few existing records could be accepted and so they had to check almost everything, either in the field, or by innumerable visits to public and private collections where they could compare earlier results with their own work. Publication of the monograph was spread over the years 1901 to 1914, with Lapworth adding a historical section in 1918. Commonly referred to as "Elles and Wood," this monumental work provided detailed descriptions of virtually every species in this distinctive, varied, and abundant group, and meticulously recorded their occurrence in successive beds; it became the basis of all subsequent research in the area for decades. In 1971, more than half a century after its publication, geologist John Challinor pointed out that the zones established by Lapworth, Elles, and Wood had remained the standard, clear proof of the thoroughness with which the work had been done. Challinor further concluded that the monograph, taken together with the work of Elles and Wood on the formations of the Welsh Borderland, probably constituted the most notable contribution by women to British geology up to the time he was writing.[96]

Although the stratigraphical value of the graptolites had been demonstrated by Lapworth in 1880,[97] the analysis of the complex evolutionary patterns the group presented was carried out by Elles. In 1922 she presented her thoughts on this subject in an important paper that summed up twenty years of work and set out her general and philosophical conclusions on the wide biological and geological significance of the graptolites. She analyzed the evolutionary history of the entire group in terms of continuous general trends and stressed their stratigraphical significance, laying the groundwork for all future work on graptolite fauna succession.[98]

She wrote more than ten later papers on graptolites, and it was as an authority in this area that she was especially remembered. Nevertheless, she also had a deep interest in general problems of land structure and particularly in Lower Paleozoic stratigraphy—approached from the paleontological point of view. Elles was in fact very much a field geologist in the tradition of the Cambridge school. Among the most notable of her stratigraphical papers were those on the Ludlow formations, Builth and Bala, classic ground since the days of Adam Sedgwick.[99] In 1921 she brought out a textbook, *The Study of Geological Maps,* based on lectures given to Cambridge honours students in geology over the years. A second edition appeared in 1931. Her presidential address to the Geology Section of the British Association in 1923 dealt with stratigraphical paleontology illustrated by modern faunal assemblages and the evolution of various fossil groups.[100]

Comparatively late in life, with her extensive body of graptolite work behind her and "an enviable reputation in Lower Palaeozoic geology,"[101] she turned to a completely new area of investigation, the metamorphic rocks of the Scottish Highlands, her favorite part of the country. That she was able, in her late fifties, to adopt new techniques, adjust her thought processes to different working procedures, and make notable contributions to a new area says much for the vigor and flexibility of her mind.[102]

Elles was an inspiring teacher with a gift for passing on her enthusiasm for her subject; several generations of Cambridge geology students, men as well as women, felt her influence strongly. For the first thirty years of her teaching career, however, she was without an official university position. From 1900 to 1903 she held a Geoffrey Research Fellowship, and then for about twenty years, starting in 1903, an assistant demonstratorship in the Sedgwick Museum. In 1926, under the Cambridge Revised Statutes, she was appointed to a university lectureship; ten years later, when she was sixty-four and soon to retire from her college posts, she became the university's first woman reader. She held the title of Reader Emeritus from 1938 until her death in 1960.

At Newnham College, her permanent home for most of her working life, she became, successively, lecturer (1904–1936), director of studies in Natural Science (1927–1936) and in Geography (1927–1933), vice principal (1925–1936, acting principal for one year), associate (1899–1923, 1936–1952), and honorary fellow (1951–1960). Although she was outspoken and at times could even be sharply critical, her solid common sense and immense capacity for work, administrative as well as academic, ensured her taking a full share of college committee responsibilities. Indeed, there were few college activities in which she did not participate. As a student she had played on the Newnham hockey team, and later, over the years, was a popular coach for Newnham teams. She was especially fond of the river, and for many years had her own canoe. Her room in Kennedy Buildings was a place where students gathered for conversation on not only scientific and social topics but all sorts of other matters of interest as well. After her retirement from her college lectureship she was elected a member of the Newnham Governing Body (1937–1951); she served on the council from 1913 to 1927 and from 1924 to 1943.

An honorary member of the Geological Societies of Edinburgh and Liverpool and a corresponding member of the Geological Society of China, Elles was one of the first group of women to be admitted to the London Geological Society, becoming a fellow in May 1919. From 1923 to 1927 she served on the council, the first woman member. She had been awarded the Lyell Fund in 1900, and following completion of the graptolite monograph in 1919 received the Murchison Medal, the first time the honor was accorded a woman. She took a D.Sc. at Trinity College Dublin in 1907.

For many years she was an active member of the British Red Cross Society, serving on its Cambridge executive committee. In 1915 she became Commandant of a small convalescent hospital for soldiers in Newnham Walk, just outside the college gates, and in 1920 received an M.B.E. for her services. During the war years she did much of her teaching and college work in the evenings.

She lived in Cambridge for almost seventy years, retiring in 1936 to a house on Barton Road, which she called "Cladah," the Gaelic for an inner harbor where boats too old to go to sea in all weathers are tied up. Cladah was just across the hockey field from Newnham and she continued to go there almost every day. She also continued to supervise undergraduates and for twenty more years occupied her research room in the Sedgwick Museum, helping other workers, and occasionally publishing papers and notes of her own at least until the mid 1940s. As late as 1950, when she was seventy-eight, she attended the Birmingham meeting of the British Association and took part in a symposium on the work of Charles Lapworth, delivering a talk on Lapworth's work in the Southern Uplands. In 1949 Cambridge awarded her an Sc.D.; at the ceremony she was presented in the Senate House, amid great acclamation, by her old friend and student T. D. Nicholas, bursar of Trinity College.

Later, "G. L. E." became something of a Cambridge character, a short, stocky figure, usually in tweeds and often with a Red Cross cap on her head, but on official occasions resplendent in scarlet doctoral gown. Those who had known her in her youth, however, remembered a striking, handsome woman, with deep blue eyes and a mass of bright, corn-colored hair. In her final years her memory began to fail and she was increasingly troubled by deafness; later her isolation was almost complete. In the autumn of 1960 arrangements were made for her to spend the winter in a nursing home in Helensburgh, Dumbartonshire, near her Scottish relatives. She died there on 18 November 1960, at age eighty-eight.

ETHEL WOOD[103] (1871–1946), a lifelong friend and colleague of Gertrude Elles, was born on 17 July 1871 at Biddenham, near Bedford, the daughter of the Rev. Henry Wood. She entered Newnham College from Bedford High School in 1891 and took the Natural Sciences Tripos examinations in 1894 and 1895, (class II, part I; class I, (geology) part II).

Gertrude Elles later recalled how Ethel was a well-known figure in her college days, as a tennis champion, an active member, on the Liberal side, of the Political Club, and most of all as an outstanding pianist.[104] Not only was she the star performer in many college concerts, but often on winter evenings she played in the darkened College Hall; people would gradually gather to listen, and then disperse quietly when she stopped. She considered becoming a professional musician, but in the end chose geology. Her connection with Newnham remained close for many years; she was an associate from 1905 until 1920.

Her joint work with Elles (in the Lake District and then in the Welsh Borderlands) began while she was still a student and continued in 1895 while she held a Bathurst research studentship. The following year, with an extension of her studentship, she went to Mason College Birmingham as assistant to Charles Lapworth, and within a short time began with Elles the preparation of the graptolite monograph. Two of her own publications from this period on Lower Paleozoic rocks were particularly notable. Each, though complete in itself, to a large extent illustrated in a limited example the type of work required in the preparation of the monograph. In the first, a 1900 paper on the Ludlow formations, she clearly demonstrated the value of graptolites in the classification of the monotonous mudstones of the Silurian rocks. The second, her 1906 paper on the Tarannon series (in Wales), virtually a small monograph on these beds, made plain for the first time their nature as the graptolite facies of the better-known Upper Llandovery horizon. Both papers included particularly fine plates prepared by the technique that she used for the monograph illustrations.[105] The outstanding quality of her work was recognized in 1904 by the award of the Wollaston Fund from the Geological Society.

Shortly after taking her D.Sc. at Birmingham in 1906 she married Gilbert Arden Shakespear, a physics lecturer. Their only child, a daughter, died in infancy. Although on marrying she formally gave up her position of research secretary to Lapworth, she nevertheless continued to collaborate fully with

Elles on the monograph. She became a fellow of the Geological Society in 1919, and the following year was awarded the Murchison Medal for her contributions to the monograph.

Throughout the First World War, as joint founder and honorary secretary of the Association of War Pensions Committee, London, she worked on behalf of disabled soldiers and their dependents, trying to secure adequate pensions, medical treatment, and suitable job training. In 1917 she was appointed to the Special Grants Committee of the Ministry of Pensions, and served on it until 1926. For many years she was also on the Birmingham and Sutton Coldfield Pensions Committee and the Birmingham Citizens Society. Her public service was recognized with the appointment of an M.B.E. in 1918 and a D.B.E. two years later. Despite the hope of her scientific colleagues that she would return to research after the war, she instead gave more and more of her time to civic matters in the City of Birmingham. In 1922 she became a Justice of the Peace for Birmingham and was much respected for her work on the bench; she was the first woman in the city to be elected to the Licensing Committee. The welfare of children and young people was her special interest, and she did a great deal of work in Birmingham's pioneering scheme for boarding out children in foster homes, herself serving as a home visitor. Women's organizations and the problems of poor and working-class girls were likewise among her major concerns. From 1929 until 1932 she was president of the Birmingham branch of the National Council of Women, and she also served for a time as president of the Birmingham and Midlands branch of the Federation of University Women.

In 1929 she and her husband moved out of the city to Caldwell Hall, a fine old house with land attached near Bromsgrove in Worcestershire, where they took up farming. She died there of cancer on 17 January 1946, at age seventy-four. It had been intended that her funeral would be a quiet one, but, as Gertrude Elles described it in her *Roll Letter* memorial, Birmingham decided otherwise. Church and graveyard were filled beyond capacity with people who came to bid her farewell, civic and university dignitaries standing in bitter winter weather alongside the children from the streets whom she had befriended over the years.

Ethel Skeat and Margaret Crosfield, the remaining two in that remarkable group of four Cambridge women geologists from the 90s, never attained the distinction of Elles and Wood, but all the same made solid contributions to their field.

ETHEL SKEAT[106] (1865–1939) was born in Cambridge, the daughter of William Walter Skeat, professor of Anglo-Saxon, and his wife, Bertha, daughter of Francis Jones of Lewisham. She went to Newnham after attending Bateman House School in Cambridge and took first class in part I of the Natural Sciences Tripos examinations in 1894. Shortly after that, first as an Arthur Hugh Clough Scholar and then with a Bathurst studentship, she spent some time on the Continent, studying paleontology for a year in Munich, under Karl von Zittel,[107] and continuing in Paris and in Switzerland.

Some of her most noteworthy work was done with Margaret Crosfield and published in the Geological Society's *Quarterly Journal.* Their first paper (1896), on the geology of the neighborhood of Carmarthen, became the basis of the Geological Survey's Carmarthen sheet memoir when that was written. A later one (1925) on the Silurian rocks of the central part of the Clwydian range in northeast Wales was another short monograph of equally high standard. Skeat also worked for a time with the Danish geologist Victor Madsen. Their paleontological investigations of Jurassic, Neocomian, and Gault boulders resulted in a substantial publication, which appeared in 1898. Six years later she brought out a summary of what was then known of the Jurassic rocks of east Greenland, extensively investigated by Madsen.[108]

For twelve years, from 1898 until 1910, she taught in secondary schools, first in Penarth, South Wales, and then in Chester. She returned to Cambridge in 1910, on her marriage to Henry Woods, Cambridge paleontologist and geological librarian. She was then forty-five. From about that time on, except for a break during the First World War, she was associated with the Cambridge Training College for Women, initially as lecturer (1911–1913), and then as honorary secretary and registrar (1919–1937). During the war she worked in the Post Office Censorship Code Department. Later she shifted her area of professional interest somewhat and published two geography monographs, *The Principles of Geography* (1923) and *The Baltic Region* (1932).

A member of the Geologists' Association from 1893, she became a fellow of the Geological Society in 1919 as soon as it opened to women; she had been awarded the Murchison Fund in 1908. Like Elles, she took a D.Sc. at Trinity College Dublin. She died on 26 January 1939.

MARGARET CROSFIELD[109] (1859–1952) was born on 7 September 1859 in Reigate, Surrey, and lived there for most of her life. She was at Newnham College from 1890 until 1893, when her studies were interrupted for a time by illness. On her return she received special permission to concentrate on geology only, without taking a Tripos, and she made full use of this privilege, attending all the advanced classes and carrying out all the fieldwork offered. The latter was then a very important part of the Cambridge course, being organized weekly throughout the term and regularly during the summer vacation. She acquired a good knowledge of many parts of the British Isles.

As well as carrying out her notable work with Ethel Skeat, she collaborated with her friend Mary Sofia Johnston on an investigation of the Wenlock limestone of Shropshire.[110] The geology of her home area in Surrey was her special interest, and she twice served as joint leader of Geologists' Association excursions there. Her short account of the geology of Surrey appeared in Charles Edgar Salmon's *Flora of Surrey* (1931).

She joined the Geologists' Association in 1892, and served on its council in 1899 and again, as librarian, from 1919 to 1923. In May 1919 she became the first woman fellow of the Geological Society. She was also a keen supporter of the Holmesdale Natural History Society, and frequently lectured to local groups. Schools, education in general, and the political rights of women were matters of special concern to her; for many years she served on the Reigate Education Committee. She died in Reigate on 13 October 1952, at age ninety-three.

Five other Cambridge-trained women—Coignou, Dale, Forster, Gardiner, and Sollas—published geology and paleontology papers listed in the bibliography. Elizabeth Dale's main interests were in botany (see chapter 1) and her single paleontological contribution was a joint report with Albert Seward on fossil ferns of the Dipteridinae family. Coignou, Gardiner, and Forster all went on to careers in education. Sollas was primarily a zoologist (see chapter 3) but published work in paleontology as well. A sixth woman with Cambridge connections whose name appears in the bibliography is Mary Caroline Hughes, wife of geologist Thomas McKenny Hughes.

CAROLINE COIGNOU[111] (1865–1932) was born in Manchester and attended Manchester High School for Girls. She went to Newnham College in 1887 on a Clothworker's scholarship, and took third class in part I of the Natural Sciences Tripos examinations in 1890. Her one geology publication, a short note on a new species of *Cyphaspis* from Carboniferous rocks in Yorkshire, appeared in the Geological Society's *Quarterly Journal* the same year. From 1890 to 1894 she was assistant mistress at Pendleton High School, and for the following six years held a similar position at her old school in Manchester. She then joined the Yorkshire West Riding Education Department and served as woman examiner (1910–1918) and later (1918–1927) as inspector of secondary schools. She died on 1 December 1932.

MARGARET GARDINER[112] (1860–1944) was one of the eight children of the historian Samuel Rawson Gardiner, a descendent of Oliver Cromwell. Her mother was Gardiner's first wife, Isabella, daughter of the theologian, Edward Irving. She went to Newnham in 1883, after attending Gower Street School and Bedford College, and took both parts of the Natural Sciences Tripos examinations (1885 and 1887), gaining second class in each. In 1884 she held a Draper's scholarship and in 1887–1888 a Bathurst studentship. From 1901 to 1914 she was a Newnham associate. Her two geological publications reporting work in stratigraphy and mineralogy appeared in the *Quarterly Journal of the Geological Society* in 1888 and 1890.

For a year or two around 1890 she taught (as assistant mistress) at St. Leonards School in St. Andrews, but in 1892 became head mistress of Withington Girls' High School, Manchester, a school started under the chairmanship of C. P. Scott, for many years owner and editor of the *Manchester Guardian.* In 1897 she opened her own school for girls, the St. Felix School, Southwold, Suffolk,[113] and in the relatively short space of twelve years, herself serving as headmistress, guided its development into an outstanding educational institution. By 1908 ill health had forced her to resign, but the school was by then established on a superb site on the coast with wide views inland and out to sea.

She lived in Horsham, Sussex, from 1915 and was active in the concerns of the community, starting a Nursing Association, a Women's Institute, and other civic groups. During the First World War she established the Queen Mary Workrooms in Cheltenham. She died in Horsham on 2 February 1944, at age eighty-five.

MARY FORSTER[114] was a student at Newnham from 1877 to 1879, holding a Jodrell scholarship in her second year. She took second class in part I of the Natural Sciences Tripos examinations in 1879. Her subsequent career was a varied one. For one year (1885–1886) she was professor of geology and demonstrator in botany at Bedford College. About this time she also worked with William Topley at the Geological Survey, and published a joint paper with him in 1887 on a geological excursion to Belgium and the Ardennes. The next fifteen years she spent in the United States, first as a lecturer at Bryn Mawr College (1888–1892), and then, until 1903, as secretary of the New York branch of the Mechanico-therapeutic and Orthopedic Zander Institute.[115] After returning to London she was a guide at the British Museum for seven years. Then in 1910 she went to India, to a post as assistant mistress at Central Hindu College, Benares. From 1914 she was principal of Gorakhpur Girls' School. She died in India on 5 August 1922.

IGERNA SOLLAS (1877–1965), lecturer in zoology at Newnham College from 1903 to 1904 and 1906 to 1913, collaborated with her father, the Oxford geologist W. J. Sollas, on a number of paleontological investigations. Their joint work included the investigation of methods for determining internal structures of fossils and ways of isolating sections for chemical analysis to determine mineral content. Of special interest was their study of a dicynodont skull by means of serial sections.[116]

MARY CAROLINE HUGHES[117] (d. 1916) was the daughter of the Rev. George F. Weston, honorary Canon of Carlisle Cathedral. She married Thomas McKenny Hughes, then Woodwardian Professor of Geology at Cambridge, in 1882; they had three sons. A skilled artist and a good linguist, she was also interested in science, and after her marriage joined Hughes in many aspects of his work, accompanying him on field excursions and helping with fossil collecting. She traveled with him in Russia, Western Europe, and North America, on occasion taking on the vital job of field cooking when the party's commissariat arrangements broke down. In 1888 she published a detailed and comprehensive account of the Mollusca of the Pleistocene gravel deposits from the Cambridge neighborhood. She also coauthored with her husband the monograph, *Cambridgeshire,* published in 1908 in the Cambridge County Geographies series.

The remaining five British women whose geological work is listed in the bibliography are Donald Longstaff, Bowdich Lee, Kelly, M'Kean, and Scott.[118]

JANE DONALD (later LONGSTAFF)[119] (1856–1935), a naturalist and specialist in invertebrate fossils, was another of the outstanding British women paleontologists of the late nineteenth and early twentieth centuries. She was the oldest of the four children of Matthewman Hodgson Donald of Carlisle, and a granddaughter, on her mother's side, of the theologian the Honorable John Henry Roper Curzon. After receiving her early education at a private school in London she studied at the Carlisle School of Art. Although she had no formal scientific training, she developed remarkably clear insight into the principles of systematic paleontology, and her work, which went on for almost half a century, was well received by her professional colleagues.

From an early age she was much interested in natural history, particularly land and freshwater Mollusca. Her first paper, "Notes on the land and fresh-water shells of Cumberland," was read at the 1881 Carlisle meeting of the Cumberland Association for the Advancement of Literature and Science. Shortly thereafter, largely at the suggestion of J. G. Goodchild, then editor of the association's *Transactions,* she began to study local fossil shells. Her paper on Carboniferous Gasteropoda from Penton and other areas, which appeared in 1885, was the first of a series of about twenty paleontological papers, all dealing with what was then the neglected field of Paleozoic Gasteropoda.[120] From her initial local investigations she soon expanded her studies into the systematic revision of various genera and families, especially the Murchisoniidae, Pleurotomariidae, and Loxonematidae. Most of her work was published in the Geological Society's *Quarterly Journal,* her last article appearing in 1933, when she was seventy-seven.

Having no connection with any institution, she lacked ready access to literature and reference materials, a formidable disadvantage. She did, however, have the necessary means and the leisure to visit the various museums in Britain and abroad where the type specimens she needed were available. Her papers were exceptional for their thoroughness and reliability and for the particularly fine illustrations, which, with her early art training, she was able to prepare herself. In 1898 the Geological Society awarded her the Murchison Fund; only Catherine Raisin had received a similar distinction previously. She was one of the first women elected to fellowship in the Geological Society when it opened to women in 1919, although she had belonged to the Geologists' Association since 1883. In 1893 she was invited to represent women geologists at the World's Congress of Geology at the Chicago World's Fair but was unable to attend.

At the age of fifty, in 1906, she married George Blundell Longstaff, a well-known entomologist and author, and with him traveled widely, visiting, among other places, North and South Africa, Australia, the West Indies, and South America, always adding to her shell collections. Over the course of several years she carried out breeding experiments with the large South African land snail, *Cochlitoma.* She was also an enthusiastic botanist and a fellow of the Linnean Society, although she does not appear to have published in this field. She died in Bath on 19 January 1935, at the age of seventy-nine. Her large collections of recent and fossil shells were presented to the British Museum.

The impressive work of the British women geologists and paleontologists of the late nineteenth and early twentieth centuries (Ogilvie Gordon, Elles, and Wood especially, but also Raisin and Donald) has as its background the long period of productive activity by women (apparent from this survey) that went back to the early 1800s. That in turn had its roots in a very widespread interest in minerals and fossils among the British upper and middle classes, women as well as men, going back even further.[121]

Reasons for the marked increase of interest in geology and paleontology among women in the early nineteenth century are not hard to find. The field was one where there was excitement and action; great discoveries were being made and a strange world of the incredibly remote past opened up to the popular imagination; women were hardly impervious to the appeal.[122] Furthermore the nature of much of the work was such that it was possible for an energetic and intelligent person to become involved without special training. William Smith, the surveyor and "amateur" geologist who pioneered the process of strata identification by fossil content (thereby laying the foundations of modern historical geology), wrote in 1818 of this widespread interest in geology. He noted particularly that in Wiltshire and the west of England, where fossils could be found fairly readily, ladies were distinguishing themselves in the work of collecting.[123]

From about the middle of the century the participation of women was further encouraged by another factor, namely the founding of a great many local and regional scientific societies and also the Geologists' Association, a nonsegregated organization with both amateur and professional members started in 1858; the latter functioned to a large extent as a national geological society for amateurs. These groups, which welcomed women in part because of the financial support they contributed with their membership fees,[124] gave a strong stimulus to regional studies. Further, they provided the important service of publishing their members' articles in the proceedings, annual reports, or transactions, which most of them brought out. Thus Elizabeth Carne's work appeared in the *Transactions of the Royal Geological Society of Cornwall,* some of Agnes Crane's papers in publications of the Brighton and Sussex Natural History Society, and Jane Donald's early articles in the *Transactions* of the Cumberland Association.

By the 1870s, when university-level training was beginning to become available to women, the most immediate factor leading those who were scientifically inclined to study the earth sciences was, arguably, the presence of the subject in the curricula of the leading girls' secondary schools of the time, and its popularity there. The fact that academic standards in average girls' schools in the late nineteenth century were very low is well-recognized, but the comparatively high quality of science instruction in a number of the best ones, those most likely to educate young women going on to advanced studies, has been less widely discussed. The latter offered instruction in a variety of science subjects, often including geology, physical geography, chemistry, and natural history.[125] The result of this particular spread of curricular offerings in the girls' schools is illustrated rather dramatically in the choice of subjects made by girls taking university entrance examinations in the late 1870s, and a report of Cambridge Local Examination results published in *Nature* in March 1879 makes interesting reading. After complaining about the generally low standard of passing in science subjects for both boys and girls, the writer presented the following remarkable information: "We do not become further consoled by the finding that 15 senior boys and 79 senior girls took zoology, 11 boys and 177 girls botany, 24 boys and 150 girls geology." Noting that "girls have no *more* right to a scientific training than boys," he went on, "Most likely, however, boys and their teachers will seek to know more of the life and past history of the globe when they find that girls can really hold their own in and enjoy these studies, and look with amazement on men for being so unwilling to learn or teach them . . ."[126] In contrast to the best of the girls' schools, the leading boys schools of the time still clung to the classical curriculum and allowed little time for science.

Lastly, in examining the reasons for the late nineteenth- and early twentieth-century flowering of geological work by British women, it may usefully be reemphasized that two-thirds of the university-trained women geologists of the time (at least as concluded from this survey) were students at Cambridge. There the Natural Sciences Tripos curriculum provided excellent opportunities and encouragement for original work.[127] Two Cambridge men geologists, Hughes and Marr, widely recognized for their inspiring teaching, had a strong influence on the women students; at a time when women were still not routinely admitted to lectures and laboratories they accepted them into theirs and encouraged independent fieldwork.

Among the four British women who were most prominent (Ogilvie Gordon, Elles, Wood, and Raisin), only Elles and Raisin held salaried professional positions, Ogilvie Gordon being an independent research worker, supported by her family (and helped by occasional small grants), and much of Wood's work being done after she had married and given up her assistantship. Despite the handicaps they faced of marginal places and limited acceptance in the scientific community (for instance, exclusion from fellowship in the London Geological Society until 1919), at least three in this group—Elles, Ogilvie Gordon, and Wood—carried out research that went well beyond the stage of data collecting and contributed to conceptual developments in the field. Furthermore their work won solid acknowledgment from their colleagues in the form of Murchison Medal awards to Elles and Wood and the Lyell Medal, as well as Austrian and Italian honors, to Ogilvie Gordon.

Two-country comparison

The pattern of steadily increasing activity by British women in geology across the nineteenth century would appear to have been peculiar to that national group. There is no evidence for a comparable degree of participation in geological work by American women amateurs over the same period; neither is there any indication of a strong interest in the earth sciences, directed toward university-level study, among American secondary school girls by the later decades of the century. Women in the United States were slower to enter the field (Figure 12–1), few choosing to take up the subject at university level until after 1900, despite the availability to them, from the 1870s onward, of courses in institutions such as the Midwestern state universities. Thus, even by the mid 1890s, when Florence Bascom, America's first professional woman geologist,

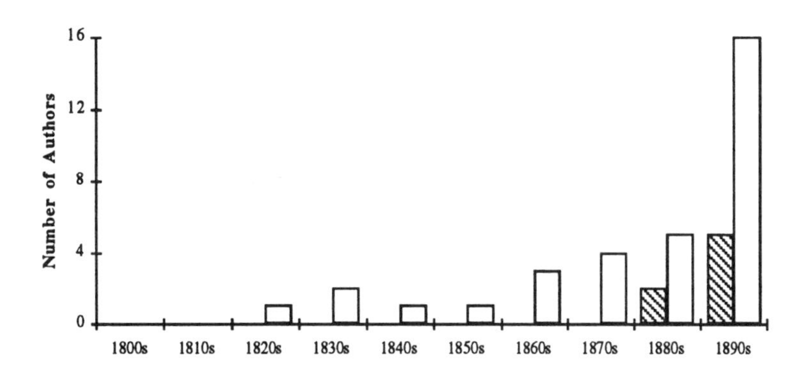

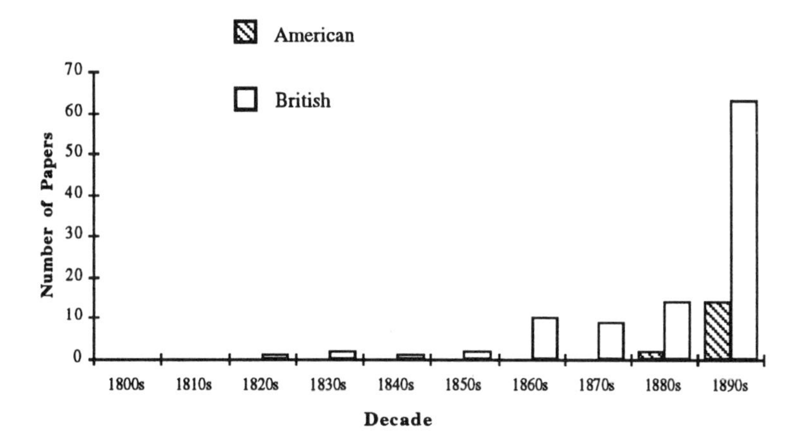

Figure 12-1. American and British authors and papers in geology, by decade, 1800–1900. Data from the Royal Society *Catalogue of Scientific Papers.*

was attempting to develop a full geology department at Bryn Mawr College, she met stiff administrative opposition; the subject was thought of as having no great appeal for women.[128] Public lectures on geology, such as those given at the Lowell Institute in Boston and in other large American cities from mid-century on, were popular with women,[129] but their interest in these did not lead to independent original work. The British pattern remains unique.[130]

Figure 12–2 summarizes the subfield distribution in the British work. (The small number of pre-1901 American authors (seven) and papers (sixteen) makes corresponding pie charts for the American contribution of limited value.) The preponderance of work in paleontology reflects the early stages of a widespread trend; women were especially active in this subfield for many decades (at least until the latter half of the present century), and indeed tended to work on the most complex and difficult of the fossil groups.[131] The substantial pre-1901 British contribution in petrology and mineralogy was largely Catherine Raisin's work, and that in tectonics Maria Ogilvie Gordon's.

Notes

1. Fielde, Hayes, and Smith were better known for their work in other areas—see chapters 2, 8, and 14, respectively. A number of American women contributed technical illustrations to nineteenth-century works in geology. Among the most notable were Orra White Hitchcock, Sarah Hall, Kate Andrews, and Harriet Huntsman. Orra Hitchcock, wife of Edward Hitchcock, the first state geologist in Massachusetts, illustrated her husband's books and reports throughout the 1830s and 1840s. Likewise, Sarah Hall, wife of James Hall, a geologist working on the state geological survey of New York in the 1840s, prepared illustrations for her husband's state report and also for his later works on paleontology. Andrews and Huntsman made drawings for geological survey reports for Ohio and Kansas, respectively. A few American women also wrote or contributed to early instructional monographs on geology. Jane Kilby Welsh's *Familiar Lessons in Geology and Mineralogy Designed for the Use of Young Persons and Lyceums* (Boston: Clapp and Hall, 2 vols.) appeared in 1832, and Miss D. W. Dodding's *First Lessons in Geology: Comprising its most Important and Interesting Facts, Simplified to the Understanding of Children* (Hartford, Conn.: H. S. Pearson)

a. Authors

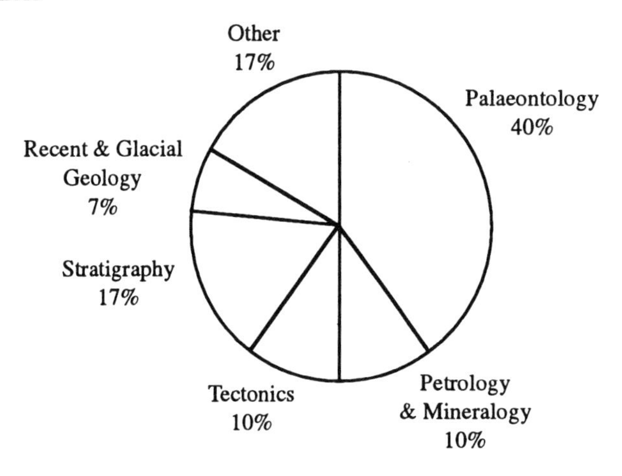

b. Papers

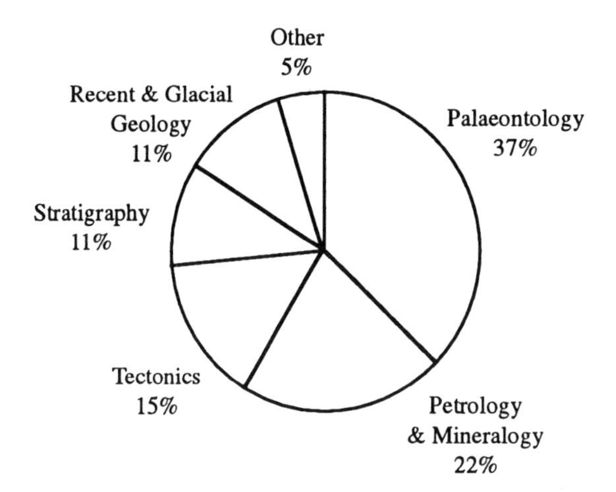

Figure 12-2. Distribution of British authors and papers in geology, by subfield, 1800–1900. (In a., an author contributing to more than one subfield is counted in each.) Data from the Royal Society *Catalogue of Scientific Papers*.

in 1847—see Michele L. Aldrich, "Women in geology," in *Women of Science. Righting the Record,* eds. G. Kass-Simon and Patricia Farnes (Bloomington, Ind.: Indiana University Press, 1990), pp. 42–71, especially pp. 44–7 and 50–1.

2. Floyd Calvin Shoemaker, *Missouri and Missourians. Land of Contrasts and People of Achievements,* 5 vols. (Chicago: Lewis Publishing Co., 1943), vol. 4, *Missouri Biography,* pp. 33–4; Jean Fahey Eberle, *The Incredible Owen Girls* (St. Louis, Mo.: Boar's Head Press, 1977); Bartlett Boder, "The three Owen sisters. Famous scientists," *Museum Graphic,* **8** (1956): 2–3 (St. Joseph Museum, St. Joseph, Missouri, ed., Roy E. Coy); Jerry D. Vineyard, "Introduction," in Luella Agnes Owen, *Cave Regions of the Ozarks and the Black Hills* (New York: Johnson Reprint Corporation, 1970), pp. v-xlii; tombstone inscriptions, Mt. Mora Cemetery, St. Joseph, Missouri; *AMS,* 1910.

3. Juliette Owen (1859–1943), a fellow of the AAAS, life member of the New York Academy of Sciences, and member or associate member of several ornithological associations including the American Ornithologists' Union, published privately a number of works on the birds of the Midwest. These included *Studies of Ornithology and Botany of Northwest Missouri,* and *Songs, Habits and Protection of Birds.* She was particularly interested in distribution of birds and in migration patterns, and is said to have contributed data for government biological surveys. She was also active in promoting national and international legislation for wildlife conservation. Mary (1849–1935) belonged to several folklore and historical organizations, both American and British, and for a time held the office of president of the Missouri Folk-lore Society. Among her writings were *Old Rabbit, the Voodoo and other Sorcerers* (London: T. Fisher Unwin, and New York: G. P. Putnam's, 1893), *The Daughter of Alouette* (London: Methuen, 1896), *Ole Rabbit's Plantation Stories as told among the Negros of the Southwest, collected from Original Sources* (Philadelphia: G. W. Jacobs, 1898), *Folk-lore of the Musquakie Indians of North America and Catalogue of the Musquakie Beadwork and other Objects in the Collection of the Folk-lore Society* (London: D. Nutt, for the Folk-lore Society, 1904), *The Sacred Council Hills, a Folk-lore Drama* (St. Joseph, Mo.: [n.p.], 1909). In 1892 she was admitted to membership in the Muskquakie Indian Tribe (Shoemaker, *Missouri and Missourians,* pp. 33–5, and, for Juliette, *AMS,* 1910).

4. Thanks to an inheritance of well-managed investments, she and her sisters were comfortably situated financially.

5. Vineyard, "Introduction," p. xix.

6. Vineyard ("Introduction," p. x) reported that many contemporary papers on loess and Quaternary geology carried references to "Miss Owen of St. Joseph." Her own loess papers included the following: "The loess at St. Joseph," *American Geologist,* **33** (1904): 223–8 (read at the AAAS meeting, St. Louis, 1904); "Evidence on the deposition of loess," *American Geologist,* **35** (1905): 291–300; "Later studies on the loess," *Pan-American Geologist,* **45** (June 1926): 377–82 (read at the AAAS meeting, Kansas City, Missouri, 1925).

7. Vineyard, "Introduction," p. ix.

8. Owen's geographical papers were "The Missouri river and its future importance to the nations of Europe," *Scottish Geographical Magazine,* **24** (1908): 588–96; "Floods in the great interior valley of North America," *Geographical Journal,* **35** (1910): 56–9 (read before Section E, British Association for the Advancement of Science, Winnipeg, 1909); "The relation of geological activity to conservation of soil and the waters of flowing streams," *Science,* **37** (1913): 459 (abstract of talk given before the AAAS, Cleveland, 1912–1913).

9. Luella A. Owen, "More concerning the Lansing skeleton," *Bibliotheca Sacra,* **73** (1903): 572–8. The Lansing find was controversial because of arguments about the nature of the deposits surrounding it (whether these were loess or river flood-plain material) and hence its age.

10. Lois Barber Arnold, *Four Lives in Science. Women's Education in the Nineteenth Century* (New York: Schocken Books, 1984), pp. 92–123; Edward H. Watson, "Bascom, Florence," *Dictionary of American Biography,* supp. 3 (New York: Scribner's, 1973), pp. 37–9; obituaries, Eleanora Bliss Knopf, "Memorial to Florence Bascom," *American Mineralogist,* **31** (1946): 168–72, Ida H. Ogilvie, "Florence Bascom, 1862–1945," *Science,* **102** (1945): 320–1; Carroll S. Rosenberg, "Bascom, Florence," in *NAW,* vol. **1,** pp. 108–10; Margaret W. Rossiter, "Geology in nineteenth-century women's education in the United States," *Journal of Geological Education,* **29** (1981): 228–32.

11. Further extraordinary action by the executive committee was required in October 1892 in order that Bascom might remain at the university for another year and complete the program she had begun.

12. Orton, president of Ohio State University in the 1870s when Bascom's father had taken the corresponding job at the University of Wisconsin, was at the time professor of geology and the state geologist of Ohio. Having successfully led the development of the university from an agricultural college to a full academic institution, his influence and prestige were considerable.

13. Miller received his Ph.D. from Johns Hopkins University in 1903. By then he also had considerable experience in geological survey work both in the Midwest and in Maryland. He joined the faculty at Lehigh University in 1907, becoming an assistant geologist on the U.S. Geological Survey the same year. For a considerable period thereafter he collaborated with Bascom in survey work and coauthored several of her most important reports.

14. See the following: (with W. B. Clark, N. H. Darton, G. N. Knapp, H. B. Kuemmel, B. L. Miller, and R. D. Salisbury) "Description of the Philadelphia district (Norristown, Germantown, Chester, and Philadelphia quadrangles), Pennsylvania-New Jersey-Delaware," *U.S. Geological Survey Folio* (1909), 24 pp.; (with W. B. Clark, N. H. Darton, G. N. Knapp, H. B. Kuemmel, and B. L. Miller) "Description of the Trenton quadrangle, New Jersey-Pennsylvania," *U.S. Geological Survey Folio* (1909), 25 pp.; (with B. L. Miller) "Description of the Elkton and Wilmington quadrangles, Maryland-Delaware-New Jersey-Pennsylvania," *U.S. Geological Survey Folio* (1920), 22 pp.; (with E. T. Wherry, G. W. Stose, and Anna I. Jonas) "Geology and mineral resources of the Quakertown-Doylestown district, Pennsylvania and New Jersey," *U.S. Geological Survey Bulletin* (1931), 62 pp.; (with G. W. Stose) "Geology and mineral resources of the Honeybrook and Phoenixville quadrangles, Pennsylvania," *U.S. Geological Survey Bulletin* (1938), 145 pp. Among Bascom's other post-1901 publications were: "Water resources of the Philadelphia district," *U.S. Geological Survey, Water Supply Paper* (1904), 75 pp.; "The resuscitation of the term Bryn Mawr gravel," *U.S. Geological Society, Professional Paper* (1925), pp. 117–19; (with G. W. Stose) "Description of the Fairfield and Gettysburg quadrangles, Pennsylvania," *U.S. Geological Survey Folio* (1929), 22 pp; (with G. W. Stose) "Description of the Coatesville and West Chester quadrangles, Pennsylvania—Delaware," *U.S. Geological Survey Folio* (1932), 14 pp.

15. Rossiter, "Geology in nineteenth-century women's education," p. 231.

16. Ibid, p. 231.

17. Knopf, "Memorial," p. 171.

18. Obituaries, Chester A. Reeds, "Memorial to Carlotta Joaquina Maury," *Proceedings of the Geological Society of America* (1939): 157–68, *NYT,* 4 January 1938, 23:3; *Who was Who in America,* vol. 1, p. 791; Edgar Wesley Owen, *Trek of the Oil Finders: a History of the Exploration for Petroleum* (Tulsa, Okla.: American Association of Petroleum Geologists, 1975), numerous references.

19. Carlotta J. Maury, "A comparison of the Oligocene of western

Europe and the southern United States," *Bulletin of American Paleontology,* no. 15 (1902), 94 pp.

20. Owen, *Trek of the Oil Finders,* p. 201.

21. Ibid., p. 208.

22. For Maury's Gulf Coast studies see Carlotta J. Maury, "Stratigraphy of the Jennings oil field, Louisiana," in G. D. Harris, "Oil and gas in Louisiana," *U.S. Geological Survey Bulletin* (1910), 192 pp., p. 56–60; "Shells from deep wells, Terrebone Parish, Louisiana, and their bearing on stratigraphy," ibid., pp. 169–73; "New Oligocene shells from Florida," *Bulletin of American Paleontology,* no. 21 (1910), 46 pp.; (with G. D. Harris and L. Reinecke) "Rock salt, its origin, geological occurrence and economic importance in the State of Louisiana, together with brief notes and references to all known salt deposits and industries of the world," *Geological Survey of Louisiana Bulletin 7* (1907), (Baton Rouge, La., 1908), 259 pp.

23. Owen, *Trek of the Oil Finders,* p. 404.

24. For Maury's Trinidad and Dominican Republic studies see, "A contribution to the paleontology of Trinidad," *Journal of the Academy of Natural Sciences of Philadelphia,* **15,** s. 2, (1912): 25–112; "Santo Domingo type sections and fossils, Part I," *Bulletin of American Paleontology,* no. 29 (1917), 240 pp., "Part II, Stratigraphy," ibid., no. 30 (1917), 43 pp; "Santo Domingan paleontological explorations," *Journal of Geology,* **26** (1918): 224–8.

25. Reeds, "Memorial," p. 161.

26. For a full annotated list of Maury's publications see ibid., pp. 162–8. Among those not already noted here are the following: "On the correlation of Porto Rican Tertiary formations with other Antillean and mainland horizons," *American Journal of Science,* **48** (1919): 209–15; "Tertiary Mollusca from Porto Rico and their zonal relations," in *Scientific Survey of Porto Rico and the Virgin Islands,* vol. 3, pt. 1, (New York: New York Academy of Sciences, 1920), pp. 1–77; "Recent Mollusca of the Gulf of Mexico and Pleistocene and Pliocene species from the Gulf States; Part I, Pelecypoda," *Bulletin of American Paleontology,* no. 34 (1920), 115 pp., "Part II, Scaphopoda, Gastropoda, Amphineura, Cephalopoda," ibid., no. 38 (1922), 142 pp.; "The recent arcas of the Panamic province," *Palaeontographica Americana,* **1** (1922): 163–208; "A further contribution to the paleontology of Trinidad (Miocene horizons)," *Bulletin of American Paleontology,* no. 42 (1925), 250 pp.; "Venezuelan stratigraphy," *American Journal of Science,* **9** (1925): 411–14; "Novas colleccoes paleontologicas do Servico Geologico do Brasil," *Servico Geologico e Mineralogico do Brasil, Boletím 33* (1929), 23 pp.; "The Soldado rock type section of Eocene," *Journal of Geology,* **37** (1929): 177–81; "Uma zona de graptolitos do Llandovery Inferior no Rio Trombetas, Estado do Para, Brasil," *Monographias do Servico Geologico e Mineralogico do Brasil,* no. 7 (1929), 53 pp.; "O Cretaceo da Parahyba do Norte," ibid., no. 8 (1930), 305 pp.; "Correlation of Antillean fossil faunas," *Science,* **72** (1930): 253–4; "Two new Dominican formational names," ibid., **73** (1931): 42–3; "Bartonian and Ludian Upper Eocene in the western hemisphere," *American Journal of Science,* **22** (1931): 375–6; "Fossil Invertebrata from northeastern Brazil," *American Museum of Natural History, Bulletin,* **67** (1934): 123–179; "*Lovenilampas,* a new echinoidean genus from the Cretaceous of Brazil," *American Museum Novitates,* no. 744 (1934), 5 pp.; "New genera and new species of fossil terrestrial Mollusca from Brazil," ibid., no. 764 (1935), 15 pp.; "The Soldado rock section," *Science,* **82** (1935): 192–3; "O Cretaceo de Sergipe," *Monographias do Servico Geologico e Mineralogico do Brasil, Boletím 11* (1936), 283 pp.; "Argillas fossilferas do Pliocenio do territorio do Acre," *Servico Geologico e Mineralogico do Brasil, Boletím 77* (1937), 34 pp.

27. Reeds, "Memorial," p. 161.

28. No biographical information about Mary Fleming has been uncovered. She discussed her observations concerning the potholes of Foster's Flats on the Niagara River at a meeting of the AAAS in Columbus, Ohio, in 1899. She was from Buffalo, New York. Mary Klem, of St. Louis, Missouri, graduated summa cum laude with a B.A. degree from Washington University, St. Louis, in 1900. Her undergraduate thesis, a study of crinoid fossils, was published as a seventeen-page paper in the *Transactions of the Academy of Sciences of St. Louis* in 1900. A member of the academy, she served as its executive secretary for many years and as its librarian in 1908 and 1909. A second geology publication by her, a substantial work entitled "A revision of the Paleozoic Palaeoechinoidea, with a synopsis of all known species," appeared in the academy's *Transactions* in 1904 (**14,** pp. 1–98). She later taught in the St. Louis public school system, continuing to do so until at least 1935. Although she brought out a biographical note on the paleontologist Gustav Hambach in 1923 (*Pan-American Geologist,* **40** (1923): 32–4), she does not appear to have published other technical papers (information from university archives, Washington University, St. Louis, including entries from the *Annual Catalog, 1899–1900* and the *General Alumni Catalog, 1917*).

29. Anon., "Familiar campus characters. Carrie A. Barbour," *Nebraska Alumnus,* (October 1926): 384; Ruth Henderson, "She is Assistant Curator," ibid., (April 1936): 14, 30; anon., "The faculty," ibid., (September 1942): 8.

30. Erwin Hinckley Barbour, "Report on the initial work of the State Geological Survey of Nebraska," *Science,* **11** (1900): 343–4.

31. Carrie Adeline Barbour, "Observations on the concretions of the Pierre shale," *Publications of the Nebraska Academy of Science,* **7** (1901): 36–8.

32. The bibliography lists papers in geology by twenty-five British women, a note by Mary Anning, not indexed by the Royal Society, having been added. In an earlier communication we gave the figure of 65 percent for the combined British and Irish fraction of the papers listed by the Royal Society (Mary R. S. Creese and Thomas M. Creese, "British women who contributed to research in the geological sciences in the nineteenth century," *British Journal for the History of Science,* **27** (1994): 23–54). The Irish contribution is not included in the present survey.

33. See n. 1. The British wife-assistants included Mary Morland Buckland (1797–1857), wife of William Buckland, Mary Ann Woodhouse Mantell (fl. 1820s and 1830s), wife of Gideon Mantell, Mary Horner Lyell (1808–1873), wife of Sir Charles Lyell, and Charlotte Hugonin Murchison (1789–1869), wife of Sir Roderick Murchison. Lady Murchison's private influence appears to have been considerable; speaking to the Edinburgh Geological Society at the time of her death, Sir Archibald Geikie reminded his audience that it was she who had induced Sir Roderick to "forsake the ordinary amusements of a retired cavalry officer, and devote himself to that branch of science in which he has so distinguished himself." Further, her work of fossil collecting on the expeditions she made with her husband, especially in the North West Highlands of Scotland, was a significant contribution to Scottish geology (*Transactions of the Edinburgh Geological Society,* **1** (1870): 265–6). Two other women geology enthusiasts of this early period who helped Buckland were Lady Mary Cole and Jane Talbot (see Creese and Creese, "British women . . . in the geological sciences").

34. Hugh S. Torrens, "Women in geology. 2—Etheldred Benett," *Open Earth,* (1983): 12–13; Earle E. Spamer, Arthur E. Bogan, and Hugh S. Torrens, "Recovery of the Etheldred Benett collection of fossils from Jurassic-Cretaceous strata of Wiltshire, England, analysis of the taxonomic nomenclature of Benett (1831), and notes and figures

of type specimens contained in the collection," *Proceedings of the Academy of Natural Sciences of Philadelphia,* **141** (1989): 115–80.

35. Etheldred Benett, "A catalogue of Wiltshire fossils," in Sir Richard Colt Hoare, *The History of Modern Wiltshire* (1822–1844), vol. 3, pt. 2 (London: J. B. and J. G. Nichols, 1831), pp. 117–26 (this catalog was followed by a short list of local plants, contributed by Etheldred's sister, Anna Maria Benett—see Spamer, Bogan, and Torrens, "Recovery of the Etheldred Benett collection," p. 130). Benett's independently published catalog was brought out by J. L. Vardy, Warminster, 1831.

36. Those who cited Benett's work included Murchison (1832), Mantell (1833, 1844, 1846), Fitton (1836), Agassiz (1842–1846, 1848), Michelin (1840–1847) and Geinitz (1846); for a listing of the many references see Spamer, Bogan, and Torrens, "Recovery of the Etheldred Benett collection," p. 134 and the reference section. Among the later authors who referred to Benett's work was A. J. Jukes-Browne of the Geological Survey; in his paper "The fossils of the Warminster Greensand" (*Geological Magazine,* **4** (1896): 261–73, especially p. 263) he pointed out the importance of her work in determining more exactly the location of sponge fossils, and quoted two paragraphs from her 1816 manuscript (Geological Society library), "Sketches of fossil Alcyonaria from the Greensand formation at Warminster Common and in the immediate vicinity of Warminister."

37. Among the exceptionally fine Benett specimens believed to have been in the Wilson acquisition, but not yet rediscovered, are those of silicified specimens of the Trigoniid *Laevitrigonia gibboza,* believed to show the internal structures of the animal. These specimens were illustrated in a plate (now in the British Museum) prepared in the 1840s, but never printed. Spamer, Bogan, and Torrens published it for the first time in their 1989 paper (Plate 1, p. 138). The illustration shows the mollusk's incredibly delicate gills and other organs. The Benett specimens are considered to be unique in the field of paleontology.

38. Spamer, Bogan, and Torrens, "Recovery of the Etheldred Benett collection."

39. Horace B. Woodward, *The History of the Geological Society of London* (London: Geological Society, 1907), pp. 118–19.

40. Benett's interests were not confined to geology. In 1833 she brought out an edition of the manuscript, *Brief Enquiry into the Antiquity, Honour and Estate of the Name and Family of Wake,* by William Wake (Warminster: [n.p.]). Wake, her great-grandfather, was an early eighteenth-century Bishop of Lincoln and Archbishop of Canterbury.

41. Hugh Torrens, "Mary Anning (1799–1847) of Lyme; 'the greatest fossilist the world ever knew'," *British Journal for the History of Science,* **28** (1995): 257–84; *DNB,* Supplement, 1901, pp. 51–2; Mrs. [Elizabeth O. Buckland] Gordon, *The Life and Correspondence of William Buckland, D.D., F.R.S.* (New York: D. Appleton, 1894), pp. 113–16; Patricia Phillips, *The Scientific Lady: A Social History of Women's Scientific Interests 1520–1918* (New York: St. Martin's Press, 1990), p. 183. See also J. M. Edmonds, "The fossil collection of the Misses Philpot of Lyme Regis," *Dorset Natural History and Archaeological Society Proceedings,* **98** (1978): 43–8; W. D. Lang, "Three letters by Mary Anning, fossilist of Lyme," ibid., **66** (1945): 169–73, and "More about Mary Anning, including a newly found letter," ibid., **71** (1950): 184–8.

42. Most of Anning's specimens went to museums in the years before incoming materials were registered (the practice started only in 1837 at the British Museum—Torrens, "Mary Anning," p. 279). This means that the process of establishing definitively the whole extent of Anning's contributions is a lengthy and tedious undertaking.

43. Woodward, *History of the Geological Society,* p. 115; Torrens, "Mary Anning."

44. Leonard G. Wilson, "The intellectual background to Charles Lyell's *Principles of Geology, 1830–1833,*" in *Toward a History of Geology,* ed. Cecil J. Schneer, (Cambridge, Mass.: MIT Press, 1969), pp. 426–43, on pp. 435–6.

45. Anning's short note in the 1839 *Magazine of Natural History* (vol. 12, p. 605) illustrates this point. It reads as follows:

> In reply to your request I beg to say that the hooked tooth is by no means new; I believe M. De la Beche described it fifteen years since in the Geological Transactions, but I am not positive; but I know that I then discovered a specimen, with about a hundred palatal teeth, as I have done several times with different specimens. I had a conversation with Agassiz on the subject; his remark was that they were the teeth by which the fish seized its prey—milling it afterwards with its palatal teeth. I am only surprised that he has not mentioned it in his work. We generally find Ichthyodorulites [defensive finbones] with them, as well as Cartilaginous bones.

46. Remark by Mary Anning to the King of Saxony on his visit to Lyme (1844), quoted by Gordon, *Life of William Buckland,* p. 115. Anning's inclusion in volume 1 of J. C. Poggendorff's classic *Biographische-literarisches Handwörterbuch zur Geschichte der exacten Wissenschaften* (Leipzig: Verlag von Johann Barth, 1863), p. 41, provides further indication of the interest of her work for European geologists.

47. Henry De la Beche, obituary notices, *Quarterly Journal of the Geological Society,* **4** (1848): xxiv–xxv.

48. See for example H. A. Forde, *The Heroine of Lyme Regis: the Story of Mary Anning, the Celebrated Fossilist* (London: Wells, Gardner & Co., 1925); *The Children's Encyclopedia,* ed. Arthur Mee (London: Educational Book Co., 1925); more recent works are Helen Brandon Bush, *Mary Anning's Treasures* (New York: McGraw Hill, 1965); Ruth van Ness Blair, *Mary's Monster* (New York: Coward, McCann, and Geoghan, 1975); Sheila Cole, *Dragon in the Cliff* (New York: Lothrop, Lee, and Shepard, 1991), and Marie Day, *Dragon in the Rocks* (Toronto: Greer de Pencier, 1992). For a full listing of works about Anning see Torrens, "Mary Anning."

49. Ibid.; see also Lang, "Three letters" and "More about Mary Anning."

50. R. J. Cleevely, R. P. Tripp, and Y. Howells, "Mrs. Elizabeth Gray (1831–1924): a passion for fossils," *Bulletin of the British Museum of Natural History* (Historical Series), **17** (1989): 167–258. I am grateful to Ron Cleevely for pointing out to me the importance of Mrs. Gray's work.

51. See for instance, Thomas Davidson, "Monograph of the British fossil Brachiopoda," *Palaeontographical Society* (*Monographs*), (1853–1883), 1–368, plus supplement (242 pp.), and "Notes on four species of Scottish Lower Silurian Brachiopoda," *Geological Magazine,* **4** (1877): 13–17; Jane Donald, "On some of the Proterozoic Gasteropoda which have been referred to as *Murchisonia* and *Pleurotomaria* with descriptions of new subgenera and species," *Quarterly Journal of the Geological Society,* **58** (1902): 313–39, and "Descriptions of Gasteropoda chiefly in Mrs. Robert Gray's collection from the Ordovician and Lower Silurian of Girvan," ibid., **80** (1924): 408–46; F. R. C. Reed, "The Lower Palaeozoic trilobites of the Girvan district, Ayrshire," *Palaeontographical Society* (*Monographs*), (1903–1935), parts 1–3, 1903–6 (184 pp.), plus 3 supplements 1914 (56 pp.), 1931, (30 pp.), 1935 (64 pp.), and "The Ordovician and Silurian Brachiopoda of the Girvan district," *Transactions of the Royal Society of Edinburgh,* **51** (1917): 795–998.

52. R. Gray, "Observations in various branches of natural history during the past summer," *Proceedings of the Glasgow Natural History Society,* **1** (1858–1869): 100–1 (1864–1865 session; specimens of univalves, bivalves, trilobites, corals and graptolites exhibited); "Specimens exhibited," ibid., **1** (1858–1869): 124 (1865–1866 session; notes on Silurian brachiopods from Girvan); "Specimens exhibited," ibid., **1** (1858–1869): 197–8 (January 1868; a rare Cystidean exhibited, *Ischadites kanigii,* from the Silurian rocks of Girvan); "Specimens exhibited," ibid., **1** (1858–1869): 229–30 (1868–1869 session; Robert Gray exhibited a specimen of *Leptaena youngiana* Davidson, a new Silurian brachiopod from Girvan. Three examples had been collected by Mrs. Gray and submitted to Thomas Davidson for naming. Specimens of *Leperditia* from Girvan were also exhibited).

53. Charles Lapworth, "Presidential Address," *Quarterly Journal of the Geological Society, Proceedings,* **59** (1903): xlvii-xlviii. The award was accepted for Mrs. Gray by Henry Woodward, women not being eligible to attend meetings of the society at the time.

54. Cleevely, Tripp, and Howells, "Mrs. Elizabeth Gray," p. 225.

55. DNB, vol. 3, 710; H. J. Rose, *A New General Biographical Dictionary* (London: B. Fellow, 1850), entry reproduced in *British Biographical Archive,* microfiche edn., ed. Paul Sieveking (London: K. G. Saur, 1984); Jane Robinson, *Wayward Women: A Guide to Women Travellers* (Oxford: Oxford University Press, 1990), pp. 44–6.

56. Maria Graham, *Journal of a Residence in Chile, during the Year 1822. And a Voyage from Chile to Brazil in 1823* (London: Longman, Hurst; John Murray, 1824). Her second book was *Journal of a Voyage to Brazil and Residence there during Part of the Years 1821, 1822, 1823* (London: Longman, Hurst; John Murray, 1824).

57. Sir Charles Lyell, *Principles of Geology* (London: John Murray, 1830–1833; New York: D. Appleton, 1857), p. 457 (N.Y. ed.).

58. See the three-part article, "On the reality of the rise of the coast of Chile in 1822, as stated by Mrs. Graham," *American Journal of Science and Arts,* **25** (1835): 239–47, on p. 240. This article includes a reprinting of Graham's 1824 account in the *Transactions of the Geological Society,* Greenough's presidential address, and Graham's response to the latter. The following brief extract (from p. 244) gives the flavor of this response (she wrote in the third person):

> Mr. Greenough mentions Mrs. Callcott's published journal, and accounts for the dead fish on the shore [*Journal,* p. 331] by an imaginary storm. Common candour would have led that gentleman to have stated that, in that very journal, it is distinctly printed, that a "delightful and calm moon-light night followed a quiet and moderately warm day" [ibid., p. 305].
>
> Mr. Greenough says, further in p. 18 of his address—"some muscles [*sic*] and oysters still adhere, she says, to the rocks on which they grew: but we do not know the nature of these rocks, whether fixed or drifted." Mrs. Callcott was ignorant that there were, or might have been, drifted rocks, until she learned it from Mr. Greenough; for much as she has been at sea, she never met with one. The rocks at Quintero, and at Valparaiso, are of grey granite, and where they lift themselves through the sand and shingle of the beach, they give the notion of bald mountain tops. At all events, they are fixed sufficiently to have caused the wreck of more than one Spanish ship of war; and when she saw them the morning after the Earthquake, that on which the wreck of the *Aquila* lay, was certainly so far above the water, that the vessel could be approached dry-shod, which had never happened before, even at the lowest tides. The beds of muscles [*sic*], of other shell-fish, and of sea-weed, were equally rocks of grey granite, fixed far below the sands of the ocean. These circumstances are stated in the published journal: but Mr. Greenough has suppressed them, and many others of the like nature, particularly the notice of some rocks and stones, that the lowest tides never left dry, but now have a passage between them and the low-water mark, sufficient to ride around without difficulty, p. 313.

59. Charles Davison, *The Founders of Modern Seismology* (Cambridge: Cambridge University Press, 1927), pp. 34–6.

60. Graham, "Reality of the rise of the coast," p. 245.

61. See, for instance, Lady Callcott, *Essays towards a History of Painting* (London: E. Moxon, 1836).

62. Nicholas Edwards, "The Hastings Collection," *Journal of the Society for the Bibliography of Natural History,* **5** (1970): 340–3; *Burke's Peerage and Baronetage* (105th ed., London: Burke's Peerage Ltd., 1975), p. 1179.

63. Searles Valentine Wood, "On the discovery of an alligator and of several new Mammalia in the Hordwell Cliff; with observations upon the geological phenomena of that locality," *London Geological Journal,* **1** (1846): 1–7, 117–22.

64. "On the tertiary beds of Hordwell, Hampshire," *Philosophical Magazine,* **6** (1853): 1–11, on p. 2.

65. Her 1851 paper in the *Bulletin de la Société Géologique de France* was quoted at length in an account of the area published by Henry Keeping thirty-two years later. (H. Keeping and E. B. Tawney, "On the section at Hordwell Cliffs, from the top of the Lower Headon to the base of the Upper Bagshot Sands," *Quarterly Journal of the Geological Society,* **39** (1883): 566–74). Keeping collated the beds of the marchioness's description with his own. A young man in his twenties when he worked with the marchioness, he later became curator, under Adam Sedgwick, of the Woodwardian Museum at Cambridge.

66. Obituary, *Journal of Botany,* **16** (1878): 64.

67. J. F. Kirk, *A Supplement to Allibone's Critical Dictionary of English Literature,* 2 vols., (Philadelphia: J. B. Lippincott, 1891), reproduced in *British Biographical Archive,* no. 386; *Burke's Landed Gentry,* 18th ed. (London: Burke's Peerage Ltd., 1972) vol. 3, p. 309; *DNB,* vol. 18, p. 107, entry for Eyton, Thomas Campbell.

68. In fact, such studies by amateurs rather than professionals produced a substantial fraction of the early knowledge of the glacial history of Britain; Eyton and Hodgson were two of the early women participants in the work—see L. Dudley Stamp, *Britain's Structure and Scenery* (London: Collins, 3d ed., 1949), p. 156.

69. *DNB,* vol. 9, pp. 135–7—entries for both Elizabeth and Joseph Carne.

70. John Altrayd Wittitterly, *Three Months Rest at Pau, in the Winter and Spring of 1859* (London: Bell and Daldy, 1860). Carne also published *Country Towns and the Place they Fill in Modern Civilization* (London: Bell and Daldy, 1868), a religious pamphlet, *England's Three Wants* (London: William Macintosh, 1871), and *The Realm of Truth* (London: H. S. King, 1873).

71. H. Cox, *Who's Who in Kent, Surrey and Sussex* (Cox's County Series. London: Horace Cox, 1911), reproduced in *British Biographical Archive,* no. 278; obituary for Edward Crane, *Geological Magazine,* **8** (1901): 286.

72. Charles Davidson, "A monograph of recent Brachiopoda," *Linnean Society Transactions,* s. 2, Zoology, **4** (1886–1888), 248 pp.

73. Obituaries, Julius Pia, "Maria Matilda Ogilvie Gordon," *Mitteilungen der Geologischen Gesellschaft in Wien,* **32** (1939): 173–86, E. B. Bailey, "Maria Ogilvie Gordon, D.B.E.," *Nature,* **144** (1939): 142–3, E. J. G[arwood], in *Quarterly Journal of the Geological Society,* **102** (1946): xl-xli, "Dame Maria Ogilvie Gordon," *Times,* 26 June

1939, 9c, and "Memorial Service: Dame Maria Ogilvie Gordon," ibid., 30 June 1939, 19d; *Who was Who,* vol. 3, p. 1018; *In Memoriam: an Obituary of Aberdeen and Vicinity for the Year 1904 with Biographical Notes and Portraits of Prominent Citizens* (Aberdeen: William Cay and Sons, 1904?), pp. 93–102, entry for the Rev. Dr. Alexander Ogilvie; information from Aberdeen Central Library; Creese and Creese, "British women . . . geological sciences," pp. 33–5; Mary R. S. Creese, "Maria Ogilvie Gordon (1864–1939)," *Earth Sciences History,* **15** (1996): 68–75.

74. Her brother Francis Ogilvie was appointed principal of Heriot Watt College in 1886, and served there for a number of years before moving on to work on government educational boards and the Geological Survey in London.

75. Permission for individual women to audit courses at the University of Munich was first granted formally by the authorities only in 1896; the first woman to take classes there subsequently was the British geology student Ethel Skeat (see below), who also studied with von Zittel—see James C. Albisetti, *Schooling German Girls and Women* (Princeton, N.J.: Princeton University Press, 1988), p. 278. Ogilvie was not the first female student to work in von Zittel's institute, the Russian Evgeniia Solomko (also a corals specialist) having listened to his lectures and worked in the Paleontological Museum about five years earlier, probably in 1886.

76. Maria Ogilvie, "Coral in the 'dolomites' of South Tyrol," *Geological Magazine,* **1** (1894): 1–10, 49–60, on pp. 2–3.

77. See Pia, "Maria Ogilvie Gordon," p. 179.

78. Ogilvie Gordon, remarks made when she accepted the London Geological Society's Lyell Medal, *Quarterly Journal of the Geological Society* (*Proceedings*), **88** (1932): lix–lx, on p. lx.

79. See Karl Alfred von Zittel, *History of Geology and Palaeontology to the End of the Nineteenth Century,* tr. Maria M. Ogilvie-Gordon (London: Walter Scott, 1901), pp. 390–1.

80. To be slightly more specific, Ogilvie Gordon postulated that in the Dolomites, in the process of the intercrossing of a later east-west thrust movement with an earlier north-south pressure system, there was produced a combination of pressure and twist (with two sets of force couples coming into play), which, in an anticline, resulted in contending spiral movements of adjacent faulted blocks. Inequalities in resistance caused these differential twisting movements to be very complex. In the course of her work she built on the classic writings of Eduard Suess (*Das Antlitz der Erde*—Prag and Wien: F. Tempsky, 1885–1909) and the contributions of several others (such as Mojsisovics, Vaĉek, Bittner, Rothpletz, Salomon, and Brögger) on the peri-Adriatic region, as well as the laboratory work on the dynamics of crust deformation by Gabriel Daubrée and others in the late 1870s.

81. Wilhelm Salomon-Calvi, "The Dolomites of South Tyrol," *Nature,* **121,** (1928): 83–5, a review of the monograph (tr., L. R. Cox).

82. Ogilvie Gordon's post-1900 papers not already mentioned include the following: "The crust-basins of southern Europe," *Verhandlungen des Internationalen Geographen Congresses,* VII, Berlin, 1899, **2** (1901): 167–80; "Monzoni and Upper Fassa," *Geological Magazine,* **9** (1902): 309–17; "The geological structure of Monzoni and Fassa," *Transactions of the Edinburgh Geological Society,* **8** (1903), Special Part, 180 pp.; "Interference-phenomena in the Alps," *Proceedings of the Geological Society,* (1905–1906): cxxxv (title only), and *Geological Magazine,* **3** (1906): 381–3; "Vorläufige Mitteilung über die Ueberschiebungsstructur im Langkofelgebiete," *Verhandlungen der Kaiserlich-königlichen Geologischen Reichsanstalt. Wein,* (1907): 263–5; "Preliminary note on overthrust-structure at Langkofel in the Dolomites," *Geological Magazine,* **4,** (1907): 408–9; "Die Ueberschiebungsmassen am Langkofel und im oberen Grödner Tal," *Verh. Geol. Reichsanst. Wein* (1909): 297–300; "Note on the Langkofel thrust-mass," *Geological Magazine,* **6** (1909): 488–90; "The thrust-masses in the western districts of the Dolomites," *Transactions of the Edinburgh Geological Society,* **9,** Special Part (1909–1910), 91 pp.; "Die Ueberschiebung am Gipfel des Sellamassivs in Gröden," *Verh. Geol. Reichsanst. Wien,* (1910): 219–30; "Geologische Profile von Grödental und Schlern," ibid., (1910): 290–4; "Ueber Lavadiskordanzen und Konglomeratbildungen in den Dolomiten Südtirols," ibid., (1911): 212–22; "Leithorizonte in der Eruptivserie des Fassa-Grödengebietes," ibid., (1913): 163–70; "Das Vorkommen von Diplopora annulatissima in Langkofelgebiete," *Verhandlungen der Geologischen Bundesanstalt. Wien,* (1925): 187–92; "Einige geologische Ergebnisse im Gebiet von Fassa und Gröden," ibid., (1925): 203–15; "Der Bau der westlichen Dolomiten," *Zeitschrift der Deutschen Geologischen Gesellschaft* (Monatsberichte), **80** (1928): 279–81; "The structure of the western Dolomites," *Quarterly Journal of the Geological Society,* **85** (1929): cxxiii–cxxvi; "Geologie des Gebietes von Pieve (Buchenstein), St. Cassian und Cortina d'Ampezzo," *Jahrbuch der Geologischen Bundesanstalt. Wien,* **79,** (1929): 357–424; "Geologie von Cortina d'Ampezzo und Cadore," ibid., **84** (1934): 59–215; (with Julius Pia) "Zur Geologie der Langkofelgruppe in den Südtiroler Dolomiten," *Alpenländischer geologischer Verein, Wien. Mitteilungen,* **32** (1940): 1–118 [Posth.].

83. Ogilvie Gordon, "The thrust-masses in the western districts," (1909–1910), p. 2.

84. See Pia, "Maria Ogilvie Gordon," p. 184.

85. Ogilvie Gordon, "Remarks made on acceptance of Lyell Medal," p. lix.

86. These were, most especially, the Scottish geologist Sir Archibald Geikie, the leading figure in British geology between 1882 and 1912, William Topley of the Geological Survey, Charles Lapworth, who did much of his research in Scotland, and Benjamin Peach and John Horne, both of whom were prominent figures in Scottish geology. Geikie secured for her a research grant of £100 from the Royal Society, a substantial amount in the 1890s and probably enough to support several seasons of work in the South Tyrol. Topley helped her with the publication of her first paper (ibid., p. lix). As far as later financial help was concerned, she occasionally found funding for specific needs; thus preparation of the many very fine large and detailed maps and sections in her 1927 monograph was paid for by a grant from the Carnegie Trustees for the Universities of Scotland.

87. See Pia, "Maria Ogilvie Gordon," p. 184.

88. Obituaries, Doris L. Reynolds, "Dr. Catherine Alice Raisin," *Nature,* **156** (1945): 327–8, E. J. G[arwood], in *Quarterly Journal of the Geological Society,* **102** (1946): xliv–xlv, L. H., in *Proceedings of the Geologists' Association,* **57** (1946): 53–4, and *Times,* 8 August 1945, 6f; Staff Records, Royal Holloway, University of London; Student Records, University College London; Office of Population Censuses and Survey, Southport, Merseyside; Mary R. S. Creese, "Raisin, Catherine Alice," *DNB,* "Missing Persons" vol., 1993, pp. 544–5.

89. Raisin also served as head of the botany department at Bedford College from 1891 to 1908, and as vice principal of the college for the three years from 1898 to 1901. The other woman department head at Bedford College at that period was Beatrice Edgell, professor of mental and moral science, appointed in 1898—see chapter 15.

90. (With T. G. Bonney) "The microscopic structure of minerals forming serpentine, and their relation to its history," *Quarterly Journal of the Geological Society,* **61** (1905): 690–714; "The formation of chert and its micro-structures in some Jurassic strata," *Proceedings of the Geologists' Association,* **18** (1903–1904): 71–82. Among Raisin's

other post-1900 publications were, "Perim Island and its relations to the area of the Red Sea," *Reports of the British Association* (1901): 640–1; "On certain altered rocks from near Bastogne and their relations to others in the district," *Quarterly Journal of the Geological Society,* **57** (1901): 55–72; "Notes on the geology of Perim Island," *Geological Magazine,* **9** (1902): 206–10; "Petrological notes on rocks from Southern Abyssinia, collected by Dr. Reginald Koettlitz," *Quarterly Journal of the Geological Society,* **59** (1903): 292–306.

91. L. H., obituary, pp. 53–4. Quoted with the permission of the Geologist's Association.

92. E. W., obituary in *Proceedings of the Geologists' Association,* **37** (1926): 229.

93. Mary C. Foley, "Visit to the British Museum (Reptilia)," ibid., **17** (1901): 117–18.

94. Obituaries, W[illiam] B[ernard] R[obinson] K[ing], ibid., **72** (1961): 168–71, O[liver] M[eredith] B[oone] B[ulman], in *Proceedings of the Geological Society* (1961): 143–5, and A. B. W., "Gertrude Lilian Elles," *Newnham College Roll Letter* (1961): 45–9; John Challinor, *The History of British Geology. A Bibliographical Study* (New York: Barnes and Nobel, 1971), p. 187; *Newnham College Register 1871–1971,* vol. 1, Staff, p. 10.

95. Lapworth at the time was so overworked that it is practically certain that only a fraction of his plans would ever have borne fruit if he had not turned the monograph work over to Wood and Elles (see G. L. E[lles], obituary for Ethel Wood, *Quarterly Journal of the Geological Society,* **102** (1946): xlvi).

96. Challinor, *History of British Geology,* pp. 144–5.

97. Charles Lapworth, "On the geological distribution of the Rhabdophora," *Annals and Magazine of Natural History,* s. 5, **3** (1879): 245–57, 449–55; **4** (1879): 331–41, 423–31; **5** (1880): 45–62; 273–85; 358–69; **6** (1880): 16–29, 185–207.

98. G. L. Elles, "The graptolite faunas of the British Isles," *Proceedings of the Geologists' Association,* **33** (1922): 168–200.

99. See the following: (with Ida L. Slater) "The highest Silurian rocks of the Ludlow district," *Quarterly Journal of the Geological Society,* **62** (1906): 195–221 (Slater was a Newnham College student, NST, Pt. I, Cl. I, 1903, Pt. II, Cl. I, geology, 1904); "The relation of the Ordovician and Silurian rocks of Conway (North Wales)," ibid., **65** (1909): 169–92; "The Bala County: its structure and succession," ibid., **78** (1922): 132–68. Among Elles's other papers on Lower Paleozoic geology and paleontology were, "A new *Azygograptus* from North Wales," *Geological Magazine,* **59** (1922): 299–301; "The age of the Hirnant beds," ibid., **59** (1922): 409–14 and **60** (1923): 559–60; "The characteristic assemblages of the graptolite zones of the British Isles," ibid., **62** (1925): 337–47; "Graptolite assemblages and the doctrine of trends," ibid., **70** (1933): 351–4; "The Lower Ordovician graptolite fauna with special reference to the Skiddaw slates," *Summary of Progress, Geological Survey for 1932,* (1933) Pt. 2, p. 94; "The classification of the Ordovician rocks," *Geological Magazine,* **74** (1937): 481–94; "Factors controlling graptolite successions and assemblages," ibid., **76,** (1939): 181–7; "Graptogonophores," ibid., **77** (1940): 283–8; "The type specimen of *Monograptus sedgwicki* (Portlock)," ibid., **79** (1942): 31–2; "The identification of graptolites," ibid., **81** (1944): 145–57; "Upper Silurian graptolite zones," ibid., **81** (1944): 257–7; "The work of Charles Lapworth in the Southern Uplands," *Advancement of Science,* **7** (1951): 435–8.

100. Gertrude L. Elles, "Presidential Address. Evolutionary palaeontology in relation to the Lower Palaeozoic rocks," *Reports of the British Association* (1923): 83–107.

101. Bulman, obituary, pp. 144–5.

102. See, for instance, "Notes on the Portsoy coastal district," *Geological Magazine,* **68** (1931): 24–34, and "The Loch na Cille boulder bed and its place in the Highland succession," *Quarterly Journal of the Geological Society,* **91** (1935): 111–44.

103. Obituaries, G. L. E[lles], ibid., **102** (1946): xlvi–xlvii, *Nature,* **157** (1946): 256–7, and "Dame Ethel Shakespear, J.P., D.Sc. (Birm)," *Newnham College Roll Letter* (1946): 47–9, anon., "Dame Ethel Shakespear," *Times,* 28 January 1946, 6e; *Newnham College Register, 1871–1971,* vol. 1, p. 112.

104. Elles, "Dame Ethel Shakespear," p. 47.

105. "The Tarannon series of Tarannon," *Quarterly Journal of the Geological Society,* **62** (1906): 644–99. Wood's 1900 paper is listed in the bibliography. She also published the note, "On graptolites from Bolivia, collected by Dr. J. W. Evans in 1901–02" (ibid., **62** (1906): 431–2).

106. Obituary, H[enry] W[oods], *Proceedings of the Geologists' Association,* **57** (1940): 114; *DNB* (1912–1921), entry for Skeat, William Walter, pp. 495–6; *Newnham College Register, 1871–1971,* vol. 1, p. 111.

107. See note 75.

108. "The Jurassic rocks of East Greenland," *Proceedings of the Geologists' Association,* **18** (1903–1904): 336–50. For Skeat's later paper with Margaret Crosfield, "The Silurian rocks of the central part of the Clwydian range," see *Quarterly Journal of the Geological Society,* **81** (1925): 170–92.

109. Obituaries, G. L. E[lles], *Proceedings of the Geological Society,* (1953): cxxxi–cxxxii, M[ary] S[ofia] J[ohnston], *Proceedings of the Geologists' Association,* **64** (1953): 62–3; *Newnham College Register,* 1871–1971, vol. 1, p. 65.

110. Margaret C. Crosfield and Mary S. Johnston "A study of Ballstone and the associated beds in the Wenlock limestone of Shropshire," *Proceedings of the Geologists' Association,* **25** (1914): 193–224.

111. *Newnham College Register, 1871–1971,* vol. 1, p. 92.

112. Ibid., p. 79; obituary, *Times,* 17 February, 1944, 7d; *DNB,* 2nd. Supplement, vol. 2, pp. 75–8, entry for Gardiner, Samuel Rawson.

113. The Newnham College Register entry for Gardiner refers to the Suffolk school she founded as the Aldeburgh School, but the *Times* obituary refers to it as St. Felix School, Southwold; it may have been renamed at some point.

114. *Newnham College Register, 1871–1971,* vol. 1, p. 60.

115. The Zander Institute, established in Stockholm by Dr. Gustaf Zander in 1865, was concerned with the promotion of medical gymnastics.

116. Igerna Sollas's post-1900 paleontological papers include the following: "Fossils in the Oxford Museum. V. On the structure and affinities of the Rhaetic plant *Naiadita,*" *Quarterly Journal of the Geological Society,* **57** (1901): 307–12; (with W. J. Sollas) "An account of the Devonian fish *Palaeospondylus gunni,* Traquair," *Proceedings of the Royal Society,* B, **72** (1903): 98–9; (with W. J. Sollas) "*Lapworthura*: a typical Brittlestar of the Silurian age, with suggestions for a new classification of the Ophiuroidea," *Philosophical Transactions of the Royal Society,* B, **202** (1911): 213–32; "On *Onychaster,* a Carboniferous brittlestar," ibid., **204** (1914): 51–62; (with W. J. Sollas) "A study of the skull of a Dicynodont by means of serial sections," ibid., **204** (1914): 201–25; (with W. J. Sollas) "On the structures of the Dicynodont skull," ibid., **207** (1916): 531–9. Igerna's elder sister Herda (Newnham College, 1897–1900, Ph.D., Heidelberg) also collaborated with their father in geology-related undertakings; her translation of the five volumes of Edward Suess's famous work, *Das Antlitz der Erde (The Face of the Earth),* came out under William Sollas's editorship over the twenty-year period 1904 to 1924.

117. See the obituary for Thomas McKenny Hughes in the *Geolog-*

ical Magazine, **54** (1917): 334–5, and also "Eminent living geologists: Thomas McKenny Hughes," ibid., **3** (1906): 1–13, especially p. 9.

118. For Bowdich Lee see chapter 4. Agnes Kelly's 1900 paper in the *Mineralogical Magazine* described work carried out in Munich. She appears to have had a connection with the crystallographer and mineralogist Sir Henry Miers, then professor of mineralogy at Oxford. Minnie M'Kean was from Edinburgh, and a member of the Edinburgh Naturalists' Society. The 1891 paper she read before the society described paleontological observations made during a visit to Suffolk. Mrs. S. Scott's paper on river action appeared in the Liverpool Geological Association's *Journal* in 1891. No later geology publications by Kelly, M'Kean, or Scott were found.

119. Obituaries, L. R. Cox, *Proceedings of the Geological Society,* **91** (1935): xcvii–xcviii, and the *Proceedings of the Geologists' Association,* **47** (1936): 97.

120. Donald Longstaff's post-1900 papers include the following: "On some Proterozoic Gasteropoda which have been referred to *Murchisonia* and *Pleurotomaria,* with descriptions of new subgenera and species," *Quarterly Journal of the Geological Society,* **58** (1902): 313–39; "Observations on some of the Loxonematidae, with descriptions of two new species," ibid., **61** (1905): 564–6; "On some Gasteropoda from the Silurian rocks of Llangadock (Carmarthenshire)," ibid., **61** (1905): 567–78; "Notes on the genera *Omospira, Lophospira* and *Turtitoma*: with descriptions of the new Proterozoic species," ibid., **62** (1906): 552–72; "Some new Lower Carboniferous Gasteropoda," ibid., **68** (1912): 295–309; "Supplementary notes on *Aclisinia,* de Koninck, and *Aclisoides,* Donald, with descriptions of new species," *Geological Magazine,* **4** (1917): 287–8; "Descriptions of Gasteropoda, chiefly in Mrs. Robert Gray's collection from the Ordovician and Lower-Silurian of Girvan," *Quarterly Journal of the Geological Society,* **80** (1924): 408–46; "A revision of the British Carboniferous Murchisoniidae with notes on their distributions and descriptions of some new species," ibid., **82** (1926): 526–55; "A review of the British Carboniferous members of the family Loxonematidae with descriptions of new forms," ibid., **89** (1933): 87–122.

121. Phillips, *Scientific Lady,* p. 108; see also Elizabeth Chambers Patterson, *Mary Somerville and the Cultivation of Science, 1815–1840* (The Hague: Martinus Nijhoff, 1983).

122. A prospectus distributed at the time of the founding of the Geologists' Association in 1858 began as follows: "There is no branch of Science which attracts so *general*—it may be said so *popular*—an interest as Geology;"—reprinted in *Proceedings of the Geologist's Association,* **7** (1880): 6–7 (italics added); see also G. S. Sweeting, "Origin and growth," in *The Geologists' Association 1858–1958: A History of the First Hundred Years,* ed. G. S. Sweeting (London: Geologists' Association, 1958), pp. 2–3, and D. V. Ager, "The Association today and tomorrow," ibid., pp. 131–3.

123. See Thomas Sheppard, *William Smith: his Maps and Memoirs* (Hull: A. Brown and Sons, 1920), p. 217.

124. Jean G. O'Connor and A. J. Meadows, "Specialization and professionalization in British geology," *Social Studies of Science,* **6** (1976): 77–89, on p. 87; see also D. E. Allen, *The Naturalist in Britain* (London: Allen Lane, 1976).

125. See Phillips, *Scientific Lady,* pp. 238–48, and Felicity Hunt, "Divided aims: the educational implications of opposing ideologies in girls' secondary schooling, 1850–1940," in *Lessons for Life: the Schooling of Girls and Women, 1850–1950,* ed. Felicity Hunt (Oxford: Basil Blackwell, 1987), pp. 3–21, especially pp. 6, 8. Phillips's discussion includes a summary of the findings of the four-year Taunton Commission of Inquiry (1864–1868) into the state of English schools.

126. "Notes," *Nature,* **19** (1879): 497 (original italics). For another similar report see ibid., **18** (1878): 423. These reports suggest that a case might be made for arguing that, at least for a time in the late 1870s, a few of the foremost girls' schools led the country in secondary education in the sciences!

127. Roy Porter, "Gentlemen and geology: the emergence of a scientific career, 1600–1920," *The Historical Journal,* **21** (1978): 809–36, especially p. 835.

128. Ogilvie, "Florence Bascom."

129. Rossiter, "Geology in nineteenth-century women's education," p. 229.

130. For a short discussion of the situation of women geologists and geology students in late nineteenth-century Russia (the country in continental Europe where contributions by women were most notable) see Creese and Creese, "British women . . . geological sciences," pp. 43–5.

131. By the mid-twentieth century the country that trained and employed most women geologists was the former Soviet Union. A 1968 survey of about 2,500 Soviet paleontologists indicated that more than half (about 1,500) were women. Further, of 480 specializing in foraminifers (small primitive protozoans) 400 were women, while of 180 studying brachiopods (larger animals whose study is less monotonous and demanding) 100 were men and 80 women (D. V. Nalivkin, *Nashi Pervye Zhenshchinye-geologi* [*Our First Women Geologists*]—Leningrad: Nauka, 1979, pp. 5–6; see also Creese and Creese, "British women . . . geological sciences," n.77). Rossiter (*Women Scientists in America,* p. 259) has pointed out that, during the increased oil exploration activity in the United States in the 1920s, several women paleontologists found oil company laboratory jobs examining the Foraminifera content of exploratory drillings (see also Owen, *Trek of the Oil Finders,* pp. 523–5).

Part 3

Social Sciences, plus Others

Chapter 13

GEOGRAPHERS, EXPLORERS, TRAVELERS, AND A HIMALAYAN CLIMBER

Less than 1 percent of the papers by women listed in the Royal Society index are on primarily geographical topics. Nine British women contributed sixteen articles, and three Americans four (although five more by Americans not listed by the Royal Society have been added to the bibliography); women in France, Germany, Austria, and Sweden together produced about six more. About two-thirds of these articles, including much of the British contribution, concerned exploration and travel, still a major area of interest to geographical societies in the late nineteenth century.

American women

Of the three American authors, one, Elizabeth Gifford Peckham, was an entomologist who probably wrote only one paper on a geographical topic, and the second, Fanny Bullock Workman, was an explorer and alpine climber; the third, Ellen Semple, one of the outstanding women geographers of her time, was one of the founders of modern academic geography in the United States.

ELLEN SEMPLE[1] (1863–1932) was born in Louisville, Kentucky, on 8 January 1863, the youngest of the five children of Alexander Bonner Semple and his second wife, Emerin (Price). Alexander Semple, a hardware merchant of Scots-Irish ancestry, accumulated a substantial amount of money during the Civil War by selling guns and other supplies to both sides, and the financial security this gave the family was the key to Ellen's freedom in later years to pursue her professional activities. Her parents separated when she was still a child and her mother was the important influence in her early life.

After some formal education in Louisville public schools, supplemented by private tuition, she followed her older sister Patty to Vassar College. Entering in 1878 at the age of fifteen, she was the youngest student there. She graduated with honors four years later and returned home to take her part in the activities of Louisville society.

Although to some extent she enjoyed this style of life, she felt the want of the intellectual stimulation she had had at Vassar, and after a trip to Europe took a position teaching ancient history and classics in a girls' school in Louisville. On a second trip to Europe in the summer of 1887 she met, at the London home of a friend, an American student who had just completed his doctoral studies at the University of Leipzig. He told her about the work of the social geographer Friedrich Ratzel, one of his teachers. Two years later, when she began working for an external M.A. degree from Vassar, she used Ratzel's *Anthropogeographie* as a guide in the sociological study she completed for her thesis.[2] Furthermore, she made up her mind to study under Ratzel, and after receiving her degree in 1891 took a year's leave from teaching and went to Leipzig.

Women were not yet admitted to the university as regular students, but she audited graduate classes and seminars not only in geography, but also in economics (taught by Wilhelm Roscher) and statistics (with August von Miaskowski); her work with Ratzel she found particularly satisfactory. He soon became her inspiration, guide, and friend; she worked hard in his courses and he repaid her efforts by his generous interest in her progress. Particularly important to her were their conversations on style in geographical writing. Despite the restrictions on women's participation in university work at the time, she would seem to have had an enjoyable experience in Leipzig.[3] She kept in close touch with Ratzel after her return to Louisville and the following year reviewed and translated a number of his works. The theory he espoused and developed, that environment plays a determining role in the historical advance of human societies, she found especially attractive; indeed it was to become the central theme in her thinking and writing for the rest of her life.

In 1893 she and her sister Patty started a girls' school, the Semple Collegiate School, where she taught history for two years. However, in 1895 she took another term in Leipzig and thereafter began to spend more and more of her time on geographical work. "The influence of the Appalachian barrier upon colonial history," the first of four early papers she published in the *Journal of School Geography,* appeared in the first issue in 1897. Four years later, with the appearance of her article on Kentucky mountain communities,[4] her scholarly reputation was established. Based on field observations made during a 350-mile horseback and camping trip she made through a little-known and inaccessible region of the eastern Kentucky highlands during the summer of 1899, the paper discussed the changes brought about by isolation and mountain environment on the traditional Anglo-Saxon lifestyle; it did much to stimulate American interest in geography. In 1903, her first book appeared, *American History and its Geographic Conditions,* much of which was also written during her summer in the Kentucky highlands. In this work she provided meaning and pattern for myriads of isolated geographical facts, and demonstrated forcefully that geographical factors are of vital significance in appraising social and economic growth. The book attracted widespread attention and remained for decades a popular text in the historical geography of the United States; a revised edition was brought out in 1933. In 1904 she published accounts of a second notable field study, an investigation of the influence of geographical environment in the lower St. Lawrence valley.[5] By then her professional standing among her colleagues was well-recognized, and she became one of the founding members of the Association of American Geographers.[6] In the same year (1904) she read papers at the International Geographical Congress in Washington, D.C., became associate editor of the *Journal of Geography,* and began to write her second monograph, *Influences of Geographical Environment; on the Bases of Ratzel's System of Anthropogeography.*

The latter work, which played a major role in the introduction of the field of social geography into the United States, was the most comprehensive of Semple's three books, and the one that aroused the greatest controversy. It had been planned by her and Ratzel as an interpretation for English-speaking readers of the principles of anthropogeography Ratzel had developed, but the final version embodied much of her own thinking. She abandoned some of the more outdated of Ratzel's ideas and modified and greatly elaborated others. Comprising seventeen chapters, the book covered such topics as "Island peoples," "Steppes and deserts," "Mountain environment," "Influence of climate on man," and "Man's relation to water." It received widespread praise on its appearance in 1911 (after Ratzel's death), a contemporary reviewer describing it as the most scholarly contribution to geographical literature produced in America to that time.[7] However, the seemingly overwhelming emphasis on environmentalism Semple put forward later came under increasing criticism. It was felt that she had greatly oversimplified the problem of understanding human movements and the development of societies. Though environment was accepted as one of the fundamental factors in the story, other considerations, such as social conflict, religion, leadership, and the all-important fact that man changes his environment (frequently for the worse) came increasingly to be seen as playing major roles, to which Semple had failed to assign sufficient significance. But as E. A. Ackerman pointed out in a 1961 assessment of *Influences,*[8] Semple's work needs to be placed in its historical context. She lived the greater part of her life in a world where the Newtonian principles of cause and effect were still the prevailing scientific concepts. Probability, randomness, and the uncertainty principle had not yet settled into the intellectual mix of the time, and an explanation of a broad range of observed or recorded social effects in terms of a single major factor, environment, fitted well with prevailing methodology and thought patterns. Nevertheless, Semple's main objective (at least as interpreted by Ackerman) was not the sweeping generalization; she sought rather to isolate a particular factor (the environment) and make possible a thorough study of its influence on the course of historical events. Whether or not she was an environmentalist is probably of less importance in any assessment of her contribution to geographical thought than the fact that her work gave impetus to the study of specific isolated factors ("isolates") as they influenced the development of historical geography in America.

In 1906, while she was writing *Influences,* she accepted the offer of a position in the new department of geography at the University of Chicago. Research and writing remained her top priority, however, and she lectured, mainly to graduate students, for one term in alternate years. Immediately after *Influences* came out in 1911 she and two Louisville women friends went on a yearlong trip around the world. During a six-week stay in Japan she visited two Japanese friends who had been her classmates at Vassar College. One had become the wife of an admiral in the Japanese navy and the other was Princess Oyama, widow of a former head of the Japanese army; these women made it possible for Semple to carry out a substantial amount of geographical work, including a 150-mile walk across the central mountain area of the country. She was especially interested in Japanese methods of intensive agriculture and presented a paper on this topic before the Royal Geographical Society in 1912. A second paper on her Japanese studies, "Japanese colonial methods," came out the following year.[9] Returning home through Europe and Britain, she stayed en route at Oxford University for a term as lecturer.

Throughout the years of the First World War she taught, either for a term or a summer session, at each of several colleges and universities in the United States (Wellesley College, 1914–1915; the University of Colorado, 1915; Western Kentucky University, 1917; Columbia University, 1918), and she also provided a course of lectures for army officers on the geography of the Italian front. In 1919 she joined the team of geographers known as The Inquiry. Under the direction of Woodrow Wilson's friend and adviser Colonel Edward House they prepared material for the American delegates to the Paris Peace Commission. Semple's contributions were manuscripts on "The Austro-Italian frontier," and "The Turkish Empire past and present." She found the war-related projects stimulating, though she objected to the rushed nature of the work, which prevented her from giving her writing its customary polish, and to the fact that women on Inquiry's research staff were not allowed to go to Europe to complete their work on the spot, although wives of high-ranking officials could accompany their husbands.

In 1921, when a graduate school in geography was established at Clark University in Worcester, Massachusetts, she joined the faculty as lecturer, the first woman to do so. Becoming professor of anthropogeography in 1923, she held the post until her death nine years later, though she alternated semesters of teaching and writing.[10] She was the only American woman geographer of her time who regularly taught in Ph.D.-granting institutions, and since Clark University and the University of Chicago, where she continued to spend some time until 1923, were the two institutions that trained the largest number of American graduate students in geography, most of the second generation of geographers in this country took courses from her. A demanding teacher remembered especially for her tightly organized lectures, she gave her time generously to the many students who asked her advice, though she supervised only two M.A. theses and two Ph.D. dissertations.

Her last major research and writing project concerned the Mediterranean region, an area in which she had had a great interest ever since she studied classical history during her undergraduate years at Vassar. On the way home from her round-the-world trip in 1911–1912 she had spent some time walking many of the routes taken by the armies of ancient Greece and Rome. The first of a series of Mediterranean articles, "Barrier boundary of the Mediterranean basin and its northern breaches as factors in history," appeared in the *Annals of the Association of American Geographers* in 1915. Her third book, *The Geography of the Mediterranean Region; its Relation to Ancient History,* was published in 1931. A substantial work of 737 pages, its second half was completed after she had been severely disabled by a heart attack and restricted to writing no more than two hours a day. She herself considered the book her best, though it never became popular with American geographers, few of whom had particular interests in the time period covered. The shift in geographical thought in the early 1930s away from the theory of environmental determinism, which was perceived as the central principle in her thinking, probably further reduced the book's interest to geographers.[11]

Semple averaged one important publication a year during her working life,[12] and her substantial contribution to the development of early twentieth-century academic geography in America was recognized by many awards and honors. Among these were the Cullum Medal of the American Geographical Society (1914), the Helen Culver Gold Medal of the Geographical Society of Chicago for "distinguished leadership and eminent achievement in geography" (1932), and election to the presidency of the Association of American Geographers (1921—she was the first woman to receive this honor). A tall and distinguished woman of great charm, she had a wide circle of friends. She died in West Palm Beach, Florida, on 8 May 1932, at the age of sixty-nine, after a lengthy period of severe illness. Her considerable private library was given to the University of Kentucky.

Very different from Ellen Semple's studies in human geography were the contributions of her contemporary, FANNY BULLOCK WORKMAN[13] (1859–1925), whose efforts were in another branch of the field entirely. Workman was one of the early women explorers in the frontier region of northwest India, and one of the first to climb in the Karakoram Range. Over the course of several years of private but well-financed, well-organized, and well-staffed expeditions, she and her husband explored the main, glacier-filled, east-west valleys of the central Karakoram and climbed several of the surrounding peaks. The geographical information they assembled was reported in a series of papers that appeared, mainly in British geography journals, between 1899 and 1912. Many included topographical or survey maps, not all of which have stood the test of time, their quality depending on whether or not good surveyors were engaged to take part in the expeditions.

Fanny Bullock was born in Worcester, Massachusetts, on 8 January 1859, the second daughter of Elvira (Hazard) and Alexander Hamilton Bullock; Alexander Bullock was the Republican governor of Massachusetts from 1866 to 1868. The family, like Semple's, was wealthy, Fanny's maternal grandfather, Augustus George Hazard, having been a successful manufacturer and merchant of gunpowder. She was taught by tutors and attended a private school in New York City, followed by two years at schools in Paris and Dresden. In 1881, when she was twenty-two, she married William Hunter Workman, a prominent Worcester doctor. Their only child, Rachel, was born three years later. William Workman gave up his medical practice in 1889 because of poor health and they moved to Europe, where, over the next several years, they spent the festival season at Bayreuth and began serious climbing in the Alps, making expeditions to the Matterhorn, Mont Blanc, and Rothorn. They also took up bicycling, which, with the arrival of the Rover Safety Cycle, had recently become a fashionable sport for women.

Throughout the 1890s the Workmans were constantly on the move; first they toured through the Mediterranean region, including North Africa, carrying twelve to twenty pounds of luggage, averaging seventy-five kilometers a day, and staying at small, out-of-the-way inns. Their two books, *Algerian Memories: a Bicycle Tour over the Atlas to the Sahara* (1895), and *Sketches Awheel in Fin de Siècle Iberia* (1897), described their trips. In 1897 they sailed for India, toured Ceylon, and the following year set out to bicycle through the subcontinent from south to north and then across the breadth of India, from Cuttack on the east to the Arabian Sea, with many side trips in the interior. When roads were impossible they took trains and often slept in railway waiting rooms or in bare dawk bungalows, the travelers' accommodation at old post stations. Their purpose was to see and photograph the remaining architecture and art of the early civilizations of India, though they appear to have had little interest in Indian society of the time and largely ignored, as much as they could, the complex and colorful life around them. Their well-illustrated book, *Through Town and Jungle: Fourteen Thousand Miles A-Wheel among the Temples and People of the Indian Plain,* appeared in 1904.

Arriving in Kashmir in the summer of 1898, they made their first climbing expedition in the Karakoram Range, going north and east from the capital, Srinagar, into Ladak, as far as the Karakoram Pass. Later the same season they attempted a second trip, at the eastern end of the Himalayas in Sikkim, but gave up after serious trouble with porters. During the cold months they toured Java, but the following June were back in Srinagar organizing their second expedition, well-staffed and well-equipped, and led by a Swiss guide, Mattia Zurbriggen, already experienced in the Karakoram. Fanny Workman was then forty and her husband fifty-two. They made three remarkable climbs of peaks in the 18,000 to 21,000-foot range, measuring the heights themselves using aneroids, and they followed the Biafo glacier to its head at the 17,500-foot Hispar Pass. Fanny Workman's 1899 paper, listed in the bibliography, describes the expedition. Her climb of the peak they named Koser Gunge she claimed as an altitude record for a woman. In 1902 and 1903 they explored the thirty-mile-long Chogo Lungma glacier in the Lesser Karakoram, and in 1906 made a circuit of the peak and glacier group of the Nun Kun massif, where Fanny's climb of Pinnacle Peak, which they measured as 23,350 feet, was claimed as another altitude record for a woman.[14]

Two years later, in 1908, they were back in the Hispar-Biafo region. This time they approached from the Gilgit side on the west, explored the Hispar glacier, and crossed the Hispar Pass to the Biafo glacier where they had climbed in 1899.[15] Their final, most ambitious, and most successful expeditions were in the eastern Karakoram in 1911 and 1912, when they explored the forty-six-mile-long Siachen glacier, one of the most inaccessible outside the polar regions. By then William Workman was in his sixties and his wife in her early fifties. They reconnoitered the southern and western approaches in 1911 and crossed the Saltoro Pass into the valley of the glacier itself. Despite the daunting prospect of having to work from a base camp twenty-five miles to the west of their objective, with very difficult terrain between, they returned the following year, with an especially well-staffed expedition. They had with them Grant Peterkin, an outstanding surveyor, and Sarjan Singh, one of the best men on the staff of the Survey of India, who would carry out the plotting for Peterkin. Instruments had been borrowed from the survey and the Royal Geographical Society. Early in the expedition they suffered a tragic loss when one of their experienced Italian porters fell into a crevasse, but nevertheless the exploration continued. The resulting map was highly professional and free from the miscalculations and topographical inaccuracies that have reduced the value of many of their earlier productions. Fanny's description of the expedition, *Two Summers in the Ice Wilds of Eastern Karakoram: the Exploration of Nineteen Hundred Square Miles of Mountain and Glacier* (1917), contained, in addition to the map, a particularly fine collection of photographs.[16]

In all, the Workmans made eight expeditions, assigning responsibilities to each other fairly and alternating the overall leadership, a point that Fanny, a strong women's rights supporter, stressed in her writings. They were not among the greatest of the Himalayan explorers, and their area of operations, a region included in the regular government survey of Kashmir begun in 1846, had previously been visited by several famous mountaineers, including Henry Godwin Austin, Sir Francis Younghusband, Sir Martin Conway, and Tom Longstaff. Though they generally obtained experienced guides, and though they took great care in organization, their inability to get along with their non-European porters frequently caused them considerable trouble. In addition, their geographical reports regularly aroused a good deal of controversy, largely because they neglected to pay attention to the work of their predecessors. Nevertheless, they were bold and persistent. Fanny Workman's achievement as a pioneering woman climber and explorer was indeed remarkable and was recognized as such in her own time with a number of medals and awards. In between expeditions she lectured extensively to learned societies throughout Europe, and she was one of the few women of her day who were invited to address the Royal Geographical Society.[17]

With the start of the First World War the Workmans' mountaineering activities came to an end. They spent the war years in the south of France, and stayed on there into the 1920s. Fanny died in Cannes, on 22 January 1925, at the age of sixty-six, after eight years of ill health. Her ashes were buried in Worcester, Massachusetts. A suffragist and a determined and uncompromising feminist throughout her life, she left large bequests from her substantial fortune to four women's colleges—Bryn Mawr, Radcliffe, Smith, and Wellesley.[18]

That few turn-of-the-century women should have carried out exploration on the scale of Fanny Workman's undertakings is hardly surprising. In addition to uncommon energy, courage, and physical endurance, such work required either institutional backing, generally beyond the reach of any would-be woman expedition leader of the time, or a private fortune comparable to Fanny's. Likewise, the lack of early American women contributors to other branches of geography is not particularly remarkable. As an academic discipline in the United States, geography in all its branches was just starting in the 1890s and during the early years of this century. Further, the ranks of the geographical profession were recruited largely from geologists, and so the first generation of American geographers worked mainly in physical geography, often geomorphology. Given the fact that only a handful of American women published original work in geology before 1901, their virtual absence from the field of physical geography is not surprising.[19] Interest in Semple's field of social and economic geography lagged in the United States, and training and direction in this area had to be sought mainly in Europe, where the field had long had closer links with history and the humanities. Semple's success is therefore especially notable. She was not the sole American geographer of her time to examine the influence of environment on human movements and social development (both Robert de Courcy Ward of Harvard and Ellsworth Huntington of Yale wrote extensively during the early years of the century on the influence of climate on man), but she was one of a very small band of pioneers who established human geography as an academic discipline and were responsible for its early development in America.[20]

British women

The nine British women whose geography papers are listed in the Royal Society *Catalogue* form the largest national group in the field. Except for Caroline Power and Miss Colthurst, they all published toward the end of the century, mainly in the 1890s.[21] Most of them would have considered themselves travelers and explorers rather than geographers. Three—Isabella Bird Bishop, Mary Kingsley (see chapter 14), and to a lesser extent Annie Taylor—were, and have remained, well-known figures among Victorian women travellers. One author whose publications are not listed by the Royal Society, the writer and commentator on geographical topics Ellen Clerke, has been added to the group. Marion Newbigin, whose pre-1901 papers were all on biological and biochemical topics, is also discussed here. She was the one professional geographer among these women, and indeed is one of the outstanding figures among turn-of-the-century British geographers.

For thirty-two years the editor of the *Scottish Geographical Magazine,* MARION NEWBIGIN[22] (1869–1934) was born in Alnwick, Northumberland. She was one of five daughters among the eight children in the family of James Newbigin, a pharmacist. All of the Newbigin girls were given the best education their father could provide for them. Marion and her sister Maude studied at the Edinburgh Association for the University Education of Women, where the curriculum (beginning about 1868) was essentially equivalent to that available to men in Edinburgh University's faculty of liberal arts.[23] Maude completed her education in Edinburgh, while Marion went on to University College Aberystwyth. She then took classes at the Extramural School of Medicine for Women in Edinburgh and at University College London. Her London B.Sc. was awarded in 1892; five years later she received a D.Sc. (London).

As a student at Edinburgh she was much influenced by the biologist Sir John Arthur Thompson, and her postgraduate research, carried out mainly in the laboratories of the Royal College of Physicians, was on biological and biochemical topics. Coloration in plants, crustaceans, and fish was her special interest. Her monograph, *Colour in Nature: a Study in Biology,* appeared in 1898, and a second book, *Life by the Seashore: an Introduction to Nature Study,* in 1901. The latter was to remain a classic for many years, a revised edition appearing as late as 1931.

For several years in the late 1890s Newbigin taught zoology at the Extra-mural School of Medicine for Women, where she succeeded Thompson. Although remembered as an excellent lecturer she never enjoyed the work, and in 1902, at the age of thirty-three, started a new career, becoming editor of the *Scottish Geographical Magazine,* at that time the most influential geographical periodical in Britain. From this position she exercised considerable influence on the development of early twentieth-century British geography; indeed, she has been described as Britain's leading geographer during the 1920s.[24] Bringing to her subject the insight of an experienced scientist, she emphasized the value of applying biological principles to human and social geography. She was one of the first to make clear the importance of dynamics, balance, and interrelationship in the constant flux of the living world, supporting a view of the ecosystem which was only beginning to come to the fore in the years before the First World War.[25]

She did not confine herself to any one subfield, however. Her editorial work kept her fully aware of developments throughout the subject as a whole, and her own interests, reflected in her many writings, were unusually broad and wide-ranging, especially for a geographer of the time. Between 1911 and her death in 1934, she wrote about twenty-five monographs. Many of these were textbooks for schools and beginning university courses, some were scholarly treatises, and some were meant for the general reader. Her style throughout was always fresh, direct, and concise. One of her first works, *Modern Geography,* in which she discussed the earth as a home and setting for man and his activities, appeared in 1911 and remained in use for many years. A Spanish edition came out in 1916. Both her *Introduction to Physical Geography,* a student standby for years, and *Man and his Conquest of Nature,* one of the earliest British works on human geography, were published in 1912; in the latter she demonstrated what now seems the obvious point that the spread of mankind throughout the globe was strongly influenced by environment.[26]

Her two pioneering works, *Ordnance Survey Maps: their Meaning and Use,* and *Animal Geography: the Faunas of the Natural Regions of the Globe,* came out in 1913. The latter was based largely on her zoology lectures at the School of Medicine for Women. The following year she published an introductory text to world geography, *The British Empire beyond the Seas,* an exposition built on a framework of seasonal climatic phenomena. Political and cultural problems of the eastern Mediterranean occupied her for many years, especially during the First World War. Her scholarly work, *Geographical Aspects of Balkan Problems in their Relation to the Great European War,* appeared in 1915, and was followed by a number of shorter articles on the same topic, mainly in the *Scottish Geographical Magazine.*[27] In 1920 she published *Aftermath: a Geographical Study of the Peace Terms,* an essay in political geography dealing with the territorial questions raised by the Versailles settlement, and a work that prompted geographers to look more closely at the new international order in Europe and the possibilities presented by the League of Nations. Her European travel book, *Frequented Ways,* appeared in 1922; in it she examined human communities of western Europe in relation to the fundamentals of geology (land forms), climate, and vegetation. An enthusiastic traveler herself, she hoped also that her pages would communicate to the reader the great pleasure the experience could bring.

She had a special interest in the "sun-lit lands" of southern Europe. Her *Mediterranean Lands* (1924), although a broad survey intended primarily for upper forms of schools and beginning university courses, was an outstandingly successful discussion of the region from the point of view of historical and human geography. The graphic picture she presented of the history of the great Mediterranean civilizations in relation to their geographical settings demonstrated very clearly the influences of natural environment on man's activity throughout the ages. In her presidential address to the geography section of the British Association in 1922 she covered similar themes, and seven years later, at the South African meeting, she presented the subject of human response to Mediterranean climate type from a worldwide perspective.[28] Her major monograph, *Southern Europe, a Regional and Economic Geography of the Mediterranean Lands* (1932), discussed the great complexity of the area, physically, economically, and socially; it served as a detailed text for the advanced student.

A last work, *Plant and Animal Geography,* which was near completion at the time of her death, was brought out by her sister Florence in 1936. This was a pioneering study in biogeography in which Newbigin, drawing on her many years of work as both biologist and geographer, set out on broad and general lines her ideas for the study of living organisms in relation to their environment, starting from basic questions about the roles of relief, climate, and soil. The book was reprinted eight times over the course of more than three decades, and a preface to the 1968 issue noted that modern quantitative and analytical work in biogeography begun in the 1950s was built on the foundations laid by Newbigin; her study remained the essential introduction to the subject for more than twenty years.[29]

Although she never held a regular geography teaching position, a great deal of her work was directly educational. For a long time she was an examiner in geography for various institutions, including the University of London, and she frequently lectured to students at Bedford College. Her influence on the training of the next generation of geographers was considerable.[30] She was best remembered, however, for her service to geography through her editorship of the *Scottish Geographical Magazine.*

She shared her home in Edinburgh with her sisters, Florence, Alice, and Hilda. Over the years, Florence contributed many of the maps and sketches in her published work. Alice, also a geographer, herself brought out a travel book, *A Wayfarer in Spain* (1926). Hilda held a post in the College of Agriculture. Later they were joined by Maude, who had spent much of her career as the deputy principal of Portsmouth Day Training College. All five of the Newbigin sisters were vigorous and enthusiastic supporters of the feminist cause. Marion died in Edinburgh on 20 July 1934.[31]

ISABELLA BIRD[32] (1831–1904), one of the best known of the British women travelers of Victorian times, was also the first woman to address a meeting of the Royal Geographical Society and one of the first women fellows of that group. She described her many long journeys, in North America, the Pacific, the Middle East and the Far East, in ten books, several of them best-sellers, published between 1856 and 1900.

Born on 15 October 1831, at Boroughbridge Hall, Yorkshire, Isabella was the elder of the two daughters of Edward Bird and his second wife, Dora, daughter of the Rev. Marmaduke Lawson of Boroughbridge. Edward Bird came from a well-to-do middle-class family with a strong tradition of service to the community and to the then energetic and influential Evangelical movement. Among his kinsmen were William Wilberforce, the antislavery campaigner, and John Bird Sumner, Bishop of Canterbury. Bird had started off in law but later took holy orders.

Since Isabella was a sickly child her doctors recommended that she spend as much time as possible out of doors, and so from an early age she rode with her father on his parish rounds in Wyton, Huntingdonshire, and rowed on the river Ouse. She was educated at home, and when barely in her teens taught in her father's Sunday school. At the age of sixteen she had her first article privately published, an essay on the rival claims of free trade and protection. From then on she contributed fairly regularly to various periodicals. In 1850, when she was nineteen, she had an operation to remove a spinal tumor, but it was only partially successful and left her a semi-invalid, suffering severely from depression and insomnia. Her doctors prescribed travel, and she spent seven months in eastern Canada and the United States. When she came home she wrote the first of her full-length books, *The Englishwoman in America* (1856), which was based on her letters to her family; despite competition from a number of similar works by other English women travelers about that time, it was reasonably successful. In 1857, when she was once again in the United States for the sake of her health, she followed a suggestion of her father and gave careful attention to the current religious revival taking place there. Her *Aspects of Religion in the United States of America,* first published as a series of articles, came out in London in 1859.

After the death of her mother in 1866 she lived with her sister Henrietta in Edinburgh and at Henrietta's cottage in Tobermory on the Isle of Mull. Although her health gradually worsened throughout the 1860s, her drive to examine critically and report on what she saw around her remained strong. She became especially interested in the material and spiritual welfare of the crofter communities of the western Highlands, toured the Outer Hebrides, and worked with Lady Gordon Cathcart on schemes for promoting emigration to Canada. In 1866 she crossed the Atlantic to find out how the settlers were getting along. Her pamphlet exposing conditions in the slums of Edinburgh, *Notes on Old Edinburgh,* appeared anonymously in 1869.

By the early 1870s she turned again to extended travel in the hope of alleviating her physical ills, and at the age of forty-one (in 1872) left for Australia and New Zealand. Rather than spending much time there, however, she embarked, on New Year's Day 1873, on a ship bound out of Auckland for the Sandwich Islands, as Hawaii was then known. A hurricane, a strike by stewards, and a desperately ill passenger whom she helped to nurse, made the voyage, to her mind, a most exciting and enjoyable one. During six months in Hawaii she explored the islands and climbed the volcanoes, traveling freely, on horseback, riding astride for the first time in her life, living for the most part in the open air, and savoring her newfound capacity for sustained physical mobility.

From Hawaii she went to San Francisco, and then, working her way east by rail and on horseback through the Sierra Nevada and the Rocky Mountains, reached the upland valley of Estes Park in Colorado. There she lodged with a settler's family for several months, climbed 14,700-foot Long's Peak, and explored the country, much of the time in the company of Jim Nugent, "Rocky Mountain Jim," an Irish-Canadian, formerly a United States government scout, but by then something of a desperado. Despite his not-infrequent bouts of drunkenness and his record of quarrels and crimes, Nugent was an attractive man, intelligent, well-read, and well-mannered. He fell in love with Isabella, and she with him, at least in some measure. She knew she could never marry him, however, his weakness for whiskey making that arrangement entirely out of the question.

During four years at home she wrote her very successful travel books, *The Hawaiian Archipelago: Six Months among the Palm Groves, Coral Reefs and Volcanoes of the Sandwich Islands* (1875) and *A Lady's Life in the Rocky Mountains* (1879). The latter, a collection of letters originally published in the magazine *Leisure Hour*, was especially popular and went through many editions; a French translation was brought out in Paris in 1888 under the title *Voyage d'une femme aux montagnes rocheuses.*

In 1878 Isabella, now a confirmed traveler, set off on the first of her trips of serious exploration, undertakings she was to continue, off and on, over the next twenty years, drawn not only by the excitement of the exotic and mysterious, but by the hope of escaping to some extent the ill health that dogged her at home. She generally chose to visit countries that were just beginning to enter a period of special interest to the West, and her subsequent reports and writings were therefore of considerable interest to geographers and ethnologists as well as the general public. She often succeeded in making close contact with the peoples she moved among, and was a careful and penetrating observer of political matters, moral and religious customs, and, not the least important, trade prospects. In Japan, the country of her first trip, she penetrated into districts hardly, if at all, known to the western world. After working her way through the wild northern interior of the main island, she reached the shores of Tsugaru Strait and crossed over to the northern island of Hokkaido. There she spent six months with the Ainu, a gentle and hospitable people to whom she was much attracted, despite the dirt and the primitive conditions they lived in.[33] From Japan she went to Hong Kong and Canton (where she saw a prison and execution ground about which she wrote a grim and graphic description), and then on to Saigon and Singapore, coming home by way of Egypt. These journeys she described in *The Golden Chersonese and the Way Thither* (1883).

In 1881, shortly after the death of her sister, Isabella married Dr. John Bishop, a gentle, humorous man, ten years her junior, who was very unlike Jim Nugent. Bishop died only five years later, however, and after that Isabella resumed her traveling. With the severing of the last of her close family ties this became almost compulsive and progressively more arduous. About the same time, she came to feel the need for some clear and socially acceptable purpose for her journeys; perhaps also she was personally beginning to find more and more comfort in religion. At any rate, she took up the missionary cause, even undergoing baptism by total immersion as a way of consecrating herself to the work. She also took the practical step of attending classes in medicine at St. Mary's Hospital, London. These preparations completed, in 1889 she set off for India with the object of founding medical missions, the instruments she considered to be the most effective pioneers of Christianity. Working closely with the Church Missionary Society, she established the Henrietta Bird Hospital for Women in Amritsar, in the foothills of northwest India, and the John Bishop Memorial Hospital in Srinagar, Kashmir. She then continued her travels in the region, going eastward as far as Leh, the capital of Ladakh, for two months of exploration that included a side trip north into the valleys of the Shyok and Nubra rivers in the Karakoram Range.[34]

Back down at the hill station of Simla she met Major Herbert Sawyer of the Intelligence Branch of the Indian Army, who was about to start on a reconnaissance of southwest Persia (now Iran). As a border state on India's northwest frontier and a buffer between the British and Russian empires, Persia was of considerable strategic interest to the British. Attempts were being made about this time to interest the Shah in the development of land and in mining schemes in southern Persia. Eager to see the region for herself, Isabella made arrangements to join Sawyer's party for a journey from Baghdad across the mountains to Tehran. The trip was done in midwinter, under terrible blizzard conditions. In the early spring the expedition went south and west into the wild, hilly Bakhtiari country, along the headwaters of the Karum and Diz rivers, finally reaching the city of Borūjerd. From there Isabella headed northwest, with only her own small party, bound for the eastern Turkish port of Trabazon on the southern shores of the Black Sea. This journey was one of the most adventurous and difficult of all her wanderings. Traversing the little-known mountainous regions and high plains of northern Iran and eastern Turkey, past the great lakes Urmia (now Daryacheh-ye Rezaiyeh) and Van, she explored the country of the wild, though often hospitable, Kurdish tribesmen, and visited scattered communities of Armenians and Nestorians, much persecuted by the Kurds. A reviewer of the book she later published on this journey wrote as follows:

> On April 30th, 1890, Mrs. Bishop started on her remarkable journey through the country of the Bakhtiaris and Luristan to Hamadan . . . and on through the district rendered notorious by the outrage of the Kurds and the oppression of the Christian races of Armenia, to Erzeroum and Trebizond. A sufficiently remarkable journey it would have been for men experienced in dealing with Orientals. The first few days' marches were through the pleasant country between Ispahan and the Kuh-i-Sukhta, on the other side of which lies Ardol, the headquarters of the Ilkhani of the Bakhtiaris . . . All will be glad to read Mrs. Bishop's evidence as to the desire felt by the Bakhtiari Khans for a prosperous trade-route through the country, and will feel with her what a regrettable fact it is that Persian jealousy has prevented this country being opened up to foreign commerce . . . Endless dangers and difficulties were met with, and the marvel is that she escaped with her life.[35]

Escape she did, however, and by early 1891 was home again in Britain. The following year she presented two lectures before the Royal Scottish Geographical Society, speaking on her Himalayan experiences at a meeting of its London branch and on her Persian journey at its anniversary meeting.[36] A member of the Scottish society since its founding in 1884, she was now elected a fellow of the Royal Geographical Society as well.[37]

She stayed home only three years, long enough to write a number of articles and her books, *Journeys in Persia and Kurdistan, including a Summer in the Upper Karun Region and a Visit to the Nestorian Rajahs* (1891), and *Among the Tibetans* (1894). In 1894, although then sixty-three, she was on her way again, this time back to the Far East. After visits to Korea and Manchuria and a brief glimpse of Siberia from Vladivostok (which she described as a cross between a city in the American West and a garrison town), she spent a summer in Japan for the sake of her health and finally set off on a journey up the Yangtze River. Seventeen days on a houseboat took her 300 miles upstream to Wanhsien, in Szechwan province, where she disembarked and began a 900-mile, three-month expedition overland by sedan chair, a conveyance from which she frequently stepped down for the exercise of walking. Going generally west in the direction of the Tibet border, she reached Chengtu, and then pushed on through the mountains as far as Somo, the region of the Miao-tze tribes, going further than any of her predecessors, and indeed as far as officialdom would allow. She stayed in the area for some weeks and made special note of the fine quality of the people's stone houses, the equal treatment accorded to women, and the distinct language, employing Tibetan characters.

By 1897 she was home once more, traveling around the country to lecture,[38] and writing more books. *Korea and her Neighbours,* which appeared in 1898, provided a considerable amount of information on internal conditions in Korea and Manchuria. It also included an appendix of tables giving precise details of Korea's trade with foreign countries, both eastern and western—a timely report, since the period was one when the western powers were focusing on the Far East in their drive for new markets and supplies of raw materials. Recognizing her as an expert on the region, the editor of the *Encyclopaedia Britannica* tapped her for a contribution on Korea to a supplement of the encyclopedia brought out in 1902. Her book was reprinted in Seoul in 1970 because it was considered a valuable analysis and criticism, by an astute contemporary eyewitness, of social institutions, living conditions, and the political situation during the last days of the old Korean empire—a clear illustration of the continuing interest and lasting value of her writings, not just to students of western women of the Victorian era, but to historians and anthropologists of the countries she visited.[39] Her description of her China journey, *The Yangtze Valley and Beyond* (1899) was a long and weighty book, profusely illustrated with her own photographs.[40] It presented a dramatic picture of the great river with its rapids, the wayside inns in Szechwan, some of them dangerous places for foreigners, and the majestic mountains and gorges of the country to the west.

After three years in Britain Isabella made one more trip, visiting Morocco and the Atlas Mountains in 1901. Although then almost seventy, she went from Tangier south to Marakkesh on horseback, covering thirty miles a day. She retired to a nursing home in Edinburgh in 1903, and died there on 7 October 1904, after fourteen months of illness, a few days before her seventy-third birthday. During the 1960s and 70s almost all her books were republished with scholarly introductions.

ANNIE TAYLOR,[41] a missionary with the China Inland Mission, was the first European to penetrate into the interior of Tibet since the Jesuits Huc and Gabet went to Lhasa in 1844–1846. She was also the first European woman ever to see the interior of that "Forbidden Land."[42] Her expedition diary was later published by the missionary William Carey as part of his *Travel and Adventure in Tibet* (1902). It recorded her seven-month, 1,300-mile journey from the northwestern border of China, west and south over incredibly difficult country as far as Na-chü, some 150 miles north of Lhasa, and then back to the Chinese border at Tatsienlu.

Annie was born in Egremont, Cumberland, on 17 October 1855, the second child in a large and well-to-do family. Her father, John Taylor, a much-traveled man and a fellow of the Royal Geographical Society, was a director of the Black Ball line of sailing ships. Her mother was of French Huguenot ancestry, the daughter of Peter Foulkes, a Brazil merchant. At some time during Annie's childhood the family moved to Kingston-upon-Thames. She was a delicate child, with a heart complaint, and except for a brief period at Clarence House School, Richmond, had very little formal education. By her early teens she had decided to be a missionary. This plan was strongly opposed by her family, but Annie bided her time. Over a period of ten years she took classes in medicine in London and visited the sick in the slums of London and Brighton. She later maintained that it was the latter experience that made possible her success in more dangerous adventures. In 1884, when she was twenty-nine, she finally sailed for Shanghai, her father paying for her passage and the equipment she needed.

Protestant missions had been active in Chinese coastal cities since the early years of the century, and for some twenty years before Annie went out they had been establishing inland stations as well, following the opening of hitherto forbidden regions in response to pressure from western trading interests. By the 1880s women missionaries were traveling alone into the interior and even taking charge of stations there. Annie spent several months in the port city of Chen-chiang, near the mouth of the Yangtze river, while she mastered the language and learned to live like a Chinese, adopting local dress and customs. Then she moved to An-ch'ing, about 100 miles upriver.

By 1887 she was posted to Lan-chou on the upper reaches of the Huang Ho, near the high country of the Tibetan plateau, and her hitherto vague interest in Tibet, an untouched and therefore specially enticing field for missionary work, strengthened from then on. Illness took her back to Shang-hai for a time, and after a period of recuperation she visited a married sister in Darjeeling, not far from the Tibet-India frontier. She crossed over into the border state of Sikkim, and despite the hostility of the Tibetan frontier authorities, managed to stay there for two years while she perfected her Tibetan. In 1891, with a Tibetan servant Pontso whom she had converted to Christianity, she obeyed the call she heard to go back to Shang-hai and enter Tibet by the long route through China. Her plan was to travel across Tibet, disguised as a Tibetan, through Lhasa to Darjeeling, "claiming the country for the Master." With a small party of four porters and a guide, little equipment but amazing determination, she traveled west and then south through wild and magnificent mountain country, seamed and cut by the headwaters of four of the mightiest rivers of China and South East Asia, the Huang-Ho, the Yangtze, the Mekong, and the Salween. The terrain was some of the most difficult in the world; routes were at best the tracks of herdsmen and traders. Having set off in September, the party had to contend with winter weather, and was continually harassed by brigands and robbed of what little they carried with them. At Na-chü, only three days' journey north of Lhasa, Annie was finally stopped by the Tibetan military authorities; one of her men with whom she had fallen out had given her away. Thanks to her remarkable courage and absolute refusal to be intimidated, neither she nor the two servants who remained with her were harmed. They had to leave Tibet, however, and made their way back to Tatsienlu on the Chinese border as directly as they could.

She returned to Britain for a time, taking Pontso with her, and lectured throughout the country about her travels. Her talks included one at a meeting of the Royal Scottish Geographical Society in December 1893. The main purpose of her lecture tour was to raise support for a Christian mission to Tibet, to be patterned on the China Inland Mission. She succeeded in gathering a group of nine missionaries, and in 1894 this Tibetan Pioneer Band went to Gan'gtok on the Sikkim-Tibet border.[43] The place was exceedingly remote and life there difficult; only Annie stayed on, living in a hut on the mountainside, which she shared with Pontso and his new wife. A year later all three moved over the 14,400-foot Jelep La pass to Yatung, where a market had been established to encourage trade between Tibet and India. Since only traders were allowed residence permits, Annie kept a small store, selling cloth, sweets, and ornaments. She stayed there for about twelve years. In 1899 she was visited by the missionary William Carey, and four years later her sister Susette made the journey across the Jelep La from Darjeeling. A brother and sister-in-law visited her in 1907, and every so often she was joined for short periods by lady missionaries. Some time before 1909 her health began to fail and she went back to England. There seems to be no record of where and when she died.

Papers on geographical topics by six other nineteenth-century British women (Mabel Bent, Mabel Rickmers, Lilly Grove, Olivia Stone, Mary Kingsley, and Ellen Clerke) are listed in the bibliography. Of these six, only Kingsley is especially well known. She could readily have been discussed here, but since the focus of much of her writing was ethnological she has been placed with the ethnologists and anthropologists.

MABEL HALL-DARE BENT[44] did her traveling with her husband, James Theodore Bent, of Baildon House in Yorkshire, an explorer and archaeologist. The daughter of Robert Westley Hall-Dare of Essex, Deputy Lieutenant for County Wexford, she was educated at home by governesses and masters. She married Bent in 1877, and over the course of twenty years, until his death in 1897, accompanied him on exploring trips he made each year to little-known localities. The Bents particularly liked the Greek islands of the southern Aegean Sea, the Kikladhes and Sporadhes (Dodecanese) groups that lie between southeastern Greece and Turkey. They also traveled in Turkey and the Persian Gulf region, Abyssinia and Mashonaland (later Rhodesia). One of their special interests was the possibility of early communication between eastern Africa and southwest Asia. They concluded that ancient remains they examined at sites in both Zimbabwe, Mashonaland, and in northern Abyssinia dated back to pre-Moslem peoples who had come over from southern Arabia in search of trade. Their last journeys were in the mountainous regions of southern Saudi Arabia, expeditions reported by Mabel Bent to the Royal Geographical Society in 1898, after her husband's death from malaria. *Arabia,* the book she published in 1900, was written mainly by her but contained much material from her husband's journals. In 1908 she brought out *Anglo-Saxons from Palestine: or, the Imperial Mystery of the Lost Tribe;* her revision of Arthur Boevey's *The Garden Tomb, Golgotha* appeared in 1931. She died on 3 July 1929.

MABEL DUFF RICKMERS spoke before the geography section of the British Association in 1899, describing her travels in Uzbekistan the previous year. Her companions were her husband, W. Rickmer Rickmers, and Dr. v. Krafft. Starting from Bukhara, they made a two-week horseback trip through steppe and loess country to the mountains of the eastern part of the Khanate, in the province of Baljuan. There they spent five months exploring the surrounding area and pushing south as far as the Oxus River (now the Amu-Darya) and the Afghan frontier. Their return was via Kitab to Samarkand. Mrs. Rickmers's major publication was a lengthy work on *The Chronology of India from Earliest Times to the Beginning of the Sixteenth Century* (1899). She also brought out two translations, both from German: P. Deussen's *The Elements of Metaphysics* (1894), and Elsa Brandström's *Among Prisoners of War in Russia and Siberia* (1929). She was a member of the Royal Asiatic Society, and a fellow of the Royal Geographical Society.

LILLY GROVE[45] (1855–1941) was French by birth, the daughter of Sigismund de Boys. With her first husband, Charles Baylee Grove, a British master mariner, she traveled widely, spending a considerable amount of time in South America. Her two papers in the *Scottish Geographical Magazine* (1893 and 1894) reported observations made in the regions of Tarapaca and Atacama in northern Chile and the archipelagos to the south, especially the large island of Chiloë and its neighbors lying in the latitude of the lower forties. She also discussed Chiloë in a talk she gave to the geography section of the British Association at its 1893 meeting. A remarkably keen and systematic observer, she presented in her papers detailed descriptions of physical features, climate, natural resources, and the occupations and customs of the inhabitants of these regions.

After Charles Grove's sudden death Lilly turned to writing as a means of supporting herself and her two teenage children. She obtained a commission from the Badminton Library to prepare a comprehensive survey of dancing,[46] and research she did for a chapter on primitive dance brought her into contact with the Scottish anthropologist James George Frazer, then at Trinity College Cambridge. She married Frazer, already well-known as the author of *The Golden Bough,* in 1896.

From then on much of her time and energy went to looking out for her husband's welfare and interests, as she saw them, and as far as she could she protected the reclusive Frazer from anything that might interrupt his scholarly work. An energetic and hardworking person, however, Lilly also continued her own writing, producing translations and adaptations of French poems and plays for use in British schools, as well as texts for courses in elementary French grammar and colloquial French. Following the outbreak of the First World War she switched her efforts to translating anti-German propaganda. A great quantity of this was being produced in France, much of it by well-known writers, to keep up spirits on the home front. Among the material she brought out were English versions of André Chiradame's *The Pan-German Plot Exposed* (1916), Philippe Millet's *Comrades in Arms* (1916) and Paul Loyson's *The Gods in Battle* (1917). Another of her major undertakings was finding and supervising French translators for each of her husband's books as these came out. She herself translated the abridged edition of *The Golden Bough* (*Le Rameau d'Or,* 1923) and also *Adonis, Attis, Orisis* (1921). Thanks in large part to her efforts, by the 1920s Frazer's books were almost as well known in France as in Britain. Among her other works was *Leaves from the Golden Bough, Culled by Lady Frazer* (1924), an illustrated, children's version of *The Golden Bough.*

The Frazers lived in Cambridge throughout much of their marriage, but Lilly, a straightforward and very forceful woman who had little experience in her early background of middle-class academic society, was never able to adjust to the expectations of the conservative, sometimes stuffy world she found there. Her difficulties would appear to have had fairly serious consequences for her husband; Frazer's biographer makes the point that not only did Lilly's "mothering" exaggerate her husband's tendency toward helplessness and unworldliness, but her abrasive personality isolated him from his colleagues more and more as the years went on.[47] She died in Cambridge, on 7 May 1941, at the age of eighty-six, a few hours after the death of her husband.

OLIVIA STONE,[48] whose short article on the Canary Islands appeared in *Nature* in 1888, also published two substantial travel books, *Norway in June . . . Accompanied by a Sketch Map, a Table of Expenses and a List of Articles Indispensable to the Traveller in Norway* (1882), and *Teneriffe and its Six Satellites: or, the Canary Islands Past and Present* (1887). Both works were essentially guidebooks for tourists of the 1880s. Stone and her husband had thoroughly explored the travel routes she recommended to her readers and recorded the local history as well.

ELLEN CLERKE[49] (1840–1906) was a regular contributor, over the course of more than twenty-five years, of articles, reviews and a column of "Notes on travel and exploration" to the *Dublin Review,* a prominent and influential Catholic periodical published in London. She was not herself a traveler, but rather an expositor and commentator who brought the latest news of geographical discoveries and related topics to a fairly general audience. Her sources included the latest books, papers presented to the Royal Geographical and regional geographical societies, articles in European geographical journals, parliamentary papers, and reports published in journals of the Catholic foreign missions, particularly *Missions Catholiques.* Although heavily influenced by her strongly conservative viewpoint on the activities of foreign missions and her unquestioning belief in the benefits brought by imperial Britain to those nonwestern societies with which relations were then being established, Clerke nevertheless presented to her readers an impressive amount of current geographical information. She contributed some thirty substantial commentaries on geographical topics to the *Dublin Review,* as well as more than sixty of her regular travel notes. She did not sign all of her articles (a not-uncommon practice at the time), but in the cases where she did not, considerable internal evidence points to her being the author. Her work was characterized by careful reading of source materials and a straightforward factual approach. Although now unquestionably dated in outlook, her essays, taken together, nevertheless provide a valuable overview of the tremendous activity in travel and geographical exploration during the final years of the Victorian era.

Ellen was the older sister of Agnes Clerke, the historian of astronomy (see chapter 10). She was born in the small town of Skibbereen in the south of Ireland, on 20 September 1840. Her father, John William Clerke, manager of the Provincial Bank in Skibbereen, had studied classics at Trinity College Dublin. Her mother was Catherine Deasy, daughter of a wealthy and influential family long resident in the neighboring town of Clonakilty. Ellen was given a sound education at home by her parents; subjects she studied included Latin and Greek, mathematics, and French. She later became fluent in both Spanish and Italian. From about 1867, because of the poor health of Agnes, the ladies of the family spent their winters in Italy, where for the four years 1873 to 1877 Ellen and Agnes lived year-round. Ellen took up the study of Italian history and literature, making good use of public libraries, especially that in Florence. They returned to Britain in 1877, settling in London, where their parents and younger brother, a practicing barrister, had already taken up residence.

Ellen's contributions to the *Dublin Review* on topics with significant geographical content began in the early 1880s with her articles based largely on reports from Catholic missions in Africa, although she regularly reinforced her discussions and geographical descriptions with material drawn from accounts of travelers and explorers. Thus, her long essay, "Madagascar past and present" (1884), was based on eight current books, from which, with considerable skill, she extracted and condensed the essential; likewise her twenty-nine page article on "Abyssinia and its people," which appeared the same year and contained a great deal of geographical information, drew on seven current books. Her special interest in Africa continued throughout the 1890s, two of her particularly interesting commentaries being "Mashunaland and its neighbours" (1894), an account of British expansion into Rhodesia and the development of the region's gold and diamond mines, and "The crisis in Rhodesia" (1896), which, along with an account of the Matabele uprising of that year, provided vivid descriptions of the country.

The Far East, and British possessions in Burma and India were also topics for her essays. She gave much attention to areas of current practical interest, particularly commercial and trading prospects and the related matter of improvements in transportation, especially the opening of canals and the building of railway lines. Her "Maritime canals," discussing Suez and Panama, came out in 1885, and the opening of the Canadian Pacific Railway occasioned her 1888 essay, "The empire route to the East." As early as 1886, in a review of worldwide petroleum resources, she rightly predicted the approach of an "age of oil." Two articles on Arctic exploration, one on Sir John Franklin's travels, were also among her contributions. Overall her work reflected an impressively wide span of interests.[50]

A sociable and outgoing person, she took an active interest in the British Association, and gave a short talk to its geography section in 1891—a discussion of aborigines of Western Australia living around the Spanish Catholic settlement of New Norcia, north of Perth. The talk is perhaps as interesting now for the light it sheds on current attitudes toward non-Europeans as for its information on aboriginal life and customs.

Ellen was not the only member of her family with geographical interests. Her first cousin Henry Hugh Deasy (1866–1947), a British Army officer, carried out exploring and survey work in Central Asia in the 1890s and was awarded the Founders Gold Medal of the Royal Geographical Society for his achievements.[51] Ellen's family and Deasy's maintained close connections, and Deasy's work would almost certainly have provided added stimulus for Ellen's efforts.

In addition to her geographical and travel writing, she authored two popular astronomy books, *Jupiter and his System* (1892) and *The Planet Venus* (1893). She was also on the permanent staff of the *Tablet,* another notable Catholic journal, conservative in its outlook on political and religious questions. Her contributions to its pages dealt mainly with current Italian and German politics, on which she kept well versed. It may well be, however, that her most lasting work was in the area of Italian history and literature. She contributed to Richard Garnett's *History of Italian Literature* (1898), translated Italian works into English, and published articles on Italian life and literature in the literary monthly, *Cornhill Magazine.* Four years before her death she brought out a novel set in Italy, *Flowers of Fire;* it contained a description of the 1872 eruption of Mount Vesuvius, which she had witnessed. She died in London on 2 March 1906, at the age of sixty-five.

Isabella Bird Bishop, Annie Taylor, and the other travelers discussed here were a few of the many British women of the latter part of Queen Victoria's reign who undertook journeys to remote and primitive countries.[52] Most were middle-class, with moderate but assured financial resources. Those who were single found in travel a way to taste independence and take advantage of opportunities for freedom of action largely beyond what they as women could expect within their own society. Both Taylor and Bird Bishop, the two most independently adventurous, were also influenced by religious concerns. For Taylor the Christian missionary cause would seem to have been the driving force for all her undertakings. In Bird Bishop the real stimulus to action most probably lay elsewhere, in her sheer restlessness and consuming passion for seeing new and exotic places, and in her quest for the relief from physical pain that travel brought her; but the missionary cause at the very least provided her with an acceptable purpose and rationale for her journeys. Considering their generally limited educational backgrounds, all these women did remarkably well in the business of collecting information and then making it available to the public through lectures and books. The contributions their fieldwork made to cultural geography and ethnology were notable; then new and expanding disciplines, these fields still had room for the self-taught amateur.

Two-country comparison

Each country at this period had only one "professional" woman geographer contributing to the literature and taking part in the academic development of the field—Semple in the United States and Newbigin in Britain. Both, however, were influential. They both had particular interests in human and social geography and pioneered the introduction of that branch of the field into academic studies in their respective countries. They also had in common special interests in the geography of the Mediterranean region and both wrote at length on the topic. Semple, however, came to her subject from a background in classical history and concentrated her efforts almost exclusively on cultural and historical geography. Newbigin, on the other hand, starting as a research biologist with strong ecological interests, ranged across the whole geographical spectrum—from physical geography and ordnance survey maps, through animal geography, to historical, cultural, and political geography.

Among the travelers and explorers, American Fanny Bullock Workman was the only one who aspired to surveying, plotting, and map production on the trips she led. The financial outlay of a Workman-style Himalayan mountaineering expedition was beyond the resources of any of the British women anyway (although another American with great enthusiasm for exploration of remote regions and the wealth to finance a large expedition was May French Sheldon—see chapter 14). The most outstanding of the British women travelers and explorers, Bird Bishop, Kingsley (chapter 14), and to a lesser extent Taylor, focused especially on social conditions, past and present, and western-nonwestern interactions; they put considerably less effort into establishing precise details of physical geography, Workman's objective in her major Himalayan expeditions.

Ellen Clerke, the prolific commentator on geographical topics, does not have an obvious American counterpart. Her success in Britain is not surprising, considering the tremendous amount of geographical exploration being carried out by the British at the time, and the great popular interest in the discoveries being made.

Notes

1. Allen D. Bushong, "Ellen Churchill Semple 1863–1932," in *Geographers: Biobibliographical Studies,* ed. T. W. Freeman (London and New York: Mansell Publishing Ltd., 1977–), vol. 8, 1984, pp. 87–94, and "Unpublished sources in biographical research: Ellen

Churchill Semple," *Canadian Geographer,* **31** (1987): 79–81; anon., "Semple, Ellen Churchill," in *NAW,* vol. 3, pp. 260–2; John K. Wright, "Miss Semple's 'Influences of Geographic Environment': notes toward a bibliobiography," *Geographical Review,* **52** (1962): 346–61; Wallace W. Atwood, "An appreciation of Ellen Churchill Semple, 1863–1932," *Journal of Geography,* **31** (1932): 267; Mildred Berman, "Sex discrimination and geography," *Professional Geographer,* **26** (1974): 8–11; Charles C. Colby, "Ellen Churchill Semple," *Annals of the Association of American Geographers,* **23,** (1933): 229–40.

2. Semple's M.A. thesis was entitled, "Slavery, a study in sociology."

3. Semple had many extracurricular interests in Leipzig and took special pleasure in the opportunities she had to attend first-class musical events. She also enjoyed her social contacts at the American church and at frequent receptions at faculty homes. The latter gave her further opportunities to get to know lecturers and fellow students (Bushong, "Unpublished sources," p. 80).

4. Ellen C. Semple, "The Anglo-Saxons of the Kentucky mountains: a study in anthropogeography," *Geographical Journal,* **17** (1901): 588–623, reprinted in *Bulletin of the American Geographical Society,* **42** (1910): 561–94.

5. Ellen C. Semple, "The influence of geographical environment on the lower St. Lawrence," *Bulletin of the American Geographical Society,* **36** (1904): 449–66; "North-shore villages of the lower St. Lawrence," in *Zu Friedrich Ratzels Gedächtnis* (Leipzig, 1904), pp. 349–60.

6. Semple's presence among the charter members of the Association of American Geographers should not be taken as an indication of that organization's ready acceptance of women. The association was small and elitist and in large measure wished to remain that way. A Ph.D. in geography was neither a required nor a sufficient qualification for membership; significant research output was what was looked for. Nonmembers could attend meetings, and many women did, particularly high school teachers who were interested in geography education. However, even well after the turn of the century, meetings at which important matters were to be discussed were not infrequently scheduled as "smokers," a name generally taken to indicate a men's meeting, and so a woman could only be present if she was willing to ignore the social conventions of the time (Allen D. Bushong, "Some aspects of the membership and meetings of the Association of American Geographers before 1949," *Professional Geographer,* **26** (1974): 435–8, on p. 437; Rossiter, *Women Scientists in America,* p. 93).

7. Anon., *Journal of Geography,* **10** (1911): 33–4, on p. 33.

8. E. A. Ackerman, "Some concepts of Ellen Churchill Semple: their meaning today," unpublished paper presented at the fifty-seventh annual meeting of the Association of American Geographers, East Lansing, Michigan, 30 August, 1961; excerpts are quoted by Bushong, "Ellen Churchill Semple," pp. 90–1.

9. See "Influences of geographic conditions upon Japanese agriculture," *Geographical Journal,* **40** (1912): 589–603; "Japanese colonial methods," *Bulletin of the American Geographical Society,* **45** (1913): 255–75.

10. Semple's salary at Clark University was substantially less than that of her male colleagues, the university trustees being of the opinion that, as a woman without dependents, she did not need a man's salary. She considered this argument to be fifty years out of date and quite inappropriate in the circumstances (Berman, "Sex discrimination," p. 8).

11. On the other hand, classical scholars took considerable interest in the work (Bushong, "Ellen Churchill Semple," p. 91).

12. Among Semple's publications not already mentioned are the following: "Mountain passes: a study in anthropogeography," *Bulletin of the American Geographical Society,* **33** (1901): 124–37, 191–203; "Geographic influences on the development of St. Louis," *Journal of Geography,* **3** (1904): 290–300; "Emphasis upon anthropogeography in schools," ibid, **3** (1904): 366–74; "Mountain peoples in relation to their soil: a study in human geography," *Geographical Teacher,* **4** (1905): 417–24, and *Journal of Geography,* **4** (1905): 417–24, "Geographical research as a field for women," in *The Fiftieth Anniversary of the Opening of Vassar College, October 10–13, 1915, a Record* (Poughkeepsie: [n.p.], 1916, pp. 70–80); "The regional geography of Turkey: a review of Banse's work," *Geographical Review,* **11** (1921): 338–50; "Alzira Ah Fi Buldan Al-Bahr Al-Mutawassit Quadiman" (["Agriculture in the ancient lands of the Mediterranean countries"], *Al-Kulliyah [Middle East Forum],* **16** (1930): 107–12; "Promontary towns of the Mediterranean," *Home Geographic Monthly,* **1** (1931): 30–5.

13. Elizabeth Knowlton, "Workman, Fanny Bullock," in *NAW,* vol. 3, pp. 672–4; Dorothy Middleton, *Victorian Lady Travellers* (London: Routledge and Kegan Paul, 1965), pp. 75–89; Luree Miller, *On Top of the World. Five Women Explorers in Tibet* (London: Paddington Press Ltd., 1976), pp. 101–29; Marion Tinling, *Women into the Unknown: a Sourcebook on Women Explorers and Travellers* (Westport, Conn.: Greenwood Press, 1989), pp. 305–11; Jane Robinson, *Wayward Women: a Guide to Women Travellers* (Oxford: Oxford University Press, 1990), pp. 30–1.

14. The height of Pinnacle Peak was later established as 22,737 feet, but it was still high enough so that Fanny Workman's climb remained a world altitude record for women for twenty-eight years, a record she held in fierce competition with her fellow American Annie Peck (1850–1935). (Peck was known for her climbing in Mexico and South America in the 1890s and the early years of this century.) For accounts of the Workmans' 1902–1906 expeditions see, for instance, Fanny Bullock Workman, "Ascent of the great Chogo Loongma glacier and other climbs in the Himalayas," *Appalachia,* **10** (1904): 241–55; "Exploration des glaciers du Kara-Korum," *Géographie,* **9** (1904): 249–56; (with William Hunter Workman) "Climbing in the north-west Himalayas," *Scottish Geographical Magazine,* **20** (1904): 47–9; "First exploration of the Hoh Lumba and Sobson glaciers. Two pioneer ascents in the Himalayas," *Geographical Journal,* **27** (1906): 129–41; "Exploration du Nun Kun," *Géographie,* **15** (1907): 93–102; "Exploration and climbing in the Nun Kun Himalaya," *Scottish Geographical Magazine,* **24** (1908): 1–14.

15. See Fanny Bullock Workman, "Further exploration in the Hunza-Nagar and the Hispar glacier," *Geographical Journal,* **32** (1908): 495–6; "The Hispar glacier. I. Its tributaries and mountains," ibid, **35** (1910): 105–15.

16. The Siachen expedition was also reported in the geographical journals. See, for example, Fanny Bullock Workman, "Some notes on my 1912 expedition to the Siachen or Rose glacier," *Geographical Journal,* **40** (1912): 615–20, and *Scottish Geographical Magazine,* **29** (1913): 13–17; "Survey of the Siachen glacier," *Bulletin of the American Geographical Society,* **44** (1912): 897–903; "The exploration of the Siachen or Rose glacier, eastern Karakoram," *Geographical Journal,* **43** (1914): 117–48, and *Géographie,* **27** (1913): 129–31; **29** (1914): 163–77.

17. William Workman was elected a fellow of the Royal Geographical Society, but Fanny never was, although she applied when fellowship was opened to women in 1913 (see also n. 37 below). She did, however, hold a membership in the Royal Asiatic Society (Dea Birkett, *Spinsters Abroad. Victorian Lady Explorers* (Oxford: Basil Blackwell, 1989), pp. 226–7). She is said to have felt she had met considerable antagonism from men scientists and mountaineers (Knowlton, "*Workman, Fanny Bullock,*" p. 674; Miller, *On Top of the World,* pp. 126–8). Middleton (*Victorian Lady Travellers,* p. 76) remarked

that the fact that Fanny, rather than her husband, was clearly the leader in many of their joint undertakings would have offended, even scandalized, the male-dominated society of British India of her time.

18. The Workmans' only daughter, Rachel, was also a woman not of the ordinary mold. Educated at Cheltenham Ladies' College (a well-known girls' boarding school in Gloucestershire) and at the Royal Holloway College for Women, London, she specialized in geology and took a London B.Sc. She then carried out postgraduate research in mineralogy and igneous petrology at Edinburgh University, at the Royal School of Mines in London, and at the Mineralogical Institute in Oslo. Her published work included studies on the rocks of Alnö in southern Norway and on the petrology of the Eildon Hills in Roxburghshire, Scotland. She was a member of the Geological Society of Sweden and one of the first group of eight women elected to fellowship in the London Geological Society in May 1919. In 1911 she married Sir Alexander MacRobert of Douneside and Cromar, director of the Cawnpore Woollen Mills Company, and a founder of the British India Corporation. She became one of the directors of the latter when her husband died in 1922. Her three sons were all killed flying, the oldest in an accident in 1938, and the two younger ones on active service in the Royal Air Force in 1941. Thereafter she presented a bomber and four fighter airplanes to the R.A.F., and gave her country house for use as an R.A.F. hospital and rest center for the duration of the Second World War. She died in 1954 and was buried in the garden of her house at Douneside (obituary, *Proceedings of the Geological Society,* no. 1529 (1955): 146).

19. For one early American woman geologist who was also interested in geography, Luella Owen, see chapter 12. Owen published studies in the early years of this century on river action and on the potential uses of rivers.

20. John Kirtland Wright, *Geography in the Making. The American Geographical Society 1851–1951* (New York: American Geographical Society, 1952), pp. 125, 171, 271. Wright points to Semple, Huntington, Ward, and Griffith Taylor as being the leaders in the development of human geography in America.

21. No biographical information has been uncovered about either Power or Colthurst. Caroline Power's letter to her uncle describing a visit to the island of Ascension in November 1834 was published in the Royal Geographical Society's *Journal* the following year. Her six-page description provides a considerable amount of information about this small, isolated, volcanic island, one of the peaks of the southern section of the Mid-Atlantic Ridge. The sole inhabitants at the time were a garrison company of Royal Marines, who were gradually developing water resources and bringing some of the land under cultivation. The island was important as a water and provisioning stop for British ships. The climate being excellent, it was also used as a place of recuperation for sick men from British stations in West Africa. Miss Colthurst's paper on the use of various scales for measuring vertical distances was read before the Royal Geographical Society by geologist George Greenough in 1849.

22. Obituaries in the *Scottish Geographical Magazine,* **50** (1934): 331–3 and the *Geographical Review,* **24** (1934): 676; *Who was Who,* vol. 3, pp. 997–8; William N. Boog Watson, "The first eight ladies," *University of Edinburgh Journal,* **23** (1967–1968): 227–34; K[enneth] C[harles] Edwards, "Geography in a University College (Nottingham)," in *British Geography, 1918–1945,* ed. Robert W. Steel (Cambridge: Cambridge University Press, 1987), pp. 90–9, on pp. 92–4; Mary R. S. Creese, "Newbigin, Marion Isabella," *DNB,* "Missing Persons" vol., 1993, pp. 492–3. See also chapter 3.

23. The women's classes were conducted by university professors or lecturers, and examinations were at the same level as those given men students. Edinburgh University did not open its degrees to women until 1893.

24. Edwards, "Geography in a University College," p. 93.

25. See H. F. F[leure], "Dr. Marion Newbigin. A Tribute," in Marion Newbigin, *Plant and Animal Geography* (London: Methuen, 1936) p. vii. See also Newbigin's paper, "The origin and maintenance of diversity in man," *Geographical Review,* **6** (1918): 411–20.

26. Newbigin's book appeared about the same time as American Ellen Semple's *Influences of Geographical Environment* (see above).

27. See, for instance, "Macedonia: the Balkan storm centre," *Scottish Geographical Magazine,* **31** (1915): 636–51; "The interrelations of Europe and Asia as exemplified in the Near East," ibid., **32** (1916): 216–27; "Balkan outlets in the present and in the future," *Geography Teacher,* **8** (1916): 333–40; "Constantinople and the Straits: the past and the future," *Scottish Geographical Magazine,* **33** (1917): 507–15; "The problem of the South Slavs (Yugoslavs)," ibid., **35** (1919): 1–15.

28. See Marion I. Newbigin, "Human geography: first principles and some applications," *British Association Reports,* (1922): 94–105, and "The Mediterranean climatic type: its world distribution and the human response," ibid., (1929): 345 (summary only). (American Ellen Semple's work, *The Geography of the Mediterranean Region,* a work focused on the historical geography of the region, appeared in 1931—see above.)

29. See Marion Cole, Preface to the 1968 reprinting of Newbigin's *Plant and Animal Geography.* The manuscript of the first (1936) edition of this work was revised and portions completed by the geographer H. J. Fleure of Manchester University. Margaret Dunlop of the Manchester Literary and Philosophical Society contributed sections on soils and climate.

30. See Marion Newbigin, "The training of the geographer: actual and ideal," *Scottish Geographical Magazine,* **41** (1925): 27–36.

31. Among Newbigin's monographs not mentioned in the text are: *Outlines of Zoology,* revision of the 4th enlarged ed. of the work by Sir John Arthur Thompson (Edinburgh: Hodder and Stoughton, 1906); *Tillers of the Ground* (London: Macmillan, 1910); *An Elementary Geography of Scotland* (Oxford: Clarendon Press, 1913); *A New Geography of Scotland for Secondary and Higher Grade Schools* (London: H. Russel, 1920); *The City of Glasgow, its Origin, Growth and Development,* hon. ed., John Gunn, ed. Marion Newbigin (Edinburgh: Royal Scottish Geographical Society, 1921); *Commercial Geography* (London: Williams and Norgate, 1924); *Canada, the Great River, the Lands and the Men* (London: Christophers, 1926); *Mediterranean Climate Type, its World Distribution and Human Response* ([n.p.], 1929); *A New Regional Geography of the World* (London: Christophers, 1929, and later eds.). Newbigin's articles not already mentioned include: "The Kingussie district, a geographical study," *Scottish Geographical Magazine,* **22** (1906): 285–315; "The Swiss Valais: a study in regional geography," ibid., **23** (1907): 169–91, 225–37; "Race and nationality," *Geographical Journal,* **50** (1917): 313–35; "Some aspects of the industrial revolution in Western Europe," *Scottish Geographical Magazine,* **34** (1918): 251–63; "The Mediterranean city state in Dalmatia," *British Association Reports,* (1921): 430 (abstract only); "Economics [of Roumania]" in *Bulgaria and Roumania,* ed. Albert E. W. Gleichen (New York: Houghton Mifflin, 1924), pp. 293–314.

32. Middleton, *Victorian Lady Travellers,* pp. 107–27; Miller, *On Top of the World,* pp. 71–99; Tinling, *Women into the Unknown,* pp. 47–55; Birkett, *Spinsters Abroad*; Robinson, *Wayward Women,* pp. 81–3; anon., "Mrs. Bishop," *Geographical Journal,* **24** (1904): 596–8; *DNB,* 2d Supplement, vol. 1, pp. 166–8.

33. Isabella Bird, *Unbeaten Tracks in Japan: an Account of Travels in the Interior, including Visits to the Aborigines of the Yezo and the*

Shrines of Nikko and Ise, 2 vols. (London: John Murray, 1880; reprint 1971; Japanese trans., *Nikon Okuchi Kiko,* Tokyo, 1973).

34. Isabella Bishop's visit to this region was some ten years before the first expedition to Kashmir of American Fanny Workman (see above).

35. J. A. D., review of Isabella Bishop's *Journey in Persia and Kurdistan, Proceedings of the Royal Geographical Society,* **14** (1892): 258–60, on p. 259.

36. Bishop also spoke on her Persian trip at the 1891 meeting of the British Association (see "The Bakhtiari country and the Karun River," *British Association Reports,* (1891): 722, title only) and at the 1892 meeting on her travels in the Himalayas ("Travels in Lesser Thibet," ibid., (1892): 812, title only).

37. The election of Isabella Bishop and several other women to fellowship in the Royal Geographical Society in 1892 did not in fact mark the permanent opening of the society to women. Dorothy Middleton described the sequence of events and decisions relating to this matter in the first chapter of her *Victorian Lady Travellers* (pp. 11–15). Initial action followed from Mrs. Bishop's address on Tibet to the recently established London branch of the Scottish Geographical Society. The meeting at which this address was given was such a brilliant event that the R. G. S. became alarmed at the success of its rival and opened its own meetings to the members, including women, of all other British geographical societies. Further, the R. G. S. council, acting on its own, agreed to the admission of women into the society on the same footing as men, and in November 1892 fifteen were accepted. This action was challenged by a group of more conservative fellows. A special general meeting of the society, packed by the conservatives, voted against the admission of women, although a referendum of the whole fellowship approved the proposal by two and a half to one. A heated debate ensued in the correspondence columns of *The Times,* legal opinion was taken, and a compromise, of sorts, finally arrived at; the twenty-two women already accepted could stay, but no more were to be admitted. Twenty years later, in 1913, the matter was again taken up, and a decision reached in favor of admitting women.

38. Isabella Bishop's 1897 talk to the Royal Geographical Society on her journeys in Szechwan (see bibliography) was the only full-scale lecture presented to the society by a woman until 1905, when Fanny Workman spoke on her travels in the Himalayas.

39. See Tae Sun Park, "Plan for reprints of books on Korea," introductory note in Isabella Bishop, *Korea and her Neighbours* (Seoul: Yonsei University Press, 1970), pp. i–ii, and Pow-key Sohn, "Foreword," ibid., pp. ii–v.

40. Bishop also published a 126-page book of photographs, with commentary, from her China trip: *Chinese Pictures; Notes on Photographs made in China* (London: Cassell, 1900).

41. Middleton, *Victorian Lady Travellers,* pp. 107–27; Miller, *On Top of the World,* pp. 47–69; Robinson, *Wayward Women,* pp. 172–3.

42. See *High Road in Tartary. An Abridged Revision of Abbé Huc's Travels in Tartary, Tibet and China during the Years 1844–56,* ed. Julie Bedier [pseud. for Sister Mary Juliana Bedier] (New York: Scribner's Sons, 1948), and [Rev.] Graham Sandberg, *The Exploration of Tibet* (London: W. Thacker, 1904).

43. See Annie R. Taylor, *The Origin of the Tibetan Pioneer Mission, together with Some Facts about Tibet* (London: Morgan and Scott, 1894; reprinted in enlarged form as *Pioneering in Tibet*). Taylor accepted only male missionaries for the work in Tibet, since she considered conditions too difficult for women. One of the men took his wife and small daughter with him.

44. *Who was Who,* vol. 3, 1941, p. 98; *DNB,* vol. 22, Supplement, p. 179, entry for Bent, James Theodore; anon., obituary for J. Theodore Bent, *Geographical Society Journal,* **9** (1897): 70–1.

45. Robert Ackerman, *J. G. Frazer. His Life and Work* (Cambridge: Cambridge University Press, 1987).

46. Lilly Grove, *Dancing* (London: Badminton Library, 1895). This work is now regarded by dance historians as a pioneering classic.

47. Ackerman, *Frazer,* p. 126.

48. Robinson, *Wayward Women,* pp. 195–6.

49. M. T. Brück, "Ellen and Agnes Clerke of Skibbereen, scholars and writers," *Seanchas Chairbre* (*Chronicles of Carbery*), **3** (1993): 23–43.

50. Ellen Clerke's *Dublin Review* articles up to 1900 are listed in *The Wellesley Index to Victorian Periodicals 1824–1900,* vol. 2., ed. Walter Edwards Houghton (Toronto: University of Toronto Press, Routledge and Kegan Paul, 1972). References for those mentioned here are as follows: "The voyage of the *Vega,* and its results," **7,** s. 3 (1882): 219–320; "Madagascar past and present," **11,** s. 3 (1884): 117–49; "Abyssinia and its peoples," **12,** s. 3 (1894): 316–45; "Maritime canals," **14,** s. 3 (1885): 1–19; "The future of petroleum," **16,** s. 3 (1886): 145–53; "The empire route to the East," **19,** s. 3 (1888): 274–93; "Sir John Franklin and the Far North," **26,** s. 3 (1891): 266–82; "Mashunaland and its neighbours," **5,** s. 4 (1894): 145–66; "The crisis in Rhodesia," **10,** s. 4 (1896): 256–70.

51. *Who was Who,* vol. 4, p. 301; information from Dr. Mary Brück. Deasy's book, *In Tibet and Chinese Turkestan,* appeared in 1901 (New York: Longman's Green; London: [n.p.]).

52. See Robinson, *Wayward Women,* for an indication of the great number of British women travelers of this period.

Chapter 14

FROM THE ZUÑI OF NEW MEXICO TO THE FANG OF GABON: EARLY CONTRIBUTIONS BY WOMEN TO ETHNOLOGY AND ANTHROPOLOGY

Articles on anthropological topics make up about 2 percent of the papers by nineteenth-century women scientists listed in the Royal Society *Catalogue.* The largest contributions came from the United States and France, with sixteen and seventeen papers, respectively; British, German, Italian, and Russian workers together produced a further twenty articles.

American women

Three of the seven Americans whose papers are listed in the *Catalogue* (Smith, Fletcher, and Stevenson) are relatively well known for their writings on native American tribal life and culture. All three held marginal positions at anthropological museums, Fletcher at Harvard's Peabody Museum, Smith and Stevenson at the Bureau of American Ethnology at the Smithsonian Institution. Two others (Babbitt and French Sheldon) were independent workers. The remaining two (Carter Cook and McGee) had major interests in other areas—see chapters 1 and 7, respectively.

ERMINNIE PLATT SMITH[1] (1836–1886) began her fieldwork in anthropology in the late 1870s. As secretary of the anthropology section of the AAAS in 1885, she has the distinction of being the first woman to hold office in that organization. Born in Marcellus, near Syracuse in western New York State on 26 April 1836, she was the ninth of the ten children of Ermina (Dodge) and Joseph Platt. Her father, whose people had been among the original settlers of the region in the eighteenth century, was a successful farmer and a Presbyterian deacon. He became the dominant influence in her early years (her mother having died when she was two) and he encouraged her interests in rocks and plants. For three years, 1850 to 1853, she studied at Emma Willard's Female Seminary in Troy, New York, where she did particularly well in languages. In 1855, at the age of nineteen, she married a wealthy Chicago lumber dealer and merchant, Simeon H. Smith, and for a number of years spent most of her time on family concerns. She managed to keep up her interests in geology, however, and assisted in a project of classifying and labeling specimens of American minerals for display in European museums. The family later moved to Jersey City, where Simeon Smith was a stockyards official and then city finance commissioner.

Erminnie traveled widely, and during a four-year stay in Germany, where her four young sons were being educated, she took the opportunity to further pursue her early interests. She studied crystallography and German literature in Strasburg and Heidelberg and completed a two-year course in mineralogy at the Mining Academy in Freiberg, Saxony. In addition she investigated the amber industry of the Baltic Coast.

After her return to Jersey City, her confidence and enthusiasm deepened by her German experience, she organized displays of mineral specimens in her home and began to offer parlor lectures on geological and cultural topics. These proved to be so popular that in 1879 she founded the Aesthetic Society of Jersey City, a group that met monthly and drew audiences of as many as 500. For the rest of her life she presided over this society and gave considerable time and energy to the preparation of the varied program it offered. She was also a member of Sorosis, a club formed in New York City in the early 1870s by a number of upper-middle class women with intellectual interests; for four years she directed its science programs. In 1879, probably with the encouragement of her cousin, Frederick Ward Putnam, curator of Harvard's Peabody Museum, she presented her first two papers at a meeting of the AAAS; one was a discussion on jade, and the other described methods for cutting and polishing gemstones. A year later her paper on amber, the result of some of her European studies, appeared in the *American Naturalist.*

Her contacts at the AAAS introduced her to work being done in American Indian ethnography, a subject she found of special interest in part because she had grown up close to the Onondaga Indian reservation near Syracuse, learned some of the language of the Indian children she played with, and occasionally attended tribal dances. In the summer of 1880, with some financial assistance from the newly established Bureau of Ethnology and introductions from Lewis Henry Morgan, an anthropologist who had pioneered ethnological work among the Iroquois, she started field studies at the Tuscarora reservation near Lewiston, New York. Subsequently, for four more summers, until 1885, she visited the scattered reservations in New York State and southeastern Canada collecting information on the history, customs, folklore, and language of the Iroquois Indians of the Six Nations, valuable work, because Iroquois life was even then changing fast. She was probably the first anthropologist to work with native "informants" in ethnographic investigations, and though in the future this method of collecting data was not always reliable (informants sometimes making up what they thought collectors would like to hear), it was an important step forward at the time. One of her collaborators, John Napoleon Brinton Hewitt, a part-Tuscarora Indian who was her interpreter and amanuensis in her earliest work, went on to carry out notable ethnographic studies of his own, continuing her research among Iroquois and Algonquin peoples and receiving notable professional recognition for his work.[2]

The first of Smith's Iroquois-language studies was published under bureau director John Wesley Powell's name in the Bureau of Ethnology's *First Annual Report, 1879–80.* A second article, entitled "Myths of the Iroquois," which appeared under her own name in the *Second Annual Report, 1880–81,* established her scholarly reputation. She held to the theory, later disputed by anthropologist Franz Boas, that sociocultural development can be understood in terms of evolutionary stages, the study being comparable to the investigation of geological strata. Iroquois mythology, a reflection of an earlier stage in man's cultural development, thus offered clues about the evolution of the human intellect.

Accounts of her continuing research were included in Powell's annual reports until 1885. In addition she regularly presented papers at annual meetings of the AAAS, sometimes speaking two or three times at a meeting, mostly on her language studies. Although it has never been published, her dictionary of the Iroquois language, a compilation of over 15,000 words, is available for research use at the Smithsonian Institution.

She worked in anthropology for only eight years, but nevertheless received considerable recognition. In addition to being the first woman to hold office in the AAAS, she was the first woman elected a fellow of the New York Academy of Sciences and was a member of the London Scientific Society. She died at age fifty of heart disease and cerebral embolism, in Jersey City, on 9 June 1886. Two years later her friends at Vassar College endowed in her name an annual student prize. Interestingly, it was for work in geology and mineralogy rather than ethnography.

ALICE FLETCHER[3] (1838–1923) began her ethnological work about the same time as Smith but had a much longer career. Over the course of four decades she carried out detailed studies of Plains Indians, which are counted among the classics of early American anthropology.

The daughter of Thomas Gilman Fletcher and his second wife, Lucia Adeline (Jenks), she was born on 15 March 1838, in Havana, Cuba. Thomas Fletcher, a New York lawyer of New Hampshire background, had graduated from Dartmouth College. By the time of Alice's birth he was suffering from tuberculosis, and despite several winters in the milder climate of the Caribbean died in 1839. His widow and two children stayed on in New York, taking a house in the new suburb of Brooklyn Heights. By the age of eight Alice was attending the newly opened Brooklyn Female Academy, later renamed the Packer Collegiate Institute. The school offered a solid curriculum, including courses in science subjects, and claimed to be as good as collegiate institutions for young men. Sometime during her preteenage years her mother remarried, and although no records of her early life have survived, scraps of information collected by her biographer indicate an extremely troubled and unhappy relationship with her stepfather; there may even have been sexual abuse.[4] When she was about eighteen she escaped from the family, taking a job as governess with a well-to-do neighbor, but the psychological scars from her early struggle, possibly a protracted one fought over several years, remained with her for life and strongly affected her ability to form close interpersonal relationships.

Her work as a governess came to an end around 1870, when she was in her early thirties. About that time she joined Sorosis, becoming the club's secretary in 1873. From that position she took a leading role in the move to found the Association for the Advancement of Women. The latter included in its membership some of the best-known and most influential women in the country, among them the astronomer Maria Mitchell and the women's activist Julia Ward Howe. About 1877, driven by financial necessity, she turned to public lecturing, using her contacts in the association to arrange talks before women's groups in the northeastern states. There was a wave of widespread public interest in American history at the time, and Fletcher offered lectures on America's European traditions and the early colonial period. When she extended her circuit westward to Ohio, Wisconsin, and Minnesota, she discovered that the most popular topics concerned the native societies of the pre-Columbian period. Her information sources for these lectures were her own reading and help she was able to get from staff members at such places as the Smithsonian Institution and the Peabody Museum. Frederick Putnam of the Peabody, anxious that public lecturers in his field be accurately informed, encouraged her to come and study at the museum. By 1880 she was a regular visitor, and from Putnam and his assistants she learned the fundamentals of archaeology and general ways of going about scientific work.

In 1881 she got a chance to do something she had for some time contemplated, namely live among Indians for an extended period in order to learn something of their way of life. She spent the winter of 1881–1882 on the Omaha reservation in Nebraska, and while there was able to help members of the tribe draft a petition to the United States Senate asking for full title to lands allotted them a decade previously but never made secure. The question of land security was causing the Indians extreme anxiety at that time when tribes were being continually moved westward to make room for white settlement. Returning to Washington in the spring of 1882, Fletcher lobbied directly for the passage of the Omaha Bill and then went back to the Midwest to continue her ethnological studies. She attended the summer dances at the Great Sioux Reservation, and, thanks to her reputation for complete honesty about her purposes and goodwill toward the Indians, she was able to witness rituals normally closed to whites. Later in the summer she again visited the Omaha, in time to celebrate the successful passage of the land allotment act she had lobbied for.

She knew how to keep careful records of her observations, and once back on the East Coast enlisted the help of Francis La Flesche, a young Indian working as a clerk at the Washington Indian Bureau, to turn these records into a publishable manuscript. Francis was the son of Joseph La Flesche, the former Omaha tribal chief Fletcher had come to know on the Nebraska Reservation. Together they produced "Five Indian ceremonies," which came out in the 1884 annual report of the Peabody Museum. A pioneering exploration of native American religious life, it was Fletcher's first major publication; like much of her later work, it relied heavily on the insight of Francis La Flesche, whose "most valuable assistance" she acknowledged.

In the spring of 1883 she was asked by the Commissioner of Indian Affairs to carry out among the Omaha the newly enacted land allotment program. As a special agent of the office she would be paid $5 a day plus expenses. Francis La Flesche was to be given leave from the Indian Bureau to accompany her as interpreter. Needing money, and seeing the assignment as an excellent way of combining ethnological studies with what she considered humanitarian work for the Indians, she took the job. Sentiment in favor of accepting the small individual landholdings was by no means unanimous throughout the tribe. However, she completed the allotment program, and over the course of the period of more than a year that she spent on the reservation also carried out a complete census of the tribe, coming to know personally every one of its 1,179 members. Her work was interrupted by a lengthy and serious illness, during which she was cared for largely by Francis La Flesche; much of her success in completing her assigned task was due to his efforts. Despite the exhausting work, ethnological concerns were not neglected. Particularly noteworthy were a number of sacred objects collected, the most important being a set of peace pipes she took back to the Peabody Museum.

On their return to the East Coast in 1884 Fletcher and La Flesche gave a joint illustrated paper on the sacred pipes of the Omaha at the Philadelphia meeting of the AAAS. The presentation was unique in the history of the anthropology section; never before had the members heard anything like the strange music produced from long, feathered pipestems by a young native man who went on to deliver a lecture on the construction of the instrument and its importance in tribal ceremonies.

Fletcher continued to work for the government, taking an assignment to prepare an exhibit on "Indian Civilization" for the World's Industrial and Centennial Cotton Exposition in New Orleans in 1885, and expanding and completing a massive report on *Indian Education and Civilization,*[5] the first part of which became, in effect, a handbook of Indian affairs. In 1887, shortly after she returned from several months of travel through Alaskan coastal settlements, she accepted another government land-allotment assignment, this time with the Winnebago tribe. She went without Francis La Flesche but accompanied by Jane Gay, a friend who was interested in photography and who was to remain with her, sharing a house in Washington, for many years. Next came more land-allotment work on the Nez Perce reservation in Idaho. Her task there was difficult because the government's plan met with almost unanimous opposition from the tribe. As in the schemes for the other reservations, individual plots were to be assigned, and much of the remaining land was to be surrendered for white settler use.

Like many liberal-minded whites of the time, Fletcher strongly and sincerely believed that the allotment arrangement was the best that could be done for the native people, a small plot with legally secure tenure being seen as preferable to a large, communally owned reservation that would inevitably continue to be cut into by white encroachment. Several years were to pass before she came to realize that the change in way of life that the schemes imposed on Indian peoples was not something to which their culture could quickly and readily adjust. As time went on the forced integration into white society that it entailed brought increasing social chaos without providing the better life for Indians that many well-intentioned whites had confidently expected. Indeed, within a short time even white settlers with a strong agricultural tradition found it difficult to make a living as subsistence farmers.

In all, Fletcher spent ten years as a special agent of the Bureau of Indian Affairs, made allotments to 4,400 Indians, some of whom, like a fraction of the Omaha, were actively seeking allotments, but the majority of whom were violently opposed. She presided over the ceding by the Omaha, Winnebago, and Nez Perce tribes of more than 620,000 acres of land from their reservations to the United States government, for sale to white settlers; the Nez Perce people lost 75 percent of their reservation.[6]

In 1890, part way through her Nez Perce work, she was awarded a lifetime research fellowship of about $1,050 annually. The money, which was to be administered by the Peabody Museum, came from Mary Copley Thaw of Pittsburgh, widow of steel magnate William Thaw, who had been impressed with Fletcher's anthropological work and helped her previously. Although the award made her an independent scholar, technically the holder of a Harvard fellowship, she was not invited to join in the scholarly activity of the male staff at the Peabody Museum. Instead, she was encouraged by Putnam to resume her work with her native informants, particularly Francis La Flesche. Putnam, probably correctly, felt that this was where she could make her best contribution to American anthropology.

Feeling somewhat rebuffed she returned to Washington, where about 1894 she took a small but pleasant house on Capitol Hill. This she shared with Jane Gay and Francis La Flesche. She was then fifty-six, Gay sixty-four, and La Flesche thirty-nine.

Her major joint monograph with La Flesche on Omaha music, prepared in collaboration with John Fillmore of the Milwaukee College of Music, appeared in the Peabody Museum papers of 1893. It was the largest collection of nonwestern music ever published in one place, ninety-two songs, all from a single tribe. Further, it was the first serious study ever made of American Indian music.

In 1895 she was elected vice president (presiding officer) of the anthropology section of the AAAS, becoming the first woman to fill such a position in any section of the organization. Her vice presidential address, entitled "The emblematic use of the tree in the Dakotan group," was a detailed discussion of the evolution of Omaha political ideas as symbolized by the sacred tree. It constituted an especially fine piece of ethnohistorical analysis and interpretation of painstakingly collected field data, and demonstrated very clearly her growing appreciation of Omaha history, ritual, and tradition. She was coming to see that the Indian was far from being a primitive man (tied to such ideas as descent from totem animals) as some of her colleagues still viewed him, and that on the contrary there was great cultural richness in his tribal ceremonies and sophistication in his artistic achievement in dance and song. Over the following years she presented a number of other notable papers, and by 1898 was widely regarded as an important contributor to studies in American anthropology. She was starred in the 1906 edition of *American Men of Science.*

By about the turn of the century she had begun to work more and more closely with the Bureau of American Ethnology, which in 1904 brought out her joint monograph with James R. Murie, "The Hako: A Pawnee ceremony." She also contributed to the bureau's *Handbook of the North American Indians,* published in two volumes in 1907 and 1910; the thirty-five entries she prepared covered topics ranging from "Adornment" to "War and war discipline." In 1903, over some opposition, she was elected president of the Anthropological Society of Washington, with which the Women's Anthropological Society of America had recently merged. She served on the council of the new American Anthropological Association, of which she was a founding member, on the editorial board of the *American Anthropologist* (1899–1916), and as president of the American Folklore Society. In addition she was a key figure in the founding and early organization of the School of American Archaeology (later the School of American Research) in Santa Fe, New Mexico. The latter was established in 1908, despite considerable resistance from many eastern anthropologists, following a great rise in interest in western regional anthropology and archaeology.

Throughout her whole career in anthropology, beginning with her first work in the early 1880s, she had depended heavily on Francis La Flesche. Their relationship was indubitably a complex one, with a primary emotional tie that went back to his concern for her during her long and serious illness and breakdown while on the Omaha reservation in 1883. During that crisis he not only provided care and affection, but led her back to her intellectual interests as well, bringing his friends and relatives to her with the songs of his people that she wanted to hear, and also making it possible for her to carry on much of her allotment-related work from her bedside.

Her status as a middle-class lady made it well nigh impossible for her to consider marriage to La Flesche, however intelligent, gifted, and talented he might be, and despite the fact that he had considerably augmented his early Presbyterian mission school education by taking a bachelor's degree in law, followed by a master's, at the National University in Washington, D.C. (1892, 1893). Further, he was seventeen years younger than she, a formidable barrier to marriage in her culture, even if not in his. For a time they settled their relationship as being that of mother and son (though informally, since he did not want to change his name) and this worked reasonably satisfactorily, although her persistent habit of giving him less than his fair share of credit in their joint work was a continuing cause of dissatisfaction to him. In the spring of 1906 he married a part-Chippewa woman, Rosa Bourassa, but the union was over in less than a year, Rosa leaving the Washington household into which Francis had apparently thought she could fit. There were other trials and vicissitudes as well throughout the course of the forty-year La Flesche-Fletcher relationship; in 1910 he left her, staying away for six months, upset about the time and attention she was giving to Edgar Hewett, one of the men she worked with in founding the Santa Fe School of American Archaeology. This decisive action by Francis, however, finally brought the denouement. Her reaction was to write to him daily, begging him to come back, and in her letters she gradually made explicit what she had never acknowledged, perhaps never even consciously recognized, namely the fact that he was the emotional center of her existence, the person beyond all others on whom she depended. She was then seventy-three, he fifty-six, and at last, after three decades, they openly acknowledged their feelings toward one another.

They had twelve more years together before Alice's death, and throughout that period the focus of their professional work shifted increasingly to his concerns. He was then collecting material relating to the Osage tribe, and Alice prepared transcriptions for him of more than 200 Osage songs. In 1911 the Bureau of American Ethnology brought out their joint 672-page monograph, *The Omaha Tribe,* an ethnohistorical work that has remained one of the classics of American Indian ethnography.[7] After 1915 only the two of them shared the Washington house, but Alice no longer worried about possible gossip this arrangement might arouse. Visitors described her at the time as pale, white-haired, and full-figured, always dressed in long, old-fashioned dresses and notably alert mentally. Francis, then middle-aged, had put on weight from his years of city living but was still a handsome man.[8] Alice died on 6 April 1923, at the age of eighty-five. She left an estate valued at $35,000. The bulk of this was in a trust, the income from which went to Francis during his lifetime; the funds then passed to the School of American Research in Santa Fe, to establish a fellowship in her name.

Francis stayed on for six more years in the Washington house and at the Bureau of American Ethnology (where he had been a staff member since 1909), and continued his ethnological work. His basic interests were fundamentally different from Fletcher's in that he wrote not for anthropologists and the public as she had done, but for later generations of American Indians. Essays of broad ethnological interest he could write if he wanted to, and he produced several that were outstanding; but what he cared about most of all were the rapidly vanishing religious ceremonies at the heart of Indian life. These he studied with the maximum possible care and thoroughness, and with all the insight that his deep understanding of the social origin and nature of religious conceptions gave him. Alice Fletcher, though clearly a competent recorder, organizer, and expositor of large amounts of anthropological information, was not one of the country's top-level anthropologists. There seems little doubt that much of the analysis and interpretation in her work originated essentially in La Flesche's mind, whether or not always formally acknowledged by joint authorship. Indeed she could be said to have been his student in many ways. La Flesche's work on *The Osage Tribe,* four volumes totaling 2,242 pages, the product of nineteen years of labor, was brought out by the American Bureau of Ethnology between 1921 and 1930. He also published *A Dictionary of the Osage Language* (1932). In 1926 the University of Nebraska gave him an honorary LL.D. He went home to the Omaha reservation in 1929 and died there three years later.

The third of America's notable early women ethnologists, MATILDA EVANS STEVENSON[9] (1849–1915), was known especially for her detailed and systematic field studies of the Indians of the Southwest, particularly the Zuñi. Her published works, which embody a considerable store of valuable information, long remained standard references for students of the pueblo cultures.

The daughter of Maria Matilda (Coxe) and Alexander H. Evans, she was born in San Augustine, Texas, on 12 May 1849, but spent much of her childhood in Washington, D.C., where her father, a lawyer and journalist, was moderately well known in the city's intellectual circles. She was taught by her mother and governesses, and then attended a private school for girls in Philadelphia. For a time she studied law with her father, working as a clerk in his law office, but she was particularly interested in rocks and minerals, took private lessons in geology and chemistry, and planned to become a mineralogist.

In 1872, shortly before her twenty-third birthday, she married James Stevenson, a government geologist. Directly after, for the six-year period between 1872 and 1878, she accompanied him on geological surveys in Colorado, Idaho, Wyoming, and Utah. The opportunity opened to her must have been almost unprecedented for an aspiring young female mineralogist in 1872. James Stevenson was executive officer on Heyden's U.S. Geological Survey of the Territories. Matilda assisted in collecting fossils and ornithological specimens on an investigation of the geysers of Yellowstone; while in Colorado and Wyoming she made a study of the Ute and Arapaho peoples, learning the basics of ethnographic technique, although she never published any of this early work.

When the Bureau of American Ethnology was founded in 1879 James Stevenson was given a staff position there also, and assigned the task of investigating archaeological remains in the Southwest. On the first of the Bureau of Ethnology expeditions (1879) the Stevenson group spent six months at Zuñi and Hopi in New Mexico and Arizona. Matilda held the post of "volunteer coadjutor" in ethnology, that is, assistant to her husband, the expedition leader. Included in the group was Frank Cushing who also became a respected Zuñi ethnologist.

Even though she was an unpaid volunteer, her position was not unimportant and her presence was fully supported by bureau director John Wesley Powell. Powell considered she complemented her husband in ethnological work, having access to information about women and children denied her husband, but which Powell considered necessary for a full picture of Indian life.[10] In fact Matilda's main interests were religion and ceremonial rituals, which she believed to be basic elements of Indian cultures. Her well-known report on child life among the Zuñis (1887) was primarily an examination of ceremonies of childhood, and her "Zuñi games" (*American Anthropologist,* 1903) an analysis of ceremonial games.

Her first publication, *The Zuñi and the Zuñians* (1881), a privately printed monograph, was the first serious ethnography of the Zuñi for a popular audience, and the first study of child-life by an American ethnologist. During the 1880s, when she and her husband were making annual trips to Zuñi, she also did much of the writing of his reports, a task he greatly disliked. The first official report to appear under her own name was her formal study of the Zuñi child in the Bureau of Ethnology's *Fifth Annual Report, 1883–84* (1887).[11]

After James Stevenson died (of Rocky Mountain fever) in 1888, Matilda was given a temporary appointment at the Bureau of Ethnology to write up his unpublished material. Her position became a permanent one in 1890. Although her salary was less than that of her male colleagues, she was the first full-time paid female ethnologist employed by a United States government agency. Almost three-quarters of a century was to elapse before another professional woman anthropologist held a full-time position at the Bureau of Ethnology.

Matilda returned to New Mexico about 1890 and began work at the Sia pueblo. "The Sia," an exhaustive study of this small community and the first major ethnography of a Rio Grande pueblo, was published in the bureau's *Eleventh Annual Report, 1889–90* (1894). She undertook further investigations at a number of pueblos but returned often to Zuñi. Over the course of the twenty-five years she worked there she formed lasting friendships with a number of well-informed people and was able to develop a genuine understanding of the society. Her obvious deep respect for the culture even gave her access to some of the rites normally closed to outsiders. She was interested in all aspects of Zuñi life, technical developments as well as religious and philosophical traditions. Topics on which she collected data included irrigation methods, clothing, dyes and pigments, and plant use, as well as art, mythology, and religion.

As the years went on she noted the profound changes that were taking place in Zuñi society. More quickly than Alice Fletcher, Stevenson realized that contact with whites was having a demoralizing effect on Indian communities. She pointed out that, although there might have been a general improvement in the material standard of living, at least initially, the integrity of the people, something far more important, was being lost. Because these social changes were so obvious to her, she felt a certain urgency in collecting information that would soon, she knew, be lost forever. Later her worries about time running out led her to press her informants too strongly for the material she sought, and her relations were sometimes strained with peoples who knew her less well than did the Zuñis.

Her major work was her 634-page monograph, "The Zuñi Indians: their mythology, esoteric fraternities and ceremonies," published in the Bureau of Ethnology's *Twenty-third Annual*

Report, 1901–02 (1904). This was followed by another substantial work, "The ethnobotany of the Zuñi Indians" (*Thirtieth Annual Report, 1908–09* (1915)), which was complemented by an herbarium she assembled of more than 200 edible, medicinal, and fetishistic plants used by the Zuñis. (The herbarium, along with a collection of sacred masks, was later contributed to the Smithsonian Institution.)

Having been refused membership in the all-male Anthropological Society of Washington, D.C., Stevenson became the prime mover in the founding, in 1885, of the Women's Anthropological Society of America. She served as the organization's first president and frequently spoke at its meetings (often on her work on the Zuñi). She was also a fellow of the AAAS and a member of the Washington Academy of Sciences. Despite many personal idiosyncracies (including a formidable temper) and a somewhat uneven relationship with Major Powell at the Bureau of Ethnology,[12] she was widely recognized as one of a small group of pioneer American ethnologists who established the tradition of impartial observation and carefully planned collection of field data.

By about 1904 she was planning and indeed had started a wide, coordinated, inter-pueblo cultural study that she saw to be necessary. She never completed this project, however. She settled near the San Ildefonso pueblo in the Rio Grande valley and expected to extend her investigations to the Tewa Indians. Although she succeeded in carrying out several shorter studies,[13] she ran into a number of difficulties, not the least of which was the fact that outsiders were by then no longer welcome in the Rio Grande pueblos. Further, she became involved in New Mexico politics and in a legal wrangle with the Bureau of Indian Affairs, distractions that took her away from her ethnological work. She also seems to have overly indulged in alcohol as she grew older. She returned to Washington, D.C., in 1915, and died there on 26 June of that year, shortly after her sixty-sixth birthday. Many of her unpublished manuscripts remain on file at the National Anthropological Archives in the Smithsonian Institution.

FRANCES BABBITT[14] (1824–1891) was especially interested in the prehistoric Ice Age peoples who gradually moved north with the retreat of the continental glaciers. Her discoveries in Minnesota of stone implements, which she correlated with the findings of geologists working on glacial deposits, were creditable contributions to the study of the distribution of Paleolithic man in North America. From time to time throughout the 1880s she spoke about her finds at meetings of the AAAS, and she set out her conclusions in one or two fairly substantial published articles.

Originally from Little Falls, Morrison County, in central Minnesota, she spent much of her life in Coldwater, southern Michigan. She was educated at the Monroe Female Seminary in Coldwater and later taught there for a number of years. From 1864 until about 1870 she served as principal of the Coldwater Female Seminary, founded by her brother-in-law, J. D. W. Fisk of Coldwater.

Much of her archaeological work was carried out in the 1870s and 1880s, when she was in her fifties and early sixties. Undoubtedly her most important discovery was a quartz "workshop" of immediate postglacial man. This she uncovered in 1879 while searching for Paleolithic remains along the glacial floodplain terraces of the Mississippi River in the township of Little Falls. Her large collection of about 1000 quartz objects from the site was examined by the experts of the time, who confirmed that many were man-made. Further, it appeared that the Little Falls tools predated similar objects collected in the eastern region of the country. When a number of specimens were exhibited at the AAAS meeting in Minneapolis in 1884 they were inspected by Frederick Putnam and accepted by him for the Peabody Museum collections.[15] With the help of Newton Horace Winchell, state geologist of Minnesota, and Warren Upham of the Minnesota Geological Survey (both experts on glacial geology and knowledgeable about Paleolithic man), Babbitt estimated that the terrace strata in which her quartzes lay had been formed about 8,000 years ago, during the retreat of the last of the ice sheets.[16] The Little Falls workshop was located near a surface quartz vein, the first source for that rock that would have become available in the area after the recession of the ice.

Babbitt also wrote on Indian cultures. Her 1887 presentation to the AAAS described the customs of the Odjibwa people, many groups of whom lived in northern Minnesota, especially in the area around Red Lake, which is still a large Indian reservation. In a long paper in the *Bulletin of the Minnesota Academy of Natural Sciences* (1881) she presented a broad discussion of pottery fragments (Sioux and Odjibwa) she had collected from Minnesota sites.

She was remembered in the Coldwater community as a woman of considerable ability and wide interests. As well as her scientific papers she wrote at least one article for Michigan newspapers on a Coldwater art collection.[17] She died in Coldwater, at the home of her brother-in-law, in July 1891.

MARY (MAY) FRENCH SHELDON[18] (1848–1936) was one of the colorful, if minor, characters in the story of African exploration. Both she and her contemporary and fellow American Fanny Bullock Workman (see chapter 13) were very much women of action on a grand scale, who, having means, energy, and initiative, embarked on expeditions of exploration in remote regions of the world at a time when such undertakings were unusual for women. Sheldon's 1891 trip to equatorial East Africa led to her being elected one of the first women fellows of the Royal Geographical Society, and her writings on the customs of East African peoples, although constituting a relatively modest contribution to the ethnological knowledge of the region, were received with interest by her contemporaries. Her second trip thirteen years later to the Belgian Congo, although it did not involve exploration of unknown country, was still an impressive undertaking for a woman at the time, and was perhaps the more important venture. Financed at least in part by a London newspaper, it was a fact-finding mission, largely to look into allegations stemming from missionary reports of unusually bad atrocities against native peoples under Belgian colonial administration.

Sheldon spent much of her life in Britain and France, but she was born in the United States. Her parents, Colonel Joseph

French and his wife Elizabeth, were wealthy and cultured southerners; the family owned cotton and tobacco plantations. She was sent to school in Europe and traveled extensively to complete her education. Her special interests were music, art, and classical literature, although she also studied geography, ethnology, and medicine. By the age of sixteen she not only was familiar with the major European cities, but had been around the world, an adventure she was to repeat three more times. Sometime before 1886 she married Eli Lemon Sheldon, an American with business interests in London and a man who had a generally liberal outlook on life. Having no children she found an outlet for her energy in literary work, managing her own publishing firm, Saxon and Company of London and New York. In 1886 she published her translation of Gustave Flaubert's *Salammbô,* a little-remembered, lesser work of the master novelist, a historical drama set in the Middle East. Her own somewhat melodramatic popular novel, *Herbert Severance,* much of which is thought to be autobiographical, appeared three years later.

Sheldon's interest in African travel almost certainly had its origins in her friendship with the celebrated Welsh-American explorer and journalist Henry Morton Stanley, whom she probably met in London in the late 1870s. But it was not until 1891, when she was forty-three, that she undertook her first expedition. Her plan was to travel by steamer from London to the east coast city of Mombasa and from there go inland to the region of Mount Kilimanjaro; she would also visit the country of the Masai, then a warlike people who did not always welcome visits from Europeans. Although as a practical goal she meant to study native customs, collect examples of native handicrafts and weapons, and accumulate material for a novel set in Africa, she also wanted to make clear to all the fact that, with sufficient resolve and strength of purpose, a woman could lead such an expedition as well as a man.

She arrived in Mombasa equipped with letters of introduction from Stanley to two highly placed officers in the Imperial British East African Company, but got no official support or encouragement for her undertaking. Stanley himself had warned her that after seeing Mombasa and its immediate surroundings she probably would not want to proceed further. However, quite undaunted, she secured the necessary passes from the Sultan of Zanzibar, signed on a caravan of about 150 porters plus reliable headmen and interpreters, and headed for Taveta, about 200 miles inland, on the border of Kenya and Tanganyika. She appears to have had little trouble with her caravan and enjoyed herself very much, traveling in a huge palanquin of wickerwork and cushions and finding the tropical heat a welcome change from London fog. The special pains she took to keep on friendly relations with the native peoples she met ensured that they in turn were civil and generous. In exchange for the many fine gifts they presented to her she gave them lengths of cloth (whose quality, she noted, they checked with some care). About fifty miles north of Taveta on the slopes of Mount Kilimanjaro she explored the crater in which lies Lake Chala, circumnavigating the lake in a copper pontoon left behind by the Hungarian explorer, Count Teleki. On her return journey she had the misfortune to fall into a river when a bridge collapsed, and throughout the rest of the journey to the coast she was ill with dysentery and fever.

Shortly after her return to England she addressed the geographical section of the British Association on her exploration of Lake Chala,[19] and in 1892 published a detailed account of her expedition, *Sultan to Sultan: Adventures among the Masai and other Tribes of East Africa.* The book recounted her many adventures and put on record her extensive observations of the customs of all the tribes she had visited. Her paper in the *Journal of the Anthropological Institute* the same year presented a summary of these observations, and included fairly extensive remarks on agricultural practices, manufacturing methods, dances, and marriage and funeral rites.[20] She was especially interested in children's activities and in the lives of native women, a number of whom she had found eager to talk with a white woman.

Following the death of her husband in 1892 she spent several years in New York. Another of her translations from French appeared in 1893.[21] She returned to London in 1902, just before considerable controversy broke out over the exploitation and maltreatment of native peoples in the Belgian-controlled Congo Free State. Through her friendship with William Thomas Stead, the influential editor of the *Pall Mall Gazette* and the *Review of Reviews,* she became interested in the matter and decided to make a personal investigation. Accepting from Stead the sum of £500 for expenses, she undertook to proceed to the Congo, carry out a careful and extensive examination, and report her findings to Stead in a series of letters, suitable for publication in the press. Matters of special concern included the use of forced native labor in general and in particular in the rubber-collecting business (then of major commercial importance), the question of free trade, the progress of missions, and whether or not the region was really a Belgian preserve or open to other Europeans and Americans.

Her trip lasted just over a year (October 1903–November 1904), and she covered a vast amount of territory, visiting innumerable trading and administrative posts. Starting from Leopoldville (now Kinshasa), she first went east along the Kasai river, then north some 400 miles to Stanley Falls on the main branch of the Congo. Generally following the Congo along the many hundred miles of the sweep of its great curve west and south, she made her way, over the course of almost nine months, back down to Leopoldville and the Atlantic coast. En route she took a number of extensive side trips into the rubber country along the Lopori and Maringa tributaries, and finally revisited the Kasai region—all told a most remarkable journey.

By February 1905 she was back in London, where she strongly defended the Belgian administration of the Congo. Although not denying that atrocities occurred, she claimed that charges that the administration countenanced abuses were false. During her trip she had attended courts in a great many of the posts she visited, and had concluded that the state "dealt fairly" with cases of misdeeds and crimes.[22] Her Congo reports were forwarded to the presidents of the United States and France, the German emperor and the British Foreign Office.[23]

Sheldon made one further visit to Africa, going to Liberia in 1905 on behalf of the Americo-Liberian Industrial Company to negotiate the concessions of a tract of land for industrial and agricultural use.

Her year of work in the Congo had given her a close and friendly connection with Belgium, and when the First World War broke out she became much involved in raising funds for Belgian relief organizations. In 1921, in recognition of her services to his country, King Albert I awarded her the distinction Chevalier de l'Ordre de la Couronne. She died in West Kensington, London, on 10 February 1936, at age eighty-eight.

This small group of late nineteenth-century American women who published work in anthropology was remarkably successful, its three best-known members—Fletcher, Smith, and Stevenson—being accorded notable professional recognition. The initial acceptance of women into the discipline would seem to have been linked to the view, current among men anthropologists from about the early 1880s onward, that women had a special role to play in collecting information on topics that were less likely to be revealed to men themselves, particularly matters concerning the lives of women and children. However, having once found an opening into the field, women did not confine themselves for long to any restricted topic but followed their particular interests and inclinations. Further, Smith, Fletcher, and Stevenson all worked through male informants. Nevertheless, the obstacles faced by the female fieldworker were not minor. Long periods of living under rough conditions were the norm and financial aid was minimal. Despite the impressive records of some of the women whose careers are sketched here, it is not surprising that only a few late nineteenth-century women took up work in anthropology.

British women

The Royal Society *Catalogue* lists nine papers on anthropological and ethnological topics by five British women (Buckland, Crane,[24] Godden, Aynsley, and Welby). Added to the bibliography are eleven more papers by three additional women—Bird Bishop and Kingsley (both of whom also wrote papers listed in the geography section the bibliography) and Layard (who has three articles listed in the botany section).

All of these women began their studies during the period that has been described as the "heroic age" of British social anthropology, the last three decades of the nineteenth century.[25] Although systematic studies in Britain of primitive cultures went back to the 1840s, and the Ethnological Society, which later merged into the Anthropological Institute, was established in 1843, the real founder of ethnology in Britain was probably E. B. Tylor, the first holder of a chair of anthropology at Oxford. Tylor's *Primitive Culture* (1871) was the first of a number of famous and extremely influential books by British anthropologists that appeared before the turn of the century (including Sir James Fraser's *The Golden Bough*). Although much of the early work was based on the collection and analysis of materials without direct firsthand investigations by the leading authorities, some field studies were also being carried out. General interest in the activities of travelers and in the exploration of nonwestern cultures was considerable (see also chapter 13).

Women participants in these undertakings who contributed the most in the way of original investigations were the travelers and ethnologists, Bird Bishop, Kingsley, and the less well-known Aynsley, Layard, whose archaeological studies began in the 1890s, and Durham, who started her ethnological work about 1900 and continued until the mid 1930s.[26] Smaller contributions came from Buckland, Godden,[27] and Welby.

ANNE WALLBANK BUCKLAND[28] (1832?–1899), one of the earliest British women to publish in the anthropological journals, was remembered especially as a popularizer of the field. She became a fellow of the Anthropological Institute soon after it opened membership to women in 1875, and for twenty-three years regularly attended its meetings, frequently presenting papers. She was also a member of the British Association, joining in 1883; the following year she traveled to Canada and read a paper (on possible prehistoric East-West contacts) before the anthropological section at the Montreal meeting.

Her most notable work was probably her well-received book *Anthropological Studies* (1891). Largely a compilation of articles she had previously published in the Anthropological Institute's *Journal* and in the *Westminster Review,* it succeeded in setting out clearly many of the current problems in the field, the methods being used in their solution, and the results so far obtained. Her special interests covered a variety of topics, ranging from Cornish and Irish prehistoric monuments, to serpent worship, neolithic surgery, and stimulants used by ancient peoples.

Among her nonanthropological publications were a travel book on Italy, *The World beyond the Esterelles* (1884), and a book on foods that included recipes from the previous century, *Our Viands* (1893). She died in West Kensington, 4 January 1899, at the age of sixty-seven. A Civil List pension, awarded for her services to anthropology, helped sustain her in her last years; it attests to the fact that her contributions were well regarded by her friends and colleagues.

HARRIET AYNSLEY[29] (1827–1898) also published early work in anthropology, notably her two books, *Some Account of the Secular and Religious Dances of Certain Primitive Peoples in Asia and Africa, together with their Survivals in Europe* (1877), and *Our Visit to Hindostan, Kashmir and Ladakh* (1879).

The daughter of the Rev. F. Manners Sutton, Harriet married the Rev. John Crugar Murray Aynsley in 1852. Until 1860 the Aynsleys lived in Rome and spent considerable time traveling in Algeria. About 1875 they made their first visit to India, fell in love with the beauty and grandeur of the subcontinent, and during the next twenty-one years traveled extensively throughout its length and breadth, from the south to the Himalayas. Harriet's third book, *An Account of a Three-month's Tour from Simla,* appeared in 1882, and *Our Tour in Southern India,* a work presenting architectural descriptions of temples, the following year. Her particular interest was the local legends and symbolism of India (an interest that earned her the most unusual distinction, for a woman, of being elected an associate of the Order of Freemasons); she published a number of scholarly articles in this area in the *Indian Antiquary.* A monograph entitled *Symbolism of the East and West,* her most extensive treatment of the topic, appeared posthumously in 1900. Her 1890 *Nature* report listed in the bibliography discussed observations made among the people of northern India.

MARIA WELBY,[30] whose article on aspects of evolution appeared in 1891, was the daughter of Augustus Henry Charles Hervey, second son of the second marquis of Bristol, and his wife Mariana Hodnett Benyon. She married Sir Charles Welby, baronet, M.P., War Office Official and Lincolnshire landowner, in 1887, and brought up a family of six children. Her 1891 paper, read at a meeting of the Anthropological Institute in 1890 when she was elected to membership, would appear to have been her only published anthropological study. A long discourse on pathways of mental evolution, it set out her conjectures on the development of the faculty of imagination and the place of superstition in human communities.

In the years before the First World War Lady Welby was much involved in the development of the Red Cross, and on the outbreak of hostilities undertook the direction of the organization's work in south Lincolnshire. She was also very active in local women's groups, working with the Women's Committee of the Board of Guardians, the Union of Women Workers and the Women's Institute movement. At the time of her death on 11 July 1920, she had been selected a delegate to the League of Nations Peace Commission.

One of the best remembered of the nineteenth-century British women who made contributions to ethnology is MARY KINGSLEY[31] (1862–1900). Still an enigmatic and controversial figure, a forceful personality of continuing interest to biographers, Kingsley was in many ways a woman of strange contradictions. She was opposed to the political emancipation of women, opposed to movements to get women admitted to the learned societies, had a sense of duty to parents, brother, and extended family that was exaggerated even by late Victorian standards, and was angered by being described as one of the "new women" of the 1890s. At the same time, she personified that "new woman." A vigorous, energetic, and amazingly courageous traveler, she was also a keen and sensitive observer who made contributions to more than one area of science. Furthermore, in her writings about West Africa and West Africans, their cultures, social and political organizations, legal systems, and the harmful influence of Europeans who were coming among them, she was not afraid to set out views that were unorthodox and at times quite controversial. Missionaries in particular she attacked for what she felt was the hypocritical self-righteousness of their attempts to destroy African culture. Her fieldwork, carried out during two short expeditions to "the Coast" in 1893 and 1895, was notable, especially her pioneering studies of West African tribal religions. Her contributions to tropical natural history were likewise considerable, and, although she disclaimed any special interest in geography and never called herself an explorer, her studies in that area anticipated the modern practice of intensive exploration of small regions; the investigation she made by canoe of the rapids in the upper reaches of the Oguooé river in Gabon and her expedition through the unexplored forest region between the Oguooé and the Remboué were both considerable achievements. When she came home she wrote two travel books, which, as well as being highly entertaining best-sellers, contained much that was new to science.

Mary Kingsley wanted very much to be taken seriously, as a person with important information to contribute to ethnology, natural history, and matters political and commercial. At the same time, however, she clung to the convenient safety net of being able to say she was "only" a woman, not the expert a white male would have been had he experience similar to hers. This modus operandi is perhaps not altogether surprising, given her lack of any formal education, her closeted early life, and the rather extreme intellectual isolation of her formative years in a family that was barely functional, an uneasy mixture of middle and working class. Possibly she also felt that she had little to gain from any opening of wider opportunities for women in general; she had already succeeded in carving out her personal niche in what was largely the white male's world of trade and exploration in West Africa and informal political maneuvering in London. She may well have preferred to maintain her special status as an exception.[32]

She was born on 13 October 1862, in Islington, Middlesex (London), the daughter of George Henry Kingsley, a society physician who spent much of his life traveling with wealthy, aristocratic patrons, first in the South Pacific and then in America. George was the brother of the well-known novelists Charles and Henry Kingsley. His wife was an innkeeper's daughter, Mary Bailey, who had been his domestic servant, probably his cook. The two were married on October 9, 1862, four days before their daughter's birth.

Mary spent much of her childhood in a dark, rather claustrophobic house in Highgate, then on the outskirts of London. With her father frequently away, her mother often ill and confined to bed, and her one brother, Charles, of little help around the house, she took on domestic responsibilities about as soon as she was physically capable, and these gradually increased. When her brother was sent away to school she lived for long periods with only her sick mother, an old servant woman and an elderly gardener for companions. Somehow she learned to read, and then proceeded to educate herself by means of her father's library, which included the classics from his boyhood, technical and scientific works, journals of learned societies, and accounts of recent exploration. Among the latter, the reports of work then going on in Africa more than anything else fired her imagination. She was deeply impressed by the journeys of Richard Burton, John Speke, and Samuel Baker in the East African lakes region in the 1850s and 60s, those of David Livingstone and Henry Morton Stanley in the 1870s, and by the experiences of the French explorers in West Africa, Paul Belloni Du Chaillu and Pierre Savorgnan de Brazza, from the 1860s to the 1880s. Livingstone's books, full of ethnological as well as geographical information, gave her an awareness of the rich complexity of African cultures.

When her parents died within a few weeks of each other in

1892, Mary, at the age of twenty-nine, was suddenly freed from the nursing and much of the domestic work that had filled her life and kept her tied down until then. That same year she did her first traveling, taking a voyage to the Canary Islands, partly for rest and recuperation. During the trip she made a brief side visit to the African mainland, and shortly after that began to make definite plans to see something more of West Africa. Later she was to rationalize her travels by emphasizing the fact that she went to Africa to extend, as a dutiful daughter, her father's uncompleted studies in anthropology. George Kingsley had traveled extensively among the islanders of the Pacific and the native peoples of North America, and over the years had collected a great deal of information on religious and law practices in nonwestern societies. Africa remained a gap in his knowledge. As the continuation of her father's work, Mary's highly unconventional journeys could be made socially acceptable and reasonable. She was also especially interested in fishes, Albert Charles Günter's *An Introduction to the Study of Fishes* (1880) having long been one of her favorite books. Thus the collecting of West African freshwater fishes became another concrete goal. She summarized her interests as "fish and fetish." Her practical route to accomplishing her stated scientific aims, however, was to go to Africa as a trader; she felt that this would give her the best chance of winning the confidence of the communities she wanted to study. Her goods would be the usual ones, cloth, beads, tobacco, and fishhooks, to be exchanged for ivory and rubber; she became an accredited agent for the Liverpool firm of Hatton and Cookson that had traded along the west coast of Africa for many years.

On her first visit to the Coast in 1893 she went out from Liverpool on the cargo ship *Lagos,* which also carried a few traders and government officials. Almost all of these men had been to Africa before and could pass along much information, although little of it was reassuring. After a number of stops at ports en route, the *Lagos* reached its destination at St. Paul de Loanda (Luanda), Angola. From there Mary went north by ship to the small enclave of Cabinda near the mouth of the Congo, where she made contact with the Fort tribe and collected fishes and insects. Continuing her way north overland, traveling rough and living on native food, she traded her way through the Congo Free State and the French-held territory that is now Congo and Gabon, as far as the port of Libreville. There by chance she came upon the British ship *La Rochelle,* which could take her home.

She wrote an account of this trip, incorporating a good deal of the ethnological material she had collected and was urged by the publisher George Macmillan to bring it out. However, after studying Frazer's *Golden Bough,* Tylor's *Primitive Culture,* and what was available of recent specialized anthropological work on West Africa, she came to feel increasingly strongly that her inquiries into native customs had been superficial, and that she needed to see and hear more before she went into print. Much of her early writing was nevertheless incorporated into her later works.

About this time she sought an introduction to Albert Günther, then keeper of the zoological department at the British Museum (Natural History), and presented him with the specimens she had brought back. Günther liked her collection, and when she told him she planned to go back to West Africa the following year he offered a certain amount of material encouragement, arranging for her to have good collecting equipment. He also commissioned her to collect freshwater fishes from the region between the Niger and the Congo and particularly from the Oguooé river. The latter was thought to contain many kinds of fishes unknown to British zoologists.

In December 1894 she left on her second trip, again sailing from Liverpool, this time in the company of Lady Ethel MacDonald, whose husband, Sir Claude MacDonald, was governor of the Niger Coast Protectorate. She spent several months with Lady MacDonald at Calabar, collecting zoological specimens in the swamps and gathering anthropological material during her talks with local people. She then proceeded south to Gabon to get the fishes she was especially interested in. Supplied with a stock of trade goods from the Hatton and Cookson store in Libreville, she went up the Oguooé river to Kangwe, near the now-famous island of Lambaréné, where she stayed with French missionaries while collecting and visiting local villages of the Fang people. Her collecting continued further upstream at the Talagouga mission, and since by then she was adept at handling an Oguooé canoe, she had complete freedom to move around the mangrove swamps as she wanted. From Talagouga she also undertook the especially risky venture of exploring by canoe the Oguooé rapids, a section of the river little known to Europeans at the time.

Going back downstream to Kangwe, she began to plan her most daring expedition—an overland trek through unmapped rainforest and swamp country north from the Oguooé to the Remboué. On the latter there was a small Hatton and Cookson bush factory, and once there she hoped to go by boat down the Remboué and its wide estuary to Libreville. The overland journey would take her through country inhabited by groups of Fang people who had had little or no contact with Europeans, and the French authorities were reluctant to give her permission to go. She even had difficulty finding Africans willing to make the journey. But eventually a party of five men, including an interpreter, was assembled. The weeklong, eighty- to 100-mile trip took them through dense forests of ebony and hardwood trees, across ravines and over steep hills; at times they waded up to their chins through leech-infested swamps. Their welcome in the isolated native villages where they spent the nights was not always as they would have wished, and much depended on their having with them someone from a previous stop who was known in the new village. They reached the Hatton and Cookson factory on the Remboué, however, and after a few days of rest Mary went down the estuary by sailing canoe to Libreville.

She made a brief visit to the island of Corsico, some twenty miles off the coast, hoping to find additional interesting fish specimens, and then took passage on a freighter bound for the German Territory of Cameroon. There, as a side trip, she climbed the 13,300-foot Mount Cameroon, the highest point along the West African coast. The third English person to make this ascent, she was undoubtedly the first white woman to do so. By late November 1895 she was back in Liverpool, where she was met on arrival by newspaper correspondents and treated as a celebrity. News of her travels had reached England before she did.

After the publication about a year later of her lively *Travels in West Africa,* taken in large part directly from her diaries, she became even more widely known and was much in demand as a public speaker. She met her fellow African traveler Henry Morton Stanley, and both Sir James Frazer and E. B. Tylor. She spent several weekends with Tylor and his wife in Oxford. Early in 1899 she brought out *West African Studies,* the work that contained the results of her ethnological research.[33] Sections of the manuscript were read and approved by Tylor and by other academics at Oxford and Cambridge. Although she gave special attention to spiritual beliefs (fetish), she covered many aspects of African tribal culture, and the general views she put forward plunged her into immediate controversy. She argued that the Negro was not a "savage" but lived by rules as precise and binding as those in European societies. Further, the white traders were not rough, lawless bullies, debauching Africans with imported gin (the latter she considered probably less damaging to the health than homemade palm wine). The missionaries, on the other hand, in trying to replace the restraints of the ancient religions with barely understood Christianity, seemed to her to be doing more harm than good. Most of these ideas were far from welcome in England, but she knew her opinions were well-founded and defended them stoutly.

Her zoological collecting had turned out very successfully; among the sixty-five species of fishes and eighteen reptiles she brought back, a number were valuable discoveries, and there were very few not of interest to the British Museum. Three new species of fishes were named after her (*Ctenopoma kingsleyae, Mormyrus kingsleyae* and *Alestes kingsleyae*) and sixteen of her captures were previously unknown modifications of known forms. Another of her fishes was one of which the museum had only one specimen; she also brought back a new snake, and a lizard for which they had been waiting for ten years.

She developed strong views on colonial policy and the administration of the West African territories. These she expressed forcefully in her main writings and in many lectures to popular audiences, chambers of commerce interested in West Africa, and scientific groups. A number of questions she argued out in the columns of the *Times,* the *Fortnightly Review,* and other major periodicals. She never shook off the notion, deeply ingrained from her earliest readings, of the basic superiority of the white man, and so did not dispute Britain's right to govern the region. But she did take issue with the methods being used, arguing that the British Crown Colony system should be replaced with "indirect rule" through the native chiefs, with as little disruption of African society as possible. European administrative activity she felt should be left largely in the hands of the traders who knew the Africans better than anyone else; trade was, after all, the real reason for Europeans being in Africa anyway. Not surprisingly these views stirred up considerable controversy, both among the public and in such quarters as the Colonial Office.

Although she fully expected to return to West Africa, Mary never did. Mindful of her family responsibilities, she kept house for her brother Charles when needed and left the country only when Charles, also given to travel, went abroad and left her free. For several years after 1895 Charles stayed at home, and, although eager to be off again, especially after her books were finished, Mary remained anchored. When she finally did leave England in 1900 her destination was South Africa rather than the Coast. With the outbreak of the second Boer War in 1899 there was a great need for nurses, and she volunteered; without any breach of the proprieties, she could count her duty to the country above even her responsibility to her brother. She hoped she might also be able to cover the war as a correspondent, and in her spare time collect some Orange River fishes for the British Museum. In the back of her mind she had a plan to return to West Africa when her war work was over, and go inland to the remote areas of northern Nigeria and Chad, regions of old and rich Islamic culture. When she arrived in Cape Town she was immediately asked to help at the desperately understaffed Simonstown hospital for Boer prisoners, then in the grip of a virulent epidemic of typhoid. After two months of grueling work there she herself succumbed to the fever. She died on 3 June 1900, at the age of thirty-seven. By her own wish she was buried at sea. The ceremony was conducted with full military honors, a party of the Fourth West Yorkshire Regiment escorting the cortege to the pier at Simonstown and a torpedo boat carrying the coffin out to sea.

The extensive contribution to studies of African life and customs that she made during the short seven years allotted her was warmly praised in the columns of the *Times* on 6 June. A second *Times* article the following day reported a special meeting of the African Trade Section of the Liverpool Chamber of Commerce at which it formally recognized the substantial benefit West African trade had derived from Mary Kingsley's writings.[34] The Mary Kingsley Hospital for the treatment of tropical disease was established in Liverpool in her memory by the Liverpool and Manchester merchants.

EDITH DURHAM[35] (1863–1944), artist, Balkan traveler, and political observer, although not a "systematic" anthropologist in the academic sense any more than was Mary Kingsley, was an authority in her own field, probably without rival. Perhaps more than any other British person of her time, she immersed herself in the life and past cultures of the Balkans, particularly Serbia and Albania. The books she wrote about her experiences in these regions during the first quarter of this century are still works of serious interest for Balkan historians.

She was born in Hanover Square, London, on 8 December 1863, the oldest of the nine children of Arthur Edward Durham, senior consulting surgeon at Guy's Hospital, and his wife, Mary, daughter of William Ellis, an economist colleague of John Stewart Mill. Her first interest was art, and after spending four years at Bedford College (1878–1882) she went to the Royal Academy of Arts. She was moderately successful professionally, with watercolor exhibitions at the academy and at various art institutes; she also illustrated the reptiles volume in the Cambridge Natural History series and a number of other books. One of her London scenes is in the Guildhall Gallery.

By the late 1890s, however, she found herself in the position of being the constant attendant of her invalid mother and in poor health herself. Her doctor's prescription for the depression that overcame her was two months' complete change every year. And so, in April 1900, at the age of thirty-six, she made her first trip, a voyage from Trieste down the Dalmatian coast on a boat of the Austrian Lloyd company. The choice was almost fortuitous, the Balkans then being rarely visited by tourists, but Durham the artist became more and more delighted by the country she saw. Just as Kingsley (also escaping from exhausting demands of family nursing) responded immediately and deeply to her first sight of the palm-fringed coastline of West Africa, so Durham was enchanted by the exotic peacock-blue of the Adriatic and the beautiful islands with their colorful, polyglot population, set against the backdrop of the mainland mountains. At Cattaro (later Kotor) she went ashore, hired a horse, and climbed up above the port into the rocky, arid, almost impassable hills, getting her first glimpse of the mountains of Montenegro and Albania, a landscape she found so magnificent that it was to her worth the journey from England just to have seen it.[36]

Once back in London, her imagination totally captivated by the seeming picturesqueness of Balkan life, she immersed herself in Balkan history and languages. Over the next few years she became fluent in both Serbian and Albanian, and, setting herself to come to some kind of understanding of the complexities of Balkan culture and politics, began to study the manners and customs of the region, one district at a time—a vast program, for which she had only eight weeks a year as far as her fieldwork was concerned. She fully intended to carry out her investigations all over the Balkans, but these plans were interrupted, first by the 1903 Macedonian revolt, quelled with exceptional brutality by the Turks, then by further troubles in 1909, the Albanian revolution against the Turks in 1911, the Balkan wars of 1912–1913, and finally the First World War. During all these upheavals the time she spent in the Balkans was fully taken up with relief work, nursing wounded and feeding refugees under extremely difficult circumstances. She became accepted (especially in Albania) as something of a national champion and was known locally as "Kraljica e Maltsorëvet" (Queen of the Highlanders). Her popularity depended, however, not so much on her effectiveness as a relief worker, but on her courage and honesty in dealing with complex and intricate problems. Her fellow countryman Aubrey Herbert, also deeply and sympathetically involved in Balkan affairs, saw her at the time as clever, aggressive, and competitive (with a Cockney accent and her hair cut short like a man), but as someone who had really accomplished a great deal for the people of the region.[37]

Over the years she came to feel more and more strongly that her early idea of studying customs as the key to understanding peoples was the right approach, and her travels, which gradually became more adventurous, acquired a serious ethnographic purpose. In 1904 she published her first major work, *Through the Lands of the Serb,* the result of four trips to Montenegro, an extended tour through the Kingdom of Serbia, and a visit to the province of Kosovo, then still part of the Ottoman Empire. Her descriptions of towns and her archaeological notes and drawings were of particular interest to western anthropologists. *The Burden of the Balkans,* a discussion of her Macedonian and Albanian experiences, followed a year later.

After the death of her mother in 1906 her freedom to travel was greater, and she continued her Balkan studies with a journey through Bosnia and Herzegovina. Montenegro remained her main base, and in 1907 she was asked to take the job of Commissioner for Montenegro at the Balkan States Exhibition at Earls Court, London. The exhibition did little to promote Montenegrin industry, then in a rudimentary state, but it did bring Edith Durham to the notice of British anthropologists. She became a fellow of the Royal Anthropological Institute the following year, later sat on the council and the executive committee, and was elected the first woman vice president.

She made another Balkan tour in 1908, a journey into the trackless, tribal areas of northern Albania that was almost unprecedented for the time. As a woman of forty-five, traveling rough with seldom more than an elderly guide and a horseboy for company, she presented no threat and was accepted where no ordinary foreign male could have made his way without a sizable armed escort. Furthermore, she spoke the language, was sympathetic to the people's concerns, and had a well-placed trust in the highlanders. *High Albania* (1909) described the communities she visited and the people she talked with.

In all her major writings Durham tried to show how no solution could be found for the Balkan question unless the vastly different cultures of the constituent states were adequately recognized. In Britain the region's problems were seen largely as a struggle of European Christian against Moslem Turk, the final stages of the long drawn-out collapse of the Ottoman Empire obscuring the details of ancient, deep-rooted regional quarrels. Durham was one of the very few who rightly interpreted the situation as a conflict of national (rather than religious) aspirations. The key problems then, as now in the struggles of the last decade of the twentieth century, were overlapping territorial claims based on transitory boundaries of the past, boundaries that had been made and remade over the centuries. The Albanians, largely a Moslem people, Durham saw as being at a disadvantage compared with their Christian neighbors, and she tried to plead their case for self-rule in her dispatches to the British press, especially the *Times,* the *Manchester Guardian,* and the *Nation,* and in essays in periodicals.[38] She also persistently lobbied the British Foreign Office. With Aubrey Herbert she founded the Anglo-Albanian Association, a small caucus of M.P.s and others concerned about Albanian interests; for a time she served as its secretary. Throughout the First World War she had to remain in Britain. She visited Tirana once again in 1921 and was welcomed by processions and singing crowds; but her health was by then frail, and she did no further traveling. She was awarded the Order of Skanderberg, and the Albanian government offered her a permanent home in its country. She chose to remain in London, however, feeling she could do more good from there by continuing to act as an independent spokesperson for Albania's interests.

During her later years she was especially active in the Anthropological Institute, contributing not only papers, but also money and much time put into administrative work. Her intimate knowledge of the Balkans was always at the disposal of her colleagues. She had assembled many fine collections of artifacts, each piece explained by an accompanying annotation. In their richness these presented a composite picture in microcosm of the complex history and ethnology of the Balkans; prehistoric motifs of the Bronze Age, still surviving beliefs from pagan times, ideas from Mithraism, Catholic and Orthodox Christianity, Rome, Byzantium, and Islam, all were abundantly represented.

Her writing continued throughout the 1920s and 1930s. As well as her many notes and articles, mainly in the journal *Man*,[39] she produced three more books, *Twenty Years of Balkan Tangle* (1920), a summary of her travels and a political commentary dealing with the years immediately after the First World War, *The Sarajevo Crime* (1925), which was purely political, and finally, *Some Tribal Origins, Laws and Customs of the Balkans* (1928), a systematic collation of her ethnological research in Albania, Montenegro, and Bosnia, much of it previously published in *Man*.[40]

She died on 15 November 1944, at her home in London, a month before her eighty-first birthday. Outspokenly critical of hypocrisy or ignorance in high places, she was remembered as a remarkably vigorous personality. In a letter to the London *Times* of 21 November, King Zog of Albania, speaking, he said, for Albania, acknowledged her efforts on behalf of his country. Noting that she had first gone to Albania when it was not much more than a geographical term, he went on to pay tribute to the help she had given over the years through her books and articles pleading Albania's case in its struggle for independence.[41] Her collections of photographs, original line drawings, and watercolors were left to the Anthropological Institute. An outstanding collection of embroideries and textiles went to the Bankfield Museum, Halifax,[42] other collections to the Pitt Rivers Museum, Oxford, and the Ethnological Museum, Cambridge.

NINA LAYARD[43] (1853?–1935), whose archaeological discoveries in East Anglia brought her a more than local reputation, was the daughter of the Rev. C. C. Layard, rector of Combe Hay, Bath; her uncle was the eminent archaeologist Sir Austen Layard, the excavator of Nineveh. She became interested in archaeology when she was in her mid-forties, following earlier work in botany, and over the course of the next twenty-five years took part in excavations not only in Suffolk and Norfolk, but in other parts of England and abroad. She was the first woman to become a fellow of the Society of Antiquaries, and for many years served as a vice president (also for a time honorary secretary) of the Suffolk Institute of Archaeology.

Beginning in the late 1890s she published a number of papers on excavations of religious sites around Ipswich, a region in which she had become especially interested after one of her brothers became a curate there. Then came several seasons of extensive work on Paleolithic sites, again mainly in the Ipswich area, following which, in 1906, she superintended the excavation of an Anglo-Saxon cemetery at Hadleigh Road, Ipswich. The latter was a particularly important piece of work and produced numerous finds, all of which were deposited in the Ipswich Museum. Among her other discoveries were Roman remains, a communal kitchen, and a Neolithic mining area. Her papers and reports, of which there were at least thirty, were presented at British Association meetings and published in both national journals and those of local institutes.[44] She also brought out two historical monographs, *Brief Sketch of the History of Ipswich School, 1477–1851* (1901) and *Seventeen Suffolk Martyrs* (1902). Her collection of verse, *Poems,* appeared in 1890. She died in August 1935 at age eighty-two.

Summary

Unlike American women ethnologists of the time who had within their own country a variety of native American communities well worth studying, British women interested in ethnological work had perforce to move beyond the frontiers of their native land. A great many did.[45] Of those mentioned here, two especially, Kingsley and Durham, are still figures of continuing interest, both having made creditable contributions to ethnological research. Like their fellow countrywoman Isabella Bird Bishop, they brought out books that constitute notable additions to the historical and anthropological records. Both were unconventional people whose exceptional vigor, force of character, and desire for the adventure of exploring new and exotic places brought them to the fore not only among anthropologists but in wider circles. Kingsley's name became familiar to the public in general, and Durham was well-known among those working on the political problems of southeastern Europe. With her more normal early family life, college-level education, and some professional success as an artist behind her, Durham was the more solidly self-confident of the two, and indeed was very much a woman of the twentieth century rather than a Victorian. Nevertheless, there are fairly strong parallels between her career and Kingsley's. Both women spoke out in support of peoples they saw as being maltreated and exploited by powerful outside interests and governments, and for both ethnological field research, while of major importance in the search for an understanding of the cultures studied, became part of a wider undertaking, one that involved a considerable amount of informal but serious political work.

Aynsley and Layard, the other two women in this group who made substantial written contributions to anthropology, were less dramatic figures than Kingsley and Durham; Aynsley's visits to "our Indian possessions" were made in a relatively conventional way, that is, in the company of her husband, and most of Layard's archaeological work was carried out near her home, where she had a family base to operate from.

Two-country comparison

A major difference between the British and American early women ethnologists and anthropologists is the fact that the British worked privately and independently without significant backing from institutions, while the Americans benefited from contacts with such organizations as Harvard's Peabody Museum or the Smithsonian Institution. These not only provided background support and a certain amount of financial help for the best known of the Americans (Smith, Fletcher, and Stevenson), but in fact men on their staffs were instrumental in introducing the women to the field in the first place. Largely as a consequence, the American women produced a number of early and important formal scholarly papers, for example Matilda Stevenson's lengthy publications in the *Reports* of the Smithsonian Institution. The British, on the other hand—Buckland, Aynsley, Kingsley, Durham, and also Bird Bishop—in their major writings necessarily aimed at a much wider popular audience. However, a considerable amount of the later British work (by Durham and Layard) appeared in scholarly journals.

Interestingly, each national group included one archaeologist, Babbitt, the American, working on Paleolithic sites in Minnesota, Layard uncovering various levels of ancient cultural remains in southeast England. Overall the best-remembered in each group—Fletcher, Stevenson, and Smith in the United States, Kingsley, Durham, Bird Bishop, and to a lesser extent Layard and Aynsley in Britain—were remarkably effective in their undertakings.

Because of the exceptionally big difference between the number of anthropology papers indexed in the Royal Society *Catalogue* and those listed in the bibliography (which gives *Catalogue* entries plus additional pre-1901 material come across) two sets of bar graphs are given below. Figure 14-1, representing only papers listed by the Royal Society, is consistent with the bar graphs presented for other fields. Figure 14-2, representing all the anthropology entries in the bibliography, is probably a truer reflection of the activities of this group of women, though no claim is made for its being a complete count of their pre-1901 journal articles. The expanded counts result largely from the inclusion of additional articles by Fletcher (United States, 1880s and 1890s) and Kingsley (Britain, 1890s).

Notes

1. Nancy Oestreich Lurie, "Smith, Erminnie Adelle Platt," in *NAW,* vol. 3, pp. 312–13, and "Women in early American anthropology," in *Pioneers of American Anthropology. The Uses of Biography,* ed. June Helm, (Seattle: University of Washington Press, 1966), pp. 31–81, on pp. 40–3; Eleanor S. Elder, "Women in early geology," *Journal of Geological Education,* **30** (1982): 287–93; *In Memoriam. Mrs. Erminnie A. Smith,* comp. Sara L. Saunders-Lee (Boston: Lee and Shepard, 1891); W[alter] H[ough], "Smith, Erminnie Adelle Platt," *DAB,* vol. 17, 1935, p. 262; Vimala Jayanti, "Erminnie Adelle Platt Smith," in *Women Anthropologists. A Biographical Dictionary,* eds. Ute Gacs, Aisha Khan, Jerrie McIntyre, Ruth Weinberg (Westport, Conn.: Greenwood Press, 1988), pp. 327–30.

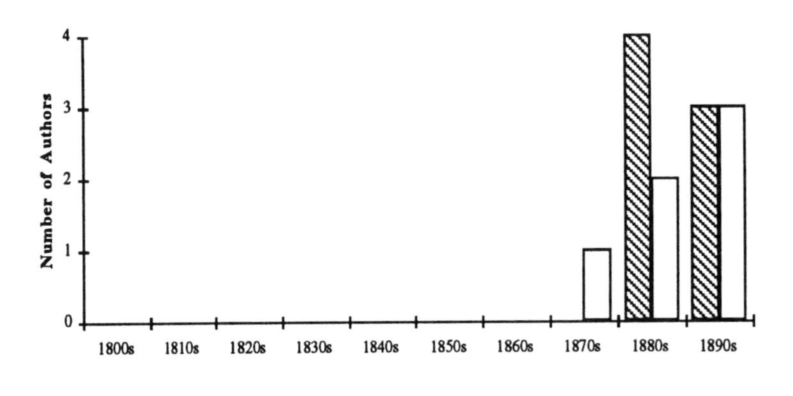

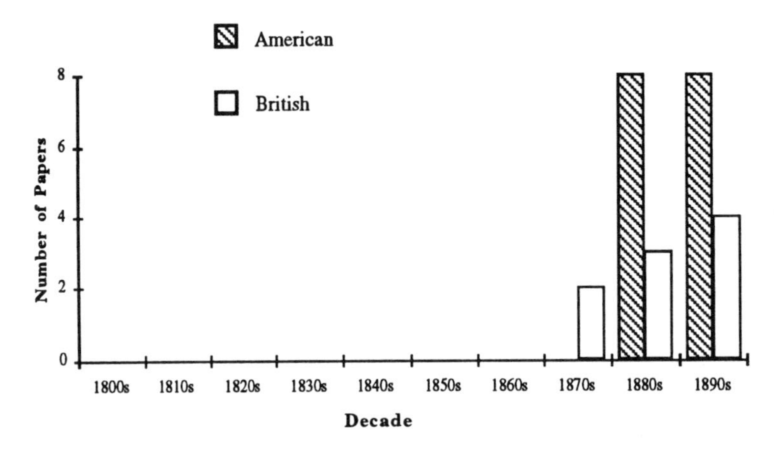

Figure 14-1. American and British authors and papers in anthropology and ethnology, by decade, 1800–1900. Data from the Royal Society *Catalogue of Scientific Papers.*

2. John R. Swanton, "John Napoleon Brinton Hewitt," *American Anthropologist,* **40** (1938): 286–90.

3. Joan Mark, *A Stranger in her Native Land. Alice Fletcher and the American Indians* (Lincoln, Neb.: University of Nebraska Press, 1988), and "Francis La Flesche: the American Indian as anthropologist," *Isis,* **73** (1982): 497–510; Lurie, "Women in anthropology," pp. 43–54; Thurman Wilkins, "Fletcher, Alice Cunningham," in *NAW,* vol. 1, pp. 630–3; obituary, "Alice Cunningham Fletcher," *American Anthropologist,* **25** (1923): 254–8.

4. Mark, *Stranger,* pp. 8–12.

5. Alice C. Fletcher, *Indian Education and Civilization,* Special Report, U.S. Bureau of Education (Washington, D.C.: U.S. Government Printing Office, 1888, 693 pp.).

6. Mark, *Stranger,* p. 200.

7. Among Fletcher's other post-1900 publications are: "Star cult among the Pawnee," *American Anthropologist,* **4** (1902): 730–6; "Pawnee star lore," *Journal of American Folklore,* **16** (1903): 10–15; "The Indian and Nature," *American Anthropologist,* **9** (1907): 440–3; "Indian music," *Bureau of American Ethnology, Bulletin 30,* 1907; "Tribal structure: a study of the Omaha and cognate tribes," in *Putnam Anniversary Volume,* ed. Franz Boas (New York: Stechert, 1909), pp. 245–76; "The child and the tribe," *Proceedings of the National Academy of Sciences,* **1** (1915): 569–74; *Indian Games and Dances with Native Songs, arranged from American Indian Ceremonial and Sports* (with music) (Boston: C. C. Birchard, 1915, and many later eds.); "The study of Indian music," *Proceedings of the National Academy of Sciences,* **1** (1915): 231–5; "The Indian and Nature: the basis of his tribal organization and rites," *Red Man,* **8** (1916): 185–8; "Prayers voiced in ancient America," *Art and Archaeology,* **9** (1920): 73–5.

8. Mark, *Stranger,* p. 345.

9. Nancy Oestreich Lurie, "Stevenson, Matilda Coxe Evans," in *NAW,* vol. 3, pp. 373–4, and "Women in anthropology," pp. 54–64; *NCAB,* vol. 20, 1929, pp. 53–4; Nancy J. Parezo, "Matilda Coxe Evans Stevenson," in Gacs, Khan, McIntyre and Weinberg, eds., *Women Anthropologists,* pp. 337–43.

10. Powell's views on this point, though still somewhat novel for the time, were fully supported by Oxford anthropologist Sir Edward Tylor, who visited the Stevensons at Zuñi. Tylor (see below), one of the most respected ethnologists of the period, also emphasized the special advantages women had in some aspects of fieldwork.

11. Alice Fletcher's article on Omaha children, "Glimpses of child-life among the Omaha tribe of Indians," appeared in the first volume of the *Journal of American Folklore* the following year (1888).

12. Stevenson had many heated arguments with bureau director Powell, who responded by dismissing her from her job at regular intervals. She in turn threatened to appeal to Congress and was duly restored to her position. On one occasion, it is said, she stormed into Powell's office and loudly and abruptly accused him of being a liar, an outburst that so aroused the Major that he had the first of the strokes that later were to cause his death (Lurie, "Women in anthropology," p. 234, n. 53).

13. "Studies of the Tewa Indians of the Rio Grande valley," *Smithsonian Miscellaneous Collections,* **60,** n. 30 (1913): 35–41; "Strange rites of the Tewa Indians," ibid., **63,** n. 8 (1914): 73–80;

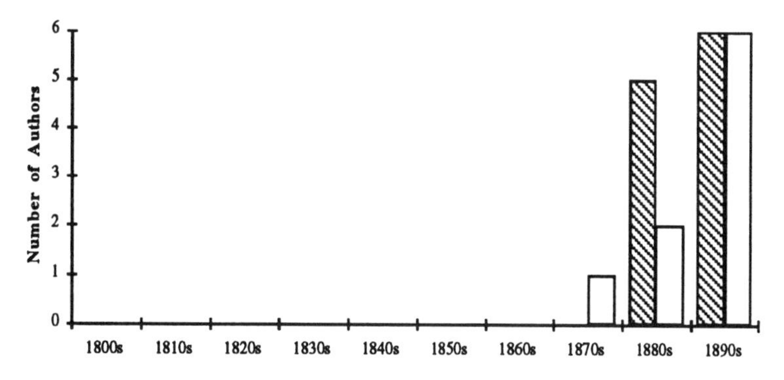

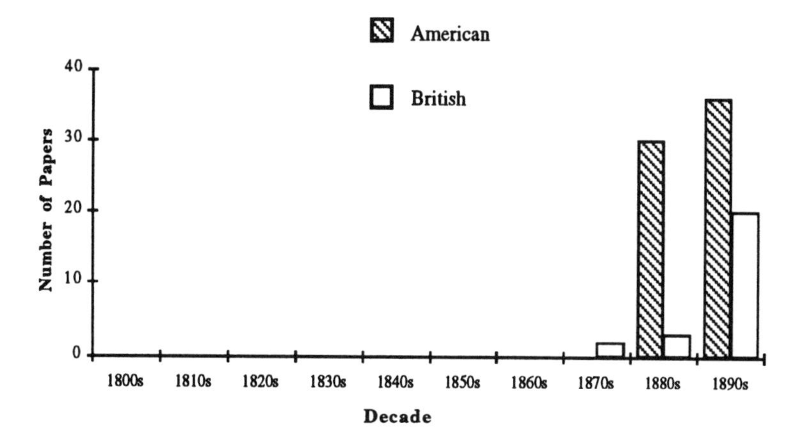

Figure 14-2. American and British authors and papers in anthropology and ethnology, by decade, 1800–1900. Data from the Royal Society *Catalogue of Scientific Papers* and the additional material listed in the bibliography.

"The Sun and Ice People among the Tewa Indians of New Mexico," ibid., **65** (1915): 73–8.

14. Anon., obituary, *Coldwater Republican,* 7 July 1891 (clipping in Branch County Library, Coldwater, Michigan).

15. Putnam later talked about Babbitt's Little Falls quartzes before the Boston Society of Natural History, comparing them with Paleolithic tools found in Delaware and Indiana (*Proceedings of the Boston Society of Natural History,* **24** (1890): 157–65; excellent photographs of several of Babbitt's specimens appear on pp. 164–5).

16. See Franc E. Babbitt, "Points concerning the Little Falls quartzes," *Proceedings of the AAAS,* (1889): 333–9, and Warren Upham, "The recession of the ice-sheet in Minnesota in its relation to the gravel deposits overlying the quartz implements found by Miss Babbitt at Little Falls, Minnesota," *Proceedings of the Boston Society of Natural History,* **23** (1888): 436–49.

17. F. E. Babbitt, "Art in Michigan," *Coldwater Republican,* 1 January 1870, reprinted from the *Detroit Post,* 20 December 1869 (clipping in Branch County Library, Coldwater, Mich.).

18. Dorothy Middleton, *Victorian Lady Travellers* (London: Routledge and Kegan Paul, 1965), pp. 90–103; François Bontinck, *Aux Origines de l'État Indépendent du Congo* (Louvain: Éditions Nauwelaerts. Publications de l'Université Lovanium de Leopoldville, 1966), Appendice II, pp. 450–8; Dea Birkett, *Spinsters Abroad. Victorian Lady Explorers* (Oxford: Basil Blackwell, 1989), pp. 24, 237, 282; Jane Robinson, *Wayward Women. A Guide to Women Travellers* (Oxford: Oxford University Press, 1990), pp. 27–8.

19. Anon., "Geographical notes: 'Mrs. French-Sheldon's visit to Lake Chala'," *Proceedings of the Royal Geographical Society,* **13** (1891): 431–2.

20. She also gave a talk entitled, "Customs of the natives of East Africa," at the International Congress of Anthropology at the 1893 Chicago World's Fair—see W. H. Holmes, "The World's Fair Congress of Anthropology," in *Selected Papers from the American Anthropologist, 1888–1920,* ed. Frederica de Laguna (Evanston, Ill.: Row, Peterson, 1960), pp. 119–30, on p. 122.

21. Félix É. Regamey, *Japan in Art and Industry,* tr. from the French by May French-Sheldon (New York: G. P. Putnam, 1893).

22. M. French-Sheldon, F.R.G.S., "The Congo State," *Times,* 3 February 1905, 8e, f.

23. Bontinck, *Aux Origines,* p. 456, n. 42. Bontinck refers to letters in the "Sheldon Papers," nine boxes of which are housed in the Ms. Division of the U.S. Library of Congress.

24. For Agnes Crane see chapter 12.

25. L. S. Hearnshaw, *A Short History of British Psychology 1840–1940* (London: Methuen, 1964), pp. 112–13.

26. For Isabella Bird Bishop see chapter 13. The one paper by Durham listed in the bibliography is in the zoology section.

27. No biographical information has been collected about Gertrude Godden, who was from Wimbledon, London. Her extensive study of frontier tribes of northeast India, presented before the Anthropological Institute in two papers in 1895 and 1898, was probably carried out with materials available in Britain. She also had broad interests in folklore. Over a period from the 1920s until about 1944 a Gertrude M. Godden brought out several books and pamphlets on current political questions, particularly the rise of Fascism and what she regarded as the threat to Britain of Communism. This may have been the same Godden.

28. F. W. Rudler, "Presidential Address," *Journal of the Royal Anthropological Institute,* **28** (1899): 325–6.

29. Robinson, *Wayward Women,* p. 80; *British Biographical Archive,* microfiche ed., no. 1055 (entry from J. F. Kirk, *A Supplement to Allibone's Critical Dictionary of English Literature*—Philadelphia: J. B. Lippincott, 1891).

30. Anon., "The late Lady Maria Welby," *Times,* 19 July 1920, 21d; *Burke's Peerage and Baronetage* (105th. ed., London: Burke's Peerage Ltd., 1975), p. 2781; *Who was Who,* vol. 3, p. 1432, entry for Welby, Sir Charles Glynne Earle.

31. Katherine Frank, *A Voyager Out. The Life of Mary Kingsley* (Boston: Houghton Mifflin, 1986); Robert D. Pearce, *Mary Kingsley: Light at the Heart of Darkness* (Oxford: Kensal Press, 1990); Middleton, *Victorian Lady Travellers,* pp. 149–76; Dea Birkett, *Mary Kingsley. Imperial Adventuress* (London: Macmillan, 1992).

32. Birkett, *Kingsley,* p. 156.

33. Kingsley brought out two other books in 1899, *The Story of West Africa,* a short historical survey of European, mainly British, trading connections with the Coast, and *Notes on Sport and Travel,* a memoir of her father together with a collection of his writings.

34. Anon., "Miss Mary Kingsley," *Times,* 6 June 1900, 8d; ibid., 7 June 1900, 9e.

35. Memorials by John L. Myres, H. J. Braunholtz and Beatrice Blackwood, "Mary Edith Durham: 8 Dec. 1863—15 Nov. 1944," *Man,* **44** (1944), Nos. 12–14; Henry Hodgkinson, "Durham, (Mary) Edith," *DNB,* "Missing Persons" Supplement, 1993, pp. 197–8; John Hodgson, "Introduction" to the 1987 ed. of Edith Durham's *High Albania* (Boston: Beacon Press), pp. xi–xvi; Robinson, *Wayward Women,* pp. 260–1; Birkett, *Spinsters Abroad; Who Was Who,* vol. 4, p. 341.

36. Edith Durham, *Through the Lands of the Serb* (London: Edward Arnold, 1904), p. 15.

37. Margaret Fitzherbert, *The Man who was Greenmantle: A Biography of Aubrey Herbert* (London: John Murray, 1983), p. 117.

38. See, for instance, M. E. Durham, "'Constitution' in North Albania," *Contemporary Review,* **94** (1908): 533–41, and "The Serb and Albanian frontiers," ibid., **95** (1909): 15–23.

39. Durham's papers on Balkan history and ethnology, published over a period of more than three decades (1904 until about 1935), include the following: "My golden sisters: a Macedonian picture," *Monthly Review,* **11** (1904): 73–81; "The story of Karageorge," *Independent Review,* **4** (1904): 58–68; "As others see it (a sketch in old Servia)," *Macmillan's Magazine,* **1** (1905): 35–44; "The western Balkans peninsula," in *Women of All Nations: a Record of their Characteristics, Habits, Manners, Customs and Influence,* eds. Thomas A. Joyce and Northcote W. Thomas (London: Cassell, 1908), pp. 672–85; "Some Montenegrin manners and customs," *Journal of the Royal Anthropological Institute,* **39** (1909): 85–96; "Balkan gypsies," *Journal of the Gypsylore Society* (Liverpool), **4** (1910): 66–70; "Albanian and Montenegrin folklore," *Folklore,* **23** (1912): 224–9; "Head-hunting in the Balkans," *Man,* **12** (1912) No. 94; "Some south Slav customs as shown in Serbian ballads and by Serbian authors," *Journal of the Royal Anthropological Institute,* **47** (1917): 435–47; "Magic," *Man,* **18** (1918) No. 57; "The Serbs as seen in their national songs," *Contemporary Review,* **117** (1920): 531–8; "Ritual nudity in Europe," *Man,* **20** (1920) No. 83; "Old beliefs and modern politics," ibid., **21** (1921) No. 41; "Cowries in the Balkans," ibid., **23** (1923) No. 11; "Head-hunting in the Balkans," ibid., **23** (1923) No. 11; "Dardania and some Balkan place-names," ibid., **23** (1923) No. 21; "A bird tradition in the west of the Balkan peninsula," ibid, **23** (1923) No. 33; "Some Balkan embroidery patterns," ibid., **23** (1923) No. 40; "Some Balkan taboos," ibid., **23** (1923) No. 52; "The seclusion of maidens from the light of the sun, and a further note on the bird tradition in the Balkans," ibid., **23** (1923) No. 61; "Some Balkan remedies for disease," ibid., **23** (1923) No. 82; "Of magic, witches and vampires in

the Balkans," ibid., **23** (1923) No. 121; "Preservation of pedigrees and commemoration of ancestors in Montenegro," ibid., **31** (1931) No. 163; "Childhood and totemism," ibid., **33** (1933) No. 45; "Jocesta's crime," ibid., **33** (1933) No. 129; "The making of a saint," ibid., **33** (1933) No. 151; "Pots and pies [cooking customs in Montenegro]," ibid., **34** (1934) No. 20; "Fear of the dead," ibid., **34** (1934) No. 157; "A Russian funeral feast," ibid., **34** (1934) No. 207; "Bride price in Albania," ibid., **35** (1935) No. 102.

40. Among Durham's works not already mentioned is her 1914 book, *The Struggle for Scutari (Turk, Slav and Albanian)* (London: Edward Arnold), an account of Albanian history between 1910 and 1914, and a description of her relief work there in the Balkan War of 1912. The book's purpose was to expose the fraud, treachery, and cold-blooded brutality of the various powers that had fought each other for possession of or influence over that little region. Her other articles include, "The blaze in the Balkans," *Monthly Review* (1903): 54–65; "The story of Essad Pasha," *Contemporary Review,* **118** (1920): 207–15; "King Nikola of Montenegro," ibid., **119** (1921): 471–7; "Croatia and Greater Serbia," ibid., **124** (1923): 590–600.

41. *Times,* 21 November 1944, 6.

42. Laura E[mily] Start, *The Durham Collection of Garments and Embroideries from Albania and Yugoslavia* (with notes by M. Edith Durham) (Halifax: Calderdale Museums, 1977, reprinted from the original ed. published in 1939 by Halifax Corporation).

43. Anon., "Miss Nina Layard," *Times,* 24 August 1935, 12b; Francis Seymour Stevenson, "In memoriam. The Rev. Edmund Farrer and Miss Nina Layard," *Proceedings of the Suffolk Institute of Archaeology,* **22** (1934–1936): 228–9; A. V. Steward, comp. and ed., *A Suffolk Bibliography* (Ipswich: Suffolk Records Society, 1979).

44. Layard's post-1900 papers on East Anglian archaeology include the following: "Notes on a human skull found in peat in the bed of the river Orwell," *Man,* **1** (1901) No. 125; "On a recent discovery of palaeolithic implements in plateau gravels of Ipswich," *Reports of the British Association* (1902): 759; "Palaeolithic implements in Ipswich," *Nature,* **66** (1902): 77; "A recent discovery of palaeolithic implements in Ipswich," *Journal of the Royal Anthropological Institute,* **33** (1903): 41–3; "Further excavations on a palaeolithic site at Ipswich," ibid., **34** (1904): 306–10, and *Reports of the British Association* (1905): 725–6, (1907): 693–4; "Some English paxes, including an example recently found in Ipswich," *Archaeological Journal,* **61** (1904): 120–30; "A winter's work on the Ipswich palaeolithic site," *Journal of the Royal Anthropological Institute,* **36** (1906): 233–6; "On a palaeolithic site in Ipswich," *Publications, Reports and Communications of the Cambridge Antiquarian Society,* **11** (1907): 493–502; "Anglo-Saxon cemetery in Ipswich," *Man,* **6** (1906) No. 101, *Archaeologica,* **60,** Pt. 2 (1907): 325–52, *Reports of the British Association* (1907): 694–5; "Anglo-Saxon cemetery, Hadleigh Road, Ipswich," *Proceedings of the Suffolk Institute of Archaeology,* **13** (1907): 1–19; "An account of a discovery at Ipswich of an Anglo-Saxon cemetery," *Proceedings of the Society of Antiquaries,* **21,** s. 2, (1905–1907): 241–7, 403; "Notes on an ancient land surface in a river terrace at Ipswich, and on palaeoliths from a gravel pit in the valley of the Lark," *Reports of the British Association* (1908): 861; "Alabaster figurines from the church of Fornham All Saints," *Proceedings of the Society of Antiquaries,* **22,** s. 2 (1907–09): 502–3; "The older series of Irish flint implements," *Man,* **9** (1909) No. 54; "Some early crucifixes, with examples from Raydon, . . . Ipswich," *Archaeological Journal,* **67** (1910): 91–7; "Animal remains from the railway cutting at Ipswich (including bones and teeth of a heavy limbed horse," *Proceedings of the Suffolk Institute of Archaeology,* **14** (1910): 59–68; (with W. E. Underwood, W. Allen Sturge, W. G. Clark, and Frank Corner) "Report of the special committee [of the Prehistorical Society of East Anglia appointed to enquire and report upon the flint implements of pre-crag man exhibited by Mr. J. Reid Moir]," *Proceedings of the Prehistoric Society of East Anglia,* **1** (1911): 24–39; "The Nayland figure-stone," *Proceedings of the Suffolk Institute of Archaeology,* **15** (1913–1915): 3–8; "Reports on discoveries in Ipswich and the neighbourhood," ibid., **15** (1913–1915): 84–6, 226–7; "Coast finds by Major Moore at Fleixstowe Ferry," *Proceedings of the Prehistoric Society of East Anglia,* **2** (1914–1918): 132–4; "Discoveries made by residents in Ipswich and the neighbourhood," *Proceedings of the Suffolk Institute of Archaeology,* **16** (1916–1918): 68–70; "Points of special interest in the Anglo-Saxon discoveries in Ipswich," ibid., **16** (1916–1918): 278–80; "Stoke bone-bed, Ipswich," *Proceedings of the Prehistoric Society of East Anglia,* **3** (1918–1922): 210–19; "Bronze crowns and a bronze head-dress from a Roman site at Cavenham Heath," *Antiquaries Journal,* **5** (1925): 258–65.

45. See Robinson, *Wayward Women.*

Chapter 15

THE FIRST GENERATION OF AMERICAN EXPERIMENTAL PSYCHOLOGISTS

Articles in psychology make up about 2 percent of the papers by women indexed by the Royal Society and almost 90 percent of them were contributed by American authors.[1] The remainder came from German and British women.

Of the twenty-four Americans whose psychology papers are listed in the bibliography, ten were associated with Wellesley College, either as students or faculty, six took Ph.D. degrees at Cornell University, and eleven taught at college level at least for a time during their careers (seven at eastern women's colleges). Most of them worked in the area of experimental psychology, a new field that then emphasized empirical investigation of elemental sensations and emotions. Several were known internationally and five in particular—Calkins of Wellesley, Washburn of Vassar, Downey of the University of Wyoming, Ladd-Franklin of Johns Hopkins and Columbia, and Martin of Stanford— received notable recognition during their lifetimes.

In the late nineteenth century the recognition of psychology as an independent discipline, separate from its ancient parent philosophy, was still relatively recent. During the 1870s and 80s Harvard psychologist William James, drawing in part on the work of Wilhelm Wundt of Leipzig in the 1860 and 70s, sought to establish a program in experimental work and formulate his own ideas on the subject. The appearance of James's *Principles of Psychology* in 1890 constituted a landmark in the development of the field in America, and the following years saw its rapid expansion.[2] James and two students of Wundt, Hugo Münsterberg (Harvard) and the Englishman Edward Titchener (Cornell), were the three men who were especially prominent in the training of early women psychologists in America.

MARY CALKINS[3] (1863–1930), a student of James and Münsterberg, was the first woman to hold the position of president in both the American Psychological Association (1905) and the American Philosophical Association (1918). She taught at Wellesley College for forty-two years.

Born in Hartford, Connecticut, 30 March 1863, Mary was the oldest of the five children of Wolcott Calkins, a Presbyterian minister with a Yale degree, whose forbears were seventeenth-century Welsh settlers in Massachusetts. His wife, Charlotte (Whiton), was descended from John and Priscilla Alden of the Plymouth Colony. Mary spent most of her childhood in Buffalo, New York, but in 1881, when she was seventeen, the family moved to Newton, Massachusetts, not far from Boston. A year later she entered Smith College as a second-year student with advanced standing. The death of her younger sister in 1883 resulted in her remaining at home for a year, but she continued her studies privately, returned to Smith in 1884, and graduated, with a concentration in classics and philosophy, in 1885. During an extended family trip to Europe the following year she spent several months in Italy and Greece, broadening her acquaintance with the classical world. Shortly after her return in 1887 her father arranged for her to meet with the president of Wellesley College, which was not far from her home, and it was agreed that she would fill an unexpected vacancy in the college's Greek department. She began to teach almost immediately, first with the rank of tutor and two years later as instructor.

Wellesley, still only fifteen years from its founding, was making plans to develop a separate department of experimental psychology (the subject was commonly taught in philosophy departments at the time). Since Calkins had studied philosophy as an undergraduate and was showing special promise as a teacher, she was asked to consider graduate training in philosophy and psychology as preparation for teaching the latter. However, there were few opportunities in the United States in 1890 for advanced training in the new discipline. Clark University in Worcester, Massachusetts, and Harvard were among the few places in New England where instruction might be had, but neither was open to women.[4] Nevertheless, Calkins's father submitted a lengthy petition to Harvard Corporation on her behalf, arguing that, since what was being requested was postgraduate and professional training for a Wellesley College faculty member, the question of coeducational instruction in general was not being raised. The plea was successful and Calkins was authorized to attend psychology courses taught by William James and Josiah Royce, with the understanding that she would not be registered as a Harvard student. At the same time, she received private instruction from Edmund Sanford at Clark University. In 1891 she returned to Wellesley as instructor of psychology in the philosophy department and proceeded to establish the college's psychology laboratory—in an attic room furnished with equipment acquired for the sum of $200, probably a fairly generous allowance for the time. The laboratory was the first in the field in an American women's college and one of the earliest in the country. Fifty-four students, a remarkable number, took the course in its first year,[5] carrying out a variety of simple experiments that allowed first-hand study of such topics as space perception and reaction time; the class also dissected a sheep's brain, a project in keeping with Calkins's close connection with physiologist/psychologist William James.

Calkins continued graduate studies at Harvard, having received further special permission to do so for the three years between 1892 and 1895, and worked under the guidance of Hugo Münsterberg. Although she requested and was given an examination that, even if informal, was equivalent to that required of a Ph.D. candidate, Harvard Corporation declined to award the degree. All the same she was acknowledged to be the strongest student the philosophy department had seen for several years. When, in 1902, with the further organization of Radcliffe College degrees, she was offered a Radcliffe Ph.D., she declined on the ground that none of her work had been done at Radcliffe. A few years later she received honorary doctorates from both Columbia (Litt.D., 1909) and Smith (LL.D., 1910).

At Wellesley she became associate professor of philosophy and psychology in 1895 and full professor three years later. She carried out a considerable amount of laboratory work during the early years of her career and produced a steady succession of papers that regularly appeared in the psychology literature under the general title, "Wellesley College Psychological Studies." Her research covered a broad range of topics typical in experimental work at the time, including color theory, space-time consciousness, child psychology, and studies on dreams. The latter she had begun with Sanford at Clark University. Her "Statistics of dreams" study (1893) on the fraction of dreams representing recent external events, and the work of her students Sarah Weed, Florence Hallam, and Emma Phinney (1896) on the relative number of good versus bad dreams, are still referred to in current works.[6]

Starting in 1892 she also brought out a number of articles on her work on memory and association, carried out in Münsterberg's laboratory at Harvard, including the research that would have been her doctoral dissertation. Her paired-associate procedure, a learning method in which the subject learns specific pair associations (numbers with colors, or words with other words—the "flash-card" principle), was considered one of her most important contributions. It augmented earlier research by Hermann Ebbinghaus in the 1880s on the formation of associations by learning lists of nonsense syllables; later refinements were added by G. E. Müller.

By about 1900 Calkins was moving away from experimental work and toward psychological theory. The first of a series of papers developing her ideas of psychology as the science of self appeared that year. However, her emphasis here on the conscious self, the whole person, as an integral unit that cannot profitably be broken down into component parts for the investigation of elemental sensations and emotions, did not fit in with current trends in the field, although it foreshadowed the work of later psychologists in the area of personality studies. She pushed her theory forcefully over the years, but it was never widely accepted by her psychologist colleagues.[7] As her career progressed she focused more and more on philosophical questions, and here also her chief interest lay in the question of the importance of self and the idea that everything that is real is ultimately mental and consequently personal.[8]

She published close to seventy papers in psychology and about thirty-seven more in philosophy. Her two textbooks, *An Introduction to Psychology* (1901), and *A First Book in Psychology* (1909), were widely used, both running to several editions. *Persistent Problems in Philosophy,* first published in 1907 and reissued five times, was probably her most important book; a second major work in philosophy, *The Good Man and the Good,* an examination of the foundations of morals, came out in 1918.

In 1927, after the presentation of an invited lecture on her theory of self-psychology to the British Psychological Association, she was elected an honorary member of that organization. The same year, a group of thirteen psychologists and philosophers, all Harvard graduates and several of them faculty members at much-respected institutions, petitioned Harvard to give her her Ph.D., pointing out the national and international reputation she held by then. (She had been starred in *American Men of Science* since the 1906 edition.) After due consideration, however, the Harvard authorities apparently failed to find any compelling reason to accede to the request.

Although not primarily an innovator or originator, Calkins was influential because of her work of systematization and exposition; her writings and lectures were models of clarity, rigor, and orderliness. An idealist and an outspoken pacifist, she was a human rights advocate who frequently voted for the socialist ticket. She retired from Wellesley in 1929 and died the following year, 26 February 1930, of inoperable cancer, shortly before her sixty-seventh birthday.

ELEANOR GAMBLE[9] (1868–1933) joined Calkins at Wellesley in 1898 and remained on the college faculty for thirty-five years until her death in 1933. Born in Cincinnati, Ohio, on 2 March 1868, she was the daughter of Joseph Gamble and his wife, Mary, daughter of Alexander T. McGill of the Princeton Theological Seminary. After graduating (A.B.) from Wellesley College in 1889 she taught Greek and philosophy for a year at the Western Female College in Oxford, Ohio, and Greek and Latin for four years at the State Normal School in Plattsburg, New York.

In 1895 she took up graduate studies in experimental psychology under the direction of Edward Titchener at Cornell University, beginning the work on the senses of taste and smell (part of Titchener's investigations of basic sensation elements) which she was to continue with considerable success for a number of years.[10] On receiving her Ph.D. (1898) she returned to Wellesley as instructor in the psychology department. She became associate professor in 1903, professor seven years later, and head of her department on Calkins's retirement in 1929. A study leave in 1906–1907 gave her a year at the University of Göttingen, where she worked with G. E. Müller on memory association; a second leave (1923–1924) she spent in Vienna studying under the guidance of the psychiatrist Theodor Reik.

She published about twenty articles in experimental psychology, a few of them reporting collaborative studies. As Calkins did, she brought out compilations of the work of her students in the "Wellesley College Psychological Studies" series; included were minor laboratory studies by undergraduates on a variety of topics, and investigations, mainly on experimental approaches to problems of memory, by about a dozen M.A. students she directed. Her interests were wide, but her work on memory and association (using Müller's methodology) reported in her 1909 monograph in the *Psychological Review* was probably her major contribution to the field. A joint study with Calkins on psychoanalytic aspects of self-psychology appeared in 1930. Her *Outline Studies in the Essentials of Psychology,* published in 1933, basically followed her lecture program and presented her attempt to integrate Titchener's sensationistic approach with Calkins's self-psychology.[11]

A small woman with wispy gray hair, she was one of the most popular and effective of Wellesley professors of her day. She taught the large introductory class in psychology (while Calkins had charge of the corresponding one in philosophy), and also conducted a course in logic. Though severely handicapped by increasing eye trouble that later made her almost blind, she was well known for her humor and liveliness. As a hobby she bred cocker spaniels and she held a membership in the Ladies' Kennel Club of Massachusetts. She and her dogs were familiar and admired figures on the Wellesley campus. She died at the age of sixty-five, on 30 August 1933, when attending a meeting of the Society of the Companions of the Holy Cross, an organization to which she had belonged for some years.

Two other women whose papers in psychology are listed in the bibliography, Miles and Andrews, also held positions on the Wellesley College staff. CAROLINE MILES[12] (1866–1951) was born in Pleasant Hill, Ohio, on 20 July 1866, the daughter of Keturah (Pickering) and Israel Miles. After taking an A.B. (1887) at Earlham College in Richmond, Indiana, she continued her studies at the University of Michigan, which awarded her a Ph.D. in 1892. Her doctoral research was a study in Kantian philosophy.[13] For a year (1891–1892) she held a fellowship in history at Bryn Mawr College. She then taught philosophy and political economy for a year at Mount Holyoke, and history and psychology for two years at Wellesley. At the same time, she continued her studies, both as a private student of Edmund Sanford at Clark University, and at Harvard, following a course remarkably similar to the one taken by Calkins. After her marriage to William Hill in 1895 she alternated periods of teaching (Bloomingdale Academy, 1910–1912, and Bethany College in West Virginia, 1912–1914?) with social work in Quaker groups, in organizations devoted to world peace, and in Chicago settlements. While in Chicago (1897–1900) she also studied at the University of Chicago, carrying out work that led to her 1898 paper on choice in the *American Journal of Psychology.* She died in Chicago, on 21 August 1951, just after her eighty-fifth birthday.

GRACE ANDREWS[14] (1875–1942), the daughter of Adrianna and George Andrews, was born in Plymouth, Massachusetts. After taking a B.A. at Wellesley in 1899 she continued her studies, specializing in English literature and philosophy (M.A., 1902). Her one publication in psychology was her "Studies of the dream consciousness" (1900). After teaching classics and English literature for a number of years in public and private schools in New York and in Wilbraham, Massachusetts, she returned to Wellesley in 1918, first as reader in the departments of philosophy and education, and later as secretary in these departments. One of her most notable undertakings was the classifying and cataloging of Wellesley's valuable collection of old textbooks, manuscripts, photographs, and other materials representative of American schoolwork from the eighteenth century on. She retired in 1941, after twenty-three years on the Wellesley staff, and died the following year, on 8 December 1942, at the age of sixty-seven.

Among the Wellesley students who contributed research reports to the pre-1901 psychological literature were Mabel Learoyd and Maude Taylor, whose jointly authored paper on mental imagery appeared in 1895.[15] MABEL LEAROYD[16] (1870–1963) had some interesting experiences in her later life. Born in Danvers, Massachusetts, on 11 December 1870, she graduated with a B.A. from Wellesley in 1894, having majored in mathematics. For thirty years (1896–1926), except for a year as secretary of the YWCA in Brooklyn, New York, and two years as principal of Northfield Bible Training School, she taught mathematics at Mount Hermon School for Boys in Mount Hermon, Massachusetts. In 1929, when she was almost sixty, she married widower Henry Franklin Cutler, for many years principal of her school and at the time the dean of the headmasters of New England. Shortly thereafter Henry Cutler retired and they went to Vienna, where he enrolled at the university as a medical student. Mabel studied German and music and presided over the American University Women's Club. After Cutler received his M.D. in 1940, he and Mabel stayed on in Vienna to allow him to complete clinics, against the advice of the consulate. In a letter to Wellesley in 1942 Mabel described their return journey across the Soviet Union and the Pacific to San Francisco. They crossed the Soviet western border ten days before the declaration of war between Germany and the USSR, but although they saw a steady succession of Soviet troop trains going west, their trip was pleasant and uneventful. At Yokohama they embarked on the next to the last steamer to sail on schedule for America; within two years it had been sunk by an American submarine. Following several years in East Northfield, Massachusetts, where Henry practiced medicine, the Cutlers retired to Lakeland, Florida. Mabel died on 9 July 1963, in her ninety-second year.

ALICE HAMLIN[17] (1869?–1934), who later carried out notable civic work in Lincoln, Nebraska, was also a Wellesley graduate. The daughter of Mary Eliza (Tenney) and Cyrus Hamlin, she was born in Constantinople, Turkey, on 20 December 1869. Her father was the founder and first president of Robert College, Constantinople. After studying at Abbott Academy in Andover, Massachusetts, she taught there for three years (1889–1892) while also continuing her studies; she took an A.B. degree at Wellesley in 1893. Her 1894 paper in the *American Journal of Psychology* on the effects of stimulae on different sense organs reported work done at Clark University summer school the same year. Going on to graduate work at Cornell under the guidance of Edward Titchener, she carried out experimental studies on the psychology of attention and distraction and received a Ph.D. in 1896. She became a member of the American Psychological Association the same year and served on its council in 1897.

Although she taught psychology at Mount Holyoke College for a year (1896–1897), she gave up full-time professional work in 1897 when she married Edgar Hinman, whom she had met as a fellow student at Cornell. Hinman was then chairman of the department of philosophy and psychology at the University of Nebraska, Lincoln. For two years Alice taught classes in advanced psychology as an assistant instructor in her husband's department, but after the birth of her daughter in 1899 she put much of her time into family concerns and civic service. As a member of the Lincoln Board of Education from 1907 to 1919 and its chairman in 1910, she played a leading role in transforming the area's somewhat backward school system into one of the most progressive in the country. For almost thirty years she also continued to teach on an occasional basis, lecturing in the School of Nursing in the University of Nebraska Medical School and at various teachers' institutes in Nebraska and South Dakota. Often she worked as a substitute for a year or two at a time. Her professional interests in her later years included the development and psychology of infants and young children. She died in Lincoln, on 28 October 1934.

Two other members of this group of early women philosopher-psychologists, Talbot and Washburn, held faculty positions at eastern women's colleges. Like Gamble and Hamlin Hinman, they both took Ph.D. degrees at Cornell.

ELLEN TALBOT[18] (1867–1968) was born in Iowa City, Iowa, on 22 November 1867, the daughter of Harriet (Bliss) and Benjamin Talbot. Her father, an educationalist, was descended from colonial settlers in Taunton, Massachusetts. After receiving her early education in public schools in Columbus, Ohio, she studied at Ohio State University (A.B., 1890) and then taught for four years in the Ohio towns of Dresden and Troy.

At Cornell University, where she held a Sage scholarship in philosophy (1895–1897) and then a Sage fellowship (1897–1898), she carried out a study of theories of the philosophers Fichte and Kant, especially as these touched on ideas of the individual consciousness.[19] She also carried out experimental work in the Cornell psychology laboratory and reported the results in papers published in 1897 and 1898, one on visual memory and another on the psychology of attention. Following two years of teaching at the Emma Willard Seminary in Troy, New York (1898–1900), and a further period of graduate study at the University of Chicago (summer, 1901), she joined the faculty of Mount Holyoke College where she remained until her retirement in 1936. For thirty-two years she was head of the department of philosophy and psychology. Shortly after a study leave at the universities of Berlin and Heidelberg (1904–1905) she published her work on Fichte in monograph form. Other shorter philosophical works appeared at intervals over the next two decades.[20] She held memberships in both the American Philosophical Association and the American Psychological Association for a number of years, but resigned from the latter in 1921, her interests having shifted more and more to philosophical questions. Throughout her life she was active in politics and supported the Democratic party. She died in Spartanburg, South Carolina, on 25 January 1968, in her one hundred and first year.

MARGARET WASHBURN[21] (1871–1939), a Vassar faculty member who was one of the prominent American psychologists of her time, was born on 25 July 1871, in Harlem, New York City, when that district was still a prosperous residential suburb. She was the only child of Elizabeth Floy (Davis) and Francis Washburn. In 1878 the family moved to Walden, New York, where Francis Washburn, by then an Episcopalian minister, had a church. Later they lived in Kingston, New York. As an only child Margaret had undisturbed leisure for her favorite pursuit of reading, and although she had some formal schooling she acquired much of her early education on her own, with encouragement from her parents. The family was comfortably off and she was taken traveling in Europe and Britain.

In 1886, at the age of fifteen, she went to Vassar College, and graduated (A.B.) in 1891, having acquired strong interests in both philosophy and science. The new field of experimental psychology she saw as combining essential elements of both these areas, and accordingly she applied for permission to study with James McKeen Cattell at Columbia University in that institution's newly established psychological laboratory. Accepted only as an auditor, however, she transferred at the end of one year, on Cattell's advice, to the Sage School of Philosophy at Cornell, where she was eligible for both a degree and a scholarship. She completed doctoral work under the guidance of Edward Titchener, receiving her degree in 1894, the first Ph.D. in psychology conferred on a woman by an American university. Her dissertation research, an examination of perception of distances and directions on skin, had the distinction of being accepted for publication in *Philosophische Studien* (1895), a journal that rarely published any work except that of the students of the editor, Wilhelm Wundt.

For six years (1894–1900) she taught philosophy, psychology, and logic at Wells College, a small women's college near enough to Cornell to allow her to return regularly for Titchener's seminars and for library work. Her translation of volume 2 of Wundt's *Ethics,* prepared in collaboration with Julia Gulliver, was brought out over the period 1897 to 1901 in addition to a number of short research papers. In 1900 she became warden of Sage College, Cornell's residence hall for women; the following year, she held in addition a lectureship in social psychology and animal psychology, both topics that were to remain special interests. She soon discovered that she had little liking for the job of supervising the personal and social lives of women students, however, and in 1902 she moved to the University of Cincinnati as assistant professor of psychology. Here also there were drawbacks. As a transplanted easterner and the only woman on the faculty she felt somewhat isolated. Within a year she accepted an invitation to join the Vassar faculty as associate professor of philosophy. She was promoted to full professor in 1908, and four years later, when a department of psychology was organized, she became head, the position she held until her retirement in 1937.

The program she developed at Vassar, where in large measure she founded the psychology laboratory, gave the college one of the best undergraduate psychology departments in the country. She never attempted to start a graduate program, having a firm belief that graduate education for women should be in a coeducational setting, but she built a strong program of experimental work for selected senior undergraduates that resulted in a series of about seventy joint publications in the *American Journal of Psychology* under the general title "Studies from the Psychology Laboratory of Vassar College." She chose the problems and the methods to be followed, turned over the experimental work and the formulation of the results to the students, and herself wrote the final drafts of resulting publications—a procedure that was uncommon among psychologists at the time. Her practice of including the names of junior assistants in the authorship of her papers was also unusual. The students examined a variety of problems, but most of the work arose from a few of Washburn's special interests—spacial perception by the different senses, memory for emotional experiences, experimental aesthetics, and detection and measurement of individual differences. There were also a few studies dealing with animal behavior. Most of the published reports were brief, but some were nevertheless notable contributions of basic data.

Washburn's best-known work, *The Animal Mind: a Textbook of Comparative Psychology* (1908) was a comprehensive survey of the field of animal behavior, based on a critical analysis of the literature. It was not the first text in the area, interest in the field having been fairly extensive since the publication of the theory of evolution (Darwin's *The Expression of the Emotions in Man and Animals* had appeared as early as 1872, and the work *Animal Intelligence* by his friend the Oxford physiologist and comparative psychologist George John Romanes in 1883). However, her approach was innovative in that she presented the problem of consciousness in animals in relation to the different senses, including space perception, memory, and problem solution. She also raised a number of questions, among them, when do ideas, as distinct from perceptions, first arise? The work gathered together the reports of a considerable amount of experimental work widely scattered throughout the literature and it had a notable influence on the development of the field. It ran to four editions (the later ones being updated as more research came out), was widely used as a text, and remained on the reading lists of university psychology courses until the 1930s.

Although she was primarily an experimentalist, Washburn gave considerable time to an examination of the role of motor phenomena in psychology. Noting that all animals, including man, are fundamentally motor organisms that react to perceptions by moving, she argued that psychology had to include movement in its basic data. Further, although she believed that consciousness exists apart from movement, she saw it as having a basis in motor processes; any explanation of consciousness, therefore, had to take into account the underlying motor phenomena. *Movement and Mental Imagery: Outlines of a Motor Theory of the Complexer Mental Processes* (1916) presented these ideas, which she developed further in later papers.[22]

An attractive woman of vivid personality, she received considerable recognition during her lifetime. Perhaps more than any other woman psychologist of the time she was accepted by male colleagues, and not only served on a number of powerful committees, but held important editorial positions. The latter included the coeditorship of the *American Journal of Psychology* (1926–1939). A member of the American Psychological Association from 1895, she was president in 1921 (the second woman to hold the position). The same year, she was vice president of the AAAS. When the select Society for Experimental Psychologists opened to women in 1929 she became one of its first women members. A further honor followed in 1931 when she became the second woman elected to the National Academy of Sciences. Her substantial contributions to her field were recognized by the dedication to her in 1927 of a commemorative volume of the *American Journal of Psychology.* That year she spoke at the international Wittenberg Conference of Feelings and Emotions (the only woman speaker). She was starred in *American Men of Science,* starting with the 1906 edition.

Washburn believed strongly in equality in education for men and women. Throughout her own career, which can scarcely be judged anything but full and exceptional by the standards of the time, she was nevertheless well aware that her professional position as a woman faculty member in a small undergraduate college remained in many ways a marginal one. In March 1937 she was completely disabled by a cerebral hemorrhage; she died on 29 October 1939, in Poughkeepsie, New York, at the age of sixty-eight.

Five of the women whose psychology papers are listed in the bibliography held faculty positions at institutions other than eastern women's colleges. These were Ladd-Franklin of Johns Hopkins and Columbia universities, Smith of the State Normal School in New Paltz, New York, Hunt of the State Normal School in Providence, Rhode Island, and the Training School for Teachers in Scranton, Pennsylvania, Schallenberger of the San Francisco State Normal School and the San Jose State Normal School, and Downey of the University of Wyoming.

CHRISTINE LADD[23] (1847–1930), one of the better remembered of America's early women scientists and the pioneer among American women psychologists, was born in Windsor, Connecticut, on 1 December 1847, the oldest of three children of Augusta (Niles) and Eliphalet Ladd. Both Ladd, a merchant, and his wife came of families prominent in early New England. From about the age of fourteen, following the death of her mother and her father's remarriage, Christine lived with her paternal grandmother in Portsmouth, New Hampshire. She got her early education at the coeducational Wesleyan Academy in Wilbraham, Massachusetts, where she took the same classes (including Greek) as boys who were preparing for Harvard.

Although lack of money meant that she had to teach for periods in public schools, she had two years at recently opened Vassar College and graduated with an A.B. in 1869. For nine years she taught science and mathematics in various high schools in Pennsylvania, Massachusetts, and New York. At the same time, she continued studies in mathematics, at Harvard with guidance from William Elwood Byerly and James Mill Pierce, and at Washington and Jefferson College in Washington, Pennsylvania, with George Vose. She also began to publish, contributing a number of notes to the newly founded Des Moines *Analyst* and solutions to set mathematical problems in the British *Educational Times.*

Having no great liking for schoolteaching she applied, in 1878, for admission to graduate study at Johns Hopkins University, recently open to men. Largely thanks to the support of British mathematician J. J. Sylvester, then on the Hopkins faculty, she was permitted to attend lectures as a special student. Sylvester had read her published work and felt she showed promise. A year later, although not given the title, she was awarded the stipend of a fellow, and proceeded to a program of study in symbolic logic under the supervision of logician and early experimental psychologist Charles Sanders Peirce, then a half-time lecturer at Hopkins. By 1882 she had fulfilled all the requirements for a doctorate in mathematics (see chapter 8), but her degree was withheld for forty-four years. Fortunately, she was still able to collect it in person when it was finally conferred at the university's semicentennial celebrations in 1926, four years before her death. Vassar awarded her an honorary LL.D. in 1887. In 1882, when she was thirty-five, she married Fabian Franklin, a member of the Johns Hopkins mathematics faculty. They had two children, of whom one, their daughter Margaret, survived to adulthood.

By the mid 1880s Ladd-Franklin had turned her attention to the theory of vision, an area in which there was much activity, particularly in Germany.[24] She began with geometric studies on the problem of binocular vision, and published a paper on an experimental method for determining the horopter in volume 1 of the *American Journal of Psychology* (1888). When her husband had a sabbatical leave in Germany in 1891–1892 she accompanied him, and, having been denied admission as a regular student to mathematics lectures at Göttingen, opted to study with psychologists Georg Müller (at Göttingen) and later with Arthur König at the University of Berlin. Both of these men were working on the theory of optics. She also managed to get access to the Berlin laboratory of the eminent physicist and physiologist Herman von Helmholtz.

At the time, following a period of intense controversy, there were two dominant theories in physiological optics, one put forward by von Helmholtz and the other by Ewald Hering. However, these being largely incompatible, several new theories, mostly hybrids of the dominant two, were also beginning to emerge. Ladd-Franklin was one of a considerable number of workers who attempted a synthesis, putting forward what she considered the strong points of each view. In a recent study by Steven Turner, Ladd-Franklin is listed as one of a "Core Set" of forty-two major participants in the Helmholtz-Hering controversy, her contributions having been cited by other participants as particularly significant.[25]

Her views, first presented at the Second International Congress of Psychology in London in 1892, assumed a photochemical model of the visual system and postulated three evolutionary stages in the development of color vision: in the most primitive stage, stimulation caused a photochemical reaction that resulted in simple light sensations (gray and white images), in the intermediate stage photochemical reactions produced blue and yellow sensations, and in the most highly developed stage further reactions produced red and green sensations in addition.

Over the next forty-five years Ladd-Franklin published several more original articles on color and other aspects of vision; *Colour and Colour Theories,* her collected works, appeared in 1929. She also contributed (by invitation) an appendix to the English translation of *Helmholtz's Treatise of Physiological Optics* (1924). Of special note were her contributions to the entry on "Vision" in volume 2 of James Mark Baldwin's *Dictionary of Philosophy and Psychology* (1901–1905), an important addition to the literature of both the fields it covered; she authored or coauthored many of its shorter articles and also served as associate editor for logic and psychology.

Later research findings in neurophysiology largely displaced the earlier theories of vision, and even by the 1920s the Helmholtz-Hering argument was beginning to die down. However, Ladd-Franklin's contributions were noteworthy. She was the only woman among the major participants in the long controversy and also the only American.[26]

Even without her doctorate she managed to hold a part-time lectureship in logic and psychology at Johns Hopkins from 1904 until 1909, and after she and her family moved to New York City in 1910 she accepted a similar position, which she kept for twenty years, at Columbia University. Both of these were marginal appointments involving the teaching of one or two courses a year, at Columbia without pay. She is said to have put considerable effort into her family and social responsibilities also, and to have been a conscientious housewife and a fine hostess.[27]

She remained active in research (in logic as well as theory of vision) until she was in her seventies, frequently presenting papers at meetings of professional groups, including the American Psychological Society (which she had joined at its second meeting in 1893), the American Philosophical Association (of which she was a charter member), and the Optical Society of America (where she was the first female member). She had been given a star in the first (1906) edition of *American Men of Science,* a mark that designated her as one of the top fifty psychologists in the country at the time. Over a number of years she took an active interest in women's issues in general, and in particular worked hard to win access for women to the "Experimentalists," the Society of Experimental Psychologists organized by Titchener in 1904. But neither her spirited correspondence with Titchener nor her persistence succeeded in breaking through his resistance, and only after his death in 1929 were women elected to the organization. She died of pneumonia, at her home on Riverside Drive, New York City, on 5 March 1930, at the age of eighty-two.

MARGARET SMITH[28] (1856–1943), the daughter of Jane (Keiver) and Cornelius Reader Smith, was born in Amherst, Nova Scotia. She graduated with a classical diploma from the State Normal School in Oswego, New York, in 1883 and spent two years in Germany, studying at Jena. From 1887 to 1896 she taught in Oswego but then returned to her studies, taking a year at the University of Göttingen. She received a Ph.D. from the University of Zurich in 1900 and continued research, with an honorary fellowship in philosophy and psychology, at Clark University in Worcester, Massachusetts, from 1900 until 1901. Thereafter she was director of psychology and geography for eight years at the State Normal School in New Platz, New York. From 1909 until about 1918 she supervised the education of a family, but before retiring returned to the Normal School for a year as instructor of languages. She later lived in Saginaw, Michigan.

Her best-known contribution to psychology was her dissertation research on rhythm and work, which, when it appeared in 1900, was considered one of the year's more notable publications.[29] *Textbook in Psychology,* her translation from the German of Johan Friedrich Herbart's monograph, appeared in 1891. She also published a number of works on pedagogical topics, with a special emphasis on the teaching of the retarded.[30] She was a member of the American Psychological Association from 1902.

HARRIET HUNT[31] of Hubbardsville, New York, was the daughter of Hannah (Lawton) and Sherebiah Hunt. Like Smith she studied at the State Normal School in Oswego, New York. Thereafter she probably taught for a time in public schools in Brooklyn, New York, and in Brookline, Massachusetts, before entering the department of pedagogy at New York University when it established its graduate program in 1890 (the first such program in the country). She received a Ped.D. (Doctor of Pedagogy) in 1891, and was one of the first two women to be awarded that degree.[32] She also studied for a year at the University of Chicago. Her special interests were the psychology of self-education and the Montessori teaching method, which she went to Rome to investigate. Her monograph, *The Psychology of Autoeducation* (1912), focused strongly on the findings of Montessori. The single paper by her listed in the bibliography reported observations in animal psychology. She taught psychology and pedagogy at the State Normal School in Providence, Rhode Island, and then served as principal of the Training School for Teachers in Scranton, Pennsylvania.

MARGARET SCHALLENBERGER[33] (1862–1951, later McNaught), who was the first woman to serve as a State Commissioner in the California Department of Education, began her career as a teacher in the San Jose Normal School in the 1880s. From the early 1890s she alternated periods of teaching with study at Stanford University. She also took courses at Cornell University, where she carried out work in experimental psychology. Her two papers on color perception in children appeared in the *American Journal of Psychology* in 1896 and 1897. After receiving an A.B. from Stanford in 1898 she went on to graduate studies in philosophy and psychology under the direction of Edward Titchener at Cornell and was awarded a Ph.D. in 1902; her dissertation reported work in child psychology.[34]

Shallenberger then joined the faculty of the San Francisco State Normal School, later moving to a similar position at the San Jose State Normal School. Her talents for committee work led to her being appointed to the position of State Commissioner of Elementary Schools. She became widely known in educational circles for her outstanding work on the National Educational Association and for her excellence as a speaker. Many of her articles on problems of teaching in elementary schools, particularly in rural areas, appeared in the *Addresses and Proceedings of the National Educational Association of the United States,* mainly in the period from 1916 to 1922. She retired in 1923, and died on 15 June 1951, in San Jose.

JUNE DOWNEY[35] (1875–1932), for twenty-five years head of the department of philosophy and psychology at the University of Wyoming, was the first woman to lead such a department at a state university, albeit at a time when the institution she served was still a small and developing one.

She was born in Laramie, Wyoming, on 13 July 1875, the oldest daughter and second child in the family of Evangeline (Owen) and Stephen Wheeler Downey. Stephen Downey, who had served as a colonel in the Union Army during the Civil War, practiced law in Washington, D.C., before moving west in 1869. He was one of the first territorial delegates to Congress from Wyoming and was largely responsible for the act creating the University of Wyoming, on whose board of trustees he later served. His wife came from a Wyoming family; her brother, William O. Owen, a pioneer surveyor of the region, led the first successful ascent of the Grand Teton in 1898. A neighboring peak, Mt. Owen, is named after him.

After receiving her early education in Laramie public schools and the university preparatory school, June attended the University of Wyoming (B.A., classics, 1895). She taught in a Laramie school for a year and then went to the University of Chicago to take up graduate studies in philosophy and psychology. On receiving an A.M. in 1898 she returned to Laramie to a position as instructor in English and philosophy at the university. A summer session at Cornell in 1901 introduced her to Edward Titchener's work in experimental psychology, and his teaching had a lasting influence on all her own future research. From about that time she taught some psychology in her philosophy classes at Wyoming. Following studies during a sabbatical year at the University of Chicago (1906–1907), where she worked under the guidance of experimental psychologist James Angell, she was awarded a Ph.D. with honors (1907). Her dissertation research on handwriting as an expression of personality was the first of a series of studies of voluntary and involuntary motor controls. She was promoted to professor of English and philosophy at Wyoming in 1905 and became head of the philosophy department in 1908; in 1915 she gave up her work in the English department and concentrated her efforts in philosophy and psychology.

Despite a heavy teaching load and occasional periods of ill health she contributed regularly to the psychology literature, and her research became widely known. This scientific prominence was appreciated by her colleagues at the University of Wyoming, where her influence in shaping the educational policies of the young institution was considerable. Known as a quiet and efficient worker, she chaired the graduate committee for a number of years (during which she saw the number of students increase from three to more than 200), and from 1908 until 1916 was also principal of the University Extension Department, a position of considerable importance in a western state university at that period. In her own department she attracted able students and gave them freedom in the laboratory to pursue projects of their own devising.

Her research interests were broad, but she continually returned to problems of imagery and aesthetics, topics related to her early work in English language and literature. These led her to the question of creativity; her book *Creative Imagination* (1929) summarized the results of thirty years of work in the area.

Her investigations of motor processes as expressed in handwriting, however, the researches she had begun as a student, were the contributions for which she was best known. Her monograph *Graphology and the Psychology of Handwriting* (1919) and a number of shorter works in the area brought her international recognition as a specialist in the study of handwriting and handedness as indicators of personality traits.[36] This line of research she developed into personality assessment in general, an area of clinical psychology to which she made important contributions. The Downey "temperament-trait" method of testing, which she developed (based mainly on handwriting), focused on nonintellectual aspects of personality at a time when most psychologists were concentrating on intelligence testing. Her procedure involved the plotting of test scores on a graph to give an individual "will-profile," which described the interrelationship of personality components and suggested that personality should be expressed in terms of a pattern. Although the measured traits were integrated into personality types (for example, aggressive, inhibited, and "hair-trigger"), variation in one trait was seen as affecting the interpretation of the entire profile—a factor that made the method extremely difficult for statisticians to use. Intelligence level was also integrated into the successful interpretation of a profile (for example, in a person of ability, "explosive" tendencies may be an advantage, while in someone less intelligent they can bring ruin and destruction). Although time-consuming to administer and marred by the fact that various parts were fairly quickly recognized as too specific to support some of the generalizations Downey drew, the method attracted immediate attention in both the United States and Britain when first published in 1919; further, it stimulated a considerable amount of work in the field of personality measurement, particularly with schoolchildren and young people.[37]

In all, Downey wrote seven books and about seventy scientific articles, including reviews, as well as contributions to literary and nontechnical magazines (poems, plays, and stories).[38] For a number of years she served as contributing editor to the *Educational Measurement Review,* as cooperating editor of the *Journal of Applied Psychology,* and as a member of the editorial staff of the international quarterly review, *Character and Personality.*

Throughout her life she remained unusually retiring, in her later years almost reclusive, maintaining contact with colleagues mainly by correspondence, and living quietly with her family in Laramie. Nevertheless, she was one of the first two women elected to the exclusive Society of Experimental Psychologists. In addition she was a charter member of the Colorado-Wyoming Academy of Sciences, a fellow of the AAAS, and from 1923 until 1925 a council member of the American Psychological Association. Her star in *American Men of Science,* coming in 1927, reflected the opinions of her colleagues about the work of her mature years. In August 1932, while attending the Third International Congress of Eugenics in New York City, she became seriously ill. She died two months later (11 October) at age fifty-seven, at her sister's home in New Jersey, following an operation for stomach cancer.

Papers in psychology by seven other women (Carman, Talbott, Carter, Sharp, Bodington, Jacobi, and Kellerman[39]) are listed in the bibliography. Ada Carman and Laura Osborne Talbott, both of whom published modest contributions in child psychology, were from Washington, D.C. Talbott and her younger sister, Mrs. Horatio King, were the founders of the Women's National Science Club; this for a time was the strongest of the separate science clubs for women, which, in response to the professional segregation of the period, were organized in the 1890s by women in the Washington area who had scientific interests.[40] Marion Carter and Stella Sharp were students at Cornell University in the 1890s. Carter's earliest journal article (1894) reported work in bacteriology. She received a B.S. in 1898, and although she does not appear to have completed the requirements for a Ph.D. she stayed on at Cornell, carrying out graduate work in philosophy and psychology. Two papers she published in 1898 discussed ideas of comparative mental development put forward by Charles Darwin and George John Romanes. Her later writings included a natural history textbook, *Nature Study with Common Things: an Elementary Laboratory Manual* (1904), a somewhat melodramatic novel entitled *The Woman with Empty Hands: the Evolution of a Suffragette* (1913), and a considerable number of educational animal stories, mainly published in magazines. The latter reflected her early interest in animal psychology. Stella Sharp (A.B., Wells College) was awarded a Ph.D. in psychology by Cornell in 1898, and stayed on as an honorary fellow the following year. Her research interests were in psychological methods.[41]

The pre-1901 publications by psychologist LILLIEN MARTIN[42] (1851–1943) which are indexed in the bibliography were in botany and plant chemistry, areas of interest to her during her fourteen years as a high school science teacher. However, her subsequent long and distinguished career was in experimental and then clinical psychology. During the 1930s and 40s she was one of the best-known women clinical psychologists in the United States.

Born in Olean, New York, on 7 July 1851, she was the oldest child in the family of Lydia (Hawes) and Russel Martin. Martin, a merchant and a graduate of Geneva College, left his family, but despite their subsequent relative poverty the children had a fairly happy life and the very best education their strong-minded mother could get for them. Lillien entered Olean Academy at age four, and when she was sixteen took her first teaching position in a girls' school run by the Episcopal church near Racine, Wisconsin, where her family then lived.

Shortly thereafter she moved to a similar post in Omaha, Nebraska. By 1876, when she was twenty-five, she had saved enough of her earnings to support herself through college. Her mother objected to her first choice, Cornell, since that institution had not yet enrolled any women, although it appeared willing to consider her application. She went to Vassar, which gave her a four-year scholarship, and graduated in 1880, having done particularly well in science.

With the help of David Starr Jordan, brother of Vassar's librarian and at that time professor of zoology at the University of Indiana, she obtained a job as teacher of botany, chemistry, and physics at a high school in Indianapolis. Nine years later, as a result of a speech delivered at an educational convention in San Francisco, she was asked to take the positions of vice principal and head of the science department in the San Francisco Girls' High School. She held this post for several years, but although very successful in administrative work her interests gradually turned to psychology. She resigned in 1894, and at the age of forty-three went to the University of Göttingen, where she studied for four years under the guidance of Georg Müller, the experimental psychologist who a few years previously had assisted Christine Ladd-Franklin with her studies. Although Martin received no degree, her research, a classical study dealing with the psycho-physics of weight lifting, was published in 1899 as a monograph coauthored with Müller; it was considered outstanding at the time.[43] While in Europe she also spent several months working in Swiss psychiatric hospitals, mainly studying problems in hypnotism.

In 1899, shortly after her return to the United States, she was invited by David Starr Jordan, by then president of recently established Stanford University, to fill a temporary post of assistant professor of psychology at Stanford. She stayed on, becoming associate professor in 1909, full professor two years later, and in 1915, a year before her retirement, executive head of the psychology department, the first woman to occupy such a position at Stanford.

A popular and successful teacher, she had charge of the general psychology class and in addition taught aesthetics, abnormal psychology, and advanced experimental psychology. She also did much of the administrative work of her then small department, and contributed to its general social well-being by regularly hosting faculty and student gatherings at her home.

Over the course of her seventeen years at Stanford she continued to publish, several of her most important works being in German. Indeed she probably had closer links with European workers than Americans.[44] She returned to Germany for several summers, going to Würtzburg in 1907, Bonn in 1908 and 1912, and Munich in 1914. One of her major contributions was the work she carried out in Oswald Külpe's laboratory in Bonn, where she collaborated in a program of experiments that demonstrated imageless thought and thus confirmed the existence in thought of components not derived from sensory images. The findings were startling to psychologists at the time, and the work brought her an honorary Ph.D. from Bonn in 1913.

After retiring from Stanford, as she was required to do at age sixty-five, she moved to San Francisco, simplified her existence by taking up her quarters in a hotel room, and began to practice as a consulting clinical psychologist. In this third career, which lasted for almost three decades, her work was pioneering. She divided her time between private practice and two mental health clinics she founded. One of these, started in 1920, was the first in the country for normal preschool children. The other was an old-age counseling center she established in 1929, where she developed a systematic approach to dealing with social problems experienced by the elderly. Indeed, although her many honors (including a star in *American Men of Science* and election to the vice presidency of the AAAS) were accorded for her earlier academic work, her most lasting contribution was probably in the area of gerontology. During her clinical career she published a number of books dealing with problems of childhood and old age, several designed for the layman. Especially notable were the two coauthored by her assistant Clare de Gruchy, *Salvaging Old Age* and *Sweeping the Cobwebs,* both of which appeared in 1930.[45]

Martin herself was an extraordinary example of a person leading a full and useful life into old age. When she was seventy-eight she visited the Soviet Union, traveling on her own. At eighty-one, having learned to drive two or three years previously when walking became difficult, she made a coast-to-coast car trip. When her handwriting became unsteady she learned touch-typing. She was a feminist, active and vocal in the suffrage movement, and, although in youth a Republican, later became a strong supporter of Franklin Roosevelt and the New Deal. She died of broncho-pneumonia, in San Francisco, on 26 March 1943, in her ninety-second year.

Of the thirteen American women psychologists discussed here who had doctoral degrees or the equivalent, nearly all had interests in experimental psychology for significant periods throughout their careers. (Among them perhaps only Talbot of Mount Holyoke was primarily a student of philosophy; Hunt and Schallenberger were largely occupied with practical problems in education.) Of the six who took their Ph.D. degrees at Cornell, at least four were students of Edward Titchener; although he declined to countenance the presence of women professionals in the experimentalists' club he started in 1904, Titchener apparently had little objection to advising women doctoral students.

This group forms one of the smaller subsets of late nineteenth-century American women scientists, but a remarkable number of its members had notably successful careers, particularly Calkins, Downey, Ladd-Franklin, Martin, and Washburn. Probably the only other groupings of American women scientists in which a comparable fraction achieved similar professional success were the medical scientists (physiologists and others), and the ethnologists. To a large extent the reasons were the same in each case: not only were these disciplines relatively new and uncrowded in the 1890s and around the turn of the century, but, in the cases of the experimental psychologists and medical scientists, the demand at expanding academic institutions for people with advanced training was still sufficiently great to allow the entry of a few well-qualified women. Thus Martin held a full faculty position at Stanford and Downey one at Wyoming, while among the medical scientists Ida Hyde taught at the University of Kansas and Alice Hamilton at Harvard. The early women psychologists had additional good fortune in that they were fairly readily accepted as members of their national professional association, the American Psychological Association, almost from its founding in 1892. Calkins and Washburn had joined by 1893 and both later held prominent leadership roles.[46]

Two-country comparison

The notable activity of American women in late nineteenth-century psychology contrasts quite markedly with the situation in Britain as it is reflected in the Royal Society's index. Only two pre-1900 psychology papers by British women were found in the *Catalogue,* one by the mathematician and educator Sophie Bryant (see chapter 8) and one by the philosophy student Constance Naden (see chapter 16). Even if the statistical studies of Alice Lee and other women assistants of Karl Pearson (see chapter 8) had been classed as work in "Darwinian psychology," the contributions of British women would still appear as relatively slight compared with the fifty-one papers published by Americans in the 1880s and 1890s.

An explanation, in broad terms, for the relative paucity of British contributions is not hard to produce. It is simply the fact that by the 1890s British psychology was to a considerable extent in a period of doldrums, whereas American work was fast coming to dominate the field—and American women were joining in the action. After having for two centuries led the world in philosophy and the study of mental processes, which laid the foundations of modern psychology, Britain, after 1870, fell behind; leadership in psychology passed first to Germany and then to the United States. Indeed, British psychology did not fully recover until after the Second World War.[47] The late nineteenth-century decline just happens to correspond to the period when women were beginning to enter the universities, and when one might have expected to see an increasing number of their research contributions.

This said, it would nevertheless be incorrect to conclude that British women were essentially absent from the scene in late nineteenth- and early twentieth-century psychology. In fact, in Hearnshaw's respected *Short History of British Psychology,* a work whose stated purpose was to bring out the broad trends in the field,[48] a remarkable number of productive and notably influential early women contributors to these broad trends are discussed. With one or two outstanding exceptions (particularly Sophie Bryant and Beatrice Edgell),[49] their professional careers began soon after the turn of the century, some ten to fifteen years later than those of most of the Americans discussed here.

It is notable, however, that mental and moral philosophy was taught at Bedford College for Women as early as the 1850s; Alexander Bain, a major figure in early British psychology, taught there from 1851 to 1854. Furthermore, in the establishment of laboratories of experimental psychology, Bedford College was one of Britain's pioneering institutions. Its laboratory opened in 1897, six years after Wellesley's and the same year that laboratories were established at Cambridge and University College London. Under the direction of Beatrice Edgell (1871–1948), who studied with Oswald Külpe in Würzburg, experimental work was quickly introduced at the Bedford College laboratory. It remained one of only three major London experimental psychology laboratories for several decades, although always smaller than that at University College. Edgell, remembered especially for her influential book *Theories of Memory* (1924), became, in 1927, the first woman professor of psychology in Britain (University of London). She was also the first woman president of the British Psychological Society. Her colleague Victoria Hazlitt (1887–1932), a pioneer in developing methods for student testing, published a well-received work on *Ability* (1926) and later carried out research on infant psychology.

Indeed, in the period before the First World War and throughout the 1920s a remarkable number of British women carried out notable work on child and educational psychology. Among them were Margaret Macmillan, Susan Isaacs, Margaret Lowenfeldt, Ruth Griffiths, Constance Simmins, and A. T. Alcock, all of whom were important enough to be included in Hearnshaw's *Short History.*

Interestingly, Britain, like the United States, had early women workers in the areas of animal psychology (Emily Mary Smith) and color vision (Mary Collins). Smith (later Lady Bartlett) was a Newnham College student (1905–1909), then a research fellow at the Cambridge University Psychological Laboratory and director of moral science at Newnham. Her survey, *The Investigation of Mind in Animals,* appeared in 1915, seven years after Margaret Washburn's *The Animal Mind.* She was one of the first British psychologists to appreciate Pavlov's work; her pioneering research on instinct and her account of homing were especially notable.[50] Dr. Mary Collins of the University Laboratory, Edinburgh, brought out her book *Colour Blindness* in 1925, by which time her American counterpart, Christine Ladd-Franklin, was nearing the end of her scholarly career. Collins also carried out a number of experimental studies.[51]

Thus, although their later start relative to their American counterparts means that their work is not listed in the pre-1900 indexes (except for papers by Bryant and Naden), British women were far from absent in research in experimental psychology within a few years of the turn of the century—the time when most of the Americans discussed here were at their most productive. Indeed, among British psychologists, especially in the areas of educational psychology and infant and child research, women would appear to have been a visible and successful group during the early years of this century, perhaps even more so than were women in American psychology.

Notes

1. A number of journals in which American psychologists published frequently were not included among those whose contents were indexed by the Royal Society. Several pre-1901 articles in such journals by Mary Calkins and Margaret Washburn have been added to the bibliography listings, increasing the count there of Calkins's publica-

tions from three to seventeen and Washburn's from two to ten. No search was made for additional pre-1901 papers by the other psychologists discussed here.

2. W. M. O'Neil, *The Beginnings of Modern Psychology,* 2d ed. (Atlantic Highlands, N.J.: Humanities Press, 1982); Thomas Hardy Leahey, *A History of Psychology. Main Currents in Psychological Thought,* 2d ed. (Englewood Cliffs, N.J.: Prentice-Hall, 1987).

3. Virginia Onderdonk, "Calkins, Mary Whiton," in *NAW,* vol. 1, pp. 278–80; Laurel Furumoto, "Mary Whiton Calkins (1863–1930)," *Psychology of Women Quarterly,* **5** (1980): 55–68; Gwendolyn Stevens and Sheldon Gardner, *The Women of Psychology,* 2 vols., *Volume 1: Pioneers and Innovators* (Cambridge, Mass.: Schenkman Publishing, 1982), pp. 77–88; Elizabeth Scarborough and Laurel Furumoto, *Untold Lives: the First Generation of American Women Psychologists* (New York: Columbia University Press, 1987), pp. 17–51; Laurel Furumoto, "Mary Whiton Calkins (1863–1930)," in Agnes N. O'Connell and Nancy F. Russo, *Women in Psychology. A Bio-bibliographic Source Book,* (Westport, Conn.: Greenwood Press, 1990), pp. 57–65.

4. Rossiter, *Women Scientists in America,* pp. 37, 329–30 n. 26.

5. Furumoto, "Mary Whiton Calkins," p. 59.

6. J. A[llan] Hobson, *The Dreaming Brain* (New York: Basic Books, 1988), p. 74.

7. Leahey, *History of Psychology,* p. 277.

8. Furumoto ("Mary Whiton Calkins," (1990 pp. 62–3) suggests that Calkins's emphasis on the significance of the self (a complex but single, unitary, continuing entity) in everyday life derived its strength at least in part from her long experience in the socially close-knit academic community that was Wellesley College; there the critical importance of the unique "self" was especially clear.

9. Alice Payne Hackett, *Wellesley. Part of the American Story* (New York: E. P. Dutton, 1949), pp. 129–30; Christian A. Ruckmick, "Eleanor Acheson McCulloch Gamble: 1868–1893," *American Journal of Psychology,* **46** (1934): 154–6; *Who was Who in America,* vol. 1, p. 437; *AMS,* 1910–1927; *WWWA,* p. 314.

10. See for instance, "Taste and smell," *Psychological Bulletin,* **8** (1911): 147–9; **10** (1913): 116–17; **12** (1915): 112–13; **13** (1916): 134–8; and "The psychology of taste and smell," ibid., **19** (1922): 297–306.

11. Gamble's post-1900 articles include, "The perception of sound direction as a conscious process," *Psychological Review,* **9** (1902): 357–73; "Attention and thoracic breathing," *American Journal of Psychology,* **16** (1905): 261–93; "A study in memorizing various materials by the reconstruction method," *Psychological Review, Monograph Series,* **10** (1909), 211 pp; "Intensity as a criterion in estimating the distance of sounds," ibid., **16** (1909): 416–26; (with L. Wilson) "A study of spacial associations in learning and in recall," ibid., **22** (1916): 41–98; "A study of three variables in memorizing," *American Journal of Psychology,* **39** (1927): 297–303; (with Mary Calkins) "The self-psychology of the psychoanalysts," *Psychological Review,* **37** (1930): 277–304.

12. *Who was Who in America,* vol. 3, p. 400; *WWWA,* p. 388.

13. Caroline Miles, "Kant's kingdom of ends," Ph.D. dissertation, University of Michigan, 1892.

14. Wellesley College archives (alumnae biographical files, 1899, Andrews).

15. No one by the name of Maude L. Taylor, the name that appears in the 1895 paper, is listed in Wellesley College records. The closest is Maud Marion Taylor, class of 1891, who received an M.L. from the University of California in 1899, married G. S. Crites in 1906, and died in 1915.

16. Wellesley College archives (alumnae biographical files, 1894, Learoyd).

17. *WWWA,* p. 391; *AMS,* 1910–1933; *NCAB,* vol. 26, p. 269 (sources differ about Hamlin's date of birth; *AMS* entries give 1869, others 1867).

18. *NCAB,* vol. 54, pp. 300–1; obituary, *NYT,* 26 January 1968: 47: 3; *WWWA,* p. 800.

19. Ellen Bliss Talbot, "The nature of Fichte's fundamental principle with special reference to its relation to the individual consciousness," Ph.D. dissertation, Cornell University, 1898. See also "The relation of the two periods of Fichte's philosophy," *Mind,* **10** (1901): 336–46.

20. See Ellen Bliss Talbot, *The Fundamental Principle of Fichte's Philosophy* (New York: Macmillan, 1906), and also "The psychology of Fichte in its relation to pragmatism," *Philosophical Review,* **16** (1907): 488–505; "Individuality and freedom," ibid., **18** (1909): 600–14; "The time process and the value of human life," ibid., **23** (1914): 634–47; **15** (1924): 17–36.

21. Edwin G. Boring, "Washburn, Margaret Floy," in *NAW,* vol. 3, pp. 546–8; Elizabeth S. Goodman, "Margaret F. Washburn (1871–1939): first woman Ph.D. in psychology," *Psychology of Women Quarterly,* **5** (1980): 69–80; *NCAB,* vol. 30, p. 28; anon., "Dr. M. F. Washburn of Vassar faculty," *NYT,* 30 October 1939: 17:4; Robert S. Woodworth, "Margaret Floy Washburn 1871–1939," *Biographical Memoirs of the National Academy of Sciences,* **25** (1949): 275–95; Scarborough and Furumoto, *Untold Lives,* pp. 91–107; Stevens and Gardner, *Women of Psychology,* vol. 1, pp. 96–105; Elizabeth Scarborough, "Margaret Floy Washburn," in O'Connell and Russo, *Women in Psychology,* pp. 342–9.

22. See, for instance, "Emotion and thought: a motor theory of their relations," in *Feelings and Emotions: the Wittenberg Symposium,* ed. M. L. Reymert (Worcester, Mass.: Clark University Press, 1928); "Purposive action," *Science,* **67** (1928): 24–8; "A system of motor psychology," in *Psychologies of 1930,* ed. C. Murchison, (Worcester, Mass.: Clark University Press, 1930).

23. Dorothea Jameson Hurvich, "Ladd-Franklin, Christine," in *NAW,* vol. 2, pp. 354–6; R. S. Woodworth, "Obituary. Christine Ladd-Franklin," *Science,* **71** (1930): 307; Scarborough and Furumoto, *Untold Lives,* pp. 109–29; Stevens and Gardner, *Women of Psychology,* vol. 1, pp. 117–21; Thomas C. Cadwallader and Joyce V. Cadwallader, "Christine Ladd-Franklin," in O'Connell and Russo, *Women in Psychology,* pp. 220–9.

24. Ladd-Franklin's graduate adviser C. S. Peirce had also carried out work in vision and optics—see C. S. Peirce, "Note on the sensation of color," *American Journal of Science,* **13** (1877): 247–51, *Philosophical Magazine,* **3** (1877): 543–7. Cadwallader and Cadwallader point out ("Christine Ladd-Franklin," p. 222), that Peirce was in fact the first experimental psychologist in the United States.

25. R. Steven Turner, *In the Eye's Mind: Vision and the Helmhlotz-Hering Controversy* (Princeton: Princeton University Press, 1994), especially p. 140.

26. Ibid., p. 141.

27. Stevens and Gardner, *Women of Psychology,* p. 118.

28. *AMS,* 1910–1927; *WWWA,* pp. 761–2.

29. See Edward B. Titchener, "The past decade in experimental psychology," *American Journal of Psychology,* **21** (1910): 404–21, on p. 419.

30. See Margaret Keiver Smith, "Sixty-two days training of a backward boy," *Psychological Clinic,* **2** (1908): 1–4, 29–47; "The training of the backward boy," ibid., **2** (1908): 134–50. Among her other works were: (with Esmond V. DeGraff) *Development Lessons.*

For Teachers (New York: A. Lovell, 1883); "The psychological and pedagogical aspects of language," *Pedagogical Seminary,* **10** (1903): 438–58; "On the reading and memorizing of meaningless syllables presented at irregular time intervals," *American Journal of Psychology,* **18** (1907): 504–13.

31. *WWWA,* p. 415 (no dates are given in this, the single biographical source found for Hunt).

32. Walter Crosby Eells, "Earned doctorates for women in the nineteenth century," *Bulletin, American Association of University Professors,* **42** (1956): 644–51.

33. [A. E. Winship], "Dr. Margaret McNaught," *Journal of Education,* **96,** (1922): 367 (I thank Gould P. Colman, Archivist, Cornell University, for bringing this to my attention); student records, Cornell University archives. Schallenberger's date of birth was probably 4 March 1862 (day and month are in doubt).

34. Margaret Schallenberger, "The growth of the child's mind," Ph.D. dissertation, Cornell University, 1902.

35. John Chynoweth Burnham, "Downey, June Etta," in *NAW,* vol. 1, pp. 514–15; Richard Stephen Uhrbrock, "June Etta Downey," *Journal of General Psychology,* **9** (1933): 351–64.

36. See "Control processes in modified handwriting: an experimental study," *Psychological Review, Monograph Series,* **9** (1908), pp. 148 (dissertation research); "Automatic phenomena of muscle reading," *Journal of Philosophy, Psychology and Scientific Methods,* **5** (1908): 650–8; "Muscle reading: a method of investigating involuntary movements and mental types," *Psychological Review,* **16** (1909): 257–301; "Judgments on the sex of handwriting," ibid., **17** (1910): 205–16; "Types of dextrality and their implications," *American Journal of Psychology,* **38** (1927): 317–67; "Back-slanted writing and sinistral tendencies," *Journal of Educational Psychology,* **23** (1932): 277–86; "Laterality of function," *Psychological Bulletin,* **30** (1933): 109–42.

37. June E. Downey, *The Will-Profile: a Tentative Scale for Measurement of the Volitional Pattern* (University of Wyoming, Department of Psychology, Bulletin, 15 (1919), 38 pp.). Downey's papers on the "will-profile" test included, "Ratings for intelligence and for will-temperament," *School and Society,* **12** (1920): 292–4; "Some volitional patterns revealed by the will-profile," *Journal of Experimental Psychology,* **3** (1920): 281–301; "Testing the will-temperament test," *School and Society,* **16** (1922): 161–8; "Jung's 'psychological types' and will-temperament patterns," *Journal of Abnormal Psychology,* **18** (1924): 345–9; (with R. S. Uhrbrock) "A non-verbal will-temperament test," *Journal of Applied Psychology,* **11** (1927): 95–105; (with R. S. Uhrbrock) "Reliability of the group will-temperament tests," *Journal of Educational Psychology,* **18** (1927): 26–39; "Observations on the validity of the group will-temperament test," ibid., **18** (1927): 592–600. For papers by other workers examining the reliability of the test see, for example, Mary Collins, "Character and temperament tests," *British Journal of Psychology,* **16** (1925–1926): 89–99, and Flora Kennedy, "Practical value of the June Downey will-temperament tests," *British Journal of Educational Psychology,* **4** (1934): 260–3 (the latter claimed to demonstrate the test's unreliability for schoolchildren).

38. Among Downey's books were a collection of poems, *The Heavenly Dykes* (Boston: R. B. Badger, 1904), written during the period when she taught English, and *Kingdom of the Mind* (New York: Macmillan, 1927), which presented the findings of experimental psychology to young readers.

39. For Jacobi and Kellerman see chapters 7 and 1, respectively. No information has been uncovered about Alice Bodington. Her four psychology papers appeared in the *American Naturalist* between 1892 and 1896. None of them included an address. She also wrote on topics in general biology and published a monograph, *Studies in Evolution and Biology* (London: Elliot Stock, 1890).

40. Rossiter, *Women Scientists in America,* p. 95.

41. Sharp's doctoral dissertation was entitled, "Individual psychology: a study in psychological method" (Cornell, 1898).

42. John Chynoweth Burnham, "Martin, Lillien Jane," in *NAW,* vol. 2, 504–5; Maud A. Merrill, "Lillien Jane Martin, 1851–1943," *American Journal of Psychology,* **56** (1943): 453–4; Norman Fenton, "Lillien Jane Martin, 1851–1943," *Psychological Review,* **50** (1943): 440–2; Scarborough and Furumoto, *Untold Lives,* pp. 189–91; Stevens and Gardner, *Women of Psychology,* vol. 1, pp. 89–96; *WWWA,* pp. 544–5 (see also chapter 1, n. 105 for a note on Martin's botanical writings).

43. Lillie J. Martin and G. E. Müller, *Zur Analyse der Unterschiedsempfindlichkeit* (Leipzig: J. A. Barth, 1899).

44. Much of Martin's work appeared as monographs. See, for instance, *Psychology of Aesthetics* (Worcester, Mass.: [n.p.], 1905); *Die Projektionsmethode und die Lokalisation visueller und anderer Vorstellungsbilder* (Leipzig: J. A. Barth, 1912); *Ein experimenteller Beitrag zur Erforschung des Unterbewussten* (Leipzig: J. A. Barth, 1915).

45. Among Martin's other later works were, *Mental Hygiene: Two Years Experience of a Clinical Psychologist* (Baltimore: Warwick and York, 1920); (with Clare de Gruchy) *Mental Training for the Preschool Child* (San Francisco: Harr Wagner, 1923); *A Handbook for Old Age Counsellors* (San Francisco: Geertz Printing, 1944).

46. After Washburn's presidency of the American Psychological Association in 1921, fifty years were to elapse before a woman was again elected to that office. By the 1920s the field had expanded considerably. However, women were by then heavily concentrated in the newer (low-paid) subfield of applied (clinical) psychology. With the tremendous increase in testing and "guidance" counseling in various public institutions, this area was developing rapidly, but its women practitioners had little contact with the academic centers where experimental and theoretical research was carried out and where the influence lay.

47. For an account of the late nineteenth-century decline in British psychology see L. S. Hearnshaw, *A Short History of British Psychology 1840–1940* (London: Methuen, 1964), pp. 120–31.

48. Ibid., p. vii.

49. Another notable exception was the early child psychologist Mary Carpenter (1807–77), sister of the physiologist and psychologist W. B. Carpenter. Mary Carpenter was the first person to establish a direct link between psychology and the study of juvenile delinquency. Her two books, *Reformatory Schools* (1851) and *Juvenile Delinquency* (1853), quickly led to legislation on juvenile problems. Carpenter was a pioneer in the field, more than two generations ahead of psychologists in general (ibid., pp. 30–1).

50. Ibid., p. 232.

51. Ibid., p. 220.

Chapter 16

OTHERS

There remain a number of women who published papers listed in the Royal Society *Catalogue* but who do not readily fit into any of the major categories discussed. Their interests and activities covered a wide range, from social work, to philosophy, to the technology of photography. Several wrote monographs; thirteen published journal articles dealing with meteorological topics.

American women

EMMA HART WILLARD[1] (1787–1870), best known as a pioneer in the development of secondary education for American girls in the 1820s and 1830s, taught science in the girls' schools she established and wrote textbooks on general science. Born on a farm at Berlin, Connecticut, on 23 February 1787, she was the sixteenth of the seventeen children of Samuel Hart, the ninth of ten by his second wife, Lydia (Hinsdale). Her father had relatively liberal views for the time and encouraged her to educate herself. In 1802, when she was fifteen, she enrolled at the Berlin Academy and two years later was herself teaching the younger children in the school. By the time she was nineteen she had been given charge of the academy over a winter term; during the next two years she taught in academies in Westfield, Massachusetts, and Middlebury, Vermont.

At age twenty-two she married John Willard, a Middlebury physician, a widower of fifty with four children. He had status and experience, and he believed in education for women. Emma continued her studies by reading the books brought into the house by her husband's nephew, who lived with them while attending Middlebury College, and so in 1814, when Dr. Willard had financial trouble, she felt herself well enough prepared to open a school for girls. Her Middlebury Female Seminary had a curriculum that was exceptional for the time; it offered instruction in both classics and the sciences, subjects that were then almost always taught only to boys.

A few years later the family moved to New York State, and shortly thereafter Emma made a strong appeal to the governor and state legislature for public financial support for the education of girls. When this effort failed to produce any results she started another private school, first in Waterford and subsequently in Troy. The Troy Female Seminary, founded in 1821, became famous. It trained no fewer than 200 women teachers before the first normal school had been established in the United States. Within ten years of its opening it had 100 boarders and 200 day students. Furthermore, it established a pattern that became characteristic of American girls' boarding schools—simple dress, an element of self-help, and, though formally nonsectarian, a generally Protestant outlook. The students trained there in the early decades of the century went on to teach all over the United States, taking with them Willard's ideas and principles. Through them she came to have a substantial influence on the intellectual tone of middle- and upper-class nineteenth-century American society.

Until his death in 1825 Dr. Willard managed the school's finances. Thereafter Emma took on the business responsibilities and ran the institution until 1838, when she turned the management over to her son (her only child) and her daughter-in-law. She then married Christopher Yates, an Albany physician, but when he demanded money from her to support his gambling, she divorced him and returned to Troy, where she spent the rest of her life as teacher, lecturer, and adviser.

Her major published works were her textbooks, principally histories of the United States and geography texts and atlases. Some were in use for decades; *Ancient Geography* ran to ten editions, the first appearing in 1822, and her *Abridged History of the United States* went through thirty reprintings and two translations (into German and Spanish) between 1831 and 1873. She also published at least two articles in medical journals. The first, entitled "Motive power of the blood," appeared in the *Boston Medical and Surgical Journal* in 1852; the other was an article on the "theory of circulation by respiration" prepared on request in 1861 for the *United States Journal of Homeopathy.* Late in life she joined the AAAS. She died in Troy, on 15 April 1870, at the age of eighty-four.

ANTOINETTE BROWN BLACKWELL[2] (1825–1921), who read a paper entitled "The comparative longevity of the sexes" to the Biological Section of the AAAS at the 1884 national meeting, was an ordained minister, social activist, lecturer, and writer with sustained interests in general science and biology. She was born in Henriette, New York, on 20 May 1825, the seventh of the ten children of Abby (Morse) and Joseph Brown. The Browns were farmers of New England background.

Antoinette attended Monroe Academy and then went to Oberlin College in Ohio, where she completed the literary course in 1847. Against opposition from parents and advisers, she proceeded to take Oberlin's theological course, but although she satisfied the requirements she was denied a degree. Nevertheless, in 1853, after overcoming stiff resistance, which came chiefly from clergymen, she was ordained as minister of the First Congregational Church in Butler and Savannah, Wayne County, New York, becoming the first woman to hold such a position in a recognized religious denomination in the United States. She left the job a year later because her increasingly liberal views were diverging from the more orthodox beliefs of the time. For a year she worked in the slums and prisons of New York City, and then in 1856 married Samuel Blackwell, brother of the pioneering woman physician Elizabeth Blackwell. Of her five surviving children, two, Edith and Ethel, became physicians.

Starting in 1869, she published a number of books on social concerns and on topics in general science and biology that interested her. They included *Studies in General Science* (1869), *The Sexes Throughout Nature* (1875), and *The Philosophy of Individuality* (1893). She was a member of the AAAS and strongly supported efforts to open science education to women. Late in her life she helped to found the All Souls' Unitarian Church in Elizabeth, New Jersey, and was its pastor emeritus until her death. She preached her last sermon in 1915 at the age of ninety, but even after that remained active in social and political work, especially the women's suffrage movement. Two books, *The Making of the Universe,* and *The Social Side of Mind and Action,* condensations of her earlier more philosophical writings, appeared in 1914 and 1915. She died in Elizabeth, on 5 November 1921, at age ninety-six.

DOROTHEA DIX[3] (1802–1887) was the oldest child of Joseph Dix, a merchant and itinerant preacher, son of an affluent and respected Boston family. His wife, Mary (Bigelow), eighteen years his senior, came from a background that gave her few advantages, social or educational. Dorothea was born on 4 April 1802 in a log cabin in the frontier village of Hampden, Maine. At the age of twelve, tired of household work that fell to her because of her mother's semi-invalid condition, she ran away and went to her widowed Dix grandmother in Boston. From there she was sent to relatives in Worcester, Massachusetts, where, at the age of fourteen, she opened and successfully operated a school for small children. In 1819 she went back to her grandmother in Boston and set about educating herself, attending whatever public lecture courses were available and reading in the city libraries.

To help support her mother after her father died in 1821 she opened a school for girls in the Dix mansion (while at the same time conducting a charity school for poor children in the coach house). In addition she continued her own studies and wrote extensively. Her elementary natural history encyclopedia *Conversations on Common Things* (1824) was a great success, running to many editions (the sixtieth appeared in 1869). Written in the conversational style then being used on both sides of the Atlantic in instructional books for young people, it presented information in simple language that children could understand. In addition to all these undertakings she took her part in church and social activities and she became engaged to a cousin. When this relationship was broken off, her health failed and she cut back her teaching, although her school continued to operate until 1835. She also kept on writing, bringing out moral and educational stories for young people. Her natural history article on spiders and butterflies appeared in the *American Journal of Science* in 1831.

Ten years later, following a period of Sunday-school teaching in the East Cambridge Massachusetts House of Correction, she embarked on a lengthy investigation of state care of the insane, who were then incarcerated as common criminals under incredibly bad conditions. Her vocal campaign for reform, coupled with remarkably successful lobbying for funds from state legislatures, eventually led to the reorganization of state mental hospitals throughout much of the United States. Among her many sociological writings are the voluminous reports she prepared for state legislatures. During a trip to Europe from 1854 to 1857 she spoke out about the need for reform in the care of the insane there also.

On the outbreak of the Civil War she volunteered her services and was appointed superintendent of Union Army nurses, a post she occupied until 1866, although she was a poor administrator and organizer. After the war, despite her advancing years, she returned to hospital and prison work, particularly in the south. In 1881 she retired to the state hospital in Trenton, New Jersey, an institution she had helped to found thirty-five years earlier. She died there on 18(?) July 1887, at age eighty-five, after several years of severe illness.

MARY HOLMES[4] (1849–1906), the daughter of the Rev. Mead Holmes and his wife Mary (Everett), was born in Chester, Ohio, on 10 April 1849. After graduating (A.B.) from Rockford Seminary in Rockford, Illinois, at the age of nineteen, she taught natural history for a year in Manitowoc, Wisconsin, and then for eleven years at Rockford Seminary. Her interests were broad. She studied music for a time around 1870 and also was active in the social work of religious organizations, holding offices in the Presbyterian Home Mission Society in Illinois. In the mid 1880s, without putting aside her social work, she took up graduate studies in zoology and paleontology at the University of Michigan (A.M., 1887; Ph.D., 1888). Her doctoral research in paleontology brought her the honor of being the first woman elected a fellow of the Geological Society of America (1889).[5] However, she did not pursue paleontological work, and indeed her only scientific journal publications were probably her two notes reporting observations and conclusions about animal instinct and intelligence, which appeared in 1884 and 1893 (see bibliography).

She was especially concerned with the situation of Southern Negro people in the years that followed Reconstruction, and from 1885 published her own monthly newsletter, *Freedman's Bulletin.* Her work in organizations of the Presbyterian church in the Synod of Illinois also continued. She was a founder and active supporter of the Mary Holmes Seminary, an industrial and general education school for Negro girls in West Point, Mississippi. From its establishment in 1892 the school had close ties with the Presbyterian church. It still exists as the

Mary Holmes College, and is now a two-year, coeducational Presbyterian college with an enrollment of about 1,000 students, offering a liberal arts and science curriculum as well as vocational training. Holmes lived in Rockford, Illinois, until her death in 1906.

MARION TALBOT[6] (1858–1948), dean of women at the University of Chicago for more than thirty years, was born on 31 July 1858 in Thun, Switzerland. The oldest of the six children of Emily (Fairbanks) and Israel Tisdale Talbot, she was of New England ancestry. Her father was the first dean of the medical school of Boston University. She grew up in Boston, where she went to private schools and the Boston Girls' High School. Her education was broadened by private tuition in Latin and Greek and a stay in Europe to learn modern languages. After attending Boston University as a special student and graduating (B.A.) in 1880, she spent some years traveling and enjoying social life. However, in 1881–1882 she joined a number of women, including chemist Ellen Swallow Richards of MIT and Alice Freeman, then acting president of Wellesley College, in the work of organizing the Association of Collegiate Alumnae, later the American Association of University Women. She was the group's first secretary and from 1895 to 1897 served as its president.

Influenced by Richards, she became interested in the new field of sanitation and enrolled at MIT in 1884. Her study of organic matter in the atmosphere, reported in the *American Meteorological Journal* in 1888, was most likely carried out at the institute. She took an S.B. in 1888 and entered the field of domestic science. In collaboration with Richards she edited *Home Sanitation; a Manual for Housekeepers,* which first appeared in 1887 and was reissued periodically over the next three decades.

In 1890 she accepted a position on the faculty of Wellesley College as instructor of domestic science, but two years later moved to Chicago to join Alice Freeman (by then Mrs. Palmer) at the new University of Chicago. Freeman had been asked to organize the university's program for women, and undertook the work on the understanding that she would be in residence for only twelve weeks of the year with Marion Talbot serving as her colleague on a full-time appointment. Talbot thus joined the University of Chicago faculty as assistant professor of sanitary science (in the department of social sciences and anthropology) and dean of undergraduate women. She became associate professor in 1895, dean of women in 1899, and professor and head of a new department of household administration in 1905.

She wrote a number of books, the first of which, *Food as a Factor in Student Life* (1894), was a technical study of the diet of undergraduate women coauthored with Ellen Richards. In *The Education of Women* (1910) she set out her response to the increasingly conservative attitude toward this issue, which was gaining strength at coeducational universities shortly after the turn of the century.[7] The work put forward her ideas about women's work in modern society and the growing need for equality in women's education. Another monograph, *The Modern Household,* coauthored with her colleague Sophonisba Breckinridge, appeared in 1912; in it she developed further her thoughts on the changing role of women and the education they needed in an increasingly technology-based society.

She was an influential force in shaping living conditions for women students at the University of Chicago, opposing the introduction of chapters of national sororities when fraternities were first brought in, leading the organization of the Women's Union in 1901, and later working to secure a clubhouse and gymnasium as well as houses for women students. With Ellen Richards she took a prominent role in the Lake Placid Conference of 1908 that led to the formation of the American Home Economics Association. In 1917 she was chosen as president of the association but declined the position because of pressure of other work. After she retired in 1925 she twice served as president of the American College for Women in Turkey (1927–1928 and 1931–1932). Cornell College, Mount Vernon, Iowa, gave her an honorary LL.D. in 1904. She died at age ninety, in Chicago, on 20 October 1948.

ZELLA ALLEN[8] was the daughter of Josiah Buffett Allen of Zanesville, Ohio, and his wife Mary Caroline (Blandy), a descendent of the famous English eighteenth century essayist Joseph Addison. After attending public schools in Zanesville and Putnam Seminary, Zella went to Mount Holyoke Seminary, from which she graduated in 1880. The following year she married Joseph Ehrman Dixson. When he died in 1885 she returned to her studies, entering Columbia University as a special student in library science while also serving as assistant librarian. From 1888 until 1890 she was both librarian and a graduate student in Denison University in Granville, Ohio (A.M., 1902). The one paper by her listed in the bibliography reported germination studies carried out in the university's botanical laboratories about 1890. With the opening of the University of Chicago in 1892, she became organizer and administrative head of the library, a post she held until 1911; she also taught library science. Editor of several literary reviews and magazines, she was also founder and proprietor of the Wisteria Cottage Press, a private press noted for publications high in artistic quality.

JULIA MCNAIR[9] (1840–1903) was born in Oswego, New York, on 1 May 1840. She was educated privately. In 1859 she married William Janes Wright, later professor of metaphysics at Westminster College, Missouri, and she spent much of her life in Fulton, Missouri. Between 1891 and 1894, when she was in her fifties, she published eight notes in *Science* on a variety of topics in natural history ranging from box turtles to water lilies. An extraordinarily prolific writer, from the late 1860s she produced more than 100 works, mainly children's stories and moral and religious tales. Many were brought out by the Philadelphia Presbyterian Board of Publications. She also wrote two books on practical housekeeping and home economics, *The Complete Home* (1879) and *Ideal Homes* (1895), both of which ran to several editions. Her popular science and natural history books included *Astronomy, the Sun and His Family* (1898), *Botany, the Story of Plant Life* (1898), and her extremely successful *Sea-side and Wayside,* which first came out in 1888 and was reissued many times, the last edition appearing in revised form in 1936. Her literary work won a medal at the Chicago World's Columbian Exposition in 1893.

MARION HOWE[10] (1867–1911, later PUGH) of Fredericksburg, Chicksaw County, in northeast Iowa, received a B.S. degree from the University of Iowa, Iowa City, in 1891. Her examination of yeasts used in bread making, probably carried out during her senior year, was reported in the *Proceedings of the Iowa Academy of Sciences* in 1892. After graduating she taught for a few years in Des Moines High School and then studied homeopathic medicine at the University of Iowa and at the Hahnemann Medical College and Hospital in Chicago. She received an H.M.D. in 1900 and after that practiced for a time in Des Moines. She died at the age of forty-four, on 30 December 1911.

MARY HINCKLEY,[11] of Milton, Massachusetts, was the daughter of Thomas Hinckley, an artist. She studied the development of frogs and toads, publishing reports of her observations in the *Proceedings* and *Memoirs* of the Boston Society of Natural History in the early 1880s.

The remaining four American women included here—Brodhead, De Riemer, Edson, and Wing—all wrote papers on meteorological topics. Jane Napier Brodhead, whose paper on the "Indian Cyclone of October, 1864" appeared in the *American Meteorological Journal,* was from New York and Aiken, South Carolina, although she lived in France for a number of years at the beginning of the present century. Her special interests were Russian history and the Catholic church in France. She brought out two monographs, *Slav and Moslem* (1894), a well-received history of Russia from the tenth century to the late nineteenth, which remains a very readable work, and *The Religious Persecution in France 1900–1906* (1907), a collection of essays on the status of the church in France, which had previously appeared in the American press. Alicia De Riemer coauthored a report with Cleveland Abbe in the *American Meteorological Journal* of 1898. She appears to have worked with climatologist Abbe at George Washington University. Helen Edson, whose paper was published in the same journal in 1895, described ice formations on Roan Mountain, Tennessee. Minerva Wing probably lived in Vermont. In 1874 she published a paper on aurora observations made at West Charlotte, Vermont; her second paper, on climate anomalies near Lake Champlain, appeared in 1885.

British women

British women whose work is listed in the bibliography but who have not been mentioned so far are the following: Constance Naden, a student of philosophy and psychology in the late 1880s; Marion Acworth, who worked with her husband on the technology of photography in the 1890s; Emma Greenland Hooker, an early experimenter with the composition of materials for encaustic art, and eight contributors of papers and notes to various journals on topics in meteorology, mainly during the last two decades of the century.

CONSTANCE NADEN[12] (1858–1889) died before reaching her thirty-second birthday, but she was considered to have shown remarkable promise in her student essay, "Induction and deduction," her major attempt at philosophical writing. She was born in Edgbaston, Birmingham, on 24 January 1858, the only child of Thomas Naden, later president of the Birmingham Architects' Association, and his wife, Caroline Anne, daughter of J. C. Woodhill of Edgbaston. Following the death of her mother two weeks after her birth, she was brought up by her maternal grandparents. Her grandfather, a retired jeweler, was an elder of the baptist church and a man of considerable literary taste. Constance attended a day school run by ladies until she was sixteen or seventeen and developed a special interest in flower painting. However, when her work was not accepted by the Birmingham Society of Artists she turned to other studies, particularly languages and philosophy. Her first book of poems, *Songs and Sonnets of Springtime,* appeared in 1881, when she was twenty-three.

From 1879 to 1881 she attended botany classes at the Birmingham and Midland Institute, became interested in science, and continued her studies at Mason College. She also joined the Birmingham Natural History Society. A young woman of ability, she won two student awards, in 1885 the Paxon Prize for an essay on local geology, and two years later the Heslop gold medal for her essay on "Induction and deduction." Her second book of poems, *A Modern Apostle, The Elixir of Life, The Story of Clarice, and Other Poems,* appeared in 1877, and her long psychological paper, "Volition," an exploration of voluntary and involuntary action, in the *Midland Naturalist* shortly after.

When her grandmother died in 1887 leaving her a fortune, she made a tour through the Middle East and India. She then bought a house in Grosvenor Square, London, and joined the Aristotelian Society and a number of benevolent groups. Confident and self-possessed as a speaker, she presented a lecture on Herbert Spenser's "Principles of Sociology" to the sociology section of Mason College in 1889. Soon after that she became ill, underwent an operation, and died within a week or two, on 23 December 1889. Her poems attracted little notice until the politician William Gladstone called attention to them in an article in which he named her among eight British women poets he considered exceptional. Her philosophical essays and other works were published in 1890 and 1891, and her collected poems reissued in 1894.[13]

MARION ACWORTH, of Cricklewood, northwest London, collaborated with her husband, Joseph Acworth, on the development of photographic techniques, particularly those relating to exposure and film development. Three of their joint papers were presented at meetings of the Royal Photographic Society and published in the society's journal in 1895. Joseph Acworth was a chemist with the Ilford Photographic Company, and later founded the very successful Imperial Dry Plate Company. Marion was an associate of the Royal College of Science.

In 1927, under the pseudonym "Neon," she published a 279-page work, *The Great Delusion,*[14] a discussion of the development and, as she saw it, the very limited success of airships and aircraft. She devoted more than half of the book to zeppelins and set out clearly her conclusion that all state support for work on airships should stop. Very critical of the entire industry, she also discussed the case against airplanes, stressing the excessive cost of the development of civil aviation, which depended on heavy government subsidy. In addition, she pointed out the inefficiency of air transportation compared with ships and railways and its unreliability because of the vagaries of weather. Finally she argued against the buildup of fleets of bombers for possible use against civilian populations in war, seeing such a course as nothing short of a reversion to barbarism; even the military effectiveness of the airplane, limited as it was by weather and its inability to achieve strike accuracy, she saw as questionable.

The book gained widespread attention, an American edition and a French translation appearing in addition to the London issue. Not surprisingly it also elicited a full and prompt public rebuttal, in the form of a work entitled *Airmen or Noahs,* by Rear Admiral Sir Murray Frazer Sueter, a pioneer of British aviation who had also worked on the development of airships. Over the course of some 440 pages, with many photographic illustrations, the admiral set out his views on aeronautics and on the influence of air power in the First World War.[15] Although Acworth was less than prescient as far as the future of airplanes was concerned, it must be admitted that she was on the mark in her opinions about the dirigibles.

EMMA HOOKER lived in Rottingdean, near Brighton. In 1786, when still Miss Greenland, she had communicated to the Society for the Encouragement of Arts, Manufactures and Commerce a recipe for preparing and using a novel composition for artwork. The material permitted imitation of ancient Greek encaustic painting, that is, inlaying with colored clays by burning into a brick or tile base. Her contribution was adjudged worthy of the "gold pallet" and her account of the process published in the society's *Transactions.* It was considered important enough to be republished in other London journals in 1808 and in the French *Annales de Chemie* in 1811, a quarter of a century after the first report.

Although details that specifically identify and quantify the key ingredients in her preparation are not given, Mrs. Hooker's procedures were, briefly, as follows. By heating and vigorous stirring, wax, gum mastiche (probably tree gum or resin), and water were combined into an emulsion. This served as a base into which were mixed the same powdered pigments as those used for preparing oil paints. Wood, canvas, cardboard, and plaster of Paris were suitable for working on; paintings were finished by brushing over with a thin layer of melted white wax. The recipe permitted considerable variation, a satisfactory emulsion base being obtainable even in the absence of either the wax or the gum mastiche. The preparation also had the advantage of keeping for years, requiring only replacement of the water when it dried out.

Of the eleven British women whose meteorological articles are listed in the bibliography, two have been discussed in other sections—Elizabeth Brown with the astronomers, Eleanor Ormerod with the entomologists. The others were Mary Baillie, Annie Baker, Mrs. Behrens, Anne Bennett, Mary Browne, Rose Mary Yeates Crawshay, W. L. Hall, Annie Ley, and Catherine Stevens. Except for Rose Mary Crawshay, wife of a prominent mid-century industrialist, none of them seem to be well remembered.

The earliest was Anne Bennett, whose paper on meteorological phenomena in connection with the climate of Berlin appeared in the *Edinburgh New Philosophical Journal* in 1853. The work was an elegant translation from the German of a long article by meteorologist Heinrich Wilhelm Dove, professor at the University of Berlin. It dealt with world climate in general and European climate in particular. Mrs. Bennett published at least two more translations, one of Goethe's *Iphigenie auf Tauris* (*Iphigenia in Taurus,* with original poems, 1851) and the other a religious work, *The Devotion to the Holy Face at St. Peter's of the Vatican and in other Celebrated Places* (1894) from the French of Pierre Désiré Janvier.

Lady Mary Baillie[16] was the oldest daughter of Stair Hathorn Stewart of Physgill, Wigtownshire. She married Sir William Baillie, Second Baronet of Polkemmet, in 1846. They had no children. Sir William was for a time in the 1840s M.P. for Linlithgowshire, the Scottish county lying to the west of Edinburgh. Lady Mary's note on remarkable sunrises on 8 and 11 December 1884 appeared in the *Proceedings of the Royal Society of Edinburgh.* She died on 7 June 1910.

ANNIE LEY was a relative (possibly a sister) of the Rev. W. Clement Ley (1840–1896) who held the benefice of Ashby Parva in Lutterworth, Leicestershire. Clement Ley was well-known and much respected for his pioneering studies of wind and cloud movement, and his contributions to early weather forecasting—see, for instance, his *Laws of the Winds Prevailing in Western Europe* (1872). Annie's meteorological observations, contributed at intervals to *Nature* during the 1880s and 90s, included reports from Ashby Parva of the atmospheric aftereffects of the 1883 Krakatoa eruption; in addition to descriptions of the spectacular skyglows then seen, she presented calculations of heights of strata producing the glow effects. These observations and calculations were incorporated into the London Royal Society's massive report, compiled from data collected from observers worldwide, *The Eruption of Krakatoa, and Subsequent Phenomena* (1888).

Annie Baker published phenological records for 1890 to 1893 in the Bristol Naturalists' Society's *Proceedings,* Miss W. L. Hall brought out notes on the meteorology of Eastbourne in the *Transactions of the Eastbourne Natural History Society* in 1882, and Mrs. Behrens wrote on rainfall in Brazil in *Symons's Monthly Meteorological Magazine* in 1889. Mary Browne contributed a description of an 1830 hurricane in the West Indies island of St. Vincent to *Timehri* (the journal of the Royal Agricultural and Commercial Society of British Guiana). Catherine Stevens and Rose Mary Yeates Crawshay each contributed short notes to *Nature* in the 1890s.

ROSE MARY YEATES[17] (1827–1907), the daughter of Mary and William Wilson Yeates of Caversham Grove, Oxfordshire, was born on 17 January 1827 at Horton Grove, four miles from Windsor. Her education, provided by governesses, was sufficiently sound to awaken broad intellectual interests ranging across literature and the sciences. In 1846, at the age of eighteen, she married Robert Crawshay, a millionaire iron-master and landowner from Merthyr Tydfil, South Wales. Crawshay, known as the "Iron King of Wales," was the third in the line of an iron dynasty that began with his grandfather Richard, a very successful London iron trader. Richard acquired the Cyfarthfa works near Merthyr Tydfil in the late eighteenth century, a time of tremendously rapid development in the industry. Under his management Cyfarthfa, by the 1790s, was by far the most productive ironworks in the world, with technology second to none in modernity.[18] By the time Robert succeeded to the business, its most prosperous period was over, but nevertheless he was an important figure in British industry, employing a workforce of 5,000.[19]

Despite her youth at the time of her marriage, Rose Mary from the first was by all accounts a capable mistress of the Crawshay family seat, Cyfarthfa Castle, an immense establishment of the turret-and-battlement style of architecture built by her father-in-law in a boom period about 1825. In addition to bringing up a family of five children and dispensing hospitality to great numbers of guests (who included eminent men of literature and science), she early became a prominent figure in Merthyr life. One of the first things she did when she arrived at Cyfarthfa was to open a soup kitchen for the poor of the surrounding villages, using food normally discarded from the castle kitchens. Three days a week, for thirty-three years, until her husband's death in 1879, she fed thirty people.

Active and energetic, she was interested in many aspects of social improvement and reform, from matters she could take up locally to the question of women's suffrage. Over the years she trained local girls at the castle for domestic service, and also ran sewing classes to teach women how to make their own clothes. Throughout the district she started reading rooms and established no fewer than seven free libraries (still uncommon at the time), supplying them with popular as well as serious literature. A member of the first Merthyr school board and an active participant in the Merthyr Debating Society, she also organized lectures and herself spoke on such subjects as women's suffrage. She stirred up much controversy among conservative religious groups in the early 1870s when she advocated euthanasia in cases of hopeless and painful illness;[20] likewise her opinion in favor of cremation drew considerable criticism.

The social project for which she became known nationally, however, was her "lady-help" scheme. In those days before much useful education was accessible to women, the country had a tremendous excess of "women of birth" who had very little means of support; without enough training to be governesses, lacking capital, and strongly discouraged from competing with men in any kind of business enterprise, life had little to offer them. Mrs. Crawshay had had ample experience with domestic help at Cyfarthfa Castle and had generally found that the better the background a woman came from the faster, more efficient, and generally better worker she proved to be. She therefore opened a London office to connect "lady-helps" with people interested in employing superior household helpers in permanent positions at good wages. Despite much initial controversy about the propriety of gentlewomen taking domestic positions, the scheme was soon recognized as both sensible and practical. It successfully placed hundreds of women who were very competent in domestic work and served a useful role for many years. Mrs. Crawshay's handbook, *Domestic Service for Gentlewomen; a Record of Experience and Success,* was brought out in 1874. The first thousand copies sold so quickly that another six thousand had to be printed; a third enlarged edition appeared in 1887. She spoke about her "experiment" at a Social Science Congress in 1874 and again the following year at the Bristol Meeting of the British Association, which she had joined in 1870. (Her B.A. talk was delivered to the section on Economic Science and Statistics, which at the time included among its interests a wide spectrum of educational and training projects.[21])

After her husband died in 1879 she turned Cyfarthfa Castle over to her oldest son. From then on she spent her winters at the Hôtel du Louvre in Mentone on the French Riviera and much of the rest of her time in London, with visits to South Wales in the summers. In 1882 she started an annual competition for the best literary essay written in English by a woman of any country. Called the "Byron-Shelley-Keats In Memoriam," and generously endowed to provide cash prizes, it continued at least until the mid 1890s and attracted considerable numbers of competitors.

Mrs Crawshay died at the age of eighty, on 2 June 1907. Cyfarthfa Castle and its 158 acres of parkland was sold to the Merthyr Corporation in 1908, and later used as a school and museum. The Cyfarthfa ironworks were closed permanently after the First World War.

Notes

1. Frederick Rudolph, "Willard, Emma Hart," in *NAW,* vol. 3, pp. 610–13; Rossiter, *Women Scientists in America,* pp. 4–9, 74, 76.

2. Barbara M. Solomon, "Blackwell, Antoinette Louise Brown," in *NAW,* vol. 1, pp. 158–61; *WWWA,* p. 104.

3. Winfred Overholser, "Dorothea Lynde Dix: a note," *Bulletin of the History of Medicine,* **9** (1941): 210–16; Helen E. Marshall, *Dorothea Dix: Forgotten Samaritan* (Chapel Hill, N.C.: University of North Carolina Press, 1937), and "Dix, Dorothea Lynde," in *NAW,* vol. 1, pp. 486–9; Harry B. Weiss, *The Pioneer Century of American Entomology* (New Brunswick, N.J.: Harry B. Weiss, 1936), pp. 127–8. Gwendolyn Stevens and Sheldon Gardner, *The Women of Psychology,* 2 vols., (Cambridge, Mass.: Schenkman, 1982), vol. 1, *Pioneers and Innovators,* pp. 51–66. (Sources are inconsistent about the date of Dix's death; it appears as both 17 and 18 July, 1887.)

4. *Who was Who in America,* vol. 1, pp. 581–2; *AMS,* 1910.

5. Holmes's Ph.D. dissertation (zoology) was entitled, "The morphology of the carinae upon the septa of rugose corals."

6. Richard J. Starr, "Talbot, Marion," in *NAW,* vol. 3, pp. 423–4; Frances L. Swain, "Our professional debt to Marion Talbot," *Journal of Home Economics,* **41** (1949): 185–6; *AMS,* 1910–1949.

7. When the University of Chicago opened in 1892 women undergraduates constituted one quarter of the student body; ten years later they accounted for half the undergraduate enrollment. Administrators tried to limit the alarming influx of women by organizing for them a separate college within the university. Talbot fought hard to

preserve full coeducation and prevent the fixing of quotas for women (Rossiter, *Women Scientists in America,* p. 109).

8. *WWWA,* p. 249.

9. *Who was Who in America,* vol. 1, p. 1385.

10. Alumni records, University of Iowa; membership lists, *Proceedings of the Iowa Academy of Sciences,* **1,** pt. 4 (1893): 5; *AMD,* 1906.

11. Burton Williamson, "Some American women in science," p. 363.

12. *DNB,* vol. 40, pp. 18–19.

13. Constance Naden, *Induction and Deduction; a Historical and Critical Sketch of Successive Philosophical Conceptions Respecting the Relations between Inductive and Deductive Thought, and Other Essays,* ed. R. Lewins (London: Bickers and Son, 1890); *Further Reliques of Constance Naden: being Essays and Tracts for our Times,* ed. George M. McCrie (London: Bickers and Son, 1891); *The Complete Poetical Works of Constance Naden* (London: Bickers and Son, 1894).

14. "Neon" [Marion Whiteford Acworth], *The Great Delusion: A Study of Aircraft in Peace and War,* with a preface by Arthur Hungerford Pollen (London: Ernest Benn, 1927).

15. Sir Murray Frazer Sueter, *Airmen or Noahs: Fair Play for our Airmen: The Great "Neon" Air Myth Exposed* (London and New York: Pitman, 1928).

16. *Burke's Peerage and Baronetage,* 105th ed., 1975, p. 156.

17. *Burke's Landed Gentry,* 18th ed., vol. 1, p. 173; *Byron-Shelley-Keats. In Memoriam. Endowed yearly Prizes. Prize Essays by Competitors. With Life Incidents of the Foundress, Rose Mary Crawshay* (Breconshire: Mrs. Crawshay, Cathedine, Bwlch, 1894; this includes the following reprinted articles: C. Wilkins, F.G.S., "Noteworthy men and women of Wales. Mrs. Rose Mary Crawshay," *Western Mail* [n.d.]; anon., "Mrs. Crawshay of Cyfarthfa," *The Queen,* 5 March 1892; Mrs. Haweis, "Interview. Mrs. Rose Mary Crawshay," *Woman's Herald,* 19 March 1892).

18. Chris Evans, *'The Labyrinth of Flames.' Work and Social Conflict in Early Industrial Merthyr Tydfil* (Cardiff: University of Wales Press, 1993), pp. 28–9.

19. *DNB,* vol. 13, p. 62; John P. Addis, *The Crawshay Dynasty: A Study in Industrial Organization and Development, 1765–1867* (Cardiff: University of Wales Press, 1957).

20. Samuel D. Williams, *Euthanasia. Reprinted from Essays by the Members of the Birmingham Speculative Club, with Preface and Thesis by Rose Mary Crawshay,* 4th ed. (London: Williams and Norgate, 1873).

21. Mrs. R. M. Crawshay, "On domestic service for gentlewomen," *Reports of the British Association* (1875): 209 (title only). The paper is printed in full in *Byron-Shelley-Keats,* Second Part, pp. 23–43.

Summary

Of the almost 1,400 papers by American women listed in the Royal Society's *Catalogue of Scientific Papers,* 390 (28 percent) were in botany and 206 (15 percent) in zoology. Contributions in entomology (13 percent), astronomy (12 percent), chemistry and biochemistry (7 percent), and the medical sciences (6 percent) were also prominent. In Britain, with a total of about 880 papers, contributions in botany again outnumbered those in any other field (211 papers, 24 percent). Papers in entomology and geology constituted 14 percent and 12 percent, respectively. Work in chemistry and biochemistry, astronomy, the medical sciences, zoology, mathematics, and physics was also noticeable (see Figure 0-3, b and d).[1]

The marked concentration of nineteenth-century activity by women in both countries in botany and entomology in part reflects the fact that these are disciplines with a component of fieldwork that can be carried out with minimal equipment and laboratory support facilities. However, it is worth emphasizing that in both countries the greater part of published research by women in these fields, and particularly in botany, was carried out by women with close ties to institutions and organizations. Thus a large fraction of the botanical work by American women was carried out under the auspices of the California Academy of Sciences and the New York Botanical Garden, or accommodated at eastern women's colleges and Midwestern state universities; in Britain the outstanding women botanists worked at the women's colleges in London and Cambridge and the British Museum (Natural History). In entomology, while a considerable amount of work by women was indeed independently conducted field collecting and life history studies, several prominent contributors had close organizational or institutional connections; for instance, in Britain Eleanor Ormerod cooperated with the Royal Entomological Society and others worked at the British Museum, while in the United States the notably productive Mary Murtfeldt collaborated over a long period with Missouri State Entomologist, later United States government entomologist, Charles Riley.

The three other fields in which women of both countries were notably active were chemistry, astronomy, and the medical sciences. In chemistry American women collaborated in a great many postgraduate research projects (usually directed by men); a few of the most outstanding of the British women were notably more successful in independent work, but as in the United States the majority made their contributions as assistants. In astronomy also the most prominent American women worked mainly as assistants and technicians (typically at major observatories), although a remarkable amount of observational research was carried out at women's colleges as well. Somewhat in contrast, published work in astronomy by women in Britain resulted, for the most part, from private studies and it included a considerable amount of commentary and review. In the medical sciences several women in each country had productive research careers, some working independently and some as full partners in larger groups.

The remaining noticeably well-populated areas of research activity for women were zoology in the United States, and geology and mathematics/physics in Britain. A major factor underlying the productivity of American women zoologists was the integration of several women as assistants and coworkers into large ongoing survey projects at important institutions, particularly the United States National Museum and the Peabody Museum (this in part paralleled the situation in astronomy). However, work by women students at a number of colleges and universities, especially Bryn Mawr and the Harvard Museum, was also significant.

The special success of late nineteenth-century British women in geology marked the culmination of many decades of considerable interest but more modest activity in that field, which would seem to have been unique to Britain. Onto this foundation was built the remarkable opportunity for thorough training in geology available in the Cambridge Natural Sciences Tripos course open to women by the 1880s. The British women's success in mathematics/physics was also largely a reflection of the opportunities for advanced study (in the form of the Mathematical Tripos) by then open to them at Cambridge. It is remarkable that the British work in mathematics and physics (taken together) constitutes 9 percent of the national total (produced by 7 percent of the authors, and equal to the chemistry/biochemistry output—Figure 0-3, c and d); the corresponding mathematics/physics output for all countries (Figure 0-2) is 6 percent.

Overall, American women concentrated more heavily in the life sciences than did the British. The notable American pre-1901 activity in experimental psychology was not matched by British women. As the bar graphs at the end of most chapters illustrate, the bulk of nineteenth-century research in all fields by women of both countries was carried out during the last two decades of the century (following the opening to women of possibilities for higher education).

Who were the most outstanding in this collection of early women research scientists, the ones who made the most impact on their fields? Among the British the following seven come to mind first: three Cambridge women, geologist Gertrude Lilian Elles and botanists Edith Rebecca Saunders and Ethel Sargant; Scottish geologist Maria Ogilvie Gordon; geographer Marion Newbigin; London botanist Margaret Benson; entomologist Eleanor Ormerod. Biochemist Ida Smedley MacLean, mathematicians Grace Chisholm Young, and Charlotte Scott, bacteriologist and biochemist Harriette Chick, and physicist Hertha Ayrton must also be ranked among those most distinguished. It is worth noting that, in addition to three of the first seven, four more in this selection of eleven (Smedley MacLean, Chisholm Young, Scott, and Ayrton) were Cambridge educated.

Among the Americans two zoologists stand out, Mary Rathbun for her tremendous productivity in systematics and taxonomy, and the unsung Julia Platt for originality. Alice Eastwood, the Canadian-American plant explorer and taxonomist, Williamina Fleming, the Scottish-American astronomer, and Midwest algologist Josephine Tilden also rank highly. Geographer Ellen Semple, psychologists Lillien Martin and Margaret Washburn, and bacteriologist Anna Williams likewise made notable contributions, as did ethnologist Matilda Coxe Stevenson, ornithologist Florence Merriam Bailey, and industrial disease specialist Alice Hamilton (although Hamilton's outstanding work was as much sociological as medical). Carlotta Maury might also be singled out for her contributions to South American paleontology.

Interestingly, Rathbun, Eastwood, Fleming, and Coxe Stevenson all lacked the advantages of formal higher education, although Rathbun's research brought her a Ph.D. when she was fifty-seven. In contrast, all except Ormerod in the especially notable British grouping were university-educated. Of the thirteen Americans mentioned above, five held academic positions, Tilden, Semple, and Martin at coeducational universities, Hamilton at Harvard, and only one, Washburn, at a women's college (there was also British-educated Scott at Bryn Mawr). The low representation of women's college faculty among the most notable research workers is hardly surprising, since funds for equipment were scarce and commitments to teaching and administration especially heavy at these institutions. Among the particularly outstanding British research workers four were faculty members at women's colleges, but two of them, Elles and Saunders, were at Cambridge where they had access to university facilities, Benson had close ties to other London institutions, and Scott pursued her mathematics at Bryn Mawr, the most research-oriented of the American women's colleges, with a mixed faculty.

What qualities and advantages did these distinguished women have that enabled them to succeed as they did? Clearly they were all well endowed with energy and drive, scientific curiosity and confidence in their own abilities; in some cases, especially among the Americans, luck in finding a position was critical. Six of the eleven British—Elles, Saunders, Smedley MacLean, Chick, Newbigin, and Scott—had institutional bases in which they worked; Sargant and Ormerod were financially independent; Ogilvie Gordon, Ayrton, and Chisholm Young (all of whom were married) relied to a large extent on family funds. Among the thirteen Americans, all except Merriam Bailey and Platt held positions that brought some income; Bailey succeeded in carrying out her work largely by joining forces with her biologist husband; Platt did her research during the decade she spent as a graduate student.

The research produced by the American and British women considered here constitutes about 66 percent of women's contributions overall during the nineteenth century, as measured by the number of papers listed in the Royal Society *Catalogue* (Figure 0-1, b). What of the other 34 percent, the work of women from other countries?

The third-largest contribution, again as measured by number of papers indexed by the Royal Society, was that produced by Russian and Polish women. It constituted about 10 percent of the total, with work in the medical sciences, chemistry, and biochemistry being especially noticeable. Further, this East European contribution, nearly all of which came in the last two decades of the century, was of almost uniformly impressive quality. Research by Italian women was also striking—in botany, astronomy, and meteorology in the earlier part of the century, but more especially in the medical sciences and zoology in the 1880s and 1890s. Indeed the Italian group was by far the most productive per person; constituting only 2 percent of the authors, it produced just over 6 percent of the papers (Figure 0-1, a and b). In France there was Marie Curie's work in atomic physics and also noticeable contributions in astronomy and the medical sciences; the combined output of botanical work by women in France, Belgium, and the Netherlands was considerable. Smaller contributions came from German and Austrian women (mainly in chemistry and zoology), from Scandinavians (mainly in botany, zoology, and chemistry), and from Irish women.

Notes

1. The major elements in the distribution pattern of British women's research uncovered in this study, namely heavy concentrations in botany, entomology, geology, and chemistry, fit well with observations of others who have examined the development of more general activity of British women in science (not just research), over longer time spans. Women's interests in entomology, geology, chemistry, and to a lesser extent botany went back beyond the nineteenth century (see Alic, *Hypatia's Heritage,* and Phillips, *Scientific Lady*).

Abbreviations

AAAS — American Association for the Advancement of Science

AAUW — American Association of University Women

AMD — *American Medical Directory.* Chicago: American Medical Association, 1906 and subsequent eds.

AMS — *American Men of Science: A Biographical Directory.* J. McKeen Cattell, ed. New York: Science Press, 1910, and subsequent eds.

DNB — *Dictionary of National Biography.* Leslie Stephen and Sidney Lee, eds. 63 vols. London: Smith, Edler, 1885–1901, and Supplements.

DSB — *Dictionary of Scientific Biography.* Charles C. Gillispie, ed. 16 vols. New York: Scribner's, 1970–1980.

JHB — John Hendley Barnhart, comp., *Biographical Notes upon Botanists.* 3 vols. Boston, G. K. Hall, 1965.

NAW — *Notable American Women,* Edward T. James, Janet Wilson James, and Paul S. Boyer, eds. 3 vols. Cambridge: Harvard University Press, 1971, and *Notable American Women: The Modern Period.* Barbara Sicherman and Carol Hurd Green, eds. Cambridge: Harvard University Press, 1980.

NCAB — *National Cyclopedia of American Biography,* vols. 1–53. New York: James T. White, 1898–1972, vol. 54, Clinton, N.J.: James T. White, 1973.

NST — Natural Sciences Tripos (Cambridge University examinations)

NYT — *New York Times*

S and F — *Women in the Scientific Search. An American Bio-bibliography 1724–1979.* Patricia Joan Siegel and Kay Thomas Finley. Metuchen, N.J.: Scarecrow Press, 1985.

WWWA — *Woman's Who's Who of America. A Biographical Dictionary of Contemporary Women of the United States and Canada. 1914–1915.* John William Leonard, ed.-in-chief. New York: American Commonwealth Co., 1914.

Bibliography of Papers by American and British Women in Scientific Periodicals, 1800–1900

Entries are from the *Catalogue of Scientific Papers* 1800–1900, compiled by the Royal Society, London (19 volumes, Cambridge: Cambridge University Press, 1867–1925). The list has been supplemented by some additional entries, marked { }. Most of these were collected from periodicals not examined by the Royal Society indexers, typically minor journals, regional journals of short run, general interest magazines, and journals in fields not covered in the *Catalogue* (such as philosophy and some areas of clinical medicine). For consistency, only the entries derived from the *Catalogue* are included in the paper counts given in the graphs that follow most chapters. For the most part, the *Catalogue's* style of presentation, nomenclature conventions, etc., have been retained. The few *Catalogue* errors found were corrected. A key to abbreviations follows the bibliography.

PART 1. LIFE SCIENCES

Bacteriology

United States

Carter, Marion Hamilton
Carter, Marion Hamilton
Chromogenic bacteria. *N. Y. Med. J.*, **59** (1894) 372

Hamilton, Alice
Hamilton, Alice
Ueber einen aus China stammenden Kapselbacillus. (Bacillus capsulatus chinensis, *nov. spec.*). *Centrbl. Bakt. (Abt. 2)*, **4** (1898) 230–6

Hansen, Emma Elsie Pammel
Pammel, Louis Hermann; Pammel, Emma
A contribution on the gases produced by certain bacteria. *Centrbl. Bakt. (Abt. 2)*, **2** (1896) 633–50

Hefferan, Mary
Hefferan, (Miss) Mary
A new chromogenic micrococcus. *Bot. Gaz.*, **30** (1900) 261–72

Hyams, Isabel F.
Hyams, (Miss) Isabel F.; Richards, (Mrs.) Ellen H.
The composition of Oscillatoria prolifica (*Greville*), O. rubescens (*De Candolle*), and its relation to the quality of water supplies. *Amer. Ass. Proc.* (1898) 234–6

Lewi, Emily
Lewi, Emily
The milk supply of New York, and the tests of its availability for infant feeding, with a review of the methods of sterilisation. *N. Y. Med. J.*, **61** (1895) 161–7

Mitchell, Charlotte
Mitchell, (Mlle.) Charlotte; Richet, Charles
De l'accoutumance des ferments aux millieux toxiques. *Paris, Soc. Biol. Mém.*, **52** (C. R.) (1900) 637–9

Peckham, Adelaide Ward
Peckham, Adelaide Ward
{A study of a case of *erysipelas genitalium,* due to the use of infected ointment. *Med. News*, **62** (1893) 148–51}
A study of the colon bacillus group, and especially of its variability in fermenting power under different conditions. *Science*, **4** (1896) 773–8
The influence of environment upon the biological processes of the various members of the colon group of bacilli. An experimental study. *J. Exper. Med.*, **2** (1897) 549–91; *Science*, **5** (1897) 981–5
Billings, John Shaw; Peckham, Adelaide Ward
The influence of certain agents in destroying the vitality of the typhoid and of the colon bacillus. *Smithsonian Rep.* (1894) 451–8; *Science*, **1** (1895) 169–74

Pennington, Mary Engle
Pennington, Mary Engle; Küsel, George C.
An experimental study of the gas-producing power of Bacillus coli communis under different conditions of environment. *Amer. Chem. Soc. J.*, **22** (1900) 556–67

Richards, Ellen Henrietta Swallow
Jordan, Edwin Oakes; Richards, (Mrs.) Ellen Henrietta
Investigations upon nitrification and the nitrifying organism. *Science*, **18** (1891) 48–52
Rolfe, George William; Richards, Ellen H.
{Reduction of nitrates by bacteria and consequent loss of nitrogen. *Tech. Quart.*, **9** (1896) 40–59}
Hyams, (Miss) Isabel F.; Richards, (Mrs.) Ellen H.
The composition of Oscillatoria prolifica (*Greville*), O. rubescens (*De Candolle*), and its relation to the quality of water supplies. *Amer. Ass. Proc.* (1898) 234–6

Williams, Anna Wessels
Williams, A. W.; Park, William Hallock
The production of diphtheria toxin. *J. Exper. Med.*, **1** (1896) 164–85

Great Britain

Chick, Harriette
Chick, Harriette
The distribution of Bacterium coli commune. *Liverpool, Thompson Yates Lab. Rep.,* **3** (Pt. 1) (1900) 1–29; **3** (Pt. 2) (1901) 117–29

Frankland, Grace Coleridge Toynbee
Frankland, (Mrs.) Grace C.
Typhoid fever epidemics in America. [1895] *Nature,* **53** (1895–6) 38–9
Bacteria and carbonated waters. *Nature,* **54** (1896) 375–6
Dr. Yersin, and plague virus. [1897] *Nature,* **55** (1896–7) 378–9
Frankland, Grace C.; Frankland, Percy F.
Studies on some new micro-organisms obtained from air. [1887] *Brit. Ass. Rep.* (1887) 745–9; *Phil. Trans. (B),* **178** (1888) 257–87
On some new and typical micro-organisms obtained from water and soil. *Roy. Soc. Proc.,* **43** (1888) 414–18; *Ztschr. Hyg.,* **6** (1889) 373–400
The nitrifying process and its specific ferment. Part I. [1890] *Phil. Trans. (B),* **181** (1889) 107–28; {*Roy. Soc. Proc.,* **47** (1890) 296–8}
Frankland, Percy Faraday; Frankland, Grace C.; Fox, Joseph J.
Contribution to the study of pure fermentations. *Brit. Ass. Rep.* (1889) 544–5
Frankland, Percy Faraday; Frew, Wm.; {Frankland, Grace C.}
A pure fermentation of mannitol and dulcitol. [With an appendix by Grace C. Frankland: "Morphological characterisation of the micro-organism causing fermentation".] *Chem. Soc. J.,* **61** (1892) 254–77

Veley, Lilian Jane Gould
Veley, Victor H.; Gould, Lilian J. (Mrs. Veley)
A bacterium in rum. *Soc. Chem. Ind. J.,* **16** (1897) 626
A bacterium living in strong spirit. *Nature,* **56** (1897) 197
The micro-organism of faulty rum. [1900] *Nature,* **61** (1899–1900) 468–9

Biochemistry

Great Britain

Newbigin, Marion Isabel
Newbigin, (Miss) Marion I.
An attempt to classify common plant pigments, with some observations on the meaning of colour in plants. *Edinb. Bot. Soc. Trans. & Proc.,* **20** (1896) 534–50
Observations on the metallic colours of the Trochilidae and the Nectarinidae. *Zool. Soc. Proc.* (1896) 238–96
The pigments of the decapod Crustacea. *J. Physiol.,* **21** (1897) 237–57
Pigments of muscle and ovary in the salmon. *Micr. Soc. J.* (1898) 526
On certain green (chlorophylloid) pigments in invertebrates. [1898] *Quart. J. Micr. Sci.,* **41** (1899) 391–431; *Edinb. R. Coll. Physns. Lab. Rep.,* **7** (1900) 391–431
On the affinities of the enterochromes. *Zool. Anz.,* **22** (1899) 325–8

Rosenheim, Mary Christine Tebb
Tebb, (Miss) M. Christine
On the transformation of maltose to dextrose. *J. Physiol.,* **15** (1894) 421–32
Note on the liver ferment. [1894] *Cambridge Phil. Soc. Proc.,* **8** (1895) 199–200
Hydrolysis of glycogen. [1898] *J. Physiol.,* **22** (1897–8) 423–32
Chemistry of reticular tissue. *J. Physiol.,* **24** (1899) x–xi
Shore, Lewis E.; Tebb, (Miss) M. Christine
On the transformation of maltose to dextrose. *J. Physiol.,* **13** (*Proc.*) (1892) xix–xx

Biology

United States

Andrews, Sara Gwendolen Foulke
Foulke, (Miss) Sara Gwendolen
Observations on Actinosphaerium eichornii. *Philad. Ac. Nat. Sci. Proc.* (1883) 125–7
Some phenomena in the life-history of Clathrulina elegans. *Philad. Ac. Nat. Sci. Proc.* (1884) 17–19
On a new species of rotifer, of the genus Apsilus. *Philad. Ac. Nat. Sci. Proc.* (1884) 37–41
Some notes on Manayunkia speciosa. *Philad. Ac. Nat. Sci. Proc.* (1884) 48–9
[Dictyophora as Apsilus vorax.] *Philad. Ac. Nat. Sci. Proc.* (1884) 51
A new species of Trachelius. *Philad. Ac. Nat. Sci. Proc.* (1884) 51–2
Manayunkia speciosa. *Science,* **3** (1884) 303–4
The reproduction of Clathrulina elegans. *Science,* **3** (1884) 435
An endoparasite of Noteus. *Amer. J. Sci.,* **30** (1885) 377–8
Andrews, Gwendolen Foulke
On a method found useful in preservation of protoplasmic spinnings. *Ztschr. Wiss. Mikr.,* **14** (1897) 447–52
Some spinning activities of protoplasm in starfish and sea-urchin eggs. *J. Morphol.,* **12** (1897) 367–89

Blackwell, Antoinette Louisa Brown
Blackwell, (Mrs.) A. Brown
The comparative longevity of the sexes. *Amer. Ass. Proc.* (1884) 515–17

Bodington, Alice
Bodington, Alice
The parasitic protozoa found in cancerous diseases. *Amer. Natlist.,* **28** (1894) 307–15

Fielde, Adele Marion
Fielde, (Miss) Adele M.
Observations on tenacity of life, and regeneration of excised parts in Lumbricus terrestris. *Philad. Ac. Nat. Sci. Proc.* (1885) 20–2
Note on the multiplication of Distoma. *Philad. Ac. Nat. Sci. Proc.* (1887) 115
Notes on freshwater rhizopods of Swatow, China. *Philad. Ac. Nat. Sci. Proc.* (1887) 122–3

Hinckley, Mary H.
Hinckley, Mary H.
Notes on the peeping frog, Hyla pickeringii, *Le Conte.* [1883] *Boston Soc. Nat. Hist. Mem.,* **3** (1878–86) 311–18
Notes on eggs and tadpoles of Hyla versicolor. [1880] *Amer. Natlist.,* **16** (1882) 636–9; *Boston Soc. Nat. Hist. Proc.,* **21** (1883) 104–7
On some differences in the mouth structure of tadpoles of the anourous batrachians found in Milton, Mass. [1882] *Boston Soc. Nat. Hist. Proc.,* **21** (1883) 307–14

Notes on the development of Rana sylvatica, *Le Conte.* [1882] *Boston Soc. Nat. Hist. Proc.,* **22** (1884) 85–95

Lewis, Graceanna
Lewis, Graceanna
{Thoughts on the structure of the animal kingdom. *Amer. Ass. Proc.* (1869) 280 [title only]}

Whitman, Emily A. Nunn
Nunn, (Miss) Emily A.
[The appearance of amoebae in an infusion of the yolk of hen's egg in Pasteur's fluid.] [1880] *Amer. Micr. J.,* **2** (1881) 19

Botany

United States

Abbott, Rosa G.
Abbott, Rosa G.
Electrical attraction of trees. *Garden & Forest,* **10** (1897) 297
Electricity in vegetation. *Garden & Forest,* **10** (1897) 337

Adamson, Margaret E.
Adamson, Margaret E.
Teratological notes on Eschscholtzia californica. *Erythea,* **7** (1899) 81–2

Atchison, Ida May Clendenin
Clendenin, Ida
Synchytrium on Stellaria media. *Bot. Gaz.,* **19** (1894) 296–7
Synchytrium on Geranium carolinianum. *Bot. Gaz.,* **20** (1895) 29–31
Lasiodiplodia, *E.* and *E., n. gen. Bot. Gaz.,* **21** (1896) 92
{Botanical teaching in secondary schools. *Amer. Ass. Proc.* (1899) 294 (title only)}

Atwater, Florence May Andrews
Andrews, Florence M.
Notes on a species of Cyathus common in lawns at Middlebury, Vermont. *Rhodora,* **2** (1900) 99–101

Arnold, Isabel S.
Arnold, Isabel S.
Notes on the flora of the Upper Chemung valley. *Torrey Bot. Club Bull.,* **15** (1888) 131–3

Austin, Rebecca (Rachel?) Merritt Smith Leonard
Austin, (Mrs.) R. M
Darlingtonia californica, *Torr.* [1878] *Bot. Gaz.,* **3** and **4** (1878–9) 70–1
Leaves of Darlingtonia californica and their two secretions. [1878] *Bot. Gaz.,* **3** and **4** (1878–9) 91
Sarcodes sanguinea. [1883] *Bot. Gaz.,* 7 and **8** (1882–3) 284–5

Bacon, Alice Elizabeth
Bacon, Alice E.
Some orchids of eastern Vermont. *Rhodora,* **2** (1900) 171–2

Beach, Alice Marie
Pammel, Louis Hermann; Beach, Alice M.
Pollination of cucurbits. [1894] *Iowa Ac. Sci. Proc.,* **2** (1895) 146–52

Beckwith, Florence E.
Beckwith, (Miss) Florence
Variation of ray-flowers in Rudbeckia hirta. [1892] *Rochester (N. Y.) Ac. Sci. Proc.,* **2** (1895) 170–1
Beckwith, (Miss) Florence; Macauley, (Miss) Mary E.
Plants of Monroe county, N.Y., and adjacent territory. [With the assistance of Joseph B. Fuller.] [1894] *Rochester (N. Y.) Ac. Sci. Proc.,* **3** (1906) 1–150

Beeler, Lora Luvernia Waters
Waters, Lora L.
Erysipheae of Riley county, Kansas. [1894] *Kan. Ac. Sci. Trans.,* **14** (1896) 200–6

Bigelow, Cassie Pearl
Bigelow, Cassie M.
Study of glands in the hop-tree. [1894] *Iowa Ac. Sci. Proc.,* **2** (1895) 138–40

Bingham, Caroline P. Lord
Bingham, Mrs. R. F.
{Common and troublesome weeds near Santa Barbara, California. *Bot. Gaz.,* **3** and **4** (1879) 226}
Flora near Santa Barbara, Cal. *Bot. Gaz.,* **12** (1887) 33–5
An American Papaver. *Bot. Gaz.,* **12** (1887) 67

Bitting, Katherine Eliza Golden
Golden, Katherine E.
Fermentation of bread. *Bot. Gaz.,* **15** (1890) 204–9
{Diseases of the sugar beet root. *Indiana Ac. Sci. Proc.* (1891) 92–7}
{The application of mathematics in botany. *Indiana Ac. Sci. Proc.* (1892) 37–41}
An auxanometer for the registration of growth of stems in thickness. *Bot. Gaz.,* **19** (1894) 113–16; {*Indiana Ac. Sci. Proc.* (1892) 46–8}
Movement of gases in rhizomes. *Amer. Ass. Proc.* (1894) 275–82
Salt as a deterrent in yeast fermentation. *Amer. Ass. Proc.* (1898) 418–20
Aspergillus oryzae. *Amer. Micr. J.,* **20** (1899) 351–60
A mold isolated from tan-bark liquors. *Amer. Ass. Proc.* (1900) 278
Arthur, J. C.; Golden, Katherine E.
{Diseases of the sugar beet root. *Bull. Indiana Agric. Expt. Stn.,* **39** (1891) 54–8}
Golden, Katherine E.; Ferris, Carleton G.
Fermentation without live cells. *Amer. Ass. Proc.* (1898) 417
Red yeasts. *Bot. Gaz.,* **25** (1898) 39–46

Blochmann, Ida May Twitchell
Twitchell, (Miss) Ida
{Felices iowenses. *Aurora,* **8** (1880) 132–7}
On the evaporation of water from leaves. *Amer. Natlist.,* **15** (1881) 385–8
Blochmann, Ida M.
California herb-lore. *Erythea,* **1** (1893) 190–1; 231–3; **2** (1894) 9–10; 39–40; 162–3

Brandegee, Mary Katharine Layne Curran
Curran, Mary K.
New species of Californian plants. [1884] *California Ac. Bull.,* **1** (1886) 12–13
List of the plants described in California, principally in the Proc. of the Cal. Acad. of Sciences, by Dr. Albert Kellogg, Dr. H. H. Behr, and Mr. H. N. Bolander; with an attempt at their identification. [With descriptions of some Californian plants collected by the writer in 1884.] [1885] *California Ac. Bull.,* **1** (1886) 128–55; 392
Botanical notes. *California Ac. Bull.,* **1** (1886) 272–5
Priority of Dr. Kellogg's genus Marah over Megarrhiza, *Torr. California Ac. Bull.,* **2** (1887) 521–4
Botanical notes. [1888] *California Ac. Proc.,* **1** (1888) 227–69; 362

Brandegee, Katharine
{Dodocatheon meadia. (Shooting Star.) *Zoe,* **1** (1890) 17–20}
{Notes on West American plants. I. *Zoe,* **1** (1890) 82, 83}
{Rhamnus californica and its allies. *Zoe,* **1** (1890) 240–4}
{Caenurus of the Hare. *Zoe,* **1** (1890) 265–8}
{The variations of Platystemon and Eschscholtzia. *Zoe,* **1** (1890) 278–82}
{Asplenium Filiz-foemina as a tree fern. *Zoe,* **1** (1890) 293–5}
{Californian Lobeliaceae. *Zoe,* **1** (1891) 373–7}
{Contributions to the knowledge of West American plants. I. *Zoe,* **2** (1891) 75–83}
{The flora of Yosemite. *Zoe,* **2** (1891) 155–67}
{Catalogue of the flowering plants and ferns growing spontaneously in the city of San Francisco. *Zoe,* **2** (1892) 334–86}
{Additions to the catalogue of San Francisco plants. *Zoe,* **3** (1892) 49, 50}
{The nomenclature of plants. *Zoe,* **3** (1892) 166–72; 258–61}
{The botanical writings of Edward L. Greene. *Zoe,* **4** (1893) 63–103}
{Sierra Nevada plants in the Coast Range. *Zoe,* **4** (1893) 168–76}
{Botanical nomenclature. *Zoe,* **4** (1893) 182–4}
{Flora of Bouldin Island. *Zoe,* **4** (1893) 211–18}
{E. L. Greene versus Asa Gray. *Zoe,* **4** (1893) 287–91}
{Botanical meetings of the annual assembly of the American Association for the Advancement of Science. *Zoe,* **4** (1893) 291–6}
{The dates of *Botany beechey, Flora boreali-americana,* and Torrey & Gray's *Flora. Zoe,* **4** (1894) 369–72}
Studies in Portulacaceae. [1894] *California Ac. Proc.,* **4** (1894) 86–91
Studies in Ceanothus. [1894] *California Ac. Proc.,* **4** (1894) 173–222
Notes on Eriogoneae. *Erythea,* **5** (1897) 79–81; 100
Notes on Cacteae. *Erythea,* **5** (1897) 111–23; 133
{Notes on Cacteae. II. *Zoe,* **5** (1900) 1–9}
{Notes on Cacteae. III. *Zoe,* **5** (1900) 31–5}

Britton, Elizabeth Gertrude Knight
Knight, Elizabeth G.
Albinism. *N. Y., Bot. Club Bull.,* **8** (1881) 125
Submersed leaves in Limnanthemum. *N. Y., Bot. Club Bull.,* **10** (1883) 34
On the fruit of Eustichium norvegicum, *Br. Eu. N. Y., Bot. Club Bull.,* **10** (1883) 99–100
Britton, (Mrs.) Elizabeth G.
Additions to the Westchester county flora. *Torrey Bot. Club Bull.,* **13** (1886) 6–7
Botanical notes in the great valley of Virginia and in the southern Alleghenies. *Torrey Bot. Club Bull.,* **13** (1886) 69–76
Plurality of embryos in Quercus alba. *Torrey Bot. Club Bull.,* **13** (1886) 95
Elongation of the inflorescence in Liquidambar. *Torrey Bot. Club Bull.,* **14** (1887) 95–6
Hypnum (Thuidium) calyptratum, *Sulliv. Torrey Bot. Club Bull.,* **15** (1888) 220
An enumeration of the plants collected by Dr. H. H. Rusby in South America, 1885–1886. III. Pteridophyta. *Torrey Bot. Club Bull.,* **15** (1888) 247–53
Ulota phyllantha in fruit from Killarney. *J. Bot.,* **26** (1888) 282
Contributions to American bryology. *Torrey Bot. Club Bull.,* **16** (1889) 106–12; **18** (1891) 49–56; **20** (1893) 393–405; **21** (1894) 1–15; 65–76; 137–59; 189–208; 343–72; **22** (1895) 36–43; 62–8; 447–58
Grimmia torquata, *Horns.,* fertile. *Rev. Bryol.,* **16** (1889) 38–9
Peristome of Grimmia torquata, *Hornsch. Rev. Bryol.,* **16** (1889) 64
Leucobryum minus, *Hampe. Torrey Bot. Club Bull.,* **19** (1892) 189–91
Notes on two of Palisot de Beauvois' species of Orthotrichum. *Rev. Bryol.,* **20** (1893) 99
The Jaeger Moss Herbarium. *Torrey Bot. Club Bull.,* **20** (1893) 335–6
[Report of the botanical exploration of southwestern Virginia during the season of 1892.] Bryophyta. Musci. [1894]; *Torrey Bot. Club Mem.,* **4** (1893–1896) 172–91
[On the collections of Mr. Miguel Bang in Bolivia.] Filices. [1895] *Torrey Bot. Club Mem.,* **4** (1893–1896) 271–3
An enumeration of the plants collected by Dr. H. H. Rusby in Bolivia, 1885–86. II. Musci. *Torrey Bot. Club Bull.,* **23** (1896) 471–99
A revision of the North American species of Ophioglossum. *Torrey Bot. Club Bull.,* **24** (1897) 545–59
Mosses of northern India. *Torrey Bot. Club Bull.,* **25** (1898) 398
A new Grimmia from Mt. Washington. *Rhodora,* **1** (1899) 148–9
A new Tertiary fossil moss. *Torrey Bot. Club Bull.,* **26** (1899) 79–81
Note on Trichostomum warnstorfii, *Limpr. Rev. Bryol.,* **27** (1900) 71
Bryological notes. *Torrey Bot. Club Bull.,* **27** (1900) 648–9

Broomell, Anna Stockton Pettit
Pettit, Anna Stockton
Arachis hypogaea, *L.* [1895] *Torrey Bot. Club Mem.,* **4** (1893–1896) 275–96

Bunting, Martha
Bunting, Martha
On the formation of cork tissue in the roots of the Rosaceae. [1897] *Amer. Natlist.,* **32** (1898) 109–10
The structure of the cork tissues in roots of some rosaceous genera. [1898] *Pennsylvania Univ. Publ. (Contrib. Bot. Lab.),* **2** (1904) 54–65

Burnett, Katharine Cleveland
Burnett, Katharine Cleveland
Notes on the influence of light on certain dorsiventral organs. *Torrey Bot. Club Bull.,* **24** (1897) 116–22

Calvert, Amelia Catherine Smith
Smith, Amelia C.
Structure and parasitism of Aphyllon uniflorum. [1898] *Amer. Natlist.,* **33** (1899) 204; *Pennsylvania Univ. Publ. (Contrib. Bot. Lab.),* **2** (1901) 111–21

Clifford, Julia Blanche
Clifford, Julia B.
Notes on some physiological properties of a Myxomycete plasmodium. *Ann. Bot.,* **11** (1897) 179–86
The Mycorhiza of Tipularia unifolia. *Torrey Bot. Club Bull.,* **26** (1899) 635–8
The Mycorhiza of Tipularia. *Amer. Ass. Proc.* (1899) 298

Cook, Alice Carter
Carter, Alice
Notes on pollination. *Bot. Gaz.,* **17** (1892) 19–22
Evolution in methods of pollination. *Bot. Gaz.,* **17** (1892) 40–6; 72–8
A sketch of the flora of the Canary Islands. *Torrey Bot. Club Bull.,* **25** (1898) 351–8

Cook, Mabel Priscilla
Cook, Mabel Priscilla
Some additions to the "Flora of Middlesex County, Massachusetts." *Rhodora,* **1** (1899) 80–2

Cooley, Grace Emily
Cooley, (Miss) Grace E.
Notes on movement of water in "Robinia pseudacacia." *Canad. Rec. Sci.,* **1** (1885) 202–7
Impressions of Alaska. *Torrey Bot. Club Bull.,* **19** (1892) 178–89
Plants collected in Alaska and Nanaimo, B. C., July and August, 1891. [With a list of mosses and lichens by Clara E. Cummings *Torrey Bot. Club Bull.,* **19** (1892) 239–49
On the reserve cellulose of the seeds of Illiaceae and of some related orders. [1895] *Boston Soc. Nat. Hist. Mem.,* **5** (1895–1904) 1–29

Cox, Mary Alice Nichols
Nichols, (Miss) Mary Alice
Achenial hairs of Compositae. *Bot. Gaz.,* **18** (1893) 378–82
Observations on the pollination of some of the Compositae. [1893] *Iowa Ac. Sci. Proc.,* **1** (Pt. 4) (1894) 100–3
Abnormal fruiting of Vaucheria. *Bot. Gaz.,* **20** (1895) 269–71
Studies in the development of the ascospores in certain Pyrenomycetes. *Bot. Gaz.,* **22** (1896) 234
The morphology and development of certain pyrenomycetous fungi. *Bot. Gaz.,* **22** (1896) 301–28
Rowlee, Willard W.; Nichols, Mary Alice
Contributions to the life-history of Symplocarpus foetidus. *Amer. Micr. Soc. Trans.,* **17** (1895) 157–64

Cross, Laura Bell Abbott
Cross, Laura B.
On the structure and pollination of the flowers of Eupatorium ageratoides and Eupatorium caelestinum. *Pennsylvania Univ. Publ.* (*Contrib. Bot. Lab.*), **1** (1897) 260–9

Cummings, Clara Eaton
Cooley, (Miss) Grace E.
Plants collected in Alaska and Nanaimo, B. C., July and August, 1891. [With a list of mosses and lichens by Clara E. Cummings.] *Torrey Bot. Club Bull.,* **19** (1892) 239–49
Cummings, Clara E.
{Cryptograms collected by Dr. C. Willard Hayes in Alaska, 1891. [An appendix to Hayes, C. W., "An expedition through the Yukon District." *Nat. Geogr. Mag.,* **4** (1892) 117–62.] *Nat. Geogr. Mag.,* **4** (1892) 160–2}

Dakin, Norra Allin
MacBride, Thomas Huston; Allin, Norra
The saprophytic fungi of eastern Iowa. The puff-balls. [1896] *Iowa Univ. Lab. Nat. Hist. Bull.,* **4** (1898) 33–6

Davidson, Alice Jane Merritt
Merritt, Alice J.
{Notes on fertilization. *Zoe,* **3** (1893) 311–12}
Notes on the pollination of some Californian mountain flowers. *Erythea,* **4** (1896) 101–3; 147–9; **5** (1897) 1–4; 15–22; 56–9; 133

Day, Mary Anna
Day, Mary A.
The local floras of New England. *Rhodora,* **1** (1899) 111–20; 138–42; 158; 174–8; 194–6; 208–11; 230; **2** (1900) 73–4
Plants from the eastern slope of Mt. Equinox. *Rhodora,* **1** (1899) 220–22

Detmers, Frederica
Detmers, Freda
{Prickly lettuce—an introduced weed. *J. Columbus Hort. Soc.,* **5** (1890) 53–4}
{A preliminary list of the rusts of Ohio. *Ohio Agric. Exp. Sta. Ann. Rep. Bull.,* **44** (1892) 133–40}

Dixson, Zella Allen
Dixson, Mrs. J. E.
Germination of Phoenix dactylifera. *Denison Univ. Sci. Lab. Bull.,* **5** (1890) 8–9

Dunn, Louise Brisbin
Dunn, Louise B.
An attempted new method of producing zygospores of Rhizopus. *Amer. Ass. Proc.* (1900) 285
Morphology of the development of the ovule in Delphinium exaltatum, *Ait. Amer. Ass. Proc.* (1900) 284; *Science,* **12** (1900) 584–5

Dunn, Luella Cushing Whitney
Whitney, Luella C.
List of Vermont Myxomycetes with notes. *Rhodora,* **1** (1899) 128–30

Eigenmann, Rosa Smith
Smith, (Miss) Rosa
Some nasturtium leaves. *Bot. Gaz.,* **10** (1884–1885) 368

Elbel, Clara Avesta Cunningham
Cunningham, Clara A.
{The effects of drought upon certain plants. *Indiana Ac. Sci. Proc.* (1896) 208–13}
A bacterial disease of the sugar beet. *Bot. Gaz.,* **28** (1899) 177–92

Ferguson, Margaret Clay
Ferguson, Margaret C.
{The development of the egg and fertilization in Pinus strobus. *Amer. Ass. Proc.* (1900) 281}

Flint, Martha Bockée
Flint, Martha Bockée
Galium verum in New York. [1885] *Bot. Gaz.,* **9** & **10** (1884–1885) 386

Fox, Henrietta G.
Fox, Henrietta G.
On the genus Cypripedium, *L.,* with reference to Minnesota species. [1895] *Minn. Bot. Stud.,* **1** (1894–1898) 423–49

Furbish, Katherine
Furbish, Kate
{A botanist's trip to "The Aroostook." [No. 2.] *Amer. Natlist.* **15** (1881) 469–70; **16** (1882) 397–9}
{Cut-leaved beech. *Amer. Natlist.* **16** (1882) 1004}
Still further notes on the flora of Rangeley Lakes, Maine. *Torrey Bot. Club Bull.,* **18** (1891) 152
Myosotis collina in Maine. *Rhodora,* **1** (1899) 76

Gloss, Mary Elgin
Gloss, Mary Elgin
Mesophyl of ferns. *Torrey Bot. Club Bull.,* **24** (1897) 432–5

Gow, Grace Darling Chester
Chester, Grace D.
Notes concerning the development of Nemaliou multifidum. *Bot. Gaz.,* **21** (1896) 340–7
Bau und Function der Spaltöffnungen auf Blumenblättern und Antheren. *Deutsch. Bot. Ges. Ber.,* **15** (1897) 420–31; (132)

Greene, Lillian Snyder
Snyder, Lillian
{The Uredineae of Tippecanoe county, Indiana. *Indiana Ac. Sci. Proc.* (1897) 216–24}
A bacteriological study of pear blight. *Amer. Ass. Proc.* (1898) 426–7

{The germ of pear blight. *Indiana Ac. Sci. Proc.* (1898) 150–6}
{The Uredineae of Madison and Nobel counties with additional specimens from Tippecanoe county. *Indiana Ac. Sci. Proc.,* (1899) 186–9}
{The Uredineae of Parke county, Indiana. *Indiana Ac. Sci. Proc.* (1900) 30 [title only]}

Gregory, Emily Lovira
Gregory, Emily L.
The pores of the libriform tissue. *Torrey Bot. Club Bull.,* **13** (1886) 197–204; 233–44
{Old and new botany. [A letter] *Bot. Gaz.,* **12** (1887) 253–4}
{Systematic botany—a correction. *Bot. Gaz.,* **12** (1887) 298}
Development of cork-wings on certain trees. *Bot. Gaz.,* **13** (1888) 249–58; 281–7; 313–18; **14** (1889) 5–10; 37–44
{Notes on some botanical reading done in the laboratory of Professor Schwendener, in Berlin, June and July, 1889. *Torrey Bot. Club. Bull.,* **16,** (1889) 297–304}
Notes on the manner of growth of the cell wall. *Torrey Bot. Club Bull.,* **17** (1890) 247–55
Abnormal growth of Spirogyra cells. *Torrey Bot. Club Bull.,* **19** (1892) 75–9
The two schools of plant physiology as at present existing in Germany and England. [1891] *Amer. Natlist.,* **26** (1892) 211–17; 279–86
Anatomy as a special department of botany. [1892] *Torrey Bot. Club Bull.,* **20** (1893) 100–7
Notes on the classifications of lichens. *Torrey Bot. Club Bull.,* **23** (1896) 359–61
What is meant by stem and leaf? *Torrey Bot. Club Bull.,* **23** (1896) 278–81
{To the editors of the *Botanical Gazette. Bot. Gaz.,* **22** (1896) 72–4}

Hagenbuck, (Mrs.) I.
Hagenbuck, (Mrs.) I.
Californian herb lore. *Erythea,* **5** (1897) 39; 97

Hansen, Emma Elsie Pammel
Pammel, Emma
A comparative study of the leaves of Lolium, Festuca, and Bromus. [1896] *Iowa Ac. Sci. Proc.,* **4** (1897) 126–31
Sirrine, Emma; Pammel, Emma
Some anatomical studies of the leaves of Sporobulus and Panicum. *Iowa Ac. Sci. Proc.,* **3** (1896) 148–59

Hewins, Nellie Pricilla
Hewins, Nellie Pricilla
A contribution to the knowledge of the organogeny of the flower and of the embryology of the Carpifoliaceae. *Amer. Ass. Proc.* (1900) 280–1

Hooker, Henrietta Edgecomb
Hooker, Henrietta E.
The germination of dodder. *Amer. Natlist.,* **22** (1888) 254
On Cuscuta gronovii. *Bot. Gaz.,* **14** (1889) 31–7

Horn, Margaretha Elise Catherine
Horn, Margaretha E. C.
The organs of attachment in Botrytis vulgaris. *Bot. Gaz.,* **22** (1896) 329–33

Horner, Charlotte N. Saunders
Horner, (Mrs.) Charlotte N. S.
Notes from Massachusetts. *Torrey Bot. Club Bull.,* **11** (1884) 8–9
Notes on the flora of South Georgetown. [1883] *Essex Inst. Bull.,* **15** (1884) 107–10
Notes on some introduced plants in Eastern Massachusetts. *Torrey Bot. Club Bull.,* **14** (1887) 219

Howard, Anne B. Townsend
Townsend, Anne B.
An hermaphrodite gametophore in Preissia commutata. *Bot. Gaz.,* **28** (1899) 360–2

Jungerich, Elizabeth Alexander Simons
Simons, Elizabeth A.
Comparative studies on the rate of circumnutation of some flowering plants. [1898] *Pennsylvania Univ. Publ. (Contrib. Bot. Lab.),* **2** (1904) 66–79
Simons, Elizabeth A.; McKenney, R. E. B.
Rapidity of circumnutation movements in relation to temperature. *Amer. Ass. Proc.* (1898) 430–1

Keener, Alice Elizabeth
Keener, Alice E.
Collinsia bicolor. *Bot. Gaz.,* **20** (1895) 232

Keller, Ida Augusta
Keller, Ida A.
The phenomenon of fertilization in the flowers of Monarda fistulosa. *Philad. Ac. Nat. Sci. Proc.* (1892) 452–4
The glandular hairs of Brasenia peltata, *Pursh. Philad. Ac. Nat. Sci. Proc.* (1893) 188–93
The jelly-like secretion of the fruit of Peltandra undulata, *Raf. Philad. Ac. Nat. Sci. Proc.* (1895) 287–90
Notes on the study of the cross-fertilization of flowers by insects. *Philad. Ac. Nat. Sci. Proc.* (1895) 555–61
Notes on underground runners. *Philad. Ac. Nat. Sci. Proc.* (1897) 161–5
Notes on plant monstrosities. *Philad. Ac. Nat. Sci. Proc.* (1897) 284–7
The growth of Viburnum lantanoides, *Michx. Philad. Ac. Nat. Sci. Proc.* (1898) 482–4
Notes on hyacinth roots. *Philad. Ac. Nat. Sci. Proc.* (1900) 438–40

Kellerman, Stella Victoria Dennis
Kellerman, Mrs. W. A.
Evolution in leaves. *Kan. Ac. Sci. Trans.,* **12** (1890) 168–73
Indications of evolution in leaves. *Science,* **18** (1891) 226–7
Some curious catnip leaves. *Science,* **19** (1892) 66–7
A series of abnormal Ailanthus leaflets. *Science,* **19** (1892) 90–1
A seeding blackberry plant. *Science,* **19** (1892) 94–5
Interesting variations of the strawberry leaf. *Bot. Gaz.,* **17** (1892) 257–8
Leaf-variation, its extent and significance. [1892] *Cincin. Soc. Nat. Hist. J.,* **16** (1893–1894) 49–53
Kellerman, W[illiam]; Kellerman, Mrs. W. A.
The Kansas forest trees identified by leaves and fruit. [1886] *Kan. Ac. Sci. Trans.,* **10** (1887) 99–111

Knowles, Etta L.
Knowles, Etta L.
Structure and distribution of resin passages of the White Pine. *Bot. Gaz.,* **11** (1886) 206–8
The "curl" of peach leaves: a study of the abnormal structure induced by Exoascus deformans. *Bot. Gaz.,* **12** (1887) 216–18
{A study of the abnormal structures induced by Ustilago zeae Mays. *J. Mycol.,* **5** (1889) No. 1, 14–18}

La Mance, Lora S.
La Mance, Lora S.
Iris hexagona. *Garden & Forest,* **8** (1895) 329

Lemmon, Sara Allen Plummer
Lemmon, Sara Allen Plummer
{The ferns of the Pacific Coast. *Pac. Rural Pr.,* **21,** No. 13 (26 March 1881) 216–17. [Reprinted, in part, in Joseph Ewan, "Bibliographical Miscellany—V. Sara Allen Plummer Lemmon and her 'Ferns of the Pacific Coast'," *Amer. Midland Natlist,* **32** (1944) 513–18]}

Lovell, (Mrs. Preston)
Lovell, Mrs. Preston
A few native orchids. *Amer. Natlist.,* **25** (1891) 248–51

Macauley, Mary Elizabeth
Beckwith, (Miss) Florence; Macauley, (Miss) Mary E.
Plants of Monroe county, N.Y., and adjacent territory. [With the assistance of Joseph B. Fuller.] [1894] *Rochester (N. Y.) Ac. Sci. Proc.,* **3** (1906) 1–150

Martin, Lillien Jane
Martin, (Miss) Lillie J.
A botanical study of the mite-gall found on the petiole of Juglans nigra, known as Erineum anomalum, *Schw.* [1884] *Amer. Ass. Proc.* (1884) 507–8; *Amer. Natlist.,* **19** (1885) 136–40
Preliminary analysis of the leaves of Juglans nigra. *Amer. J. Pharm.,* **58** (1886) 468–74
{Plan for laboratory work in chemical botany. *Amer. Ass. Proc.* (1886) 258(Abs.)}

McComb, Amanda
McComb, Amanda
The development of the karyokinetic spindle in vegetative cells of higher plants. *Torrey Bot. Club Bull.,* **27** (1900) 451–9

McCormick, Florence Anna
McCormick, Florence A.
History of the leaf of Pinus virginiana. *Amer. Micr. J.,* **20** (1899) 33–8

McEwen, Marion C.
McEwen, Marion C.
The comparative anatomy of Corema alba and Corema conradii. *Torrey Bot. Club Bull.,* **21** (1894) 277–85

McFadden, Effie B.
McFadden, Effie B.
The development of the antheridium of Targionia hypophylla. *Torrey Bot. Club Bull.,* **23** (1896) 242–4

McGee, Emma Rachel
McGee, Emma R.
Some Nebraska plants. *Bot. Gaz.,* **13** (1888) 301

Merry, Martha
Merry, Martha
The identity of Podosphaera minor, *Howe,* and Microsphaera fulvo-fulcra, *Cooke. Bot. Gaz.,* **12** (1887) 189–91

Millington, Lucy Ann Bishop
Millington, (Mrs.) Lucy A.
[Botanical notes. Arceuthobium. Variations. Epilobium.] *N. Y., Bot. Club Bull.,* **2** (1871) 42; 47–8; **3** (1872) 55–6; {43–4} ; **5** (1874) 9; **10** (1883) 24
{Sphagnum. *Old and New,* **6** (1872) 237–43}
{Ferns. *Old and New,* **8** (1873) 694–8}
{Winter in the Adirondaks. *Old and New,* **11** (1875) 201–6}
{Thoreau and Wilson Flagg. *Old and New,* **11** (1875) 460–4}

Mitchell, Ann Maria
Mitchell, Ann Maria
The white blackberry. *Rhodora,* **1** (1899) 205–6

Mulford, Anna Isabella
Mulford, A. Isabel
Notes upon the northwestern and Rocky Mountain flora. *Bot. Gaz.,* **19** (1894) 117–20
A study of the agaves of the United States. *St. Louis, Bot. Gard. Rep.* (1896) 47–100

Murtfeldt, Mary Esther
Murtfeldt, (Miss) Mary E.
Floral eccentricities. *Bot. Gaz.,* **14** (1889) 18

Myers, Alice Ward Hess
Hess, Alice Ward; Vandivert, Harriet
Basidiomycetae of central Iowa. [1899] *Iowa Ac. Sci. Proc.,* **7** (1900) 183–7

Newell, Jane Hancox
Newell, Jane H.
The flowers of the horsechestnut. *Bot. Gaz.,* **18** (1893) 107–9

Nichols, Susan Percival
Rowlee, Willard W.; Nichols, Susie P.
The taxonomic value of the staminate flowers of some of the oaks. *Bot. Gaz.,* **29** (1900) 353–6

Northup, Alice Belle Rich
Northup, John I.; Northrop, Alice B.
Plant notes from Tadousac and Temiscouata county, Canada. *Torrey Bot. Club Bull.,* **17** (1890) 27–32; iv

Norton, Florence May Lyon
Lyon, Florence May
Dehiscence of the sporangium of Adiantum pedatum. *Torrey Bot. Club Bull.,* **14** (1887) 180–3
A contribution to the life history of Euphorbia corollata. *Bot. Gaz.,* **25** (1898) 418–26

Nott, Edith Sumner Byxbee
Byxbee, Edith Sumner
The development of the karyokinetic spindle in the pollen-mother-cells of Lavatera. [1900] *California Ac. Proc.* (*Bot.*), **2** (1904) 63–82

Osband, Lucy A.
Osband, Lucy A.
A freak of inflorescence. *Science,* **23** (1894) 92
Abnormal plant growths. *Amer. Natlist.,* **28** (1894) 706

Owen, Maria Louise Tallant
Owen, Maria L.
{Nantucket plants. *N. Y., Bot. Club Bull.,* **6** (1879) 330}
Notes on Corema conradii. *Torrey Bot. Club Bull.,* **11** (1884) 117
Trillium cernuum, *L. Bot. Gaz.,* **19** (1894) 337–8
Tillaea simplex. *Bot. Gaz.,* **20** (1895) 80–1
The Connecticut Valley Botanical Society. *Rhodora,* **1** (1899) 95–6

Paine, Harriet Eliza
Paine, (Miss) Harriet E.
Groveland plants not reported by Mr. Robinson in country flora. [1883] *Essex Inst. Bull.,* **14** (1884) 134
Plants shown at the meeting in Groveland, Mass., Aug. 1883. [1883] *Essex Inst. Bull.,* **15** (1884) 133

Parsons, Mary Elizabeth
Parsons, Mary Elizabeth
{The ferns of Talmalpais. *Zoe,* **2** (1891) 129–33}

Patterson, Flora Wambaugh
Patterson, (Mrs.) Flora W.
Species of Taphrina parasitic on Populus. *Amer. Ass. Proc.* (1894) 293–94

A study of North American parasitic Exoascaceae. [1895] *Iowa Univ. Lab. Nat. Hist. Bull.,* **3** (1894–6) No. 3, 89–135
An Exoascus upon Alnus leaves. *Amer. Ass. Proc.* (1895) 192
New species of fungi. *Torrey Bot. Club Bull.,* **27** (1900) 282–6
{Some woody fungi. *Asa Gray Bull.* **8** (1900) 13–19}

Peacock, Bertha Leon De Graffe
De Graffe, Bertha L.
Opuntia vulgaris, *Mill. Amer. J. Pharm.,* **68** (1896) 169–77

Pease, Mrs. F. S.
Pease, (Mrs. F. S.)
The honey-plant. *Amer. Ass. Proc.* (1887) 277

Peirce, Mary Frances
Peirce, Mary F.
Note on Sarracenia variolaris. *Torrey Bot. Club Bull.,* **14** (1887) 229

Pennington, Mary Engle
Pennington, Mary Engle
A chemico-physiological study of Spirogyra nitida. *Pennsylvania Univ. Publ. (Contrib. Bot. Lab.),* **1** (1897) 203–59

Perkins, Janet Russel
Perkins, Janet R.
Beiträge zur Kenntnis der Monimiaceae. 1. Ueber die Gliederung der Gatterungen der Mollinedieae. *Engler, Bot. Jbüch.,* **25** (1898) 547–77
Monographie der Gattung Mollinedia. *Engler, Bot. Jbüch.,* **27** (1900) 636–83

Perrett, Fanny J.
Perrett, (Miss) Fanny J.; [Bessey, C. E.]
[The histology of the asparagus.] [1881] *Bot. Gaz.,* **5** and **6** (1880–1881) 294–5

Plumer, Mary N.
Plumer, Mary N.
Dissemination of seeds. [1881] *Essex Inst. Bull.,* **13** (1882) 121–47

Price, Sarah (Sadie) Frances
Price, Sadie F.
A rare fern. *Garden & Forest,* **6** (1893) 99–100
Cave plants. *Garden & Forest,* **6** (1893) 403

Putnam, Bessie Lucina
Putnam, Bessie L.
A double may-weed. *Garden & Forest,* **7** (1894) 439
{Poison ivy. *Garden & Forest,* **8** (1895) 249}
A day-blooming Cereus grandiflorus. *Bot. Gaz.,* **20** (1895) 462–3
{Hamamelis virginiana. *Bot. Gaz.,* **21** (1896) 170}
{Determination of sex in Arisaema triphyllum. *Asa Gray Bull.,* **6** (1898) 50–2}
{A white form of Carduus arvensis. *Asa Gray Bull.,* **7** (1899) 37}

Reed, Minnie
Reed, Minnie
Cross and self-fertilization. *Bot. Gaz.,* **17** (1892) 330
Some notes on condensed vegetation in western Kansas. [1892] *Kan. Ac. Sci. Trans.,* **13** (1893) 91–6
A peculiar malformation of an ovary and placenta on Begonia rubra-grandiflora. *Bot. Gaz.,* **19** (1894) 298
Cross fertilization of petunias. *Bot. Gaz.,* **19** (1894) 336–7
Long-continued blooming of "Malvastrum coccineum." [1894] *Kan. Ac. Sci. Trans.,* **14** (1896) 132
Ferns of Wyandotte County. [1894] *Kan. Ac. Sci. Trans.,* **14** (1896) 150–1
Kansas mosses. [1894] *Kan. Ac. Sci. Trans.,* **14** (1896) 152–99

Reynolds, Mary Collins
Reynolds, (Miss) Mary C.
Florida ferns. [1877] *N. Y., Bot. Club Bull.,* **6** (1875–1879) 175–6; **7** (1880) 89–90
Some Florida ferns. [1879–1881] *Bot. Gaz.,* **3** and **4** (1878–1879) 139; 177
Notes from St. Augustine, Florida. [1879–1881] *Bot. Gaz.,* **3** and **4** (1878–1879) 227
Some notes of rare ferns [1879–1881] *Bot. Gaz.* **5** (1880) 42
Baptisia calycosa, W. M. Canby. *Bot. Gaz.,* **5** (1880) 89–90
New localities for some Florida plants. *Bot. Gaz.,* **6** (1881) 158–9
Queer places for ferns. *Bot. Gaz.,* **6** (1881) 161–2

Robbins, Mary Caroline Pike
Robbins, (Mrs.) Mary C.
A remarkable group of pines. *Garden & Forest,* **8** (1895) 332–3
The Dalles of the St. Croix, Wisconsin and Minnesota. *Garden & Forest,* **10** (1897) 330–1
The St. Croix River. *Garden & Forest,* **10** (1897) 421–2; 431–2

Saunders, Mary T.
Saunders, (Miss) Mary T.
The flora of Colonial days. [1895] *Essex Inst. Bull.,* **27** (1897) 74–88

Schively, Adeline Frances
Schively, Adeline Frances
Contributions to the life history of Amphicarpaea monoica. [1897] *Pennsylvania Univ. Publ. (Contrib. Bot. Lab.),* **1** (1897) 270–363
Recent experiments and observations on fruit production in Amphicarpaea monoica. [1897] *Amer. Natlist.,* **32** (1898) 109
Recent observations on Amphicarpaea monoica. [1898] *Pennsylvania Univ. Publ. (Contrib. Bot. Lab.),* **2** (1904) 20–30

Schuck, Mary C. Rolfs
Rolfs, Mary C.
Notes on the pollination of some Liliaceae and a few other plants. [1893] *Iowa Ac. Sci. Proc.,* **1** (Pt. 4) (1894) 98–100

Searing, Anna H.
Searing, Anna H.
{The life history of some fungi. *Rochester (N. Y.) Ac. Sci. Proc.,* **1** (1889–91) 56 (title only)}
The flora of Long Pond. [1894] *Rochester (N. Y.) Ac. Sci. Proc.,* **2** (1895) 297–300

Seavey, Fanny Copley
Seavey, Fanny Copley
{The Rose Garden on the wooded island, Jackson Park, Chicago. *Garden & Forest,* **8** (1895) 328}
Proposed enlargement of the Missouri Botanical Garden. *Garden & Forest,* **10** (1897) 213–14

Sirrine, Emma F.
Sirrine, Emma
Structure of the seed coats of Polygonaceae. [1894] *Iowa Ac. Sci. Proc.,* **2** (1895) 128–35
A study of the leaf anatomy of some species of the genus Bromus. [1896] *Iowa Ac. Sci. Proc.,* **4** (1897) 119–25; 241
Sirrine, Emma; Pammel, Emma
Some anatomical studies of the leaves of Sporobulus and Panicum. *Iowa Ac. Sci. Proc.,* **3** (1896) 148–59

Slosson, Annie Trumbull
Slosson, (Mrs.) Annie Trumbull
Subularia aquatica. *Torrey Bot. Club Bull.,* **11** (1884) 118–19
Prolification in phleum. *Torrey Bot. Club Bull.,* **11** (1884) 120

Slosson, (Mrs.) A. L.
Slosson, (Mrs.) A. L.
Personal observations upon the flora of Kansas. *Kan. Ac. Sci. Trans.*, **11** (1889) 19–22
A partial list of plants found in Cherokee County, Texas. *Kan. Ac. Sci. Trans.*, **12** (1890) 62–3

Smith, Arma Anna
Smith, Arma Anna
The development of the cystocarp of Griffithsia bornetiana. *Bot. Gaz.*, **22** (1896) 35–47
Abortive flower buds of Trillium. *Bot. Gaz.*, **22** (1896) 402–3

Smith, Frances Grace
Smith, (Miss) F. Grace
A peculiar case of contact irritability. *Torrey Bot. Club Bull.*, **27** (1900) 190–4
Distribution of red color in vegetative parts in the New England flora. [1899] *Science*, **11** (1900) 301

Smyth, Lumina Cotton Riddle
Riddle, Lumina Cotton
The embryology of Alyssum. *Bot. Gaz.*, **26** (1898) 314–24
{Further studies in plant embryology [*Syaphylea trifolia* L.]. *Ohio State Ac. Sci. Ann. Rep.*, 7 (1899) 45}
{Some abnormal plant specimens. *Ohio State Ac. Sci. Ann. Rep.*, 7 (1899) 45–6}

Snow, Julia Warner
Snow, Julia W.
Pseudo-Pleurococcus, *nov. gen. Ann. Bot.*, **13** (1899) 189–95
Ulvella americana. *Bot. Gaz.*, **27** (1899) 309–14

Spalding, Effie Almira Southworth
Southworth, (Miss) Effie A.
Structure, development, and distribution of stomata in Equisetum arvense. *Amer. Natlist.*, **18** (1884) 1041–2
Development of stomata of the oat. *Amer. Natlist.*, **19** (1885) 710–11
Leaf-blight and cracking of the pear. Entomosporium maculatum, *Lév. [U. S.] Comm. Agr. Rep.* (1888) 357–64
Leaf-spot of the rose. Cercospora rosaecola. *Pass. [U. S.] Comm. Agr. Rep.* (1888) 364–5
Septosporium on grape leaves. *[U. S.] Comm. Agr. Rep.* (1888) 381–3
A disease of the sycamore. Gloeosporium nervisequum, *Sacc. [U. S.] Comm. Agr. Rep.* (1888) 387–9
Gloeosporium nervisequum (*Fckl.*), *Sacc. J. Mycol.*, **5** (1889) 51–2
A new hollyhock disease. [1890] *J. Mycol.*, **6** (1891) 45–50
Anthracnose of cotton. *J. Mycol.*, **6** (1891) 100–5
Additional observations on anthracnose of the hollyhock. *J. Mycol.*, **6** (1891) 115–16
Ripe rot of grapes and apples. *J. Mycol.*, **6** (1891) 164–73
Notes on some curious fungi. *Torrey Bot. Club Bull.*, **18** (1891) 303–4
Galloway, Beverly T.; Southworth, (Miss) Effie A.
Treatment of apple scab. *J. Mycol.*, **5** (1889) 210–14
Preliminary notes on a new and destructive oat disease. [1890] *J. Mycol.*, **6** (1891) 72–3

Stabler, Louise Merritt
Stabler, Louise Merritt
An economical maple. *Torrey Bot. Club Bull.*, **18** (1891) 304–5

Stanford, Mary Emma Olson
Olson, Mary E.
Acrospermum urceolatum, a new discomycetous parasite of Selaginella rupestris. *Bot. Gaz.*, **23** (1897) 367–71
Observations on Gigartina. [1899] *Minn. Bot. Stud.*, **2** (1898–1902) 154–68

Taplin, Emily Louise
Taplin, Emily Louise
Two rare orchids. *Garden & Forest*, **1** (1888) 281–2

Thomas, Hannah
Pammel, L. H.; Burnip, J. R.; Thomas, Hannah
Some studies on the seeds and fruits of Berberidaceae. [1897] *Iowa Ac. Sci. Proc.*, **5** (1898) 209–23

Thompson, Caroline Burling
Thompson, (Miss) Caroline B.
The structure and development of internal phloem in Gelsemium sempervirens, *Ait.* [1897] *Amer. Natlist.*, **32** (1898) 110; *Amer. J. Pharm.*, **71** (1899) 422–33; *Pennsylvania Univ. Publ.* (*Contrib. Bot. Lab.*), **2** (1904) 41–53

Tilden, Josephine Elizabeth
Tilden, Josephine E.
Note on the development of a filamentous form of Protoccus in entomostracan appendages. *Bot. Gaz.*, **19** (1894) 334–5
List of fresh-water algae collected in Minnesota during 1893-[97]. [1894–1898] *Minn. Bot. Stud.*, **1** (1894–1898) 25–31; 228–37; 597–600
On the morphology of hepatic elaters, with special reference to branching elaters of Concephalus conicus. [1894] *Minn. Bot. Stud.*, **1** (1894–1898) 43–53
A contribution to the bibliography of American algae. [1895] *Minn. Bot. Stud.*, **1** (1894–1898) 295–421
A contribution to the life-history of Pilinia diluta, *Wood*, and Stigeoclonium flagelliferum, *Kg.* [1896] *Minn. Bot. Stud.*, **1** (1894–1898) 601–35
A new Oscillatoria from California. *Torrey Bot. Club Bull.*, **23** (1896) 58–69
Some new species of Minnesota algae which live in a calacareous or silliceous matrix. *Bot. Gaz.*, **23** (1897) 95–104
On some algal stalactites of the Yellowstone National Park. *Bot. Gaz.*, **24** (1897) 194–9
Observations on some West American thermal algae. *Bot. Gaz.*, **25** (1898) 89–105
List of fresh-water algae collected in Minnesota during 1893-[97]. [1894–98] *Minn. Bot. Stud.*, **2** (1898–1902) 25–9
{The study of algae. *Plant World*, **1** (1898) 148–50}
{The study of algae in high schools. *Plant World*, **2** (1899): 59–63}

Trask, Luella Blanche Engles
Trask, (Mrs.) Blanche
San Clemente island. *Erythea*, **5** (1897) 30
Field notes from Santa Catalina island. *Erythea*, **7** (1899) 135–46

Treat, Mary Allen Davis
Treat, (Mrs.) Mary
{Pine barren plants. *Amer. Ent. and Bot.*, **2** (1870) 318–19}
Drosera (Sundew) as a fly-catcher. *Amer. J. Sci.*, **2** (1871) 463–4
Observations on the Sundew: Drosera filiformis. *Amer. Natlist.*, **7** (1873) 705–8
Plants that eat animals. *Amer. Natlist.*, **9** (1875) 658–62

{Is the valve of Utricularia sensitive?, *Harper's New Mon. Mag.*, **52** (1876) 382–7}
Argiope riparia, *var.* multiconcha. *Amer. Natlist.*, **21** (1887) 1122
April in the pine barrens. *Garden & Forest*, **1** (1888) 124
The pine barrens in May. *Garden & Forest*, **1** (1888) 182
Among the pines in June. *Garden & Forest*, **1** (1888) 243
The pines in July. *Garden & Forest*, **1** (1888) 290–1
August in the pines. *Garden & Forest*, **1** (1888) 362–3
The pines in October. *Garden & Forest*, **1** (1888) 435
The pines in mid-November. *Garden & Forest*, **1** (1888) 494–5
Christmas in the pines. *Garden & Forest*, **1** (1888) 518–19
Botanical names. *Garden & Forest*, **3** (1890) 206–7
Evergreens in the New Jersey pine region. *Garden & Forest*, **3** (1890) 546–7
Weeds in southern New Jersey. *Garden & Forest*, **5** (1892) 292; **10** (1897) 313–14
Water-plants in southern New Jersey. *Garden & Forest*, **5** (1892) 363
Winter-blooming plants in the pines. *Garden & Forest*, **7** (1894) 102
Wayside plants in the pines. *Garden & Forest*, **7** (1894) 302–3
Troublesome grasses in southern New Jersey. *Garden & Forest*, **8** (1895) 103–4
The pines in a dry summer. *Garden & Forest*, **8** (1895) 362–3
Cruelty of Asclepias. *Garden & Forest*, **10** (1897) 341

Vail, Anna Murray
Vail, (Miss) Anna Murray
The Alleghenies of Virginia in June. *Garden & Forest*, **3** (1890) 367–8; 391–2
Notes on the spring flora of southwestern Virginia. [With annotations by N. L. Britton, and a list of mosses by E. G. Britton.] *Torrey Bot. Club Mem.*, **2** (1890–1891) 27–53
An undescribed Desmodium from Texas and Mexico. *Torrey Bot. Club Bull.*, **18** (1891) 120
A preliminary list of the species of the genus Meibomia, *Heist.*, occurring in the United States and British America. *Torrey Bot. Club Bull.*, **19** (1892) 107–18
Notes on the flora of Smythe county, Virginia. *Garden & Forest*, **5** (1892) 364; 375–6; 388–9; 424; 437–8
Albinos among orchids. *Garden & Forest*, **5** (1892) 395
A study of the genus Psoralea in America. *Torrey Bot. Club Bull.*, **21** (1894) 91–119
A revision of the North American species of the genus Cracca. *Torrey Bot. Club Bull.*, **22** (1895) 25–36
A preliminary list of the North American species of Malpighiaceae and Zygophyllaceae. *Torrey Bot. Club Bull.*, **22** (1895) 228–31
Two undescribed species of Rhynchosia. *Torrey Bot. Club Bull.*, **22** (1895) 458–9
A study of the genus Galactica in North America. *Torrey Bot. Club Bull.*, **22** (1895) 500–11
The June flora of a Long Island swamp. *Garden & Forest*, **8** (1895) 282–3
The collection of funeral wreaths and offerings in the Museum of Egyptian Antiquities at Giseh. *Garden & Forest*, **8** (1895) 312–13
The sacred lotus in Egypt. *Garden & Forest*, **8** (1895) 378
Studies in the Leguminosae. *Torrey Bot. Club Bull.*, **23** (1896) 139–41; **24** (1897) 14–18; **26** (1899) 106–17
An undescribed species of Kallstroemia from New Mexico. *Torrey Bot. Club Bull.*, **24** (1897) 206–7
Studies in the Asclepiadaceae. *Torrey Bot. Club Bull.*, **24** (1897) 305–10; **25** (1898) 30–9; 171–82; **26** (1899) 423–31
Notes on Covillea and Fagonia. *Torrey Bot. Club Bull.*, **26** (1899) 301–2
Morong, Thomas; Britton, Nathaniel Lord; Vail, (Miss) Anna Murray
An enumeration of the plants collected by Dr. Thomas Morong in Paraguay, 1888–1890. [1892] *N. Y. Ac. Ann.*, **7** (1892–1894) 45–280
Small, John K.; Vail, (Miss) Anna Murray
Report of the botanical exploration of southwestern Virginia during the season of 1892. [1893–94] *Torrey Bot. Club Mem.*, **4** (1893–1896) 93–201

Vandivert, Harriet A.
Hess, Alice Ward; Vandivert, Harriet
Basidiomycetae of central Iowa. [1899] *Iowa Ac. Sci. Proc.*, **7** (1900) 183–7

Webster, Helen Maria Noyes
Noyes, Helen M.
The ferns of Alstead, New Hampshire. *Rhodora*, **2** (1900) 181–5

Williams, Clara Louise
Williams, Clara L.
The origin of the karyokinetic spindle in Passiflora calerulea, *Linn.* [1899] *California Ac. Proc. (Bot.)*, **1** (1904) 189–206

Wilson, Kate Eastman
Wilson, Kate Eastman
Double flowers of the Epigaea repens. *Bot. Gaz.*, **15** (1890) 19–20

Wilson, Lucy Langdon Williams
Wilson, Lucy L.W.
Observations on the American squawroot (Conopholis americana, *Wallr.*). [1897] *Amer. Natlist.*, **32** (1898) 108
Observations on Conopholis americana. [1898] *Pennsylvania Univ. Publ. (Contrib. Bot. Lab.)*, **2** (1904) 1–19

Windsor, Margaret Fursman Boynton
Boynton, Margaret Fursman
Observations upon the dissemination of seeds. *Bot. Gaz.*, **20** (1895) 502–3

Wright, Julia McNair
Wright, Julia McNair
Notes on water lilies, etc. *Science*, 23 (1894) 39–40

United States—Canada

Eastwood, Alice
Eastwood, Alice
{Common shrubs of southwest Colorado. *Zoe*, **2** (1891) 102–4}
{The fertilization of geraniums. *Zoe*, **2** (1891) 112}
{Mariposa lilies of Colorado. *Zoe*, **2** (1892) 201–3}
{Additions to the flora of Colorado. I. *Zoe*, **2** (1892) 226–33}
{The loco weeds. *Zoe*, **3** (1892) 53–8}
{Notes on some species of the genus Oenothera. *Zoe*, **3** (1892) 248–52}
{A trip through southeastern Utah. *Zoe*, **3** (1893) 354–61}
{Notes on some Colorado plants. *Zoe*, **4** (1893) 2–12}
{Additions to the flora of Colorado. II. *Zoe*, **4** (1893) 16–20}
{List of plants collected in southeastern Utah, with notes and descriptions of new species. *Zoe*, **4** (1893) 113–27}
{Field notes at San Emidio. *Zoe*, **4** (1893) 144–7}
{Botanical notes. *Zoe*, **4** (1893) 286–7}
{Gilia superba. Phacelia nudicaulis. *Zoe*, **4** (1893) 296}
Argemone hispida. *J. Bot.*, **33** (1895) 376–7
Observations on the habits of Nemophila. *Erythea*, **3** (1895) 151–3

Two species of Aquilegia from the Upper Sonoran zone of Colorado and Utah. *California Ac. Proc.,* **4** (1895) 559–62

New localities for West American plants. *Erythea,* **4** (1896) 32; 200

Pelargonium anceps, *Ait. Erythea,* **4** (1896) 34

New localities for two introduced plants. *Erythea,* **4** (1896) 34–5

Trillium sessile. *Erythea,* **4** (1896) 71

On Dr. D. Prain's account of the genus Argemone. *Erythea,* **4** (1896) 93–6

Arbutus menziesii in San Francisco County. *Erythea,* **4** (1896) 99

New stations for two introduced plants. *Erythea,* **4** (1896) 99

The Alpine flora of Mt. Shasta. *Erythea,* **4** (1896) 136–42

Geranium parviflorum, *Willd. Erythea,* **4** (1896) 145

Scolymus hispanicus, *L. Erythea,* **4** (1896) 145

New stations for two plants. *Erythea,* **4** (1896) 151

On heteromorphic organs of Sequoia sempervirens, *Endl. California Ac. Proc.,* **5** (1896) 170–6

Report on a collection of plants from San Juan county, in southeastern Utah. [1896] *California Ac. Proc.,* **6** (1897) 270–329

Descriptions of some new species of Californian plants. [1896] *California Ac. Proc.,* **6** (1897) 422–30

The Coniferae of the Santa Lucia mountains. *Erythea,* **5** (1897) 71–4

Ferns of the Yosemite and the neighboring Sierras. *Erythea,* **6** (1898) 14–15

The plant inhabitants of Nob Hill, San Francisco. *Erythea,* **6** (1898) 61–7

Notes on the flora of Marin county, California. *Erythea,* **6** (1898) 72–5; 117–18

Is Xerophyllum tenax a septennial? [With note by Willis Linn Jepson.] *Erythea,* **6** (1898) 75–6

Pyrola minor, *L.,* in California. *Erythea,* **6** (1898) 93

Plants in flower in November and December, 1897. *Erythea,* **6** (1898) 114–15

New localitites for rare Californian plants. *Erythea,* **7** (1899) 76–7

Introduced plants in Placer county, California. *Erythea,* **7** (1899) 150

Sedge used in Indian basket-making. *Erythea,* **7** (1899) 150

Migratory plants in Alameda county. *Erythea,* **7** (1899) [189]; 175–6

Studies in the herbarium and the field. [1897–98] *California Ac. Proc.* (Bot.), **1** (1904) 71–146

{Notes on Cupressus macnabiana. *Zoe,* **5** (1900) 11–13}

{Aquilegia exima. *Zoe,* **5** (1900) 28–30}

{Short articles. *Zoe,* **5** (1900) 35–7}

{Some plants of Mendocino county new to the flora of California. *Zoe,* **5** (1900) 58–60}

{Rediscovery of Thermopsis macrophylla. *Zoe,* **5** (1900) 76–8}

{New species of California plants. *Zoe,* **5** (1900) 80–90}

Great Britain

Auld, Helen P.

Harvey-Gibson, R. J.; Auld, Helen P.

L.M.B.C. Memoirs. No. IV. Codium. *Liverpool Biol. Soc. Proc. & Trans.,* **14** (1900) 326–43

Barnard, Alicia Mildred

Barnard, (Miss) A. M.

Thrincia tuberosa, *DC.* [1885] *Norf. Norw. Nat. Soc. Trans.,* **4** (1889) 143

Bateson, Anna

Bateson, (Miss) Anna

The effect of cross-fertilization on inconspicuous flowers. [1888] *Ann. Bot.,* **1** (1887–1888) 255–61

On the change of shape exhibited by turgescent pith in water. [1889] *Ann. Bot.,* **4** (1889–1891) 117–25

Bateson, (Miss) Anna; Darwin, Francis

The effect of stimulation on turgescent vegetable tissues. [1887] *Linn. Soc. J. (Bot.),* **24** (1888) 1–27

On a method of studying geotropism. [1888] *Ann. Bot.,* **2** (1888–1889) 65–8

On the change of shape in turgescent pith. *Cambridge Phil. Soc. Proc.,* **6** (1889) 358–9

Bateson, William; Bateson, (Miss) Anna

On variations in the floral symmetry of certain plants having irregular corollas. *Linn. Soc. J. (Bot.),* **28** (1891) 386–424

Becker, Lydia Ernestine

Becker, Lydia E.

On alteration in the structure of Lychnis dioica, observed in connection with the development of a parasitic fungus. *Brit. Ass. Rep.,* **39**(Sect.) (1869) 106; *J. Bot.,* **7** (1869) 291–2

Benson, Margaret Jane

Benson, (Miss) Margaret

Contributions to the embryology of the Amentiferae. [1893] *Linn. Soc. Trans. (Bot.),* **3** (1888–1894) 409–24

On the phylogenic position of the chalazogamic Amentiferae. *Brit. Ass. Rep.* (1894) 687

Birley, Caroline

Copland, (Miss) L.; Birley, (Miss) Caroline

Notes on the flora of the Faeroes. *J. Bot.,* **29** (1891) 179–83

Calvert, Sara Agnes

Calvert, Agnes

The laticiferous tissue in the stem of Hevea brasiliensis. [1887] *Ann. Bot.,* **1** (1887–1888) 75–7

Calvert, Agnes; Boodle, Leonard A.

On laticiferous tissue in the pith of Manihot glaziovii, and on the presence of nuclei in this tissue. [1887] *Ann. Bot.,* **1** (1887–1888) 55–62

Copland, L.

Copland, (Miss) L.; Birley, (Miss) Caroline

Notes on the flora of the Faeroes. *J. Bot.,* **29** (1891) 179–83

Dale, Elizabeth

Dale, (Miss) Elizabeth

Intumescences of Hibiscus vitifolius (*L.*) *Brit. Ass. Rep.* (1899) 930–1; (1900) 940

On certain outgrowths (intumescences) on the green parts of Hibiscus vitifolius, *Linn.* [1899] *Cambridge Phil. Soc. Proc.,* **10** (1900) 192–209

Ward, H. Marshall; Dale, (Miss) Elizabeth

On Craterostigma pumilum, *Hochst,* a rare plant from Somali-Land. [1898] *Linn. Soc. Trans.* (*Bot.*), **5** (1895–1901) 343–55

Dawson, Maria

Dawson, (Miss) Maria

On the structure of an ancient paper. *Ann. Bot.,* **12** (1898) 111–15

"Nitragin" and the nodules of leguminous plants. [1898] *Phil. Trans. (B),* **192** (1900) 1–28; {*Roy. Soc. Proc.,* **66** (1900) 63–5}

Further observations on the nature and functions of the nodules of leguminous plants. *Phil. Trans. (B),* **193** (1900) 51–67

On the biology of Poronia punctata (*L.*). *Ann. Bot.,* **14** (1900) 245–62

On the anatomical characters of the substance "Indian Soap." [1900] *Canad. Inst. Trans.,* **7** (1904) 1–6

Dodd, (Miss)
Dodd, (Miss)
The wild flowers of Mona. [1887] *Yn Lioar Manninagh,* **1,** Pt. 1 (1894) 97–8

Edmonds, (Miss)
Edmonds, (Miss)
Aster salignus, *Willd.* ? found at Derwentwater. *J. Bot.,* 7 (1869) 139

Ewart, Mary Frances
Ewart, (Miss) Mary Frances
On the staminal hairs of Thesium. *Ann. Bot.,* **6** (1892) 271–90
Notes on abnormal Cypripedium flowers. [1892] *Linn. Soc. J. (Bot.),* **30** (1895) 45–50
On the leaf glands of Ipomoea paniculata. *Ann. Bot.,* **9** (1895) 275–88

Fingland, (Mrs.)
Elliot, George Francis Scott; Fingland, (Mrs.)
Note on raspberry roots. [1898] *Glasgow Nat. Hist. Soc. Proc. & Trans.,* **5** (1900) 205–7

Fry, Agnes
Fry, Agnes
The unfolding of wood-sorrel leaves. *J. Bot.,* **29** (1891) 301–3
Position of boughs in summer and winter. [1896] *Nature,* **55** (1896–7) 198

Gaye, Selina
Gaye, Selina
Some facts about arums. *Midland Natlist.,* **8** (1885) 301–5

Gepp, Ethel Sarel Barton
Barton, Ethel Sarel
A systematic and structural account of the genus Turbinaria, *Lamx.* [1891] *Linn. Soc. Trans. (Bot.),* **3** (1888–1894) 215–26
On the occurrence of galls in Rhodymenia palmata, *Grev. J. Bot.,* **29** (1891) 65–8
A provisional list of the marine algae of the Cape of Good Hope. *J. Bot.,* **31** (1893) 53–6; 81–4; 110–14; 138–44; 171–7; 202–10
Notes on Bryopsis. *J. Bot.,* **33** (1895) 161–2
Cape algae. *J. Bot.,* **34** (1896) 193–8; 458–61
Welwitsch's African marine algae. *J. Bot.,* **35** (1897) 369–74
On the structure and development of Soranthera, *Post. et Rupr.* [1898] *Linn. Soc. J. (Bot.),* **33** (1897–1898) 479–86
On the fruit of Chnoospora fastigiata, *J. Ag.* [1898] *Linn. Soc. J. (Bot.),* **33** (1897–1898) 507–8
On Notheia anomala, *Harv. et Bail.* [1899] *Linn. Soc. J. (Bot.),* **34** (1898–1900) 417–25
On the forms, with a new species, of Halimeda from Funafuti. [1900] *Linn. Soc. J. (Bot.),* **34** (1898–1900) 479–82
Murray, George Robert Milne; Barton, Ethel Sarel
On the structure and systematic position of Chantransia; with a description of a new species. [1890] *Linn. Soc. J. (Bot.),* **28** (1891) 209–16

Gifford, Isabella
Gifford, Isabella
Observations on the marine flora of Somerset. *Somersetsh. Archaeol. Soc. Proc.,* **4** (1853) 116–23
Notices of the rare and most remarkable plants in the neighbourhoods of Dunster, Blue Anchor, Minehead, etc. *Somersetsh. Archaeol. Soc. Proc.,* **4** (1855) 131–7
{The true tetraspores of Seirospora griffithsiana. *J. Bot.* (1871) 113}

Gowan, Jane
Seward, Albert C.; Gowan, (Miss) J.
The maidenhair tree (Ginko biloba, *L.*). *Ann. Bot.,* **14** (1900) 111–54; 320

Gregg, Mary Kirby
Kirby, (Miss) M.
{Revivifying property of the Leicestershire Udora. *Phytologist,* **3** (1848) 30}
Note on the "Flora of Leicestershire" with addenda thereto. *Phytologist,* **3** (1848) 179–80

Griffiths, Frances Elizabeth
Griffiths, Mrs. A. B.
On degenerated specimens of Tulipa sylvestris. [1887] *Edinb. Roy. Soc. Proc.,* **14** (1888) 349–51
Griffiths, Arthur Bower; Griffiths, Mrs. A. B.
Investigations on the influence of certain rays of the solar spectrum on root-absorption and on the growth of plants. [1887] *Edinb. Roy. Soc. Proc.,* **14** (1888) 125–9

Hall, Kate Marion
Jennings, A. Vaughan; Hall, Kate M.
Notes on the structure of Tmesipteris. [1891] *Irish Ac. Proc.,* **2** (1891–1893) 1–18

Hodgson, Elizabeth
Hodgson, (Miss) E.
North or Lake Lancashire; a sketch of its botany, geology and physical geography. *J. Bot.,* **3** (1874) 268–77; 296–305

Hottinger, (Miss)
Hottinger, (Miss)
On the teaching of botany. [1890] *Leicester Soc. Trans.,* **2** (1889–1892) 183–7
Notes on the alpine flora. [1892] *Leicester Soc. Trans.,* **3** (1895) 107–12
Notes on the flora of Australia. [1895] *Leicester Soc. Trans.,* **4** (1898) 37–42

Huie, Lily H.
Huie, Lily H.
On some protein crystalloids and their probable relation to the nutrition of the pollen tube. [1895] *Cellule,* **11** (1895) 81–92.
Changes in the tentacle of Drosera rotundifolia, produced by feeding with egg albumen. *Brit. Ass. Rep.* (1896) 1014–15
Changes in the cell-organs of Drosera rotundifolia, produced by feeding with egg albumen. *Quart. J. Micr. Sci.,* **39** (1897) 387–425
Preliminary note on changes in the gland cells of Drosera produced by various food materials. *Brit. Ass. Rep.* (1898) 1066–7; *J. Physiol.,* **23** (1898–1899) vi-vii; [38]
Further study of cytological changes produced in Drosera. *Quart. J. Micr. Sci.,* **42** (1899) 203–22

Hunter, Anne
Hunter, Anne
Description of an Agaric (Agaricus caperatus), new to the British flora. *Berwick. Nat. Club Hist.,* **2** (1842–1849) 174–5; *Ann. Nat. Hist.,* **28** (1846) 474–5
Notice of Morchella semi-libera. [1867] *Berwick. Nat. Club Hist.,* **5** (1868) 359

Ibbetson, Agnes Thomson
Ibbetson, Agnes
On the impregnation of the seed, and first shooting of the nerve of life, in the embryo of plants. *Nicholson, J.,* **23** (1809) 161–9
On the perspiration of plants. *Nicholson, J.,* **23** (1809) 169–73

On the formation of the winter leaf-bud, and of leaves. *Nicholson, J.*, **23** (1809) 293–300
On the stem of trees, with an attempt to discover the cause of motion of plants. *Nicholson, J.*, **23** (1809) 334–50
On the supposed perspiration of plants. *Nicholson, J.*, **23** (1809) 351–4
Remaining proof of the cause of motion in plants explained; and what is called the sleep of plants shown to be relaxation only. *Nicholson, J.*, **24** (1809) 114–24
On the effects produced by the grafting and budding of trees. *Nicholson, J.*, **24** (1809) 337–46
On the defects of grafting and budding. *Nicholson, J.*, **24** (1809) 346–57
On the structure and growth of seeds. *Nicholson, J.*, **27** (1810) 1–17
On the structure and classification of seeds. *Nicholson, J.*, **27** (1810) 174–84
On the method of Jussieu. *Nicholson, J.*, **28** (1811) 98–105
The beautiful tint of flowers acquired by the same means that paint the rainbow. *Nicholson, J.*, **28** (1811) 170–80
On the interior of plants. *Nicholson, J.*, **28** (1811) 254–66; **29** (1811) 1–12
Description of firs, illustrated by dissections. *Nicholson, J.*, **29** (1811) 202–12
On the motion of the flower of the barberry. *Nicholson, J.*, **29** (1811) 213–14
Farther observations on the fructification of the firs. *Nicholson, J.*, **29** (1811) 295–8
On the hairs of plants. *Nicholson, J.*, **30** (1812) 1–9
Of the mechanical powers in the leaf stalks of various plants. *Nicholson, J.*, **30** (1812) 179–83
On the mechanism of leaves. *Nicholson, J.*, **31** (1812) 1–4
On the mechanism of flowers. *Nicholson, J.*, **31** (1812) 81–7
On the different sorts of wood, with some remarks on the work of Du Thouars. *Nicholson, J.*, **31** (1812) 161–8
On fresh-water plants. *Nicholson, J.*, **31** (1812) 241–8
On the fructification of the plants of the class Cryptogamia. *Nicholson, J.*, **32** (1812) 1–13
Letter on the structure of the water lily. *Nicholson, J.*, **32** (1812) 137–8
On the dissection of flowers. *Nicholson, J.*, **32** (1812) 169–76
On the interior buds of all plants. *Nicholson, J.*, **33** (1812) 1–10
On the secret and open nectaries of various flowers. *Nicholson, J.*, **33** (1812) 171–9
On the growth or increase of trees. *Nicholson, J.*, **33** (1812) 241–51
On the roots of trees. *Nicholson, J.*, **33** (1812) 334–44
On the formation of the seeds of plants, and other objects. *Nicholson, J.*, **35** (1813) 19–29
On the wood and bark of trees much magnified. *Nicholson, J.*, **35** (1813) 87–94
The seeds of all plants first formed in the roots. *Nicholson, J.*, **36** (1813) 34–45
Letter showing that the spiral wire is the cause of all motions in plants. *Nicholson, J.*, **36** (1813) 266–77
On the use of air-vessels in plants. *J. de Phys.*, **78** (1814) 453–61; *Tilloch, Phil. Mag.*, **43** (1814) 81–8
Explanation of the cuticle of leaves. *Tilloch, Phil. Mag.*, **44** (1814) 161–71
On the nourishment produced to the plant by its leaves. *Tilloch, Phil. Mag.*, **45** (1815) 3–15
On the phaenomena attending the roots of plants in snowy weather. *Tilloch, Phil. Mag.*, **45** (1815) 177–8
A paper proving that the embryos of the seeds are formed in the root alone. *Tilloch, Phil. Mag.*, **45** (1815) 183–8
On the phaenomena of vegetation. *Tilloch, Phil. Mag.*, **45** (1815) 321–9
Experiments introductory to an attempt to exhibit the comparative anatomy of animals and vegetables. *Tilloch, Phil. Mag.*, **46** (1815) 46–8
Comparative anatomy; or a slight attempt to draw up a comparison between animal and vegetable life. *Tilloch, Phil. Mag.*, **46** (1815) 81–100
On the anatomy of vegetables; intended to substitute many important truths in phytology. *Tilloch, Phil. Mag.*, **48** (1816) 96–111
On the physiology of vegetables. *Tilloch, Phil. Mag.*, **48** (1816) 173–89; 401–8; **49** (1817) 125–32; **50** (1817) 341–8
A new view of vegetable life. *Tilloch, Phil. Mag.*, **48** (1816) 278–86
{On the adapting of plants to the soil, and not the soil to plants. *Lett. and Papers on Agr.*, **14** (1816) 136–59}
On the death of plants. *Thomson, Ann. Phil.*, **11** (1818) 252–62
{On the injurious effect of burying weeds. *Thomson, Ann. Phil.*, **12** (1818) 87–91}
On the seeds of plants. *Tilloch, Phil. Mag.*, **51** (1818) 404–11
On the fructification of seeds. *Tilloch, Phil. Mag.*, **52** (1818) 81–8
On the action of lime upon animal and vegetable substances. *Thomson, Ann. Phil.*, **14** (1819) 125–9
On the physiology of botany. *Tilloch, Phil. Mag.*, **56** (1820) 3–9
On the flower-buds of trees passing through the wood, as noticed by Cicero and Pliny. *Tilloch, Phil. Mag.*, **59** (1822) 3–8
On the perspiration alleged to take place in plants. *Tilloch, Phil. Mag.*, **59** (1822) 243–8
On the pollen of flowers. *Tilloch, Phil. Mag.*, **60** (1822) 56–60

James, Cecely Louisa Grey-Egerton, Countess of Selkirk
Selkirk, (Countess of)
The peach leaf-curl fungus. [1900] *Hortic. Soc. J.*, **25** (1900–1901) 163

Jex-Blake, L. Eleanor
Jex-Blake, L. Eleanor
Variations in plants of the herb Paris. *Nature*, **62** (1900) 174

Kent, Elizabeth
Kent, (Miss)
Considerations on botany, as a study for young people, intended as an introduction to a series of papers illustrative of the Linnaean system of plants. *Mag. Nat. Hist.*, **1** (1829) 124–35
An introductory view of the Linnean system of plants. *Mag. Nat. Hist.*, **1** (1829) 228–38; 429–36; **2** (1829) 155–64; **3** (1830) 52–62; 134–42; 350–61

Kingsley, Mary Henrietta
Kingsley, Mary H.
{Gardening in West Africa. *Climate* (April 1900) 77–87}

Layard, Antonia (Nina) Frances
Layard, (Miss) Nina F.
On reversion. *Brit. Ass. Rep.* (1890) 973–4
On the arrangement of the buds in Lemna minor. *Brit. Ass. Rep.* (1892) 747–8
Remarks on the roots of the Lemna and the reversing of the fronds in Lemnatrisulea. *Brit. Ass. Rep.* (1893) 803–4

Lee, Sarah Wallis Bowdich
Bowdich, (Mrs.) E.
On the natural order of plants, Dicotyledoneae anonaceae. *Mag. Nat. Hist.*, **1** (1829) 438–41

Lister, Gulielma
Lister, (Miss) G.
On the origin of the placentas in the tribe Alsineae of the order Caryophylleae. [1883] *Linn. Soc. J. (Bot.),* **20** (1884) 423–9

M'Inroy, (Miss)
M'Inroy, (Miss)
Notice of mosses found near Blair-Athole, Perthshire. [1864] *Edinb., Bot. Soc. Trans.,* **8** (1866) 75–6
Notice of the mosses found in the neighbourhood of The Burn, near Brechin, Forforshire. [1864] *Edinb., Bot. Soc. Trans.,* **8** (1866) 109–10

Merrifield, Mary Philadelphia Watkins
Merrifield, (Mrs.) Mary Philadelphia
List of marine algae found at Brighton and its vicinity, with observations on a few of the most remarkable plants. *Phytologist,* **6** (1862–1863) 513–26
Observations on the fruit of Nitophyllum versicolor. [1874] *Linn. Soc. J., (Bot.)* **14** (1875) 421–3; 632
Recent additions to the British marine flora. *J. Bot.,* **5** (1876) 147–8
Gulf-weed. *Nature,* **18** (1878) 708–11

Nuttall, Gertrude Clarke
Clarke, (Miss) Gertrude
On lichens. [1893] *Leicester Soc. Trans.,* **3** (1895) 267–74
Nuttall, Mrs. C. D.
Symbiosis in plant life. [1894] *Leicester Soc. Trans.,* **4** (1898) 43–9

Ogilvie-Farquharson, Marion Sarah Ridley
Farquharson, (Mrs.)
Notes on the mosses of North Scotland. [1886] *Scott. Natlist.,* **8** (1885–1886) 381
Ferns and mosses of the Alford district. [1890] *Scott. Natlist.,* **10** (1889–1890) 193–8

Pertz, Dorothea Frances Matilda
Pertz, (Miss) Dorothea F. M.
Note on Pleurococcus. *Brit. Ass. Rep.* (1897) 864
On the gravitation stimulus in relation to position. *Ann. Bot.,* **13** (1899) 620
Darwin, Francis; Pertz, (Miss) Dorothea F. M.
On the artificial production of rhythm in plants. *Brit. Ass. Rep.* (1891) 695; *Ann. Bot.,* **6** (1892) 245–64
On rectipetality and on a modification of the klinostat. [1891] *Cambridge Phil. Soc. Proc.,* **7** (1892) 141–2
On the effect of water currents on the assimilation of aquatic plants. [1896] *Cambridge Phil. Soc. Proc.,* **9** (1898) 76–90
On the injection of the intercellular spaces occurring in the leaves of Elodea during recovery from plasmolysis. [1897] *Cambridge Phil. Soc. Proc.,* **9** (1898) 272
Experiments on the periodic movement of plants. *Cambridge Phil. Soc. Proc.,* **10** (1900) 259
Bateson, William; Pertz, (Miss) Dorothea F. M.
Note on the inheritance of variation in the corolla of Veronica buxbaummi. [1897] *Cambridge Phil. Soc. Proc.,* **10** (1900) 78–92

Rathbone, Mary May
Rathbone, May
Colouring of plants. [1899] *Nature,* **59** (1898–1899) 342

Riley, Margaretta (Meta) Hopper
Riley, Meta
{Polypodium dryopteris and calcareum. *Phytologist,* **1** (1841–1844) 94}

Roberts, May
Roberts, May
The mosses of Upper Dovey. *J. Bot.,* **34** (1896) 330–4; **35** (1897) 492–3

Russell, Anna Worsley
Russell, Anna
{Note on a list of Newbury plants. *Phytologist,* **3** (1850) 716}
{Botanical notes [on the fern Pteris aquiliana]. *Phytologist,* **1** (1856) 390}
{Sonchus palustris. *Phytologist,* **2** (1857) 279}
List of some of the rarer fungi found near Kenilworth. *J. Bot.,* **6** (1868) 90–1
Note on Hygrophorus calyptraeformis. *J. Bot.,* 7 (1869) 116

Sargant, Ethel
Sargant, (Miss) Ethel
Some details of the first nuclear divison in the pollen mother cells of Lilium martagon, *L. Micr. Soc. J.* (1895) 283–7
Direct nuclear division in the embryo-sac of Lilium martagon. *Ann. Bot.,* **10** (1896) 107–8
On the heterotype divisions of Lilium martagon. *Brit. Ass. Rep.* (1896) 1021
The formation of the sexual nuclei in Lilium martagon.[I. Oögenesis.] *Ann. Bot.,* **10** (1896) 445–77
Photomicrography as an aid to research. [1898] *Holmesdale Nat. Hist. Club Proc.* (1896–1898) 67–70
The formation of the sexual nuclei in Lilium martagon.[II. Spermatogenesis.] *Ann. Bot.,* **11** (1897) 187–224
A new type of transition from stem to root in the vascular system of seedlings. *Ann. Bot.,* **14** (1900) 633–8
On a fourth type of transition from stem to root-structure occurring in certain monocotyledonous seedlings. *Brit. Ass. Rep.* (1900) 937–8
On the presence of two vermiform nuclei in the fertilized embryo-sac of Lilium martagon. [1899] *Roy. Soc. Proc.,* **65** (1900) 163–5
Recent work on the results of fertilization in angiosperms. *Ann. Bot.,* **14** (1900) 689–712
Scott, Dukinfield H.; Sargant, (Miss) Ethel
On the pitchers of Dischidia rafflesiana (*Wall.*) *Ann. Bot.,* 7 (1893) 243–69
Scott, (Mrs.) Rina; Sargant, (Miss) Ethel
On the development of Arum maculatum from the seed. *Ann. Bot.,* **12** (1898) 399–414

Saunders, Edith Rebecca
Saunders, (Miss) Edith Rebecca
On the structure and function of the septal glands in Kniphofia. [1890] *Ann. Bot.,* **5** (1890–1891) 11–25
On a discontinuous variation occurring in Biscutella laevigata. [1897] *Roy. Soc. Proc.,* **62** (1898) 11–26

Saunders, Helen
Saunders, Helen
A list of plants growing wild in the parish of South Molton, and in the neighbouring parishes of North Molton, Filleigh, Chittlehampton, George Nympton, Satterleigh, King's Nympton, and Bishop's Nympton. *Devon. Ass. Trans.,* **26** (1894) 451–66
Botanical notes. *Devon. Ass. Trans.,* **30** (1898) 198–202

Scott, Henderina (Rina) Victoria Klaassen
Scott, (Mrs.) Rina; Sargant, (Miss) Ethel
On the developement of Arum maculatum from the seed. *Ann. Bot.,* **12** (1898) 399–414

Scrivenor, (Mrs.)
Scrivenor, (Mrs.)
Method of preserving flowers. *J. Bot.,* **7** (1869) 341–3

Selby, Ada
Selby, (Miss) Ada
List of flowering plants observed in Hertfordshire during the [years 1883–1885]. [1884–1886] *Herts. Nat. Hist. Soc. Trans.,* **3** (1886) 101–2; **4** (1888) 118

Smith, Anna Maria
Smith, Anna Maria
Flora von Fiume. [Mit einer Einleitung von J. A. Knapp.] [1878] *Wien Zool. Bot. Verh.,* **28** (Abh.) (1879) 335–86

Smith, Annie Lorrain
Smith, (Miss) Annie Lorrain
On the development of the cystocarps in Callophyllis laciniata., *Kütz.* [1890] *Linn. Soc. J. (Bot.),* **28** (1891) 205–8
East African fungi. *J. Bot.,* **33** (1895) 340–4
On the anatomy of a plant from Senegambia. [1893] *Linn. Soc. J. (Bot.),* **30** (1895) 155–7
Nomenclature of British Pyrenomycetes. *J. Bot.,* **34** (1896) 358–9
Microscopic fungi new to, or rare in, Britain. *J. Bot.,* **35** (1897) 7–8; 100
British mycology. *Brit. Mycol. Soc. Trans.* (1897–1901) 68–75
Fungi new to Britain. [1899–1900] *Brit. Mycol. Soc. Trans.* (1897–1901) 113–16; 150–8
Supplement to Welwitsch's African fungi. *J. Bot.,* **36** (1898) 177–80
New or rare British fungi. *J. Bot.,* **36** (1898) 180–2
Some new microscopic fungi. *Micr. Soc. J.* (1900) 422–4
Carruthers, William; Smith, Annie Lorrain
On a disease in turnips caused by bacteria. *Agr. Soc. J.,* **11** (1900) 738–41

Stackhouse, Emily
Stackhouse, (Miss) Emily
List of Musci, natives of Cornwall. *Cornwall, J. Roy. Instit.,* **3** (1865) 58–62
Rare plants in the neighbourhood of Truro. [1866] *Cornwall, J. Roy. Instit.,* **2** (1866) 245–50

Tansley, Edith Chick
Chick, Edith
On the vascular system of the hypocotyl and embryo of Ricinus communis, *L.* [1899] *Edinb. Roy. Soc. Proc.,* **22** (1900) 652–72

Thomas, Ethel Nancy Miles (Mrs. Hyndman)
Thomas, Ethel N.
On the presence of vermiform nuclei in a dicotyledon. *Ann. Bot.,* **14** (1900) 318–19
Double fertilization in a dicotyledon, Caltha palustris. *Ann. Bot.,* **14** (1900) 527–35

Thomas, Millicent
Thomas, (Miss) Millicent
On the alpine flora of Clova. [1900] *Perthsh. Soc. Sci. Trans. & Proc.,* **3** (1903) 60–9

Thomas, Rose Haig
Thomas, Rose Haig
Attractive characters in fungi. [1890] *Nature,* **43** (1891) 79–80

Tindall, Ella M.
Tindall, Ella M.
Fossombronia mittenii, *n. sp. J. Bot.,* **36** (1898) 44–5

Twining, Elizabeth
Twining, (Miss)
On the comparison of the flora of Britain with that of other countries. *Brit. Ass. Rep.* (1847) Pt.2, 87–8

Vickers, Anna
Vickers, (Mlle.) A.
Contribution à la flore algologique des Canaries. *Ann. Sci. Nat. (Bot.),* **4** (1896) 293–303

Walker, A. W. Paton
Walker, (Mrs.)
Journal of an ascent to the summit of Adam's Peak, Ceylon. *Hooker, Comp. Bot. Mag.,* **1** (1835) 3–14
Journal of a tour in Ceylon. *Hooker's J. Bot.,* **2** (1840) 223–56

Warham, Amy E.
Harvey-Gibson, R. J.; Warham, (Miss) Amy E.
Note on the stinging hairs of Urtica dioica. *Liverpool Biol. Soc. Proc. & Trans.,* **4** (1890) 91–4

Warren, Elizabeth Andrew
Warren, (Miss)
On the recent botanical discoveries in Cornwall. *Cornwall, Polyt. Soc. Trans.* (1842) 24–5
Marine algae found on the Falmouth shores. *Cornwall, Polyt. Soc. Trans.* (1849) 31–7

White, M. Buchanan
White, (Miss) M. Buchanan
Strobilomyces strobilaceus in Perthshire. *Ann. Scott. Nat. Hist.* (1892) 274

Whitting, Frances G.
Murray, George R. M.; Whitting, Frances G.
New Peridinaceae from the Atlantic. [1899] *Linn. Soc. Trans. (Bot.),* **5** (1895–1901) 321–42

Wright, (Mrs.)
Wright, (Mrs.)
Notes on the meres of Shropshire. *Edinb., Bot. Soc. Trans.,* **11** (1873) 280–1
Notes on some rare plants gathered near Mentone. [1872] *Edinb., Bot. Soc. Trans.,* **11** (1873) 394–5

Conchology

United States

Bradshaw, M. F.
Bradshaw, (Mrs.) M. F.
Haminea virescens. [1895] *Nautilus,* **8** (1894–1895) 100–1
Megatebennus bimaculatus. [1895] *Nautilus,* **8** (1894–1895) 112–13
Orange, California. [1898] *Nautilus,* **12** (1898–1899) 9

Campbell, E. D. G.
Campbell, (Mrs.) E. D. G.
Marine shells on the southern California coast. [1895] *Nautilus,* **10** (1896–1897) 56–7

Drake, Marie
Drake, (Mrs.) Marie
Marine shells of Puget Sound. [1894] *Nautilus,* **9** (1895–1896) 38–42

King, E. H.
King, (Mrs.) E. H.
Collecting in Monterey Bay. [1896] *Nautilus,* **11** (1897–1898) 23–4

Monks, Sarah Preston
Monks, Sarah P.
San Pedro as a collecting ground. [1893] *Nautilus,* **7** (1893–1894) 74–7

Oldroyd, Ida Mary Shepard
Shepard, Ida M.
{With a dredge. *Nautilus,* **9** (1895–1896) 71–2}

Price, Sarah (Sadie) Frances
Price, Sadie F.
Mollusca of southern Kentucky. [1900] *Nautilus,* **14** (1900–1901) 75–9

Soper, C.
Soper, (Miss) C.
My Snailery. [1895] *Nautilus,* **10** (1896–1897) 113–15

Wentworth, D. J.
Wentworth, D. J. (Mrs. Edwin P.)
Marine shells on the coast of Maine. [1893] *Nautilus,* **9** (1895–1896) 34–5
Along the Damariscotta. [1895] *Nautilus,* **9** (1895–1896) 140–3
Purpura lapillus, *L.* [1896] *Nautilus,* **11** (1897–1898) 57–8

White, G. W.
White, (Mrs.) G. W.
Collecting in Southern California. [1894] *Nautilus,* **9** (1895–1896) 102–4

Williamson, Martha Burton Woodhead
Williamson, (Mrs.) M. Burton
{Collecting Chitons on the Pacific Coast. *Nautilus,* **4** (1890–1891) 32–3}
On Clementia subdiaphana, *Cpr.,* in San Pedro Bay. [1893] *Nautilus,* **6** (1892–1893) 116
An annotated list of the shells of San Pedro Bay and vicinity. With a description of two new species by W. H. Dall. [1892] *U. S. Mus. Proc.,* **15** (1893) 179–220
Edible mollusks of Southern California. [1893] *Nautilus,* **7** (1893–1894) 27–9
Beach shell collecting in connection with a study of oceanic phenomena. [1893] *Nautilus,* **7** (1893–1894) 41–2
Abalone or Haliotis shells of the Californian coast. *Amer. Natlist.,* **23** (1894) 849–58
{Conchological researches in San Pedro Bay and vicinity, including the Alamitos oyster fishery. *Hist. Soc. Southern Calif., Ann. Pub.,* **3** (1894) 10–15}
The aestivation of snails in Southern California. *Amer. Natlist.,* **29** (1895) 1106–8
An interrogation in regard to Septifer bifurcatus, *Rve.,* and Mytilus bifurcatus, *Conr.* [1898] *Nautilus,* **12** (1898–1899) 67–9
Helix (Epiphragmophora) Kelletti, *Forbes,* and its habitat. [1899] *J. Malacol.,* **7** (1900) 87–8
Aestivation of Epiphragmophora traskii in Southern California. [1900] *Nautilus,* **14** (1900–1901) 13–15

Great Britain

Carphin, Janet
Carphin, Janet
Helix hispida, *var.* sinistrorsum, in Berwickshire. *Ann. Scott. Nat. Hist.* (1895) 254

Longstaff, Mary Jane Donald
Donald, (Miss) Jane
Notes on the land and fresh-water shells of Cumberland [and Westmorland]. *Cumberland Ass. Trans.,* **7** (1882) 51–60; **9** (1885) 217–19

Thew, (Mrs. Edward)
Thew, (Mrs. Edward)
List of the shells found on the shore between Alnmouth and Amble Pier. *Berwick. Nat. Club Hist.,* **15** (1897) 309–12

Entomology

United States

Aaron, C. B.
Aaron, (Mrs.) C. B.
The bee moth. *Philad., Ent. News,* **8** (1897) 189–90

Beach, Alice Marie
Beach, Alice M.
Additions to the known species of Iowa Ichneumonida. [1893] *Iowa Ac. Sci. Proc.,* **1** (Pt. 4) (1894) 128–9
Some bred parasitic Hymenoptera in the Iowa Agricultural College collection. [1894] *Iowa Ac. Sci. Proc.,* **2** (1895) 92–4
Contributions to a knowledge of the Thripidae of Iowa. *Iowa Ac. Sci. Proc.,* **3** (1896) 214–27

Casad, Jessie E.
Cockerell, Theodore D. A.; Casad, (Miss) Jessie E.
New species of Mutillidae. *Philad., Ent. News,* **5** (1894) 293–6
Descriptions of new Hymenoptera. [With notes by Wm. J. Fox] *Amer. Ent. Soc. Trans.,* **22** (1895) 297–300

Chapman, Bertha L.
Chapman, Bertha L.
Two new species of Trichodectes (Mallophaga). *Philad., Ent. News,* **8** (1897) 185–7
Kellogg, Vernon Lyman; Chapman, Bertha L.
[New Mallophaga. III.] Mallophaga from birds of California. *California Ac. Occ. Pap.,* **6** (1899) 53–143

Clarke, Cora Huidekoper
Clarke, Cora H.
Description of two interesting houses made by native Caddis-fly larvae. [1882] *Boston Soc. Nat. Hist. Proc.,* **22** (1884) 67–71
Caddis-worms of Stony Brook. [1891] *Psyche,* **6** (1893) 153–8

Cockerell, Wilmatte Porter
Cockerell, Theodore D. A.; Porter, (Miss) Wilmatte (Mrs. Cockerell)
Contributions from the New Mexico Biological station. VIII. The New Mexico bees of the genus Bombus. [VII. Observations on bees, with descriptions of new genera and species.] *Ann. Mag. Nat. Hist.,* **4** (1899) 386–93; 403–21
A new variety of Argynnis nitocris. *Philad., Ent. News,* **11** (1900) 622

Dickens, Bertha Sarah Kimball
Kimball, Bertha S.
Conorhinus sanguisugus, its habits and life-history. [1894] *Kan. Ac. Sci. Trans.,* **14** (1896) 128–31

Dimmock, Anna Katherine Hofmann
Dimmock, Anna Katherine
Asymmetry of the nervous system in the larva of Harpyia. [1882] *Psyche,* **3** (1880–1882) 340–1

Sexual attraction in Prionus. [1884] *Psyche,* **4** (1890) 159
The insects of Betula in North America. [1885] *Psyche,* **4** (1890) 239–43; 271–6

Dix, Dorothea Lynde
Dix, (Miss) D. L.
Notice of the Aranea aculeata, the Phalaena antigua, and some species of the Papilio. *Silliman, J.,* **19** (1831) 61–3

Eigenmann, Rosa Smith
Smith, (Miss) Rosa
Insect-life among spider eggs. *Amer. Natlist.,* **18** (1884) 77

Eliot, Ida Mitchell
Soule, Caroline G.; Eliot, Ida M.
{Notes on an undetermined Lepidopterous larva. *Canad. Ent.,* **18** (1886) 124–5}
Notes on the early stages of some Heterocera. [1889] *Psyche,* **5** (1888–1890) 259–69
{Smerinthus astylus. *Psyche,* **6** (1891–1893) 31}
Hemaris diffinis: from larvae sent from Missouri. [1891] *Psyche,* **6** (1891–1893) 142–5

Fernald, Maria Elizabeth Smith
Fernald, Mrs. C. H.
Notes on Sphingidae captured at Orono, Maine, and vicinity. *Canad. Ent.,* **16** (1884) 20–2
List of Zygaenidae and Bombycidae taken at Orono, Maine, and vicinity. *Canad. Ent.,* **16** (1884) 57–8
List of Geometridae captured at Orono, Maine, and vicinity. *Canad. Ent.,* **16** (1884) 129–30
Northern localities for southern butterflies. *Amer. Natlist.,* **18** (1884) 77
Notes on Papilio turnus and Pyrameis cardui. *Canad. Ent.,* **18** (1886) 50–1

Fielde, Adele Marion
Fielde, (Miss) Adele M.
The meng-leng. *Science,* **4** (1884) 159
Fishing lines and ligatures from the silk-glands of lepidopterous larvae. *Philad. Ac. Nat. Sci. Proc.* (1886) 298–9
On an aquatic larva and its case. *Philad. Ac. Nat. Sci. Proc.* (1887) 293–4
Notes on an aquatic insect, or insect-larva, having jointed dorsal appendages. *Philad. Ac. Nat. Sci. Proc.* (1888) 129–30
On an insect-larva habitation. *Philad. Ac. Nat. Sci. Proc.* (1888) 176–7
Portable ant nests. [1900] *Woods Holl Mar. Biol. Lab. Bull.,* **2** (1901) 81–5

Harward, Winnie
Harward, Winnie
Insectivorous habits of lizards. *U. S. Div. Ent. Bull.,* **22** (1900) 96–7

Highfield, Laurene
Highfield, Laurene
A curious hemipteron. *Amer. Natlist.,* **28** (1894) 283

Hitchcock, Fanny Rysam Mulford
Hitchcock, Fanny R. M.
Occurrence of early stages of Blepharocera. *Amer. Natlist.,* **20** (1886) 651–2
Notes on the larvae of Amblystoma. [1888] *N. Y. Ac. Trans.,* **7** (1887–1889) 255–8

Jordan, Alice M.
Jordan, (Miss) Alice M.
Life history of Papilio zolicaon. *Canad. Ent.,* **26** (1894) 257–8

King, Helen Selina
King, Helen Selina (Mrs. V. O.)
Phosphorescent insects: their metamorphoses. *Amer. Natlist.,* **12** (1878) 354–8
The fireflies and their phosphorescent phenomena. *Amer. Natlist.,* **12** (1878) 662–5
Life history of Pleotomus pallens, *Lec.* [1880] *Psyche,* **3** (1886) 51–3
Thyridopteryx ephemeraeformis, *Haworth.* Its habits and metamorphosis. [1881] *Psyche,* **3** (1880–1882) 241–3
Internal organization of Hesperia ethlius, *Cram.,* as observed in the living animal. [1882] *Psyche,* **3** (1880–1882) 322–4

Meek, Elizabeth B.
Meek, Elizabeth B.
Some variations in Lucanus placidus, statistically examined. [1900] *Science,* **13** (1901) 375

Middleton, Nettie
Middleton, (Miss) Nettie
Larvae of butterflies. *Ill. Insects Rep.,* (**10**) (1881) 73–98
A new species of Aphis, of the genus Colopha [C. eragrostidis]. [1878] *Ill. Lab. Nat. Hist. Bull.,* **1** (No. 2) (1884) 17

Mills, Helen
Mills, (Miss) Helen; Snow, W. A.
The destructive diplosis of the Monterey pine. [With preparatory note by Vernon L. Kellogg.] *Philad., Ent. News,* **11** (1900) 489–94

Monks, Sarah Preston
Monks, Sarah P.
{Curious habit of a dragon fly. *Amer. Natlist.,* **15** (1881) 141}
{Aestivation of California Mason spiders. *Hist. Soc. Southern Calif., Ann. Pub.,* **1** (1888) 18–22}
{Trap-door spiders. *Hist. Soc. Southern Calif., Ann. Pub.,* **1** (1888) 28–36}

Morris, Margaretta Hare
Morris, (Miss) Margaretta
On the Cecidomyia destructor, or Hessian Fly. *Silliman, J.,* **40** (1841) 381–2; *Amer. Phil. Soc. Trans.,* **8** (1843) 49–52
Observations on the development of the Hessian Fly. *Philad. Ac. Nat. Sci. Proc.,* **1** (1841–1843) 66–8
[On the larvae of Cicada septemdecim.] *Philad. Ac. Nat. Sci. Proc.,* **3** (1846–1847) 132–4
On the habits of Cecidomyia culmicola. *Philad. Ac. Nat. Sci. Proc.,* **4** (1848–1849) 194
On the Seventeen-years Locust (in a letter to Charles Girard). *Boston Soc. Nat. Hist. Proc.,* **4** (1851) 110

Morton, Emily L.
Morton, (Miss) Emily L.
Notes on Danais archippus, *Fabr. Canad. Ent.,* **20** (1888) 226–8
Notes from New Windsor. *Philad., Ent. News,* **3** (1892) 1–3
{Edwards, W. H.
On the history and preparatory stages of Fenescia tarquinius, *Fabr.* [The article includes extensive quotations from observations sent to Edwards by Emily Morton.] *Canad. Ent.,* **18** (1886) 141–53}
Dyar, Harrison G.; Morton, Emily L.
The life-histories of the New York slug caterpillars. *N. Y. Ent. Soc. J.,* **3** (1895) 145–57; **4** (1896) 1–9; 167–90

Murtfeldt, Mary Esther
Murtfeldt, (Miss) Mary Esther
Notes on Attelabus bipustulaties, *Fabr. Canad. Ent.,* **4** (1872) 143–5

The larvae of Depressaria dubitella and Gelechia rubensella. *Canad. Ent.*, **6** (1874) 221–2

Larva of Anaphora agrotipennella. *Canad. Ent.*, **8** (1876) 185–6

An experiment with a stinging larva [Lagoa opercularis]. *Canad. Ent.*, **8** (1876) 201–2

{Forest-tree borers. *Kansas City Rev. Sci. Ind.*, **2** (1878–9) 91–5}

New species of Tineidae: [Gelechia chambersella, G. (formosella) vernella, G. (cinerella) inconspicuella, G. beneficentelli; Lithocolletis gregariella (? L. desmodiella, *Clemens*), *nn. spp.*] *Canad. Ent.*, **13** (1881) 242–6; **15** (1883) 138–9

Attacus cinctus, *Tepper. Canad. Ent.*, **16** (1884) 131–2

Larval longevity of certain Coleophorae. [1885] *Entomologica Amer.*, **1** (1885–6) 222–4

A fragrant butterfly (Callidryas eubule ♂). [Habits of Hypoprepia packardii, *Grote.* The grape-berry moth (Eudemis botrana, *s. v.*) Xylocopa and Megachile cutting flowers.] [1881–1882] *Psyche*, **3** (1886) 198; 243–4; 276; 343

Vernal habit of Apatura. [1886] *Entomologica Amer.*, **2** (1886–1887) 180–1

Notes from Missouri for the season of 1886. *U. S. Div. Ent. Bull.*, **13** (1887) 59–65

Traces of maternal affection in Eutillia sinuata, *Fabr.* [1887] *Entomologica Amer.*, **3** (1887–1888) 177–8

Entomological notes of the season of 1888. *[U. S.] Comm. Agr. Rep.* (1888) 133–9

Life-history of Graptodera foliacea, *Lec.* [1888] *U. S. Div. Ent. Insect Life*, **1** (1888–1889) 74–6

The carnivorous habits of tree crickets. [1889] *U. S. Div. Ent. Insect Life*, **2** (1889–1890) 130–2

An interesting tineid. (Menesta melanella, *n. sp.*) [1890] *U. S. Div. Ent. Insect Life*, **2** (1889–1890) 303–5

Entomological notes from Missouri for the season of 1889. *U. S. Div. Ent. Bull.*, **22** (1890) 73–84

Some experiences in rearing insects. *Canad. Ent.*, **22** (1890) 220–5

Entomological notes for the season of 1890[-92]. *U. S. Div. Ent. Bull.*, **23** (1891) 45–56; **26** (1892) 36–44; **30** (1893) 49–56

Longevity and vitality of Argas and Trombidium. *Canad. Ent.*, **23** (1891) 248–9

The use of grape bags by a paper-making wasp. [1891] *U. S. Div. Ent. Insect Life*, **4** (1892) 192–3

Hominivorous habits of the screw worm in St. Louis. [1891] *U. S. Div. Ent. Insect Life*, **4** (1892) 200–1

The web-worm tiger (Plochionus timidus, *Hald.*). *Canad. Ent.*, **24** (1892) 279–82

The Osage-orange pyralid. (Loxostege maclurae, *n. sp.*, *Riley.*) *U. S. Div. Ent. Insect Life*, **5** (1893) 155–7; 370

The cheese or meat skipper (Piophila casei). [1893] *U. S. Div. Ent. Insect Life*, **6** (1894) 170–5

Entomological memoranda for 1893. *U. S. Div. Ent. Insect Life*, **6** (1894) 257–9

Habits of Stibadium spumosum, *Gr. U. S. Div. Ent. Insect Life*, **6** (1894) 301–2

Acorn insects: primary and secondary. *U. S. Div. Ent. Insect Life*, **6** (1894) 318–24

Notes on the insects of Missouri for 1893. *U. S. Div. Ent. Bull.*, **32** (1894) 37–45

A new pyralid. *Canad. Ent.*, **29** (1897) 71–2

[Indiana caves and their fauna. By W. S. Blatchley.] A cave-inhabiting moth. *Indiana Dept. Geol. Ann. Rep.*, **21** (1897) 191–2

New Tineidae, with life-histories. *Canad. Ent.*, **32** (1900) 161–6

Palmer, Mary T.

Palmer, Mary T.

A trap-door spider at work. *Science*, **7** (1886) 240–1

Peckham, Elizabeth Gifford

Peckham, (Mrs.) Elizabeth Gifford

Protective resemblances in spiders. [1889] *Wisconsin Nat. Hist. Soc. Occ. Pap.*, **1** (1889–1890) 61–113

Peckham, George W.; Peckham, Elizabeth G.

Genera of the family Attidae: with a partial synonymy. [1883] *Wisconsin Ac. Trans.*, **6** (1885) 255–342

On some new genera and species of Attidae from the eastern part of Guatemala. [1885] *Wisconsin Nat. Hist. Soc. Proc.* (1885–1889) 62–86

Some observations on the special senses of wasps. [1887] *Wisconsin Nat. Hist. Soc. Proc.* (1885–1889) 91–132

Some observations on the mental powers of spiders. *J. Morphol.*, **1** (1887) 383–419

The duration of memory in wasps. *Amer. Natlist.*, **21** (1887) 1038–40

Attidae of North America. *Wisconsin Ac. Trans.*, **7** (1889) 1–104

Observations on sexual selection in spiders of the family Attidae. [1889] *Wisconsin Nat. Hist. Soc. Occ. Pap.*, **1** (1889–1890) 3–60

Additional observations on sexual selection in spiders of the family Attidae, with some remarks on Mr. Wallace's theory of sexual ornamentation. [1890] *Wisconsin Nat. Hist. Soc. Occ. Pap.*, **1** (1889–1890) 115–41

Ant-like spiders of the family Attidae. [1892] *Wisconsin Nat. Hist. Soc. Occ. Pap.*, **2** (1892–1895) 1–83

Spiders of the Marptusa group of the family Attidae. [1894] *Wisconsin Nat. Hist. Soc. Occ. Pap.*, **2** (1892–1895) 85–156

Spiders of the Homalattus group of the family Attidae. [1895] *Wisconsin Nat. Hist. Soc. Occ. Pap.*, **2** (1892–1895) 159–83

On the spiders of the family Attidae of the island of Saint Vincent. *Zool. Soc. Proc.* (1893) 692–704

Notes on the habits of Trypoxyllon rubrocinctum and Trypoxyllon albopilosum. [1895] *Psyche*, **7** (1894–1896) 303–6

The sense of sight in spiders, with some observations on the color sense. *Wisconsin Ac. Trans.*, **10** (1895) 231–61

Spiders of the family Attidae from Central America and Mexico. *Wisconsin Nat. Hist. Soc. Occ. Pap.*, **3** (1896) 1–101

On the instincts and habits of the solitary wasps. *Wisconsin Geol. Nat. Hist. Surv. Bull.*, **2** (1898) iv + 245 pp.

Instinct or reason? *Amer. Natlist.*, **34** (1900) 817–18

Pellenes, and some other genera of the family Attidae. *Wisconsin Nat. Hist. Soc. Bull.*, **1** (1900) 195–233

Peckham, George W.; Peckham, Elizabeth G.; Wheeler, William Morton

Spiders of the subfamily Lyssomanae. *Wisconsin Ac. Trans.*, **7** (1889) 221–56

Pigeon, Emily Adella Smith

Smith, Emily A.

Report on the noxious insects of northern Illinois. *Ill. Insects Rep.*, (7) (1878) 107–31

The maple-tree bark-louse. *Amer. Natlist.*, **12** (1878) 655–61

Modes of spreading and means of extinguishing the maple-tree bark-louse. *Amer. Natlist.*, **12** (1878) 808–9

{Biological and other notes on Pseudococcus aceris. *N. Amer. Ent.*, **1** (1880) 73–87; *Amer. Ent.*, **3** (1880) 220–1}

Sargent, Annie Bell
Sargent, (Miss) Annie Bell
Some observations on the hunting spider, Lycosa vulpina. [1897] *Philad., Ent. News,* **9** (1898) 131–4
Preliminary notes on the rate of growth and on the development of instincts in spiders. *Philad. Ac. Nat. Sci. Proc.* (1900) 395–411

Sheldon, Jennie Maria Arms
Hyatt, Alpheus; Arms, (Miss) J. M.
A general survey of the modes of development in insects and their meaning. [1891] *Psyche,* **6** (1893) 37–44

Slater, Florence Wells
Slater, Florence Wells
The egg-carrying habit of Zaitha. *Amer. Natlist.,* **33** (1899) 931–3

Slosson, Annie Trumbull
Slosson, (Mrs.) Annie Trumbull
Southern form of E[cpantheria] scribonia, *Stoll.* [1888] *Entomologica Amer.,* **3** (1887–1888) 212
A new species of Euphanessa. *Entomologica Amer.,* **5** (1889) 7
A new Spilosoma. *Entomologica Amer.,* **5** (1889) 40
Phragmatobia assimilans, *Walker. Entomologica Amer.,* **5** (1889) 85–6
Larvae of Seirarctia Echo. *Entomologica Amer.,* **6** (1890) 8–9
Cressonia hyperbola, *n. var. Entomologica Amer.,* **6** (1890) 59
Varina ornata, *Neum. Entomologica Amer.,* **6** (1890) 136
The home of Seirarctia Echo. *Entomologica Amer.,* **6** (1890) 153–5
May moths in northern New Hampshire. *Philad., Ent. News,* **1** (1890) 17–19; 47
Winter collecting in Florida. *Philad., Ent. News,* **1** (1890) 81–3; 101–2
Phragmatobia assimilans, *Walker. Philad., Ent. News,* **2** (1891) 2–3
Phragmatobia assimilans, *n. var.* franconia. *Philad., Ent. News,* **2** (1891) 41
A long-lived basket maker. *Philad., Ent. News,* **3** (1892) 49–51
A new Arctia. *Philad., Ent. News,* **3** (1892) 257–8
A new Dasylophia from Florida. *Canad. Ent.,* **24** (1892) 139
Collecting on Mt. Washington. *Philad., Ent. News,* **4** (1893) 249–52; 287–92
{Common *versus* proper. *N. Y. Ent. Soc. J.,* **1** (1893) 1–5}
{Spring collecting in northern Florida. *N. Y. Ent. Soc. J.,* **1** (1893) 147–52}
List of insects taken in Alpine region of Mt. Washington. *Philad., Ent. News,* **5** (1894) 1–6; 271–4; **6** (1895) 4–7; 316–21; **7** (1896) 262–5; **8** (1897) 237–40; , **9** (1898) 251–3; **11** (1900) 319–23
Hyparpax, *var.* tyria, *n. var. Philad., Ent. News,* **5** (1894) 198
Coleoptera of Lake Worth, Florida. *Canad. Ent.,* **27** (1895) 9–10
Collecting at Lake Worth, Fla. *Philad., Ent. News,* **6** (1895) 133–6
The season on Mt. Washington. *Philad., Ent. News,* **6** (1895) 276–80
More about the red bug. *Philad., Ent. News,* **7** (1896) 40–1
Singular habit of a cecidomyid. *Philad., Ent. News,* **7** (1896) 238
List of the Araneae taken in Franconia, New Hampshire. *N. Y. Ent. Soc. J.,* **6** (1898) 247–9
A new Cossonus. *Canad. Ent.,* **31** (1899) 193
Collecting on Biscayne Bay. *Philad., Ent. News,* **10** (1899) 94–6; 124–6

Smallwood, Mabel E.
Smallwood, Mabel E.
Statistical studies on sand fleas. *Science,* **12** (1900) 373

Snyder, Mrs. Arthur J.
Snyder, Mrs. A. J.
Trypeta solidaginis. *Canad. Ent.,* **30** (1898) 99–100

Soule, Caroline Gray
Soule, Caroline G.
Description of the larva of Sphinx luscitiosa. [1888] *Psyche,* **5** (1888–1890) 85–6
Notes on the larval stages of Samia cynthia. [1888] *Psyche,* **5** (1888–1890) 135–6
Description of eggs and larva of Apatelodes torrefacta. [1889] *Psyche,* **5** (1888–1890) 148–9
The march of Hyperchiria io. [1891] *Psyche,* **6** (1891–1893) 15
Halisidota caryae. [1891] *Psyche,* **6** (1891–1893) 158–60
Heteropacha rileyana. [1891] *Psyche,* **6** (1891–1893) 193–4
The early stages of Nerice bidentata. [1892] *Psyche,* **6** (1891–1893) 276–7
Early stages of Spilosoma latipennis. [1894] *Psyche,* **7** (1894–1896) 71–2
Polygamy of Actias luna and Callosamia promethea. [1894] *Psyche,* **7** (1894–1896) 167
Uncertainty of the duration of any stage in the life-history of moths. [1895] *Psyche,* 7 (1894–1896) 191
Description of some of the larval stages of Amphion nessus. [1895] *Psyche,* 7 (1894–1896) 212–13
Life-history of Deilephila lineata. *Psyche,* 7 (1894–1896) 458–60
Early stages of Triptogon modesta. *Psyche,* **8** (1897–1899) 308–10
Color-variation in larvae of Papilio polyxenes, and other notes. *Psyche,* **8** (1897–1899) 435–6
The "cocoons" or "cases" of some burrowing caterpillars. [1900] *Psyche,* **9** (1900–1902) 7–8
Soule, Caroline G.; Eliot, Ida M.
{Notes on an undetermined Lepidopterous larva. *Canad. Ent.,* **18** (1886) 124–5}
Notes on the early stages of some Heterocera. [1889] *Psyche,* **5** (1888–1890) 259–69
{Smerinthus astylus. *Psyche,* **6** (1891–1893) 31}
Hemaris diffinis: from larvae sent from Missouri. [1891] *Psyche,* **6** (1891–1893) 142–5

Treat, Mary Allen Davis
Treat, (Mrs.) Mary
{White grub fungus. *Amer. Ent.,* **2** (1869–1870) 53}
{The tomato worm. *Amer. Ent.,* **2** (1869–1870) 87}
{Parasitic mites on the house fly. *Amer. Ent.,* **2** (1869–1870) 87}
{Polyphemus moth. *Amer. Ent.,* **2** (1869–1870) 88}
{Plant lice and their enemies. *Amer. Ent.,* **2** (1869–1870) 141–3}
{My raspberry and verbena moths and what came of them. *Amer. Ent. Bot.,* **2** (1870) 203–5}
{Pupa of the girdled sphinx. *Amer. Ent. Bot.,* **2** (1870) 241}
{To kill the pea weevil. *Amer. Ent. Bot.,* **2** (1870) 241}
Controlling sex in butterflies. *Amer. Natlist.,* 7 (1873) 129–32; *Entomologist,* **6** (1873) 372–5
{The enemies of the oak. *Amer. Agriculturist.* **33** (1874) 344}
The habits of a tarantula. *Amer. Natlist.,* **13** (1879) 485–9
{Notes on the slave-making ant (F. sanguinea). *Amer. Natlist.,* **13** (1879) 707–8}
{Notes on harvesting ants in New Jersey. *Amer. Ent.,* **1,** n.s. (1880) 225–6}
Behavior of Dolomedes tenebrosus. *Science,* **3** (1884) 217–8
{Curculio and injury to cherries. *Prairie Farmer* (1888) 538}
Insect enemies of the pitch pine. *Garden & Forest,* **4** (1891) 62–3

{Some injurious insects of the orchard and garden. *N. Y. Ent. Soc. J.*, **1** (1893) 16–20}

Wadsworth, Mattie
Wadsworth, (Miss) Mattie
List of the dragonflies (Odonata) taken at Manchester, Kennebec Co., Me., in 1888 and 1889, etc. *Philad., Ent. News,* **1** (1890) 36–7; 55–7; 80; **2** (1891) 11–12; **3** (1892) 8–9; **5** (1894) 132; **9** (1898) 111

Warner, Hattie H.
Warner, (Miss) Hattie H.
Kentucky butterflies. *Canad. Ent.*, **26** (1894) 289–91

Wright, Julia McNair
Wright, Julia McNair
A water-beetle. *Science,* **17** (1891) 134–5
A wasp study. *Science,* **20** (1892) 220–1

Great Britain

Alderson, E. Maude
Alderson, (Miss) E. Maude
{Collecting in the neighbourhood of Worksop in 1893. *Entomologist,* **27** (1894) 140–2}

Bazett, E. C.
Bazett, (Mrs.) E. C.
Notes on rearing Stauropus fagi. *Ent. Month. Mag.,* **36** (1900) 275–7

Bewsher, Eva M. A.
Bewsher, Eva M. A.
Ventilating bees. *Nature,* **39** (1889) 224

Chawner, Ethel Frances
Chawner, (Miss) E. F.
Notes on saw-flies. *Entomologist,* **27** (1894) 175
Life-history of Tenthredopsis microcephala. *Entomologist,* **28** (1895) 168–9

Cooper, Charlotte
Cooper, Charlotte
Curious particulars respecting bees. *Tilloch, Phil. Mag.,* **13** (1802) 181–2

Cottingham, M. L.
Cottingham, (Miss) M. L.
{Captures in Argyllshire. *Entomologist,* **27** (1894) 223}
{Collecting in Argyllshire from June 10th, 1894. *Entomologist,* **28** (1895) 20–1}

Edwards, A. D.
Edwards, (Miss) A. D.
{Clostera anachoreta. *Entomologist,* **27** (1894) 176}

Fountaine, Margaret Elizabeth
Fountaine, Margaret E.
Notes on the butterflies of Sicily. *Entomologist,* **30** (1897) 4–11
{White female of Colias chrysotheme near Vienna. (Field report) *Entomologist,* **30** (1897) 296–7}

Fowler, Alice
Fowler, Alice
Death's head moth at Inverbroom, west Ross-shire. *Ann. Scott. Nat. Hist.* (1900) 125

Fraser, Jane
Fraser, Jane
About some Samoan butterflies. *Ent. Month. Mag.,* **30** (1894) 146–9
In an old orange garden. *Ent. Month. Mag.,* **31** (1895) 13–15

Hopley, Catherine Cooper
Hopley, (Miss) Catherine C.
Note on Phosphoenus hemipterus. *Ent. Month. Mag.,* **5** (1868–1869) 70
{Jumping beans and jumping eggs. *Entomologist,* **28** (1895) 52–3}
{Those "jumping eggs." *Entomologist,* **28** (1895) 159–60}

Huie, Lily H.
Huie, Lily H.
Note on the life-history of a weevil (Hypera plantaginis, *De Geer*). *Ann. Scott. Nat. Hist.* (1894) 117

Kimber, Mary
Kimber, Mary
The Macro-lepidoptera of the Newbury district. *Newb. Field Club Trans.,* **4** (1895) 20–34

Miller, Elizabeth
Miller, (Miss) Elizabeth
On a cluster of cocoons of Aphomia sociella. [1896] *Ent. Record,* **7** (1895–1896) 237

Nicholl, Mary de la Beche.
Nicholl, (Mrs.) Mary de la B.
The butterflies of Aragon. *Ent. Soc. Trans.* (1897) 427–34
Butterfly hunting in Dalmatia, Montenegro, Bosnia, and Hercegovina. *Ent. Record,* **11** (1899) 1–8
Bulgarian butterflies. *Ent. Record,* **12** (1900) 29–34; 64–9

Ormerod, Eleanor Anne
Ormerod, (Miss) Eleanor A.
Notes on the development of Volucella bombylans, parasitical in the nests of Carder-bees; with observations on the development of the tubular head appendages of its pupa. *Ent. Month. Mag.,* **10** (1873–1874) 196–200
Life history of Meligethes. [1874] *Ent. Month. Mag.,* **11** (1874–1875) 46–52
Turkey oak-galls. *Entomologist,* **10** (1877) 42–3
Phytoptus of the birch-knots. *Entomologist,* **10** (1877) 83–6
Oak-galls: Aphilothrix corticis, *L. Entomologist,* **10** (1877) 165–6
Workings of Hylesinus fraxini. *Entomologist,* **10** (1877) 183–7
The Colorado beetle. *Entomologist,* **10** (1877) 217–20
Turnip and cabbage-gall weevil, Ceutorhynchus sulcicollis. *Entomologist,* **10** (1877) 246–9
Notes on the egg and development of the Phytoptus. *Entomologist,* **10** (1877) 280–3
On the development of galls of Cecidomyia ulmariae. *Entomologist,* **11** (1878) 12–14
Considerations on abnormal gall-growth. *Entomologist,* **11** (1878) 82–7
Turkey oak-galls. *Entomologist,* **11** (1878) 201–4
Notes on Psylliodes chrysocephala. *Entomologist,* **11** (1878) 217–20
The prevention of insect injury by the use of phenol preparations. *Ent. Soc. Trans.* (1878) 333–5
Notes on leaf-galls on Parinarium curatellifolium. [1878] *Ent. Month. Mag.,* **15** (1878–1879) 97–9
On an undetermined oak-gall. [1879] *Ent. Month. Mag.,* **15** (1878–1879) 197–8
Cane-borers. *Ent. Soc. Trans. (Proc.)* (1879) xxxvi-xl
Sitophilus granarius. *Entomologist,* **12** (1879) 51–4
Egg of Calycophthora avellanae. *Entomologist,* **12** (1879) 110–11

Considerations as to effects of temperature on insect development. *Entomologist,* **12** (1879) 137–42
Undescribed oak galls. *Entomologist,* **12** (1879) 193–4
Observations of the effects of low temperatures on larvae. *Ent. Soc. Trans.* (1879) 127–30
Sugar-cane borers of British Guiana. *Ent. Soc. Trans. (Proc.)* (1879) xxxiii–xxxvi
Cane-borers. *Ent. Soc. Trans. (Proc.)* (1880) xv-xix
Effects of warmth and surrounding atmospheric conditions on silk-worm larvae. *Entomologist,* **15** (1882) 127–9
Observations on the development of Sitones lineatus. *Ent. Soc. Trans. (Proc.)* (1882) xiv-xvi
Report on wireworm. *Agr. Soc. J.,* **19** (1883) 104–43
Reports of the honorary consulting entomologist. *Agr. Soc. J.,* **20** (1884) 324–31; 698–709; **21** (1885) 322–9; **22** (1886) 311–15; **23** (1887) 310–15; **24** (1888) 289–96; **25** (1889) 329–43; **1** (1890) 170–84; 407–12; 837–47; **2** (1891) 168–71; 389–98; 595–6; 849–54; **3** (1892) 132–5; 365–71
Observations on the Oestridae, commonly known as bot flies. [1884] *Agr. Stud. Gaz.,* **2** (1884–1886) 5–9
Observations on the development of ox warble, and warble maggot. *Agr. Soc. J.,* **21** (1885) 490–9
The recent appearance of the Hessian fly, Cecidomyia destructor (*Say*), in Great Britain. *Agr. Soc. J.,* **22** (1886) 721–7
Cecidomyia destructor, *Say,* in Great Britain. [1886] *Ent. Soc. Trans.* (1887) 1–6
Mustard beetles. *Agr. Soc. J.,* **23** (1887) 273–84
The Hessian fly. *Entomologist,* **20** (1887) 262–4; *Nature,* **36** (1887) 439
Parasites of the "Hessian fly" (Cecidomyia destructor, *Say*). *Entomologist,* **20** (1887) 317–18
The Hessian fly and its introduction into Britain. [1889] *Herts. Nat. Hist. Soc. Trans.,* **5** (1890) 168–76
The diamond-back moth. *Agr. Soc. J.,* **2** (1891) 596–630
A few remarks on insect prevalence during the summer of 1893. *Nature,* **48** (1893) 394–5
Abundance of caterpillars of the antler moth, Charaeas graminis, *Linn.,* in the south of Scotland. *Ent. Month. Mag.,* **30** (1894) 169–71
Lamellicorn beetles on pasturage in the Argentine territories. *Entomologist,* **27** (1894) 229–32
The "turnip mud-beetle." [1894] *Agr. Stud. Gaz.,* **7** (1894–1896) 37–8
Hippobosca equina, *Linn.,* at Ystalyfera, Glamorganshire. *Entomologist,* **31** (1898) 225–6
Indian "forest flies." Hippobosca (aegyptiaca ?), *Macq.* [1895] *Indian Mus. Notes,* **4** (1900) 79–80

Pasley, L. M. S.
Pasley, (Miss) L. M. S.
Singular instance of parasitism. *Ent. Month. Mag.,* **1** (1864–1865) 281

Redmayne, Mary B.
Redmayne, (Mrs.) Mary B.
Listrodromus quinqueguttus, *Grav.,* bred from Cyaniris argiolus. *Ent. Record,* **12** (1900) 164

Ricardo, Gertrude
Ricardo, (Miss) Gertrude
Notes on the Pangoninae of the family Tabanidae in the British Museum Collection. *Ann. Mag. Nat. Hist.,* **5** (1900) 97–121; 167–82
Description of five new species of Pangoninae from South America. *Ann. Mag. Nat. Hist.,* **6** (1900) 291–4
Notes on Diptera from South Africa (Tabanidae and Asilidae). *Ann. Mag. Nat. Hist.,* **6** (1900) 161–78

Robson, Mrs. Samuel
Robson, Mrs. S.
Notes on Argynnis niphe, *Linnaeus,* a nymphalid butterfly. [1893] *Bombay Nat. Hist. Soc. J.,* **8** (1893–1894) 151–2
Notes on Callerebia nirmala, *Moore,* a satyrid butterfly. [1894] *Bombay Nat. Hist. Soc. J.,* **8** (1893–1894) 551–3
Life-history of Rapala schistacea, *Moore,* a lycaenid butterfly. [1895] *Bombay Nat. Hist. Soc. J.,* **9** (1894–1895) 337
Life-history of Athyma opalina, *Kollar,* a nymphaline butterfly. [1895] *Bombay Nat. Hist. Soc. J.,* **9** (1894–1895) 338–9
Life-history of Camena cleobis, *Godart,* a lycaenid butterfly. [1895] *Bombay Nat. Hist. Soc. J.,* **9** (1894–1895) 339–40
Description of the larva of Papilio cloanthus, *Westwood.* [1895] *Bombay Nat. Hist. Soc. J.,* **9** (1894–1895) 497
Life-history of Papilio glycerion, *Westwood.* [1895] *Bombay Nat. Hist. Soc. J.,* **9** (1894–1895) 497–8

Sanders, Cora Brooking
Poulton, Edward B.; Sanders, Cora B.
An experimental enquiry into the struggle for existence in certain common insects. *Brit. Ass. Rep.* (1898) 906–9

Sharpe, Emily Mary Bowdler
Sharpe, (Miss) Emily Mary
Descriptions of new species of East-African butterflies. *Ann. Mag. Nat. Hist.,* **5** (1890) 335–6
Further descriptions of butterflies and moths collected by Mr. F. J. Jackson in Eastern Africa. *Ann. Mag. Nat. Hist.,* **5** (1890) 440–3
On some new species of African Lycaenidae in the collection of Philip Crowley, Esq. *Ann. Mag. Nat. Hist.,* **6** (1890) 103–6
Descriptions of some new species of African butterflies in the collection of Captain G. E. Shelley. *Ann. Mag. Nat. Hist.,* **6** (1890) 346–50
On a collection of Lepidoptera made by Mr. Edmund Reynolds on the rivers Tocantins and Araguaya and in the province of Goyaz, Brazil. *Zool. Soc. Proc.* (1890) 552–77
Descriptions of new butterflies collected by Mr. F. J. Jackson, F.Z.S., in British East Africa, during his recent expedition. *Zool. Soc. Proc.* (1891) 187–94; 633–8
Descriptions of some new species of Lepidoptera collected by Mr. Herbert Ward at Bangala, on the Congo. *Ann. Mag. Nat. Hist.,* **7** (1891) 130–5
Descriptions of two new species of Lycaenidae from West Africa, in the collection of Mr. Philip Crowley. *Ann. Mag. Nat. Hist.,* **8** (1891) 240–1
On a collection of Lepidoptera from Bangala. *Iris,* **4** (1891) 53–60
Descriptions of new species of butterflies from the island of St. Thomas, West Africa. *Zool. Soc. Proc.* (1893) 553–8
List of butterflies collected by Captain J. W. Pringle, R.E., on the march from Teita to Uganda, in British East Africa. *Zool. Soc. Proc.* (1894) 334–53
Descriptions of three new species of Lepidoptera from East Africa. *Ann. Mag. Nat. Hist.,* **17** (1896) 125–7
Descriptions of two new species of Lepidoptera collected by Dr. W. J. Ansorge in East Africa. *Ann. Mag. Nat. Hist.,* **18** (1896) 158–9
List of Lepidoptera collected in Somali-land by Mrs. E. Lort Phillips. *Zool. Soc. Proc.* (1896) 523–9
List of Lepidoptera obtained by Dr. A. Donaldson Smith during

his recent expedition to Lake Rudolf. *Zool. Soc. Proc.* (1896) 530–7
A list of the lepidopterous insects collected on the Red Sea, in the neighbourhood of Suakim, by Mr. Alfred J. Cholmeley. *Zool. Soc. Proc.* (1897) 775–7
Descriptions of some new species of Acraeidae collected by Mr. F. J. Jackson at Ntebi, Uganda. *Ann. Mag. Nat. Hist.*, **19** (1897) 581–2
On a collection of lepidopterous insects from San Domingo. [With field-notes by the collector, Dr. Cuthbert Christy.] *Zool. Soc. Proc.* (1898) 362–9
A list of the lepidopterous insects collected by Mrs. Lort Phillips in Somaliland. *Zool. Soc. Proc.* (1898) 369–72
Descriptions of two new butterflies collected by Major E. M. Woodward in Nandi, Equatorial Africa. *Ann. Mag. Nat. Hist.*, **3** (1899) 243–4
Descriptions of two new moths collected by Dr. Christy on the Upper Niger. *Ann. Mag. Nat. Hist.*, **3** (1899) 371–3
On a collection of butterflies from the Bahamas. *Zool. Soc. Proc.* (1900) 197–203

Smee, Elizabeth Mary
Smee, (Miss) Elizabeth Mary
{Habits of the larva of Phryganea. *Zool. Soc. Proc.* (1863) 78–80}

Sotheby, R. M.
Sotheby, (Miss) R. M.
Macrolepidoptera of Eastbourne and the neighbourhood. [1882] *Eastbourne Nat. Hist. Soc. Trans.*, **1** Pt. 3(4?) (1883) 13–18

Thomas, M. K.
Thomas, (Mrs.) M. K.
Descriptions of three new species of the genus Iletica (Cantharidae) in the collection of the British Museum. *Ann. Mag. Nat. Hist.*, **12** (1893) 138–40
Note on Deridea, *Westwood* (Lyttidae), with the description of a new species. *Ann. Mag. Nat. Hist.*, **19** (1897) 389–90
Descriptions of five new species of Mylabrinae (Lyttidae) in the collection of the British Museum. *Ann. Mag. Nat. Hist.*, **19** (1897) 501–3
Descriptions of two new species of Mylabrinae collected during Capt. Bottego's last expedition. *Genova Mus. Civ. Ann.*, **39** (1898) 555–6

Thomas, Rose Haig
Thomas, Rose Haig
Protective mimicry. *Nature*, **46** (1892) 612
Migration of a water-beetle. *Nature*, **52** (1895) 223
A luminous centipede. [1895] *Nature*, **53** (1895–1896) 131

Veley, Lilian Jane Gould
Gould, Lilian J.
Experiments in 1890 and 1891 on the colour-relation between certain lepidopterous larvae and their surroundings, together with some other observations on lepidopterous larvae. *Ent. Soc. Trans.* (1892) 215–46

Williams, Juliet N.
Williams, Juliet N.
Carnivorous caterpillars. *Nature*, **46** (1892) 128

Wollaston, Edith Shepherd
Wollaston, Mrs. Thomas Vernon
Notes on the Lepidoptera of St. Helena, with descriptions of new species. *Ann. Mag. Nat. Hist.*, **3** (1879) 219–33; 329–43; 415–41

Medicine

United States

Baldwin, Helen
Baldwin, Helen
An experimental study of oxaluria with special reference to its fermentative origin. [1900] *J. Exper. Med.*, **5** (1900–1901) 27–46

Bloom, Selina
Bloom, Selina
Ueber die Retrochlorioidealblutungen nach Staarextractionen. *Arch. f. Ophthalm.*, **46** (1898) 184–235
Bloom, Selina; Garten, Siegfried
Vergleichende Untersuchung der Sehschärfe des hell- und des dunkeladaptiren Auges. *Pflüger, Arch. Physiol.*, **72** (1898) 372–408

Bryson, Louise Fiske
Bryson, Louise Fiske
Preliminary note on the study of exophthalmic goitre. *N. Y. Med. J.*, **50** (1889) 656–8
The present epidemic of influenza. [With discussion.] *N. Y. Med. J.*, **51** (1890) 120–4; 156–8

Cushier, Elizabeth M.
Cushier, Elizabeth M.
{A case of puerperal albuminuria, with uraemic symptoms, treated by jaborandi. *Med. Rec.*, **15** (1879) 174–6}
{A case of epithelioma of the vulva succeeding a long-standing pruritus; operation; cure. *Med. Rec.*, **16** (1879) 440}
{Epithelioma of the leg. *N. Y. Med. J.*, **31** (1880) 635–7}
{Epithelioma of the leg; amputation. *Med. Rec.*, **17** (1880) 436–8}
A case of myxoedema. *Arch. Med.*, **8** (1882) 203–18
{Removal (by Dr. Thomas) of a submucus fibroid from the cavity of the uterus with contracted os uteri. *N. Y. Med. J.*, **36** (1882) 472–5}
{Cyst of the broad ligament twisted upon its pedicle, simulating enlargement of the uterus. *Philad. Med. Times*, **14** (1883–4) 588}
{Cyst of the parovarium simulating pregnancy. *Med. Rec.*, **25** (1884) 501}
{Cystic ovaries; Battey's operation. *Med. Rec.*, **25** (1884) 705}
{A parovarian cyst with twisted pedicle, attended with persistent uterine haemorrhage. *N. Y. Med. J.*, **40** (1884) 181–3}
{Dermoid cyst of the ovary. *Med. Rec.*, **28** (1885) 498}
{Prolapse of the uterus; hysterroraphy. *Int. J. Surg.*, **2** (1889) 136}
{Curetting the uterus for retained placenta. *Int. J. Surg.*, **2** (1889) 214}
{Fibro-sarcomata of the uterus and liver. *Proc. N. Y. Path. Soc.*, (1889), (1890) 2–5}
{Crural hernia in the foetus. *Med. Rec.*, **41** (1892) 471; *Proc. N. Y. Path. Soc.*, 1891, (1892) 69}
{Venous thrombosis as a cause of sudden death in the puerperium. *Amer. Med.-Surg. Bull.*, **8** (1895) 1519–21}

Davenport, Gertrude Crotty
Crotty, Gertrude
{Some statistics relating to the health of college women. *Kan. Ac. Sci. Trans.*, **13** (1893) 33–7}

Davis, Josephine Griffith
Griffith-Davis, J.
Notes on methyl violet: its uses in malignant diseases. *N. Y. Med. J.*, **64** (1896) 482–5

The determination of sex at will. [1899] *N. Y. Med. J.*, **71** (1900) 256–8

Day, Mary Hannah Gage

Day, Mary Gage

{On some unpromising gynecological cases. *Amer. J. Surg. Gyn.*, **11** (1899) 141–2}

Hinds, Clara Bliss

Hinds, Clara Bliss

Child growth. *Amer. Natlist.*, **20** (1886) 745–8

Jacobi, Mary Cortina Putnam

Putnam, Mary C.

Note on a case of human nosencephalian monster who lived 29 hours. *Arch. Sci. Pract. Med.*, **1** (1873) 342–50; 446–52

Jacobi, Mary Putnam

{Provisional report of the effect of quinine upon cerebral circulation. *Arch. Med.*, **1** (1879) 33–43}

{Aphasis due to thrombosis of small twigs supplying the speech region. *Med. Rec.*, **15** (1879) 425}

{Contributions to sphygmography. *Arch. Med.*, **2** (1879) 51–62}

{Note on the cause of sudden death during the operation of thoracentesis. *Med. Rec.*, **16** (1879) 139, 331}

{Fatty degeneration of the placenta. *Med. Rec.*, **16** (1879) 162}

{Generalized tuberculosis. *Med. Rec.*, **16** (1879) 570}

{Case of tubercular meningitis, with measurements of cranial temperatures. *J. Nerv. Ment. Dis.*, n.s. **5** (1880) 51–6}

{Cystic calculi; spasmodic action of the gall-bladder. *Med. Rec.*, **17** (1880) 238}

{Suppurative fibrinous pleuritis supervening upon bronchopneumonia in a child. *Med. Rec.*, **17** (1880) 271}

{Case of uterine fibroid treated with ergotine injections and finally removed by means of Thomas's scoop. *Amer. J. Med. Sci.*, n.s. **79** (1880) 422–6}

{Case of nocturnal rotary spasm. *J. Nerv. Ment. Dis.*, n.s. **5** (1880) 390–402}

{Théorie de la menstruation. [Translated by R. Fauquez, from "The question of rest for women during menstruation."] *Rev. méd-chir. d. mal. d. femmes*, **2** (1880) 411, 459, 512, 572–82}

{Case of facial and palatine paralysis, and loss of equilibrium, produced by a fall on the head. *Indep. Pract.*, **2** (1881) 69–74}

{Scarlatinous nephritis. *Med. Rec.*, **19** (1881) 353–5}

{A case of microcephalus. *Med. Rec.*, **19** (1881) 645–50}

{Inaugural address [on medical education] at the opening of the Women's Medical College of the New York Infirmary, October 1, 1880. *Chicago M. J. Exam.*, **42** (1881) 561–85}

{The prophylaxis of insanity. *Arch Med.*, **6** (1881) 120–35}

{Case of trephining of sternum for osteomyelitis. *Amer. J. Obst.*, **14** (1881) 981–92}

{Specialism in medicine [edit.]. *Arch. Med.*, **7** (1881) 87–97}

{Shall women practice medicine? *N. Amer. Rev.*, **134** (1882) 52–75}

{Hysterical locomotor ataxia. *Arch. Med.*, **9** (1883) 88–93}

{Opening lecture on diseases of children, at the Post-Graduate Medical School, New York. *Boston Med. Surg. J.*, **108** (1883) 121, 145}

{Spina bifida complicated by hydrocephalus. *Med. Rec.*, **23** (1883) 358}

{An address delivered at the commencement of the Woman's Medical College of the New York Infirmary, May 30, 1883. *Arch. Med.*, **10** (1883) 59–71}

{Two cases of aspiration of dermoid cysts followed by inflammation. *Amer. J. Obst.*, **16** (1883) 1160–70}

{Two peculiar cases of typhoid fever; one at the age of six months; one beginning with pneumonia, and with heart failure conspicuous; effect of digitalis. *Arch. Med.*, **11** (1884) 30–72}.

{Some considerations on endometritis. *Boston Med. Surg. J.*, **110** (1884) 468–70}

Studies in endometritis. *N. Y. Med. J.*, **39** (1884) 557–558; *Amer. J. Obst.*, **18** (1885) {36, 113} 262, 376, 519, 596, 802, 915

{Menstrual subinvolution or metritis of the non-parturient uterus. (Studies in endometritis.) *Amer. J. Obst.*, **18** (1885) 915–32}

{Case of aortic stenosis followed by left hemiplegia in a child of six. *Arch. Pediat.*, **2** (1885) 710–19}

{Pseudo-hypertropic paralysis. *Syst. Pract. Med.* **4** (1886) 557–80}

{Pulmonary consolidation; enlarged bronchial glands; pressure upon one of the recurrent laryngeal nerves; heart clot. *Med. Rec.*, **29** (1886) 401}

{The ovarian complication of endometritis. (Studies in endometritis.) *Amer. J. Obst.*, **19** (1886) 352–86}

{Profuse epistaxis; dilation and fatty degeneration of the heart. *N. Y. Med. J.*, **43** (1886) 669}

{Dilation and fatty degeneration of the heart; blood changes of pernicious malarial fever. *Med. Rec.*, **30** (1886) 106}

{Infantile and spinal paralysis. *Syst. Pract. Med.*, **5** (1886) 1113}

{Some considerations on hysteria. *Med. Rec.*, **30** (1886) 365, 329, 396}

{The indication of quinine in pneumonia. *N. Y. Med. J.*, **45** (1887) 589–93, 620–4}

{Dermoid cyst by inclusion. *Proc. N. Y. Path. Soc.*, 1887, (1888) 137}

{Notes on uterine versions and flections. *Amer. J. Obst.*, **21** (1888) 225–38}

{Case of uterine fibroid treated by Apostoli's method; enucleation of the tumor. *Amer. J. Obst.*, **21** (1888) 806–15}

{Tumor of the pons. *Med. News*, **53** (1888) 708}

Remarks on occulus indicus. *N. Y. Med. J.*, **48** (1888) 29–35

{Case of the post-epileptic hysteria; effect of inhalation of compressed air; phenomena of transfer. *J. Nerv. Ment. Dis.*, **15** (1888) 442–5}

{Case of probable tumor of the pons. *J. Nerv. Ment. Dis.*, **16** (1889) 115–29}

{Acute mania after operations. *Med. Rec.*, **35** (1889) 446–8}

{Case of cirrhosis of liver with splenic tumor. *Arch. Pediat.*, **6** (1889) 273–82}

{The practical study of biology. *Boston Med. Surg. J.*, **120** (1889) 631}

{Intra-uterine therapeutics. *Amer. J. Obst.*, **22** (1889) 449, 598, 697}

{Stomach washing. *Rep. Proc. Alum. Assoc. Woman's Med. Coll. Penn.* (1889) 50–3}

{The use of electricity in gynaecology. *Rep. Proc. Alum. Assoc. Woman's Med. Coll. Penn.* (1889) 60–73}

{The higher education of women. *Med. News*, **56** (1890) 75–7; *Science*, **18** (1891) 295}

{Case of typhoid fever treated by cold baths. *Times and Reg.*, **7** (1890) 34–6}

{Remarks upon empyema. *Med. News*, **56** (1890) 117, 136, 170}

{Female physicians for insane women. *Med. Rec.*, **37** (1890) 543}

{Hysterical fever. *Rep. Proc. Alum. Assoc. Woman's Med. Coll. Penn.*, **15** (1890) 87–95}

Functional disturbances of heart and pulse. *N. Y. Med. J.*, **54** (1891) 725

{A case of infantile pneumonia traversing three dangerous crises. *Arch. Pediat.*, **9** (1892) 118–28}

{A case of myelitis with vertebral tumor. *Int. Med. Mag.*, **1** (1892) 449–59}
{Urethral irritation. *Coll. and Clin. Rec.*, **13** (1892) 291–5; *Amer. Lancet*, n.s. **17** (1892) 1–4}
{Permanent drainage in the treatment of endometritis. *Woman's Med. J.*, **1** (1893) 1–3}
{Diseased placenta. *Med. Rec.*, **48** (1895) 711; *Woman's Med. J.*, **4** (1895) 339–41}
{The Stephenson wave. *Amer. J. Obst.*, **32** (1895) 90–2}
{Case of absent uterus; with considerations on the significance of hermaphroditism. *Amer. J. Obst.*, **32** (1895) 510–42}
{Ataxia in a child. *Med. Rec.*, **51** (1897) 761–5}
{Considerations on Flechsig's Gehirn und Seele. *J. Nerv. Ment. Dis.*, **24** (1897) 747–68 [discussion, 778–82]}
{A suggestion in regard to suggestive therapeutics. *N. Y. Med. J.*, **67** (1898) 485–9}
{The blood count in anemia and certain nervous affections. *Med. Rec.*, **53** (1898) 933–5}
{Varieties of nephritis. *Woman's Med. J.*, **8** (1898) 191–6}

Jacobi, Mary Putnam; White, Victoria A.
On the use of the cold pack followed by massage in the treatment of aneamia. *Arch. Med.*, **3** (1880) 296–323; **4** (1880) 51–72; 163–90

Jacobi, Mary Putnam (*et alii*)
A discussion on diphtheria. [1884] *N. Y. Med. J.*, **41** (1885) 49–53

Jacobi, Mary Putnam; Kydd, Mary M.
{Experiments on urinary toxicity. *Med. Rec.*, **52** (1897) 653–9}

Jones, Mary A. Dixon

Jones, (Mrs.) Mary A. Dixon
Another hitherto undescribed disease of the ovaries. Anomalous menstrual bodies. *N. Y. Med. J.*, **51** (1890) 511–16; 542–7
Carcinoma on the floor of the pelvis. Two discoveries in cancerous disease. *Amer. Micr. Soc. Trans.*, **20** (1899) 165–76

Kydd, Mary Mitchell

Jacobi, Mary Putnam; Kydd, Mary M.
{Experiments on urinary toxicity. *Med. Rec.*, **52** (1897) 653–9}

McNutt, Sarah J.

McNutt, S. J. (*et alii*)
Intra-cranial haemorrhage in children. *N. Y. Med. J.*, **41** (1885) 104–6; 135–9

Mergler, Marie Josepha

Mergler, M. J.
The pathology of leukaemia. [*With discussion.*] *N. Y. Med. J.*, **46** (1887) 693–4
{Report of two cases of abdominal section. *Chicago Med. Rec.*, **3** (1892) 481–4 [discussion, 497]}
{A case of acquired sterility. *Chicago Clin. Rev.*, **1** (1892–1893) 23–6}
{What are the indications for the removal of uterine appendages. *Med. Surg. Rep.*, **69** (1893) 117–20}
{Fibroid tumors of the uterus; the complications. *Trans. Illinois Med. Soc.*, **44** (1894) 429–39}
{Report of cases of abdominal section. *Chicago Med. Rec.*, **10** (1896) 402–6}
{Report of two cases of extensive skin-grafting. *Woman's Med. J.*, **6** (1897) 77–80}

Moody, Kate Cameron

Moody, Kate Cameron
Reflex pain. *Brit. J. Dental Sci.*, **28** (1885) 788–94

Moody, Mary Blair

Moody, Mary B.
{A case of hysterectomy with peculiar features. *Med. Press West. N. Y.*, **3** (1888) 54–7}

Mosher, Eliza M.

Mosher, Eliza M.
{A case of purulent inflammation of the inner ear. *Amer. J. Otol.*, **2** (1880) 110–13}
{Health of criminal women. *Boston Med. Surg. J.*, **107** (1882) 316}
Pulmonary gangrene. *N. Y. Med. J.*, **42** (1885) 233–5
{Utilization of the outgoing air in the replacement of the uterus by the knee-chest position. *Amer. J. Obst.*, **20** (1887) 1027}
{A critical study of the biceps cruris muscle as it relates to disease in and around the knee joint. *Ann. Surg.*, **14** (1891) 356–69}
{The influence of habitual posture on the symmetry and health of the body. *Brooklyn Med. J.*, **6** (1892) 393–414}
{Habits of posture a cause of deformity and displacement of the uterus. *N. Y. J. Gyn. Obst.*, **3** (1893) 962–77}
{A posture model. *Brooklyn Med. J.*, **9** (1895) 30–4}
{Habitual postures of school-children. *Ann. Hyg.*, **10** (1895) 1–10}
{Flat chest; produced by habits of posture; its prevention and correction. *Brooklyn Med. J.*, **10** (1896) 355–63}

Niles, Mary W.

Niles, Mary W.
Bubonic plague in Canton. *N. Y. Med. J.*, **60** (1894) 467–8

Peckham, Grace

Peckham, Grace
Rhythmical myoclonus. *Arch. Med.*, **9** (1883) 97–117
Metallo-therapy, theoretically and practically considered. *Arch. Med.*, **10** (1883) 155–75; 283–301
Mirror-writing and other pathological chirography of nervous origin. *N. Y. Med. J.*, **43** (1886) 104–5
{Tumors of the clitoris. *Amer. J. Obst.*, **24** (1891) 1153–72}

Post, Sarah E.

Post, S. E.
A pseudo negative trace. *Arch. Med.*, **11** (1884) 1–11
Iodoform in diabetes. A report of two cases showing coincident diminution of sugar and of urea together with an increase in weight; also and more especially showing that toxic doses of the drug are unnecessary and, further, that they are prejudicial to its therapeutic action. *Arch. Med.*, **11** (1884) 116–44

Robinovitch, Louise G.

Robinovitch, Louise G.
On the reduction of fever, particularly in typhoid. The comparative value of antipyretics and the cold-water treatment. *N. Y. Med. J.*, **55** (1892) 320–2
{On fever and its reduction; bleeding in some cases of meningitis. *N. Y. Med. J.*, **63** (1896) 353}
{On morphinism. *N. Y. Med. J.*, **69** (1899) 298–302}

Robinson, Ethel Blackwell

Blackwell, Ethel
Tests on women of the red coloring matter of blood. *N. Y. Med. J.*, **56** (1892) 570–3
Tests on the haemoglobin of pregnant women. *N. Y. Med. J.*, **61** (1895) 776–8

Sargent, Elizabeth R. C.

Sargent, Elizabeth
Cocaine in glaucoma. *Arch. Ophthalm.*, **16** (1887) 205–6

Sherwood, Mary
Sherwood, Mary
Polyneuritis recurrens. *Virchow, Arch.,* **123** (1891) 166–82

White, Victoria A.
Jacobi, Mary Putnam; White, Victoria A.
On the use of the cold pack followed by massage in the treatment of anemia. *Arch. Med.,* **3** (1880) 296–323; **4** (1880) 51–72; 163–90

Willard, Emma Hart
Willard, Emma
Motive power of the blood. *Boston, Med. Surg. J.,* **46** (1852) 336–41

United States—Great Britain

Latham, Vida Annette
Latham, Vida A.
{A brief study of a case of elephantiasis and its history. *Amer. Soc. Micr. Proc.,* **14** (1892) 133–40}
{The value of differential diagnosis in dentistry. *Amer. Med. Assoc. J.,* **26** (1896) 722–5}
{Report on a case of leukaemia. *Chicago Path. Soc. Trans.,* **1** (1896) 38–40}
{Dental septicemia of the antrum: two cases with obscure symptoms. *Amer. Med. Assoc. J.,* **32** (1899) 121}

Great Britain

Corthorn, Alice Mary
Corthorn, Alice M.
Plague in monkeys and squirrels. *Indian Med. Gaz.,* **34** (1899) 81

Fowke, Fanny
Fowke, (Miss) Fanny
The physical condition of pauper children boarded out under the Local Government Orders, 1870 and amended, 1889. [With discussion] *Int. Cong. Hyg. Trans.,* **10** (1891) 329–36

Frankland, Grace C. Toynbee
Frankland, (Mrs.) Grace C.
The toxicity of eel serum, and further studies on immunity. *Nature,* **58** (1898) 369–71

Kingsley, Mary Henrietta
Kingsley, Mary H.
{Nursing in West Africa. *Chamber's J.* (June 1900) 369–96}

Nightingale, Florence
Nightingale, (Miss) Florence
Village sanitation in India. *Congr. Int. Hyg. C. R.,* **2** (1894) 580–3

Owen, Sara S.
Owen, Sara S.
Unconscious bias in walking. *Nature,* **29** (1884) 336–7

Sharpe, Margaret Mary
Sharpe, Margaret M.
The X-ray treatment of skin diseases. [1899] *Fortschr. Röntgenstr.,* **3** (1899–1900) 197

Shove, Edith
Rémy, C.; Shove, (Miss) E.
Expériences à propos des lésions du pancréas chez les diabétiques. *Paris, Soc. Biol. Mém.,* **34** (C.R.) (1882) 598–603

Sime, Jessie A.
Sime, Jessie A.
The worsted test for colour vision. *Nature,* **56** (1897) 516

Wood, Catherine Jane
Wood, Catherine Jane
The progress of nursing during the Victorian era. *Practitioner,* **58** (1897) 709–16

Microscopy

United States

Bishop, Frances Lewis
Bishop, Fanny Lewis
Kaufmann's method for the staining of tubercle bacilli. *N. Y. Med. J.,* **57** (1893) 458

Claypole, Agnes Mary
Claypole, (Miss) Agnes Mary
A new method for securing paraffin sections to the slide or coverglass. *Amer. Micr. Soc. Proc.,* **16** (1894) 65–7

Davenport, Gertrude Crotty
Crotty, Gertrude
Methods of collecting, cleaning, and mounting diatoms. *Kan. Ac. Sci. Trans.,* **12** (1890) 81–3

Detmers, Frederica
Detmers, (Miss) Freda
The comparative size of blood corpuscles of man and domestic animals. *Amer. Soc. Micr. Proc.* (1887) 216–32

Gage, Susanna Stuart Phelps
Gage, (Mrs.) Susanna S. Phelps
Staining and permanent preservation of histological elements isolated by means of caustic potash (KOH) or nitric acid (HNO_3). *Amer. Soc. Micr. Proc.,* **11** (1899) 34–45

United States—Great Britain

Latham, Vida Annette
Latham, (Miss) Vida A.
Mounting mosses. *Micr. Soc. J.* (1887) 843–4
A decayed nut. *Amer. Micr. J.,* **9** (1888) 173–4
{Histology of the teeth; notes on methods of preparation. *J. Micr. Nat. Sci.,* n.s. **2** (1889) 137–52}
Short notes in practical biology. Amoeba. *Amer. Micr. J.,* **10** (1889) 151–5
Alcoholic method of mounting Bryozoa. *Micr. Soc. J.* (1890) 681
The use of stains, especially with reference to their value for differential diagnosis. *Amer. Soc. Micr. Proc.,* **13** (1891) 94–109
A brief account of the microscopical anatomy in a case of chrome lead poisoning. *Amer. Soc. Micr. Proc.,* **13** (1891) 110–15
Balsam mounting. *Micr. Soc. J.* (1892) 159–61
A plea for the study of re-agents in micro-work. *Amer. Micr. Soc. Proc.,* **15** (1893) 209–11
{Preparing sections of teeth for histology and bacteriology. *J. Micr.,* s. 3, **2** (1893) 241; s. 3, **3** (1893) 25}
The question of correct naming and use of micro-reagents. *Amer. Micr. Soc. Trans.,* **17** (1895) 350–8
What is the best method of teaching microscopical science in medical schools? *Amer. Micr. Soc. Trans.,* **18** (1896) 311–20
[Reaction of normal and diabetic blood to methylen-blue.] *Micr. Soc. J.* (1899) 352
The reaction of diabetic blood to some of the anilin dyes. *Amer. Micr. Soc. Trans.,* **21** (1900) 31–40

Great Britain

Hart, Mrs. Ernest
Hart, Mrs. Ernest
On the micrometric numeration of the blood-corpuscles and the estimation of their haemoglobin. *Quart. J. Micr. Sci.,* **21** (1881) 132–45
Note on the formation of fibrine. *Quart. J. Micr. Sci.,* **22** (1882) 255–9

Hoggan, Frances Elizabeth Morgan
Hoggan, Frances Elizabeth
On a new process of histological staining. [1876] *Quekett Micr. Club J.,* **4** (1874–1877) 180–1

Natural History

United States

Kellerman, Stella Victoria Dennis
Kellerman, Mrs. W. A.
Snake story. *Science,* **21** (1893) 36–7

Moody, Mary Blair
Moody, Mary B.
Singing of birds. *Science,* **21** (1893) 264

Wright, Julia McNair
Wright, Julia McNair
Fiddler-crabs. *Amer. Natlist.,* **21** (1887) 415–18
The food of moles. *Science,* **17** (1891) 121
Traumatic hypnotism. *Science,* **19** (1892) 66
Notes on the food of the box tortoise. *Science,* **19** (1892) 95
Hypnotism among the lower animals. *Science,* **19** (1892) 95–6

Great Britain

Brown, (Mrs.)
Brown, (Mrs.)
The wild animals of Palestine. *Dumfr. Gallow. Soc. Trans.,* **14** (1898) 68–72

Carphin, Janet
Carphin, Janet
Argulus foliaceus in the Edinburgh district. *Ann. Scott. Nat. Hist.* (1895) 255

Crane, Agnes
Crane, (Miss) Agnes
The origin of birds. *Brighton Nat. Hist. Soc. Rep.* (1894) 26–32
An arrangement in horns and hoofs. *Brighton Nat. Hist. Soc. Rep.* (1896) 16–19

Gough, Frances
Gough, (Mrs.) Frances
Sea anemones. [1887] *Penzance Soc. Trans.,* **2** (1884–1888) 275–81

Harvey, Mrs. M.
Harvey, Mrs. M.
Some account of a particular variety of bull (Bos taurus), now exhibiting in London. *Mag. Nat. Hist.,* **1** (1828) 113–14

Hatch, E. D. W.
Hatch, (Mrs.) E. D. W.
The "trading-rat." *J. Sci.,* **22** (1885) 656–60

Lee, Sarah Wallis Bowdich
Bowdich, (Mrs.) S.
{Anecdotes of a tamed panther. *Mag. Nat. Hist.,* **1** (1828) 108–12}
{Anecdotes of a Diana monkey. *Mag. Nat. Hist.,* **2** (1829) 9–13}
{Attachments formed by animals. *Mag. Nat. Hist.,* **2** (1829) 62}
{Some account of the progress of natural history, during the year 1828, as reported to the Academy of Sciences at Paris by the Baron Cuvier. *Mag. Nat. Hist.,* **2** (1829) 409–28}
Lee, Mrs. R. (formerly Mrs. Bowdich)
{Some details respecting the Garden of Plants and the National Museum at Paris. *Mag. Nat. Hist.,* **3** (1830) 22–6}
{Two poodles from Milan. [Foreign notices. France.] *Mag. Nat. Hist.,* **3** (1830) 290–1}

Martin, (Miss)
Martin, (Miss)
A month in the Canaries. [1892] *Holmesdale Nat. Hist. Club Proc.* (1890–1892) 88–96

Power, Jeannette de Villepreux
Power, Jeannette
Observations on the habits of the common Marten. (Martes foina.) *Ann. Nat. Hist.,* **20** (1857) 416–22

Sprague, Miss
Sprague, (Miss)
Microscopic life. [1897] *Edinb. Nat. Soc. Trans.,* **3** (1898) 304–12

Thynne, Anna
Thynne, (Mrs.)
On the increase of Madrepores; with notes by P. H. Gosse. *Ann. Nat. Hist.,* **3** (1859) 449–61

W., E.
W., (Miss) E.
Quelques observations de Miss E. W. sur les animaux mollusques. Extraites d'une lettre à M. Defrance, Avril 1822. *J. de Phys.,* **95** (1822) 387–92

Yate, Mary Theodosia McCausland
Yate, M. T. (Mrs. A. C.)
Polecats as pets. [1898] *Bombay Nat. Hist. Soc. J.,* **11** (1897–1888) 737–39

Ornithology

United States

Bailey, Florence Augusta Merriam
Merriam, Florence A.
Was he a philanthropist? *Auk,* **7** (1890) 404–7
Interesting nesting site of a winter wren (Troglodytes hiemalis). *Auk,* **7** (1890) 407
Nesting habits of Phainopepla nitens in California. *Auk,* **13** (1896) 38–43
Notes on some of the birds of southern California. *Auk,* **13** (1896) 115–24

Ball, Helen A.
Ball, Helen A.
Peculiar nest of a chipping sparrow. *Auk,* **12** (1895) 305
{Pinicola enucleator at Worcester, Massachusetts. *Auk,* **13** (1896) 259}
{A mockingbird at Worcester, Massachusetts. *Auk,* **14** (1897) 324}

Bates, Abby Frances Caldwell
Bates, Abby F. C.
A swallow roost at Waterville, Maine. *Auk,* **12** (1895) 48–57

Berry, Mabel C.
Berry, Mabel C.
First occurrence of the blue grosbeak in New Hampshire. *Auk,* **13** (1896) 342–3

Boyce, Caroline
Boyce, Caroline
The robin [Turdus migratorius]. *Amer. Natlist.,* **8** (1874) 203–8

Bruce, Mary Emily
Bruce, Mary Emily
A month with the goldfinches. *Auk,* **15** (1898) 239–43

Eckstorm, Fannie Pearson Hardy
Hardy, Fannie P.
Testimony of some early voyagers on the great auk. *Auk,* **5** (1888) 380–4

Foote, F. Huberta
Foote, F. Huberta
The Carolina wren (Thryothorus ludovicianus) at Inwood-on-Hudson, New York City. *Auk,* **14** (1897) 224
A provident nuthatch. *Auk,* **16** (1899) 283

Furness, Caroline Ellen
Furness, Caroline E.
The pine grosbeak at Poughkeepsie, N. Y. *Auk,* **13** (1896) 175

Hine, Jane L.
Hine, Jane L.
Observations on the ruby-throated hummingbird. *Auk,* **11** (1894) 253–4

Hyatt, Mary
Hyatt, Mary
Bird-music in August. *Science,* **21** (1893) 4–5
Familiarity of certain wood birds. *Science,* **23** (1894) 58

Lewis, Graceanna
Lewis, (Miss) Grace Anna
{On the plumage of terns. *Amer. Ass. Proc.* (1869) 280 [title only]}
Remarks on the fluids contained in the bulbs of feathers of living birds. [1869] *Essex Inst. Bull.,* **1** (1870) 126–7
Note on the Lyre bird. *Amer. Natlist.,* **4** (1871) 321–31
Symmetrical figures in birds' feathers. *Amer. Natlist.,* **5** (1871) 675–8
On the genus Hyliota. *Philad. Ac. Nat. Sci. Proc.* (1883) 128–30

Miller, Mary Mann
Miller, Mary Mann
Birds feeding on hairy caterpillars. *Auk,* **16** (1899) 362

Miller, Olive Thorne
Miller, Olive Thorne
The eccentricities of a pair of robins. *Science,* **22** (1893) 188–9
{Captive wild birds. *Auk,* **14** (1897) 251}

Millington, Lucy Ann Bishop
Millington, L. A.
An albino humming-bird. *Zoologist,* **3** (1868) 1343; *Amer. Natlist.,* **2** (1869) 110

Treat, Mary Allen Davis
Treat, (Mrs.) Mary
{Our mocking bird. *Forest and Stream,* **8** (1877) 112–13}
The great crested flycatcher. *Amer. Natlist.,* **15** (1881) 601–4
{Winter birds in the pines. *Garden & Forest,* **6** (1893) 39}
Autumn birds in the pines. *Garden & Forest,* **9** (1896) 452–3

Tyler, Martha G.
Tyler, Martha G.
The barn owl (Strix pratinocola) in northern Vermont. *Auk,* **11** (1894) 253

Watson, Amelia Montague
Watson, Amelia M.
Taming a chipping sparrow (Spizella socialis). *Auk,* **11** (1894) 256–7

Webster, Ellen E.
Webster, Ellen E.
Ring-billed gull in New Hampshire. *Auk,* **17** (1900) 169
White-winged crossbills and Brünnich's murres in central New Hampshire. *Auk,* **17** (1900) 175–6

Whipple, Jennie May
Whipple, (Miss) J. M.
Hawking. *Ornith. Ool.,* **10** (1885) 89

Woodworth, Nelly Hart
Woodworth, Nelly Hart
Bird notes from St. Albans, Vermont. *Auk,* **12** (1895) 311–12

Wright, Julia McNair
Wright, Julia McNair
The shrike. *Science,* **17** (1891) 217
Mocking-birds and their young. *Science,* **17** (1891) 361
Red birds and a grosbeak. *Science,* **22** (1893) 134

Great Britain

Barnard, (Mrs.)
Barnard, (Mrs.)
Migration of birds. [1879] *Rugby, Nat. Hist. Soc. Rep.* (1880) 39–42

Greet, C. H.
Greet, (Miss) C. H.
Curious nesting place of the great tit. *Berwick. Nat. Club Hist.,* [**11**] (1887) 245

Hubbard, Emma
Hubbard, Emma
Late appearance of the cuckoo. *Nature,* **50** (1894) 338

Ley, Annie
Ley, Annie
The habits of the cuckoo. [1896] *Nature,* **53** (1895–1896) 223

Saxby, Jessie Margaret Edmonston
Saxby, Jessie M. E.
The food of the great skua (Stercorarius catarrhactes, *L.*) *Ann. Scott. Nat. Hist.* (1892) 201–2
Gossip about gulls. [1896] *Edinb. Nat. Soc. Trans.,* **3** (1898) 158–68

Traill, Adelaide L.
Traill, Adelaide L.
Fulmar petrel (Fulmarus glacialis) breeding in Papa Stour, Shetland. *Ann. Scott. Nat. Hist.* (1893) 184
Wryneck at the island of Foula, Shetland. *Ann. Scott. Nat. Hist.* (1898) 182

Warrender, Eleanor
Warrender, (Miss)
Finches mobbing a hawk. *Berwick. Nat. Club Hist.,* **14** (1894) 400

Physiology, Neurology, Anatomy, and Pathology

United States

Ballantyne, Bertha L.

Hough, Theodore; Ballantyne, Bertha L.

Preliminary note on the effects of changes in external temperature on the circulation of blood in the skin. *Boston Soc. Med. Sci. J.,* **3** (1899) 330–4

Barker, Alice Hopkins Albro

Chittenden, R. H.; Albro, (Miss) Alice H.

The influence of bile and bile salts on pancreatic proteolysis. *Amer. J. Physiol.,* **1** (1898) 307–35

Bunting, Martha

Bunting, Martha

Ueber die Bedeutung der Otolithenorgane für die geotropischen Functionen von Astacus fluviatilis. *Pflüger, Arch. Physiol.,* **54** (1893) 531–7

On the origin of the sex-cells in Hydractinia and Podocoryne; and the development of Hydractinia. *J. Morphol.,* **9** (1894) 203–36

Cone, Claribel

Cone, Claribel

On a polymorphous cerebral tumor (alveolar glioma ?) containing tubercules and tubercule bacilli. *N. Y. Med. J.,* **69** (1899) 331–6; 361–6; 403–5

Cooke, Elizabeth W.

Cooke, (Miss) Elizabeth

Experiments upon the osmotic properties of the living frog's muscle. [1898] *J. Physiol.,* **23** (1898–1899) 137–49

Howell, William Henry; Cooke, (Miss) Elizabeth

Action of the inorganic salts of serum, milk, gastric juice, etc., upon the isolated working heart, with remarks upon the causation of the heart-beat. *J. Physiol.,* **14** (1893) 198–220

Day, Mary Hannah Gage

Day, Mary Gage

Experimental demonstration of the toxicity of the "loco-weed" (Astragalus mollissimus and Oxytropis lamberti). *N. Y. Med. J.,* **49** (1889) 237–8

DeWitt, Lydia Maria Adams

Huber, G. Carl; DeWitt, Mrs.

Endings of sensory and motor nerves in the "muscle spindles" of voluntary muscle with demonstration of preparations. *Science,* **5** (1897) 908–9

The innervation of motor tissues, with especial reference to nerve-endings in the sensory muscle-spindles. *Brit. Ass. Rep.* (1897) 810–11

Eubank, Marion D.

Hall, Winfield S.; Eubank, Marion D.

The regeneration of blood. *J. Exper. Med.,* **1** (1896) 656–76

Gage, Susanna Stuart Phelps

Gage, (Mrs.) Susanna S. Phelps

Ending and relation of the muscular fibres in the muscles of minute animals (mouse, mole, bat, and English sparrow). *Amer. Soc. Micr. Proc.* (1887) 207–8

{Form, endings, and relation of striated muscular fibers in the muscles of minute animals (mouse, shrew, bat and English sparrow). [Parts 1 and 2.] *Microscope,* **8** (1888) 225–37; 257–72}

The intramuscular endings of fibers in the skeletal muscles of the domestic and laboratory animals. *Amer. Soc. Micr. Proc.,* **12** (1890) 132–9

A preliminary account of the brain of Diemyctylus viridescens, based upon sections made through the entire head. *Amer. Ass. Proc.* (1892) 197

A reference model. *Amer. Micr. Soc. Proc.,* **14** (1892) 154–5

Comparative morphology of the brain of the soft-shelled turtle (Amyda mutica) and the English sparrow (Passer domesticus). *Amer. Micr. Soc. Trans.,* **17** (1895) 185–238

Modifications of the brain during growth. *Amer. Natlist.,* **30** (1896) 836–7; {*Science,* **4** (1896) 602–3}

The brain of the embryo soft-shelled turtle. *Amer. Micr. Soc. Trans.,* **18** (1896) 282–6

A series of specimens illustrating the development of the chick. [1897] *Science,* **7** (1898) 226–7

Notes on the chick's brain. *Amer. Ass. Proc.* (1899) 256

Gage, Simon Henry; Gage, (Mrs.) Susanna S. Phelps

Aquatic respiration in soft-shelled turtles (Amyda mutica and Aspidonectes spinifer); a contribution to the physiology of respiration in vertebrates. *Amer. Ass. Proc.* (1885) 316–18; *Amer. Natlist.,* **20** (1886) 233–6

{Amoeboid movements of the cell-nucleus in Necturus. *Science,* **7** (1886) 35}

Combined aerial and aquatic respiration. *Science,* **7** (1886) 394–5

Pharyngeal respiratory movements of adult amphibia under water. *Science,* **7** (1886) 395

Changes in the ciliated areas of the alimentary canal of the amphibia during development, and the relation to the mode of respiration. *Amer. Ass. Proc.* (1890) 337–8

Griffith, Mary

Griffith, (Mrs.) Mary

Observations on the vision of the retina. *Froriep, Notizen,* **40** (1834) 177–81; *Phil. Mag.,* **4** (1834) 43–6

Observations on the spectra of the eye and the seat of vision. *Phil. Mag.,* **5** (1834) 192–6; *Poggend. Annal.,* **33** (1834) 477–9

On the halo or fringe which surrounds all bodies. *Silliman, J.,* **38** (1840) 22–32

Hamilton, Alice

Hamilton, Alice

A peculiar form of fibrosarcoma of the brain. *J. Exper. Med.,* **4** (1899) 597–608

On the presence of new elastic fibres in tumors. [1900] *J. Exper. Med.,* **5** (1900–1) 131–8

Thomas, H. M.; Hamilton, Alice

The clinical course and pathological histology of a case of neuroglioma of the brain. *J. Exper. Med.,* **2** (1897) 635–56

Hyde, Ida Henrietta Heidenheimer

Hyde, (Miss) Ida H.

Notes on the hearts of certain mammals. *Amer. Natlist.,* **25** (1891) 861–3

Entwicklungsgeschichte einiger Scyphomedusen. *Ztschr. Wiss. Zool.,* **58** (1894) 531–65

The nervous mechanism of the respiratory movements in Limulus polyphemus. *J. Morphol.,* **9** (1894) 431–48

Beobachtungen über die Secretion der sogenannten Speicheldrüsen von Octopus macropus. *Ztschr. Biol.,* **35** (1897) 459–77

The effect of distension of the ventricle on the flow of blood through the walls of the heart. *Amer. J. Physiol.,* **1** (1898) 215–24

Collateral circulation in the cat after ligation of the postcava. *Kan. Univ. Quarterly,* **9** (1900) 167–71
Ewald, J. Rich.; Hyde, Ida H.
{Zur physiologie des labyrinths. IV. Mittheilung die Beziehungen des Grosshirns zum Tonuslabyrinth. *Pflüger, Arch. Physiol.,* **60** (1895) 492–508}

Katz, Louise
Katz, Louise
Histolysis of muscle in the transforming toad. [Bufo lentiginosus.] *Amer. Ass. Proc.* (1900) 233–4

Kendall, A. Josephine
Kendall, A. Josephine; Luchsinger, Balthasar
Zur Innervation der Gefässe. *Pflüger, Arch. Physiol.,* **13** (1876) 197–212
Zur Theorie der Secretionen. *Pflüger, Arch. Physiol.,* **13** (1876) 212

Latimer, Caroline Wormeley
Latimer, Caroline W.
On the modification of rigor mortis resulting from previous fatigue of the muscle, in cold-blooded animals. [1898] *Amer. J. Physiol.,* **2** (1899) 29–46
Latimer, Caroline W.; Warren, Joseph W.
On the presence of the amylolytic ferment and its zymogen in the salivary glands. *J. Exper. Med.,* **2** (1897) 465–73

Moore, Anne
Moore, Anne
Dinophilus gardineri (*sp. nov.*). [1899] *Woods Holl Mar. Biol. Lab. Bull.,* **1** (1900) 15–18
Further evidence of the poisonous effects of a pure NaCl solution. [1900] *Amer. J. Physiol.,* **4** (1900) 386–96

Pollard, Myra E.
Sewall, Henry; Pollard, Myra E.
On the relations of diaphragmatic and costal respiration, with particular reference to phonation. *J. Physiol.,* **11** (1890) 159–78

Schively, Mary A.
Schively, Mary A.
Ueber die Abhängigkeit der Herzthätigkeit einiger Seethiere von der Concentration des Seewassers. [1893] *Pflüger, Arch. Physiol.,* **55** (1894) 307–18
Structure and development of Spirorbis borealis. *Philad. Ac. Nat. Sci. Proc.* (1897) 153–60

Stone, Ellen Appleton
Stone, Ellen A.
Some observations on the physiological function of the pyloric caeca of Asterias vulgaris. *Amer. Natlist.,* **31** (1897) 1035–41

Towle, Elizabeth Williams
Towle, Elizabeth W.
A study in the heliotropism of Cypridopsis. *Amer. J. Physiol.,* **3** (1900) 345–65

Welch, Jeannette Cora
Welch, Jeannette C.
On the measurement of mental activity through muscular activity and the determination of a constant of attention. *Amer. J. Physiol.,* **1** (1898) 283–306

Whitman, Emily A. Nunn
Nunn, Emily A.
The structural changes in the epidermis of the frog, brought about by poisoning with arsenic and antimony. [1878] *J. Physiol.,* **1** (1878–1879) 247–56

United States—Great Britain

Latham, Vida Annette
Latham, Vida A.
{Disease of the maxillary bones and their periosteum. Amer. Med. Assoc. J., 21 (1893) 994–1001}
{Sarcoma of the antrum. *Ohio Dent. J.,* **1** (1897) 415–17}
{The technic and pathology of the peridental membrane. *Amer. Med. Assoc. J.,* **28** (1897) 69–71}
{Five cases of sarcoma of the head and neck. *Amer. Med. Assoc. J.,* **30** (1898) 1098–1102)}

Great Britain

Alcock, Rachel
Alcock, (Miss) Rachel
The digestive processes of Ammocoetes. [1891] *Cambridge Phil. Soc. Proc.,* **7** (1892) 252–5
The peripheral distribution of the cranial nerves of Ammocoetes. [1898] *J. Anat. Physiol.,* **33** (1899) 131–53
On proteid digestion in Ammocoetes. *J. Anat. Physiol.,* **33** (1899) 612–37

Bidder, Marion Greenwood
Greenwood, (Miss) Marion
On the gastric glands of the pig. [1884] *J. Physiol.,* **5** (1884–1885) 195–208; vii–ix
On the digestive process in some rhizopods. *J. Physiol.,* **7** (1886) 253–73; **8** (1887) 263–87
On digestion in Hydra; with some observations on the structure of the endoderm. *J. Physiol.,* **9** (1888) 317–44
On the action of nicotin upon certain invertebrates. *J. Physiol.,* **11** (1890) 573–605
On retractile cilia in the intestine of Lumbricus terrestris. *J. Physiol.,* **13** (1892) 239–59
On the constitution and mode of formation of "food vacuoles" in infusoria, as illustrated by the history of the processes of digestion in Carchesium polypinum. [1893–1894] *Phil. Trans. (B),* **185** (1895) 355–83
On structural change in the resting nuclei of protozoa. Part I. The macro nucleus of Carchesium polypinum. *J. Physiol.,* **20** (1896) 427–54
Greenwood, (Miss) Marion; Saunders, (Miss) E. R.
On the rôle of acid in protozoan digestion. *J. Physiol.,* **16** (1894) 441–67

Brinck, Julia
Brinck, (Frl.) Julia
On the nutrition of skeletal muscle. *J. Physiol.,* **10** (1889) ix–x
Ueber synthetische Wirkung lebender Zellen. *Ztschr. Biol.,* **25** (1889) 453–73
Brinck, (Frl.) Julia; Kronecker, Hugo
Ueber synthetische Wirkung lebender Zellen. *Arch. Anat. Physiol. (Physiol. Abth.)* (1887) 347–9; *Bern Mitth.* (1887) xviii–xxii

Buchanan, Florence
Buchanan, Florence
The efficiency of the contraction of veratrinised muscle. [1899] *J. Physiol.,* **25** (1899–1900) 137–56

Divine, Julia

Divine, Julia

Ueber die Atmung des Krötenherzens. *Arch. Anat. Physiol. (Physiol. Abth.)* (1898) 533–4; *Schweiz. Natf. Ges. Verh.* (1898) 124–5; *J. Physiol.,* **23** (1898–1899) [12]-[13]

Kronecker, Hugo; Divine, (Mlle.)

[Respiration of the heart of the tortoise.] *Nature,* **58** (1898) 481

Eves, Florence Elizabeth

Eves, (Miss) Florence

Some experiments on the liver ferment. [1885] *J. Physiol.,* **5** (1884–1885) 342–51

On some experiments on the liver ferment. [1884] *Cambridge Phil. Soc. Proc.,* **5** (1886) 182–3

Langley, John Newport; Eves, Florence

On certain conditions which influence the amylolytic action of saliva. [1883] *J. Physiol.,* **4** (1883–1884) 18–28

Flemming, E. E.

Flemming, (Mrs.) E. E.

Absence of the left internal carotid. *J. Anat. Physiol.,* **29** (1895) xxiii–xxxiv

Malposition of the colon. *J. Anat. Physiol.,* **31** (1897) xxxii–xxxiv

Hoggan, Frances Elizabeth Morgan

Hoggan, George; Hoggan, Frances Elizabeth

On the origin of the lymphatics. *Brit. Ass. Rep., (Sect.)* (1875) 165–7

Lymphatics and their origin in muscular tissues. *Roy. Soc. Proc.,* 25 (1877) 550–1

On the minute structure and relationships of the lymphatics of the mammalian skin, and on the ultimate distribution of nerves to the epidermis and subepidermic lymphatics. [1877] *Roy. Soc. Proc.,* **26** (1878) 289–90

Étude sur les lymphatiques de la peau. *Robin, J. Anat.,* **15** (1879) 50–69

Étude sur les lymphatiques des muscles striés. *Robin, J. Anat.,* **15** (1879) 584–611

Des lymphatiques du périchondre. [Note préliminaire.] *Paris, Ac. Sci. C. R.,* **89** (1879) 320–2

On the development and retrogression of the fat-cell. *Micr. Soc. J.,* **2** (1879) 353–80

Étude sur le rôle des lymphatiques de la peau dans l'infection cancéreuse. *Arch. de Physiol.,* 7 (1880) 284–306

Note sur les lymphatiques des muscles striés. [1879] *Paris, Soc. Biol. Mém.,* 1, C.R. (1880) 197–9

On the development and retrogression of blood-vessels. *Micr. Soc. J.,* **3** (1880) 568–84

On the lymphatics of cartilage or of the perichondrium. [1880] *J. Anat. Physiol.,* **15** (1881) 121–36

On the comparative anatomy of the lymphatics of the mammalian urinary bladder. *J. Anat. Physiol.,* **15** (1881) 355–77

On the lymphatics of the pancreas. *J. Anat. Physiol.,* **15** (1881) 475–95

Zur pathologischen Histologie der schmerzhaften subcutanen Geschwulst. *Virchow, Arch.,* **83** (1881) 233–42

Étude sur les changements subis par le système nerveux dans la lèpre. *Arch. de Physiol.,* **10** (1882) 83–127; 233–65

De la dégénération et de la régénération du cylindre-axe et des autres élémentes des fibres nerveuses dans les lésions non traumatiques. [1880] *Robin, J. Anat.,* **18** (1882) 27–59

On the comparative anatomy of the lymphatics of the uterus. [1881] *J. Anat. Physiol.,* **16** (1882) 50–89

Étude sur les terminaisons nerveuses dans la peau. *Robin, J. Anat.,* **19** (1883) 377–98

On some cutaneous nerve-terminations in mammals. [1882] *Linn. Soc. J. (Zool.),* **16** (1883) 546–93

The lymphatics of the walls of the larger blood-vessels and lymphatics. [1882] *J. Anat. Physiol.,* **17** (1883) 1–23

The lymphatics of periosteum. *J. Anat. Physiol.,* **17** (1883) 308–28

Forked nerve endings on hairs. *J. Anat. Physiol.,* **27** (1893) 224–31

Newbigin, Marion Isabel

Paton, D. N.; {Boyd, F. D.; Dunlop, J. C.; Gillespie, A. L.; Gulland, G. L.; Greig, E. D. W.; Newbigin, Marion Isabel}

The physiology of salmon in fresh water. [1898] *J. Physiol.,* **22** (1897–1898) 355–6

Saunders, Edith Rebecca

Greenwood, (Miss) Marion; Saunders, (Miss) E. R.

On the rôle of acid in protozoan digestion. *J. Physiol.,* **16** (1894) 441–67

Southall, Gertrude

Southall, (Miss) Gertrude; Haycraft, John Berry

Note on an amylolytic ferment found in the gastric mucous membrane of the pig. *J. Anat. Physiol.,* **23** (1889) 452–4

Sowton, S. C. M.

Sowton, (Miss) S. C. M.

Galvanometric records of the decline of the current of injury in medullated nerve and of the changes in its response to periodic stimulation. [1898] *Nature,* **58** (1898) 484; *J. Physiol.,* **23** (1898–1899) [36]-[37]

On the reflex electrical effects in mixed nerve and in the anterior and posterior roots. *Roy. Soc. Proc.,* **64** (1899) 353–9

Observations on the electromotive phenomena of nonmedullated nerve. *Roy. Soc. Proc.,* **66** (1900) 379–89

Waller, Augustus D.; Sowton, (Miss) S. C. M.

Action of carbonic dioxide on voluntary and on cardiac muscle. *J. Physiol.,* **20** (1896) xvi-xvii

[Functional activity of nerve cells. Report of the Committee . . . appointed to investigate the changes which are associated with the functional activity of nerve cells and their peripheral extensions. Appendix.] IV. On the action of reagents upon isolated nerve. *Brit. Ass. Rep.* (1897) 518–20

[The action on isolated nerve of muscarine, choline and neurine.] *Nature,* **58** (1898) 484

Action upon isolated nerve of muscarine, choline and neurine. [1898] *J. Physiol.,* **23** (1898–1899) [35]-[36]

Stoney, Florence Ada

Stoney, (Miss) F. A.

Oesophagus with two well-marked diverticula. [1899] *J. Anat. Physiol.,* **34** (1900) ii–iii

Zoology

United States

Bennett, Ethelwyn Foote

Foote, Ethelwyn

The extrabranchial cartilages of the elasmobranchs. *Anat. Anz.,* **13** (1897) 305–8

Brace, Edith Minerva

Brace, (Miss) Edith M.

Notes on Aelosoma tenebrarum. *Amer. Ass. Proc.,* (1898) 363; {*J. Morphol.* **17** (1901) 177–84}

Bush, Katharine Jeannette
Bush, Katharine Jeannette
Additions to the shallow-water Mollusca of Cape Hatteras, N. C., dredged by the U. S. Fish Commission steamer *Albatross* in 1883 and 1884. [1885] *Connecticut Ac. Trans.,* **6** (1882–1885) 453–80
Catalogue of Mollusca and Echinodermata, dredged on the coast of Labrador by the expedition under the direction of Mr. W. A. Stearns, in 1882. [1883] *U. S. Mus. Proc.,* **6** (1884) 236–49
List of the shallow-water Mollusca dredged off Cape Hatteras by the *Albatross* in 1883. *U. S. Fish Comm. Rep.,* **11** (1885) 579–90
List of deep-water Mollusca dredged by the United States Fish Commission steamer *Fish Hawk,* in 1880, 1881, and 1882, with their range in depth. *U. S. Fish Comm. Rep.,* **11** (1885) 701–27
Reports on the results of dredging, ...by the U. S. Coast Survey steamer *Blake.* XXXIV. Report on the Mollusca dredged by the *Blake* in 1880, including descriptions of several new species. *Harvard Mus. Zool. Bull.,* **23** (1892–1893) 199–244
Descriptions of new species of Turbonilla of the western Atlantic fauna, with notes on those previously known. *Philad. Ac. Nat. Sci. Proc.* (1899) 145–77
Revision of the marine gastropods referred to Cyclostrema, Adeorbis, Vitrinella, and related genera; with descriptions of some new genera and species belonging to the Atlantic fauna of America. [1897] *Connecticut Ac. Trans.,* **10** (1899–1900) 97–144; 672
Verrill, Addison E.; Bush, (Miss) Katharine Jeannette
Revision of the genera of Ledidae and Nuculidae of the Atlantic coast of the United States. *Amer. J. Sci.,* **3** (1897) 51–63
Revision of the deep-water Mollusca of the Atlantic coast of North America, with descriptions of new genera and species. Pt. I. Bivalvia. *U. S. Mus. Proc.,* **20** (1898) 775–901; xii
Additions to the marine Mollusca of the Bermudas. [1900] *Connecticut Ac. Trans.,* **10** (1899–1900) 513–44; 672

Byrnes, Esther Fussell
Byrnes, Esther Fussell
The maturation and fertilization of the eggs of Limax. [1896] *Science,* **5** (1897) 391
Experimental studies on the development of limb-muscles in Amphibia. *J. Morphol.,* **14** (1898) 105–40
On the regeneration of limbs in frogs after the extirpation of limb-rudiments. [1898] *Anat. Anz.,* **15** (1899) 104–7
The maturation and fertilization of the egg of Limax agrestis (*Linné*). [1899] *J. Morphol.,* **16** (1900) 201–36

Calvert, Amelia Catherine Smith
Smith, Amelia C.
Multiple canals in the spinal cord of a chick embryo. [1898] *Anat. Anz.,* **15** (1899) 59–60

Clapp, Cornelia Maria
Clapp, Cornelia M.
Some points in the development of the toad fish (Batrachus tau). *J. Morphol.,* **5** (1891) 494–501
Relation of the axis of the embryo to the first cleavage plane. *Wood's Holl Mar. Biol. Lab. Lect.* (1898) 139–51
The lateral line system of Batrachus tau. [1898] *J. Morphol.,* **15** (1899) 223–64

Claypole, Agnes Mary
Claypole, (Miss) Agnes Mary
The enteron of the Cayuga Lake lamprey. *Amer. Micr. Soc. Proc.,* **16** (1894) 125–64
The appendages of an insect embryo. *Canad. Ent.,* **28** (1896) 289; *Science,* **4** (1896) 603–4
Some points on cleavage among arthropods. *Amer. Micr. Soc. Trans.,* **19** (1897) 74–82
The embryology and oögenesis of Anurida maritima (*Guér.*). *J. Morphol.,* **14** (1898) 219–300
The embryology of the Apterygota. *Zool. Bull.,* **2** (1899) 69–76

Claypole, Edith Jane
Claypole, (Miss) Edith Jane
The action of leucocytes toward foreign material. *Amer. Micr. J.,* **14** (1893) 334–6
Blood coagulation. *Amer. Micr. J.,* **14** (1893) 355–6
Blood corpuscles. *Amer. Micr. J.,* **14** (1893) 356–7
The blood of Necturus and Cryptobranchus. (Necturus maculatus, Cryptobranchus alleghaniensis.) *Amer. Micr. Soc. Proc.,* **15** (1893) 39–76
The action of leucocytes toward foreign substances. *Amer. Natlist.,* **28** (1894) 316–25
Notes on the comparative histology of blood and muscle. *Amer. Micr. Soc. Trans.,* **18** (1896) 49–70
The comparative histology of the digestive tract. *Amer. Micr. Soc. Trans.,* **19** (1897) 83–92
{An interesting abnormality. *J. App. Micr.,* **2** (1899) 261–3}

Cockerell, Wilmatte Porter
Cockerell, Theodore D. A.; Porter, (Miss) Wilmatte
A new crayfish from New Mexico. *Philad. Ac. Nat. Sci. Proc.* (1900) 434–5

Congdon, Edna M.
Ritter, William E.; Congdon, (Miss) Edna M.
On the inhibition by artificial section of the normal fission plane in Stenostoma. [1900] *California Ac. Proc. (Zool.),* **2** (1904) 365–76

Crocker, Gulielma R.
Ritter, William E.; Crocker, Gulielma R.
Papers from the Harriman Alaska Expedition. III. Multiplication of rays and bilateral symmetry in the 20-rayed star-fish, Pycnopodia helianthoides (*Stimpson*). *Washington Ac. Sci. Proc.,* **2** (1900) 247–74

Davenport, Gertrude Crotty
Davenport, Gertrude C.
Agassiz's work on the embryology of the turtle. *Amer. Natlist.,* **32** (1898) 187–8
Variation in the sea anemone, Sagartia luciae. [1899] *Science,* **11** (1900) 253
Variation in the madreporic body and stone canal of Asterias vulgaris. [1900] *Science,* **13** (1901) 374–5

Eigenmann, Rosa Smith
Smith, (Miss) Rosa
On the occurrence of a species of Cremnobates at San Diego, Calif. (*U. S. Nat. Mus. Proc.,* **3,** 1880, 147–9) *Smithsonian Miscell. Coll.,* **22** (1882) Art. 1
Description of a new gobiod fish, (Othonops eos), from San Diego, Calif. (*U. S. Nat. Mus. Proc.,* **4,** 1881, 19–21). *Smithsonian Miscell. Coll.,* **22** (1882) Art. 2
Description of a new species of Gobiesox, (G. rhessodon), from San Diego, Calif. (*U. S. Nat. Mus. Proc.,* **4,** 1881, 140–1). *Smithsonian Miscell. Coll.,* **22** (1882) Art. 2
Description of a new species of Uranidea (U. rhothea) from Spokane River, Washington territory. [1882] *U. S. Mus. Proc.,* **5** (1883) 347–8

On the life coloration of the young of Pomacentrus rubicundus. *U. S. Mus. Proc.,* **5** (1883) 652–3
The life colors of Cremnobates integripinnis. [1883] *U. S. Mus. Proc.,* **6** (1884) 216–17
Note on the occurrence of Gasterosteus williamsoni, *Grd.,* in an artesian well at San Bernadino, Cal. [1883] *U. S. Mus. Proc.,* **6** (1884) 217
Notes on the fishes of Todos Santos Bay, Lower California. [1883] *U. S. Mus. Proc.,* **6** (1884) 232–6
Notes on fishes collected at San Cristobal Bay, Lower California, by Mr. Charles H. Townsend. *U. S. Mus. Proc.,* 7 (1885) 551–3; [viii]
Description of a new species of Squalius. *California Ac. Bull.,* **1** (1886) [No.1, 1884] 3–4
On Tetraodon setosus, a new species allied to T. meleagris, *Lacép.* [1886] *California Ac. Bull.,* **2** (1887) 155–6
On the occurrence of a new species of Rhinoptera (R. encenadae) in Todos Santos Bay, Lower California. [1886] *U. S. Mus. Proc.,* **9** (1887) 220
{Note on Typhlogobius californiensis. *Zoe,* **1** (1890) 181–2}
New Californian fishes. *Amer. Natlist.,* **25** (1891) 153–6
Description of a new species of Catostomus (C. rex) from Oregon. *Amer. Natlist.,* **25** (1891) 667–8
Description of a new species of Euprotomicrus. [1890] *California Ac. Proc.,* **35** (1893) 35

Smith, (Miss) Rosa; Swain, Joseph
Notes on a collection of fishes from Johnston's Island, including descriptions of 5 new species: [Ophichthys stypurus; Upeneus velifer, U. preorbitalis; Julis verticalis, J. clepsydralis]. [1882] *U. S. Mus. Proc.,* **5** (1883) 119–43

Eigenmann, Carl H.; Smith, (Miss) Rosa (Mrs. Eigenmann)
A revision of the edentulous genera of Curimatinae. [1889] *N. Y. Ac. Ann.,* **4** (1887–1889) 409–40
American Nematognathi. *Amer. Natlist.,* **22** (1888) 647–9
A list of the American species of Gobiidae and Callionymidae, with notes on the specimens contained in the Museum of Comparative Zoology, at Cambridge, Massachusetts. [1888] *California Ac. Proc.,* **1** (1889) 51–78
Notes on some Californian fishes, with descriptions of two new species. *U. S. Mus. Proc.,* **11** (1889) 463–6
Preliminary notes on South American Nematognathi. [1888–1889] *California Ac. Proc.,* **1** (1889) 119–72; 361; **2** (1890) 28–56
Description of a new species of Cyprinodon. *California Ac. Proc.,* **1** (1889) 270
A review of the Erythrininae. [1889] *California Ac. Proc.,* **2** (1890) 100–16
A revision of the South American Nematognathi, or cat-fishes. *California Ac. Occ. Pap.,* **1** (1890) 508 pp.
Cottus beldingii, *sp. nov. Amer. Natlist.,* **25** (1891) 1132
A catalogue of the fishes of the Pacific coast of America, north of Cerros Island. [1892] *N. Y. Ac. Ann.,* **6** (1891–1892) 349–58
A catalogue of the freshwater fishes of South America. [1891] *U. S. Mus. Proc.,* **14** (1892) 1–81
New fishes from western Canada. *Amer. Natlist.,* **26** (1892) 961–4
Additions to the fauna of San Diego. [1890] *California Ac. Proc.,* **3** (1893) 1–24
Description of a new species of Sebastodes. [1890] *California Ac. Proc.,* **3** (1893) 36–8
Preliminary descriptions of new fishes from the North-West. *Amer. Natlist.,* **27** (1893) 151–4; 592

Field, Eva H.
Field, (Miss) Eva H.
A contribution to the study of malignant growths in the lower animals. *Amer. Micr. Soc. Proc.,* **16** (1894) 223–33

Foot, Katharine
Foot, Katharine
Preliminary note on the maturation and fertilization of the egg of Allolobophora foetida. *J. Morphol.,* **9** (1894) 475–84
The centrosomes of the fertilized egg of Allolobophora foetida. *Wood's Holl Mar. Biol. Lab. Lect.* (1896–1897) 45–57
Centrosome and archoplasm. *Science,* **5** (1897) 231
Yolk-nucleus and polar rings. [1896] *J. Morphol.,* **12** (1897) 1–16
The origin of the cleavage centrosomes. *J. Morphol.,* **12** (1897) 809–14
The cocoons and eggs of Allolobophora foetida. *J. Morphol.,* **14** (1898) 481–506

Foot, Katharine; Strobell, Ella Church
Further notes on the egg of Allolobophora foetida. [1898] *Zool. Bull.,* **2** (1899) 129–50
Photographs of the egg of Allolobophora foetida. *J. Morphol.,* **16** (1900) 601–18

Goddard, Martha Freeman
Goddard, Martha Freeman
On the second abdominal segment in a few Libellulidae. *Amer. Phil. Soc. Proc.,* **35** (1896) 205–12

Green, Isabella M.
Green, (Miss) Isabella M.
The peritoneal epithelium in amphibia. *Amer. Natlist.,* **30** (1896) 944–5; *Science,* **4** (1896) 604
The peritoneal epithelium of some Ithaca amphibia (Necturus, Amblystoma, Desmognathus, and Diemyctylus). *Amer. Micr. Soc. Trans.,* **18** (1896) 76–106

Greene, Flora Hartley
Hartley, Flora
Description of a new species of wood-rat from Arizona. [1894] *California Ac. Proc.,* **4** (1895) 157–60
Notes on a specimen of Alepisaurus aesculapius, *Bean,* from the coast of San Luis Obispo county, California. [1895] *California Ac. Proc.,* **5** (1896) 49–50

Gregory, Emily Ray
Gregory, Emily Ray
Origin of the pronephric duct in selachians. [1897] *Science,* **5** (1897) 1000; *Zool. Bull.,* **1** (1898) 123–9
The pronephros in Testudinata. *Science,* **7** (1898) 576
Observations on the development of the excretory system in turtles. *Zool. Jbüch. (Anat.),* **13** (1900) 683–714

Hall, Mary Alice Bowers
Bowers, Mary A.
Peripheral distribution of the cranial nerves of Spelerpes bilineatus. [1900] *Amer. Ac. Proc.,* **36** (1901) 177–93

Hazen, Annah Putnam
Hazen, Annah Putnam
The regeneration of a head instead of a tail in an earthworm. *Anat. Anz.,* **16** (1899) 536–41

Morgan, T. H.; Hazen, Annah Putnam
The gastrulation of Amphioxus. *J. Morphol.,* **16** (1900) 569–600

Patten, Wm.; Hazen, Annah Putnam
The development of the coxal gland, branchial cartilages, and gen-

ital ducts of Limulus polyphemus. *J. Morphol.,* **16** (1900) 459–502

Hefferan, Mary
Hefferan, (Miss) Mary
Variations in jaws of Nereis limbata. *Amer. Ass. Proc.* (1900) 206–7
Variation in the teeth of Nereis. *Woods Holl Mar. Biol. Lab. Bull.,* **2** (1901) 129–43

Hempstead, Marguerite
Hempstead, Marguerite
Development of the lungs in the frogs, Rana catesbiana, R. silvatica, and R. virescens. *Amer. Ass. Proc.* (1900) 242–3

Henchman, Annie Parker
Henchman, Annie P.
The origin and development of the central nervous system in Limax maximus. *Harvard Mus. Zool. Bull.,* **20** (1890–1891) 169–208
The eyes of Limax maximus. [1896] *Science,* **5** (1897) 428–9

Hitchcock, Fanny Rysam Mulford
Hitchcock, Fanny R. M.
Preliminary paper on structure of Alosa sapidissima. *Amer. Ass. Proc.* (1887) 259–60
On the homologies of Edestus. *Amer. Ass. Proc.* (1887) 260–1
Preliminary notes on the osteology of Alosa sapidissima. *Amer. Natlist.,* **21** (1887) 1032–3
Further notes on the osteology of the shad (Alosa sapidissima). [1881] *N. Y. Ac. Ann.,* **4** (1887–1889) 225–8

Horning, Jennie
Eigenmann, Carl H.; Horning, J.
A review of the Chaetodontidae of North America. [1887] *N. Y. Ac. Ann.,* **4** (1887–1889) 1–18

Hughes, Elizabeth G.
Jordan, David Starr; Hughes, Elizabeth G.
A review of the genera and species of Julidinae found in American waters. [1886] *U. S. Mus. Proc.,* **9** (1887) 56–70
A review of the species of the genus Prionotus. [1886] *U. S. Mus. Proc.,* **9** (1887) 327–38
Eigenmann, Carl H.; Hughes, Elizabeth G.
A review of the North American species of the genera Lagodon, Archosargus, and Diplodus. [1887] *U. S. Mus. Proc.,* **10** (1888) 65–74

Kenyon, Josephine Hemenway
Hemenway, Josephine
The structure of the eye of Scutigera (Cermatia) forceps. *Woods Holl Mar. Biol. Lab. Bull.,* **1** (1900) 205–13

Key, Wilhelmine Marie Enteman
Enteman, (Miss) Minnie Marie
The unpaired ectodermal structures of the Antennata. *Zool. Bull.,* **2** (1899) 275–82
Variation in Daphnia hyalina. *Science,* **12** (1900) 229
Variations in the crest of Daphnia hyalina. *Amer. Natlist.,* **34** (1900) 879–90
[On the behaviour of Polistes.] [1900] *Science,* **13** (1901) 112–13

King, Helen Dean
King, Helen Dean
Regeneration in Asteris vulgaris. *Arch. EntwMech.,* **7** (1898) 351–63; **9** (1900) 724–37

Langdon, Fanny E.
Langdon, Fanny E.
The sense organs of Lumbricus agricola, *Hoffm.* [1893] *Anat. Anz.,* **10** (1895) 114–17; *J. Morphol.,* **11** (1895) 193–234
The peripheral nervous system of Neries virens. [1896] *Science,* **5** (1897) 427–8

Langenbeck, Clara
Langenbeck, Clara
Formation of the germ layers in the amphipod Microdeutopus gryllotalpa, *Costa. J. Morphol.,* **14** (1898) 301–36

Merrill, Harriet Bell
Merrill, Harriet Bell
The structure and affinities of Bunops scutifrons, *Birge. Wisconsin Ac. Trans.,* **9** (1893) 319–42
Preliminary note on the eye of the leech. *Zool. Anz.,* **17** (1894) 286–8

Monks, Sarah Preston
Monks, Sarah P.
The columella and staples in some North American turtles. *Amer. Phil. Soc. Proc.,* **17** (1878) 335–7
{The spotted salamander. *Amer. Natlist.,* **14** (1880) 371–4}
A partial biography of the green lizard [Anolis principalis]. *Amer. Natlist.,* **15** (1881) 96–9

Morgan, Lilian Vaughan Sampson
Sampson, Lilian V.
Die Muskulatur von Chiton. *Jena. Ztschr.,* **28** (1894) 460–78; *J. Morphol.,* **11** (1895) 595–628
Unusual modes of breeding and development among Anura. *Amer. Natlist.,* **34** (1900) 687–715

Mosenthal, Johanna Kroeber
Kroeber, Johanna
An experimental demonstration of the regeneration of the pharynx of Allolobophora from endoderm. [1900] *Woods Holl Mar. Biol. Lab. Bull.,* **2** (1901) 105–10

Mozley, Annie Elizabeth
Mozley, (Miss) Annie E.
List of the Kansas snakes in the museum of the Kansas State University. [1877] *Kan. Ac. Sci. Trans.,* **6** (1878) 34–5

Nickerson, Margaret Lewis
Lewis, Margaret
Centrosome and sphere in certain of the nerve cells of an invertebrate. *Anat. Anz.,* **12** (1896) 291–9
Epidermal sense organs in certain polychaetes. [1896] *Science,* **5** (1897) 428
A method of removing cuticula from marine annelids. *Zool. Bull.,* **1** (1898) 243
Studies of the central and peripheral nervous systems of two polychaete annelids. *Amer. Ac. Proc.,* **33** (1898) 223–68
Clymene producta, *n. sp.* [1897] *Boston Soc. Nat. Hist. Proc.,* **28** (1899) 111–15
Nickerson, Margaret Lewis
Intracellular differentiations in gland cells of Phascolosoma gouldii. [1898] *Science,* **9** (1899) 365–6
Intracellular canals in the skin of Phascolosoma. [1899] *Zool. Jbüch. (Anat.),* **13** (1900) 191–6

Peebles, Florence
Peebles, Florence
Experimental studies on Hydra. *Arch. EntwMech.,* **5** (1897) 749–819
Some experiments on the primitive streak of the chick. *Arch. EntwMech.,* **7** (1898) 405–29
The effect of temperature on the regeneration of Hydra. [1898] *Zool. Bull.,* **2** (1899) 125–8

Experiments in regeneration and in grafting of Hydrozoa. *Arch. EntwMech.*, **10** (1900) 435–88

Pennington, Mary Engle
Ryder, John A.; Pennington, Mary Engle
Non-sexual conjugation of the nuclei of the adjacent cells of an epithelium. *Anat. Anz.*, **9** (1894) 759–64

Perkins, Helen
Davenport, Charles B.; Perkins, Helen
A contribution to the study of geotaxis in the higher animals. [1897] *J. Physiol.*, **22** (1897–1898) 99–110

Phelps, Jessie
Phelps, (Miss) Jessie
The development of the adhesive organ of Amia. *Science,* **9** (1899) 366
Reighard, J. E.; Phelps, (Miss) Jessie; Mast, [Samuel Ottmar]
The development of the adhesive organ and hypophysis in Amia. [1899] *Science,* **11** (1900) 251

Platt, Julia Barlow
Platt, Julia B.
Studies on the primitive axial segmentation of the chick. *Harvard Mus. Zool. Bull.*, **17** (1888–1889) 171–90
The anterior head-cavities of Acanthis. (Preliminary notice.) *Zool. Anz.*, **13** (1890) 239
A contribution to the morphology of the vertebrate head, based on a study of Acanthias vulgaris. *J. Morphol.*, **5** (1891) 79–112
Further contribution to the morphology of the vertebrate head. *Anat. Anz.*, **6** (1891) 251–65
Fibres connecting the central nervous system and chorda in Amphioxus. *Anat. Anz.*, **7** (1892) 282–4
Ectodermic origin of the cartilages of the head. *Anat. Anz.*, **8** (1893) 506–9
Ontogenetic differentiations of the ectoderm in Necturus. [Preliminary notice.] [1893] *Anat. Anz.*, **9** (1894) 51–6
Ontogenetische Differenzirung des Ektoderms in Necturus. *Arch. Mikr. Anat.*, **43** (1894) 911–66
Ontogenetic differentiations of the ectoderm in Necturus. Study II. On the development of the peripheral nervous system. *Quart. J. Micr. Sci.*, **38** (1896) 485–547
The development of the thyroid gland and of the suprapericardial bodies in Necturus. *Anat. Anz.*, **11** (1896) 557–67
The development of the cartilaginous skull and of the branchial and hypoglossal musculature in Necturus. [1897] *Morphol. Jbuch.*, **25** (1898) 377–464
On the specific gravity of Spirostomum, Paramaecium, and the tadpole in relation to the problem of geotaxis. *Amer. Natlist.*, **33** (1899) 31–8

Randolph, Harriet
Randolph, Harriet
The regeneration of the tail in Lumbriculus. [Preliminary communication.] *Zool. Anz.*, **14** (1891) 154–6
Ein Beitrag zur Kenntniss der Tubificiden. *Jena. Ztschr.*, **27** (1892) 463–76; *Zürich Vrtljschr.*, **37** (1892) 145–7
The regeneration of the tail in Lumbriculus. *J. Morphol.*, **7** (1892) 317–44
Observations and experiments on regeneration in planarians. *Arch. EntwMech.*, **5** (1897) 352–72
Chloretone (acetonchloroform): an anaesthetic and macerating agent for lower animals. *Zool. Anz.*, **23** (1900) 436–9

Rathbun, Mary Jane
Rathbun, Mary J.
Catalogue of the crabs of the family Periceridae, in the U. S. National Museum. [With extract from the unpublished report of Dr. William Stimpson, on the Crustacea of the North Pacific Exploring Expedition, 1853 to 1856.] [1892] *U. S. Mus. Proc.*, **15** (1893) 231–77
Catalogue of the crabs of the family Maiidae in the U. S. National Museum. [With extract from an unpublished report of Dr. William Stimpson, on the Crustacea of the North Pacific Exploring Expedition 1853 to 1856.] [1893] *U. S. Mus. Proc.*, **16** (1894) 63–103
Scientific results of explorations by the U. S. Fish Commission steamer *Albatross*. No. XXIV. Descriptions of new genera and species of crabs from the west coast of America and the Sandwich Islands. [1893] *U. S. Mus. Proc.*, **16** (1894) 223–60
Descriptions of new species of American freshwater crabs. [1893] *U. S. Mus. Proc.*, **16** (1894) 649–61
Descriptions of two new species of crabs from the western Indian Ocean, presented to the National Museum by Dr. W. L. Abbott. [1894] *U. S. Mus. Proc.*, **17** (1895) 21–4
Description of a new genus and two new species of African freshwater crabs. [1894] *U. S. Mus. Proc.*, **17** (1895) 25–7
Notes on the crabs of the family Inachidae in the United States National Museum. [1894] *U. S. Mus. Proc.*, **17** (1895) 43–75
Descriptions of a new genus and four new species of crabs from the Antillean region. [1894] *U. S. Mus. Proc.*, **17** (1895) 83–6
The genus Callinectes. [With observations upon the habits of Callinectes sapidus by John D. Mitchell, Benjamin Harrison and Willard Nye, jun.] *U. S. Mus. Proc.*, **18** (1896) 349–75
Description of two new species of freshwater crabs from Costa Rica. *U. S. Mus. Proc.*, **18** (1896) 377–9
Description of a new genus and four new species of crabs from the West Indies. [1896] *U. S. Mus. Proc.*, **19** (1897) 141–4
Descriptions de nouvelles espèces de crabes d'eau douce, appartenant aux collections du Muséum d'Histoire Naturelle de Paris. *Paris, Mus. Hist. Nat. Bull.*, **3** (1897) 58–62
Synopsis of the American Sesarmae, with description of a new species. *Washington Biol. Soc. Proc.*, **11** (1897) 89–92; 112
Synopsis of the American species of Palicus philippi (=Cymopolia, *Roux*), with descriptions of six new species. *Washington Biol. Soc. Proc.*, **11** (1897) 93–9
Synopsis of the American species of Ethusa, with description of a new species. *Washington Biol. Soc. Proc.*, **11** (1897) 109–10
Description of a new species of Cancer from Lower California, and additional note on Sesarma. *Washington Biol. Soc. Proc.*, **11** (1897) 111–12
The African swimming crabs of the genus Callinectes. *Washington Biol. Soc. Proc.*, **11** (1897) 149–51
A revision of the nomenclature of the Brachyura. *Washington Biol. Soc. Proc.*, **11** (1897) 153–67
Descriptions of three new species of fresh-water crabs of the genus Potamon. *Washington Biol. Soc. Proc.*, **12** (1898) 27–30
The Brachyura of the Biological Expedition to the Florida Keys and the Bahamas in 1893. *Iowa Univ. Lab. Nat. Hist. Bull.*, **4** (1898) 250–94
A contribution to a knowledge of the fresh-water crabs of America. The Pseudothelphusinae. [1898] *U. S. Mus. Proc.*, **21** (1899) 507–37
Jamaica Crustacea. *Jamaica Inst. J.*, **2** (1899) 628–9

Notes on the Crustacea of the Tres Marias Islands. *U. S. N. Amer. Fauna,* No. 14 (1899) 73–5

The Brachyura collected by the U. S. Fish Commission steamer *Albatross,* on the voyage from Norfolk, Virginia, to San Francisco, California, 1887–88. [1898] *U. S. Mus. Proc.,* **21** (1899) 567–616

Results of the Branner-Agassiz Expedition to Brazil. I. The decapod and stomatopod Crustacea. *Washington Ac. Sci. Proc.,* **2** (1900) 133–56

Synopses of North-American invertebrates. VII. The cyclometopous or cancroid crabs of North America. [X. The oxyrhynchous and oxystomatous crabs of North America. XI. The catometopous or grapsoid crabs of North America.] *Amer. Natlist.,* **34** (1900) 131–43; 503–20; 583–92

The decapod crustaceans of West Africa. *U. S. Mus. Proc.,* **22** (1900) 271–316

Benedict, James E.; Rathbun, Mary J.

The genus Panopeus. [1891] *U. S. Mus. Proc.,* **14** (1892) 355–85

Robertson, Alice

Robertson, Alice

Studies in Pacific Coast Entoprocta. [1900] *California Ac. Proc. (Zool.),* **2** (1897–1901) 323–48

Papers from the Harriman Alaska Expedition. VI. The Bryozoa. *Washington Ac. Sci. Proc.,* **2** (1900) 315–40

Ross, Mary J.

Ross, (Miss) Mary J.

Special structural features in the air-sacks of birds. *Amer. Micr. Soc. Trans.,* **20** (1899) 29–40

Rucker, Augusta

Rucker, Augusta

A description of the male of Peripatus eisenii, *Wheeler. Woods Holl Mar. Biol. Lab. Bull.,* **1** (1900) 251–9

Searle, Harriet Richardson

Richardson, Harriet

Description of a new species of Sphaeroma. *Washington Biol. Soc. Proc.,* **11** (1897) 105–7

Description of a new genus and species of Sphaeromidae from Alaskan waters. *Washington Biol. Soc. Proc.,* **11** (1897) 181–3

Description of a new crustacean of the genus Sphaeroma from a warm spring in New Mexico. [1897] *U. S. Mus. Proc.,* **20** (1898) 465–6

Description of a new parasite isopod of the genus Aega from the southern coast of the United States. *Washington Biol. Soc. Proc.,* **12** (1898) 39–40

Description of four new species of Rocinela, with a synopsis of the genus. *Amer. Phil. Soc. Proc.,* **37** (1898) 8–17

Key to the isopods of the Pacific coast of North America, with descriptions of twenty-two new species. *U. S. Mus. Proc.,* **21** (1899) 815–69

Description of a new species of Idotea from Hakodate Bay, Japan. *U. S. Mus. Proc.,* **22** (1900) 131–4

Results of the Branner-Agassiz Expedition to Brazil. II. The isopod Crustacea. *Washington Ac. Sci. Proc.,* **2** (1900) 157–9

Synopses of North American invertebrates: VIII. The Isopoda, Part I. Chelifera, Flabellifera, Valvifera. [Part II. Asellota, Oniscoidea, Epicaridea.] *Amer. Natlist.,* **34** (1900) 207–30; 295–309

Seelye, Anne Ide Barrows

Barrows, Anne Ide

Respiration of Desmognathus. *Anat. Anz.,* **18** (1900) 461–4

Strobell, Ella Church

Foot, Katharine; Strobell, Ella Church

Further notes on the egg of Allolobophora foetida. [1898] *Zool. Bull.,* **2** (1899) 129–50

Photographs of the egg of Allolobophora foetida. *J. Morphol.,* **16** (1900) 601–18

Sturges, Mary Mathews

Sturges, Mary M.

Preliminary note on Distomum patellare, *n. sp.* [1897] *Zool. Bull.,* **1** (1898) 57–69

Polymorphic nuclei in embryonic germ-cells. [1898] *Science,* **9** (1899) 183

Tayler, Louise

Tayler, Louise

The striped muscle fibre: a few points in its comparative histology. *Amer. Micr. J.,* **18** (1897) 73–9

Our present knowledge of the kidney worm (Sclerostoma pinguicola) of swine. *U. S. Bur. Anim. Ind. Rep.,* **16** (1900) 612–37

Thompson, Caroline Burling

Thompson, (Miss) Caroline B.

Preliminary description of Zygeupolia litoralis, a new genus and new species of heteronemertean. *Zool. Anz.,* **23** (1900) 151–3

Carinoma tremaphoros, a new mesonemertean species. *Zool. Anz.,* **23** (1900) 627–30

Wallace, Louise Baird

Wallace, Louise B.

The structure and development of the axillary gland of Batrachus. *J. Morphol.,* **8** (1893) 563–8

The germ-ring in the egg of the toadfish (Batrachus tau). [1898] *J. Morphol.,* **15** (1899) 9–16

The accessory chromosome in the spider. A preliminary notice. *Anat. Anz.,* **18** (1900) 327–9

Whitman, Emily A. Nunn

Nunn, Emily A.

{The development of the ctenophore Mnemiopsis. *Johns Hopkins University Circular,* **1** (1879–1882) 16 [title only]}

The Naples zoological station. *Science,* **1** (1883) 479–81; 507–10

On the development of the enamel of the teeth of vertebrates. [1882] *Roy. Soc. Proc.,* **34** (1883) 156–65

{The zoological station at Naples. *The Century,* n.s. **10** (1886) 791–9}

Willcox, Mary Alice

Willcox, M. A.

{Anatomy of the grasshopper. *Observer* (Portland, Conn.), 7 (1896) 184–92}

{Anatomy of the May beetle. *Observer* (Portland, Conn.), 7 (1896) 365–73}

{Directions for the practical study of the grasshopper. *Observer* (Portland, Conn.), 7 (1896) 374–8}

Zur Anatomie von Acmaea fragilis, *Chemnitz. Jena. Ztschr.,* **32** (1898) 411–56

Notes on the occipital region of the trout, Trutta fario. *Zool. Bull.,* **2** (1899) 151–4

Hermaphroditism among the Docoglossa. *Science,* **12** (1900) 230–1

Notes on the anatomy of Acmaea testudinalis, *Müller.* [1899] *Science,* **11** (1900) 171

Great Britain

Abbott, Elizabeth Caroline Jane
Abbott, (Miss) E. C.; Gadow, Hans Friedrich
On the evolution of the vertebral column of fishes. [1894] *Phil. Trans.,* (B) **186** (1895) 163–221

Beck, Emma J.
Beck, (Miss) E. J.
Description of the muscular and endoskeletal systems of Scorpio. [1883] *Zool. Soc. Trans.,* **11** (1885) 339–60

Browne, Margaret Robinson
Robinson, (Miss) Margaret
On the nauplius eye persisting in some decapods. *Quart. J. Micr. Sci.,* **33** (1892) 283–7

Buchanan, Florence
Buchanan, (Miss) Florence
On the ancestral development of the respiratory organs in the decapodus Crustacea. *Quart. J. Micr. Sci.,* **29** (1889) 451–67
Hekaterobranchus Shrubsolii. A new genus and species of the family Spionidae. *Quart. J. Micr. Sci.,* **31** (1890) 175–200
Report on the occupation of the table [at the Laboratory of the Marine Biological Association at Plymouth]. *Brit. Ass. Rep.* (1892) 356–60
Report on polychaets collected during the Royal Dublin Society's Survey off the west coast of Ireland. Part I. Deep-water forms. [1893] *Dublin Soc. Sci. Proc.,* **8** (1893–8) 169–79
Peculiarities in the segmentation of certain polychaetes. *Quart. J. Micr. Sci.,* **34** (1893) 529–44
Note on the worm associated with Lophohelia prolifera. [1897] *Dublin Soc. Sci. Proc.,* **8** (1893–8) 432–3
A polynoid with branchiae (Eupolydontes cornishii). *Quart. J. Micr. Sci.,* **35** (1894) 433–50
On a blood-forming organ in the larva of Magelona. *Brit. Ass. Rep.* (1895) 469–70

Collcutt, Margaret C.
Collcutt, Margaret C.
On the structure of Hydractinia echinata. [1897] *Quart. J. Micr. Sci.,* **40** (1898) 77–99

Crane, Agnes
Crane, (Miss) Agnes
Notes on the habits of the Manatees (Manatus australis) in captivity in the Brighton Aquarium. *Zool. Soc. Proc.* (1881) 456–60
On a brachiopod of the genus Atretia, named in *Ms.* by the late Dr. T. Davidson. *Zool. Soc. Proc.* (1886) 181–4

Durham, Mary Edith
Durham, (Miss) M. Edith
Notes on the mode of feeding of the egg-eating snake (Dasypeltis scabra.) *Zool. Soc. Proc.* (1896) 715

Fedarb, Sophie M.
Fedarb, Sophie M.
On some earthworms from India. [1896] *Bombay Nat. Hist. Soc. J.,* **11** (1897–1898) 431–7
On some earthworms from British India. [1896] *Zool. Soc. Proc.* (1898) 445–50
Beddard, F. E.; Fedarb, Sophie M.
On some Perichaetae from the Eastern Archipelago collected by Mr. Everett. *Ann. Mag. Nat. Hist.,* **16** (1895) 69–73
Notes upon two earthworms, Perichaeta biserialis and Trichochaeta hesperidum. *Zool. Soc. Proc.* (1899) 803–9

Heath, Alice
Heath, (Miss) Alice
On the structure of the polycarp and the endocarp in the Tunicata. *Liverpool Lit. Phil. Soc. Proc.,* **37** (1883) 185–93
Notes on a tract of modified ectoderm in Crania anomala and Lingula anatina. [1887] *Liverpool Biol. Soc. Proc.,* **2** (1888) 95–104

Hopley, Catherine Cooper
Hopley, Catherine C.
The glottis of snakes. *Amer. Natlist.,* **18** (1884) 732–3
Observations on a remarkable development in the mud-fish. *Amer. Natlist.,* **25** (1891) 487–9

Johnson, Alice
Johnson, (Miss) Alice
On the development of the pelvic girdle of the hind limb in the chick. *Cambridge Phil. Soc. Proc.,* **4** (1883) 328–31; *Quart. J. Micr. Sci.,* **23** (1883) 399–411
On the changes and ultimate fate of the blastopore in the newt (Triton cristatus). *Roy. Soc. Proc.,* **37** (1884) 65–6
On the fate of the blastopore and the presence of a primitive streak in the newt (Triton cristatus) *Quart. J. Micr. Sci.,* **24** (1884) 659–72
Johnson, (Miss) Alice; Sheldon, (Miss) Lilian
Notes on the development of the newt (Triton cristatus). *Quart. J. Micr. Sci.,* **26** (1886) 573–89
On the development of the cranial nerves of the newt. *Roy. Soc. Proc.,* **40** (1886) 94–5

Kirkaldy, Jane Willis
Kirkaldy, (Miss) J. W.
On the head kidney of Myxine. *Quart. J. Micr. Sci.,* **35** (1894) 353–9
On the species of Amphioxus. *Brit. Ass. Rep.* (1894) 685–6
A revision of the genera and species of the Branchiostomidae. *Quart. J. Micr. Sci.,* **37** (1895) 303–23

Maclagan, Nellie
Maclagan, Nellie
List of edible British fishes, with their English, Latin, French, Italian and German synonyms. *Edinb., Fish. Brd. Rep.,* **2** (1884) 74–7

Musgrave, Edith M. Pratt
Pratt, Edith M.
Contribution to our knowledge of the marine fauna of the Falkland Islands. *Manchester Lit. Phil. Soc. Mem. & Proc.,* **42** (1898) No. 13, 26 pp.
The Entomostrasca of Lake Bassenthwaite. [With an introductory note by S. J. Hickson, M. A., F. R. S.] *Ann. Mag. Nat. Hist.,* **2** (1898) 467–76

Newbigin, Marion Isabel
Newbigin, (Miss) Marion I.
On British species of Siphonostoma. *Ann. Mag. Nat. Hist.,* **5** (1900) 190–5
Paton, Diarmid Noel; Newbigin, (Miss) Marion Isabel
Further investigations on the life-history of the salmon in fresh water. [1899] *Edinb. R. Coll. Physns. Lab. Rep.,* **7** (1900) 1–11; *Edinb., Fish. Brd. Rep.,* **18** (Pt. 2) (1900) 78–85; *Edinb. Roy. Soc. Proc.,* **23** (1902) 44–5

Ormerod, Eleanor Anne
Ormerod, (Miss) Eleanor A.
Observations on the cutaneous exhudation of the Triton cristatus or Great Water-newt. [1872] *Linn. Soc. J. (Zool.),* **11** (1873) 493–6

Palethorpe, Fanny D.
Palethorpe, Fanny D.; Wilson, Charlotte
Preliminary paper on a collection of simple ascidians from Australian seas. *Liverpool Biol. Soc. Proc.,* **1** (1887) 63–6

Pollard, E. C.
Pollard, (Miss) E. C.
A new sporozoon in Amphioxus. *Quart. J. Micr. Sci.,* **34** (1893) 311–16
Notes on the Peripatus of Dominica. [1893] *Quart. J. Micr. Sci.,* **35** (1894) 285–93

Power, Jeannette de Villepreux
Power, Jeannette
Osservazioni fisiche sopra il polpo dell' Argonauta argo. *Catania Acc. Gioen. Atti,* **12** (1837) 129–48; *Mag. Nat. Hist.,* **3** (1839) 101–6; 149–54
Experiments made with a view of ascertaining how far certain marine testaceous animals possess the power of renewing parts which may have been removed. {Translated by James Power, Esq.} *Froriep, Notizen,* **6** (1838) col. 209–11; *Mag. Nat. Hist.,* **2** (1838) 63–5
Further experiments and observations on the Argonauta argo. *Brit. Ass. Rep.,* 1844 (Pt. 2) (1844) 74–7; *Wiegmann, Archiv,* **11** (1845) 369–83
Observations on the habits of various marine animals:- 1. On the food and digestion of the Bulla lignaria. 2. On the nourishment and digestion of the Asterias (Astrorpecten) aurianticus. 3. Observations upon Octopus vulgaris and Pinna nobilis. *Ann. Nat. Hist.,* **20** (1857) 334–6

Sallitt, Jessie Amelia
Sallitt, Jessie A.
On the chlorophyll corpuscles of some infusoria. *Quart. J. Micr. Sci.,* **24** (1884) 165–70

Sheldon, Lilian
Sheldon, (Miss) Lilian
Note on the ciliated pit of ascidians and its relation to the nerve-ganglion and the so-called hypophysial gland; and an account of the anatomy of Cynthia rustica(?). [1887] *Quart. J. Micr. Sci.,* **28** (1888) 131–48
On the development of Peripatus novae-zealandiae. [1887–1888] *Quart. J. Micr. Sci.,* **28** (1888) 205–37; **29** (1889) 283–93
Notes on the anatomy of Peripatus capensis and Peripatus novae-zealandiae. *Quart. J. Micr. Sci.,* **28** (1888) 495–9
The maturation of the ovum in the Cape and New Zealand species of Peripatus. [1889] *Quart. J. Micr. Sci.,* **30** (1890) 1–29
Johnson, (Miss) Alice; Sheldon, (Miss) Lilian
Notes on the development of the newt (Triton cristatus). *Quart. J. Micr. Sci.,* **26** (1886) 573–89
On the development of the cranial nerves of the newt. *Roy. Soc. Proc.,* **40** (1886) 94–5

Staveley, E. F.
Staveley, (Miss) E. F.
Notes on the form of the comb (Pecten) in different Andrenidae and Apidae, and on the alar hooks of the species of Sphecodes and Halictus. *Zool. Soc. Proc.* (1862) 118–23
Observations on the neuration of the hind wings of hymenopterous insects, and on the hooks which join the fore and hind wings together in flight. [1860] *Linn. Soc. Trans.,* **23** (1862) 125–38
Note on the presence of teeth on the maxillae of Spiders. *Zool. Soc. Proc.* (1865) 673–4; *Ann. Mag. Nat. Hist.,* **17** (1866) 399–400

Thorneley, Laura Roscoe
Thorneley, (Miss) Laura Roscoe
Supplementary report upon the hydroid zoophytes of the L.M.B.C. district. *Liverpool Biol. Soc. Proc. & Trans.,* **8** (1894) 140–7
[Notes on Rockall Island and Bank . . . vi. Reports.] On the Polyzoa. [On the Hydrozoa.] [1896] *Irish Ac. Trans.,* **31** (1896–1901) 79–80; 81

Veley, Lilian Jane Gould
Gould, Lilian J.
Notes on the minute structure of Pelomyxa palustris (*Greeff*). [With addendum by M. D. Hill] *Quart. J. Micr. Sci.,* **36** (1894) 295–306

Warham, Amy E.
Warham, (Miss) Amy E.
On variations in the dorsal tubercle of Ascidia virginia. [1892] *Liverpool Biol. Soc. Proc. & Trans.,* **7** (1893) 98–9

Wilson, Charlotte
Palethorpe, Fanny D.; Wilson, Charlotte
Preliminary paper on a collection of simple ascidians from Australian seas. *Liverpool Biol. Soc. Proc.,* **1** (1887) 63–6

Part 2. Mathematical, Physical, and Earth Sciences

Astronomy

United States

Babbitt, Mary Edith Tarbox
Tarbox, Mary E.
Observations of minor planets, made at the Vassar College Observatory. [1897] *Astr. J.,* **18** (1898) 8

Bardwell, Elisabeth Miller
Bardwell, (Miss) Elisabeth M.
"A star shower." *Sidereal Messenger,* **5** (1886) 29
Mount Holyoke Observatory, sunspot observations. *Sidereal Messenger,* **10** (1891) 37
Meteor in daytime. *Popular Astr.,* **1** (1894) 192
Leonid meteors observed at Mount Holyoke College Observatory. *Popular Astr.,* **7** (1899) 49–50
Bardwell, (Miss) Elisabeth M.; Mack, (Miss) I. G.
Observations of the solar eclipse of March 16th, 1885, at the John B. Williston Observatory, South Hadley, Mass. *Sidereal Messenger,* **4** (1885) 126

Byrd, Mary Emma
Byrd, Mary E.
First observations of the sun and moon. *Popular Astr.,* **1** (1894) 216–21; 252–7
Observations of Encke's Comet, made at Smith College Observatory, Northampton, Mass., with the 11-inch refractor and filar micrometer. *Astr. J.,* **15** (1895) 103
Byrd, Mary E.; Whitney, Mary W.
Longitude of Smith College Observatory. *Harvard Astr. Obs. Ann.,* **29** (1893) 35–63

Davidson, Ellinor Fauntleroy
Davidson, (Mrs.) E.
Observations on the transit of Venus at Nagasaki. *Amer. J. Sci.,* **9** (1875) 235–6

Davidson, (Mrs.) E.; Davidson, G.
Observations of meteors, 13–14 Nov. 1869, at Santa Barbara, California. *Astr. Soc. Month. Not.*, **30** (1870) 67–8

Davis, Elizabeth Brown
Davis, (Mrs.) Elizabeth Brown
Elements of Comet 1888 I. *Sidereal Messenger*, **7** (1888) 352
Elements of the Comet *e* 1891 (Barnard). *Astr. J.*, **11** (1892) 119

Drake, Eleanor B.
Drake, Eleanor B.
A brilliant meteor. *Popular Astr.*, **7** (1899) 543–4

Furness, Caroline Ellen
Furness, Caroline E.
Sunspot observations. *Sidereal Messenger*, **10** (1891) 150
The scientific value of a total lunar eclipse. [1895] *Popular Astr.*, **3** (1896) 109–14
Tycho Brahe. *Popular Astr.*, **3** (1896) 221–6
Observations of Comet *b* 1898, made at the Observatory of Vassar College. [1898] *Astr. J.*, **19** (1899) 39
Whitney, Mary W.; Furness, Caroline E.
Observations of minor planet (372), made at the Vassar College Observatory. *Astr. J.*, **15** (1895) 162
Observations of minor planets, made at the Vassar College Observatory. *Astr. J.*, **16** (1896) 47; **17** (1897) 77; 144; 181–2; **18** (1898) 109; 178–9
Observations of Comet *a* 1896, made at the Vassar College Observatory with the 12-inch equatorial. *Astr. J.*, **16** (1896) 72
Observations of Comets *a* and *b* 1896, made at the Vassar College Observatory. *Astr. J.*, **16** (1896) 103
Observations of Comet *b* 1896 (Swift), made at the Vassar College Observatory. *Astr. J.*, **16** (1896) 119; **17** (1897) 136
Observations of minor planets and comets, made at the Vassar College Observatory. [1896] *Astr. J.*, **17** (1897) 37–8
Observations of Comet *b* 1897, made at the Vassar College Observatory. [1897] *Astr. J.*, **18** (1898) 80
Observations of minor planets and Comet *i* 1898, made at the Observatory of Vassar College with the 12-inch refractor. *Astr. J.*, **19** (1899) 187
Observations of minor planets, made at Vassar College Observatory with the 12-inch refractor. [1899] *Astr. J.*, **20** (1900) 29–30
Observations of comets and minor planets, made at the Vassar College Observatory. *Astr. J.*, **20** (1900) 159–60
Observations of minor planet 1899 EY made at the Vassar College Observatory. *Astr. Nachr.*, **152** (1900) 171–4
Observations of minor planets made with the 12-inch equatorial of the Vassar College Observatory, Poughkeepsie, N. Y. *Astr. Nachr.*, **154** (1901) 139–42

Harpham, Florence Ellen
Harpham, (Miss) Flora E.
Elements of Comet 1892 I. (Swift March 6). *Astr. & Astrophys.*, **11** (1892) 625
Gale's Comet in an inch-and-a-half glass. *Popular Astr.*, **1** (1894) 472
Hints for a small telescope. [1894] *Popular Astr.*, **2** (1895) 106–7
The Rutherfurd photographs. *Popular Astr.*, **8** (1900) 129–36
Sivaslian, A. G.; Harpham, (Miss) Flora E.
Elements and ephemeris of Comet *b* 1892 (Swift). *Astr. & Astrophys.*, **11** (1892) 344; 444
Wilson, Herbert C.; Harpham, (Miss) Flora E.
Ephemeris of Comet *a* 1892 (Swift). *Astr. & Astrophys.*, **11** (1892) 536

Hayes, Ellen Amanda
Hayes, Ellen
Sun-heat and orbital eccentricity. *Science*, **21** (1893) 227–8

Hobe, Adelaide M.
Hobe, A.; Kuno, Y.; Phipps, S. C.; Sprague, Roger
Elements and Ephemeris of Comet *e*, 1899 (Giacobini). [1899] *Astr. Soc. Pacific Publ.*, **11** (1899) 190–1; *Astr. J.*, **20** (1900) 124

Howe, Fannie McClurg Shattuck
Howe, Mrs. Herbert A.
Observations [of the transit of Mercury, 1894 November 9, 10. Made at the Chamberlin Observatory, University Park, Colorado.] *Astr. J.*, **14** (1895) 164

Iliff, Edna
Iliff, Edna; Waterbury, Lottie
Groups of time-stars for amateur astronomers. [1894] *Popular Astr.*, **2** (1895) 74–8

Mack, Isabella Graham
Bardwell, (Miss) Elisabeth M.; Mack, (Miss) I. G.
Observations of the solar eclipse of March 16th, 1885, at the John B. Williston Observatory, South Hadley, Mass. *Sidereal Messenger*, **4** (1885) 126

Maury, Antonia Caetana de Paiva Pereira
Maury, (Miss) Antonia C. de P. P.
Spectra of bright stars photographed with the 11-inch Draper telescope as a part of the Henry Draper Memorial. [Under the direction of Edward C. Pickering, Director of the Observatory.] *Harvard Astr. Obs. Ann.*, **28** (Pt.1) (1897) 1–128
The K lines of β Aurigae. *Astrophys. J.*, **8** (1898) 173–5; *Science*, **8** (1898) 457
Corona seen after totality. *Popular Astr.*, **8** (1900) 401–2

Mitchell, Maria
Mitchell, (Miss) Maria
Observations and elements of Miss Mitchell's comet. *Astr. Soc. Month. Not.*, **8** (1847–1848) 9–11; 130–1
Minima of Algol. Gould, *Astr. J.*, **5** (1858) 7
Observations on some of the double stars. *Amer. J. Sci.*, **36** (1863) 38–40
On Jupiter and its satellites. *Amer. J. Sci.*, **1** (1871) 393–5
Notes on observations on Jupiter and its satellites. *Amer. J. Sci.*, **5** (1873) 454–6; **15** (1878) 38–41
Notes on the satellites of Saturn. *Amer. J. Sci.*, **17** (1879) 430–2

Moulton, Mary Etta
Moulton, (Miss) Mary Etta
Total solar eclipse in India, Jan. 22, 1898. *Popular Astr.*, **6** (1898) 121–3
The Leonids in India. *Popular Astr.*, **8** (1900) 104–5

O'Halloran, Rose
O'Halloran, (Miss) Rose
"When the moon runs highest and runs lowest." *Astr. Soc. Pacific Publ.*, **7** (1895) 265–72
Observations of Mira Ceti. [1895–1899]. *Astr. Soc. Pacific Publ.*, **8** (1896) 79–81; **10** (1898) 103–4; **11** (1899) 80–2
The meteors of the 13th November. [1895] *Popular Astr.*, **3** (1896) 213
Observations of *o* Ceti. *Popular Astr.*, **3** (1896) 384; 440; **6** (1898) 56; 255; **7** (1899) 223
Observations with a four-inch telescope of the recent maxima of the variable stars R and S Scorpii. *Astr. Soc. Pacific Publ.*, **8** (1896) 254

Maximum of *o* Ceti (Mira), 1896–97. [1897] *Astr. Soc. Pacific Publ.,* **9** (1897) 86–7; *Popular Astr.,* **5** (1898) 55
The Leonids. *Popular Astr.,* **4** (1897) 453
The sun-spot minimum. *Popular Astr.,* **6** (1898) 478–9
Observations of the maximum of Mira (*o* Ceti) 1898. *Popular Astr.,* **6** (1898) 586–7
The red stars V Hydrae and 277 of Birmingham's Catalogue. *Astr. Soc. Pacific Publ.,* **10** (1898) 105; *Popular Astr.,* **6** (1898) 254
The surface of the Sun. *Astr. Soc. Pacific Publ.,* **10** (1898) 222–4
Cyclonic motion in a sun-spot. *Observatory, London,* **22** (1899) 274
Maxima of *o* Ceti and other variables (1899). *Astr. Soc. Pacific Publ.,* **11** (1899) 246–7
The Philippine firmament. *Popular Astr.,* **7** (1899) 201–5
The recent cyclonic sunspot. *Popular Astr.,* **7** (1899) 278
The solar cyclones. *Popular Astr.,* **7** (1899) 331
Maximum of *o* Ceti (Mira) in 1899. *Popular Astr.,* **7** (1899) 544
Observations of variable stars. *Astr. Soc. Pacific Publ.,* **12** (1900) 24–5
Variables. *Popular Astr.,* **8** (1900) 112
The total eclipse of the Sun at New Orleans. *Popular Astr.,* **8** (1900) 349–50

Palmer, Margaretta
Palmer, Margaretta
Determination of the orbit of Comet 1847 VI. [1893] *Yale Univ. Obs. Trans.,* **1** (1887–1904) 183–207

Sanborn, Alice J.
Sanborn, (Miss) A. J.
Investigation of the inequality of the pivots of the Fauth transit of the Washburn Observatory. [1883] *Sidereal Messenger,* **2** (1883–1884) 225
{Determination of the thread intervals. *Washburn Obs. Publ.,* **2** (1884) 36–8}

Silliman, Hepsa Ely
Silliman, Mrs. G. S.
On the origin of aerolites. *Edinb. New Phil. J.,* **16** (1862) 227–48

Sparks, Marion E.
Sparks, Marion E.
Brilliant meteor. *Popular Astr.,* **5** (1898) 558–9

Updegraff, Alice Maxwell Lamb
Lamb, (Miss) Alice Maxwell
The change of latitude in the observations made at Willet's Point, New York. *Sidereal Messenger,* **4** (1885) 129–33
Observations at Willet's Point. *Sidereal Messenger,* **5** (1886) 179–80
An index to certain classes of stars contained in the Greenwich Catalogues, reduced to 1875.0. *Washburn Obs. Publ.,* **5** (1887) 23*; 116–230
Schreiben betr. die Bemerkungen zu dem "Index [to . . . the Greenwich Catalogues"]. *Astr. Nachr.,* **119** (1888) 191–2
Updegraff, Milton; Lamb, (Miss) Alice Maxwell (Mrs. Updegraff)
Some observations of the 303 stars between $5^{h.}$ and $12^{h.}$ of right ascension. *Washburn Obs. Publ.,* **4** (1886) 1*-24*
Observations with the meridian circle [made at the Washburn Observatory]. *Washburn Obs. Publ.,* **6** (1890), Pt. 1, 1–21

Wagner, Mary S.
Wagner, Mary S.
Double-star measures, made at the observatory of the University of Minnesota. [1897] *Astr. J.,* **18** (1898) 13
Whitney, Mary W.; Wagner, Mary S.
Observations of Comet *g* 1892 (Brooks), made at the Vassar College Observatory. [1893] *Astr. J.,* **12** (1893) 188; 188; **13** (1894) 7; 7

Waterbury, Lottie
Iliff, Edna; Waterbury, Lottie
Groups of time-stars for amateur astronomers. [1894] *Popular Astr.,* **2** (1895) 74–8

Wentworth, F. Gertrude
Wentworth, (Miss) F. Gertrude
Elements of Comet *a* 1892 (Swift). [1892] *Astr. J.,* **12** (1893) 72
Ephemeris of Comet *a* 1892 (Swift). [1892] *Astr. J.,* **12** (1893) 80; 96; 126; 149
Ephemeris of Comet *e* 1892 (Barnard). *Astr. J.,* **12** (1893) 104
On the parallax of β Cygni. *Astr. J.,* **12** (1893) 191

Whiting, Sarah Frances
Whiting, S. F.
The Whitin Observatory of Wellesley College. *Popular Astr.,* **8** (1900) 482–4

Whitney, Mary Watson
Whitney, Mary W.
Double star S 503. *Sidereal Messenger,* **10** (1891) 300
Observations of Comet *a* 1890, made at the Vassar College Observatory with the 12-inch equatorial. [1890] *Astr. J.,* **10** (1891) 55
Observations of Nova Aurigae at Vassar College Observatory. *Astr. & Astrophys.,* **11** (1892) 461–2
Filar-micrometer observations of Comet *a* 1892 (Swift), made with the 12-inch equatorial of Vassar College Observatory. [1892] *Astr. J.,* **12** (1893) 48; 93
Observations of Comet *f* 1892, made at Vassar College Observatory. [1892] *Astr. J.,* **12** (1893) 133; 149
Filar-micrometer observations of Comet *g* 1892, made at Vassar College Observatory. *Astr. J.,* **12** (1893) 175
Some recent markings on Jupiter. *Astr. & Astrophys.,* **12** (1893) 22–3
Observations of Comet 1892 III (Holmes), made at the Vassar College Observatory. [1893] *Astr. J.,* **13** (1894) 38–9
Lunar eclipse of 1895 March 10. *Astr. J.,* **15** (1895) 38
Observations of Comet *c* 1895 (Perrine), made at the Vassar College Observatory. *Astr. J.,* **16** (1896) 38
Total lunar eclipse of September 3, 1895 observed at Poughkeepsie, N. Y. *Astr. Nachr.,* **139** (1896) 79–80
Determination of the positions of comets. [1896] *Popular Astr.,* **4** (1897) 177–81
Observation[s] of Comet *g* 1896 (Perrine), made at the Vassar College Observatory. [1896–97] *Astr. J.,* **17** (1897) 56; 72; 92
Problem of solar motion. [1897] *Popular Astr.,* **5** (1898) 309–15
[Observations of Comet *b* 1898 (Perrine),] at Vassar College Observatory. [1898] *Astr. J.,* **19** (1899) 8
Observations of Leonids. *Astr. J.,* **20** (1900) 148
Observations of the total solar eclipse by the Vassar College party. *Astrophys. J.,* **122** (1900) 96
Occultations observed during the lunar eclipse of Dec. 16, 1899. *Astr. Nachr.,* **152** (1900) 261–2
The eclipse of May 28th. *Science,* **11** (1900) 992–3
Byrd, Mary E.; Whitney, Mary W.
Longitude of Smith College Observatory. *Harvard Astr. Obs. Ann.,* **29** (1893) 35–63
Whitney, Mary W.; Wagner, Mary S.
Observations of Comet *g* 1892 (Brooks), made at the Vassar College Observatory. [1893] *Astr. J.,* **12** (1893) 188; 188; **13** (1894) 7; 7

Whitney, Mary W.; Furness, Caroline E.
Observations of minor planet (372), made at the Vassar College Observatory. *Astr. J.,* **15** (1895) 162
Observations of minor planets, made at the Vassar College Observatory. *Astr. J.,* **16** (1896) 47; **17** (1897) 77; 144; 181–2; **18** (1898) 109; 178–9
Observations of Comet *a* 1896, made at the Vassar College Observatory with the 12-inch equatorial. *Astr. J.,* **16** (1896) 72
Observations of Comets *a* and *b* 1896, made at the Vassar College Observatory. *Astr. J.,* **16** (1896) 103
Observations of Comet *b* 1896 (Swift), made at the Vassar College Observatory. *Astr. J.,* **16** (1896) 119; **17** (1897) 136
Observations of minor planets and comets, made at the Vassar College Observatory. [1896] *Astr. J.,* **17** (1897) 37–8
Observations of Comet *b* 1897, made at the Vassar College Observatory. [1897] *Astr. J.,* **18** (1898) 80
Observations of minor planets and Comet *i* 1898, made at the Observatory of Vassar College with the 12-inch refractor. *Astr. J.,* **19** (1899) 187
Observations of minor planets, made at Vassar College Observatory with the 12-inch refractor. [1899] *Astr. J.,* **20** (1900) 29–30
Observations of comets and minor planets, made at the Vassar College Observatory. *Astr. J.,* **20** (1900) 159–60
Observations of minor planet 1899 EY made at the Vassar College Observatory. *Astr. Nachr.,* **152** (1900) 171–4
Observations of minor planets made with the 12-inch equatorial of the Vassar College Observatory, Poughkeepsie, N. Y. *Astr. Nachr.,* **154** (1901) 139–42
Whitney, Mary W.; Everett, Alice
Observations of minor planets, made at the observatory of Vassar College with the 12-inch refractor. [1899] *Astr. J.,* **20** (1900) 47
Observations of minor planets and Comet *a* 1899, made at the observatory of Vassar College with the 12-inch refractor. [1899] *Astr. J.,* **20** (1900) 76–7

Willard, Charlotte Richards
Willard, (Miss) Charlotte R.
Time and time signals. [1893] *Popular Astr.,* **1** (1894) 60–2
Willard, (Miss) Charlotte; Wilson, Herbert C.
Carleton College sunspot observations. *Sidereal Messenger,* **9** (1890) 136; 230; 280; 325; 418; 468; **10** (1891) 37; 102; 150
Wilson, Herbert C.; Willard, (Miss) Charlotte R.
Ephemeris of comet 1892 I (Swift). *Astr. & Astrophys.,* **12** (1893) 184

Winlock, Anna
Rogers, William A.; Winlock, Anna
A catalogue of 130 polar stars for the epoch of 1875.0, resulting from all the available observations made between 1860 and 1885, and reduced to the system of the Catalogue of Publication XIV of the Astronomische Gesellschaft. [1886] *Amer. Ac. Mem.,* **11** (1888) 227–99

Young, Anne Sewell
Young, Anne Sewell
Elliptic elements of Comet *g* 1896. *Astr. J.,* **17** (1897) 192
The Leonids. *Popular Astr.,* **4** (1897) 498–9
The Sun's heat. *Popular Astr.,* **6** (1898) 145–8
Wilson, Herbert C.; Young, Anne Sewell
Ephemeris of Comet *g* 1896 (Perrine). *Popular Astr.,* **4** (1897) 450

United States—Great Britain

Fleming, Williamina Paton Stevens
Fleming, (Mrs.) M.
Spectra of δ and μ Centauri. *Astr. Nachr.,* **123** (1890) 383–4
New variable in Caelum. *Astr. Nachr.,* **124** (1890) 175–6
Stars having peculiar spectra. [Including new variables in Triangulus and Hydra, in Aquarius, Delphinus and Camelopardalus, in Sagittarius, in Lacerta, a group of the fifth type in Cepheus, and variables in Centaurus, Lupus, Pavo and Microscopium.] *Astr. Nachr.,* **125** (1890) 155–6; 363–4; **126** (1891) 117–20; 165–6; **127** (1891) 5–6; **128** (1891) 11–14; 121–2; 403–4; **135** (1894) 195–8; **137** (1895) 71–4; **138** (1895) 175–80
Two new variable stars near the cluster 5 M Librae. (N. G. C. 5904) *Astr. Nachr.,* **125** (1890) 157–8
New variable star in Scorpius RA. $16^h 48^m$. 4 Decl. -44° 57' (1900). *Astr. Nachr.,* **125** (1890) 361–4
New variable star in Sagittarius RA. $20^h 9^m$. 4 Decl. -39° 29' (1900). *Astr. Nachr.,* **125** (1890) 365–6
New variable star in Camelopardalis. *Sidereal Messenger,* **10** (1891) 152
New planetary nebula. (DM.-12° 1172 magn. 9. 2.) *Sidereal Messenger,* **10** (1891) 240
Objects of interest on spectrum plates and two new variable stars in Perseus. *Astr. Nachr.,* **126** (1891) 163–6
Stars having peculiar spectra. New variable stars in Perseus, Triangulum and Hydra, [and in Aquarius and Delphinus]. *Sidereal Messenger,* **10** (1891) 7–8; 106
Harvard College Observatory astronomical expedition to Peru. *Astr. Soc. Pacific Publ.,* **4** (1892) 58–62
New variable stars in the southern sky. *Astr. Nachr.,* **130** (1892) 125–6
Stars having peculiar spectra. *Astr. & Astrophys.,* **11** (1892) 418–19; 765–7; **12** (1893) 170; 546–7; 810–11; **13** (1894) 501–3
{A field for woman's work in astronomy. *Astr. & Astrophys.,* **12** (1893) 683–9}
Entdeckung eines neuen Sterns im Sternbilde Norma. *Astr. Nachr.,* **134** (1894) 59–60
Two new variable stars. *Astr. & Astrophys.,* **13** (1894) 195; *Popular Astr.,* **1** (1894) 324
Stars having peculiar spectra. Eleven new variable stars. [Eight new variable stars in Cetus, Vela, Centaurus, Lupus, Scorpio, Aquila and Pegasus.] *Astrophys. J.,* **1** (1895) 411–15; **2** (1895) 354–59
Seven new variable stars. *Astrophys. J.,* **2** (1895) 198–201
Stars of the fifth type in the Magellanic clouds. *Astrophys. J.,* **8** (1898) 232; *Science,* **8** (1898) 452
Classification of the spectra of variable stars of long period. *Astrophys. J.,* **8** (1898) 233; *Science,* **8** (1898) 455

Proctor, Mary
Proctor, Mary
Total eclipse of the Sun. [1896] *Popular Astr.,* **4** (1897) 221–2
Evenings with the stars. [1896–97] *Popular Astr.,* **4** (1897) 321–7; 374–80; 436–43; 515–20; 562–9; **5** (1898) 37–45; 263–7; 321–4; 382–5
Heavens for June, July and August. [1897] *Popular Astr.,* **5** (1898) 97–105; 152–9; 207–14

Great Britain

Ashley, Mary

Ashley, Mary

The eclipse of the moon. [1877, Aug. 23] *Observatory, London,* **1** (1878) 177

{Hyginus N. [1880] *Selenographical J.,* **3** (1878–1882) 36}

{Notes [on lunar features]. [1880] *Selenographical J.,* **3** (1878–1882) 36–8}

Brown, Elizabeth

Brown, (Miss) Elizabeth

{Observations of proper motions in certain sunspots. *Liverpool Astr. Soc. Abs. Proc.,* **2** (1883–1884) 18–19}

{A short review of the sunspots of 1882 and 1883. *Liverpool Astr. Soc. Abs. Proc.,* **2** (1883–1884) 62–4}

{An equatored sunspot. *Liverpool Astr. Soc. J.,* **3** (1884) 92}

{Report. Solar Section Director. *Liverpool Astr. Soc. J.,* **4** (1885) 2}

Variable orange stars. *Observatory, London,* **18** (1885) 200–1

{Remarkable sunspots. *Liverpool Astr. Soc. J.,* **5** (1886) 50–2}

{The search for Vulcan. *Liverpool Astr. Soc. J.,* **5** (1886) 146}

{Short note on Solar Section observations. *Liverpool Astr. Soc. J.,* **6** (1887–1888) 133}

{Aurorae and sunspots. *Liverpool Astr. Soc. J.,* **7** (1888–1889) 52–4}

{Solar section. *Brit. Astr. Ass. J.,* **1** (1890–1891) 58–60}

{A few notes on the sun-spots of June 1891. *Brit. Astr. Ass. J.,* **1** (1890–1891) 546–8}

{A few hints to beginners in solar observation. *Astr. Soc. Pacific Publ.,* **3** (1891) 172–5}

{Section for the observation of the Sun. Report. *Brit. Astr. Ass. Mem.* **1** (1891) 99–138; **2** (1892) 53–105; **3** (1895) 49–118; **4** (1896) 43–106; **5** (1897) 83–122; **6** (1898) 143–74; **7** (1899) 17–46}

{Note on the recent large groups of sun-spots. *Brit. Astr. Ass. J.,* **2** (1891–1892) 210–11}

{Note on the sun-spots of December 24–26, 1892. *Brit. Astr. Ass. J.,* **3** (1892–1893) 186–7}

{Some typical forms of sun-spots. *Brit. Astr. Ass. J.,* **3** (1892–1893) 272–3}

{Large sun-spots of August 1893. *Brit. Astr. Ass. J.,* **3** (1892–1893) 494–6}

Anthélie. *Astronomie* (1893) 278

{A few questions with relation to sun-spots. *Brit. Astr. Ass. J.,* **4** (1893–1894) 202}

{Note on a peculiar feature of the large sun-spot of February 1894. *Brit. Astr. Ass. J.,* **4** (1893–1894) 300}

{A visit to the Madrid observatory. *Brit. Astr. Ass. J.,* **4** (1893–1894) 361–2}

{Growth and decay of sun-spots. *Brit. Astr. Ass. J.,* **5** (1894–1895) 460–1}

{Some curious changes in sun-spots. *Brit. Astr. Ass. J.,* **5** (1894–1895) 513–4}

{The sun-spots of the past year (1894, October 1 to 1895, September 30). *Brit. Astr. Ass. J.,* **6** (1895–1896) 15–16}

{The colours of stars. *Brit. Astr. Ass. J.,* **6** (1895–1896) 32–3}

{Growth and decay of sun-spots. *Brit. Astr. Ass. J.,* **6** (1895–1896) 170–1}

{Summary of sun-spots for the half year ending March 31, 1896. *Brit. Astr. Ass. J.,* **6** (1895–1896) 251–2}

{A brief review of the sun-spots of the past year. *Brit. Astr. Ass. J.,* **7** (1896–1897) 50–3}

{The large sun-spot of January 1897. *Brit. Astr. Ass. J.,* **7** (1896–1897) 204–5}

{The "black holes" in sun-spots. *Brit. Astr. Ass. J.,* **7** (1896–1897) 206}

{Summary of sun-spots for the half-year ending March 31, 1897. *Brit. Astr. Ass. J.,* **7** (1896–1897) 454–5}

Observations of variable and suspected variable stars. [1896] *Brit. Astr. Ass. Mem.,* **5** (1897) 28–9

{Remarkable group of sun-spots in December 1897. *Brit. Astr. Ass. J.,* **8** (1897–1898) 186–7}

{Summary of sun-spots (for the half year ending March 31, 1898). *Brit. Astr. Ass. J.,* **8** (1898–1898) 276–7}

{Summary of sun-spots for the half year ending September 30, 1898. *Brit. Astr. Ass. J.,* **9** (1898–1899) 24}

Fry, Isabel

Fry, Isabel

[A brilliant meteor.] *Nature,* **36** (1887) 30

Jones, S. Grace

Jones, (Miss) S. Grace

Meteor seen at Kensworth. [1898] *Herts. Nat. Hist. Soc. Trans.,* **10** (1901) 68

Lehmann, Susanna

Lehmann, Susanna

A bright meteor. [1898] *Nature,* **57** (1897–1898) 271

Somerville, Mary Fairfax Greig

Somerville, Mary

{Art. VII.—Ueber den Halleyschen Cometen. Von Littrow. Wein, 1835. 2. Ueber den Halleyschen Cometen. Von Professor von Encke. Berliner Jahrbuch, 1835, etc. *Quarterly Rev.* **55** (1835) 195–223}

Young, Grace Emily Chisholm

Chisholm, (Miss) Grace

A meteor. *Nature,* **46** (1892) 490

Great Britain—Ireland

Clerke, Agnes Mary

Clerke, (Miss) Agnes M.

{The great southern comet of 1880. *Fraser's Magazine for Town and Country,* **21,** n.s. (1881) 224–34}

{The recent aurora. *Nature,* **26** (1882) 549}

The Pleaides. *Nature,* **33** (1886) 561–4

Homeric astronomy. *Nature,* **35** (1887) 585–8; 607–8

Globular star clusters. *Nature,* **38** (1888) 365–7

Variable double stars. *Observatory, London,* **11** (1888) 188–92

Southern star spectra. *Observatory, London,* **11** (1888) 429–32

Irregular star clusters. [1888] *Nature,* **39** (1889) 13–14

An historical and descriptive list of some double stars suspected to vary in light. [1888] *Nature,* **39** (1889) 55–8

Photographic star-gauging. *Nature,* **40** (1889) 344–6

Some southern red stars. *Observatory, London,* **12** (1889) 134–6

Stellar spectroscopy at the Lick Observatory. *Observatory, London,* **13** (1890) 46–9

Dark stars. *Observatory, London,* **13** (1890) 235–8

New double stars. [1889] *Nature,* **41** (1890) 132–3

Rigel and the Great Nebula. *Observatory, London,* **13** (1890) 313–16; 347–8

The system of Zeta Cancri. *Astr. Soc. Pacific Publ.,* **2** (1890) 188–91

{The rotation periods of Mercury and Venus. [Critical review]. *Brit. Astr. Assoc. J.,* **1** (1890–1891) 20–5}
A southern observatory. [1889] *Smithsonian Rep.* (1891) 115–26
Recent photographs of the Annular Nebula in Lyra. *Nature,* **43** (1891) 419–20
The Sun's motion in space. *Nature,* **44** (1891) 572–4
Nova Aurigae and its suggestions. *Observatory, London,* **15** (1892) 286–9
Nova Aurigae. *Observatory, London,* **15** (1892) 334–9
The new star in Auriga. *Astr. & Astrophys.,* **11** (1892) 504–13
The system of Algol. *Science,* **19** (1892) 298–9
The distribution of the stars. *Astr. & Astrophys.,* **12** (1893) 515–20
A new method in astronomy. *Observatory, London,* **17** (1894) 234–6
Some anomalous sidereal spectra. *Observatory, London,* **18** (1895) 193–6
Five short-period variables. *Observatory, London,* **19** (1896) 115–16
A new class of variable stars. *Observatory, London,* **20** (1897) 52–5
Is star strewn space infinite? *Popular Astr.,* **4** (1897) 431–4
Les progrès de l'astronomie depuis soixante ans. [Tr.] [1897] *Ciel et Terre,* **18** (1897–1898) 107–15; 140–6
On the spectra of certain nebulae. *Science,* **8** (1898) 454
Notes on the Wolf-Rayet stars. *Observatory, London,* **22** (1899) 52–5
Helium in long-period variables. *Observatory, London,* **22** (1899) 198–9
The spectrum of β Cygni. *Observatory, London,* **22** (1899) 387–9
Some remarkable spectroscopic binaries. *Observatory, London,* **23** (1900) 127–9

Everett, Alice
Everett, (Miss) Alice
{The total lunar eclipse of November 15, 1891. *Brit. Astr. Ass. J.,* **2** (1891–1892) 88}
{On the photographic magnitude of Nova Aurigae. *Brit. Astr. Ass. J.,* **2** (1891–1892) 276–8}
Note on the binary I Leonis. R. A. $11^{h.}$ $18^{m.}$ $27^{s.}$, Dec. 11° $6^{1}.5$, 1895. ($3^{m.}$ 9 and $7^{m.}$ 1.) *Astr. Soc. Month. Not.,* **55** (1895) 440–1
Galactic longitude and latitude of poles of binary-star orbits. *Astr. Soc. Month. Not.,* **56** (1896) 462–5
Whitney, Mary W.; Everett, Alice
Observations of minor planets and Comet *a* 1899, made at the Observatory of Vassar College with the 12-inch refractor. [1899] *Astr. J.,* **20** (1900) 76–7
Observations of minor planets, made at the Observatory of Vassar College with the 12-inch refractor. [1899] *Astr. J.,* **20** (1900) 47

Huggins, Margaret Lindsay Murray (Lady)
Huggins, Margaret L. (Lady)
The astrolabe. A summary. *Astr. & Astrophys.,* **13** (1894) 793–801
{". . . teach me how to name the . . . light." *Astrophys. J.,* **8** (1898) 54}
Huggins, (Sir) William; Huggins, Margaret L. (Lady)
On the spectrum, visible and photographic, of the Great Nebula in Orion. [1889] *Roy. Soc. Proc.,* **46** (1890) 40–60
Note on the photographic spectra of Uranus and Saturn. [1889] *Roy. Soc. Proc.,* **46** (1890) 231–3
On a re-determination of the principal line in the spectrum of the Nebula in Orion, and on the character of the line. [1890] *Roy. Soc. Proc.,* **48** (1891) 202–13
Note on the photographic spectrum of the Great Nebula in Orion. [1890] *Roy. Soc. Proc.,* **48** (1891) 213–16
On a new group of lines in the photographic spectrum of Sirius. [1890] *Roy. Soc. Proc.,* **48** (1891) 216–17
On Wolf and Rayet's bright-line stars in Cygnus. [1890] *Roy. Soc. Proc.,* **49** (1891) 33–46
Preliminary note on Nova Aurigae. *Astr. Nachr.,* **129** (1892) 107; *Roy. Soc. Proc.,* **50** (1892) 465–6
On Nova Aurigae. *Astr. & Astrophys.,* **11** (1892) 571–81; *Roy. Soc. Proc.,* **51** (1892) 486–95
On the bright bands in the present spectrum of Nova Aurigae. [1893] *Roy. Soc. Proc.,* **54** (1894) 30–6
On the relative behaviour of the H and K lines of the spectrum of calcium. *Roy. Soc. Proc.,* **61** (1897) 433–40
Spectroscopic notes. *Astrophys. J.,* **6** (1897) 322–7

Chemistry

United States

Atkinson, Elizabeth Allen
Atkinson, (Miss) Elizabeth Allen
I. Metal separation by means of hydrobromic acid gas. II. Indium in tungsten minerals. *Amer. Chem. Soc. J.,* **20** (1898) 797–813
Atkinson, (Miss) Elizabeth Allen; Smith, Edgar Fahs
The separation of iron from beryllium. *Amer. Chem. Soc. J.,* **17** (1895) 688–9

Balcom, Lillian Lynn
Orndorff, William Ridgely; Balcom, (Miss) L. L.
The polymeric modifications of propionic aldehyde: parapropionic and metapropionic aldehydes. *Amer. Chem. J.,* **16** (1894) 645–50

Barker, Alice Hopkins Albro
Chittenden, R. H.; Albro, (Miss) Alice H.
The formation of melanins or melanin-like pigments from proteid substances. *Amer. J. Physiol.,* **2** (1899) 291–305

Bills, Harriet V.
Bills, Harriet V.; Vaughan, V. C.
Estimation of lime in the shell and in the interior of the egg, before and after incubation. *J. Physiol,* **1** (1879), 434–6

Bouton, Rosa
Avery, Samuel; Bouton, (Miss) Rosa
On phenylglutaric acid and its derivatives. *Amer. Chem. J.,* **20** (1898) 509–15

Bragg, Charlotte Almira
Richards, Ellen H.; Bragg, Lottie A.
{The distribution of phosphorus and nitrogen in the products of modern milling. *Tech. Quart,* **3** (1890) 246–52}

Breed, Mary Bidwell
Breed, Mary Bidwell
On phenolphthalein and methyl-orange as indicators. [With discussion.] *Franklin Inst. J.,* **135** (1893) 312–16
Keiser, Edward H.; Breed, Mary Bidwell
The atomic weight of palladium. *Amer. Chem. J.,* **16** (1894) 20–8
The action of magnesium upon the vapors of alcohols and a new method of preparing allylene. [1894] *Franklin Inst. J.,* **139** (1895) 304–9

Cheney, Margaret Swan
Cheney, Margaret S.; Richards, Ellen Swallow
A new and ready method for the estimation of nickel in pyrrhotites and mattes. *Amer. J. Sci.,* **14** (1877) 178–81; *Chem. News,* **36** (1877) 161–2

Clarke, Mary Eva
Kastle, Joseph Hoeing; Clark, Mary E.
On the effect of various solvents on the allotropic change of mercuric iodide. *Amer. Chem. J.,* **22** (1899) 473–84

Cook, Ellen Parmelee
Cook, Ellen P.
Ueber die optische Drehrichtung der Asparaginsäure in wässerigen Lösungen. *Berlin, Chem. Ges. Ber.,* **30** (1897) 294–7

Day, Mary Hannah Gage
Day, Mary Gage
The separation of the poison of the "loco-weed". *N. Y. Med. J.,* **50** (1889) 604–5

Doan, Martha
Doan, (Miss) Martha
Index to the literature of thallium, 1861–96. [1899] *Smithsonian Miscell. Coll.,* **41** (1902) Art.1, 26 pp.
Dennis, Louis Monroe; Doan, (Miss) Martha; Gill, Adam Capen
Some new compounds of thallium. *Amer. Chem. Soc. J.,* **18** (1896) 970–7

Fairbanks, Charlotte
Fairbanks, (Miss) Charlotte
An iodometric method for the determination of phosphorus in iron. *Amer. J. Sci.,* **2** (1896) 181–5; *Ztschr. Anorg. Chem.,* **13** (1897) 117–20
Gooch, Frank Austin; Fairbanks, (Miss) Charlotte
The estimation of the halogens in mixed silver salts. *Amer. J. Sci.,* **50** (1895) 27–32; *Ztschr. Anorg. Chem.,* **9** (1895) 349–55
The iodometric estimation of molybdic acid. *Amer. J. Sci.,* **2** (1896) 156–62; *Ztschr. Anorg. Chem.,* **13** (1897) 101–9

Fossler, Mary Louise
Avery, Samuel; Fossler, Mary L.
On α-methyl-β-phenylglutaric acid. *Amer. Chem. J.,* **20** (1898) 516–18

Foster, Julia Catherine Morris
Morris, Julia Catherine; Gooch, Frank Austin
The iodometric estimation of arsenic acid. *Amer. J. Sci.,* **10** (1900) 151–7; *Ztschr. Anorg. Chem.,* **25** (1900) 227–35

Gibson, Harriet Winfield
Vulté, Hermann T.; Gibson, Harriet Winfield
The chemistry of corn oil. *Amer. Chem. Soc. J.,* **22** (1900) 453–67

Hitchcock, Fanny Rysom Mulford
Hitchcock, Fanny R. M.
The tungstates and molybdates of the rare earths. [1894] *Amer. Chem. Soc. J.,* **17** (1895) 483–94; 520–37
Introductory note on the reduction of metallic oxides at high temperatures. *Amer. Chem. Soc. J.,* **20** (1898) 232–3
Notes on the atomic mass of tungsten. [1897] *N. Y. Ac. Trans.,* **16** (1898) 332–4

Hooker, Mary Elizabeth Owens
Owens, Mary E.; Clarke, Frank Wigglesworth
On a new variety of tetrahedrite. [1880] *Berlin, Chem. Ges. Ber.,* **13** (1880) 1786–7; *Amer. Chem. J.,* **2** (1880–1881) 173
Some new salts of uranium. [1880] *Amer. Chem. J.,* **2** (1880–1881) 331; *Berlin, Chem. Ges. Ber.,* **14** (1881) 35–6
Some new compounds of platinum. [1881] *Amer. Chem. J.,* **3** (1881–1882) 350–1; *Chem. News,* **45** (1882) 62–3
Japp, Francis Robert; Owens, (Miss) Mary E.
On condensation compounds of benzil with ethyl alcohol. *Chem. Soc. J.,* **47** (1885) 90–4; 938; *Berlin, Chem. Ges. Ber.,* **18** (1885) 174–8; *Amer. Chem. J.,* **7** (1885–1886) 16–21

Jeffreys, Elizabeth
Jeffreys, Elizabeth
Ueber die Darstellung der höheren Amine der aliphatischen Reihe. Pentadecylamin. *Berlin, Chem. Ges. Ber.,* **30** (1897) 898–901
On undecylamine and pentadecylamine and the preparation of the higher amines of the aliphatic series. *Amer. Chem. J.,* **22** (1899) 14–44

Johnson, Katherine Crane
Crane, (Miss) Kate
The cohesion figures of oils as tests for their identity and purity. [1874] {*Amer. J. Pharm.,* **46** (1874) 406–9} ; *Pharm. J.,* **5** (1875) 242–4

Keller, Ida Augusta
Keller, Ida A.
The coloring matter of the aril of Celastrus scandens. *Amer. J. Pharm.,* **68** (1896) 183–6; *Philad. Ac. Nat. Sci. Proc.* (1896) 212–18

Kline, Fannie Talbot Littleton
Littleton, Fannie T.
Remarkable molecular change in silver amalgam. *Chem. Soc. J.,* **67** (1895) 239–42
Note on the heat of formation of the silver amalgam, $Ag_2 Hg_8$. [1896] *Chem. Soc. Proc.,* **12** (1897) 220–1
On the conditions affecting the volumetric determination of starch by means of a solution of iodine. *Amer. Chem. Soc. J.,* **19** (1897) 44–9

Linton, Laura Alberta
Linton, Laura A.
On the technical analysis of asphaltum. [1894–1895] *Amer. Chem. Soc. J.,* **16** (1894) 809–22; **18** (1896) 275–9
Peckham, S. F.; Linton, Laura A.
On Trinidad pitch. *Amer. J. Sci.,* **1** (1896) 193–207

Lloyd, Rachel Abbie Holloway
Lloyd, Rachel
Ueber die Umwandlung höherer Homologen des Benzolphenols in primäre und secundäre Amine. *Berlin, Chem. Ges. Ber.,* **20** (1887) 1254–65
On the conversion of some of the homologues of benzo-phenol into primary and secondary amines. [1888] *Nebraska Univ. Stud.,* **1** (1888–1892) 97–118
Lloyd, Rachel; Mabery, Charles F.
On the diiodbromacrylic and chlorbromacrylic acids. *Amer. Ac. Proc.,* **16** (1881) 235–40; *Amer. Chem. J.,* **3** (1881–1882) 124–9
Dibromiodacrylic and chlorbromiodacrylic acids. *Amer. Ac. Proc.,* **17** (1882) 94–103; *Amer. Chem. J.,* **4** (1882–1883) 92–100
On the formation and constitution of chlordibromacrylic acid. *Amer. Ass. Proc.* (1883) 161–4
On α- and β-chlordibromacrylic acids. [1884] *Amer. Ac. Proc.,* **19** (1884) 281–9; *Amer. Chem. J.,* **6** (1884–1885) 157–65

Lucas, Minnie Brooks Yoder
Yoder, Minnie B.; Stillman, J.M.
On the combination of anhydrous ammonia and aluminium chloride. *Amer. Chem. J.*, **17** (1895) 748–53

MacDonald, Margaret Baxter
Kohler, Elmer Peter; MacDonald, Margaret B.
Disulphones and ketosulphones. *Amer. Chem. J.*, **22** (1899) 219–26
{Action of sulfonic chloride on the metallic derivatives of the ethyl salts of ketonic acid. *Amer. Chem. J.*, **22** (1899) 227–39}

Mason, Elizabeth Spaulding
Richards, Ellen H.; Mason, Elizabeth S.
{The effect of heat upon the digestibility of gluten. *Tech. Quart.*, 7 (1894) 63–5}

Michael, Helen Cecilia De Silver Abbott
Abbott, (Miss) Helen C. De S.
{Some observations on the nutritive value of condiments. *Polyclinic* (1883)}
Preliminary analysis of the bark of Fouquieria splendens. *Amer. Ass. Proc.*, **33** (1884) 190–8; {*Amer. J. Pharm.*, **58** (1885) 81–8}
A chemical study of Yucca angustifolia. [1885] {*Amer. Ass. Proc. (Abs.)*, **34** (1885) 125–6}; *Amer. Phil. Soc. Trans.*, **16** (1890) 254–84
Certain chemical constituents of plants considered in relation to their morphology and evolution. *Bot. Gaz.*, **11** (1886) 270–2
On haematoxylin in the bark of Saraca indica. *Philad. Ac. Nat. Sci. Proc.* (1886) 352–4
{Preliminary analysis of a Honduran plant named Chichipate. *Amer. Ass. Proc.*, **35** (1886) 161}
Comparative chemistry of higher and lower plants. *Amer. Natlist.*, **21** (1887) 719–30; 800–10
Plant analysis as an applied science. *Franklin Inst. J.*, **124** (1887) 1–33
{Plant chemistry as illustrated in the production of sugar from sorghum. *Philad. Coll. Pharm. Alum. Ass. Proc.* (Feb. 1887)}
The chemical basis of plant forms. *Franklin Inst. J.*, **124** (1887) 161–85
[Untersuchung über Alloisomerie. II.] Zur Kenntniss der Addition von Brom und Chlor zu fester Crotonsäure. *J. Prakt. Chem.*, **46** (1892) 273–85
Zur constitution des Phloretins. *Berlin, Chem. Ges. Ber.*, **27** (1894) 2686–9
A review of recent synthetic work in the class of carbohydrates. [1895] *Franklin Inst. J.*, **142** (1896) 217–40
Abbott, (Miss) Helen C. De S.; Trimble, Henry
On the occurrence of solid hydrocarbons in plants. *Amer. Chem. J.*, **10** (1888) 439–40; *Amer. J. Pharm.*, **60** (1888) 321–4; *Amer. Phil. Soc. Proc.*, **25** (March 1888) 124–5; *Berlin, Chem. Ges. Ber.*, **21** (1888) 2598–9
Abbott, (Miss) Helen C. De S.; Jeanprêtre, John
Ueber eine neue Bildungsweise von aromatischen Nitrilen. *Berlin, Chem. Ges. Ber.*, **25** (1892) 1615–19
Zur Kenntniss der Mandelsäure und ihres Nitrils. *Berlin, Chem. Ges. Ber.*, **25** (1892) 1678–84

Ordway, Evelyn M. Walton
Walton, Evelyn M.
Liquefaction and cold produced by the mutual reaction of solid substances. *Amer. J. Sci.*, **22** (1881) 206–13; *Phil. Mag.*, **12** (1881) 290–8

Paetow, Lily Gavit Kollock
Kollock, Lily G.
Electrolytic determinations and separations. *Amer. Chem. Soc. J.*, **21** (1899) 911–28

Palmer, Alice Williams
Palmer, Alice W.; Richards, Ellen Swallow
Notes on antimony tannate. *Amer. Ass. Proc.* (1878) 150–6; *Amer. J. Sci.*, **16** (1878) 196–8; 361–4

Palmer, Helen Cooley
Cooley, Helen
Comparison of methods of testing indigo. *J. Anal. Chem.*, **2** (1888) 129–37

Peacock, Bertha Leon De Graffe
De Graffe, Bertha L.
The tannins of some Ericaceae. *Amer. J. Pharm.*, **68** (1896) 313–21

Peirce, Gertrude Klein
Peirce, Gertrude K.; Smith, Edgar F.
On the products obtained by the nitration of metachlorsalicylic acid. [1879] *Amer. Chem. J.*, **1** (1879–1880) 176–81; *Berlin, Chem. Ges. Ber.*, **13** (1880) 34–6

Pennington, Mary Engle
Pennington, Mary Engle
Derivatives of columbium and tantalum. [1895] *Amer. Chem. Soc. J.*, **18** (1896) 38–67
Pennington, Mary Engle; Smith, Edgar Fahs
The atomic mass of tungsten. *Amer. Phil. Soc. Proc.*, **33** (1894) 332–6; *Ztschr. Anorg. Chem.*, **8** (1895) 198–204

Perkins, Mary Antoinette Rice
Bailey, E. H. S.; Rice, Mary A.
On the composition of the water from a mineral spring in the vicinity of the Great Spirit Spring, Mitchell County, Kansas. [1893] *Kan. Ac. Sci. Trans.*, **14** (1896) 40–1

Phelps, Martha Austin
Austin, Martha
On the estimation of manganese separated as the carbonate. *Amer. J. Sci.*, **5** (1898) 382–4; *Ztschr. Anorg. Chem.*, **17** (1898) 272–5
The double ammonium phosphates of beryllium, zinc, and cadmium in analysis. [1899] *Amer. J. Sci.*, **8** (1899) 206–16; *Ztschr. Anorg. Chem.*, **22** (1900) 207–20
The constitution of the ammonium magnesium arseniate of analysis. *Amer. J. Sci.*, **9** (1900) 55–61; *Ztschr. Anorg. Chem.*, 23 (1900) 146–54
Gooch, Frank Austin; Austin, Martha
The estimation of manganese as the sulphate and as the oxide. *Amer. J. Sci.*, **5** (1898) 209–14; *Ztschr. Anorg. Chem.*, **17** (1898) 264–91
On the condition of oxidation of manganese precipitated by the chlorate process. *Amer. J. Sci.*, **5** (1898) 260–8; *Ztschr. Anorg. Chem.*, **17** (1898) 253–63
On the determination of manganese as the pyrophosphate. *Amer. J. Sci.*, **6** (1898) 233–43; *Ztschr. Anorg. Chem.*, **18** (1898) 339–51
The constitution of the ammonium magnesium phosphate of analysis. [1899] *Amer. J. Sci.*, **7** (1899) 187–98; *Ztschr. Anorg. Chem.*, **20** (1899) 121–36; **22** (1900) 163

Placeway, Lola Ann
Bennett, Alfred Allen; Placeway, (Miss) Lola Ann
The quantitative determination of the three halogens, chlorine, bromine, and iodine, in mixtures of their binary compounds. *Amer. Chem. Soc. J.*, **18** (1896) 688–92

Reed, Mary F.

Reed, (Miss) Mary F.

Study of the quantitative effect of temperature in the reaction of oxalic acid with potassic permanganate. *Amer. Chemist,* **5** (1875) 358–9

Richards, Ellen Henrietta Swallow

Swallow, Ellen H.

Analysis of samarskite from a new locality. *Boston Soc. Nat. Hist. Proc.,* **17** (1875) 424–8; *Amer. J. Sci.,* **14** (1877) 71 [part only]

On the occurrence of boracic acid in mineral water. *Boston Soc. Nat. Hist. Proc.,* **17** (1875) 428–30

Notes on the chemical composition of some of the mineral species accompanying the lead ore of Newburyport. *Boston Soc. Nat. Hist. Proc.,* **17** (1875) 462–5

Richards, (Mrs.) Ellen H.

{Notes on some reactions of titanium. *Trans. Amer. Inst. Min. Eng.,* **9** (1882–1883) 90–1}

Note on the determination of carbon monoxide. [1885] *Amer. Chem. J.,* 7 (1885–1886) 143–4

{A delicate test for alum in potable water. *Tech. Quart.,* **4** (1891) 94}

Carbon dioxide as a measure of the efficiency of ventilation. *Amer. Chem. Soc. J.,* **15** (1893) 572–4

The hardness of water and methods by which it is determined. *Massachusetts Brd. of Health Rep.,* **27** (1896) 433–42

Cheney, Margaret S.; Richards, Ellen Swallow

A new and ready method for the estimation of nickel in pyrrhotites and mattes. *Amer. J. Sci.,* **14** (1877) 178–81; *Chem. News,* **36** (1877) 161–2

Palmer, Alice W.; Richards, Ellen Swallow

Notes on antimony tannate. *Amer. Ass. Proc.* (1878) 150–6; *Amer. J. Sci.,* **16** (1878) 196–8; 361–4

Richards, Ellen H.; Bragg, Lottie A.

{The distribution of phosphorus and nitrogen in the products of modern milling. *Tech. Quart.,* **3** (1890) 246–52}

Richards, Ellen H.; Mason, Elizabeth S.

{The effect of heat upon the digestibility of gluten. *Tech. Quart.,* **7** (1894) 63–5}

Richards, (Mrs.) Ellen H.; Ellms, Joseph W.

The coloring matter of natural waters; its source, composition, and quantitative measurement. *Amer. Ass. Proc.* (1895) 87–9; *Amer. Chem. Soc. J.,* **18** (1896) 68–81

Richards, Ellen H.; Hopkins, Arthur T.

{The normal chlorine of the water supplies of Jamaica. *Tech. Quart.,* **11** (1898) 227–40}

Roberts, Charlotte Fitch

Roberts, Charlotte F.

On the reduction of nitric acid by ferrous salts. *Amer. J. Sci.,* **46** (1893) 126–34

On the estimation of chlorates and nitrates and of nitrites and nitrates in one operation. *Amer. J. Sci.,* **46** (1893) 231–5

On the blue iodide of starch. *Amer. J. Sci.,* **47** (1894) 422–9

The action of reducing agents on iodic acid. *Amer. J. Sci.,* **48** (1894) 151–8

The standardization of potassium permanganate in iron analysis. *Amer. J. Sci.,* **48** (1894) 290–2

Stallo, Helena

Stallo, Helena; Clarke, Frank Wigglesworth

Die Constitution der Antimontartarate. *Berlin, Chem. Ges. Ber.,* **13** (1880) 1787–96; *Amer. Chem. J.,* **2** (1880–1881) 319–29

Stickney, Delia

Stickney, Delia

The reduction of copper sulphide. *Amer. Chem. J.,* **18** (1896) 502–4

Talbot, Marion

Talbot, (Miss) Marion

Organic matter in the atmosphere. [1888] *Amer. Meteorol. J.,* **5** (1888–1889) 318–21

Titcomb, Lillian Emeline Ray

Sanford, Fernando; Ray, Lillian E.

On a possible change of weight in chemical reactions. *Phys. Rev.,* **5** (1897) 247–53; 7 (1898) 236–8

Walker, Cora

Walker, Cora

Chemical composition of two silver ornaments from Inca graves at Chimbote, Peru. *Chem. Soc. J.,* **67** (1895) 242–45

Great Britain

Aston, Emily Alicia

Aston, (Miss) Emily

Notes on some compounds of the oxides of silver and lead. *Chem. Soc. J.,* **59** (1891) 1093–5

On an Alpine nickel-bearing serpentine, with fulgurites; with petrographical notes by T. G. Bonney. *Geol. Soc. Quart. J.,* **52** (1896) 452–9

Note on a portion of the Nubian desert southeast of Korosko. III. Water analyses. *Geol. Soc. Quart. J.,* **53** (1897) 374

Aston, (Miss) Emily; Pickering, Spencer Percival Umfreville

On multiple sulphates. *Chem. Soc. J.,* **49** (1886) 123–30

Ramsay, William; Aston, (Miss) Emily

Atomic weight of boron. *Brit. Ass. Rep.* (1892) 687–9; *Chem. Soc. J.* (1893) 207–17

The molecular surface-energy of mixtures of nonassociating liquids. *Roy. Soc. Proc.,* **56** (1894) 182–91; *Ztschr. Physikal. Chem.,* **15** (1894) 89–97

The molecular surface energy of the esters, showing its variation with chemical constitution. *Roy. Soc. Proc.,* **56** (1894) 162–70; *Ztschr. Physikal. Chem.,* **15** (1894) 98–105

The molecular formula of some liquids, as determined by their molecular surface energy. *Chem. Soc. J.,* **65** (1894) 167–73

Walker, James; Aston, (Miss) Emily

Affinity of weak bases. *Chem. Soc. J.* (1895) 576–86

Aston, (Miss) Emily; Collie, J. Norman

Oxidation products of α,γ-dimethyl-α'-chloropyridine. *Chem. Soc. J.,* **71** (Pt.2) (1897) 653–7

Aston, (Miss) Emily ; Newton, (Miss)

A note on the estimation of zinc oxide. *Chem. News,* **75** (1897) 133–4

Dutoit, Paul; Aston, (Miss) Emily; Friderich, Louis

Sur la conductibilité de solutions d'électrolytes dans quelques dissovents organiques polymérisés. *Arch. Sci. Phys. Nat.,* **5** (1896) 287

Dutoit, Paul; Aston, (Miss) Emily

Relation entre la polymérisation des corps liquides et leur pouvoir dissociant sur les électrolytes. *Paris, Ac. Sci. C. R.,* **125** (1897) 240–3

Guye, Philippe A.; Aston, (Miss) Emily

Influence de la température sur le pouvoir rotatoire. *Paris, Ac. Sci. C. R,* **124** (1897) 194–7

Influence de la température sur le pouvoir rotatoire des liquides. *Paris, Ac. Sci. C. R.,* **125** (1897) 819–21
Variations du pouvoir rotatoire de l'alcool amylique avec la température. *Arch. Sci. Phys. Nat.,* **4** (1897) 592
Sur le pouvoir rotatoire de l'acide valérique actif. *Paris, Ac. Sci. C. R.,* **130** (1900) 585–8

Boole, Lucy Everest
Dunstan, W. R.; Boole, (Miss) Lucy Everest
Chemical observations on tartar emetic. [1888] *Pharm. J.,* **19** (1889) 385–7
An enquiry into the nature of the vesicating constituent of croton oil. [1895] *Roy. Soc. Proc.,* **59** (1895) 237–49

Buchanan, Margaret E.
Buchanan, (Miss) M. E.
Note on the reaction of hydrocyanic acid. [1887] *Pharm. J.,* **18** (1888) 512

Dougal, Margaret Douie
Dougal, (Mrs.) Margaret Douie
On the teaching of inorganic chemistry. *Chem. News,* **68** (1893) 223–5; 237–8; 247–8; 260–1
A specimen of early Scottish iron. *Chem. Soc. J.,* **65** (1894) 744–50
Effect of heat on aqueous solutions of chrome alum. *Chem. Soc. J.,* **69** (1896) 1526–30

Earp, Annie Gertrude
Earp, (Miss) A. G.
Notes on the effect of the replacement of oxygen by sulphur on the boiling- and melting-points of compounds. *Phil. Mag.,* **35** (1893) 458–62

Evans, Clare de Brereton
Evans, Clare de Brereton
Researches on tertiary benzenoid amines. [1895–1896] *Chem. Soc. Proc.,* **11** (1896) 235–6; **12** (1897) 234–5
Studies on the chemistry of nitrogen. Enantiomorphous forms of ethylpropylpiperidonium iodide. *Chem. Soc. J.,* **71** (Pt.1) (1897) 522–6

Farrer, E. M.
Farrer, (Miss) E. M.; Pickering, Spencer Umfreville
Hydration of salts. *Chem. News,* **53** (1886) 279

Field, Elizabeth Eleanor
Field, (Miss) Eleanor
Chromic acid. *Chem. Soc. J.,* **61** (1892) 405–8
Note on the interactions of alkali haloids with lead and bismuth haloids. *Chem. Soc. J.,* **63** (1893) 540–7

Fortey, Emily C.
Fortey, Emily C.
Hexamethylene from American and Galician petroleum. *Chem. Soc. J.,* **73** (Pt.2) (1898) 932–49
Hexanaphthene and its derivatives. Preliminary note. [1897] *Chem. Soc. Proc.,* **13** (1898) 161–2
Action of light and oxygen on dibenzyl ketone. *Chem. Soc. J.,* **75** (Pt.2) (1899) 871–3
Richardson, Arthur; Fortey, Emily C.
Action of light on amyl alcohol. *Chem. Soc. J.,* **69** (1896) 1349–52
Note on the action of light on ether. *Chem. Soc. J.,* **69** (1896) 1352–5
Action of light on amyl alcohol. *Chem. Soc. Proc.,* **12** (1897) 164–5
Note on the action of light on ether. *Chem. Soc. Proc.,* **12** (1897) 165–6
Young, Sydney; Fortey, Emily C.
The vapour pressures, specific volumes and critical constants of hexamethylene. *Chem. Soc. J.,* **75** (Pt.2) (1899) 873–83
Note on the refraction and magnetic rotation of hexamethylene, chlorohexamethylene, and dichlorohexamethylene. *Chem. Soc. J.,* **77** (Pt.1) (1900) 372–4
Vapour pressures, specific volumes, and critical constants of diisopropyl and diisobutyl. *Chem. Soc. J.,* **77** (Pt.2) (1900) 1126–44

Green, Wilhelmina M.
Green, Wilhelmina M.
On the infusion of tea. *Chem. News,* **52** (1885) 229–31

Halcrow, Lucy
Halcrow, (Miss) Lucy; Frankland, Edward
On the action of air upon peaty water. *Chem. Soc. J.,* **37** (1880) 506–17

Hall, Lucy
Hall, (Miss) L.; Collie, J. Norman
Production of some nitro- and amido-oxylutidines. Part II. *Chem. Soc. J.,* **73** (Pt.1) (1898) 235–41
Kipping, Frederic Stanley; Hall, (Miss) L.
Synthesis of phenoketoheptamethylene. [1899] *Chem. Soc. Proc.,* **15** (1900) 173–4

Johnston, Etta J.
Johnston, (Miss) Etta J.; Carnelley, Thomas
Effect of floor-deafening on the sanitary condition of dwelling houses. *Roy. Soc. Proc.,* **45** (1889) 346–51
The relation of physiological action to atomic weight. *Nature,* **41** (1890) 189–90

Johnston, Margaret Neill
Kennedy, (Miss) Catherine Lucy; Johnston, (Miss) Margaret Neill
Contributions to the history of the mineral waters of Yorkshire. Analysis of the water of the Boston Spa, Thorp Arch. *Chem. Soc. J.,* **38** (1881) 515–16 [Part IV of a paper by T. E. Thorpe]

Judson, Winifred
Walker, James Wallace; Judson, Winifred
Reduction of bromic acid and the law of mass action. *Brit. Ass. Rep.* (1897) 613–16; {*Chem. Soc. Proc.* **13** (1898) 64}; *Chem. Soc. J.,* **73** (1898) 410–22

Kennedy, Catherine Lucy
Kennedy, (Miss) Catherine Lucy; Johnston, (Miss) Margaret Neill
Contributions to the history of the mineral waters of Yorkshire. Analysis of the water of the Boston Spa, Thorp Arch. *Chem. Soc. J.,* **38** (1881) 515–16 [Part IV of a paper by T. E. Thorpe]

Lloyd, Emily Jane
Lloyd, Emily J.
Osmotic pressure: with special reference to the work of Rudorff and Raoult. *Chem. News,* **60** (1889) 284–6; 297–9

Marshall, Dorothy
Marshall, (Miss) Dorothy
Sur la chaleur de vaporisation de l'acide formique. *Paris, Ac. Sci. C. R.,* **122** (1896) 1333–5
On the heats of vaporization of liquids at their boiling-points. *Phil. Mag.,* **43** (1897) 27–32
Griffiths, Ernest Howard; Marshall, (Miss) Dorothy
The latent heat of evaporation of benzene. [1895] *London Phys. Soc. Proc.,* **14** (1896) 16–56; *Phil. Mag.,* **41** (1896) 1–37
Marshall, (Miss) Dorothy; Ramsay, William
A method of comparing directly the heats of evaporation of differ-

ent liquids at their boiling-points. [1895] *London Phys. Soc. Proc.,* **14** (1896) 57–72; *Phil. Mag.,* **41** (1896) 38–52

Marshall, Margaret Henrietta
Marshall, Margaret H.
Experiments and investigations as to the influence exerted over some minerals by animal and vegetable matter, under certain conditions. *Edinb. Roy. Soc. Proc.,* **2** (1851) 58–61

McKillop, Margaret Seward
Pendlebury, William H.; Seward, (Miss) Margaret
An investigation of a case of gradual chemical change: the interaction of hydrogen chloride and chlorate in presence of potassium iodide. [1888] *Roy. Soc. Proc.,* **45** (1889) 396–423
The interaction of hydrogen chloride and potassium chlorate. [1893] *Chem. Soc. Proc.,* **9** (1895) 211–12

MacLean, Ida Smedley
Smedley, Ida
Benzylanilinesulphonic acids. [1900] *Chem. Soc. Proc.,* **16** (1901) 160

Mills, Mildred May Gostling
Fenton, Henry J. Horstman; Gostling, (Miss) Mildred
Action of hydrogen bromide in presence of ether on carbohydrates and certain organic acids. *Chem. Soc. J.,* **73** (Pt.1) (1898) 554–8
Bromomethylfurfuraldehyde. *Chem. Soc. J.,* **75** (Pt.1) (1899) 423–33; **75** (Pt.2) (1899) 1219

Newton, Lucy
Aston, (Miss) Emily; Newton, (Miss)
A note on the estimation of zinc oxide. *Chem. News,* **75** (1897) 133–4

Passavant, Laura M.
Passavant, Laura M.
On the specific volume of chloral. *Chem. Soc. J.,* **39** (1881) 53–7

Walker, Annie Purcell Sedgwick (Lady)
Ruhemann, Siegfried; Sedgwick, (Miss) A.P.
Weiteres über den Dicarboxyglutaconsäureester. *Berlin, Chem. Ges. Ber.,* **28** (1895) 822–5
Sedgwick, (Miss) A.P.; Collie, J. Norman
Some oxypyridine derivatives. *Chem. Soc. J.,* **67** (1895) 399–413

Walter, Lavinia Edna
Walter, L. Edna
Note on thio-derivatives from sulphanilic acid. [1895] *Chem. Soc. Proc.,* **11** (1896) 141–2

Whiteley, Martha Annie
Whiteley, (Miss) Martha Annie
The oxime of mesoxamide and some allied compounds. *Chem. Soc. J.,* **77** (1900) Pt. 2, 1040–6
Crompton, Holland; Whiteley, (Miss) Martha Annie
The melting points of mixtures. *Chem. Soc. J.,* **67** (1895) 327–37

Williams, Katherine Isabella
Williams, (Miss) Katherine I.
The composition of cooked vegetables. *Chem. Soc. J.,* **61** (1892) 226–41
The composition of cooked fish. *Chem. Soc. J.,* **71** (Pt.2) (1897) 649–53
Williams, (Miss) Katherine I.; Ramsay, William
Note as to the existence of an allotropic modification of nitrogen. *Chem. Soc. Proc.,* **2** (1886) 223–5
The estimation of free oxygen in water. *Chem. Soc. J.,* **49** (1886) 751–61; [877]; *Chem. Soc. Proc.,* **2** (1886) 223

Geology

United States

Barbour, Carrie Adeline
Barbour, Carrie Adeline
{Some methods of collecting, preparing and mounting fossils. *Nebraska Ac. Sci., Pub. 6* (1898) 258–62}
{Report of the work of the Morrill geological expeditions of the University of Nebraska. *Science,* **11** (1900) 856–8}

Bascom, Florence
Bascom, Florence
The structures, origin and nomenclature of the acid volcanic rocks of South Mountain. *J. Geol.,* **1** (1893) 813–32
A pre-tertiary nepheline-bearing rock. *J. Geol.,* **4** (1896) 160–5
Perido-steatite and diabase. *Philad. Ac. Nat. Sci. Proc.* (1896) 219–20
The ancient volcanic rocks of South Mountain, Pennsylvania. *U. S. Geol. Surv. Bull.,* No. **136** (1896) 124 pp.
Aporhyolite of South Mountain, Pennsylvania. [1896] *Amer. Geol. Soc. Bull.,* **8** (1897) 393–6
The relation of the streams in the neighborhood of Philadelphia to the Bryn Mawr gravel. *Amer. Geologist,* **19** (1897) 50–7
On some dykes in the vicinity of Johns Bay, Maine. *Amer. Geologist,* **23** (1899) 275–80
Volcanics of Neponset valley, Massachusetts. *Amer. Geol. Soc. Bull.,* **11** (1900) 115–26
Bascom, Florence; Dale, T. N.
{Note on the dyke rocks in the slate belt of eastern New York and western Vermont. *U. S. Geol. Survey, Ann. Rep.,* **19,** pt. 3 (1899) 223–6}

Fielde, Adele Marion
Fielde, (Miss) Adele M.
Notes on the geology of China. *Philad. Ac. Nat. Sci. Proc.* (1887) 30–1

Fleming, Mary A.
Fleming, (Miss) Mary A.
The pot-holes of Foster's Flats (now called Niagara Glen) on the Niagara river. *Amer. Ass. Proc.* (1899) 226–7

Hayes, Ellen Amanda
Hayes, Ellen
The temperature of the Earth's crust. *Science,* **3** (1896) 518–19

Klem, Mary Jeanette
Klem, Mary
The development of Agaricocrinus. *St. Louis Ac. Trans.,* **10** (1900) 167–84

Maury, Carlotta Joaquina
Maury, C. Joaquina; Harris, G. D.
{Geology of Chiapas, Tabasco and the peninsula of Yucatan. (Translation and brief summary from the original by Carlos Sapper, "La Geograpfía Física y la Geología de la Peninsula de Yucatán", Boletín del Instituto Geologico de Mexico, Número 3, 57 pp.) *J. Geol.,* **4** (1896) 938–47}

Owen, Luella Agnes
Owen, (Miss) Luella Agnes
{Cavernes américaines. *Spelunca,* **2** (1896) 8–13}
Marble Cave (Missouri) et Wind Cave (Dakota). *Spelunca,* **3** (1897) 22–31
La Caverne de Cristal. *Spelunca,* **4** (1898) 77–81

Les cavernes de Ha Ha Tonka. *Spelunca,* **5** (1899) 16–20
{The bluffs of the Missouri river. *Verhandlunge des VII Internationalen Geographen-Kongresses in Berlin, Pt. 2,* (1899) 686–90}

Smith, Erminnie Adelle Platt
Smith, Erminnie A.
Concerning amber. [1879] *Amer. Natlist.,* **14** (1880) 179–80
{Monograph on jade, illustrated with fine specimens in all different varieties. *Amer. Ass. Proc.* (1879) 523–5 [abstract]}
{The great Oberstein industry; method of coloring, cutting and polishing agates and secondary gems. *Amer. Ass. Proc.* (1879) 248 [title only]}

Great Britain

Anning, Mary
Anning, (Miss) Mary
{Note on the supposed frontal spine of the genus Hybodus. *Mag. Nat. Hist.,* **12** (1839) 605}

Carne, Elizabeth Catherine Thomas
Carne, (Miss) E.
On the evidence to be derived from cliff boulders, with regard to a former condition of the land and sea in the Land's End district. *Cornwall Geol. Soc. Trans.,* 7 (1847–1860) 368–78
On transition and metamorphosis of the rocks in Land's End district. [1875] [Posth.] *Cornwall Geol. Soc. Trans.,* **9** (1878) 1–21
Enquiry into the nature of the forces that have acted on the formation and elevation of the Land's End granite. [1874] [Posth.] *Cornwall Geol. Soc. Trans.,* **9** (1978) 132–51.

Coignou, Carloine Pauline Marie
Coignou, (Miss)
On a new species of Cyphaspis from the Carboniferous rocks of Yorkshire. *Geol. Soc. Quart. J.,* **46** (1890) 421–2

Crane, Agnes
Crane, (Miss) Agnes
On certain genera of living fishes and their fossil affinities. *Geol. Mag.,* **4** (1877) 209–19; *Brighton Nat. Hist. Soc. Proc.,* **24** (1878) 44–58
The general history of the Cephalopoda, recent and fossil. *Geol. Mag.,* **5** (1878) 487–99
New classifications of the Brachiopoda. *Geol. Mag.,* **10** (1893) 318–23; 384
The generic evolution of the Palaeozoic Brachiopoda. *Science,* **21** (1893) 72–4
The evolution of the Brachiopoda. A sequel to Dr. Thomas Davidson's, "What is a brachiopod." [1893] *Geol. Mag.,* **2** (1895) 65–7; 103–16

Crosfield, Margaret Chorley
Crosfield, (Miss) Margaret C.
The Tremadoc slates. [1896] *Holmesdale Nat. Hist. Club Proc.* (1896–1898) 11–22
Redhill and Merstham. [1897] *Holmesdale Nat. Hist. Club Proc.* (1896–1898) 65–6
Excursion to Reigate. [1899] *Geol. Ass. Proc.,* **16** (1900) 162–3
Crosfield, (Miss) Margaret C.; Skeat, (Miss) Ethel G.
On the geology of the neighbourhood of Carmarthen. *Geol. Soc. Quart. J.,* **52** (1896) 523–41

Dale, Elizabeth
Seward, Albert C.; Dale, (Miss) Elizabeth
On the structure and affinities of Dipteris conjugata, *Reinw.,* with note on the geological history of the Dipteridinae. *Brit. Ass. Rep.* (1900) 946

Elles, Gertrude Lilian
Elles, (Miss) Gertrude L.
The subgenera Petalograptus and Cephalograptus. *Geol. Soc. Quart. J.,* **53** (1897) 186–212
The graptolite fauna of the Skiddaw slates. *Geol. Soc. Quart. J.,* **54** (1898) 463–539
The occurrence of Placoparia in the Skiddaw slates. *Geol. Mag.,* **5** (1898) 141
The zonal classification of the Wenlock shales of the Welsh borderland. [1899] *Geol. Soc. Quart. J.,* **56** (1900) 370–413
Elles, (Miss) Gertrude L.; Wood, (Miss) Ethel M. R.
Woodwardian Museum notes. Supplementary notes on the Drygill shales. *Geol. Mag.,* **2** (1895) 216–49
On the Llandovery and associated rocks of Conway (North Wales). *Geol. Soc. Quart. J.,* **52** (1896) 273–88

Eyton, Elizabeth Charlotte
Eyton, (Miss) Charlotte
On an ancient coastline in North Wales. *Geol. Mag.,* **3** (1866) 289–91
On an old lake-basin in Shropshire. *Geol. Mag.,* **4** (1867) 1–2
On the glacio-marine denudation of certain districts. *Geol. Mag.,* **4** (1867) 545–9
The drift-beds of Lhandrillo Bay, Denbigshire. *Geol. Mag.,* **5** (1868) 349–53
On the Pleistocene deposits of North Shropshire. *Geol. Mag.,* 7 (1870) 106–13
On the age and geological position of the blue clay of the western counties. *Geol. Mag.,* 7 (1870) 545–7

Foley, Mary Cecilia
Foley, Mary C.
On enclosures of glass in a basalt near Bertrich, in the Eifel. *Geol. Mag.,* **3** (1896) 242–5

Forster, Mary
Topley, William; Forster, (Miss) Mary
Excursion to Belgium and the French Ardennes: Brussels, Givet, Dinant, Namur, Grotto of Han, etc. [1885] *Geol. Ass. Proc.,* **9** (1887) 261–86

Gardiner, Margaret Isabella
Gardiner, (Miss) Margaret I.
The greensand bed at the base of the Thanet Island. *Geol. Soc. Quart. J.,* **44** (1888) 755–60
Contact-alteration near New Galloway. *Geol. Soc. Quart. J.,* **46** (1890) 569–80

Gordon, Maria Matilda Ogilvie (later **Dame Maria Ogilvie Gordon**)
Ogilvie, (Miss) Maria M. (Mrs. Gordon)
Landslips in the St. Cassian strata of S. Tyrol. *Brit. Ass. Rep.* (1892) 721–2
Preliminary note on the sequence and fossils of the Upper Triassic strata of the neighbourhood of St. Cassian, Tyrol. *Geol. Mag.,* **9** (1892) 145–7
Contributions to the geology of the Wengen and St. Cassian strata in the southern Tyrol. [1892] *Geol. Soc. Quart. J.,* **49** (1893) 1–77
Coral in the "dolomites" of South Tyrol. *Geol. Mag.,* **1** (1894) 1–10; 49–60
The "Gemmi" disaster. *Nature,* **52** (1895) 573–5

Recent work on the madreporarian skeleton. [1896] *Nature,* **55** (1896–1897) 126–7

The classification of Madreporaria. [1897] *Nature,* **55** (1896–1897) 280–4

Die Korallen der Stramberger Schichten. [1896–1897] *Palaeontographica, Suppl.* **2** (1897) Abth.7, 73–282; i-iv

Microscopic and systematic study of madreporarian types of corals. [1895] *Phil. Trans.* (B), **187** (1897) 83–345

Sigmoidal curves. *Brit. Ass. Rep.* (1899) 754–5

The torsion-structure of the Dolomites. [1898] *Geol. Soc. Quart. J.,* **55** (1899) 560–633

Torsion-structure in the Alps. *Nature,* **60** (1899) 443–6

Similar geological structure in South Tyrol and the Isle of Man. [1900] *Nature,* **61** (1899–1900) 490

On the fauna of the Upper Cassian zone in Falzarego valley, South Tyrol. *Geol. Mag.,* **7** (1900) 337–49

Rock-structures in the Isle of Man and in South Tyrol. *Nature,* **62** (1900) 7

The origin of land-forms through crust-torsion. *Geogr. J.,* **16** (1900) 457–69

Ueber die Obere Cassianer Zone an der Falzarego-Strasse (Südtirol). *Wien, Geol. Verh.* (1900) 306–22

Graham, Maria Dundas (later **Lady Callcott**)

Graham, Maria

An account of some effects of the late earthquakes in Chile. [1824] *Geol. Soc. Trans.,* **1** (1824) 413–15

On the reality of the rise of the coast of Chile, in 1822. *Silliman, J.,* **28** (1835) 236–47

Hastings, Barbara, Marchioness of (Baroness Grey de Ruthyn)

Hastings (Marchioness of)

On the freshwater Eocene beds of Hordle Cliff, Hants. *Brit. Ass. Rep.* (1847) Pt.2, 63–4; *Roma, Corrisp. Scient.,* **1** (1848) 133

Descriptions géologiques des falaises d'Hordle, et sur la côte de Hampshire, en Angleterre. *Paris, Soc. Géol. Bull.,* **9** (1851–2) 191–203

On the tertiary beds of Hordwell, Hampshire. *Phil. Mag.,* **6** (1853) 1–11

Hodgson, Elizabeth

Hodgson, (Miss) E.

On a deposit containing diatomaceae, leaves, etc., in the iron-ore mines near Ulverstone. *Geol. Soc. Quart. J.,* **19** (1863) 19–31

On Helix, and perforated limestone. *Geologist,* **7** (1864) 42–4

On the glacial drift of Furness, Lancashire. *Geologist,* **7** (1864) 209–17

The moulded limestones of Furness. *Geol. Mag.,* **4** (1867) 401–6

The south coast of Furness. *Geol. Mag.,* **6** (1869) 286–7

On the situation of the iron-ore fossils in the Water Blain mines, South Cumberland. *Geol. Mag.,* **7** (1870) 113–15

The granite-drift of Furness. *Geol. Mag.,* **7** (1870) 328–39

Visits to trap dykes. [Reminiscences] [*Posth.*] *Barrow Field Club Rep.,* **2** (1878) 89–94

Hughes, Mary Caroline Weston

Hughes, (Mrs. Thomas McKenney)

On the Mollusca of the Pleistocene gravels in the neighbourhood of Cambridge. *Geol. Mag.,* **5** (1888) 193–207

Kelly, Agnes

Kelly, Agnes

Conchite, a new form of calcium carbonate. [1900] *Min. Mag.,* **12** (1900) 363–70; *München Ak. Sber.,* **30** (1901) 187–94

Lee, Sarah Wallis Bowdich

Lee, (Mrs.)

Notice of a fossil Nautilus found in the sandstone of the Isle of Sheppey. *Mag. Nat. Hist.,* **4** (1831) 137–8

Longstaff, Mary Jane Donald

Donald, (Miss) Jane

Notes on some carboniferous Gasteropoda from Penton and elsewhere. [1884] *Cumberland Ass. Trans.,* **9** (1885) 127–36

Some additional notes on the land and freshwater shells of Cumberland and Westmorland. [1884–1886] *Cumberland Ass. Trans.,* **9** (1885) 217–19; **11** (1886) 150–1

Notes upon some Carboniferous species of Murchisonia in our public museums. *Geol. Soc. Quart. J.,* **43** (1887) 617–31

Descriptions of some new species of Carboniferous Gasteropoda. *Geol. Soc. Quart. J.,* **45** (1889) 619–25

Notes on some new and little-known species of Carboniferous Murchisonia. *Geol. Soc. Quart. J.,* **48** (1892) 562–75

Notes on the genus Murchisonia and its allies; with a revision of the British Carboniferous species, and descriptions of some new forms. *Geol. Soc. Quart. J.,* **51** (1895) 210–34

Observations on the genus Aclisina, *De Koninck,* with descriptions of British species and of some other Carboniferous Gasteropoda. [1897] *Geol. Soc. Quart. J.,* **54** (1898) 45–72

Remarks on the genera Ectomaria, *Koken,* and Hormotoma, *Salter,* with descriptions of British species. *Geol. Soc. Quart. J.,* **55** (1899) 251–72

On some recent Gasteropoda referred to the family Turritellidae and their supposed relationship to the Murchisoniidae. [1900] *London Malacol. Soc. Proc.,* **4** (1901) 47–55

M'Kean, Minnie

M'Kean, (Miss) Minnie

The fossils of the Red Crag and chalk pits, Suffolk. [1887] *Edinb. Nat. Soc. Trans.,* **2** (1891) 51–4

Raisin, Catherine Alice

Raisin, (Miss) Catherine A.

Notes on the metamorphic rocks of South Devon. *Geol. Soc. Quart. J.,* **43** (1887) 715–33

The metamorphic rocks of South Devon. *Geol. Mag.,* **5** (1888) 190–1

On some rock specimens from Somali Land. *Geol. Mag.,* **5** (1888) 414–18

On some rock-specimens from Socotra. *Geol. Mag.,* **5** (1888) 504–7

Devonian greenstones and chlorite schists of South Devon. *Geol. Mag.,* **6** (1889) 265–9

On some nodular felstones of the Lleyn. *Geol. Soc. Quart. J.,* **45** (1889) 247–69

On the lower limit of the Cambrian series in N. W. Caernarvonshire. *Geol. Soc. Quart. J.,* **47** (1891) 329–40

The so-called serpentines of the Lleyn. *Geol. Mag.,* **9** (1892) 408–13

Contributions to the geology of Africa. *Geol. Mag.,* **10** (1893) 436–43

Variolite of the Lleyn, and associated volcanic rocks. *Geol. Soc. Quart. J.,* **49** (1893) 145–63

Note on a portion of the Nubian desert south-east of Korosko. II. Petrology. *Geol. Soc. Quart. J.,* **53** (1897) 364–73

On a hornblende-picrite from the Zmutthal (Canton Valais). *Geol. Mag.,* **4** (1897) 202–5

On the nature and origin of the Rauenthal serpentine. *Geol. Soc. Quart. J.*, **53** (1897) 246–68

Dendritic patterns caused by evaporation. *Nature*, **58** (1898) 224

On certain structures formed in the drying of a fluid with particles in suspension. *Roy. Soc. Proc.*, **63** (1898) 217–27

Bonney, (Rev.) Thomas George; Raisin, (Miss) Catherine A.

Report on some rock-specimens from the Kimberley diamond-mines. *Geol. Mag.*, **8** (1891) 412–15

On the so-called spilites of Jersey. *Geol. Mag.*, **10** (1893) 59–64

On rocks and minerals collected by Mr. W. M. Conway in the Karakoram Himalayas. *Roy. Soc. Proc.*, **55** (1894) 468–87

On the relations of some of the older fragmental rocks in north-western Caernarvonshire. *Geol. Soc. Quart. J.*, **50** (1894) 578–601

On varieties of serpentine and associated rocks in Anglesey. *Geol. Soc. Quart. J.*, **55** (1899) 276–302

Stone, (Sir) J. B.; Bonney, (Rev.) Thomas George; Raisin, (Miss) Catherine A.

Notes on the diamond-bearing rock of Kimberley, South Africa. *Geol. Mag.*, **2** (1895) 492–502

Scott, S.

Scott, (Mrs.) S.

The work of a river. *Liverpool Geol. Ass. J.*, **11** (1891) 27–30

Sollas, Igerna Brünhilda Johnson

Sollas, Igerna B. J.

On Naiadita from the Upper Rhaetic (bed K of Wilson's section) of Redland, near Bristol. *Brit. Ass. Rep.* (1900) 752–3

Wood, Ethel Mary Reader (later **Dame Ethel Shakespear**)

Wood, (Miss) Ethel M. R.

The Lower Ludlow formation and its graptolite fauna. *Geol. Soc. Quart. J.*, **56** (1900) 415–91

Elles, (Miss) Gertrude L.; Wood, (Miss) Ethel M. R.

Woodwardian Museum notes. Supplementary notes on the Drygill shales. *Geol. Mag.*, **2** (1895) 216–49

On the Llandovery and associated rocks of Conway (North Wales). *Geol. Soc. Quart. J.*, **52** (1896) 273–88

Woods, Ethel Gertrude Skeat

Crosfield, (Miss) Margaret C.; Skeat, (Miss) Ethel G.

On the geology of the neighbourhood of Carmarthen. *Geol. Soc. Quart. J.*, **52** (1896) 523–41

Skeat, (Miss) Ethel G.; Madsen, Victor

On Jurassic, Neocomian and Gault boulders found in Denmark. *Danmarks Geol. Unders.*, Raekke 2, **8** (1898) 213 pp.

Mathematics, Applied Mathematics and Statistics

United States

Fitch, Annie Louise MacKinnon

MacKinnon, (Miss) Annie L.

Concomitant binary forms in terms of the roots. [1895–1898] *Ann. Math.*, **9** (1894–1895) 95–157; **12** (1898–1899) 95–109

Growe, Bessie E.

Growe, (Miss) Bessie E.

On new canonical forms of the binary quintic and sextic. *Kan. Univ. Quarterly*, **6** (1897) 201–4

The reduction of binary quantics to canonical forms by linear transformation. [1900] *N. Y., Amer. Math. Soc. Bull.*, **7** (1901) 3; 12–13

Hayes, Ellen Amanda

Hayes, Ellen

Mean values. *Science*, **21** (1893) 333

Ladd-Franklin, Christine

Ladd, (Miss) Christine

Quaternions. *Des Moines, Analyst*, **4** (1877) 172–4

On some properties of four circles inscribed in one another and circumscribed about another. *Des Moines, Analyst*, **5** (1878) 116–17

The polynomial theorem. *Des Moines, Analyst*, **5** (1878) 145–7

The Pascal hexagram. *Amer. J. Math.*, **2** (1879) 1–12

On De Morgan's extension of the algebraic processes. *Amer. J. Math.*, **3** (1880) 210–25

The nine-line conic. *Des Moines, Analyst*, **7** (1880) 147–9

On segments made on lines by curves. *Amer. J. Math.*, **4** (1881) 272

{On the so-called d'Alembert-Carnot geometrical paradox. *Messenger Math.* **15** (1885–1886) 36–7}

On some characteristics of symbolic logic. *Amer. J. Psychol.*, **2** (1889–1890) 542–67

{Some proposed reforms in common logic. *Mind*, **15** (1890) 75–88}

{Dr. Hillebrand's syllogistic scheme. *Mind*, n.s. **1** (1892) 527–30}

Newson, Mary Frances Winston

Winston, Mary Frances

{Eine Bemerkungen zur Theorie der hypergeometrischen Funktion. *Math. Ann.*, **46** (1895) 159–60}

Schottenfels, Ida May

Schottenfels, Ida May

On groups of order 8!/2. *N. Y., Amer. Math. Soc. Bull.*, **6** (1900) 440–3

Two non-isomorphic simple groups of the same order 20,160. [1898] *Ann. Math.*, **1** (1900) 147–52

Wood, Ruth Goulding

Wood, (Miss) R. G.

The collineations of space which transform a non-degenerate quadric surface into itself. [1900] *N. Y., Amer. Math. Soc. Bull.*, **7** (1901) 202; 207

United States—Great Britain

Maddison, Ada Isabel

Maddison, Isabel

On certain factors of the *c*- and *p*-discriminants and their relation to fixed points on the family of curves. *Quart. J. Math.*, **26** (1893) 307–21

On singular solutions of differential equations of the first order and the geometrical properties of certain invariants and covariants of their complete primitives. *Quart. J. Math.*, **28** (1896) 311–74

Scott, Charlotte Angas

Scott, (Miss) Charlotte Angas

The binomial equation $x^p - 1 = 0$. *Amer. J. Math.*, **8** (1886) 261–4

On the higher singularities of plane curves. *Amer. J. Math.*, **14** (1892) 301–25

The nature and effect of singularities of plane algebraic curves. *Amer. J. Math.*, **15** (1893) 221–43

On plane cubics. [1893] *Phil. Trans. (A)*, **185** (1895) 247–77

Note on adjoint curves. *Quart. J. Math.*, **28** (1896) 377–81

Note on equianharmonic cubics. *Messenger Math.*, **255** (1896) 180–5

On Cayley's theory of the absolute. *N. Y., Amer. Math. Soc. Bull.*, **3** (1897) 235–46

Sur la transformation des courbes planes. *Ass. Franç. C. R.*, (Pt. 2) (1897) 50–9

Note on linear systems of curves. *Nieuw Arch. Wisk.*, **3** (1898) 243–52

On the intersection of plane curves. *N. Y., Amer. Math. Soc. Bull.*, **4** (1898) 260–73

Studies in the transformation of plane algebraic curves. Pt. I *Quart. J. Math.*, **29** (1898) 329–81 {Pt. II. **32** (1901) 309–39}

A proof of Noether's fundamental theorem. *Math. Ann.*, **52** (1899) 593–7

Arithmetical note. *Science*, **12** (1900) 648–9

On Von Staudt's Geometrie der Lage. *Math. Gaz.*, **1** (1900) 307–14; 323–31; 363–70

The status of imaginaries in pure geometry. [1899] *N. Y., Amer. Math. Soc. Bull.*, **6** (1900) 163–8

On a memoir by Ricardo De Paolis: [Le trasformazioni piane doppie]. [1900] *N. Y., Amer. Math. Soc. Bull.*, **7** (1901) 24–38

Great Britain

Ayrton, Phoebe Sarah Marks

Marks, (Miss) Sarah

The uses of a line-divider. [1885] *Phil. Mag.*, **19** (1885) 280–5; *London Phys. Soc. Proc.*, **7** (1886) 1–6

Barwell, Mildred Emily

Barwell, (Miss) M. E.

The conformal representation of a pentagon on a half plane. *London Math. Soc. Proc.*, **29** (1898) 695–706

Beeton, Mary

Beeton, (Miss) Mary; Pearson, Karl

Data for the problem of evolution in man. II. A first study of the inheritance of longevity and the selective death-rate in man. [1899] *Roy. Soc. Proc.*, **65** (1900) 290–305

Beeton, (Miss) Mary; Yule, G. U.; Pearson, Karl

Data for the problem of evolution in man. V. On the correlation between duration of life and the number of offspring. [1900] *Roy. Soc. Proc.*, **67** (1901) 159–79

Bryant, Sophie Willcock

Bryant, (Mrs.) Sophie

The position of mathematics in the school education of girls. *Educ. Times*, **31** (1878) 145–9

On the failure of the attempt to deduce inductive principles from the mathematical theory of probabilities. *Phil. Mag.*, **17** (1884) 510–18

On the ideal geometrical form of natural cell-structure. [1885] *London Math. Soc. Proc.*, **16** (1884–1885) 311–15

{On the nature and functions of a complete symbolic language. *Mind*, **13** (1888) 118–208}

An example in "correlation of averages" for four variables. *Phil. Mag.*, **36** (1893) 372–7

Byron, Augusta Ada, Countess of Lovelace

Lovelace, Augusta Ada (Countess of)

{Notes upon the Memoir, "Sketch of the analytical engine invented by Charles Babbage," by L. F. Menabrea of Turin. [Notes by the translator.] *Taylor, Scientif. Mem.*, **3** (1843) 666–731}

Note. Although the above entry does not appear in the Royal Society *Catalogue*, the following erroneous one does: Lovelace, (Lady), "Notes to a translation of Mitscherlich's Memoir on the 'Chemical reactions produced by bodies which act only by contact,'" *Taylor, Scientific Mem.*, **4** (1846) 1–15; the article occupying these journal pages does not include notes by Lovelace.

Fawcett, Cicely Deborah

Fawcett, (Miss) Cicely D.; Pearson, Karl

Mathematical contributions to the theory of evolution. On the inheritance of the cephalic index. *Roy. Soc. Proc.*, **62** (1898) 413–17

Pearson, Karl; Lee, Alice; Warren, Ernest; Fry, Agnes; Fawcett, Cicely D., et al.

[Mathematical contributions to the theory of evolution. IX.] On the principle of homotyposis and its relation to heredity, to the variability of the individual, and to that of the race. Part I. Homotyposis in the vegetable kingdom. *Phil. Trans.* (A), **197** (1901) 285–379

Fawcett, Philippa Garrett

Fawcett, (Miss) P.G.

Note on the motion of solids in a liquid. *Quart. J. Math.*, **26** (1893) 231–58

Fry, Agnes

Pearson, Karl; Lee, Alice; Warren, Ernest; Fry, Agnes; Fawcett, Cicely D., et al.

[Mathematical contributions to the theory of evolution. IX.] On the principle of homotyposis and its relation to heredity, to the variability of the individual, and to that of the race. Part I. Homotyposis in the vegetable kingdom. *Phil. Trans.* (A), **197** (1901) 285–379

Hardcastle, Frances

Hardcastle, (Miss) Frances

A theorem concerning the special systems of point-groups on a particular type of base curve. [1897] *London Math. Soc. Proc.*, **29** (1898) 132–9

Some observations on the modern theory of point-groups. *N. Y., Amer. Math. Soc. Bull.*, **4** (1898) 390–402

Report on the present state of the theory of point-groups. Part I. *Brit. Ass. Rep.* (1900) 121–31

Hudson, Hilda Phoebe

Hudson, Hilda

{Simple proof of Euclid II 9 and 10. *Nature*, **45** (1891) 189}

Lee, Alice Elizabeth

Lee, (Miss) Alice; Pearson, Karl

Mathematical contributions to the theory of evolution. On the relative variation and correlation in civilised and uncivilised races. *Roy. Soc. Proc.*, **61** (1897) 343–57

On the distribution of frequency (variation and correlation) of the barometric height at divers stations. [1897] *Phil. Trans.* (A), **190** (1898) 423–67

[Mathematical contributions to the theory of evolution. VI.] II. On the inheritance of fertility in mankind. [VIII. On the inheritance of characters not capable of exact quantitative measurement. Part I. Introductory. Part II. On the inheritance of coat-colour in horses. Part III. On the inheritance of eye-colour in man.] [1898–1900] *Phil. Trans.* (A), **192** (1899) 279–90; 195; **195** (1901) 79–150

On the vibrations in the field round a theoretical Hertzian oscillator. [1899] *Phil. Trans.* (A), **193** (1900) 159–88

Data for the problem of evolution in man. VI. A first study of the correlation of the human skull. [1900] *Phil. Trans.* (A), **196** (1901) 225–64

Pearson, Karl; Lee, Alice; Warren, Ernest; Fry, Agnes; Fawcett, Cicely D. et al.
[Mathematical contributions to the theory of evolution. IX.] On the principle of homotyposis and its relation to heredity, to the variability of the individual, and to that of the race. Part I. Homotyposis in the vegetable kingdom. *Phil. Trans.* (A), **197** (1901) 285–379

Stott, Alicia Boole
Stott, Alicia Boole
On certain series of sections of the regular four-dimensional hypersolids. [1900] *Amsterdam, Ak. Verh.* (*Sect.* 1), 7 (1901) No. 3, 21 pp.

Whiteley, Martha Annie
Whiteley, (Miss) Martha Annie; Pearson, Karl
Data for the problem of evolution in man. I. A first study of the variability and correlation of the hand. [1899] *Roy. Soc. Proc.,* **65** (1900) 126–51

Young, Grace Emily Chisholm
Chisholm, (Miss) Grace (Mrs. W. H. Young)
On the curve, $y = 1/\{x^2 + \sin^2\psi\}^{3/2}$, and its connection with an astronomical problem. *Astr. Soc. Month. Not.,* **57** (1897) 379–87; 501
Sulla varietà razionale normale M_3^4 di S_6 rappresentante della trigonometria sferica. [1899] *Torino Acc. Sci. Atti,* **34** (1898) 429–38

Meteorology

United States

Brodhead, Jane Milliken Napier
Brodhead, (Mrs.) Jane Napier
Indian cyclone of October, 1864. [1888] *Amer. Meteorol. J.,* **5** (1888–1889) 209–13

De Riemer, Alicia
De Riemer, (Miss) Alicia; Abbe, Cleveland
The average frequency of days of hail during 1893–97. *U. S. Monthly Weath. Rev.,* **26** (1898) 546–7

Edson, Helen R.
Edson, (Mrs.) Helen R.
Ice needles on Roan Mountain, Tennessee. [1895] *Amer. Meteorol. J.,* **11** (1894–1895) 387–9

Wing, Minerva E.
Wing, Minerva E.
Observations of the aurora borealis made at West Charlotte, Vt., lat. 14° 19′, long. 73° 15′ W. from Greenwich. *Amer. J. Sci.,* **8** (1874) 157
Anomalies of climate near Lake Champlain. [1885] *Amer. Meteorol. J.,* **2** (1885–1886) 55

Great Britain

Baillie, Mary Stewart (Lady)
Baillie, (Lady) Mary
[Remarkable sunrises on December 8th and 11th, 1884.] *Edinb. Roy. Soc. Proc.,* **13** (1886) 25

Baker, Annie
Baker, (Miss) Annie (et alii)
Phenological records for 1890[-93]. *Bristol Nat. Soc. Proc.,* **6** (1891) 278–86; **7** (1894) 4–12; 64–70; 145–52

Behrens, (Mrs.)
Behrens, (Mrs.)
Brazilian rainfall. [1888] *Symons, Meteorol. Mag.,* **23** (1889) 59

Bennett, Anne Ramsden
Bennett, Anne R.
Meteorological phenomena in connection with the climate of Berlin. *Edinb. New Phil. J.,* **54** (1853) 155–62; 214–29

Brown, Elizabeth
Brown, (Miss) Elizabeth
The aurora of 1881, January 31. *Observatory, London,* **4** (1881) 92
[The remarkable sunsets.] [1883] *Nature,* **29** (1884) 132
The sky-glows. *Nature,* **30** (1884) 607
Aurorae. *Nature,* **31** (1885) 458
Remarkable rime and mist. *Nature,* **39** (1889) 342
Unusual darkness. *Meteorol. Soc. Quart. J.,* **24** (1898) 166

Browne, Mary
Browne, Mary
The hurricane of 1830 in St. Vincent; by an eye-witness. *Timehri,* **5** (1886) 54–78

Crawshay, Rose Mary Yeates
Crawshay, Rose Mary
A double moon. *Nature,* **45** (1892) 224

Hall, W. L.
Hall, (Miss) W. L.
Notes on the meteorology of Eastbourne. [Reply to Mr. Bretton] *Eastbourne Nat. Hist. Soc. Trans.,* **1** (Pt.2) (1882) 41–3

Ley, Annie
Ley, Annie
[The remarkable sunsets.] [1883] *Nature,* **29** (1884) 103; 130
A remarkable rime. *Nature,* **39** (1889) 270
[Remarkable rime and mist.] *Nature,* **39** (1889) 342
Sun pillar. *Nature,* **45** (1892) 484

Ormerod, Eleanor Anne
Ormerod, (Miss) Eleanor A.
Water storage. *Symons, Meteorol. Mag.,* **20** (1885) 140
Heavy rain at St. Albans, July 12th. [1889] *Symons, Meteorol. Mag.,* **24** (1890) 96

Stevens, Catherine O.
Stevens, Catherine O.
A curious rainbow. *Nature,* **54** (1896) 211
Solar halo of July 3rd. *Nature,* **58** (1898) 224

Physics

United States

Bruère, Alice H.
Bruère, (Miss) Alice H.
A comparison of two concave Rowland gratings. *Phys. Rev.,* **3** (1896) 301–5
On the polarization of light reflected from hard rubber. *Phys. Rev.,* **6** (1898) 140–52

Crehore, Mary L.
Nichols, Edward L.; Crehore, Mary L.
Studies of the lime-light. [1894] *Phys. Rev.,* **2** (1895) 161–9

Foote, Eunice
Foote, Eunice (Mrs. Elisha)
Circumstances affecting the heat of the Sun's rays. *Silliman, J.,* **22** (1856) 382–3

On a new source of electrical excitation. *Amer. Ass. Proc.* (1857) 123–6; *Silliman, J.,* **24** (1857) 386–7

Maltby, Margaret Eliza

Maltby, (Miss) Margaret E.

Methode zur Bestimmung grosser elektrolytischer Widerstände. *Ztschr. Physikal. Chem.,* **18** (1895) 133–58

Methode zur Bestimmung der Periode electrischer Schwingungen. *Ann. Phys. Chem.,* **61** (1897) 553–77

Cross, Charles R.; Maltby, Margaret E.

On the least number of vibrations necessary to determine pitch. *Amer. Ac. Proc.,* **27** (1893) 222–35; {*Tech. Quart.,* **5** (1892) 213–28}

Kohlrausch, Friedrich, W. G.; Maltby, (Miss) Margaret E.

Das elektrische Leitvermögen wässeriger Lösungen von Alkali-Chloriden und Nitraten. *Berlin Ak. Sber.* (1899) 665–71; 779; *Berlin, Phys. Reichsanst. Abh.,* **3** (1900) 155–227

Morrison, Caroline Willard Baldwin

Baldwin, Caroline Willard

A photographic study of arc spectra. *Phys. Rev.,* **3** (1896) 370–80; 448–75

Noyes, Mary Chilton

Noyes, Mary Chilton

The influence of heat and the electric current upon Young's modulus for a piano wire. *Phys. Rev.,* **2** (1895) 277–97

The influence of heat, of the electric current, and of magnetization upon Young's modulus. *Phys. Rev.,* **3** (1896) 432–47

Sabine, Annie W.

Sabine, Annie W.

The strength of the microphone current as influenced by variations in normal pressure and mass of the electrodes. *Amer. Ac. Proc.,* **24** (1889) 90–3

Cross, Charles R.; Sabine, Annie W.

Researches on microphone currents. *Amer. Ac. Proc.,* **24** (1889) 94–104

Spencer, Mary Cass

Nichols, Edward L.; Spencer, Mary C.

The influence of temperature upon the transparency of solutions. *Phys. Rev.,* **2** (1895) 344–60

Stone, Isabelle

Stone, Isabelle

On the electrical resistance of thin films. *Phys. Rev.,* **6** (1898) 1–16

Great Britain

Ayrton, Phoebe Sarah Marks

Marks, (Miss) Sarah (Mrs. Ayrton)

The electric arc. *Electrician,* **34** (1895) 335–9; 364–8; 399–401; 471–5; 541–5; 610–16; **35** (1895) 418–21; 635–9; 743–8; **36** (1896) 36–9; 225–8; 539–42

[The E. M. F. etc. of electric arc.] *Nature,* **52** (1895) 535

On the relations between arc curves and crater ratios with cored positive carbons. *Brit. Ass. Rep.* (1897) 575–7

The drop of potential at the carbons of the electric arc. *Brit. Ass. Rep.* (1898) 805–7

The hissing of the electric arc. [With discussion] *Inst. Elect. Engin. J.,* **28** (1899) 400–35; 438–50; {*Elect. Rev.,* **44** (1899) 526–8; 614–16; 657–8; *Nature,* **60** 282–6; 302–5}

The light emitted by the continous-current arc. *Electrician,* **45** (1900) 921–4; 966–7

Bryant, Ella Mary

Bryant, Ella Mary

On the thermal condition of iron, steel and copper when acting as boiler-plate. *Inst. Civ. Engin. Proc.,* **132** (1898) 274–87

Chambers, Jesse Mary

Chambers, (Miss) J. M.

A simple way of explaining Clerk Maxwell's electromagnetic theory of light. *Phil. Mag.,* **21** (1886) 162–3

On the Weberian theory of diamagnetism, and on the rotation of the plane of polarisation of light by magnetic and diamagnetic substances. *Electrician,* **17** (1886) 27–9

On the possible connection of Hall's phenomenon with the rotation of the plane of polarisation of light. *Electrician,* **17** (1886) 69

On pyroelectric and doubly refracting crystals, dielectric action and Kerr's phenomenon. *Electrician,* **17** (1886) 193–4

Crosse, Cornelia Augusta Hewitt Berkeley

Crosse, (Mrs.)

On the apparent mechanical action accompanying electrical transfer. *Brit. Ass. Rep.* (1855) Pt.2, 55

Fawcett, Philippa Garrett

Fawcett, (Miss) Philippa G.

The electric strength of mixtures of nitrogen and hydrogen. *Roy. Soc. Proc.,* **56** (1894) 263–71

Hubbard, Emma

Hubbard, Emma

Bullet-proof shields. *Nature,* **50** (1894) 148

Klaassen, Helen G.

Klaassen, (Miss) Helen G.

The effect of temperature on the conductivity of solutions of sulphuric acid. [1891] *Cambridge Phil. Soc. Proc.,* **7** (1892) 137–41

Change of phase on reflection at the surface of highly absorbing media. *Phil. Mag.,* **44** (1897) 349–55

Ewing, James Alfred; Klaassen, (Miss) Helen G.

The dissipation of energy through reversals of magnetism in the core of a transformer. *Electrician,* **28** (1892) 111

Magnetic qualities of iron. [1893] *Phil. Trans. (A),* **184** (1894) 985–1039

Sidgwick, Eleanor Mildred Balfour

Rayleigh, John William Strutt (Lord); Sidgwick, Eleanor Mildred

On the specific resistance of mercury. [1882] *Roy. Soc. Proc.,* **34** (1883) 27–8; *Phil. Trans.,* **174** (1884) 173–85

Experiments by the method of Lorenz, for the further determination of the absolute value of the British Association unit of resistance with an appendix on the determination of the pitch of a standard tuning fork. [1882] *Chem. News,* **47** (1883) 27; *Roy. Soc. Proc.,* **34** (1883) 438–9; *Phil. Trans.,* **174** (1884) 295–322

On the electro-chemical equivalent of silver, and on the absolute electromotive force of Clark cells. [1884] *Roy. Soc. Proc.,* **36** (1884) 448–50; **37** (1884) 142–6; *Phil. Trans.,* **175** (1885) 411–60

Somerville, Mary Fairfax Greig

Somerville, Mary

On the magnetizing power of the more refrangible solar rays. *Annal. de Chimie,* **31** (1826) 393–400; *Edinb. J. Sci.,* **4** (1826) 328–32; *Froriep, Notizen,* **14** (1826) col. 49–52; *Phil. Trans.* (1826) Pt. 2, 132–9; *Poggend. Annal.,* **6** (1826) 493–6; *Quetelet, Corresp. Math.,* **2** (1826) 161–7

Expériences sur la transmission des rayons chimiques du spectre solaire, à travers différents millieux. *Bibl. Univ.,* **5** (1836) 391–4;

Paris, Ac. Sci. C. R., **3** (1836) 473–6; *Poggend. Annal.,* **39** (1836) 219–24; *Edinb. New Phil. J.,* **22** (1837) 180–3
On the action of the rays of the spectrum upon vegetable juices. [1845] *Phil. Trans.* (1846) 111–21

Technology

United States

Pugh, Marion Adell Howe
Howe, Minnie
Some experiments for the purpose of determining the active principles of bread making. [1891] *Iowa Ac. Sci. Proc.,* **1** (1892) Pt. 2, 64

Richards, Ellen Henrietta Swallow
Richards, Ellen Henrietta Swallow
{Notes on a naptha process for cleaning wool. *Bull. Nat. Ass. Wool Manuf.,* **9** (2) (1879) 96–101}
An apparatus for determining the liability of oils to spontaneous combustion. *J. Anal. Chem.,* **6** (1892) 269–71

Great Britain

Acworth, Marion Whiteford
Acworth, J. J.; Acworth, (Mrs.) Marion W.
Exposure and development. A rejoinder to Mr. Sterry's communication. *Photogr. J.,* **19** (1895) 361–6
Notes on the Hurter and Driffield system of speed testing. [With discussion] *Photogr. J.,* **19** (1895) 208–21
Exposure and development relatively considered. [1895] *Photogr. J.,* **20** (1896) 48–9

Hooker, Emma Jane Greenland
Hooker, Emma Jane
Method of preparing and applying a composition for painting in imitation of the ancient Grecian manner, called encaustic painting. {*Soc. Arts Trans.,* **5** (1787) 104–10; **10** (1792) 168–73} ; *Nicholson, J.,* **21** (1808) 81–5; *Tilloch, Phil. Mag.,* **32** (1808) 120–3; *Annal. de Chimie,* **77** (1811) 161–7

PART 3. SOCIAL SCIENCES

Anthropology and Ethnology

United States

Babbitt, Frances E.
Babbitt, (Miss) Franc E.
{Red Lake notes. *Minnesota Ac. Nat. Sci. Bull.,* **2** (1881) 86–101}
Exhibition and description of some palaeolithic quartz implements from central Minnesota. *Amer. Ass. Proc.* (1884) 593–9
Vestiges of glacial man in Minnesota. *Amer. Natlist.,* **18** (1884) 594–605; 697–708
Illustrative notes concerning the Minnesota Odjibwas. *Amer. Ass. Proc.* (1887) 303–7
Points concerning the Little Falls quartzes. *Amer. Ass. Proc.* (1889) 333–9
{Ancient quartz workers and their quarries in Minnesota. [Posth.] *Amer. Antiq. Orient. J.,* **3** (1895) 18–23}

Cook, Alice Carter
Cook, Alice Carter
{The aborigines of the Canary Islands. *Amer. Anthrop.,* **2** (1900) 451–93}

Fletcher, Alice Cunningham
Fletcher, Alice C.
{The Sun dance of the Ogalalla Sioux. *Amer. Ass. Proc.* (1882) 580–4}
{Among the Omahas. *Woman's J.,* **13** (1882) 46–7}
{On Indian education and self-support. *Century Mag.,* **26** (1883) 312–15}
{Five Indian ceremonies. *Peabody Mus. Amer. Arch. Ethnol., Rep. XVI* (1883) 260–333}
{Observations of the laws and privileges of the gens in Indian society. *Amer. Ass. Proc.* (1883) 395–6 [abstract]}
{Symbolic earth formations of the Winnebagoes. *Amer. Ass. Proc.* (1883) 396–7}
Observations upon the usage, symbolism and influence of the sacred pipes of fellowship among the Omahas. *Amer. Ass. Proc.* (1884) 615–17
{Lands in severality to Indians: illustrated by experiences with the Omaha tribe. *Amer. Ass. Proc.* (1884) 654–65}
{An evening in camp among the Omahas. *Science,* **6,** (1885) 88–90}
{An average day in camp among the Sioux. *Science,* **6,** (1885) 285–7}
{Land and education for the Indian. *Southern Workman,* **14** (1885) 6}
{Composite portraits of American Indians. *Science,* **7** (1886) 408}
{The problem of the Omahas. *Southern Workman,* **15** (1886) 55}
{The supernatural among the Omaha tribe of Indians. *Amer. Soc. Psychical Res. Proc.,* **1** (1887) 3–18}
{On the preservation of archaeologic monuments. *Amer. Ass. Proc.,* (1887) 317}
{Glimpses of child-life among the Omaha tribe. *J. Amer. Folkl.,* **1** (1888) 115–23}
{Joseph La Flesche. *J. Amer. Folkl.,* **2** (1889) 11}
{Leaves from my Omaha notebook. *J. Amer. Folkl.,* **2** (1889) 219–26}
The phonetic alphabet of the Winnebago Indians. *Amer. Ass. Proc.* (1889) 354–7; {*J. Amer. Folkl.,* **3** (1890) 299–301}
The Nez Perce country. *Amer. Ass. Proc.* (1891) 357
{The Indian Messiah. *J. Amer. Folkl.,* **4** (1891) 57–60}
{Hal-thu-ska Society of the Omaha tribe. *J. Amer. Folkl.,* **5** (1892) 135–44}
{Music as found in certain North American Indian tribes. *Music Rev.,* **2** (1893) 534–8; *Music,* **4** (1893) 457–67}
{Personal studies of Indian life: politics and "pipe-dancing." *Century Mag.,* **45** (1893) 441–5}
{Love songs among the Omaha Indians. *Memoirs Int. Congr. Anthrop., Chicago* (1894) 153–7}
{The religion of the North American Indians. *World's Congr. Religions, Addr.* (1894) 541–5}
{Indian songs: personal studies of Indian life. *Century Mag.,* **47** (1894) 421–31}
{Some aspects of Indian music and its study. *Archaeologist,* **2** (1894) 195–234}
{Hunting customs of the Omahas. *Century Mag.,* **47** (1895) 691–702}
The sacred pole of the Omaha tribe. *Amer. Ass. Proc.* (1895) 270–80; {*Amer. Antiquarian,* **17** (1895) 257–68}

{Indian songs and music. *Amer. Ass. Proc.* (1895) 281–4}
{Tribal life among the Omahas. *Century Mag.,* **51** (1896) 450–1}
Address [to the Anthrop. Sect.]. The emblematic use of the tree in the Dakotan group. *Amer. Ass. Proc.* (1896) 191–209
A study from the Omaha tribe: the import of the totem. *Amer. Ass. Proc.* (1897) 325–34; {*Science,* **7** (1898) 296–304}
{The Indian at the Trans-Mississippi Exposition. *Southern Workman,* **27** (1898) 216–17}
{Indian songs and music. *J. Amer. Folkl.,* **11** (1898) 85–104}
{The significance of the garment. *Amer. Ass. Proc.* (1898) 471–2}
{The significance of the scalp-lock: a study of Omaha ritual. *Anthrop. Inst. J.,* **27** (1898) 436–50}
{Flotsam and jetsam from aboriginal America. *Southern Workman,* **28** (1899) 12–14}
{The Indian woman and her problem. *Southern Workman,* **28** (1899) 172–6}
{Indian speech. *Southern Workman,* **28** (1899) 426–8}
{A Pawnee ritual used when changing a man's name. *Amer. Anthrop.,* **1** (1988) 82–97}
{Giving thanks: a Pawnee ceremony. *J. Amer. Folkl.,* **13** (1900) 261–6}
{Indian characteristics. *Southern Workman,* **29** (1900) 202–5}
{The old man's love song: an Indian story. *Music,* **18** (1900) 137}
{The Osage Indians in France. *Amer. Anthrop.,* **2** (1900) 395–400}
The hako: a Pawnee ceremony. *Smithsonian Inst. Bur. Ethnol. Rep.,* (Pt. 2) (1900–1901) 5–372

Fletcher, Alice C.; Stevenson, T. E.
{Report of the committee on the preservation of archaeologic remains on public lands. *Amer. Ass. Proc.* (1888) 35–7}

Fletcher, Alice C.; La Flesche, Francis; Fillmore, John C.
{A study of Omaha Indian music. *Peabody Mus. Amer. Arch. Ethnol.,* **1** (1893) 237–87}

McGee, Anita Newcomb

McGee, Anita Newcomb
Notes on American communities. *Amer. Ass. Proc.* (1888) 322–3
The evolution of a sect. *Amer. Ass. Proc.* (1890) 376
An experiment in human stirpiculture. *Amer. Ass. Proc.* (1891) 357–8

Sheldon, Mary (May) French

French-Sheldon, (Mrs.)
Customs among the natives of East Africa, from Teita to Kilimegalia, with special reference to their women and children. *Anthrop. Inst. J.,* **21** (1892) 358–90

Smith, Eriminnie Adele Platt

Smith, Erminnie A.
{On the Iroquois language. *Amer. Ass. Proc.* (1880) 736 [title only]}
{Myths of the Iroquois. *Smithsonian Inst. Bur. Ethnol. Rep.* (1880–1881) 47–116}
{Comparative differences in the Iroquois group of dialects. *Amer. Ass. Proc.* (1881) 315–19 [abstract]}
{Animal myths. *Amer. Ass. Proc.* (1881) 321–3 [abstract]}
{A few deductions from a "Dictionary of the Tuscarora dialect." *Amer. Ass. Proc.* (1882) 595 [title only]}
{Life among the Mohawks in the Catholic missions of Quebec province. *Amer. Ass. Proc..* (1883) 398–9 [abstract]}
{Studies in the Iroquois concerning the verb "to be" and its substitute. *Amer. Ass. Proc.* (1883) 399–402 [abstract]}
{Accidents or mode signs of verbs in the Iroquois dialects. *Amer. Ass. Proc.* (1883) 402–3 [abstract]}
{Disputed points concerning Iroquois pronouns. *Amer. Ass. Proc.* (1884) 606–9 [abstract]}
{Formation of Iroquois words. *Amer. Ass. Proc.* (1884) 617 [title only]}
{Etymology of the Iroquois word Rha-wen-ni-yu. *Amer. Ass. Proc.* (1884) 617 [title only]}
{Artificial wampum. *Science,* **5** (1885) 3–4}
{The significance of flora to the Iroquois. *Amer. Ass. Proc.* (1885) 404–11}

Stevenson, Matilda Coxe Evans

Stevenson, (Mrs.) Matilda Coxe
{The Cliff-Dwellers of the New Mexico canyons. *Kansas City Review,* **6** (1883) 636–9}
{The religious life of the Zuñi child. *Smithsonian Inst. Bur. Ethnol. Rep.* (1883–1884) 533–55}
{Zuñi religion. *Science,* **11** (1888) 136–7}
The Sia. *Smithsonian Inst. Bur. Ethnol. Rep.* (1889–1890) 3–157
{Tusayan legends of the Snake and Flute people. *Amer. Ass. Proc.* (1893) 258–70}
{A chapter in Zuñi mythology. *Memoirs Int. Congr. Anthrop., Chicago* (1894) 312–19}
{Zuñi ancestral gods and masks. *Amer. Anthrop.,* **11** (1898) 468–97}

Fletcher, Alice C; Stevenson, T[illy] E.
{Report of the committee on the preservation of archaeological remains on public lands. *Amer. Ass. Proc.* (1888) 35–7}

Great Britain

Aynsley, Harriet Georgina Maria Sutton

Murray-Aynsley, (Mrs.) Harriet G. M.
Thought and breathing. *Nature,* **41** (1890) 441

Bishop, Isabella Lucy Bird

Bishop, (Miss) Isabella L.
{Shadow of the Kurd. *Rev. of Revs.,* **59** (1891) 642–54, 819–35}
{Marriage system of Tibet. *Rev. of Revs.,* **7** (1893) 471}

Buckland, Anne Wallbank

Buckland, (Miss) A. W.
Ethnological hints afforded by the stimulants in use among savages and among the ancients. [1878] *Anthrop. Inst. J.,* **8** (1879) 239–53
Notes on some Cornish and Irish pre-historic monuments. [1879] *Anthrop. Inst. J.,* **9** (1880) 146–65
Facts suggestive of prehistoric intercourse between East and West. [1884] {*Brit. Ass. Rep.* (1884) 916–17}; *Anthrop. Inst. J.,* **14** (1885) 222–32
Distribution of animals and plants by ocean currents. *Nature,* **38** (1888) 245
{On tattooing. *Anthrop. Inst. J.,* **17** (1888) 318–27}

Crane, Agnes

Crane, (Miss) Agnes
The origin of speech and development of language. *Brighton Nat. Hist. Soc. Rep.* (1889) 20–30; 30–43
{Discovery of Mexican feather-work in Madrid. *Science,* **21** (1893) 11}

Godden, Gertrude H.

Godden, Gertrude H.
Nägä and other frontier tribes of north-east India. [With discussion] [1896–1897] *Anthrop. Inst. J.,* **26** (1897) 161–201; *Anthrop. Inst. J.,* **27** (1898) 2–51

Kingsley, Mary Henrietta
Kingsley, Mary H.
{The Negro future. *Spectator* (7 Dec. 1895)}
{The development of dodos. *National Rev.* (March 1896) 66–79}
{Black ghosts. *Cornhill Mag.* (June 1896) 79–92}
{Two days' African entertainment. *Cornhill Mag.* (March 1897) 354–9}
{The fetish view of the human soul. *Folklore,* **8** (1897) 138–51}
{African religion and law. [Hibbert lecture at Oxford] *National Rev.,* **29** (1897) 122–39}
{Fishing in West Africa. *National Rev.,* **29** (1897) 213–27}
{West Africa from an ethnologist's point of view. *Liverpool Geogr. Soc.* (11 Nov. 1897) 58–73}
{The laws and nature of property among the peoples of the true Negro stock. *Brit. Ass. Rep.* (1898) 1018–19}
{The forms of apparition in West Africa. *Psychical Res. Soc. J.,* **14** (1899) 331–42}
{West Africa from an ethnological point of view. *Imperial Inst. J.* (April 1900)}

Layard, Antonia (Nina) Frances
Layard, Nina F.
{Underground Ipswich [Blackfriar's excavations]. *East Anglian Daily Times* (28 Sept. 1898)}
{Remarks on Woolsey's cottage and the priory of St. Peter and St. Paul, Ipswich. *Archaeol. J.,* **56** (1899) 211–15}
{Original researches on the sites of religious houses in Ipswich. *Archaeol. J.,* **56** (1899) 232–8; *Suffolk Inst. Archaeol. Proc.,* **10** (1898–1900) 183–8}

Welby, Maria Louisa Helen Hervey (Lady)
Welby, (Lady)
An apparent paradox in human evolution. [With discussion.] [1890] *Anthrop. Inst. J.,* **20** (1891) 304–29

Geography

United States

Peckham, Elizabeth Maria Gifford
Gifford, Elizabeth M.; Peckham, Geo. W.
Temperature of Pine, Beaver and Okanchee Lakes, Waukesha County, Wisconsin, at different depths, extending from May to December, 1879; also particulars of depths of Pine Lake. *Wisconsin Ac. Trans.,* **5** (1882) 273–5

Semple, Ellen Churchill
Semple, Ellen Churchill
{The American Mediterranean and the interoceanic canal. *Vassar Misc. Mon.* **24** (1894) 54–65}
{The influence of the Appalachian barrier upon colonial history. *J. School Geogr.,* **1** (1897) 33–41}
{Some geographic causes determining the location of cities. *J. School Geogr.,* **1** (1897) 225–31}
{The Indians of southeastern Alaska in relation to their environment. *J. School Geogr.,* **2** (1898) 206–15}
{A comparative study of the Atlantic and Pacific oceans. *J. School Geogr.,* **3** (1899) 121–9; 172–80}
The development of the Hanse towns in relation to their geographical environment. *Amer. Geogr. Soc. Bull.,* **31** (1899) 236–55
{Louisville, a study in economic geography. *J. School Geogr.,* **4** (1900) 361–70}
{A new departure in social settlements. *Ann. Amer. Ac. Pol. Soc. Sci.,* **15** (1900) 301–4}

Workman, Fanny Bullock
Workman, (Mrs.) Fanny Bullock
Ascent of the Biafo glacier and Hispar pass: two pioneer ascents in the Karakoram. *Scott. Geogr. Mag.,* **15** (1899) 523–6
Dans les neiges du Baltistan. [1900] *Paris, Club Alpin Franç. Annu.,* **27** (1901) 320–55; {*Scott. Geogr. Mag.,* **17** (1901) 74–86 [under the title, "Amid the snows of Baltistan"]}

Great Britain

Bent, Mabel Virginia Anna Hall-Dare
Bent, (Mrs. J. Theodore)
Exploration in the Yafei and Fadhli countries. *Geogr. J.,* **12** (1898) 41–63
Sokotra. *Brit. Ass. Rep.* (1898) 943; *Scott. Geogr. Mag.,* **14** (1898) 629–36

Bishop, Isabella Lucy Bird
Bishop, (Mrs.) Isabella L.
{The Bakhtiari country. *Roy. Geogr. Soc. Proc.,* **13** (1891) 633–5}
Anniversary address. The Upper Karun region and the Bakhtiari Lurs. [1891] *Scott. Geogr. Mag.,* **8** (1892) 1–14
A journey through Lesser Tibet. *Scott. Geogr. Mag.,* **8** (1892) 513–28
A journey in western Sze-Chuan. *Geogr. J.,* **10** (1897) 19–50
The valley of the Yangtze. *Brit. Ass. Rep.* (1898) 940

Colthurst, (Miss)
Colthurst, (Miss)
Comparative view of the various standards commonly used to express vertical distances. *Geogr. Soc. J.,* **19** (1849) 192

Grove, Elisabeth (Lilly) de Boys (later **Lady Frazer**)
Grove, (Mrs.) Lilly
Deserts of Atacama and Tarapaca. *Scott. Geogr. Mag.,* **9** (1893) 57–65
On the islands of Chilöe. *Brit. Ass. Rep.* (1893) 833
The islands of Chilöe and the south of Chili. *Scott. Geogr. Mag.,* **10** (1894) 113–19

Kingsley, Mary Henrietta
Kingsley, (Miss) Mary H.
Travels in West Africa. *Geogr. J.,* **7** (1896) 95–6
Travels on the western coast of Equatorial Africa. *Scott. Geogr. Mag.,* **12** (1896) 113–24
{The ascent of Cameroons Peak and travels in French Congo. *Liverpool Geogr. Soc.* (19 March 1896) 36–52}
{The throne of thunder. *National Rev.* (May 1896) 357–74}

Power, Caroline
Power, Caroline
On the island of Ascension. *Geogr. Soc. J.,* **5** (1835) 256–62

Rickmers, C. Mabel Duff
Rickmers, (Mrs.)
Travels in East Bokhara. *Brit. Ass. Rep.* (1899) 806

Stone, Olivia M.
Stone, Olivia M.
The Canary Islands. [1887] *Nature,* **37** (1888) 201–2

Taylor, Annie Royle
Taylor, Annie R.
[Her] experiences in Tibet. [1893] *Scott. Geogr. Mag.,* **10** (1894) 1–8

Great Britain—Ireland

Clerke, Ellen Mary
Clerke, (Miss) E. M.
{On the aborigines of Western Australia. *Brit. Ass. Rep.* (1891) 716–17}

Psychology

United States

Andrews, Grace Allerton
Andrews, Grace A.
Studies of the dream consciousness. [1900] *Amer. J. Psychol.,* **12** (1900–1901) 131–4

Bodington, Alice
Bodington, Alice
Mental evolution in man and the lower animals. *Amer. Natlist.,* **26** (1892) 482–92; 543–606
Insanity in royal families. A study in heredity. *Amer. Natlist.,* **29** (1895) 118–29; 408
A study in morbid psychology, with some reflections. *Amer. Natlist.,* **30** (1896) 510–18; 599–605
Mental action during sleep, or sub-conscious reasoning. *Amer. Natlist.,* **30** (1896) 849–54

Calkins, Mary Whiton
Calkins, Mary Whiton
{A suggested classification of cases of association. *Phil. Rev.,* **1** (1892) 389–402}
{Experimental psychology at Wellesley College. *Amer. J. Psychol.,* **5** (1892–1893) 260–71}
Statistics of dreams. *Amer. J. Psychol.,* **5** (1892–1893) 311–43
A statistical study of pseudo-chromesthesia and of mental forms. *Amer. J. Psychol.,* **5** (1892–1893) 439–64
{Assocation I. *Psychol. Rev.,* **1** (1894) 476–83}
{Wellesley College psychological studies. *Educ. Rev.,* **8** (1894) 269–86; *Pedagog. Seminary,* **3** (1895) 319–41; *Psychol. Rev.,* **2** (1895) 363–7; 7 (1900) 580–91}
Synaesthesia. *Amer. J. Psychol.,* **7** (1895–1896) 90–107
{Association: An essay analytic and experimental. *Psychol. Rev. Monograph Supp.,* No. 2 (1896) 1–56}
{Association II. *Psychol. Rev.,* **3** (1896) 32–49}
{Community of ideas of men and women. *Psychol. Rev.,* **3** (1896) 426–30}
{Short studies in memory and association from the Wellesley laboratory. *Psychol. Rev.,* **5** (1898) 451–62}
{Attributes of sensation. *Psychol. Rev.,* **6** (1899) 506–14}
{Psychology as science of selves. *Phil. Rev.,* **9** (1900) 490–501}
{Wellesley College psychological studies. *Psychol. Rev.* 7 (1900) 580–91}

Carman, Ada
Carman, Ada
Pain and strength measurements of 1,507 school children in Saginaw, Michigan. [1899] *Amer. J. Psychol.,* **10** (1898–1899) 392–8

Carter, Marion Hamilton
Carter, Marion Hamilton
Darwin's idea of mental development. [1898] *Amer. J. Psychol.,* **9** (1897–1898) 534–59
Romanes' idea of mental development. [1898] *Amer. J. Psychol.,* **11** (1899–1900) 101–18

Cutler, Mabel Woodbury Learoyd
Learoyd, Mabel W.; Taylor, Maude L.
The "continued story." *Amer. J. Psychol.,* 7 (1895–1896) 86–90

Downey, June Etta
Downey, June E.
A musical experiment. [1897] *Amer. J. Psychol.,* **9** (1897–1898) 63–9

Gamble, Eleanor Acheson McCulloch
Gamble, Eleanor Acheson McCulloch
The applicability of Weber's law to smell. [1898] *Amer. J. Psychol.,* **10** (1898–1899) 82–142

Hallam, Florence M.
Weed, Sarah C.; Hallam, Florence M.; Phinney, Emma D.
A study of the dream-consciousness. [1896] *Amer. J. Psychol.,* 7 (1895–1896) 405–11

Hill, Caroline Miles
Miles, Caroline
A study of individual psychology. [1894] *Amer. J. Psychol.,* **6** (1893–1894) 534–58
Hill, Caroline Miles
On choice. [1898] *Amer. J. Psychol.,* **9** (1897–1898) 587–90

Hinman, Alice Julia Hamlin
Hamlin, Alice J.
On the least observable interval between stimuli addressed to disparate senses and to different organs of the same sense. [1894] *Amer. J. Psychol.,* **6** (1893–1894) 564–75
Attention and distraction. [1896] *Amer. J. Psychol.,* **8** (1896–1897) 3–36

Holmes, Mary Emilée
Holmes, Mary E.
Intelligence of the cat. *Amer. Natlist.,* **18** (1884) 95–7
Is it instinct or intelligence? *Science,* **21** (1893) 11–12

Hunt, Harriet E.
Hunt, Hattie E.
Observations on newly hatched chicks. [1897] *Amer. J. Psychol.,* **9** (1897–1898) 125–7

Jacobi, Mary Cortina Putnam
Jacobi, Mary Putnam
Note on aphasis with reference to loss of nouns. *N. Y. Med. J.,* **45** (1887) 217
{Note on the special liability to loss of nouns in aphasis. *J. Nerv. Ment. Dis.,* **14** (1887) 94–110}
The place for the study of language in a curriculum of education. [1888] *Amer. J. Psychol.,* **2** (1889–1890) 91–140

Kellerman, Stella Victoria Dennis
Kellerman, Mrs. W. A.
Epidemic forms of mental or nervous diseases or disorders. *Science,* **21** (1893) 305

Ladd-Franklin, Christine
Ladd, Christine (Mrs. Franklin)
Illusive memory. *Science,* **3** (1884) 434–5
A method for the experimental determination of the horopter. [1887] *Amer. J. Psychol.,* **1** (1888–1889) 99–111
An unknown organ of sense. *Science,* **14** (1889) 183–5
A new theory of light sensation. [1892] *Johns Hopkins Univ. Circ.,* **12** (1892–1893) 108–10; *Ztschr. Psychol.,* **4** (1893) 211–21; *Nature,* **49** (1893–1894) 394

{Hering's theory of color vision. *Mind,* **2** (1893) 517}
Theory of color sensation. *Science,* **22** (1893) 80–1
Color vision. *Science,* **22** (1893) 135; **7** (1898) 773–6; **8** (1898) 329–30
The normal night-blindness of the fovea. [1894] *Amer. J. Psychol.,* **6** (1893–1894) 630; *Science,* **1** (1895) 115–16
The positions of retinal images. [1896] *Nature,* **53** (1895–1896) 341
An optical illusion. *Science,* **3** (1896) 247–75
The new theory of the light-sense. *Amer. Ass. Proc.* (1898) 473–4
Phosphorescence in deep-sea animals. *Science,* **11** (1900) 954
The problem of color. *Science,* **12** (1900) 408–10

McNaught, Margaret Everitt Schallenberger
Schallenberger, Margaret
Discussion. Professor Baldwin's method of studying the color-perception of children. [1897] *Amer. J. Psychol.,* **8** (1896–1897) 560–76; **9** (1897–1898) 62

Phinney, Emma D.
Weed, Sarah C.; Hallam, Florence M.; Phinney, Emma D.
A study of the dream-consciousness. [1896] *Amer. J. Psychol.,* **7** (1895–1896) 405–11

Sharp, Stella Emily
Sharp, Stella Emily
Individual psychology: a study in psychological method. [1899] *Amer. J. Psychol.,* **10** (1898–1899) 329–91

Smith, Margaret Keiver
Smith, Margaret Keiver
Rhythmus und Arbeit. *Phil. Stud.,* **16** (1900) 71–133; 197–305

Talbot, Ellen Bliss
Talbot, Ellen Bliss
An attempt to train the visual memory. [1897] *Amer. J. Psychol.,* **8** (1896–1897) 414–17
Darlington, L.; Talbot, E. B.
A study of certain methods of distracting the attention. Distraction by musical sounds. The effect of pitch upon attention. [1898] *Amer. J. Psychol.,* **9** (1897–1898) 332–45
{The relation between human consciousness and its ideal as conceived by Kant and Fichte. *Kantstudien,* **4** (1899) 286–310}

Talbott, Laura Osborne
Talbott, Laura Osborne
The psychical study of child-life. *Amer. Ass. Proc.* (1891) 358–61
A few psychological inquiries. *Amer. Ass. Proc.* (1892) 294–6

Taylor, Maude L.
Learoyd, Mabel W.; Taylor, Maude L.
The "continued story." *Amer. J. Psychol.,* **7** (1895–1896) 86–90

Washburn, Margaret Floy
Washburn, Margaret Floy
Some apparatus for cutaneous stimulation. [With note by E. B. Titchener.] [1894] *Amer. J. Psychol.,* **6** (1893–1894) 422–4; 523
{The perception of distance in the inverted landscape. *Mind,* **3** (1894) 438–40}
Ueber den Einfluss der Gesichtsassociationen auf die Raumwahrnehmungen der Haut. *Phil. Stud.,* **11** (1895) 190–225
{The intensive statement of particular and negative propositions. *Phil. Rev.,* **5** (1896) 403–5}
{The process of recognition. *Phil. Rev.,* **6** (1897) 267–74}
{The psychology of deductive logic. *Mind,* **7** (1898) 523–30}
{Subjective colors and the after-image: their significance for the theory of attention. *Mind,* **8** (1899) 25–34}
{After-images. *Psychol. Rev.,* **6** (1899) 653}
{Recent discussions of imitation. *Phil. Rev.,* **8** (1899) 101–4}
{The color changes of the white light after-image, central and peripheral. *Psychol. Rev.,* **7** (1900) 39–46}

Weed, Sarah C.
Weed, Sarah C.; Hallam, Florence M.; Phinney, Emma D.
A study of the dream-consciousness. [1896] *Amer. J. Psychol.,* **7** (1895–1896) 405–11

Great Britain

Bryant, Sophie Willock
Bryant, (Mrs.) Sophie
Experiments in testing the character of school children. [With discussion.] [1885] *Anthrop. Inst. J.,* **15** (1886) 338–51
{Antipathy and sympathy. *Mind,* **4** (1895) 365–70}
{The double effect of mental stimulae: a contrast of types. *Mind,* **9** (1900) 305–18}

Naden, Constance Caroline Woodhill
Naden, Constance C. W.
Volition. *Midland Natlist.,* **11** (1888) 53–7; 109–14

Periodical Title Abbreviations Key

Abbreviations follow the Royal Society *Catalogue* usage. For periodicals not included in the *Catalogue* but added to the bibliography, a similar style of title abbreviation has been adopted.

Agr. Soc. J.	The Journal of the Royal Agricultural Society of England. London.
Agr. Stud. Gaz.	Agricultural Students' Gazette. Cirencester.
Amer. Ac. Mem.	Memoirs of the American Academy of Arts and Sciences. Cambridge, Mass.
Amer. Ac. Proc.	Proceedings of the American Academy of Arts and Sciences. Boston.
Amer. Agriculturalist	American Agriculturalist. New York.
Amer. Anthrop.	American Anthropologist. Washington, D.C., New York, & Lancaster, Pa.
Amer. Antiq. Orient. J.	American Antiquarian and Oriental Journal. Chicago.
Amer. Ass. Proc.	Proceedings of the American Association for the Advancement of Science. Washington, D.C., & Salem, Mass.

Amer. Chem. J.	American Chemical Journal. Baltimore.
Amer. Chem. Soc. J.	The Journal of the American Chemical Society. New York & Easton, Pa.
Amer. Chemist	The American Chemist, a Monthly Journal of Theoretical Chemistry. New York.
Amer. Ent.	American Entomologist. St. Louis & New York.
Amer. Ent. Bot.	American Entomologist and Botanist. St. Louis & New York.
Amer. Ent. Soc. Trans.	Transactions of the American Entomological Society. Philadelphia.
Amer. Geol. Soc. Bull.	Bulletin of the Geological Society of America. New York & Rochester, N.Y.
Amer. Geologist	The American Geologist. Minneapolis, Minn.
Amer. J. Math.	American Journal of Mathematics. Baltimore.
Amer. J. Med. Sci.	American Journal of Medical Sciences. Philadelphia.
Amer. J. Obst.	American Journal of Obstetrics and Diseases of Women and Children. New York.
Amer. J. Otol.	American Journal of Otology. Boston.
Amer. J. Pharm.	The American Journal of Pharmacy. Philadelphia.
Amer. J. Physiol.	The American Journal of Physiology. Boston.
Amer. J. Psychol.	The American Journal of Psychology. Baltimore & Worcester, Mass.
Amer. J. Sci.	The American Journal of Science. New Haven, Conn.
Amer. J. Surg. Gyn.	American Journal of Surgery and Gynecology. New York.
Amer. Med. Assoc. J.	Journal of the American Medical Association. Chicago.
Amer. Med. Surg. Bull.	American Medico-Surgical Bulletin. New York.
Amer. Meteorol. J.	American Meteorological Journal. Detroit, Ann Arbor, Mich., Boston, New York, Chicago, & London.
Amer. Micr. J.	The American Monthly Microscopical Journal. Washington, D.C.
Amer. Micr. Soc. Proc.	Proceedings of the American Microscopical Society. Washington, D.C., & Ithaca, N.Y.
Amer. Micr. Soc. Trans.	Transactions of the American Microscopical Society. Lincoln, Neb., & New York.
Amer. Midland Natlist.	American Midland Naturalist. Lafayette, Ind.
Amer. Natlist.	The American Naturalist. Salem, Mass., Philadelphia, & New York.
Amer. Phil. Soc. Proc.	Proceedings of the American Philosophical Society. Philadelphia.
Amer. Phil. Soc. Trans.	Transactions of the American Philosophical Society. Philadelphia.
Amer. Soc. Micr. Proc.	Proceedings of the American Society of Microscopists. [n.p.]
Amer. Soc. Psychical Res. Proc.	Proceedings of the American Society of Psychical Research. Boston.
Amsterdam, Ak. Verh. (Sect. 1)	Verhandelingen der Koninklijke Akademie van Wetenschappen. Amsterdam.
Anat. Anz.	Anatomischer Anzeiger. Jena.
Ann. Amer. Ac. Pol. Soc. Sci.	Annals of the American Academy of Political and Social Sciences. Philadelphia.
Ann. Bot.	Annals of Botany. London.
Ann. Hyg.	Annals of Hygiene. Philadelphia.
Ann. Mag. Nat. Hist.	The Annals and Magazine of Natural History, including Zoology, Botany, and Geology. London.
Ann. Mat.	Annali di Matematica pura ed applicata . . . Milano.
Ann. Math.	Annals of Mathematics. Charlottesville, Va., etc.
Ann. Nat. Hist.	Annals of Natural History, or Magazine of Zoology, Botany, and Biology. London.
Ann. Phys. Chem.	Annalen der Physik und Chemie. Leipzig.
Ann. Sci. Nat. (Bot.)	Annales des Science Naturelles. Botanique. Paris.
Ann. Scott. Nat. Hist.	The Annals of Scottish Natural History. Edinburgh.
Ann. Surg.	Annals of Surgery. St. Louis & Philadelphia.
Annal. de Chimie	Annales de Chimie. Paris.
Anthrop. Inst. J.	The Journal of the Anthropological Institute of Great Britain and Ireland. London.
Arch. Anat. Physiol. (Physiol. Abth.)	Archiv für Anatomie und Physiologie. Physiologische Abtheilung. Archiv für Physiologie. Leipzig.
Arch. de Physiol.	Archives de Physiologie Normale et Pathologique. Paris.
Arch. f. Ophthalm.	Albrecht von Graefe's Archiv für Ophthalmologie. Berlin & Leipzig.
Arch. EntwMech.	Archiv für Entwickelungsmechanik der Organismen. Leipzig.
Arch. Med.	Archives of Medicine. New York.
Arch. Mikr. Anat.	Archiv für Mikroskopische Anatomie [und Entwickelungsgeschichte]. Bonn.
Arch. Ophthalm.	Archives of Ophthalmology. New York.
Arch. Pediat.	Archives of Pediatrics. Jersey City, N.J., Philadelphia, & New York.
Arch. Sci. Phys. Nat.	Bibliotheque Universelle. Archives des Sciences Physiques et Naturelles. Genève.

Arch. Sci. Pract. Med.	Archives of Scientific and Practical Medicine. New York.
Archaeologist	Archaeologist. Waterloo, Ind. & Columbus, Ohio.
Archaeol. J.	Archaeological Journal. London.
Asa Gray Bull.	Asa Gray Bulletin. New York.
Ass. Franç. C. R.	Association Française pour l'Avancement des Sciences. Comptes Rendus. Paris.
Astr. & Astrophys.	Astronomy and Astro-Physics. Northfield, Minn.
Astr. J.	The Astronomical Journal. Boston.
Astr. Nachr.	Astronomische Nachrichten. Kiel.
Astr. Soc. Month. Not.	Monthly Notices of the [Royal] Astronomical Society. London.
Astr. Soc. Pacific Publ.	Publications of the Astronomical Society of the Pacific. San Francisco.
Astronomie	L'Astronomie. Paris.
Astrophys. J.	The Astrophysical Journal. Chicago.
Auk	The Auk. Boston & New York.
Aurora	The Aurora. Ames, Iowa.
Barrow Field Club Rep.	Barrow Naturalists' Field Club and Literary and Scientific Association. Annual Report and Proceedings. Barrow-in-Furness & Kendal.
Berlin Ak. Sber.	Sitzungsberichte der Königlich Preussischen Akademie der Wissenschaften zu Berlin. Berlin.
Berlin, Chem. Ges. Ber.	Berichte der Deutschen Chemischen Gesellschaft. Berlin.
Berlin, Phys. Reichsanst. Abh.	Wissenschaftliche Abhandlungen der Physikalisch-Technischen Reichsanstalt. Berlin.
Bern Mitth.	Mittheilungen der Naturforschenden Gesellschaft in Bern. Bern.
Berwick. Nat. Club Hist.	History of the Berwickshire Naturalists' Club. Alnwick.
Bibl. Univ.	Bibliothéque Universelle des Sciences, Belles-Lettres, et Arts. Partie des Sciences. Genève.
Bombay Nat. Hist. Soc. J.	The Journal of the Bombay Natural History Society. Bombay.
Boston Med. Surg. J.	Boston Medical and Surgical Journal. Boston.
Boston Soc. Med. Sci. J.	Journal of the Boston Society of Medical Sciences. Boston.
Boston Soc. Nat. Hist. Mem.	Memoirs read before the Boston Society of Natural History. Boston.
Boston Soc. Nat. Hist. Proc.	Proceedings of the Boston Society of Natural History. Boston.
Bot. Gaz.	The Botanical Gazette. Crawfordsville, Ind., & Indianapolis.
Brighton Nat. Hist. Soc. Proc.	Annual Reports and Abstracts of Proceedings of the Brighton and Sussex Natural History Society. Brighton.
Brighton Nat. Hist. Soc. Rep.	Brighton and Sussex Natural History and Philosophical Society. Annual Report together with Abstracts of Papers. Brighton.
Bristol Nat. Soc. Proc.	Proceedings of the Bristol Naturalists' Society. Bristol.
Brit. Ass. Rep.	Reports of the British Association for the Advancement of Science. London.
Brit. Astr. Ass. J.	Journal of the British Astronomical Association. London.
Brit. Astr. Ass. Mem.	Memoirs of the British Astronomical Association. London.
Brit. J. Dental Sci.	British Journal of Dental Science. London.
Brit. Mycol. Soc. Trans.	The British Mycological Society. Transactions. Worcester.
Brooklyn Med. J.	Brooklyn Medical Journal. New York.
Bull. Nat. Ass. Wool Manuf.	Bulletin of the National Association of Wool Manufacturers. Boston.
California Ac. Bull.	Bulletin of the California Academy of Sciences. San Francisco.
California Ac. Occ. Pap.	Occasional Papers of the California Academy of Sciences. San Francisco.
California Ac. Proc.	Proceedings of the California Academy of Sciences. San Francisco.
California Ac. Proc. (Bot.)	Proceedings of the California Academy of Sciences. Botany. San Francisco.
California Ac. Proc. (Zool.)	Proceedings of the California Academy of Sciences. Zoology. San Francisco.
Cambridge Phil. Soc. Proc.	Proceedings of the Cambridge Philosophical Society. Cambridge.
Canad. Ent.	The Canadian Entomologist. London, Ontario.
Canad. Inst. Trans.	Transactions of the Canadian Institute. Toronto.
Canad. Rec. Sci.	The Canadian Record of Science. Montreal.
Catania Acc. Gioen. Att.	Atti dell' Accademia Gioenia di Scienze Naturali in Catania. Catania.
Cellule	La Cellule. Recueil de Cytologie et d'Histologie générale. Liers, Gent, & Leuven.
Century	The Century. New York.
Century Mag.	Century Magazine. [Title varies.] New York.
Chamber's J.	Chamber's Journal. Edinburgh & London.
Chem. News	The Chemical News and Journal of Physical Science. London.
Chem. Soc. J.	Journal of the Chemical Society. London.

Chem. Soc. Proc. Proceedings of the Chemical Society. London.
Chicago Clin. Rev. Chicago Clinical Review. Chicago.
Chicago M. J. Exam. Chicago Medical Journal and Examiner. Chicago.
Chicago Med. Rec. Chicago Medical Record. Chicago.
Chicago Path. Soc. Trans. Transactions of the Chicago Pathological Society. Chicago.
Ciel et Terre Ciel et Terre. Bruxelles.
Cincin. Soc. Nat. Hist. J. The Journal of the Cincinnati Society of Natural History. Cincinnati.
Climate Climate. A Quarterly Journal of Health and Travel. London.
Coll. and Clin. Rec. [Dunglison's] College and Clinical Record. Philadelphia.
Congr. Int. Hyg. C. R. . . . Congrès International d'Hygiène et de Démographie . . . Compte[s] Rendu[s et Mémoires]. Budapest.
Connecticut Ac. Trans. Transactions of the Connecticut Academy of Arts and Sciences. New Haven.
Cornhill Mag. Cornhill Magazine. London.
Cornwall Geol. Soc. Trans. Transactions of the Royal Geological Society of Cornwall. Penzance.
Cornwall, J. Roy. Instit. Journal of the Royal Institution of Cornwall. Truro.
Cornwall, Polyt. Soc. Trans. Reports and Transactions of the Royal Polytechnic Society of Cornwall. Falmouth.
Cumberland Ass. Trans. Transactions of the Cumberland and Westmorland Association for the Advancement of Literature and Science. Carlisle.
Danmarks Geol. Unders. Danmarks Geologiske Undersøgelse. København.
Denison Univ. Sci. Lab. Bull. Bulletin of [the Scientific Laboratories of] Denison University. Granville, Ohio.
Des Moines, Analyst The Analyst: a monthly Journal of pure and applied Mathematics. Des Moines, Iowa.
Deutsch. Bot. Ges. Ber. Berichte der Deutschen Botanischen Gesellschaft. Berlin.
Devon. Ass. Trans. Report and Transactions of the Devonshire Association for the Advancement of Science, Literature, and Art. Plymouth.
Dublin Soc. Sci. Proc. The Scientific Proceedings of the Royal Dublin Society. Dublin.
Dublin Soc. Sci. Trans. The Scientific Transactions of the Royal Dublin Society. Dublin.
Dumfr. Gallow. Soc. Trans. The Transactions and Journal of Proceedings of the Dumfriesshire and Galloway Natural History and Antiquarian Society. Dumfries.
East Anglian Daily Times East Anglian Daily Times. Ipswich.
Eastbourne Nat. Hist. Soc. Trans. Transactions of the Eastbourne Natural History Society. Eastbourne.
Edinb., Bot. Soc. Trans. Transactions of the Botanical Society. Edinburgh.
Edinb. Bot. Soc. Trans. & Proc. Transactions and Proceedings of the Botanical Society of Edinburgh. Edinburgh.
Edinb., Fish. Brd. Rep. Annual Report of the Fishery Board for Scotland. Edinburgh.
Edinb. J. Sci. The Edinburgh Journal of Science. Edinburgh.
Edinb. Nat. Soc. Trans. Transactions of the Edinburgh Field Naturalists' and Microscopical Society. Edinburgh.
Edinb. New Phil. J. The Edinburgh New Philosophical Journal. Edinburgh.
Edinb. R. Coll. Physns. Lab. Rep. Reports from the Laboratory of the Royal College of Physicians, Edinburgh. Edinburgh & London.
Edinb. Roy. Soc. Proc. Proceedings of the Royal Society of Edinburgh. Edinburgh.
Educ. Rev. Educational Review. New York.
Educ. Times The Educational Times and Journal of the College of Preceptors. London.
Elect. Rev. The Electrical Review. London.
Electrician The Electrician. London.
Engler, Bot. Jbüch. Botanische Jahrbücher für Systematik, Pflanzengeschichte und Pflanzengeographie herausgegeben von A. Engler. Leipzig.
Ent. Month. Mag. The Entomologist's Monthly Magazine. London.
Ent. Record The Entomologist's Record and Journal of Variation. London.
Ent. Soc. Trans. The Transactions of the Entomological Society of London. London.
Entomologica Amer. Entomologica Americana. Brooklyn, N.Y.
Entomologist The Entomologist. London.
Erythea Erythea. Berkeley, Calif.
Essex Inst. Bull. Bulletin of the Essex Institute. Salem, Mass.
Folklore Folklore. London.
Forest and Stream Forest and Stream. New York.
Fortschr. Röntgenstr. Forschritte auf dem Gebiete der Röntgenstrahlen. Hamburg.
Franklin Inst. J. The Journal of the Franklin Institute devoted to Science and the Mechanic Arts. Philadelphia.

Froriep, Notizen	Notizen aus dem Gebiete der Natur- und Heilkunde; von L. von Froriep. Erfurt & Weimar.
Gard. Chron.	The Gardener's Chronicle. London.
Garden & Forest	Garden and Forest. New York.
Genova Mus. Civ. Ann.	Annali del Museo Civico di Storia Naturale di Genova. Genova.
Geogr. J.	The Geographical Journal. London.
Geogr. Soc. J.	Journal of the Royal Geographical Society of London. London.
Geol. Ass. Proc.	Proceedings of the Geologists' Association. London.
Geol. Mag.	The Geological Magazine. London.
Geol. Soc. Quart. J.	The Quarterly Journal of the Geological Society of London. London.
Geol. Soc. Trans.	Transactions of the Geological Society of London. London.
Geologist	The Geologist. London.
Glasgow Nat. Hist. Soc. Proc. & Trans.	Transactions of the Natural History Society of Glasgow. Glasgow.
Harper's New Mon. Mag.	Harper's New Monthly Magazine. New York.
Harvard Astr. Obs. Ann.	Annals of the Astronomical Observatory of Harvard College. Cambridge, Mass., Karlsruhe, & Waterville, Me.
Harvard Mus. Zool. Bull.	Bulletin of the Museum of Comparative Zoology at Harvard College. Cambridge, Mass.
Herts. Nat. Hist. Soc. Trans.	Transactions of the Hertfordshire Natural History Society and Field Club. London.
Hist. Soc. Southern Calif., Ann. Pub.	Annual Publication of the Historical Society of Southern California. Los Angeles.
Holmesdale Nat. Hist. Club Proc.	Proceedings of the Holmesdale Natural History Club. London.
Hooker, Comp. Bot. Mag.	Companion to the Botanical Magazine; by W. J. Hooker. London.
Hooker's J. Bot.	Hooker's Journal of Botany. London.
Hortic. Soc. J.	The Journal of the Royal Horticultural Society. London.
Ill. Insects Rep.	. . . Report of the State Entomologist. . . . on the Noxious and Beneficial Insects of the State of Illinois. Bloomington, Ill.
Ill. Lab. Nat. Hist. Bull.	Bulletin of the Illinois State Laboratory of Natural History. Bloomington, Ill.
Imperial Inst. J.	Imperial Institute Journal. London.
Indep. Pract.	Independent Practitioner. Baltimore & Buffalo.
Indian Med. Gaz.	The Indian Medical Gazette. Calcutta.
Indian Mus. Notes	Indian Museum Notes. Calcutta.
Indiana Ac. Sci. Proc.	Proceedings of the Indiana Academy of Sciences. Indianapolis, Ind.
Indiana Agric. Expt. Stn. Bull.	Indiana Agricultural Experiment Station. Bulletin. Lafayette, Ind.
Indiana Dept. Geol. Ann. Rep.	Indiana, Department of Geology and Natural History Reports. Indianapolis.
Inst. Civ. Engin. Proc.	Minutes of Proceedings of the Institution of Civil Engineers. London.
Inst. Elect. Engin. J.	Journal of the Institution of Electrical Engineers. London & New York.
Int. Cong. Hyg. Trans.	Transactions of the Seventh International Congress of Hygiene and Demography. London.
Int. J. Surg.	International Journal of Surgery. New York.
Int. Med. Mag.	International Medical Magazine. Philadelphia.
Iowa Ac. Sci. Proc.	Proceedings of the Iowa Academy of Sciences. Des Moines, Iowa.
Iowa Univ. Lab. Nat. Hist. Bull.	Bulletin from the Laboratories of Natural History of the State University of Iowa. Iowa City, Iowa.
Iris	Iris. Dresden.
Irish Ac. Proc.	Proceedings of the Royal Irish Academy. [Science.] Dublin.
Irish Ac. Trans.	The Transactions of the Royal Irish Academy. Science. Dublin.
Irish Natlist.	The Irish Naturalist. Dublin, Belfast, & London.
J. Amer. Folkl.	Journal of American Folklore. Boston, New York, & Lancaster, Pa.
J. Anal. Chem.	The Journal of Analytical Chemistry. Easton, Pa.
J. Anat. Physiol.	The Journal of Anatomy and Physiology, normal and pathological. London, Cambridge, & Edinburgh.
J. App. Micr.	Journal of Applied Microscopy and Laboratory Methods. Rochester, N.Y.
J. Bot.	The Journal of Botany, British and Foreign. London.
J. Columbus Hort. Soc.	Columbus Horticultural Society Journal. Columbus, Ohio.
J. Conch.	The Journal of Conchology. Leeds.
J. Exper. Med.	The Journal of Experimental Medicine. New York.
J. Geol. (Chicago)	The Journal of Geology. Chicago.
J. Malacol.	The Journal of Malacology. London & Berlin.
J. Micr. Nat. Sci.	Journal of Microscopy and Natural Science. London.

J. Morphol. The Journal of Morphology. Boston.
J. Mycol. The Journal of Mycology. [U.S. Dept. of Agriculture.] Washington, D.C.
J. Nerv. Ment. Dis. Journal of Nervous and Mental Disease. Chicago.
J. Physiol. The Journal of Physiology. Cambridge & London.
J. Prakt. Chem. Journal für praktische Chemie. Leipzig.
J. School Geogr. Journal of School Geography. Lancaster, Pa.
J. Sci. The Journal of Science and Annals of Astronomy, Biology, Geology, Industrial Arts, Manufactures, and Technology. London.
J. de Phys. Journal de Physique, de Chimie, et de l'Histoire naturelle. Paris.
Jamaica Inst. J. Journal of the Institute of Jamaica. Kingston, Jamaica.
Jena. Ztschr. Jenaische Zeitschrift für Naturwissenschaft. Jena.
Kan. Ac. Sci. Trans. Transactions of the . . . . annual Meeting of the Kansas Academy of Science. Topeka, Kansas.
Kan. Univ. Quarterly The Kansas University Quarterly. Lawrence, Kansas.
Kansas City Rev. Sci. Ind. Kansas City Review of Science & Industry. Kansas City.
Kantstudien Kantstudien. Hamburg & Berlin.
Leicester Soc. Trans. Transactions of the Leicester Literary and Philosophical Society. Leicester.
Lett. and Papers on Agr. Letters and Papers on Agriculture, Planting, etc. Selected from the Correspondence of the Bath and West of England Society. Bath.
Linn. Soc. J. (Bot.) The Journal of the Linnean Society. Botany. London.
Linn. Soc. J. (Zool.) The Journal of the Linnean Society. Zoology. London.
Linn. Soc. Trans. The Transactions of the Linnean Society of London. London.
Linn. Soc. Trans. (Bot.) The Transactions of the Linnean Society of London. Botany. London.
Liverpool Astr. Soc. Abs. Proc. Liverpool Astronomical Society. Abstract of Proceedings. Liverpool.
Liverpool Astr. Soc. J. Journal of the Liverpool Astronomical Society. Liverpool.
Liverpool Biol. Soc. Proc. Proceedings of the Liverpool Biological Society. Liverpool.
Liverpool Biol. Soc. Proc. & Trans. Proceedings and Transactions of the Liverpool Biological Society. Liverpool.
Liverpool Geogr. Soc. Liverpool Geographical Society Transactions. Liverpool.
Liverpool Geol. Ass. J. Liverpool Geological Association. Journal. Liverpool.
Liverpool Lit. Phil. Soc. Proc. Proceedings of the Literary and Philosophical Society of Liverpool. London & Liverpool.
Liverpool, Thompson Yates Lab. Rep. The Thompson Yates Laboratories Report. Liverpool.
London Malacol. Soc. Proc. Proceedings of the Malacological Society of London. London.
London Math. Soc. Proc. Proceedings of the London Mathematical Society. London.
London Phys. Soc. Proc. Proceedings of the Physical Society of London. London.
Mag. Nat. Hist. The Magazine of Natural History, and Journal of Zoology, Botany, Mineralogy, Geology, and Meteorology. London.
Manchester Lit. Phil. Soc. Mem. & Proc. Memoirs and Proceedings of the Manchester Literary and Philosophical Society. Manchester.
Massachusetts Brd. of Health Rep. Annual Report of the State Board of Health of Massachusetts. Boston.
Math. Ann. Mathematische Annalen. Leipzig.
Math. Gaz. The Mathematical Gazette. London.
Med. News Medical News. Philadelphia & New York.
Med. Press West. N.Y. Medical Press of Western New York. Buffalo, N.Y.
Med. Rec. Medical Record. New York.
Med. Surg. Rep. Medical and Surgical Reporter. Burlington, N.J., & Philadelphia.
Memoirs Int. Cong. Anthrop., Chicago Memoirs of the International Congress of Anthropology. [Chicago]
Messenger Math. The Messenger of Mathematics. London & Cambridge.
Meteorol. Soc. Quart. J. Quarterly Journal of the Royal Meteorological Society. London.
Micr. Soc. J. Journal of the Royal Microscopical Society. London & Edinburgh.
Microscope The Microscope. Ann Arbor, Mich.
Midland Natlist. The Midland Naturalist. Birmingham.
Min. Mag. The Mineralogical Magazine and Journal of the Mineralogical Society. London.
Mind Mind. London.
Minn. Bot. Stud. Geological and Natural History Survey of Minnesota. Minnesota Botanical Studies. Minneapolis.
Minnesota Ac. Nat. Sci. Bull. Minnesota Academy of [Natural] Sciences. Bulletin. Minneapolis.
Morphol. Jbuch. Morphologisches Jahrbuch. Leipzig.

München Ak. Sber. Sitzungsberichte der mathematisch-physikalischen Classe der k. b. Akademie der Wissenschaften zu München. München.
Music Music. Chicago.
Music Rev. Music Review. Chicago.
N. Amer. Ent. North American Entomologist. Buffalo, N.Y.
N. Amer. Rev. North American Review. Boston & New York.
N. Y. Ac. Ann. Annals of the New York Academy of Sciences. New York.
N. Y. Ac. Trans. Transactions of the New York Academy of Sciences. New York.
N. Y., Amer. Math. Soc. Bull. Bulletin of the American Mathematical Society. New York & Lancaster, Pa.
N. Y., Bot. Club Bull. Bulletin of the Torrey Botanical Club. New York.
N. Y. Ent. Soc. J. Journal of the New York Entomological Society. New York.
N. Y. J. Gyn. Obst. New York Journal of Gynecology and Obstetrics. New York.
N. Y. Med. J. The New York Medical Journal. New York.
Nat. Geogr. Mag. National Geographic Magazine. Washington.
National Rev. National Review. London.
Nautilus The Nautilus. Philadelphia.
Nebraska Ac. Sci., Pub. Publications of the Nebraska Academy of Science. Lincoln, Neb.
Nebraska Univ. Stud. University Studies. Lincoln, Neb.
Newb. Field Club Trans. Transactions of the Newbury District Field Club. Newbury.
Nicholson, J. Journal of Natural Philosophy, Chemistry, and the Arts; by W. Nicholson. London.
Nieuw Arch. Wisk. Nieuw Archief voor Wiskunde. Amsterdam.
Norf. Norw. Nat. Soc. Trans. Transactions of the Norfolk and Norwich Naturalists' Society. Norwich.
Observatory, London The Observatory. London.
Observer Observer. Portland, Conn.
Ohio Agric. Exp. Sta. Ann. Rep. Bull. Ohio Agricultural Experiment Station. Annual Report. Columbus, Ohio.
Ohio Dent. J. Ohio Dental Journal. Columbus, Ohio.
Ohio State Ac. Sci. Ann. Rep. Ohio State Academy of Science. Annual Report. Columbus, Ohio.
Old and New Old and New. Boston & New York.
Ornith. Ool. The Ornithologist and Oologist. Pawtucket, R.I., Boston, Mass., & Hyde Park, Mass.
Pac. Rural Pr. Pacific Rural Press. San Francisco.
Palaeontographica Palaeontographica. Cassel & Stuttgart.
Paris, Ac. Sci. C. R. Comptes Rendus hebdomadaires des Séances de l'Académie des Sciences. Paris.
Paris, Club Alpin Franç. Annu. Annuaire du Club Alpin Français. Paris.
Paris, Mus. Hist. Nat. Bull. Bulletin du Muséum d'Histoire Naturelle. Paris.
Paris, Soc. Biol. Mém. Comptes Rendus [hebdomadaires] des Séances et Mémoires de la Société de Biologie. Paris.
Paris, Soc. Géol. Bull. Bulletin de la Société Géologique de France. Paris.
Peabody Mus. Amer. Arch. Ethnol., Papers Peabody Museum of American Archeology and Ethnology. Papers. New Haven, Conn.
Peabody Mus. Amer. Arch. Ethnol. Rep. Peabody Museum of American Archeology and Ethnology. Annual Report. New Haven, Conn.
Pedagog. Seminary Pedagogical Seminary and Journal of Genetic Psychology. Worcester, Mass. & Provincetown, Mass.
Pennsylvania Univ. Publ. (Contrib. Bot. Lab.) Publications of the University of Pennsylvania. Contributions fromthe Botanical Laboratory. Philadelphia.
Penzance Soc. Trans. Transactions of the Penzance Natural History and Antiquarian Society. Plymouth.
Perthsh. Soc. Sci. Trans. & Proc. Transactions and Proceedings of the Perthshire Society of Natural Science. Perth.
Pflüger, Arch. Physiol. Archiv für die gesammte Physiologie des Menschen und der Thiere. Herausgegeben von Dr. E. F. W. Pflüger. Bonn.
Pharm. J. The Pharmaceutical Journal. London.
Phil. Mag. The London and Edinburgh Philosophical Magazine and Journal of Science. London.
Phil. Rev. Philosophical Review. Boston.
Phil. Stud. Philosophische Studien. Leipzig.
Phil. Trans. Philosophical Transactions of the Royal Society of London. London.
Philad. Ac. Nat. Sci. Proc. Proceedings of the Academy of Natural Sciences of Philadelphia. Philadelphia.
Philad. Coll. Pharm. Alum. Ass. Proc. Proceedings of the Alumni Association of the Philadelphia College of Pharmacy. Philadelphia.
Philad., Ent. News Entomological News. Philadelphia.
Philad. Med. Times Philadelphia Medical Times. Philadelphia.

Photogr. J.	The Photographic Journal. London.
Phys. Rev.	The Physical Review. New York, London, & Berlin.
Phytologist	The Phytologist. London.
Plant World	Plant World. Binghampton, N.Y.
Poggend. Annal.	Annalen der Physik und Chemie; von J. C. Poggendorff. Leipzig.
Polyclinic	The Polyclinic. Philadelphia.
Popular Astr.	Popular Astronomy. Northfield, Minn.
Practitioner	The Practitioner. London, Paris, New York, & Melbourne.
Prairie Farmer	Prairie Farmer. Chicago.
Proc. N.Y. Path. Soc.	Proceedings of the New York Pathological Society. New York.
Psyche	Psyche. Cambridge, Mass.
Psychical Res. Soc. J.	Journal of the Society for Psychical Research. London.
Psychol. Rev.	Psychological Review. New York, etc.
Psychol. Rev. Monograph Suppl.	Psychological Review. Monograph Supplement. New York, etc.
Quart. J. Math.	The Quarterly Journal of pure and applied Mathematics. London.
Quart. J. Micr. Sci.	Quarterly Journal of Microscopical Science. London.
Quarterly Rev.	Quarterly Review. London.
Quekett Micr. Club J.	The Journal of the Quekett Microscopical Club. London.
Quetelet., Corresp. Math.	Correspondance Mathématique et Physique. Gent & Bruxelles.
Rep. Proc. Alum. Assoc. Woman's Med. Coll. Penn.	Reports of Proceedings of the Alumnae Association of the Woman's Medical College of Pennsylvania. Philadelphia.
Rev. Bryol.	Revue Bryologique. Bulletin bimestriel consacré à l'Étude des Mousses et des Hépatiques. Paris & Caen.
Rev. méd-chir. d. mal. d. femmes	Revue médico-chirurgicale des maladies des femmes. Paris.
Rev. of Revs.	Review of Reviews. London.
Rhodora	Rhodora. Journal of the New England Botanical Club. Boston, Mass. & Providence, R.I.
Robin, J. Anat.	Journal de l'Anatomie et de la Physiologie Normales et Pathologiques de l'Homme et des Animaux. Paris.
Rochester (N.Y.) Ac. Sci. Proc.	Proceedings of the Rochester Academy of Science. Rochester, N.Y.
Roma, Corrisp. Scient.	Corrispondenza Scientifica in Roma. Roma.
Roy. Geogr. Soc. Proc.	Proceedings of the Royal Geographical Society. London.
Roy. Soc. Proc.	Proceedings of the Royal Society of London. London.
Rugby, Nat. Hist. Soc. Rep.	Reports of the Rugby School Natural History Society. Rugby.
Schweiz. Natf. Ges. Verh.	Verhandlungen der Schweizerischen Naturforschenden Gesellschaft. Luzern.
Science	Science. Cambridge, Mass. & New York.
Scott. Geogr. Mag.	The Scottish Geographical Magazine. Edinburgh.
Scott. Natlist.	The Scottish Naturalist. Perth.
Selenographical J.	Selenographical Journal. London.
Sidereal Messenger	The Sidereal Messenger. Northfield, Minn.
Silliman, J.	The American Journal of Science and Arts; by B. Silliman. New Haven, Conn.
Smithsonian Inst. Bur. Ethnol. Rep.	Annual Report of the Bureau of Ethnology to the Secretary of the Smithsonian Institution. Washington.
Smithsonian Miscell. Coll.	Smithsonian Miscellaneous Collections. Washington.
Smithsonian Rep.	Annual Report of the Board of Regents of the Smithsonian Institution. Washington.
Soc. Arts. Trans.	Transactions of the Society of Arts. London.
Soc. Chem. Ind. J.	The Journal of the Society of Chemical Industry. Manchester & London.
Somersetsh. Archaeol. Soc. Proc.	Proceedings of the Somersetshire Archaeological and Natural History Society. Taunton.
Southern Workman	Southern Workman. Hampton, Va.
Spectator	Spectator. London.
Spelunca, Paris	Spelunca. Bulletin de la Société de Spéléologie. Paris.
St. Louis Ac. Trans.	The Transactions of the Academy of Science of St. Louis. St. Louis, Mo.
St. Louis, Bot. Gard. Rep.	Missouri Botanical Garden. [Reports.] St. Louis, Mo.
Suffolk Inst. Archaeol. Proc.	Proceedings of the Suffolk Institute of Archaeology. Ipswich.
Symons, Meteorol. Mag.	Symon's Monthly Meteorological Magazine. London.
Syst. Pract. Med.	System of Practical Medicine. Philadelphia.
Taylor, Scientif. Mem.	Scientific Memoirs; by R. Taylor, J. Tybdall, and W. Fran. London.

Tech. Quart.	Technology Quarterly. Boston.
Thomson, Ann. Phil.	Annals of Philosophy, or Magazine of Chemistry, Mineralogy, Mechanics, and the Arts; by T. Thomson. London.
Tilloch, Phil. Mag.	The Philosophical Magazine (Ed. A. Tilloch.) London.
Timehri	Timehri [Journal of the Royal Agricultural and Commercial Society of British Guiana]. Demerara, British Guiana.
Times and Reg.	Times and Register. Philadelphia.
Torino Acc. Sci. Atti	Atti della R. Accademia delle Scienze di Torino. Torino.
Torrey Bot. Club Bull.	Bulletin of the Torrey Botanical Club. New York.
Torrey Bot. Club Mem.	Memoirs of the Torrey Botanical Club. New York.
Trans. Amer. Inst. Min. Eng.	Transactions of the American Institute of Mining Engineers. New York.
Trans. Illinois Med. Soc.	Transactions. Illinois Medical Society. Chicago.
U.S. Bur. Anim. Ind. Rep.	U.S. Department of Agriculture. . . . Annual Report of the Bureau of Animal Industry. Washington.
[U.S.] Comm. Agr. Rep.	Report of the Commissioner of Agriculture. Washington.
U.S. Div. Ent. Bull.	U.S. Department of Agriculture. Division of Entomology. Bulletin. Washington.
U.S. Div. Ent. Insect Life	U.S. Department of Agriculture. Division of Entomology. Insect Life. Washington.
U.S. Fish Comm. Rep.	United States Commission of Fish and Fisheries. Report of the Commissioner. Washington.
U.S. Geol. Surv. Ann. Rep.	Annual Report of the United States Geological Survey. Washington.
U.S. Geol. Surv. Bull.	Bulletin of the United States Geological Survey. Washington.
U.S. Monthly Weath. Rev.	United States of America: Department of Agriculture. Monthly Weather Review and Annual Summary. Washington.
U.S. Mus. Proc.	Proceedings of the United States National Museum. Washington.
U.S. N. Amer. Fauna	U.S. Department of Agriculture. Division of Biological Survey. North American Fauna. Washington.
Vassar Misc. Mon.	Vassar Miscellany Monthly. Poughkeepsie, N.Y.
Verh. Internat. Geogr. Kongr. Berlin, VII.	Verhandlunge des VII Internationalen Geographen-Kongresses in Berlin. Berlin.
Virchow, Arch.	Archiv für Pathologische Anatomie und Physiologie und für Klinische Medicin. Herausgegeben von Rudolf Virchow. Berlin.
Washburn Obs. Publ.	Publications of the Washburn Observatory of the University of Wisconsin. Madison, Wis.
Washington Ac. Sci. Proc.	Proceedings of the Washington Academy of Sciences. Washington.
Washington Biol. Soc. Proc.	Proceedings of the Biological Society of Washington. Washington.
Wiegmann, Archiv	Archiv für Naturgeschichte; von Ar. Fr. Aug. Wiegmann. Berlin.
Wien, Geol. Verh.	Verhandlungen der kaiserlich-königlichen Geologischen Reichsanstalt. Wien.
Wien, Zool. Bot. Verh.	Verhandlungen der kaiserlich-königlichen Zoologisch-Botanischen Gesellschaft in Wien. Wien.
Wisconsin Ac. Trans.	Transactions of the Wisconsin Academy of Science, Arts, and Letters. Madison, Wis.
Wisconsin Geol. Nat. Hist. Surv. Bull.	Wisconsin Geological and Natural History Survey. Bulletin. Madison, Wis.
Wisconsin Nat. Hist. Soc. Occ. Pap.	Occasional Papers of the Natural History Society of Wisconsin. Milwaukee, Wis.
Wisconsin Nat. Hist. Soc. Proc.	Proceedings of the Natural History Society of Wisconsin. Milwaukee, Wis.
Woman's J.	Woman's Journal. Boston & Chicago.
Woman's Med. J.	Woman's Medical Journal. Cincinnati & Washington.
Woods Holl Mar. Biol. Lab. Bull.	Biological Bulletin. [Marine Biological Laboratory, Woods Holl, Mass.] Boston.
Wood's Holl Mar. Biol. Lab. Lect.	Biological Lectures delivered at the Marine Biological Laboratory, Wood's Holl, [Mass.]. Boston. [From 1877 until 1897 Woods Hole, Massachusetts, was officially known as Wood's Holl.]
World's Congr. Religions. Addr.	World's Congress of Religions. Addresses and Papers. Philadelphia.
Yale Univ. Obs. Trans.	Transactions of the Astronomical Observatory of Yale University. New Haven, Conn.
Yn Lioar Manninagh	Yn Lioar Manninagh. Douglas, Isle of Man.
Zoe	Zoe. San Francisco.
Zool. Anz.	Zoologischer Anzeiger. Leipzig.
Zool. Bull.	Zoological Bulletin. [Continued as Woods Holl Mar. Biol. Lab. Bull.] Boston.
Zool. Jbüch. (Anat.)	Zoologische Jahrbücher. Abteilung für Anatomie und Ontonogenie der Thiere. Jena.
Zool. Soc. Proc.	Proceedings of the Zoological Society of London. London.
Zool. Soc. Trans.	Transactions of the Zoological Society of London. London.

Zoologist	The Zoologist. London.
Ztschr. Anorg. Chem.	Zeitschrift für anorganische Chemie. Hamburg & Leipzig.
Ztschr. Biol.	Zeitschrift für Biologie. München & Leipzig.
Ztschr. Hyg.	Zeitschrift für Hygiene. Leipzig.
Ztschr. Physikal. Chem.	Zeitschrift für Physikalische Chemie, Stochiometrie und Verwandtschaftlslehre. Leipzig.
Ztschr. Wiss. Mikr.	Zeitschrift für wissenschaftliche Mikroskopie und für mikroskopische Technik. Leipzig.
Ztschr. Wiss. Zool.	Zeitschrift für wissenschaftliche Zoologie. Leipzig.
Zürich Vrtljschr.	Vierteljahrsschrift der naturforschenden Gesellschaft in Zürich. Zürich.

Selected Bibliography

Abir-Am, Pnina G., and Outram, Dorinda, eds. *Uneasy Careers and Intimate Lives: Women in Science 1787–1979.* New Brunswick, N.J.: Rutgers University Press, 1987.

Albisetti, James C. *Schooling German Girls and Women: Secondary and Higher Education in the Nineteenth Century.* Princeton, N.J.: Princeton University Press, 1988.

Alic, Margaret. *Hypatia's Heritage: A History of Women in Science from Antiquity to the Late Nineteenth Century.* London: The Women's Press, 1986.

Allen, David Elliston. *The Naturalist in Britain: A Social History.* London: Allen Lane, 1976.

———. "The women members of the Botanical Society of London, 1836–1856." *British Journal for the History of Science,* **13** (1980): 240–54.

Bell, E. Moberly. *Storming the Citadel: The Rise of the Woman Doctor.* London: Constable and Co., 1953.

Birkett, Dea. *Spinsters Abroad: Victorian Lady Explorers.* Oxford: Basil Blackwell, 1989.

———. *Mary Kingsley: Imperial Adventuress.* London: Macmillan, 1992.

Blake, Catriona. *The Charge of the Parasols: Women's Entry to the Medical Profession.* London: The Women's Press, 1990.

Bonner, Thomas Neville. *To the Ends of the Earth: Women's Search for Education in Medicine.* Cambridge, Mass.: Harvard University Press, 1992.

Brück, Mary T. "Ellen and Agnes Clerke of Skibbereen, scholars and writers." *Seanchas Chairbre,* **3** (1993): 23–43.

———. "Agnes Mary Clerke, chronicler of astronomy." *Quarterly Journal of the Royal Astronomical Society,* **35** (1994): 59–79.

Caine, Barbara. *Victorian Feminists.* Oxford: Oxford University Press, 1992.

Chaff, Sandra L.; Hainbach, Ruth; Fenichel, Carol; Woodside, Nina B., comps. and eds. *Women in Medicine: A Bibliography of the Literature on Women Physicians.* Metuchen, N.J.: Scarecrow Press, 1977.

Challinor, John. *The History of British Geology: A Bibliographical Study.* New York: Barnes and Noble, 1971.

Chick, Harriette; Hume, Margaret; Macfarlane, Marjorie. *War on Disease: A History of the Lister Institute.* London: André Deutsch, 1971.

Clause, Bonnie Tochter. "The Wistar rat as the right choice: establishing mammalian standards and the ideal of a standardized mammal." *Journal of the History of Biology,* **26** (1993): 329–49.

Cleevely, R. J.; Tripp, R. P.; Howells, Y. "Mrs. Elizabeth Gray (1831–1924): a passion for fossils." *Bulletin. British Museum (Natural History) Historical Series,* **17** (1989): 167–258.

Creese, Mary R. S. "British women of the nineteenth and early twentieth centuries who contributed to research in the chemical sciences." *British Journal for the History of Science,* **24** (1991): 275–305.

———. and Creese, Thomas M. "British women who contributed to research in the geological sciences in the nineteenth century." Ibid., **27** (1994): 23–54.

Dakin, Susanna Bryant. *The Perennial Adventure: A Tribute to Alice Eastwood: 1859–1953.* San Francisco: California Academy of Sciences, 1954.

Desmond, Ray. *Dictionary of British and Irish Botanists and Horticulturalists, including Plant Collectors and Botanical Artists.* London: Taylor and Francis, 1977.

Eberle, Jean Fahey. *The Incredible Owen Girls.* St. Louis, Mo.: Boar's Head Press, 1977.

Eells, Walter Crosby. "Earned doctorates for women in the nineteenth century," *American Association of University Professors. Bulletin,* **42** (1956): 644–51.

Fountaine, Margaret. *Love among the Butterflies: The Travels and Adventures of a Victorian Lady,* ed. W. F. Cater. London: Collins, 1980.

———. *Butterflies and Late Loves: The Further Travels and Adventures of a Victorian Lady,* ed. W. F. Cater. London: Collins, 1986.

Frank, Katherine. *A Voyager Out: The Life of Mary Kingsley.* Boston: Houghton Mifflin, 1986.

Gacs, Ute; Kahn, Aisha; McIntyre, Jerrie; Weinberg, Ruth, eds. *Women Anthropologists: A Biographical Dictionary.* Westport, Conn.: Greenwood Press, 1988.

Gage, Andrew Thomas, and Stearn, William Thomas. *A Bicentenary History of the Linnean Society of London.* London: Linnean Society/Academic Press, 1988.

Geison, Gerald L. *Michael Foster and the Cambridge School of Physiology: The Scientific Enterprise in Late Victorian Society.* Princeton, N.J.: Princeton University Press, 1978.

———, ed. *Physiology in the American Context 1850–1940.* Bethesda, Md.: American Physiological Society, 1987.

Girton College Register, 1869–1946. Cambridge: printed for Girton College, 1948.

Grattan-Guiness, I. "A mathematical union. William Henry and Grace Chisholm Young." *Annals of Science,* **29** (1972): 105–86.

Grinstein, Louise S., and Campbell, Paul J., eds. *Women of Mathematics: A Biobibliographic Sourcebook.* Westport, Conn.: Greenwood Press, 1987.

———; Rose, Rose K; Rafailovich, Miriam H., eds. *Women in Chemistry and Physics: A Biobibliographic Sourcebook.* Westport, Conn.; Greenwood Press, 1993.

Hackett, Alice Payne. *Wellesley: Part of the American Story.* New York: E. P. Dutton, 1949.

Hamilton, Alice. *Exploring the Dangerous Trades: The Autobiography of Alice Hamilton, M.D.* Boston: Little, Brown and Co., 1943; Boston: Northeastern University Press, 1985 (reprint).

Hawksworth, D. L., and Seaward, M. R. D. *Lichenology in the British Isles 1568–1975: An Historical and Bibliographical Survey.* Richmond: Richmond Publishing Co., 1977.

Hearnshaw, L. S. *A Short History of British Psychology 1840–1940.* London: Methuen, 1964.

Jones, Bessie Zaban, and Boyd, Lyle Clifford. *The Harvard College Observatory: The First Four Directorships, 1839–1919.* Cambridge: Harvard University Press, 1971.

Kass-Simon, G., and Farnes, Patricia, eds. *Women of Science. Righting the Record.* Bloomington, Ind.: Indiana University Press, 1990.

Keenan, Katherine. "Lilian Vaughan Morgan (1870–1952): her life and work." *American Zoologist,* **23** (1983): 867–76.

Kessell, Edward L., ed. *A Century of Progress in the Natural Sciences, 1853–1953.* San Francisco: California Academy of Sciences, 1955; New York: Arno Press, 1974 (reprint).

Kidwell, Peggy Aldrich. "Women astronomers in Britain, 1780–1930." *Isis,* **75** (1984): 534–46.

Kirby, Mary (Mrs. Gregg). *Leaflets from my Life: a Narrative Autobiography.* London: Simpkin, Marshall and Co., 1887.

Kofalk, Harriet. *No Woman Tenderfoot: Florence Merriam Bailey, Pioneer Naturalist.* College Station, Tex.: Texas A and M University Press, 1989.

Kohler, Robert E. *From Medical Chemistry to Biochemistry: The Making of a Biomedical Discipline.* Cambridge: Cambridge University Press, 1982.

Leahey, Thomas Hardy. *A History of Psychology: Main Currents in Psychological Thought.* 2d. ed. Englewood Cliffs, N.J.: Prentice-Hall, 1987.

Lewis, Jane. *Women and Social Action in Victorian and Edwardian England.* Stanford, Calif.: Stanford University Press, 1991.

Lillie, Frank R. *The Woods Hole Marine Biological Laboratory.* Chicago: University of Chicago Press, 1944.

Lindberg, David R. "Mary Alice Willcox, 1856–1953." *Annual Report: Western Society of Malacologists,* **10** (1977): 16–17.

Love, Rosaleen. "'Alice in eugenics-land'; feminism and eugenics in the scientific careers of Alice Lee and Ethel Elderton." *Annals of Science,* **36** (1979): 145–8.

MacHale, Desmond. *George Boole: His Life and Work.* Dublin: Boole Press, 1985.

MacLeod, Roy, and Moseley, Russell. "Fathers and daughters: reflections on women, science and Victorian Cambridge." *History of Education,* **8** (1979): 321–33.

Mallis, Arnold. *American Entomologists.* New Brunswick, N.J.: Rutgers University Press, 1971.

Maltby, Margaret E., comp. *History of the Fellowships Awarded by the American Association of University Women, 1888–1929, with the Vitas of the Fellows.* Washington, D.C.: American Association of University Women, [1929].

Mark, Joan. *A Stranger in her Native Land: Alice Fletcher and the American Indians.* Lincoln, Neb.: University of Nebraska Press, 1988.

Mason, Joan. "Hertha Ayrton (1854–1923) and the admission of women to the Royal Society of London." *Notes and Records. Royal Society of London,* **45** (1991): 201–20.

McWilliams-Tullberg, Rita. *Women at Cambridge: A Men's University—Though of a Mixed Type.* London: Victor Gollancz, 1975.

Middleton, Dorothy. *Victorian Lady Travellers.* London: Routledge and Kegan Paul, 1965.

Miles, Wyndham D., and Gould, Robert F., eds. *American Chemists and Chemical Engineers.* Vol. 2. Guilford, Conn.: Gould Books, 1994.

Miller, Luree. *On Top of the World: Five Women Explorers in Tibet.* London: Paddington Press, 1976.

Moldow, Gloria. *Women Doctors in Gilded-age Washington: Race, Gender and Professionalization.* Urbana, Ill.: University of Illinois Press, 1987.

Morton, A. G. *History of Botanical Science: An Account of the Development of Botany from Ancient Times to the Present Day.* London: Academic Press, 1981.

Newnham College Register, 1871–1971. Vol., 1, 1871–1923. 2d. ed. Cambridge: Newnham College, 1979.

O'Connell, Agnes N., and Russo, Nancy Felipe, eds. *Women in Psychology: A Bio-Bibliographic Sourcebook.* Westport, Conn.: Greenwood Press, 1990.

O'Connor, Jean G., and Meadows, A. J. "Specialization and professionalization in British geology." *Social Studies in Science,* **6** (1976): 77–89.

Ogilvie, Marilyn Bailey. *Women in Science: Antiquity through the Nineteenth Century: A Biographical Dictionary with Annotated Bibliography.* Cambridge, Mass.: MIT Press, 1986.

Osborn, Herbert. *Fragments of Entomological History: Including Some Personal Recollections of Men and Events.* Columbus, Ohio: [n.p.], 1937.

Overfield, Richard A. *Science with Practice: Charles E. Bessey and the Maturing of American Botany.* Ames, Iowa: Iowa State University Press, 1993.

Patterson, Elizabeth Chambers. *Mary Somerville and the Cultivation of Science, 1815–1840.* The Hague: Martinus Nijhoff, 1983.

Pearce, Robert D. *Mary Kingsley: Light at the Heart of Darkness.* Oxford: Kensal Press, 1990.

Phillips, Ann, ed. *A Newnham Anthology.* Cambridge: Cambridge University Press, 1979.

Phillips, Patricia. *The Scientific Lady: A Social History of Women's Scientific Interests, 1520–1918.* New York: St. Martin's Press, 1990.

Remington, Jeanne E. "Katharine Jeannette Bush: Peabody's mysterious zoologist." *Discovery,* **12,** n.3 (1977): 3–8.

Robinson, Jane. *Wayward Women: A Guide to Women Travellers.* Oxford: Oxford University Press, 1990.

Robinson, Lisa Mae. "The electrochemical school of Edgar Fahs Smith, 1878–1913." Ph.D. dissertation, University of Pennsylvania, 1986.

Rossiter, Margaret W. *Women Scientists in America: Struggles and Strategies to 1940.* Baltimore, Md.: Johns Hopkins University Press, 1982.

Sarjeant, William Antony S. *Geologists and the History of Geology: An International Bibliography from the Origins to 1978.* New York: Arno Press, 1980.

Scarborough, Elizabeth, and Furumoto, Laurel. *Untold Lives: The First Generation of American Women Psychologists.* New York: Columbia University Press, 1987.

Schäfer, Edward Albert. *History of the Physiological Society during its First Fifty Years, 1876–1926.* London: Cambridge University Press, 1927.

Scringeour, R.M., ed. *The North London Collegiate School 1850–1950: A Hundred Years of Girls' Education.* London: Oxford University Press, 1950.

Sharp, Evelyn. *Hertha Ayrton, 1854–1923: A Memoir.* London: Edward Arnold, 1926.

Sidgwick, Ethel. *Mrs. Henry Sidgwick: A Memoir by her Niece.* London: Sidgwick and Jackson, 1938.

Siegel, Patricia Joan, and Finley, Kay Thomas. *Women in the Scientific Search: An American Biobibliography, 1724–1979,* Metuchen, N.J.: Scarecrow Press, 1985.

Sieveking, Paul, ed. *British Biographical Archive.* Microfiche edn. New York: K. G. Saur, 1984.

Smith, Beatrice Scheer. "Lucy Bishop Millington, nineteenth-century botanist: her life and letters to Charles Horton Peck, State Botanist of New York." *Huntia,* **8** (1992): 111–53.

Smith, F. B. *Florence Nightingale: Reputation and Power.* London: Croom Helm, 1982.

Somerville, Martha. *Personal Recollections from Early Life to Old Age of Mary Somerville. With Selections from her Correspondence.* Boston: Roberts Brothers: 1876.

Stein, Dorothy. *Ada: A Life and a Legacy.* Cambridge, Mass.: MIT Press, 1985.

Stevens, Gwendolyn, and Gardner, Sheldon. *The Women of Psychology. Volume 1: Pioneers and Innovators.* 2 vols. Cambridge, Mass.: Schenkman Publishing, 1982.

Stevens, Helen Norton. *Memorial Biography of Adele M. Fielde: Humanitarian.* New York: Fielde Memorial Committee, 1918.

Stuckey, Ronald L. *Women Botanists from Ohio Born before 1900: With Reference Calendars from 1776 to 2028.* Columbus, Ohio: RLS Creations, 1992.

Sweeting, G. S., ed. *The Geologists' Association 1858–1958: A History of the First Hundred Years.* London: Geologists' Association, 1958.

Thomas, Onfel. *Frances Elizabeth Hoggan 1843–1927.* Newport, Monmouthshire: R. H. Johns, 1971.

Tomlinson, Margaret. *Three Generations in the Honiton Lace Trade: A Family History.* Exeter: Devon Print Group, 1983.

Torrens, Hugh S. "Mary Anning (1799–1847) of Lyme; 'the greatest fossilist the world ever knew'," *British Journal for the History of Science,* **28** (1995): 257–84.

———. "Women in geology. 2—Etheldred Benett." *Open Earth* (1983): 12–13.

Tuke, Margaret J. *A History of Bedford College for Women 1849–1937.* London: Oxford University Press, 1939.

Vardi, Winifred I. (Mrs. E. W. Candler), comp. *King Edward VI High School for Girls, Birmingham, 1883–1925.* London: Ernest Benn, 1928.

Wallace, Robert, ed. *Eleanor Ormerod, LL.D., Economic Entomologist: Autobiography and Correspondence.* New York: E. P. Dutton, 1904.

Warner, Deborah Jean. *Graceanna Lewis, Scientist and Humanitarian.* Washington, D.C.: Smithsonian Institution Press, 1979.

Weiss, Harry B. "Mrs Mary Treat, 1830–1923, early New Jersey naturalist." *Proceedings of the New Jersey Historical Society,* **73** (1955): 258–73.

Welsh, Lilian, M.D., LL.D. *Reminiscences of Thirty Years in Baltimore.* Baltimore: Norman, Remington Co., 1925.

Williamson, M[artha] Burton. "Some American women in science." *Chautauquan,* **28,** n.s.**19** (1898–1899): 161–8, 361–8, 465–73.

Wilson, Carol Green. *Alice Eastwood's Wonderland: The Adventures of a Botanist.* San Francisco: California Academy of Sciences, 1955.

Woodward, Horace B. *The History of the Geological Society of London.* London: Longman's, Green, 1908.

Zottoli, Steven J., and Seyfarth, Ernst-August. "Julia B. Platt (1857–1935): pioneer comparative embryologist and neurologist." *Brain, Behaviour and Evolution,* **43** (1994): 92–106.

Index

About the Author and Contributor

Mary Creese (née Weir) was born in Orkney, Scotland, in 1935. After almost thirty years as a research chemist she turned to the subject of women's contributions to scientific work, particularly in the nineteenth century. Since 1990 she has published more than 20 articles on early women scientists. She holds a B.Sc. degree from the University of Glasgow and a Ph.D. from the University of California, Berkeley (both in chemistry) and coauthored some 20 technical publications. She has lived in Kansas since 1964, nearly all of the time working at the University of Kansas, most recently as an associate at the Hall Center for the Humanities. She is married to Thomas Creese and they have two daughters.

Thomas Creese was born in New Jersey in 1934. He holds degrees in mathematics from Massachusetts Institute of Technology and the University of California, Berkeley. Since 1964 he has been a member of the Department of Mathematics at the University of Kansas and has coauthored both a textbook and a research monograph. He and Mary Creese have collaborated on projects in computational chemistry and history of women in science.